tor Optimization

ries in Vector Optimization contains publications in various fields of opti-
on with vector-valued objective functions, such as multiobjective optimiza-
nulti criteria decision making, set optimization, vector-valued game theory
order areas to financial mathematics, biosystems, semidefinite programming
ultiobjective control theory. Studies of continuous, discrete, combinatorial
ochastic multiobjective models in interesting fields of operations research
o included. The series covers mathematical theory, methods and applications
nomics and engineering. These publications being written in English are
rily monographs and multiple author works containing current advances in
fields.

information about this series at
www.springer.com/series/8175

Vector Optimization

Series Editor:

Johannes Jahn
University of Erlangen-Nürnberg
Department of Mathematics
Cauerstr. 11
81058 Erlangen
Germany
johannes.jahn@fau.de

Akhtar A. Khan • Christiane Tammer •
Constantin Zălinescu

Set-valued Optimization

An Introduction with Applications

Springer

Akhtar A. Khan
Rochester Institute of Technology School
of Mathematical Sciences
Rochester
New York
USA

Christiane Tammer
Halle
Germany

Constantin Zălinescu
University "Al. I. Cuza" Iasi
Faculty of Mathematics
Iasi
Romania

ISSN 1867-8971 ISSN 1867-898X (electronic)
ISBN 978-3-642-54264-0 ISBN 978-3-642-54265-7 (eBook)
DOI 10.1007/978-3-642-54265-7
Springer Heidelberg New York Dordrecht London

Library of Congress Control Number: 2014951215

Printed on acid-free paper

Springer is part of Springer Science+Business Media (www.springer.com)

With our deepest appreciation, we dedicate this book to the 80th birthday of Prof. A. Göpfert

Acknowledgments

We take this opportunity to express our most sincere gratitude to Truong Bao, Mihail Barbosu, Patricia Clark, Marius Durea, Joydeep Dutta, Rosalind Elster, Franco Giannessi, Mark Gockenbach, Joachim Gwinner, Alfred Göpfert, Frank Heyde, Jonas Ide, Johannes Jahn, Kathrin Klamroth, Elisabeth Köbis, Andreas Löhne, Sophia Maggelakis, Juan Enrique Martìnez-Legaz, Antonino Maugeri, Douglas Meadows, Dumitru Motreanu, Boris Mordukhovich, Zuhair Nashed, Fabio Raciti, Miguel Sama, Anita Schöbel, Abul Hasan Siddiqi, Radu Strugariu, Doug Ward, and Petra Weidner for the joint work and for their support.

We would also like to thank Rosalind Elster, Alfred Göpfert, Miguel Sama, and Doug Ward for carefully reading the manuscript and many valuable suggestions. Moreover, we are grateful to Andreas Löhne and Brian Winkler for drawing some of the figures for this book.

I (AAK) take an immense pleasure in expressing my gratitude and deep love to my mother Shahida Begum who, since my birth, alone bore all the hardships of life and made uncountable sacrifices for me. It is also a pleasure to acknowledge the support of my late uncle and his family. I thank my loving wife Basca and my adorable daughters Sophie and Amelie for all their love and support. In addition, I would like to thank Mary Hoeft for her boundless motherly love.

We are extremely thankful to Mr. Christian Rauscher of Springer-Verlag for his limitless patience and unfailing support throughout this process.

Finally, all the authors are grateful to their coauthors for all their support, understanding, and hard work over this long and fruitful collaboration. We also acknowledge that Sects. 2.3, 2.4 (partially), 3.1 and 3.2; Chapters 6, 7, 9, and 10 were written by CZ and the remaining parts were prepared by AAK and CT. The work of A.A. Khan was partially supported by a grant from the Simons Foundation

(#210443 to Akhtar Khan). The research of C. Zălinescu was supported partially by the grant PN-II-ID-PCE-2011-3-0084, CNCS-UEFISCDI, Romania.

Rochester, NY, USA — Akhtar A. Khan
Halle, Germany — Christiane Tammer
Iasi, Romania — Constantin Zălinescu

List of Symbols

$\mathbb{R}$	The set of real numbers
$\overline{\mathbb{R}}$	$\overline{\mathbb{R}} := \mathbb{R} \cup \{-\infty\} \cup \{+\infty\}$
$\mathbb{R}_+$	Set of non-negative real numbers
$(a,b),]a,b[, (a,b],]a,b], [a,b), [a,b[, [a,b]$	Intervals in $\overline{\mathbb{R}}$
$\mathbb{P}$	Set of positive real numbers
$\mathbb{N}, \mathbb{N}^*$	Set of non-negative integers, $\mathbb{N}^* = \mathbb{N} \setminus \{0\}$
$X, Y, Z, \ldots$	Real linear spaces or topological linear spaces
$\|\|\cdot\|\|, \|\|\cdot\|\|_*$	Norm in X, norm in X^*
$Y^\bullet$	$Y^\bullet := Y \cup \{-\infty\} \cup \{+\infty\}$ (extended space)
C, K	Cones (in Y, Z)
$C^+, C^- := -C^+$	Positive and negative dual cones of C
$C^\#$	Quasi-interior of the dual cone C^+
$\overset{\circ}{B}_X, B_X$	Open and closed unit balls in the n.v.s. X
$B(k^0, \varepsilon)$	Closed ball with the center k^0 and the radius ε
$F: X \rightrightarrows Y$	Set-valued mapping
$f: X \to Y^\bullet$	Single-valued map, function
φ_{A,k^0}	$\varphi_{A,k^0}(y) := \inf\{t \in \mathbb{R} \mid y \in tk^0 - A\}$ $(y \in Y)$
$\mathcal{N}_X(x)$	The set of neighborhoods of $x \in X$ in the t.v.s. (X, τ)
$\mathcal{N}_X$	The set of balanced neighborhoods of $0 \in X$ in the t.v.s. (X, τ)

$\operatorname{Im} f = \{f(x) \mid x \in X\}$	Image (range) of the function $f : X \to Y$
$\operatorname{dom} f$	Domain of the single-valued map $f : X \to Y^\bullet$
$\operatorname{dom} F$	Domain of the set-valued map $F : S \rightrightarrows Y$
$\operatorname{Dom} F$	Domain of the set-valued map $F : S \rightrightarrows Y^\bullet$
$f^{-1}(B) = \{x \in X \mid f(x) \in B\}$	Inverse image of the set $B \subset Y$ of the function $f : X \to Y$
$F^{-1}(B) := \{x \in X \mid F(x) \cap B \neq \emptyset\}$	Inverse image by F of the set $B \subset F$
$F^{+1}(B) := \{x \in X \mid F(x) \subset B\}$	Another type of inverse image by F of the set $B \subset F$
(r, s), $(r, s]$, $[r, s)$, $[r, s]$	Intervals determined by $r, s \in \overline{\mathbb{R}}$, $r \le s$
$\hat{N}_\varepsilon(x; \Omega)$	ε-Normals
$\hat{N}(x; \Omega)$	Fréchet normal cone to Ω at $x \in \Omega$
$N(x; \Omega)$	(Basic, limiting, Mordukhovich) *normal cone* to Ω at x
$\operatorname{graph} F$	Graph of F
$D^* F(\overline{x}, \overline{y})$	(Basic, normal, Mordukhovich) *coderivative* of F at $(\overline{x}, \overline{y})$
$\operatorname{epi} f$	Epigraph of $f : X \to \overline{\mathbb{R}}$
$\partial F(\overline{x}, \overline{y})(y^*)$	(Limiting, basic, normal) subdifferential of F at $(\overline{x}, \overline{y})$ for $y^* \in Y^*$
$\mathscr{E}_F : X \rightrightarrows Y$	Epigraphical multifunction of F, $\operatorname{graph} \mathscr{E}_F = \operatorname{epi} F$
$\operatorname{int} A$, $\operatorname{r-int} A$	Interior and relative interior of the set A
$\operatorname{cl} A$	Closure of the set A
$\operatorname{bd} A$, $\operatorname{r-bd} A$	Boundary and relative boundary of the set A
$\operatorname{qi} A$, $\operatorname{qri} A$	Quasi-interior and quasi-relative interior of the set A
$\operatorname{span} A$, $\overline{\operatorname{span}} A$	Linear and closed linear hulls of the set A
$\operatorname{aff} A$, $\overline{\operatorname{aff}} A$	Affine and closed affine hulls of the set A
$\operatorname{cone} A$, $\overline{\operatorname{cone}} A$	Conic and closed conic hulls of the set A
$\operatorname{conv} A$, $\overline{\operatorname{conv}} A$	Convex and closed convex hulls of the set A
$\operatorname{core} A$	Algebraic interior (or core) of the set A

$\operatorname{icr} A$	Relative algebraic interior (or intrinsic core) of the set A
$\operatorname{sic} A$	Strong intrinsic core of the set A
A_∞	Asymptotic cone of the nonempty set A
$RC(A)$	Recession cone of the nonempty set A
$\mathscr{P}(Y)$	Power set of Y
$\mathscr{P}_C(Y)$	Family of all C-proper subsets of Y
$\operatorname{Min}(A, C)$	Set of minimal elements of A with respect to C
$\operatorname{Max}(A, C)$	Set of maximal elements of A with respect to C
$(\ldots)\text{-PMin}(A, C)$	Set of $(\ldots)$ properly minimal elements of A with respect to C
$\operatorname{WMin}(A, C)$	Set of weakly minimal elements of A with respect to C
$\operatorname{Min}(F(S), \preceq)$	Set of minimal solutions of the set-valued problem (SP)
$\operatorname{QMin}(A, C)$	Q-minimal point of A with respect to C
$T(S, x)$	Contingent cone of S at x
$C(S, x)$	Clarke's tangent cone of S at x
$A(S, x)$	Adjacent cone of S at x
$DF(x, y)$	Contingent derivative of F at $(x, y) \in \operatorname{graph}(F)$
$D_A F(x, y)$	Adjacent derivative of F at $(x, y) \in \operatorname{graph}(F)$
$D_L F(x, y)$	Lower Dini derivative of F at $(x, y) \in \operatorname{graph}(F)$
$D_E F(x, y)$	Contingent epiderivative of F at $(x, y) \in \operatorname{graph}(F)$
P_C	Metric projection mapping
Pr_X	$\operatorname{Pr}_X : X \times Y \to X$, $\operatorname{Pr}_X(x, y) = x$
f^*, f^{**}	Conjugate and biconjugate of f
σ_A	Support function of the set A
ι_A	Indicator function of the set A
$d(x, S), d_S(x)$	Distance from x to S
$(x_n), \{x_n\}$ $(x_n \in X, n \in \mathbb{N})$	Sequences in X
$\operatorname{Cl}_+ A := \operatorname{Cl}_+^C A$	Upper closure (with respect to C) of A
$\inf A$	Infimum of $A \subseteq Y$, $(Y, \leq)$ partially ordered set
$\sup A$	Supremum of $A \subseteq Y$, $(Y, \leq)$ partially ordered set

$\operatorname{Inf} A$	Infimal set of $A \subseteq Y^{\bullet}$ with respect to C
$\mathscr{F}$	Hyperspace of upper closed sets
$\mathscr{I}$	Family of all self-infimal subsets
$\preccurlyeq_{\mathscr{F}}$	Order relation in $\mathscr{F}$
$\oplus_{\mathscr{F}}$	Addition in $\mathscr{F}$
$\odot_{\mathscr{F}}$	Multiplication by non-negative real numbers in $\mathscr{F}$
$\preccurlyeq_{\mathscr{I}}$	Order relation in $\mathscr{I}$
$\oplus_{\mathscr{I}}$	Addition in $\mathscr{I}$
$\odot_{\mathscr{I}}$	Multiplication by non-negative real numbers in $\mathscr{I}$
$\langle y^*, \cdot \rangle$, $(y^*, \cdot)$, $y^*(\cdot)$	Linear continuous functional $y^* : Y \longrightarrow \mathbb{R}$
$\mathscr{L}(X, Y)$	Set of linear continuous mappings from X to Y
$\mathscr{M}(\mathscr{P}_C(Y), \mathbb{R})$	Set of all functions from $\mathscr{P}_C(Y)$ to $\mathbb{R}$
$L : Y \rightrightarrows Y$	Level-set mapping
t.v.s.	Real topological vector (linear) space
n.v.s.	Real normed (vector) space
l.c.s.	Locally convex topological vector space
l.s.c.	Lower semicontinuous
u.s.c.	Upper semicontinuous
w.l.o.g.	Without loss of generality

Contents

List of Figures

Chapter 1
Introduction

Set-valued optimization is a vibrant and expanding branch of applied mathematics that deals with optimization problems where the objective map and/or the constraint maps are set-valued maps acting between abstract spaces. Since the notion of set-valued maps subsumes single-valued maps, set-valued optimization provides an important generalization and unification of scalar as well as vector optimization problems. Therefore, this relatively new discipline has justifiably attracted a great deal of attention in recent years.

Set-valued optimization benefits enormously from the recently-developed techniques of nonsmooth and variational analysis, and it not only provides elegant proofs of known results in scalar, vector, and nonsmooth optimization, but it also gives a new spectrum of powerful techniques to these vital branches of applied mathematics. Even more importantly, there are many research domains that directly lead to optimization problems which can most satisfactorily be studied in the unified framework provided by set-valued optimization. For instance, duality principles in vector optimization, gap functions for vector variational inequalities, inverse problems for partial differential equations and variational inequalities, fuzzy optimization, image processing, viability theory, and mathematical economics all lead to optimization problems that can be conveniently cast as set-valued optimization problems. Moreover, since set-valued maps appear naturally in many branches of pure and applied mathematics, set-valued optimization has the evident potential to remain an important and active research topic in both the near and foreseeable future.

1.1 Motivating Examples

In the following, we collect some examples that depict some of the natural occurrences of set-valued optimization problems. We begin with the following example emerging in game theory:

A.A. Khan et al., *Set-valued Optimization*, Vector Optimization,
DOI 10.1007/978-3-642-54265-7_1

Example 1.1.1 (Game Theory).

In this book we deal with set order relations like those defined below. Let $C \subset Y$ be a proper convex and pointed cone in a linear space Y. Furthermore, let $A, B \subset Y$ be arbitrarily chosen sets. Then the **lower set less order relation** $\preceq_C^l$ is defined by (see Definition 2.6.9)

$$A \preceq_C^l B \iff A + C \supset B \iff \forall b \in B \; \exists a \in A : a \leq_C b.$$

Moreover, we will also use the **upper set less order relation** $\preceq_C^u$ defined by (see Definition 2.6.9)

$$A \preceq_C^u B \iff A \subset B - C \iff \forall a \in A \; \exists b \in B : a \leq_C b.$$

These set order relations can be applied to problems in **game theory**. We consider a **vector-valued two-person game**. Let M and N be nonempty sets, Y a linear space and let $f : M \times N \longrightarrow Y$ be a single-valued map. Suppose that **Player 1** chooses $a \in M$ and **Player 2** next chooses $b \in N$. Then $f(a,b)$ describes the loss for Player 1.

Consider a set-valued map $F : M \rightrightarrows Y$ defined by

$$F(a) := \{f(a,b) \mid b \in N\}.$$

In the case that Player 2 is cooperative toward Player 1, Player 1 may choose the **lower set less order relation** $\preceq_C^l$ and look for $\preceq_C^l$-minimal solutions $\overline{a} \in M$ in the sense of Definition 2.6.19, i.e., Player 1 is looking for $\overline{a} \in M$ with

$$F(a) \preceq_C^l F(\overline{a}) \quad \text{for some} \quad a \in M \Longrightarrow F(\overline{a}) \preceq_C^l F(a).$$

This means that the vector-valued game can be formulated as a set-valued optimization problem, and Player 1 has to solve the following:

$$\preceq_C^l -\text{minimize} \quad F(a), \quad \text{subject to} \quad a \in M. \qquad (SP- \preceq_C^l)$$

When Player 2 is non-cooperative (Player 2 wants Player 1 to lose), Player 1 should choose the **upper set less order relation** $\preceq_C^u$ and look for $\preceq_C^u$-minimal solutions $\overline{a} \in M$ in the sense of Definition 2.6.19. Thus, Player 1 is looking for $\overline{a} \in M$ with

$$F(a) \preceq_C^u F(\overline{a}) \quad \text{for some} \quad a \in M \Longrightarrow F(\overline{a}) \preceq_C^u F(a).$$

Then the vector-valued game can again be formulated as a set-valued optimization where Player 1 is now looking to solve the problem:

$$\preceq_C^u -\text{minimize} \quad F(a), \quad \text{subject to} \quad a \in M. \qquad (SP- \preceq_C^u)$$

Further interesting applications of similar set-valued approaches are given in the paper by Fernandez, Hinojosa and Puerto [183].

To further illustrate the wide range of set-valued optimization problems, we now examine their connection with duality principles in vector optimization and gap functions for vector variational inequalities.

Example 1.1.2 (Duality Principles). Consider the following vector optimization problem:

$$\text{minimize}_C\ f(x) \qquad \text{subject to} \quad x \in S = \{x \in M \mid -g(x) \in K\} \tag{P}$$

where $M \subset \mathbb{R}^n$, $C \subset \mathbb{R}^p$ and $K \subset \mathbb{R}^m$ are all pointed closed convex cones, and where $f : \mathbb{R}^n \to \mathbb{R}^p$ and $g : \mathbb{R}^n \to \mathbb{R}^m$ are single-valued mappings. Let $\mathscr{L}$ be the family of $p \times m$ matrices Λ such that $\Lambda K \subset C$ and let $L : M \times \mathscr{L} \to \mathbb{R}^p$ be the **vector-valued Lagrangian** given by $L(x, \Lambda) = f(x) + \Lambda g(x)$. We define a set-valued map $\Phi : \mathscr{L} \rightrightarrows \mathbb{R}^p$ as follows:

$$\Phi(\Lambda) = \text{Min}(\{L(x, \Lambda) \mid x \in M\}, C).$$

Then the dual problem associated with (P) is the following set optimization problem:

$$\text{maximize}_C\ \Phi(\Lambda) \quad \text{subject to} \quad \Lambda \in \mathscr{L}. \tag{D}$$

We derive duality statements for vector optimization problems using a set-valued approach in detail in Sect. 15.1. Specifically, we discuss set-valued approaches to duality for linear vector optimization and give some applications in mathematical finance.

Example 1.1.3 (Gap Functions). Let X and Y be normed spaces and let $\mathscr{L}(X, Y)$ be the set of all linear continuous operators from X to Y. Let $Q \subset X$ be a nonempty closed convex set, let $Q_0 \subset X$ be a pointed closed convex cone and let $\Gamma : X \to Y$ be a given map.

Consider the following vector variational inequality (VVI, for short) : Find $\bar{x} \in Q$ such that

$$\langle \Gamma(\bar{x}), x - \bar{x} \rangle \notin -Q_0 \backslash \{0\} \quad \text{for every} \quad x \in Q.$$

We define a set-valued mapping $\Phi : X \rightrightarrows Y$ by

$$\Phi(x) = \text{Max}(\{\langle \Gamma(\bar{x}), x - Q \rangle,\ x \in Q\}, Q_0).$$

Then the above VVI is equivalent to the following set optimization problem:

$$\text{minimize}_{Q_0}\ \Phi(x) \quad \text{subject to} \quad x \in Q.$$

Robust optimization has become very popular in the last few years, possibly because of its many applications and its close approximation to real-world scenarios. Usually when solving an optimization problem, all parameters are given. Instead, it might happen that only an estimated value or a set of possible parameter values are available. This is especially true in problems of traffic optimization, scheduling, portfolio optimization, network flow and network design.

On the other hand, many real-world optimization problems instead require the minimization of multiple conflicting objectives, e.g. the maximization of expected return versus the minimization of risk in portfolio optimization, the minimization of production time versus the minimization of manufacturing equipment cost, or the maximization of tumor control versus the minimization of normal tissue complication in radiotherapy treatment design.

Example 1.1.4 (Robust Optimization). Robust optimization can be applied to problems where a solution is required that hedges against the worst case that may happen. For example, consider the problem of placing a landing pad for rescue helicopters at a ski resort. The location of the landing pad should be chosen in such a way that the flight time to all the resort's protected slopes is minimized in the worst case, even though flight times are uncertain due to unknown future weather conditions. Similarly, in determining the flight schedule of an airline, one would want to be able to serve as many passengers as possible in a cost-effective manner, even though the exact number of passengers will not be known when the schedule is fixed.

A concept of robustness for vector-valued optimization problems was introduced and studied in a paper by Ehrgott, Ide and Schöbel [159]. Using set order relations, we derive and explore new concepts of robustness in Sect. 15.4.

Example 1.1.5 (Stochastic/Fuzzy Programming).

Another rich source for set optimization problems is the situation in mathematical programming when the objective and the constraint functions cannot be assigned exact values, but are allowed to vary in a specified range. This leads to set-valued data, for example, those found in stochastic and fuzzy programming. In fuzzy programming, this fact is usually characterized by a membership function, but in the general setting of set optimization, the use of membership functions can be avoided as whole sets can be taken into account.

We also note that set optimization makes several other interesting appearances in sensitivity analysis for vector optimization problems.

1.2 Book Structure

The organization of this book is as follows:

Chapter 2 begins with an introduction to order relations and cone properties and then gives a detailed overview of solution concepts in vector-valued as well as set-valued optimization. We introduce and discuss the following solution concepts for set-valued optimization problems:

- solution concepts based on a vector approach,
- solution concepts based on a set approach,
- solution concepts based on lattice structure.

Furthermore, we present the embedding approach of Kuroiwa and show how it is possible to transform a set-valued optimization problem into a vector optimization problem. Solution concepts for set-valued optimization problems with respect to abstract preference relations and for set-valued problems with variable order structure are also studied. Subsequently, we introduce approximate solutions of set-valued optimization problems, and finally, the relationships between the different solution concepts are studied.

In Chap. 3, we present continuity notions for set-valued mappings and corresponding properties under certain convexity assumptions. Additionally, we introduce Lipschitz properties for both single-valued and set-valued maps as well as recall some concepts of generalized differentiability and corresponding calculus rules.

Chapter 4 is a thorough introduction to tangent cones of first-order and tangent sets and tangent cones of second-order. This chapter includes definitions, properties, and examples of some of the most commonly used first-order tangent cones and second-order tangent sets and cones. Specifically, we study six of the most commonly used first-order tangent cones and six of their modified analogues as well as give numerous equivalent characterizations for a selection of them. We also study the notion of the local cone approximation from an abstract point of view, illustrating not only the key features of tangent cones, but also suggesting new ways for their generation. Furthermore, we investigate convex cones which lie between the contingent cone and the Clarke's tangent cone: two of the most popular tangent cones of first-order. In this discussion, we also study recession tangent cones, the Treiman tangent cone, the prototangent cone, and the quasi-strict tangent cone. We go on to study six of the most commonly used second-order tangent sets and a related notion of second-order generalized tangent sets. We examine, in sufficient detail, second-order asymptotic tangent cones and second-order projective tangent cones as well as collect their intrinsic features. Useful connections among various tangent cones and tangent sets are also thoroughly investigated. A brief description of some higher-order tangent cones and higher-order tangent sets is likewise provided.

In Chap. 5 we introduce nonlinear scalarization methods, which are techniques that are important from both a theoretical and computational point of view. We introduce several different scalarizing functionals and discuss their properties; particularly we examine their monotonicity, continuity, Lipschitz continuity, sublinearity, and convexity. Using these nonlinear functionals, we provide nonconvex separation theorems and also derive necessary optimality conditions for solutions of set-valued optimization problems (Sect. 12.8). Generally, these functionals prove useful in a variety of proofs, but especially in the proof of minimal point theorems in Chap. 10. Continuing, we study characterizations of the solutions of set-valued optimization problems using nonlinear scalarizing functionals. Finally, we present the extremal principle of Kruger and Mordukhovich and discuss its relationship to the separation properties of nonconvex sets. This extremal principle is later applied in Sect. 12.9 to derive a subdifferential variational principle for set-valued mappings, and again in Sect. 12.11 to prove a first order necessary condition for fully-localized minimizers of set-valued optimization problems.

In Chap. 6, we present a generalization of the Hahn–Banach–Kantorovich extension theorem to K-convex set-valued maps as well as Yang's extension theorem. We also give classical separation theorems for convex sets, the core convex topology on a linear space, and a criterion for the convexity of the cone generated by a set.

To each type of efficiency for optimization problems, it is possible to associate notions of a conjugate and subdifferential for either vector valued functions or set-valued maps. Thus Chap. 7 studies the conjugate and subdifferential corresponding to the notion of strong efficiency as well as the subdifferentials corresponding to weak and Henig-type efficiencies. For the strong conjugate and subdifferential, we establish results similar to the convex scalar case. For the other types of subdifferential, we establish formulas for the subdifferentials of the sum and composition of both functions and set-valued maps.

Chapter 8 contains presentations of duality assertions for set-valued optimization problems in infinite dimensional spaces where the solution concept is based on a vector, set, or lattice approach. We give conjugate duality statements for problems where the solution concept is based on a vector approach. The notions of conjugate maps and subdifferentials as well as a perturbation approach used for deriving these duality assertions are also all provided. A Lagrange duality for set-valued problems based on the vector approach is also given. Furthermore, we consider set-valued optimization problems where the solution concept is given by a set order relation (introduced by Kuroiwa) and derive several corresponding saddle point assertions. For set-valued problems where the solution concept is based on a lattice structure, we present duality theorems that are based on the consequent use of the infimum and supremum (in the sense of greatest lower and least upper bounds with respect to a partial ordering). We derive conjugate duality assertions as well as Lagrange duality statements. A comparison of different approaches to duality in set-valued optimization is then given at the end of the chapter.

In Chap. 9 we establish several existence results for minimal points with respect to transitive relations and then apply them in topological vector spaces for quasi-orders generated by convex cones. We continue with the presentation of several types of convex cones along with related notions of compactness. We end the chapter with some existence results for vector and set optimization problems.

Chapter 10 presents existence results for the minimal points of subsets of the Cartesian product of a complete metric space and a topological vector space with respect to order relations determined by generalized set-valued metrics. Such results are in turn useful for deriving EVP type results for vector and scalar functions. We then derive EVP results of Ha's type as well as an EVP result for bi-set-valued maps.

We end the chapter with an application to error bounds for set-valued optimization problems.

Chapter 11 is devoted to the derivatives of set-valued maps, which are motivated by the geometric interpretation of the classical derivative for single-valued maps as a local approximation of its graph. We study the contingent derivative, the contingent epiderivative, and examine several analogues of these notions such as the adjacent derivative, the Clarke derivative, the lower Dini derivative, and the radial derivative. We also investigate their second-order analogues: the second-order contingent derivative, the second-order adjacent derivative, the second-order lower Dini derivative, and their epiderivative counterparts. We give several calculus rules for the first-order derivatives and epiderivatives along with their second-order extensions. The existence of contingent epiderivatives is a vital issue in set-valued optimization and we subsequently present a comprehensive study pertaining to the existence of contingent epiderivatives, their generalized analogues, and their various characterizations. We also study other important issues such as proto-differentiability and semi-differentiability along with several other related notions. Numerous additional concepts which play a critical role in the differentiability of set-valued maps are also explored in depth.

In Chap. 12, we give a wide variety of optimality conditions in set-valued optimization. To be specific, we give necessary first-order and second-order optimality conditions in set-valued optimization by a direct approach using graphical derivatives and epiderivatives. We also necessary first-order and second-order optimality conditions in set-valued optimization by the Dubovitskii-Milyutin approach using graphical derivatives and epiderivatives. Sufficient optimality conditions using graphical derivatives and epiderivatives are also given. A brief discussion of optimality conditions for Q-minimizers using graphical derivatives and coderivatives is also made. We describe the necessary optimality conditions for set-valued optimization problems in Asplund spaces using the limiting subdifferential of Mordukhovich, where the solution concept is based on a vector approach. The given results do not assume convexity concerning the objective set-valued map or feasible set, nor do they assume that the ordering cone of the image space has a nonempty interior. In order to prove the optimality conditions, we use the nonlinear scalarization method presented in Sect. 5.2. Furthermore, we derive necessary conditions for Q-minimizers of the set-valued optimization problem in Banach spaces using Clarke's coderivative and Ioffe's approximate coderivative. In the proof of the necessary optimality condition, we use the scalarizing functional by Hiriart-Urruty (introduced in Sect. 5.3). Moreover, we also detail necessary conditions for approximate solutions of a set-valued optimization problem in Asplund spaces using the extremal principle of variational analysis given in Sect. 5.5. Using a directional derivative of the set-valued map, we develop necessary and sufficient conditions for solutions described by the set approach. We also present necessary conditions for those solutions introduced in Sect. 2.6.5 that are related to applications in welfare economics (see Sect. 15.3). The main tool for the proof of these necessary conditions is the extended extremal principle given in Theorem 5.5.10. Due to its importance for the derivation of algorithms, we also consider set-valued optimization problems from the point of view of the stability of *KKT* points. Corresponding results for a vector approach in finite dimensional spaces are likewise given.

Chapter 13 addresses sensitivity analysis for a variety of problems. Here we study first-order and second-order sensitivity analysis of set-valued optimization problems. We introduce several set-valued perturbation maps and investigate their differentiability properties and give related results. Sensitivity analysis is done by using both the graphical derivatives such as the contingent derivatives and the lower Dini derivatives as well as by using the coderivatives. In this chapter, we will also study differentiability properties of set-valued gap functions associated to vector variational inequalities and sensitivity analysis for vector variational inequalities. For vector variational inequalities, both the first-order as well as the second-order results are given.

In Chap. 14, we introduce several solution procedures for set-valued optimization problems. Section 14.1 gives a Newton method for solving general set-valued optimization problems. We also present an algorithm for a special class of set-valued optimization problems where the objective map is polyhedral, convex, and set-valued (Sect. 14.2). We further discuss the noteworthy relationship between multiobjective optimization problems with uncertainties and set-valued optimization and then go on to present solution methods for these special set-valued optimization problems (see Sect. 15.4 and particularly Sect. 15.4.4).

Chapter 15 develops several applications of set-valued optimization. We use the solution concept based on a lattice approach to derive a useful dual problem to a given vector optimization problem and subsequently to obtain corresponding duality assertions. This approach is characterized by an embedding of the image space of the vector optimization problem into a complete lattice. The results are discussed for the linear case, and then some applications of the duality results to mathematical finance are given. Specifically, we study a Markowitz-type bicriterial portfolio optimization problem where the expected return of the portfolio should be maximized and the risk of the portfolio, measured by the Conditional Value at Risk, should be minimized. Moreover, we give an interpretation of the corresponding dual problem in the financial sense. In addition, we apply a variational approach and necessary conditions for set-valued optimization problems to derive extensions of the second fundamental theorem of welfare economics to nonconvex economies with general preference relations.

Finally, the last chapter deals with uncertainty in vector optimization, a topic which is of obvious importance in many applications. Most real-world optimization problems are contaminated with uncertain data and some parameters are either not known or not known exactly. In this chapter we apply solution concepts for set-valued optimization problems based on a set approach, particularly making use of the set order relations introduced by Kuroiwa in order to study robustness for vector-valued optimization problems. The presented algorithms can be considered as algorithms for solving special classes of set-valued optimization problems where the solution concept is given by different set order relations.

1.3 Useful Notation

We conclude this chapter by introducing some notation:

Let X, Y be arbitrary nonempty sets. A function $F : X \to 2^Y$ is called a **set-valued map**, and is denoted by $F : X \rightrightarrows Y$. So, if $F : X \rightrightarrows Y$ is a set-valued map, the image of F at $x \in X$ is a (possibly empty) subset $F(x)$ of Y. The **domain** of F is

$$\operatorname{dom} F := \{x \in X \mid F(x) \neq \emptyset\},$$

while its **image** is

$$\operatorname{Im} F := \{y \in Y \mid \exists x \in X \ : \ y \in F(x)\}.$$

The set-valued map $F : X \rightrightarrows Y$ is usually identified with its graph,

$$\operatorname{graph} F := \{(x, y) \in X \times Y \mid y \in F(x)\}.$$

In this way, with each set-valued map one associates a relation and vice versa. It is obvious that

$$\operatorname{dom} F = \operatorname{Pr}_X(\operatorname{graph} F)$$

and

$$\operatorname{Im} F = \operatorname{Pr}_Y(\operatorname{graph} F),$$

where $\operatorname{Pr}_X(\operatorname{graph} F)$ denotes the projection of graph F onto X and $\operatorname{Pr}_Y(\operatorname{graph} F)$ the projection of graph F onto Y. The image by F of the set $A \subset X$ is

$$F(A) := \bigcup_{x \in A} F(x);$$

so, $\operatorname{Im} F = F(X)$. The inverse image by F of the set $B \subset F$ is

$$F^{-1}(B) := \{x \in X \mid F(x) \cap B \neq \emptyset\}.$$

In fact, $F^{-1}(B)$ is the image of B given by the inverse set-valued map $F^{-1} : Y \rightrightarrows X$ defined by

$$F^{-1}(y) := \{x \in X \mid y \in F(x)\};$$

so $F^{-1}(y) = F^{-1}(\{y\})$. Of course, $\operatorname{dom} F^{-1} = \operatorname{Im} F$ and $\operatorname{Im} F^{-1} = \operatorname{dom} F$. We shall also use another type of inverse image: $F^{+1}(B) := \{x \in X \mid F(x) \subset B\}$.

When $\Delta : X \rightrightarrows W$ is another set-valued map, the **composition** of Δ and F is the set-valued map $\Delta \circ F : X \rightrightarrows W$ defined by $(\Delta \circ F)(x) := \{z \in W \mid \exists\, y \in F(x)$ with $z \in \Delta(y)\}$; note that

$$\operatorname{graph}(\Delta \circ F) = \operatorname{Pr}_{X\times W}\big(\operatorname{graph} F \times W \cap X \times \operatorname{graph} \Delta\big).$$

When Y is a linear space and $F, F_1, F_2 : X \rightrightarrows Y$, we define the sum $F_1 + F_2$ and multiplication by a scalar γF as the set-valued maps $F_1 + F_2, \gamma F : X \rightrightarrows Y$ defined by

$$(F_1 + F_2)(x) := F_1(x) + F_2(x),$$

$$(\gamma F)(x) := \gamma \cdot F(x)$$

with the usual conventions that $A + \emptyset := \emptyset + A := \emptyset$, $\gamma \cdot \emptyset = \emptyset$ for $A \subset Y$.

Furthermore, we introduce some useful notions and notation related to vector-valued functions. To Y, quasiordered by the proper convex cone C, we adjoin the greatest and the smallest elements $\infty := +\infty$, $-\infty$ ($\notin Y$), thereby obtaining $Y^\bullet := Y \cup \{-\infty, +\infty\}$. We consider that $y + \infty = \infty$, $\lambda \cdot \infty = \infty$, and $-\infty \leq_C y \leq_C \infty$ for all $y \in Y^\bullet$ and $\lambda \in \mathbb{R}_+$. Of course, if $f : X \to Y^\bullet$, the **domain** of f is

$$\operatorname{dom} f := \{x \in X \mid f(x) \in Y \cup \{-\infty\}\},$$

the **sublevel** and **strict sublevel sets of** f **of height** $y \in Y$ are $\operatorname{lev}_f(y) := \{x \in X \mid f(x) \leq_C y\}$ and $\operatorname{lev}_f^<(y) := \{x \in X \mid y - f(x) \in \operatorname{int} C\}$, and the **epigraph** of f is $\operatorname{epi} f := \{(x, y) \in X \times Y \mid f(x) \leq_C y\}$. With such an f we associate the set-valued map $F_{f,C} : X \rightrightarrows Y$ whose graph is $\operatorname{epi} f$. So, $F_{f,C}(x) = f(x) + C$ for every $x \in \operatorname{dom} F_{f,C} = \operatorname{dom} f$, $\operatorname{epi} f = \operatorname{epi} F_{f,C}$, $\operatorname{lev}_f(y) = \operatorname{lev}_{F_{f,C}}(y)$, and $\operatorname{lev}_f^<(y) = \operatorname{lev}_{F_{f,C}}^<(y)$ for every $y \in Y$; in particular, $F_{f,C} = \big(F_{f,C}\big)_C$.

Chapter 2
Order Relations and Ordering Cones

In this chapter, first, we give an introduction to order relations and cone properties. Then we present a detailed overview of solution concepts in vector-valued as well as set-valued optimization. We introduce and discuss the following solution concepts for set-valued optimization problems:

- solution concepts based on vector approach,
- solution concepts based on set approach,
- solution concepts based on lattice structure.

Furthermore, we present the embedding approach by Kuroiwa and show how it is possible to transform a set-valued optimization problem into a vector optimization problem using this embedding approach. Solution concepts for set-valued optimization problems with respect to abstract preference relations and for set-valued problems with variable order structure are studied. Moreover, we introduce approximate solutions of set-valued optimization problems. Finally, relationships between different solution concepts are studied.

2.1 Order Relations

In this section, our objective is to study some useful order relations. We begin by recalling that given a nonempty set M, by $M \times M$ we represent the set of ordered pairs of elements of M, that is,

$$M \times M := \{(x_1, x_2) \mid x_1, x_2 \in M\}.$$

The following definition gives the notion of an order relation.

Definition 2.1.1. Let M be a nonempty set and let $\mathscr{R}$ be a nonempty subset of $M \times M$. Then $\mathscr{R}$ is called an **order relation** (or a **binary relation**) on M and the

A.A. Khan et al., *Set-valued Optimization*, Vector Optimization,
DOI 10.1007/978-3-642-54265-7_2

pair $(M, \mathscr{R})$ is called a set M with **order relation** $\mathscr{R}$. The containment $(x_1, x_2) \in \mathscr{R}$ will be denoted by $x_1 \mathscr{R} x_2$. The order relation $\mathscr{R}$ is called:

(a) **reflexive** if for every $x \in M$, we have $x \mathscr{R} x$;
(b) **transitive** if for all $x_1, x_2, x_3 \in M$, the relations $x_1 \mathscr{R} x_2$ and $x_2 \mathscr{R} x_3$ imply that $x_1 \mathscr{R} x_3$;
(c) **antisymmetric** if for all $x_1, x_2 \in M$, the relations $x_1 \mathscr{R} x_2$ and $x_2 \mathscr{R} x_1$ imply that $x_1 = x_2$.

Moreover, an order relation $\mathscr{R}$ is called a **preorder** on M if $\mathscr{R}$ is transitive, a **quasiorder** if $\mathscr{R}$ is reflexive and transitive and a **partial order** on M if $\mathscr{R}$ is reflexive, transitive, and antisymmetric. In all the three cases, the containment $(x_1, x_2) \in \mathscr{R}$ is denoted by $x_1 \leq_{\mathscr{R}} x_2$, or simply by $x_1 \leq x_2$ if there is no risk of confusion. The binary relation $\mathscr{R}$ is called a **linear** or **total order** if $\mathscr{R}$ is a partial order and any two elements of M are **comparable**, that is

(d) for all $x_1, x_2 \in M$ either $x_1 \leq_{\mathscr{R}} x_2$ or $x_2 \leq_{\mathscr{R}} x_1$.

Furthermore, if each nonempty subset M' of M has a first element x' (meaning that $x' \in M'$ and $x' \leq_{\mathscr{R}} x \ \forall x \in M'$), then M is called **well-ordered**.

We recall Zermelo's theorem: For every nonempty set M there exists a partial order $\mathscr{R}$ on M such that $(M, \mathscr{R})$ is well-ordered.

An illustrative example of a relation is $\Delta_M := \{(x, x) \mid x \in M\}$ which is reflexive, transitive, and antisymmetric, but it satisfies (d) only when M is a singleton.

We recall that the **inverse** of the relation $\mathscr{R} \subset M \times M$ is the relation

$$\mathscr{R}^{-1} := \{(x_1, x_2) \in M \times M \mid (x_2, x_1) \in \mathscr{R}\},$$

and if $\mathscr{S}$ is a relation on M, then the **composition** of $\mathscr{R}$ and $\mathscr{S}$ is the relation

$$\mathscr{S} \circ \mathscr{R} := \{(x_1, x_3) \mid \exists x_2 \in M \mid (x_1, x_2) \in \mathscr{R},\ (x_2, x_3) \in \mathscr{S}\}.$$

Using these two notations, the conditions (a), (b), (c), and (d) are equivalent to $\Delta_M \subset \mathscr{R}, \mathscr{R} \circ \mathscr{R} \subset \mathscr{R}, \mathscr{R} \cap \mathscr{R}^{-1} \subset \Delta_M$ and $\mathscr{R} \cup \mathscr{R}^{-1} = M \times M$, respectively.

Definition 2.1.2. Let $\mathscr{R}$ be an order relation on the nonempty set M and let $M_0 \subset M$ be nonempty. An element $x_0 \in M_0$ is called a **maximal (minimal) element** of M_0 **relative to** $\mathscr{R}$ if for every $x \in M_0$,

$$x_0 \mathscr{R} x \Rightarrow x \mathscr{R} x_0 \qquad (x \mathscr{R} x_0 \Rightarrow x_0 \mathscr{R} x). \tag{2.1}$$

The collection of all maximal (minimal) elements of M_0 with respect to (w.r.t. for short) $\mathscr{R}$ is denoted by $\mathrm{Max}(M_0, \mathscr{R})$ ($\mathrm{Min}(M_0, \mathscr{R})$).

Note that x_0 is a maximal element of M_0 w.r.t. $\mathscr{R}$ if and only if x_0 is a minimal element of M_0 w.r.t. $\mathscr{R}^{-1}$, and hence $\mathrm{Max}(M_0, \mathscr{R}) = \mathrm{Min}(M_0, \mathscr{R}^{-1})$.

Remark 2.1.3. 1. If the order relation $\mathscr{R}$ in Definition 2.1.2 is antisymmetric, then $x_0 \in M_0$ is maximal (minimal) if and only if for every $x \in M_0$

$$x_0 \mathscr{R} x \Rightarrow x = x_0 \quad (x \mathscr{R} x_0 \Rightarrow x_0 = x). \tag{2.2}$$

2. If $\mathscr{R}$ is an order relation on M and $\emptyset \neq M_0 \subset M$, then $\mathscr{R}_0 := \mathscr{R} \cap (M_0 \times M_0)$ is an order relation on M_0. In such a situation, the set M_0 will always be endowed with the order structure $\mathscr{R}_0$ if not stated explicitly otherwise. If $\mathscr{R}$ is a preorder (partial order, linear order) on M, then $\mathscr{R}_0$ is a preorder (partial order, linear order) on M_0. Therefore, x_0 is a maximal (minimal) element of M_0 relative to $\mathscr{R}$ iff x_0 is a maximal (minimal) element of M_0 relative to $\mathscr{R}_0$.

In the following, we give some examples to illustrate the above notions.

Example 2.1.4. (1) Assume that X is a nonempty set and $M := \mathscr{P}(X)$ represents the collection of subsets of X. Then the order relation $\mathscr{R} := \{(A, B) \in M \times M \mid A \subset B\}$ is a partial order on M. However, if X contains at least two elements, then $\mathscr{R}$ is not a linear order.

(2) Assume that $\mathbb{N}$ is the set of non-negative integers and

$$\mathscr{R}_{\mathbb{N}} := \{(n_1, n_2) \in \mathbb{N} \times \mathbb{N} \mid \exists\, p \in \mathbb{N} \; : \; n_2 = n_1 + p\}.$$

Then $\mathbb{N}$ is well-ordered by $\mathscr{R}_{\mathbb{N}}$. Note that $\mathscr{R}_{\mathbb{N}}$ defines the usual order relation on $\mathbb{N}$, and $n_1 \mathscr{R}_{\mathbb{N}} n_2$ will always be denoted by $n_1 \leq n_2$ or, equivalently, $n_2 \geq n_1$.

(3) Let $\mathbb{R}$ be the set of real numbers and let $\mathbb{R}_+ := [0, \infty[$ be the set of non-negative real numbers. The usual order relation on $\mathbb{R}$ is defined by

$$\mathscr{R}_1 := \{(x_1, x_2) \in \mathbb{R} \times \mathbb{R} \mid \exists\, y \in \mathbb{R}_+ \; : \; x_2 = x_1 + y\}.$$

Then $\mathscr{R}_1$ is a linear order on $\mathbb{R}$, but $\mathbb{R}$ is not well-ordered by $\mathscr{R}_1$. In the following, the fact $x_1 \mathscr{R}_1 x_2$ will always be denoted by $x_1 \leq x_2$ or, equivalently, $x_2 \geq x_1$.

(4) Given $n \in \mathbb{N}$, $n \geq 2$, we consider the binary relation $\mathscr{R}_n$ on $\mathbb{R}^n$ defined by

$$\mathscr{R}_n := \{(x, y) \in \mathbb{R}^n \times \mathbb{R}^n \mid \forall\, i \in \overline{1, n} \; : \; x_i \leq y_i\},$$

where $x = (x_1, \ldots, x_n)$, $y = (y_1, \ldots, y_n)$ and $\overline{1, n} := \{i \in \mathbb{N} \mid 1 \leq i \leq n\}$. Then $\mathscr{R}_n$ is a partial order on $\mathbb{R}^n$, but $\mathscr{R}_n$ is not a linear order. For example, the elements e_1 and e_2 are not comparable w.r.t. $\mathscr{R}_n$, where $e_i := (0, \ldots, 0, 1, 0, \ldots, 0) \in \mathbb{R}^n$. As usual, by e_i we denote the vector whose entries are all 0 except the ith one, which is 1).

Remark 2.1.5. Every well-ordered subset W of $\mathbb{R}$ (equipped with its usual partial order defined above) is at most countable. Indeed, every element $y \in W$, except the greatest element w of W (provided that it exists), has a successor $s(y) \in W$. Clearly, if $y, y' \in W$, $y < y'$, then $s(y) \leq y'$. Therefore, fixing $q_y \in \mathbb{Q}$ such that

$y < q_y < s(y)$ for $y \in W \setminus \{w\}$, we get an injective function from $W \setminus \{w\}$ into $\mathbb{Q}$, and so W is at most countable.

We emphasize that even when $\mathscr{R}$ is a partial order on M, a nonempty subset M_0 of M may have zero, one, or several maximal elements, but if $\mathscr{R}$ is a linear order, then every subset has at most one maximal (minimal) element.

Definition 2.1.6. Let $\emptyset \neq M_0 \subset M$ and let $\mathscr{R}$ be an order relation on M. Then:

1. M_0 is **lower (upper) bounded** (w.r.t. $\mathscr{R}$) if there exists $a \in M$ such that $a\mathscr{R}x$ ($x\mathscr{R}a$) for every $x \in M_0$. In this case, the element a is called a **lower (upper) bound** of M_0 (w.r.t. $\mathscr{R}$).
2. If, moreover, $\mathscr{R}$ is a partial order, we say that $a \in M$ is the **infimum (supremum)** of M_0 if a is a lower (upper) bound of M_0 and for any lower (upper) bound a' of M_0 we have that $a'\mathscr{R}a$ ($a\mathscr{R}a'$).

In set-valued optimization, the existence of maximal elements w.r.t. order relations is an important problem. For this, the following **Zorn's lemma** (or Zorn's axiom) plays a crucial role.

Axiom 2.1.7 (Zorn) *Let $(M, \leq)$ be a reflexively preordered set. If every nonempty totally ordered subset of M is upper bounded, then M has maximal elements.*

We recall that given a linear space X, a nonempty set $M \subset X$ is **affine** (or a **linear manifold**) if $\lambda x + (1-\lambda)y \in M$ for all $x, y \in M$ and $\lambda \in \mathbb{R}$. A nonempty set C of X is called **convex** if $[x, y] := \{\lambda x + (1-\lambda)y \mid \lambda \in [0,1]\} \subset C$ for all $x, y \in C$. By convention the empty set $\emptyset$ is considered to be affine and convex. It is obvious that a linear subspace is affine and an affine set is convex. Moreover, any intersection of linear subspaces, affine sets, or convex sets is a linear subspace, an affine set, or a convex set, respectively. These properties allow us to introduce the **linear hull**, the **affine hull**, and the **convex hull** of a nonempty set $A \subset X$ as being, respectively,

$$\operatorname{lin} A := \bigcap \{Y \subset X \mid A \subset Y,\ Y \text{ linear subspace}\},$$

$$\operatorname{aff} A := \bigcap \{M \subset X \mid A \subset M,\ M \text{ linear manifold}\},$$

$$\operatorname{conv} A := \bigcap \{C \subset X \mid A \subset C,\ C \text{ convex set}\}.$$

Clearly, for $X = \mathbb{R}^n$ and $\mathscr{R} = \mathscr{R}_n$ (from Example 2.1.4 (4)), we have

$$\forall\, x_1, x_2 \in X,\ \forall\, \lambda \in \mathbb{R}\ :\ x_1\mathscr{R}x_2,\ 0 \leq \lambda \Rightarrow \lambda x_1 \mathscr{R} \lambda x_2, \tag{2.3}$$

$$\forall\, x_1, x_2, x \in X\ :\ x_1\mathscr{R}x_2 \Rightarrow (x_1 + x)\mathscr{R}(x_2 + x). \tag{2.4}$$

It is easy to find examples of relations satisfying (2.4). In fact, a nonempty relation $\mathscr{R}$ on the linear space X satisfies (2.4) if and only if there exists $\emptyset \neq D \subset X$ such that $\mathscr{R} = \mathscr{R}_D$, where

$$\mathscr{R}_D := \{(x_1, x_2) \in X \times X \mid x_2 - x_1 \in D\}.$$

Moreover, $\mathscr{R}_D$ is reflexive if and only if $0 \in D$, and $\mathscr{R}_D$ is transitive if and only if $D + D \subset D$.

Definition 2.1.8. Let $\mathscr{R}$ be an order relation on the linear space X; we say that $\mathscr{R}$ is **compatible** with the linear structure of X if (2.3) and (2.4) hold.

In linear spaces, a large number of relations $\mathscr{R}$ can be defined by cones which are compatible with the linear structure of the space. For this we first give the following:

Definition 2.1.9. A nonempty set $C \subset X$ is a **cone** if for every $x \in C$ and for every $\lambda \in \mathbb{R}_+$ we have $\lambda x \in C$. Clearly, if C is a cone, then $0 \in C$. The cone C is called

(a) **convex** if for all $x_1, x_2 \in C$ we have $x_1 + x_2 \in C$,
(b) **nontrivial** or **proper** if $C \neq \{0\}$ and $C \neq X$,
(c) **reproducing** if $C - C = X$,
(d) **pointed** if $C \cap (-C) = \{0\}$.

Clearly, the cone C satisfies condition (b) in the definition above iff, C is a convex set.

In the following, we collect a few examples of cones.

Example 2.1.10. (1) Let

$$\mathbb{R}^n_+ := \{x \in \mathbb{R}^n \mid x_i \geq 0 \; \forall\, i \in \overline{1,n}\} = \{x \in \mathbb{R}^n \mid (0, x) \in \mathscr{R}_n\}. \tag{2.5}$$

$\mathbb{R}^n_+$ is obviously a cone in the linear space $\mathbb{R}^n$, which fulfills all the conditions of Definition 2.1.9.

(2) Let $C[0,1]$ be the linear space of all real functions defined and continuous on the interval $[0,1] \subset \mathbb{R}$. Addition and multiplication by scalars are defined, as usual, by

$$(x + y)(t) = x(t) + y(t), \quad (\lambda x)(t) = \lambda x(t) \quad \forall\, t \in [0,1]$$

for $x, y \in C[0,1]$ and $\lambda \in \mathbb{R}$. Then

$$C_+[0,1] := \{x \in C[0,1] \mid x(t) \geq 0 \; \forall\, t \in [0,1]\} \tag{2.6}$$

is a convex, nontrivial, pointed, and reproducing cone in $C[0,1]$. Note that the set

$$Q := \{x \in C_+[0,1] \mid x \text{ is nondecreasing}\} \tag{2.7}$$

is also a convex, nontrivial, and pointed cone in the space $C[0,1]$, but it doesn't satisfy condition (c) from Definition 2.1.9: $Q - Q$ is the proper linear subspace of all functions with bounded variation of $C[0,1]$.

(3) Consider the set $C \subset \mathbb{R}^n$ defined by

$$\begin{aligned} C := \{x = (x_1, \ldots, x_n)^T \in \mathbb{R}^n \mid x_1 > 0, \text{ or} \\ x_1 = 0,\ x_2 > 0, \text{ or} \\ \cdots \\ x_1 = \cdots = x_{n-1} = 0,\ x_n > 0, \text{ or} \\ x = 0\}. \end{aligned}$$

Then the cone C satisfies all the conditions of Definition 2.1.9.

In the following result, we characterize compatibility between linear and order relations:

Theorem 2.1.11. *Let X be a linear space and let C be a cone in X. Then the relation*

$$\mathscr{R}_C := \{(x_1, x_2) \in X \times X \mid x_2 - x_1 \in C\} \tag{2.8}$$

is reflexive and satisfies (2.3) and (2.4). Moreover, C is convex if and only if $\mathscr{R}_C$ is transitive, and, respectively, C is pointed if and only if $\mathscr{R}_C$ is antisymmetric. Conversely, if $\mathscr{R}$ is a reflexive relation on X satisfying (2.3) and (2.4), then $C := \{x \in X \mid 0\mathscr{R}x\}$ is a cone and $\mathscr{R} = \mathscr{R}_C$.

Proof. See [214, Theorem 2.1.13]. □

The above result shows that when $\emptyset \neq C \subset X$, the relation $\mathscr{R}_C$ defined by (2.8) is a reflexive preorder iff C is a convex cone, and $\mathscr{R}_C$ is a partial order iff C is a pointed convex cone.

We note that $\mathscr{R}_{\mathbb{R}^n_+} = \mathscr{R}_n$ (defined in Example 2.1.4 (4)), while the relation $\mathscr{R}_C$ with $C \subset \mathbb{R}^n$ defined in Example 2.1.10 (3) is a linear order, called the **lexicographic order** on $\mathbb{R}^n$.

Let Y be a linear topological space, partially ordered by a proper pointed convex closed cone $C \subset Y$.

Denote this order by "$\leq_C$". Its ordering relation is described by

$$y_1 \leq_C y_2 \quad \text{if and only if} \quad y_2 - y_1 \in C \quad \text{for all } y_1, y_2 \in Y. \tag{2.9}$$

In the sequel, we omit the subscript C as no confusion occurs.

As usual, we denote by

$$C^+ := \{y^* \in Y^* \mid y^*(y) \geq 0\ \forall y \in C\}$$

the continuous positive dual cone of C, and by

$$C^\# := \{y^* \in C^+ \mid y^*(y) > 0\ \forall\, y \in C \setminus \{0\}\}$$

the quasi-interior of C^+.

We recall that the **interior** and the **closure** of the subset A of the topological space (X, τ) are defined, respectively, by

$$\operatorname{int} A := \bigcup\{D \subset X \mid D \subset A,\ D \text{ open}\},$$

$$\operatorname{cl} A := \overline{A} := \bigcap\{B \subset X \mid A \subset B,\ B \text{ closed}\}.$$

Clearly, $\operatorname{int} A$ is open and $\operatorname{cl} A$ is closed.

2.2 Cone Properties Related to the Topology and the Order

We discuss now the connections between **topology** and **order**. Unlike the notion of an ordered linear space (i.e., a linear space equipped with a compatible reflexive preorder), the notion of an ordered topological linear space does not demand for any direct relation between the order and the involved topology. However, because a compatible reflexive preorder on a linear space is defined by a convex cone, it is customary to ask that the cone defining the order be closed, have nonempty interior, or be normal. Before introducing the notion of a normal cone, we recall that a nonempty set A of the linear space X is **full** with respect to the convex cone $C \subset X$ if $A = [A]_C$, where

$$[A]_C := (A + C) \cap (A - C).$$

Note that $[A]_C$ is full w.r.t. C for every set $A \subset X$.

Definition 2.2.1. Let (X, τ) be a t.v.s. and let $C \subset X$ be a convex cone. Then C is called **normal** (relative to τ) if the origin $0 \in X$ has a neighborhood base formed by full sets w.r.t. C.

In the next result we give several characterizations of normal cones. We are using the notation $\mathcal{N}_X$ for the set of balanced neighborhoods of $0 \in X$ in the t.v.s. (X, τ).

Theorem 2.2.2. *Let (X, τ) be a topological linear space and let $C \subset X$ be a convex cone. Then the following statements are equivalent:*

(i) *C is normal,*
(ii) $\forall\, V \in \mathcal{N}_X,\ \exists\, U \in \mathcal{N}_X\ :\ [U]_C \subset V$,
(iii) *for all nets $(x_i)_{i \in I}, (y_i)_{i \in I} \subset X$ such that $0 \leq_C x_i \leq_C y_i$ for every $i \in I$ one has $(y_i) \to 0 \Rightarrow (x_i) \to 0$,*
(iv) *$\operatorname{cl} C$ is normal.*

Proof. See [214, Theorem 2.1.22]. □

The following corollary is immediate.

Corollary 2.2.3. *Let (X, τ) be a Hausdorff t.v.s. and let $C \subset X$ be a convex cone. If C is normal, then* $\operatorname{cl} C$ *is pointed, and so C is pointed, too.*

Let (X, τ) be a Hausdorff t.v.s. partially ordered by the convex cone C. We say that a net $(x_i)_{i\in I} \subset X$ is nonincreasing if

$$\forall\, i, j \in I \;:\; j \succeq i \Rightarrow x_j \leq_C x_i. \tag{2.10}$$

Given $\emptyset \neq A \subset X$, we say that A is **lower bounded with respect to** C if A is lower bounded with respect to $\mathscr{R}_C$ (see Definition 2.1.6). Similarly, $a \in X$ is a lower bound (infimum) of A w.r.t. C if a is so for $\mathscr{R}_C$. Hence $a \in X$ is a lower bound of A w.r.t. C if $a \leq_C x$ for every $x \in A$. An element a is the infimum of A w.r.t. C if a is a lower bound and for any lower bound a' of A we have that $a' \leq_C a$. The infimum of A w.r.t. C will be denoted by $\inf_C A$ when it exists.

Proposition 2.2.4. *Let (X, τ) be a Hausdorff t.v.s. partially ordered by the closed convex cone C. If the net $(x_i)_{i\in I} \subset X$ is nonincreasing and convergent to $x \in X$, then $\{x_i \mid i \in I\}$ is bounded below and $x = \inf\{x_i \mid i \in I\}$.*

Proof. See [214, Proposition 2.1.24]. □

We emphasize that in ordered topological linear spaces, the classical result concerning the bounded monotone sequences is not generally true. We consider the linear space ℓ^∞ of all bounded sequences $x = (x^k)_{k\geq 1} \subset \mathbb{R}$ endowed with the norm $\|x\| = \sup\{|x^k| \mid k = 1, 2, \ldots\}$. In ℓ^∞ we consider the "usual" partial order generated by the cone $\ell^\infty_+ := \{x \in l^\infty \mid x^k \geq 0 \;\forall\, k \geq 1\}$; ℓ^∞_+ is a pointed closed convex cone (even reproducing and with nonempty interior).

Example 2.2.5 (Peressini [475, p. 91]). The sequence $\{x_n\} \subset \ell^\infty$, defined by (for n fixed)

$$x_n^k = \begin{cases} -1 & \text{if } 1 \leq k \leq n, \\ 0 & \text{if } k > n, \end{cases}$$

is nonincreasing w.r.t. C, and $\inf\{x_n \mid n \geq 1\} = e' := -e$ where $e = (1, 1, 1, \ldots) \in \ell^\infty$. But $\|x_n - e'\| = 1$ for every $n \geq 1$. Consequently, $\{x_n\}_{n\geq 1}$ does not converge to its infimum.

We also recall that a cone C that partially orders a Hausdorff linear topological space (X, τ) is said to be **Daniell** if any nonincreasing net having a lower bound τ-converges to its infimum (see Jahn [292, p.29], Luc [402, p. 47], Borwein [67]).

In the following, let us recall some useful notions of cones which play an important role in proving existence results for solutions of optimization problems in infinite dimensional spaces.

Definition 2.2.6. Let Y be a Hausdorff topological vector space and $C \subset Y$ a proper convex cone.

Fig. 2.1 Cone properties

$$\begin{array}{ccc}
C \text{ well-based} & \overset{C=\operatorname{cl} C,\ Y=\mathbb{R}^n}{\Longleftarrow} & C \text{ pointed} \\
\Updownarrow\ Y \text{ normed} & & \Uparrow \\
C \text{ nuclear} & \Longrightarrow & C \text{ normal} \\
\Downarrow\ C \text{ complete} & & \\
C \text{ Daniell} & \Longleftarrow & C \text{ compact base.}
\end{array}$$

(i) C is **based** if there exists a nonempty convex subset B of C such that $C = \mathbb{R}_+ B$ (where $\mathbb{R}_+ B := \{\lambda b \mid b \in B \text{ and } \lambda \geq 0\}$) and $0 \notin \operatorname{cl} B$; the set B is called a **base** for C.

(ii) C is called **well-based** if C has a bounded base.

(iii) Let the topology of Y be defined by a family $\mathscr{P}$ of seminorms. C is called **supernormal** or **nuclear** if for each $p \in \mathscr{P}$ there exists $y^* \in Y^*$, such that $p(y) \leq \langle y, y^* \rangle$ for all $y \in C$; it holds $y^* \in C^+$ in this case.

(iv) C is said to be **Daniell** if any nonincreasing net having a lower bound converges to its infimum.

(iv) C is said to be **regular** if any decreasing (increasing) net which has a lower bound (upper bound) is convergent.

In Fig. 2.1 we give an overview of such additional cone properties and corresponding relations for the case that Y is a Banach space, C a proper and convex cone in Y.

The following result gives useful information for cones with bases:

Theorem 2.2.7. *Let X be a Hausdorff locally convex space and $C \subset X$ a proper convex cone. Then C has a base if and only if $C^{\#} \neq \emptyset$.*

Proof. See [214, Theorem 2.2.12]. □

In the following, we collect a few examples of Daniell cones.

Example 2.2.8. 1. We recall that if $(x_\alpha)_{\alpha \in A}$ is a net which is increasing (decreasing) in a topological vector space (Y, τ) ordered by a closed convex cone C and if $\overline{x}$ is a cluster point of (x_α), then $\overline{x} = \sup_{\alpha \in A} x_\alpha$ ($\overline{x} = \inf_{\alpha \in A} x_\alpha$) (see Peressini [475, Proposition 3.1]). Therefore, any regular cone is Daniell.

2. If $(Y, ||\cdot||)$ is a Banach lattice, that is, Y is a Banach space, vector lattice and the norm is absolute, i.e., $||x|| = ||\,|x|\,||$ for any $x \in Y$, then the cone $Y_+ = \{y \in Y \mid y \geq 0\}$ is Daniell if Y has weakly compact intervals.

3. A convex cone with a weakly compact base is a Daniell cone.

The following result connects some useful cones.

Proposition 2.2.9 (Isac [280]). *Let $(Y, \mathscr{P})$ be a Hausdorff locally convex space and $C \subset Y$ a proper convex cone. Then*

$$C \text{ well-based} \Longrightarrow C \text{ nuclear} \Longrightarrow C \text{ normal.}$$

If Y is a normed space, then

$$C \text{ nuclear} \implies C \text{ well-based.}$$

Remark 2.2.10. Among the classical Banach spaces their usual positive cones are well-based only in l^1 and $L^1(\Omega)$ (but l^1 is not an Asplund space (see Definition 3.5.3)).

Let Y be a topological vector space over $\mathbb{R}$. Assume (Y, C) is simultaneously a vector lattice with the lattice operations $x \mapsto x^+$, $x \mapsto x^-$, $x \mapsto |x|$, $(x, y) \mapsto \sup\{x, y\}$ and $(x, y) \mapsto \inf\{x, y\}$.

Definition 2.2.11. A set $A \subset Y$ is called **solid**, if $x \in A$ and $|y| \leq |x|$ implies $y \in A$. The space Y is called **locally solid**, if it possesses a neighborhood of 0 consisting of solid sets.

Lemma 2.2.12. *The following properties are equivalent:*

(i) *Y is locally solid.*
(ii) *C is normal, and the lattice operations are continuous.*

In order to derive optimality conditions or duality statements in general spaces (cf. Chaps. 8, 12), the ordering cone is often required to have a nonempty interior. Therefore, in the following, we give some examples of convex cones with nonempty interior.

Example 2.2.13. 1. Any closed convex cone C in the Euclidean space $(\mathbb{R}^n, \langle\cdot,\cdot\rangle)$ such that C is self-adjoint (i.e., $C = C^+$) has a nonempty interior.

2. Consider the space of continuous functions $C[a,b]$ with the norm $||x|| = \sup\{|x(t)| \mid t \in [a,b]\}$. Then the cone of positive functions in $C[a,b]$

$$C[a,b]_+ := \{x \in C[a,b] \mid \forall t \in [a,b] \ : \ x(t) \geq 0\}$$

has a nonempty interior.

3. Let $Y = l^2(\mathbb{N}^*, \mathbb{R})$ with the well-known structure of a Hilbert space. The convex cone

$$C_{l^2} := \{x = \{x_i\}_{i \geq 1} \mid x_1 \geq 0 \text{ and } \sum_{i=2}^{\infty} x_i^2 \leq x_1^2\}$$

has a nonempty interior

$$\text{int } C_{l^2} := \{x = \{x_i\}_{i \geq 1} \mid x_1 > 0 \text{ and } \sum_{i=2}^{\infty} x_i^2 < x_1^2\}.$$

4. Let l^∞ be the space of bounded sequences of real numbers, equipped with the norm $||x|| = \sup_{n \in \mathbb{N}}\{|x_n|\}$. The cone

$$l^\infty{}_+ := \{x = \{x_n\}_{n\in\mathbb{N}} \mid x_n \geq 0 \text{ for any } n \in \mathbb{N}\}$$

has a nonempty interior (cf. Peressini [475], p. 186).

5. Let $C^1[a,b]$ be the real vector space formed by all real continuously differentiable functions defined on $[a,b]$ $(a, b \in \mathbb{R}, a < b)$, equipped with the norm

$$||f||_1 := \{\int_a^b (f(t))^2 \, \mathrm{d}t + \int_a^b (f'(t))^2 \, \mathrm{d}t\}^{1/2}$$

for any $f \in C^1[a,b]$. Using a Sobolev's imbedding theorem, we can show that the natural ordering cone

$$C^1[a,b]_+ := \{f \in C^1[a,b] \mid f \geq 0\}$$

has a nonempty interior. The proof is based on some technical details (cf. da Silva [532]).

6. About the locally convex spaces, we put in evidence the following result. If (Y, τ) is a real locally convex space, then for every closed convex pointed cone $C \subset Y$, with nonempty interior, there exists a continuous norm $|| \cdot ||$ on Y such that C has a nonempty interior in the normed space $(Y, || \cdot ||)$.

Proof. Take $y_0 \in \operatorname{int} C$ and $A := (y_0 - C) \cap (C - y_0)$. Then A is a closed convex and balanced set with $0 \in \operatorname{int} A$ such that the Minkowski functional $p_A : Y \to \mathbb{R}$ defined by

$$p_A(y) := \inf\{t > 0 \mid y \in tA\}$$

is a seminorm. Because $\operatorname{int} A = \operatorname{core} A = \{y \in Y \mid p_A(y) < 1\} \subset A = \{y \in Y \mid p_A(y) \leq 1\}$ (see Proposition 6.2.1), p_A is also continuous. Take $y \in Y$ with $p_A(y) = 0$. Then $y \in n^{-1}A$ for every $n \geq 1$, whence $n^{-1}y_0 \pm y \in C$ for such n. It follows that $\pm y \in \operatorname{cl} C = C$, and so $y = 0$. Hence $\|\cdot\| := p_A$ is a norm and $A = B_{(Y,\|\cdot\|)}$, and so $y_0 \in \operatorname{int}_{\|\cdot\|} C$. □

Finally, we give an example of a normed (vector) space (n.v.s.) where the natural ordering cone has a nonempty interior as well as the Daniell property.

Example 2.2.14 (see Jahn [293]). Consider the real linear space $L^\infty(\Omega)$ of all (equivalence classes of) essentially bounded functions $f : \Omega \to \mathbb{R}$ $(\emptyset \neq \Omega \subset \mathbb{R}^n)$ measurable with the norm $|| \cdot ||_{L^\infty(\Omega)}$ given by

$$||f||_{L^\infty(\Omega)} := \operatorname{ess\,sup}_{x\in\Omega}\{|f(x)|\} \text{ for all } f \in L^\infty(\Omega).$$

The ordering cone

$$L^\infty(\Omega)_+ := \{f \in L^\infty(\Omega) \mid f(x) \geq 0 \text{ almost everywhere on } \Omega\}$$

has a nonempty interior and is weak* Daniell.

2.3 Convexity Notions for Sets and Set-Valued Maps

Throughout this section X, Y are real topological vector spaces.

Definition 2.3.1. Let $A \subseteq X$ be a nonempty set. We say that A is **α-convex**, where $\alpha \in]0,1[$, if $\alpha x + (1-\alpha)y \in A$ for all $x, y \in A$. The set A is **mid-convex** if A is $\frac{1}{2}$-convex. The set A is **nearly convex** if A is α-convex for some $\alpha \in]0,1[$. The set A is **closely convex** if $\operatorname{cl} A$ is convex. The empty set is α-convex for all $\alpha \in]0,1[$ and closely convex (and so nearly convex).

Of course, A is convex if and only if A is α-convex for every $\alpha \in]0,1[$. Moreover, if $T : X \to Y$ is a linear operator and $A \subseteq X$, $B \subseteq Y$ are α-convex (nearly convex, convex), then $T(A)$ and $T^{-1}(B)$ are α-convex (nearly convex, convex), too.

Some properties of nearly convex sets are mentioned in the next result (see [214, Proposition 2.4.3, Corollary 2.4.4]).

Proposition 2.3.2. *Let $A \subseteq X$ be a nonempty nearly convex set. Then*

(i) $\operatorname{cl} A$ *is convex.*
(ii) *If* $x \in \operatorname{icr} A$ *and* $y \in A$, *then* $[x, y] \subseteq A$. *Moreover, if* $x \in \operatorname{int} A$ *and* $y \in A$, *then* $[x, y[\subseteq \operatorname{int} A$.
(iii) *If* $\operatorname{int} A \neq \emptyset$, *then* $\operatorname{int} A$ *is convex and* $\operatorname{icr} A = \operatorname{int} A$.
(iv) *If A is open or closed, then A is convex.*

Definition 2.3.3. Let $C \subseteq Y$ be a convex cone. We say that $A \subseteq Y$ is **C-α-convex** if $A + C$ is α-convex; A is **nearly C-convex** if $A + C$ is nearly convex; A is **closely C-convex** if $A + C$ is closely convex. Moreover, A is **closely c-C-convex** (nearly C-subconvexlike in [601]) if $\operatorname{cl}(\mathbb{P}(A + C))$ is convex; A is **ic-C-convex** (see [517]) if $\operatorname{int}(\mathbb{P}(A + C))$ is convex and $\mathbb{P}(A + C) \subseteq \operatorname{cl}(\operatorname{int}(\mathbb{P}(A + C)))$.

The next result, stated essentially in [79, Lemma 2.5], proves to be useful in the following sections.

Lemma 2.3.4. *Assume that $C \subseteq Y$ is a convex cone with* $\operatorname{int} C \neq \emptyset$ *and let $A \subseteq Y$. Then*

$$\operatorname{cl}(A + C) = \operatorname{cl}(\operatorname{cl} A + C) = \operatorname{cl}(A + \operatorname{int} C), \tag{2.11}$$

$$A + \operatorname{int} C = \operatorname{cl} A + \operatorname{int} C = \operatorname{int}(A + C) = \operatorname{int}(\operatorname{cl} A + C) = \operatorname{int}(\operatorname{cl}(A + C)). \tag{2.12}$$

Therefore, $\operatorname{cl}(A + C)$ *is convex iff* $A + \operatorname{int} C$ *is convex.*

Proof. The equalities in (2.11) follow immediately from the known relation

$$\operatorname{cl}(A + B) = \operatorname{cl}(\operatorname{cl} A + B) = \operatorname{cl}(\operatorname{cl} A + \operatorname{cl} B), \tag{2.13}$$

valid for all subsets $A, B \subseteq Y$, and the fact that $\operatorname{cl} C = \operatorname{cl}(\operatorname{int} C)$.

Note that $A + \operatorname{int} C$ is open being the union $\cup_{a \in A}(a + \operatorname{int} C)$ of open sets. The inclusions $A + \operatorname{int} C \subseteq \operatorname{cl} A + \operatorname{int} C$ and $\operatorname{int}(A + C) \subseteq \operatorname{int}(\operatorname{cl}(A + C))$ are obvious. Take $y \in \operatorname{cl} A$ and $k \in \operatorname{int} C$. Since $k - \operatorname{int} C \in \mathscr{N}_Y(0)$, we have that $A \cap (y + k - \operatorname{int} C) \neq \emptyset$, whence $y + k \in A + \operatorname{int} C$; hence $\operatorname{cl} A + \operatorname{int} C \subseteq A + \operatorname{int} C$, and so the first equality in (2.12) is true.

The inclusion $A + \operatorname{int} C \subseteq \operatorname{int}(A + C)$ is obvious because $A + \operatorname{int} C$ is open. Fix $k^0 \in \operatorname{int} C$ and take $y \in \operatorname{int}(A + C)$. Then there exists $\alpha > 0$ such that $y - \alpha k^0 \in A + C$, whence $y \in A + C + \operatorname{int} C = A + \operatorname{int} C$. It follows that $\operatorname{int}(A + C) \subseteq A + \operatorname{int} C$, and so the second equality in (2.12) is true. The third equality in (2.12) follows immediately from the first two equalities.

Clearly, $\operatorname{cl}(A + C) = \operatorname{cl}(A + C) + C$; using the first three equalities in (2.12) we get

$$\begin{aligned}\operatorname{int}(\operatorname{cl}(A + C)) &= \operatorname{int}(\operatorname{cl}(A + C) + C) = \operatorname{cl}(A + C) + \operatorname{int} C \\ &= (A + C) + \operatorname{int} C = A + \operatorname{int} C.\end{aligned}$$

If $\operatorname{cl}(A + C)$ is convex then $A + \operatorname{int} C = \operatorname{int}(\operatorname{cl}(A + C))$ is convex. Conversely, if $A + \operatorname{int} C$ is convex, then $\operatorname{cl}(A + C) = \operatorname{cl}(A + \operatorname{int} C)$ is convex. The proof is complete. □

It is worth observing that

$$\mathbb{P}(A + C) = \mathbb{P}A + C, \quad \mathbb{P}(A + \operatorname{int} C) = \mathbb{P}A + \operatorname{int} C. \tag{2.14}$$

Moreover, if $\operatorname{int} C \neq \emptyset$, using (2.14) and (2.12) we get

$$\operatorname{int}(\mathbb{P}(A + C)) = \mathbb{P}A + \operatorname{int} C = \operatorname{int}(\operatorname{cl}(\mathbb{P}(A + C))). \tag{2.15}$$

In the next result we establish some relationships between the C-convexity notions above.

Proposition 2.3.5. *Let $A \subseteq Y$. The following assertions hold:*

(i) *Let $\alpha \in \,]0, 1[$. Then A is C-α-convex iff $\alpha A + (1 - \alpha)A \subseteq A + C$.*
(ii) *A is closely C-convex iff $\lambda A + (1 - \lambda)A \subseteq \operatorname{cl}(A + C)$ for all $\lambda \in \,]0, 1[$.*
(iii) *A is closely C-convex and* $\operatorname{int} C \neq \emptyset$ *iff*

$$\exists k \in \operatorname{int} C,\ \forall \alpha > 0,\ \forall \lambda \in \,]0, 1[\,: \alpha k + \lambda A + (1 - \lambda)A \subseteq A + C. \tag{2.16}$$

(iv) *If A is nearly C-convex then A is closely C-convex.*
(v) *If A is closely C-convex then A is closely c-C-convex.*
(vi) *A is ic-C-convex iff A is closely c-C-convex and*

$$\operatorname{int}(\operatorname{cl}(\mathbb{P}(A + C))) = \operatorname{int}(\mathbb{P}(A + C)) \neq \emptyset. \tag{2.17}$$

(vii) *Assume that* $\operatorname{int} C \neq \emptyset$. *Then A is ic-C-convex iff A is closely c-C-convex.*

Proof. All the assertions a clearly true if A is empty. Therefore, we assume that $A \neq \emptyset$.

(i) The assertion is (almost) evident.

(ii) Assume that A is closely C-convex and take $\lambda \in]0,1[$. Since $\operatorname{cl}(A+C)$ is convex we get

$$\lambda A + (1-\lambda)A \subseteq \lambda \operatorname{cl}(A+C) + (1-\lambda)\operatorname{cl}(A+C) = \operatorname{cl}(A+C).$$

Conversely, assume that $\lambda A + (1-\lambda)A \subseteq \operatorname{cl}(A+C)$ for all $\lambda \in]0,1[$. Taking $\lambda \in]0,1[$, we have that $\lambda(A+C) + (1-\lambda)(A+C) \subseteq \operatorname{cl}(A+C) + C = \operatorname{cl}(A+C)$. Using (2.13) we get $\lambda \operatorname{cl}(A+C) + (1-\lambda)\operatorname{cl}(A+C) \subseteq \operatorname{cl}(A+C)$, and so $\operatorname{cl}(A+C)$ is convex.

(iii) Assume that (2.16) holds; then clearly $\operatorname{int} C \neq \emptyset$. Take $y_1, y_2 \in A$ and $\lambda \in]0,1[$. Then $n^{-1}k + \lambda y_1 + (1-\lambda)y_2 \in A + C$ for every $n \in \mathbb{N}^*$; taking the limit we get $\lambda y_1 + (1-\lambda)y_2 \in \operatorname{cl}(A+C)$, and so A is closely C-convex.

Assume now that A is closely C-convex and $\operatorname{int} C \neq \emptyset$. Consider $k \in \operatorname{int} C$. Take $\alpha > 0$, $y_1, y_2 \in A$ and $\lambda \in]0,1[$. Then $\lambda y_1 + (1-\lambda)y_2 \in \operatorname{cl}(A+C)$. Then using (2.12),

$$\begin{aligned}\alpha k + \lambda y_1 + (1-\lambda)y_2 &\in \operatorname{cl}(A+C)\\ &+ \operatorname{int} C = (A+C) + \operatorname{int} C = A + \operatorname{int} C \subseteq A + C.\end{aligned}$$

(iv) Assume that A is nearly C-convex. Then, using (i), $A+C$ is nearly convex. Then, by Proposition 2.3.2 (i) we obtain that $\operatorname{cl}(A+C)$ is convex, that is A is closely C-convex.

(v) Clearly, $\operatorname{cl}(\mathbb{P}(A+C)) = \operatorname{cl}(\mathbb{P}(\operatorname{cl}(A+C)))$. Since $\operatorname{cl}(A+C)$ is convex, from the preceding relation we obtain that $\operatorname{cl}(\mathbb{P}(A+C))$ is convex, that is A is c-C-convex.

(vi) Assume that A is ic-C-convex. From the definition of the ic-C-convexity we have that $B := \operatorname{int}(\mathbb{P}(A+C))$ is nonempty and convex, and $\mathbb{P}(A+C) \subseteq \operatorname{cl} B$. It follows that $\operatorname{int}(\operatorname{cl} B) = B \subseteq \mathbb{P}(A+C)$ and $\operatorname{cl}(\mathbb{P}(A+C)) \supseteq \operatorname{cl} B$, and so $\operatorname{cl}(\mathbb{P}(A+C)) = \operatorname{cl} B$ is convex. Therefore, A is closely c-C-convex. Moreover, since B is open, convex and nonempty, we have that $\operatorname{int}(\operatorname{cl}(\mathbb{P}(A+C))) = \operatorname{int}(\operatorname{cl} B) = B$. Therefore, (2.17) holds.

Assume now that A is closely c-C-convex and (2.17) holds. Then $C := \operatorname{cl}(\mathbb{P}(A+C))$ is convex $\operatorname{int} C = \operatorname{int}(\mathbb{P}(A+C)) =: B \neq \emptyset$. Then clearly B is convex and $\operatorname{cl} B = \operatorname{cl}(\operatorname{int} C) = C \supseteq \mathbb{P}(A+C)$. Hence A is ic-C-convex.

(vii) Let $\operatorname{int} C \neq \emptyset$. Then (2.15) holds and $\operatorname{int}(\mathbb{P}(A+C)) \neq \emptyset$. The conclusion follows using (vi). The proof is complete. □

Note that Proposition 2.3.5 (vii) is stated in [601, Theorem 3.1], while the fact that A is closely c-C-convex if A is ic-C-convex in Proposition 2.3.5 (vi) is proved in [601, Theorem 3.2]

Let $F : X \rightrightarrows Y$. We say that F is **α-convex (mid-convex, nearly convex, convex)** if graph F is α-convex (mid-convex, nearly convex, convex). It is obvious that if F is α-convex (mid-convex, nearly convex, convex), so are dom F, Im F, and $F(x)$ for every $x \in X$. It is easy to see that F is α-convex if and only if

$$\forall\, x, x' \in \operatorname{dom} F : \alpha F(x) + (1-\alpha)F(x') \subseteq F\left(\alpha x + (1-\alpha)x'\right);$$

in the relation above $x, x' \in \operatorname{dom} F$ can be replaced by $x, x' \in X$.

To $F : X \rightrightarrows Y$ we associate the set-valued maps $\operatorname{cl} F$, $\operatorname{conv} F$, $\overline{\operatorname{conv}} F : X \rightrightarrows Y$ defined by

$$\begin{aligned}(\operatorname{cl} F)(x) &:= \operatorname{cl}[F(x)], \ (\operatorname{conv} F)(x) := \operatorname{conv}[F(x)],\\ (\overline{\operatorname{conv}} F)(x) &:= \overline{\operatorname{conv}}[F(x)] \ (x \in X).\end{aligned}$$

It is almost obvious that $\operatorname{cl} F$, $\operatorname{conv} F$ and $\overline{\operatorname{conv}} F$ are α-convex (mid-convex, nearly convex, convex) if F is α-convex (mid-convex, nearly convex, convex).

To $F : X \rightrightarrows Y$ and $y^* \in Y^*$, where Y^* is the topological dual of Y, we also associate

$$\phi_{y^*} := \phi^F_{y^*} : X \to \overline{\mathbb{R}}, \quad \phi_{y^*}(x) := \inf\{\langle y, y^*\rangle \mid y \in F(x)\} \quad (x \in X), \tag{2.18}$$

where, as usual, $\inf \emptyset := +\infty$; then $\operatorname{dom} \phi_{y^*} = \operatorname{dom} F$ for every $y^* \in Y^*$ and $\phi_0 = \iota_{\operatorname{dom} F}$. Clearly, $\phi^F_{y^*} = \phi^{\operatorname{cl} F}_{y^*} = \phi^{\operatorname{conv} F}_{y^*} = \phi^{\overline{\operatorname{conv}} F}_{y^*}$ for every $y^* \in Y^*$. The function ϕ_{y^*} (but with sup instead of inf) was introduced in [138], and used (for example) in [422, 423, 518], too.

Proposition 2.3.6. *Let $F : X \rightrightarrows Y$.*

(i) *If F is convex then ϕ_{y^*} is convex for every $y^* \in Y^*$.*
(ii) *Assume that Y is a locally convex space. If ϕ_{y^*} is convex for every $y^* \in Y^*$ then $\overline{\operatorname{conv}} F$ is convex.*

Proof. (i) Consider $x, x' \in \operatorname{dom} \phi_{y^*}$ and $\alpha \in\,]0,1[$. Take $\gamma, \gamma' \in \mathbb{R}$ such that $\phi_{y^*}(x) < \gamma$, $\phi_{y^*}(x') < \gamma'$. Then there exist $y \in F(x)$, $y' \in F(x')$ such that $\langle y, y^*\rangle < \gamma$, $\langle y', y^*\rangle < \gamma'$. Then $\alpha y + (1-\alpha)y' \in F(\alpha x + (1-\alpha)x')$, and so

$$\begin{aligned}\phi_{y^*}\left(\alpha x + (1-\alpha)x'\right) &\leq \left\langle \alpha y + (1-\alpha)y', y^*\right\rangle = \alpha\, \langle y, y^*\rangle + (1-\alpha)\left\langle y', y^*\right\rangle\\ &< \alpha\gamma + (1-\alpha)\gamma'.\end{aligned}$$

Letting $\gamma \to \phi_{y^*}(x)$, $\gamma' \to \phi_{y^*}(x')$ we get $\phi_{y^*}(\alpha x + (1-\alpha)x') \leq \alpha\phi_{y^*}(x) + (1-\alpha)\phi_{y^*}(x')$. Hence ϕ_{y^*} is convex.

(ii) Since $\phi^F_{y^*} = \phi^{\overline{\operatorname{conv}} F}_{y^*}$ for every $y^* \in Y^*$, we may (and do) assume that $F = \overline{\operatorname{conv}} F$. We have that $\operatorname{dom} F = \operatorname{dom} \phi_0$ is convex, $\phi_0 = \iota_{\operatorname{dom} F}$ being convex. Assume that F is not convex. Then there exist $x, x' \in \operatorname{dom} F$, $y \in F(x)$, $y' \in$

$F(x')$ and $\alpha \in]0,1[$ such that $z := \alpha y + (1-\alpha)y' \notin F(\alpha x + (1-\alpha)x') =: A$. Since A is a nonempty closed convex set, there exists $y^* \in Y^*$ such that $\langle z, y^* \rangle < \inf\{\langle v, y^* \rangle \mid v \in A\} = \phi_{y^*}(\alpha x + (1-\alpha)x')$. Since

$$\alpha\phi_{y^*}(x) + (1-\alpha)\phi_{y^*}(x') \leq \alpha \langle y, y^* \rangle + (1-\alpha)\langle y', y^* \rangle = \langle z, y^* \rangle$$
$$< \phi_{y^*}\left(\alpha x + (1-\alpha)x'\right),$$

we get the contradiction that ϕ_{y^*} is not convex. Hence F is convex. □

Let $C \subseteq Y$ be a convex cone. We say that F is **C-α-convex (C-mid-convex, C-nearly convex, C-convex)** if the set-valued map

$$F_C : X \rightrightarrows Y, \quad F_C(x) := F(x) + C,$$

is α-convex (mid-convex, nearly convex, convex). Of course, F is C-α-convex if and only if

$$\forall\, x, x' \in \operatorname{dom} F : \alpha F(x) + (1-\alpha)F(x') \subseteq F\left(\alpha x + (1-\alpha)x'\right) + C.$$

Note that sometimes graph F_C is denoted by $\operatorname{epi}_C F$, or simply epi F, and is called the **epigraph** of F.

Corollary 2.3.7. *Let $F : X \rightrightarrows Y$ and $C \subset Y$ be a convex cone.*

(i) *If F is C-convex then ϕ_{y^*} is convex for every $y^* \in C^+$.*
(ii) *Assume that Y is a locally convex space and ϕ_{y^*} is convex for every $y^* \in C^+$. Then $\overline{\operatorname{conv}} F_C$ is convex; in particular, if $F(x) + C$ is closed and convex for every $x \in X$, then F is C-convex.*

Proof. Of course, F is C-convex if and only if F_C is convex. Let us set $\tilde{\phi}_{y^*} := \phi_{y^*}^{\overline{\operatorname{conv}} F}$. Note that $\tilde{\phi}_{y^*} = \phi_{y^*}^F = \phi_{y^*}$ for $y^* \in C^+$, while for $y^* \in Y^* \setminus C^+$, $\tilde{\phi}_{y^*}(x) = +\infty$ for $x \in \operatorname{dom} F$ and $\tilde{\phi}_{y^*}(x) = -\infty$ for $x \notin \operatorname{dom} F$. Hence $\tilde{\phi}_{y^*}$ is convex for every $y^* \in Y^*$ if and only if ϕ_{y^*} is convex for every $y^* \in C^+$. The conclusion follows applying Proposition 2.3.6 to F_C. □

Of course, Proposition 2.3.6 can be obtained from Corollary 2.3.7 taking $C = \{0\}$. Corollary 2.3.7 can be found, essentially, in [138, Proposition 1.6] and [518, Lemma 3].

The **sublevel set of F of height** y (w.r.t. C) is the set

$$\operatorname{lev}_F(y) := \{x \in X \mid F(x) \cap (y - C) \neq \emptyset\};$$

when $\operatorname{int} C \neq \emptyset$ we also consider the **strict sublevel set of F of height** y (w.r.t. C) defined by

$$\operatorname{lev}_F^<(y) := \{x \in X \mid F(x) \cap (y - \operatorname{int} C) \neq \emptyset\}.$$

In this way we get the **sublevel** and **strict sublevel set-valued maps** lev_F, $\operatorname{lev}_F^<$: $Y \rightrightarrows X$.

We say that F is **C-α-quasiconvex (C-mid-quasiconvex, C-nearly quasiconvex, C-quasiconvex)** if for every $z \in Y$ the sublevel set $\operatorname{lev}_F(z)$ is α-convex (mid-convex, nearly convex, convex). An equivalent definition of C-α-quasiconvexity is that

$$\forall\, x, x' \in \operatorname{dom} F : (F(x) + C) \cap (F(x') + C) \subseteq F\left(\alpha x + (1-\alpha)x'\right) + C.$$

Notice that F is C-α-quasiconvex whenever

$$\begin{aligned}\forall x, x' \in \operatorname{dom} F : F(x) &\subseteq F\left(\alpha x + (1-\alpha)x'\right) + C \text{ or}\\ F(x') &\subseteq F\left(\alpha x + (1-\alpha)x'\right) + C.\end{aligned}$$

Note also that F is C-α-quasiconvex (C-mid-quasiconvex, C-nearly quasiconvex, C-quasiconvex) whenever F is C-α-convex (C-mid-convex, C-nearly convex, C-convex).

The set-valued map F is **C-convexlike** if

$$\begin{aligned}\forall x_1, x_2 &\in X,\ \forall y_1 \in F(x_1),\ \forall y_2 \in F(x_2),\ \forall \lambda \in]0,1[,\\ \exists x_3 &\in X : \lambda y_1 + (1-\lambda) y_2 \in F(x_3) + C,\end{aligned}$$

or, equivalently, $F(X) + C$ is convex, that is $F(X)$ is C-convex. Of course, if F is C-convex then F is C-convexlike.

Li and Chen [387] (see also [602]) say that F is **C-subconvexlike** if

$$\exists k \in \operatorname{int} C,\ \forall \alpha > 0,\ \forall x, x' \in X,\ \forall \lambda \in]0,1[: \alpha k + \lambda F(x) + (1-\lambda) F(x') \subseteq F(X) + C.$$

Using Proposition 2.3.5 (iii), F is C-subconvexlike iff $\operatorname{int} C \neq \emptyset$ and $F(X)$ is closely C-convex.

We say that $f : X \to Y^\bullet$ is **C-α-convex (C-mid-convex, C-nearly convex, C-convex, C-α-quasiconvex, C-mid-quasiconvex, C-nearly quasiconvex, C-quasiconvex)** if the set-valued map $F_{f,C}$ is C-α-convex (C-mid-convex, C-nearly convex, C-convex, C-α-quasiconvex, C-mid-quasiconvex, C-nearly quasiconvex, C-quasiconvex); in particular, f is C-convex if and only if

$$\forall\, x, x' \in X,\ \forall\, \alpha \in [0,1] : f\left(\alpha x + (1-\alpha)x'\right) \leq_C \alpha f(x) + (1-\alpha) f(x').$$

If f is C-α-convex (C-mid-convex, C-nearly convex, C-convex), then $\operatorname{dom} f$ is so, and f is C-α-quasiconvex (C-mid-quasiconvex, C-nearly quasiconvex, C-quasiconvex).

2.4 Solution Concepts in Vector Optimization

In this section, we first recall concepts of Pareto minimal points, weakly and properly minimal points, then we introduce the concept of Q-minimal points and establish relations among them.

Unless otherwise mentioned, in the following we consider a linear topological space Y, partially ordered by a proper pointed convex closed cone C and a nonempty set $A \subset Y$.

We introduce the following sets of *Pareto minimal points* (*Pareto maximal points*, respectively) of A with respect to C:

Definition 2.4.1 (Pareto Minimal (Maximal) Points). Consider

$$\mathrm{Min}(A, C) := \{\overline{y} \in A \mid A \cap (\overline{y} - C) = \{\overline{y}\}\}. \tag{2.19}$$

An element $\overline{y} \in \mathrm{Min}(A, C)$ is called a **Pareto minimal point** of A with respect to C.

Furthermore, consider

$$\mathrm{Max}(A, C) := \{\overline{y} \in A \mid A \cap (\overline{y} + C) = \{\overline{y}\}\}. \tag{2.20}$$

An element $\overline{y} \in \mathrm{Max}(A, C)$ is called a **Pareto maximal point** of A with respect to C.

Moreover, in order to describe weak minimality we will study the following solution concept in Y. Many solution procedures for vector optimization problems generate weakly minimal elements.

Definition 2.4.2 (Weakly Minimal (Maximal) Points). Suppose that $\mathrm{int}\, C \neq \emptyset$. Consider

$$\mathrm{WMin}(A, C) := \{\overline{y} \in A \mid A \cap (\overline{y} - \mathrm{int}\, C) = \emptyset\}. \tag{2.21}$$

An element $\overline{y} \in \mathrm{WMin}(A, C)$ is called a **weakly minimal point** of A with respect to C. Furthermore, consider

$$\mathrm{WMax}(A, C) := \{\overline{y} \in A \mid A \cap (\overline{y} + \mathrm{int}\, C) = \emptyset\}. \tag{2.22}$$

An element $\overline{y} \in \mathrm{WMax}(A, C)$ is called a **weakly maximal point** of A with respect to C.

Moreover, we introduce the concept of strongly minimal points:

Definition 2.4.3. Consider

$$\mathrm{StrMin}(A, C) := \{\overline{y} \in A \mid A \subseteq \overline{y} + C\}. \tag{2.23}$$

An element $\overline{y} \in \text{StrMin}(A, C)$ is called a **strong minimal point** of A with respect to C.

In the following, we introduce different concepts of properly minimal points. Properly minimal points are important in the proofs of many theoretical assertions because corresponding scalarizing functionals (see Sect. 5.1) are strictly C-monotone. The first concept for proper minimality in the following definition (cf. Ha [228]) is based on scalarization by means of (strictly C-monotone) functionals $y^* \in C^\#$.

Definition 2.4.4 (Properly Minimal Points).

(a) Suppose that $C^\# \neq \emptyset$ and consider

$$\text{S-PMin}(A, C) := \{\overline{y} \in A \mid \exists y^* \in C^\#, \;\; \forall y \in A : \; y^*(\overline{y}) \leq y^*(y)\}. \tag{2.24}$$

An element $\overline{y} \in \text{S-PMin}(A, C)$ is called a **S-properly minimal point** of A w.r.t. C.

(b) Let

$$\text{Hu-PMin}(A, C) := \{\overline{y} \in A \mid (\text{cl conv cone}[(A - \overline{y}) \cup C]) \cap (-C) = \{0\}\}.$$

An element $\overline{y} \in \text{Hu-PMin}(A, C)$ is called a **Hurwicz properly minimal point** of A w.r.t. C.

(c) Assume that Y is a n.v.s.; $\overline{y}$ is a **Hartley properly minimal point** of A w.r.t. C ($\overline{y} \in \text{Ha-PMin}(A, C)$) if $\overline{y} \in \text{Min}(A, C)$ and there exists a constant $M > 0$ such that, whenever there is $\lambda \in C^+$ with $\lambda(y - \overline{y}) > 0$ for some $y \in A$, one can find $\mu \in C^+$ with

$$\lambda(y - \overline{y})/\|\lambda\| \leq -M(\mu(y - \overline{y})/\|\mu\|).$$

(d) Consider

$$\text{Be-PMin}(A, C) := \{\overline{y} \in A \mid \text{cl cone}[(A - \overline{y}) + C] \cap (-C) = \{0\}\}.$$

$\overline{y} \in \text{Be-PMin}(A, C)$ is called a **Benson properly minimal point** of A w.r.t. C.

(e) Consider

$$\text{Bo-PMin}(A, C) := \{\overline{y} \in A \mid \text{cl cone}(A - \overline{y}) \cap (-C) = \{0\}\}.$$

$\overline{y} \in \text{Bo-PMin}(A, C)$ is called a **Borwein properly minimal point** of A w.r.t. C.

(f) Consider

GHe-PMin(A, C) := $\{\overline{y} \in A \mid \exists$ proper convex pointed cone D with $C \setminus \{0\} \subset \operatorname{int} D$ such that $(A - \overline{y}) \cap (-\operatorname{int} D) = \emptyset\}$.
$\overline{y} \in$ GHe-PMin(A, C) is called a **Henig global properly minimal point** of A w.r.t. C.

(g) Suppose that Y is a n.v.s., C has a base Θ and consider

He-PMin(A, C) := $\{\overline{y} \in A \mid \exists\, \varepsilon > 0$ such that $\operatorname{cl}\operatorname{cone}(A - \overline{y}) \cap (-\Theta + \varepsilon B_Y) = \emptyset\}$.

$\overline{y} \in$ He-PMin(A, C) is called a **Henig properly minimal point** of A w.r.t. to C.

(h) Assume that Y is a n.v.s. Consider

Sup-PMin(A, C) := $\{\overline{y} \in A \mid \exists\, \rho > 0$ such that $\operatorname{cl}\operatorname{cone}(A - \overline{y}) \cap (B_Y - C) \subseteq \rho B_Y\}$.

$\overline{y} \in$ Sup-PMin(A, C) is called a **super efficient point** of A w.r.t. C.

For the notions of minimal points in Definition 2.4.4, we refer to [292, 293, 402] and [228]. The concepts of Henig proper minimality and Henig global proper minimality have been presented in [242]. The above definition of Henig properly minimal points can be found in [71, 627]; see also [228, 242, 618, 619]. For an equivalent definition of Henig properly minimal points by means of a functional from $C^{\#}$ the reader is referred to [619]. We note that positive proper minimality has been introduced by Hurwicz [11], and super efficiency has been introduced by Borwein and Zhuang [71]. We refer the reader to [219] for a survey and materials on proper efficiency.

In the sequel, when speaking of weakly minimal points (resp. S-properly minimal points) we mean that $\operatorname{int} C$ (resp. $C^{\#}$) is nonempty, when speaking of Henig minimal points we mean that C has a base Θ and when speaking that C has a bounded base we mean that Θ is bounded.

Let $B \subset Y$ be a convex set such that $0 \notin \operatorname{cl} B$ (that is B is a base for cone B); we set

$$\mathcal{N}_Y^B := \mathcal{N}^B := \{V \in \mathcal{N}_Y \mid V \text{ convex}, V \cap B = \emptyset\}; \tag{2.25}$$

clearly, $\mathcal{N}_Y^B \neq \emptyset$. For $V \in \mathcal{N}^B$ we set

$$P_V^B := \operatorname{cone}(B + V);$$

then P_V^B is a proper convex cone with $\operatorname{int} P_V^B = \mathbb{P}(B + \operatorname{int} V) \neq \emptyset$.

Similar to the case of normed vector spaces, if Θ is a base of C, we set

$$\text{He-PMin}(A,\Theta) := \{y \in A \mid \exists V \in \mathscr{N}_Y^{\Theta} : [\operatorname{cl}\operatorname{cone}(A-y)] \cap (V-\Theta) = \emptyset\}.$$

Observe that

$$\text{He-PMin}(A,\Theta) = \{y \in A \mid \exists V \in \mathscr{N}_Y^{\Theta} : (A-y) \cap \left(-P_V^{\Theta}\right) = \{0\}\} \tag{2.26}$$

$$= \{y \in A \mid \exists V \in \mathscr{N}_Y^{\Theta} : (A-y) \cap \left(-\operatorname{int} P_V^{\Theta}\right) = \emptyset\}. \tag{2.27}$$

Indeed, $[\operatorname{cl}\operatorname{cone}(A-y)] \cap (V-\Theta) = \emptyset \Rightarrow (A-y) \cap \operatorname{cone}(V-\Theta) = \{0\}$ and $(A-y) \cap \operatorname{cone}(V-\Theta) = \{0\} \Rightarrow [\operatorname{cl}\operatorname{cone}(A-y)] \cap (\operatorname{int} V-\Theta) = \emptyset$.

Moreover, the super efficiency in the case in which Y is a locally convex space is defined by

$\text{Sup-PMin}(A,C) := \{y \in A \mid \forall V \in \mathscr{N}_Y, \exists U \in \mathscr{N}_Y : \operatorname{cl}\operatorname{cone}(A-y) \cap (U-C) \subseteq V\}$.

The following two Propositions 2.4.5 and 2.4.6 are shown under weaker assumptions concerning the cone $C \subset Y$ (Y a linear topological space), namely that C is a proper convex cone, for the corresponding solution concepts.

Proposition 2.4.5. *Let $A \subset Y$ be nonempty. Then*

(i) $\text{StrMin}(A,C) \subseteq \bigcap\{\arg\min_A y^* \mid y^* \in C^+ \setminus 0\}$, *with equality if Y is a l.c.s. and C is closed.*

(ii) $\text{S-PMin}(A,C) = \bigcup\{\arg\min_A y^* \mid y^* \in C^{\#}\} = \text{S-PMin}(A+C,C)$.

(iii) *If* $\operatorname{int} C \neq \emptyset$, *then* $\text{Min}(A,C) \subseteq \text{WMin}(A,C)$ *and*

$$\text{WMin}(A,C) = A \cap \text{WMin}(A+C,C) = A \cap \operatorname{bd}(A+C) \tag{2.28}$$

$$\supseteq \bigcup\{\arg\min_A y^* \mid y^* \in C^+ \setminus \{0\}\}, \tag{2.29}$$

with equality if A is closely C-convex.

(iv) *We have that* $\text{GHe-PMin}(A,C) \subseteq \text{Min}(A,C)$ *and*

$$\text{GHe-PMin}(A,C) = \bigcup\{\text{WMin}(A,D) \mid D \in \mathscr{D}_C\}$$

$$= \text{GHe-PMin}(A+C,C) \supseteq \text{S-PMin}(A,C) \tag{2.30}$$

with equality if A is closely C-convex, where

$$\mathscr{D}_C := \{D \subset Y \mid D \text{ proper pointed convex cone with } C \setminus \{0\} \subset \operatorname{int} D\}. \tag{2.31}$$

(v) *Assume that Θ is a base of C. Then*

$$\text{He-PMin}(A,\Theta) = \bigcup\{\text{WMin}(A,D) \mid D \in \mathscr{D}_\Theta\} = \text{He-PMin}(A+C,\Theta) \tag{2.32}$$

$$\supseteq \bigcup \{\arg\min_A y^* \mid y^* \in Y^*,\ \inf y^*(\Theta) > 0\}, \tag{2.33}$$

with equality if A is closely C-convex, where

$$\mathscr{D}_\Theta := \{P_V^\Theta \mid V \in \mathscr{N}_Y^\Theta\}. \tag{2.34}$$

(vi) *If Θ is a base of C then $\mathscr{D}_\Theta \subseteq \mathscr{D}_C$; consequently,* He-PMin$(A,\Theta) \subseteq$ GHe-PMin(A,C).

Proof. (i) Let $\overline{y} \in \mathrm{StrMin}(A,C)$, that is $\overline{y} \in A \subseteq \overline{y} + C$. Then clearly $\mathrm{StrMin}(A,C) = \{\overline{y}\}$ and the inclusion holds. Assume that Y is a l.c.s. and C is closed, and take $\overline{y} \in \bigcap\{\arg\min_A y^* \mid y^* \in C^+\}$. Then $\overline{y} \in A$ and $\langle y - \overline{y}, y^*\rangle \geq 0$ for all $y \in A$ and $y^* \in C^+$. By the bipolar theorem we get $y - \overline{y} \in C^{++} = \mathrm{cl}\, C = C$, whence $A \subseteq \overline{y} + C$.

(ii) The first equality is given by the definition of S-PMin(A,C). The inclusion S-PMin$(A,C) \subseteq$ S-PMin$(A+C,C)$ is obvious. If $\overline{y} \in$ S-PMin$(A+C,C)$, then $\overline{y} = \overline{a} + \overline{c}$ for some $\overline{a} \in A$, $\overline{c} \in C$, and there exists $\overline{y}^* \in C^\#$ such that $\langle \overline{a} + \overline{c}, \overline{y}^*\rangle \leq \langle y, \overline{y}^*\rangle$ for all $y \in A + C$. In particular $\langle \overline{a} + \overline{c}, \overline{y}^*\rangle \leq \langle \overline{a}, \overline{y}^*\rangle$, whence $\langle \overline{c}, \overline{y}^*\rangle \leq 0$. It follows that $\overline{c} = 0$, and so $\overline{y} = \overline{a} \in A$. Consequently, $\overline{y} \in$ S-PMin(A,C).

(iii) Assume that $\mathrm{int}\, C \neq \emptyset$. Taking into account that $\mathrm{int}\, C \subseteq C \setminus \{0\}$, the inclusion $\mathrm{Min}(A,C) \subseteq \mathrm{WMin}(A,C)$ follows.
For $y \in Y$, using Lemma 2.3.4, we have

$$(A - y) \cap (-\mathrm{int}\, C) = \emptyset \Leftrightarrow y \notin A + \mathrm{int}\, C \Leftrightarrow y \notin A + C + \mathrm{int}\, C \Leftrightarrow y \notin \mathrm{int}(A+C).$$

It follows that

$$\begin{aligned} \overline{y} \in \mathrm{WMin}(A,C) &\Leftrightarrow [\overline{y} \in A,\ (A + C - \overline{y}) \cap (-\mathrm{int}\, C) = \emptyset] \\ &\Leftrightarrow \overline{y} \in A \cap \mathrm{WMin}(A,C) \\ &\Leftrightarrow [\overline{y} \in A,\ \overline{y} \notin \mathrm{int}(A+C)] \Leftrightarrow \overline{y} \in A \cap \mathrm{bd}(A+C). \end{aligned}$$

Hence the equalities in (2.28) hold.

Take $\overline{y} \in \arg\min_A y^*$ for some $y^* \in C^+ \setminus \{0\}$. Because $\mathrm{int}\, C \subseteq \{y \in Y \mid \langle y, y^*\rangle > 0\}$, we obtain that $(A - \overline{y}) \cap (-\mathrm{int}\, C) = \emptyset$, and so $\overline{y} \in \mathrm{WMin}(A,C)$. Hence the inclusion in (2.28) holds.

Assume now that A is closely C-convex and $\overline{y} \in \mathrm{WMin}(A,C)$ [$\subseteq \mathrm{WMin}(A + C, C)$]. Then $(A + C) \cap (\overline{y} - \mathrm{int}\, C) = \emptyset$; it follows that $\mathrm{cl}(A + C) \cap (\overline{y} - \mathrm{int}\, C) = \emptyset$. Because $\mathrm{cl}(A + C)$ is convex, by a separation theorem (see [293, Theorem 3.16]) there exists $y^* \in Y^* \setminus \{0\}$ such that $\langle y + v, y^*\rangle \geq \langle \overline{y} - v', y^*\rangle$ for all $y \in A$ and $v, v' \in C$. It follows that $y^* \in C^+$ and $\langle y - \overline{y}, y^*\rangle \geq 0$ for every $y \in A$, and so $y^* \in C^+ \setminus \{0\}$ and $\overline{y} \in \arg\min_A y^*$.

(iv) The first equality in (2.30) is just the definition of GHe-PMin(A, C).

Take $\overline{y} \in$ GHe-PMin(A, C); then $\overline{y} \in$ WMin(A, D) for some $D \in \mathscr{D}_C$, and so $(A - \overline{y}) \cap (-\operatorname{int} D) = \emptyset$. Because $C + \operatorname{int} D = \operatorname{int} D$ (for every $D \in \mathscr{D}_C$), we clearly have that GHe-PMin$(A, C) \subseteq$ GHe-PMin$(A + C, C)$.

Take $\overline{y} \in$ GHe-PMin$(A + C, C)$; then $\overline{y} = y + v$ for some $y \in A$, $v \in C$. Assuming that $v \neq 0$, we get the contradiction $v \in (C \setminus \{0\}) \cap (A + C - \overline{y}) \subseteq (-\operatorname{int} D) \cap (A + C - \overline{y}) = \emptyset$. Hence $\overline{y} = y \in A$.

Take $\overline{y} \in$ S-PMin(A, C). Then there exists $y^* \in C^\#$ with $0 \leq \langle y - \overline{y}, y^* \rangle \leq \langle y + v - \overline{y}, y^* \rangle$ for all $y \in A$, $v \in C$. Take $D := \{0\} \cup \{y \in Y \mid \langle y, y^* \rangle > 0\}$. Then D is a pointed convex cone with $\operatorname{int} D = \{y \in Y \mid \langle y, y^* \rangle > 0\}$. It follows that $(A + C - \overline{y}) \cap (-\operatorname{int} D) = \emptyset$, and so $\overline{y} \in$ GHe-PMin(A, C). Hence S-PMin$(A, C) \subseteq$ GHe-PMin(A, C).

Assume that A is closely C-convex and take $\overline{y} \in$ GHe-PMin(A, C). Then there exists $D \in \mathscr{D}_C$ such that $(A - \overline{y}) \cap (-\operatorname{int} D) = \emptyset$. It follows that $\overline{y} \in A \cap$ WMin(A, D). Since A is closely C-convex and $C \subseteq D$, A is closely D-convex. From (iii) we get $y^* \in D^+ \setminus \{0\}$ such that $\overline{y} \in \arg\min_A y^*$. Because $D^+ \setminus \{0\} \subseteq C^\#$, we obtain that $\overline{y} \in$ S-PMin(A, C).

(v) The first equality in (2.32) is given in (2.27). For the equality He-PMin$(A, \Theta) =$ He-PMin$(A + C, \Theta)$ use a similar argument to that used in (iv) (possibly taking into account that $\mathscr{D}_\Theta \subseteq \mathscr{D}_C$).

Take $\overline{y} \in \arg\min_A y^*$ for some $y^* \in Y^*$ with $2\gamma := \inf y^*(\Theta) > 0$ and set $V := \{y \in Y \mid |\langle y, y^* \rangle| < \gamma\} \in \mathscr{N}_Y^\Theta$. Clearly, $D := P_V^\Theta \in \mathscr{D}_\Theta$. Because $y^* \in D^+ \setminus \{0\}$, from (iii) we get $\overline{y} \in$ WMin(A, D), and so $\overline{y} \in$ He-PMin(A, Θ). Hence the inclusion in (2.33) holds.

The proof of the equality in (2.33) for A closely C-convex is similar to the proof of the corresponding equality in (2.30).

(vi) The arguments used at the beginning of the proof of (v) show that $\mathscr{D}_\Theta \subseteq \mathscr{D}_C$. □

Proposition 2.4.6. *Let $A \subset Y$ be nonempty.*

(i) *One has*

$$\begin{aligned} \text{S-PMin}(A, C) &\subseteq \text{Hu-PMin}(A + C, C) \subseteq \text{Hu-PMin}(A, C) \subseteq \text{Be-PMin}(A, C) \\ &= \text{Be-PMin}(A + C, C) \end{aligned} \tag{2.35}$$

with Hu-PMin$(A, C) =$ Be-PMin(A, C) *if A is closely C-convex, and*

$$\text{Be-PMin}(A, C) = \text{Bo-PMin}(A + C, C) \subseteq \text{Bo-PMin}(A, C) \subseteq \text{Min}(A, C), \tag{2.36}$$

$$\text{GHe-PMin}(A, C) \subseteq \text{Be-PMin}(A, C). \tag{2.37}$$

Moreover, if Y has the property that for any closed convex cone $K \subseteq Y$ there exists $y^* \in Y^*$ such that $\langle y, y^* \rangle > 0$ for every $y \in K \setminus (-K)$ (for example if Y is a separable normed vector space) then S-PMin$(A, C) =$ Hu-PMin(A, C).

(ii) *Assume that* Θ *is a base of* C *and* Y *is a locally convex space. Then* $\text{Sup-PMin}(A, C) \subseteq \text{He-PMin}(A, \Theta)$. *Moreover, if* Θ *is bounded, then* $\text{Sup-PMin}(A, C) = \text{He-PMin}(A, \Theta)$.

(iii) *Assume that* Θ *is a compact base of* C *and* Y *is a locally convex space. Then*

$$\text{S-PMin}(A, C) = \text{Hu-PMin}(A, C) \quad \textit{and} \quad \text{He-PMin}(A, \Theta) = \text{GHe-PMin}(A, C).$$

Moreover, if A *is closely* C*-convex, then* $\text{S-PMin}(A, C) = \text{Be-PMin}(A, C)$, *while if* A *is closely convex then* $\text{S-PMin}(A, C) = \text{Bo-PMin}(A, C)$.

(iv) *Assume that* $\text{StrMin}(A, C) \neq \emptyset$. *Then* $\text{StrMin}(A, C) = \text{Min}(A, C)$. *If* C *is closed (or, more generally,* $\operatorname{cl} C \cap (-C) = \{0\}$*), then*

$$\text{StrMin}(A, C) = \text{Hu-PMin}(A, C) = \text{Hu-PMin}(A + C, C); \tag{2.38}$$

if $C^{\#} \neq \emptyset$ *then*

$$\text{StrMin}(A, C) = \text{S-PMin}(A, C) = \text{GHe-PMin}(A, C), \tag{2.39}$$

and $\text{StrMin}(A, C) = \text{He-PMin}(A, \Theta)$ *for every base* Θ *of* C *if, furthermore,* Y *is a locally convex space.*

Proof. (i) Take $\overline{y} \in \text{S-PMin}(A, C)$. Then $\overline{y} \in A$ and there exists $\overline{y}^* \in C^{\#}$ such that $\langle \overline{y}, \overline{y}^* \rangle \leq \langle y, \overline{y}^* \rangle$ for all $y \in A$. It follows that $0 \leq \langle y, \overline{y}^* \rangle$ for all $y \in E_1 := \operatorname{cl} \operatorname{conv} \operatorname{cone}(A + C - \overline{y})$, and so $E_1 \cap (-C) = \{0\}$, because $\overline{y}^* \in C^{\#}$. Hence the first inclusion in (2.35) holds.

Take $\overline{y} \in \text{Hu-PMin}(A+C, C)$ $(\subseteq A+C)$. Hence $\overline{y} = y+v$ with $y \in A$, $v \in C$. Then $-v \in (A + C - \overline{y}) \cap (-C) \subseteq (\operatorname{cl} \operatorname{conv} \operatorname{cone}(A + C - \overline{y})) \cap (-C) = \{0\}$. Therefore, $v = 0$, and so $\overline{y} \in A$. It follows that $\overline{y} \in \text{Hu-PMin}(A, C)$.

Take $\overline{y} \in \text{Hu-PMin}(A, C)$. Then $(A - \overline{y}) \cup C \subseteq E_2 := \operatorname{cl} \operatorname{conv} \operatorname{cone}[(A - \overline{y}) \cup C]$, whence $(A - \overline{y}) + C \subseteq E_2$. It follows that $F := \operatorname{cl} \operatorname{cone}(A + C - \overline{y}) \subseteq E_2$, and so $\overline{y} \in \text{Be-PMin}(A, C)$. Assuming that A is closely C-convex and $\overline{y} \in \text{Be-PMin}(A, C)$, we have that F is a convex cone and $F \cap (-C) = \{0\}$. Since $(A - \overline{y}) \cup C \subseteq A + C - \overline{y}$, it follows that $E_2 \subseteq F$, and so $\overline{y} \in \text{Hu-PMin}(A, C)$.

The equalities in (2.35) and (2.36) follow directly from the definitions of the corresponding sets.

Take $\overline{y} \in \text{Bo-PMin}(A + C, C)$. Then $\overline{y} = y + v$ for some $y \in A$, $v \in C$. Then $-v \in (A + C - \overline{y}) \cap (-C) \subseteq (-C) \cap \operatorname{cl} \operatorname{cone}(A + C - \overline{y}) = \{0\}$. Hence $\overline{y} \in A$, and so $\overline{y} \in \text{Bo-PMin}(A, C)$.

Take $\overline{y} \in \text{Bo-PMin}(A, C)$. Since $\operatorname{cl} \operatorname{cone}(A - \overline{y}) \supseteq A - \overline{y}$, it follows that $\overline{y} \in \text{Min}(A, C)$.

Take $\overline{y} \in \text{GHe-PMin}(A, C) = \text{GHe-PMin}(A + C, C)$. Then there exists $D \in \mathscr{D}_C$ such that $(A + C - \overline{y}) \cap (-\operatorname{int} D) = \emptyset$, and so $[\mathbb{R}_+(A + C - \overline{y})] \cap (-\operatorname{int} D) = \emptyset$, whence $[\operatorname{cl} \operatorname{cone}(A + C - \overline{y})] \cap (-\operatorname{int} D) = \emptyset$. Hence $\overline{y} \in \text{Be-PMin}(A, C)$. Therefore, (2.37) holds.

Assume now that for any closed convex cone $K \subset Y$ there exists $y^* \in Y^*$ such that $\langle y, y^* \rangle > 0$ for every $y \in K \setminus (-K)$, and take $\overline{y} \in \text{Hu-PMin}(A, C)$. Then $E_3 \cap (-C) = \{0\}$, or, equivalently, $(C \setminus \{0\}) \cap (-E_3) = \emptyset$, where $E_3 := \operatorname{cl} \operatorname{conv} \operatorname{cone} [(A - \overline{y}) \cup C]$. Then there exists $y^* \in Y^*$ such that $\langle y, y^* \rangle > 0$ for every $y \in E_3 \setminus (-E_3)$. Because $C \subseteq E_3$, if $y \in C \setminus \{0\}$ then $y \in E_3 \setminus (-E_3)$, and so $\langle y, y^* \rangle > 0$. Hence $y^* \in C^{\#}$. Therefore, $\overline{y} \in \text{S-PMin}(A, C)$.

(ii) Assume that Θ is a base of C and Y is a locally convex space. Because Θ is a base of C, there exists $V_0 \in \mathcal{N}_Y^c$ such that $(2V_0) \cap \Theta = \emptyset$, or, equivalently, $V_0 \cap (V_0 - \Theta) = \emptyset$. Take $\overline{y} \in \text{Sup-PMin}(A, C)$, that is $\overline{y} \in A$ and for every $V \in \mathcal{N}_Y$ there exists $U \in \mathcal{N}_Y$ such that $K \cap (U - C) \subseteq V$. Therefore, there exists $U_0 \in \mathcal{N}_Y^c$ such that $U_0 \subseteq V_0$ and $[K \cap (U_0 - \Theta)] \subseteq K \cap (U_0 - C) \subseteq V_0$. It follows that $K \cap (U_0 - \Theta) \subseteq V_0 \cap (U_0 - \Theta) \subseteq V_0 \cap (V_0 - \Theta) = \emptyset$. Hence $\overline{y} \in \text{He-PMin}(A, \Theta)$.

Assume, moreover, that Θ is bounded and take $\overline{y} \in \text{He-PMin}(A, \Theta)$. Then there exists $V_0 \in \mathcal{N}_Y$ such that $K \cap (V_0 - \Theta) = \emptyset$. Suppose that $\overline{y} \notin \text{Sup-PMin}(A, C)$. Then there exists $V_1 \in \mathcal{N}_Y^c$ such that for every $U \in \mathcal{N}_Y^c$ one has $K \cap (U - C) \nsubseteq V_0$. Hence, for every $U \in \mathcal{N}_Y^c$ there exists $y_U \in U$, $t_U \in \mathbb{R}_+$, $z_U \in \Theta$ such that $y_U - t_U z_U \in K \setminus V_0$. Clearly, $t_U > 0$ for every $U \in \mathcal{N}_Y^c$ with $U \subseteq V_0$; otherwise we get the contradiction $y_U \in (K \setminus V_0) \cap V_0 = \emptyset$. Taking $p := p_{V_0}$ the Minkowski functional of V_0, p is a continuous seminorm with $\operatorname{int} V_0 = \{y \in Y \mid p(y) < 1\}$ (see Proposition 6.2.1). Clearly, $t_U^{-1} y_U - z_U \in K$. Because $K \cap (V_0 - \Theta) = \emptyset$ we get $t_U^{-1} y_U \notin V_0$ for $U \subseteq V_0$, and so $t_U^{-1} p(y_U) = p(t_U^{-1} y_U) \geq 1$ for such U. It follows that $t_U \leq p(y_U)$ for $U \subseteq V_0$. Hence $(t_U)_{U \in \mathcal{N}_Y^c} \to 0$. Since $(y_U)_{U \in \mathcal{N}_Y^c} \to 0$ and $(z_U)_{U \in \mathcal{N}_Y^c}$ is bounded, it follows that $(y_U - t_U z_U)_{U \in \mathcal{N}_Y^c} \to 0$, contradicting the fact that $y_U - t_U z_U \in K \setminus V_0$ for every U. This contradiction shows that $\overline{y} \in \text{Sup-PMin}(A, C)$.

(iii) Assume that Θ is a compact base of C and Y is a locally convex space.

Take $\overline{y} \in \text{Hu-PMin}(A, C)$, that is $\overline{y} \in A$ and $E_4 \cap (-C) = \{0\}$, where $E_4 := \operatorname{cl} \operatorname{conv} \operatorname{cone} \left[(A - \overline{y}) \cup C\right]$. It follows that $E_4 \cap (-\Theta) = \emptyset$. Using a separation theorem, there exist $y^* \in Y^*$ and $\alpha, \beta \in \mathbb{R}$ such that

$$\forall y \in (A - \overline{y}) \cup C,\ v' \in \Theta,\ t \in \mathbb{R}_+ : t \langle y, y^* \rangle \geq \alpha > \beta \geq \left\langle -v', y^* \right\rangle. \tag{2.40}$$

It follows that $\alpha \leq 0$, $2\gamma := \inf_\Theta y^* > 0$, whence $y^* \in C^{\#}$, and $\langle y - \overline{y}, y^* \rangle \geq 0$ for every $y \in A$. Hence $\overline{y} \in \text{S-PMin}(A, C)$. Taking into account (2.35) we obtain that $\text{S-PMin}(A, C) = \text{Hu-PMin}(A, C)$.

Take $\overline{y} \in \text{GHe-PMin}(A, C)$. There exists $D \in \mathcal{D}_C$ such that $(A - \overline{y}) \cap (-\operatorname{int} D) = \emptyset$. Clearly, $\Theta \subseteq \operatorname{int} D$. Since Θ is compact and Y is a l.c.s., there exists $V \in \mathcal{N}_Y^\Theta$ such that $\Theta + V \subseteq \operatorname{int} D$. Taking $D' := P_V^\Theta$ we have that $D' \in \mathcal{D}_\Theta$ and $\overline{y} \in \text{WMin}(A, D')$, and so $\overline{y} \in \text{He-PMin}(A, \Theta)$. Hence $\text{He-PMin}(A, \Theta) = \text{GHe-PMin}(A, C)$.

Assume now that A is closely C-convex and take $\overline{y} \in \text{Be-PMin}(A, C)$, that is $\overline{y} \in A$ and $\operatorname{cl}\operatorname{cone}(A + C - \overline{y}) \cap (-C) = \{0\}$. Since A is closely C-convex, we obtain that $\operatorname{cl}\operatorname{cone}(A + C - \overline{y})$ is a closed convex cone. From our hypothesis we have that $\operatorname{cl}\operatorname{cone}(A + C - \overline{y}) \cap (-\Theta) = \emptyset$. Using a separation theorem, there exist $y^* \in Y^*$ and $\alpha, \beta \in \mathbb{R}$ such that (2.40) holds with $y \in A + C - \overline{y}$ instead of $y \in (A - \overline{y}) \cup C$. It follows that $\alpha \leq 0$, $2\gamma := \inf_\Theta y^* > 0$, whence $y^* \in C^\#$, and $\langle y - \overline{y}, y^* \rangle \geq 0$ for every $y \in A$. Hence $\overline{y} \in \text{S-PMin}(A, C)$. Hence $\text{Be-PMin}(A, C) = \text{S-PMin}(A, C)$.

Assume now that A is closely convex and take $\overline{y} \in \text{Bo-PMin}(A, C)$, that is $\overline{y} \in A$ and

$$\operatorname{cl}\operatorname{cone}(A - \overline{y}) \cap (-C) = \{0\},$$

whence $\operatorname{cl}\operatorname{cone}(A - \overline{y}) \cap (-\Theta) = \emptyset$. Since A is closely convex, $\operatorname{cl}\operatorname{cone}(A - \overline{y})$ is convex. Using a separation theorem, as above, we get $\overline{y} \in \text{S-PMin}(A, C)$. Hence $\text{S-PMin}(A, C) = \text{Bo-PMin}(A, C)$.

(iv) Assume now that $\text{StrMin}(A, C) \neq \emptyset$. Because C is pointed we have that $\text{StrMin}(A, C) = \{\overline{y}\}$ for some $\overline{y} \in A$; hence $A \subseteq \overline{y} + C$. Clearly, $(A - \overline{y}) \cap (-C) \subseteq C \cap (-C) = \{0\}$, and so $\overline{y} \in \text{Min}(A, C)$. Conversely, take $y \in \text{Min}(A, C)$; then $\{0\} = (A - y) \cap (-C) \ni \overline{y} - y$, and so $y = \overline{y} \in \text{StrMin}(A, C)$. Therefore, $\text{StrMin}(A, C) = \text{Min}(A, C)$.

Since $A - \overline{y} \subseteq C$, we have that $(A - \overline{y}) \cup C = (A + C - \overline{y}) \cup C = C$.

Assume, moreover, that C is closed; then

$$\operatorname{cl}\operatorname{conv}\operatorname{cone}[(A + C - \overline{y}) \cup C] = C,$$

and so $\overline{y} \in \text{Hu-PMin}(A, C)$. Then (2.38) follows from $\text{Min}(A, C) = \{\overline{y}\}$ and relations (2.35), (2.36).

Assume that $C^\# \neq \emptyset$. Because $\overline{y} \in A \subseteq \overline{y} + C$, clearly, $\arg\min_A y^* = \{\overline{y}\}$ for every $y^* \in C^\#$. Hence $\text{S-PMin}(A, C) = \{\overline{y}\}$. The second equality in (2.39) follows using Proposition 2.4.5 (iv).

Assume that Y is a l.c.s. and Θ is a base of C. Since $0 \notin \operatorname{cl}\Theta$, there exists $y^* \in Y^*$ with $\inf_\Theta y^* > 0$. Hence $y^* \in C^\#$, and so $\arg\min_A y^* = \{\overline{y}\}$. Using Proposition 2.4.5 (v) we have that $\overline{y} \in \text{S-PMin}(A, C)$. Using (2.39) and Proposition 2.4.5 (vi) we get $\text{He-PMin}(A, \Theta) = \{\overline{y}\}$. □

It is worth observing that for $y^* \in Y^* \setminus \{0\}$ and $C := K_{y^*} := \{0\} \cup \{y \in Y \mid \langle y, y^* \rangle > 0\}$ we have that $C^\# = \mathbb{P}y^*$ and $\text{S-PMin}(A, C) = \text{Min}(A, C) = \arg\min_A y^*$, and so all the efficiency sets used in Propositions 2.4.5, 2.4.6, excepting StrMin, reduce to $\arg\min_A y^*$. Moreover, $\text{StrMin}(A, C) \neq \emptyset$ if and only if y^* has a unique minimum point on A.

The inclusions in assertions (i), and (iii), (iv) of Proposition 2.4.5, as well as the corresponding equalities for C closed or A convex, respectively, are established in [175, Theorem 3.1]. The most part of the inclusions in assertions (i) and (iii) of Proposition 2.4.6, as well as the mentioned equalities, can be found [392]; note that the hypothesis that Θ is compact can be replaced by the fact that Θ is weakly compact (for this just apply the corresponding result for the weak topology on Y). Assertion (ii) of Proposition 2.4.6 can be found in [618], while the equality $\text{S-PMin}(A, C) = \text{Hu-PMin}(A, C)$ in Proposition 2.4.6 (i) is obtained in [271, Theorem V.2.4]; the case of separable normed vector spaces in (i) is based on the following extension of the Krein–Rutman theorem obtained by Hurwicz [271, Lemma V.2.2].

Theorem 2.4.7. *Let $(Y, \|\cdot\|)$ be a separable normed vector space and $C \subseteq Y$ be a closed convex cone. Then there exists $y^* \in Y^*$ such that $\langle y, y^* \rangle > 0$ for every $y \in C \setminus (-C)$.*

Proof. If $C \setminus (-C) = \emptyset$ (that is C is a linear space) we can take $y^* = 0$. So, assume that $C \setminus (-C) \neq \emptyset$. Because $C^{++} = C$, C^+ is not trivial, too. The set $C_1 := C^+ \cap U_{Y^*}$ is a weakly* closed subset of U_{Y^*} (hence C_1 is w^*-compact). Because Y is separable, the weak* topology on U_{Y^*} is metrizable (see [108, Theorem V.5.1]), and so C_1 (being w^*-compact) is w^*-separable. Let $A = \{y_1^*, y_2^*, \ldots\} \subseteq C_1$ be w^*-dense in C_1. Take

$$\overline{y}^* := \sum_{k=1}^{\infty} \frac{1}{2^k} y_k^*;$$

the series is strongly convergent because it is absolutely convergent and Y^* is a Banach space. Clearly, $\overline{y}^* \in C_1 \subseteq C^+$. Assume that there exists $\overline{y} \in C \setminus (-C)$ such that $\langle \overline{y}, \overline{y}^* \rangle = 0$. Because $\langle \overline{y}, y_k^* \rangle \geq 0$ for every $k \geq 1$, we obtain that $\langle \overline{y}, y^* \rangle = 0$ for every $y^* \in A$. With the set A being w^*-dense in C_1, we obtain that $\langle \overline{y}, y^* \rangle = 0$ for every $y^* \in C_1$, and so $\langle -\overline{y}, y^* \rangle = 0 \geq 0$ for every $y^* \in C^+ = \mathbb{R}_+ C_1$. Therefore, we get the contradiction $-\overline{y} \in C^{++} = C$. □

In order to describe weak and proper minimality in a unified way, we use the notation of Q-minimal points (compare Ha [228]).

Definition 2.4.8. Assume that $D \subset Y$ is a proper cone with nonempty interior and put $Q := \operatorname{int} D$. We say that $\overline{y}$ is a **Q-minimal point** of A ($\overline{y} \in \text{QMin}(A, C)$) if

$$A \cap (\overline{y} - Q) = \emptyset$$

or, equivalently,

$$(A - \overline{y}) \cap (-Q) = \emptyset.$$

In the paper by Gerstewitz and Iwanow [197] properly minimal elements are defined using a set $Q \subset Y$ with $0 \in \operatorname{bd} Q$ and $\operatorname{cl} Q + (C \setminus \{0\}) \subset \operatorname{int} Q$. This approach is related to the well-known concept of *dilating cone* (or a *dilation*) of C:

Definition 2.4.9. Suppose that $D \subset Y$ is a proper cone with nonempty interior and put $Q := \operatorname{int} D$. Q is said to be a *dilation* of C, or *dilating* C if it contains $C \setminus \{0\}$.

Remark 2.4.10. Makarov and Rachkovski [409] studied more detailed some concepts of proper efficiency and introduced the notion of $\mathscr{B}$-efficiency, i.e., efficiency w.r.t. a family of dilations of C. Namely, given $\mathscr{B} \in \mathscr{F}(C)$, where $\mathscr{F}(C)$ is the class of families of dilations of C, $\overline{y}$ is said to be a *$\mathscr{B}$-minimal point* of A ($\overline{y} \in \mathscr{B}\,\mathrm{Min}(A, C)$) if there exists $B \in \mathscr{B}$ such that

$$(A - \overline{y}) \cap (-B) = \emptyset.$$

It has been established that Borwein proper efficiency, Henig global proper efficiency, Henig proper efficiency, super efficiency and Hartley proper efficiency are $\mathscr{B}$-efficiency with $\mathscr{B}$ being appropriately chosen family of dilating cones. The reader will see that in contrast with $\mathscr{B}$-efficiency, the concept introduced in Definition 2.4.8 includes not only some concepts of proper efficiency among which are these ones considered in [409] but also the concepts of strong efficiency and weak efficiency.

In order to study the relationships between weakly / properly minimal points and Q-minimal points let Y be a n.v.s. and Θ as before a base of C. Setting

$$\delta := \delta_\Theta := d(0, \Theta) = \inf\{\|\theta\| \mid \theta \in \Theta\} > 0,$$

for each $0 < \eta < \delta$, we can associate to C a convex, pointed and open set V_η, defined by

$$V_\eta := \mathrm{cone}(\Theta + \eta \mathring{B}_Y). \tag{2.41}$$

For each scalar $\varepsilon > 0$, we also associate to C an open set $C(\varepsilon)$

$$C(\varepsilon) := \{y \in Y \mid d(y, C) < \varepsilon d(y, -C)\}.$$

We are going to show that the weakly / properly minimal points introduced in Definitions 2.4.2 and 2.4.4 are in fact Q-minimal points (Definition 2.4.8) with Q being appropriately chosen sets. The following result is shown in [228].

Theorem 2.4.11. (a) $\overline{y} \in \mathrm{WMin}(A, C)$ *iff* $\overline{y} \in \mathrm{QMin}(A, C)$ *with* $Q = \mathrm{int} C$.
(b) $\overline{y} \in \text{S-PMin}(A, C)$ *iff* $\overline{y} \in \mathrm{QMin}(A, C)$ *with* $Q = \{y \in Y \mid y^*(y) > 0\}$ *and* $y^* \in C^\#$.
(c) $\overline{y} \in \text{Hu-PMin}(A, C)$ *iff* $\overline{y} \in \mathrm{QMin}(A, C)$, *with* $Q = Y \setminus -\mathrm{cl\,conv\,cone}[(A - \overline{y}) \cup C]$.
(d) $\overline{y} \in \text{Be-PMin}(A, C)$ *iff* $\overline{y} \in \mathrm{QMin}(A, C)$, *with* $Q = Y \setminus -\mathrm{cl\,cone}[(A - \overline{y}) + C]$.
(e) $\overline{y} \in \text{Ha-PMin}(A, C)$ *iff* $\overline{y} \in \mathrm{QMin}(A, C)$ *with* $Q = C(\varepsilon)$ *for some* $\varepsilon > 0$.
(f) $\overline{y} \in \text{Bo-PMin}(A, C)$ *iff* $\overline{y} \in \mathrm{QMin}(A, C)$, *with Q being some dilation of* C.
(g) $\overline{y} \in \text{GHe-PMin}(A, C)$ *iff* $\overline{y} \in \mathrm{QMin}(A, C)$, *with* $Q = \mathrm{int}\, D$, *being some dilation of C, where D is a proper pointed convex cone in Y.*

(h) *(supposing that Y is a n.v.s.)* $\overline{y} \in$ He-PMin(A, C) *iff* $\overline{y} \in$ QMin(A, C) *with* $Q = V_\eta$ *and* η *is some scalar satisfying* $0 < \eta < \delta = d(0, \Theta)$.

(i) *(supposing that Y is a n.v.s. and C has a bounded base Θ)* $\overline{y} \in$ Sup-PMin(A) *iff* $\overline{y} \in$ QMin(A, C) *with* $Q = V_\eta$ *and* η *is some scalar satisfying* $0 < \eta < \delta = d(0, \Theta)$.

Proof. Using Definitions 2.4.4 and 2.4.8 one can easily prove the assertions (a)–(d) and (g). The assertions (e)–(f) are formulated in a slightly different form as established by Makarov and Rachkovski [409].

We prove now the assertion (h), namely, we show that $\overline{y} \in$ He-PMin(A, C) iff there is a scalar η with $0 < \eta < \delta$ such that

$$(A - \overline{y}) \cap (-V_\eta) = \emptyset. \tag{2.42}$$

Recall that by definition, $\overline{y} \in$ He-PMin(A, C) iff

$$\text{cl cone}(A - \overline{y}) \cap (\Theta + \varepsilon B_Y) = \emptyset. \tag{2.43}$$

It is also known [619] that $\overline{y} \in$ He-PMin(A, C) iff

$$(A - \overline{y}) \cap (-\overline{S}_n) = \{0\} \tag{2.44}$$

for some integer $n \in \mathbb{N}$, where $\overline{S}_n = \text{cl cone}(\Theta + \delta/(2n)B_Y)$. Now, suppose that $\overline{y} \in$ He-PMin(A, C). Then (2.43) holds. Without loss of generality we can assume that $0 < \varepsilon < \delta$. We show that (2.43) holds with $\eta = \varepsilon$. Suppose to the contrary that there is $y' \in A - \overline{y}$ such that $y' \in -V_\varepsilon$. Clearly, $y' \in \text{cl cone}(A - \overline{y}) \cap (-\text{cl cone}(\Theta + \varepsilon B_Y))$. On the other hand, as $0 < \eta = \varepsilon < \delta$ and by the definition of δ, $0 \notin V_\varepsilon$. Hence $y' \neq 0$. This is a contradiction to (2.43). Next, suppose that (2.42) holds for some η. Let n be an integer satisfying $n - 1 > \delta/(2\eta)$ or $\delta/(2n - 2) < \eta$. By (2.42) we have

$$(A - \overline{y}) \cap (-V_{\delta/(2n-2)}) \subseteq (A - \overline{y}) \cap (-V_\eta) = \emptyset.$$

Then $(A - \overline{y}) \cap (-V_{\delta/(2n-2)} \cup \{0\}) = \{0\}$. On the other hand, [619, Lemma 2.1] states that if $(A - \overline{y}) \cap (-V_{\delta/(2n-2)} \cup \{0\}) = \{0\}$, then $(A - \overline{y}) \cap (-\overline{S}_n) = \{0\}$. Thus, (2.44) holds and therefore, $\overline{y} \in$ He-PMin(A, C), as it was to be shown.

To complete the proof note that the last assertion (i) of this theorem follows from (h) and the assertion (ii) in Proposition 2.4.6. □

Remark 2.4.12. The assertion (h) in the above theorem is inspired by the definition of Henig properly minimal points for sets in locally convex spaces given by Gong in [213]. One can deduce that any Henig properly minimal point is a global Henig properly minimal point.

Furthermore, we mention solution concepts for vector optimization problems introduced by ElMaghri–Laghdir[175] where it is not supposed that the ordering cone $C \subset Y$ is pointed and closed. This concept is based on a generalization of the concept of dilations of a cone $C \subset Y$ (see Definition 2.4.9). In the text below we follow the presentation in [175]. We assume that X, Y, Z are topological vector spaces and S is a subset of X. For the following notations and results the **lineality** of C, defined by

$$l(C) := C \cap (-C),$$

is very important. Of course, C is pointed if $l(C) = \{0\}$.

Unlike the assumptions made before we assume in the sequel in this section that Y is ordered by the proper convex cone C (C is proper if $C \neq l(C)$, or equivalently, C is not a linear subspace of Y).

Furthermore, using $l(C)$ we introduce

$$C^{\&} := \{y^* \in Y^* \mid \langle y, y^* \rangle > 0 \; \forall y \in C \setminus l(C)\} = (C \setminus l(C))^{\#}.$$

Hence, if C is pointed then $C^{\&} = C^{\#}$. Note that Theorem 2.4.7 gives sufficient conditions for $C^{\&} \neq \emptyset$.

We use the notations $y \leqq y'$ if $y' - y \in C$; $y < y'$ if $y' - y \in \operatorname{int} C$ and furthermore, $y \leq y'$ if $y' - y \in C \setminus l(C)$.

In the following we consider a proper vector-valued objective function $f : X \to Y^\bullet$, $S \subseteq X$ and use the notation

$$f(S) := \{f(x) \mid x \in S \cap \operatorname{dom} f\} \subset Y.$$

Consider now the vector optimization problem

$$\text{minimze } f(x) \text{ subject to } x \in S. \tag{VP}$$

Using the lineality of C and without assuming the pointedness and closedness of C we introduce the following solution concepts for (VP). These concepts are extensions of the solution concepts introduced in Definitions 2.4.1, 2.4.2 and 2.4.4.

Definition 2.4.13. $\overline{x} \in S \cap \operatorname{dom} f$ is

- **strongly l(C)-minimal** if $f(\overline{x}) \leqq f(x)$ for all $x \in S$, or equivalently

$$f(S) \subseteq f(\overline{x}) + C,$$

- **Pareto l(C)-minimal** if $f(x) \leqq f(\overline{x}) \Rightarrow f(\overline{x}) \leqq f(x)$ for all $x \in S$, or equivalently

$$f(S) \cap (f(\overline{x}) - (C \setminus l(C))) = \emptyset,$$

- **weakly l(C)-minimal** if $f(x) \not< f(\overline{x})$ for all $x \in S$, or equivalently

$$f(S) \cap (f(\overline{x}) - \operatorname{int} C) = \emptyset,$$

- **l(C)-properly minimal** if there exists $D \subset Y$ a proper convex cone with $C \setminus l(C) \subseteq \operatorname{int} D$ such that $\overline{x}$ is efficient with respect to D, $f(S) \cap (f(\overline{x}) - D \setminus l(D)) = \emptyset$.

Definition 2.4.14. Suppose that $D \subset Y$ is a proper convex cone with nonempty interior. Put $Q := \operatorname{int} D$. Q is said to be a **generalized dilation** of C or **generalized dilating** C if it contains $C \setminus l(C)$.

Set

$$\mathcal{Q}_C := \{D \subset Y \mid D \text{ is a proper convex cone with } C \setminus l(C) \subset \operatorname{int} D\}.$$

Lemma 2.4.15. *Let $C \subset Y$ be a proper convex cone. Then $\mathcal{Q}_C \neq \emptyset$ if and only if $C^{\&} \neq \emptyset$. Moreover, if $C^{\&} \neq \emptyset$ then*

$$C + \operatorname{int} D = \operatorname{int} D \quad \forall D \in \mathcal{Q}_C \tag{2.45}$$

and

$$C^{\&} = \cup_{D \in \mathcal{Q}_C} (D^+ \setminus \{0\}). \tag{2.46}$$

Proof. Assume that $\mathcal{Q}_C \neq \emptyset$ and take $D \in \mathcal{Q}_C$. Since D is proper, $\operatorname{int} D \neq Y$. Take $q_0 \in Y \setminus \operatorname{int} D$. By a separation theorem there exists $y^* \in Y^*$ such that $\langle q_0, y^* \rangle \leq 0 < \langle y, y^* \rangle$ for all $y \in \operatorname{int} D$. It follows that $y^* \in C^{\&}$.

Conversely, assume that $C^{\&} \neq \emptyset$ and take $y^* \in C^{\&}$. Consider $D := \{y \in Y \mid \langle y, y^* \rangle \geq 0\}$. From the very definition of $C^{\&}$ we have that $C \setminus l(C) \subset \{y \in Y \mid \langle y, y^* \rangle > 0\} = \operatorname{int} D$.

Let us prove (2.45). First note that $C \subseteq \operatorname{cl}(C \setminus \operatorname{cl} C)$. Indeed, there exists $k_0 \in C \setminus l(C)$. Let $k \in C$. Then $k + \lambda k_0 \in C + (C \setminus l(C)) = C \setminus l(C)$ for every $\lambda > 0$. The claim follows taking the limit for $\lambda \to 0$. Then, taking $D \in \mathcal{Q}_C$ and using repeatedly [79, Lemma 2.5] we get

$$\operatorname{int} D \subseteq C + \operatorname{int} D = \operatorname{cl}(C \setminus l(C)) + \operatorname{int} D = (C \setminus l(C)) + \operatorname{int} D \subseteq \operatorname{int} D,$$

and so (2.45) holds.

Take now $D \in \mathcal{Q}_C$; then $C \setminus l(C) \subseteq \operatorname{int} D$, and so $C^{\&} = (C \setminus l(C))^{\#} \supseteq (\operatorname{int} D)^{\#} = D^+ \setminus \{0\}$. Hence the inclusion $\supseteq$ holds in (2.46). Let now $y^* \in C^{\&}$. Then $D := \{y \in Y \mid \langle y, y^* \rangle \geq 0\} \in \mathcal{Q}_C$ and $y^* \in D^+ \setminus \{0\}$. It follows that (2.46) holds. □

For $A \subset Y$ we set

$$E_s(A) := E_s^C(A) := \{\overline{a} \in A \mid A \subseteq \overline{a} + C\},$$

$$E_e(A) := E_e^C(A) := \{\overline{a} \in A \mid A \cap (\overline{a} - (C \setminus l(C))) = \emptyset\},$$

$$E_w(A) := E_w^C(A) := \{\overline{a} \in A \mid A \cap (\overline{a} - \operatorname{int} C) = \emptyset\},$$

$$E_p(A) := E_p^C(A) := \cup \left\{E_e^D(A) \mid D \in \mathcal{Q}_C\right\}.$$

Since $\operatorname{int} C \subseteq C \setminus l(C) \subseteq \operatorname{int} D$, we have that

$$E_p(A) \subseteq E_e(A) \subseteq E_w(A), \tag{2.47}$$

the last inclusion makes sense for $\operatorname{int} C \neq \emptyset$. It is worth to observe that

$$E_p(A) = \cup \left\{E_w^D(A) \mid D \in \mathcal{Q}_C\right\}. \tag{2.48}$$

Indeed, the inclusion $\subseteq$ in (2.48) follows from the last inclusion in (2.47). Take $\overline{a} \in E_w^D(A)$ for some $D \in \mathcal{Q}_C$ and consider $D' := \{0\} \cup \operatorname{int} D$. Then $D' \in \mathcal{Q}_C$ and $\overline{a} \in E_w^{D'}(A)$ because $D' \subseteq D$. Since $E_w^{D'}(A) = E_e^{D'}(A)$, we obtain that $\overline{a} \in E_p(A)$.

Lemma 2.4.16. *$E_\sigma(A) = A \cap E_\sigma(A+C)$ for $\sigma \in \{s, e, w, p\}$. Moreover, if $l(C) = \{0\}$, that is C is pointed, then $E_\sigma(A) = E_\sigma(A + C)$ for $\sigma \in \{s, e, p\}$. Generally, $E_w(A) = A \cap \operatorname{bd}(A + C)$ (for $\operatorname{int} C \neq \emptyset$), and so $E_w(A)$ and $E_w(A + C)$ ($= \operatorname{bd}(A + C)$) might be different (take $A = \{0\}$).*

Proof. For $\sigma = s$ use the fact that $A \subseteq a + C \Rightarrow A \subseteq A + C \subseteq a + C$. For $\sigma = e$ use the fact that $C \setminus l(C) = C + (C \setminus l(C))$, while for $\sigma = w$ use the fact that $C + \operatorname{int} C = \operatorname{int} C$.

For $\sigma = p$ one uses (2.45). If $a \in E_p(A)$ then there exists $D \in \mathcal{Q}_C$ such that $A \cap (a - \operatorname{int} D) = \emptyset$. Hence $0 \notin A - a + \operatorname{int} D = (A + C) - a + \operatorname{int} D$ by (2.45), and so $a \in E_p(A + C)$. Conversely, let $a \in A \cap E_p(A + C)$. Then $0 \notin (A + C) - a + \operatorname{int} D = A - a + \operatorname{int} D$, and so $a \in E_p(A)$.

Assume that $l(C) = \{0\}$. We have to prove that $E_\sigma(A + C) \subseteq A$ for $\sigma \in \{s, e, p\}$. Assume that $a + k \in E_\sigma(A + C)$ for some $a \in A$ and $k \in C \setminus \{0\}$. For $\sigma = s$ we get $a + k \leqq a$, whence the contradiction $k \in -C$. For $\sigma = e$ we get the contradiction $a \in (A + C) \cap (a + k - (C \setminus \{0\})$. For $\sigma = p$, we have that $E_p(A) \subseteq E_e(A) \subseteq A$. □

Generally, $E_s(A) = \emptyset$.

Lemma 2.4.17. *If $E_s(A) \neq \emptyset$ then $E_s(A) = E_e(A)$. Moreover, if $C^{\&} \neq \emptyset$ then $E_s(A) = E_p(A) = E_e(A)$.*

Proof. Because $0 \notin C \setminus l(C) = C + (C \setminus l(C))$ we have that

$$C \cap (-(C \setminus l(C))) = \emptyset.$$

Fix $\overline{a} \in E_s(A)$. Let $a \in E_s(A)$. Then $A \subseteq a + C$, and so

$$A \cap (a - (C \setminus l(C))) \subseteq (a + C) \cap (a - (C \setminus l(C))) = a + [C \cap (-(C \setminus l(C)))] = \emptyset.$$

Hence $a \in E_e(A)$, and so $E_s(A) \subseteq E_e(A)$.

Take now $a \in E_e(A) \subseteq A \subseteq \overline{a} + C$. Hence $a = \overline{a} + k$ for some $k \in C$. If $k \notin l(C)$ then $\overline{a} \in A \cap ((a - (C \setminus l(C)))$, contradicting the fact that $a \in E_e(A)$. Hence $k \in l(C)$, and so $A \subseteq \overline{a} + C = a - k + C \subseteq a + C$, which shows that $a \in E_s(A)$.

Assume, moreover, that $C^{\&} \neq \emptyset$, and take $y^* \in C^{\&}$. It follows that $D := \{y \in Y \mid \langle y, y^* \rangle \geq 0\} \in \mathcal{Q}_C$. Using (2.45) we have that $C \cap (-\operatorname{int} D) = \emptyset$, and so

$$A \cap (\overline{a} - \operatorname{int} D) \subseteq (\overline{a} + C) \cap (\overline{a} - \operatorname{int} D) = \overline{a} + (C \cap (-\operatorname{int} D)) = \emptyset;$$

therefore, $\overline{a} \in E_p A$. Hence $E_s(A) \subseteq E_p(A)$. Since always $E_p(A) \subseteq E_e(A)$ we get the conclusion. □

The corresponding sets of solutions for (VP) are denoted by $E_s(f, S)$, $E_e(f, S)$, $E_w(f, S)$, $E_p(f, S)$, respectively. More precisely,

$$E_\sigma(f, S) := \{\overline{x} \in S \cap \operatorname{dom} f \mid f(\overline{x}) \in E_\sigma(f(S \cap \operatorname{dom} f))\}, \quad \sigma \in \{s, e, w, p\}.$$

2.5 Vector Optimization Problems with Variable Ordering Structure

Yu introduced in [606] nondominated solutions of vector optimization problems with variable ordering structure based on general domination set mappings, compare also Chen, Huang, Yang [91]. Vector optimization problems with variable ordering structure are studied intensively by Eichfelder in [162, 164–166], Eichfelder, Ha [168] where corresponding solution concepts, characterizations by scalarization methods, optimality conditions and numerical procedures are presented. Eichfelder [163, 166] gives a very detailed overview on solution concepts for vector optimization problems with variable ordering structure and presents a complete characterization of these solution concepts (see also Eichfelder, Kasimbeyli [170] and Eichfelder, Gerlach [167]).

Let X and Y be Banach spaces, $\emptyset \neq S \subset X$, $f : X \to Y$ and let $C : X \rightrightarrows Y$ be a set-valued map such that for each $x \in X$, $C(x)$ is a nonempty convex set with $0 \in \operatorname{bd} C(x)$.

We consider the following vector optimization problems with variable ordering structure

$$\text{v-minimize} \quad f(x) \quad \text{subject to } x \in S.$$

The solution concept for this problem is given in the following definition (compare [91, Definition 1.15]):

Definition 2.5.1 (v-Minimal Points, Weakly v-Minimal Points). Let $C : X \rightrightarrows Y$ be a set-valued map with $0 \in \operatorname{bd} C(x)$, $C(x)$ a convex set for all $x \in X$, $S \subset X$ and $f : S \to Y$. An element $\overline{x} \in S$ is said to be a **v-minimal point** of f w.r.t. $C(\cdot)$ if

$$(f(S) - f(\overline{x})) \cap (-(C(\overline{x}) \setminus \{0\})) = \emptyset.$$

The set of all $f(x)$ with x a v-minimal point of f w.r.t. $C(\cdot)$ is denoted by $\operatorname{Min}(f(S), C(\cdot))$.

Suppose that for all $x \in X$, $\operatorname{int} C(x) \neq \emptyset$. An element $\overline{x} \in S$ is said to be a **weakly v-minimal point** of f w.r.t. $C(\cdot)$ if

$$(f(S) - f(\overline{x})) \cap (-\operatorname{int} C(\overline{x})) = \emptyset.$$

The set of all $f(x)$ with x a weakly v-minimal point of f w.r.t. $C(\cdot)$ is denoted by $\operatorname{WMin}(f(S), C(\cdot))$.

Remark 2.5.2. For further solution concepts, especially nondominated elements, of vector optimization problems with variable ordering structure see Eichfelder [166].

In the case that we ask for v-minimal points of f w.r.t. $C(\cdot)$ the vector optimization problem with variable ordering structure is given by

$$\operatorname{Min}(f(S), C(\cdot)). \tag{VP_v}$$

When we are looking for weakly v-minimal points of f w.r.t. $C(\cdot)$ we study the problem

$$\operatorname{WMin}(f(S), C(\cdot)).$$

The following relationships between v-minimal solutions of the vector optimization problem (VP_v) and solutions of suitable scalarized problems are shown by Chen, Huang and Yang [91, Theorem 2.18].

Theorem 2.5.3. *Consider the vector optimization problem with variable ordering structure (VP_v), where $C : X \rightrightarrows Y$ is a set-valued map such that for each $x \in X$, $C(x)$ is a convex subset of Y with $0 \in \operatorname{bd} C(x)$ and $\operatorname{int} C(x) \neq \emptyset$. Then:*

(a) *Let $\overline{x} \in S$. Suppose that there exists $y^* \in Y^*$ with $y^*(y) > 0$ for all $y \in C(\overline{x}) \setminus \{0\}$ such that $\overline{x} \in S$ is a minimal solution of the scalar optimization problem*

$$\min_{x \in S} y^*(f(x)). \tag{P_{y^*}}$$

Then $\overline{x}$ is a v-minimal point of f w.r.t. $C(\cdot)$ concerning the vector optimization problem (VP_v).

(b) *Let $f(S)$ be a convex subset of Y and $\overline{x}$ a v-minimal point of the problem (VP_v). Then there exists $y^* \in Y^*$ satisfying $y^*(y) > 0$ for all $y \in \operatorname{int} C(\overline{x})$, such that $\overline{x}$ is a minimal solution of the scalar optimization problem (P_{y^*}).*

Remark 2.5.4. Characterizations of solutions of general vector optimization problems with variable ordering structure by means of nonlinear scalarizing functionals are given by Eichfelder in [163, 164, 166] and by Eichfelder, Ha [168].

2.6 Solution Concepts in Set-Valued Optimization

Unless otherwise mentioned, let Y be a linear topological space, partially ordered by a proper pointed convex closed cone C. Let $\mathscr{P}(Y) = 2^Y$ be the **power set** of Y.

We consider a set-valued optimization problem with a general geometric constraint:

$$\text{minimize} \quad F(x) \quad \text{subject to} \quad x \in S, \tag{SP}$$

where S is a subset of X, X is a linear space and the cost mapping $F : S \rightrightarrows Y$ is a set-valued mapping. As already introduced, we use the notations $F(S) = \bigcup_{x \in S} F(x)$ and $\operatorname{dom} F = \{x \in S \mid F(x) \neq \emptyset\}$.

In Sects. 2.6.1, 2.6.2 and 2.6.3 we introduce different solution concepts for the problem (*SP*). Furthermore, in Sect. 2.6.4 we present the embedding approach by Kuroiwa [353, 354, 357], in Sect. 2.6.5 we discuss solution concepts with respect to abstract preference relations by Bao and Mordukhovich [28], in Sect. 2.6.6 we introduce solution concepts for set-valued optimization problems with variable ordering structure, in Sect. 2.6.7 we study approximate solutions of set-valued optimization problems and finally, in Sect. 2.7 we discuss relations between the solution concepts.

2.6.1 Solution Concepts Based on Vector Approach

First, we introduce a solution concept where "minimization" in (*SP*) is to be understood with respect to the partial order $\leq_C$ defined in (2.9). In contrast to single-valued functions, for every $\overline{x} \in \operatorname{dom} F$ there are many distinct values $\overline{y} \in Y$ such that $\overline{y} \in F(\overline{x})$. Hence, in the first approach, when studying minimizers of a set-valued mapping, we fix one element $\overline{y} \in F(\overline{x})$, and formulate the following solution concept based on the concept of Pareto minimality introduced in Definition 2.4.1.

Definition 2.6.1 (Minimizer of (SP)). Let $\overline{x} \in S$ and $(\overline{x}, \overline{y}) \in \operatorname{graph} F$. The pair $(\overline{x}, \overline{y}) \in \operatorname{graph} F$ is called a **minimizer** of the problem (*SP*) if $\overline{y} \in \operatorname{Min}\big(F(S), C\big)$.

Furthermore, also the other notions of weakly / properly minimal points for sets (see Definitions 2.4.2 and 2.4.4) naturally induce corresponding notions of *weak / proper minimizers* to the corresponding set optimization problems (see Ha [228]).

Definition 2.6.2 (Weak Minimizer of (SP)). Let $\overline{x} \in S$ and $(\overline{x}, \overline{y}) \in \operatorname{graph} F$. The pair $(\overline{x}, \overline{y}) \in \operatorname{graph} F$ is called a **weak minimizer** of the problem (*SP*) if $\overline{y} \in \operatorname{WMin}\big(F(S), C\big)$.

Let $D \subset Y$ be as before (see Definition 2.4.8) a proper cone with nonempty interior and $Q := \operatorname{int} D$.

Definition 2.6.3 (Q-Minimizer of (SP)). Consider the set-valued optimization problem (SP). Let $\overline{x} \in S$ and $(\overline{x}, \overline{y}) \in \operatorname{graph} F$. We say that $(\overline{x}, \overline{y})$ is an **S-proper minimizer** (**Hurwicz proper minimizer**, **Hartley proper minimizer**, **Benson proper minimizer**, **Borwein proper minimizer**, **Henig global proper minimizer**, **Henig proper minimizer**, **super minimizer** and **Q-minimizer**, respectively) of (SP) if $\overline{y}$ is an S-properly minimal (Hurwicz properly minimal, Hartley properly minimal, Benson properly minimal, Borwein properly minimal, Henig global properly minimal, Henig properly minimal, super efficient and Q-minimal, respectively) point of $F(S)$, i.e., $\overline{y} \in$ S-PMin$(F(S), C)$ ($\overline{y} \in$ Hu-PMin$(F(S), C)$, $\overline{y} \in$ Ha-PMin$(F(S), C)$, $\overline{y} \in$ Be-PMin$(F(S), C)$,
$\overline{y} \in$ Bo-PMin$(F(S), C)$, $\overline{y} \in$ GHe-PMin$(F(S), C)$, $\overline{y} \in$ He-PMin$(F(S), C)$, $\overline{y} \in$ Sup-PMin$(F(S), C)$, $\overline{y} \in$ Q-Min$(F(S), C)$, respectively).

Moreover, especially in Chaps. 8 and 15 we study set-valued optimization problems, where the set-valued objective map $F : X \rightrightarrows Y$ is to be maximized over the feasible set $S \subseteq X$ (X is a linear space)

$$\text{maximize} \quad F(x) \quad \text{subject to} \quad x \in S. \tag{SP_{max}}$$

Analogously to Definitions 2.6.1 and 2.6.2 we now introduce maximizers and weak maximizer of (SP_{max}).

Definition 2.6.4 (Maximizer of (SP_{max})). Let $\overline{x} \in S$ and $(\overline{x}, \overline{y}) \in \operatorname{graph} F$. The pair $(\overline{x}, \overline{y}) \in \operatorname{graph} F$ is called a **maximizer** of the problem (SP_{max}) if $\overline{y} \in \operatorname{Max}\big(F(S), C\big)$.

Definition 2.6.5 (Weak Maximizer of (SP_{max})). Consider the set-valued optimization problem (SP_{max}). Let $\overline{x} \in S$ and $(\overline{x}, \overline{y}) \in \operatorname{graph} F$. We say that $(\overline{x}, \overline{y})$ is a **weak maximizer** of (SP_{max}) if $\overline{y}$ is a weakly maximal point of $F(S)$, i.e., $\overline{y} \in \operatorname{WMax}(F(S), C)$.

Furthermore, we consider set-valued optimization problems with a special structure concerning the restrictions, namely inequality restrictions: Let X, Y, Z be real locally convex Hausdorff spaces, Y, Z be ordered by proper pointed closed convex cones C, K, respectively, $F : X \rightrightarrows Y$ and $G : X \rightrightarrows Z$. Under these assumptions we study a set-valued optimization problem of the following form (see Tasset [570]):

$$\text{minimize} \quad F(x) \quad \text{subject to} \quad x \in S, \tag{SP_T}$$

where $M \subset X$ is a set satisfying $M \subseteq \operatorname{dom} F \cap \operatorname{dom} G$ and

$$S := \{x \in M \mid G(x) \cap (-K) \neq \emptyset\}. \tag{2.49}$$

For set-valued problems (SP_T) with a feasible set S given by (2.49) we derive duality assertions in Sect. 8.1 using the following solution concept with respect to the quasi(-relative) interior of a cone $C \subset Y$.

Let $B \subset Y$ be a nonempty convex set; the **quasi interior of** B is

$$\operatorname{qi} B := \{y \in B \mid \operatorname{cl}(\mathbb{R}_+(B - y)) = Y\}$$

and the **quasi-relative interior of** B is

$$\operatorname{qri} B := \{y \in B \mid \operatorname{cl}(\mathbb{R}_+(B - y)) \text{ is a linear space}\}.$$

Because $\operatorname{cl}(\mathbb{R}_+(B - y)) \subset \operatorname{cl}\operatorname{aff} B - y$, we have that $\operatorname{cl}\operatorname{aff} B = Y$ whenever $\operatorname{qi} B \neq \emptyset$. In fact we have

$$0 \in \operatorname{qi}(B - B) \iff \operatorname{cl}\operatorname{aff} B = Y \Longrightarrow \operatorname{qi} B = \operatorname{qri} B. \tag{2.50}$$

It is worth to observe that for $y_0 \in B$ we have that

$$y_0 \notin \operatorname{qri} B \iff \exists y^* \in Y^* : \inf y^*(B) \geq \langle y_0, y^* \rangle < \sup y^*(B); \tag{2.51}$$

in particular,

$$y_0 \in B \setminus \operatorname{qri} B \Longrightarrow \exists y^* \in Y^* \setminus \{0\} : \inf y^*(B) = \langle y_0, y^* \rangle. \tag{2.52}$$

Note that in the above implications we do not assume that $\operatorname{qri} B \neq \emptyset$. Note that (2.51) covers [72, Theorem 2.7].

Observe also that for $B = C$ a convex cone,

$$y \in \operatorname{qi} C, \; y^* \in C^+ \setminus \{0\} \Longrightarrow \langle y, y^* \rangle > 0. \tag{2.53}$$

Indeed, if $\langle y, y^* \rangle = 0$ then $\langle y' - y, y^* \rangle \geq 0$ for every $y' \in C$, and so $\langle y'', z^* \rangle \geq 0$ for every $y'' \in \operatorname{cl}(\mathbb{R}_+(C - y)) = Y$. We get so the contradiction $y^* = 0$.

Using these notations Tasset [570] introduced the following solution concept for the set-valued problem (SP_T) with restrictions given by (2.49).

Definition 2.6.6 (Quasi-Weak Minimizer of (SP_T)). Assume $\operatorname{qi} C \neq \emptyset$ and consider the set-valued optimization problem (SP_T) with restrictions given by (2.49). Let $\overline{x} \in S$ and $(\overline{x}, \overline{y}) \in \operatorname{graph} F$. The pair $(\overline{x}, \overline{y})$ is called a **quasi-weak minimizer** of the problem (SP_T) with restrictions given by (2.49) if $F(S) \cap (\overline{y} - \operatorname{qi} C) = \emptyset$, and we denote this by $\overline{y} \in \operatorname{Min}(F(S), \operatorname{qi} C)$.

2.6.2 Solution Concepts Based on Set Approach

Although the concept of a minimizer of the set-valued problem (*SP*) given in Definitions 2.6.1 and 2.6.3 is of mathematical interest, it cannot be often used in practice. It is important to mention that a minimizer $(\overline{x}, \overline{y})$ depends on only certain special element $\overline{y}$ of $F(\overline{x})$ and other elements of $F(\overline{x})$ are ignored. In other words, an element $\overline{x} \in S$ for that there exists at least one element $\overline{y} \in F(\overline{x})$ which is a Pareto minimal point (Definition 2.4.1) of the image set of F even if there exist many bad elements in $F(\overline{x})$, is a solution of the set-valued optimization problem (*SP*). For this reason, the solution concepts introduced in Sect. 2.6.1 are sometimes improper.

In order to avoid this drawback it is necessary to work with practically relevant order relations for sets. This leads to solution concepts for set-valued optimization problems based on comparisons among values of the set-valued objective map F.

First, we will introduce several order relations that are used in order to formulate corresponding solution concepts for the set-valued problem (*SP*). The **set less order relation** $\preceq^s_C$ is introduced independently by Young [605] and Nishnianidze [443] (cf. Eichfelder, Jahn [169]) for the comparison of sets:

Definition 2.6.7 (Set Less Order Relation).

Let $C \subset Y$ be a proper closed convex and pointed cone. Furthermore, let $A, B \in \mathscr{P}(Y)$ be arbitrarily chosen nonempty sets. Then the set less order relation is defined by

$$A \preceq^s_C B :\Longleftrightarrow A \subseteq B - C \text{ and } A + C \supseteq B.$$

Remark 2.6.8. Of course, we have

$$A \subseteq B - C \Longleftrightarrow \forall a \in A \; \exists b \in B : a \leq_C b$$

and

$$A + C \supseteq B \Longleftrightarrow \forall b \in B \; \exists a \in A : a \leq_C b.$$

Kuroiwa [347, 349, 351] introduced the following order relations:

Definition 2.6.9 (Lower (Upper) Set Less Order Relation). Let $C \subset Y$ be a proper closed convex and pointed cone. Furthermore, let $A, B \in \mathscr{P}(Y)$ be arbitrary nonempty sets. Then the lower set less order relation $\preceq^l_C$ is defined by

$$A \preceq^l_C B :\Longleftrightarrow A + C \supseteq B$$

and the upper set less order relation $\preceq^u_C$ is defined by

$$A \preceq^u_C B :\Longleftrightarrow A \subseteq B - C.$$

The lower set less order relation $\preceq^l_C$ is illustrated in Fig. 2.2 and the upper set less order relation $\preceq^u_C$ in Fig. 2.3.

Remark 2.6.10. There is the following relationship between the lower set less order relation $\preceq^l_C$ and the upper set less order relation $\preceq^u_C$:

$$A \preceq^l_C B :\Longleftrightarrow A+C \supseteq B \Longleftrightarrow B \subseteq A-(-C) \Longleftrightarrow: B \preceq^u_{-C} A \Longleftrightarrow (-B) \preceq^u_C (-A).$$

Remark 2.6.11. It is easy to see that $A \preceq^l_C B$ is equivalent to

$$A + C \supseteq B + C.$$

Furthermore, $A \preceq^u_C B$ is equivalent to

$$A - C \subseteq B - C.$$

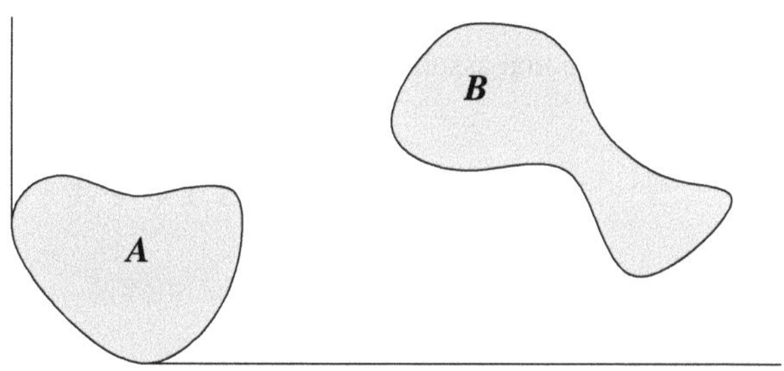

Fig. 2.2 Lower set less order relation $\preccurlyeq^l_C$

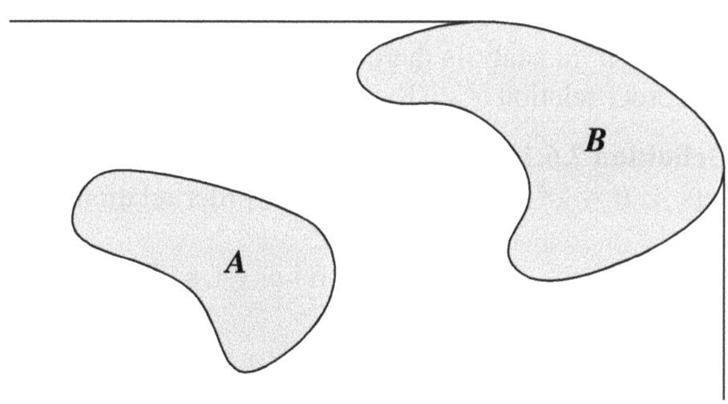

Fig. 2.3 Upper set less order relation $\preccurlyeq^u_C$

It is important to mention that

$$A \preceq^l_C B \text{ and } B \preceq^l_C A \Longleftrightarrow A + C = B + C.$$

Under our assumption that C is a pointed closed convex cone it holds $\mathrm{Min}(A + C, C) = \mathrm{Min}(A, C)$ and $\mathrm{Min}(B + C, C) = \mathrm{Min}(B, C)$ such that we get

$$A \preceq^l_C B \text{ and } B \preceq^l_C A \Longrightarrow \mathrm{Min}(A, C) = \mathrm{Min}(B, C).$$

Under the additional assumptions $A \subseteq \mathrm{Min}(A, C) + C$ and $B \subseteq \mathrm{Min}(B, C) + C$ (domination property, see [47, 400]) we have

$$\mathrm{Min}(A, C) = \mathrm{Min}(B, C) \Longleftrightarrow A + C = B + C$$

and so

$$A \preceq^l_C B \text{ and } B \preceq^l_C A \Longleftrightarrow \mathrm{Min}(A, C) = \mathrm{Min}(B, C).$$

Similarly,

$$A \preceq^u_C B \text{ and } B \preceq^u_C A \Longleftrightarrow A - C = B - C$$

and because of $\mathrm{Max}(A - C, C) = \mathrm{Max}(A, C)$ and $\mathrm{Max}(B - C, C) = \mathrm{Max}(B, C)$ it holds

$$A \preceq^u_C B \text{ and } B \preceq^u_C A \Longrightarrow \mathrm{Max}(A, C) = \mathrm{Max}(B, C).$$

Under the additional assumption $A \subseteq \mathrm{Max}(A, C) - C$ and $B \subseteq \mathrm{Max}(B, C) - C$ it holds

$$\mathrm{Max}(A, C) = \mathrm{Max}(B, C) \Longleftrightarrow A - C = B - C$$

and so

$$A \preceq^u_C B \text{ and } B \preceq^u_C A \Longleftrightarrow \mathrm{Max}(A, C) = \mathrm{Max}(B, C).$$

In interval analysis there are even more order relations in use, like the certainly less order relation $\preceq^c_C$ (Kuroiwa [347–349, 351], compare Eichfelder, Jahn [169]):

Definition 2.6.12 (Certainly Less Order Relation $\preceq^c_C$). For arbitrary nonempty sets $A, B \in \mathscr{P}(Y)$ the certainly less order relation $\preceq^c_C$ is defined by

$$A \preceq^c_C B :\Longleftrightarrow (A = B) \text{ or } (A \neq B, \forall a \in A\ \forall b \in B : a \leq_C b).$$

An illustration of the certainly less order relation $\preceq^c_C$ is given in Fig. 2.4.

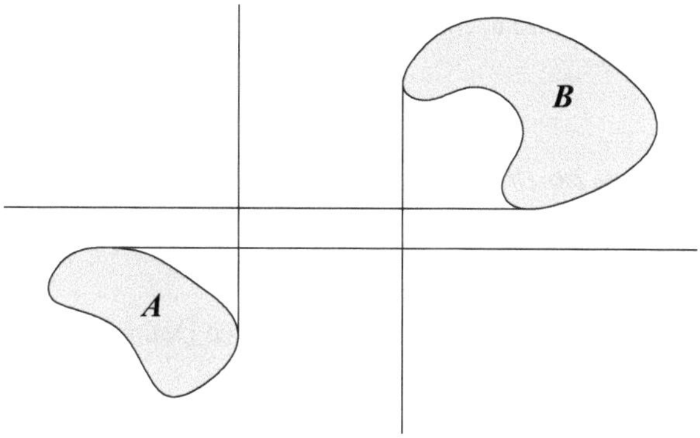

Fig. 2.4 Certainly less order relation $\preccurlyeq_C^c$

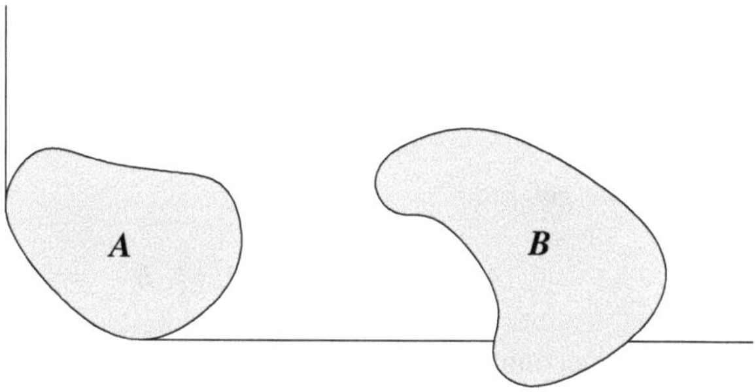

Fig. 2.5 Possibly less order relation $\preccurlyeq_C^p$

Moreover, the possibly less order relation $\preceq_C^p$ (Kuroiwa [348, 349, 351]) is given in the following definition:

Definition 2.6.13 (Possibly Less Order Relation $\preceq_C^p$). For arbitrary nonempty sets $A, B \in \mathscr{P}(Y)$ the possibly less order relation $\preceq_C^p$ is defined by

$$A \preceq_C^p B :\Longleftrightarrow (\exists a \in A, \exists b \in B : a \leq_C b).$$

The possibly less order relation $\preceq_C^p$ is illustrated in Fig. 2.5.

Remark 2.6.14. It is clear that $A \preceq_C^c B$ implies

$$\exists a \in A \quad \text{such that} \quad \forall b \in B : a \leq_C b. \tag{2.54}$$

Moreover, (2.54) implies $A \preceq_C^l B$ (see Definition 2.6.9) such that

$$A \preceq_C^c B \Longrightarrow A \preceq_C^l B.$$

Furthermore, $A \preceq_C^l B$ implies

$$\exists a \in A, \ \exists b \in B \text{ such that } a \leq_C b. \tag{2.55}$$

Taking into account Definition 2.6.13, we have

$$A \preceq_C^c B \quad \Longrightarrow \quad A \preceq_C^l B \quad \Longrightarrow \quad A \preceq_C^p B. \tag{2.56}$$

Remark 2.6.15. The relation $A \preceq_C^c B$ implies

$$\exists b \in B \quad \text{such that} \quad \forall a \in A : a \leq_C b. \tag{2.57}$$

Moreover, (2.57) implies $A \preceq_C^u B$ (see Definition 2.6.9) such that

$$A \preceq_C^c B \Longrightarrow A \preceq_C^u B.$$

Furthermore, $A \preceq_C^u B$ implies

$$\exists a \in A,\ \exists b \in B \text{ such that } a \leq_C b, \tag{2.58}$$

such that we get

$$A \preceq_C^c B \quad \Longrightarrow \quad A \preceq_C^u B \quad \Longrightarrow \quad A \preceq_C^p B$$

taking into account Definition 2.6.13.

Furthermore, the minmax less order relation $\preceq_C^m$ is introduced for sets A, B belonging to

$$\mathscr{F} := \{A \in \mathscr{P}(Y) \mid \operatorname{Min}(A, C) \neq \emptyset \text{ and } \operatorname{Max}(A, C) \neq \emptyset\}.$$

Note that for instance in a topological real linear space Y for every compact set in $\mathscr{P}(Y)$ minimal and maximal elements exist.

Definition 2.6.16 (Minmax Less Order Relation). Let A, B be sets belonging to $\mathscr{F}$. Then the minmax less order relation $\preceq_C^m$ is defined by

$$A \preceq_C^m B :\Longleftrightarrow \operatorname{Min}(A, C) \preceq_C^s \operatorname{Min}(B, C) \text{ and } \operatorname{Max}(A, C) \preceq_C^s \operatorname{Max}(B, C).$$

The minmax certainly less order relation $\preceq_C^{mc}$ is introduced in the next definition:

Definition 2.6.17 (Minmax Certainly Less Order Relation). For arbitrary $A, B \in \mathscr{F}$ the *minmax* certainly less order relation $\preceq_C^{mc}$ is given by

$$\begin{aligned} A \preceq_C^{mc} B :\Longleftrightarrow (A = B)\ or\ (A \neq B, &\operatorname{Min}(A, C) \preceq_C^c \operatorname{Min}(B, C) \\ &\text{and } \operatorname{Max}(A, C) \preceq_C^c \operatorname{Max}(B, C)). \end{aligned}$$

Finally, we introduce the minmax certainly nondominated order relation $\prec_C^{mn}$ (see Jahn, Ha [295]).

Definition 2.6.18 (Minmax Certainly Nondominated Order Relation). For arbitrary nonempty $A, B \in \mathscr{P}(Y)$ the *minmax certainly nondominated order relation* $\preceq_C^{mn}$ is defined by

$$A \preceq_C^{mn} B :\Longleftrightarrow (A = B) \; or \; (A \neq B, \mathrm{Max}(A, C) \preceq_C^{s} \mathrm{Min}(B, C)).$$

The set less order relation $\preceq_C^s$ and the order relations $\preceq_C^l$, $\preceq_C^u$, $\preceq_C^m$, $\preceq_C^{mc}$ and $\preceq_C^{mn}$ are preorders. If $\preceq_C$ denotes one of these order relations, then we can define optimal solutions with respect to the preorder $\preceq_C$ and the corresponding set-valued optimization problem is given by

$$\preceq_C -\text{minimize} \quad F(x), \quad \text{subject to} \quad x \in S, \qquad (SP - \preceq_C)$$

where we assume again (compare (SP)) that Y is a linear topological space, partially ordered by a proper pointed convex closed cone C, S is a subset of X, X is a linear space, $F : X \rightrightarrows Y$.

Definition 2.6.19 (Minimal Solutions of $(SP - \preceq_C)$ w.r.t. the Preorder $\preceq_C$). An element $\overline{x} \in S$ is called a minimal solution of problem $(SP - \preceq_C)$ w.r.t. the preorder $\preceq_C$ if

$$F(x) \preceq_C F(\overline{x}) \quad \text{for some} \quad x \in S \Longrightarrow F(\overline{x}) \preceq_C F(x).$$

Remark 2.6.20. When we use the set relation $\preceq_C^l$ introduced in Definition 2.6.9 in the formulation of the solution concept, i.e., when we study the set-valued optimization problem $(SP - \preceq_C^l)$, we observe that this solution concept is based on comparisons among sets of minimal points of values of F (see Definition 2.4.1). Furthermore, considering the upper set less order relation $\preceq_C^u$ (Definition 2.6.9), i.e., considering the problem $(SP - \preceq_C^u)$ we recognize that this solution concept is based on comparisons of maximal points of values of F (see Definition 2.4.1).

When $\overline{x} \in S$ is a minimal solution of problem $(SP - \preceq_C^l)$ there does not exist $x \in S$ such that $F(x)$ is strictly smaller than $F(\overline{x})$ with respect to the set order $\preceq_C^l$.

In the following we give three examples (see Kuroiwa [347]) of set-valued optimization problems in order to illustrate the different solution concepts introduced in Definitions 2.6.1 and 2.6.19.

Example 2.6.21. Consider the set-valued optimization problem

$$\text{minimize} \quad F_1(x), \quad \text{subject to} \quad x \in S,$$

with $X = \mathbb{R}$, $Y = \mathbb{R}^2$, $C = \mathbb{R}^2_+$, $S = [0, 1]$ and $F_1 : S \rightrightarrows Y$ is given by

$$F_1(x) := \begin{cases} [(1, 0), (0, 1)] & \text{if } x = 0 \\ [(1 - x, x), (1, 1)] & \text{if } x \in (0, 1], \end{cases}$$

where $[(a, b), (c, d)]$ is the line segment between (a, b) and (c, d).

Only the element $\overline{x} = 0$ is a minimal solution in the sense of Definition 2.6.19 w.r.t. $\preceq^l_C$. However, all elements $(\overline{x}, \overline{y}) \in \text{graph } F_1$ with $\overline{x} \in [0,1]$, $\overline{y} = (1 - \overline{x}, \overline{x})$ for $\overline{x} \in (0,1]$ and $\overline{y} = (1,0)$ for $\overline{x} = 0$ are minimizers of the set-valued optimization problem in the sense of Definition 2.6.1. This example shows that the solution concept with respect to the set relation $\preceq^l_C$ (see Definition 2.6.19) is more natural and useful than the concept of minimizers introduced in Definition 2.6.1.

Example 2.6.22. Now we discuss the set-valued optimization problem

$$\text{minimize} \quad F_2(x), \quad \text{subject to} \quad x \in S,$$

with $X = \mathbb{R}$, $Y = \mathbb{R}^2$, $C = \mathbb{R}^2_+$, $S = [0,1]$ and $F_2 : S \rightrightarrows Y$ is given by

$$F_2(x) := \begin{cases} \left[(1, \frac{1}{3}), (\frac{1}{3}, 1)\right] & \text{if } x = 0 \\ [(1-x, x), (1,1)] & \text{if } x \in (0,1]. \end{cases}$$

The set of minimal solutions in the sense of Definition 2.6.19 w.r.t. $\preceq^l_C$ is the interval [0, 1], but the set of minimizers in the sense of Definition 2.6.1 is given by

$$\{(\overline{x}, \overline{y}) \in \text{graph } F_2 \mid \overline{x} \in (0,1],\ \overline{y} = (1 - \overline{x}, \overline{x})\}.$$

Here we observe that $\overline{x} = 0$ is a $\preceq^l_C$-minimal solution but the set $F_2(\overline{x})$ $(\overline{x} = 0)$ has no Pareto minimal points.

Example 2.6.23. In this example we are looking for minimal solutions of a set-valued optimization problem with respect to the set relation $\preceq^u_C$ introduced in Definition 2.6.9.

$$\preceq^u_C -\text{minimize} \quad F_3(x), \quad \text{subject to} \quad x \in S, \qquad (SP - \preceq^u_C)$$

with $X = \mathbb{R}$, $Y = \mathbb{R}^2$, $C = \mathbb{R}^2_+$, $S = [0,1]$ and $F_3 : S \rightrightarrows Y$ is given by

$$F_3(x) := \begin{cases} [[(1,1), (2,2)]] & \text{if } x = 0 \\ [[(0,0), (3,3)]] & \text{if } x \in (0,1], \end{cases}$$

where $[[(a,b),(c,d)]] := \{(y_1, y_2) \mid a \leq y_1 \leq c,\ b \leq y_2 \leq d\}$.

Then a minimal solutions of $(SP - \preceq^u_C)$ in the sense of Definition 2.6.19 is only $\overline{x} = 0$. On the other hand, $x \in (0,1]$ are not minimal solutions of $(SP - \preceq^u_C)$ in the sense of Definition 2.6.19, but for all $\overline{x} \in (0,1]$ there are $\overline{y} \in F_3(\overline{x})$ such that $(\overline{x}, \overline{y})$ are minimizers in the sense of Definition 2.6.1.

Further relationships between different solution concepts in set-valued optimization are discussed in Sect. 2.7.

Applications of solution concepts based on set approach introduced in this section are given in Sect. 15.4 concerning robustness, in Sect. 1.1 concerning game theory. Furthermore, in Sect. 8.2 we present duality assertions for the primal problem $(SP - \preceq^l_C)$.

2.6.3 Solution Concepts Based on Lattice Structure

We recall in this section the concept of an **infimal set** (resp. **supremal set**), which is due to Nieuwenhuis [442], was extended by Tanino [563], and slightly modified with respect to the elements $\pm\infty$ by Löhne and Tammer [397]. We will shortly discuss the role of the space of self-infimal sets, which was shown in [397] to be a complete lattice. As we will see in Sect. 15.1, this complete lattice is useful for applications of set-valued approaches in the theory of vector optimization, especially in duality theory.

First, we recall the definitions of lower and upper bounds as well as the infimum and supremum of a subset of a partially ordered set. Consider a partially ordered set $(Y, \leq)$ and $A \subseteq Y$. As already introduced in Definition 2.1.6, an element $l \in Y$ is called a **lower bound** of A if $l \leq y$ for all $y \in A$. Furthermore, an element $u \in Y$ is called an **upper bound** of A if $u \geq y$ for all $y \in A$. Using lower and upper bounds, the infimum and supremum for a subset A of a partially ordered set $(Y, \leq)$ is defined in Definition 2.1.6. An element $\bar{l} \in Y$ is called greatest lower bound or **infimum** of $A \subseteq Y$ if $\bar{l}$ is a lower bound of A and for every other lower bound l of A it holds $l \leq \bar{l}$. If the infimum of A exists we use the notation $\bar{l} = \inf A$ for it. Analogously, we define the least upper bound or **supremum** of $A \subseteq Y$ and denote it by $\sup A$.

Based on the definition of the infimum and supremum we introduce the notion of a complete lattice that is important for the approach in this section.

Definition 2.6.24. A partially ordered set $(Y, \leq)$ is called a **complete lattice** if the infimum and supremum exist for every subset $A \subseteq Y$.

A characterization of a complete lattice based on the existence of the infimum of subsets $A \subseteq Y$ is given by Löhne [395, Proposition 1.6].

Proposition 2.6.25. *A partially ordered set $(Y, \leq)$ is a complete lattice if and only if the infimum exists for every subset $A \subseteq Y$.*

Let us give some examples for complete lattices (compare [395]).

Example 2.6.26. It is well known that $\overline{\mathbb{R}} = \mathbb{R} \cup \{\pm\infty\}$ equipped with the natural order relation $\leq$ provide a complete lattice.

Example 2.6.27. Consider a nonempty set Y and let $\mathscr{P}(Y) = 2^Y$ be the power set of Y. $(\mathscr{P}(Y), \supseteq)$ is a complete lattice. The infimum and supremum of a nonempty subset $\mathscr{A} \subseteq \mathscr{P}(Y)$ are described by

$$\inf \mathscr{A} = \bigcup_{A \in \mathscr{A}} A, \qquad \sup \mathscr{A} = \bigcap_{A \in \mathscr{A}} A. \tag{2.59}$$

If $\mathscr{A}$ is empty, we put $\sup \mathscr{A} = Y$ and $\inf \mathscr{A} = \emptyset$. $Y \in \mathscr{P}(Y)$ is the least element and $\emptyset \in \mathscr{P}(Y)$ is the greatest element in $(\mathscr{P}(Y), \supseteq)$.

Example 2.6.28. Consider a linear space Y and let $\mathscr{C}(Y)$ be the family of all convex subsets of Y. $(\mathscr{C}(Y), \supseteq)$ provides a complete lattice. The infimum and supremum of a nonempty subset $\mathscr{A} \subseteq \mathscr{C}(Y)$ is described by

$$\inf \mathscr{A} = \operatorname{conv} \bigcup_{A \in \mathscr{A}} A, \qquad \sup \mathscr{A} = \bigcap_{A \in \mathscr{A}} A. \tag{2.60}$$

If $\mathscr{A}$ is empty, we put again $\sup \mathscr{A} = Y$ and $\inf \mathscr{A} = \emptyset$.

Example 2.6.29. Consider a topological space Y and let $\mathscr{F}(Y)$ be the family of all closed subsets of Y. $(\mathscr{F}(Y), \supseteq)$ provides a complete lattice. By

$$\inf \mathscr{A} = \operatorname{cl} \bigcup_{A \in \mathscr{A}} A, \qquad \sup \mathscr{A} = \bigcap_{A \in \mathscr{A}} A \tag{2.61}$$

the infimum and supremum of a nonempty subset $\mathscr{A} \subseteq \mathscr{F}(Y)$ are given.

If $\mathscr{A}$ is empty, we put again $\sup \mathscr{A} = Y$ and $\inf \mathscr{A} = \emptyset$.

Results concerning the infimum and supremum in the space of upper closed sets are given in Proposition 2.6.40.

In the sequel, in this section we assume that $(Y, \leq)$ is a partially ordered linear topological space, where the order is induced by a proper pointed convex cone C satisfying $\emptyset \neq \operatorname{int} C \neq Y$. Here we do not assume that C is closed. However, in Sect. 15.1 we will give a reformulation of a vector optimization problem as $\mathscr{I}$-valued problem, where the closedness of C is important (compare Proposition 2.17 in [395] and Sect. 2.7). We write $y \leq y'$ iff $y' - y \in C$ and $y < y'$ iff $y' - y \in \operatorname{int} C$. We denote by $Y^\bullet := Y \cup \{-\infty\} \cup \{+\infty\}$ the **extended space**, where the ordering is extended by the convention

$$\forall y \in Y: \ -\infty \leq y \leq +\infty.$$

The linear operations on $Y^\bullet$ are extended by the following calculus rules in analogy to that ones stated for the extended real space $\overline{\mathbb{R}}$:

$$\begin{array}{ll} 0 \cdot (+\infty) = 0, & 0 \cdot (-\infty) = 0, \\ \forall \alpha > 0 & : \alpha \cdot (+\infty) = +\infty, \\ \forall \alpha > 0 & : \alpha \cdot (-\infty) = -\infty, \\ \forall y \in Y^\bullet & : y + (+\infty) = +\infty + y = +\infty, \\ \forall y \in Y \cup \{-\infty\} & : y + (-\infty) = -\infty + y = -\infty. \end{array}$$

The extended space $Y^\bullet$ is not a linear space.

In the following definition we introduce the upper closure of $A \subseteq Y^\bullet$ (see [395, 397]) that is important for the formulation of the solution concept.

Definition 2.6.30. The **upper closure** (with respect to C) of $A \subseteq Y^\bullet$ is defined to be the set

$$\mathrm{Cl}_+ A := \begin{cases} Y & \text{if} \quad -\infty \in A \\ \emptyset & \text{if} \quad A = \{+\infty\} \\ \{y \in Y \mid \{y\} + \operatorname{int} C \subseteq A \setminus \{+\infty\} + \operatorname{int} C\} & \text{otherwise.} \end{cases}$$

We have [395, Proposition 1.40]

$$\mathrm{Cl}_+ A := \begin{cases} Y & \text{if} \quad -\infty \in A \\ \emptyset & \text{if} \quad A = \{+\infty\} \\ \mathrm{cl}\,((A \setminus \{+\infty\}) + C) & \text{otherwise.} \end{cases} \tag{2.62}$$

As introduced in Definition 2.4.2, the set of *weakly minimal points* of a subset $A \subseteq Y$ (with respect to C) is defined by

$$\mathrm{WMin}(A, C) = \{y \in A \mid A \cap (\{y\} - \operatorname{int} C) = \emptyset\},$$

and the set of *weakly maximal points* of A is defined by

$$\mathrm{WMax}(A, C) = \{y \in A \mid A \cap (\{y\} + \operatorname{int} C) = \emptyset\}.$$

If $A \neq \emptyset$, $A \subseteq Y$ we have [442, Theorem I-18]

$$\mathrm{WMin}(\mathrm{Cl}_+ A, C) = \emptyset \iff A + \operatorname{int} C = Y \iff \mathrm{Cl}_+ A = Y.$$

In order to formulate set-valued optimization problems where the solution concept is based on the lattice structure we introduce the notion of an infimal set for a subset of $Y^\bullet$ (see Nieuwenhuis [442], Tanino [563, 566] and Löhne, Tammer [397]).

Definition 2.6.31 (Infimal Set). The **infimal set** of $A \subseteq Y^\bullet$ (with respect to C) is defined by

$$\mathrm{Inf}\, A := \begin{cases} \mathrm{WMin}(\mathrm{Cl}_+ A, C) & \text{if } \emptyset \neq \mathrm{Cl}_+ A \neq Y \\ \{-\infty\} & \text{if } \mathrm{Cl}_+ A = Y \\ \{+\infty\} & \text{if } \mathrm{Cl}_+ A = \emptyset. \end{cases}$$

Remark 2.6.32. We see that the infimal set of a nonempty set $A \subset Y$ (with respect to C) coincides with the set of weakly minimal elements of the set $\mathrm{cl}\,(A + C)$ with respect to C ($\mathrm{WMin}(\mathrm{cl}\,(A + C), C)$), if $\mathrm{cl}\,(A + C) \neq Y$. Note that if $A \subset Y$ then $\mathrm{WMin}(A, C) = A \cap \mathrm{Inf}\, A$.

By our conventions, $\mathrm{Inf}\, A$ is always a nonempty set. Clearly, if $-\infty$ belongs to A, we have $\mathrm{Inf}\, A = \{-\infty\}$, in particular, $\mathrm{Inf}\{-\infty\} = \{-\infty\}$. Furthermore, it holds $\mathrm{Inf}\, \emptyset = \mathrm{Inf}\{+\infty\} = \{+\infty\}$ and $\mathrm{Cl}_+ A = \mathrm{Cl}_+(A \cup \{+\infty\})$ and hence $\mathrm{Inf}\, A = \mathrm{Inf}(A \cup \{+\infty\})$ for all $A \subseteq Y^\bullet$.

Considering the set of weakly maximal points $\mathrm{WMax}(A, C)$ of a set $A \subset Y$ with respect to C (see Definition 2.4.2) we define analogously the **lower closure** $\mathrm{Cl}_- A$ and the **supremal set** $\mathrm{Sup}\, A$ of a set $A \subseteq Y^\bullet$. It holds

$$\mathrm{Sup}\, A = -\mathrm{Inf}(-A).$$

The following assertions were proved by Nieuwenhuis [442] and, in an extended form, by Tanino [563].

Proposition 2.6.33. *For $A, B \subseteq Y$ with $\emptyset \neq \mathrm{Cl}_+ A \neq Y$ and $\emptyset \neq \mathrm{Cl}_+ B \neq Y$ it holds*

(i) $\mathrm{Inf}\, A = \{y \in Y \mid y \notin A + \mathrm{int}\, C,\ \{y\} + \mathrm{int}\, C \subseteq A + \mathrm{int}\, C\}$,
(ii) $A + \mathrm{int}\, C = B + \mathrm{int}\, C \iff \mathrm{Inf}\, A = \mathrm{Inf}\, B$,
(iii) $A + \mathrm{int}\, C = \mathrm{Inf}\, A + \mathrm{int}\, C$,
(iv) $\mathrm{Cl}_+ A = \mathrm{Inf}\, A \cup (\mathrm{Inf}\, A + \mathrm{int}\, C)$,
(v) $\mathrm{Inf}\, A$, $(\mathrm{Inf}\, A - \mathrm{int}\, C)$ *and* $(\mathrm{Inf}\, A + \mathrm{int}\, C)$ *are disjoint,*
(vi) $\mathrm{Inf}\, A \cup (\mathrm{Inf}\, A - \mathrm{int}\, C) \cup (\mathrm{Inf}\, A + \mathrm{int}\, C) = Y$.

Proposition 2.6.34. *For $A \subseteq Y^\bullet$ it holds*

(i) $\mathrm{Inf}\,\mathrm{Inf}\, A = \mathrm{Inf}\, A$, $\mathrm{Cl}_+\mathrm{Cl}_+ A = \mathrm{Cl}_+ A$, $\mathrm{Inf}\,\mathrm{Cl}_+ A = \mathrm{Inf}\, A$, $\mathrm{Cl}_+\,\mathrm{Inf}\, A = \mathrm{Cl}_+ A$,
(ii) $\mathrm{Inf}(\mathrm{Inf}\, A + \mathrm{Inf}\, B) = \mathrm{Inf}(A + B)$,
(iii) $\alpha\, \mathrm{Inf}\, A = \mathrm{Inf}(\alpha A)$ *for* $\alpha > 0$.

Proposition 2.6.35. *Let $A_i \subset Y^\bullet$ for $i \in I$, where I is an arbitrary index set. Then it holds*

(i) $\mathrm{Cl}_+ \bigcup_{i \in I} A_i = \mathrm{Cl}_+ \bigcup_{i \in I} \mathrm{Cl}_+ A_i$,
(ii) $\mathrm{Inf} \bigcup_{i \in I} A_i = \mathrm{Inf} \bigcup_{i \in I} \mathrm{Inf}\, A_i$.

Proof. (i) As $\mathrm{Cl}_+ \{+\infty\} = \emptyset$ and $\mathrm{Cl}_+ A = \mathrm{Cl}_+(A \setminus \{+\infty\})$ we can assume that $+\infty \notin \bigcup_{i \in I} A_i$. We also assume $\{-\infty\} \notin \bigcup_{i \in I} A_i$, because the statement is otherwise obvious.

So we have

$$\mathrm{Cl}_+ \bigcup_{i \in I} A_i = \mathrm{cl}\left(\bigcup_{i \in I} A_i + C\right) = \mathrm{cl}\left(\bigcup_{i \in I} \mathrm{cl}(A_i + C)\right) + C = \mathrm{Cl}_+ \bigcup_{i \in I} \mathrm{Cl}_+ A_i.$$

(ii) Follows from (i) and Proposition 2.6.33 (iv). □

In order to formulate set-valued optimization problems in an appropriate form for deriving duality assertions (see Sects. 8.3 and 15.1) we introduce in the following definitions the hyperspaces of upper closed sets and self-infimal sets.

Using the upper closure of $A \subseteq Y$ (Definition 2.6.30) we introduce the hyperspace of upper closed sets (compare Löhne, Tammer [397]).

Definition 2.6.36 (Hyperspace of Upper Closed Sets).
The family $\mathscr{F} := \mathscr{F}_C(Y)$ of all sets $A \subseteq Y$ with

$$\mathrm{Cl}_+ A = A$$

is called the **hyperspace of upper closed sets**.

In $\mathscr{F}$ we introduce an addition $\oplus_{\mathscr{F}} : \mathscr{F} \times \mathscr{F} \to \mathscr{F}$, a multiplication by non-negative real numbers $\odot_{\mathscr{F}} : \mathbb{R}_+ \times \mathscr{F} \to \mathscr{F}$ and an order relation $\preccurlyeq_{\mathscr{F}}$ as follows:

$$A \oplus_{\mathscr{F}} B := \mathrm{cl}\,(A + B),$$

$$\alpha \odot_{\mathscr{F}} A := \mathrm{Cl}_+(\alpha \cdot A)$$

$$A \preccurlyeq_{\mathscr{F}} B : \Longleftrightarrow \; A \supseteq B.$$

We use the rule $0 \cdot \emptyset = \{0\}$. This implies $0 \odot_{\mathscr{F}} \emptyset = \mathrm{Cl}_+\{0\} = \mathrm{cl}\, C$.
Furthermore, using the infimal set of $A \subseteq Y^\bullet$ (Definition 2.6.31) we introduce the hyperspace of self-infimal sets (compare [397]).

Definition 2.6.37 (Hyperspace of Self-Infimal Sets). The family $\mathscr{I} := \mathscr{I}_C(Y^\bullet)$ of all self-infimal subsets of $Y^\bullet$, i.e., all sets $A \subseteq Y^\bullet$ satisfying

$$\mathrm{Inf}\, A = A$$

is called **hyperspace of self-infimal sets**.

In $\mathscr{I}$ we introduce an addition $\oplus_{\mathscr{I}} : \mathscr{I} \times \mathscr{I} \to \mathscr{I}$, a multiplication by non-negative real numbers $\odot_{\mathscr{I}} : \mathbb{R}_+ \times \mathscr{I} \to \mathscr{I}$ and an order relation $\preccurlyeq_{\mathscr{I}}$ as follows:

$$A \oplus_{\mathscr{I}} B := \mathrm{Inf}(A + B),$$

$$\alpha \odot_{\mathscr{I}} A := \mathrm{Inf}(\alpha \cdot A)$$

$$A \preccurlyeq_{\mathscr{I}} B : \Longleftrightarrow \; \mathrm{Cl}_+ A \supseteq \mathrm{Cl}_+ B.$$

Note that the definition of $\oplus_{\mathscr{I}}$ is based on the inf-addition in $Y^\bullet$. As a consequence we obtain $\{-\infty\} \oplus_{\mathscr{I}} \{+\infty\} = \{+\infty\}$. Of course, for all $A \in \mathscr{I}$ we get $0 \odot_{\mathscr{I}} A = \mathrm{Inf}\{0\} = \mathrm{bd}\, C$. In the space of self-supremal sets the sup-addition in $Y^\bullet$ is the underlying operation (Fig. 2.6).

Lemma 2.6.38. *For $A, B \in \mathscr{I}$ with $\emptyset \neq \mathrm{Cl}_+ A \neq Y$ and $\emptyset \neq \mathrm{Cl}_+ B \neq Y$ we have*

$$A \preccurlyeq_{\mathscr{I}} B \quad \Longleftrightarrow \quad A \cap (B + \mathrm{int}\, C) = \emptyset.$$

Proof. If $\mathrm{Cl}_+ A = \emptyset$ or $\mathrm{Cl}_+ B = \emptyset$ the proof is immediate, such that we assume that $\mathrm{Cl}_+ A \neq \emptyset$ and $\mathrm{Cl}_+ B \neq \emptyset$.

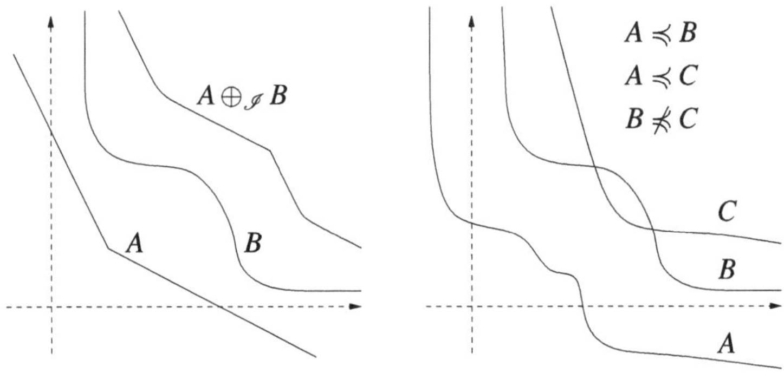

Fig. 2.6 The addition and the ordering in $\mathscr{I}$ for $C = \mathbb{R}^2_+$

Suppose that $A \preccurlyeq_{\mathscr{I}} B$. Taking into account the definition of $\preccurlyeq_{\mathscr{I}}$ it holds $\operatorname{Cl}_+ B \subseteq \operatorname{Cl}_+ A$. By Proposition 2.6.33 (iv) we have $\operatorname{Cl}_+ B = B \cup (B + \operatorname{int} C)$. This yields $(B \cup (B + \operatorname{int} C)) \cap (A - \operatorname{int} C) = \emptyset$ because of $\operatorname{Cl}_+ B \subseteq \operatorname{Cl}_+ A$. Therefore $A \cap (B + \operatorname{int} C) = \emptyset$.

Conversely, if $A \cap (B + \operatorname{int} C) = \emptyset$ then $B + \operatorname{int} C \subset A + \operatorname{int} C$. Hence $\operatorname{Cl}_+ B \subset \operatorname{Cl}_+ A$ and so $A \preccurlyeq_{\mathscr{I}} B$. □

Proposition 2.6.39. *The spaces* $(\mathscr{F}, \oplus_{\mathscr{F}}, \odot_{\mathscr{F}}, \supset)$ *and* $(\mathscr{I}, \oplus_{\mathscr{I}}, \odot_{\mathscr{I}}, \preccurlyeq_{\mathscr{I}})$ *are isomorphic and isotone. The corresponding bijection is given by*

$$j : \mathscr{F} \to \mathscr{I}, \quad j(\cdot) = \operatorname{Inf}(\cdot), \quad j^{-1}(\cdot) = \operatorname{Cl}_+(\cdot).$$

Proof. By Proposition 2.6.34 (i), j is a bijection between $\mathscr{F}$ and $\mathscr{I}$.

For $A_1, A_2 \in \mathscr{F}$, we have $j(A_1) \oplus_{\mathscr{I}} j(A_2) = j(A_1 \oplus A_2)$. This follows from Proposition 2.6.34 (ii).

Similarly, we can easily verify that for $\alpha \geq 0$ and $A, B \in \mathscr{F}$ we have

$$\alpha \odot_{\mathscr{I}} j(A) = j(\alpha \odot_{\mathscr{F}} A) \quad \text{and} \quad A \supseteq B \iff j(A) \preccurlyeq j(B).$$ □

Proposition 2.6.40. $(\mathscr{F}, \supseteq)$ *and* $(\mathscr{I}, \preccurlyeq)$ *are complete lattices. For nonempty subsets* $\mathscr{A} \subseteq \mathscr{F}$ *and* $\mathscr{B} \subseteq \mathscr{I}$ *the infimum and supremum can be expressed by*

$$\inf \mathscr{A} = \operatorname{cl} \bigcup_{A \in \mathscr{A}} A, \qquad \sup \mathscr{A} = \bigcap_{A \in \mathscr{A}} A,$$

$$\inf \mathscr{B} = \operatorname{Inf} \bigcup_{B \in \mathscr{B}} \operatorname{Cl}_+ B, \qquad \sup \mathscr{B} = \operatorname{Inf} \bigcap_{B \in \mathscr{B}} \operatorname{Cl}_+ B.$$

Proof. For the space $(\mathscr{F}, \supseteq)$ the statements are obvious and for $(\mathscr{I}, \preccurlyeq)$ they follow from Proposition 2.6.39. □

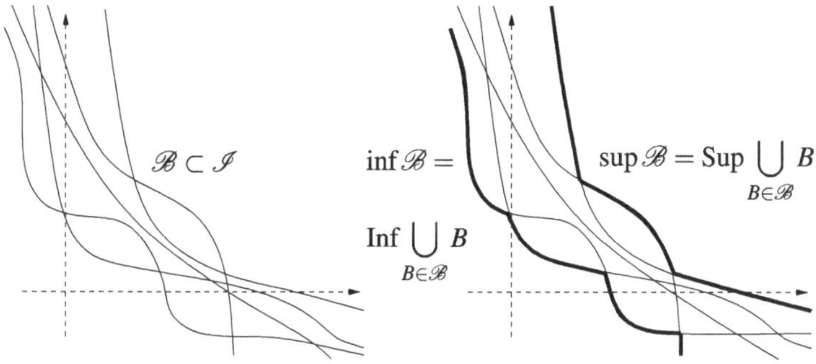

Fig. 2.7 The infimum and supremum in $\mathscr{I}$ for $C = \mathbb{R}^2_+$

As usual, if $\mathscr{A} \subset \mathscr{F}$ and $\mathscr{B} \subset \mathscr{I}$ are empty we define the infimum (supremum) to be the largest (smallest) element in the corresponding complete lattice, i.e., $\inf \mathscr{A} = \emptyset$, $\sup \mathscr{A} = Y$, $\inf \mathscr{B} = \{+\infty\}$ and $\sup \mathscr{B} = \{-\infty\}$.

It follows the main result of this section, which shows that the infimum as well as the supremum in $\mathscr{I}$ can be expressed in terms that frequently are used in vector optimization (compare [442], [145, 563, 566]), but up to now not in the framework of complete lattices (see Fig. 2.7).

Theorem 2.6.41. *For nonempty sets* $\mathscr{B} \subset \mathscr{I}$ *it holds*

$$\inf \mathscr{B} = \operatorname{Inf} \bigcup_{B \in \mathscr{B}} B, \qquad \sup \mathscr{B} = \operatorname{Sup} \bigcup_{B \in \mathscr{B}} B.$$

Proof. (i) It holds $\inf \mathscr{B} = \operatorname{Inf} \bigcup_{B \in \mathscr{B}} \operatorname{Cl}_+ B = \operatorname{Inf} \operatorname{Cl}_+ \bigcup_{B \in \mathscr{B}} \operatorname{Cl}_+ B = \operatorname{Inf} \operatorname{Cl}_+ \bigcup_{B \in \mathscr{B}} B = \operatorname{Inf} \bigcup_{B \in \mathscr{B}} B$.

(ii) We have to show that

$$\operatorname{Sup} \bigcup_{B \in \mathscr{B}} B = \operatorname{Inf} \bigcap_{B \in \mathscr{B}} \operatorname{Cl}_+ B.$$

Then the assertion follows with Proposition 2.6.40.

a) If $\{+\infty\} \in \mathscr{B}$ we have $+\infty \in \bigcup_{B \in \mathscr{B}} B$ and hence $\operatorname{Sup} \bigcup_{B \in \mathscr{B}} B = \{+\infty\}$. On the other hand, since $\operatorname{Cl}_+ \{+\infty\} = \emptyset$, we have $\operatorname{Inf} \bigcap_{B \in \mathscr{B}} \operatorname{Cl}_+ B = \operatorname{Inf} \emptyset = \{+\infty\}$.
b) Let $\{+\infty\} \notin \mathscr{B}$ but $\{-\infty\} \in \mathscr{B}$. If $\{-\infty\}$ is the only element in $\mathscr{B}$ the assertion is obvious, otherwise we can omit this element without changing the expressions.
c) Let $\{+\infty\} \notin \mathscr{B}$ and $\{-\infty\} \notin \mathscr{B}$. Then, $B \subseteq Y$ and $\emptyset \neq \operatorname{Cl}_+ B \neq Y$ for all $B \in \mathscr{B}$, i.e., we can use the statements of Proposition 2.6.33. Define the sets

$$V := \bigcup_{B \in \mathscr{B}} (B - \operatorname{int} C) = \left(\bigcup_{B \in \mathscr{B}} B \right) - \operatorname{int} C$$

and

$$W := \bigcap_{B\in\mathscr{B}} \mathrm{Cl}_+ B.$$

We show that $V\cap W = \emptyset$ and $V\cup W = Y$. Assume there exists some $y \in V\cap W$. Hence there is some $\bar{B} \in \mathscr{B}$ such that $y \in \big(\bar{B}-\operatorname{int} C\big)\cap \mathrm{Cl}_+\bar{B} = \emptyset$, a contradiction. Let $y \in Y \setminus W$ (we have $W \neq Y$, because otherwise it holds $\mathrm{Cl}_+ B = Y$ for all $B \in \mathscr{B}$ and hence $\{-\infty\} \in \mathscr{B}$). Then there exists some $\bar{B} \in \mathscr{B}$ such that $y \notin \mathrm{Cl}_+\bar{B}$. By Proposition 2.6.33 (iv), (vi) we obtain $y \in \bar{B} - \operatorname{int} C \subset V$.

If $\mathrm{Cl}_- V = Y$ we have $W = \emptyset$, hence $\operatorname{Sup}\bigcup_{B\in\mathscr{B}} B = \operatorname{Sup} V = \{+\infty\} = \operatorname{Inf}\emptyset = \operatorname{Inf} W$. Otherwise, we have $\emptyset \neq \mathrm{Cl}_- V \neq Y$ and $\emptyset \neq \mathrm{Cl}_+ W \neq Y$. By Proposition 2.6.33, we obtain

$$\begin{aligned}\operatorname{Sup}\bigcup_{B\in\mathscr{B}} B &= \{y \in Y \mid y \notin V,\ \{y\} - \operatorname{int} C \subset V\}\\ &= \{y \in Y \mid y \in W, (\{y\} - \operatorname{int} C) \cap W = \emptyset\}\\ &= \operatorname{WMin}(W, C) = \operatorname{WMin}(\mathrm{Cl}_+ W, C) = \operatorname{Inf} W\end{aligned}$$

and so the proof is completed. □

We will see in Sect. 15.1 that the infimum/supremum in $\mathscr{I}$ is closely related to the solution concepts of vector optimization because the infimal/supremal set is closely related to the set of weakly minimal/maximal elements (see also Sect. 2.7).

We next show some calculus rules in the hyperspace $\mathscr{I}$ of self-infimal sets given by Löhne, Tammer [398] and Löhne [395, Proposition 1.56].

Proposition 2.6.42. *For subsets $\mathscr{A}, \mathscr{B} \subseteq \mathscr{I}$ we have*

(i) $\inf \mathscr{A} \oplus_{\mathscr{I}} \mathscr{B} = \inf \mathscr{A} \oplus_{\mathscr{I}} \inf \mathscr{B}$
(ii) $\sup \mathscr{A} \oplus_{\mathscr{I}} \mathscr{B} \preccurlyeq \sup \mathscr{A} \oplus_{\mathscr{I}} \sup \mathscr{B}$.

Proof. (i) If $\mathscr{A} = \emptyset$, it holds $\inf \mathscr{A} \oplus_{\mathscr{I}} \mathscr{B} = \inf \mathscr{A} = \{+\infty\}$ and so

$$\inf \mathscr{A} \oplus_{\mathscr{I}} \mathscr{B} = \inf \mathscr{A} \oplus_{\mathscr{I}} \inf \mathscr{B} = \{+\infty\}.$$

Otherwise, we have

$$\begin{aligned}\inf \mathscr{A} \oplus_{\mathscr{I}} \mathscr{B} &= \operatorname{Inf} \bigcup_{A\in\mathscr{A}, B\in\mathscr{B}} A \oplus_{\mathscr{I}} B = \operatorname{Inf} \bigcup_{A\in\mathscr{A}, B\in\mathscr{B}} A + B\\ &= \operatorname{Inf}(\bigcup_{A\in\mathscr{A}} A + \bigcup_{B\in\mathscr{B}} B) = \operatorname{Inf} \bigcup_{A\in\mathscr{A}} A \oplus_{\mathscr{I}} \operatorname{Inf} \bigcup_{B\in\mathscr{B}} B\\ &= \inf \mathscr{A} \oplus_{\mathscr{I}} \inf \mathscr{B}. \end{aligned} \tag{2.63}$$

(ii) For all $A \in \mathscr{A}$ and $B \in \mathscr{B}$ we have $A \oplus_{\mathscr{I}} B \preccurlyeq \sup \mathscr{A} \oplus_{\mathscr{I}} \sup \mathscr{B}$. As $\mathscr{I}$ is a complete lattice we can take the supremum on the left-hand side of the inequality. This yields (ii).

□

In the same manner like $\mathscr{F}$ and $\mathscr{I}$ we define the space $\mathscr{F}^{\diamond}$ of lower closed subsets of Y and the space $\mathscr{S}$ of self-supremal subsets of $Y^{\bullet}$, where we underlie the sup-addition in $Y^{\bullet}$ in the latter case.

Using the *infimal set* given by Definition 2.6.31 we formulate set-valued optimization problems based on lattice structure:

For an arbitrary set S, consider the following set-valued optimization problem

$$\operatorname{Inf} \bigcup_{x \in S} F(x) \tag{SP_Y}$$

with a set-valued objective map $F : S \rightrightarrows Y^{\bullet}$. We denote the **domain** of $F : S \rightrightarrows Y^{\bullet}$ by $\operatorname{Dom} F := \{x \in S \mid F(x) \neq \emptyset \text{ and } F(x) \neq \{+\infty\}\}$.

Next, we introduce a solution concept for set-valued optimization problems in the hyperspace of self-infimal sets $(\mathscr{I}, \oplus_{\mathscr{I}}, \odot_{\mathscr{I}, \preccurlyeq_{\mathscr{I}}})$ (see Definition 2.6.37 and Definition 2.6.31 for infimal sets).

Definition 2.6.43 (Infimal Set of (SP_Y)). The set $\bar{P} := \operatorname{Inf} \bigcup_{x \in S} F(x)$ is the solution set of the set-valued optimization problem (SP_Y).

Since $\operatorname{Inf} \bigcup_{x \in S} F(x) = \operatorname{Inf} \bigcup_{x \in S} \operatorname{Inf} F(x)$, (SP_Y) can be expressed as an $\mathscr{I}$-valued problem; without loss of generality we can assume that the sets $F(x)$ are self-infimal, i.e., $F : S \to \mathscr{I}$. Thus we consider the following problem.

$$\bar{P} = \operatorname{Inf} \bigcup_{x \in S} F(x) = \inf_{x \in S} F(x). \tag{$SP- \preceq_{\mathscr{I}}$}$$

Furthermore, in Sect. 8.3.1 we derive duality assertions for $\mathscr{F}$-valued optimization problems. We consider the space $(\mathscr{F}, \supseteq)$ (see Definition 2.6.36), where the order relation is given by

$$A \preccurlyeq_{\mathscr{F}} B :\Longleftrightarrow A \supseteq B$$

for subsets A, B of $\mathscr{F}$.

In order to formulate the $\mathscr{F}$-valued optimization problem we study an objective map $F : X \to \mathscr{F}$, i.e., the objective function values of F are subsets of the hyperspace of upper closed sets introduced in Definition 2.6.36. Using these notations we study the $\mathscr{F}$-valued problem

$$\bar{P} := \inf_{x \in X} P(x) = \operatorname{cl} \bigcup_{x \in X} F(x). \tag{$SP- \preceq_{\mathscr{F}}$}$$

Remark 2.6.44. In Sect. 8.3 we will show duality assertions for $\mathscr{F}$-valued ($\mathscr{I}$-valued, respectively) primal problems and corresponding dual problems based on conjugation as well as Lagrangian technique. Furthermore, in Sect. 15.1 we will use the lattice approach for deriving duality assertions for vector optimization problems.

Remark 2.6.45. In the special case of single-valued functions $F = f : S \longrightarrow Y$ the problems $(SP\text{--} \preceq_{\mathscr{I}})$ and $(SP\text{--} \preceq_{\mathscr{F}})$ coincide.

In Sect. 14.2 we will present an algorithm for solving set-valued optimization problems where the objective map has a polyhedral convex graph. There we will need the following notions and assertions concerning the infimum and supremum in a subspace of the hyperspace of upper closed sets $(\mathscr{F}, \oplus_{\mathscr{F}}, \odot_{\mathscr{F}}, \supseteq)$ (see Definition 2.6.36), see Löhne [395].

Definition 2.6.46. The **subspace of all closed convex subsets** of an extended partially ordered linear topological space $Y^\bullet$ is given by

$$\mathscr{F}_{\text{conv}} := \{A \subset \mathscr{F} \mid \forall \lambda \in (0,1) \; : \; \lambda \odot A \; \oplus \; (1-\lambda) \odot A = A\}.$$

The space $\mathscr{F}_{\text{conv}}$ can be characterized using the convex hull (compare Löhne [395, Proposition 1.59]).

Proposition 2.6.47. *Assume that Y is a linear topological space ordered by a proper pointed convex cone $C \subset Y$ with $\operatorname{int} C \neq \emptyset$ and let $\mathscr{F} = \mathscr{F}_C(Y)$. Then*

$$\mathscr{F}_{\text{conv}} = \{A \subseteq Y \mid \text{Cl}_+ \operatorname{conv} A = A\}.$$

Proof. Taking into account

$$A = \text{Cl}_+ \operatorname{conv} A \iff A = \operatorname{conv} A \wedge A = \text{Cl}_+ A \iff$$

$$\forall \lambda \in [0,1] : A = \text{Cl}_+(\lambda A + (1-\lambda)A) \iff$$

$$\forall \lambda \in [0,1] : A = \lambda \odot A \; \oplus \; (1-\lambda) \odot A$$

□

Proposition 2.6.48. *$(\mathscr{F}_{\text{conv}}, \supseteq)$ is a complete lattice. For nonempty subsets $\mathscr{A} \subseteq \mathscr{F}_{\text{conv}}$ the infimum and supremum can be expressed by*

$$\inf \mathscr{A} = \operatorname{cl} \operatorname{conv} \bigcup_{A \in \mathscr{A}} A, \qquad \sup \mathscr{A} = \bigcap_{A \in \mathscr{A}} A.$$

Proof. For all $A \in \mathscr{A}$, where $\mathscr{A}$ is a nonempty subset of $\mathscr{F}_{conv}$ we get with Proposition 2.6.47 $A = \text{Cl}_+ \operatorname{conv} A$. This yields

$$\operatorname{cl} \operatorname{conv} \bigcup_{A \in \mathscr{A}} A = \operatorname{cl} \operatorname{conv} \bigcup_{A \in \mathscr{A}} \text{Cl}_+ \operatorname{conv} A = \operatorname{cl} \operatorname{conv} \bigcup_{A \in \mathscr{A}} \operatorname{cl} \operatorname{conv}(A + C)$$

$$= \operatorname{cl}\operatorname{conv} \bigcup_{A\in\mathscr{A}} (A + C) = \operatorname{cl}\big(\big(\operatorname{conv} \bigcup_{A\in\mathscr{A}} A\big) + C\big)$$

$$= \operatorname{Cl}_{+} \operatorname{conv} \bigcup_{A\in\mathscr{A}} A.$$

So we get $\operatorname{cl}\operatorname{conv} \bigcup_{A\in\mathscr{A}} A \in \mathscr{F}_{conv}$. Since $\bigcap_{A\in\mathscr{A}} A$ is convex and upper closed, we can conclude $\bigcap_{A\in\mathscr{A}} A \in \mathscr{F}_{conv}$. So we get the assertions of the proposition. □

2.6.4 *The Embedding Approach by Kuroiwa*

An important approach for deriving optimality conditions and algorithms for solving set-valued optimization problems is based on the introduction of an **embedding space** into which the set-valued optimization problem is embedded (see Kuroiwa [353, 354]). With this approach Kuroiwa [354] defines notions of directional derivatives for set-valued maps and derives corresponding necessary and sufficient optimality conditions (compare Sect. 12.10). In this section we present results given by Kuroiwa in [353, 354, 357].

Let Y be a n.v.s., let C be a proper closed convex pointed cone in Y with $\operatorname{int} C \neq \emptyset$ and $\operatorname{int} C^{+} \neq \emptyset$.

The set relation $\preceq_C^{cl}$ discussed in this section is defined as follows (cf. Definition 2.6.9): For $A,\ B \subset Y$,

$$A \preceq_C^{cl} B :\Longleftrightarrow \operatorname{cl}(A + C) \supset B. \tag{2.64}$$

We consider minimal solutions with respect to the quasi-order $\preceq_C^{cl}$ given in (2.64) in the sense of Definition 2.6.19, and the corresponding set-valued optimization problem is given by

$$\preceq_C^{cl} -\text{minimize} \quad F(x), \quad \text{subject to} \quad x \in S, \qquad (SP_A - \preceq_C^{cl})$$

where $F : S \rightrightarrows Y$ is a set-valued objective mapping and S is a set.

We are looking for minimal solutions of $(SP_A - \preceq_C^{cl})$ in the sense of Definition 2.6.19, i.e., for elements $\overline{x} \in S$ with

$$F(x) \preceq_C^{cl} F(\overline{x}) \quad \text{for some} \quad x \in S \Longrightarrow F(\overline{x}) \preceq_C^{cl} F(x).$$

As already mentioned, a subset A of Y is said to be C**-convex** if $A+C$ is convex. Furthermore, $A \subseteq Y$ is said to be C^{+}**-bounded** if $\langle y^*, A\rangle$ is bounded from below for any $y^* \in C^{+}$, where C^{+} is the positive dual cone of C.

Let $\mathscr{G}$ be the **family of all nonempty C-convex and C^{+}-bounded subsets of** Y. In the following, we introduce a process of construction of a normed space $\mathscr{V}$ into which $\mathscr{G}$ is embedded. This approach goes back to Kuroiwa and Nuriya [357].

At first, we introduce an **equivalence relation** $\equiv$ on $\mathcal{G}^2$: For all (A, B), $(D, E) \in \mathcal{G}^2$,

$$(A, B) \equiv (D, E) :\Longleftrightarrow \mathrm{cl}(A + E + C) = \mathrm{cl}(B + D + C).$$

The **quotient space** $\mathcal{G}^2/\equiv$ is denoted by $\mathcal{V}$, where

$$\mathcal{V} := \{[A, B] | (A, B) \in \mathcal{G}^2\},$$

where $[A, B] := \{(D, E) \in \mathcal{G}^2 | (A, B) \equiv (D, E)\}$.

Furthermore, we define addition and scalar multiplication on the quotient space $\mathcal{V}$ as follows:

$$[A, B] + [D, E] := [A + D, B + E],$$

$$\lambda \cdot [A, B] := \begin{cases} [\lambda A, \lambda B] & \text{if } \lambda \geq 0 \\ [(-\lambda) B, (-\lambda) A] & \text{if } \lambda < 0. \end{cases}$$

Then $(\mathcal{V}, +, \cdot)$ is a vector space over $\mathbb{R}$.

In order to introduce an order relation on $\mathcal{V}$ we define the following subset of $\mathcal{V}$ by

$$\mu(C) := \{[A, B] \in \mathcal{V} | B \preceq_C^{cl} A\}. \tag{2.65}$$

It is easy to see that $\mu(C)$ is a pointed convex cone in $\mathcal{V}$.

Using the pointed convex cone $\mu(C)$, we define an **order relation** $\preceq_{\mu(C)}$ on $\mathcal{V}$ as follows:

$$[A, B] \preceq_{\mu(C)} [D, E] :\Longleftrightarrow [D, E] - [A, B] \in \mu(C).$$

Then, $(\mathcal{V}, +, \cdot, \preceq_{\mu(C)})$ is an ordered vector space over $\mathbb{R}$.

Let a function φ from $\mathcal{G}$ to $\mathcal{V}$ be given by

$$\varphi(A) := [A, \{0\}] \quad \text{for all} \quad A \in \mathcal{G},$$

then

$$A \preceq_C^{cl} B \Leftrightarrow \varphi(A) \preceq_{\mu(C)} \varphi(B),$$

for any $A, B \in \mathcal{G}$.

By using this function φ, the set optimization problem $(SP_A - \preceq_C^{cl})$ can be transformed into a **vector optimization problem** in the following sense: If F is a map from S to $\mathscr{G}$, then $\overline{x} \in S$ is a minimal solution of $(SP_A - \preceq_C^{cl})$ if and only if

$$\varphi \circ F(S) \cap (\varphi \circ F(\overline{x}) - \mu(C)) = \{\varphi \circ F(\overline{x})\}. \tag{2.66}$$

Formula (2.66) means that $\varphi \circ F(\overline{x})$ is a Pareto minimal point of $\varphi \circ F(S)$ with respect to $\mu(C)$, i.e., $\varphi \circ F(\overline{x}) \in \text{Min}(\varphi \circ F(S), \mu(C))$ (see Definition 2.4.1).

Finally, we introduce a norm $|\cdot|$ in $\mathscr{G}^2/\equiv$. Consider $c \in \text{int}\, C$ and a weak* compact base $W := \{y^* \in C^+ \mid \langle y^*, c\rangle = 1\}$ of C^+, then for each $[A, B] \in \mathscr{V}$,

$$|\,[A, B]\,| := \sup_{y^* \in W} |\inf\langle y^*, A\rangle - \inf\langle y^*, B\rangle|,$$

is well-defined. Furthermore, let

$$\mathscr{V}(W) := \{[A, B] \in \mathscr{V} \mid |\,[A, B]\,| < +\infty\},$$

then $(\mathscr{V}(W), |\cdot|)$ is a normed vector space, and $\mu(C)$ is closed in $(\mathscr{V}(W), |\cdot|)$.

In Sect. 12.10 we derive necessary and sufficient conditions for solutions of $(SP_A - \preceq_C^{cl})$ using this embedding approach.

2.6.5 Solution Concepts with Respect to Abstract Preference Relations

In this section we present a solution concept for set-valued optimization problems with geometric constraints useful in welfare economics (introduced by Bao and Mordukhovich [28]).

$$\text{minimize} \qquad F(x) \quad \text{subject to} \quad x \in S, \tag{SP_A}$$

where the cost mapping $F : X \rightrightarrows Y$ is a set-valued mapping, X and Y are Banach spaces and S is a subset of X.

The "minimization" in (SP_A) is understood with respect to a certain preference relation on Y. This general (abstract) preference relation on Y is defined as follows (see Sect. 2.1 or [431, Subsection 5.3.1]): For a given nonempty subset $\mathscr{R} \subset Y \times Y$, one says that y^1 is preferred to y^2 (we write $y^1 \prec y^2$) if $(y^1, y^2) \in \mathscr{R}$.

Following Bao and Mordukhovich [28] we will study a preference on Y directly in terms of a given preference mapping $L : Y \rightrightarrows Y$ instead of a preference $\prec$ on Y described via a subset $\mathscr{R} \subset Y \times Y$. The **level-set mapping** $L : Y \rightrightarrows Y$ associated to the preference relation $\prec$ is defined by

$$L(y) := \{u \in Y \mid u \prec y\}. \tag{2.67}$$

This means that $u \in Y$ is preferred to y if $u \in L(y)$.

An abstract preference $\prec$ has to satisfy some requirements in order to be useful in optimization and applications, especially in economics and engineering. In [431, Definition 5.55] three properties imposed in order to postulate the notion of **closed preference relations** as follows:

- The preference relation $\prec$ is **nonreflexive**, this means that $(y, y) \notin \mathcal{R}$ for all $y \in Y$;
- Given some $\overline{y} \in Y$ (a local minimizer in the sequel), the preference $\prec$ is **locally satiated** around $\overline{y}$ in the sense that $y \in \operatorname{cl} L(y)$ for all y in some neighborhood of $\overline{y}$.
- The preference $\prec$ is **almost transitive** meaning that

$$[u \in L(y),\ v \in \operatorname{cl} L(u)] \Longrightarrow v \in L(y). \tag{2.68}$$

Especially, in the study of vector optimization problems the almost transitivity property is widely used (under different names) ; see, e.g., [41, 431, 433, 624, 625] and the references therein. However, the almost transitivity property turns out to be rather restrictive, in contrast to the first two properties of $\prec$ formulated above. In particular, for the Pareto minimality (compare Definition 2.4.1) defined by

$$y^1 \prec y^2 :\Longleftrightarrow y^2 - y^1 \in C \setminus \{0\} \tag{2.69}$$

via an ordering cone $C \subset Y$, the preference $\prec$ is almost transitive if and only if C is convex and pointed; see [431, Proposition 5.56]. For the lexicographical order on $\mathbb{R}^q$ and other natural preference relations important in vector optimization and its applications including those to welfare economics (see Sect. 15.3) this property of the ordering cone is not fulfilled.

Bao and Mordukhovich [28] developed an approach to preference relations in set-valued optimization that is motivated by applications to models in welfare economics and allows to avoid the almost transitivity condition (2.68), which does not usually hold for many economies. They required the most natural local satiation property of the preference mapping L formulated at the reference optimal solution. The following solution concept is introduced by Bao and Mordukhovich [28, Definition 3.1]:

Definition 2.6.49 (Fully Localized Minimizers for Constrained Set-Valued Optimization Problems). Consider the problem (SP_A). Let $(\overline{x}, \overline{y}) \in \operatorname{graph} F$ and $\overline{x} \in S$. Then

- $(\overline{x}, \overline{y})$ is called a **fully localized weak minimizer for** (SP_A) if there exist neighborhoods U of $\overline{x}$ and V of $\overline{y}$ such that there is no $y \in F(S \cap U) \cap V$ preferred to $\overline{y}$, i.e.,

$$F(S \cap U) \cap L(\overline{y}) \cap V = \emptyset. \tag{2.70}$$

- $(\overline{x}, \overline{y})$ is called a **fully localized minimizer for** (SP_A) if there exist neighborhoods U of $\overline{x}$ and V of $\overline{y}$ such that there is no $y \in F(S \cap U) \cap V$ with $y \neq \overline{y}$ and $y \in \operatorname{cl} L(\overline{y})$, i.e.,

$$F(S \cap U) \cap \operatorname{cl} L(\overline{y}) \cap V = \{\overline{y}\}. \tag{2.71}$$

- $(\overline{x}, \overline{y})$ is called a **fully localized strong minimizer for** (SP_A) if there exist neighborhoods U of $\overline{x}$ and V of $\overline{y}$ such that there is no $(x, y) \in \operatorname{graph} F \cap (U \times V)$ with $(x, y) \neq (\overline{x}, \overline{y})$ satisfying $x \in S$ and $y \in \operatorname{cl} L(\overline{y})$, i.e.,

$$\operatorname{graph} F \cap (S \times \operatorname{cl} L(\overline{y})) \cap (U \times V) = \{(\overline{x}, \overline{y})\}. \tag{2.72}$$

In Sect. 12.11, Theorem 12.11.3 we derive first order necessary conditions for fully localized minimizers. Furthermore, we study applications in welfare economics in Sect. 15.3.

Remark 2.6.50. It is easy to see that each fully localized strong minimizer for (SP_A) is also a fully localized minimizer for (SP_A). Furthermore, each fully localized minimizer for (SP_A) is also a fully localized weak minimizer for (SP_A). In the case $S = X$ in Definition 2.6.49, we speak about the corresponding fully localized minimizers for the mapping F.

Remark 2.6.51. For all of the concepts in Definition 2.6.49 (see Bao and Mordukhovich [28, Definition 3.1]) the underlying feature is that one introduces the image localization of minimizers in constructions (2.70), (2.71), (2.72). These concepts introduced in [28, Definition 3.1] are different from the concepts discussed before even for minimal points and weakly minimal points of single-valued objectives $F := f : X \longrightarrow Y$ and allow to study local Pareto-type optimal allocations of welfare economics (see Sect. 15.3). The concept of (global or local) strong minimizers was first time introduced in [28, Definition 3.1] for set-valued optimization problems; it is related to Khan's notion of strong Pareto optimal allocations for models of welfare economics (compare Khan [324]) and the corresponding relationships established in Sect. 15.3.

Example 2.6.52. In this example we will see that a fully localized strong minimizer may not provide a partially localized (i.e., with $V = Y$) minimum or weak minimum in (2.70) and (2.71). To see this, we consider a set-valued mapping $F : \mathbb{R} \rightrightarrows \mathbb{R}$ defined by

$$F(x) := \begin{cases} \{-x\} & \text{if } x < 0 \\ \{0, 1\} & \text{if } x = 0 \\ \{x + 1\} & \text{if } x > 0, \end{cases}$$

with respect to the usual order on $\mathbb{R}$ generating the level sets $L(y) = (-\infty, y)$, it is easy to see that condition (2.72) is fulfilled at $(0, 1)$ with $U = (-\frac{1}{2}, \frac{1}{2})$ and $V = (\frac{1}{2}, \frac{3}{2})$ but conditions (2.70) and (2.71) do not hold with $V = \mathbb{R}$.

Example 2.6.53. Now, we will see that the localized minimizers and weak minimizers in Definition 2.6.49 are identical in the case of scalar set-valued optimization with $Y = \mathbb{R}$ and $L(y) = (-\infty, y)$, but they may be quite different in the vector-valued case. For example, if we consider an objective map $F : \mathbb{R} \rightrightarrows \mathbb{R}^2$ with $F(x) \equiv \mathbb{R}^2 \setminus \operatorname{int} \mathbb{R}^2_-$ and the usual weak preference on $\mathbb{R}^2$ with the level sets $L(y) = y - \operatorname{int} \mathbb{R}^2_+$ (see Definition 2.6.2), we have that $(0, 0) \in \mathbb{R} \times \mathbb{R}^2$ is a localized weak minimizer for F, but it is not a localized minimizer for this mapping. Note finally that localized strong minimizers reduce to standard isolate minimizers for scalar single-valued optimization problems.

2.6.6 Set-Valued Optimization Problems with Variable Ordering Structure

In the book by Chen, Huang and Yang [91] set-valued optimization problems with variable ordering structure are introduced, where the solution concept is related to the solution concept for vector optimization problems with variable ordering structure given in Definition 2.5.1.

Let X and Y be Banach spaces , $S \subset X$ be nonempty. Furthermore, let $C : X \rightrightarrows Y$ be a cone-valued mapping. We assume that for every $x \in X$, the set $C(x)$ is a proper closed convex cone with nonempty interior $\operatorname{int} C(x)$.

We consider a set-valued objective mapping $F : X \rightrightarrows Y$ and a set-valued optimization problem with variable ordering structure

$$\text{v-minimize } F(x) \qquad \text{subject to} \quad x \in S, \qquad (SP_v)$$

where "v-minimize" stands for problems with variable ordering structure with respect to a cone-valued mapping $C : X \rightrightarrows Y$ in the following sense:

Definition 2.6.54 ((Weak) v-Minimizer of (SP_v)). Let $\overline{x} \in S$ and $\overline{y} \in F(\overline{x})$.

(a) The pair $(\overline{x}, \overline{y})$ is called a **v-minimizer** of (SP_v) if

$$(F(S) - \overline{y}) \cap (-C(\overline{x})) = \{0\}.$$

(b) The pair $(\overline{x}, \overline{y})$ is called a **weak v-minimizer** of (SP_v) if

$$(F(S) - \overline{y}) \cap (-\operatorname{int} C(\overline{x})) = \emptyset.$$

In the following we will show that a set-valued optimization problem (SP_v) can be transformed into an equivalent vector-optimization problem in the sense that their v-minimal solution pairs are identical.

Definition 2.6.55. A cone-valued map $C : X \rightrightarrows Y$ is **pointed** on $S \subset X$ if the cone $\cup_{x \in S} C(x)$ is pointed, i.e.,

$$(\cup_{x\in S} C(x)) \cap (-\cup_{x\in S} C(x)) = \{0\}.$$

Remark 2.6.56. A cone-valued map $C : X \rightrightarrows Y$ is pointed on S if and only if

$$\forall x_1, x_2 \in S: \qquad C(x_1) \cap (-C(x_2)) = \{0\}.$$

For deriving the relationships between v-minimizers of a set-valued problem with variable ordering structure (SP_v) in the sense of Definition 2.6.54 and v-minimal points of a corresponding vector optimization problem with an objective function $f : X \to Y$ (see Definition 2.5.1) we need a certain monotonicity property concerning the set-valued map C with respect to f.

Definition 2.6.57. Let $f : X \to Y$ be a vector-valued function and $C : X \rightrightarrows Y$ be a cone-valued mapping. The cone-valued mapping C is called **weakly f-monotone**, if for all $x_1, x_2 \in X$, $c_1 \in C(x_1)$,

$$f(x_1) - f(x_2) \in c_1 + C(x_2) \quad \Longrightarrow C(x_2) \subset C(x_1).$$

The following relationships between v-minimal points of a vector optimization problem (VP_v) and v-minimizers of a set-valued optimization problem (SP_v) are shown by Chen, Huang and Yang [91, Proposition 2.63].

Proposition 2.6.58. *Let $S \subset X$ and $C : X \rightrightarrows Y$ be a pointed cone-valued mapping on S. Furthermore, let $f : X \to Y$ be a vector-valued function and $F : X \rightrightarrows Y$ be given by*

$$F(x) = f(x) + C(x) \quad (x \in X). \tag{2.73}$$

(a) *Suppose that C is weakly f-monotone. If $\overline{x} \in S$ is a v-minimal point of the vector optimization problem (VP_v):*

$$\text{Min}(f(S), C(\cdot)),$$

then $(\overline{x}, f(\overline{x}))$ is a v-minimizer of the set-valued optimization problem (SP_v):

$$v\text{-minimize } F(x) \quad \text{subject to} \quad x \in S.$$

(b) *If $(\overline{x}, \overline{y})$ is a v-minimizer of the set-valued optimization problem (SP_v), then $\overline{x}$ is a v-minimal point of the vector optimization problem (VP_v) and $\overline{y} = f(\overline{x})$.*

Proof. First, we show that (a) holds. Consider a v-minimal point $\overline{x} \in S$ of the problem (VP_v). Then

$$\forall x \in S: \quad f(x) - f(\overline{x}) \notin -C(\overline{x}) \setminus \{0\}.$$

This yields

$$\forall x \in S: \quad (f(x) - f(\overline{x}) + C(x)) \cap (-C(\overline{x}) \setminus \{0\}) = \emptyset. \tag{2.74}$$

Indeed, if there exists $\tilde{x} \in S$ with

$$(f(\tilde{x}) - f(\overline{x}) + C(\tilde{x})) \cap (-C(\overline{x}) \setminus \{0\}) \neq \emptyset,$$

then there exists $\overline{c} \in C(\overline{x})$ and $\overline{c} \neq 0$ such that

$$-\overline{c} \in f(\tilde{x}) - f(\overline{x}) + C(\tilde{x}).$$

This means

$$f(\overline{x}) - f(\tilde{x}) \in \overline{c} + C(\tilde{x}).$$

Taking into account the weak f-monotonicity of C, we get

$$f(\tilde{x}) - f(\overline{x}) \in -\overline{c} - C(\overline{x}).$$

Furthermore, since C is pointed and $\overline{c} \neq 0$, it follows that

$$f(\tilde{x}) - f(\overline{x}) \in -C(\overline{x}) \setminus \{0\},$$

in contradiction to the assumption that $\overline{x}$ is a v-minimal point of the problem (VP_v).

Taking into account (2.73) and (2.74) we get

$$\forall y \in F(x),\ x \in S: \quad y - f(\overline{x}) \notin -C(\overline{x}) \setminus \{0\}.$$

Hence, $(\overline{x}, f(\overline{x}))$ is a v-minimizer of the set-valued optimization problem (SP_v).

Now, we will prove (b). Let us assume that $(\overline{x}, \overline{y})$ is a v-minimizer of the set-valued optimization problem (SP_v). Then,

$$\overline{y} \in F(\overline{x}) = f(\overline{x}) + C(\overline{x})$$

and

$$\forall y \in F(S): \qquad y - \overline{y} \notin -C(\overline{x}) \setminus \{0\}. \tag{2.75}$$

Of course, it holds $\overline{y} = f(\overline{x})$. We have to show that $\overline{x}$ is a v-minimal point of the vector optimization problem (VP_v). Contrarily, suppose that $\overline{x}$ is not a v-minimal point of the vector optimization problem (VP_v). Then, for some element $\tilde{x} \in S \setminus \{\overline{x}\}$,

$$f(\tilde{x}) - f(\overline{x}) \in -C(\overline{x}) \setminus \{0\}.$$

So we get

$$f(\tilde{x}) - \overline{y} \in -C(\overline{x}) \setminus \{0\},$$

because of $f(\overline{x}) = \overline{y}$, in contradiction to (2.75). Hence, $\overline{x}$ is a v-minimal point of the vector optimization problem (VP_v). □

Remark 2.6.59. In the book by Chen, Huang, Yang [91, Theorem 2.64] necessary conditions for **weak v-minimizers** of (SP_v) are shown using the contingent derivative.

Remark 2.6.60. Necessary and sufficient optimality conditions in form of the Fermat rule for **nondominated** solutions of unconstrained set-valued optimization problems with variable ordering structure and the Lagrange multiplier rule for the constrained set-valued problems with variable ordering structure are given by Eichfelder and Ha in [168].

2.6.7 *Approximate Solutions of Set-Valued Optimization Problems*

In this section we introduce a concept of approximate solutions in set-valued optimization. Approximate solutions are of interest from the theoretical as well as computational point of view. Especially, in order to formulate set-valued versions of Ekeland's variational principle (compare Chap. 10) and a subdifferential variational principle for set-valued mappings (see Sect. 12.9) one is dealing with approximate solutions.

We consider a set-valued optimization problem:

$$\text{minimize} \qquad F(x) \quad \text{subject to} \quad x \in X, \tag{SP}$$

where X is a linear space, Y is a linear topological space, $C \subset Y$ is a proper closed convex cone and the cost mapping $F : X \rightrightarrows Y$ is a set-valued mapping.

The following concepts for approximate solutions of the set-valued problem (SP) was given by Bao and Mordukhovich [27, Definition 3.4] and is related to minimizers introduced in Definition 2.6.1.

Definition 2.6.61 (Approximate Minimizers of Set-Valued Optimization Problems). Let $\overline{x} \in X$ and $(\overline{x}, \overline{y}) \in \operatorname{graph} F$. Then:

(i) Consider $\varepsilon > 0$ and $k^0 \in C \setminus \{0\}$. The pair $(\overline{x}, \overline{y}) \in \operatorname{graph} F$ is called an **approximate εk^0-minimizer** for F if

$$y + \varepsilon k^0 \notin \overline{y} - C \qquad \text{for all } \; y \in F(x) \text{ with } x \neq \overline{x}.$$

(ii) Consider $\varepsilon > 0$ and $k^0 \in C \setminus \{0\}$. The pair $(\overline{x}, \overline{y}) \in \operatorname{graph} F$ is called a **strict approximate εk^0-minimizer** for F if there is a number $0 < \tilde{\varepsilon} < \varepsilon$ such that $(\overline{x}, \overline{y})$ is an approximate $\tilde{\varepsilon} k^0$-minimizer of this mapping.

In Sect. 12.9, Theorem 12.9.1, we will show necessary conditions for strict approximate εk^0-minimizers of F.

2.7 Relationships Between Solution Concepts

In this section we study the relationships between different solution concepts in set-valued optimization. Furthermore, we discuss the special case that the objective map is single-valued.

Let Y be a linear topological space, partially ordered by a proper pointed convex closed cone C, X a linear space, S a subset of X and $F : X \rightrightarrows Y$. We consider the set-valued optimization problem (SP):

$$\text{minimize} \qquad F(x) \quad \text{subject to} \quad x \in S. \tag{2.76}$$

In the formulation of the solution concepts based on set approach the underlying space is a linear topological space Y whereas the extended space $Y^\bullet := Y \cup \{-\infty\} \cup \{+\infty\}$ is considered in the formulation of the solution concepts based on lattice approach in order to work with infimum and supremum.

Remark 2.7.1. The differences between the solution concepts based on set-approach in Definition 2.6.19 and the solution concepts based on vector approach in Definition 2.6.1 are already discussed in Examples 2.6.21, 2.6.22 and 2.6.23.

Remark 2.7.2. In the special case of single-valued functions $F = f : X \longrightarrow Y$ the concept of minimizers of the set-valued problem (SP) (see Definition 2.6.1) coincides with the solution concept for Pareto minimal points of $f(S)$ with respect to C introduced in Definition 2.4.1: $(\overline{x}, f(\overline{x})) \in \operatorname{graph} f$ is a minimizer in the sense of Definition 2.6.1 if and only if $f(\overline{x})$ is a Pareto minimal point of $f(S)$ with respect to C, i.e., $f(\overline{x}) \in \operatorname{Min}(f(S), C)$.

Remark 2.7.3. In the special case of single-valued functions $F = f : X \longrightarrow Y$ the concept of minimal solutions of the problem $(SP - \preceq)$ (see Definition 2.6.19) w.r.t. the order relations introduced in Definition 2.6.9 coincides with the solution concept for Pareto minimal points given in Definition 2.4.1.

In the following we consider a linear topological space Y, a linear space X, $S \subseteq X$, a set-valued map $F : X \rightrightarrows Y$, $F(S) = \bigcup_{x \in S} F(x) \neq \emptyset$ and a proper pointed closed convex cone $C \subset Y$ with $\operatorname{int} C \neq \emptyset$. The relationship between the infimal set of $F(S)$ (Definition 2.6.31) and weak minimizers of $F(S)$ in the sense of Definition 2.6.2 is given in the next proposition.

Proposition 2.7.4. *Under the assumption that* $F(S) = \mathrm{cl}(F(S) + C)$ *we get*

$$\mathrm{WMin}(F(S), C) = \mathrm{Inf}\, F(S).$$

Proof. Taking into account the assumption $F(S) = \mathrm{cl}(F(S)+C)$ and the definition of the infimal set we get

$$\mathrm{WMin}(F(S), C) = \mathrm{WMin}(\mathrm{cl}(F(S)+C), C) = \mathrm{WMin}(\mathrm{Cl}_+ F(S), C) = \mathrm{Inf}\, F(S).$$

The proof is completed. □

The assertion of Proposition 2.7.4 says that the solution concept for $\mathscr{I}$-valued problems coincides with the set $\mathrm{WMin}(F(S), C)$ in Definition 2.6.3 for weak minimizers.

Furthermore, if we assume that $F(S) + C$ is closed we get the following assertion.

Proposition 2.7.5. *Under the assumption that* $F(S) + C$ *is closed we get*

$$\mathrm{Inf}\, F(S) = \mathrm{WMin}(F(S) + C, C).$$

Proof. Because of the closedness of $F(S) + C$ we get

$$\mathrm{WMin}(F(S) + C, C) = \mathrm{WMin}(\mathrm{cl}(F(S) + C), C) = \mathrm{Inf}\, F(S),$$

taking into account Definition 2.6.31. □

Corollary 2.7.6. *Assuming that* $F(S) + C$ *is closed and* $\mathrm{WMin}(F(S) + C, C) = \mathrm{WMin}(F(S), C)$ *we get*

$$\mathrm{Inf}\, F(S) = \mathrm{WMin}(F(S), C).$$

In Sect. 15.1 we will use methods of set-valued optimization for deriving duality assertions for vector optimization problems. The relationships between vector optimization problems and $\mathscr{I}$-valued problems are discussed by Löhne and Tammer [397] and in a comprehensive and detailed way by Löhne [395].

Let Y be a linear topological space, partially ordered by a proper pointed convex closed cone C, X a linear space, S a subset of X and $f : X \longrightarrow Y^\bullet$ a vector-valued function. We consider the vector optimization problem

$$\mathrm{Min}(f(S), C). \tag{VOP}$$

In Sect. 15.1 we will see that it is very useful to assign to (VOP) a corresponding $\mathscr{I}$-valued problem such that one can use the complete lattice structure of $(\mathscr{I}, \preceq := \preceq_{\mathscr{I}})$, where $(\mathscr{I}, \preceq)$ is defined with respect to the ordering cone C of the vector optimization problem.

For a given vector-valued function $f : X \longrightarrow Y^\bullet$ we put

$$\overline{f} : X \longrightarrow \mathscr{I}, \qquad \overline{f}(x) := \mathrm{Inf}\{f(x)\}$$

and assign to (VOP) the $\mathscr{I}$-valued problem

$$\preceq -\text{minimize} \quad \overline{f} \quad \text{subject to} \quad x \in S. \qquad (VOP_{\mathscr{I}})$$

Problem $(VOP_{\mathscr{I}})$ is said to be the $\mathscr{I}$-extension of the vector optimization problem (VOP) (see Löhne [395]). The lattice extension of the vector optimization problem (VOP) allows us to handle the problem in the framework of complete lattices. For this extension it is important that the ordering cone C is closed as we will see in the proof of the following proposition.

The following assertion is shown by Löhne [395, Proposition 2.17].

Proposition 2.7.7. *For all $x, u \in X$ it holds*

$$f(x) \leq_C f(u) \quad \Longleftrightarrow \quad \overline{f}(x) \preceq \overline{f}(u).$$

Proof. Consider $y = f(x)$ and $z = f(u)$. Let $\mathrm{Inf}\{y\} \preceq \mathrm{Inf}\{z\}$, then $\mathrm{Cl}_+\{y\} \supseteq \mathrm{Cl}_+\{z\}$. With (2.62) we can conclude $z \in \mathrm{cl}(\{z\} + C) \subseteq \mathrm{cl}(\{y\} + C)$. Because of the assumption that C is closed, we get $z \in \{y\} + C$. This means $y \leq_C z$. The opposite inclusion is obvious. □

As a direct consequence of Proposition 2.7.7 we get corresponding assertions concerning the solutions of (VOP) and $(VOP_{\mathscr{I}})$ (see Löhne [395, Proposition 2.18]).

Finally, it is important to mention the following references. In the paper by Hernández, Jiménez, Novo [244], Benson proper efficiency in set-valued optimization is discussed. Hernández, Jiménez, Novo study in [245] weak and proper efficiency in set-valued optimization. Flores-Bazán, Hernández characterize efficiency without linear structure in [189]. Moreover, Hernández, Rodríguez-Marín, Sama describe solutions of set-valued optimization problems [253]. Furthermore, in Hernández, Rodríguez-Marín [250] certain existence results for solutions of set optimization problems are derived.

Chapter 3
Continuity and Differentiability

In this chapter we present continuity notions for set-valued mappings and corresponding properties under convexity assumptions. Furthermore, we introduce Lipschitz properties for single-valued and set-valued maps. Concepts of generalized differentiability and corresponding calculus rules are recalled.

3.1 Continuity Notions for Set-Valued Maps

In this section X and Y are separated (in the sense of Hausdorff) topological spaces and $F : X \rightrightarrows Y$ is a set-valued map. When mentioned explicitly, Y is a separated topological vector space (s.t.v.s.). The presentation of this section follows that in [214, Section 1.2.5] where one can find the proofs for the results which are only stated.

Definition 3.1.1. Let $x_0 \in X$. We say that

(a) F is **upper continuous** (u.c.) **at** x_0 if

$$\forall\, D \subseteq Y,\ D \text{ open},\ F(x_0) \subseteq D,\ \exists\, U \in \mathscr{N}_X(x_0),\ \forall\, x \in U\ :\ F(x) \subseteq D, \tag{3.1}$$

i.e., $F^{+1}(D)$ is a neighborhood of x_0 for each open set $D \subseteq Y$ such that $F(x_0) \subseteq D$;

(b) F is **lower continuous** (l.c.) **at** x_0 if

$$\forall\, D \subseteq Y,\ D \text{ open},\ F(x_0) \cap D \neq \emptyset,\ \exists\, U \in \mathscr{N}_X(x_0),\ \forall\, x \in U\ :\ F(x) \cap D \neq \emptyset, \tag{3.2}$$

i.e., $F^{-1}(D)$ is a neighborhood of x_0 for each open set $D \subseteq Y$ such that $F(x_0) \cap D \neq \emptyset$.

A.A. Khan et al., *Set-valued Optimization*, Vector Optimization,
DOI 10.1007/978-3-642-54265-7_3

(c) F is **continuous at** x_0 if F is u.c. and l.c. at x_0.
(d) F is **upper continuous (lower continuous, continuous)** if F is so at every $x \in X$;
(e) F is **lower continuous at** $(x_0, y_0) \in X \times Y$ if

$$\forall\, V \in \mathcal{N}_Y(y_0),\ \exists\, U \in \mathcal{N}_X(x_0),\ \forall\, x \in U\ :\ F(x) \cap V \neq \emptyset.$$

It follows from the definition that $x_0 \in \operatorname{int}(\operatorname{dom} F)$ and $y_0 \in \operatorname{cl}(F(x_0))$ if F is lower continuous at (x_0, y_0) and F is lower continuous at $x_0 \in \operatorname{dom} F$ if and only if F is lower continuous at every (x_0, y) with $y \in F(x_0)$; moreover, F is lower continuous at every $x_0 \in X \setminus \operatorname{dom} F$. If $x_0 \in X \setminus \operatorname{dom} F$, then F is u.c. at x_0 if and only if $x_0 \in \operatorname{int}(X \setminus \operatorname{dom} F)$. So, if F is u.c. then $\operatorname{dom} F$ is closed, while if F is l.c. then $\operatorname{dom} F$ is open.

The next result is an immediate consequence of the definitions.

Proposition 3.1.2. (i) *F is upper continuous if and only if $F^{+1}(D)$ is open for every open set $D \subseteq Y$;*
(ii) *F is lower continuous if and only if $F^{-1}(D)$ is open for every open set $D \subseteq Y$.*

The **limit inferior of** F **at** $x_0 \in X$ is defined by

$$\liminf_{x \to x_0} F(x) := \{y \in Y \mid \forall\, V \in \mathcal{N}_Y(y),\ \exists\, U \in \mathcal{N}_X(x_0), \forall\, x \in U^{\bullet}\ :\ F(x) \cap V \neq \emptyset\},$$

while the **limit superior of** F **at** $x_0 \in X$ is defined by

$$\begin{aligned}\limsup_{x \to x_0} F(x) :=& \{y \in Y \mid \forall\, V \in \mathcal{N}_Y(y),\ \forall\, U \in \mathcal{N}_X(x_0), \exists\, x \in U^{\bullet}\ :\ F(x) \cap V \neq \emptyset\} \\ =& \bigcap\nolimits_{U \in \mathcal{N}_X(x_0)} \operatorname{cl}\left(F(U^{\bullet})\right),\end{aligned}$$

where $U^{\bullet} := U \setminus \{x_0\}$ for $U \in \mathcal{N}_X(x_0)$.

Note that x_0 is an accumulation point of $\operatorname{dom} F$ if $\limsup_{x \to x_0} F(x) \neq \emptyset$, while $x_0 \in \operatorname{int}(\operatorname{dom} F \cup \{x_0\})$ if $\liminf_{x \to x_0} F(x) \neq \emptyset$; if x_0 is an isolated point of X (i.e., $\{x_0\} \in \mathcal{N}_X(x_0)$), then $\liminf_{x \to x_0} F(x) = Y$. Of course, if x_0 is an accumulation point of $\operatorname{dom} F$, then

$$\liminf_{x \to x_0} F(x) \subseteq \limsup_{x \to x_0} F(x),$$

both of them being closed sets.

Sometimes in the definitions of $\liminf_{x \to x_0} F(x)$ and $\limsup_{x \to x_0} F(x)$ one takes $x \in \operatorname{dom} F$. Note that this situation reduces to the preceding one by considering the restriction of F at $\{x_0\} \cup \operatorname{dom} F$.

Recall first that if (X, d) is a metric space and $A, B \subseteq X$, then the **excess of** A **over** B is

$$e(A, B) = \sup_{x \in A} \operatorname{dist}(x, B) \text{ if } A, B \neq \emptyset, \quad e(\emptyset, B) = 0, \quad e(A, \emptyset) = \infty \text{ if } A \neq \emptyset,$$

where $\operatorname{dist}(x, A) := \inf_{a \in A} d(x, a)$; in particular, $\operatorname{dist}(x, \emptyset) = \infty$. It is simple to show that for a nonempty and compact set $A \subseteq X$ and an open set $D \subseteq X$, if $A \subseteq D$, then there exists $\varepsilon > 0$ such that $A_\varepsilon := \{x \in X \mid \operatorname{dist}(x, A) < \varepsilon\} = \{x \mid A \cap B(x, \varepsilon) \neq \emptyset\} \subseteq D$, where $B(x, \varepsilon) := \{x' \in X \mid d(x, x') < \varepsilon\}$.

In particular cases for X or/and Y one has useful characterizations for the elements of $\liminf_{x \to x_0} F(x)$ and $\limsup_{x \to x_0} F(x)$.

Proposition 3.1.3. *Let $x_0 \in X$ and $y \in Y$.*

(i) $y \in \liminf_{x \to x_0} F(x)$ *if and only if*

$$\forall\, X \setminus \{x_0\} \supseteq (x_i)_{i \in I} \to x_0,\ \exists\, (x_{\varphi(j)})_{j \in J},\ \exists\, Y \supseteq (y_j)_{j \in J} \to y,$$
$$\forall\, j \in J\ :\ y_j \in F(x_{\varphi(j)}),$$

and $y \in \limsup_{x \to x_0} F(x)$ *if and only if*

$$\exists\, X \setminus \{x_0\} \supseteq (x_i)_{i \in I} \to x_0,\ \exists\, Y \supseteq (y_i)_{i \in I} \to y,\ \forall\, i \in I\ :\ y_i \in F(x_i).$$

(ii) *If X and Y are first countable, then* $y \in \liminf_{x \to x_0} F(x)$ *if and only if*

$$\forall X \setminus \{x_0\} \supseteq (x_n) \to x_0,\ \exists Y \supseteq (y_n) \to y,\ \exists n_0 \in \mathbb{N},\ \forall\, n \geq n_0\ :\ y_n \in F(x_n),$$

and $y \in \limsup_{x \to x_0} F(x)$ *if and only if*

$$\exists\, X \setminus \{x_0\} \supseteq (x_n) \to x_0,\ \exists\, Y \supseteq (y_n) \to y,\ \forall\, n \in \mathbb{N}\ :\ y_n \in F(x_n).$$

(iii) *If (Y, ρ) is a metric space, then*

$$y \in \liminf_{x \to x_0} F(x) \Leftrightarrow \lim_{x \to x_0} \operatorname{dist}(y, F(x)) = 0,$$
$$y \in \limsup_{x \to x_0} F(x) \Leftrightarrow \liminf_{x \to x_0} \operatorname{dist}(y, F(x)) = 0.$$

In (i) above $(x_i)_{i \in I}$, $(y_j)_{j \in J}$, $(y_i)_{i \in I}$ denote nets, while $(x_{\varphi(j)})_{j \in J}$ denotes a subnet of $(x_i)_{i \in I}$, that is $\varphi : J \to I$ is such that for every $i \in I$ there exists $j_i \in J$ with $\varphi(j) \geq i$ for all $j \geq j_i$.

We have the following characterization of upper continuity in a special case.

Proposition 3.1.4. *Suppose that (Y, ρ) is a metric space and $F(x_0)$ is compact. Then F is u.c. at x_0 if and only if* $\lim_{x \to x_0} e(F(x), F(x_0)) = 0$.

In the next results we provide other characterizations for upper and lower continuity at a point.

Proposition 3.1.5. *Let X, Y be first countable (in particular, let X, Y be metric spaces) and $x \in \operatorname{dom} F$. The following assertions are equivalent:*

(i) *the set-valued map F is u.c. at x;*
(ii) *for every closed set $E \subseteq Y$ with $F(x) \cap E = \emptyset$, there exists $U \in \mathcal{N}_X(x)$ such that $F(U) \cap E = \emptyset$;*
(iii) *for every closed set $E \subseteq Y$ and every sequence $X \supseteq (x_n) \to x$ with $F(x_n) \cap E \neq \emptyset$ for $n \in \mathbb{N}$, we have $F(x) \cap E \neq \emptyset$;*
(iv) *for every open set $D \subseteq Y$ with $F(x) \subseteq D$, and every sequence $X \supseteq (x_n) \to x$ there exists $n_D \in \mathbb{N}$ such that $F(x_n) \subseteq D$ for $n \geq n_D$;*
(v) *for all sequences $(x_n) \subseteq X$, $(y_n) \subseteq Y$ with $(x_n) \to x$ and $y_n \in F(x_n) \setminus F(x)$ for $n \in \mathbb{N}$, there exists a subsequence $(y_{n_k}) \to y \in F(x)$.*

If X, Y are not first countable, the conditions (i)–(iv) *remain equivalent by replacing sequences by nets.*

Proposition 3.1.6. *Let $(x_0, y_0) \in X \times Y$. The following statements are equivalent:*

(i) *the set-valued map F is lower continuous at (x_0, y_0);*
(ii) *for every net $(x_i)_{i \in I} \to x_0$ there exist a subnet $(x_{\varphi(j)})_{j \in J}$ of (x_i) and a net $(y_j)_{j \in J} \to y_0$ such that $y_j \in F(x_{\varphi(j)})$ for $j \in J$.*
(iii) $y_0 \in \liminf_{x \to x_0} F(x) \cap \operatorname{cl} F(x_0)$.

Suppose now that X, Y are first countable. Then (i) *is equivalent to*

(iv) *for every sequence $X \supseteq (x_n) \to x_0$ there exist a sequence $Y \supseteq (y_n) \to y_0$ and $n_0 \in \mathbb{N}$ such that $y_n \in F(x_n)$ for every $n \geq n_0$.*

From the preceding result we obtain that

$$F \text{ is lower continuous at } x_0 \Leftrightarrow F(x_0) \subseteq \liminf_{x \to x_0} F(x).$$

Definition 3.1.7. We say that

(i) F is **closed** if $\operatorname{graph} F$ is a closed subset of $X \times Y$;
(ii) F is **closed at** $x \in X$ if for every net $((x_i, y_i))_{i \in I} \subseteq \operatorname{graph} F$ converging to (x, y) we have that $y \in F(x)$;
(iii) F is **closed-valued** if $F(x)$ is closed for every $x \in X$;
(iv) F is **compact at** $x \in X$ if for every net $((x_i, y_i))_{i \in I} \subseteq \operatorname{graph} F$ with $(x_i) \to x$, there exists a subnet $(y_{\varphi(j)})_{j \in J}$ converging to some $y \in F(x)$.

Of course, F is closed if and only if F is closed at every $x \in X$; moreover, if F is closed, then F is closed-valued. If F is compact at x, then F is closed at x; moreover, if F is closed (compact) at x, then $F(x)$ is closed (compact). Note also that F is compact at $x \in X \setminus \operatorname{dom} F$ if and only if $x \in \operatorname{int}(X \setminus \operatorname{dom} F)$, but F may be closed at some $x \in \operatorname{cl}(\operatorname{dom} F) \setminus \operatorname{dom} F$.

Proposition 3.1.8. *Assume that $\emptyset \neq A \subseteq \operatorname{dom} F$ and F is u.c. at every $x \in A$.*

(i) *If A is compact and $F(x)$ is compact for every $x \in A$, then $F(A)$ is compact.*

(ii) *If A is connected and $F(x)$ is connected for every $x \in A$, then $F(A)$ is connected.*

Recall that the nonempty subset A of the topological space (X, τ) is **connected** if $A \cap D_1 \cap D_2 \neq \emptyset$ whenever D_1, D_2 are open subsets of X such that $A \cap D_1 \neq \emptyset$, $A \cap D_2 \neq \emptyset$ and $A \subseteq D_1 \cup D_2$. Recall that the connected subsets of $\mathbb{R}$ are the intervals, while in a topological vector space every convex set is connected.

When Y is a metric space or a topological vector space, Proposition 3.1.8(i) follows also from the next result.

Proposition 3.1.9. *Let $x \in X$. The following assertions hold:*

(i) *F is compact at x if and only if $F(x)$ is compact and F is u.c. at x.*
(ii) *If (Y, ρ) is a metric space, then F is compact at x if and only if $\lim_{x' \to x} e(F(x'), F(x)) = 0$ and $F(x)$ is compact.*
(iii) *Let Y be first countable. If F is compact at x, then for every sequence $((x_n, y_n))_{n \in \mathbb{N}} \subseteq \operatorname{graph} F$ with $(x_n) \to x$, there exists a subsequence $(y_{n_k}) \to y \in F(x)$. If X is first countable and Y is a metric space, the converse is true.*

Related to closedness of set-valued maps we have the following result.

Proposition 3.1.10. *Let $x \in X$. The following assertions hold:*

(i) *F is closed at x if and only if for every $y \in Y \setminus F(x)$ there exist $U \in \mathscr{N}_X(x)$ and $V \in \mathscr{N}_Y(y)$ such that $F(U) \cap V = \emptyset$.*
(ii) *F is closed at x if and only if $\limsup_{x' \to x} F(x') \subseteq F(x) = \operatorname{cl} F(x)$.*
(iii) *Suppose that X, Y are first countable. Then F is closed at x if and only if for every sequence $\operatorname{graph} F \supseteq ((x_n, y_n))_{n \in \mathbb{N}} \to (x, y)$ one has $y \in F(x)$.*
(iv) *If F is u.c. at x and $F(x)$ is compact, then F is closed at x.*
(v) *If Y is regular (in particular, if Y is a metric space or a topological vector space), F is u.c. at x, and $F(x)$ is closed, then F is closed at x.*

The next result gives sufficient conditions for the upper continuity of the intersection of two set-valued maps.

Proposition 3.1.11. *Consider $F_1, F_2 : X \rightrightarrows Y$ and $F(x) = F_1(x) \cap F_2(x)$ for every $x \in X$. Then F is u.c. at $x_0 \in X$ if one of the following conditions holds:*

(i) *F_1 is closed at x_0, F_2 is u.c. at x_0, and $F_2(x_0)$ is compact;*
(ii) *Y is normal (in particular, Y is a metric space), F_1, F_2 are u.c. at x_0, and $F_1(x_0)$, $F_2(x_0)$ are closed.*

Recall that $\mathscr{N}_Y$ denotes the class of balanced neighborhoods of the origin of the t.v.s. Y.

Definition 3.1.12. Let Y be a topological vector space and $x_0 \in X$. We say that

(a) F is **Hausdorff upper continuous** (H-u.c.) **at** x_0 if

$$\forall\, V \in \mathscr{N}_Y,\ \exists\, U \in \mathscr{N}_X(x_0),\ \forall\, x \in U \ :\ F(x) \subseteq F(x_0) + V. \tag{3.3}$$

(b) F is **Hausdorff lower continuous** (H-l.c.) **at** x_0 if

$$\forall\, V \in \mathcal{N}_Y,\ \exists\, U \in \mathcal{N}_X(x_0),\ \forall\, x \in U\ :\ F(x_0) \subseteq F(x) + V. \tag{3.4}$$

(c) F is **Hausdorff continuous at** x_0 if F is H-u.c. and H-l.c. at x_0.
(d) F is **Hausdorff upper continuous (Hausdorff lower continuous, Hausdorff continuous)** if F is so at every $x \in X$.

The above definition can be given when Y is a metric space, too; just replace $V \in \mathcal{N}_Y$ by $\varepsilon > 0$ and $F(x_0) + V$ by $(F(x_0))_\varepsilon$. Of course, if Y is a metric space, then F is H-u.c. at x_0 if and only if $\lim_{x \to x_0} e\,(F(x), F(x_0)) = 0$, and F is H-lower continuous at x_0 if and only if $\lim_{x \to x_0} e\,(F(x_0), F(x)) = 0$.

Concerning the continuity of the sum of set-valued maps and of the multiplication by scalars, we have the following result.

Proposition 3.1.13. *Let Y be a topological vector space, $F, F_1, F_2 : X \rightrightarrows Y$, $x_0 \in X$, and $\alpha \in \mathbb{R}$.*

(i) *If F is u.c. (resp. lower continuous, H-u.c., H-l.c) at x_0, then αF is u.c. (resp. l.c., H-u.c., H-l.c.) at x_0.*
(ii) *If F_1 and F_2 are l.c. (resp. H-u.c., H-l.c.) at x_0, then $F_1 + F_2$ is l.c. (resp. H-u.c., H-l.c.) at x_0.*

Note that we have not a similar result to Proposition 3.1.13(ii) for upper continuity. Also note that if F is u.c. at x_0, then F is H-u.c. at x_0; the converse implication is true when $F(x_0)$ is compact. On the other hand, if F is H-l.c. at x_0, then F is l.c. at x_0, the converse being true if $F(x_0)$ is compact.

We can characterize Hausdorff upper and lower continuities by using nets, and even sequences when X and Y are first countable.

Proposition 3.1.14. *Suppose that Y is a topological vector space and $x \in X$. Then:*

(i) *F is H-lower continuous at x if and only if for all nets $(x_i)_{i\in I} \subseteq X$ with $(x_i) \to x$ and $(\overline{y}_i)_{i\in I} \subseteq F(x)$ there exist a subnet $(x_{\varphi(j)})$ and a net $(y_j)_{j\in J}$ such that $y_j - \overline{y}_{\varphi(j)} \to 0$ and $y_j \in F(x_{\varphi(j)})$ for all $j \in J$;*
(ii) *F is H-u.c. at x if and only if for every net $((x_i, y_i))_{i\in I} \subseteq \operatorname{graph} F$ with $(x_i) \to x$ there exists a subnet $(y_{\varphi(j)})_{j\in J}$ and a net $(\overline{y}_j)_{j\in J} \subseteq F(x)$ such that $y_{\varphi(j)} - \overline{y}_j \to 0$.*

Suppose now that X and Y are first countable.

(iii) *F is H-lower continuous at x if and only if for every sequence $(x_n)_{n\in\mathbb{N}} \subseteq X$ with $(x_n) \to x$ and every sequence $(\overline{y}_n) \subseteq F(x)$ there exists a sequence (y_n) such that $y_n - \overline{y}_n \to 0$ and $y_n \in F(x_n)$ for all $n \geq n_0$;*
(iv) *F is H-u.c. at x if and only if for every sequence $((x_n, y_n))_{n\in\mathbb{N}} \subseteq \operatorname{graph} F$ with $(x_n) \to x$ there exists a sequence $(\overline{y}_n) \subseteq F(x)$ such that $y_n - \overline{y}_n \to 0$.*

Proposition 3.1.15. *Suppose that F is upper continuous and closed-valued. Then* graph F *is closed. The same conclusion holds if Y is a topological vector space and F is H-u.c. instead of being u.c.*

Note that requiring the upper continuity of F only on dom F is not sufficient for the conclusion of the preceding result.

Let $(A_n)_{n\in\mathbb{N}} \subseteq \mathscr{P}(Y)$. Recall that

$$\liminf_{n\to\infty} A_n := \{y \in Y \mid \exists\,(y_n) \to y \text{ such that } y_n \in A_n \text{ for } n \geq n_0\},$$

$$\limsup_{n\to\infty} A_n := \{y \in Y \mid \exists\,(y_{n_k}) \to y \text{ such that } y_{n_k} \in A_{n_k} \text{ for } k \in \mathbb{N}\};$$

of course, $\liminf_{n\to\infty} A_n \subseteq \limsup_{n\to\infty} A_n$. We say that (A_n) **converges in the sense of Kuratowski–Painlevé** to $A \subseteq Y$ if $\limsup_{n\to\infty} A_n \subseteq A \subseteq \liminf_{n\to\infty} A_n$.

Having $A, A_n \subseteq Y$, $n \in \mathbb{N}$, and taking $X := \mathbb{N} \cup \{\infty\} \subseteq \overline{\mathbb{R}}$ (X endowed with the topology induced by that of $\overline{\mathbb{R}}$) we may consider the set-valued map $F : X \rightrightarrows Y$ defined by $F(n) = A_n$ for $n \in \mathbb{N}$ and $F(\infty) = A$. When Y is first countable, $\liminf_{n\to\infty} A_n$ is exactly $\liminf_{n\to\infty} F(n)$, and $\limsup_{n\to\infty} A_n$ is $\limsup_{n\to\infty} F(n)$.

Definition 3.1.16. Suppose that Y is a separated topological vector space and $C \subseteq Y$ is a convex cone. We say that F is **C-u.c.,C-l.c.,H-C-u.c.**, or **H-C-l.c. at** $x_0 \in X$ if relation (3.1), (3.2), (3.3) or (3.4) holds with $F(x) \subseteq D + C$, $F(x) \cap (D - C) \neq \emptyset$, $F(x) \subseteq F(x_0) + V + C$, $F(x_0) \subseteq F(x) + V + C$ instead of $F(x) \subseteq D$, $F(x) \cap D \neq \emptyset$, $F(x) \subseteq F(x_0) + V$, $F(x_0) \subseteq F(x) + V$, respectively; similarly for F **C-l.c. at** (x_0, y_0). F is **C-continuous** (**C-Hausdorff continuous**) **at** x_0 if F is C-u.c. and C-l.c. (H-C-u.c. and H-C-l.c.) at x_0.

Remark 3.1.17. As in Proposition 3.1.13 one can prove that having the set-valued maps $F, F_1, F_2 : X \rightrightarrows Y$ that are C-l.c. (resp. C-H-u.c., C-H-l.c.) at $x_0 \in X$, Y being a t.v.s., $C \subseteq Y$ a convex cone, and $\alpha \in \mathbb{R}_+$, then $F_1 + F_2$ and αF are C-l.c. (resp. C-H-u.c., C-H-l.c.) at x_0.

When $P \subseteq C$ is another convex cone, if F is P-l.c. (P-u.c., H-P-l.c., or H-P-u.c.) at x_0, then F is C-l.c. (C-u.c., H-C-l.c., or H-C-u.c.) at x_0. Note also that F is C-l.c., H-C-u.c., or H-C-l.c. if and only if F_C is l.c., H-u.c., or H-l.c., respectively, but such an equivalence is not true for upper continuity. In fact, we have that F is C-u.c. at x_0 if F_C is u.c. at x_0, but the converse is not true even if $F = F_C$.

Definition 3.1.18. When Y is a topological vector space, we say that F is **uniformly C-l.c. at** x_0 **on** A if $A \subseteq F(x_0)$ and

$$\forall\, V \in \mathscr{N}_Y,\ \exists\, U \in \mathscr{N}_X(x_0),\ \forall\, x \in U \ :\ A \subseteq F(x) + V + C.$$

Hence F is uniformly C-l.c. at x_0 on $F(x_0)$ exactly when F is C-H-l.c. at x_0.

Remark 3.1.19. Clearly, if F is C-u.c. (resp. C-H-l.c.) at $x_0 \in X$ then F is C-H-u.c. (resp. C-l.c.) at x_0. If F is C-l.c. at (x_0, y) for every $y \in A \subseteq F(x_0)$ and A is compact, then F is uniformly C-l.c. at x_0 on A.

Proposition 3.1.20. *If $F(x_0)$ is compact, then F is C-u.c. (resp. C-H-l.c.) at $x_0 \in X$ provided F is C-H-u.c. (resp. C-l.c.) at x_0.*

Proof. The case $x_0 \notin \operatorname{dom} F$ is trivial. So let $x_0 \in \operatorname{dom} F$ be such that $F(x_0)$ is compact.

First assume that F is C-H-u.c. at x_0. Consider $D \subseteq Y$ an open set with $F(x_0) \subseteq D$. Then there exists $V \in \mathscr{N}_Y$ such that $F(x_0) + V \subseteq D$. Since F is C-H-u.c. at x_0, there exists $U \in \mathscr{N}_X(x_0)$ such that $F(x) \subseteq F(x_0) + V + C \subseteq D + C$ for every $x \in U$, and so F is C-u.c. at x_0.

Assume now that F is C-l.c. at x_0 and take $V \in \mathscr{N}_Y$. There exists $V_0 \in \mathscr{N}_Y$ such that $V_0 + V_0 \subseteq V$. Since $F(x_0)$ is compact, there exists a finite set $Y_0 \subseteq F(x_0)$ such that $F(x_0) \subseteq Y_0 + V_0$. Since F is C-l.c. at x_0, for any $y \in Y_0$ there exists $U_y \in \mathscr{N}(x_0)$ such that $F(x) \cap (y + V_0 - C) \neq \emptyset$. Take $U := \bigcap_{y \in Y_0} U_y$; then $U \in \mathscr{N}(x_0)$. Take $x \in U$ and $y \in F(x_0)$. Then $y = y_0 + v$ for some $y_0 \in Y_0$ and $v \in V_0$. Since $y_0 \in Y_0$ and $x \in U \subseteq U_{y_0}$, we have that $F(x) \cap (y_0 + V_0 - C) \neq \emptyset$, that is $y_0 + v' - c \in F(x)$ for some $v' \in V_0$ and $c \in C$, whence $y = y_0 + v \in F(x) - V_0 + C + V_0 \subseteq F(x) + V + C$. Hence F is C-H-l.c. at x_0. □

With any set-valued map $F : X \rightrightarrows Y$ we associate its closure $\overline{F} : X \rightrightarrows Y$ defined by $\overline{F}(x) := \operatorname{cl}(F(x))$ for $x \in X$. Of course, $\operatorname{dom} F = \operatorname{dom} \overline{F}$. Continuity properties of F and $\overline{F}$ are deeply related.

Proposition 3.1.21. *Let $x \in X$ and $y \in Y$. The following assertions hold:*

(i) *F is l.c. at (x, y) if and only if $\overline{F}$ is l.c. at (x, y).*
(ii) *F is l.c. at x if and only if $\overline{F}$ is l.c. at x.*
(iii) *If Y is a topological vector space and $C \subseteq Y$ is a convex cone, then F is H-C-u.c. (H-C-l.c.) at x if and only if $\overline{F}$ is H-C-u.c. (H-C-l.c.) at x.*
(iv) *If F is u.c. at x and either Y is normal or Y is regular and* $\operatorname{cl} F(x)$ *is compact, then $\overline{F}$ is u.c. at x.*
(v) *If F is u.c. at x and either Y is regular or* $\operatorname{cl} F(x)$ *is compact, then $\overline{F}$ is closed at x.*
(vi) *F is closed at x if and only if $F(x)$ is closed and $\overline{F}$ is closed at x.*
(vii) *If Y is a metric space, then F is compact at x if and only if $F(x)$ is closed and $\overline{F}$ is compact at x.*

Note that if $\overline{F}$ is u.c. at x, it does not follow that F is u.c. at x (but the implication is true if $F(x)$ is closed). Take $F : \mathbb{R} \rightrightarrows \mathbb{R}$, $F(x) =]x, x+1[$; F is not u.c. at 0, but $\overline{F}$ is.

From now on (in this section), Y is a separated topological vector space and $C \subseteq Y$ is a convex cone. Recall that the convex cone C determines a quasi-order $\leq_C$ defined by $y_1 \leq_C y_2$ if $y_2 - y_1 \in C$.

Definition 3.1.22. We say that

(i) F is C-**lower semicontinuous** (C-l.s.c. for short) if $\operatorname{lev}_F(y) := \{x \in X \mid F(x) \cap (y - C) \neq \emptyset\}$ is closed for every $y \in Y$;
(ii) F is C-**upper semicontinuous** (C-u.s.c. for short) if $\operatorname{lev}_F^<(y) := \{x \in X \mid F(x) \cap (y - \operatorname{int} C) \neq \emptyset\}$ is open in X for every $y \in Y$.

Note that $\operatorname{lev}_F(y) = (F_C)^{-1}(y)$ and $\operatorname{lev}_F^<(y) = (F_{\operatorname{int} C})^{-1}(y)$ for every $y \in Y$, so that F is C-l.s.c. iff $(F_C)^{-1}$ is closed-valued, and F is C-u.s.c. iff $(F_{\operatorname{int} C})^{-1}$ is open valued. Moreover, if $y_1 \leq_C y_2$, then $\operatorname{lev}_F(y_1) \subseteq \operatorname{lev}_F(y_2)$ and $\operatorname{lev}_F^<(y_1) \subseteq \operatorname{lev}_F^<(y_2)$. Furthermore, when $\operatorname{int} C \neq \emptyset$ we have that $\operatorname{int}\left(\operatorname{dom}(\operatorname{lev}_F)\right) = \operatorname{dom}(\operatorname{lev}_F^<)$.

The following result holds.

Proposition 3.1.23. *The following assertions hold:*

(i) *If F is C-u.c. and C is closed then F is C-lower semicontinuous.*
(ii) *If* epi F *is closed, then F is C-lower semicontinuous.*
(iii) *Suppose that Y is a locally convex space, C is closed,* $\operatorname{int} C \neq \emptyset$, *$F$ is C-lower semicontinuous, and either $F(x)$ is weakly compact or $F(x) + C$ is closed for every $x \in X$. Then* epi F *is closed. Moreover, if $y_0, y_1 \in Y$ are such that $y_1 - y_0 \in \operatorname{int} C$ and $\operatorname{lev}_F(y_1)$ is compact, then $\operatorname{lev}_F(y_0)$ is compact and lev_F is upper continuous at y_0.*

Proof. (i) Take $y \in Y$ and $x \in X \setminus \operatorname{lev}_F(y)$. Then $F(x) \cap (y - C) = \emptyset$ (or, equivalently, $F(x) \subseteq Y \setminus (y - C)$). Since C is closed and F is C-u.c. at x, there exists $U \in \mathscr{N}(x)$ such that $F(x') \cap (y - C) = \emptyset$ for every $x' \in U$. This means that $U \subseteq X \setminus \operatorname{lev}_F(y)$, whence $\operatorname{lev}_F(y)$ is closed.

For the proof of (ii) and (iii) see [214, Proposition 2.5.22]. □

Proposition 3.1.24. *Consider the set-valued map $F : X \rightrightarrows Y$.*

(i) *Suppose that C is closed and $P \subseteq C$ is another convex cone. If F is P-upper continuous, then F is C-lower semicontinuous.*
(ii) *Suppose that* $\operatorname{int} C \neq \emptyset$; *then F is C-lower continuous if and only if F is C-upper semicontinuous.*

From the preceding result we obtain that F is C-l.s.c. and $-C$-l.s.c. when F is upper continuous (and C is closed).

Proposition 3.1.25. *Suppose that* $\operatorname{int} C \neq \emptyset$ *and* $y \in \operatorname{dom}(\operatorname{lev}_F)$. *If* lev_F *is lower continuous at y, then* $\operatorname{lev}_F(y) \subseteq \operatorname{cl}\left(\operatorname{lev}_F^<(y)\right)$. *Conversely, if X is a topological vector space (or a metric space),* $\operatorname{lev}_F(y)$ *is totally bounded, and* $\operatorname{lev}_F(y) \subseteq \operatorname{cl}\left(\operatorname{lev}_F^<(y)\right)$, *then* lev_F *is Hausdorff lower continuous at y. Moreover, if* $\operatorname{lev}_F(y)$ *is compact, then* lev_F *is lower continuous at y.*

Definition 3.1.26. We say that $F : X \rightrightarrows Y$ is **weakly C-upper bounded** on a set $A \subseteq X$ if there exists $y \in Y$ such that $A \subseteq \operatorname{lev}_F(y)$.

Proposition 3.1.27. *Consider the following assertions:*

(i) $\operatorname{int}(\operatorname{epi} F) \neq \emptyset$,
(ii) *there exists* $x_0 \in \operatorname{int}(\operatorname{dom} F)$ *such that* F *is weakly* C*-upper bounded on a neighborhood of* x_0,
(iii) *there exists* $(x_0, y_0) \in \operatorname{graph} F$ *such that* F_C *is l.c. at* (x_0, y_0).

Then (i) $\Rightarrow$ (ii); *moreover, if* $\operatorname{int} C \neq \emptyset$ *then* (iii) $\Rightarrow$ (ii) $\Rightarrow$ (i).

Proof. (i) $\Rightarrow$ (ii) Take $(x_0, y_0) \in \operatorname{int}(\operatorname{epi} F)$. Then there exists $U \in \mathscr{N}_X(x_0)$ and $V \in \mathscr{N}_Y$ such that $U \times (y_0 + V) \subseteq \operatorname{epi} F$; clearly $x_0 \in \operatorname{int}(\operatorname{dom} F)$ and $U \times \{y_0\} \subseteq \operatorname{epi} F$. It follows that $U \subseteq \operatorname{lev}_F(y_0)$, and so (ii) holds.

Assume now that $\operatorname{int} C \neq \emptyset$.

(iii) $\Rightarrow$ (ii) Take $y \in \operatorname{int} C$; then $V := (C - y) \cap (y - C) \in \mathscr{N}_Y$. Because F_C is l.c. at (x_0, y_0), there exists $U \in \mathscr{N}_X(x_0)$ such that $(y_0 + V) \cap (F(x) + C) \neq \emptyset$, or, equivalently, $y_0 \in F(x) + V + C$ for every $x \in U$. We obtain that $y_0 + y \in F(x) + C$ for every $x \in U$, and so $U \subseteq \operatorname{lev}_F(y_0 + y)$. Therefore, (ii) holds.

(ii) $\Rightarrow$ (i) There exist $y \in Y$ and $U \in \mathscr{N}_X(x_0)$ such that $U \subseteq \operatorname{lev}_F(y)$. Then $U \times \{y\} \subseteq \operatorname{epi} F$, whence $U \times (y + C) \subseteq \operatorname{epi} F$. It follows that $\operatorname{int}(\operatorname{epi} F) \neq \emptyset$. □

The continuity properties for **vector-valued functions** and those of set-valued maps are related. We give first some continuity notions for extended vector-valued functions.

Definition 3.1.28. Let $f : X \to Y^\bullet$ be proper and $x_0 \in X$.

(i) f is C**-lower continuous** (C-l.c.) **at** x_0 if

$$\forall y \in Y,\ y \leq f(x_0),\ \forall V \in \mathscr{N}_Y,\ \exists U \in \mathscr{N}_X(x_0),$$
$$\forall x \in U \cap \operatorname{dom} f : f(x) \in y + V + C; \tag{3.5}$$

f is C**-lower continuous** (C-l.c.) if f is C-l.c. at any $x_0 \in X$.

(ii) f is C**-upper continuous** (C-u.c.) **at** $x_0 \in \operatorname{dom} f$ if

$$\forall\, V \in \mathscr{N}_Y,\ \exists\, U \in \mathscr{N}_X(x_0),\ \forall\, x \in U\ :\ f(x) \in f(x_0) + V - C;$$

f is C**-upper continuous** (C-u.c.) if f is C-u.c. at any $x_0 \in \operatorname{dom} f$.

(iii) f is C**-lower semicontinuous** (C-l.s.c.) if $\operatorname{lev}_f(y)$ is closed for every $y \in Y$.
(iv) f is C**-upper semicontinuous** (C-u.s.c.) if $\operatorname{lev}_f^<(y)$ is open for every $y \in Y$.

Note that for $x_0 \in \operatorname{dom} f$ condition (3.5) is equivalent to

$$\forall V \in \mathscr{N}_Y,\ \exists U \in \mathscr{N}_X(x_0),\ \forall x \in U \cap \operatorname{dom} f : f(x) \in f(x_0) + V + C, \tag{3.6}$$

while for $f(x_0) = +\infty$ (3.5) is equivalent to

$$\forall y \in Y,\ \forall V \in \mathscr{N}_Y,\ \exists U \in \mathscr{N}_X(x_0),\ \forall x \in U \cap \operatorname{dom} f : f(x) \in y + V + C. \tag{3.7}$$

Recall that $F_{f,C} : X \rightrightarrows Y$ is the set-valued map associated to $f : X \to Y^\bullet$, that is, $\operatorname{graph} F_{f,C} = \{(x, y) \in X \times Y \mid f(x) \leq_C y\}$.

Consider $x_0 \in \operatorname{dom} f$. If f is C-u.c. at x_0, then $x_0 \in \operatorname{int}(\operatorname{dom} f)$. On the other hand, if $F_{f,C}$ is closed at x_0, then C is closed; if C is closed and $F_{f,C}$ is u.c. at x_0, then $F_{f,C}$ is closed at x_0. Moreover, f is $\{0\}$-u.c. at $x_0 \Leftrightarrow f$ is continuous at $x_0 \Leftrightarrow x_0 \in \operatorname{int}(\operatorname{dom} f)$ and f is $\{0\}$-l.c. at x_0.

Proposition 3.1.29. *Let $f : X \to Y^\bullet$ be proper and $x_0 \in \operatorname{dom} f$. Then:*

(i) *f is C-l.c. at $x_0 \Leftrightarrow F_{f,C}$ is C-u.c. at $x_0 \Leftrightarrow F_{f,C}$ is H-u.c. at x_0.*
(ii) *f is C-u.c. at $x_0 \Leftrightarrow F_{f,C}$ is l.c. at $x_0 \Leftrightarrow F_{f,C}$ is H-l.c. at x_0.*

When $Y = \mathbb{R}$ and $C = \mathbb{R}_+$ we have more refined statements; $F_f := F_{f,\mathbb{R}_+}$ in this case.

Proposition 3.1.30. *Let $f : X \to \overline{\mathbb{R}}$ with the associated set-valued map $F_f : X \rightrightarrows \mathbb{R}$ whose graph is $\operatorname{epi} f$.*

(i) *Suppose that $x_0 \in \operatorname{dom} f$ $(= \operatorname{dom} F_f)$. Then f is l.c. at $x_0 \Leftrightarrow F_f$ is u.c. at x_0 $\Leftrightarrow F_f$ is H-u.c. at x_0.*
(ii) *Suppose that $x_0 \in X$. Then F_f is H-l.c. at $x_0 \Rightarrow f$ is u.c. at $x_0 \Leftrightarrow F_f$ is l.c. at x_0; moreover, if either $f(x_0) > -\infty$ or X is a topological vector space and f is convex, then f is u.c. at $x_0 \Leftrightarrow F_f$ is H-l.c. at x_0.*

Proposition 3.1.31. (i) *If $f : X \to Y^\bullet$ is a proper C-l.c. function and C is a proper closed convex cone, then $\operatorname{epi} f$ is closed.*
(ii) *Let $f_i : X_i \to Y^\bullet$ $(i \in \{1, 2\})$ be proper and C-l.c., where (X_i, τ_i) are topological spaces and $f : X_1 \times X_2$ be defined by $f(x_1, x_2) := f_1(x_1) + f_2(x_2)$. Then f is C-l.c.*
(iii) *Let $f : (Z, \sigma) \to Y^\bullet$ be a proper C-l.c. function and $g : (X, \tau) \to (Z, \sigma)$ be continuous. Then $f \circ g$ is C-l.c.*
(iv) *Let Z be a topological vector space ordered by the convex cone P, $f : Z \to Y^\bullet$ be proper, C-l.c. and increasing such that for every $y \in Y$ there exists $z \in Z$ with $f(z) \geq y$, and $g : (X, \tau) \to Z^\bullet$ be proper and P-l.c. Then $f \circ g$ is C-l.c., where $f(+\infty) := +\infty$.*

Proof. (i) Let $(x, y) \in X \times Y \setminus \operatorname{epi} f$. We have the following three situations: a) $x \notin \operatorname{cl}(\operatorname{dom} f)$, b) $x \in \operatorname{cl}(\operatorname{dom} f) \setminus \operatorname{dom} f$, and c) $x \in \operatorname{dom} f$.

a) In this case there exists $U \in \mathscr{N}_X(x)$ such that $U \cap \operatorname{dom} f = \emptyset$, and so $(U \times Y) \cap \operatorname{epi} f = \emptyset$. Hence $(x, y) \notin \operatorname{cl}(\operatorname{epi} f)$.
b) In this case $f(x) = +\infty$. Because $Y \neq C = \cap\{C + V \mid V \in \mathscr{N}_Y\}$, there exist $y' \in Y$ and $V' \in \mathscr{N}_Y$ such that $y' \notin C + V'$. Then, there exists $V \in \mathscr{N}_Y$ such that $V + V \subseteq V'$. Using (3.7), there exists $U \in \mathscr{N}_X(x)$ such that $f(x') \in y - y' + V + C$ for every $x' \in U \cap \operatorname{dom} f$. Then $(U \times (y + V)) \cap \operatorname{epi} f = \emptyset$. In the contrary case, there exists $x' \in U$ such that $f(x') \in y + V - C$. Since $f(x') \in y - y' + V + C$, we get $y' \in V - V + C + C \subseteq V' + C$, a contradiction. Hence $(x, y) \notin \operatorname{cl}(\operatorname{epi} f)$.

c) In this case we have that $y \notin f(x) + C$. Since C is closed, as above, there exists $V \in \mathcal{N}_Y$ such that $(y + V) \cap (f(x) + V + C) = \emptyset$. Using (3.6), there exists $U \in \mathcal{N}_X(x)$ such that $f(x') \in f(x) + V + C$ for every $x' \in U \cap \operatorname{dom} f$. It follows that $(U \times (y + V)) \cap \operatorname{epi} f = \emptyset$, and so $(x, y) \notin \operatorname{cl}(\operatorname{epi} f)$.

(ii) Let first $x = (x_1, x_2) \in \operatorname{dom} f = \operatorname{dom} f_1 \times \operatorname{dom} f_2$ and $V \in \mathcal{N}_Y$; there exists $V' \in \mathcal{N}_Y$ such that $V' + V' \subseteq V$. Since f_i is C-l.c. at x_i, there exists $U_i \in \mathcal{N}_{X_i}(x_i)$ such that $f_i(x_i') \in f_i(x_i) + V' + C$ for every $x_i' \in U_i$ $(i \in \{1, 2\})$. Therefore, for $(x_1', x_2') \in (U_1 \times U_2) \cap \operatorname{dom} f$ we have that

$$\begin{aligned} f(x_1', x_2') &= f_1(x_1') + f_2(x_2') \in f_1(x_1) + V' + C + f_2(x_2) \\ &\quad + V' + C \subseteq f(x_1, x_2) + V + C. \end{aligned}$$

Hence f is C-l.c. at x.

Let now $x \in \operatorname{cl}(\operatorname{dom} f) \setminus \operatorname{dom} f = (\operatorname{cl}(\operatorname{dom} f_1) \times \operatorname{cl}(\operatorname{dom} f_2)) \setminus (\operatorname{dom} f_1 \times \operatorname{dom} f_2)$. We can have one of the following situations: a) $x_1 \in \operatorname{cl}(\operatorname{dom} f_1) \setminus \operatorname{dom} f_1$ and $x_2 \in \operatorname{cl}(\operatorname{dom} f_2) \setminus \operatorname{dom} f_2$, b) $x_1 \in \operatorname{cl}(\operatorname{dom} f_1) \setminus \operatorname{dom} f_1$ and $x_2 \in \operatorname{dom} f_2$, c) $x_1 \in \operatorname{dom} f_1$ and $x_2 \in \operatorname{cl}(\operatorname{dom} f_2) \setminus \operatorname{dom} f_2$.

a) Let $y \in Y$, $V \in \mathcal{N}_Y$, and take $V' \in \mathcal{N}_Y$ such that $V' + V' \subseteq V$. Then for $i \in \{1, 2\}$,

$$\exists U_i \in \mathcal{N}_{X_i}(x_i),\ \forall x_i' \in U_i \cap \operatorname{dom} f_i : f_i(x_i') \in \tfrac{1}{2}y + V' + C,$$

and so

$$\begin{aligned} \exists U :&= U_1 \times U_2 \in \mathcal{N}_{X_1 \times X_2}(x),\ \forall x' \\ &= (x_1', x_2') \in U \cap \operatorname{dom} f : f(x') \in y + V + C. \end{aligned} \tag{3.8}$$

Hence f is C-l.c. at x.

b) Let $y \in Y$, $V \in \mathcal{N}_Y$, and take $V' \in \mathcal{N}_Y$ such that $V' + V' \subseteq V$. Then

$$\exists U_1 \in \mathcal{N}_{X_1}(x_1),\ \forall x_1' \in U_1 \cap \operatorname{dom} f_1 : f_1(x_1') \in y - f_2(x_2) + V' + C,$$
$$\exists U_2 \in \mathcal{N}_{X_2}(x_2),\ \forall x_2' \in U_2 \cap \operatorname{dom} f_2 : f_2(x_2') \in f_2(x_2) + V' + C.$$

It follows that (3.8) holds, and so f is C-l.c. at x.

c) This case follows from b) interchanging 1 and 2.

(iii) We have that $\operatorname{dom}(f \circ g) = g^{-1}(\operatorname{dom} f)$. Let $x \in X$ and set $z := g(x)$. If $f(z) = -\infty$, it is nothing to do.

Assume that $f(z) \in Y$ and take $V \in \mathcal{N}_Y$. Because f is C-l.c. at z, there exists $W \in \mathcal{N}_Z(z)$ such that $f(z') \in f(z) + V + C$ for every $z' \in W \cap \operatorname{dom} f$. By the continuity of g at x, there exists $U \in \mathcal{N}_X(x)$ such that $g(x') \in W$ for every $x' \in U$. It follows that $(f \circ g)(x') = f(g(x')) \in f(g(x)) + V + C$ for every $x' \in U \cap \operatorname{dom} f \circ g$. Hence $f \circ g$ is C-l.c. at x in this case.

Assume now that $f(z) = \infty$ and take $y \in Y$, $V \in \mathcal{N}_Y$. Because f is C-l.c. at z, there exists $W \in \mathcal{N}_Z(z)$ such that $f(z') \in y + V + C$ for every $z' \in W \cap \operatorname{dom} f$. Continuing as in the previous case, we get $U \in \mathcal{N}_X(x)$ such that $g(x') \in W$ for every $x' \in U$. It follows that $(f \circ g)(x') = f(g(x')) \in y + V + C$ for every $x' \in U \cap \operatorname{dom}(f \circ g)$. Hence $f \circ g$ is C-l.c. at x in this case, too.

(iv) Consider $x_0 \in X$ and set $y_0 := (f \circ g)(x_0) \in Y \cup \{+\infty\}$, $z_0 := g(x_0) \in Z \cup \{+\infty\}$. Assume first that $z_0 \in Z$. Let $V \in \mathcal{N}_Y$ and $y \in Y$, with $y = y_0$ if $y_0 \in Y$. Because f is C-l.c. at z_0, there exists $W \in \mathcal{N}_Z$ such that $f(z) \in y + V + C$ for every $z \in (z_0 + W) \cap \operatorname{dom} f$. Because g is P-l.c. at x_0, there exists $U \in \mathcal{N}_X(x_0)$ such that $g(x) \in z_0 + W + P$ for every $x \in U \cap \operatorname{dom} g$. Take $x \in U \cap \operatorname{dom}(f \circ g) \subseteq U \cap \operatorname{dom} g$. Then $g(x) \geq_P z$ for some $z \in z_0 + W$, and so $f(g(x)) \geq_C f(z)$ because f is increasing. It follows that $z \in (z_0 + W) \cap \operatorname{dom} f$ and so $f(g(x)) \in f(z) + C \subseteq y + V + C + C = y + V + C$. Therefore, $f \circ g$ is C-l.c. at x_0.

Assume now that $z_0 = +\infty$; consequently, $y_0 = +\infty$. Let $V \in \mathcal{N}_Y$ and $y \in Y$. By hypothesis, there exists $z \in Z$ such that $f(z) \geq y$. Because f is C-l.c. at z, there exists $W \in \mathcal{N}_Z$ such that $f(z') \in y + V + C$ for every $z' \in (z + W) \cap \operatorname{dom} f$. Because g is P-l.c. at x_0 and $z \leq_P g(x_0) = +\infty$, there exists $U \in \mathcal{N}_X(x_0)$ such that $g(x) \in z + W + P$ for every $x \in U \cap \operatorname{dom} g$. Take $x \in U \cap \operatorname{dom}(f \circ g) \subseteq U \cap \operatorname{dom} g$. Then $g(x) \geq_P z'$ for some $z' \in z + W$, and so $f(g(x)) \geq_C f(z')$. It follows that $z' \in (z + W) \cap \operatorname{dom} f$ and so $f(g(x)) \in f(z') + C \subseteq y + V + C + C = y + V + C$. Therefore, $f \circ g$ is C-l.c. at x_0. Hence $f \circ g$ is C-l.c. □

We are now interested in continuity properties of the composition of two set-valued maps and the composition of a function with a set-valued map. So consider another topological space U, the functions $f : X \times U \to Y$, $g : X \to Y$, and the set-valued maps $\Lambda : U \rightrightarrows X$, $F : X \times U \rightrightarrows Y$ and $G : X \rightrightarrows Y$; we associate the set-valued maps $f\Lambda, g\Lambda, F\Lambda, G\Lambda : U \rightrightarrows Y$ defined by $(f\Lambda)(u) := f(\Lambda(u) \times \{u\})$, $(g\Lambda)(u) := g(\Lambda(u))$, $(F\Lambda)(u) = F(\Lambda(u) \times \{u\})$, and $(G\Lambda)(u) := G(\Lambda(u))$.

Proposition 3.1.32. *Let f, g and Λ, F, G be as above and $u_0 \in \operatorname{dom} \Lambda$.*

(i) *If Λ is u.c. at u_0 and g is C-l.c. on $\Lambda(u_0)$, then $g\Lambda$ is C-u.c. at u_0. Moreover, if $\Lambda(u_0)$ is compact, then $g\Lambda$ is H-C-u.c. at u_0.*
(ii) *If Λ is u.c. at u_0, F is C-u.c. at (x, u_0) for all $x \in \Lambda(u_0)$, and $\Lambda(u_0)$ is compact, then $F\Lambda$ is C-u.c. at u_0. If f is C-l.c. on $\Lambda(u_0)$, then $f\Lambda$ is H-C-u.c. at u_0.*
(iii) *If Λ is l.c. at $(x_0, u_0) \in \operatorname{graph} \Lambda$ and F is C-l.c. at $(x_0, u_0; y_0)$, then $F\Lambda$ is C-l.c. at (u_0, y_0). In particular, if f is C-u.c. at (x_0, u_0), then $f\Lambda$ is C-l.c. at (u_0, y_0), where $y_0 = f(u_0, x_0)$.*
(iv) *If Λ is compact at u_0 and F is closed at (x, u_0) for all $x \in \Lambda(u_0)$, then $F\Lambda$ is closed at u_0.*
(v) *If Λ is compact at every $u \in U$ and G is C-l.s.c., then $G\Lambda$ is C-l.s.c., too.*

Suppose now that X is a topological vector space, too.

(vi) *If Λ is H-u.c. at u_0 and f is equi-C-l.c. on $\Lambda(u_0) \times \{u_0\}$, then $f\Lambda$ is H-C-u.c. at u_0.*
(vii) *If Λ is H-l.c. at u_0 and f is equi-C-u.c. on $\Lambda(u_0) \times \{u_0\}$, then $f\Lambda$ is H-C-l.c. at u_0.*

Detailed references for the most part of the results in this section are mentioned at the end of [214, Section 2.5]; we mention only some missing references in [214, Section 2.5] and those for the results not mentioned there. So, Propositions 3.1.23, 3.1.24, 3.1.25 are stated in [184, Proposition 2.3] and [185, Proposition 3.1]. Proposition 3.1.27 is essentially established in Song [535]. The continuity notions for vector-valued functions are in accordance with those of Penot and Théra [473]. The statement of Proposition 3.1.31 is adapted from [611].

3.2 Continuity Properties of Set-Valued Maps Under Convexity Assumptions

Throughout this section X and Y are topological vector spaces, $C \subseteq Y$ is a convex cone and $\Gamma : X \rightrightarrows Y$ is a set-valued map.

Proposition 3.2.1. *Let $x_0 \in \operatorname{dom} \Gamma$.*

(i) *Assume that* (a) *Γ is nearly convex, and* (b) *there exist $U \in \mathscr{N}_X(x_0)$ and $B \subseteq Y$ a bounded set such that $\Gamma(x) \cap (B - C) \neq \emptyset$ for every $x \in U$. Then* (c) *Γ is C-lower continuous at x_0.*
(ii) *Conversely, assume that* (c) *and* (d) *$B - C \in \mathscr{N}_Y(0)$ for some bounded set $B \subseteq Y$ hold. Then* (b) *holds.*

Proof. (i) Because Γ is nearly convex, there exists $\alpha \in]0,1[$ such that Γ is C-α-convex. Let $\Lambda^\alpha := \bigcup_{n\geq 0} \Lambda_n^\alpha$, where

$$\Lambda_0^\alpha := \{0,1\}, \quad \Lambda_{n+1}^\alpha := \{\alpha t + (1-\alpha)s \mid t,s \in \Lambda_n^\alpha\} \; \forall\, n \geq 0.$$

As seen in [214, Lemma 2.4.2], one has that $\operatorname{cl} \Lambda^\alpha = [0,1]$.

Let $y_0 \in \Gamma(x_0)$ and $V \in \mathscr{N}_Y(0)$. Since B is bounded, $B - y_0$ is also bounded, and so there is some $\lambda \in \Lambda^\alpha \setminus \{0,1\}$ such that $\lambda(B - y_0) \subseteq V$. Fix $u \in U' := U - x_0 \in \mathscr{N}_X(0)$; because $\Gamma(x_0 + u) \cap (B - C) \neq \emptyset$, there is some $b \in B$ and $v \in C$ such that $b - v \in \Gamma(x_0 + u)$. By the α-convexity of Γ we obtain that

$$(1-\lambda)y_0 + \lambda(b - v) \in (1-\lambda)\Gamma(x_0) + \lambda\Gamma(x_0 + u) \subseteq \Gamma(x_0 + \lambda u) + C,$$

whence $y_0 + \lambda(b - y_0) \in (y_0 + V) \cap (\Gamma(x_0 + \lambda u) + C)$. Hence $(y_0 + V) \cap (\Gamma(x) + C) \neq \emptyset$ for every $x \in x_0 + \lambda U'$ $(\in \mathscr{N}(x_0))$, and so Γ is C-lower continuous at x_0.

(ii) By (d) we have that $V := B - C \in \mathscr{N}_Y(0)$. Take $y_0 \in \Gamma(x_0)$. Because Γ is C-lower continuous at x_0, there exists $U \in \mathscr{N}(x_0)$ such that $(\Gamma(x) + C) \cap (y_0 + V) \neq \emptyset$ for every $x \in U$. It follows that $\Gamma(x) \cap (y_0 + B - C) \neq \emptyset$ for $x \in U$, which proves that Γ verifies condition (b) ($y_0 + B$ being bounded). The proof is complete. □

Corollary 3.2.2. *Assume that Γ is nearly convex. Then* (ii) $\Rightarrow$ (iii) *in Proposition 3.1.27 holds. Therefore, if* $\operatorname{int} C \neq \emptyset$, *then conditions* (i), (ii) *and* (iii) *in Proposition 3.1.27 are equivalent.*

Proof. If (ii) of Proposition 3.1.27 holds, then condition (b) in Proposition 3.2.1 holds, and so Γ is C-lower continuous at x_0 by the latter result. Taking $y_0 \in F(x_0)$, (iii) in Proposition 3.2.1 holds. The proof is complete. □

Note that the condition (d) in Proposition 3.2.1 is verified if $\operatorname{int} C \neq \emptyset$ or if Y is a normed vector space. Moreover, if $\operatorname{int} C \neq \emptyset$ then condition (b) in Proposition 3.2.1 is equivalent to the weak C-upper boundedness of Γ on a neighborhood of x_0.

As in [25] and [26], we say that $E \subseteq Y$ is **quasi (C-lower) bounded** if there exists a bounded set $B \subseteq Y$ such that $E \subseteq B + C$;

Theorem 3.2.3. *Let Γ be C-nearly convex. Suppose that Γ is C-lower continuous at* $(x_0, y_0) \in \operatorname{graph} \Gamma$. *Then Γ is C-lower continuous at every* $x \in \operatorname{int}(\operatorname{dom} \Gamma) = \operatorname{core}(\operatorname{dom} \Gamma)$. *Moreover, Γ is H-C-lower continuous at every* $x \in \operatorname{int}(\operatorname{dom} \Gamma)$ *for which $\Gamma(x)$ is quasi-bounded.*

The next result is related to the boundedness condition of the preceding theorem.

Proposition 3.2.4. *Let Γ be C-nearly convex. Suppose that* $x_0 \in \operatorname{int}(\operatorname{dom} \Gamma)$ *is such that $\Gamma(x_0)$ is quasi-bounded. Then $\Gamma(x)$ is quasi-bounded for every* $x \in \operatorname{dom} \Gamma$.

As an application of the preceding result we obtain that $f : X \to \overline{\mathbb{R}}$ is finite-valued on $\operatorname{dom} f$ whenever f is nearly convex and finite-valued at $x_0 \in \operatorname{int}(\operatorname{dom} f)$.

Proposition 3.2.5. *Assume that Γ is C-lower continuous at* $(x_0, y_0) \in \operatorname{graph} \Gamma$ *and C-nearly convex. If $\Gamma(x_0)$ is quasi-bounded, then Γ is H-C-u.c. at x_0.*

The preceding results lead to the following important theorem.

Theorem 3.2.6. *Let Γ be a C-nearly convex. Suppose that Γ is C-lower continuous at* $(x_0, y_0) \in \operatorname{graph} \Gamma$ *and $\Gamma(x_0)$ is quasi-bounded. Then Γ is H-C-continuous on* $\operatorname{int}(\operatorname{dom} \Gamma)$.

The next result corresponds to another well-known result of convex analysis. It is an immediate consequence of Proposition 3.2.1 and Theorem 3.2.6.

Corollary 3.2.7. *Let Γ be a C-nearly convex. Suppose that there exist* $x_0 \in \operatorname{int}(\operatorname{dom} \Gamma)$, *$B \subseteq Y$ a bounded set, and $U \in \mathscr{N}_X$ such that $\Gamma(x_0) \subseteq B + C$ and $\Gamma(x_0 + u) \cap (B - C) \neq \emptyset$ for every $u \in U$. Then Γ is H-C-continuous on* $\operatorname{int}(\operatorname{dom} \Gamma)$.

From Corollary 3.2.7, also taking into account Proposition 3.1.30, we obtain that $f : X \to \overline{\mathbb{R}}$ is (finite-valued and) continuous on $\operatorname{int}(\operatorname{dom} f)$ whenever f is nearly convex, $f(x_0) \in \mathbb{R}$, and f is bounded above on a neighborhood of x_0.

Theorem 3.2.8. *Let Γ be nearly convex. Assume that Γ is lower continuous at some $(x_0, y_0) \in \operatorname{graph} \Gamma$. Then Γ is closed at any $x \in \operatorname{int}(\operatorname{dom} \Gamma)$ with $\Gamma(x)$ closed.*

Proof. Using Theorem 3.2.3 we have that Γ is lower continuous at any $x \in \operatorname{int}(\operatorname{dom} \Gamma)$. Because Γ is nearly convex, there exists $\alpha \in]0, 1/2]$ such that Γ is α-convex.

Fix $x \in \operatorname{int}(\operatorname{dom} \Gamma)$ such that $\Gamma(x)$ is closed. Take $y_1 \in Y \setminus \Gamma(x)$. We claim that there exists $y_2 \in \Gamma(x)$ such that $\alpha y_1 + (1-\alpha)y_2 \notin \Gamma(x)$. In the contrary case for every $z \in \Gamma(x)$ one has $\alpha y_1 + (1-\alpha)z \in \Gamma(x)$. Inductively we obtain that $(1-(1-\alpha)^n)y_1 + (1-\alpha)^n z \in \Gamma(x)$ for every $n \geq 1$, and so $y_1 \in \operatorname{cl} \Gamma(x) = \Gamma(x)$, a contradiction. Consider $y_2 \in \Gamma(x)$ with $\alpha y_1 + (1-\alpha)y_2 \notin \Gamma(x)$. Because $\Gamma(x)$ is closed, there exists $V \in \mathscr{N}_Y(0)$ such that

$$[\alpha y_1 + (1-\alpha)y_2 + \alpha V + (1-\alpha)V] \cap \Gamma(x) = \emptyset. \tag{3.9}$$

Because Γ is lower continuous at (x, y_2), there exists $U \in \mathscr{N}_X$ such that $\Gamma(x') \cap (y_2 + V) \neq \emptyset$ for every $x' \in x + U$. We claim that

$$[(x + U) \times (y_1 + V)] \cap \operatorname{graph} \Gamma = \emptyset, \tag{3.10}$$

and so $(x, y_1) \notin \operatorname{cl}(\operatorname{graph} \Gamma)$. Otherwise, let $(\hat{x}, \hat{y}) \in [(x + U) \times (y_1 + V)] \cap \operatorname{graph} \Gamma$ and take $x' := (1-\alpha)^{-1}(x - \alpha\hat{x})$. Since $\hat{x} \in x + U$, we have that $x' \in \frac{1}{1-\alpha}x - \frac{\alpha}{1-\alpha}x - \frac{\alpha}{1-\alpha}U \subseteq x + U$. Therefore, there exists $y' \in \Gamma(x') \cap (y_2 + V)$. On one hand, because Γ is α-convex, we obtain that $\alpha(\hat{x}, \hat{y}) + (1-\alpha)(x', y') \in \operatorname{graph} \Gamma$, and so $\alpha\hat{y} + (1-\alpha)y' \in \Gamma(\alpha\hat{x} + (1-\alpha)x') = \Gamma(x)$. On the other hand, $\alpha\hat{y} + (1-\alpha)y' \in \alpha(y_1 + V) + (1-\alpha)(y_2 + V)$, contradicting (3.9). Therefore, (3.10) holds.

Let now $((x_i, y_i))_{i \in I} \subseteq \operatorname{graph} \Gamma$ converge to (x, y); clearly, $(x, y) \in \operatorname{cl}(\operatorname{graph} \Gamma)$. Assuming that $y \notin \Gamma(x)$, by the argument above, we obtain the contradiction $(x, y) \notin \operatorname{cl}(\operatorname{graph} \Gamma)$. Hence $y \in \Gamma(x)$, and so Γ is closed at x. □

In fact, semicontinuous nearly convex set-valued maps are very close to convex set-valued maps.

Proposition 3.2.9. *Suppose that $\Gamma : X \rightrightarrows Y$ is closed-valued and nearly convex. If* $\operatorname{dom} \Gamma$ *is convex (for example, if* $\operatorname{dom} \Gamma$ *is closed or open) and Γ is H-u.c. at every point of* $\operatorname{dom} \Gamma$*, then Γ is convex.*

From Theorem 3.2.6 one gets the following characterization of the continuity of nearly convex proper functions.

Corollary 3.2.10. *Let $f : X \to Y^\bullet$ be a nearly convex proper function and $x_0 \in \operatorname{int}(\operatorname{dom} f)$. Suppose that C is normal. Then: f is C-u.c. at $x_0 \Leftrightarrow \Gamma_f$ is lower continuous at $x_0 \Leftrightarrow \Gamma_f$ is H-continuous at every $x \in \operatorname{int}(\operatorname{dom} f) \Leftrightarrow f$ is continuous on $\operatorname{int}(\operatorname{dom} f) \Leftrightarrow f$ is continuous at x_0.*

Another characterization of the continuity of a convex function is provided in the next result.

Proposition 3.2.11. *Let C be normal and $f : X \to Y^\bullet$ be a proper convex operator. Then:*

(i) *f is continuous at $x_0 \in \operatorname{dom} f$ iff*

$$\exists z \in Y,\ \forall V \in \mathscr{N}_Y,\ \exists U \in \mathscr{N}_X,\ \forall x \in x_0 + U : f(x) \in z + V - C. \tag{3.11}$$

Consequently, if $g : X \to Y^\bullet$ is another proper convex operator such that $g \leq f$ and f is continuous at $x_0 \in \operatorname{dom} f$, then g is continuous at x_0, too.

(ii) *If*

$$\exists\, B \subseteq Y,\ B \text{ bounded},\ \exists U \in \mathscr{N}_X\ :\ f(x_0 + U) \subseteq B - C, \tag{3.12}$$

then f is continuous at x_0. If $\operatorname{int} C \neq \emptyset$ and f is continuous at $x_0 \in \operatorname{dom} f$ then (3.12) holds (even with B a singleton).

Proof. (i) Let us set

$$\mathscr{N} := \{V \in \mathscr{N}_Y \mid V = (V + C) \cap (V - C)\};$$

because C is normal, $\mathscr{N}$ is a base of neighborhoods of the origin of Y.

Assume that f is continuous at $x_0 \in \operatorname{int}(\operatorname{dom} f)$. Then

$$\forall\, V \in \mathscr{N}_Y,\ \exists\, U \in \mathscr{N}_X\ :\ f(x_0 + U) \subseteq f(x_0) + V,$$

and so, taking $z := f(x_0)$, (3.11) holds. Conversely, assume that (3.11) holds. Replacing if necessary f by $\tilde{f}$ defined by $\tilde{f}(x) := f(x_0 + x) - f(x_0)$, we assume that $x_0 = 0$ and $f(0) = 0$. Consider $V \in \mathscr{N}_Y$; there exists $V_1 \in \mathscr{N}$ such that $V_1 + V_1 \subseteq V$. Since $V_1 \in \mathscr{N}_Y$, there exists $\lambda \in\ ]0, 1]$ such that $\lambda z \in V_1$. From (3.11) there exists $U \in \mathscr{N}_X$ such that $f(x) \in z + V_1 - C$ for every $x \in x_0 + U$. Taking into account that V_1 is balanced, f is convex and $f(0) = 0$, for $x \in U$ we have that $f(\lambda x) \leq \lambda f(x)$, and so

$$f(\lambda x) \in \lambda f(x) - C \subseteq \lambda z + \lambda V_1 - C \subseteq V_1 + V_1 - C \subseteq V - C;$$

moreover,

$$0 = f(0) = f\left(\tfrac{1}{2}\lambda x + \tfrac{1}{2}(-\lambda x)\right) \leq \tfrac{1}{2} f(\lambda x) + \tfrac{1}{2} f(-\lambda x),$$

and so

$$f(\lambda x) \in -f(-\lambda x) + C \subseteq C - V + C = V + C$$

because $-\lambda x \in U$. Therefore, $f(u) \in (V - C) \cap (V + C) = V$ for every $u \in \lambda U$, and so f is continuous at $0 = x_0$.

(ii) Assume that (3.12) holds. As above we assume that $x_0 = 0$ and $f(0) = 0$. Therefore, there exist $U_0 \in \mathscr{N}_X$ and $B \subseteq Y$ bounded such that $f(U_0) \subseteq B - C$. Consider $V \in \mathscr{N}$. Since B is bounded, there exists $\lambda \in]0,1]$ such that $\lambda B \subseteq V$. Taking into account that V is balanced, f is convex and $f(0) = 0$, for $x \in U_0$ we have that $f(\lambda x) \leq \lambda f(x)$, and so $f(\lambda x) \in \lambda f(x) - C \subseteq \lambda B - \lambda C - C \subseteq V - C$. Therefore, (3.11) holds with $z := 0$.

Assume that $\operatorname{int} C \neq \emptyset$ and take $z \in \operatorname{int} C$; then $z - C$ is a neighborhood of $0 \in Y$. Since f is continuous at x_0, there exists $U \in \mathscr{N}_X$ such that $f(x_0 + U) \subseteq f(x_0) + z - C$. Therefore, (3.12) holds with $B := \{f(x_0) + z\}$. □

Recall that the Minkowski (gauge) functional associated to a subset A of a linear space E is the functional

$$\mu_A : E \to \overline{\mathbb{R}}, \quad \mu_A(x) := \begin{cases} \inf\{t > 0 \mid x \in tA\} & \text{if } x \neq 0, \\ 0 & \text{if } x = 0. \end{cases}$$

It is clear that μ_A is positively homogeneous and $x \in tA$ for all $t > \mu_A(x)$ whenever $A = [0,1]A$. Moreover, μ_A is convex (hence sublinear) if A is convex. In the case in which A is an absorbing set, μ_A is finite-valued and it coincides with p_A as introduced in Proposition 6.2.1.

Lemma 3.2.12. *Let X, Y be topological vector spaces, $f : X \to Y^\bullet$ be a proper convex function and $x_0 \in \operatorname{dom} f$. Suppose that for some $U \in \mathscr{N}_X$ and some balanced set $V \subseteq Y$ one has*

$$\forall\, x \in x_0 + U \;:\; f(x) \in f(x_0) + V - C. \tag{3.13}$$

Then

$$\forall\, x \in x_0 + U \;:\; \mu_{[V]_C}(f(x) - f(x_0)) \leq p_U(x - x_0). \tag{3.14}$$

Proof. Replacing f by $g : X \to Y^\bullet$, $g(x) = f(x_0 + x) - f(x_0)$, we may suppose that $x_0 = 0$ and $f(x_0) = 0$. Therefore $f(x) \in V - C$ for every $x \in U$. Since $0 = f\left(\frac{1}{2}x' + \frac{1}{2}(-x')\right) \leq \frac{1}{2}f(x') + \frac{1}{2}f(-x')$ for every $x' \in X$, we obtain that $f(x) \in -f(-x) + C \subseteq -V + C + C = V + C$ for $x \in U$, and so $f(x) \in [V]_C$ for all $x \in U$.

From Proposition 6.2.1 we have that $\{x \in X \mid p_U(x) < 1\} \subseteq U \subseteq \{x \in X \mid p_U(x) \leq 1\}$. Let $x \in U$ ($\Leftrightarrow -x \in U$) and set $1 \geq t := p_U(x) \geq 0$. Take $s := 1$ if $t = 1$ and $s \in]t,1]$ if $t < 1$. Then $x' := s^{-1}x \in U$. Since f

is convex and $s \leq 1$, we obtain that $f(x) \leq sf(x') + (1-s)f(0) \in s[V]_C$. It follows that $f(\pm x) \in s[V]_C - C$. Since $f(x) \in -f(-x) + C$, we obtain that $f(x) \in [s[V]_C]_C = s[V]_C$. Therefore, $\mu_{[V]_C}(f(x)) \leq t = p_U(x)$, and so (3.14) holds. □

Under slightly stronger conditions than those in Lemma 3.2.12 we have a stronger conclusion.

Proposition 3.2.13. *Let X, Y be topological vector spaces, $f : X \to Y^\bullet$ be a proper convex function and $x_0 \in \operatorname{dom} f$. Suppose there exist a closed and convex $U \in \mathcal{N}_X$ and a balanced convex set $V \subseteq Y$ such that condition (3.13) is satisfied. Then for every $\rho \in]0,1[$,*

$$\forall\, x, x' \in x_0 + \rho U \ : \mu_{[V]_C}\left(f(x) - f(x')\right) \leq \frac{1+\rho}{1-\rho} p_U(x - x'). \tag{3.15}$$

Proof. As in the proof of Lemma 3.2.12, we may assume that $x_0 = 0$ and $f(x_0) = 0$. Of course, by Proposition 6.2.1 and Remark 6.2.6, p_U is a continuous semi-norm with $U = \{x \mid p_U(x) \leq 1\}$ and $\operatorname{int} U = \{x \mid p_U(x) < 1\}$. Let $\rho \in]0,1[$ and consider $x, x' \in \rho U \subseteq \operatorname{int} U$. Observe that $|p_U(tx + (1-t)x') - p_U(x')| \leq |t|\, p_U(x - x')$, and so

$$p_U\left(tx + (1-t)x'\right) \leq p_U(x') + |t|\, p_U(x - x') \leq \rho + |t|\, p_U(x - x') \quad \forall t \in \mathbb{R}.$$

Let us set

$$A := \{t \in \mathbb{R} \mid z_t := tx + (1-t)x' \in U\}, \quad t_0 := \sup A.$$

Clearly, A is an interval and $t_0 \in]1, \infty]$ (because $x \in \operatorname{int} U$); moreover, $t_0 = \infty$ if $p_U(x - x') = 0$. We claim that $t_0 \geq \frac{1-\rho}{p_U(x-x')}$ if $p_U(x - x') > 0$. Indeed, let $\varphi : \mathbb{R} \to \mathbb{R}$, $\varphi(t) = p_U(tx + (1-t)x')$. Since p_U is convex and continuous, so is φ. Moreover, $\varphi(t) \geq tp_U(x - x') - p_U(-x')$, and so $\lim_{t\to\infty} \varphi(t) = \infty$. As $\varphi(1) = p_U(x) \leq \rho < 1$, there exists $\bar{t} > 1$ such that $\varphi(\bar{t}) = 1$. It follows that $\varphi(t) < 1$ for $t \in [1, \bar{t})$, and so $[1, \bar{t}) \subseteq A$. Since $p_U(z_{\bar{t}}) = 1$ and $\bar{t}(x - x') = z_{\bar{t}} - x'$, we get $\bar{t}\, p_U(x - x') = p_U(z_{\bar{t}} - x') \geq p_U(z_{\bar{t}}) - p_U(x') \geq 1 - \rho$, and so $t_0 \geq \frac{1-\rho}{p_U(x-x')}$ as claimed. For $t \in]1, t_0[$ we have that $x = (1 - t^{-1})x' + t^{-1}z_t$, and so $f(x) \leq (1 - t^{-1})f(x') + t^{-1} f(z_t)$. Using Lemma 3.2.12, for all $s \in]1, \infty[$, $t \in]1, t_0[$ and $\rho' > \rho$ we have

$$\begin{aligned} f(x) \ - f(x') &\in t^{-1}\left(f(z_t) - f(x')\right) - C \subseteq t^{-1}\left(s[V]_C + \rho'[V]_C\right) - C \\ &\subseteq t^{-1}(s + \rho')[V]_C - C, \end{aligned}$$

and so, because $\alpha[V]_C \subseteq \beta[V]_C$ for $0 < \alpha < \beta$, $f(x) - f(x') \in t^{-1}(s+\rho')[V]_C - C$ for all $t \in]0, t_0[$, $s > 1$, $\rho' > \rho$. It follows that

$$\forall t > \frac{1+\rho}{1-\rho} p_U(x-x') : f(x) - f(x') \in t[V]_C - C.$$

Changing the roles of x and x', and taking into account that $[[V]_C]_C = [V]_C$, we get

$$\forall t > \frac{1+\rho}{1-\rho} p_U(x-x') : f(x) - f(x') \in t[V]_C; \tag{3.16}$$

therefore, (3.15) holds. □

In the case in which X, Y are normed vector spaces Proposition 3.2.13 can be used to get the local lipschitzianity of the proper convex function $f : X \to Y^{\bullet}$ (see Theorem 3.3.4).

The results presented without proofs are quoted from the book [214]; see there the bibliographical comments. Proposition 3.2.1 extends [260, Proposition 3.8]. Theorem 3.2.8 is a slight generalization of Theorem 1 in [486]. The proof of Proposition 3.2.11 follows that given in [609]. The function μ_A is introduced and studied for general sets in [474]. Lemma 3.2.12, Proposition 3.2.13 and Theorem 3.3.4 are slight generalizations of Proposition 2.3 and Corollary 2.4 (c), (d) in [67]; their proofs mimics the proofs of Lemma 2.2.8, Theorem 2.2.11 and Corollary 2.2.12 in [614].

3.3 Lipschitz Properties for Single-Valued and Set-Valued Maps

It is well known from scalar optimization that one needs certain assumptions concerning the objective function as well as the feasible set, especially in order to show the stability of solutions with respect to perturbations or rates of convergence in numerical procedures (compare Klatte, Kummer [329]). Under the assumption of classical Lipschitz continuity (see Definition 3.3.1) it is possible to show such results. An important characteristic of the classical Lipschitz continuity (Definition 3.3.1) in comparison with the general continuity for single-valued functions is a *linear rate* of continuity quantified by some modulus (Lipschitz constant L).

Under certain Lipschitz properties characterizations of well-posedness and sensitivity analysis for constraint systems as well as variational systems are given in Chap. 4 of the book by Mordukhovich [430]. Furthermore, Lipschitz continuity is important for deriving calculus rules in generalized differentiation (see Sects. 3.4, 3.5).

Definition 3.3.1. Consider a map $f : S \subseteq X \to Y$, where X and Y are n.v.s.. The mapping f is **Lipschitz continuous around** $\overline{x} \in X$ (or Lipschitzian around $\overline{x} \in X$) if there is a neighborhood U of $\overline{x}$ and a constant $L \geq 0$ such that

$$||f(x) - f(u)|| \leq L||x - u|| \quad \text{for all} \quad x, u \in S \cap U. \tag{3.17}$$

Remark 3.3.2. In Definition 3.3.1 we can take $f : X \to Y^\bullet$ and $S = \operatorname{dom} f$.

It is known that a proper convex function $f : \mathbb{R} \to \overline{\mathbb{R}}$ is locally Lipschitzian on int(dom f) (see [614, Theorem 2.1.5]). In [407, Theorem 3.1] a corresponding result is shown for C-convex vector-valued functions in the finite dimensional setting. Recall that a vector-valued function $f : S \to Y$, $S \subset X$ convex, is called C-convex (C is a proper convex cone in Y), if for every $x, y \in S$, $\lambda \in [0, 1]$, one has

$$f(\lambda x + (1 - \lambda)y) \in \lambda f(x) + (1 - \lambda) f(y) - C.$$

Theorem 3.3.3. *Suppose that $C \subset \mathbb{R}^q$ is a proper convex cone, $\operatorname{cl} C$ is pointed, $S \subseteq \mathbb{R}^n$ is a proper convex set. Furthermore, assume that $f : S \to \mathbb{R}^q$ is C-convex. Then f is locally Lipschitzian on the relative interior of S.*

A more general result is given by Borwein [67, Corollary 2.4] for normed vector spaces X, Y and a normal cone $C \subset Y$; we give a proof using Proposition 3.2.13.

Theorem 3.3.4. *Let X, Y be normed vector spaces, $f : X \to Y^\bullet$ be a proper convex function. Assume that f is continuous at some $x_0 \in \operatorname{dom} f$ and the cone C is normal. Then f is locally Lipschitz on* int(dom f)*. Moreover, if*

$$\forall x \in x_0 + RU_X : \|f(x) - f(x_0)\| \leq M, \tag{3.18}$$

for some $R, M > 0$, then

$$\forall x, x' \in x_0 + rU_X : \left\|f(x) - f(x')\right\| \leq \frac{M\gamma}{R} \cdot \frac{R + r}{R - r} \cdot \left\|x - x'\right\|,$$

where $r \in]0, R[$ and $\gamma := \sup\{\|y\| \mid y \in [U_Y]_C\}$.

Proof. Assume that (3.18) holds. Take $U := RU_X \in \mathcal{N}_X$, $V := MU_Y \in \mathcal{N}_Y$ and $\rho := r/R \in]0, 1[$. Clearly, (3.13) holds. Observe that $[V]_C \subseteq M\gamma U_Y$, $p_U = R^{-1} \|\cdot\|$ and $\mu_{[V]_C} = p_{[V]_C} \geq p_{M\gamma U_Y} = (M\gamma)^{-1} \|\cdot\|$. Using Proposition 3.2.13, for $x, x' \in x_0 + rU$ we get

$$\begin{aligned}(M\gamma)^{-1} \left\|f(x) - f(x')\right\| &\leq \mu_{[V]_C}\left(f(x) - f(x')\right) \leq \frac{1 + \rho}{1 - \rho} p_U(x - x') \\ &= \frac{1 + \rho}{1 - \rho} R^{-1} \left\|x - x'\right\|.\end{aligned}$$

Hence (3.16) holds.

Because the continuity of f at some point of dom f implies the continuity of f on int(dom f) (see Corollary 3.2.10), the conclusion follows. □

In order to show a subdifferential chain rule (Lemma 3.5.14) or the relationship between the coderivative of a vector-valued function and the subdifferential of its scalarization ([430, Theorem 3.28], compare Sect. 3.5) we assume that a scalar function is strictly Lipschitzian in the following sense:

Definition 3.3.5. Suppose that X and Y are n.v.s. and $f : X \to Y$ is a single-valued map. f is called **strictly Lipschitzian** at $\overline{x}$ if f is Lipschitzian around $\overline{x}$ and there exists a neighborhood V of the origin in X such that the sequence

$$y_k := \frac{f(x_k + t_k v) - f(x_k)}{t_k}, \quad k \in \mathbb{N},$$

contains a norm convergent subsequence whenever $v \in V$, $x_k \to \overline{x}$, and $t_k \downarrow 0$.

Remark 3.3.6. If Y is finite-dimensional, the class of strictly Lipschitzian mappings reduces to the class of locally Lipschitzian mappings $f : X \to \mathbb{R}^q$. This is not the case for $\dim Y = +\infty$, see [430, Example 1.35]. Every at $\overline{x}$ strictly differentiable map is strictly Lipschitzian at this point.

In the following we study natural extensions of Lipschitz continuity to set-valued mappings. These Lipschitzian properties will be used for deriving generalized differential calculus and optimality conditions (see Chap. 12). Especially in Sect. 12.8, we need an extension of Lipschitzian continuity in order to prove necessary conditions for minimizers of the set-valued optimization problem (SP).

In the following we present several notions of Lipschitz-like set-valued mappings. We study set-valued mappings $F : X \rightrightarrows Y$, where X and Y are n.v.s., with $\operatorname{dom} F \neq \emptyset$.

Definition 3.3.7. Consider nonempty subsets $U \subset X$ and $V \subset Y$. A set-valued map $F : X \rightrightarrows Y$ is called **Lipschitz-like on** U **relative to** V if there exist a number $L \geq 0$ such that for every $x, u \in U$ one has:

$$F(x) \cap V \subset F(u) + L \|x - u\| B_Y,$$

where B_Y denotes the closed unit ball of Y.

Remark 3.3.8. If $F(U) \cap V = \emptyset$ then F is Lipschitz-like on U relative to V. If $F : X \rightrightarrows Y$ is Lipschitz-like on U relative to V and $F(U) \cap V \neq \emptyset$, then $U \subset \operatorname{dom} F$.

Definition 3.3.9. Consider a set-valued map $F : X \rightrightarrows Y$ and a point $(\overline{x}, \overline{y}) \in \operatorname{graph} F$. F is called **Lipschitz-like around** $(\overline{x}, \overline{y})$ **with modulus** $L \geq 0$ if there are neighborhoods U of $\overline{x}$ and V of $\overline{y}$ s.t. for every $x, u \in U$ one has:

$$F(x) \cap V \subset F(u) + L \|x - u\| B_Y.$$

The infimum of all such moduli L is called the **exact Lipschitzian bound of** F **around** $(\overline{x}, \overline{y})$ and is denoted by $\operatorname{Lip} F(\overline{x}, \overline{y})$.

Remark 3.3.10. The local Lipschitz-like property is also known as the **pseudo-Lipschitzian property** or the **Aubin property** of multifunctions.

The case $V = Y$ in Definitions 3.3.7 and 3.3.9 leads us to the following definition:

Definition 3.3.11. Consider a nonempty set $U \subset X$.

(i) A set-valued map $F : X \rightrightarrows Y$ is called **Lipschitz continuous on** U if there exist a number $L \geq 0$ such that for every $x, u \in U$ one has:

$$F(x) \subset F(u) + L \, \|x - u\| \, B_Y.$$

(ii) F is called **Lipschitzian around** $\overline{x}$ **with the exact bound** $\operatorname{Lip} F(\overline{x})$ if $V = Y$ in Definition 3.3.9.

Remark 3.3.12. If $F : X \rightrightarrows Y$ is Lipschitz on U one has $U \cap \operatorname{dom} F = \emptyset$ or $U \subset \operatorname{dom} F$.

Remark 3.3.13. It is easy to see that for single-valued mappings $f : S \subseteq X \to Y$ all the properties in Definitions 3.3.9 and 3.3.11 reduce to the classical Lipschitz continuity given in Definition 3.3.1.

Remark 3.3.14. The local Lipschitz-like property of a set-valued mapping $F : X \rightrightarrows Y$ (Definition 3.3.9) can be viewed as a localization of the Lipschitzian behavior not only relative to a point of the domain but also relative to a particular point of the image $\overline{y} \in F(\overline{x})$.

Remark 3.3.14 leads us to the following characterization in terms of the local Lipschitz continuity of the scalar distance function $d(u, A) := \inf_{v \in A} ||u - v||$ $(u \in Y, A \subset Y)$, where we use the convention $\inf \emptyset := +\infty$, to the moving set $F(x)$ with respect to both variables (x, y) (see Mordukhovich [430, Theorem 1.41]).

Theorem 3.3.15 (Scalarization of the Lipschitz-Like Property). *Consider a set-valued mapping $F : X \rightrightarrows Y$ and $(\overline{x}, \overline{y}) \in \operatorname{graph} F$. The following properties are equivalent:*

(a) *F is Lipschitz-like around $(\overline{x}, \overline{y})$.*
(b) *The scalar function $\mu : X \times Y \to \overline{\mathbb{R}}$ defined by (with the convention $\inf \emptyset = +\infty$)*

$$\mu(x, y) := d(y, F(x)) := \inf_{v \in F(x)} ||y - v||$$

is Lipschitzian around $(\overline{x}, \overline{y})$.

Remark 3.3.16. The local properties in Definitions 3.3.9 and 3.3.11 are stable / robust with respect to small perturbations of the reference points. These properties hold for F if and only if they hold for the mapping $\overline{F} : X \rightrightarrows Y$ with $\overline{F}(x) := \operatorname{cl} F(x)$.

Remark 3.3.17. From Definition 3.3.11, (i), it follows that the Lipschitz continuity of F on U is equivalent to

$$\forall x, u \in U: \qquad \text{haus}(F(x), F(u)) \leq L||x - u||,$$

where $\text{haus}(A_1, A_2)$ is the **Hausdorff distance** (or Pompieu-Hausdorff distance) between $A_1, A_2 \subset Y$ that is defined by

$$\text{haus}(A_1, A_2) := \inf\{\mu \geq 0 \mid A_1 \subset A_2 + \mu B_Y,\ A_2 \subset A_1 + \mu B_Y\}.$$

It is important to mention that the Hausdorff distance furnishes a **metric** on the space of all nonempty and compact (even closed and bounded) subsets of Y. Thus, if a set-valued map $F: X \rightrightarrows Y$ is compact-valued, its Lipschitz continuity in the sense of Definition 3.3.11 is equivalent to the Lipschitz continuity of a single-valued map $x \to F(x)$ (see Definition 3.3.1) from X to the space of all nonempty, compact subsets of Y equipped with the Haussdorff metric.

In the following we explain some relationships between the local Lipschitzian (Definition 3.3.11, (ii)) and Lipschitz-like properties (Definition 3.3.7) of multifunctions. It follows directly from the definitions that, if F is Lipschitzian around $\overline{x} \in \text{dom}\, F$, then it is Lipschitz-like around $(\overline{x}, \overline{y})$ for every $\overline{y} \in F(\overline{x})$ with

$$\text{Lip}\, F(\overline{x}) \geq \sup\{\text{Lip}\, F(\overline{x}, \overline{y}) \mid \overline{y} \in F(\overline{x})\}. \tag{3.19}$$

The converse of this assertion holds with the equality in (3.19) under some additional assumptions concerning F. The following theorem is shown by Mordukhovich [430, Theorem 1.42] under the assumption that $F: X \rightrightarrows Y$ is **compact around** $\overline{x} \in \text{dom}\, F$, i.e., there exist a neighborhood U of $\overline{x}$ such that $\text{cl}\, F(U)$ is compact, that is $F(U)$ is relatively compact. Furthermore, we assume that F is **closed at** $\overline{x}$, i.e., for every $y \notin F(\overline{x})$ there are neighborhoods U of $\overline{x}$ and V of y such that $F(x) \cap V = \emptyset$ for all $x \in U$. F is **closed at** $\overline{x}$ in this sense if F is closed at $\overline{x}$ in the sense of [214, Definition 2.5.7 (ii)]. F is closed at $\overline{x}$ implies that F is closed-valued at $\overline{x}$, i.e., $F(\overline{x})$ is closed.

In the following theorem we discuss the assumptions for the equivalence of the properties that F is Lipschitzian around $\overline{x}$ (Definition 3.3.11, (ii)) and that F is Lipschitz-like around $(\overline{x}, \overline{y})$ (Definition 3.3.9).

Theorem 3.3.18 (Lipschitz Continuity of Locally Compact Multifunctions). *Let $F: X \rightrightarrows Y$ be closed at $\overline{x} \in \text{dom}\, F$ and compact around this point.*

Then F is Lipschitzian around $\overline{x}$ if and only if it is Lipschitz-like around $(\overline{x}, \overline{y})$ for every $\overline{y} \in F(\overline{x})$. In this case we have

$$\text{Lip}\, F(\overline{x}) = \max\{\text{Lip}\, F(\overline{x}, \overline{y}) \mid \overline{y} \in F(\overline{x})\} < +\infty.$$

The proof of this result is given by Mordukhovich [430, Theorem 1.42].

Let us now consider the case of a mapping $F\colon X \rightrightarrows Y$ between Asplund spaces (see Definition 3.5.3) with the range space Y *partially ordered* by a proper, closed, convex and pointed cone $C \subset Y$. Recall that the **epigraph** of F with respect to the ordering cone C is given by

$$\operatorname{epi} F = \{(x, y) \in X \times Y \mid y \in F(x) + C\},$$

and the **epigraphical multifunction** of F denoting by $\mathscr{E}_F\colon X \rightrightarrows Y$ is defined by

$$\mathscr{E}_F(x) := \{y \in Y \mid y \in F(x) + C\} \quad \text{with} \quad \operatorname{graph} \mathscr{E}_F = \operatorname{epi} F. \tag{3.20}$$

In Sect. 12.8, Theorem 12.8.9 we will show necessary conditions for minimizers of set-valued optimization problems under the assumption that the set-valued mapping $F : X \rightrightarrows Y$ is epigraphically Lipschitz-like in the following sense:

Definition 3.3.19. A set-valued mapping $F : X \rightrightarrows Y$ is **epigraphically Lipschitz-like** (ELL) around $(\overline{x}, \overline{y})$ with modulus $L \geq 0$ if there are neighborhoods U of $\overline{x}$ and V of $\overline{y}$ such that

$$\mathscr{E}_F(x) \cap V \subset \mathscr{E}_F(u) + L\|x - u\| B_Y \quad \text{for all} \quad x, u \in U,$$

where B_Y stands for the closed unit ball of Y.

In other words, F is ELL at $(\overline{x}, \overline{z})$ if its epigraphical multifunction $\mathscr{E}_F$ is Lipschitz-like at that point (Definition 3.3.9). This property is known also as the Aubin property, the pseudo Lipschitzian property; cf. [431, 499]. It agrees with the classical local Lipschitzian behavior in the case of single-valued functions, and reduces to the standard Hausdorff of local Lipschitzian property for set-valued mappings when $V = Y$.

For deriving necessary optimality conditions using the approximate subdifferential by Ioffe (see Definition 3.6.1) we need the following Lipschitz property:

Definition 3.3.20. A function $f : X \to Y$ is said to be **strongly compactly Lipschitzian** at $\overline{x}$ if there exist a multifunction $R : X \rightrightarrows \operatorname{Comp}(Y)$, where $\operatorname{Comp}(Y)$ is the collection of all nonempty $\|\cdot\|$-compact subsets of Y, and a function $r : X \times X \to \mathbb{R}_+$ satisfying the following properties:

(i) $\lim\limits_{\substack{x \to \overline{x} \\ h \to 0}} r(x, h) = 0$.

(ii) There is $\mu > 0$ such that for all $h \in \mu B_X$, $x \in \overline{x} + \mu B_X$, and all $t \in (0, \mu)$, one has

$$t^{-1}\big(f(x + th) - f(x)\big) \in R(h) + \|h\| r(x, th) B_X.$$

(iii) $R(0) = \{0\}$ and R is upper semicontinuous.

The strongly compactly Lipschitzian property agrees with the usual Lipschitzian continuity when Y is finite dimensional.

3.4 Clarke's Normal Cone and Subdifferential

In this section we recall the concept of the Clarke subdifferential (see [100]) assuming that X is a Banach space. Suppose that $f : X \to \overline{\mathbb{R}}$ is Lipschitzian around $x \in X$ with $f(x) \in \mathbb{R}$.

Definition 3.4.1. The **Clarke generalized subdifferential** of f at x is the set

$$\partial_C f(x) := \{x^* \in X^* \mid \forall v \in X \ : \ \langle x^*, v\rangle \leq f^0(x; v)\}, \tag{3.21}$$

where $f^0(x; v)$ is the **generalized directional derivative** of f at x in the direction $v \in X$

$$f^0(x; v) := \limsup_{y \to x,\ t \to +0} \frac{f(y + tv) - f(y)}{t}.$$

Definition 3.4.2. The **Clarke normal cone** to S at $x \in S$ is described via the weak* closure of the cone spanned on the generalized gradient of the distance function

$$N_C(x; S) = \mathrm{cl}^{w^*} \cup_{\lambda \geq 0} \lambda \partial_C d(x; S).$$

Now, assume that f is l.s.c. on X. The **Clarke subdifferential** of f at x is the set

$$\partial_C f(x) := \{x^* \in X^* \mid (x^*, -1) \in N_C((x, f(x)); \mathrm{epi}\, f)\}. \tag{3.22}$$

Remark 3.4.3. If f is Lipschitzian around $x \in X$ with $f(x) \in \mathbb{R}$ the definitions of Clarke's subdifferential given by (3.21) and (3.22) are equivalent (see [430, Section 1.4.3]).

The **indicator function** of a set $S \subset X$ is the function $\iota_S : X \to \overline{\mathbb{R}}$ defined by $\iota_S(x) := 0$ for $x \in S$ and $\iota_S(x) := +\infty$ for $x \in X \setminus S$.

The above normal cone to a set can also be equivalently defined using indicator function ι_S associated with this set as follows

$$N_C(x; S) = \partial_C \iota_S(x).$$

The following result is given by Clarke [100, Proposition 2.3.1].

Proposition 3.4.4. *Consider $f : X \to \overline{\mathbb{R}}$. Assume that f is l.s.c. on its domain, $x \in X$ with $f(x) \in \mathbb{R}$. Then, if f is Lipschitz around x we have*

$$\partial_C(-f)(x) = -\partial_C fx).$$

Moreover, we introduce the **Clarke coderivative** for a set-valued mapping $F : X \rightrightarrows Y$ acting between Banach spaces X and Y:

Definition 3.4.5. Let $F : X \rightrightarrows Y$ and $(x, y) \in \operatorname{graph} F$. Assuming that F is closed, the *Clarke coderivative* $D_C^* F(x, y)$ of F at (x, y) is a set-valued map from Y^* to X^* defined by

$$D_C^* F(x, y)(y^*) := \{x^* \in X^* \mid (x^*, -y^*) \in N_C((x, y); \operatorname{graph} F)\}.$$

3.5 Limiting Cones and Generalized Differentiability

We will study generalized differentiability properties in Asplund spaces (see Definition 3.5.3), a special class of Banach spaces, because it is possible to show fundamental results like extremal principles (see Sect. 5.5) in these spaces (see [342, 343, 430, 432]). Furthermore, several calculus rules in generalized differentiation theory presented in this section are working in Asplund spaces. This is important for deriving optimality conditions based on limiting subdifferentials in Sect. 12.8.

We begin by recalling the following:

Definition 3.5.1. For a set-valued map $F : X \rightrightarrows X^*$ acting between a Banach space X and its dual X^* we denote the sequential **Kuratowski-Painlevé upper limit** with respect to the norm topology on X and weak* topology on X^* by

$$\operatorname*{Lim\,sup}_{x \to x_0} F(x),$$

which is given as

$$\operatorname*{Lim\,sup}_{x \to x_0} F(x) := \{x^* \in X^* \mid \exists \text{ sequences } x_k \to x_0, \text{ and } x_k^* \xrightarrow{w^*} x^*, \\ \text{with } \; x_k^* \in F(x_k), \quad \forall k = 1, 2, \ldots\},$$

where $\xrightarrow{w^*}$ denotes convergence in the weak-star topology of X^*.

Let X be a Banach space and Ω be a subset of X. The **Fréchet normal cone** to Ω at $x \in \Omega$ is defined by

$$\hat{N}(x; \Omega) := \left\{x^* \in X^* \;\middle|\; \limsup_{u \xrightarrow{\Omega} x} \frac{\langle x^*, u - x\rangle}{\|u - x\|} \le 0\right\}, \tag{3.23}$$

where $u \xrightarrow{\Omega} x$ means $u \to x$ with $u \in \Omega$. This definition can be extended:

Definition 3.5.2 (Generalized Normals). Let $\Omega \subset X$ be a nonempty subset of a Banach space X. Consider $x \in \Omega$ and $\varepsilon \geq 0$. Then

$$\hat{N}_\varepsilon(x;\Omega) := \left\{x^* \in X^* \;\middle|\; \limsup_{u \xrightarrow{\Omega} x} \frac{\langle x^*, u - x\rangle}{\|u - x\|} \leq \varepsilon\right\} \tag{3.24}$$

defines the **set of ε-normals** to Ω at x.

If $x \notin \Omega$ we put $\hat{N}_\varepsilon(x;\Omega) := \emptyset$ for all $\varepsilon \geq 0$.

In the sequel, we present definitions and properties of the basic generalized differential constructions held in the Asplund space setting and enjoying a full calculus (see Mordukhovich [425], [426], [427] and Kruger, Mordukhovich [342]).

Definition 3.5.3. A Banach space X is called **Asplund space** if every convex continuous function $\phi : U \to \mathbb{R}$ defined on an open convex subset U of X is Fréchet differentiable on a dense subset of U.

Remark 3.5.4. The class of Asplund spaces is quite broad including every reflexive Banach space and every Banach space with a separable dual; in particular, c_0 and l^p, $L^p[0,1]$ for $1 < p < +\infty$ are Asplund spaces, but l^1 and l^∞ are not Asplund spaces. It has been comprehensively investigated in geometric theory of Banach spaces, and largely employed in variational analysis; see, e.g., [66, 430, 431].

In the following we suppose that Ω is a subset of an Asplund space X.

Definition 3.5.5. Consider $\overline{x} \in \Omega$. Ω is said to be **closed** around $\overline{x} \in \Omega$ if there is a neighborhood U of $\overline{x}$ such that $\Omega \cap U$ is a closed set.

Given $\overline{x} \in \Omega$. Assume that Ω is closed around $\overline{x} \in \Omega$.

Definition 3.5.6. The (**basic, limiting, Mordukhovich**) **normal cone** to Ω at $\overline{x} \in \Omega$ is defined by

$$\begin{aligned} N(\overline{x};\Omega) &:= \operatorname{Lim\,sup}_{x \to \overline{x}} \hat{N}(x;\Omega) \\ &= \left\{x^* \in X^* \;\middle|\; \exists\, x_k \xrightarrow{\Omega} \overline{x},\ x_k^* \xrightarrow{w^*} x^* \text{ with } x_k^* \in \hat{N}(x_k,\Omega)\right\}, \end{aligned} \tag{3.25}$$

where Lim sup stands for the sequential Painlevé-Kuratowski outer limit of Fréchet normal cones to Ω at x as x tends to $\overline{x}$.

If $\overline{x} \notin \Omega$ we put $N(\overline{x};\Omega) := \emptyset$.

Note that, in contrast to (3.24), the basic normal cone (3.25) is often *nonconvex* enjoying nevertheless *full calculus*, and that both the cones (3.25) and (3.23) reduce to the normal cone of convex analysis when Ω is convex.

We define the (basic, limiting, Mordukhovich) subdifferential of an extended-real-valued function through basic normals to its epigraph.

Definition 3.5.7. Consider a function $f : X \to \overline{\mathbb{R}}$ and a point $x^0 \in X$ with $|f(x^0)| < +\infty$. The **(basic, limiting, Mordukhovich) subdifferential** of f at x^0 is defined by

$$\partial f(x^0) := \{x^* \in X^* \mid (x^*, -1) \in N((x^0, f(x^0)); \text{epi } f)\}.$$

The elements of $\partial f(x^0)$ are called **(basic, limiting) subgradients** of f at x^0.

We put $\partial f(x^0) := \emptyset$ if $|f(x^0)| = +\infty$.

We recall the notions of coderivatives and subdifferentials for set-valued mappings (see [24–27, 426, 427]). Given a set-valued mapping $F: X \rightrightarrows Y$ with its graph

$$\text{graph } F = \big\{(x, y) \in X \times Y \big| \; y \in F(x)\big\}$$

between Asplund spaces X and Y.

Definition 3.5.8. The **(basic, normal, Mordukhovich) coderivative** mapping $D^*F(\overline{x}, \overline{y}): Y^* \rightrightarrows X^*$ of F at $(\overline{x}, \overline{y}) \in \text{graph } F$ is defined by

$$D^*F(\overline{x}, \overline{y})(y^*) := \big\{x^* \in X^* \big| \; (x^*, -y^*) \in N\big((\overline{x}, \overline{y}); \text{graph } F\big)\big\}, \tag{3.26}$$

We omit $\overline{y} = f(\overline{x})$ in (3.26) if $F = f: X \to Y$ is single-valued. If $f: X \to Y$ happens to be *strictly differentiable* at $\overline{x}$ (which is automatic when it is C^1 around this point), then

$$D^* f(\overline{x})(y^*) = \big\{\nabla f(\overline{x})^* y^*\big\} \quad \text{for all} \quad y^* \in Y^*.$$

Furthermore, we introduce the Fréchet coderivative mapping of F at $(\overline{x}, \overline{y}) \in \text{graph } F$ (see Mordukhovich [426], [427] and Bao, Mordukhovich [27]) using the Fréchet normal cone given by (3.23).

Definition 3.5.9. The **Fréchet coderivative** mapping $\hat{D}^*F(\overline{x}, \overline{y}): Y^* \rightrightarrows X^*$ of F at $(\overline{x}, \overline{y}) \in \text{graph } F$ is defined by

$$\hat{D}^*F(\overline{x}, \overline{y})(y^*) := \big\{x^* \in X^* \big| \; (x^*, -y^*) \in \hat{N}\big((\overline{x}, \overline{y}); \text{graph } F\big)\big\}, \tag{3.27}$$

Remark 3.5.10. It is important to mention (see Bao, Mordukhovich [27]) that both coderivative set-valued mappings $D^*F(\overline{x}, \overline{y}): Y^* \rightrightarrows X^*$ and $\hat{D}^*F(\overline{x}, \overline{y}): Y^* \rightrightarrows X^*$ are positively homogeneous functions of $y^* \in Y^*$ reducing to the single-valued adjoint derivative linear operator

$$\hat{D}^* f(\overline{x}) = \{\nabla f(\overline{x})^* y^*\}, \qquad D^* f(\overline{x}) = \{\nabla f(\overline{x})^* y^*\} \text{ for all } y^* \in Y^*, \tag{3.28}$$

provided that f is Fréchet differentiable at $\overline{x}$ for the first equality and strictly differentiable at this point for the second one, i.e.,

$$\lim_{x,u \to \overline{x}} \frac{f(x) - f(u) - \langle \nabla f(\overline{x}), u - x \rangle}{||x - u||} = 0,$$

which is automatic when $f \in C^1$ around $\overline{x}$.

Using coderivatives of the epigraphical multifunction (see (3.20)), we can define appropriate extensions of the subdifferential notion from extended-real-valued function to vector-valued and set-valued mappings with values in partially ordered spaces. The following notions of subdifferentials for set-valued mappings $F : X \rightrightarrows Y$ acting between Asplund spaces X and Y are first introduced in [25] and then further developed in [24, 26, 27]; cf. the epi-coderivatives for set-valued mappings in [621] and references therein.

Definition 3.5.11. Given $(\overline{x}, \overline{y}) \in \operatorname{graph} F$, the **(limiting, basic, normal) subdifferential** of F at $(\overline{x}, \overline{y})$ in *direction* $y^* \in Y^*$ is defined by

$$\partial F(\overline{x}, \overline{y})(y^*) := D^* \mathscr{E}_F(\overline{x}, \overline{y})(y^*), \tag{3.29}$$

where $\mathscr{E}_F$ is the epigraphical multifunction of F (see (3.20)).

As pointed in [27], we always have the following implication

$$D^* \mathscr{E}_F(\overline{x}, \overline{y})(y^*) \neq \emptyset \quad \Longrightarrow \quad -y^* \in N(0; C). \tag{3.30}$$

Consequently, the requirement that $-y^* \in N(0; C)$ is redundant for the definitions of the subdifferential for set-valued mappings in [24–27]; in particular, [25, Definition 2.1]. As usual, when $F = f : X \to Y$ is a single-valued vector-valued function, we omit $\overline{y}$ from the notation of subdifferentials. The subdifferential of f at $\overline{x}$ in (3.29) becomes

$$\partial f(\overline{x})(y^*) = D^* \mathscr{E}_f(\overline{x})(y^*),$$

and when $F = f : X \to (-\infty, \infty]$ is a lower semicontinuous extended-real-valued function, the subdifferential (3.29) with $\|y^*\|_* = 1$ agrees with the (basic, limiting, Mordukhovich) subdifferential; see [430] and Definition 3.5.7. By [430, Theorem 3.28], the relationship between the coderivative of a vector-valued function and the subdifferential of its scalarization

$$D^* f(\overline{x})(y^*) = \partial \langle y^*, f \rangle(\overline{x}) \quad \text{with} \quad \langle y^*, f \rangle(x) := \big\langle y^*, f(x) \big\rangle \tag{3.31}$$

holds provided that f is strictly Lipschitzian at $\overline{x}$ in the sense of Definition 3.3.5.

Furthermore, we will use the Fréchet subdifferential F at $(\overline{x}, \overline{y}) \in \operatorname{graph} F$ in Sect. 12.9, Theorem 12.9.1, for deriving necessary conditions for approximate solutions of a set-valued optimization problem.

Definition 3.5.12. For $(\overline{x}, \overline{y}) \in \operatorname{graph} F$, the **Fréchet subdifferential** of F at $(\overline{x}, \overline{y})$ in *direction* $y^* \in Y^*$ is defined by

$$\hat{\partial} F(\overline{x}, \overline{y}) := \{x^* \in X^* \mid x^* \in \hat{D}^* \mathscr{E}_F(\overline{x}, \overline{y})(y^*), \ -y^* \in N(0; C), ||y^*||_* = 1\}. \tag{3.32}$$

Observe that the condition $-y^* \in N(0; C)$ in (3.32) is not a restriction; in fact, it automatically follows from the inclusion $x^* \in \hat{D}^* \mathscr{E}_F(\overline{x}, \overline{y})(y^*)$.

For the sake of compact exposition, we will not recall all the calculus rules for normal cones and coderivatives. However, we will refer to them in Mordukhovich's books [430, 431].

We need the following calculus rules given by Mordukhovich in [430] for proving necessary optimality conditions in Sect. 12.8. The next lemma follows from [430, Theorem 3.36] taking into account [430, Theorem 1.26 and Corollary 1.81].

Lemma 3.5.13. *Let X be an Asplund space and let $x_0 \in X$. Let $f_i : X \to \overline{\mathbb{R}}$, $i = 1, 2$, be proper lower-semicontinuous functions and one of these is Lipschitzian around x_0. Then one has*

$$\partial(f_1 + f_2)(x_0) \subseteq \partial f_1(x_0) + \partial f_2(x_0).$$

The next result is shown in [430, Theorem 3.41 and Corollary 3.43]:

Lemma 3.5.14. *Assume that X and Y are Asplund spaces. Let $f : X \to Y$ be strictly Lipschitzian at x_0 and $\phi : Y \to \overline{\mathbb{R}}$ be Lipschitzian around $f(x_0)$. Then one has*

$$\partial(\phi \circ f)(x_0) \subseteq \bigcup_{y^* \in \partial\phi(f(x_0))} \partial\langle y^*, f\rangle(x_0).$$

3.6 Approximate Cones and Generalized Differentiability

Assume that X is a Banach space and $f : X \to \mathbb{R} \cup \{+\infty\}$ is l.s.c. on X and $x \in X$ with $f(x) \in \mathbb{R}$. Ioffe [275, 277] (compare Mordukhovich [425] for the finite dimensional case) introduced the **approximate subdifferential** of f at x:

Definition 3.6.1. The **approximate subdifferential** of f at x is the set

$$\partial_A f(x) := \cap_{L \in \mathscr{F}} \operatorname*{Lim\,sup}_{(\varepsilon, y) \to (+0, x)} \partial_\varepsilon^- f_{y+L}(y),$$

where $\mathscr{F}$ is the collection of all finite dimensional subspaces of X, $f_{y+L}(u) = f(u)$ if $u \in y + L$ and $f_{y+L}(u) = +\infty$ otherwise, and furthermore, for $\varepsilon \geq 0$

$$\begin{aligned} \partial_\varepsilon^- f_{y+L}(y) : &= \{x^* \in X^* \mid \forall v \in X \ : \ \langle x^*, v\rangle \\ &\leq \varepsilon\|v\| + \liminf_{t \to +0} t^{-1}[f_{y+L}(y + tv) - f_{y+L}(y)]\}. \end{aligned}$$

The **approximate (Ioffe) normal cone** to $S \subset X$ at $x \in S$ is given by

$$N_A(x; S) = \cup_{\lambda > 0} \lambda \partial_A d(x; S).$$

The above normal cone to a set can also be equivalently defined using the indicator function ι_S associated with this set as follows

$$N_A(x, S) = \partial_A \iota_S(x).$$

Definition 3.6.2. Let $F : X \rightrightarrows Y$, X and Y are Banach spaces and $(x, y) \in \operatorname{graph} F$. Assuming that F is closed, the **approximate (Ioffe) coderivative** $D_A^* F(x, y)$ of F at (x, y) is a set-valued map from Y^* to X^* defined by

$$D_A^* F(x, y)(y^*) := \{x^* \in X^* \mid (x^*, -y^*) \in N_A((x, y); \operatorname{graph} F)\}.$$

Among all the properties of the above subdifferentials and coderivative, let us recall those ones which will be used in the sequel. The following proposition is shown for Banach spaces X and Y (cf. [228, Proposition 21.2]).

Proposition 3.6.3. *Let f and g be functions from X to $\mathbb{R} \cup \{+\infty\}$ which are l.s.c. on their domains, let F be a set-valued closed map from X to Y. We assume that $x \in \operatorname{dom} f$ or $x \in \operatorname{dom} F$.*

(i) $\partial_A f(x) \subset \partial_C f(x)$ *and these subdifferentials coincide with the subdifferential of convex analysis* $(\partial f(x) = \{x^* \in X^* \mid \forall v \in X : \langle x^*, v - x \rangle \leq f(v) - f(x)\})$ *if f is convex and Lipschitz around x.*
(ii) *If f attains a local minimum at x, then* $0 \in \partial_A f(x)$.
(iii) $\partial_A(f + g)(x) \subset \partial_A f(x) + \partial_A g(x)$ *for* $x \in \operatorname{dom} f \cap \operatorname{dom} g$ *if f is Lipschitz around x.*
(iv) $D_A^* F(x, y)(y^*) \subseteq D_C^* F(x, y)(y^*)$ *for* $(x, y) \in \operatorname{graph} F$ *and* $y^* \in Y^*$.

Chapter 4
Tangent Cones and Tangent Sets

Tangent cones of first-order and tangent cones and tangent sets of higher-order play a very important role in set-valued optimization. For instance, derivatives and epiderivatives of set-valued maps are commonly defined by taking tangent cones and tangent sets of graphs and epigraphs of set-valued maps. Moreover, properties of tangent cones and tangent sets are quite decisive in giving calculus rules for derivatives and epiderivatives of set-valued maps. Furthermore, optimality conditions in set-valued optimization are also most conveniently expressed by using tangent cones and tangent sets. Sensitivity analysis, constraints qualifications, and many other issues in set-valued optimization heavily rely on tangent cones and tangent sets.

This chapter is a thorough introduction to tangent cones of first-order and tangent sets and tangent cones of second-order. We collect definitions, properties, and examples of some of the most commonly used tangent cones of first-order and tangent sets and tangent cones of second-order. To be specific, we study six of the most commonly used tangent cones of first-order, six of their modified analogues, and give numerous equivalent characterizations of some of them. We also study the notion of the local cone approximation from an abstract point of view. This notion not only illustrates the key features of the tangent cones, but it also suggests new ways of generating tangent cones. We investigate convex cones which lie in between the contingent cone and the Clarke's tangent cone, which are two of the most popular tangent cones of first-order. In this discussion, we study recession tangent cones, the Treiman tangent cone, the prototangent cone, and the quasi-strict tangent cone. We study six most commonly used second-order tangent sets and a related notion of second-order generalized tangent sets. We study, in sufficient detail, second-order asymptotic tangent cones and second-order projective tangent cones and collect their intrinsic features. Useful connections among various tangent cones and tangent sets are thoroughly investigated. A brief description of some higher-order tangent cones and higher-order tangent sets is also given. The results of this chapter will play an important role in several other chapters of this book.

A.A. Khan et al., *Set-valued Optimization*, Vector Optimization,
DOI 10.1007/978-3-642-54265-7_4

For simplicity in the presentation, we have chosen to work in the setting of a normed space or a Banach space. However, almost all the results are either directly valid in more general spaces or have their analogous statements valid in general spaces.

Finally, we recall that given a normed space Z and an element $z \in Z$, by $U(z)$, we denote an open ball centered at z and by $U_\varepsilon(z)$, we denote the open ball of radius ε centered at z. By B_Z (or simply B), we represent the unit ball in the space Z. We also recall that $\mathbb{P} := \{t \in \mathbb{R} \mid \ t > 0\}$, and, if not specified otherwise, the notation $\{t_n\} \downarrow 0$ means that $\{t_n\} \subset \mathbb{P}$ and $t_n \to 0$. Moreover, given a nonempty set S of a normed space Z, the notion $x \overset{S}{\to} \bar{x}$ means that $x \to \bar{x}$ with $x \in S$.

4.1 First-Order Tangent Cones

A large number of first-order tangent cones have been proposed in the literature. For instance, Ioffe [276], Penot [463], and Ward [592] present a unified framework for defining a large number of tangent cones of first-order. An interesting survey article that collects many tangent cones of first-order is by Palata [449], see also [140, 307]. However, not all of the known tangent cones of first-order have been very useful. It turns out the tangent cones with most useful properties are actually the most commonly used ones. In this section, we study only the most commonly used tangent cones of first-order. To be specific, we will study the radial tangent cone, the feasible tangent cone, the contingent cone, the interiorly contingent cone, the adjacent cone, and the interiorly adjacent cone. We give useful properties of these six cones and give various characterizations of some of them. We collect numerous examples from the literature to illustrate these notions. This section follows the exposition of the excellent papers by Elster and Thierfelder [178] and Giorgi and Guerraggio [208], and the monograph by Bazarra and Shetty [38], among others. A few examples are taken from [38]. Some related results can be found in [131, 254, 255, 338, 362, 363].

4.1.1 The Radial Tangent Cone and the Feasible Tangent Cone

We focus on the following two notions of tangent cones of first-order.

Definition 4.1.1. Let X be a normed space, let S be a subset of X, and let $\bar{x} \in X$.

1. The **radial tangent cone** $R(S, \bar{x})$ of S at $\bar{x}$ is the set of all $x \in X$ for which there exists a sequence $\{t_n\} \subset \mathbb{P}$ with $t_n \downarrow 0$ such that $\bar{x} + t_n x \in S$ for every $n \in \mathbb{N}$.

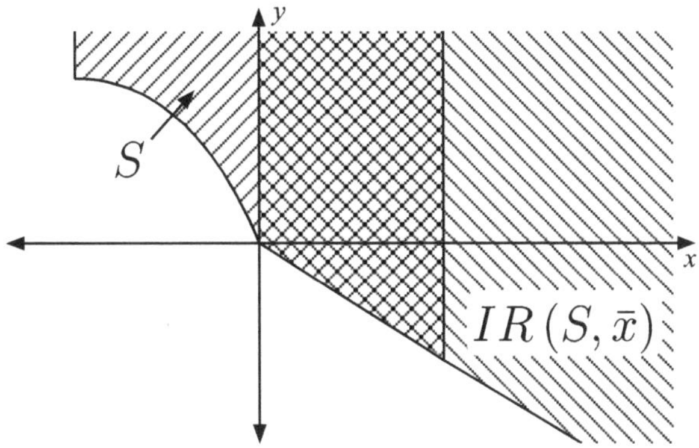

Fig. 4.1 Feasible tangent cone in Example 4.1.6

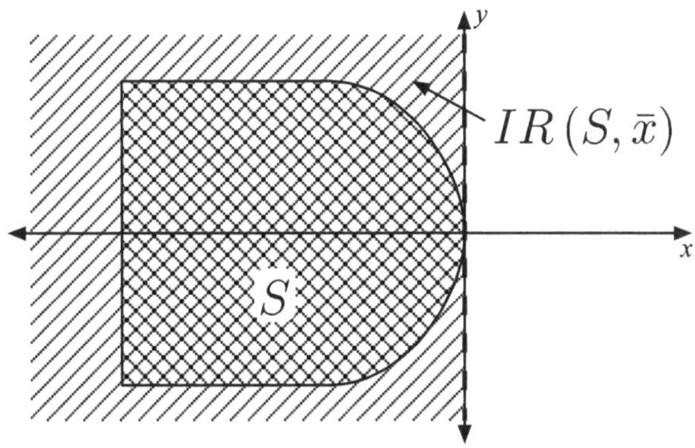

Fig. 4.2 Feasible tangent cone in Example 4.1.9

2. The **feasible tangent cone** $IR(S,\bar{x})$ of S at $\bar{x}$ is the set of all $x \in X$ such that for every $\{t_n\} \subset \mathbb{P}$ with $t_n \downarrow 0$, there exists an index $m \in \mathbb{N}$ satisfying that $\bar{x} + t_n x \in S$ for every $n \geq m$.

Remark 4.1.2. The radial tangent cone and the feasible tangent cone are the simplest examples of first-order tangent cones. Clearly these cones are independent of the topology of X.The feasible tangent cone is also known as the **cone of feasible directions**, the **cone of admissible directions**, the **cone of linear tangent vectors**, the **cone of inner tangent vectors**, the **cone of radial interior displacements**, the **weak feasible tangent cone**, and the **radial cone** (see [85, 178, 584]).

The following result gives properties of the radial tangent cone and the feasible tangent cone.

Theorem 4.1.3. *Let X be a normed space, let S, S_1, S_2 be subsets of X, and let $\bar{x}, \bar{x}_1, \bar{x}_2 \in X$. Then:*

1. $R(S,\bar{x}) = X \backslash IR(X \backslash S, \bar{x})$
2. $IR(S,\bar{x}) = X \backslash R(X \backslash S, \bar{x})$
3. $IR(S,\bar{x}) \subseteq R(S,\bar{x})$

4. $\bar{x} \in S \iff 0 \in R(S, \bar{x}) \iff 0 \in IR(S, \bar{x})$
5. $S_1 \subseteq S_2 \implies R(S_1, \bar{x}) \subseteq R(S_2, \bar{x})$
6. $S_1 \subseteq S_2 \implies IR(S_1, \bar{x}) \subseteq IR(S_2, \bar{x})$
7. $R(S_1 \cup S_2, \bar{x}) = R(S_1, \bar{x}) \cup R(S_2, \bar{x})$
8. $IR(S_1 \cap S_2, \bar{x}) = IR(S_1, \bar{x}) \cap IR(S_2, \bar{x})$
9. $R(S_1 \times S_2, (\bar{x}_1, \bar{x}_2)) \subseteq R(S_1, \bar{x}_1) \times R(S_2, \bar{x}_2)$
10. $IR(S_1 \times S_2, (\bar{x}_1, \bar{x}_2)) = IR(S_1, \bar{x}_1) \times IR(S_2, \bar{x}_2)$.

Proof. A proof can be deduced from the arguments used in the proof of Theorem 4.2.10. □

We collect a few examples of the radial tangent cone and the feasible tangent cone.

Example 4.1.4. Let $S_1 = \mathbb{Q}$ and let $S_2 = \{\mathbb{R} \backslash \mathbb{Q}\} \cup \{0\}$. Clearly $S_1 \cap S_2 = \{0\}$ and we obtain that $R(S_1 \cap S_2, 0) = \{0\}$, $R(S_1, 0) = \mathbb{R}$, and $R(S_2, 0) = \mathbb{R}$. This reveals the fact that the intersection formula $R(S_1 \cap S_2, \bar{x}) = R(S_1, \bar{x}) \cap R(S_2, \bar{x})$, is, in general, false. □

Example 4.1.5. For $S \subset \mathbb{R}^2$ given by $S = \{(x, y) |\ y = -x^3\}$, the feasible tangent cone $IR(S, \bar{x})$ of S at $\bar{x} = (0, 0)$ is the cone given by $IR(S, \bar{x}) = \{(0, 0)\}$. Note that if we exclude $(0, 0)$ from the set S, then the feasible tangent cone is empty. □

Example 4.1.6. For $S \subset \mathbb{R}^2$ given by $S = \{(x, y) |\ -1 \leq x \leq 1,\ y \geq x^{1/3},\ y \geq -x\}$, the feasible tangent cone $IR(S, \bar{x})$ of S at $\bar{x} = (0, 0)$ is given by $IR(S, \bar{x}) = \{(x, y) |\ x \geq 0,\ y \geq x\}$. □

Example 4.1.7. For $S \subset \mathbb{R}^2$ given by $S = \{(x, y) |\ y \geq |x|\}$, the feasible tangent cone $IR(S, \bar{x})$ of S at $\bar{x} = (0, 0)$ is given by $IR(S, \bar{x}) = \{(x, y) |\ y \geq |x|\}$. □

Example 4.1.8. For $S \subset \mathbb{R}^2$ given by $S = \{(x, y) |\ y > x^2\}$, the feasible tangent cone $IR(S, \bar{x})$ of S at $\bar{x} = (0, 0)$ is given by $IR(S, \bar{x}) = \{(x, y) |\ y > 0\}$. □

Example 4.1.9. For $S \subset \mathbb{R}^2$ given by $S = \{(x, y) |\ 0 \leq x \leq 2,\ 0 \leq y \leq 3\} \bigcup \{(x, y) |\ (x - 2)^2 + (y - 1)^2 \leq 1\}$, the feasible tangent cone $IR(S, \bar{x})$ of S at $\bar{x} = (3, 1)$ is given by $IR(S, \bar{x}) = \{(x, y) |\ x < 0\} \cup \{(0, 0)\}$. □

Remark 4.1.10. Examples 4.1.7 and 4.1.8 reveal an important fact: The feasible tangent cone, in general, is neither closed nor open. This is also true for the radial tangent cone.

4.1.2 The Contingent Cone and the Interiorly Contingent Cone

In this subsection, we study the following two notions of tangent cones of first-order.

Definition 4.1.11. Let X be a normed space, let S be a subset of X, and let $\bar{x} \in X$.

1. The **contingent cone** $T(S, \bar{x})$ of S at $\bar{x}$ is the set of all $x \in X$ for which there are sequences $\{t_n\} \subset \mathbb{P}$ and $\{x_n\} \subset X$ with $t_n \downarrow 0$ and $x_n \to x$ satisfying that $\bar{x} + t_n x_n \in S$ for every $n \in \mathbb{N}$.
2. The **interiorly contingent cone** $IT(S, \bar{x})$ of S at $\bar{x}$ is the set of all $x \in X$ such that for all sequences $\{t_n\} \subset \mathbb{P}$ and $\{x_n\} \subset X$ with $t_n \downarrow 0$ and $x_n \to x$, there exists an integer $m \in \mathbb{N}$ satisfying that $\bar{x} + t_n x_n \in S$ for every $n \geq m$.

The notion of the contingent cone is commonly attributed to G. Bouligand (see [75]) who coined the phrase *contingent* to generalize the setting when there is one and only one tangent half-line at every point of a given set. Independently of [75], F. Severi (see [527]) also proposed the notion of the contingent cone. Recently, in an interesting paper [143], S. Dolecki and G. Greco trace the evolution of the concepts of differentiability, tangency and related notions (such as total differential, contingent, paratangent cones, etc.) to Giuseppe Peano (see also [49, 142]). The sequential definition of the interiorly contingent cone, which is also known as the **cone of interior displacements**, can be traced back to the monograph by Flett [187, Section 4.5]. Penot [463] also gives the above sequential definition of the interiorly contingent cone. The contingent cone has been recognized under numerous names such as the **Bouligand tangent cone**, the **tangent cone**, the **cone of adherent displacements**, the **outer tangent cone**, among others, see [53, 262].

The contingent cone has been used extensively in optimization theory and variational analysis and numerous characterizations for this notion have been proposed in the literature. In our next result, we show the equivalence among some of the most commonly used definitions of the contingent cone. We follow the scheme of the excellent paper by Giorgi and Guerraggio [208] and the monograph by Bazarra and Shetty [38], where further details can be found.

The following is an equivalence of different characterizations of the contingent cone.

Theorem 4.1.12. *Let X be a normed space, let S be a subset of X, and let $\bar{x} \in \operatorname{cl} S$. Then, for $i \in \{1, \ldots, 16\}$, we have $T(S, \bar{x}) = T_i(S, \bar{x})$, where $T_i(S, \bar{x})$ are given as follows:*

$$T_1(S, \bar{x}) = \{x \in X \mid \exists \{t_n\} \subset \mathbb{P} \ \exists \{x_n\} \subset S \text{ such that } x_n \to \bar{x},\ t_n (x_n - \bar{x}) \to x\}$$
$$T_2(S, \bar{x}) = \left\{x \in X \mid \exists \{t_n\} \subset \mathbb{P} \ \exists \{x_n\} \subset S \text{ such that } t_n \downarrow 0,\ t_n^{-1}(x_n - \bar{x}) \to x\right\}$$
$$T_3(S, \bar{x}) = \{x \in X \mid \forall t > 0 \ \exists s \in (0, t) \ \exists \hat{x} \in U_t(0) \text{ such that } \bar{x} + s(x + \hat{x}) \in S\}$$
$$T_4(S, \bar{x}) = \{x \in X \mid \forall U(x) \ \forall t > 0 \ \exists s \in (0, t) \ \exists \hat{x} \in U(x) \text{ such that } \bar{x} + s\hat{x} \in S\}$$
$$T_5(S, \bar{x}) = \{x \in X \mid \forall U(x) \ \forall t > 0 \ \exists s \in (0, t) \text{ such that } (\bar{x} + sU(x)) \cap S \neq \emptyset\}$$
$$T_6(S, \bar{x}) = \left\{ x \in X \mid \forall U(x) \ \forall U(0) \text{ we have } \left(\bar{x} + U(0) \bigcap \left(\bigcup_{t>0} tU(x)\right)\right) \bigcap S \neq \emptyset \right\}$$
$$T_7(S, \bar{x}) = \left\{ x \in X \mid \forall U(x) \ \forall U(\bar{x}) \text{ we have } \left(\bar{x} + \bigcup_{t>0} tU(x)\right) \bigcap U(\bar{x}) \bigcap S \neq \emptyset \right\}$$

$$T_8(S,\bar{x}) = \bigcap_{n=1}^{\infty} \text{cl cone}\,(S \cap U_{1/n}(\bar{x}) - \bar{x})$$

$$T_9(S,\bar{x}) = \bigcap_{U(\bar{x})} \text{cl cone}\,(S \cap U(\bar{x}) - \bar{x})$$

$$T_{10}(S,\bar{x}) = \Big\{x \in X \mid \exists\, t > 0\ \exists\, \{x_n\} \subset S \text{ with } x_n \neq \bar{x},\ x_n \to \bar{x} \text{ such that } \frac{x_n - \bar{x}}{\|x_n - \bar{x}\|} \to tx\Big\} \bigcup \{0\}$$

$$T_{11}(S,\bar{x}) = \Big\{x \in X \mid \exists\, \{x_n\} \text{ with } \bar{x} + x_n \in S,\ x_n \to 0,\ \exists\, t > 0 \text{ such that } \frac{x_n}{\|x_n\|} \to tx\Big\} \bigcup \{0\}$$

$$T_{12}(S,\bar{x}) = \limsup_{t \downarrow 0} \frac{S - \bar{x}}{t}$$

$$T_{13}(S,\bar{x}) = \Big\{x \in X \mid \liminf_{t \downarrow 0} \frac{d(\bar{x} + tx, S)}{t} = 0\Big\}$$

$$T_{14}(S,\bar{x}) = \Big\{x \in X \mid \exists\, z(\cdot) : \mathbb{R} \to X \text{ with } \lim_{r \to 0} z(r) = x \text{ such that } \forall\, t > 0\ \exists\ s \in (0,t) \text{ with } \bar{x} + sz(s) \in S\Big\}$$

$$T_{15}(S,\bar{x}) = \Big\{x \in X \mid \exists\, t_n \downarrow 0\ \exists\ \{x_n\} \subset S \text{ such that } x_n = \bar{x} + t_n x + t_n \varepsilon(t_n) \text{ where } \lim_{t \to 0} \varepsilon(t) = 0\Big\}$$

$$T_{16}(S,\bar{x}) = \Big\{x \in X \mid \forall\, \varepsilon > 0\ \exists\, \hat{x} \in U_\varepsilon(\bar{x}) \cap S \text{ such that } \Big\| \frac{\hat{x} - \bar{x}}{\|\hat{x} - \bar{x}\|} - \frac{x}{\|x\|} \Big\| < \varepsilon\Big\} \bigcup \{0\}.$$

Proof. We divide the proof into several parts.

1. We begin by showing that $T(S,\bar{x}) = T_1(S,\bar{x})$. Let $x \in T(S,\bar{x})$ be arbitrary. Then there are sequences $\{t_n\} \subset \mathbb{P}$ and $\{x_n\} \subset X$ such that $t_n \downarrow 0$, $x_n \to x$ and $\bar{x} + t_n x_n \in S$. Define $s_n = \frac{1}{t_n}$ and $\hat{x}_n = \bar{x} + t_n x_n$. Then $\{s_n\} \subset \mathbb{P}$, $\{\hat{x}_n\} \subset S$ with $\hat{x}_n \to \bar{x}$ and $s_n\,(\hat{x}_n - \bar{x}) = x_n \to x$. Therefore, $x \in T_1(S,\bar{x})$ and hence we have shown that $T(S,\bar{x}) \subseteq T_1(S,\bar{x})$. For the converse, let $x \in T_1(S,\bar{x})$ be arbitrary. Then there are sequences $\{t_n\} \subset \mathbb{P}$ and $\{x_n\} \subset S$ such that $x_n \to \bar{x}$ and $t_n(x_n - \bar{x}) \to x$. Clearly $t_n \to \infty$. By setting $s_n = \frac{1}{t_n}$ and $\hat{x}_n = t_n(x_n - \bar{x})$, we obtain that $\hat{x}_n \to x$ with $\bar{x} + s_n\hat{x}_n \in S$. Consequently, $x \in T(S,\bar{x})$.
2. Since the identity $T_1(S,\bar{x}) = T_2(S,\bar{x})$ is trivial, we have obtained that $T(S,\bar{x}) = T_2(S,\bar{x})$.

3. In view of the fact that

$$\limsup_{t\downarrow 0} \frac{S-\bar{x}}{t} = \bigcap_{U(0),s>0} \bigcup_{t\in(0,s)} \left[\frac{S-\bar{x}}{t} + U(0)\right],$$

the equivalence among $T_4(S,\bar{x})$, $T_5(S,\bar{x})$ and $T_{12}(S,\bar{x})$ is evident.
4. The equivalence among $T_{10}(S,\bar{x})$, $T_{11}(S,\bar{x})$ and $T_{16}(S,\bar{x})$ is also easy to prove.
5. Noticing that the identity $T_8(S,\bar{x}) = T_9(S,\bar{x})$ is trivial, we proceed to prove that $T_1(S,\bar{x}) = T_9(S,\bar{x})$. Let $x \in T_1(S,\bar{x})$ be arbitrary. Then, there exist $\{t_n\} \subset \mathbb{P}$ and $\{x_n\} \subset S$ such that $x_n \to \bar{x}$ and $t_n(x_n - \bar{x}) \to x$. Due to the convergence $x_n \to \bar{x}$, there exists $n_1 \in \mathbb{N}$ such that $x_n \in S \cap U(\bar{x})$ for every $n > n_1$. Consequently, $t_n(x_n - \bar{x}) \in \text{cone}\,(S \cap U(\bar{x}) - \bar{x})$ for every $n > n_1$, which ensures that $x = \lim_{n\to\infty} t_n(x_n - \bar{x}) \in \text{cl cone}(S \cap U(\bar{x}) - \bar{x})$. Since this argument is valid for every $U(\bar{x})$, we have $T_1(S,\bar{x}) \subseteq \bigcap_{U(\bar{x})} \text{cl cone}(S \cap U(\bar{x}) - \bar{x})$. For the converse, let $x \in \bigcap_{U(\bar{x})} \text{cl cone}(S \cap U(\bar{x}) - \bar{x})$. Given $n \in \mathbb{N}$, we choose $U_{1/n}(\bar{x})$. Then the above containment implies that x is the limit of elements $t_m(x_m - \bar{x})$, where $t_m > 0$ and $x_m \in S \cap U_{1/n}(\bar{x})$. We choose m_n with $\|x - t_{m_n}(x_{m_n} - \bar{x})\| < n^{-1}$. By varying n, we construct the sequences $\{t_{m_n}\} \subset \mathbb{P}$, $\{x_{m_n}\} \subset S$ such that $t_{m_n}(x_{m_n} - \bar{x}) \to x$. Therefore, $x \in T_1(S,\bar{x})$.
6. We claim that $T_1(S,\bar{x}) = T_{15}(S,\bar{x})$. Let $x \in T_1(S,\bar{x})$. Then there are sequences $\{t_n\} \subset \mathbb{P}$, $\{x_n\} \subset S$ such that $x_n \to \bar{x}$ and $t_n(x_n - \bar{x}) \to x$. In other words, $x = t_n(x_n - \bar{x}) - \varepsilon\left(t_n^{-1}\right)$, where $\varepsilon : \mathbb{R} \to X$ is a function such that $\varepsilon(t) \to 0$ as $t \to 0$. Since $x_n \to \bar{x}$ and $\lim_{n\to\infty} (t_n(x_n - \bar{x}))$ exists, we deduce that $t_n \to \infty$. Defining $s_n = 1/t_n$, we obtain that $x_n = \bar{x} + s_n x + s_n \varepsilon(s_n)$, ensuring $T_1(S,\bar{x}) \subseteq T_{15}(S,\bar{x})$. For the converse, assume that there exists $\{x_n\} \subset S$ such that $x_n = \bar{x} + s_n x + s_n\varepsilon(s_n)$, which implies that $s_n^{-1}(x_n - \bar{x}) \to x$ and by setting $t_n = \frac{1}{s_n}$, we deduce that $x \in T_1(S,\bar{x})$.
7. We next claim that $T(S,\bar{x}) \subseteq T_6(S,\bar{x})$. Let $x \in T(S,\bar{x})$ be arbitrary. Then there exists $\{t_n\} \subset \mathbb{P}$ and $\{x_n\} \subset S$ such that $t_n \downarrow 0$, $x_n \to x$ and $\bar{x} + t_n x_n \in S$. Since $t_n \downarrow 0$, there exists an index $m \in \mathbb{N}$ such that $t_n x_n \in U(0)$ for every $n \geq m$. Furthermore, since $x_n \to x$, we also have that $t_n x_n \in t_n U(x)$ for some $U(x)$. Therefore, for suitable $n \in \mathbb{N}$, we have $\bar{x} + t_n x_n \in \left(\bar{x} + U(0) \bigcap \left(\bigcup_{t>0} tU(x)\right)\right) \bigcap S$. In other words, $\left(\bar{x} + U(0) \bigcap \left(\bigcup_{t>0} tU(x)\right)\right) \bigcap S \neq \emptyset$, which confirms that $T(S,\bar{x}) \subseteq T_6(S,\bar{x})$.
8. Let us now prove that $T_6(S,\bar{x}) \subseteq T_{16}(S,\bar{x})$. Let $x \in T_6(S,\bar{x})$ be arbitrary. Then for every $U_\varepsilon(0)$, we can find $0 \neq \hat{x} \in U(0)$ such that $\bar{x} + \hat{x} = \check{x} \in S$. It suffices to show that

$$\left\| \frac{\check{x} - \bar{x}}{\|\check{x} - \bar{x}\|} - \frac{x}{\|x\|} \right\| < \varepsilon.$$

Since $\hat{x} \in \bigcup_{t>0} tU(x)$, for every $U(x)$ or $\hat{x} = t(x+z)$, with z sufficiently small. Therefore,

$$\left\| \frac{\breve{x}-\bar{x}}{\|\breve{x}-\bar{x}\|} - \frac{\hat{x}}{\|x\|} \right\| = \left\| \frac{\|\hat{x}\|\|x\| - x\|\hat{x}\|}{\|\hat{x}\|\|x\|} \right\| = \lambda \left\| \frac{x\|x\| + z\|x\| - x\|x+s\|}{\|x+z\|\|x\|\|} \right\|,$$

which can be made smaller than ε.

9. We now claim that $T_{16}(S,\bar{x} \subseteq T_{13}(S,\bar{x})$. Let $x \in T_{16}(S,\bar{x})$ be arbitrary. Then $\tilde{x} = tx$ with $t \in \mathbb{R}$ and $\|\tilde{x}\| = 1$ also belongs to $T_{16}(S,\bar{x})$. Therefore there exists a sequence $\{z_n\} \subset S$ with $z_n \to \bar{x}$ such that $\lim_{n\to\infty} \left\| \|z_n - \bar{x}\|^{-1}(z_n - \bar{x}) - \tilde{x} \right\| = 0$, or $\lim_{\|h\|\to 0} \inf \left\| \|h\|^{-1}(z-\bar{x}) - \tilde{x} \right\| = 0$. Since $\lim_{\|h\|\to 0} \inf h^{-1}\|\bar{x} + \|h\|\tilde{x} - z\| = 0$, we obtain that $\tilde{x} \in T_{13}(S,\bar{x})$, implying that $x \in T_{13}(S,\bar{x})$.
10. We next show that $T_{13}(S,\bar{x}) \subseteq T_1(S,\bar{x})$. Let $x \in T_{13}(S,\bar{x})$ be arbitrary. Then for every $\{t_n\} \subset \mathbb{P}$ with $t_n \downarrow 0$, we have $\inf_{\hat{x}\in S} h_n^{-1}\|\bar{x} + h_n x - \hat{x}\| < t_n$. Therefore, there exists $\{\hat{x}_n\} \subset S$ such that $h_n^{-1}\|\bar{x} + h_n x - \hat{x}_n\| < t_n$. From, $\lim_{n\to\infty} h_n^{-1}(\bar{x} + h_n x - \hat{x}_n) = 0$, or its equivalent expression $h_n^{-1}(\bar{x} + h_n x - \hat{x}_n) = o(1)$, it follows that $x = o(1) + h_n^{-1}(\hat{x}_n - \bar{x})$, and therefore, $\lim_{n\to\infty} h_n^{-1}(\hat{x}_n - \bar{x}) = x$ with $\hat{x}_n \to \bar{x}$. Consequently, $T_{13}(S,\bar{x}) \subseteq T_1(S,\bar{x})$.
11. We claim that $T_{15}(S,\bar{x}) \subseteq T_3(S,\bar{x})$. Let $x \in T_{15}(S,\bar{x})$ be arbitrary. Then there are sequences $\{t_n\} \subset \mathbb{P}$ and $\{x_n\}$ such that $\bar{x} + t_n x + t_n x_n \in S$. Therefore, for every $\delta > 0$, there are scalars $n_1 > 0$ and $n_2 > 0$ such that for every n with $n \geq \max\{n_1, n_2\}$, we have $|t_n| < \delta$ and $\|x_n\| < \delta$. Consequently, for $\bar{n} = \max\{n_1, n_2\}$, $t_n = \lambda$ and $x_{\bar{n}} = z$ we obtain $\bar{x} + \lambda x + \lambda z \in S$ with $\lambda \in (0,\delta)$ and $\|z\| < \delta$. This ensures that $x \in T_3(S,\bar{x})$.
12. Next we show that $T_3(S,\bar{x}) \subseteq T_4(S,\bar{x})$. Let $x \in T_3(S,\bar{x})$ be arbitrary. Let $U_\eta(x)$ be a neighborhood and let $\varepsilon > 0$ be fixed but arbitrary. For $\delta = \min\{\eta, \varepsilon\}$, there is a positive number $t < \delta \leq \varepsilon$ and an element $z \in U_\delta(0) \subset U_\eta(0)$ for which $\bar{x} + tx + tz = \bar{x} + t(x+z) \in S$. Therefore, $x \in T_4(S,\bar{x})$.
13. Our next claim is that $T_{12}(S,\bar{x}) \subseteq T_{10}(S,\bar{x})$. Let $x \in T_{12}(S,\bar{x})$ be arbitrary. We consider a sequence of neighborhoods of 0 and a sequence $\{\lambda_n\}$ such that $\lambda_n \to 0$. We construct sequences $\{x_n\} \subset S$, $\{t_n\} \to 0$, and $\{z_n\} \to 0$ such that $x = t_n^{-1}(x_n - \bar{x}) + z_n$, implying that $t_n^{-1}(x_n - \bar{x}) = x - z_n \to x$, which proves that $x \in T_{10}(S,\bar{x})$.
14. We claim that $T_{11}(S,\bar{x}) \subseteq T_{14}(S,\bar{x})$. Let $x \in T_{11}(S,\bar{x})$ be arbitrary. Then there exists $\{x_n\} \to 0$ for which $x_n = \alpha x\|x_n\| + o(\|x_n\|)$. By hypothesis, we have $\bar{x} + x_n \in S$ and hence $\bar{x} + \alpha x\|x_n\| + o(\|x_n\|) \in S$. Then, for every $\lambda > 0$ and $t \in (0,\lambda)$ $t = \alpha\|x_n\|$ with n large enough, for which $\bar{x} + t\phi(t) \in S$ with $\phi(t)$ given by $[x + o(1)]$.
15. Finally, the inclusion $T_{11}(S,\bar{x}) \subseteq T(S,\bar{x})$ is easy to prove. □

Remark 4.1.13. The above characterizations have their own advantages. For instance, it follows immediately from the definition of $T_1(S,\bar{x})$ that if $\bar{x}$ is an isolated point then $T_1(S,\bar{x}) = \{0\}$. In fact, in this case, as required by $T_1(S,\bar{x})$, $x_n = \bar{x}$ is the only sequence in S converging to $\bar{x}$. Furthermore, the cone $T_5(S,\bar{x})$

reduces to the empty-set when $\bar{x}$ is an isolated point of S. Therefore, to be able to provide a reasonable comparison, following the approach of [207], the element $\{0\}$ has been added to this cone. We remark that $T_1(S,\bar{x})$ was used in [1, 34, 36, 37, 262], among others. Characterizations $T_2(S,\bar{x})$ and $T_{15}(S,\bar{x})$ were proposed by Bazaraa, Goode, and Nashed [36]. Characterizations $T_8(S,\bar{x})$ (or $T_9(S,\bar{x})$) is due to Varaiya [582]. The monographs [254] and [255] by M. R. Hestenes contains $T_{10}(S,\bar{x})$. R. T. Rockafellar in [491] used $T(S,\bar{x})$ and $T_5(S,\bar{x})$. Definition of $T_{11}(S,\bar{x})$ is given in Canon et. al [83]. Hiriart-Urruty [262] also uses $T_6(S,\bar{x})$. Federer [182] contain $T_{16}(S,\bar{x})$. A more detailed remark on the use of these characterizations can be found in [207]. Some other works dealing with various aspects of contingent cones are [58, 59, 64, 113, 119, 131, 192, 261, 308, 309, 413, 435, 436, 453, 463–466, 468–470, 501, 519, 571, 577–579, 592].

The following result from [178] gives characterizations of the interiorly contingent cone.

Theorem 4.1.14. *Let X be a normed space, let S be a subset of X, and let $\bar{x} \in \operatorname{cl} S$. Then, for $i \in \{1,2\}$, we have $IT(S,\bar{x}) = IT_i(S,\bar{x})$, where $IT_i(S,\bar{x})$ are given as follows:*

$$IT_1(S,\bar{x}) = \{x \in X \mid \exists\, U(x)\ \exists\, t > 0\ \forall\, s \in (0,t)\ \forall\, z \in U(x) \text{ we have } \bar{x} + sz \in S\}$$

$$IT_2(S,\bar{x}) = \left\{ x \in X \mid \exists\, U(\bar{x})\ \exists\, U(x) \text{ such that } \left(\bar{x} + \bigcup_{t>0} tU(x)\right) \bigcap U(\bar{x}) \subseteq S \right\}.$$

Proof. The equivalence between $IT_1(S,\bar{x})$ and $IT(S,\bar{x})$ is given in Flett [187, Section 4.5] and is easy to show. Here we will only prove the equivalence between $IT_1(S,\bar{x})$ and $IT_2(S,\bar{x})$. Let $x \in IT_1(S,\bar{x})$ be arbitrary. Then there are a neighborhood $U(x)$ and a scalar $t > 0$ such that $\bar{x} + \bigcup_{s\in(0,t)} sU(x) \subset S$. We can find neighborhoods $U(\bar{x})$ and $V(x) \subset U(x)$ such that $U(\bar{x}) \bigcap (\bar{x} + tV(x)) = \emptyset$ which implies that $\left(\bar{x} + \bigcup_{s>0} sV(x)\right) \bigcap U(\bar{x}) \subset \bar{x} + \bigcup_{s\in(0,t)} sU(x)$ and, in particular, $\left(\bar{x} + \bigcup_{s>0} sV(x)\right) \bigcap U(\bar{x}) \subset S$, confirming that $x \in IT_2(S,\bar{x})$. We have shown that $IT_1(S,\bar{x}) \subseteq IT_2(S,\bar{x})$.

For the converse, let $x \in IT_2(S,\bar{x})$. Then there are neighborhoods $U(x)$ and $U(\bar{x})$ satisfying $\left(\bar{x} + \bigcup_{t>0} tU(x)\right) \bigcap U(\bar{x}) \subset S$. Since there exist a scalar $t > 0$ and a neighborhood $V(x) \subset U(x)$ such that for every $s \in (0,t)$, we have $\bar{x} + sV(x) \subset U(\bar{x})$, and consequently, for every $s \in (0,t)$, we have $\bar{x} + sV(x) \subset S$, which ensures that $x \in IT_1(S,\bar{x})$. The proof is complete. □

We next collect some properties of the contingent cone and the interiorly contingent cone.

Theorem 4.1.15. *Let X be a normed space, let S, S_1, S_2 be subsets of X, and let $\bar{x}$, $\bar{x}_1$, $\bar{x}_2 \in X$. Then:*

1. $IT(S,\bar{x}) = X \backslash T(X \backslash S,\bar{x})$
2. $T(S,\bar{x}) = X \backslash IT(X \backslash S,\bar{x})$
3. $IT(S,\bar{x})$ *is an open cone*

4. *$T(S,\bar{x})$ is a closed cone*
5. $T(\mathrm{cl}\, S,\bar{x}) = T(S,\bar{x})$
6. $IT(\mathrm{int}\, S,\bar{x}) = IT(S,\bar{x})$
7. $\bar{x} \in \mathrm{cl}\, S \iff 0 \in T(S,\bar{x})$
8. $x \in \mathrm{int}\, S \iff 0 \in IT(S,\bar{x})$
9. $IT(S,\bar{x}) \subseteq IR(S,\bar{x}) \subseteq R(S,\bar{x}) \subseteq T(S,\bar{x})$
10. $S_1 \subseteq S_2 \implies IT(S_1,\bar{x}) \subseteq IT(S_2,\bar{x})$
11. $S_1 \subseteq S_2 \implies T(S_1,\bar{x}) \subseteq T(S_2,\bar{x})$
12. $IT(S_1 \cap S_2,\bar{x}) = IT(S_1,\bar{x}) \cap IT(S_2,\bar{x})$
13. $T(S_1 \cup S_2,\bar{x}) = T(S_1,\bar{x}) \cup T(S_2,\bar{x})$
14. $IT(S_1 \times S_2,(\bar{x}_1,\bar{x}_2)) = IT(S_1,\bar{x}_1) \times IT(S_2,\bar{x}_2)$
15. $T(S_1 \times S_2,(\bar{x}_1,\bar{x}_2)) \subseteq T(S_1,\bar{x}_1) \times T(S_2,\bar{x}_2)$.

Proof. The proof can be derived from the arguments supplied in the proof of Theorem 4.2.10. □

We now give a few examples of the contingent cone and the interiorly contingent cone.

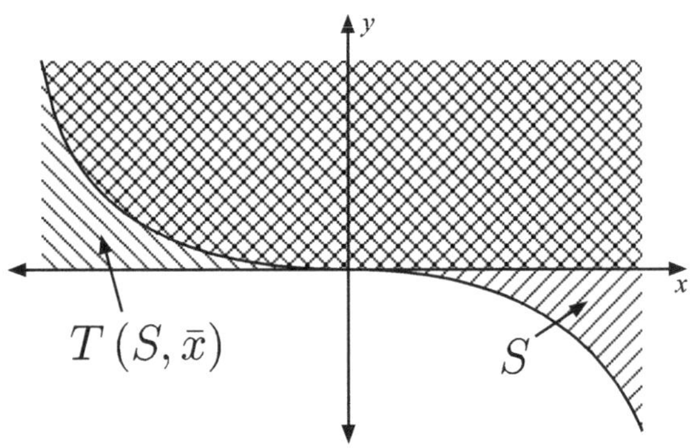

Fig. 4.3 $T(S,\bar{x})$ for Example 4.1.16

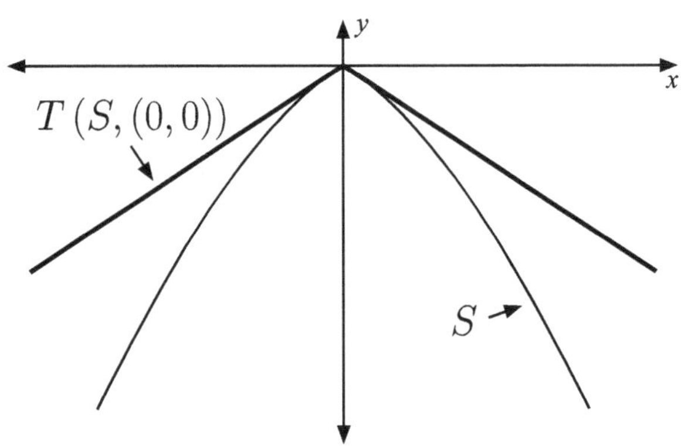

Fig. 4.4 $T(S,\bar{x})$ for Example 4.1.17

Example 4.1.16. For $S \subset \mathbb{R}^2$ given by $S = \{(x, y) \in | \ y \geq -x^3\}$, the contingent cone $T(S, \bar{x})$ of S at $\bar{x} = (0, 0)$ is given by $T(S, \bar{x}) = \{(x, y) | \ y \geq 0\}$. □

Example 4.1.17. For $S \subset \mathbb{R}^2$ given by $S = \{(x, y) | \ x^2 + |x| + y = 0\}$, the contingent cone $T(S, \bar{x})$ of S at $\bar{x} = (0, 0)$ is given by $T(S, (0, 0)) = \{(x, y) | \ y + |x| = 0\}$. □

Example 4.1.18. For the sets $S_1, S_2 \subset \mathbb{R}$ given by

$$S_1 := \{x \in \mathbb{R} | \ x = 2^{-2n}, \ n \in \mathbb{N}\} \cup \{0\},$$

$$S_2 := \{x \in \mathbb{R} | \ x = 2^{-2n+1}, \ n \in \mathbb{N}\} \cup \{0\},$$

we have $T(S_1, 0) = T(S_2, 0) = \mathbb{R}_+$, whereas, $(1, 1) \notin T(S_1 \times S_2, (0, 0))$, which shows that the inclusion $T(S_1 \times S_2, (\bar{x}_1, \bar{x}_2)) \supset T(S_1, \bar{x}_1) \times T(S_2, \bar{x}_2)$, is, in general, false (see [60]). □

Example 4.1.19. For the sets $S_1, S_2 \subset \mathbb{R}^2$ given by

$$S_1 = \left\{(x, y) | \ y \geq x^2\right\},$$

$$S_2 = \left\{(x, y) | \ y \leq -x^2\right\},$$

we have

$$T(S_1 \cap S_2, (0, 0)) = \{(0, 0)\},$$

$$T(S_1, (0, 0)) \cap T(S_2, (0, 0)) = \{(x, y) \in \mathbb{R}^2 | \ y = 0\},$$

showing that the intersection formula $T(S_1 \cap S_2, \bar{x}) = T(S_1, \bar{x}) \cap T_2(S, \bar{x})$, is, in general, false. □

The following definition gives a useful way of extending the scope of tangent cones:

Definition 4.1.20. Let X be a reflexive Banach space, let $S \subset X$, and let $\bar{x} \in S$. The **weak contingent cone** $T_w(S, \bar{x})$ of S at $\bar{x}$ is the collection of all $x \in S$ such that there are a sequence $\{x_n\} \subset S$ with $x_n \to \bar{x}$ and a sequence $\{t_n\} \subset \mathbb{P}$ such that $t_n (x_n - \bar{x})$ converges weakly to x.

Remark 4.1.21. The notion of weak contingent cone was proposed by Bazaraa, Goode, Nashed and Shetty [37]. Borwein [59] conducted a thorough study of an extension of the above notion where he employed a general topology that subsumed the weak as well as the weak* topology.

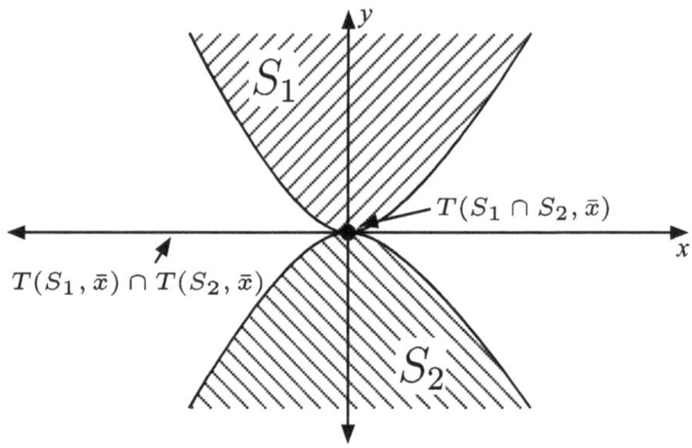

Fig. 4.5 Failure of the intersection formula $T(S_1 \cap S_2, \bar{x}) = T(S_1, \bar{x}) \cap T_2(S, \bar{x})$ in Example 4.1.19

4.1.3 The Adjacent Cone and the Interiorly Adjacent Cone

This subsection deals with the following two notions of tangent cones.

Definition 4.1.22. Let X be a normed space, let S be a subset of X, and let $\bar{x} \in S$.

1. The **adjacent cone** $A(S, \bar{x})$ of S at $\bar{x}$ is the set of all $x \in X$ such that for every sequence $\{t_n\} \subset \mathbb{P}$ with $t_n \downarrow 0$, there exists $\{x_n\} \subset X$ with $x_n \to x$ satisfying $\bar{x} + t_n x_n \in S$ for every $n \in \mathbb{N}$.
2. The **interiorly adjacent cone** $IA(S, \bar{x})$ of S at $\bar{x}$ is the set of all $x \in X$ for which there exists $\{t_n\} \subset \mathbb{P}$ with $t_n \downarrow 0$ such that for every $\{x_n\} \subset X$ with $x_n \to x$, there exists an index $m \in \mathbb{N}$ satisfying $\bar{x} + t_n x_n \in S$ for every $n \geq m$.

Remark 4.1.23. The adjacent cone has also been referred to as **derivable cone**, **intermediate cone**, **cone of attainable directions**, **Nagumo cone**, and **Ursescu cone**, among others. The interiorly adjacent cone is also known as the **cone of quasi-interior directions**, see [268].

The following result gives some of the most commonly used definitions of the adjacent cone:

Theorem 4.1.24. *Let X be a normed space, let S be a subset of X, and let $\bar{x} \in \operatorname{cl} S$. For $i \in \{1, \ldots, 5\}$, we have $A(S, \bar{x}) = A_i(S, \bar{x})$, where $A_i(S, \bar{x})$ are given as follows:*

$$A_1(S, \bar{x}) = \Big\{x \in X \mid \exists\, \varepsilon : \mathbb{R}_+ \to X,\ \lim_{t \to 0} \varepsilon(t) = 0,\ \exists\, t > 0 \text{ such that } \bar{x} + sx + s\varepsilon(s) \in S,\ \forall\, s \in (0, t)\Big\}$$

$$A_2(S, \bar{x}) = \big\{x \in X \mid \exists\, \phi : \mathbb{R}_+ \to X,\ \exists\, t > 0 \text{ such that } \phi(0) = \bar{x},\ \phi'(0) = x,\ \phi(s) \in S,\ \forall\, s \in (0, t)\big\}$$

$$A_3(S, \bar{x}) = \{x \in X \mid \forall\, U(x)\ \exists\, t > 0\ \forall\, s \in (0, t)\ \exists\, \hat{x} \in U(x) \text{ such that } \bar{x} + s\hat{x} \in S\}$$

$$A_4(S,\bar{x}) = \left\{x \in X \mid \lim_{t\to 0} \frac{d(\bar{x}+tx,S)}{t} = 0\right\}$$
$$A_5(S,\bar{x}) = \liminf_{t\downarrow 0} \frac{S-\bar{x}}{t}.$$

Proof. The proof, which is based on [38, 207], is divided into several parts.

1. We begin by showing that $A_1(S,\bar{x}) = A_2(S,\bar{x})$. Let $x \in A_2(S,\bar{x})$ be arbitrary. Then from the expression $\phi'(0) = x$, we obtain that $\lim_{s\downarrow 0} s^{-1}(\phi(s)-\bar{x}) = x$, which implies that $\phi(s)-\bar{x} = sx + s\varepsilon(s)$, where $\varepsilon(s) \to 0$ as $s \to 0$. In other words, $\phi(s) = \bar{x} + sx + s\varepsilon(s)$ with $\phi(s) \in S$ for every $s \in (0,t]$, with suitable t. Therefore $x \in A_1(S,\bar{x})$. For the converse, assume that $x \in A_1(S,\bar{x})$. Then we have $\phi(s) = \bar{x} + sx + s\varepsilon(s)$, for the defined s and ε. Therefore, $\bar{x} = \phi(0)$ and $\lim_{s\to 0} s^{-1}(\phi(s)-\bar{x}) = x + \lim_{s\to 0}\varepsilon(s) = x$, and consequently $x \in A_2(S,\bar{x})$.
2. The identity $A(S,\bar{x}) = A_2(S,\bar{x})$ is immediate.
3. The proofs for $A_1(S,\bar{x}) \subseteq A_3(S,\bar{x})$ and $A_3(S,\bar{x}) \subseteq A_4(S,\bar{x})$ follow from the definition.
4. We next prove that $A_4(S,\bar{x}) \subseteq A_2(S,\bar{x})$. For $x \in A_4(S,\bar{x})$, we have $\lim_{h\to 0^+} h^{-1}d(\bar{x}+hx,S) = 0$. Consequently, for every $\varepsilon > 0$ there exists $\delta > 0$ such that $h^{-1}(d(\bar{x}+hx,S)) < \varepsilon$ for every $h < \delta$. Therefore, for every $h \in (0,\delta)$, there is an element $y \in S$ such that $\left\|h^{-1}(\bar{x}+hx-y(h))\right\| < \varepsilon$ or equivalently $\lim_{h\to 0^+} h^{-1}(\bar{x}+hx-y(h)) = 0$. It follows that $\bar{x}+hx+o(h) = y(h) \in S$. By taking $\phi(h) = \bar{x}+hx+o(h)$ continuously extended to $[0,\delta)$, we get that $x \in A_2(S,\bar{x})$. □

The characterizations given above have been known for quite some time. Recently, in an interesting paper, Jiménez and Novo [306] proposed two new characterizations of the adjacent cone by using the notion of **slowly convergent sequences**. In the following, by Ω we represent the set of all strictly decreasing sequences $\{t_n\}$ of positive real numbers converging to 0 such that $\frac{t_n}{t_{n-1}} \to 1$. In [306], the following two characterizations of the adjacent cone were proposed:

Theorem 4.1.25. *Let X be a normed space, let S be a subset of X, and let $\bar{x} \in \operatorname{cl} S$. For $i \in \{6,7\}$, we have $A(S,\bar{x}) = A_i(S,\bar{x})$, where $A_i(S,\bar{x})$ are given as follows:*

$$A_6(S,\bar{x}) := \{x \in X \mid \exists\, \{t_n\} \subset \Omega,\ \exists\, \{x_n\} \to 0,\ \textit{such that } \bar{x}+t_nx+t_nx_n \in S, \textit{ for every } n \in \mathbb{N}\}$$
$$A_7(S,\bar{x}) := \{x \in X \mid \exists\, \{t_n\}\} \subset \Omega,\ \exists\, \{x_n\} \to x,\ \textit{such that } \bar{x}+t_nx_n \in S, \textit{ for every } n \in \mathbb{N}\}.$$

Proof. See Jiménez and Novo [306]. □

We note that the interiorly adjacent cone can equivalently be expressed as follows:

$$IA(S,\bar{x}) = \{x \in X \mid \exists\, U(x)\ \forall\, t > 0\ \exists\, s \in (0,t)\ \forall\, \hat{x} \in U(x),\ \textit{we have }\ \bar{x}+s\hat{x} \in S\}.$$

The following theorem collects some of the main features of the adjacent cone and the interiorly adjacent cone.

Theorem 4.1.26. *Let X be a normed space, let S, S_1, S_2 be subsets of X, and let $\bar{x}, \bar{x}_1, \bar{x}_2 \in X$. Then:*

1. $A(S,\bar{x}) = X \backslash IA(X \backslash S, \bar{x})$
2. $IA(S,\bar{x}) = X \backslash A(X \backslash S, \bar{x})$
3. $A(S,\bar{x})$ *is a closed cone*
4. $IA(S,\bar{x})$ *is an open cone*
5. $A(\operatorname{cl} S, \bar{x}) = A(S,\bar{x})$
6. $IA(\operatorname{int} S, x) = IA(S,\bar{x})$
7. $x \in \operatorname{cl} S \iff 0 \in A(S,\bar{x})$
8. $x \in \operatorname{int} S \iff 0 \in IA(S,\bar{x})$
9. $IR(S,\bar{x}) \subseteq A(S,\bar{x}) \subseteq T(S,\bar{x})$
10. $IT(S,\bar{x}) \subseteq IA(S,\bar{x}) \subseteq R(S,\bar{x})$
11. $S_1 \subseteq S_2 \implies A(S_1,\bar{x}) \subseteq A(S_2,\bar{x})$
12. $S_1 \subseteq S_2 \implies IA(S_1,\bar{x}) \subseteq IA(S_2,\bar{x})$
13. $A(S_1 \times S_2, (\bar{x}_1, \bar{x}_2)) = A(S_1, \bar{x}_1) \times A(S_2, \bar{x}_2)$
14. $A(S_1 \times S_2, (\bar{x}_1, \bar{x}_2)) = A(S_1, \bar{x}_1) \times A(S_2, \bar{x}_2)$
15. $IA(S_1 \times S_2, (\bar{x}_1, \bar{x}_2)) \subseteq IA(S_1, \bar{x}_1) \times IA(S_2, \bar{x}_2)$.

Proof. The proof can be derived from the proof of Theorem 4.2.10. □

We now give a few examples of the adjacent cone and the interior adjacent cone.

Example 4.1.27. For $S \subset \mathbb{R}^2$ given by $S = \{(x,y) | \; y = x^3\}$, the adjacent cone $A(S,\bar{x})$ of S at $\bar{x} = (0,0)$ is given by $A(S,\bar{x}) = \{(x,y) | \; y = 0\}$. □

Example 4.1.28. For $S \subset \mathbb{R}^2$ given by $S = \{(x,y) | \; x \geq 0, \; y \geq x^3\} \bigcup \{(x,y) | \; x \leq 0, \; y \geq -x\}$, the adjacent cone $A(S,\bar{x})$ of S at $\bar{x} = (0,0)$ is given by $A(S,\bar{x}) = \{(x,y) | \; y \geq \max\{0,-x\}\}$. □

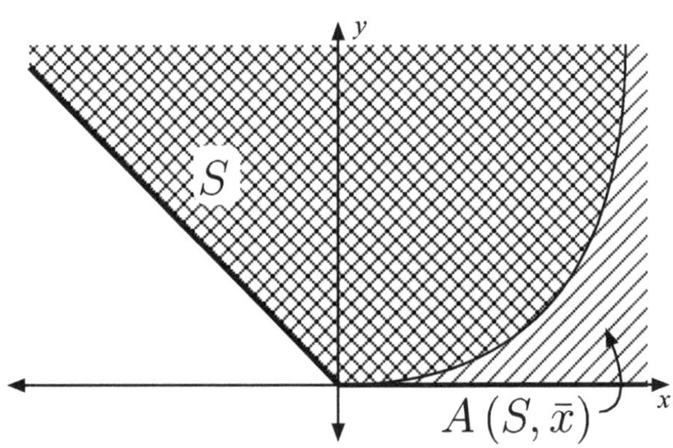

Fig. 4.6 $A(S,\bar{x})$ for Example 4.1.28

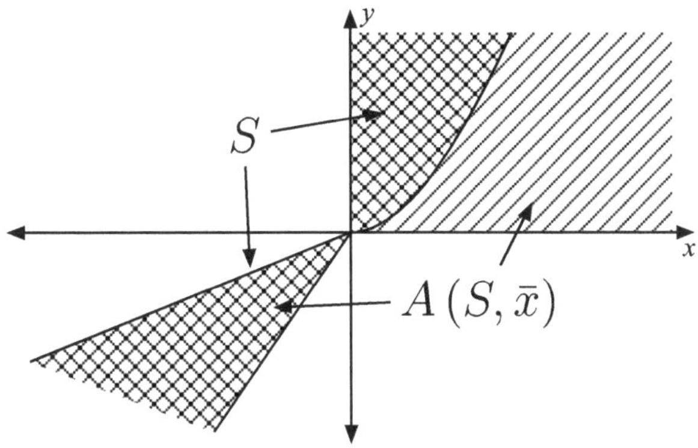

Fig. 4.7 $A(S,\bar{x})$ for Example 4.1.30

Example 4.1.29. For $S \subset \mathbb{R}^2$ given by $S = \{(x,y) \mid y = x\sin(x^{-1}),\ x \neq 0\} \cup \{(0,0)\}$, the adjacent cone $A(S,\bar{x})$ of S at $\bar{x} = (0,0)$ is given by $A(S,\bar{x}) = \{(x,y) \mid |y| \leq |x|\}$. □

Example 4.1.30. For $S \subset \mathbb{R}^2$ defined by

$$S = \{(x,y) \mid x \geq 0,\ y \geq x^2\} \bigcup \left\{(x,y) \mid x \leq 0,\ \frac{2}{3}x \leq y \leq -\frac{1}{3}x\right\},$$

the adjacent cone $A(S,\bar{x})$ of S at $\bar{x} = (0,0)$ is given by

$$A(S,\bar{x}) = \{(x,y) \mid x \geq 0,\ y \geq 0\} \bigcup \left\{(x,y) \mid x \leq 0,\ -\frac{2}{3}x \leq y \leq -\frac{1}{3}x\right\}.$$

4.2 Modified First-Order Tangent Cones

All the first-order tangent cones discussed so far in this chapter are, in general, non-convex. A quite easy fix, as done in some of the earlier works, where convexity of the tangent cone was required, is to consider the so-called **pseudo tangent cone** $\Pi(S,\bar{x})$ by taking the convex hull of the contingent cone $T(S,\bar{x})$ of S at $\bar{x}$:

$$\Pi(S,\bar{x}) = \operatorname{conv}(T(S,\bar{x})).$$

It turns out that the convexity achieved through the above procedure of taking the convex hull of the contingent cone is not optimal, as the pseudo tangent cone is not easily analytically tractable. Furthermore, it often gives too large an approximation of the underlying set. In the following, we discuss another procedure of getting convex approximations of sets by considering the modified analogues of the cones studied above. These cones are always convex. The starting point of this idea was the influential work of F. H. Clarke who introduced the notion of the so-called Clarke's cone (see Clarke [100]). As we will see below, the Clarke's cone is a modified version of the adjacent cone. The foregoing discussion is based on the excellent paper [178].

4.2.1 The Modified Radial and the Modified Feasible Tangent Cones

In this section, we deal with the following two modified tangent cones of first-order.

Definition 4.2.1. Let X be a normed space, let S be a subset of X, and let $\bar{x} \in X$.

1. The **modified radial tangent cone** $R_{\text{M}}(S,\bar{x})$ of S at $\bar{x}$ is the collection of all $x \in X$ such that for every neighborhood $V(\bar{x})$ and for every scalar $t > 0$, there exist a scalar $s \in (0,t)$ and an element $\hat{x} \in V(\bar{x}) \cap S \cup \{\bar{x}\}$ satisfying that $\hat{x} + sx \in S$.
2. The **modified feasible tangent cone** $IR_{\text{M}}(S,\bar{x})$ of S at $\bar{x}$ is the collection of all $x \in X$ for which there exist a neighborhood $V(\bar{x})$ and a scalar $t > 0$ such that for every $s \in (0,t)$ and for every $\hat{x} \in V(\bar{x}) \cap S \cup \{\bar{x}\}$, we have $\hat{x} + sx \in S$.

Remark 4.2.2. In [178], $IR_{\text{M}}(S,\bar{x})$ is referred to as the **modified cone of feasible directions**. This cone also appears in the works of R. T. Rockafellar (see [492,493]).

The following result collects some of the useful features of the modified radial tangent cone and the modified feasible tangent cone.

Theorem 4.2.3. *Let X be a normed space, let S, S_1, S_2 be subsets of X, and let $\bar{x}$, $\bar{x}_1$, $\bar{x}_2 \in X$. Then:*

1. $R_M(S,\bar{x}) = X \backslash IR_M(X \backslash S,\bar{x})$
2. $IR_M(S,\bar{x}) = X \backslash R_M(X \backslash S,\bar{x})$
3. $IR_M(S,\bar{x})$ *is a convex cone*
4. $R_M(S,\bar{x})$ *is the complement of a convex cone*
5. $IR_M(S,\bar{x}) \subseteq IR(S,\bar{x}) \subseteq R(S,\bar{x}) \subseteq R_M(S,\bar{x})$
6. $\bar{x} \in S \Longleftrightarrow 0 \in IR_M(S,\bar{x}) \Longleftrightarrow 0 \in R_M(S,\bar{x})$
7. $IR_M(S_1 \cap S_2,\bar{x}) \supseteq IR_M(S_1,\bar{x}) \cap IR_M(S_2,\bar{x})$
8. $R_M(S_1 \cup S_2,\bar{x}) \subseteq R_M(S_1,\bar{x}) \cup R(S_2,\bar{x})$
9. $IR_M(S_1 \times S_2,(\bar{x}_1,\bar{x}_2)) = IR_M(S_1,\bar{x}_1) \times IR_M(S_2,\bar{x}_2)$
10. $R_M(S_1 \times S_2,(\bar{x}_1,\bar{x}_2)) \subseteq R_M(S_1,\bar{x}_1) \times R_M(S_2,\bar{x}_2)$.

Proof. The proof can be derived from the proof of Theorem 4.2.10. □

4.2.2 The Modified Contingent and the Modified Interiorly Contingent Cones

Here we study the following two modified tangent cones of first-order.

Definition 4.2.4. Let X be a normed space, let S be a subset of X, and let $\bar{x} \in X$.

1. The **modified contingent cone** $T_{\mathrm{M}}(S, \bar{x})$ of S at $\bar{x}$ is the collection of all $x \in X$ such that for every neighborhood $V(\bar{x})$, every neighborhood $V(x)$, and every scalar $t > 0$, there are a scalar $s \in (0, t)$, an element $\hat{x} \in V(\bar{x}) \cap S \cup \{\bar{x}\}$ and an element $\check{x} \in V(x)$ so that we have $\hat{x} + s\check{x} \in S$.
2. The **modified interiorly contingent cone** $IT_{\mathrm{M}}(S, \bar{x})$ of S at $\bar{x}$ is the collection of all $x \in X$ for which there are a neighborhood $V(\bar{x})$, a neighborhood $V(x)$, and a scalar $t > 0$ such that for every $s \in (0, t)$ for every $\check{x} \in V(x)$ and for every $\hat{x} \in V(\bar{x}) \cap S \cup \{\bar{x}\}$, we have $\hat{x} + s\check{x} \in S$.

Remark 4.2.5. In [178], $IT_{\mathrm{M}}(S, \bar{x})$ is referred to as the **modified cone of interior displacements**. The cone $IT_{\mathrm{M}}(S, \bar{x})$ is also known as the **cone of epi Lipschitzian directions** (see [492, 493]), and the **hypertangent cone** (see [524]).

The following sequential definition of the hypertangent cone has been used quite often:

$$IT_M(S, \bar{x}) = \{x \in X \mid \forall \{x_n\} \xrightarrow{S} \bar{x},\ \forall \{t_n\} \downarrow 0,\ \forall \{z_n\} \to x,\ \text{we have}$$
$$x_n + t_n z_n \in S, \text{for large } n\}. \tag{4.1}$$

The following result collects some of the useful features of the modified contingent cone and of the modified interiorly contingent cone.

Theorem 4.2.6. *Let X be a normed space, let S, S_1, S_2 be subsets of X, and let $\bar{x}$, $\bar{x}_1$, $\bar{x}_2 \in X$. Then:*

1. $IT_M(S, \bar{x}) = X \backslash T_M(X \backslash S, \bar{x})$
2. $T_M(S, \bar{x}) = X \backslash IT_M(X \backslash S, \bar{x})$
3. $T_M(S, \bar{x})$ *is a closed cone*
4. $IT_M(S, \bar{x})$ *is an open cone*
5. $IT_M(S, \bar{x})$ *is a convex cone*
6. $T_M(S, \bar{x})$ *is the complement of a convex cone*
7. $IT_M(S, \bar{x}) \subseteq IR_M(S, \bar{x}) \subseteq IR(S, \bar{x}) \subseteq R(S, \bar{x}) \subseteq R_M(S, \bar{x}) \subseteq T_M(S, \bar{x})$
8. $IT_M(S, \bar{x}) \subseteq IT(S, \bar{x}) \subseteq IR(S, \bar{x}) \subseteq R(S, \bar{x}) \subseteq T(S, \bar{x}) \subseteq T_M(S, \bar{x})$
9. $\bar{x} \in \operatorname{cl} S \iff 0 \in T_M(S, \bar{x})$
10. $x \in \operatorname{int} S \iff 0 \in IT_M(S, \bar{x})$
11. $IT_M(\operatorname{int} S, \bar{x}) \supseteq IT_M(S, \bar{x})$
12. $T_M(\operatorname{cl} S, \bar{x}) \subseteq T_M(S, \bar{x})$
13. $IT_M(S_1 \cap S_2, \bar{x}) \supseteq IT_M(S_1, \bar{x}) \cap IT_M(S_2, \bar{x})$
14. $T_M(S_1 \cup S_2, \bar{x}) \subseteq T_M(S_1, \bar{x}) \cup T_M(S_2, \bar{x})$
15. $IT_M(S_1 \times S_2, (\bar{x}_1, \bar{x}_2)) = IT_M(S_1, \bar{x}_1) \times IT_M(S_2, \bar{x}_2)$
16. $T_M(S_1 \times S_2, (\bar{x}_1, \bar{x}_2)) \subseteq T_M(S_1, \bar{x}_1) \times T_M(S_2, \bar{x}_2)$.

Proof. The proof can be derived from the proof of Theorem 4.2.10. □

4.2.3 The Modified Adjacent and the Modified Interiorly Adjacent Cones

Now we investigate the following modified tangent cones of first-order.

Definition 4.2.7. Let X be a normed space, let S be a subset of X, and let $\bar{x} \in X$.

1. The **modified adjacent cone** $A_{\mathrm{M}}(S,\bar{x})$ of S at $\bar{x}$ is the collection of all $x \in X$ such that for every neighborhood $V(x)$, there are a neighborhood $V(\bar{x})$ and a scalar $t > 0$ satisfying that for every scalar $s \in (0,t)$ and for every element $\hat{x} \in V(\bar{x}) \cap S \cup \{\bar{x}\}$, there exists $\check{x} \in V(x)$ with $\hat{x} + s\check{x} \in S$.
2. The **modified interiorly adjacent cone** $IA_{\mathrm{M}}(S,\bar{x})$ of S at $\bar{x}$ is the collection of all $x \in X$ for which there exists a neighborhood $V(x)$ so that for every neighborhood $V(\bar{x})$ and every scalar $t > 0$, there are a scalar $s \in (0,t)$ and an element $\hat{x} \in V(\bar{x}) \cap S \cup \{\bar{x}\}$ such that for every $\check{x} \in V(x)$, we have $\hat{x} + s\check{x} \in S$.

The following result gives some characterizations of the above notion.

Theorem 4.2.8. *Let X be a normed space, let S be a subset of X, and let $\bar{x} \in \operatorname{cl} S$. For $i \in \{1,\ldots,7\}$, we have $A_M(S,\bar{x}) = C_i(S,\bar{x})$, where $C_i(S,\bar{x})$ are given as follows:*

$$C_1(S,\bar{x}) = \left\{ x \in X \mid \limsup_{\hat{x}\to\bar{x},\, t\downarrow 0} \frac{d(\hat{x}+tx,S) - d(\hat{x},S)}{t} = 0 \right\}$$

$$C_2(S,\bar{x}) = \{x \in X \mid \forall \{x_n\} \to \bar{x},\ \textit{with } \{x_n\} \subset \operatorname{cl} S,\ \forall \{t_n\} \downarrow 0,\ \exists \{z_n\} \to x \ \textit{such that } x_n + t_n z_n \in S\}$$

$$C_3(S,\bar{x}) = \left\{ x \in X \mid \lim_{t\to 0} \left[\liminf_{\lambda\to 0} \frac{d(\bar{x}+h+\lambda x), S}{\lambda} \right] \right\}$$

$$C_4(S,\bar{x}) = \bigcap_{U(0)} \bigcup_{\lambda\downarrow 0,\, U(\bar{x})} \bigcap_{x\in S\cap U(\bar{x}),\, t\in(0,\lambda)} \left(\frac{S-\bar{x}}{t} - U(\bar{x}) \right)$$

$$C_5(S,\bar{x} = \liminf_{t\downarrow 0,\, x \xrightarrow{S} \bar{x}} \frac{S-\bar{x}}{t}$$

$$C_6(S,\bar{x}) = \{x \in X \mid \forall\, \varepsilon > 0,\ \exists\, \alpha,\beta > 0,\ \forall\, y \in S \cap U_\alpha(\bar{x}),\ \forall h \in (0,\beta),\ \exists\, x \in U_\varepsilon(x) \ \textit{with } y + hx \in S\}$$

$$C_7(S,\bar{x}) = \left\{ x \in X \mid \lim_{\lambda\downarrow 0,\, \hat{x} \xrightarrow{S} \bar{x}} \frac{d(\hat{x}+\lambda x,S)}{\lambda} = 0 \right\}.$$

Proof. A proof of this result can be found in the excellent paper Giorgi and Guerraggio [208]. □

Remark 4.2.9. Definition 4.2.7 was proposed by [178]. However, Theorem 4.2.8 shows that when $\bar{x} \in \operatorname{cl} S$, $A_{\mathrm{M}}(S,\bar{x})$ coincides with the celebrated **Clarke's tangent cone** proposed by F. H. Clarke and used extensively in optimization and variational analysis. In fact, Giorgi and Guerraggio [208] show that if $\bar{x} \in \operatorname{cl} S$, then $A_{\mathrm{M}}(S,\bar{x})$ and $C_6(S,\bar{x})$ are the same, however, if $\bar{x} \notin \operatorname{cl} S$, then they are two different cones.

In fact, if $\bar{x} \notin \mathrm{cl}\,(S)$, then $C_6(S, \bar{x})$ coincides with X whereas $A_{\mathrm{M}}(S, \bar{x}) = \emptyset$. The notion $C_1(S, \bar{x})$, which was originally introduced by F. H. Clarke in a finite-dimensional setting, was adopted in general spaces by Hiriart-Urruty [262], who also proposed $C_2(S, \bar{x})$ and proved its equivalence with $C_1(S, \bar{x})$.

In the following $A_{\mathrm{M}}(S, \bar{x})$ and $IA_{\mathrm{M}}(S, \bar{x})$ will be denoted by $C(S, \bar{x})$ and $IC(S, \bar{x})$, respectively.

We give some features of the Clarke's tangent cone and the modified interiorly adjacent cone.

Theorem 4.2.10. *Let X be a normed space, let S, S_1, S_2 be subsets of X, and let $\bar{x}, \bar{x}_1, \bar{x}_2 \in X$. Then:*

1. $IC(S, \bar{x}) = X \backslash C(X \backslash S, \bar{x})$
2. $C(S, \bar{x}) = X \backslash IC(X \backslash S, \bar{x})$
3. $C(S, \bar{x})$ *is a closed cone*
4. $IC(S, \bar{x})$ *is an open cone*
5. $C(S, \bar{x})$ *is a convex cone*
6. $IC(S, \bar{x})$ *is the complement of a convex cone*
7. $\bar{x} \in \mathrm{cl}\, S \Longleftrightarrow 0 \in C(S, \bar{x})$
8. $\bar{x} \in \mathrm{int}\, S \Longleftrightarrow 0 \in IC(S, \bar{x})$
9. $IC(\mathrm{int}\, S, \bar{x}) = IC(S, \bar{x})$
10. $C(\mathrm{cl}\, S, \bar{x}) = C(S, \bar{x})$
11. $IR_M(S, \bar{x}) \subseteq C(S, \bar{x}) \subseteq A(S, \bar{x})$
12. $IA(S, \bar{x}) \subseteq IC(S, \bar{x}) \subseteq R_M(S, \bar{x})$
13. $C(S_1 \times S_2, (\bar{x}_1, \bar{x}_2)) = C(S_1, \bar{x}_1) \times C(S_2, \bar{x}_2)$
14. $IC(S_1 \times S_2, (\bar{x}_1, \bar{x}_2)) \subseteq IC(S_1, \bar{x}_1) \times IC(S_2, \bar{x}_2)$.

Proof. The assertions (1) through (4) follow directly from the definition. We will prove (5). Since $C(S, \bar{x})$ is a cone, to prove that it is convex, it suffices to show that for arbitrarily chosen elements $x_1, x_2 \in C(S, \bar{x})$, we have $x_1 + x_2 \in C(S, \bar{x})$. Let $U(x_1 + x_2)$ be an arbitrary neighborhood of $x_1 + x_2$. We will show that there are a neighborhood $V(\bar{x})$ and a scalar $t > 0$ such that for every element $\hat{x} \in V(\bar{x}) \cap S \cup \{\bar{x}\}$ and for every scalar $s \in (0, t)$, there exists $\check{x} \in U(x_1 + x_2)$ with $\hat{x} + s\check{x} \in S$, or equivalently,

$$\frac{S - \hat{x}}{s} \cap U(x_1 + x_2) \neq \emptyset. \tag{4.2}$$

Let $U(x_1)$ and $U(x_2)$ be neighborhoods of x_1 and x_2 such that

$$U(x_1) + U(x_2) \subseteq U(x_1 + x_2). \tag{4.3}$$

Since $x_1 \in C(S, \bar{x})$, there are a neighborhood $V_1(\bar{x})$ and a scalar $t_1 > 0$ such that

$$\frac{S - \hat{x}_1}{s_1} \cap U(x_1) \neq \emptyset \quad \text{for every } \hat{x}_1 \in V_1(\bar{x}) \cap S \cup \{\bar{x}\} \quad \text{for every } s_1 \in (0, t_1). \tag{4.4}$$

Analogously, since $x_2 \in C(S, \bar{x})$, there are a neighborhood $V_2(\bar{x})$ and a scalar $t_2 > 0$ such that

$$\frac{S - \hat{x}_2}{s_2} \cap U(x_2) \neq \emptyset \quad \text{for every } \hat{x}_2 \in V_2(\bar{x}) \cap S \cup \{\bar{x}\} \quad \text{for every } s_2 \in (0, t_2). \tag{4.5}$$

We choose a neighborhood $V(\bar{x})$ and a scalar $t > 0$ such that

$$V(\bar{x}) \subset V_1(\bar{x}), \tag{4.6a}$$

$$t \leq \min\{t_1, t_2\}, \tag{4.6b}$$

$$V(\bar{x}) + sU(x_1) \subset V_2(\bar{x}) \quad \text{for every } s \in (0, t). \tag{4.6c}$$

Then, for every $\hat{x} \in V(\bar{x}) \cap S \cup \{\bar{x}\}$ and for every $s \in (0, t)$, there exists a vector $\tilde{x}_1 \in U(x_1)$ such that $\hat{x} + s\tilde{x}_1 \in S$ and $\hat{x} + s\tilde{x}_1 \in V_2(x)$. Using (4.5), we confirm that there exists a vector $\tilde{x}_2 \in U(x_2)$ such that $\hat{x} + s\tilde{x}_1 + s\tilde{x}_2 \in S$. This, in view of (4.3), implies that $\tilde{x}_1 + \tilde{x}_2 \in U(x_1 + x_2)$, confirming that (4.2) holds. Therefore, $x_1 + x_2 \in C(S, \bar{x})$. This establishes the convexity of the cone $C(S, \bar{x})$.

Having proved (5), the assertion (6) is a direct consequence.

We next prove (7). Let $\bar{x} \in \text{cl } S$ be arbitrary. Then, for every $t > 0$, we have

$$S \cap (\bar{x} + tU(0)) \neq \emptyset, \tag{4.7}$$

where $U(0)$ is arbitrary. We set $t = 1$ and $V(\bar{x}) = \bar{x} + U(0)$, then pick $\hat{x} \in V(\bar{x}) \cap S \cup \{\bar{x}\}$ and $s \in (0, t)$.

Note that if $\hat{x} = \bar{x}$, then by (4.7), we have $s^{-1}(S - \bar{x}) \cap U(0) \neq \emptyset$. On the other hand, if $\bar{x} \neq \hat{x}$, then $\hat{x} \in S$ and we have $0 \in s^{-1}(S - \bar{x}) \cap U(0) \neq \emptyset$, confirming that $0 \in C(S, \bar{x})$. Therefore, we have shown that $\bar{x} \in \text{cl } S$ implies that $0 \in C(S, \bar{x})$. For the converse, we assume that $0 \in C(S, \bar{x})$. Then, for every $U(0)$, there exists a scalar $s > 0$ such that $s^{-1}(S - \hat{x}) \cap U(0) \neq \emptyset$, which confirms that $x \in \text{cl } S$.

A proof of (8) is a consequence of the duality relationship given in (1) and (2).

The proof of (9) and (10) are similar, so we will only prove (10). Let $x \in C(\text{cl } S, \bar{x})$ be arbitrary. Then for each neighborhood $U(x)$ there are a neighborhood $V(\bar{x})$ and a scalar $t > 0$ such that

$$(\hat{x} + sU(x)) \cap \text{cl } S \neq \emptyset \quad \text{for every } \hat{x} \in V(\bar{x}) \cap \text{cl } S \cup \{\bar{x}\}, \ \text{ for every } s \in (0, t).$$

Since the set $U(x)$ is open, the above expression implies that $(\hat{x} + sU(x)) \cap S \neq \emptyset$ for every $\hat{x} \in V(\bar{x}) \cap S \cup \{\bar{x}\}$, and for every $s \in (0, t)$, confirming that $x \in C(S, \bar{x})$. Therefore, we have shown that $C(\text{cl } S, \bar{x}) \subset C(S, \bar{x})$. To prove the converse inclusion, we pick an arbitrary $x \in C(S, \bar{x})$. Let $U(x)$ be a given neighborhood of x. Then, for $U_1(x) = 2^{-1}(U(x) - x) + x$, there are a neighborhood $V(\bar{x})$ and a scalar $t > 0$ such that $\hat{x} + sU_1(x) \cap S \neq \emptyset$ for every $\hat{x} \in V(\bar{x}) \cap \cup\{\bar{x}\}$ and for every $s \in (0, t)$.

We choose $\hat{x} \in V(\bar{x}) \cap \text{cl } S \cup \{\bar{x}\}$ and $s \in (0, t)$. Then, for each $U(0)$, there is an element $\check{x}$ such that $\check{x} \in (U(0) + \hat{x}) \cap (V(\bar{x}) \cap S \cup \{\bar{x}\})$ and $(\check{x} + sU(x)) \cap S \neq \emptyset$, which implies that $(\hat{x} + sU_1(x) + U(0)) \cap S \neq \emptyset$ for every $U(0)$, and $(\hat{x} + sU(x)) \cap \text{cl } S \supset \text{cl } (\hat{x} + sU_1(x)) \cap \text{cl } S \neq \emptyset$, confirming that $x \in C(\text{cl } S, \bar{x})$.

Properties (11) and (12) follow from the definitions of the tangent cones involved.

We next prove (13). Let $(x_1, x_2) \in C(S_1 \times S_2, (\bar{x}_1, \bar{x}_2))$ be arbitrary. Therefore, for every neighborhood $U(x_1)$ and $U(x_2)$, there are neighborhoods $V(\bar{x}_1)$ and $V(\bar{x}_2)$ and a scalar $t > 0$ satisfying that for every scalar $s \in (0, t)$ and for every element $(\hat{x}_1, \hat{x}_2) \in V(\bar{x}_1) \times V(\bar{x}_2) \cap S \cup \{(\bar{x}_1, \bar{x}_2)\}$ there exist $\check{x}_1 \in U(x_1)$ and $\check{x}_2 \in U(x_2)$ with $(\hat{x}_1, \hat{x}_2) + s(\check{x}_1, \check{x}_2) \in S_1 \times S_2$. In other words, for every neighborhood $U(x_1)$ and $U(x_2)$ there are neighborhoods $V(\bar{x}_1)$ and $V(\bar{x}_2)$ and a scalar $t > 0$ satisfying that, for every scalar $s \in (0, t)$ and for every element $\hat{x}_1 \in V_1(\bar{x}_1) \cap S_1 \cup \bar{x}_1$ and $\hat{x}_2 \in V_2(\bar{x}_2) \cap S_2 \cup \bar{x}_2$ there exist $\check{x}_1 \in U(x_1)$ and $\check{x}_2 \in U(x_2)$ satisfying that $\hat{x}_1 + s\check{x}_1 \in S_1$ and $\hat{x}_2 + s\check{x}_2 \in S_2$, confirming that $(x_1, x_2) \in C(S_1, \bar{x}_1) \times C(S_2, \bar{x}_2)$. This proves that $C(S_1 \times S_2, (\bar{x}_1, \bar{x}_2)) \subseteq C(S_1, \bar{x}_1) \times C(S_2, \bar{x}_2)$. The converse inclusion follows by reversing the arguments.

Finally, analogous arguments lead to a proof of property (14). □

In the following, we give a few simple examples of the Clarke tangent cone.

Example 4.2.11. For $S \subset \mathbb{R}^2$ given by $S = \{(x, y) | \, x^2 + |x| + y = 0\}$, the Clarke's tangent cone of S at $(0, 0)$ is given by $C(S, (0, 0)) = \{(0, 0)\}$. □

Remark 4.2.12. The above example depicts a situation in which the Clarke's tangent cone is trivially convex and the contingent cone is nonconvex (See Example 4.1.17).

Example 4.2.13. For $S \subset \mathbb{R}^2$ given by $S = \{(x, y) | \; y - \log(|x| + 1) \leq 0\}$, the Clarke's tangent cone of S at $(0, 0)$ is given by $C(S, (0, 0)) = \{(x, y) | \; y + |x| \leq 0\}$. □

The convexity of the Clarke's tangent cone is obtained by varying the direction x. Unfortunately, as a consequence, the isotone property is lost, as shown below:

Example 4.2.14. For $X = \mathbb{R}^2$, we define two sets S_1 and S_2 in X by $S_1 = \left\{(x, y) \in \mathbb{R}^2 | \; y \geq x^2\right\}$ and $S_2 = \left\{(x, y) \in \mathbb{R}^2 | \; y - x^2 \sin\left(\frac{1}{x}\right) \geq 0\right\} \bigcup \{(0, 0)\}$. Then $S_1 \subset S_2$, and $C(S_1, (0, 0)) = \mathbb{R} \times \mathbb{R}_+$, and $C(S_2, (0, 0)) = \{(x, y) | \; y \geq |x|\}$, showing that $C(S_1, (0, 0)) \not\subset C(S_2, (0, 0))$. □

4.3 Miscellaneous Properties of First-Order Tangent Cones

In this section, we give some relationships among the contingent cone, the adjacent cone, and the Clarke's cone which are among the most commonly used tangent cones. We also give some formulae based on a combination of different cones, and give a few additional examples.

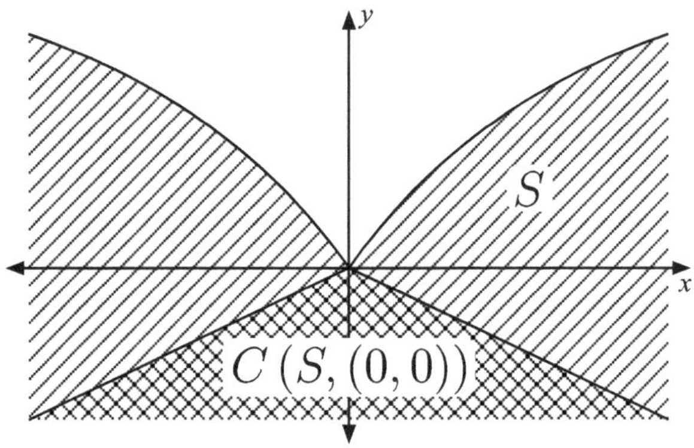

Fig. 4.8 The Clarke's tangent cone in Example 4.2.13

We begin with the following interesting example (see [267]).

Example 4.3.1. For $S \subset \mathbb{R}^2$ defined by $S = S_1 \cup S_2 \cup S_3$, where

$$S_1 = \left\{(x_1, x_2) \in \mathbb{R}^2 |\ (x_1+1)^2 + x_2^2 = 1,\ x_1 \in [-2, 0],\ x_2 \le 0\right\},$$
$$S_2 = \left\{(x_1, x_2) \in \mathbb{R}^2 |\ -2 \le x_1 \le 0\right\},$$
$$S_3 = \left\{(x_1^{-1}, 0) \in \mathbb{R}^2 |\ x_1 \in \mathbb{N}^*\right\},$$

we have

$$T(S, (0,0)) = \left\{(x_1, x_2) \in \mathbb{R}^2 |\ x_1 = 0,\ x_2 \le 0\right\} \bigcup \left\{(x_1, x_2) \in \mathbb{R}^2 |\ x_2 = 0\right\},$$
$$A(S, (0,0)) = \left\{(x_1, x_2) \in \mathbb{R}^2 |\ x_1 = 0,\ x_2 \le 0\right\} \bigcup \left\{(x_1, x_2) \in \mathbb{R}^2 |\ x_1 \le 0,\ x_2 = 0\right\},$$
$$C(S, (0,0)) = \{(0,0)\},$$

which shows that all the three cones, in general, are different. □

To obtain the desired features of these cone the following notions have been introduced:

Definition 4.3.2. Let X be a normed space, let $S \subset X$ be nonempty, and let $\bar{x} \in S$. The set S is said to be **tangentially regular** at $\bar{x}$, if $T(S, \bar{x}) = C(S, \bar{x})$, that is, the contingent cone and the Clarke's tangent cone coincide. Moreover, the set S is said to be **derivable** at $\bar{x}$, if $T(S, \bar{x}) = A(S, \bar{x})$, that is, the contingent cone and the adjacent cone coincide.

Example 4.3.3. For a set $S \subset \mathbb{R}^2$ given by $S := \{(x, y) \in \mathbb{R}^2 |\ x^2 + y^2 \le 2|x|\}$, we have $T(S, (0,0)) = \mathbb{R}^2$, however, $C(S, (0,0)) = \{(0,0)\}$. This shows that the tangential regularity concept is not related to the convexity of the contingent cone (see [60]). □

We have the following useful result which is quite easy to prove:

Proposition 4.3.4. *Let X and Y be normed spaces, let S be a subset of X, and let $\bar{x} \in S$ be arbitrary. Assume that $\ell : X \to Y$ is a linear and continuous map. Then the following inclusions hold:*

$$\ell(T(S,\bar{x})) \subset T(\ell(S),\ell(\bar{x})), \tag{4.8}$$

$$\ell(A(S,\bar{x})) \subset A(\ell(S),\ell(\bar{x})). \tag{4.9}$$

Our next result is a collection of intersection formulas relating a few cones (see [577]):

Theorem 4.3.5. *Let X be a normed space, let S_1 and S_2 be two nonempty subsets of X, and let $\bar{x} \in \operatorname{cl} S_1 \cap \operatorname{cl} S_2$. Then the following relationships hold:*

1. $IT(S_1,\bar{x}) \cap IA(S_2,\bar{x}) \subseteq IA(S_1 \cap S_2,\bar{x})$
2. $IT(S_1,\bar{x}) \cap A(S_2,\bar{x}) \subseteq A(S_1 \cap S_2,\bar{x})$
3. $IT(S_1,\bar{x}) \cap T(S_2,\bar{x}) \subseteq T(S_1 \cap S_2,\bar{x})$
4. $IA(S_1,\bar{x}) \cap A(S_2,\bar{x}) \subseteq T(S_1 \cap S_2,\bar{x})$.

Proof. The proof can be obtained by using the definition of the cones involved. □

We have another simple result connecting the three cones:

Theorem 4.3.6. *Let X be a normed space, let S be a nonempty subset of X, and let $\bar{x} \in \operatorname{cl} S$. Then the following inclusions hold:*

1. $T(S,\bar{x}) + C(S,\bar{x}) \subset T(S,\bar{x})$
2. $A(S,\bar{x}) + C(S,\bar{x}) \subset A(S,\bar{x})$
3. $IT_M(S,\bar{x}) + C(S,\bar{x}) \subset IT_M(S,\bar{x})$.

Proof. Let $x \in T(S,\bar{x})$ and $z \in C(S,\bar{x})$ be arbitrary. Then, by the definition of the contingent cone, there are sequences $\{t_n\} \subset \mathbb{P}$ and $\{x_n\} \subset X$ with $t_n \downarrow 0$ and $x_n \to z$ such that $y_n := \bar{x} + t_n x_n \in S$ for every $n \in \mathbb{N}$. Clearly, $z_n \to \bar{x}$. We recall that due to the containment $z \in C(S,\bar{x})$ for every $\{y_n\} \subset X$ with $y_n \stackrel{S}{\to} x$ and for every $\{s_n\} \subset \mathbb{P}$ with $s_n \to 0$, there exists $\{z_n\}$ with $z_n \to z$ such that $y_n + s_n z_n \in S$ for every $n \in \mathbb{N}$. This implies that $\bar{x} + t_n(x_n + z_n) = y_n + t_n z_n \in S$, confirming that $x + z \in T(S,\bar{x})$. This proves the first inclusion. The proof of the second and the third inclusions can be given analogously (see also Theorem 4.12.8). □

We have the following useful consequence of the above result:

Corollary 4.3.7. *Let X be a normed space, let $S \subset X$, and let $\bar{x} \in S$. If $IT_M(S,\bar{x}) \neq \emptyset$, then*

$$IT_M(S,\bar{x}) = \operatorname{int}(C(S,\bar{x}))\,.$$

Remark 4.3.8. A set S is called **epi-Lipschitzian** at $\bar{x}$, if $IT_M(S,\bar{x}) \neq \emptyset$. Rockafellar [492] showed that in a finite dimensional setting any set S is epi-Lipschitzian if and only if $\operatorname{int}(C(S,\bar{x})) \neq \emptyset$. He also gave an example to show that in general spaces, the condition $\operatorname{int}(C(S,\bar{x})) \neq \emptyset$ does not imply that S is epi-Lipschitzian.

Several authors have studied the relationship between the Clarke's tangent cone and limit inferior of contingent cones at neighboring points. An important contribution was made by Penot [461] (see also Cornet [113], Wang [589]) who showed that for closed sets in finite dimensional setting the Clarke's cone is equal to the limit inferior of contingent cones. Treiman [574] extended this result for epi-Lipschitzian closed sets in Banach spaces. Many related results and their consequences can also be found in [18, 267].

In the following, we give a result (without proof) by Borwein and Strojwas [70, Theorem 4.1]:

Theorem 4.3.9. *Let X be a locally convex Hausdorff space, let $S \subset X$ be compactly epi-Lipschitzian at $\bar{x} \in S$. Suppose that one of the following assumptions holds:*

1. *S is tangentially regular at $\bar{x}$.*
2. *S is closed relative to some neighborhood of $\bar{x}$ and X is a Banach space.*
3. *S is weakly compact relative to some neighborhood of $\bar{x}$ and X is a normed space.*

Then:

$$\liminf_{x \xrightarrow{S} \bar{x}} T(S,x) = C(S,\bar{x}).$$

4.4 First-Order Tangent Cones on Convex Sets

On convex sets, the tangent cones of first-order have simpler structure and nice properties. The highlight of this section is the relationship among the tangent cones on convex sets.

4.4.1 Connections Among First-Order Tangent Cones on Convex Sets

We recall that in Sect. 4.1 and Sect. 4.2, we studied the following tangent cones:

$$\begin{aligned} R(S,\bar{x}) &= \{x \in X \mid \forall\, t > 0\ \exists\, s \in (0,t) \text{ such that } \bar{x} + sx \in S\}, \\ IR(S,\bar{x}) &= \{x \in X \mid \exists\, t > 0\ \forall\, s \in (0,t) \text{ we have } \bar{x} + sx \in S\}, \end{aligned}$$

$$
\begin{aligned}
T(S,\bar{x}) &= \{x \in X \mid \forall\, U(x)\ \forall\, t > 0\ \exists\, s \in (0,t)\ \exists\, \hat{x} \in U(x) \text{ such that} \\
&\qquad \bar{x} + s\hat{x} \in S\}, \\
IT(S,\bar{x}) &= \{x \in X \mid \exists\, U(x)\ \exists\, t > 0\ \forall\, s \in (0,t)\ \forall\, z \in U(x) \text{ we have } \bar{x} + sz \in S\}, \\
A(S,\bar{x}) &= \{x \in X \mid \forall\, U(x)\ \exists\, t > 0\ \forall\, s \in (0,t)\ \exists\, \hat{x} \in U(x) \text{ such that} \\
&\qquad \bar{x} + s\hat{x} \in S\}, \\
IA(S,\bar{x}) &= \{x \in X \mid \exists\, U(x)\ \forall\, t > 0\ \exists\, s \in (0,t)\ \forall\, \hat{x} \in U(x) \text{ we have} \\
&\qquad \bar{x} + s\hat{x} \in S\}, \\
R_{\mathrm{M}}(S,\bar{x}) &= \{x \in X \mid \forall\, V(\bar{x})\ \forall\, t > 0\ \exists\, s \in (0,t)\ \exists\, \hat{x} \in V(\bar{x}) \cap S \cup \{\bar{x}\} \text{ such that} \\
&\qquad \hat{x} + sx \in S\}, \\
IR_{\mathrm{M}}(S,\bar{x}) &= \{x \in X \mid \exists\, V(\bar{x})\ \exists\, t > 0\ \forall\, s \in (0,t)\ \forall\, \hat{x} \in V(\bar{x}) \cap S \cup \{\bar{x}\} \text{ we have} \\
&\qquad \hat{x} + sx \in S\}, \\
T_{\mathrm{M}}(S,\bar{x}) &= \{x \in X \mid \forall\, V(x)\ \forall\, V(\bar{x})\ \forall\, t > 0\ \exists\, s \in (0,t) \exists\, \hat{x} \in V(\bar{x}) \cap S \cup \{\bar{x}\} \\
&\qquad \exists\, \breve{x} \in V(x) :\ \hat{x} + s\breve{x} \in S\}, \\
IT_{\mathrm{M}}(S,\bar{x}) &= \{x \in X \mid \exists\, V(x)\ \exists\, V(\bar{x})\ \exists\, t > 0\ \forall\, s \in (0,t)\ \forall\, \hat{x} \in V(\bar{x}) \cap S \cup \{\bar{x}\}\ \forall \\
&\qquad \breve{x} \in V(x) \text{ we get } \hat{x} + s\breve{x} \in S\}, \\
C(S,\bar{x}) &= \{x \in X \mid \forall\, U(x)\ \exists\, V(\bar{x})\ \exists\, t > 0\ \forall\, s \in (0,t)\ \forall\, \hat{x} \in V(\bar{x}) \cap S \cup \{\bar{x}\}\ \exists \\
&\qquad \breve{x} \in U(x) :\ \hat{x} + s\breve{x} \in S\}, \\
IC(S,\bar{x}) &= \{x \in X \mid \exists\, U(x)\ \forall\, V(\bar{x})\ \forall\, t > 0\ \exists\, s \in (0,t)\ \exists\, \hat{x} \in V(\bar{x}) \cap S \cup \{\bar{x}\}\ \forall \\
&\qquad \breve{x} \in V(x) :\ \hat{x} + s\breve{x} \in S\}.
\end{aligned}
$$

The following relationships hold among the above cones for $\bar{x} \in \mathrm{cl}\,(S) \subset X$:

$$
\begin{array}{ccccccc}
IT_{\mathrm{M}}(S,\bar{x}) & \subseteq & IT(S,\bar{x}) & \subseteq & IA(S,\bar{x}) & \subseteq & IC(S,\bar{x}) \\
\cap\!\!\!| & & \cap\!\!\!| & & \cap\!\!\!| & & \cap\!\!\!| \\
IR_{\mathrm{M}}(S,\bar{x}) & \subseteq & IR(S,\bar{x}) & \subseteq & R(S,\bar{x}) & \subseteq & R_{\mathrm{M}}(S,\bar{x}) \\
\cap\!\!\!| & & \cap\!\!\!| & & \cap\!\!\!| & & \cap\!\!\!| \\
C(S,\bar{x}) & \subseteq & A(S,\bar{x}) & \subseteq & T(S,\bar{x}) & \subseteq & T_{\mathrm{M}}(S,\bar{x}),
\end{array}
$$

where the cones in the first row are open and the cones in the third row are closed.

It turns out that this relationship simplifies significantly on convex sets. The following result specifies the connections among various tangent cones of first-order.

Theorem 4.4.1. *Let X be a normed space, let $S \subset X$ be convex, and let $x \in \operatorname{cl} S$. Then:*

1. $IT_M(S,\bar{x}) = IT(S,\bar{x}) = IA(S,\bar{x}) = \operatorname{cone}(\operatorname{int} S - x)$.
2. $IR(S,\bar{x}) = R(S,\bar{x}) = \operatorname{cone}(S - \bar{x})$.
3. $C(S,\bar{x}) = A(S,\bar{x}) = T(S,\bar{x}) = \operatorname{cl}\operatorname{cone}(S - \bar{x})$.

Proof. We begin by showing that when the set S is convex, the following inclusions hold

$$IT(S,\bar{x}) = IA(S,\bar{x}), \tag{4.10}$$

$$IR(S,\bar{x}) = R(S,\bar{x}), \tag{4.11}$$

$$A(S,\bar{x}) = T(S,\bar{x}). \tag{4.12}$$

Let $x \in IA(S,\bar{x})$ be arbitrary. Then there are a neighborhood $U(x)$ and a scalar $t > 0$ such that $\bar{x}+tU(x) \subseteq S$. Since the set S is convex, it follows that $\bar{x}+sU(x) \subset S$ for every $s \in (0,t)$ which confirms that $x \in IT(S,\bar{x})$. The converse inclusion follows from the properties mentioned above. This proves (4.10). Identity (4.11) can be shown is an analogous manner.

Let $x \in T(S,\bar{x}) = T(\operatorname{cl} S,\bar{x})$. Then for each neighborhood $U(x)$, there exists a scalar $t > 0$ such that $(\bar{x} + tU(x)) \cap \operatorname{cl} S \neq \emptyset$, which, in view of the convexity, of the set S, yields $(\bar{x} + sU(x)) \cap \operatorname{cl} S \neq \emptyset$ for every $s \in (0,t)$, proving that $x \in A(\operatorname{cl} S,\bar{x}) = A(S,\bar{x})$. Since $A(S,\bar{x}) \subseteq T(S,\bar{x})$, we proved (4.12).

Next we claim that

$$IT(S,\bar{x}) = IT_{\mathrm{M}}(S,\bar{x}), \tag{4.13}$$

$$A(S,\bar{x}) = C(S,\bar{x}). \tag{4.14}$$

Let $x \in IT(S,\bar{x})$. Then, there are a neighborhood $U(x)$ and a scalar $t > 0$ such that for every $s \in (0,t]$, we have $\bar{x} + sU(x) \subseteq S$. For some $\theta \in (0,1)$, we define $V_1(\bar{x}) := \bar{x} + t\theta(U(x) - x)$, and $U_1(x) := x + (1-\theta)(U(x) - x)$. Since $U(x)$ is convex, we have

$$\begin{aligned} V_1(\bar{x}) + tU_1(x) &= \bar{x} + t\theta(U(x) - x) + tx + t(1-\theta)(U(x) - x) \\ &\subseteq \bar{x} + tx + t(U(x) - x) = \bar{x} + tU(x) \subseteq S, \end{aligned}$$

and using the convexity of the set S, we have $V_1(\bar{x}) \cap S \bigcup \{\bar{x}\} + sU_1(x) \subset S$ for every, $s \in (0,t]$. Therefore, $y \in IC(S,\bar{x})$, proving that $IT(S,\bar{x}) \subset IT_{\mathrm{M}}(S,\bar{x})$. Due to $IT(S,\bar{x}) \supset IT_{\mathrm{M}}(S,\bar{x})$, we have proven (4.13).

For (4.14), we assume that $y \in A(S,\bar{x})$. Let $U(x)$ be an arbitrary neighborhood of x. For some $\theta \in (0,1)$, we define $U_1(x) := x + (1-\theta)(U(x) - x)$. By the definition of cone $A(S,\bar{x})$, there exists a scalar $t > 0$ such that $(x+sU_1(x)) \cap S \neq \emptyset$ for every $s \in (0,t]$. We define a neighborhood $V_1(\bar{x})$ by $V_1(\bar{x}) := x + s\theta(U(x) - x)$. Then, for every $\hat{x} \in V_1(\bar{x})$, we have

$$(\hat{x} + tU(x)) \cap S \supseteq (\hat{x} + t\theta(U(x) - x) + tx + t(1-\theta)(U(x) - x))$$
$$\cap S \supseteq (\bar{x} + tU_1(x)) \cap S \neq \emptyset,$$

and using the convexity of the set S once again, we deduce that $(\hat{x} + sU(x)) \cap S \neq \emptyset$ for every $\hat{x} \in V_1(\bar{x}) \cap S \cup \{\bar{x}\}$, and for every $s \in (0,t]$, which proves that $x \in IC(S,\bar{x})$. We have shown that $IA(S,\bar{x}) \subset IC(S,\bar{x})$, and since the converse is always true, we established (4.14).

Finally, for $\bar{x} \in \text{cl } S$, we claim that

$$IR(S,\bar{x}) = \text{cone}\,(S - \bar{x}), \tag{4.15}$$
$$IT(S,\bar{x}) = \text{cone}\,(\text{int } S - \bar{x}), \tag{4.16}$$
$$T(S,\bar{x}) = \text{cl}\,\text{cone}\,(S - \bar{x}). \tag{4.17}$$

The identity (4.15) is a consequence of the definition. Therefore, we have

$$IT(S,\bar{x}) = IT(\text{int } S,\bar{x}) \subset IR(\text{int } S,\bar{x}) = \text{cone}(\text{int } S - \bar{x}).$$

Let $x \in \text{cone}(\text{int } S - \bar{x})$. Then there exists a scalar $t > 0$ and an element $\hat{x} \in \text{int } S$ such that $x = t(\hat{x} - \bar{x})$. Therefore, for $U = t(\text{int } S - \bar{x})$, which is an open neighborhood of x, we have $\bar{x} + sU = \bar{x} + st(\text{int } S - \bar{x}) \subseteq S$ for every $s \in \left(0, t^{-1}\right]$, confirming that $x \in IT(S,\bar{x})$. Therefore, we have shown that $IT(S,\bar{x}) \supset \text{cone}(\text{int } S - \bar{x})$, and this proves (4.16).

For (4.17), we note that due to (4.15), we get $T(S,\bar{x}) \supseteq \text{cl}\,IR(S,\bar{x}) = \text{cl}\,\text{cone}\,(S - \bar{x})$. To prove the converse, we assume that $x \in T(S,\bar{x})$. Then, for any neighborhood $U(x)$, there exists a scalar $t > 0$ such that $(\bar{x} + tU(x)) \cap S \neq \emptyset$, implying $U(x) \cap \text{cone}(S - \bar{x}) \neq \emptyset$, and hence $x \in \text{cl}\,\text{cone}\,(S - \bar{x})$. □

Let us now recall another notion of a cone:

Definition 4.4.2. Let X be a normed space, let $S \subset X$, and let $\bar{x} \in S$. The **closed radial cone** $\bar{R}(S,\bar{x})$ is the collection of all $x \in X$ for which there are sequences $\{t_n\} \subset \mathbb{P}$ and $\{x_n\} \subset X$ such that $\bar{x} + t_n x_n \in S$.

Remark 4.4.3. Taa [548] traces this notion back to Bouligand [74]. It is clear that $T(S,\bar{x}) \subset \bar{R}(S,\bar{x})$. Furthermore, $T(S,\bar{x}) = \bar{R}(S,\bar{x})$, if the set S is convex, justifying its name (see (4.11) and (4.12)).

Our objective now is to study cones generated by taking a conical hull of set. The importance of such cones and their connection to tangent cones is evident from Theorem 4.4.1.

Let X be a linear topological space and let K be a convex cone with nonempty interior. Let $\bar{x} \in X$ be a point. By $K(\bar{x})$, we denote the closure of the convex hull of $K + \bar{x}$, that is,

$$K(\bar{x}) = \text{cl}\,\text{cone}\,(K + \bar{x}) = \text{cl}\,\{\lambda\,(k + \bar{x} | \, k \in K,\ \lambda \geq 0)\}.$$

Clearly, cone $(K + \bar{x}) \subset K + \{\lambda \bar{x} | \lambda \geq 0\}$. In particular, if $\bar{x} \in -K$, then the equality holds, or

$$K(\bar{x}) = \text{cl}\ (K + \{\lambda \bar{x} | \lambda \geq 0\}) \quad \text{whenever } \bar{x} \in -K. \tag{4.18}$$

We can repeat the above construction. With another element $\bar{z} \in X$, we denote by $K(\bar{x}, \bar{z})$ the closure of the conical hull of $K(\bar{x}) + \bar{z}$, that is,

$$K(\bar{x}, \bar{z}) = \text{cl}\,\text{cone}\ (K(\bar{x}) + \bar{z}) = \text{cl}\ \{\lambda\ (k(\bar{x}) + \bar{z}) \mid k(\bar{x}) \in K(\bar{x}),\ \lambda \geq 0\}.$$

We have the following result (see [45]):

Proposition 4.4.4. *Let $\bar{x} \in -K$ and let $\bar{z} \in -K(\bar{x})$. Then*

$$K(\bar{x}, \bar{z}) = \text{cl}\ (K + \{\lambda \bar{x} + \mu \bar{z} | \lambda \geq 0,\ \mu \geq 0\}).$$

Proof. We set $\{\bar{x}\} := \{\lambda \bar{x} | \lambda \geq 0, \}$ and $\{\bar{z}\} := \{\lambda \bar{z} | \lambda \geq 0\}$, and apply (4.18) twice to obtain that

$$K(\bar{x}, \bar{z}) = \text{cl}\ (K(\bar{x}) + \bar{z}) = \text{cl}\ (\text{cl}\ (K + \bar{x}) + \bar{z}). \tag{4.19}$$

Now, setting $A := K + \{\bar{x}\} + \{\bar{z}\}$, we have $\text{int}(A) \subset \text{cl}\,(K + \{\bar{x}\}) + \{\bar{z}\} \subset \text{cl}\,(A)$. Noting that A is convex, and by the fact that K is solid, A is solid as well. Using $\text{cl}\,(\text{int}(A)) = \text{cl}\,(A)$, we deduce that $\text{cl}\,(\text{cl}\,(K + \{\bar{x}\}) + \{\bar{z}\}) = \text{cl}\,(K + \{\bar{x}\} + \{\bar{z}\})$. This proves (4.19) and the proof is complete. □

We have the following important result based on the above calculation (see [45]):

Proposition 4.4.5. *Let W be a topological space and let $g : W \to \mathbb{R}^n$. We set $K = \mathbb{R}^n_+$ and define $I(\bar{x}) := \{i \in \{1, 2, \ldots, n\} | g_i(\bar{x}) = 0)\}$ and $J(\bar{x}, d) := \{i \in I(\bar{x}) | g_i'(\bar{x})(d) = 0\}$, where $d \in W$. Then:*

1. *If $g(\bar{x}) \in -K$, then*

$$K(g(\bar{x})) = \Pi_{i=1}^n S_i \quad \text{where } S_i = \begin{cases} \mathbb{R}_+ & \text{if } i \in I(\bar{x}) \\ \mathbb{R} & \text{otherwise.} \end{cases}$$

2. *If $g(\bar{x}) \in -K$, and $g'(\bar{x})(d) \in -K(g(\bar{x}))$, then*

$$K(g(\bar{x}), g(\bar{x})d) = \Pi_{i=1}^n S_i \quad \text{where } S_i = \begin{cases} \mathbb{R}_+ & \text{if } i \in J(\bar{x}, d) \\ \mathbb{R} & \text{otherwise.} \end{cases}$$

The first part of this result says that $g'(\bar{x})(d) \in -K(g(\bar{x}))$ if and only if $g_i'(\bar{x})(d) \leq 0$ for $i \in I(\bar{x})$.

We conclude this section by another result concerning the interior of the conical hulls:

Proposition 4.4.6. *Let $\bar{x} \in -K$ and let $\bar{z} \in -K(\bar{x})$. Then,*

$$\operatorname{int}(K(\bar{x})) = \operatorname{int}(K) + \{\lambda\bar{x} \mid \lambda \geq 0\}, \tag{4.20}$$

$$\operatorname{int}(K(\bar{x}, \bar{z})) = \operatorname{int}(K) + \{\lambda\bar{x} + \mu\bar{z} \mid \lambda \geq 0,\ \mu \geq 0\}. \tag{4.21}$$

Proof. By (4.18), cone $(K + \bar{x}) = K + \{\lambda\bar{x} \mid \lambda \geq 0\}$. It is known that $\operatorname{int}(K(\bar{x})) = \operatorname{int}(K) + \{\lambda\bar{x} \mid \lambda \geq 0\}$, see [628, Theorem 2.2]. Therefore, (4.20) follows from the observation that $\operatorname{int}\operatorname{cl} A = \operatorname{int} A$, where A is a solid convex set. Using (4.20) we deduce (4.21). The proof is complete. □

4.4.2 Properties of First-Order Tangent Cones on Convex Sets

We begin by recalling that given a nonempty set S of a normed space X with X^* as the topological dual of X, the normal cone to S at $\bar{x} \in S$ is given by:

$$N(\bar{x}, S) = (S - \bar{x})^{-} = \{x^* \in X^* \mid \langle x^*, x - \bar{x}\rangle \leq 0, \text{ for every } x \in S\}.$$

The following well-known result computes the contingent cone on a constraint set.

Theorem 4.4.7. *Let X and Y be Banach spaces, let $L \subset X$ and $M \subset Y$ be closed and convex, let $A : X \to Y$ be linear and continuous, and let $K := \{x \in L \mid A(x) \in M\} = L \cap A^{-1}(M)$ be nonempty, that is, $0 \in A(L) - M$. Then for any $\bar{x} \in K$, we have*

$$T(K, \bar{x}) \subseteq T(L, \bar{x}) \cap A^{-1}(T(M, A(\bar{x}))),$$

$$N(\bar{x}, K) \supset N(\bar{x}, L) + A^* N(A(\bar{x}), M).$$

If we assume that

$$0 \in \operatorname{int}(M - A(L)), \tag{4.22}$$

then for every $\bar{x} \in K = A^{-1}(M) \cap L$, the following intersection formulas hold:

$$T(K, \bar{x}) = T(L, \bar{x}) \cap A^{-1}(T(M, A(\bar{x}))),$$

$$N(\bar{x}, K) = N(\bar{x}, L) + A^* N(A(\bar{x}), M).$$

Proof. See [18]). □

The following two corollaries are immediate:

Corollary 4.4.8. *Let X and Y be Banach spaces, let $L \subset X$ be closed and convex, and let $A : X \to Y$ be linear and continuous. Then for any $\bar{y} \in \operatorname{int}(A(L))$ and for any $\bar{x} \in L \cap A^{-1}(\bar{y})$, we have*

$$T(L \cap A^{-1}(\bar{y}), \bar{x}) = T(L, \bar{x}) \cap \operatorname{Ker} A;$$

$$N(\bar{x}, L \cap A^{-1}(\bar{y})) = N(\bar{x}, L) + \operatorname{Im} A^*.$$

Corollary 4.4.9. *Let X and Y be Banach spaces and let $K \subset X$ and $L \subset Y$ be closed and convex. If $0 \in \operatorname{int}(K - L)$, then for any $\bar{x} \in K \cap L$, we have*

$$T(K \cap L, \bar{x}) = T(K, \bar{x}) \cap T(L, \bar{x}),$$

$$N(\bar{x}, K \cap L) = N(\bar{x}, K) + N(\bar{x}, L).$$

The constraint qualification (4.22) plays an important role in the proof. However, there are instances when this condition is violated but the intersection formula remains valid. This situation is depicted in the following example given by Křivan [341]:

Example 4.4.10. For $X = Y = \mathbb{R}^2$, let K be a triangle with vertices $(0,0)$, $(1,1)$ and $(-1,1)$. By choosing $L = -K$ and $A(x) = x$, we observe that $0 \notin \operatorname{int}(K - L)$. However, the intersection formula holds:

$$T(K, (0,0)) \cap T(L, (0,0)) = T(K \cap L, (0,0)).$$

In the following, our objective is to give an extension of Theorem 4.4.7 by relaxing the condition (4.22). This extension is motivated by an interesting work of Křivan [341], who provided a sufficient condition in terms of the support functions for the contingent cone to the intersection of two closed and convex subsets of a Hilbert space to be equal to the intersection of contingent cones to these sets. He also showed that for Hilbert spaces the surjectivity condition, which is essential to prove the same formula in the context of reflexive Banach spaces, follows from his condition (see Aubin and Cellina [15] and Aubin and Ekeland [16]).

We recall the following result:

Theorem 4.4.11. *Let X, Y be separated locally convex spaces, $f \in \Lambda(X)$, $g \in \Lambda(Y)$, and $A \in \mathscr{L}(X, Y)$ be such that $A(\operatorname{dom} f) \cap \operatorname{dom} g \neq \emptyset$. If*

$$(f + g \circ A)^*(x^*) = \min\{f^*(x^* - A^* y^*) + g^*(y^*) \mid y^* \in Y^*\}, \quad \forall x^* \in X^*, \tag{4.23}$$

then

$$\partial(f + g \circ A)(x) = \partial f(x) + A^*(\partial g(Ax)), \quad \forall x \in X. \tag{4.24}$$

Proof. The proof can be found in Hirriart-Urruty [263]. We provide a proof for completeness.

Clearly, $f + g \circ A \in \Lambda(X)$. The inclusion $\supset$ in (4.24) holds always.

Let $x^* \in \partial(f + g \circ A)(x)$. Then $x \in \operatorname{dom}(f + g \circ A) = \operatorname{dom} f \cap A^{-1}(\operatorname{dom} g)$ and

$$(f + g \circ A)(x) + (f + g \circ A)^*(x^*) = \langle x, x^* \rangle .$$

From (4.23) we get $y^* \in Y^*$ $(f + g \circ A)^*(x^*) = f^*(x^* - A^* y^*) + g^*(y^*)$. It follows that

$$\left[f(x) + f^*(x^* - A^* y^*) - \left\langle x, x^* - A^* y^* \right\rangle\right] + \left[g(Ax) + g^*(y^*) - \left\langle Ax, y^* \right\rangle\right] = 0.$$

Since $f(x) + f^*(x^* - A^* y^*) - \langle x, x^* - A^* y^* \rangle \geq 0$ and $g(Ax) + g^*(y^*) - \langle Ax, y^* \rangle \geq 0$ (by the Young–Fenchel inequality), we obtain that $f(x) + f^*(x^* - A^* y^*) - \langle x, x^* - A^* y^* \rangle = 0$ and $g(Ax) + g^*(y^*) - \langle Ax, y^* \rangle = 0$, and so $x^* - A^* y^* \in \partial f(x)$ and $y^* \in \partial g(Ax)$, whence $x^* = x^* - A^* y^* + A^* y^* \in \partial f(x) + A^* (\partial g(Ax))$. The conclusion follows. □

We have the following extension of Theorem 4.4.7:

Theorem 4.4.12. *Let X and Y be normed vector spaces, $f \in \Lambda(X)$, $g \in \Lambda(Y)$ and $A \in \mathscr{L}(X, Y)$ be such that $A(\operatorname{dom} f) \cap \operatorname{dom} g \neq \emptyset$. If there exists $c > 0$ such that*

$$(f + g \circ A)^*(x^*) = \inf\{f^*(x^* - A^* y^*) + g^*(y^*) \mid y^* \in Y^*, \ \|y^*\| \\ \leq c \|x^*\|\}, \ \forall\, x^* \in X^*, \tag{4.25}$$

then (4.24) holds.

Proof. It is sufficient to observe that (4.23) holds. Because f^*, g^* are w^*-lsc and A^* is w^*-w^* continuous we obtain that the mapping $Y^* \ni y^* \mapsto f^*(x^* - A^* y^*) + g^*(y^*) \in \bar{\mathbb{R}}$ is w^*-lsc. Since the set $U := \{y^* \in Y^* \mid \|y^*\| \leq c \|x^*\|\}$ is w^*-compact, there exists some $\bar{y}^* \in U \subset Y^*$ such that

$$\inf\{f^*(x^* - A^* y^*) + g^*(y^*) \mid y^* \in Y^*, \ \|y^*\| \leq c \|x^*\|\} \\ = f^*(x^* - A^* \bar{y}^*) + g^*(\bar{y}^*). \tag{4.26}$$

Since one has always

$$(f + g \circ A)^*(x^*) \leq \inf\{f^*(x^* - A^* y^*) + g^*(y^*) \mid y^* \in Y^*\} \quad \forall x^* \in X^*,$$

from (4.25) and (4.26) we get that (4.23) holds. The conclusion follows using Theorem 4.4.11. □

Recall that in a normed space X with X^* as its dual, the support function of $K \subset X$ is given by

$$\sigma_K(x^*) := \sup_{x \in K} \langle x^*, x \rangle \quad x^* \in X^*.$$

The following result of Křivan [341] is a direct consequence of the above result:

Theorem 4.4.13. *Let X, Y be normed vector space, $L \subset X$, $M \subset Y$ be convex sets and $A \in \mathscr{L}(X, Y)$ such that $L \cap A^{-1}(M) \neq \emptyset$. Assume that there exists $c > 0$ such that*

$$\sigma_{L \cap A^{-1}(M)}(x^*) = \inf \{\sigma_L(x^* - A^* y^*) + \sigma_M(y^*) \mid y^* \in Y^*, \\ \|y^*\| \le c\,\|x^*\|\}, \quad \forall x^* \in X^*. \tag{4.27}$$

Then the following identities hold:

$$N(x, L \cap A^{-1}(M)) = N(x, L) + A^* \left(N(Ax, M)\right), \quad \forall x \in L \cap A^{-1}(M), \tag{4.28}$$

$$T(L \cap A^{-1}(M), x) = T(L, x) \cap A^{-1} \left(T(M, Ax)\right), \quad \forall x \in L \cap A^{-1}(M). \tag{4.29}$$

Proof. Take $f := \iota_L$ and $g := \iota_M$. Then $f \in \Lambda(X)$, $g \in \Lambda(Y)$ and $A(\operatorname{dom} f) \cap \operatorname{dom} g = A(L) \cap M \neq \emptyset$. Then $f + g \circ A = \iota_{L \cap A^{-1}(M)}$, $f^* = \sigma_L$, $g^* = \sigma_M$, $(f + g \circ A)^* = \sigma_{L \cap A^{-1}(M)}$, $\partial f(x) = N(x, L)$ for $x \in L$, $\partial g(y) = N(y, M)$ for $y \in M$ and $\partial(f + g \circ A)(x) = N(x, L \cap A^{-1}(M))$ for $x \in L \cap A^{-1}(M)$. Applying Theorem 4.4.12 we get (4.28).

The relation (4.29) follows immediately from (4.28) taking into account that for (convex) cones $K_1, K_2 \subset U$ one has $(K_1 + K_2)^- = K_1^- \cap K_2^-$, while for the (convex) cone $K \subset U$ and $A \in \mathscr{L}(U, V)$ one has $(A(K))^- = A^{*-1}(K^-)$ without any supplementary condition, where U, V are separated locally convex spaces. □

The following useful intersection formula is a direct consequence of the above result.

Theorem 4.4.14. *Let X be a normed space and let $L \subset X$ and $K \subset Y$ be two nonempty, closed, and convex sets. Assume that there exists a constant $c > 0$ such that for every $x^* \in X^*$ we have*

$$\sigma_{L \cap A^{-1}(K)}(x^*) = \inf \left\{\sigma_L\left(x^* - z^*\right) + \sigma_K(z^*) \middle| \, z^* \in Y^*, \ \|z^*\| \le c\|x^*\|\right\}. \tag{4.30}$$

Then, for every $\bar{x} \in L \cap K$, the following intersection formula holds:

$$T(L \cap K, \bar{x}) = T(L, \bar{x}) \cap T(K, \bar{x}).$$

Example 4.4.15. This example is a continuation of Example 4.4.10. That is, we have $X = \mathbb{R}^2$, and K is a triangle with vertices $(0,0)$, $(1,1)$ and $(-1,1)$, and $L = -K$. Since

$$B((0,0),1) \subset [N((0,0),K) \cap B((0,0),1)] + [N((0,0),L) \cap B((0,0),1)],$$

it follows that for every $x^* \in X$, there exists $x_1^* \in (N(0,K) \cap B((0,0),1))$ and $x_2^* \in (N(0,L) \cap B((0,0),1))$ such that $x^* = \|x^*\|x_1^* + \|x^*\|x_2^*$. We set $u^* = \|x^*\|x_1^*$. Since $u^* \in N(0,K)$ and $x^* - u^* \in N(0,L)$, we deduce that

$$\inf\{\sigma_K(w^*) + \sigma_L(x^* - w^*) |\ w^* \in X,\ \|w^*\| \le \|x^*\|\} \le \sigma_K u^* - \sigma_L(x^* - u^*) = 0.$$

Furthermore, it follows from the convexity of the support function $\sigma_{K \cap L}$ that

$$\sigma_{K\cap L}(x^*) \le \sigma_{K\cap L}(u^*) + \sigma_{K\cap L}(x^* - u^*) \le \sigma_K(u^*) + \sigma_L(x^* - u^*).$$

Since $\sigma_{K\cap L}(x^*) = 0$ for every $x^* \in X$, it follows that

$$0 = \sigma_{K\cap L}(x^*) = \inf\{\sigma_K(v^*) + \sigma_L(x^* - v^*) |\ v^* \in X,\ \|v^*\| \le \|x^*\|\},$$

confirming that (4.30) holds, and therefore, $T(L \cap K, (0,0)) = T(L,(0,0)) \cap T(K,(0,0))$. □

The following result, whose proof can be found in Křivan [341], connects the two constraint qualification conditions, namely, (4.22) and (4.27) in a Hilbert space setting.

Proposition 4.4.16. *Let X and Y be Hilbert spaces and let $L \subset X$ and $K \subset Y$ be nonempty, closed, and convex with $L \subset B(0,p)$ for some $p > 0$. Let $A : X \to Y$ be a continuous and linear map. Assume that* (4.22) *holds. Then, there exists $c > 0$ such that* (4.30) *is satisfied.*

Conditions ensuring (4.27) in a finite-dimensional setting are also given by Křivan [341].

The following result, whose proof can be found in [16, Page 174] and [18, Page 141], gives some additional properties of the contingent cone on a convex set. Clearly, these stronger properties are a byproduct of the convexity of the underlying sets. In the following, we set $T(S,\bar{x}) = N(\bar{x},S)^-$.

Theorem 4.4.17. *Let $X, X_1, \ldots, X_n$, and Y be Banach spaces, let $S, S_1, S_2, \ldots, S_n$ be closed, convex sets, and let $\bar{x}, \bar{x}_1, \bar{x}_2, \ldots \bar{x}_n$ be given elements. Then the following properties hold:*

1. *If $\bar{x}_1 \in S_1 \subset X$ and $\bar{x}_2 \in S_2 \subset X$, then*

$$T(S_1 + S_2, \bar{x}_1 + \bar{x}_2) = \mathrm{cl}\,(T(S_1,\bar{x}_1) + T(S_2,\bar{x}_2)),$$
$$N(\bar{x}_1 + \bar{x}_2, S_1 + S_2) = N(\bar{x}_1, S_1) \cap N(\bar{x}_2, S_2).$$

2. *If* $A : X \to Y$ *is linear and continuous and* $\bar{x} \in S \subset X$, *then*

$$T(A(S), A(\bar{x})) = \mathrm{cl}\,(A(T(S, \bar{x})))$$
$$N(A(\bar{x}), A(S)) = (A^*)^{-1}(N(\bar{x}, S)).$$

3. *If* $\bar{x}_i \in S_i \subset X_i$ *for* $i \in \{1, \ldots, n\}$, *are convex, then*

$$T(S_1 \times S_2 \times \ldots \times S_n, (\bar{x}_1, \bar{x}_2, \ldots, \bar{x}_n)) = T(S_1, \bar{x}_1) \times T(S_2, \bar{x}_2) \times \ldots \times T(S_n, \bar{x}_n),$$
$$N((\bar{x}_1, \bar{x}_2, \ldots, \bar{x}_n), S_1 \times S_2 \times \ldots \times S_n) = N(\bar{x}_1, S_1) \times N(\bar{x}_2, S_2) \times \ldots \times N(\bar{x}_n, S_n).$$

4. *Assume that for closed and convex sets* $S_i \subset X$, *for* $i \in \{1, \ldots, n\}$ *and for* $\bar{x} \in \cap_{i=1}^n S_i$, *there exists* $k > 0$ *such that for every* x_i *with* $\|x_i\| \leq k$, *we have* $\cap_{i=1}^n (S_i - x_i) \neq \emptyset$. *Then*

$$T(S_1 \cap S_2 \cap \ldots \cap S_n, \bar{x}) = T(S_1, \bar{x}) \cap \ldots \cap T(S_n, \bar{x})$$
$$N(\bar{x}, S_1 \cap S_2 \cap \ldots \cap S_n) = N(\bar{x}, S_1) + \ldots + N(\bar{x}, S_n).$$

We note that Bazaraa, Goode and Nashed [36] showed that if S_1 and S_2 are closed and convex subsets of $\mathbb{R}^n$ such that $\mathrm{ri}(S_1) \cap \mathrm{ri}(S_2) \neq \emptyset$, then for every $\bar{x} \in \mathrm{cl}\, S_1 \cap \mathrm{cl}\, S_2$, we have

$$T(S_1, \bar{x}) \cap T(S_2, \bar{x}) = T(S_1 \cap S_2, \bar{x}). \tag{4.31}$$

We now collect a few characteristic examples of the contingent cones:

Example 4.4.18. Let X be a Hilbert space and let $S \subset X$ be a closed and convex cone. Then

$$N(\bar{x}, S) = S^- \cap \{\bar{x}\}^\perp.$$

Consequently, $x \in T(S, \bar{x})$ if and only if

$$\langle s, x \rangle \leq 0 \quad \text{for every } s \in S^- \text{ with } \langle s, \bar{x} \rangle = 0.$$

In particular, if S is a closed subspace, then $T(S, \bar{x}) = S$ and $N(\bar{x}, S) = S^\perp$. □

Example 4.4.19. Let $\Sigma = \{x \in \mathbb{R}^n_+ \mid \sum_{i=1}^n x_i = 1\}$ be the simplex and let $I(\bar{x}) = \{i \in \{1, \ldots, n\}$ with $\bar{x}_i = 0\}$. Then

$$x \in T(\mathbb{R}^n_+, \bar{x}) \iff x_i \geq 0 \quad \text{for every } i \in I(\bar{x}),$$
$$x \in T(\Sigma_n, \bar{x},) \iff x_i \geq 0 \quad \text{for every } i \in I(\bar{x}) \text{ and } \sum_{i=1}^n x_i = 0.$$

4.5 First-Order Local Cone Approximation

Having seen several interesting examples of tangent cones of first-order, our goal in this section is to study an axiomatic approach for local cone approximation of sets. The primary objective of this approach to single out the common features shared by the most frequently used tangent cones. This approach also gives ways to construct new local cone approximations from the known ones. This section is based on papers by K. H. Elster and J. Thierfelder (see [179–181]), and it also benefits from Castellani and Pappalardo [85], Dolecki [140, 141], and Ward [592].

For the following notion of abstract local cone approximation, we recall that the recession cone S_∞ of a set S is the collection of all z such that for every $r \geq 0$, we have $S + rz \subset S$.

Definition 4.5.1. Given a normed space X, a set-valued map $K : 2^X \times X \to 2^X$ is called a **local cone approximation**, if it associates to each set $S \subseteq X$ and each point $\bar{x} \in X$, a cone $K(S, \bar{x}) \subseteq X$ satisfying the following properties:

1. $K(S, \bar{x}) = K(S - \bar{x}, 0)$,
2. $K(S, \bar{x}) = K(S \cap U(\bar{x}), \bar{x})$ for every $U(\bar{x})$,
3. $K(S, \bar{x}) = \emptyset$ if $\bar{x} \notin \operatorname{cl} S$,
4. $K(S, \bar{x}) = X$ if $x \in \operatorname{int} S$,
5. $\ell(K(S, \bar{x})) = K(\ell(S), \ell(\bar{x}))$, where $\ell : X \to X$ is a linear homomorphism,
6. $S_\infty \subset (K(S, \bar{x}))_\infty$, where S_∞ and $(K(S, \bar{x}))_\infty$ are the recession cones of S and $K(S, \bar{x})$.

Remark 4.5.2. In the above definition, property (1) demands the invariance of the cone approximation with respect to contemporaneous shifting of the set S and the point $\bar{x}$. In other words, the cone approximation $K(S, \bar{x})$ and an approximation of $S - \bar{x}$ at the origin are identical. Properties (2), (3), and (4) depict the local nature of the local cone approximation. Axiom (5) shows the invariance of the cone approximation with respect to a linear homomorphism such as rotation or reflection. Condition (6) gives a relationship between the recession cone of S and $K(S, \bar{x})$. This property is crucial in deriving some conclusions regarding the so-called K-epiderivatives.

The following theorem confirms that the notion of local cone approximation is quite rich:

Theorem 4.5.3. *Let X be a normed space, let S be a nonempty subset of X, and let $\bar{x} \in X$. Then, the radial tangent cone, the feasible tangent cone, the contingent cone, the interiorly contingent cone, the adjacent cone, the interiorly adjacent cone, the modified radial tangent cone, the modified feasible tangent cone, the modified contingent cone, the modified interiorly contingent cone, the Clarke's cone, and the modified interiorly adjacent cone are all local cone approximations.*

Proof. We begin by noting that the first four properties either follow from the definition or can be obtained from the features of these cones given in the previous sections. Therefore, it suffices to prove the last two properties. We will give a proof of these two properties only for the Clarke's cone $C(S, \bar{x})$. Similar arguments can be used to develop detailed proofs for other cones.

We begin by showing that $\ell(C(S, \bar{x})) = C(\ell(S), \ell(\bar{x}))$. Assume that $x \in \ell(C(S, \bar{x}))$ is arbitrary. This implies that $\ell^{-1}(x) \in C(S, \bar{x})$. Let $U(x)$ be an arbitrary neighborhood of x. Then, due to the continuity of ℓ, there is a neighborhood $U(\ell^{-1}(x))$ of $\ell^{-1}(x)$ such that $\ell(U(\ell^{-1}(x))) \subset U(x)$. Since $\ell^{-1}(x) \in C(S, \bar{x})$, there are a neighborhood $V(\bar{x})$ and a scalar $t > 0$ such that for every $s \in (0, t)$ and for every $\hat{x} \in V(\bar{x}) \cap S \cup \bar{x}$, we have $\frac{S-\hat{x}}{s} \cap U(\ell^{-1}(x)) \neq \emptyset$.

Also, there are a neighborhood $V(\ell(\bar{x}))$ and a scalar $t > 0$ such that for every $s \in (0, t)$ and for every $\hat{x} \in V(\ell(\bar{x})) \cap \ell(S) \cup \{\ell(\bar{x})\}$, we have

$$\frac{S - \ell^{-1}(\hat{x})}{s} \cap U(\ell^{-1}(x)) \neq \emptyset.$$

Using $\ell(U(\ell^{-1}(x))) \subset U(x)$, we obtain

$$\frac{\ell(S) - \hat{x}}{s} \cap U(x) \supseteq \frac{\ell(S) - \hat{x}}{s} \cap \ell(U(\ell^{-1}(x))) \neq \emptyset,$$

which ensures that $x \in C(\ell(S), \ell(\bar{x}))$ and hence proving the inclusion $\ell(C(S, \bar{x})) \subseteq C(\ell(S), \ell(\bar{x}))$.

For the converse, let $x \in C(\ell(S), \ell(\bar{x}))$ be arbitrary. We claim that $\ell^{-1}(x) \in C(S, \bar{x})$. Due to the continuity of ℓ^{-1}, for a neighborhood $U(\ell^{-1}(x))$, there exists a neighborhood $U(x)$ such that

$$\ell^{-1}(U(x)) \subseteq U(\ell^{-1}(x)). \tag{4.32}$$

Using the definition of the Clarke's tangent cone, there are a neighborhood $V(\ell(\bar{x}))$ and a scalar $t > 0$ such that for every $s \in (0, t)$ and for every $\hat{x} \in V(\ell(\bar{x})) \cap \ell(S) \cup \{\ell(\bar{x})\}$, we have

$$\frac{\ell(S) - \hat{x}}{s} \cap U(x) \neq \emptyset.$$

Also, there are a neighborhood $V(\bar{x})$ and a scalar $t > 0$ such that for every $s \in (0, t)$ and for every $\hat{x} \in V(\bar{x}) \cap S \cup \{\bar{x}\}$, we have

$$\frac{\ell(S) - \ell(\hat{x})}{s} \cap U(x) \neq \emptyset,$$

which, by using (4.32), we obtain

$$\frac{S - \hat{x}}{s} \cap U(\ell^{-1}(x)) \supseteq \frac{S - \hat{x}}{s} \cap \ell^{-1}(U(x)) \neq \emptyset$$

confirming that $\ell^{-1}(x) \in C(S, \bar{x})$.

Finally, we proceed to prove the inclusion $S_\infty \subset (C(S,\bar{x}))_\infty$. We begin by choosing $z \in S_\infty$. Therefore, for every $r \geq 0$, we have $S + rz \subseteq S$. Let $x \in C(S,\bar{x})$ be arbitrary. Then for every neighborhood $U(x)$, there exists a neighborhood $V(\bar{x})$ and a scalar $t > 0$ such that for every $\hat{x} \in V(\bar{x}) \cap S \cup \{\bar{x}\}$ and for every $s \in (0,t)$, we have

$$\frac{S - \hat{x}}{s} \cap U(x) \neq \emptyset.$$

For $x + z$, by taking $U(x+z) := U(x) + z$, we obtain

$$\begin{aligned}
\frac{S - \hat{x}}{s} \cap U(x+z) &= \left[\left(\frac{S - \hat{x}}{s} - z\right) \cap U(x)\right] + z, \\
&= \left[\left(\frac{S - sz - \hat{x}}{s}\right) \cap U(x)\right] + z, \\
&\supseteq \left[\frac{S - \hat{x}}{s} \cap U(x)\right] + z, \\
&\neq \emptyset
\end{aligned}$$

which proves that $x + z \in C(S,\bar{x})$ and hence $z \in (C(S,\bar{x}))_\infty$. □

The following result shows that all the conditions of the local cone approximation are vital.

Theorem 4.5.4. *For each axiom of the local cone approximation given in Definition 4.5.1, there exists a local cone approximation which violates this axiom and satisfies all the other axioms. In other words, none of the axioms can be expressed by the others.*

Proof. We will construct local cone approximations violating one axiom and verifying all the others. Let X be a normed space, let $S \subset X$, and $\bar{x} \in X$. Let $K(S,\bar{x})$ be a local cone approximation.

1. The local cone approximation defined by $K_1(S,\bar{x}) := K(S,\bar{x}) + \bar{x}$ violates the first property $K(S - \bar{x}, 0) = K(S,\bar{x})$ but satisfies all the other conditions.
2. The local cone approximation defined by

$$K_2(S,\bar{x}) := \begin{cases} \operatorname{cone}(S - \bar{x}) & \text{if } x \in \operatorname{int} S, \\ \emptyset & \text{if } x \in X \setminus \operatorname{int} S, \end{cases}$$

   generally violates the second property $K(S,\bar{x}) = K(S \cap U(\bar{x}), \bar{x})$ but satisfies all the other axioms.
3. The local cone approximation defined by $K_3(S,\bar{x}) := X$ violates the third property but satisfies all the other axioms.

4. The local cone approximation defined by $K_4(S, \bar{x}) := \emptyset$ violates the fourth property but satisfies all the other axioms.
5. Let A be a nonempty cone. The local cone approximation defined by $K_5(S, \bar{x}) := K(S, \bar{x}) + A$ generally violates the fifth axiom but satisfies all the other conditions. In fact,

$$K_5(\ell(S), \ell(\bar{x})) = K(\ell(S), \ell(\bar{x})) + A = \ell(K(S, \bar{x})) + A,$$
$$\ell(K_5(S, \bar{x})) = \ell(K(S, \bar{x})) + A) = \ell(K(S, \bar{x})) + \ell(A).$$

   Clearly, the first four axioms hold. The proof of the sixth axiom is as follows: Let $z \in S_\infty$. Then, we have $(K(S, \bar{x}))_\infty$. Furthermore, $K_5(S, \bar{x}) + z = K(S, \bar{x}) + A + z \subseteq K(S, \bar{x}) + A = K_5(S, \bar{x})$.
6. The local cone approximation defined by

$$K_6(S, \bar{x}) = \begin{cases} \emptyset \text{ if} & x \notin \operatorname{cl} S, \\ X \text{ if} & x \in \operatorname{int} S, \\ \{0\} \text{ elsewhere}, \end{cases}$$

   violates the sixth property for the case when $X = \mathbb{R}^2$, $S = \mathbb{R}^2_+$ and $\bar{x} = (0, 0)$. In fact, we have

$$K_6(S, (0, 0)) = \{(0, 0)\},$$
$$(K_6(S, (0, 0)))_\infty = \{(0, 0)\},$$

   and this completes the proof. □

The following results gives ways of constructing new notions of local cone approximations:

Theorem 4.5.5. *Let X be a normed space, let S be a nonempty subset of X, and let $\bar{x} \in X$. Let I be a finite index set. Let $K(S, \bar{x})$ and $K_i(S, \bar{x})$, with $i \in I$, be local cone approximations. Then each of the following relationships define a local cone approximation:*

1. $\operatorname{int} K(S, \bar{x})$
2. $\operatorname{cl} K(S, \bar{x})$
3. $\operatorname{conv} K(S, \bar{x})$
4. $X \backslash K(X \backslash S, \bar{x})$
5. $\bigcup_{i \in I} K_i(S, \bar{x})$
6. $\bigcap_{i \in I} K_i(S, \bar{x})$
7. $\Sigma_{i \in I} K_i(S, \bar{x})$.

Proof. We begin with a proof of part (1). The first four properties of Definition 4.5.1 being easy to verify, we give a proof for the last two properties. Let ℓ be a linear homomorphism. Then:

$$\text{int}\, K(\ell(S), \ell(\bar{x})) = \text{int}\, \ell(K(S, \bar{x})) \subset \ell(\text{int}\, K(S, \bar{x})).$$

Furthermore, since ℓ^{-1} is continuous, we have

$$\text{int}\, K(S, \bar{x}) = \text{int}\, \ell^{-1}(K(\ell(S), \ell(\bar{x}))) \subset \ell^{-1}(\text{int}\, K(\ell(S), \ell(\bar{x}))),$$

implying that $\ell(\text{int}\, K(S, \bar{x})) \subset \text{int}\ K(\ell(S), \ell(\bar{x}))$, and the property is proved.

For the sixth property of Definition 4.5.1, it suffices to note that $(K(S, \bar{x}))_\infty \subset (\text{int}\, K(S, x))_\infty$.

The proofs of the statements that $\text{cl}\, K(S, \bar{x})$ and $\text{conv}\, K(S, \bar{x})$ are local cone approximations can be made by analogous arguments.

For showing that $X \backslash K(X \backslash S, \bar{x})$ is a local cone approximation, we only prove the sixth axiom of Definition 4.5.1. Due to the identity $S_\infty = -(X \backslash S)_\infty$, we infer that

$$S_\infty = -(X \backslash S)_\infty \subset -(K(X \backslash S, \bar{x}))_\infty = (X \backslash K(X \backslash S, \bar{x}))_\infty.$$

The proof for the other properties is also easy to prove. □

Before any further advancement, we recall the following definition:

Definition 4.5.6. Given a normed space X, a local cone approximation $K : 2^X \times X \to X$ is called **isotone**, if for each set $A \subset B \subset X$ and for each $\bar{x} \in X$, we have $K(A, \bar{x}) \subset K(B, \bar{x})$.

The following result gives a useful implication of an isotone local cone approximation.

Theorem 4.5.7. *Let X be a normed space, let S be a subset of X, and let $\bar{x} \in X$. Given a local cone approximation $K(S, \bar{x})$, the local cone approximation $M(S, \bar{x}) := X \backslash K(X \backslash S, \bar{x})$ is isotone and forms a* ***disjunction preserving*** *pair with $K(\cdot, \cdot)$. That is, for every $A, B, \subset X$ and every $\bar{x} \in X$ with $A \cap B = \emptyset$, it holds that $K(A, \bar{x}) \cap M(B, \bar{x}) = \emptyset$.*

4.6 Convex Subcones of the Contingent Cone

As we have observed in the previous sections, two of the most commonly used tangent cones, namely the contingent cone and the Clarke's tangent cone, have their own merits. The contingent cone is isotone, a property that plays a vital role in deriving necessary optimality conditions and has numerous other useful implications. On the other hand, the convexity of the Clarke's tangent cone is a remarkably useful analytical tool. As it has already been stated, convexity on the contingent cone can be imposed by taking the convex hull of the contingent cone. However, this process results in a cone that is often too large to give much useful information as a local approximation tool. Therefore, it is of importance to ask whether there are closed and convex cones that lie between the Clarke's tangent cone and the contingent cone. To answer this and study related questions, we begin by investigating a local cone approximation $K(\cdot,\cdot)$ which is convex-valued and satisfies the following chain of inclusions

$$C(S,\bar{x}) \subset K(S,\bar{x}) \subset T(S,\bar{x}), \tag{4.33}$$

where S is a nonempty subset of a normed space X and $\bar{x} \in X$.

Convex and closed cones $K(\cdot,\cdot)$ satisfying (4.33) have been known to play an important role in nonsmooth analysis and optimization and we anticipate their use in set-valued optimization.

An important question pertaining to the above discussion is whether we can find a cone that contains the most important features of the contingent cone and the Clarke's cone, namely, it is isotone and convex both.

The following impossibility theorem of Ward [590] gives a negative answer to this question:

Theorem 4.6.1. *There is no mapping $K : 2^{\mathbb{R}^n} \times \mathbb{R}^n \to 2^{\mathbb{R}^n}$ that has all the following properties:*

1. *K is isotone*
2. *K is convex*
3. *$K(S,\bar{x}) \subset T(S,\bar{x})$ for every $S \subset \mathbb{R}^n$ and $\bar{x} \in \operatorname{cl} S$*
4. *$S \subset K(S,\bar{x})$, whenever S is a one dimensional subspace of $\mathbb{R}^n$ and $\bar{x} \in S$.*

Proof. Consider the subsets of $\mathbb{R}^n$ defined by $S_1 = \mathbb{R} \times \{0\}$ and $S_2 = \{0\} \times \mathbb{R}$. Note that if K possessed the third property, then $K(S_1 \cup S_2, (0,0)) \subset S_1 \cup S_2$. On the other hand, if K is isotone and possessed the fourth property, then $S_1 \cup S_2 \subset K(S_1, 0) \cup K(S_2, 0) = K(S_1 \cup S_2, (0,0))$. Therefore, $K(S_1 \cup S_2, (0,0)) = S_1 \cup S_2$, if K is isotone and satisfies the last two properties. Evidently, such a mapping is bound to fail to be convex. □

Having obtained a negative answer, our next genuine step is to ask whether there are closed and convex cones satisfying (4.33) which yield more accurate approximation of a set than the Clarke's tangent cone and at the same time are also equipped with its analytical strengths? In the following, we will show that an answer to this question is affirmative. We intend to discuss a few cones that satisfy (4.33) and have significantly useful properties. Our treatment of this subject is based on D. Ward's seminal contributions [590, 591, 593, 597].

We begin by noticing that a simple way to obtain a closed and convex cone is by taking the recession cone of a closed cone. Given a nonempty subset S of a normed space X and an element $\bar{x} \in \operatorname{cl} S$, we define the **recession contingent cone** and the **recession adjacent cone** as follows:

$$T_\infty(S,\bar{x}) = \{x \in X \mid T(S,\bar{x}) + x \subset T(S,\bar{x})\}, \tag{4.34}$$

$$A_\infty(S,\bar{x}) = \{x \in X \mid A(S,\bar{x}) + x \subset A(S,\bar{x})\}. \tag{4.35}$$

It follows from the above definition that

$$C(S,\bar{x}) \subset T_\infty(S,\bar{x}) \subset T(S,\bar{x}),$$
$$C(S,\bar{x}) \subset A_\infty(S,\bar{x}) \subset A(S,\bar{x}),$$

and it can be shown that the above chain of inclusions are, in general, strict (see [414]).

The following example shows that despite of the well-known inclusion $A(S,\bar{x}) \subset T(S,\bar{x})$, the inclusion $A_\infty(S,\bar{x}) \subset T_\infty(S,\bar{x})$ is, in general, false.

Example 4.6.2. Define $f : \mathbb{R} \to \mathbb{R}$ by

$$f(x) = \begin{cases} 0 & \text{if } x = 0, \\ -2^{-(n+1)} & \text{if } 2^{-(n+1)} \le |x| < 2^{-n}, \ n = 0, \pm 1, \pm 2, \ldots. \end{cases}$$

Then, for $S = \operatorname{epi}(f)$ and $\bar{x} = (0,0)$, we have

$$T(S,\bar{x}) = \{(x,y) \mid y \ge -|x|\},$$
$$A(S,\bar{x}) = \left\{(x,y) \mid y \ge -\frac{|x|}{2}\right\},$$

and consequently

$$T_\infty(S,\bar{x}) = \{(x,y) \mid y \ge |x|\},$$
$$A_\infty(S,\bar{x}) = \left\{(x,y) \mid y \ge \frac{|x|}{2}\right\}.$$

In the following, we give the two cones, both introduced by Penot [463].

Definition 4.6.3. Let X be a normed space, let S be a subset of X, and let $\bar{x} \in S$.

1. The **prototangent cone** $P(S,\bar{x})$ of S at $\bar{x}$ is the collection of all $x \in X$ such that for every sequence $(x_n,t_n) \to (\bar{x},0_+)$ with $\{x_n\} \subset S$ and $t_n^{-1}(x_n - \bar{x})$ convergent, there exists $\{z_n\} \subset X$ converging to x with $x_n + t_n z_n \in S$ for every $n \in \mathbb{N}$.
2. The **interiorly prototangent cone** $P(S,\bar{x})$ of S at $\bar{x}$ is the collection of all $x \in X$ such that for every sequence $(x_n,t_n) \to (\bar{x},0_+)$ with $\{x_n\} \subset S$ and $t_n^{-1}(x_n - \bar{x})$ convergent, for every $\{z_n\} \subset X$ converging to x, we have $x_n + t_n z_n \in S$, for sufficiently large $n \in \mathbb{N}$.

It follows from the above definition that

$$C(S,\bar{x}) \subset P(S,\bar{x}) \subset A_\infty(S,\bar{x}) \subset A(S,\bar{x}) \subset T(S,\bar{x}),$$

and, if, $A(S,\bar{x}) = T(S,\bar{x})$, then

$$P(S,\bar{x}) = A_\infty(S,\bar{x}) = T_\infty(S,\bar{x}).$$

The following result, which follows directly from the definitions of the involved tangent cones, gives some useful information:

Proposition 4.6.4. *Let X and Y be normed spaces, let $S_1 \subset X$ and $S_2 \subset Y$ be two subsets, and let $\bar{x} \in \operatorname{cl} S_1$ and $\bar{y} \in \operatorname{cl} S_2$. Then*

$$A_\infty(S_1,\bar{x}) \times A_\infty(S_2,\bar{y}) = A_\infty(S_1 \times S_2,(\bar{x},\bar{y})),$$
$$P(S_1,\bar{x}) \times P(S_2,\bar{y}) \subset P(S_1 \times S_2,(\bar{x},\bar{y})).$$

Note that often product formulas are given by combining two tangent cones. For instance, it has been known that the following inclusion

$$T(S_1,\bar{x}) \times A(S_2,\bar{y}) \subset T(S_1 \times S_2,(\bar{x},\bar{y})), \tag{4.36}$$

holds. The following example shows that $T_\infty(S,\bar{x})$ cannot be combined with any other cone to produce an analogue of such formulas. This deficiency of $T_\infty(S,\bar{x})$ limits is usefulness.

Example 4.6.5. For the two sets S_1 and S_2 given by

$$S_1 = \{x \in \mathbb{R} |\ x = 2^{-2n},\ n = 1,2,3,\ldots\} \cup \{0\},$$
$$S_2 = \{x \in \mathbb{R} |\ x = 2^{-2n+1},\ n = 1,2,3,\ldots\} \cup \{0\},$$

we have

$$T(S_1,0) = T(S_2,0) = \mathbb{R}_+,$$
$$A(S_1,0) = A(S_2,0) = \{0\},$$
$$T_\infty(S_1 \times S_2,(0,0)) = \{(0,0)\},$$

and hence, the inclusion $T_\infty(S_1,0) \times K(S_2,0) \subset T_\infty(S_1 \times S_2,(0,0))$ fails for $K = T, T_\infty, A, A_\infty, P$ or C. □

The following two results shed some light on the behaviour of the tangent cones under linear and continuous maps.

Proposition 4.6.6. *Let X and Y be normed spaces, let $S \subset X$ with $\bar{x} \in \operatorname{cl} S$, and let $L : X \to Y$ be a linear and continuous map. Then the following implications hold:*

$$A(L(S), L(\bar{x})) \subset L(A(S,\bar{x})) \quad \Longrightarrow \quad L(A_\infty(S,\bar{x})) \subset A_\infty(L(S), L(\bar{x})),$$
$$T(L(S), L(\bar{x})) \subset L(T(S,\bar{x})) \quad \Longrightarrow \quad L(T_\infty(S,\bar{x})) \subset T_\infty(L(S), L(\bar{x})).$$

Proof. For $x \in A_\infty(S,\bar{x})$, we set $z = L(x)$. Let $v \in A(L(S), L(\bar{x}))$ be arbitrary. By hypothesis $v = L(u)$ for some $u \in A(S,\bar{x})$. Therefore, $z + v = L(x+u)$, and since $x + u \in A(S,\bar{x})$, we get $z + v \in L(A(S,\bar{x})) \subset A(L(S), L(\bar{x}))$. Consequently, $z = A_\infty(L(S), L(\bar{x}))$ and the first implication is established. The second implication can be proved in an analogous fashion. □

Proposition 4.6.7. *Let X and Y be normed spaces, let $S \subset X$, let $\bar{x} \in \operatorname{cl} S$, and let $L : X \to Y$ be a linear and continuous map. Assume that the following condition holds: For every $(v_n, t_n) \to (L(\bar{x}), 0_+)$ such that $\{v_n\} \subset L(S)$ and $t_n^{-1}(v_n - L(\bar{x}))$ converges, there are $\{u_n\} \subset S$ with $v_n = L(u_n)$ and $t_n^{-1}(u_n - \bar{x})$ that are convergent. Then the following inclusions hold:*

$$L(P(S,\bar{x})) \subset P(L(S), L(\bar{x})),$$
$$L(K_\infty(S,\bar{x})) \subset K_\infty(L(S), L(\bar{x})).$$

Proof. Let $z \in L(P(S,\bar{x}))$ be arbitrary. Then $z = L(x)$ with $x \in P(S,\bar{x})$. Assume that $(z_n, t_n) \to (L(\bar{x}), 0_+)$, with $\{z_n\} \subset L(S)$ and $t_n^{-1}(z_n - L(\bar{x}))$, is convergent. By the hypothesis, there exists $\{x_n\} \subset S$ with $z_n = L(x_n)$ and convergent $t_n^{-1}(x_n - \bar{x})$. Then there exists $u_n \to x$ such that $x_n + t_n u_n \in S$. Hence $L(u_n) \to L(x)$ and $z_n + t_n L(u_n) = L(x_n + t_n u_n) \in L(S)$. Consequently, $z \in P(L(S), L(\bar{x}))$ and the first inclusion is verified. The second inclusion follows the same lines of argument. □

In the following, we study another closed convex cone and its "interiorly" counterpart.

Definition 4.6.8. Let X be a Banach space, and let S be a subset of X that is closed near $\bar{x} \in S$.

1. The **Treiman tangent cone** $E(S,\bar{x})$ to S at $\bar{x}$ is the set of all $x \in T(S,\bar{x})$ such that for every $\varepsilon, \lambda > 0$, there exists $U(\bar{x})$ such that for all $\hat{x} \in [U(\bar{x}) \cap S] \setminus \{\bar{x}\}$, we have
$$S \cap \{\hat{x} + (0,\lambda] \|\hat{x} - \bar{x}\| U_\varepsilon(x)\} \neq \emptyset. \tag{4.37}$$
2. The **interiorly Treiman tangent cone** $IE(S,\bar{x})$ to S at $\bar{x}$ is the set of all $x \in T(S,\bar{x})$ for which there are $\varepsilon, \lambda > 0$ such that for every $t \in (0,\lambda)$, there exists $U(\bar{x})$ satisfying that for every $\hat{x} \in [U(\bar{x}) \cap S] \setminus \{\bar{x}\}$, we have $\hat{x} + t\|\hat{x} - \bar{x}\| U_\varepsilon(x) \subset S$.

The Treiman tangent cone was introduced by Treiman [545] whereas the interiorly Tangent cone was introduced by Ward [593]. J. Treiman gave several equivalent characterizations of the Treiman cone. In particular, he showed that (4.37) can be replaced by the following condition:

$$S \cap \{\hat{x} + \lambda\|\hat{x} - \bar{x}\|U_\varepsilon(x)\} \neq \emptyset. \tag{4.38}$$

We now give two characterizations of the Trieman tangent cone proposed by Ward [593]:

Proposition 4.6.9. *Let X be a Banach space and let $S \subset X$ be closed near $\bar{x} \in$ bd(S). Then*

$$E(S,\bar{x}) = \{x \in T(S,\bar{x})|\ \forall\varepsilon,\ \lambda > 0,\ \exists\, U(\bar{x}) \textit{ such that } (4.37) \textit{ holds}$$
$$\forall\hat{x} \in [U(\bar{x}) \cap \mathrm{bd}(S)] \setminus \{\bar{x}\}\}. \tag{4.39}$$

Proof. Let T be the set on the right hand side of (4.39). Since S is closed near $\bar{x}$, there exists $U(\bar{x})$ such that $U(\bar{x}) \cap \mathrm{bd}(S) \subset U(\bar{x}) \cap S$. It then becomes evident that $E(S,\bar{x}) \subset T$. For the converse, let $x \in X \setminus E(S,\bar{x})$. Then there exists $\varepsilon, \lambda > 0$ such that for every $U(\bar{x})$, there exists $\hat{x} \in (U(\bar{x}) \cap S)\setminus\{\bar{x}\}$ satisfying that

$$S \cap \{\hat{x} + (0,\lambda]\|\hat{x} - \bar{x}\|U_\varepsilon(x)\} = \emptyset.$$

Clearly such an element $\hat{x}$ must belong to bd(S), so $x \in X\setminus\{T\}$. This implies that $T \subset E(S,\bar{x})$ and (4.39) holds. The proof is complete. □

The following sequential version of $E(S,\bar{x})$ follows easily from Proposition 4.6.9.

Proposition 4.6.10. *Let X be a Banach space and let $S \subset X$ be closed near $\bar{x} \in S$. Then the Trieman tangent cone $E(S,\bar{x})$ is the collection of all $z \in T(S,\bar{x})$ such that for all $\{x_n\} \subset$ bd$(S) \setminus \{\bar{x}\}$ with $x_n \to \bar{x}$, there exists a subsequence $\{x_m\}$ and there exist sequences $\{t_m\} \subset \mathbb{P}$ and $\{z_m\} \subset X$ with $t_m \downarrow 0$, $z_m \to z$ and $x_m + t_m\|x_n - \bar{x}\|z_m \in S$.*

Before any advancement, we also recall the following version of prototangent cone proposed by Michel and Penot [420] (see [545] and Definition 4.6.3).

Definition 4.6.11. Let X be a normed space, let S be a subset of X, and let $\bar{x} \in S$. The **P-tangent cone** $PM(S,\bar{x})$ of S at $\bar{x}$ is the collection of all $x \in X$ such that for every sequence $(x_n,t_n) \to (\bar{x},0_+)$ with $\{x_n\} \subset S$ and $t_n^{-1}(x_n - \bar{x})$ convergent, there exists a subsequence $\{x_m\}$ of $\{x_n\}$ and a sequence $\{z_m\}$ converging to x with $x_m + t_m z_m \in S$.

We have the following results concerning the properties of the above cones (see [593]).

Theorem 4.6.12. *Let X be a Banach space, let S, S_1, S_2 be subsets of X which are closed near $\bar{x} \in S \cap S_1 \cap S_2$. Then*

1. *$E(S, \bar{x})$ is a closed, convex cone*
2. *$E(S, \bar{x})$ is an open convex cone*
3. *$C(S, \bar{x}) \subset E(S, \bar{x}) \subset PM(S, \bar{x}) \subset T(S, \bar{x})$*
4. *$IC(S, \bar{x}) \subset IE(S, \bar{x}) \subset E(S, \bar{x})$*
5. *$E(S, \bar{x}) = PM(S, \bar{x})$ provided that $X = \mathbb{R}^n$*
6. *$E(S_1, \bar{x}) \cap IE(S_2, \bar{x}) \subset E(S_1 \cap S_2, \bar{x})$*
7. *$E(S, \bar{x}) + IE(S, \bar{x}) \subset IE(S, \bar{x})$*
8. *$E(S, \bar{x} = \text{cl } IE(S, \bar{x}))$ provided that $IE(S, \bar{x}) \neq \emptyset$*
9. *$IE(S, \bar{x} = \text{int } E(S, \bar{x}))$ provided that $IE(S, \bar{x}) \neq \emptyset$*
10. *$E(S_1, \bar{x}) \cap E(S_2, \bar{x}) \subset E(S_1 \cap S_2, \bar{x})$ provided that $E(S_1, \bar{x}) \cap IE(S_2, \bar{x}) \neq \emptyset$.*

Proof. A proof of the first property can be found in [545] whereas the next four properties follow from the definition. We remark that the convexity of $IE(S, \bar{x})$ directly follows from the inclusion $E(S, \bar{x}) + IE(S, \bar{x}) \subset IE(S, \bar{x})$ proved in the following.

We give a proof for the inclusion $E(S_1, \bar{x}) \cap IE(S_2, \bar{x}) \subset E(S_1 \cap S_2, \bar{x})$. Let $x \in E(S_1, \bar{x}) \cap IE(S_2, \bar{x})$ be arbitrary. Then $x \in A(S_1, \bar{x}) \cap A(S_2, \bar{x})$ due to the fact that $E(\cdot, \cdot) \subset A(\cdot, \cdot)$ (see [593]). Therefore, $x \in A(S_1 \cap S_2, \bar{x})$ implying that $x \in T(S_1 \cap S_2, \bar{x})$. Now let $\varepsilon > 0$ and $\lambda > 0$ be given. Then there exist $\lambda_2 \in (0, \lambda)$ and $\varepsilon_2 \in (0, \varepsilon)$ such that for each $t \in (0, \lambda_2)$, there exists $U_t(\bar{x})$ satisfying that for all $\hat{x} \in (U_t(\bar{x}) \cap S_2) \backslash \{\bar{x}\}$, we have

$$\hat{x} + t\|\hat{x} - \bar{x}\| U_{\varepsilon_2}(x) \subset S_2. \tag{4.40}$$

Let $t \in (0, \lambda_2)$ be arbitrary. By (4.38), there exists $U(\bar{x})$ with $U(\bar{x}) \subset U_t(\bar{x})$ such that for each $\hat{x} \in (U(\bar{x}) \cap S_1) \backslash \{\bar{x}\}$, there exists $z \in U_{\varepsilon_2}(x)$ satisfying

$$\hat{x} + \lambda\|\hat{x} - \bar{x}\| z \in S_1. \tag{4.41}$$

Now, let $\hat{x} \in (U(\bar{x}) \cap S_1 \cap S_2) \backslash \{\bar{x}\}$. Then there exists $z \in U_{\varepsilon_2}(x)$ satisfying the above equation. In addition $\hat{x} + t\|\hat{x} - \bar{x}\| z \subset S_2$ by (4.40). Therefore, $x \in E(S_1 \cap S_2, \bar{x})$.

For the inclusion $E(S, \bar{x}) + IE(S, \bar{x}) \subset IE(S, \bar{x})$, let $x \in E(S, \bar{x})$ and $z \in IE(S, \bar{x})$. Without any loss of generality, we assume that x and z are both nonzero. It is easy to show that $x + z \in K(S, \bar{x})$. Since $z \in IE(S, \bar{x})$, there exists $\varepsilon \in (0, \min\{\|x\|, \|z\|\})$ and $\lambda > 0$ such that for each $t \in (0, \lambda)$, there exists $U(\bar{x})$ with $\hat{x} + t\|\hat{x} + \bar{x}\|(x + U_\varepsilon(0)) \subset S$, for every $\hat{x} \in (U(\bar{x}) \cap S) \backslash \{\bar{x}\}$. Choose $\lambda_1 \in (0, \lambda)$ such that

$$\lambda_1 < \min\left\{\left(\|x\| + \frac{\varepsilon}{4}\right)^{-1}, \left[\left(\|x\| + \frac{\varepsilon}{4}\right)(\|z\| + \varepsilon)\right]^{-1} \frac{\varepsilon}{2}\right\}.$$

Then, for each $t \in (0, \lambda_1)$, there exists $U_t(\bar{x})$ such that for every $\hat{x} \in (U_t(\bar{x}) \cap S)\backslash\{\bar{x}\}$, we have $\hat{x} + t\|\hat{x} - \bar{x}\|(z + U_\varepsilon(0)) \subset S$. Moreover, there exists $U(\bar{x})$ with $U(\bar{x}) \subset U_t(\bar{x})$ such that for every $\hat{x} \in (U(\bar{x}) \cap S)\backslash\{\bar{x}\}$ we have $\hat{x} + t\|\hat{x} - \bar{x}\|(x + U_\varepsilon(0)) \subset U_t(\bar{x})$. Since $x \in E(S, \bar{x})$, there exists $U_1(\bar{x})$ with $U_1(\bar{x}) \subset U(\bar{x})$ and $\left\{\hat{x} + t\|\hat{x} - \bar{x}\|\left(x + U_{\varepsilon/4}(0)\right)\right\} \cap S \neq \emptyset$, for every $\hat{x} \in (U_1(\bar{x}) \cap S)\backslash\{\bar{x}\}$.

Now let $t \in (0, \lambda_1)$, $\hat{x} \in (U_1(\bar{x}) \cap S)\backslash\{\bar{x}\}$, and $w \in U_{\varepsilon/4}(0)$. There exists $w_1 \in U_{\varepsilon/4}(0)$ such that $\tilde{x} := \hat{x} + t\|\hat{x} - \bar{x}\|(x + w_1) \in S \cap U_t(\bar{x})$. We also have $\tilde{x} \neq \bar{x}$, since

$$\|\tilde{x} - \bar{x}\| \geq \|\hat{x} - \bar{x}\| - \lambda_1\|\hat{x} - \bar{x}\|(\|x\| + \|w_1\|) = \|\hat{x} - \bar{x}\|(1 - \lambda_1)(\|x\| + \|w_1\|) > 0$$

by our choice of λ_1. Therefore, $\tilde{x} + t\|\tilde{x} - \bar{x}\|(z + U_\varepsilon(0)) \subset S$, or in other words,

$$\hat{x} + t\|\hat{x} - \bar{x}\|(x + w_1) + t\|\tilde{x} + \bar{x}\|(z + U_\varepsilon(0)) \subset S.$$

Finally, we define u such that $\|\tilde{x} - \bar{x}\|(z + u) = \|\hat{x} - \bar{x}\|(z + w - w_1)$ which implies that

$$u = \|\tilde{x} - \bar{x}\|^{-1}\{(\|\hat{x} - \bar{x}\| - \|\tilde{x} - \bar{x}\|)x + \|\hat{x} - \bar{x}\|(w - w_1)\}.$$

To finish the proof, it suffices to show that $\|u\| < \varepsilon$. For this, we first note that

$$|\,\|\hat{x} - \bar{x}\| - \|\tilde{x} - \bar{x}\|\,| \leq \|\tilde{x} - \hat{x}\| \leq t\|\hat{x} - \bar{x}\|(\|x\| + \|w_1\|).$$

Then by our choice of λ_1, we have

$$\begin{aligned}\|u\| &\leq \|\tilde{x} - \bar{x}\|^{-1}\|\hat{x} - \bar{x}\|\{t\|z\|(\|x\| + \|w_1\|) + \|w - w_1\|\}\\ &\leq \left[1 - \lambda_1\left(\|x\| + \frac{\varepsilon}{4}\right)\right]^{-1}\left\{\lambda_1\left(\|x\| + \frac{\varepsilon}{4}\right)\|z\| + \frac{\varepsilon}{2}\right\} < \varepsilon,\end{aligned}$$

and we conclude that $x + z \in IE(S, \bar{x})$. This at once confirms the eighth and the ninth properties.

For the final property, let $x \in E(S_1, \bar{x}) \cap E(S_2, \bar{x})$ and $y \in E(S_1, \bar{x}) \cap IE(S_2, \bar{x})$ be arbitrary. By using the convexity of cones, we get $x + ty \in E(S_1, \bar{x}) \cap IE(S_2, \bar{x})$ for every $t > 0$. Then, $x + ty \in E(S_1 \cap S_2, \bar{x})$ for every $t > 0$. Since the cone $E(\cdot, \cdot)$ is closed, we deduce that $x \in E(S_1 \cap S_2, \bar{x})$. □

The following example shows that $E(S, \bar{x}) = PM(S, \bar{x})$ is false in infinite dimensions (see [545]):

Example 4.6.13. Let X be an infinite dimensional Banach space and let $\{x_0, x_1, x_2 \ldots\}$ be a sequence of elements of norm 1 such that for all m, n we have $\|x_n - x_m\| > \delta > 0$. Define

$$S := [0, \infty) \cdot \{x_0\} \cup \cup_{n=1}^{\infty} \left\{\frac{x_n}{n}\right\}.$$

Then S is a closed set such that

$$T(S,0) = PM(S,0) = [0,\infty)\cdot\{x_0\},$$
$$E(S,0) = C(S,0) = \{0\}.$$

We give another example to show that the Treiman tangent cone may be strictly larger than the Clarke's tangent cone.

Example 4.6.14. For $S \subset \mathbb{R}^2$ defined by $S = \{(x,y)|\ y \le 0\} \cup \{(x,y)|\ y = x^2\}$, we have

$$E(S,(0,0)) = PM(S,(0,0)) = \{(x,y)|\ y \le 0\},$$
$$C(S,(0,0)) = \{(x,0)|\ x \in \mathbb{R}\}.$$

The following result gives a useful property of the Treiman tangent cone:

Theorem 4.6.15. *Let X and Y be Banach spaces and let $L : X \to Y$ be a linear and continuous mapping. Let $S \subset X$ be closed near $\bar{x} \in S$. Assume that for each sequence $\{w_n\} \subset \mathrm{bd}\,(L(S))\backslash\{L(\bar{x})\}$ with $w_n \to L(\bar{x})$, there are a subsequence $\{w_m\}$ and a sequence $\{x_m\} \subset \mathrm{bd}(S)\backslash\{\bar{x}\}$ such that $L(x_m) = w_m$ with $x_m \to \bar{x}$ and $\{\|x_m - \bar{x}\|/\|w_m - L(\bar{x})\|\}$ is bounded above. Then:*

$$L(E(S,\bar{x})) \subset E(L(S),L(\bar{x})).$$

Proof. Let $z \in L(E(S,\bar{x}))$ with $z = L(x)$ with $x \in E(S,\bar{x})$. Then $x \in T(S,\bar{x})$ and hence $z \in T(L(S),L(\bar{x}))$. Let $\{w_n\}$ be a sequence in $\mathrm{bd}(L(S))\backslash\{L(\bar{x})\}$ that converges to $L(\bar{x})$. By hypothesis, there are a subsequence $\{w_m\}$ and a sequence $\{x_m\} \subset \mathrm{bd}(S)\backslash\{\bar{x}\}$ such that $L(x_m) = w_m$, $x_m \to \bar{x}$, and the sequence $\{\|x_m - \bar{x}\|/\|w_m - L(\bar{x})\|\}$ is bounded from above. Since $x \in E(S,\bar{x})$, the sequence $\{x_n\}$ may be chosen so that there exist sequences $\{t_m\}$ and $\{z_m\}$ with $t_m \downarrow 0$, $z_m \to x$, and $x_m + t_m\|x_m - \bar{x}\|z_m \in S$. Then $w_m + t_m\|x_m - \bar{x}\|L(z_m) \in L(S)$ and $L(z_m) \to z$, since L is linear and continuous. We define

$$s_m := t_m \frac{\|x_m - \bar{x}\|}{\|w_m - L(\bar{x})\|}.$$

Since $t_m \to 0$ and $\{\|x_m - \bar{x}\|/\|w_m - L(\bar{x})\|\}$ is bounded, we deduce that $s_m \to 0$ and that $w_m + s_m\|w_m - L(\bar{x})\|L(z_m) \in S$ which confirms that $z \in E(L(S),L(\bar{x}))$ and the proof is complete. □

We conclude this section by discussing the notion of quasi-strict tangent cone of Penot [463].

Definition 4.6.16. Let X be a normed space, let S be a subset of X, and let $\bar{x} \in S$.

1. The **quasi-strict tangent cone** $Q(S,\bar{x})$ of S at $\bar{x}$ is the collection of all $x \in X$ such that for every sequence $(x_n,t_n) \to (\bar{x},0_+)$ with $\{x_n\} \subset S$ and $\lim_{n\to\infty} t_n^{-1}(x_n - \bar{x}) \in A(S,\bar{x})$, there exists $\{z_n\} \subset X$ converging to x with $x_n + t_n z_n \in S$ for every $n \in \mathbb{N}$.
2. The **interiorly quasi-strict tangent cone** $IQ(S,\bar{x})$ of S at $\bar{x}$ is the collection of all $x \in X$ such that for every sequence $(x_n,t_n) \to (\bar{x},0_+)$ with $\{x_n\} \subset S$ and $\lim_{n\to\infty} t_n^{-1}(x_n - \bar{x}) \in A(S,\bar{x})$, for every $\{z_n\} \subset X$ converging to x, we have $x_n + t_n z_n \in S$, for sufficiently large $n \in \mathbb{N}$.

We have the following results concerning the properties of the above cone (see [463, 593]).

Theorem 4.6.17. *Let X be a Banach space, let S, S_1, S_2 be subsets of X which are closed near $\bar{x} \in S \cap S_1 \cap S_2$. Then*

1. *$Q(S,\bar{x})$ is a closed, convex cone*
2. *$C(S,\bar{x}) \subset E(S,\bar{x}) \subset Q(S,\bar{x}) \subset A(S,\bar{x}) \subset T(S,\bar{x})$*
3. *$Q(S_1,\bar{x}) \cap Q(S_2,\bar{x}) \subset Q(S_1 \cap S_2,\bar{x})$ provided that $Q(S_1,\bar{x}) \cap IQ(S_2,\bar{x}) \neq \emptyset$*
4. *$A(S_1,\bar{x}) \cap A(S_2,\bar{x}) = A(S_1 \cap S_2,\bar{x})$ provided that $Q(S_1,\bar{x}) \cap IQ(S_2,\bar{x}) \neq \emptyset$.*

Proof. The proof is very similar to arguments used above and can be found in [463, 593]. □

We also state the following analogue of Theorem 4.6.15.

Theorem 4.6.18. *Let X and Y be Banach spaces, let $L : X \to Y$ be a linear and continuous mapping, let $S \subset X$, and let $\bar{x} \in S$. Assume that $A(L(S), L(\bar{x})) \subset L(A(S,\bar{x}))$. Then $L(Q(S,\bar{x})) \subset Q(L(S), L(\bar{x}))$.*

Proof. See [590, 593]. □

4.7 First-Order Inversion Theorems and Intersection Formulas

We now proceed to give inversion theorems which play a fundamental role in deriving calculus rules for the derivatives of set-valued maps. For this, we first collect a few definitions and results:

Definition 4.7.1. Let X be a Banach space and let $S \subset X$ be nonempty.

1. The set S is said to be closed near $x \in \operatorname{cl} S$, if $U_\varepsilon(x) \cap S$ is closed for some $\varepsilon > 0$.
2. If S is closed near $\bar{x}$, it is called **epi-Lipschitz-like** at $\bar{x}$, if there are $\tau > 0$, a convex set Q with polar set Q^o weak* locally compact, and $\beta > 0$ such that $S \cap U_\tau(x) + tQ \subseteq S$ for every $t \in (0,\beta)$.

Remark 4.7.2. Note that if X is finite-dimensional, then any locally closed set is epi-Lipschitz-like by setting $Q = \{0\}$. See [62, 63] for an excellent introduction to epi-Lipschitz-like sets.

Definition 4.7.3. Let X and Y be normed spaces and let $g : X \to Y$ be a given map.

1. The map g is called **strictly differentiable** at $\bar{x} \in X$, if there is a continuous linear map $Dg(\bar{x}) : X \to Y$ such that

$$\lim_{(x,y,t)\to(\bar{x},\bar{y},0_+)} \frac{g(x+ty)-g(x)}{t} = Dg(\bar{x})(\bar{y}), \quad \text{for every } \bar{y} \in X.$$

2. The map g is called **Hadamard differentiable** at $\bar{x} \in X$, if

$$\lim_{(y,t)\to(\bar{y},0_+)} \frac{g(\bar{x}+ty)-g(\bar{x})}{t} = Dg(\bar{x})(\bar{y}), \quad \text{for every } \bar{y} \in X.$$

The following important result is a special case of Borwein and Strojwas [65, Theorem 4.1]. This result, which unifies the finite-dimensional and Banach space case treated separately by Borwein [60], is taken from Ward [590].

Theorem 4.7.4. *Let X and Y be Banach spaces, and let $g : X \to Y$ be strictly differentiable on $U_\lambda(\bar{x})$, for some $\lambda > 0$, where $\bar{x} \in g^{-1}(0) \cap S$ and $S \subset X$ is epi-Lipschitz -like at $\bar{x}$. Suppose*

$$Dg(x)C(S,\bar{x}) = Y. \tag{4.42}$$

Then there are $k > 0$ and $\delta > 0$ such that for each $x \in S \cap U_\delta(\bar{x})$, there exists $z \in S \cap g^{-1}(0)$ with

$$\|x - z\| \leq k\|g(x)\|.$$

With the above preparation, we give the following inversion theorem taken from [590]:

Theorem 4.7.5. *Let X and Y be Banach spaces, and let $g : X \to Y$ be strictly differentiable on $U_\lambda(\bar{x})$ for some $\lambda > 0$, where $\bar{x} \in g^{-1}(0) \cap S$ and $S \subset X$ is epi-Lipschitz -like at $\bar{x}$. Assume that* (4.42) *holds. Then the following inclusions hold:*

$$C(S,\bar{x}) \cap Dg(\bar{x})^{-1}(0) \subseteq C(S \cap g^{-1}(0),\bar{x}), \tag{4.43}$$

$$A(S,\bar{x}) \cap Dg(\bar{x})^{-1}(0) \subseteq A(S \cap g^{-1}(0),\bar{x}), \tag{4.44}$$

$$T(S,\bar{x}) \cap Dg(\bar{x})^{-1}(0) \subseteq T(S \cap g^{-1}(0),\bar{x}), \tag{4.45}$$

$$P(S,\bar{x}) \cap Dg(\bar{x})^{-1}(0) \subseteq P(S \cap g^{-1}(0),\bar{x}), \tag{4.46}$$

$$T_\infty(S,\bar{x}) \cap Dg(\bar{x})^{-1}(0) \subseteq T_\infty(S \cap g^{-1}(0),\bar{x}), \tag{4.47}$$

$$A_\infty(S,\bar{x}) \cap Dg(\bar{x})^{-1}(0) \subseteq A_\infty(S \cap g^{-1}(0),\bar{x}). \tag{4.48}$$

Proof. The inclusions for the Clarke's tangent cone and the contingent cone are given in [60]. We prove the inclusion for A_∞ and the remaining inclusions can be proved by using similar arguments. Let $x \in A_\infty(S,\bar{x}) \cap Dg(\bar{x})^{-1}(0)$ and $z \in A(S \cap g^{-1}(0),\bar{x})$ be chosen arbitrarily. It is sufficient to show that $x + z \in A(S \cap g^{-1}(0),\bar{x})$. Since $A(\cdot,\cdot)$ is isotone, $z \in A(S,\bar{x})$ and $z \in A(g^{-1}(0),\bar{x}) \subset Dg(\bar{x})^{-1}(0)$. Therefore, for an arbitrary sequence $\{t_n\} \subset \mathbb{P}$ with $t_n \to 0$, there exists $\{w_n\}$ with $w_n \to (x+z)$ and $\bar{x} + t_n w_n \in S$. Since g is strictly differentiable at $\bar{x}$, we have

$$\frac{g(\bar{x} + t_n w_n) - g(\bar{x})}{t_n} \to Dg(\bar{x})(x+z) = 0.$$

Now, Theorem 4.7.4 implies that there exists $u_n \in S \cap g^{-1}(0)$ with $t_n^{-1}(u_n - \bar{x} - t_n w_n) \to 0$. By setting $v_n = t_n^{-1}(u_n - \bar{x})$, we note that $v_n \to (x+z)$ and $\bar{x} + t_n v_n = u_n \in S \cap g^{-1}(0)$. Consequently $x + z \in A(S \cap g^{-1}(0),\bar{x})$ and the proof is complete. □

The following inversion, which is recently given by Durea and Strugariu [155], is based on the notion of metric subregularity which is an extension of the metric regularity, see [10, 146, 155, 622]. We remark that a detailed study of such results under the notion of metric regularity have been carried out by Borwein [60], Cominetti [103] and Jourani [309], among others.

Theorem 4.7.6. *Let X and Y be Banach spaces, let $S \subset X$ and $Q \subset Y$ be closed sets, let $g : X \to Y$ be a continuously Fréchet differentiable map, and let $\bar{x} \in S \cap g^{-1}(Q)$. Assume that $f : X \times Y \to Y$ defined by $f(x,y) := g(x) - y$ is metrically subregular at $(\bar{x}, g(\bar{x}), 0)$ with respect to $S \times Q$ in the following sense: There exist a constant $k > 0$ and neighborhoods $U_\tau(\bar{x})$ and $U_\tau(g(\bar{x}))$ such that for every $(x,y) \in [U_\tau(\bar{x}) \times U_\tau(g(\bar{x}))] \cap S \times Q$, we have*

$$d\left((x,y), f^{-1}(0) \cap (S \times Q)\right) \le k\|g(x) - y\|. \tag{4.49}$$

Then the following formulas hold:

$$T(S,\bar{x}) \cap Dg(\bar{x})^{-1}(A(Q,g(\bar{x}))) \subset T\left(S \cap g^{-1}(Q),\bar{x}\right),$$

$$A(S,\bar{x}) \cap Dg(\bar{x})^{-1}(T(Q,g(\bar{x}))) \subset T\left(S \cap g^{-1}(Q),\bar{x}\right),$$

$$A(S,\bar{x}) \cap Dg(\bar{x})^{-1}(A(Q,g(\bar{x}))) = A\left(S \cap g^{-1}(Q),\bar{x}\right).$$

Proof. In view of the metric subregularity hypothesis, there are a number $k > 0$ and neighborhoods $U_\tau(\bar{x})$ and $U_\tau(g(\bar{x}))$ such that for every $(x,y) \in [U_\tau(\bar{x}) \times U_\tau(g(\bar{x}))] \cap S \times Q$, (4.49) holds.

We choose $x \in T(S,\bar{x}) \cap Dg(\bar{x})^{-1}(A(Q,g(\bar{x})))$ so that $x \in T(S,\bar{x})$ and $Dg(\bar{x})(x) \in A(Q,g(\bar{x}))$. By using the definitions of the contingent cone and the adjacent cone, there are sequences $\{t_n\} \subset \mathbb{P}$, $\{x_n\} \subset X$ and $\{z_n\} \subset Y$ such that $t_n \downarrow 0$, $x_n \to x$ and $z_n \to Dg(\bar{x})(x)$ with $\bar{x} + t_n x_n \in S$ and $g(\bar{x}) + t_n z_n \in Q$. For sufficiently large $n \in \mathbb{N}$, (4.49) with $(x,y) = (\bar{x} + t_n x_n, g(\bar{x}) + t_n z_n)$ ensures that there exists $(p_n, q_n) \in S \times Q$ with $g(p_n) = q_n$ such that

$$\|(\bar{x} + t_n x_n, g(\bar{x}) + t_n z_n) - (p_n, q_n)\| < k\|g(\bar{x} + t_n x_n) - g(\bar{x}) - t_n z_n\| + t_n^2.$$

Consequently, for sufficiently large $n \in \mathbb{N}$, we have $p_n \in S \cap g^{-1}(Q)$ and

$$\|\bar{x} + t_n x_n - p_n\| < k\|g(\bar{x} + t_n x_n) - g(\bar{x}) - t_n z_n\| + t_n^2$$

which implies that

$$\left\| \frac{p_n - \bar{x}}{t_n} + x_n \right\| < k \left\| \frac{g(\bar{x} + t_n x_n) - g(\bar{x})}{t_n} - z_n \right\| + t_n.$$

Noticing that

$$\frac{g(\bar{x} + t_n x_n) - g(\bar{x})}{t_n} \to Dg(\bar{x})(x),$$

we deduce that $w_n := t_n^{-1}(p_n - \bar{x}) \to x$ and the proof of the first inclusion is complete. The remaining parts follow by using the properties of the contingent cone and the adjacent cone. □

For the sake of completeness, we also state the following inversion theorem from [18, Page 147]. As it can be seen that (4.50) is used to prove that the inverse of a certain map is pseudo-Lipschitz and hence it can be obtained by the arguments used in the proof of Theorem 4.7.6.

Theorem 4.7.7. *Let X and Y be Banach spaces, let $S \subset X$ and $Q \subset Y$ be closed sets, let $g : X \to Y$ be a continuously differentiable around $\bar{x} \in S \cap g^{-1}(Q)$. Assume that either Y is finite-dimensional and*

$$Dg(\bar{x})(C(S,\bar{x})) - C(Q,g(\bar{x})) = Y, \tag{4.50}$$

or, there are $c > 0$, $\alpha \in [0,1[$ and $\tau > 0$ such that for every $x \in S \cap U_\tau(\bar{x})$ and for every $y \in Q \cap U_\tau(g(\bar{x}))$,

$$B_Y \subset Dg(x)\,(A(S,x) \cap cB_X) - T(Q,y) + \alpha B_Y.$$

Then the following relations hold:

$$A(S,\bar{x}) \cap Dg(\bar{x})^{-1}(T(Q,g(\bar{x}))) \subset T\left(S \cap g^{-1}(Q),\bar{x}\right),$$
$$C(S,\bar{x}) \cap Dg(\bar{x})^{-1}(C(Q,g(\bar{x}))) \subset C\left(S \cap g^{-1}(Q),\bar{x}\right),$$
$$A(S,\bar{x}) \cap Dg(\bar{x})^{-1}(A(Q,g(\bar{x}))) = A\left(S \cap g^{-1}(Q),\bar{x}\right).$$

The following two corollaries of the above result are immediate:

Corollary 4.7.8. *Let X and Y be Banach spaces, let $Q \subset Y$ be closed, let $g : X \to Y$ be a continuously differentiable around $\bar{x} \in g^{-1}(Q)$. Assume that either Y is finite-dimensional and*

$$\operatorname{Im}(Dg(\bar{x})) + C(Q,g(\bar{x})) = Y,$$

or, there are $c > 0$, $\alpha \in [0,1[$ and $\tau > 0$ such that for every $x \in U_\tau(\bar{x})$ and for every $y \in Q \cap U_\tau(g(\bar{x}))$,

$$B_Y \subset c\,Dg(x)(B_X) + T(Q,y) + \alpha B_Y.$$

Then the following relations hold:

$$Dg(\bar{x})^{-1}(T(Q,g(\bar{x}))) = T\left(g^{-1}(Q),\bar{x}\right),$$
$$Dg(\bar{x})^{-1}(A(Q,g(\bar{x}))) = A\left(g^{-1}(Q),\bar{x}\right),$$
$$Dg(\bar{x})^{-1}(C(Q,g(\bar{x}))) \subset C\left(g^{-1}(Q),\bar{x}\right).$$

Corollary 4.7.9. *Let X be a Banach space, let $S_1, S_2 \subset X$ be closed sets, and let $\bar{x} \in S_1 \cap S_2$. Assume that either X is finite-dimensional and*

$$C(S_1,\bar{x}) - C(S_2,\bar{x}) = X$$

or there are $c > 0$, $\alpha \in [0,1[$ and $\tau > 0$ such that for every $x \in S_1 \cap U_\tau(\bar{x})$ and for every $y \in S_2 \cap U_\tau(\bar{x})$,

$$B_Y \subset (A(S_1,x) \cap cB_X) - T(S_2,y) + \alpha B_Y.$$

Then the following relations hold:

$$A(S_1,\bar{x}) \cap T(S_2,\bar{x}) \subset T(S_1 \cap S_2,\bar{x}),$$
$$C(S_1,\bar{x}) \cap C(S_2,\bar{x}) \subset C(S_1 \cap S_2,\bar{x}),$$
$$A(S_1,\bar{x}) \cap A(S_2,\bar{x}) = A(S_1 \cap S_2,\bar{x}).$$

In finite dimensions, an inversion theorem can be given involving the basic normal cone (see [596, Theorem 2.4]):

Theorem 4.7.10. *Let $S \subset \mathbb{R}^n$ be locally closed at $\bar{x} \in \mathbb{R}^n$ and let $g : \mathbb{R}^n \to \mathbb{R}^m$ be strictly differentiable at $\bar{x} \in S \cap g^{-1}(0)$. If*

$$0 \in y^* Dg(\bar{x}) + N(S, \bar{x}) \quad \Longrightarrow \quad y^* = 0, \tag{4.51}$$

then the following formulas hold:

$$T(S, \bar{x}) \cap Dg(\bar{x})^{-1}(0) = T\left(S \cap g^{-1}(0), \bar{x}\right),$$
$$A(S, \bar{x}) \cap Dg(\bar{x})^{-1}(0) = A\left(S \cap g^{-1}(0), \bar{x}\right).$$

4.8 Expressions of the Contingent Cone on Some Constraint Sets

In this section, we give explicit formulas for the computation of the contingent cone on certain constraint sets. Similar results can be derived by using the inversion theorem studied in the previous sections, however, by taking into account the specific structure of the constraint sets, the conditions can be significantly relaxed.

Let X be a Hilbert space, let $I_1 = \{1, 2, \ldots, m\}$ and $I_2 = \{m+1, m+2, \ldots, s\}$ be index sets, and let $h_i : X \to \mathbb{R}$ with $i \in I_1 \cup I_2$ be maps defining the sets of equality and inequality constraints by

$$S_i = \{x \in X | \, h_i(x) = 0\} \quad \text{for } i \in I_1,$$
$$S_j = \{x \in X | \, h_j(x) \geq 0\} \quad \text{for } j \in I_2,$$

and, in a combined form,

$$S := \bigcap_{i=1}^{s} S_i = \left\{x \in X | \; h_i(x) = 0, \text{ for } i \in I_1, \text{ and } h_j(x) \geq 0, \text{ for } j \in I_2\right\}. \tag{4.52}$$

We aim to give formulas for the contingent cone for the constraint set $P \cap S$ at any $\bar{x} \in P \cap S$, where $P \subset X$ is an abstract constraint set. For this, we first derive expressions for the contingent cone for S in terms of the sets S_i with $i \in I_1 \cup I_2$ and then extend the formulas to the set $S \cap P$. This section is based on the interesting work by Di [134] where more details can be found.

We will recall some notions of derivatives and auxiliary results. We begin with the following:

Definition 4.8.1. Let X be a Hilbert space, and let $g : X \to \mathbb{R}$ be a given function. We say that the map g admits first-order Fréchet derivative at $\bar{x}$, if there exists an element $\nabla g(\bar{x}) \in X$ such that

$$\lim_{t \to 0} \frac{g(\bar{x} + tx) - g(\bar{x})}{t} = \langle \nabla g(\bar{x}), x \rangle, \tag{4.53}$$

and the limit is uniform with respect to x in bounded sets.

By Riesz's theorem, first-order Fréchet derivatives can be seen as a linear map from X to $\mathbb{R}$.

We now give a formula for the contingent cone to a single equality and inequality constraint:

Proposition 4.8.2. *Let X be a Hilbert space and let $h : X \to \mathbb{R}$ be continuous around $\bar{x} \in X$ and differentiable at $\bar{x}$. Assume that $h(\bar{x}) = 0$ and $\nabla h(\bar{x}) \neq 0$. Then:*

1. *For $S = \{x \in X \mid h(x) = 0\}$, we have*

$$T(S, \bar{x}) = \{x \in X \mid \langle \nabla h(\bar{x}), x \rangle = 0\}.$$

2. *For $S = \{x \in X \mid h(x) \geq 0\}$, we have*

$$T(S, \bar{x}) = \{x \in X \mid \langle \nabla h(\bar{x}), x \rangle \geq 0\}.$$

We also recall the following notion:

Definition 4.8.3. Let X be a Hilbert space, let $S \subset X$ be closed, and let $\bar{x} \in S$. The **strict normal cone** $\hat{N}(\bar{x}, S)$ of S at $\bar{x}$ is the collection of all elements $u \in X$ such that

$$\limsup_{z \to \bar{x},\, \bar{x} \neq z \in S} \langle u, z - \bar{x} \rangle \leq 0. \tag{4.54}$$

The strict normal cone is known to have the characterization given by: $\hat{N}(\bar{x}, S) = -T(S, \bar{x})^{+}$.

We next recall the following alternative theorem:

Lemma 4.8.4. *Let X be a Hilbert space and let $A \in \mathscr{L}(\mathbb{R}^m, X)$. Assume that Γ_1 and Γ_2 are two closed, convex cones in $\mathbb{R}^m$ such that $-\Gamma_2^* \subset \Gamma_1$ and $\operatorname{int}(\Gamma_2) \neq \emptyset$. Then exactly one of the following systems has a solution:*

1. *$-A^T A x \in \operatorname{int}(\Gamma_2)$ for $x \in \Gamma_1$ such that $|x| \leq 1$;*
2. *$Ap = 0 \in X$, for $0 \neq p \in \Gamma_2^*$.*

Recall that an m-**dimensional regular box** in $\mathbb{R}^m$ is a nonempty, closed, polyhedron with 2^m vertices and $2m$ faces such that it is linearly homomorphic to the standard box $[0, 1]^m$.

We now recall the following generalized mean-value theorem that uses the above notion:

Lemma 4.8.5. *Let $g = (g_1, \ldots, g_m)$ be maps given on an open set containing an m-dimensional regular box $\mathbb{B}$ in $\mathbb{R}^n$. Let Π_i^1 and Π_i^2 be its i-th pair of opposite sides. If for each $i = 1, 2, \ldots, m$, $g_i(x) > a_i$ for $x \in \Pi_1^1$ and $g_i(x) < a_i$ for $x \in \Pi_1^2$, then there exists $z \in \operatorname{int}(\mathbb{B})$ with $g(z) = a := (a_1, a_2, \ldots, a_m)$.*

The following fundamental theorem on systems of linear equations will also be used shortly:

Lemma 4.8.6. *Let X be a Hilbert space, let $x, a_i \in X$ and let $b_i \in \mathbb{R}$, where $i \in \{0, 1, \ldots, s\}$. Then for an inequality $\langle a_0, x\rangle \leq b_0$ to be a consequence of a consistent system*

$$\langle a_i, x\rangle = b_i \quad \text{for } i \in I_1 \quad \langle a_i, x\rangle \leq b_i \quad \text{for } i \in I_2, \tag{4.55}$$

it is necessary and sufficient to have the following condition: There exists $\lambda = (\lambda_1, \ldots, \lambda_s)$, with $\lambda_i \geq 0$ for $i \in I_2$ such that $a_0 = \sum_{i=1}^{s} \lambda_i a_i$, and $\sum_{i=1}^{s} \lambda_i b_i \leq b_0$.

Proof. See Zeidler [617]. □

In order to compute the contingent cone to the constraint set S at $\bar{x} \in S$, it is convenient to divide the inequality constraints into active and non-active types as follows:

$$I_2^{=} = \{i \in I_2 |\ h_i(\bar{x}) = 0\},$$
$$I_2^{>} = \{i \in I_2 |\ h_i(\bar{x}) > 0\}.$$

We also define

$$\Lambda := \{\lambda \in \mathbb{R}^s |\ \lambda_i \geq 0, \text{ for } i \in I_2^{=} \text{ and } \lambda_i = 0 \text{ for } i \in I_2^{>}\}, \tag{4.56}$$

and note that

$$\Lambda^* = \{\lambda \in \mathbb{R}^s |\ \lambda_i = 0, \text{ for } i \in I_1 \text{ and } \lambda_i \geq 0 \text{ for } i \in I_2^{=}\}. \tag{4.57}$$

Finally, we recall the following technical result (see [134]):

Lemma 4.8.7. *Let $v_1, v_2, \ldots, v_s$ be elements in a Hilbert space X. If there is no $0 \neq \lambda \in \Lambda$ such that*

$$\sum_{i=1}^{s} \lambda_i v_i = 0, \tag{4.58}$$

then there exists an m-dimensional regular box $\mathbb{B}$ in $\mathbb{R}^s$, a linearly bijective image of $[0, 1]^m$, with the jth pair of opposite sides Π_j^1 and Π_j^2, such that for $j \in I_1$, we have

$$\left\langle v_j, \sum_{i=1}^{s} \alpha_i v_i \right\rangle > 0 \quad for\, \alpha \in \Pi_j^1,$$

$$\left\langle v_j, \sum_{i=1}^{s} \alpha_i v_i \right\rangle < 0 \quad for\, \alpha \in \Pi_j^2,$$

and for $j \in I_2^=$, *we have* $\langle v_j, \sum_{i=1}^{s} \alpha_i v_i \rangle > 0$, *for* α *in all sides of* $\mathbb{B}$, *where* $\alpha = (\alpha_1, \alpha_2, \ldots, \alpha_s) \in \mathbb{R}^s$.

We are now ready to give the following formula, originally proved by Di [134] (see [135, 136]).

Theorem 4.8.8. *Let* X *be a Hilbert space, let* $S \subset X$ *be given by* (4.52), *and let* $\bar{x} \in S$. *Assume that the following* ***Mangasarian-Fromovitz constraint qualification*** *holds: Let* $h = (h_1, h_2, \ldots, h_s)$ *be continuous around* $\bar{x}$ *and Fréchet differentiable at* $\bar{x}$. *The only* $\lambda \in \Lambda$, *where* Λ *is given by* (4.56), *such that* $\sum_{i=1}^{s} \lambda_i \nabla h_i(\bar{x}) = 0$ *is* $\lambda = 0$. *Then the following formula holds:*

$$T(S, \bar{x}) = \bigcap_{i=1}^{s} T(S_i, \bar{x}) = \left\{x \in X \,|\, \nabla h^T(\bar{x})(x) \in \Lambda^*\right\} = \nabla h(\bar{x})^{-1} \Lambda^*$$

$$= \left\{ x \in X \;\middle|\; \begin{array}{ll} \langle \nabla h_i(\bar{x}), x \rangle = 0, & for\ i \in I_1 \\ \langle \nabla h_i(\bar{x}), x \rangle \geq 0, & for\ i \in I_2^= \end{array} \right\},$$

where Λ^* *is given by* (4.57) *and* $\nabla h(\bar{x})^{-1} \Lambda^*$ *is an inverse image of* $\nabla h(\bar{x}) = (\nabla h_1(\bar{x}), \ldots, \nabla h_s(\bar{x}))$.

Proof. It suffices to show that $\bigcap_{i=1}^{s} T(S_i, \bar{x}) \subset T(S, \bar{x})$, as the converse inclusion always holds and the other formulas follow from Proposition 4.8.2.

Let $x \in \bigcap_{i=1}^{s} T(S_i, \bar{x})$ be arbitrary. Then, using Proposition 4.8.2, it follows that

$$\langle \nabla h_i(\bar{x}), x \rangle = 0 \quad \text{for } i \in I_1,$$

$$\langle \nabla h_i(\bar{x}), x \rangle \geq 0 \quad \text{for } i \in I_2^=.$$

By the Mangasarian-Fromovitz constraint qualification and Lemma 4.8.7, there exists an m-dimensional regular box $\mathbb{B}$ in $\mathbb{R}^s$ with m pairs of opposite sides $\left\{\left(\Pi_j^1, \Pi_j^2\right) \mid j = 1, \ldots, m\right\}$ such that for $j \in I_1$, we have

$$\left\langle \nabla h_j(\bar{x}), \sum_{i=1}^{s} \alpha_i \nabla h_i(\bar{x}) \right\rangle > 0 \quad \text{for } \alpha \in \Pi_j^1; \tag{4.59a}$$

$$\left\langle \nabla h_j(\bar{x}), \sum_{i=1}^{s} \alpha_i \nabla h_i(\bar{x}) \right\rangle < 0 \quad \text{for } \alpha \in \Pi_j^2, \tag{4.59b}$$

and for $j \in I_2^{=}$, we have

$$\left\langle \nabla h_j(\bar{x}), \sum_{i=1}^{s} \alpha_i \nabla h_i(\bar{x}) \right\rangle > 0 \quad \text{for } \alpha \text{ in all sides of the box } \mathbb{B}. \tag{4.60}$$

For $n \in \mathbb{N}$, we define

$$x_{n,t}(\alpha) = \bar{x} + t\left(x + \frac{1}{n}\sum_{i=1}^{s} \alpha_i \nabla h_i(\bar{x})\right), \tag{4.61}$$

and by using the Fréchet differentiability of functions at $\bar{x}$, for any $j \in I_1$ or $j \in I_2^{=}$, we have

$$\begin{aligned} h_j(x_{n,t}(\alpha)) &= h_j(\bar{x}) + t\left\langle \nabla h_j(\bar{x}), x + \frac{1}{n}\sum_{i=1}^{s} \alpha_i \nabla h_i(\bar{x}) \right\rangle \\ &\quad + o\left(t\left(x + \frac{1}{n}\sum_{i=1}^{s} \alpha_i \nabla h_i(\bar{x})\right)\right) \\ &= t\left\langle \nabla h_j(\bar{x}), x + \frac{1}{n}\sum_{i=1}^{s} \alpha_i \nabla h_i(\bar{x}) \right\rangle + o(t). \end{aligned}$$

Furthermore, for an inactive constraint, that is, for $j \in I_2^{>}$ or j such that $h_j(\bar{x}) > 0$, it follows from the continuity of h_j around $\bar{x}$ that for sufficiently small t, we have $h_j(x_{n,t}(\alpha)) > 0$ for $\alpha \in \mathbb{B}$. Also from (4.60) and (4.61), we can choose a sufficiently small t, denoted by t_n, such that for all $j \in I_1$, we have

$$h_j(x_{n,t_n}(\alpha)) > 0 \quad \text{for } \alpha \in \Pi_j^1,$$
$$h_j(x_{n,t_n}(\alpha)) < 0 \quad \text{for } \alpha \in \Pi_j^2,$$

and for all $j \in I_2^{=}$, we have

$$h_j(x_{n,t_n}(\alpha)) \geq 0 \quad \text{for } \alpha \in \mathbb{B}.$$

By using Lemma 4.8.5, there exists $\hat{\alpha}_n \in \mathbb{B}$ such that

$$h_j(x_{n,t_n}(\hat{\alpha}_n)) = 0 \quad \text{for } j \in I_1 \tag{4.62a}$$

$$h_j(x_{n,t_n}(\hat{\alpha}_n)) \geq 0 \quad \text{for } j \in I_2, \tag{4.62b}$$

which implies that $x_n := x_{n,t_n}(\hat{\alpha}_n) \in S$ and since $t_n^{-1}(x_n - \bar{x}) \to x$, we deduce that $x \in T(S, \bar{x})$. □

Remark 4.8.9. Ledzewicz and Walczak [370] proved the above result in a Banach space setting by imposing an additional condition that h is Lipschitz around $\bar{x}$. See also [371].

Since the strict normal cone is the polar of the contingent cone, we have the following result:

Corollary 4.8.10. *Under the assumptions of Theorem 4.8.8, we have*

$$\hat{N}(\bar{x}, S) = \left\{ -\sum_{i=1}^{s} \lambda_i \nabla h_i(\bar{x}) \mid \lambda \in \Lambda \right\} = -\nabla h(\bar{x})\Lambda.$$

Having obtained expressions for the contingent cone on the set of equality and inequality constraints, we now proceed to incorporate an abstract constraint set:

Theorem 4.8.11. *Let X be a finite-dimensional Hilbert space, let $P \subset X$ be closed and convex, let $S \subset X$ be given by* (4.52), *and $\bar{x} \in P \cap S$. Assume that the following constraint qualification holds: Let $h = (h_1, h_2, \ldots, h_s)$ be continuous around $\bar{x}$ and Fréchet differentiable at $\bar{x}$. Assume that there is no $0 \neq \lambda \in \Lambda$, where Λ is given by* (4.56), *such that $\sum_{i=1}^{s} \lambda_i \nabla h_i(\bar{x}) \in \hat{N}(\bar{x}, P)$. Then:*

$$T(S \cap P, \bar{x}) = \bigcap_{i=1}^{s} T(S_i, \bar{x}) \cap T(P, \bar{x}). \tag{4.63}$$

Proof. Let $x \in \bigcap_{i=1}^{s} T(S_i, \bar{x}) \cap T(P, \bar{x})$ be arbitrary. Then there exist sequences $\{x_n\} \subset P$ and $\{t_n\} \subset \mathbb{P}$ such that $t_n \downarrow 0$, $z_n = t_n^{-1}(x_n - \bar{x}) \to x$, and

$$\langle \nabla h_i(\bar{x}), x \rangle = 0 \quad \text{for } i \in I_1$$

$$\langle \nabla h_i(\bar{x}), x \rangle \geq 0 \quad \text{for } i \in I_2^{=}.$$

We define $\beta_n := \|z_n - x\|$ and note that $\beta_n \to 0$ and $x_n = \bar{x} + t_n x + t_n \beta_n \tilde{x}_n$ for some $\tilde{x}_n$ with $|\tilde{x}_n| \leq 1$. Similar to the proof of Theorem 4.8.8, there exists an m-dimensional regular box $\mathbb{B}$ in $\mathbb{R}^s$ with m pairs of opposite sides

$\left\{\left(\Pi_j^1, \Pi_j^2\right) \mid j = 1, \ldots, m\right\}$ and with 2^m vertices such that $j \in I_1$, we have

$$\left\langle \nabla h_j(\bar{x}), \sum_{i=1}^{s} \alpha_i \nabla h_i(\bar{x}) \right\rangle > 0 \quad \text{for } \alpha \in \Pi_j^1; \tag{4.64a}$$

$$\left\langle \nabla h_j(\bar{x}), \sum_{i=1}^{s} \alpha_i \nabla h_i(\bar{x}) \right\rangle < 0 \quad \text{for } \alpha \in \Pi_j^2, \tag{4.64b}$$

and for $j \in I_2^{=}$, we have

$$\left\langle \nabla h_j(\bar{x}), \sum_{i=1}^{s} \alpha_i \nabla h_i(\bar{x}) \right\rangle > 0 \quad \text{for } \alpha \text{ in all sides of the box } \mathbb{B}. \tag{4.65}$$

We define

$$W = \left\{ w \in X \;\middle|\; |w| \le 1, \begin{array}{l} \langle w, \nabla h_j(\bar{x})\rangle = 0 \text{ for } j \in I_1 \\ \langle w, \nabla h_j(\bar{x})\rangle \ge 0 \text{ for } j \in I_2^{=} \end{array} \right\}. \tag{4.66}$$

Given $n \in \mathbb{N}$, $\alpha \in \mathbb{B}$, $w \in W$, and $t > 0$, we define

$$x_{n,t}(\alpha, w) = \bar{x} + t \cdot t_n \left(x + \beta_n^{\frac{1}{2}} \sum_{i=1}^{s} \alpha_i \nabla h_i(\bar{x}) + \beta_n^{\frac{1}{3}} w \right), \tag{4.67}$$

and by using the Fréchet differentiability of functions h_i at $\bar{x}$, for any $j \in I_1$ or $j \in I_2$, we have

$$\begin{aligned} h_j(x_{n,t}(\alpha, w)) = {} & h_j(\bar{x}) + t \cdot t_n \langle \nabla h_j(\bar{x}), x\rangle + t \cdot t_n \beta_n^{\frac{1}{2}} \left\langle \nabla h_j(\bar{x}), \sum_{i=1}^{s} \alpha_i \nabla h_i(\bar{x}) \right\rangle \\ & + t \cdot t_n \langle \nabla h_j(\bar{x}), w\rangle + o\left(t \cdot t_n \left| x + \beta_n^{\frac{1}{2}} \sum_{i=1}^{s} \alpha_i \nabla h_i(\bar{x}) + \beta_n^{\frac{1}{3}} w \right| \right), \end{aligned}$$

which is positive for $j \in I_2^{>}$, $\alpha \in \mathbb{B}$ and $w \in W$ when t is small enough, say $t \le s_n$, because $h_j(\bar{x}) > 0$; and is positive for $j \in I_2^{=}$, $\alpha \in \mathbb{B}$ and $w \in W$ when t is small enough, say $t \le s_n$, because $h_j(\bar{x}) = 0$ and $\langle \nabla h_j(\bar{x}), w\rangle \ge 0$, and equals

$$t \cdot t_n \beta_n^{\frac{1}{2}} \left\langle \nabla h_j(\bar{x}), \sum_{i=1}^{s} \alpha_i \nabla h_i(\bar{x}) \right\rangle + o\left(t \cdot t_n \left| x + \beta_n^{\frac{1}{2}} \sum_{i=1}^{s} \alpha_i \nabla h_i(\bar{x}) + \beta_n^{\frac{1}{3}} w \right| \right),$$

for $j \in I_1$, $\alpha \in \mathbb{B}$, $w \in W$ and for all t, because $h_j(\bar{x}) = 0$, $\langle \nabla h_j(\bar{x}), x\rangle = 0$ and $\langle \nabla h_j(\bar{x}), w\rangle = 0$. Therefore, for $j \in I_1$, we have

$$h_j(x_{n,s_n}(\alpha, w)) > 0 \quad \text{for } \alpha \in \Pi_j^1, \text{ and } w \in W, \tag{4.68a}$$

$$h_j(x_{n,s_n}(\alpha, w)) < 0 \quad \text{for } \alpha \in \Pi_j^2 \text{ and } w \in W, \tag{4.68b}$$

and for all $j \in I_2$, we have

$$h_j(x_{n,s_n}(\alpha, w)) \geq 0 \quad \text{for } \alpha \in \mathbb{B}, \text{ and } w \in W. \tag{4.69}$$

Assuming that there exists an $\tilde{\alpha} \in \mathbb{B}$ such that the compact, convex set

$$S_{n,s_n}(\tilde{\alpha}) = \{x_{n,s_n}(\alpha, w))| \ w \in W\}$$

is disjoint with P, then by the separation theorem, we know that there exists a hyperplane strongly separating $S_{n,s_n}(\tilde{\alpha})$ from P, that is,

$$\langle u_n, y\rangle \leq \langle u_n, x_{n,s_n}(\tilde{\alpha}, w)\rangle, \tag{4.70}$$

for every $w \in W$ and $y \in P$ where $u_n \neq 0$. Without any loss of generality, we assume that $\|u_n\| = 1$ and $u_n \to u$. This means that in the limit, $\langle u, y - \bar{x}\rangle \leq 0$ for every $y \in P$, confirming that $u \in \hat{N}(\bar{x}, P)$. On the other hand, given $w \in W$ such that $\|w\| = 1$, we define $z_n(s_n) := s_n \cdot x_n + (1 - s_n)\bar{x}$, $v_n(w) := x_{n,s_n}(\tilde{\alpha}, w) - z_n(s_n)$, and get $z_n(s_n) = \bar{x} + s_n(x_n - \bar{x}) = \bar{x} + s_n t_n v_n$ with

$$v_n(w) = s_n t_n \left(x - v_n + \beta_k^{\frac{1}{2}} \sum_{i=1}^{s} \tilde{\alpha}_i \nabla h(\bar{x}) + \beta_n^{\frac{1}{3}} w \right)$$

and by (4.70), we have

$$\left\langle u_n, \frac{v_n(w)}{|v_n(w)|} \right\rangle \geq 0, \tag{4.71}$$

Since $\|x - v_n\| = \beta_n$, we deduce that $\frac{v_n(w)}{\|v_n(w)\|} \to 0$. Therefore, by taking the limit in (4.71), we have $\langle u, w\rangle \geq 0$, implying that $u \in W^*$. By employing Lemma 4.8.6, there exists a nonzero vector $\lambda \in \Lambda$ such that $\sum_{i=1}^{s} \lambda_i \nabla h_i(\bar{x}) = u \in \hat{N}(\bar{x}, P)$, which, however, contradicts the imposed constraint qualification, and hence for any $\alpha \in \mathbb{B}$, we must have $S_{n,s_n} \cap P \neq \emptyset$. Therefore, for any vertex $\alpha_i \in \mathbb{B}$ there exists $w_i \in W$ such that $z_{n,i} = x_{n,s_n}(\alpha_i, w_i) \in P$. Now, $z_{n,1}, \ldots z_{n,2^m}$ form a new m-dimensional regular box $\tilde{\mathbb{B}}$. Therefore, we apply the generalized mean value theorem to $\tilde{\mathbb{B}}$ which is a subset of P since P is convex. From (4.69) and (4.70), we can apply generalized mean value theorem to $\tilde{\mathbb{B}}$. Consequently, there exists $\hat{x}_n \in \tilde{\mathbb{B}} \cap P$ such that

$$h_j(\hat{x}_n) = 0 \quad \text{for } j \in I_1 \text{ and } h_j(\hat{x}_n) \geq 0, \ \text{for} j \in I_2, \qquad .$$

that is, $\hat{x}_n \in P \cap S$. It can be shown that $(t_n s_n)^{-1}(\hat{x}_n - \bar{x}) \to x$ which confirms that $x \in T(P \cap S, \bar{x})$ because $\hat{x}_n \in \mathbb{B}_n$. Since the converse inclusion is always true, the proof is complete. □

Since the Clarke normal cone and the strict normal cone on a convex set P coincide, we get:

Corollary 4.8.12. *Under the assumptions of Theorem 4.8.11, we have*

$$\hat{N}(\bar{x}, P \cap S) = -\nabla h(\bar{x}) + \hat{N}(\bar{x}, P). \tag{4.72}$$

Proof. We will show that the right side is closed. Assume that $z_n := x_n + y_n$ with $x_n \in -\nabla h(\bar{x})\Lambda$ and $y_n \in \hat{N}(\bar{x}, P)$, and $z_n \to \bar{z}$. Since $-\nabla h(\bar{x})\Lambda$ is a closed set in a finite-dimensional subspace, if $\{x_n\}$ is bounded, then we can assume that $x_n \to \tilde{x} \in -\nabla h(\bar{x})\Lambda$, and hence $y_n = z_n - x_n \to \bar{z} - \tilde{x} \in \tilde{N}(\bar{x}, P)$, that is, $\bar{z}$ belong to the right side of the equality. If x_n is not bounded, then we can assume that $\|x_n\| \to \infty$ and $\dfrac{x_n}{\|x_n\|} \to \tilde{x} \neq 0$. Therefore, the limit of

$$\frac{z_n}{\|x_n\|} = \frac{x_n}{\|x_n\|} + \frac{y_n}{\|x_n\|}$$

contradicts the basic constraint qualification and the proof is complete. □

4.9 Second-Order Tangent Sets

In this section, we study second-order tangent sets, which are natural extensions of some of the tangent cones of first-order investigated in the previous sections. In set-valued optimization, second-order derivatives and second-order epiderivatives of set-valued maps are defined by using second-order tangent sets and second-order tangent cones. These second-order notions are also fundamental in giving second-order optimality conditions and second-order sensitivity analysis, among others. We note that Hoffmann and Kornstaed [265] were among the first ones to conduct a systematic study of second-order local approximation (see also [192]). The seminal contributions of Penot [465, 466, 468] and Ward [594, 596]) and the monographs by Bonnans and Shapiro [53] and Rockafellar and Wets [499] serves as an excellent introduction to this subject. The survey article by Giorgi, Jiménez, and Novo [209] provided a good guideline for arranging the properties of the second-order tangent sets. Some arguments that we use can also be found in Castellani [84], Castellani and Pappalardo [86]. See also [104–107].

4.9.1 Second-Order Radial Tangent Set and Second-Order Feasible Tangent Set

In this subsection, we focus on the following two notions of second-order tangent sets:

Definition 4.9.1. Let X be a normed space, let S be a subset of X, and let $(\bar{x}, \bar{u}) \in X \times X$.

1. The **second-order radial tangent set** $R^2(S, \bar{x}, \bar{u})$ of S at $(\bar{x}, \bar{u})$ is the set of all $x \in X$ for which there exists a sequence $\{t_n\} \subset \mathbb{P}$ with $t_n \downarrow 0$ such that $\bar{x} + t_n\bar{u} + t_n^2 x/2 \in S$ for every $n \in \mathbb{N}$.
2. The **second-order feasible tangent set** $IR^2(S, \bar{x}, \bar{u})$ of S at $(\bar{x}, \bar{u})$ is the set of all $x \in X$ such that for every $\{t_n\} \subset \mathbb{P}$ with $t_n \downarrow 0$, there is an $m \in \mathbb{N}$ satisfying that $\bar{x} + t_n\bar{u} + t_n^2 x/2 \in S$ for every $n \geq m$.

   An equivalent formulation of the above two notions is as follows:

$$R^2(S, \bar{x}, \bar{u}) = \{x \in X \mid \forall\, t > 0\ \exists\, s \in (0, t) \text{ such that } \bar{x} + s\bar{u} + s^2 x/2 \in S\},$$

$$IR^2(S, \bar{x}, \bar{u}) = \{x \in X \mid \exists\, t > 0\ \forall\, s \in (0, t) \text{ we have } \bar{x} + s\bar{u} + s^2 x/2 \in S\}.$$

The following theorem collects some properties of the second-order radial tangent set and the second-order feasible tangent set.

Theorem 4.9.2. *Let X be a normed space, let S, S_1, S_2 be nonempty subsets of X, let $\bar{u} \in X$, and let $\bar{x} \in \operatorname{cl} S \cap \operatorname{cl} S_1 \cap \operatorname{cl} S_2$. Then the following properties hold:*

1. $R^2(S, \bar{x}, 0) = R(S, \bar{x})$
2. $IR^2(S, \bar{x}, 0) = IR(S, \bar{x})$
3. $IR^2(S, \bar{x}, 0) \subset R^2(S, \bar{x}, 0)$
4. $IR^2(X \setminus S, \bar{x}, \bar{u}) = X \setminus R^2(S, \bar{x}, \bar{u})$
5. $0 \in IR^2(S, \bar{x}, \bar{u}) \iff \bar{u} \in IR(S, \bar{x})$
6. $0 \in R^2(S, \bar{x}, \bar{u}) \iff \bar{u} \in R(S, \bar{x})$
7. $S_1 \subseteq S_2 \implies R^2(S_1, \bar{x}, \bar{u}) \subseteq R^2(S_2, \bar{x}, \bar{u})$
8. $S_1 \subseteq S_2 \implies IR^2(S_1, \bar{x}, \bar{u}) \subseteq IR^2(S_2, \bar{x}, \bar{u})$
9. $R^2(S_1 \cup S_2, \bar{x}, \bar{u}) = R^2(S_1, \bar{x}, \bar{u}) \cup R^2(S_1, \bar{x}, \bar{u})$
10. $IR^2(S_1 \cap S_2, \bar{x}, \bar{u}) = IR^2(S_1, \bar{x}, \bar{u}) \cap IR^2(S_1, \bar{x}, \bar{u})$.

Proof. The proof is based on arguments similar to the one given for the first-order analogues. □

4.9.2 Second-Order Contingent Set and Second-Order Interiorly Contingent Set

Our focus is on the following notions of second-order tangent sets.

Definition 4.9.3. Let X be a normed space, let S be a subset of X, and let $(\bar{x}, \bar{u}) \in X \times X$.

1. The **second-order contingent set** $T^2(S, \bar{x}, \bar{u})$ of S at $(\bar{x}, \bar{u})$ is the set of all $x \in X$ for which there are a sequence $\{t_n\} \subset \mathbb{P}$ with $t_n \downarrow 0$ and a sequence $\{x_n\} \subset X$ with $x_n \to x$ such that $\bar{x} + t_n\bar{u} + t_n^2 x_n/2 \in S$ for every $n \in \mathbb{N}$.
2. The **second-order interiorly contingent set** $IT^2(S, \bar{x}, \bar{u})$ of S at $(\bar{x}, \bar{u})$ is the set of all $x \in X$ such that for every sequence $\{t_n\} \subset \mathbb{P}$ with $t_n \downarrow 0$ and every sequence $\{x_n\} \subset X$ with $x_n \to x$ there exists an index $m \in \mathbb{N}$ satisfying that $\bar{x} + t_n\bar{u} + t_n^2 x_n/2 \in S$ for every $n \geq m$.

An equivalent formulation of the above two notions reads:

$$\begin{aligned} T^2(S, \bar{x}, \bar{u}) &= \{x \in X \mid \forall\, U(x)\ \forall\, t > 0\ \exists\, s \in (0, t)\ \exists\ z \in U(x) \text{ with} \\ &\qquad \bar{x} + t\bar{u} + t^2 z/2 \in S\}, \\ IT^2(S, \bar{x}, \bar{u}) &= \{x \in X \mid \exists\, U(x)\ \exists\, t > 0\ \forall\, s \in (0, t)\ \forall\, z \in U(x) \text{ we have} \\ &\qquad \bar{x} + s\bar{u} + s^2 z/2 \in S\}. \end{aligned}$$

We now give other equivalent characterizations of the second-order contingent set.

Theorem 4.9.4. *Let X be a normed space, let S be subset of X, let $\bar{u} \in X$, and let $\bar{x} \in \operatorname{cl} S$. Then, for $i \in \{1, \ldots, 5\}$, we have $T^2(S, \bar{x}, \bar{u}) = T_i^2(S, \bar{x}, \bar{u})$, where $T_i^2(S, \bar{x}, \bar{u})$ are given as follows:*

$$\begin{aligned} T_1^2(S, \bar{x}, \bar{u}) &= \left\{x \in X \mid \exists\, \{t_n\} \downarrow 0\ \exists\, \{x_n\} \subset S \text{ such that } \frac{x_n - \bar{x} - t_n\bar{u}}{\frac{1}{2}t_n^2} \to x\right\}, \\ T_2^2(S, \bar{x}, \bar{u}) &= \left\{x \in X \mid \exists\, \{t_n\} \downarrow 0 \text{ such that } d(\bar{x} + t_n\bar{u} + t_n^2 x/2, S) = o(t_n^2)\right\}, \\ T_3^2(S, \bar{x}, \bar{u}) &= \left\{x \in X \mid \exists\, \{x_n\} \subset S\ \exists\, \{t_n\} \downarrow 0 \text{ such that } x_n = \bar{x} + t_n\bar{u} + t_n^2 x/2 + o(t_n^2)\right\}, \\ T_4^2(S, \bar{x}, \bar{u}) &= \left\{x \in X \mid \liminf_{t \downarrow 0} \frac{d(\bar{x} + t\bar{u} + t^2 x/2, S)}{t^2} = 0\right\}, \\ T_5^2(S, \bar{x}, \bar{u}) &= \limsup_{t \downarrow 0} \frac{S - \bar{x} - t\bar{u}}{\frac{1}{2}t^2}. \end{aligned}$$

Proof. The proof is based on standard arguments and is omitted. □

The following result collects some properties of the second-order contingent sets and the second-order interiorly contingent sets. (In the proofs, we drop the factor $1/2$, see Sect. 4.10.)

Theorem 4.9.5. *Let X be a normed space, let S, S_1, S_2 be nonempty subsets of X, let $\bar{u} \in X$, and let $\bar{x} \in \operatorname{cl} S \cap \operatorname{cl} S_1 \cap \operatorname{cl} S_2$. Then the following properties hold:*

1. $T^2(S, \bar{x}, 0) = T(S, \bar{x})$
2. $IT^2(S, \bar{x}, 0) = IT(S, \bar{x})$

3. $T^2(X\backslash S, \bar{x}, \bar{u}) = X\backslash IT^2(S, \bar{x}, \bar{u})$
4. $0 \in T^2(S, \bar{x}, \bar{u}) \implies \bar{u} \in T(S, \bar{x})$
5. $0 \in IT^2(S, \bar{x}, \bar{u}) \impliedby \bar{u} \in IT(S, \bar{x})$
6. $T^2(S, \bar{x}, \bar{u})$ *is closed*
7. $T^2(S, \bar{x}, \bar{u}) = T^2(\text{cl } S, \bar{x}, \bar{u})$
8. $IT^2(S, \bar{x}, \bar{u})$ *is open*
9. $IT^2(S, \bar{x}, \bar{u}) = IT^2(\text{int } S, \bar{x}, \bar{u})$
10. $T^2(S, \bar{x}, \bar{u}) \neq \emptyset \implies \bar{u} \in T(S, \bar{x})$ *and hence* $\bar{u} \notin T(S, \bar{x}) \implies T^2(S, \bar{x}, \bar{u}) = \emptyset$
11. $IT^2(S, \bar{x}, \bar{u}) \neq X \implies \bar{u} \notin IT(S, \bar{x})$ *and hence* $\bar{u} \in IT(S, \bar{x}) \implies IT^2(S, \bar{x}, \bar{u}) = X$
12. $IT^2(S, \bar{x}, \bar{u}) \subseteq IR^2(S, \bar{x}, \bar{u}) \subseteq R^2(S, \bar{x}, \bar{u}) \subseteq T^2(S, \bar{x}, \bar{u})$
13. $S_1 \subseteq S_2 \implies T^2(S_1, \bar{x}, \bar{u}) \subseteq T^2(S_2, \bar{x}, \bar{u})$
14. $S_1 \subseteq S_2 \implies IT^2(S_1, \bar{x}, \bar{u}) \subseteq IT^2(S_2, \bar{x}, \bar{u})$
15. $T^2(S_1 \cup S_2, \bar{x}, \bar{u}) = T^2(S_1, \bar{x}, \bar{u}) \cup T^2(S_2, \bar{x}, \bar{u})$
16. $IT^2(S_1 \cap S_2, \bar{x}, \bar{u}) = IT^2(S_1, \bar{x}, \bar{u}) \cap IT^2(S_2, \bar{x}, \bar{u})$
17. $T^2(S, \bar{x}, \bar{u}) + t\bar{u} \subset T^2(S, \bar{x}, \bar{u})$, *for every* $t \in \mathbb{R}$
18. $IT^2(S, \bar{x}, \bar{u}) + t\bar{u} \subset IT^2(S, \bar{x}, \bar{u})$, *for every* $t \in \mathbb{R}$.

Proof. The proof is based on arguments similar to the one given for the first-order analogues. For the sake of completeness, we give proofs of a few properties. The first five properties are entirely analogous to the first-order case. We note that the closedness of $T^2(S, \bar{x}, \bar{u})$ follows from the openness of $IT^2(S, \bar{x}, \bar{u})$ and the duality relationship between them. To prove that $IT^2(S, \bar{x}, \bar{u})$ is open, we choose an arbitrary $x \in IT^2(S, \bar{x}, \bar{u})$. Let $\hat{x} \in U_\varepsilon(x)$ with $\varepsilon > 0$. Due to the containment $x \in IT^2(S, \bar{x}, \bar{u})$, there exists $t > 0$ such that for each $s \in (0, t)$ and for each $\tilde{x} \in U_t(x)$, we have $\bar{x} + s\bar{u} + s^2\tilde{x} \in S$. For $\varepsilon < t$, we choose $\bar{\varepsilon} < t - \varepsilon$ which implies that $U_{\bar{\varepsilon}}(\hat{x}) \subset U_t(x)$ and for each $s \in (0, \bar{\varepsilon})$, we have $\bar{x} + s\bar{u} + s^2 U_{\bar{\varepsilon}}(\hat{x}) \subset S$. This confirms that $U_{\bar{\varepsilon}}(x) \subset IT^2(S, \bar{x}, \bar{u})$.

It is evident that if $\bar{u} \notin T(S, \bar{x})$ then $T^2(S, \bar{x}, \bar{u}) = \emptyset$. This fact further implies that if $\bar{u} \in IT(S, \bar{x})$ then $IT^2(S, \bar{x}, \bar{u}) = X$. In fact, this follows from the chain of implications that $\bar{u} \in IT(S, \bar{x})$ implies $\bar{u} \notin T(X\backslash S, \bar{x})$ which further implies that $T^2(X\backslash S, \bar{x}, \bar{u}) = \emptyset$ and hence $IT^2(S, \bar{x}, \bar{u}) = X$. Proofs of some the properties can also be deduced from the proof of Theorem 4.10.3 □

The following example shows that the inclusions stated above are, in general, strict.

Example 4.9.6. For $S \subset \mathbb{R}^2$ given by $S = \{(x, y) |\ y > 0\} \bigcup \bigcup_{n \in \mathbb{N}} \{(\frac{1}{n}, 0)\} \bigcup \{(0, 0)\}$, we have

$$IT^2(S, (1, 0), (0, 0)) = \text{int } S;$$

$$R^2(S, (1, 0), (0, 0)) = \text{int } S \cup \{(0, 0)\};$$

$$IR^2(S,(1,0),(0,0)) = \operatorname{int} S \cup \{(x,0) \mid x \geq 0\};$$
$$T^2(S,(1,0),(0,0)) = \operatorname{cl} S.$$

We give a simple example (see [53]).

Example 4.9.7. For the subsets S_1 and S_2 given by

$$S_1 := \bigcup_{n=0}^{\infty} [2^{-(2n+1)}, 2^{-2n}] \bigcup \{0\};$$

$$S_2 := \bigcup_{n=0}^{\infty} [2^{-2n}, 2^{-(2n-1)}] \bigcup \{0\},$$

we have $S := S_1 \cup S_2 = [0,1]$ and $T^2(S_1,0,1) = T^2(S_2,0,1) = T^2(S,0,1) = \mathbb{R}$.
□

In contrast with first-order tangent cones, the second-order analogues given here, are not, in general, cones or even connected sets (see Example 4.10.8) and they could be empty.

We conclude this subsection by the following simple example (see [53]):

Example 4.9.8. Let X be a normed space and let $S \subset X$ be closed and convex. Then

$$T^2(S \times \mathbb{R}_+, (\bar{x},0),(h,k)) = \begin{cases} T^2(S,\bar{x},h) \times \mathbb{R}, & \text{if } k > 0 \\ T^2(S,\bar{x},h) \times \mathbb{R}_+, & \text{if } k = 0 \\ \emptyset, & \text{if } k < 0. \end{cases}$$

4.9.3 *Second-Order Adjacent Set and Second-Order Interiorly Adjacent Set*

Our focus here is on the following two notions of second-order tangent sets.

Definition 4.9.9. Let X be a normed space, let S be a subset of X, and let $(\bar{x},\bar{u}) \in X \times X$.

1. The **second-order adjacent set** $A^2(S,\bar{x},\bar{u})$ of S at $(\bar{x},\bar{u})$ is the set of all $x \in X$ such that for every sequence $\{t_n\} \subset \mathbb{P}$ with $t_n \downarrow 0$, there exists a sequence $\{x_n\} \subset S$ with $x_n \to x$ satisfying that $\bar{x} + t_n\bar{u} + t_n^2 x_n/2 \in S$ for every $n \in \mathbb{N}$.
2. The **second-order interiorly adjacent set** $IA^2(S,\bar{x},\bar{u})$ of S at $(\bar{x},\bar{u})$ is the set of all $x \in X$ such that there exists a sequence $\{t_n\} \subset \mathbb{P}$ with $t_n \downarrow 0$ so that for every sequence $\{x_n\} \subset S$ with $x_n \to x$ there exists an index $m \in \mathbb{N}$ satisfying that $\bar{x} + t_n\bar{u} + t_n^2 x_n/2 \in S$ for every $n \geq m$.

An equivalent formulation of the above second-order tangent sets reads:

$$\begin{aligned}A^2(S,\bar{x},\bar{u}) &= \{x \in X \mid \forall\, U(x)\ \exists\, t>0\ \forall\, s \in (0,t)\ \exists\ z \in U(x) \text{ with}\\ &\qquad \bar{x}+t\bar{u}+t^2z/2 \in S\},\\ IA^2(S,\bar{x},\bar{u}) &= \{x \in X \mid \exists\, U(x)\ \forall t>0\ \exists\, s \in (0,t)\ \forall\, z \in U(x) \text{ we have}\\ &\qquad \bar{x}+s\bar{u}+s^2z/2 \in S\}.\end{aligned}$$

We now give other equivalent characterizations of the second-order adjacent set.

Theorem 4.9.10. *Let X be a normed space, let S be nonempty subset of X, let $\bar{u} \in X$, and let $\bar{x} \in \operatorname{cl} S$. Then for $i \in \{1,2,3\}$, we have $A^2(S,\bar{x},\bar{u}) = A_i^2(S,\bar{x},\bar{u})$, where $A_i^2(S,\bar{x},\bar{u})$ are given as follows:*

$$\begin{aligned}A_1^2(S,\bar{x},\bar{u}) &= \left\{x \in X \mid \forall\, \{t_n\} \downarrow 0\ \exists\, \{x_n\} \subset S \text{ such that } \frac{x_n-\bar{x}-t_n\bar{u}}{\frac{1}{2}t_n^2} \to x\right\}\\ A_2^2(S,\bar{x},\bar{u}) &= \left\{x \in X \mid \limsup_{t\downarrow 0} \frac{d(\bar{x}+t\bar{u}+t^2x/2,S)}{t^2} = 0\right\}\\ A_3^2(S,\bar{x},\bar{u}) &= \liminf_{t\downarrow 0} \frac{S-\bar{x}-t\bar{u}}{\frac{1}{2}t^2}.\end{aligned}$$

Proof. The proof is based on standard arguments and hence omitted. □

The following result collects some properties of the second-order adjacent sets and second-order interiorly adjacent sets.

Theorem 4.9.11. *Let X be a normed space, let S, S_1, S_2 be nonempty subsets of X, let $\bar{u} \in X$, and let $\bar{x} \in \operatorname{cl} S \cap \operatorname{cl} S_1 \cap \operatorname{cl} S_2$. Then the following properties hold:*

1. $A^2(S,\bar{x},0) = A(S,\bar{x})$
2. $IA^2(S,\bar{x},0) = IA(S,\bar{x})$
3. $A^2(X\backslash S,\bar{x},\bar{u}) = X\backslash IA^2(S,\bar{x},\bar{u})$
4. $0 \in A^2(S,\bar{x},\bar{u}) \implies \bar{u} \in A(S,\bar{x})$
5. $0 \in IA^2(S,\bar{x},\bar{u}) \impliedby \bar{u} \in IA(S,\bar{x})$
6. $A^2(S,\bar{x},\bar{u})$ *is closed*
7. $A^2(S,\bar{x},\bar{u}) = A^2(\operatorname{cl} S,\bar{x},\bar{u})$
8. $IA^2(S,\bar{x},\bar{u})$ *is open*
9. $IA^2(S,\bar{x},\bar{u}) = IA^2(\operatorname{int} S,\bar{x},\bar{u})$
10. $A^2(S,\bar{x},\bar{u}) \neq \emptyset \implies \bar{u} \in A(S,\bar{x})$ *and hence* $\bar{u} \notin A(S,\bar{x}) \implies A^2(S,\bar{x},\bar{u}) = \emptyset$
11. $IA^2(S,\bar{x},\bar{u}) \neq X \implies \bar{u} \notin IA(S,\bar{x})$ *and hence* $\bar{u} \in IA(S,\bar{x}) \implies IA^2(S,\bar{x},\bar{u}) = X$
12. $IT^2(S,\bar{x},\bar{u}) \subseteq IR^2(S,\bar{x},\bar{u}) \subseteq A^2(S,\bar{x},\bar{u}) \subseteq T^2(S,\bar{x},\bar{u})$
13. $IT^2(S,\bar{x},\bar{u}) \subseteq IA^2(S,\bar{x},\bar{u}) \subseteq R^2(S,\bar{x},\bar{u}) \subseteq T^2(S,\bar{x},\bar{u})$
14. $S_1 \subseteq S_2 \implies A^2(S_1,\bar{x},\bar{u}) \subseteq A^2(S_2,\bar{x},\bar{u})$
15. $S_1 \subseteq S_2 \implies IA^2(S_1,\bar{x},\bar{u}) \subseteq IA^2(S_2,\bar{x},\bar{u})$

16. $IA^2(S_1 \cup S_2, \bar{x}, \bar{u}) = IA^2(S_1, \bar{x}, \bar{u}) \cup IA^2(S_2, \bar{x}, \bar{u})$
17. $A^2(S_1 \cup S_2, \bar{x}, \bar{u}) \supset A^2(S_1, \bar{x}, \bar{u}) \cup A^2(S_2, \bar{x}, \bar{u})$
18. $A^2(S, \bar{x}, \bar{u}) + t\bar{u} \subset A^2(S, \bar{x}, \bar{u})$, *for every* $t \in \mathbb{R}$
19. $IA^2(S, \bar{x}, \bar{u}) + t\bar{u} \subset IA^2(S, \bar{x}, \bar{u})$, *for every* $t \in \mathbb{R}$.

The following is an analogue of Example 4.9.7 (see [53]).

Example 4.9.12. For the subsets S_1 and S_2 given by

$$S_1 := \bigcup_{n=0}^{\infty} [2^{-2n+1}, 2^{-2n}] \bigcup \{0\};$$

$$S_2 := \bigcup_{n=0}^{\infty} [2^{-(2n-1)}, 2^{-2n}] \bigcup \{0\},$$

we have $S := S_1 \cup S_2 = [0, 1]$, $A^2(S_1, 0, 1) = A^2(S_2, 0, 1) = \emptyset$, and $A^2(S, 0, 1) = \mathbb{R}$, showing that the inclusion $A^2(S_1 \cup S_2, \bar{x}, \bar{u}) \supset A^2(S_1, \bar{x}, \bar{u}) \cup A^2(S_2, \bar{x}, \bar{u})$ is, in general, strict.

The following example, given by Penot [467], shows that $T^2(\cdot, \cdot, \cdot)$ is larger than $A^2(\cdot, \cdot, \cdot)$.

Example 4.9.13. Let $\{r_n\} \subset \mathbb{P}$ be a sequence such that for some $c \in (0, 1)$, $r_{n+1} \leq c r_n$, for each $n \in \mathbb{N}$ and let $S \subset \mathbb{R}^2$ be given by $S := \{(x, y) | \ y \geq (r_n + r_{n+1}x - r_n r_{n+1}$ for every $n \in \mathbb{N})\}$. Then, it turns out that $(0, 2) \in T^2(S, (0, 0), (1, 0))$, however, $(0, 2) \notin A^2(S, (0, 0), (1, 0))$.

4.10 Generalized Second-Order Tangent Sets

In this section, we study an extension of the second-order contingent sets proposed by Cambini, Martein, and Vlach [82]. This generalization not only gives an interesting insight into the second-order contingent set but it has also served as a motivation to introduce the important notion of second-order asymptotic tangent cones which will be studied in the next section. In this section, which is based on [82], we mainly focus on the generalized contingent sets of second-order, a notion that corresponds to the second-order contingent set. However, analogues of other second-order tangent sets can also be defined and studied.

For later use, we formulate the following condition: Let $\{\alpha_n\} \subset \mathbb{P}$, $\{\beta_n\} \subset \mathbb{P}$, and $\{x_n\} \subset X$ be arbitrary sequences and k be a nonnegative number satisfying

$$\alpha_n \to \infty, \quad \beta_n \to \infty, \quad \frac{\beta_n}{\alpha_n} \to k, \tag{4.73a}$$

$$x_n \to \bar{x} \quad \alpha_n(x_n - \bar{x}) \to \bar{u} \quad \beta_n[\alpha_n(x_n - \bar{x}) - \bar{u}] \to x. \tag{4.73b}$$

We begin with the following:

Definition 4.10.1. Let X be a normed space, let $S \subset X$, let $\bar{u} \in X$, let $\bar{x} \in S$, and let $k \geq 0$.

1. The **second-order indexed contingent set** $T_k^2(S, \bar{x}, \bar{u})$ of S at $(\bar{x}, \bar{u})$ is the set of all x such that there are sequences $\{\alpha_n\} \subset \mathbb{P}$, $\{\beta_n\} \subset \mathbb{P}$, and $\{x_n\} \subset S$ satisfying (4.73). The **generalized second-order contingent set** $\hat{T}^2(S, \bar{x}, \bar{u})$ of S at $(\bar{x}, \bar{u})$ is then defined by

$$\hat{T}^2(S, \bar{x}, \bar{u}) := \bigcup_{k>0} T_k^2(S, \bar{x}, \bar{u}).$$

2. The **second-order indexed adjacent set** $A_k^2(S, \bar{x}, \bar{u})$ of S at $(\bar{x}, \bar{u})$ is the set of all x such that for all sequences $\{\alpha_n\} \subset \mathbb{P}$ and $\{\beta_n\} \subset \mathbb{P}$ satisfying (4.73a), there exists $\{x_n\} \subset S$ satisfying (4.73b). The **generalized second-order adjacent set** $\hat{A}^2(S, \bar{x}, \bar{u})$ of S at $(\bar{x}, \bar{u})$ is then defined by

$$\hat{A}^2(S, \bar{x}, \bar{u}) := \bigcup_{k>0} A_k^2(S, \bar{x}, \bar{u}).$$

Remark 4.10.2. We emphasize that in the above definition, the sequences $\{\alpha_n\}$ and $\{\beta_n\}$ depend on k. Furthermore, to define the second-order indexed contingent set $T_\infty^2(S, \bar{x}, \bar{u})$ of S at $(\bar{x}, \bar{u})$, we replace the requirement that $\frac{\beta_n}{\alpha_n} \to \infty$ by $\frac{\alpha_n}{\beta_n} \to 0$. The same argument applies to other sets.

We now give basic properties of the second-order indexed contingent set and the generalized second-order contingent set. Similar results for other generalized adjacent sets can be given.

Theorem 4.10.3. *Let X be a normed space, let S be a subset of X, let $\bar{u} \in X$, and let $\bar{x} \in S$. Then:*

1. *For each $k > 0$, we have*

$$T^2(S, \bar{x}, \bar{u}) = \frac{2}{k} T_k^2(S, \bar{x}, \bar{u}).$$

2. *Let $t \in \mathbb{R}$ and $k \geq 0$ be arbitrary. Then $x + t\bar{u} \in T_k^2(S, \bar{x}, \bar{u})$, whenever $x \in T_k^2(S, \bar{x}, \bar{u})$.*
3. *Assume that X is finite-dimensional. Then $0 \in T_0^2(S, \bar{x}, \bar{u})$ Moreover, $T^2(S, \bar{x}, \bar{u}) = \emptyset$ implies that there exists $\bar{u} \in T_0^2(S, \bar{x}, \bar{u})$ such that $\bar{u} \neq 0$ and $\langle \bar{u}, d \rangle = 0$.*
4. *$T_\infty^2(S, \bar{x}, \bar{u}) \neq \emptyset$ if and only if $0 \in T^2(S, \bar{x}, \bar{u})$.*

Proof. We begin with a proof of the first part. Let $x \in T^2(S, \bar{x}, \bar{u})$ be arbitrary. Then there exist sequences $\{\gamma_n\} \subset \mathbb{P}$ and $\{z_n\} \subset X$ such that $\gamma_n \downarrow 0$, $z_n \to x$, and $\bar{x} + \gamma_n \bar{u} + \frac{1}{2}\gamma_n^2 z_n \in S$ for every $n \in \mathbb{N}$. Let $k > 0$ be a given number. We define

$\alpha_n := \frac{1}{\gamma_n}$, $\beta_n := \frac{k}{\gamma_n}$, and $x_n := \bar{x} + \gamma_n \bar{u} + \frac{1}{2}\gamma_n^2 z_n$, and note that the sequences $\{\alpha_n\} \subset \mathbb{P}$ and $\{\beta_n\} \subset \mathbb{P}$ satisfy (4.73a). Furthermore,

$$\alpha_n(x_n - \bar{x}) = \frac{1}{\gamma_n}\left(\gamma_n \bar{u} + \frac{1}{2}\gamma_n^2 z_n\right) = \left(\bar{u} + \frac{1}{2}\gamma_n z_n\right) \to \bar{u},$$

$$\beta_n\left[\alpha_n(x_n - \bar{x}) - \bar{u}\right] = \frac{k}{\gamma_n}\left(\bar{u} + \frac{1}{2}\gamma_n z_n - \bar{u}\right) = \frac{k}{2} z_n \to \frac{k}{2} x,$$

which means that $\frac{k}{2}x \in T_k^2(S, \bar{x}, \bar{u})$ and hence proving the inclusion $T^2(S, \bar{x}, \bar{u}) \subseteq \frac{2}{k}T_k^2(S, \bar{x}, \bar{u})$.

For the converse, let $z \in \frac{2}{k}T_k^2(S, \bar{x}, \bar{u})$, where $k > 0$, be arbitrary. Then there exists $x \in T_k^2(S, \bar{x}, \bar{u})$ with $z = \dfrac{2}{k}x$ confirming that there exist sequences $\{\alpha_n\} \subset \mathbb{P}$, $\{\beta_n\} \subset \mathbb{P}$, and $\{x_n\} \subset S$ such that (4.73) holds. We define sequences $\{\gamma_n\} \subset \mathbb{P}$ and $\{z_n\} \subset \mathbb{P}$ by $\gamma_n := \frac{1}{\alpha_n}$, and $z_n := 2\alpha_n\left[\alpha_n(x_n - \bar{x}) - \bar{u}\right]$, and note that $\{\gamma_n\} \subset \mathbb{P}$, $\gamma_n \downarrow 0$ and $\bar{x} + \gamma_n \bar{u} + \frac{1}{2}\gamma_n^2 z_n =: x_n \in S$. Since $\frac{\alpha_n}{\beta_n} \to \frac{1}{k}$, and $\beta_n\left[\alpha_n(x_n - \bar{x}) - \bar{u}\right] \to x$, we obtain $z_n = \frac{2\alpha_n}{\beta_n}\left[\alpha_n(x_n - \bar{x}) - \bar{u}\right] \to \frac{2}{k}x = z$, giving $x \in T^2(S, \bar{x}, \bar{u})$. Therefore, $T^2(S, \bar{x}, \bar{u}) \supset \frac{2}{k}T_k^2(S, \bar{x}, \bar{u})$ and the proof of the first part is complete.

Let us now proceed to prove the second part. For an arbitrary $k > 0$ and $t \in \mathbb{R}$, let $x \in T_k^2(S, \bar{x}, \bar{u})$. Then there are sequences $\{\alpha_n\} \subset \mathbb{P}$, $\{\beta_n\} \subset \mathbb{P}$, and $\{x_n\} \subset S$ such that (4.73) holds. We define sequences $\{\hat{\alpha}_n\}$, $\{\hat{\beta}_n\}$ and $\{\hat{x}_n\}$ by $\hat{\alpha}_n := \alpha_n + \frac{t}{k}$, $\hat{\beta}_n := \beta_n$, and $\hat{x}_n := x_n$, and note that, without any loss of generality, we can assume that $\{\hat{\alpha}_n\} \subset \mathbb{P}$ and $\{\hat{\beta}_n\} \subset \mathbb{P}$, as they are positive for sufficiently large n. Clearly, $\hat{\alpha}_n \to \infty$, $\hat{\beta}_n \to \infty$, $\frac{\hat{\beta}_n}{\hat{\alpha}_n} \to k$, and $\hat{x}_n \to \bar{x}$. Furthermore, we have

$$\hat{\alpha}_n(\hat{x}_n - \bar{x}) = \left(\alpha_n(x_n - \bar{x}) + \frac{t}{k}(x_n - \bar{x})\right) \to \bar{u},$$

and

$$\begin{aligned}\hat{\beta}_n\left[\hat{\alpha}_n(\hat{x}_n - \bar{x}) - \bar{u}\right] &= \beta_n\left[\alpha_n(x_n - \bar{x}) - \bar{u} + \frac{t}{k}(x_n - \bar{x})\right] \\ &= \beta_n\left[\alpha_n(x_n - \bar{x}) - \bar{u}\right] + \frac{t}{k}\beta_n(x_n - \bar{x}),\end{aligned}$$

and by $\beta_n(x_n - \bar{x}) = \frac{\beta_n}{\alpha_n}\alpha_n(x_n - \bar{x}) \to k\bar{u}$, we get $\hat{\beta}_n\left[\hat{\alpha}_n(\hat{x}_n - \bar{x}) - \bar{u}\right] \to (x + t\bar{u})$ proving $x + t\bar{u} \in T_k^2(S, \bar{x}, \bar{u})$.

We now take $k = 0$. Then for $x \in T_0^2(S, \bar{x}, \bar{u})$, there are sequences $\{\alpha_n\} \subset \mathbb{P}$, $\{\beta_n\} \subset \mathbb{P}$, and $\{x_n\} \subset S$ such that (4.73) holds with $k = 0$. We define sequences $\{\hat{\alpha}_n\}$, $\{\hat{\beta}_n\}$, and $\{\hat{x}_n\}$ by $\hat{\alpha}_n := \alpha_n + \frac{t\alpha_n}{\beta_n}$, $\hat{\beta}_n := \beta_n$, $\hat{x}_n := x_n$, and assume, without any loss of generality, that $\{\hat{\alpha}_n\} \subset \mathbb{P}$ and $\{\hat{\beta}_n\} \subset \mathbb{P}$ as they are positive for sufficiently large n. Clearly, $\hat{\alpha}_n \to \infty$, $\hat{\beta}_n \to \infty$, $\frac{\hat{\beta}_n}{\hat{\alpha}_n} \to 0$, and $\hat{x}_n \to \bar{x}$, and

$$\hat{\alpha}_n(\hat{x}_n - \bar{x}) = \alpha_n(x_n - \bar{x}) + \frac{t\alpha_n}{\beta_n}(x_n - \bar{x}),$$

$$\hat{\beta}_n\,[\hat{\alpha}_n(\hat{x}_n - \bar{x}) - \bar{u}] = \beta_n\,[\alpha_n(x_n - \bar{x}) - \bar{u}] + t\alpha_n(x_n - \bar{x}),$$

which, due to the fact that $\frac{t}{\beta_n} \to 0$, implies that $\hat{\alpha}_n(\hat{x}_n - \bar{x}) \to \bar{u}$ and $\hat{\beta}_n\,[\hat{\alpha}_n(\hat{x}_n - \bar{x}) - \bar{u}] \to (x + t\bar{u})$, and hence $x + t\bar{u} \in T_0^2(S, \bar{x}, \bar{u})$. This completes the proof of the second part.

For the third part, we assume that we assume that X is finite-dimensional. Let $T^2(S, \bar{x}, \bar{u}) \neq \emptyset$. Then for $x \in T^2(S, \bar{x}, \bar{u})$, there are sequences $\{\alpha_n\} \subset \mathbb{P}$, $\{\beta_n\} \subset \mathbb{P}$, and $\{x_n\} \subset S$ such that

$$\frac{\beta_n}{\alpha_n} \to 2 \quad \alpha_n(x_n - \bar{x}) \to \bar{u} \quad \beta_n\,[\alpha_n(x_n - \bar{x}) - \bar{u}] \to x.$$

We define $\{\hat{\alpha}_n\} \subset \mathbb{P}$, $\{\hat{\beta}_n\} \subset \mathbb{P}$, and $\{\hat{x}_n\} \subset S$ by $\hat{\alpha}_n := \alpha_n$, $\hat{\beta}_n := \sqrt{\alpha}_n$, and $\hat{x}_n := x_n$ and note that

$$\frac{\hat{\beta}_n}{\hat{\alpha}_n} \to 0 \quad \hat{\alpha}_n(\hat{x}_n - \bar{x}) \to \bar{u} \quad \hat{\beta}_n\,[\hat{\alpha}_n(\hat{x}_n - \bar{x} - \bar{u})] = \sqrt{\alpha_n}\,[\alpha_n(x_n - \bar{x}) - \bar{u}] \to 0,$$

which confirms that $0 \in T_0^2(S, \bar{x}, \bar{u})$.

Now assume that $T^2(S, \bar{x}, \bar{u}) = \emptyset$. Then, for an arbitrary $0 \neq \bar{u} \in T(S, \bar{x})$, there are sequences $\{\gamma_n\} \subset \mathbb{P}$ with $\gamma_n \to \infty$ and $\{x_n\} \subset S$ with $x_n \neq \bar{x}$ such that $\gamma_n(x_n - \bar{x}) \to \bar{u}$. For $\alpha_n := \frac{\|\bar{u}\|}{\|x_n - \bar{x}\|}$, we have $\alpha_n(x_n - \bar{x}) \to \bar{u}$. We now define $z_n := \alpha_n(x_n - \bar{x}) - \bar{u}$ and $\beta_n := \frac{1}{\|z_n\|}$. Without loss of generality, we may assume that the sequence $2\alpha_n z_n$ is not bounded because otherwise we could find its convergent sequences. The limit of such a subsequence would belong to $T^2(S, \bar{x}, \bar{u})$, which would contradict our assumption. It follows that, without loss of generality, we may assume that $\alpha_n \|z_n\| \to \infty$. Therefore, $\frac{\beta_n}{\alpha_n} = \frac{1}{\alpha_n \|z_n\|} \to 0$, and we conclude that $\beta_n z_n = \frac{z_n}{\|z_n\|}$ has a subsequence that converges to some $z \in T_0^2(S, \bar{x}, \bar{u})$ with $\|z\| = 1$. From the identity $z_n + \bar{u} = \alpha_n(x_n - \bar{x})$, we have $\|z_n\|^2 + \|\bar{u}\|^2 + 2\langle z_n, \bar{u}\rangle = \alpha_n^2 \|x_n - \bar{x}\| = \|\bar{u}\|^2$. Therefore, $\|z_n\|^2 + 2\langle z_n, \bar{u}\rangle = 0$ and taking the limit appropriately, we obtain $\langle z, \bar{u}\rangle = 0$. (See also [209] for the alternative arguments given for this proof.)

It remains to show that $0 \in T_0^2(S, \bar{x}, \bar{u})$ for the present case. For this, we define, $\hat{\beta}_n := \sqrt{\beta_n}$, $\hat{\alpha}_n := \alpha_n$, $\hat{x}_n := x_n$. Then,

$$\frac{\hat{\beta}_n}{\hat{\alpha}_n} = \frac{\sqrt{\beta_n}}{\alpha_n} = \frac{1}{\alpha_n \sqrt{z_n}} = \frac{\sqrt{\|z_n\|}}{\alpha_n \|z_n\|} \to 0$$

$$\hat{\beta}_n \left[\hat{\alpha}_n(x_n - \bar{x}) - \bar{u}\right] = \sqrt{\beta_n} z_n = \frac{\beta_n z_n}{\sqrt{\beta_n}} \to 0.$$

Let us now prove the last part. We begin with assuming that $0 \in T^2(S, \bar{x}, \bar{u})$. It follows that there exists $\{\alpha_n\} \subset \mathbb{P}$, $\{\beta_n\} \subset \mathbb{P}$, and $\{x_n\} \subset S$ such that

$$\alpha_n \to \infty \quad \beta_n \to \infty \quad \frac{\beta_n}{\alpha_n} \to 2 \quad \alpha_n(x_n - \bar{x}) \to \bar{u} \quad \beta_n(\alpha_n(x_n - \bar{x}) - \bar{u}) \to 0.$$

Define $z_n := \alpha_n(x_n - \bar{x}) - \bar{u}$, $\hat{\beta}_n := \frac{1}{\|z_n\|}$, and $\hat{\alpha}_n := \alpha_n \quad \hat{x}_n = x_n$. Since $\alpha_n z_n = \frac{\alpha_n}{\beta_n}\beta_n z_n \to 0$, we have $\frac{\hat{\beta}_n}{\hat{\alpha}_n} = \frac{1}{\alpha_n \|z_n\|} \to \infty$, and hence from $\hat{\beta}_n \left[\hat{\alpha}_n(\hat{x}_n) - \bar{u}\right] = \frac{z_n}{\|z_n\|} \to z$, we conclude that $z \in T^2_\infty(S, \bar{x}, \bar{u})$.

Now let $z \in T^2_\infty(S, \bar{x}, \bar{u})$. Then there are sequences $\{\alpha_n\} \subset \mathbb{P}$, $\{\beta_n\} \subset \mathbb{P}$ and $\{x_n\} \subset S$ such that

$$\alpha_n \to \infty \quad \beta_n \to \infty \quad \frac{\beta_n}{\alpha_n} \to \infty \quad \alpha_n(x_n - \bar{x}) \to \bar{u} \quad \beta_n(z_n) \to z,$$

where $z_n = \alpha_n(x_n - \bar{x}) - \bar{u}$. Clearly, $\beta_n z_n = \frac{\beta_n}{2\alpha_n}(2\alpha_n z_n) \to z$. Furthermore, since $\frac{\beta_n}{2\alpha_n} \to \infty$, we obtain that $2\alpha_n z_n \to 0 \in T^2(S, \bar{x}, \bar{u})$. The proof is complete. □

As a consequence of the above result, we have:

Corollary 4.10.4. *Let X be a normed space, let S be a subset of X, let $\bar{u} \in X$, and let $\bar{x} \in S$. Then:*

$$T^2_2(S, \bar{x}, \bar{u}) = T^2(S, \bar{x}, \bar{u}),$$

$$T^2(S, \bar{x}, \bar{u}) \subseteq \hat{T}^2(S, \bar{x}, \bar{u}),$$

$$\hat{T}^2(S, \bar{x}, \bar{u}) = \bigcup_{k>0} \frac{k}{2} T^2(S, \bar{x}, \bar{u}).$$

In the following, we give a few examples to illustrate the notion of the generalized contingent sets. Our first example shows that the inclusion $T^2(S, \bar{x}, \bar{u}) \subseteq \hat{T}^2(S, \bar{x}, \bar{u})$ is, in general, strict.

Example 4.10.5. For $X = \mathbb{R}^2$, we define $S = \{(x_1, x_2) \in \mathbb{R}^2 | \ x_2 = x_1^2\}$. We set $\bar{x} = (0, 0)$, and $\bar{u} = (1, 0)$. For $x \in T^2(S, \bar{x}, \bar{u})$, there are sequences $\{\gamma_n\} \subset \mathbb{P}$ and $\{x_n\} \subset X$ with $x_n = (x_{n1}, x_{n2})$ such that $\gamma_n \to 0$, $x_n \to x$, and $\bar{x} + \gamma_n \bar{u} + \frac{1}{2}\gamma_n^2 x_n \in S$, where the containment $\bar{x} + \gamma_n \bar{u} + \frac{1}{2}\gamma_n^2 x_n \in S$ implies that $\frac{1}{2}\gamma_n^2 x_{n2} = \left(\gamma_n + \frac{1}{2}\gamma_n^2 x_{n1}\right)^2$, and hence $x_{n2} = 2\left(1 + \frac{1}{2}\gamma_n x_{n1}\right)^2 \to 2$,

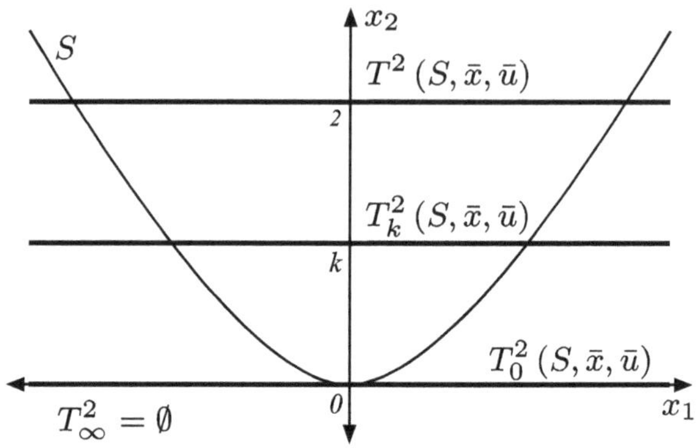

Fig. 4.9 Contingent sets in Example 4.10.5

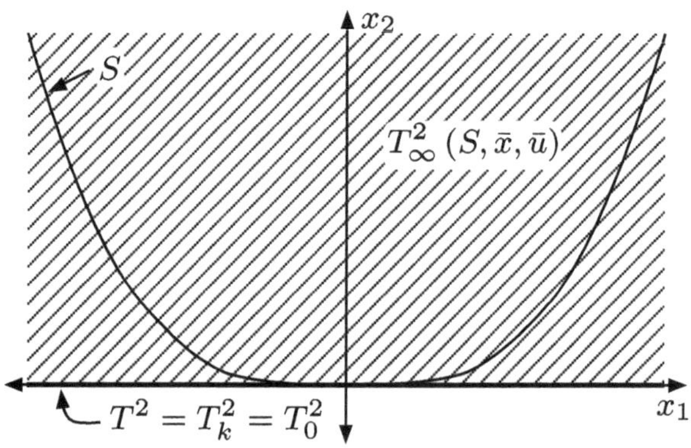

Fig. 4.10 Contingent sets in Example 4.10.7

because $\{x_{n1}\}$ is convergent and $\gamma_n \to 0$. We have shown that $T^2(S,\bar{x},\bar{u}) \subseteq \{(x_1,x_2) \in \mathbb{R}^2 | \ x_2 = 2, \ x_1 \in \mathbb{R}\}$. To complete the argument, it suffices to show that $x = (x_1,x_2) = (0,1)$ belongs to $\hat{T}^2(S,\bar{x},\bar{u})$. For this, let $\{\alpha_n\} \subset \mathbb{P}$ be an arbitrary sequence such that $\alpha_n \to \infty$. Set $\beta_n = \alpha_n$ and define $\{x_n\} \subset \mathbb{R}^2$ by $x_{n1} = \frac{1}{\alpha_n}$, and $x_{n2} = \frac{1}{\alpha_n^2}$, so that $\{x_n\} \subset S$ with $x_n \to \bar{x}$. Moreover,

$$\begin{aligned}
\alpha_n(x_{n1} - \bar{x}_1) &= \alpha_n x_{n1} = 1,\\
\alpha_n(x_{n2} - \bar{x}_2) &= \alpha_n x_{n2} = \frac{1}{\alpha_n},\\
\beta_n\left[\alpha_n(x_{n1} - \bar{x}_1) - \bar{u}_1\right] &= \beta_n \cdot 0 = 0,\\
\beta_n\left[\alpha_n(x_{n2} - \bar{x}_2) - \bar{u}_2\right] &= \frac{\beta_n}{\alpha_n} - 0 = 1,
\end{aligned}$$

confirming that $x \in T_1^2(S,\bar{x},\bar{u}) \subseteq \hat{T}^2(S,\bar{x},\bar{u})$. In fact, it can shown that

$$\begin{aligned} T^2(S,\bar{x},\bar{u}) &= \{(x_1,2) |\ x_1 \in \mathbb{R}\}, \\ T_k^2(S,\bar{x},\bar{u}) &= \{(x_1,k) |\ x_1 \in \mathbb{R}\}, \\ T_0^2(S,\bar{x},\bar{u}) &= \{(x_1,0) |\ x_1 \in \mathbb{R}\}, \\ T_\infty^2(S,\bar{x},\bar{u}) &= \emptyset. \end{aligned}$$

Example 4.10.6. For $S := \{(x_1,x_2) \in \mathbb{R}^2 |\ x_2 = x_1\sqrt{x_1},\ x_1 \geq 0\}$, we set $\bar{x} = (0,0)$, $\bar{u} = (1,0)$, and $k > 0$. Then,

$$\begin{aligned} T^2(S,\bar{x},\bar{u}) &= \emptyset, \\ T_k^2(S,\bar{x},\bar{u}) &= \emptyset, \\ T_0^2(S,\bar{x},\bar{u}) &= \{(x_1,x_2) |\ x_2 \geq 0\}, \\ T_\infty^2(S,\bar{x},\bar{u}) &= \emptyset. \end{aligned}$$

Example 4.10.7. For $S := \{(x_1,x_2) \in \mathbb{R}^2 |\ x_2 = x_1^4\}$, we set $\bar{x} = (0,0)$, $\bar{u} = (1,0)$, and $k > 0$. Then,

$$\begin{aligned} T^2(S,\bar{x},\bar{u}) &= \{(x_1,x_2) |\ x_2 = 0\}, \\ T_k^2(S,\bar{x},\bar{u}) &= \{(x_1,x_2) |\ x_2 = 0\}, \\ T_0^2(S,\bar{x},\bar{u}) &= \{(x_1,x_2) |\ x_2 = 0\}, \\ T_\infty^2(S,\bar{x},\bar{u}) &= \{(x_1,x_2) |\ x_2 \geq 0\}. \end{aligned}$$

The next example shows that the second-order contingent set is not even a connected set:

Example 4.10.8. Define a set $S \subset \mathbb{R}^2$ by $S := \{(x,y) |\ \left(y \geq x^2\right) \cup \left(-x^2 \geq y\right)$ for $x < 0\}$. Then for $\bar{x} = (0,0)$, we have $T(S,(0,0)) = \mathbb{R}^2$ and for $\bar{u}_\alpha = (-\alpha,0)$ with $\alpha \in \mathbb{R}+$, we have

$$T_1^2(S,\bar{x},\bar{u}_\alpha) = \{(x,y) |\ \left(y \geq \alpha^2\right) \cup \left(-y \geq \alpha^2\right)\}.$$

4.11 Second-Order Asymptotic Tangent Cones

All the second-order tangent sets studied in the previous section are, in general, not cones. In this section, we collect a related notion of second-order asymptotic tangent cones, which, as evident from the name, are always cones. We remark that probably,

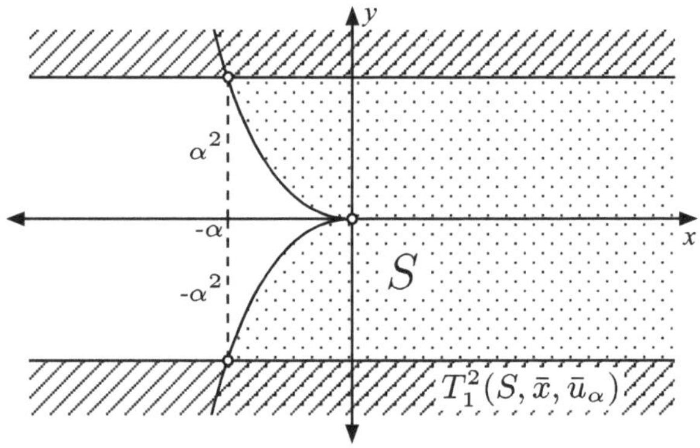

Fig. 4.11 Non-connected contingent set of second-order for Example 4.10.8

Ledzewicz and Schaettler [364] were the first ones to introduce second-order notions of local approximation which were actually cones. In fact, there are a few other authors who introduced second-order conical notions of local approximation. Some of these notions will be studied in later part of this chapter.

4.11.1 Second-Order Asymptotic Feasible Tangent Cone and Second-Order Asymptotic Radial Tangent Cone

We begin with the following:

Definition 4.11.1. Let X be a normed space, let S be a subset of X, and let $(\bar{x},\bar{u}) \in X \times X$.

1. The **second-order asymptotic radial tangent cone** $\mathbb{R}^2(S,\bar{x},\bar{u})$ of S at $(\bar{x},\bar{u})$ is the set of all $x \in X$ for which there exist sequences $\{(s_n,t_n)\} \subset \mathbb{P}\times\mathbb{P}$ with $(s_n,t_n) \downarrow 0$ and $\frac{s_n}{t_n} \downarrow 0$ such that $\bar{x} + s_n\bar{u} + s_nt_nx \in S$ for every $n \in \mathbb{N}$.
2. The **second-order asymptotic feasible tangent cone** $\mathbb{IR}^2(S,\bar{x},\bar{u})$ of S at $(\bar{x},\bar{u})$ is the set of all $x \in X$ such that for every sequence $\{(s_n,t_n)\} \subset \mathbb{P}\times\mathbb{P}$ with $(s_n,t_n) \downarrow 0$ and $\frac{s_n}{t_n} \downarrow 0$, there exists an index $m \in \mathbb{N}$ satisfying that $\bar{x} + s_n\bar{u} + s_nt_nx \in S$ for every $n \geq m$.

   Equivalently,

$$\mathbb{R}^2(S,\bar{x},\bar{u}) = \{x \in X \mid \forall\, t>0\ \exists\, s,r \in (0,t) \text{ with } \frac{s}{r} < t \text{ such that } \bar{x}+s\bar{u}+srx \in S\},$$

$$\mathbb{IR}^2(S,\bar{x},\bar{u}) = \{x \in X \mid \exists\, t>0\ \forall\, s,r \in (0,t) \text{ with } \frac{s}{r} < t \text{ we have } \bar{x}+s\bar{u}+srx \in S\}.$$

The following theorem collects some properties of the second-order asymptotic radial tangent cone and the second-order asymptotic feasible tangent cone.

Theorem 4.11.2. *Let X be a normed space, let S, S_1, S_2 be nonempty subsets of X, let $\bar{u} \in X$, and let $\bar{x} \in \operatorname{cl} S \cap \operatorname{cl} S_1 \cap \operatorname{cl} S_2$. Then the following properties hold:*

1. $\mathbb{R}^2(S, \bar{x}, 0) = R(S, \bar{x})$
2. $\mathbb{IR}(S, \bar{x}, 0) = IR(S, \bar{x})$
3. $\mathbb{IR}^2(X \backslash S, \bar{x}, \bar{u}) = X \backslash \mathbb{R}^2(S, \bar{x}, \bar{u})$
4. $0 \in \mathbb{IR}^2(S, \bar{x}, \bar{u}) \iff \bar{u} \in IR(S, \bar{x})$
5. $0 \in \mathbb{R}^2(S, \bar{x}, \bar{u}) \iff \bar{u} \in R(S, \bar{x})$
6. $S_1 \subseteq S_2 \implies \mathbb{R}^2(S_1, \bar{x}, \bar{u}) \subseteq \mathbb{R}^2(S_2, \bar{x}, \bar{u})$
7. $S_1 \subseteq S_2 \implies \mathbb{IR}^2(S_1, \bar{x}, \bar{u}) \subseteq \mathbb{IR}^2(S_2, \bar{x}, \bar{u})$
8. $\mathbb{R}^2(S_1 \cup S_2, \bar{x}, \bar{u}) = \mathbb{R}^2(S_2, \bar{x}, \bar{u}) \cup \mathbb{R}^2(S_1, \bar{x}, \bar{u})$
9. $\mathbb{IR}^2(S_1 \cap S_2, \bar{x}, \bar{u}) = \mathbb{IR}^2(S_1, \bar{x}, \bar{u}) \cap \mathbb{IR}^2(S_2, \bar{x}, \bar{u})$.

Proof. The proof is a direct consequence of the definition of the cones involved. □

4.11.2 Second-Order Asymptotic Contingent Cone and Second-Order Asymptotic Interiorly Contingent Cone

Our focus is on the following two notions of second-order asymptotic tangent sets.

Definition 4.11.3. Let X be a normed space, let S be a subset of X, and let $(\bar{x}, \bar{u}) \in X \times X$.

1. The **second-order asymptotic contingent cone** $\mathbb{T}^2(S, \bar{x}, \bar{u})$ of S at $(\bar{x}, \bar{u})$ is the set of all $x \in X$ for which there are a sequence $\{(s_n, t_n)\} \subset \mathbb{P} \times \mathbb{P}$ with $(s_n, t_n) \downarrow 0$ and $\frac{s_n}{t_n} \downarrow 0$ and a sequence $\{x_n\} \subset X$ with $x_n \to x$ satisfying that $\bar{x} + s_n \bar{u} + s_n t_n x_n \in S$ for every $n \in \mathbb{N}$.
2. The **second-order asymptotic interiorly contingent cone** $\mathbb{IT}^2(S, \bar{x}, \bar{u})$ of S at $(\bar{x}, \bar{u})$ is the set of all $x \in X$ such that for every sequence $\{(s_n, t_n)\} \subset \mathbb{P} \times \mathbb{P}$ with $(s_n, t_n) \downarrow 0$ and $\frac{s_n}{t_n} \downarrow 0$ and every sequence $\{x_n\} \subset X$ with $x_n \to x$, there exists an index $m \in \mathbb{N}$ satisfying that $\bar{x} + s_n \bar{u} + s_n t_n x_n \in S$ for every $n \geq m$.

   Equivalently,

$$\mathbb{T}^2(S, \bar{x}, \bar{u}) = \{x \in X \mid \forall\, U(x)\ \forall\, t > 0\ \exists\, r, s \in (0, t) \text{ with } \frac{s}{r} < t\ \exists\, z \in U(x) \text{ with } \bar{x} + s\bar{u} + srz \in S\},$$

$$\mathbb{IT}^2(S, \bar{x}, \bar{u}) = \{x \in X \mid \exists\, U(x)\ \exists\, t > 0\ \forall\, s, r \in (0, t)\ ) \text{ with } \frac{s}{r} < t\ \forall\, z \in U(x) \text{ we have } \bar{x} + s\bar{u} + srz \in S\}.$$

Remark 4.11.4. The second-order asymptotic contingent cone is exactly the object $T_0(S, \bar{x}, \bar{u})$ which was proposed in Definition 4.10.1. Consequently, some of its properties which are given in Theorem 4.11.6, follow from Theorem 4.10.3.

The following theorem collects some of the other commonly used characterizations of the second-order asymptotic contingent cone.

Theorem 4.11.5. *Let X be a normed space, let S be subset of X, let $\bar{u} \in X$, and let $\bar{x} \in \operatorname{cl} S$. Then, for $i \in \{1, \ldots, 5\}$, we have $\mathbb{T}^2(S, \bar{x}, \bar{u}) = \mathbb{T}_i^2(S, \bar{x}, \bar{u})$, where $\mathbb{T}_i^2(S, \bar{x}, \bar{u})$ are given by*

$$\mathbb{T}_1^2(S, \bar{x}, \bar{u}) = \left\{ x \in X \mid \exists\ (s_n, t_n) \downarrow (0,0) \text{ with } \frac{s_n}{t_n} \downarrow 0,\ \exists\ \{x_n\} \subset S \text{ such that } \frac{x_n - x - s_n \bar{u}}{s_n t_n} \to x \right\}$$

$$\mathbb{T}_2^2(S, \bar{x}, \bar{u}) = \left\{ x \in X \mid \liminf_{(s,t)\downarrow(0,0),\ \frac{s}{t}\downarrow 0} \frac{d(\bar{x} + s\bar{u} + stx, S)}{st} = 0 \right\}$$

$$\mathbb{T}_3^2(S, \bar{x}, \bar{u}) = \left\{ x \in X \mid \exists\ (s_n, t_n) \downarrow (0,0),\ \frac{s_n}{t_n} \downarrow 0,\ \exists\ \{x_n\} \to x,\ \bar{x} + s_n \bar{u} + \frac{t_n^2}{s_n} x_n \in S \right\}$$

$$\mathbb{T}_4^2(S, \bar{x}, \bar{u}) = \left\{ x \in X \mid \exists\ (\alpha_n, \beta_n) \to (\infty, \infty),\ \frac{\beta_n}{\alpha_n} \downarrow 0,\ \exists\ \{x_n\} \subset S, \text{ such that } \beta_n(\alpha_n(x_n - \bar{x}) - \bar{u}) \to x \right\}$$

$$\mathbb{T}_5^2(S, \bar{x}, \bar{u}) = \limsup_{(s,t)\downarrow(0,0),\ \frac{s}{t}\downarrow 0} \frac{S - \bar{x} - s\bar{u}}{st}.$$

Proof. The proof is based on standard arguments. □

The following theorem collects some properties of the second-order asymptotic contingent cone and second-order asymptotic interiorly contingent cone.

Theorem 4.11.6. *Let X be a normed space, let S, S_1, S_2 be nonempty subsets of X, let $\bar{u} \in X$, and let $\bar{x} \in \operatorname{cl} S \cap \operatorname{cl} S_1 \cap \operatorname{cl} S_2$. Then the following properties hold:*

1. $\mathbb{T}^2(S, \bar{x}, 0) = T(S, \bar{x})$
2. $\mathbb{IT}^2(S, \bar{x}, 0) = IT(S, \bar{x})$
3. $\mathbb{T}^2(X \setminus S, \bar{x}, \bar{u}) = X \setminus \mathbb{IT}^2(S, \bar{x}, \bar{u})$
4. $0 \in \mathbb{T}^2(S, \bar{x}, \bar{u}) \implies \bar{u} \in T(S, \bar{x})$
5. $0 \in \mathbb{IT}^2(S, \bar{x}, \bar{u}) \impliedby \bar{u} \in IT(S, \bar{x})$
6. $\mathbb{T}^2(S, \bar{x}, \bar{u})$ *is a closed cone*
7. $\mathbb{T}^2(S, \bar{x}, \bar{u}) = \mathbb{T}^2(\operatorname{cl} S, \bar{x}, \bar{u})$

8. $\mathbb{IT}^2(S, \bar{x}, \bar{u})$ *is an open cone*
9. $\mathbb{IT}^2(S, \bar{x}, \bar{u}) = \mathbb{IT}^2(\text{int } S, \bar{x}, \bar{u})$
10. $\mathbb{T}^2(S, \bar{x}, \bar{u}) \neq \emptyset \implies \bar{u} \in T(S, \bar{x})$ *and hence* $\bar{u} \notin T(S, \bar{x}) \implies \mathbb{T}^2(S, \bar{x}, \bar{u}) = \emptyset$
11. $\mathbb{IT}^2(S, \bar{x}, \bar{u}) \neq X \implies \bar{u} \notin IT(S, \bar{x})$ *and hence* $\bar{u} \in IT(S, \bar{x}) \implies \mathbb{T}^2(S, \bar{x}, \bar{u}) = X$
12. $S_1 \subseteq S_2 \implies \mathbb{T}^2(S_1, \bar{x}, \bar{u}) \subseteq \mathbb{T}^2(S_2, \bar{x}, \bar{u})$
13. $S_1 \subseteq S_2 \implies \mathbb{IT}^2(S_1, \bar{x}, \bar{u}) \subseteq \mathbb{IT}^2(S_2, \bar{x}, \bar{u})$
14. $\mathbb{IT}^2(S_1 \cap S_2, \bar{x}, \bar{u}) = \mathbb{IT}^2(S_2, \bar{x}, \bar{u}) \cap \mathbb{IT}^2(S_1, \bar{x}, \bar{u})$
15. $\mathbb{T}^2(S_1 \cup S_2, \bar{x}, \bar{u}) = \mathbb{T}^2(S_1, \bar{x}, \bar{u}) \cup \mathbb{T}^2(S_2, \bar{x}, \bar{u})$
16. $\mathbb{T}^2(S, \bar{x}, \bar{u}) + t\bar{u} \subset \mathbb{T}^2(S, \bar{x}, \bar{u})$, *for every* $t \in \mathbb{R}$
17. $\mathbb{IT}^2(S, \bar{x}, \bar{u}) + t\bar{u} \subset \mathbb{IT}^2(S, \bar{x}, \bar{u})$, *for every* $t \in \mathbb{R}$.

Proof. The proof is based on standard arguments used earlier in this chapter. □

Remark 4.11.7. The authors in [209] give a nice proof of the fact that, in a finite dimensional setting, $0 \in \mathbb{T}^2(S, \bar{x}, \bar{u})$ is equivalent to $\bar{u} \in T(S, \bar{x})$.

The following example, originally given in [209] shows that, in general, $T^2(S, \bar{x}, \bar{u})$ and $\mathbb{T}^2(S, \bar{x}, \bar{u})$ are different objects:

Example 4.11.8. For $S \subset \mathbb{R}^2$ given by $S = \{(x, y) \in \mathbb{R}^2 |\ y \geq x^2\} \cup \{(x, y) \in \mathbb{R}^2 | y = -x^2\}$, and for $\bar{x} = (0, 0)$ and $\bar{u} = (1, 0)$, we have

$$T^2(S, \bar{x}, \bar{u}) = \{(x, y) \in \mathbb{R}^2 |\ y \geq 2 \cup y = -2\},$$
$$\mathbb{T}^2(S, \bar{x}, \bar{u}) = \{(x, y) \in \mathbb{R}^2 |\ y \geq 0\}.$$

4.11.3 *Second-Order Asymptotic Adjacent Cone and Second-Order Asymptotic Interiorly Adjacent Cone*

Our focus is on the following two notions:

Definition 4.11.9. Let X be a normed space, let S be a subset of X, and let $(\bar{x}, \bar{u}) \in X \times X$.

1. The **second-order asymptotic adjacent cone** $\mathbb{A}^2(S, \bar{x}, \bar{u})$ of S at $(\bar{x}, \bar{u})$ is the set of all $x \in X$ such that for every sequence $\{(s_n, t_n)\} \subset \mathbb{P} \times \mathbb{P}$ with $(s_n, t_n) \downarrow 0$ and $\frac{s_n}{t_n} \downarrow 0$, there exists a sequence $\{x_n\} \subset X$ with $x_n \to x$ satisfying that $\bar{x} + s_n \bar{u} + s_n t_n x_n \in S$ for every $n \in \mathbb{N}$.
2. The **second-order asymptotic interiorly adjacent cone** $\mathbb{IA}^2(S, \bar{x}, \bar{u})$ of S at $(\bar{x}, \bar{u})$ is the set of all $x \in X$ such that there exists a sequence $\{(s_n, t_n)\} \subset \mathbb{P} \times \mathbb{P}$ with $(s_n, t_n) \downarrow 0$ and $\frac{s_n}{t_n} \downarrow 0$ so that for every sequence $\{x_n\} \subset X$ with $x_n \to x$, there exists an index $m \in \mathbb{N}$ satisfying that $\bar{x} + s_n \bar{u} + s_n t_n x_n \in S$ for every $n \geq m$.

Equivalently,

$$\begin{aligned}\mathbb{A}^2(S,\bar{x},\bar{u}) &= \{x \in X \mid \forall\ U(x)\ \exists\ t>0\ \forall\ s,r \in (0,t)\ )\ \text{such that}\\ &\qquad \frac{s}{r} < t\ \exists\ z \in U(x) \text{ with } \bar{x}+s\bar{u}+srz \in S\},\\ \mathbb{IA}^2(S,\bar{x},\bar{u}) &= \{x \in X \mid \exists\ U(x)\ \exists\ t>0\ \forall\ s,r \in (0,t)\ )\ \text{with}\\ &\qquad \frac{s}{r} < t\ \forall\ z \in U(x) \text{ we have } \bar{x}+s\bar{u}+srz \in S\}.\end{aligned}$$

The following result collects some of the other commonly used definitions of the second-order asymptotic adjacent cones.

Theorem 4.11.10. *Let X be a normed space, let S be a subset of X, let $\bar{u} \in X$, and let $\bar{x} \in cl\ S$. Then for $i \in \{1,\ldots,5\}$, we have $\mathbb{A}^2(S,\bar{x},\bar{u}) = \mathbb{A}_i^2(S,\bar{x},\bar{u})$, where $\mathbb{A}_i(S,\bar{x},\bar{u})$ are given as follows:*

$$\mathbb{A}_1^2(S,\bar{x},\bar{u}) = \left\{x \in X \mid \forall\ \ (s_n,t_n) \downarrow (0,0) \text{ with } \frac{s_n}{t_n} \downarrow 0,\ \exists\ \{x_n\} \subset S \text{ such that}\right.$$
$$\left.\frac{x_n - x - s_n\bar{u}}{s_nt_n} \to x\right\}$$

$$\mathbb{A}_2^2(S,\bar{x},\bar{u}) = \left\{\{x \in X \mid \limsup_{(s,t)\downarrow(0,0),\ \frac{s}{t}\downarrow 0} \frac{d(\bar{x}+t_n\bar{u}+stx,S)}{st} = 0\right\}$$

$$\mathbb{A}_3^2(S,\bar{x},\bar{u}) = \{x \in X \mid \forall\ (s_n,t_n) \downarrow (0,0) \text{ with } \frac{s_n}{t_n} \downarrow 0\ \exists\{x_n\} \to x \text{ such that}$$
$$\bar{x}+s_n\bar{u}+s_n^{-1}t_n^2x_n \in S\}$$

$$\mathbb{A}_4^2(S,\bar{x},\bar{u}) = \left\{x \in X \mid \forall\ (\alpha_n,\beta_n) \to (\infty,\infty) \text{ with } \frac{\beta_n}{\alpha_n} \downarrow 0\ \exists\{x_n\} \subset S \text{ such that}\right.$$
$$\left.\beta_n(\alpha_n(x_n-\bar{x})-\bar{u}) \to x\right\}$$

$$\mathbb{A}_5^2(S,\bar{x},\bar{u}) = \liminf_{(s,t)\downarrow(0,0),\frac{s}{t}\downarrow 0} \frac{S-\bar{x}-s\bar{u}}{st}.$$

The following result collects some properties of the second-order asymptotic adjacent cones and the second-order asymptotic interiorly adjacent cones.

Theorem 4.11.11. *Let X be a normed space, let S, S_1, S_2 be nonempty subsets of X, let $\bar{u} \in X$, and let $\bar{x} \in \mathrm{cl}\,S \cap \mathrm{cl}\,S_1 \cap \mathrm{cl}\,S_2$. Then the following properties hold:*

1. $\mathbb{A}^2(S,\bar{x},0) = A(S,\bar{x})$
2. $\mathbb{IA}^2(S,\bar{x},0) = IA(S,\bar{x})$
3. $\mathbb{A}^2(X\backslash S,\bar{x},\bar{u}) = X\backslash\mathbb{IA}^2(S,\bar{x},\bar{u})$

4. $0 \in \mathbb{A}^2(S, \bar{x}, \bar{u}) \implies \bar{u} \in A(S, \bar{x})$
5. $0 \in \mathbb{IA}^2(S, \bar{x}, \bar{u}) \impliedby \bar{u} \in IA(S, \bar{x})$
6. $\mathbb{A}^2(A, \bar{x}, \bar{u})$ *is a closed cone*
7. $\mathbb{A}^2(S, \bar{x}, \bar{u}) = \mathbb{A}^2(\text{cl } S, \bar{x}, \bar{u})$
8. $\mathbb{IA}^2(S, \bar{x}, \bar{u})$ *is an open cone*
9. $\mathbb{IA}^2(S, \bar{x}, \bar{u}) = \mathbb{IA}^2(\text{int } S, \bar{x}, \bar{u})$
10. $\mathbb{A}^2(S, \bar{x}, \bar{u}) \neq \emptyset \implies \bar{u} \in A(S, \bar{x})$ *and hence* $\bar{u} \notin A(S, \bar{x}) \implies \mathbb{A}^2(S, \bar{x}, \bar{u}) = \emptyset$.
11. $\mathbb{IA}^2(S, \bar{x}, \bar{u}) \neq X \implies \bar{u} \notin IT(S, \bar{x})$ *and hence* $\mathbb{IA}^2(S, \bar{x}, \bar{u}) \neq X \implies \mathbb{IA}^2(S, \bar{x}, \bar{u}) = X$.
12. $S_1 \subseteq S_2 \implies \mathbb{A}^2(S_1, \bar{x}, \bar{u}) \subseteq \mathbb{A}^2(S_2, \bar{x}, \bar{u})$
13. $S_1 \subseteq S_2 \implies \mathbb{IA}^2(S_1, \bar{x}, \bar{u}) \subseteq \mathbb{IA}^2(S_2, \bar{x}, \bar{u})$
14. $\mathbb{A}^2(S, \bar{x}, \bar{u}) + t\bar{u} \subset \mathbb{A}^2(S, \bar{x}, \bar{u})$, *for every* $t \in \mathbb{R}$
15. $\mathbb{IA}^2(S, \bar{x}, \bar{u}) + t\bar{u} \subset \mathbb{IA}^2(S, \bar{x}, \bar{u})$, *for every* $t \in \mathbb{R}$.

Proof. The proof is based on standard arguments used earlier in this chapter. □

In the above, we have studied second-order extensions of the first-order tangent cones. In fact, it is possible to modify some of these definitions to obtain second-order versions of the first-order modified tangent cones. We explain this by giving the following definition of the second-order Clarke's tangent set given by Ward [595]:

Definition 4.11.12. Let X be a normed space, let $S \subset X$, and let $(\bar{x}, \bar{u}) \in S \times X$. The **second-order Clarke's tangent cone** $C^2(S, \bar{x}, \bar{u})$ is the set of all $x \in X$ such that for every $\{t_n\} \subset \mathbb{P}$ with $t_n \downarrow 0$ and for every $\{z_n\} \subset S$ with $z_n \to \bar{x}$, there exists a sequence $\{x_n\} \subset X$ such that $x_n \to x$ and $z_n + t_n\bar{u} + t_n^2 x_n \in S$ for every $n \in \mathbb{N}$.

Clearly, it is possible to obtain variants of the other first-order modified tangent cones.

4.12 Miscellaneous Properties of Second-Order Tangent Sets and Second-Order Asymptotic Tangent Cones

In this section, we collect useful properties of the second-order tangent cones and the second-order tangent sets studied in the previous sections.

We begin with the following result that shows that, on convex sets, the relationship among some of the second-order tangent sets and second-order tangent cones is quite elegant:

Theorem 4.12.1. *Let X be a normed space, let $S \subset X$ be convex, let $\bar{x} \in S$, and let $\bar{u} \in T(S, \bar{x})$. Then:*

1. $T^2(S, \bar{x}, \bar{u}) + T(T(S, \bar{x}), \bar{u}) \subseteq T^2(S, \bar{x}, \bar{u}) \subseteq T(T(S, \bar{x}), \bar{u})$
2. $A^2(S, \bar{x}, \bar{u}) + T(T(S, \bar{x}), \bar{u}) \subseteq A^2(S, \bar{x}, \bar{u}) \subseteq T(T(S, \bar{x}), \bar{u})$

3. $\mathbb{T}^2(S,\bar{x},\bar{u}) + T(T(S,\bar{x}),\bar{u}) \subseteq \mathbb{T}^2(S,\bar{x},\bar{u})$
4. $A(A(S,\bar{x}),\bar{u}) = T(T(S,\bar{x}),\bar{u}) = \text{cl cone}\,[\text{cone}\,(S-\bar{x}) - \bar{u}]$
5. $A^2(S,\bar{x},\bar{u}) \subseteq A(A(S,\bar{x}),\bar{u})$
6. $T^2(S,\bar{x},\bar{u}) = T(T(S,\bar{x}),\bar{u})$ *if* $0 \in T^2(S,\bar{x},\bar{u})$
7. $A^2(S,\bar{x},\bar{u}) = T^2(S,\bar{x},\bar{u}) = T(T(S,\bar{x}),\bar{u})$ *if* $0 \in A^2(S,\bar{x},\bar{u})$
8. $\mathbb{T}^2(S,\bar{x},\bar{u}) = \text{cl cone}\,[\text{cone}\,(S-\bar{x}) - \bar{u}]$, *provided that* $\mathbb{T}^2(S,\bar{x},\bar{u}) \neq \emptyset$
9. $T^2(S,\bar{x},\bar{u}) \subseteq \mathbb{T}^2(S,\bar{x},\bar{u})$, *provided that* $\mathbb{T}^2(S,\bar{x},\bar{u}) \neq \emptyset$
10. $\alpha T^2(S,\bar{x},\bar{u}) + \beta T^2(S,\bar{x},\bar{u}) \subseteq T^2(S,\bar{x},\bar{u})$, *for any* $\alpha \geq 0$ *and* $\beta \geq 0$
11. $T(T(S,\bar{x}),\bar{u}) = \text{cl}\,(T(S,\bar{x}) + \text{Lin}\{\bar{u}\})$ *where* $\text{Lin}\{\bar{u}\} := \{t\bar{u} |\ t \in \mathbb{R}\}$, *the linear space generated by* $\bar{u}$.

Proof. Let $x \in T^2(S,\bar{x},\bar{u})$ be arbitrary. Then there are sequences $\{t_n\} \subset \mathbb{P}$ and $\{x_n\} \subset X$ with $t_n \downarrow 0$ and $x_n \to x$ such that $\bar{x} + t_n\bar{u} + \frac{1}{2}t_n^2 x_n \in S$ for every $n \in \mathbb{N}$. Next we choose an element $z \in S$ and positive scalars $\alpha, \beta \in \mathbb{P}$ arbitrarily. Since S is convex, for sufficiently large $n \in \mathbb{N}$, we have $\bar{x} + \beta t_n(z - \bar{x}) \in S$. Using the convexity of S once again, for sufficiently large $n \in \mathbb{N}$, we also have

$$\left(1 - \frac{\alpha t_n}{2}\right)\left[\bar{x} + t_n\bar{u} + \frac{1}{2}t_n^2 x_n\right] + \left(\frac{\alpha t_n}{2}\right)[\bar{x} + \beta t_n(z - \bar{x})] \in S$$

which implies that

$$\bar{x} + t_n\bar{u} + \frac{1}{2}t_n^2\,[x_n + \alpha\,[\beta(z-\bar{x}) - \bar{u}] + \alpha t_n x_n] \in S$$

and hence $x + \alpha\,[\beta(z-\bar{x}) - \bar{u}] \in T^2(S,\bar{x},\bar{u})$. Since S is convex, we have $T(S,\bar{x}) = \text{cl}\,\left(\cup_{\beta>0}\beta(S-\bar{x})\right)$, and this, in view of the closedness of $T^2(S,\bar{x},\bar{u})$, ensures that $x + \alpha\,[T(S,\bar{x}) - \bar{u}] \subseteq T^2(S,\bar{x},\bar{u})$.

Furthermore, since $T(S,\bar{x})$ is also convex, by repeating the same arguments again, we deduce that $x + T(T(S,\bar{x}),\bar{u}) \subseteq T^2(S,\bar{x},\bar{u})$ proving that

$$T^2(S,\bar{x},\bar{u}) + T(T(S,\bar{x}),\bar{u}) \subseteq T^2(S,\bar{x},\bar{u}).$$

For the second inclusion, we note that, since $S \subset \bar{x} + T(S,\bar{x})$, we have

$$d\left(\bar{x} + t_n\bar{u} + \frac{1}{2}t_n^2 x, S\right) \geq d\left(t_n\bar{u} + \frac{1}{2}t_n^2 x, T(S,\bar{x})\right),$$

and hence $x \in T^2(S,\bar{x},\bar{u})$ implies that $d\left(\bar{u} + \frac{t_n}{2}x, T(S,\bar{x})\right) = o(t_n)$, confirming that $\frac{x}{2} \in T(T(S,\bar{x}),\bar{u})$. Hence $x \in T(T(S,\bar{x}),\bar{u})$ due to the fact that $T(T(S,\bar{x}),\bar{u})$ is a cone.

All the remaining parts can be proved by similar arguments. For instance, for $T^2(S,\bar{x},\bar{u}) = T(T(S,\bar{x}),\bar{u})$, we note that if $0 \in T^2(S,\bar{x},\bar{u})$, then we have the inclusion $T(T(S,\bar{x}),\bar{u}) \subseteq T^2(S,\bar{x},\bar{u})$. Since the converse inclusion holds, we easily deduce the identity. Finally, the last part follows from the fact if K is a closed and convex cone and $\bar{w} \in K$, then $T(K,\bar{w}) = \text{cl}\,(K + \text{Lin}\{\bar{w}\})$ (see [53]). The proof is complete. □

Remark 4.12.2. The second inclusion was given by Cominetti [103]. The first two inclusions, the sixth, the seventh, and the last part is given in [53]. See also [209, 304, 305]. The second to last part is given by Di [134]. We note that the last part relates nicely Proposition 4.4.4.

It is natural to consider second-order analogues by taking tangent cones of tangent cones. The following interesting example, originally given by Jourani [309], shows that, in general, the second-order tangent sets are different from the tangent cone of a tangent cone.

Example 4.12.3. For $S = \big\{(x,y) \subset \mathbb{R}^2 |\ x \neq -1,\ (x+1)y = 1\big\}$, we set $\bar{x} = (0,1)$ and $\bar{u} = (1,-1)$. Then,

$$A^2(S,\bar{x},\bar{u}) = \{(x,y) \in \mathbb{R}^2 |\ y = -x + 1\},$$
$$A(A(S,\bar{x}),\bar{u}) = \{(x,y) \in \mathbb{R}^2 |\ y = -x\}.$$

Example 4.12.4. For $S = \big\{(x,y) \in \mathbb{R}^2 |\ x > -1,\ (x+1)y \geq 1\big\}$, we set $\bar{x} = (0,1)$ and $\bar{u} = (1,-1)$. Then,

$$A^2(S,\bar{x},\bar{u}) = \{(x,y) \in \mathbb{R}^2 |\ y \geq -x + 1\},$$
$$A(A(S,\bar{x}),\bar{u}) = \{(x,y) \in \mathbb{R}^2 |\ y \geq -x\}.$$

Remark 4.12.5. If the underlying set S is convex, then $A^2(S,\bar{x},\bar{u})$ is convex. However, $T^2(S,\bar{x},\bar{u})$ may fail to be convex even for a convex set S. See [53, Example 3.3.1]

Our next result is a collection of intersection formulas relating a few cones:

Theorem 4.12.6. *Let X be a normed space, let S_1 and S_2 be two nonempty subsets of X, let $\bar{x} \in \text{cl}\, S_1 \cap \text{cl}\, S_2$, and let $\bar{u} \in X$. Then the following relationships hold:*

1. $IT^2(S_1,\bar{x},\bar{u}) \cap IA^2(S_2,\bar{x},\bar{u}) \subseteq IA^2(S_1 \cap S_2,\bar{x},\bar{u})$
2. $IT^2(S_1,\bar{x},\bar{u}) \cap A^2(S_2,\bar{x},\bar{u}) \subseteq A^2(S_1 \cap S_2,\bar{x},\bar{u})$
3. $IT^2(S_1,\bar{x},\bar{u}) \cap T^2(S_2,\bar{x},\bar{u}) \subseteq T^2(S_1 \cap S_2,\bar{x},\bar{u})$
4. $IA^2(S_1,\bar{x},\bar{u}) \cap A^2(S_2,\bar{x},\bar{u}) \subseteq T^2(S_1 \cap S_2,\bar{x},\bar{u})$.

Proof. The proof can be obtained by using the definition of the sets involved. □

The following result collects a useful feature of the second-order contingent cone and the second-order adjacent cone (see [594]).

Theorem 4.12.7. *Let X be a normed space, let S be a subset of X, and let $(\bar{x}, \bar{u}) \in S \times X$. Let $\ell : X \to Y$ be linear and continuous. Then the following inclusions hold:*

$$\ell(T^2(S, \bar{x}, \bar{u})) \subseteq T^2(\ell(S), \ell(\bar{x}), \ell(\bar{u})), \tag{4.74}$$

$$\ell(A^2(S, \bar{x}, \bar{u})) \subseteq A^2(\ell(S), \ell(\bar{x}), A(\bar{u})). \tag{4.75}$$

Proof. Following directly from the definitions of the involved sets. □

We have the following auxiliary result which will be used shortly (see Ward [594]):

Theorem 4.12.8. *Let X be a normed space, let S be a subset of X, let $\bar{x} \in S$ and let $\bar{u} \in X$. Then the following inclusions hold:*

$$T^2(S, \bar{x}, \bar{u}) + C(S, \bar{x}) \subseteq T^2(S, \bar{x}, \bar{u})$$

$$A^2(S, \bar{x}, \bar{u}) + C(S, \bar{x}) \subseteq A^2(S, \bar{x}, \bar{u})$$

$$A^2(S, \bar{x}, \bar{u}) + IT_M(S, \bar{x}) \subseteq IT^2(S, \bar{x}, \bar{u}).$$

Proof. Let $x \in T^2(S, \bar{x}, \bar{u})$ and $y \in C(S, \bar{x})$ be arbitrary. Then, due to the containment $x \in T^2(S, \bar{x}, \bar{u})$, there are sequences $\{t_n\} \subset \mathbb{P}$ and $\{x_n\} \subset X$ such that $t_n \downarrow 0$, $x_n \to x$ and $z_n := \bar{x} + t_n\bar{u} + \frac{t_n^2}{2}x \in S$ for every $n \in \mathbb{N}$. Since $\{z_n\} \subset S$ and $z_n \to \bar{x}$, the containment $y \in C(S, \bar{x})$ ensures that there exists a sequence $\{y_n\} \subset Y$ with $y_n \to y$ such that for every sequence $\{s_n\} \subset \mathbb{P}$ with $s_n \downarrow 0$, we have $z_n + s_n y_n \in S$ for every $n \in \mathbb{N}$. Since, $\bar{x} + t_n\bar{u} + \frac{t_n^2}{2}(x_n + y_n) = z_n + \frac{t_n^2}{2}y_n \in S$, we at once deduce that $x + y \in T^2(S, \bar{x}, \bar{u})$. This proves the first inclusion. The proof of the second inclusion is obtained by similar arguments.

For the final inclusion, let $x \in A^2(S, \bar{x}, \bar{u})$ and $y \in IT_M(S, \bar{x})$ be arbitrary. We set $z := x + y$. Let $\{t_n\} \subset \mathbb{P}$ with $t_n \downarrow 0$ and $\{z_n\} \subset X$ with $z_n \to z$ be arbitrary sequences. Since $x \in A^2(S, \bar{x}, \bar{u})$, there exists a sequence $\{y_n\} \subset X$ with $y_n \to x$ such that $x_n := \bar{x} + t_n\bar{u} + t_n^2 y_n/2 \in S$ for every $n \in \mathbb{N}$. We set $u_n = z_n - y_n$ and note that $u_n \to y$, and since $y \in IT_M(S, \bar{x})$, we have

$$\bar{x} + t_n\bar{u} + \frac{t_n^2}{2}z_n = \bar{x} + t_n\bar{u} + \frac{t_n^2}{2}(y_n + u_n) = x_n + \frac{t_n^2}{2}u_n \in S$$

for n sufficiently large. Consequently $x + y \in IT^2(S, \bar{x}, \bar{u})$. □

We now state a useful consequence of the above result.

Proposition 4.12.9. *Let X be a normed space, let S be a subset of X, let $\bar{x} \in S$ and let $\bar{u} \in X$. Assume that $IT_M(S, \bar{x}) \neq \emptyset$ and $A^2(S, \bar{x}, \bar{u}) \neq \emptyset$. Then*

$$A^2(S, \bar{x}, \bar{u}) = \operatorname{cl}\left(IT^2(S, \bar{x}, \bar{u})\right), \tag{4.76}$$

$$IT^2(S, \bar{x}, \bar{u}) = \operatorname{int}\left(A^2(S, \bar{x}, \bar{u})\right). \tag{4.77}$$

The following intersection formula for second-order contingent sets is due to Ward [594]:

Theorem 4.12.10. *Let X be a normed space, let $\{S_1, S_2, \ldots, S_n\}$ be a family of nonempty subsets of X, let $\bar{x} \in \bigcap_{i=1}^{n} S_i$, and let $\bar{u} \in X$. Assume that*

$$C(S_1, \bar{x}) \cap IT_M(S_2, \bar{x}) \cap \cdots \cap IT_M(S_n, \bar{x}) \neq \emptyset. \tag{4.78}$$

Then the following intersection formulas hold:

$$T^2(S_1, \bar{x}, \bar{u}) \cap A^2(S_2, \bar{x}, \bar{u}) \cap \cdots \cap A^2(S_n, \bar{x}, \bar{u}) \subseteq T^2(S_1 \cap S_2 \cap \cdots \cap S_n, \bar{x}, \bar{u}); \tag{4.79}$$

$$A^2(S_1, \bar{x}, \bar{u}) \cap A^2(S_2, \bar{x}, \bar{u}) \cap \cdots \cap A^2(S_n, \bar{x}, \bar{u}) = A^2(S_1 \cap S_2 \cap \cdots \cap S_n, \bar{x}, \bar{u}). \tag{4.80}$$

Furthermore, if $A^2(S_i, \bar{x}, \bar{u}) = T^2(S_i, \bar{x}, \bar{u})$ for every $i = 2, 3, \ldots, n$, then

$$T^2(S_1, \bar{x}, \bar{u}) \cap T^2(S_2, \bar{x}, \bar{u}) \cap \cdots \cap T^2(S_n, \bar{x}, \bar{u}) = T^2(S_1 \cap S_2 \cap \cdots \cap S_n, \bar{x}, \bar{u}). \tag{4.81}$$

Proof. To prove (4.79), we first note that the following inclusion is always true:

$$T^2(S_1, \bar{x}, \bar{u}) \cap IT^2(S_2, \bar{x}, \bar{u}) \cap \cdots \cap IT^2(S_n, \bar{x}, \bar{u}) \subseteq T^2(S_1 \cap S_2 \cap \cdots \cap S_n, \bar{x}, \bar{u}). \tag{4.82}$$

Let $x \in T^2(S_1, \bar{x}, \bar{u}) \cap A^2(S_2, \bar{x}, \bar{u}) \cap \cdots \cap A^2(S_n, \bar{x}, \bar{u})$ and $y \in C(S_1, \bar{x}) \cap IT_M(S_2, \bar{x}) \cap \cdots \cap IT_M(S_n, \bar{x})$ be arbitrary. By using Theorem 4.12.8 and the fact that $C(S_1, \bar{x})$ and $IT_M(S_i, \bar{x})$ are cones, we deduce that for every $t > 0$,

$$x + ty \in T^2(S_1, \bar{x}, \bar{u}) \cap IT^2(S_2, \bar{x}, \bar{u}) \cap \cdots \cap IT^2(S_n, \bar{x}, \bar{u}),$$

and by using (4.82), we get

$$x + ty \in T^2(S_1 \cap S_2 \cap \cdots \cap S_n, \bar{x}, \bar{u}),$$

which confirms that

$$y \in \operatorname{cl}\left(T^2(S_1 \cap S_2 \cap \cdots \cap S_n, \bar{x}, \bar{u})\right) = T^2(S_1 \cap S_2 \cap \cdots \cap S_n, \bar{x}, \bar{u}).$$

Inclusion (4.79) and equality (4.80) follow from analogous arguments. □

In the above result, (4.78) played a crucial role involving convex cones. In the following, we state an analogue of the above result for the case when the underlying sets are closed and convex and all but one are solid. The proof of this result relies on the Robinson's constraint qualification (see [53, Proposition 3.36]) and hence makes good use of the imposed conditions:

Theorem 4.12.11. *Let X be a Banach space and let $\{S_1, S_2, \ldots, S_n\}$ be a family of closed and convex subsets of X. Assume that there exists $\bar{z} \in S_1$ such that $\bar{z} \in \operatorname{int}(S_i)$, for $i \in \{2, \ldots, n\}$. Assume that $\bar{x} \in S_1 \cap S_2 \cap \cdots \cap S_n$ and $\bar{u} \in T(S_1, \bar{x}) \cap T(S_2, \bar{x}) \cap \cdots \cap T(S_n, \bar{x})$. Then:*

$$T(S_1 \cap S_2 \cap \cdots \cap S_n, \bar{x}) = T(S_1, \bar{x}) \cap T(S_2, \bar{x}) \cap \cdots \cap T(S_n, \bar{x}), \tag{4.83}$$

$$T^2(S_1 \cap S_2 \cap \cdots \cap S_n, \bar{x}, \bar{u}) \subset T^2(S_1, \bar{x}, \bar{u}) \cap T^2(S_2, \bar{x}, \bar{u}) \cap \cdots \cap T^2(S_n, \bar{x}, \bar{u}), \tag{4.84}$$

$$A^2(S_1 \cap S_2 \cap \cdots \cap S_n, \bar{x}, \bar{u}) = A^2(S_1, \bar{x}, \bar{u}) \cap A^2(S_2, \bar{x}, \bar{u}) \cap \cdots \cap A^2(S_n, \bar{x}, \bar{u}). \tag{4.85}$$

4.13 Second-Order Inversion Theorems

In this section, we give some second-order inversion theorems which will play a fundamental role in deriving calculus rules for the second-order graphical and epigraphical derivatives. In the following, we give two results. The first result, Theorem 4.13.1, is by Cominetti [103, Theorem 3.1] where the proof shows the importance of the metric regularity. The second result is taken from [155], where a more general notion of metric subregularity is used.

We begin with the following result due to Cominetti [103].

Theorem 4.13.1. *Let X and Y be Banach spaces, let $S \subset X$ and $Q \subset Y$ be nonempty, closed and convex, let $g : X \to Y$ is continuous on S and twice differentiable at $\bar{x} \in S$ with $g(\bar{x}) \in Q$, and let $\bar{u} \in X$. Assume that the following Robinson's constraint qualification holds:*

$$0 \in \operatorname{core}\{Dg(\bar{x})(S - \bar{x}) - (Q - g(\bar{x}))\}. \tag{4.86}$$

Then the following identities hold:

$$A(S \cap g^{-1}(Q), \bar{x}) = A(S, \bar{x}) \cap Dg(\bar{x})^{-1} \left[A(Q, g(\bar{x}))\right], \tag{4.87}$$

$$A^2(S \cap g^{-1}(Q), \bar{x}, \bar{u}) = A^2(S, \bar{x}, \bar{u}) \cap Dg(\bar{x})^{-1} \left[A^2(Q, g(\bar{x}), Dg(\bar{x})(\bar{u})) - D^2 g(\bar{x})(\bar{u}, \bar{u})\right]. \tag{4.88}$$

Proof. Since (4.87) follows from (4.88), it is sufficient to prove the latter. The map g being continuously differentiable, is Lipschitz continuous near $\bar{x}$, and hence for some $k \geq 0$ and every x near $\bar{x}$, we have

$$d(g(x), Q) \leq k\, d(x, g^{-1}(Q)) \leq k\, d(x, S \cap g^{-1}(Q)).$$

Then for an arbitrary $x \in A^2(S \cap g^{-1}(Q), \bar{x}, \bar{u})$, we have

$$d\left(g\left(\bar{x} + t\bar{u} + \frac{1}{2}t^2 x\right), Q\right) = o(t^2).$$

By using the Taylor expansion, we have

$$g\left(\bar{x} + t\bar{u} + \frac{1}{2}t^2 x\right) = g(\bar{x}) + tDg(\bar{x})(\bar{u}) + \frac{1}{2}t^2\left[Dg(\bar{x})(x) + D^2 g(\bar{x})(\bar{u}, \bar{u})\right] + o(t^2),$$

implying that

$$Dg(\bar{x})(x) + D^2 g(\bar{x})(\bar{u}, \bar{u}) \in A^2(Q, g(\bar{x}), Dg(\bar{x})(\bar{u})),$$

which, when combined with the inclusion $A^2(S \cap g^{-1}(Q), \bar{x}, \bar{u}) \subset A^2(S, \bar{x}, \bar{u})$, yields

$$A^2(S \cap g^{-1}(Q), \bar{x}, \bar{u}) \subseteq A^2(S, \bar{x}, \bar{u}) \cap Dg(\bar{x})^{-1}$$
$$\left[A^2(Q, g(\bar{x}), Dg(\bar{x})(\bar{u})) - D^2 g(\bar{x})(\bar{u}, \bar{u})\right].$$

For the converse inclusion, we note that for some $k \geq 0$ and all $x \in S$ near $\bar{x}$, we have (see [103, Corolary 2.2])

$$d(x, S \cap g^{-1}(Q)) \leq k d(g(x), Q). \tag{4.89}$$

Now, if $x \in A^2(S, \bar{x}, \bar{u}) \cap Dg(\bar{x})^{-1}\left[A^2(Q, g(\bar{x}), Dg(\bar{x})(\bar{u})) - D^2 g(\bar{x})(\bar{u}, \bar{u})\right]$, then

$$d\left(g(\bar{x}) + tDg(\bar{x})(\bar{u}) + \frac{1}{2}t^2\left[Dg(\bar{x})(x) + D^2 g(\bar{x})(\bar{u}, \bar{u})\right], Q\right) \sim o(t^2)$$

and hence by choosing $x_t \to x$ with $\bar{x} + t\bar{u} + \frac{1}{2}t_n^2 x_t \in S$, we get

$$d\left(g\left(\bar{x} + t\bar{u} + \frac{1}{2}t^2 x_t\right), Q\right) \sim o(t^2).$$

Therefore, by using (4.89), we deduce that

$$d\left(\bar{x} + t\bar{u} + \frac{1}{2}t_n^2 x_t, S \cap g^{-1}(Q)\right) \sim o(t^2),$$

which gives that $x \in A^2(S \cap g^{-1}(Q), \bar{x}, \bar{u})$. The proof is complete. □

We remark that Robinson's constraint qualification $0 \in \text{core}\{Dg(\bar{x})(S - \bar{x}) - (Q - g(\bar{x}))\}$ can equivalently be written as

$$Dg(\bar{x})(IR(S, \bar{x})) - IR(Q, g(\bar{x})) = Y,$$

where $IR(A, \bar{z}) = \cup_{t>0} t(A - \bar{z})$ is the feasible tangent cone to A at $\bar{z}$. If either Y is finite dimensional or $\text{int}(Q) \neq \emptyset$, this condition is equivalent to the condition

$$\left[Dg(\bar{x})^*\right]^{-1}(N(\bar{x}, S)) \cap (-N(g(\bar{x}), Q)) = \{0\},$$

where $N(\cdot, \cdot)$ is the classical normal cone.

For the case of mathematical programming where $g : \mathbb{R}^n \to \mathbb{R}^p \times \mathbb{R}^q$ with

$$\begin{aligned} g &= (g_1, g_2, \ldots, g_p, h_1, \ldots, h_q), \\ S &= \mathbb{R}^n, \\ Q &= \mathbb{R}^p_- \times \{0_{\mathbb{R}^q}\}, \end{aligned}$$

we have

$$S \cap g^{-1}((y, z) + Q) = \left\{x \in \mathbb{R}^n \mid g_i(x) \leq y_i \quad h_j(x) = z_j\right\},$$

and Robinson's constraint qualification becomes the Mangasarian-Fromowitz constraint qualification. That is:

1. $\{\nabla h_j(\bar{x})\}_{j=1}^p$ is linearly independent.
2. There exists $v \in \mathbb{R}^n$ such that

$$\begin{aligned} \langle \nabla h_j(\bar{x}), v\rangle &= 0 \quad \text{for every } j, \\ \langle \nabla g_j(\bar{x}), v\rangle &< 0 \quad \text{for all } i \text{ such that } g_i(\bar{x}) = 0. \end{aligned}$$

In [53], an excellent discussion of various equivalent formulations of these constraint qualifications and their limitations can be found.

Furthermore, under the setting of Theorem 4.13.1, with $g : X \to Y$ twice continuously differentiable, for the particular case when $S = X$ and Q is a closed and convex, under the following Robinson's constraint qualification

$$0 \in \text{int}\left[g(\bar{x}) + Dg(\bar{x})X - Q\right], \tag{4.90}$$

Bonnans and Shapiro [53, Page 167] gave the following formulas:

$$T^2(g^{-1}(Q), \bar{x}, \bar{u}) = Dg(\bar{x})^{-1}\left[T^2(Q, g(\bar{x}), Dg(\bar{x})(\bar{u})) - D^2g(\bar{x})(\bar{u}, \bar{u})\right], \tag{4.91}$$

$$A^2(g^{-1}(Q), \bar{x}, \bar{u}) = Dg(\bar{x})^{-1}\left[A^2(Q, g(\bar{x}), Dg(\bar{x})(\bar{u})) - D^2g(\bar{x})(\bar{u}, \bar{u})\right]. \tag{4.92}$$

Several particular cases of Theorem 4.13.1 and formulas (4.91)–(4.91) have been given in the literature. For instance, Theorem 12.9 of the book by Ben-Israel, Ben-Tal, and Zlobec [42] gives (4.88) for the case $S = X$ and $Q = \{0\}$. For this case, (4.88) reads:

$$A^2(g^{-1}(0), \bar{x}, \bar{u}) = \{x \in X \mid Dg(\bar{x})(x) + D^2g(\bar{x})(\bar{u}, \bar{u}) = 0\}.$$

Furthermore, Kawasaki [317] proved analogous result for the case when $S = X$ and $Q = \{0\} \times K$, where K is a closed, convex, and solid cone. For the particular case when $S = X$ and Q is a convex polyhedra, it was shown by Rockafellar [494] that

$$A^2(g^{-1}(Q), \bar{x}, \bar{u}) = Dg(\bar{x})^{-1}\left[A(A(Q, g(\bar{x})), Dg(\bar{x})(\bar{u})) - D^2g(\bar{x})(\bar{u}, \bar{u})\right],$$

which is equivalent to the given result as

$$A(A(Q, g(\bar{x})), Dg(\bar{x})(\bar{u})) = A^2(Q, g(\bar{x}), Dg(\bar{x})(\bar{u}))$$

when Q is a polyhedra.

We conclude this discussion by a remark that the above results also generalizes the following well-known inversion theorem (see [524]):

Theorem 4.13.2. *Let X and Y be Banach spaces, let Q be a nonempty, closed, and convex subset of Y, let $g : X \to Y$ be continuous on X and continuously differentiable at $\bar{x} \in g^{-1}(Q)$. Assume that Robinson's constraint qualification* (4.90) *holds. Then*

$$T(g^{-1}(Q), \bar{x}) = Dg(\bar{x})^{-1}(T(Q, \bar{x})). \tag{4.93}$$

We note that the above result subsumes the following classical tangent space theorem:

Theorem 4.13.3. *Let X and Y be Banach spaces, let $g : X \to Y$ be continuous on X and continuously differentiable at $\bar{x} \in g^{-1}(0)$. If $Dg(\bar{x})$ is surjective, then*

$$T(\mathrm{Ker}(g), \bar{x}) = \mathrm{Ker}(Dg(\bar{x})).$$

It is evident from the proof of Theorem 4.13.1 that the imposed conditions make a good use of the metric regularity. In our next result, we give an inversion theorem under the weaker notion of metric subregularity (see [155]).

Theorem 4.13.4. *Let X and Y be Banach spaces, let $S \subset X$ and $Q \subset Y$ be closed sets, let $g : X \to Y$ be a twice continuously differentiable map, and let $\bar{x} \in S \cap g^{-1}(Q)$. Assume that $f : X \times Y \to Y$ defined by $f(x, y) = g(x) - y$ is metrically subregular at $(\bar{x}, g(\bar{x}), 0)$ with respect to $S \times Q$ in the following sense: There exist a constant $k > 0$ and neighborhoods $U_\tau(\bar{x})$ and $U_\tau(g(\bar{x}))$ such that for* *every* $(x, y) \in [U_\tau(\bar{x}) \times U_\tau(g(\bar{x}))] \cap S \times Q$, (4.49) *holds. Then for any* $\bar{u} \in X$, *we have*

$$T^2(S, \bar{x}, \bar{u}) \cap Dg(\bar{x})^{-1}\left(A^2\left(Q, g(\bar{x}), Dg(\bar{x})(\bar{u})\right) - \frac{1}{2}D^2g(\bar{x})(\bar{u}, \bar{u})\right)$$
$$\subset T^2\left(S \cap g^{-1}(Q), \bar{x}, \bar{u}\right),$$
$$A^2(S, \bar{x}, \bar{u}) \cap Dg(\bar{x})^{-1}\left(T^2\left(Q, g(\bar{x}), Dg(\bar{x})(\bar{u})\right) - \frac{1}{2}D^2g(\bar{x})(\bar{u}, \bar{u})\right)$$
$$\subset T^2\left(S \cap g^{-1}(Q), \bar{x}, \bar{u}\right),$$
$$A^2(S, \bar{x}, \bar{u}) \cap Dg(\bar{x})^{-1}\left(A^2\left(Q, g(\bar{x}), Dg(\bar{x})(\bar{u})\right) - \frac{1}{2}D^2g(\bar{x})(\bar{u}, \bar{u})\right)$$
$$\subset A^2\left(S \cap g^{-1}(Q), \bar{x}, \bar{u}\right).$$

Proof. It follows from the metric subregularity hypothesis that there are a number $k > 0$ and neighborhoods $U_\tau(\bar{x})$ and $U_\tau(g(\bar{x}))$ such that for every $(x, y) \in [U_\tau(\bar{x}) \times U_\tau(g(\bar{x}))] \cap S \times Q$, we have

$$d\left((x, y), f^{-1}(0) \cap (S \times Q)\right) \leq \mu \|g(x) - y\|. \tag{4.94}$$

Let

$$x \in T^2(S, \bar{x}, \bar{u}) \cap Dg(\bar{x})^{-1}\left(A^2\left(Q, g(\bar{x}), Dg(\bar{x})(\bar{u})\right) - \frac{1}{2}D^2g(\bar{x})(\bar{u}, \bar{u})\right)$$

be arbitrary. Therefore, $x \in T^2(S, \bar{x}, \bar{u})$ and

$$Dg(\bar{x})(x) + \frac{1}{2}D^2g(\bar{x})(\bar{u}, \bar{u}) \in A^2(Q, g(\bar{x}), Dg(\bar{x})(\bar{u})).$$

By using the definitions of the second-order adjacent set and the second-order contingent set, there are sequences $\{t_n\}$, $\{x_n\}$, and $\{z_n\}$ such that $t_n \downarrow 0$, $x_n \to x$, with $\bar{x} + t_n\bar{u} + t_n^2 x_n \in S$, and

$$z_n \to \left(Dg(\bar{x})(x) + \frac{1}{2}D^2g(\bar{x})(\bar{u}, \bar{u})\right)$$

with $g(\bar{x}) + t_n Dg(\bar{x})(\bar{u}) + t_n^2 z_n \in Q$.

Equation (4.94) with

$$(x, y) = (\bar{x} + t_n\bar{u} + t_n^2 x_n, g(\bar{x}) + t_n Dg(\bar{x})(\bar{u}) + t_n^2 z_n)$$

ensures that for sufficiently large $n \in \mathbb{N}$, there exists $(p_n, q_n) \in S \times Q$ with $g(p_n) = q_n$ such that

$$\begin{aligned}&\|(\bar{x} + t_n\bar{u} + t_n^2 x_n, g(\bar{x}) + t_n Dg(\bar{x})(\bar{u}) + t_n^2 z_n) - (p_n, q_n)\|\\&\quad < k\|g(\bar{x} + t_n\bar{u} + t_n^2 x_n) - g(\bar{x}) - t_n Dg(\bar{x})(\bar{u}) - t_n^2 z_n\| + t_n^3.\end{aligned}$$

Consequently, $p_n \in S \cap g^{-1}(Q)$ and

$$\left\|\frac{p_n - \bar{x} - t_n\bar{u}}{t_n^2} - x_n\right\| < k\left\|\frac{g(\bar{x} + t_n\bar{u} + t_n^2 x_n) - g(\bar{x}) - t_n Dg(\bar{x})(\bar{u})}{t_n^2} - z_n\right\| + t_n.$$

Since

$$\frac{g(\bar{x} + t_n\bar{u} + t_n^2 x_n) - g(\bar{x}) - t_n Dg(\bar{x})(\bar{u})}{t_n^2} \to Dg(\bar{x})(x) + \frac{1}{2}D^2 g(\bar{x})(\bar{u}, \bar{u}),$$

we deduce that

$$w_n = \frac{p_n - \bar{x} - t_n\bar{u}}{t_n^2} \to x$$

with

$$\bar{x} + t_n\bar{u} + t_n^2 w_n = p_n \in S \cap g^{-1}(Q)$$

and hence the proof of the first inclusion is complete. The remaining parts follow by using the properties of the second-order contingent set and the second-order adjacent set. □

4.14 Expressions of the Second-Order Contingent Set on Specific Constraints

In this section, we give explicit formulas for the computation of the second-order contingent set on certain constraint sets. This section also contains some characteristic examples of second-order contingent sets. We begin with the following definition proposed by Di [134].

Definition 4.14.1. Given a Hilbert space X, a map $g : X \to \mathbb{R}$ is said to admit a second-order Fréchet derivative at $\bar{x}$ if there exists an element $\nabla^2 g(\bar{x}) \in \mathscr{L}(X, X)$ such that

$$\lim_{t\downarrow 0} \frac{g(\bar{x} + tx) - g(\bar{x}) - t\langle \nabla g(\bar{x}), x\rangle}{(t^2/2)} = \langle x, \nabla^2 g(\bar{x})(x)\rangle, \tag{4.95}$$

and the limit is uniform with respect to x in bounded sets, where $\nabla g(\bar{x})$ is the first-order derivative.

Remark 4.14.2. As noted by Di [134], the above notion of second-order Fréchet differentiability is not a local property, and it does not imply that the function is strictly differentiable at that point, nor continuously differentiable around the point, nor Lipschitz around the point.

For the second-order contingent set, we have the following analogue of Proposition 4.8.2:

Proposition 4.14.3. *Let X be a Hilbert space and let $h : X \to \mathbb{R}$ be continuous around $\bar{x} \in X$ and differentiable at $\bar{x}$. Let $h(\bar{x}) = 0$, let $\nabla h(\bar{x}) \neq 0$, and let h be second-order Fréchet differentiable at $\bar{x}$.*

1. *For $S = \{x \in X \mid h(x) = 0\}$ and $\bar{u} \in T(S, \bar{x})$, we have*

$$T^2(S, \bar{x}, \bar{u}) = \left\{x \in X \mid \left\langle \bar{u}, \nabla^2 h(\bar{x})(\bar{u}) \right\rangle + \langle \nabla h(\bar{x}), x \rangle = 0 \right\}.$$

2. *For $S = \{x \in X \mid h(x) \geq 0\}$ and $\bar{u} \in T(S, \bar{x})$, we have*

$$T^2(S, \bar{x}, \bar{u}) = \left\{x \in X \mid \left\langle \bar{u}, \nabla^2 h(\bar{x})(\bar{u}) \right\rangle + \langle \nabla h(\bar{x}), x \rangle \geq 0 \right\}.$$

Proof. We will only prove the case of the equality constraint. Let $x \in T^2(S, \bar{x}, \bar{u})$. Then there exist $\{x_n\} \subset S$ and $\{t_n\} \downarrow 0$ such that $x_n = \bar{x} + t_n \bar{u} + \frac{t_n^2}{2} x + o(t_n^2)$. By the differentiability of h at $\bar{x}$, we get

$$\begin{aligned}
0 &= h(x_n) - h(\bar{x}) \\
&= \left\langle \nabla h(\bar{x}), t_n \bar{u} + \frac{t^2}{2} x + o(t_n^2) \right\rangle \\
&\quad + \frac{1}{2} \left\langle t_n \bar{u} + \frac{t_n^2}{2} x + o(t_n^2), \nabla^2 h(\bar{x}) \left(t_n \bar{u} + \frac{t_n^2}{2} x + o\left(t_n^2\right) \right) \right\rangle \\
&\quad + o\left(t_n \bar{u} + \frac{t_n^2}{2} x + o(t_n^2) \right)^2 \\
&= \frac{t_n^2}{2} \left(\langle \nabla h(\bar{x}), x \rangle + \left\langle \bar{u}, \nabla^2 h(\bar{x})(\bar{u}) \right\rangle \right) + o(t_n^2),
\end{aligned}$$

which confirms that $\langle \nabla h(\bar{x}), x \rangle + \left\langle \bar{u}, \nabla^2 h(\bar{x})(\bar{u}) \right\rangle = 0$ and hence

$$T^2(S, \bar{x}, \bar{u}) \subseteq \left\{x \in X \mid \left\langle \bar{u}, \nabla^2 h(\bar{x})(\bar{u}) \right\rangle + \langle \nabla h(\bar{x}), x \rangle = 0 \right\}.$$

For the converse, assume that for some x we have

$$\langle \nabla h(\bar{x}), x \rangle + \left\langle \bar{u}, \nabla^2 h(\bar{x})(\bar{u}) \right\rangle = 0.$$

Then, for any $n \in \mathbb{N}$, we define

$$y_n(t) := \bar{x} + t\left(\bar{u} + \frac{t}{2}x + \frac{t}{n}\nabla h(\bar{x})\right),$$

$$z_n(t) := \bar{x} + t\left(\bar{u} + \frac{t}{2}x - \frac{t}{n}\nabla h(\bar{x})\right).$$

Since h has the second-order Fréchet derivative at $\bar{x}$, we have

$$\begin{aligned}
h(y_n(t)) - h(\bar{x}) &= t\left\langle \nabla h(\bar{x}), \bar{u} + \frac{t}{2}x + \frac{t}{n}\nabla h(\bar{x})\right\rangle \\
&\quad + \frac{t_n^2}{2}\left\langle \bar{u} + \frac{t}{2}x + \frac{t}{n}\nabla h(\bar{x}), \nabla^2 h(\bar{x})\left(\bar{u} + \frac{t}{2}x + \frac{t}{n}\nabla h(\bar{x})\right)\right\rangle \\
&\quad + o\left(t\bar{u} + \frac{t^2}{2}x + \frac{t^2}{n}\nabla h(\bar{x})\right)^2 \\
&= \frac{t^2}{n}\langle \nabla h(\bar{x}), \nabla h(\bar{x})\rangle \\
&\quad + \frac{t^2}{2}\left(\langle \nabla h(\bar{x}), x\rangle + \langle \bar{u}, \nabla^2 h(\bar{x})(\bar{u})\rangle\right) + o(t^2), \\
&= \frac{t^2}{n}\langle \nabla h(\bar{x}), \nabla h(\bar{x})\rangle + o(t^2), \qquad (4.96)
\end{aligned}$$

and analogously

$$h(z_n(t)) - h(\bar{x}) = -\frac{t^2}{n}\langle \nabla h(\bar{x}), \nabla h(\bar{x})\rangle + o(t^2). \qquad (4.97)$$

By employing the uniform convergence in Fréchet differentiability, there exists a t, denoted by t_n, with $t_n \downarrow 0$ such that (4.96) is positive and (4.97) is negative. Therefore, by the intermediate value theorem, along the line segment between $y_n(t_n)$ and $z_n(t_n)$, there exists x_n such that $h(x_n) - h(\bar{x}) = 0$, confirming that $x_n \in S$. Clearly, we also have

$$\frac{x_n - \bar{x} - t_n\bar{u}}{\frac{1}{2}t_n^2} \to x$$

which ensures that $x \in T^2(S, \bar{x}, \bar{u})$. The proof is complete. □

In the following result, we give explicit expressions for the second-order contingent set on the set of equality and inequality constraints:

Theorem 4.14.4. *Let X be a Hilbert space, let $S \subset X$ be the constraint defined by* (4.52)*, and $\bar{x} \in S$ be arbitrary. Assume that the following* ***Mangasarian-Fromovitz***

constraint qualification *holds: Let* $h = (h_1, h_2, \ldots, h_s)$ *be continuous around* $\bar{x}$ *and Fréchet differentiable at* $\bar{x}$. *The only* $\lambda \in \Lambda$, *where* Λ *is given by* (4.56), *such that* $\sum_{i=1}^{s} \lambda_i \nabla h_i(\bar{x}) = 0$, *is* $\lambda = 0$. *Assume that* $h_i : X \to \mathbb{R}$ *admits second-order Fréchet derivatives at* $\bar{x}$ *for* $i \in \{1, 2, \ldots, s\}$ *and* $\bar{u} \in T(S, \bar{x})$. *Then the following formula holds:*

$$\begin{aligned} T^2(S, \bar{x}, \bar{u}) &= \bigcap_{i=1}^{s} T(S_i, \bar{x}, \bar{u}) \\ &= \left\{x \in X \mid \left\langle \bar{u}, \nabla^2 h(\bar{x})(\bar{u}) \right\rangle + \langle x, \nabla h(\bar{x}) \rangle \in \Lambda^* \right\} \\ &= \left\{ x \in X \;\middle|\; \begin{array}{ll} \left\langle \bar{u}, \nabla^2 h_i(\bar{x})(\bar{u}) \right\rangle + \left\langle x, \nabla^2 h_i(\bar{x}) \right\rangle = 0, & \text{for } i \in I_1 \\ \left\langle \bar{u}, \nabla^2 h_i(\bar{x})(\bar{u}) \right\rangle + \left\langle x, \nabla^2 h_i(\bar{x}) \right\rangle \geq 0, & \text{for } i \in I_2^{=} \end{array} \right\}. \end{aligned}$$

Proof. It suffices to show that

$$\bigcap_{i=1}^{s} T(S_i, \bar{x}, \bar{u}) \subseteq T(S, \bar{x}, \bar{u}),$$

as the converse inclusion holds universally. Let $x \in \bigcap_{i=1}^{s} T(S_i, \bar{x}, \bar{u})$ be arbitrary. For $n \in \mathbb{N}$, we define

$$x_{n,t}(\alpha) = \bar{x} + t \left(\bar{u} + \frac{t}{2} x + \frac{t}{n} \sum_{i=1}^{s} \alpha_i \nabla h_i(\bar{x}) \right), \tag{4.98}$$

and by using the Frećhet differentiability of h_i at $\bar{x}$, we get

$$\begin{aligned} h_j(x_{n,t}(\alpha)) = h_j(\bar{x}) &+ t \left\langle \nabla h_j(\bar{x}), \bar{u} + \frac{t}{2} x + \frac{t}{n} \sum_{i=1}^{s} \alpha_i \nabla h_i(\bar{x}) \right\rangle \\ &+ \frac{t^2}{2} \left\langle \bar{u} + \frac{t}{2} x + \frac{t}{n} \sum_{i=1}^{s} \alpha_i \nabla h_i(\bar{x}), \nabla^2 h_j(\bar{x}) \right. \\ &\times \left. \left(\bar{u} + \frac{t}{2} x + \frac{t}{n} \sum_{i=1}^{s} \nabla \alpha_i h_i(\bar{x}) \right) \right\rangle \\ &+ o \left(t \left(\bar{u} + \frac{t^2}{2} x + \frac{t}{n} \sum_{i=1}^{s} \alpha_i \nabla h_i(\bar{x}) \right) \right)^2 \end{aligned}$$

$$= h_j(\bar{x}) + t\left\langle \nabla h_j(\bar{x}), \bar{u}\right\rangle + \frac{t^2}{n}\left\langle \nabla h(\bar{x}), \sum_{i=1}^{s} \alpha_i \nabla h(\bar{x})\right\rangle$$

$$+ \frac{t^2}{2}\left(\left\langle \nabla h_j(\bar{x}), x\right\rangle + \left\langle \bar{u}, \nabla^2 h_j(\bar{x})(\bar{u})\right\rangle\right) + o(t^2).$$

Consequently,

$$h_j(x_{n,t}(\alpha)) = \frac{t^2}{n}\left\langle \nabla h_j(\bar{x}), \sum_{i=1}^{s} \alpha_i \nabla h_i(\bar{x})\right\rangle + o(t^2)$$

for $j \in I_i$ because $h_j(\bar{x}) = 0$, $\langle \nabla h_j(\bar{x}), \bar{u}\rangle = 0$, and $\langle \nabla h_j(\bar{x}), x\rangle + \langle \bar{u}, \nabla^2 h_j(\bar{x})(\bar{u})\rangle$ for such j. Moreover,

$$h_j(x_{n,t}(\alpha)) \geq \frac{t^2}{n}\left\langle \nabla h_j(\bar{x}), \sum_{i=1}^{s} \alpha_i \nabla h_i(\bar{x})\right\rangle + o(t^2)$$

for $j \in I_2^{=}$ because $h_j(\bar{x}) \geq 0$, $\langle \nabla h_j(\bar{x}), \bar{u}\rangle \geq 0$, and $\langle \nabla h_j(\bar{x}), x\rangle + \langle \bar{u}, \nabla^2 h_j(\bar{x})(\bar{u})\rangle \geq 0$ for such j, and $h_j(x_{n,t}(\alpha)) > 0$ for $j \in I_2^{>}$ if t is sufficiently small. By the same reasoning as in the proof of Theorem 4.8.8, there exist t_n and α_n such that $\{t_n\} \downarrow 0$ and $x_{n,t}(\alpha_n)$ satisfies (4.62). Therefore, $x_n := x_{n,t}(\alpha_n) \in S$. It can be checked that

$$\frac{x_n - \bar{x} - t_n \bar{u}}{\frac{t_n^2}{2}} \to x,$$

and hence it follows that $x \in T^2(S, \bar{x}, \bar{u})$. The proof is complete. □

The following result is an analogous of Theorem 4.14.4 for the second-order contingent sets and can be proved by employing similar arguments (see Di [134]):

Theorem 4.14.5. *Assume that the conditions of Theorem 4.8.11 hold. In addition, assume that $h_i : X \to \mathbb{R}$ admits second-order Fréchet derivatives at $\bar{x}$ for $i \in \{1, \ldots, s\}$ and $\bar{u} \in T(P \cap S, \bar{x})$. Then:*

$$T^2(P \cap S, \bar{x}, \bar{u}) = \bigcap_{i=1}^{s} T(S_i, \bar{x}, \bar{u}) \bigcap T^2(P, \bar{x}, \bar{u})$$

$$= \left\{x \in T^2(P, \bar{x}, \bar{u}) \mid \langle \bar{u}, \nabla^2 h(\bar{x})\rangle + \langle x, \nabla h(\bar{x})\rangle \in \Lambda^*\right\}. \tag{4.99}$$

4.15 Miscellaneous Second-Order Tangent Cones

In this section, we study a few other notions of second-order tangent cones.

4.15.1 *Second-Order Tangent Cones of Ledzewicz and Schaettler*

Probably, Ledzewicz and Schaettler [364] were the first to introduce second-order notions of local approximation which were actually cones. We recall the two notions proposed in [364].

Definition 4.15.1. Let X be a Banach space, let $S \subset X$, let $\bar{x} \in S$, and let $\bar{u} \in X$.

1. The **second-order feasible cone** $FC^2(S, \bar{x}, \bar{u})$ of S at $\bar{x}$ in the direction $\bar{u}$ is the collection of all pairs $(x, \gamma) \in X \times \mathbb{R}_+$ for which there exist an $\varepsilon_0 > 0$ and a neighborhood $V(x)$ such that for every $\varepsilon \in (0, \varepsilon_0)$, we have $\bar{x} + \varepsilon\sqrt{\gamma} + \varepsilon^2 V(\bar{x}) \subset S$.
2. The **second-order incident cone** $TC^2(S, \bar{x}, \bar{u})$ of S at $\bar{x}$ in the direction $\bar{u}$ is the collection of all pairs $(x, \gamma) \in X \times \mathbb{R}_+$ for which there exists an $\varepsilon_0 > 0$ such that for every $\varepsilon \in (0, \varepsilon_0)$, there exists an element $r(\varepsilon) \in X$ of $o(\varepsilon^2)$ satisfying that $\bar{x} + \varepsilon\sqrt{\gamma} + \varepsilon^2 x + r(\varepsilon) \in S$.

Remark 4.15.2. In [364], the second-order incident cone is called the second-order tangent cone. It is evident that $TC^2(S, \bar{x}, \bar{u})$ is a cone, that is, for any $x \in TC^2(S, \bar{x}, \bar{u})$ and any $\lambda > 0$, it holds that $\lambda x \in TC^2(S, \bar{x}, \bar{u})$. The second-order feasible cone, $FC^2(S, \bar{x}, \bar{u})$, on the other hand, is an open cone. It is also easy to see that if $\bar{u} \in IT(S, \bar{x})$, then $FC^2(S, \bar{x}, \bar{u}) = X \times (0, \infty)$. The interiorly contingent cone, studied earlier in this chapter, is called the first-order feasible cone in [364].

We now give a few examples to illustrate the above notions (see [364, 365]):

Example 4.15.3. For $S \subset \mathbb{R}^2$ given by $S := \{(x, y) |\ x \geq 0,\ x^2 \leq y \leq 2x^2\}$, we have $IT(S, (0, 0)) = \emptyset$, and $FC^2(S, (0, 0), (1, 0)) = \{((x, y), \gamma) \in \mathbb{R}^3 |\ 0 < \gamma < y < 2\gamma\}$. □

Example 4.15.4. For $S \subset \mathbb{R}^2$ given by $S := \{(x, y) |\ y \geq |x|\}$, we obtain that $IT(S, (0, 0)) = \operatorname{int} S$, and $FC^2(S, (0, 0), (1, 1)) = \{((x, y), \gamma) \in \mathbb{R}^3 |\ \gamma > 0, x < y\}$. □

Example 4.15.5. For $S \subset \mathbb{R}^2$ given by $S := \{(x, y) |\ x^2 + y^2 = 1\}$, a simple computation leads to $TC^2(S, (1, 0), (0, 1)) = \{((x, y), \gamma) \in \mathbb{R}^3 |\ \gamma > 0, \gamma + 2x = 0\}$. □

We now specify the above cones and their duals on some specific sets. For proofs see [364].

Proposition 4.15.6. *Let X be a Banach space, let $g : X \to \mathbb{R}$ be twice Fréchet differentiable at $\bar{x}$ with $g'(\bar{x}) \neq 0$. Let $\bar{u}$ satisfy $g'(\bar{x})(\bar{u}) = 0$. Then, for $S := \{x \in X |\ g(x) \leq 0\}$, the second-order feasible cone $FC^2(S, \bar{x}, \bar{u})$ and its dual $FC^2(S, \bar{x}, \bar{u})^+$ are given by:*

$$FC^2(S, \bar{x}, \bar{u}) = \left\{(x, \gamma) \in X \times \mathbb{R}_+ |\ g'(\bar{x})(x) + \frac{1}{2}\gamma g''(\bar{x})(\bar{u}, \bar{u}) < 0\right\},$$

$$FC^2(S, \bar{x}, \bar{u})^+ = \Big\{(\lambda, \mu) \in X^* \times \mathbb{R}^*_+ |\ \exists\, \alpha_1 \leq 0\ \alpha_2 \geq 0 \text{ such that}$$

$$\begin{pmatrix} \lambda \\ \mu \end{pmatrix} = \begin{pmatrix} g'(\bar{x})^* & 0 \\ \frac{1}{2} g''(\bar{x})(\bar{u}, \bar{u}) & 1 \end{pmatrix} \begin{pmatrix} \alpha_1 \\ \alpha_2 \end{pmatrix}\Big\}.$$

Let X and Y be Banach spaces, let $g : X \to Y$ be a map, and let $S := \{x \in X |\ g(x) = 0\}$. If g is continuously Fréchet differentiable in a neighborhood of $\bar{x} \in S$, and if the derivative $g'(\bar{x})$ is onto, that is, g is the so-called regular map at $\bar{x}$, then by the classical Lusternik theorem, we have

$$T(S, \bar{x}) = \big\{x \in X |\ g'(\bar{x})(x) = 0\big\} = \text{Ker}\left(g'(\bar{x})\right). \tag{4.100}$$

Among many generalization of the Lusternik theorem, the one given by Avakov [20] deals with non-regular operators that are twice Fréchet differentiable at $\bar{x}$. Assuming that $\text{Im}(g'(\bar{x}))$ is closed in Y, for $\bar{u} \in X$, we define a linear map $G(\bar{x}, \bar{u}) : X \to Y \times Y \backslash \text{Im}(g'(\bar{x}))$ by the formula

$$x \to G(\bar{x}, \bar{u})(x) = \left(g'(\bar{x})(\bar{u}), \pi\left(g''(\bar{x})(\bar{u}, x)\right)\right), \tag{4.101}$$

where $\pi : Y \to Y \backslash \text{Im}(g'(\bar{x}))$ is the quotient map. Then, under the conditions that $G(\bar{x}, \bar{u})(\bar{u}) = 0$ and $\text{Im}(G(\bar{x}, \bar{u})) = \text{Im}(g'(\bar{x})) \times Y \backslash Im\,(g'(\bar{x}))$, it holds that $\bar{u} \in T(S, \bar{x})$.

Using the above result, the following characterization of the second-order tangent cone has been proved by Ledzewicz and Schaettler [364].

Proposition 4.15.7. *Let X and Y be Banach spaces, let $g : X \to Y$ be three times Fréchet differentiable at $\bar{x} \in S$, let $\text{Im}(g'(\bar{x}))$ be closed, let $g'(\bar{x})(\bar{u}) = 0$ and $g''(\bar{x})(\bar{u}, \bar{u}) \in (Im)(g'(\bar{x}))$, and let $\text{Im}(G(\bar{x}, \bar{u})) = \text{Im}\,(g'(\bar{x})) \times Y \backslash \text{Im}(g'(\bar{x}))$. Then:*

$$TC^2(S, \bar{x}, \bar{u}) = \Big\{(x, \gamma) \in X \times \mathbb{R}_+ |\ g'(\bar{x})(x) + \frac{1}{2}\gamma\, g''(\bar{x})(\bar{u}, \bar{u}) = 0,$$

$$g''(\bar{x})(\bar{u}, x) + \frac{1}{6}\gamma g'''(\bar{x})(\bar{u}, \bar{u}, \bar{u}) \in \text{Im}(g'(\bar{x}))\Big\},$$

$$TC^2(S, \bar{x}, \bar{u})^+ = \Big\{(\lambda, \mu) \in X^* \times \mathbb{R}^* |\ \exists\, y_1^* \in \text{Im}(g'(\bar{x}))^*,\ y_2^* \in \text{Im}(g'(\bar{x}))^\perp \text{ and}$$

$$r \geq 0 \text{ such that } \lambda = g'(\bar{x})^* y_1^* + (g''(\bar{x})(\bar{u}, \cdot))^* y_2^*$$

$$\mu = \left\langle y_1^*, \frac{1}{2} g''(\bar{x})(\bar{u}, \bar{u}) \right\rangle + \left\langle y_2^*, \frac{1}{2} g''(\bar{x})(\bar{u}, \bar{u}) \right\rangle$$
$$+ \left\langle y_2^*, \frac{1}{6} g'''(\bar{x})(\bar{u}, \bar{u}, \bar{u}) \right\rangle + r \Big\} .$$

4.15.2 Projective Tangent Cones of Second-Order

Our focus in this section is on the following notion of projective second-order tangent sets introduced by Penot [468] in a higher-order setting. See also [465, 466, 469].

Definition 4.15.8. Let X be a normed space, let S be a subset of X, let $\bar{u} \in X$, and let $\bar{x} \in \operatorname{cl} S$.

1. The **second-order projective contingent cone** $\mathbb{P}^2(S, \bar{x}, \bar{u})$ of S at $(\bar{x}, \bar{u})$ is the set of all $(x, t) \in X \times \mathbb{R}$ such that there are a sequence $\{(s_n, t_n)\} \subset \mathbb{P} \times \mathbb{P}$ with $(s_n, t_n) \to (0_+, t)$ and $\dfrac{s_n}{t_n} \downarrow 0$ and a sequence $\{x_n\} \subset S$ with $x_n \to x$ satisfying that, for every $n \in \mathbb{N}$, we have $\bar{x} + s_n \bar{u} + \frac{s_n^2}{2t_n} x_n \in S$.
2. The **second-order projective adjacent cone** $\hat{\mathbb{P}}^2(S, \bar{x}, \bar{u})$ of S at $(\bar{x}, \bar{u})$ is the set of all $(x, t) \in X \times \mathbb{R}$ such that for every sequence $\{(s_n, t_n)\} \subset \mathbb{P} \times \mathbb{P}$ with $(s_n, t_n) \to (0_+, t)$ and $\dfrac{s_n}{t_n} \downarrow 0$, there exists a sequence $\{x_n\} \subset S$ with $x_n \to x$ satisfying that for every $n \in \mathbb{N}$, we have $\bar{x} + s_n \bar{u} + \frac{s_n^2}{2t_n} x_n \in S$.

Remark 4.15.9. In the above definitions, if instead of the convergence $x_n \to x$, we assume that x_n converges weakly to x, we obtain the notions of the **second-order weak projective contingent cone** and the **second-order weak projective adjacent cone**, denoted by $\mathbb{P}^2_{\mathrm{w}}(S, \bar{x}, \bar{u})$ and $\hat{\mathbb{P}}^2_{\mathrm{w}}(S, \bar{x}, \bar{u})$, respectively.

We recall that the projective space $P(X)$ associated with the space X is the set of equivalence classes of pairs $(x, s) \in X \times \mathbb{R}_+$ for the relation

$$(x, s) \sim (\hat{x}, \hat{s}) \text{ if } (\hat{x}, \hat{s}) = (tx, ts) \text{ for some } t > 0.$$

We shall denote by $\mathscr{P}$ the quotient mapping $\mathscr{P} : X \times \mathbb{R} \to P(X)$ and call it the projective projection. Evidently, $P(X)$ can be decomposed as the union

$$P(X) = X_1 \cup X_0,$$

where X_1 is the image of $X \times \{1\}$ and X_0 is the image of $X \times \{0\}$.

Therefore, $\mathbb{P}^2(S, \bar{x}, \bar{u})$ can be split into two parts, namely the union of $\mathscr{P}(T^2(S, \bar{x}, \bar{u})) \times \{1\}$ and $\mathscr{P}(\mathbb{T}^2(S, \bar{x}, \bar{u})) \times \{0\}$ where

$$T^2(S,\bar{x},\bar{u}) = \Big\{x \in X \mid \exists\, \{t_n\} \downarrow 0,\ \exists\, \{x_n\} \to x,\ \text{so that}$$
$$\bar{x} + t_n\bar{u} + \frac{t_n^2}{2}x_n \in S,\ \forall\, n \in \mathbb{N}\Big\},$$
$$\mathbb{T}^2(S,\bar{x},\bar{u}) = \Big\{x \in X \mid \exists\, \{(t_n, r_n\} \downarrow (0,0),\ \text{with } \frac{t_n}{r_n} \to 0\ \exists\, \{x_n\} \to x,\ \text{with}$$
$$\bar{x} + t_n\bar{u} + \frac{t_n^2}{2r_n}x_n \in S,\ \forall\, n \in \mathbb{N}\Big\},$$

are the second-order contingent set and the second-order asymptotic contingent cone studied earlier in this chapter.

It has been noted that the second-order contingent set can be empty, and, seemingly to remedy this situation, has been one of the motivations behind the second-order projective contingent cone. The following result clarifies this statement:

Theorem 4.15.10. *Let X be a reflexive Banach space, let S be a subset of X, let $\bar{x} \in \operatorname{cl} S$ and let $\bar{u} \in T(S,\bar{x})$. Then $\mathbb{P}_w^2(S,\bar{x},\bar{u})$ is nonempty.*

Proof. We show that either $T^2(S,\bar{x},\bar{u})$ or $\mathbb{T}^2(S,\bar{x},\bar{u})$ is nonempty where these cones are considered to be equipped with the weak topology. Due to the containment $\bar{u} \in T(S,\bar{x})$, there exists a sequence $\{t_n\} \in \mathbb{P}$ with $t_n \downarrow 0$ such that the sequence $\{s_n\}$ given by $s_n := t_n^{-1}d(\bar{x} + t_n\bar{u})$ converges to 0. Since $0 \in T^2(S,\bar{x},\bar{u})$ if $s_n = 0$ for infinitely many n, we may assume that $s_n > 0$ and set $r_n = \frac{t_n}{2s_n}$, and $w_n = s_n^{-1}t_n^{-1}(z_n - \bar{x} - t_n\bar{u})$, where $z_n \in S$ is such that $\|\bar{x} + t_n\bar{u} - z_n\| \leq 2s_nt_n$. Then $r_n^{-1}t_n = 2s_n \to 0$ and $\bar{x} + t_n\bar{u} + \frac{t_n^2}{2r_n}w_n = z_n \in S$.

Taking a subsequence if necessary, we may assume that $\{r_n\} \to r \in [0,\infty]$ and $\{w_n\}$ has a weak limit u in $2B_X$. If $r = \infty$, setting $\bar{w}_n = r_n^{-1}w_n$, we get $\{\bar{w}_n\} \to 0$ and $0 \in T^2S,\bar{x},\bar{u})$. If $r \in \mathbb{P}$, the same choice shows that $r^{-1}w \in T^2(\bar{x},\bar{u})$. Finally, if $r = 0$, we have $w \in \mathbb{T}^2(S,\bar{x},\bar{u})$. □

In the following, we collect some of the useful properties of the projective contingent cones of second-order and for their proof we refer the reader to the original contribution of Penot [468].

Proposition 4.15.11. *Let X and Y be reflexive Banach spaces, let $g : X \to Y$ be a mapping of class C^2 on an open subset S_0 of X. Let S be a subset of X, let $\bar{x} \in S_0 \cap S$ and let C be a subset of Y with $g(S) \subset C$. Then for each $\bar{u} \in X$, and $(w,r) \in \mathbb{P}^2(S,\bar{x},\bar{u})$, we have*

$$(g^{-1}(\bar{x})(w) + rg''(\bar{x})(\bar{u},\bar{u})) \in \mathbb{P}^2(C, g(\bar{x}), g'(\bar{x})(\bar{u})).$$

Proposition 4.15.12. *Let X be a reflexive Banach space, let S be a convex subset of X, let $\bar{x} \in S$, and let $\bar{u} \in T(S,\bar{x})$. Then, for any $z \in T(T(S,\bar{x}),\bar{u})$, $(w,r) \in \mathbb{P}^2(S,\bar{x},\bar{u})$, we have $(w+z,r) \in \mathbb{P}^2(S,\bar{x},\bar{u})$.*

For many other results on the above cones, the reader is referred to Penot [468].

4.15.3 Second-Order Tangent Cone of N. Pavel

We now study a notion that was introduced by Pavel [455] and later modified by Pavel and Ursescu [457], Pavel, Huang, and Kim [456], and by Constantin [104–106].

Definition 4.15.13. Let X be a normed space, let $S \subset X$, and let $\bar{x} \in S$. The **second-order derivable cone** $T^2(S, \bar{x})$ is the collection of all $x \in X$ such that there exists $\bar{u} \in X$ with

$$\lim_{t \downarrow 0} \frac{1}{t^2} d\left(\bar{x} + t\bar{u} + \frac{t^2}{2} x, S\right) = 0. \tag{4.102}$$

It follows that (4.102) is equivalent to that $\bar{u} \in T(S, \bar{x})$ and for some $\varepsilon(t) = o(t)$, we have

$$\bar{x} + t\bar{u} + \frac{t^2}{2}(x + \varepsilon(t)) \in S. \tag{4.103}$$

We have the following simple observation:

Proposition 4.15.14. *Let X be a normed space, let $S \subset X$, and let $\bar{x} \in S$. Then, $T^2(S, \bar{x})$ is a cone.*

Proof. Let $x \in T^2(S, \bar{x})$ and let $\lambda > 0$. By taking $t = \sqrt{\lambda} t$ in (4.102), we get

$$\lim_{t \downarrow 0} \frac{1}{t^2} d\left(\bar{x} + t\left(\sqrt{\lambda}\bar{u}\right) + \frac{t^2}{2}(\lambda x), S\right) = 0, \tag{4.104}$$

which at once ensures that $\lambda x \in T^2(S, \bar{x})$. □

We have the following characteristic example (see [456]):

Example 4.15.15. Let X be a Hilbert space and let $S := \{x \in X \mid |x| = r\}$, and let $\bar{x} \in S$. Then the second-order derivable cone is given by $T^2(S, \bar{x}) = \{x \in X \mid \exists\, \bar{u} \in X \text{ such that } \langle \bar{u}, \bar{x}\rangle = 0,\ |\bar{u}|^2 + \langle \bar{x}, x\rangle = 0\}$. □

We have the following useful result (see [456]):

Proposition 4.15.16. *Let X be a normed space, let Y be a finite dimensional space, and let $g : X \to Y$ be a twice Fréchet differentiable function. Then, for S defined by $S := \{x \in X \mid g(x) = 0\}$, the second-order derivable cone at $\bar{x} \in S$ is given by*

$$\begin{aligned} T^2(S, \bar{x}) = \big\{ x \in X \mid \exists\, \bar{u} \text{ with } g(\bar{x}) = 0,\ \langle g'(\bar{x}), \bar{u}\rangle = 0, \\ \langle g''(\bar{x})(\bar{u}), \bar{u}\rangle + \langle g'(\bar{x}), x\rangle = 0 \big\}. \end{aligned}$$

4.15.4 Connections Among the Second-Order Tangent Cones

As noted by Penot [468], by setting $t = s\sqrt{r}$, we note that for $r > 0$, $(w, r) \in TC^2(S, \bar{x}, \bar{u})$ if and only if $(w, r) \in \hat{\mathbb{P}}^2(S, \bar{x}, \bar{u})$ if and only if $r^{-1}w \in A^2(S, \bar{x}, \bar{u}) := \liminf_{t\to 0^+} 2t^{-2}(S - \bar{x} - t\bar{u})$. However, for $(w, 0) \in TC^2(S, \bar{x}, \bar{u})$ if and only if $w \in A(S, z) := \liminf_{t\to 0^+} t^{-1}(S - \bar{x})$, the first-order adjacent cone, and there is no relation with the case $(w, 0) \in \hat{\mathbb{P}}^2(S, \bar{x}, \bar{u})$.

It is also evident that $TC^2(S, \bar{x}, \bar{u}) \subset T^2(S, \bar{x}) \times \mathbb{R}_+$. In fact, if $(x, \gamma) \in TC^2(S, \bar{x}, \bar{u})$, then we have $\bar{x} + t\sqrt{\gamma}\bar{u} + t^2(x + \varepsilon(t))$ where $\varepsilon(t) \to 0$ as $t \to 0$. Then, by noting that

$$\bar{x} + t\sqrt{\gamma}\bar{u} + t^2(x + \varepsilon(t)) = \bar{x} + t\sqrt{\gamma}\bar{u} + \frac{t^2}{2}(2x + 2\varepsilon(t)) \in S,$$

we deduce that $2x \in T^2(S, \bar{x})$, and since $T^2(S, \bar{x})$ is a cone, we conclude that $x \in T^2(S, \bar{x})$.

4.16 Second-Order Local Approximation

This section is devoted to a second-order local approximation of sets, which is a natural generalization of the first-order local cone approximation studied in Sect. 4.5. Our exposition of this section is based on the work by Castellani and Pappalardo [86].

We begin by introducing the following notion of second-order local approximation:

Definition 4.16.1. Given a normed space X, a set-valued map $K^2 : 2^X \times X \times X \to 2^X$ is called a **local approximation of second-order** if it associates, to each set $S \subseteq X$ and each pair $(\bar{x}, \bar{u}) \in X \times X$, a set $K^2(S, \bar{x}, \bar{u}) \subseteq X$ such that

$$t^2 K^2(S, \bar{x}, \bar{u}) = K^2(S, \bar{x}, t\bar{u})$$

for every $t \geq 0$ and satisfies the following properties:

1. $K^2(S, \bar{x}, \bar{u}) = K^2(S - \bar{x}, 0, \bar{u})$
2. $K^2(S, \bar{x}, \bar{u}) = K^2(S \cap U(\bar{x}), \bar{x}, \bar{u})$, for every $U(\bar{x})$
3. $K^2(S, \bar{x}, \bar{u}) = \emptyset$ if $\bar{x} \notin \operatorname{cl} S$
4. $K^2(S, \bar{x}, \bar{u}) = X$ if $\bar{x} \in \operatorname{int} S$.
5. $\ell(K^2(S, \bar{x}, \bar{u})) = K^2(\ell(S), \ell(\bar{x}), \ell(\bar{u}))$ for any linear homomorphism $\ell : X \to X$
6. $S_\infty \subset \left(K^2(S, \bar{x}, \bar{u})\right)_\infty$ for every $\bar{x} \in \operatorname{cl} S$, and for every $\bar{u} \in \operatorname{cl} K^2(S, \bar{x}, 0)$.

Evidently, all the axioms of the above definition are satisfied when

$$K^2(S, \bar{x}, \bar{u}) = K(S, \bar{x}),$$

where $K(S, \bar{x})$ is a local cone approximation of first-order introduced in Sect. 4.5.

By following the arguments given for the local cone approximation of first-order, it can be shown that the above axioms are independent or each other. That is, no axiom can be expressed by the others.

Furthermore, it can be shown that the second-order radial tangent set, the second-order interiorly radial tangent set, the second-order contingent set, the second-order interiorly contingent set, the second-order adjacent set, and the second-order interiorly adjacent set are all local approximations of second-order. We prove this statement only for the second-order contingent set in the following result (for simplicity, we drop the factor $1/2$):

Theorem 4.16.2. *Let X be a normed space, let S be a subset of X, let $\bar{x} \in S$ and $\bar{u} \in X$. The second-order contingent set $T^2(S, \bar{x}, \bar{u})$ is a local approximation of second-order in the sense of Definition 4.16.1.*

Proof. The first four axioms of Definition 4.16.1 and the fact that $t^2 T^2(S, \bar{x}, \bar{u}) = T^2(S, \bar{x}, t\bar{u})$ for every $t \geq 0$, are quite easy to verify and can also be seen from the properties of the second-order contingent set stated earlier. Therefore, we will only verify the last two axioms. We begin with a proof of the fifth axiom. Let $x \in T^2(\ell(S), \ell(\bar{x}), \ell(\bar{u}))$ be arbitrary, where $\ell : X \to X$ is a linear homomorphism. Then for every $t > 0$, there are $\hat{x} \in U_t(x)$ and $s \in (0, t)$ such that

$$\ell(\bar{x}) + s\ell(\bar{u}) + s^2\hat{x} \in \ell(S),$$

which, due to the continuity of ℓ^{-1}, is equivalent to the containment that

$$\bar{x} + s\bar{u} + s^2\ell^{-1}(\hat{x}) \in S.$$

By using the continuity of ℓ^{-1} once again, we deduce that for each $t > 0$ there are $\tilde{x} \in U_t(\ell(x))$ and $s \in (0, t)$ such that $\bar{x} + s\bar{u} + s^2\tilde{x} \in S$. This confirms that $x \in \ell(T^2(S, \bar{x}, \bar{u}))$ and hence we have shown that $T^2(\ell(S), \ell(\bar{u}), \ell(\bar{x})) \subset \ell(T^2(S, \bar{x}, \bar{u}))$. The converse inclusion is obtained by simply employing the used arguments in a reverse order.

For the final axiom, let $z \in S_\infty$ and $x \in T^2(S, \bar{x}, \bar{u})$ be arbitrary. We will need to show that for every $t > 0$ there are $s \in (0, t)$ and $\bar{z}$ with $\|\bar{z}\| \leq t$ such that $\bar{x} + s\bar{u} + s^2(x + z + \bar{z}) \in S$. We clearly have the existence of $s \in (0, t)$ and $\bar{z}$ with $\|z\| \leq t$ such that $\bar{x} + s\bar{u} + s^2(x + \bar{z}) \in S$ which implies that $(\bar{x} + s\bar{u} + s^2(x + \bar{z})) + s^2 z \in S$. The proof is complete. □

The following result gives a mechanism to construct new notions of local approximation of second-order from the known ones:

Theorem 4.16.3. *Let X be a normed space, let S be a subset of X, and let $(\bar{x}, \bar{u}) \in X \times X$. Let $K^2(S, \bar{x}, \bar{u})$ be a local approximation of second-order in the sense of Definition 4.16.1. Then $L^2(S, \bar{x}, \bar{u})$ defined by the following relationship is also a local approximation of second-order in the sense of Definition 4.16.1:*

$$L^2(S, \bar{x}, \bar{u}) = X \backslash K^2(X \backslash S, \bar{x}, \bar{u}).$$

Furthermore, if $K^2(\cdot, \cdot, \cdot)$ is isotone, then for every $S_1, S_2 \subset X$ and every $(\bar{x}, \bar{u}) \in X \times X$, the following implication holds:

$$S_1 \cap S_2 = \emptyset \implies K^2(S_1, \bar{x}, \bar{u}) \cap L^2(S_2, \bar{x}, \bar{u}) = \emptyset. \tag{4.105}$$

Proof. We will show that the six axioms of Definition 4.16.1 hold for $L^2(S, \bar{x}, \bar{u})$.

1. The first axiom holds due to the fact that

$$L^2(S - \bar{x}, 0, \bar{u}) = X \backslash K^2(X \backslash S - \bar{x}, 0, \bar{u}) = X \backslash K^2(X \backslash S, \bar{x}, \bar{u}) = L^2(S, \bar{x}, \bar{u}).$$

2. For the second axiom, we note that for an arbitrary $U(\bar{x})$, we have

$$\begin{aligned} L^2(S \cap U(\bar{x}), \bar{x}, \bar{u}) &= X \backslash K^2(X \backslash (S \cap U(\bar{x}), \bar{x}, \bar{u}) \\ &= X \backslash K^2(X \backslash S \cup (X \backslash U(\bar{x})), \bar{x}, \bar{u}) \\ &= X \backslash \left[K^2((X \backslash S \cap U(\bar{x})) \cup ((X \backslash U(\bar{x})) \cap U(\bar{x})), \bar{x}, \bar{u})\right] \\ &= X \backslash K^2(X \backslash S \cap U(\bar{x}), \bar{x}, \bar{u}) \\ &= X \backslash K^2(Q \backslash S, \bar{x}, \bar{u}) \\ &= L^2(S, \bar{x}, \bar{u}). \end{aligned}$$

3. Note that if $\bar{x} \in \operatorname{int} S$ then $\bar{x} \notin \operatorname{cl}(X \backslash S)$ and consequently

$$L^2(S, \bar{x}, \bar{u}) = X \backslash K^2(X \backslash S, \bar{x}, \bar{u}) = X \backslash \emptyset = X.$$

4. The proof for this axiom is analogous to the proof of the third axiom.
5. For a linear homomorphism $\ell : X \to X$, we have

$$\begin{aligned} L^2(\ell(S), \ell(\bar{x}), \ell(\bar{u})) &= X \backslash K^2(X \backslash \ell(S), \ell(\bar{x}), \ell(\bar{u})) \\ &= \ell(X) \backslash (K^2(\ell(X \backslash S), \bar{x}, \bar{u})) \\ &= \ell(K^2(X \backslash S, \bar{x}, \bar{u})) \\ &= \ell(L^2(S, \bar{x}, \bar{u})). \end{aligned}$$

6. Since $(X\backslash S)_\infty = -S_\infty$, we have

$$S_\infty = -(X\backslash S)_\infty \subset -\left(K^2(X\backslash S,\bar{x},\bar{u})\right)_\infty$$
$$= \left(X\backslash K^2(X\backslash S,\bar{x},\bar{u})\right)_\infty = \left(L^2(S,\bar{x},\bar{u})\right)_\infty.$$

Therefore, all the six axioms have been verified and the proof is complete. □

Note that if $L^2(S,\bar{x},\bar{u})$ is built from $K^2(S,\bar{x},\bar{u})$ using (4.105) then we have

$$X\backslash L^2(S,\bar{x},\bar{u}) = X\backslash K^2(X\backslash S,\bar{x},\bar{u}).$$

4.17 Higher-Order Tangent Cones and Tangent Sets

In the previous sections, we witnessed numerous second-order tangent cones and second-order tangent sets. These objects are natural extensions of tangent cones of first-order. It turns out that all the second-order tangent cones and the second-order tangent sets were either initially defined as higher-order approximation tools or have a natural higher-order extension. See Constantin [104–106], Furukawa and Yoshinaga [192], Hoffmann and Kornstaedt[265], Ledzewicz and Schättler [365, 365], Penot [468] and the cited references therein. In recent years, numerous authors have focused on higher-order tangent sets and used them to define higher-order derivatives and epiderivatives of set-valued maps. Higher-order tangent sets have also been used to give higher-order optimality conditions and higher-order sensitivity analysis. See Anh, Khanh, and Tung [8], Li, Sun, and Zhu [382], Wang, Li, and Teo [586], and the cited references therein. Some interesting results for nonsmooth optimization problems can be found in Ginchev, Guerraggio, and Rocca [205, 206]. Although higher-order derivatives, epiderivatives, optimality conditions, and sensitivity analysis are of interest, they typically follow the same pattern of their second-order counterparts, and, in general, all the results which hold for second-order study have their higher-order analogues. Therefore, in this work, we will not dwell on the higher-order tangent sets and tangent cones but, for the sake of completeness, we collect the definitions of higher-order extensions of the most commonly used second-order tangent sets.

Definition 4.17.1. Let X be a normed space, let S be a nonempty subset of X, let $\bar{x} \in \mathrm{cl}\,(S)$, and let $\bar{u}_1, \bar{u}_2, \ldots, \bar{u}_{m-1}$ be elements of X.

1. The **m-th order radial tangent set** $R^m(S,\bar{x},\bar{u},\ldots,\bar{u}_{m-1})$ of S at $(\bar{x},\bar{u},\ldots,\bar{u}_{m-1})$ is the set of all $x \in X$ for which there exists a sequence $\{t_n\} \subset \mathbb{P}$ with $t_n \downarrow 0$ such that

$$\bar{x} + t_n\bar{u} + t_n^2\bar{u}_2 + \cdots + t_n^{m-1}\bar{u}_{m-1} + t_n^m x \in S \quad \text{for every } n \in \mathbb{N}.$$

2. The **m-th order feasible tangent set** $IR^m(S, \bar{x}, \bar{u}, \dots, \bar{u}_{m-1})$ of S at $(\bar{x}, \bar{u}, \dots, \bar{u}_{m-1})$ is the set of all $x \in X$ such that for every $\{t_n\} \subset \mathbb{P}$ with $t_n \downarrow 0$, there is an $m \in \mathbb{N}$ satisfying that

$$\bar{x} + t_n \bar{u} + t_n^2 \bar{u}_2 + \cdots + t_n^{m-1} \bar{u}_{m-1} + t_n^m x \in S \quad \text{for every } n \geq m.$$

3. The **m-th-order contingent set** $T^m(S, \bar{x}, \bar{u}, \dots, \bar{u}_{m-1})$ of S at $(\bar{x}, \bar{u}, \dots, \bar{u}_{m-1})$ is the set of all $x \in X$ for which there are a sequence $\{t_n\} \subset \mathbb{P}$ with $t_n \downarrow 0$ and a sequence $\{x_n\} \subset X$ with $x_n \to x$ such that

$$\bar{x} + t_n \bar{u} + t_n^2 \bar{u}_2 + \cdots + t_n^{m-1} \bar{u}_{m-1} + t_n^m x_n \in S \quad \text{for every } n \in \mathbb{N}.$$

4. The **m-th-order interiorly contingent set** $IT^m S, \bar{x}, \bar{u}, \dots, \bar{u}_{m-1})$ of S at $(\bar{x}, \bar{u}, \dots, \bar{u}_{m-1})$ is the set of all $x \in X$ such that for every sequence $\{t_n\} \subset \mathbb{P}$ with $t_n \downarrow 0$ and every sequence $\{x_n\} \subset X$ with $x_n \to x$ there exists an index $m \in \mathbb{N}$ satisfying that

$$\bar{x} + t_n \bar{u} + t_n^2 \bar{u}_2 + \cdots + t_n^{m-1} \bar{u}_{m-1} + t_n^m x_n \in S \quad \text{for every } n \geq m.$$

5. The **m-th order adjacent set** $A^m(S, \bar{x}, \bar{u}, \dots, \bar{u}_{m-1})$ of S at $(\bar{x}, \bar{u}, \dots, \bar{u}_{m-1})$ is the set of all $x \in X$ such that for every sequence $\{t_n\} \subset \mathbb{P}$ with $t_n \downarrow 0$, there exists a sequence $\{x_n\} \subset S$ with $x_n \to x$ satisfying that

$$\bar{x} + t_n \bar{u} + t_n^2 \bar{u}_2 + \cdots + t_n^{m-1} \bar{u}_{m-1} + t_n^m x_n \in S \quad \text{for every } n \in \mathbb{N}.$$

6. The **m-th order interiorly adjacent set** $IA^m(S, \bar{x}, \bar{u}, \dots, \bar{u}_{m-1})$ of S at $(\bar{x}, \bar{u}, \dots, \bar{u}_{m-1})$ is the set of all $x \in X$ such that there exists a sequence $\{t_n\} \subset \mathbb{P}$ with $t_n \downarrow 0$ so that for every sequence $\{x_n\} \subset S$ with $x_n \to x$ there exists an index $m \in \mathbb{N}$ satisfying that

$$\bar{x} + t_n \bar{u} + t_n^2 \bar{u}_2 + \cdots + t_n^{m-1} \bar{u}_{m-1} + t_n^m x_n \in S \quad \text{for every } n \geq m.$$

Clearly, when $m = 2$, we recover the second-order tangent sets (without the factor $1/2$). It is evident that the properties discussed for the second-order objects can be extended to the higher-order tangent sets.

Chapter 5
Nonconvex Separation Theorems

In this chapter we introduce nonlinear scalarization methods that are very important from the theoretical as well as computational point of view. We introduce different scalarizing functionals and discuss their properties, especially monotonicity, continuity, Lipschitz continuity, sublinearity, convexity. Using these nonlinear functionals we show nonconvex separation theorems. These nonlinear functionals are used for deriving necessary optimality conditions for solutions of set-valued optimization problems in Sect. 12.8 and in different proofs, especially in the proof minimal point theorems in Chap. 10. Moreover, we study characterizations of solutions of set-valued optimization problems by means of nonlinear scalarizing functionals. Finally, we present the extremal principle by Kruger and Mordukhovich and discuss its relationship to separation properties of nonconvex sets. This extremal principle will be applied in Sect. 12.9 for deriving a subdifferential variational principle for set-valued mappings and in Sect. 12.11 in order to prove a first order necessary condition for fully localized minimizers of set-valued optimization problems.

5.1 Separating Functions and Examples

Separating functionals play an important role in many fields of mathematics; especially, in order to prove minimal point theorems (see Chap. 10) and optimality conditions (see Chap. 12). In this section we introduce an axiomatic approach for scalarization by means of (in general nonlinear) functionals. We consider a linear topological space Y, $k^0 \in Y \setminus \{0\}$, a proper set $D \subset Y$ and a **separating functional** $\varphi : Y \to \mathbb{R} \cup \{\pm\infty\}$ having some of the following properties:

($A_{sf}1$) The function φ is **D-monotone**, i.e., $y, w \in Y$, $y \in w - D$ implies $\varphi(y) \le \varphi(w)$.

($A_{sf}1'$) The function φ is **strictly C-monotone**, i.e., $y, w \in Y$, $y \in w - (D \setminus \{0\})$ implies $\varphi(y) < \varphi(w)$.

A.A. Khan et al., *Set-valued Optimization*, Vector Optimization,
DOI 10.1007/978-3-642-54265-7_5

$(A_{sf}2)$ The function φ is **convex**, i.e., for all $y^1, y^2 \in Y$, $\lambda \in [0,1]$ it holds

$$\varphi(\lambda y^1 + (1-\lambda)y^2) \leq \lambda\varphi(y^1) + (1-\lambda)\varphi(y^2).$$

$(A_{sf}2')$ The function φ is **sublinear**, i.e., φ is subadditive (for all $y^1, y^2 \in Y$ it holds

$$\varphi(y^1 + y^2) \leq \varphi(y^1) + \varphi(y^2))$$

and positively homogenous (for all $y \in Y$ and $\lambda \geq 0$ it holds $\varphi(\lambda y) = \lambda\varphi(y)$).

$(A_{sf}2'')$ The function φ is **linear**.

$(A_{sf}3)$ The function φ enjoys the property

$$\forall s \in \mathbb{R},\ \forall y \in Y : \ \varphi\left(y + sk^0\right) = \varphi(y) + s. \tag{5.1}$$

$(A_{sf}4)$ The function φ is **lower continuous**.

$(A_{sf}4')$ The function φ is **continuous**.

Examples for functions satisfying the axioms given above are listed in the following:

Example 5.1.1. Assume that A is a proper closed subset of Y. Let $k^0 \in Y \setminus \{0\}$ be such that $A - [0, +\infty)k^0 \subset A$. Consider the function $\varphi : Y \to \mathbb{R} \cup \{\pm\infty\}$, defined by

$$\varphi_{A,k^0}(y) := \inf\left\{t \in \mathbb{R} \mid y \in tk^0 + A\right\}. \tag{5.2}$$

We use the convention $\inf \emptyset = +\infty$. Then it holds $\operatorname{dom} \varphi_{A,k^0} = \mathbb{R}k^0 + A$ (see [214, Theorem 2.3.1]).

If $A = -D$ is a *proper closed convex cone* and $k^0 \in \operatorname{int}\ D$, the function (5.2) fulfills $(A_{sf}1)$, $(A_{sf}2')$ $(A_{sf}3)$ and $(A_{sf}4')$ (see [214, Corollary 2.3.5] and Corollary 5.2.8 of this section).

If A is a *proper closed convex subset* of Y with $A - [0, +\infty)k^0 \subset A$ and $D \subset Y$ with $A - D \subset A$, the functional (5.2) fulfills $(A_{sf}1)$, $(A_{sf}2)$, $(A_{sf}3)$ and $(A_{sf}4)$. These properties of the functional (5.2) are shown in [214, Theorem 2.3.1] (see Theorems 5.2.3, 5.2.6).

Example 5.1.2. The scalarization method by Pascoletti and Serafini [454] for a vector optimization problem

$$\text{minimize} \qquad f(x) \quad \text{subject to} \quad x \in S, \tag{VP}$$

(where $Y = \mathbb{R}^m$, $D = \mathbb{R}^m_+$, $S \subset \mathbb{R}^n$, $f(x) = (f_1(x), \ldots, f_m(x))^T$ and $f_i : S \to \mathbb{R}$ for all $i = 1, \ldots, m$) given by

$$\min\ t \tag{5.3}$$

subject to the constraints

$$f(x) \in a + tr - D,$$

$$x \in S,$$

$$t \in \mathbb{R},$$

with parameters $a, r \in \mathbb{R}^m$, $r \in \operatorname{int} \mathbb{R}^m_+$, satisfies the axioms $(A_{sf}1)$, $(A_{sf}2)$, $(A_{sf}3)$, $(A_{sf}4')$.

Example 5.1.3. The following functional was introduced by Hiriart-Urruty [262]: For a nonempty subset A of a Banach space Y with $A \neq Y$, the **oriented distance function** $\Delta_A : Y \to \mathbb{R}$ is given as

$$\Delta_A(y) := d(y, A) - d(y, Y \setminus A).$$

Zaffaroni has shown (see [607, Proposition 3.2]):

(i) Δ_A is Lipschitz of rank 1.
(ii) If A is convex, then Δ_A is convex and if A is a cone, then Δ_A is positively homogeneous.
(iii) Assume that $A = -D$ is a proper closed convex cone. If $y^1, y^2 \in Y$ with $y^1 - y^2 \in A$, then $\Delta_A(y^1) \leq \Delta_A(y^2)$.

The functional Δ_A satisfies the axioms $(A_{sf}1)$, $(A_{sf}2')$, $(A_{sf}4')$ if $A = -D$ is a proper closed convex cone. Further results concerning the functional Δ_A are given in Sect. 5.3.

Example 5.1.4. Certain nonlinear functionals are used in **Mathematical Finance** in order to express a risk measure (for example a valuation of risky investments) with respect to an acceptance set $B \subset Y$. Let Y be a linear space of random variables, and let Ω be a set of elementary events (a set of all possible states of the future). Then a future payment of an investment is a random variable $y : \Omega \to \mathbb{R}$. Positive payments in the future are wins, negative ones are losses. If no investment is being done, then y takes on the value zero. In order to valuate such an investment, we need to valuate random variables by comparing them. To do that, we introduce an ordering relation that is induced by a set $B \subseteq Y$. Artzner, Delbaen, Eber and Heath proposed in [12] axioms for a cone $B \subseteq Y$ of random variables that represent acceptable investments:

(A1) $\{y \in Y \mid y(\omega) \geq 0\ (\omega \in \Omega)\} \subset B, \quad B \cap \{y \in Y \mid y(\omega) < 0\ (\omega \in \Omega)\} = \emptyset,$
(A2) $B + B \subset B.$

Cones $B \subset Y$ satisfying the axioms (A1) and (A2) of acceptable investments can be used in order to introduce a preference relation on Y. The decision maker prefers y_1 to y_2 (changing from y_2 to y_1 is an acceptable risk) if and only if $y_1 - y_2$ is an element of B, i.e.,

$$y_1 \geq_B y_2 \Longleftrightarrow y_1 - y_2 \in B.$$

Artzner, Delbaen, Eber and Heath [12] (compare Heyde [256]) introduced **coherent risk measures** $\mu : Y \to \overline{\mathbb{R}}$. In the papers by Artzner, Delbaen, Eber and Heath [12] and Rockafellar, Uryasev and Zabarankin [488, 498] (compare also Rockafellar, Uryasev [496, 497]) the following properties of coherent risk measures μ are supposed:

$(P1)$ $\mu(y + tk^0) = \mu(y) - t$,
$(P2)$ $\mu(0) = 0$ and $\mu(\lambda y) = \lambda\mu(y)$ for all $y \in Y$ and $\lambda > 0$,
$(P3)$ $\mu(y^1 + y^2) \leq \mu(y^1) + \mu(y^2)$ for all $y^1, y^2 \in Y$,
$(P4)$ $\mu(y^1) \leq \mu(y^2)$ if $y^1 \geq_B y^2$.

The sublevel set $L_\mu(0) = B$ of μ to the level 0 is a convex cone and corresponds to the acceptance set B. It can be shown (see [498]) that a coherent risk measure admits a representation as

$$\mu(y) = \inf\{t \in \mathbb{R} \mid y + tk^0 \in B\}. \tag{5.4}$$

Furthermore, Heyde [256] has shown that a coherent risk measure can be identified with the functional $\varphi_{-B,k^0}(-y)$ (see (5.2)) by

$$\varphi_{-B,k^0}(y) = \mu(-y).$$

We get corresponding properties $(A_{sf}1)$, $(A_{sf}2')$, $(A_{sf}3)$ for the functional $\mu(-y)$ like in Example 5.1.1 for the functional φ_{A,k^0} depending from the properties of the set $A = -B$, i.e., of the acceptance set B in Mathematical Finance.

Examples for coherent risk measures are the **conditional value at risk** (see [191, Section 4.4, Definition 4.43]) and the **worst-case risk measure** (see [191, Example 4.8]).

Example 5.1.5 (Value at Risk). Let Ω be a fixed set of scenarios. A financial position is described by a mapping $x : \Omega \to \mathbb{R}$ and x belongs to a given class $\mathscr{X}$ of financial positions. Here $x(\omega)$ is the discounted net worth of the position at the end of the trading period if the scenario $\omega \in \Omega$ is realized. Now, the aim is to quantify the risk of x by some real number $\rho(x)$. Assume that $\mathscr{X}$ is the linear space of bounded measurable functions (containing the constants) on some **measurable space** $(\Omega, \mathscr{S})$. Furthermore, let P be a **probability measure** on $(\Omega, \mathscr{S})$. A position x is considered to be acceptable if the probability of a loss is bounded by a given level $\lambda \in (0, 1)$, i.e., if $P[x < 0] \leq \lambda$. The corresponding monetary risk measure $V@R_\lambda$, defined by

$$V@R_\lambda(x) := \inf\{m \in \mathbb{R} \mid P(m + x < 0) \leq \lambda\}$$

is called *Value at Risk* at level λ. $V@R_\lambda$ is the smallest amount of capital which, if added to x and invested in the risk-free asset, keeps the probability of a negative outcome below the level λ. The Value at Risk only controls the probability of a loss. It is important to mention that the Value at Risk does not capture the size of such a loss if it occurs.

$V@R_\lambda$ is positively homogeneous but in general it is not convex (see Föllmer and Schied [191], Example 4.11), this means that $(A_{sf}2)$ and $(A_{sf}2')$ are not fulfilled.

Example 5.1.6 (Worst-Case Risk Measure). Consider the **worst-case risk measure** ρ_{max} defined by

$$\rho_{max}(x) := -\inf_{\omega\in\Omega} x(\omega) \text{ for all } x \in \mathscr{X},$$

where Ω is a fixed set of scenarios, $x : \Omega \to \mathbb{R}$, x belongs to a given class $\mathscr{X}$ of financial positions and $\mathscr{X}$ is the linear space of bounded measurable functions on some measurable space $(\Omega, \mathscr{S})$. The value $\rho_{max}(x)$ is the least upper bound for the potential loss that can occur in each scenario. ρ_{max} is a **coherent risk measure** (see Föllmer and Schied [191, Example 4.8]) such that we get the properties mentioned in Example 5.1.4.

5.2 Nonlinear Separation

In order to show minimal-point theorems and corresponding variational principles in Chap. 10 or optimality conditions in Chap. 12 we use a scalarization method by means of certain nonlinear functionals. In this section we discuss useful properties of these functionals (cf. Göpfert, Riahi, Tammer, Zălinescu [214] and Tammer, Zălinescu [560]).

5.2.1 Construction of Scalarizing Functionals

Having a nonempty subset A of a real linear space Y and an element $k^0 \neq 0$ of Y, Gerstewitz [196] introduced the function (5.5) (see Fig. 5.1) as scalarizing functional in vector optimization. In the papers by Gerstewitz, Iwanow [197], Gerth, Weidner [198] and Göpfert, Tammer, Zălinescu [215] the essential properties for the usage of the functional (5.5) as scalarizing functional in vector optimization are derived, especially monotonicity properties. The construction in (5.5) was mentioned by Krasnosel'ski [340] (see Rubinov [511]) in the context of operator theory. Furthermore, the functional (5.5) is used by Pascoletti, Serafini [454] as scalarizing functional in multicriteria optimization.

$$\varphi_A := \varphi_{A,k^0} : Y \to \overline{\mathbb{R}}, \quad \varphi_{A,k^0}(y) := \inf\{t \in \mathbb{R} \mid y \in tk^0 + A\}, \tag{5.5}$$

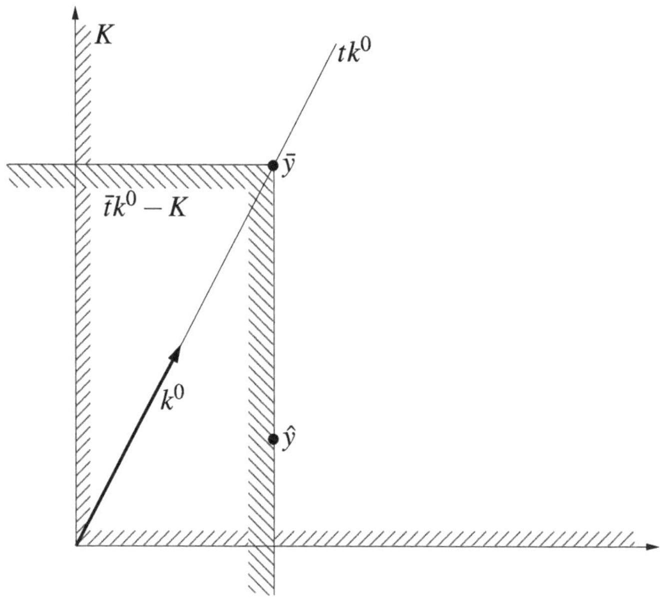

Fig. 5.1 Level sets of the function φ_{A,k^0} from (5.5), where $A = -K = -\mathbb{R}^2_+$ and $k^0 \in \operatorname{int} K$ hold

where, as usual, $\inf \emptyset := +\infty$ (and $\sup \emptyset := -\infty$); we use also the convention $(+\infty) + (-\infty) := +\infty$.

This function was used by many authors, mainly for scalarization of vector optimization problems. Luenberger [408, Definition 4.1] considered

$$\sigma(g; y) := \inf\{\xi \in \mathbb{R} \mid y - \xi g \in \mathscr{Y}\},$$

the corresponding function being called the **shortage function** associated to the production possibility set $\mathscr{Y} \subset \mathbb{R}^m$ and $g \in \mathbb{R}^m_+ \setminus \{0\}$. The case when $g = (1, \ldots, 1)$ was introduced earlier by Bonnisseau and Cornet [54]. A similar function is introduced in [408, Definition 2.1] under the name of **benefit function**.

More recently such a function was considered in the context of mathematical finance beginning with Artzner et. al. [12]; see Heyde [256] (see Example 5.1.4). Under the name of topical function such functions were studied by Singer and his collaborators (see [510]). We discuss many important properties of φ_{A,k^0} in Sect. 5.2.2. Moreover, we study local continuity properties in Sect. 5.2.4. Bonnisseau and Crettez [55] obtained local Lipschitz properties for φ_{A,k^0} (called Luenberger shortage function in [55]) in a very special case, more general results are given by Tammer and Zălinescu [560]. Of course, φ_{A,k^0} is a continuous sublinear functional if A is a proper closed convex cone and $k^0 \in \operatorname{int} A$ (see Corollary 5.2.8) and so φ_{A,k^0} is Lipschitz continuous. Such Lipschitz properties of φ_{A,k^0} are of interest also in the case when $A \subset Y$ is an arbitrary (convex) set and the interior of the usual ordering cone in Y is empty like in mathematical finance where the acceptance sets are in function spaces as L_p and the corresponding risk measures are formulated by means of φ_{A,k^0} (see e.g. Föllmer and Schied [190, 191]).

5.2.2 Properties of Scalarization Functions

Throughout this section Y is a separated locally convex space and Y^* is its topological dual, $C \subset Y$ is a proper closed convex cone, $k^0 \in C \setminus (-C)$ and $A \subset Y$ is a nonempty set.

The **recession cone** of the nonempty set $A \subset X$ is the set

$$A_\infty := \{u \in X \mid x + tu \in A \ \forall x \in A, \ \forall t \in \mathbb{R}_+\}.$$

It follows easily that A_∞ is a convex cone; A_∞ is also closed when A is closed. If A is a closed convex set then $A_\infty = \bigcap_{t \in \mathbb{P}} t(A - a)$, where $\mathbb{P} :=]0, +\infty[$ and $a \in A$ (A_∞ does not depend on $a \in A$).

Furthermore, we assume that A satisfies the following condition (see also [55]):

(A1) A is closed, satisfies the **free-disposal** assumption $A - C = A$, and $A \neq Y$.

We shall use also the (stronger) condition:

(A2) A is closed, satisfies the **strong free-disposal** assumption $A - (C \setminus \{0\}) = \operatorname{int} A$, and $A \neq Y$.

Because $A - C = A \cup (A - (C \setminus \{0\}))$, we have that $(A2) \Rightarrow (A1)$. Moreover, the condition $A - (C \setminus \{0\}) = \operatorname{int} A$ is equivalent to $A - (C \setminus \{0\}) \subset \operatorname{int} A$.

Remark 5.2.1. Assume that the nonempty set A satisfies assumption $(A2)$. Then C is pointed, that is, $C \cap (-C) = \{0\}$, and $A - \mathbb{P}k^0 \subset \operatorname{int} A$ for $k^0 \in C \setminus \{0\}$.

The last assertion is obvious. For the first one, assume that $k \in C \cap (-C) \setminus \{0\}$. Take $a \in \operatorname{bd} A$ $(\subset A)$; such an a exists because $A \neq Y$. Then $a' := a - k \in \operatorname{int} A \subset A$, and so $a = a' - (-k) \in \operatorname{int} A$, a contradiction.

Remark 5.2.2. When A satisfies condition (A1) or (A2) with respect to C and $k^0 \in C \setminus (-C)$ then A satisfies condition (A1) or (A2), respectively, with respect to $\mathbb{R}_+ k^0$. In fact in many situations it is sufficient to take $C = \mathbb{R}_+ k^0$ for some $k^0 \in Y \setminus \{0\}$. In such a situation (A1) [resp. (A2)] means that A is a closed proper subset of Y and $A - \mathbb{R}_+ k^0 = A$ [resp. $A - \mathbb{P}k^0 \subset \operatorname{int} A$].

The free-disposal condition $A = A - C$ shows that $C \subset -A_\infty$. As observed above A_∞ is also closed because A is closed. Hence $-A_\infty$ is the largest closed convex cone C verifying the free-disposal assumption $A = A - C$.

The **aim of this section** is to find a suitable functional $\varphi : Y \to \mathbb{R}$ and conditions such that two given nonempty subsets A and H of Y can be separated by φ.

To $A \subset Y$ satisfying (A1) and $k^0 \in C \setminus (-C)$ we associate the function φ_{A,k^0} defined in (5.5). We consider the set

$$A' := \{(y,t) \in Y \times \mathbb{R} \mid y \in tk^0 + A\}.$$

The assumption on A shows that A' is of **epigraph type**, i.e. if $(y,t) \in A'$ and $t' \geq t$, then $(y,t') \in A'$. Indeed, if $y \in tk^0 + A$ and $t' \geq t$, since

$$tk^0 + A = t'k^0 + A - (t'-t)k^0 \subset t'k^0 + A,$$

(because of (A1)) we obtain that $(y,t') \in A'$. Also observe that $A' = T^{-1}(A)$, where $T : Y \times \mathbb{R} \to Y$ is the continuous linear operator defined by $T(y,t) := tk^0 + y$. So, if A is closed (convex, cone), then A' is closed (convex, cone). Obviously, the domain of φ_A is the set $\mathbb{R}k^0 + A$ and $A' \subset \operatorname{epi}\varphi_A \subset \operatorname{cl} A'$ (because A' is of epigraph type), from which it follows that $A' = \operatorname{epi}\varphi_A$ if A is closed, and so φ_A is a lower semicontinuous (l.s.c.) function.

In the next results we collect several useful properties of φ_A (compare Göpfert, Riahi, Tammer, Zălinescu [214] and Tammer, Zălinescu [560]).

Theorem 5.2.3. *Assume that $C \subset Y$ is a proper closed convex cone, $k^0 \in C\setminus(-C)$ and $A \subset Y$ is a nonempty set. Furthermore, suppose*

(A1) *A is closed, satisfies the free-disposal assumption $A - C = A$, and $A \neq Y$.*

Then φ_A (defined in (5.5)) is l.s.c., $\operatorname{dom}\varphi_A = \mathbb{R}k^0 + A$,

$$\{y \in Y \mid \varphi_A(y) \leq \lambda\} = \lambda k^0 + A \quad \forall\, \lambda \in \mathbb{R}, \tag{5.6}$$

and

$$\varphi_A(y + \lambda k^0) = \varphi_A(y) + \lambda \quad \forall\, y \in Y,\ \forall\, \lambda \in \mathbb{R}. \tag{5.7}$$

Moreover,

(a) *φ_A is convex if and only if A is convex; $\varphi_A(\lambda y) = \lambda\varphi_A(y)$ for all $\lambda > 0$ and $y \in Y$ if and only if A is a cone.*
(b) *φ_A is proper if and only if A does not contain lines parallel to k^0, i.e.,*

$$\forall\, y \in Y,\ \exists\, t \in \mathbb{R} \,:\, y + tk^0 \notin A. \tag{5.8}$$

(c) *φ_A is finite-valued if and only if A does not contain lines parallel to k^0 and*

$$\mathbb{R}k^0 + A = Y. \tag{5.9}$$

(d) *Let $B \subset Y$; φ_A is B-monotone if and only if $A - B \subset A$.*
(e) *φ_A is subadditive if and only if $A + A \subset A$.*

Proof. We have already observed that $\operatorname{dom}\varphi_A = \mathbb{R}k^0 + A$ and φ_A is l.s.c. when A is closed. From the definition of φ_A the inclusion $\supset$ in (5.6) is obvious, while the converse inclusion is immediate, taking into account the closedness of A. Formula (5.7) follows easily from (5.6).

(a) Since the operator T defined above is onto and $\operatorname{epi}\varphi_A = T^{-1}(A)$, we have that $\operatorname{epi}\varphi_A$ is convex (cone) if and only if $A = T(\operatorname{epi}\varphi_A)$ is so. The conclusion follows.

(b) We have

$$\varphi_A(y) = -\infty \Leftrightarrow y \in tk^0 + A \ \forall t \in \mathbb{R} \Leftrightarrow \{y + tk^0 \mid t \in \mathbb{R}\} \subset A.$$

The conclusion follows.

(c) The conclusion follows from (b) and the fact that $\operatorname{dom}\varphi_A = \mathbb{R}k^0 + A$.

(d) Suppose first that $A - B \subset A$ and take $y_1, y_2 \in Y$ with $y_2 - y_1 \in B$. Let $t \in \mathbb{R}$ be such that $y_2 \in tk^0 + A$. Then $y_1 \in y_2 - B \subset tk^0 + (A - B) \subset tk^0 + A$, and so $\varphi_A(y_1) \leq t$. Hence $\varphi_A(y_1) \leq \varphi_A(y_2)$. Assume now that φ_A is B-monotone and take $y \in A$ and $b \in B$. From (5.6) we have that $\varphi_A(y) \leq 0$. Since $y - (y - b) \in B$, we obtain that $\varphi_A(y - b) \leq \varphi_A(y) \leq 0$, and so, using again (5.6), we obtain that $y - b \in A$.

(e) Suppose first that $A + A \subset A$ and take $y_1, y_2 \in Y$. Let $t_i \in \mathbb{R}$ be such that $y_i \in t_i k^0 + A$ for $i \in \{1, 2\}$. Then $y_1 + y_2 \in (t_1 + t_2)k^0 + (A + A) \subset (t_1 + t_2)k^0 + A$, and so $\varphi_A(y_1 + y_2) \leq t_1 + t_2$. It follows that $\varphi_A(y_1 + y_2) \leq \varphi_A(y_1) + \varphi_A(y_2)$. Assume now that φ_A is subadditive and take $y_1, y_2 \in A$. From (5.6) we have that $\varphi_A(y_1), \varphi_A(y_2) \leq 0$. Since φ_A is subadditive, we obtain that $\varphi_A(y_1 + y_2) \leq \varphi_A(y_1) + \varphi_A(y_2) \leq 0$, and so, using again (5.6), we obtain that $y_1 + y_2 \in A$. □

Remark 5.2.4. Under assumption (A1), from Theorem 5.2.3 we get that φ_A is l.s.c.,

$$A = \{y \in Y \mid \varphi_A(y) \leq 0\}, \quad \operatorname{int} A \subset \{y \in Y \mid \varphi_A(y) < 0\}, \tag{5.10}$$

and so

$$\operatorname{bd} A = A \setminus \operatorname{int} A \supset \{y \in Y \mid \varphi_A(y) = 0\}. \tag{5.11}$$

In general the inclusion in (5.11) is strict.

Example 5.2.5. Consider $C := \mathbb{R}^2_+$, $k^0 := (1, 0)$ and

$$A := (]-\infty, 0] \times]-\infty, 0]) \cup ([0, +\infty[\times]-\infty, -1]).$$

Then $\varphi_A(u, v) = -\infty$ for $v \leq -1$, $\varphi_A(u, v) = u$ for $v \in (-1, 0]$ and $\varphi_A(u, v) = +\infty$ for $v > 0$. In particular, $\varphi_A(0, -1) = -\infty$ and $(0, -1) \in \operatorname{bd} A$ (see Fig. 5.2).

Theorem 5.2.6. *Assume that $C \subset Y$ is a proper closed convex cone, $k^0 \in C \setminus (-C)$ and $A \subset Y$ is a nonempty set. Furthermore, suppose*

(A2) *A is closed, satisfies the strong free-disposal assumption $A - (C \setminus \{0\}) = \operatorname{int} A$, and $A \neq Y$.*

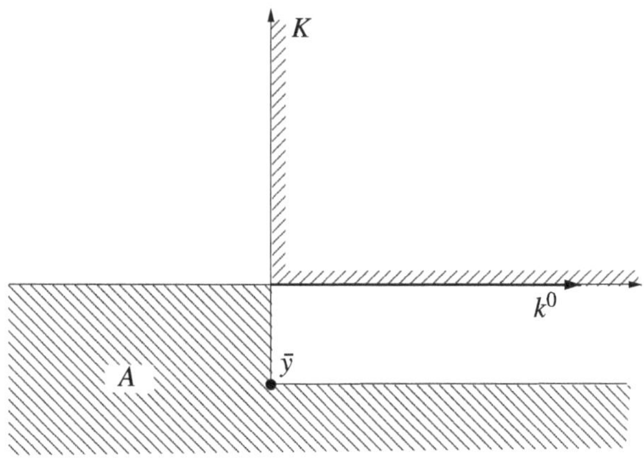

Fig. 5.2 $\overline{y} \in \text{bd } A$ with $\varphi_A(\overline{y}) = -\infty$ in Example 5.2.5

Then (5.6), (5.7), (a), (b), (c) from Theorem 5.2.3 hold, and moreover

(f) φ_A *is continuous and*

$$\{y \in Y \mid \varphi_A(y) < \lambda\} = \lambda k^0 + \text{int } A, \quad \forall\, \lambda \in \mathbb{R}, \tag{5.12}$$

$$\{y \in Y \mid \varphi_A(y) = \lambda\} = \lambda k^0 + \text{bd } A, \quad \forall\, \lambda \in \mathbb{R}. \tag{5.13}$$

(g) *If* φ_A *is proper, then*

$$\varphi_A \text{ is } B\text{-monotone} \Leftrightarrow A - B \subset A \Leftrightarrow \text{bd } A - B \subset A.$$

Moreover, if φ_A *is finite-valued, then*

$$\varphi_A \text{ strictly } B\text{-monotone} \Leftrightarrow A - (B \setminus \{0\}) \subset \text{int } A \Leftrightarrow \text{bd } A - (B \setminus \{0\}) \subset \text{int } A.$$

(h) *Assume that* φ_A *is proper; then*

$$\varphi_A \text{ is subadditive} \Leftrightarrow A + A \subset A \Leftrightarrow \text{bd } A + \text{bd } A \subset A.$$

Proof. Suppose now that (A2) holds.

(f) Let $\lambda \in \mathbb{R}$ and take $y \in \lambda k^0 + \text{int } A$. Since $y - \lambda k^0 \in \text{int } A$, there exists $\varepsilon > 0$ such that $y - \lambda k^0 + \varepsilon k^0 \in A$. Therefore $\varphi_A(y) \leq \lambda - \varepsilon < \lambda$, which shows that the inclusion $\supseteq$ always holds in (5.12). Let $\lambda \in \mathbb{R}$ and $y \in Y$ be such that $\varphi_A(y) < \lambda$. There exists $t \in \mathbb{R}$, $t < \lambda$, such that $y \in tk^0 + A$. It follows with (A2) that $y \in \lambda k^0 + A - (\lambda - t)k^0 \subset \lambda k^0 + \text{int } A$. Therefore (5.12) holds, and so φ_A is upper semicontinuous. Because φ_A is also lower semicontinuous, we have that φ_A is continuous. From (5.6) and (5.12) we obtain immediately that (5.13) holds.

(g) Let us prove the second part, the first one being similar to that of (and partially proved in) Theorem 5.2.3 (d). So, let φ_A be finite-valued.

Assume that φ_A is strictly B-monotone and take $y \in A$ and $b \in -B \setminus \{0\}$. From (5.6) we have that $\varphi_A(y) \le 0$, and so, by hypothesis, $\varphi_A(y-b) < 0$. Using (5.12) we obtain that $y - b \in \operatorname{int} A$. Assume now that $\operatorname{bd} A - (B \setminus \{0\}) \subset \operatorname{int} A$. Consider $y_1, y_2 \in Y$ with $y_2 - y_1 \in B \setminus \{0\}$. From (5.13) we have that $y_2 \in \varphi_A(y_2)k^0 + \operatorname{bd} A$, and so $y_1 \in \varphi_A(y_2)k^0 + \operatorname{bd} A - (B \setminus \{0\}) \subset \varphi_A(y_2)k^0 + \operatorname{int} A$. From (5.12) we obtain that $\varphi_A(y_1) < \varphi_A(y_2)$. The remaining implication is obvious.

(h) Let φ_A be proper. One has to prove $\operatorname{bd} A + \operatorname{bd} A \subset A \Rightarrow \varphi_A$ is subadditive. Consider $y_1, y_2 \in Y$. If $\{y_1, y_2\} \not\subset \operatorname{dom} \varphi_A$, there is nothing to prove; hence let $y_1, y_2 \in \operatorname{dom} \varphi_A$. Then, by (5.13), $y_i \in \varphi_A(y_i)k^0 + \operatorname{bd} A$ for $i \in \{1, 2\}$, and so $y_1 + y_2 \in (\varphi_A(y_1) + \varphi_A(y_2))k^0 + (\operatorname{bd} A + \operatorname{bd} A) \subset (\varphi_A(y_1) + \varphi_A(y_2))k^0 + A$. Therefore $\varphi_A(y_1 + y_2) \le \varphi_A(y_1) + \varphi_A(y_2)$. □

When $k^0 \in \operatorname{int} C$ we get an additional important property of φ_A (see also Theorem 5.2.12).

Corollary 5.2.7. *Assume that $C \subset Y$ is a proper closed convex cone, $k^0 \in \operatorname{int} C$ and $A \subset Y$ satisfies condition (A1). Then φ_A is finite-valued and continuous.*

Proof. Because $k^0 \in \operatorname{int} C$ we have that $\mathbb{R}k^0 + C = Y$. From Theorem 5.2.3 (c) it follows that

$$\operatorname{dom} \varphi_A = A + \mathbb{R}k^0 = A - C + \mathbb{R}k^0 = A + Y = Y.$$

Assuming that φ_A is not proper, from Theorem 5.2.3 (c) we get $y + \mathbb{R}k^0 \subset A$ for some $y \in Y$. Then $Y = y + \mathbb{R}k^0 - C \subset A - C = A$, a contradiction. Hence φ_A is finite-valued.

Moreover, we have that $A - \mathbb{P}k^0 \subset A - \operatorname{int} C \subset \operatorname{int}(A - C) = \operatorname{int} A$. Applying Theorem 5.2.6 (f) for C replaced by $\mathbb{R}_+k^0$ we obtain that φ_A is continuous. □

From the preceding results we get the following particular case.

Corollary 5.2.8. *Let $C \subset Y$ be a proper closed convex cone and $k^0 \in \operatorname{int} C$. Then*

$$\varphi_C : Y \to \mathbb{R}, \qquad \varphi_C(y) := \inf\{t \in \mathbb{R} \mid y \in tk^0 - C\}$$

is a well-defined continuous sublinear function such that for every $\lambda \in \mathbb{R}$,

$$\{y \in Y \mid \varphi_C(y) \le \lambda\} = \lambda k^0 - C, \quad \{y \in Y \mid \varphi_C(y) < \lambda\} = \lambda k^0 - \operatorname{int} C.$$

Moreover, φ_C is strictly $(\operatorname{int} C)$-monotone.

Proof. The assertions follow using Theorem 5.2.6 and Corollary 5.2.7 applied for $A := -C$. For the last part note that $C + \operatorname{int} C = \operatorname{int} C$. □

Now all preliminaries are done, and we can prove the following nonconvex separation theorem.

Theorem 5.2.9 (Nonconvex Separation Theorem). *Let $A \subset Y$ be a closed proper set with nonempty interior, $H \subset Y$ a nonempty set such that $H \cap \operatorname{int} A = \emptyset$. Let $C \subset Y$ be a proper closed convex cone and $k^0 \in \operatorname{int} C$. Furthermore, assume*

(A2) *A is closed, satisfies the strong free-disposal assumption $A - (C \setminus \{0\}) = \operatorname{int} A$, and $A \neq Y$.*

Then φ_A defined by (5.5) is a finite-valued continuous function such that

$$\varphi_A(x) \geq 0 > \varphi_A(y) \quad \forall\, x \in H,\ \forall\, y \in \operatorname{int} A; \tag{5.14}$$

moreover, $\varphi_A(x) > 0$ for every $x \in \operatorname{int} H$.

Proof. By Corollary 5.2.7 φ_A is a finite-valued continuous function. By Theorem 5.2.6 (f) we have that $\operatorname{int} A = \{y \in Y \mid \varphi_A(y) < 0\}$, and so (5.14) obviously holds.

Take $y \in \operatorname{int} H$; then there exists $t > 0$ such that $y - tk^0 \in H$. From (5.7) and (5.12) we obtain that $0 \leq \varphi_A(y - tk^0) = \varphi_A(y) - t$, whence $\varphi_A(y) > 0$. □

Of course, if we impose additional conditions on A, we have additional properties of the separating functional φ_A (see Theorems 5.2.3 and 5.2.6).

5.2.3 Continuity Properties

If A is a proper closed subset of Y (hence $\emptyset \neq A \neq Y$) and $A - \mathbb{P}k^0 \subset \operatorname{int} A$, applying Theorem 5.2.6 for $C := \mathbb{R}_+k^0$ we obtain that φ_A is continuous (on Y) and (5.13) holds. In the next result we characterize the continuity of φ_A at a point $y_0 \in Y$ (compare Tammer and Zălinescu [559, 560, 614]).

Proposition 5.2.10. *Assume that $C \subset Y$ is a proper closed convex cone, $k^0 \in C \setminus (-C)$ and $A \subset Y$ is a nonempty set satisfying condition (A1). Then the function φ_A is (upper semi-) continuous at $y_0 \in Y$ if and only if $y_0 -]\varphi_A(y_0), +\infty[\cdot k^0 \subset \operatorname{int} A$.*

Proof. If $\varphi_A(y_0) = +\infty$ it is clear that φ_A is upper semicontinuous at y_0 and the inclusion holds. So let $\varphi_A(y_0) < +\infty$.

Assume first that φ_A is upper semicontinuous at y_0. Let $\lambda \in]\varphi_A(y_0), +\infty[$. Then there exists a neighbourhood V of y_0 such that $\varphi_A(y) < \lambda$ for every $y \in V$. It follows that for $y \in V$ we have $y \in \lambda k^0 + A$, that is, $V \subset \lambda k^0 + A$. Hence $y_0 \in \lambda k^0 + \operatorname{int} A$, whence $y_0 - \lambda k^0 \in \operatorname{int} A$.

Assume now that $y_0 -]\varphi_A(y_0), +\infty[\cdot k^0 \subset \operatorname{int} A$ and take $\varphi_A(y) < \lambda < +\infty$. Then, by our hypothesis, $V := \lambda k^0 + A$ is a neighbourhood of y_0 and from the definition of φ_A we have that $\varphi_A(y) \leq \lambda$ for every $y \in V$. Hence φ_A is upper semicontinuous at y_0. □

Corollary 5.2.11. *Under the hypotheses of Proposition 5.2.10 assume that φ_A is continuous at $y_0 \in \operatorname{bd} A$. Then $\varphi_A(y_0) = 0$.*

Proof. Of course, $\varphi_A(y_0) \leq 0$. If $\varphi_A(y_0) < 0$, from the preceding proposition we obtain the contradiction $y_0 = y_0 - 0k^0 \in \operatorname{int} A$. □

5.2.4 Lipschitz Properties

The primary goal of this section is to study local Lipschitz properties of the functional φ_{A,k^0} under as weak as possible assumptions concerning the subset $A \subset Y$ and $k^0 \in Y$ (compare Tammer and Zălinescu [560]).

When A is a convex set, as noticed above, φ_A is convex. In such a situation from the continuity of φ_A at a point in the interior of its domain one obtains the local Lipschitz continuity of φ_A on the interior of its domain (if the function is proper). Moreover, when $A = -C$ and $k^0 \in \operatorname{int} C$ then (it is well known that) φ_A is a continuous sublinear function, and so φ_A is Lipschitz continuous.

Recently in the case $Y = \mathbb{R}^m$ and for $C = \mathbb{R}^m_+$ Bonnisseau–Crettez [55] obtained the Lipschitz continuity of φ_A around a point $y \in \operatorname{bd} A$ when $-k^0$ is in the interior of the Clarke tangent cone of A at y. The (global) Lipschitz continuity of φ_A can be related to a result of Gorokhovik–Gorokhovik [217] established in normed vector spaces as we shall see in the sequel.

Theorem 5.2.12. *Assume that $C \subset Y$ is a proper closed convex cone, $k^0 \in C \setminus (-C)$ and $A \subset Y$ is a nonempty set satisfying condition (A1).*

(i) *One has*

$$\varphi_A(y) \leq \varphi_A(y') + \varphi_{-C}(y - y') \quad \forall y, y' \in Y. \tag{5.15}$$

(ii) *If $k^0 \in \operatorname{int} C$ then φ_A is finite-valued and Lipschitz on Y.*

Proof. (i) By Theorem 5.2.3 (applied for A and $A := -C$, respectively) we have that φ_A and φ_{-C} are lower semicontinuous functions, φ_{-C} being sublinear and proper.

Let $y, y' \in Y$. If $\varphi_A(y') = +\infty$ or $\varphi_{-C}(y - y') = +\infty$ it is nothing to prove. In the contrary case let $t, s \in \mathbb{R}$ be such that $y - y' \in tk^0 - C$ and $y' \in sk^0 + A$. Then, taking into account assumption $(A1)$

$$y \in tk^0 - C + sk^0 + A = (t + s)k^0 + (A - C) = (t + s)k^0 + A.$$

It follows that $\varphi_A(y) \leq t + s$. Passing to infimum with respect to t and s satisfying the preceding relations we get (5.15).

(ii) Assume that $k^0 \in \operatorname{int} C$. Let $V \subset Y$ be a symmetric closed and convex neighbourhood of 0 such that $k^0 + V \subset C$ and let $p_V : Y \to \mathbb{R}$ be the Minkowski functional associated to V; then p_V is a continuous seminorm and

$V = \{y \in Y \mid p_V(y) \leq 1\}$. Let $y \in Y$ and $t > 0$ such that $y \in tV$. Then $t^{-1}y \in V \subset k^0 - C$, whence $y \in tk^0 - C$. Hence $\varphi_{-C}(y) \leq t$. Therefore, $\varphi_{-C}(y) \leq p_V(y)$. This inequality confirms that $(\mathbb{R}k^0 - C =)\ \mathrm{dom}\,\varphi_{-C} = Y$. Moreover, since φ_{-C} is sublinear we get $\varphi_{-C}(y) \leq \varphi_{-C}(y') + p_V(y - y')$ and so

$$\left|\varphi_{-C}(y) - \varphi_{-C}(y')\right| \leq p_V(y - y') \quad \forall y, y' \in Y, \tag{5.16}$$

that is, φ_{-C} is Lipschitz.

By Corollary 5.2.7 we have that φ_A is finite-valued (and continuous). From (5.15) we have that $\varphi_A(y) - \varphi_A(y') \leq \varphi_{-C}(y - y') \leq p_V(y - y')$, whence (interchanging y and y')

$$\left|\varphi_A(y) - \varphi_A(y')\right| \leq p_V(y - y') \quad \forall y, y' \in Y. \tag{5.17}$$

Hence φ_A is Lipschitz continuous (on Y). □

Note that the condition $A - (C \setminus \{0\}) \subset \mathrm{int}\, A$ does not imply that φ_A is proper.

Example 5.2.13. Take $A := \{(x, y) \in \mathbb{R}^2 \mid y \geq -|x|^{-1}\}$, with the convention $0^{-1} := +\infty$, and $C := \mathbb{R}_+ k^0$ with $k^0 := (0, -1)$. Then $A - (C \setminus \{0\}) = \mathrm{int}\, A$ and $\varphi_A(0, 1) = -\infty$.

Note that, with our notation, [55, Proposition 7] asserts that φ_{A,k^0} is finite and locally Lipschitz provided $Y = \mathbb{R}^n$, $C = \mathbb{R}^n_+$ and $k^0 \in \mathrm{int}\, C$, which is much less than the conclusion of Theorem 5.2.12(ii).

Of course, in the conditions of Theorem 5.2.12(ii) we have that $-k^0 \in \mathrm{int}\, A_\infty$ because $C \subset -A_\infty$. In fact we have also a converse of Theorem 5.2.12(ii).

Proposition 5.2.14. *Assume that $C \subset Y$ is a proper closed convex cone, $k^0 \in C \setminus (-C)$ and $A \subset Y$ is a nonempty set satisfying condition (A1). If φ_A is finite-valued and Lipschitz then $-k^0 \in \mathrm{int}\, A_\infty$.*

Proof. By hypothesis there exists a closed convex and symmetric neighbourhood V of 0 such that (5.17) holds. We have that $A = \{y \in Y \mid \varphi_A(y) \leq 0\}$. Let $y \in A$, $v \in V$ and $\alpha \geq 0$. Then

$$\varphi_A(y + \alpha(v - k^0)) \leq \varphi_A(y + \alpha v) - \alpha \leq \varphi_A(y) + \alpha p_V(v) - \alpha \leq 0$$

because $V = \{y \in Y \mid p_V(y) \leq 1\}$. Hence $V - k^0 \subset A_\infty$, which shows that $-k^0 \in \mathrm{int}\, A_\infty$. □

Corollary 5.2.15. *Under the assumptions of Proposition 5.2.14, the function φ_A is finite-valued and Lipschitz if and only if $-k^0 \in \mathrm{int}\, A_\infty$.*

Proof. The necessity is given by Proposition 5.2.14. Assume that $-k^0 \in \mathrm{int}\, A_\infty$. Taking $C := -A_\infty$, using Theorem 5.2.12(ii) we obtain that φ_A is finite-valued and Lipschitz. □

If $\operatorname{int} C \neq \emptyset$ and $k^0 \notin \operatorname{int} C$, φ_{-C} is not finite-valued, and so it is not Lipschitz. One may ask if the restriction of φ_{-C} at its domain is Lipschitz. The next examples show that both situations are possible.

Example 5.2.16. Take $C = \mathbb{R}^2_+$ and $k^0 = (1, 0)$. We have that $\varphi_{-C}(y_1, y_2) = y_1$ for $y_2 \leq 0$, $\varphi_{-C}(y_1, y_2) = +\infty$ for $y_2 > 0$, and so $\varphi_{-C}|_{\operatorname{dom} \varphi_{-C}}$ is Lipschitz.

Example 5.2.17. Take $C := \{(u, v, w) \in \mathbb{R}^3 \mid v, w \geq 0,\ u^2 \leq vw\}$ and $k^0 := (0, 0, 1)$; then

$$\varphi_{-C}(x, y, z) = \begin{cases} +\infty & \text{if } y > 0 \text{ or } [y = 0 \text{ and } x \neq 0], \\ z & \text{if } x = y = 0, \\ z - x^2/y & \text{if } y < 0. \end{cases}$$

It is clear that the restriction of φ_{-C} at its domain is not continuous at $(0, 0, 0) \in \operatorname{dom} \varphi_{-C}$ and the restriction of φ_{-C} at the interior of its domain is not Lipschitz. However, φ_{-C} is locally Lipschitz on the interior of its domain.

The last property mentioned in the previous example is a general one for φ_A when A is convex.

Proposition 5.2.18. *Let A be a proper closed subset of Y and $k^0 \in Y \setminus \{0\}$ be such that $A - \mathbb{R}_+ k^0 = A$. If A is convex, has nonempty interior, and does not contain any line parallel with k^0 (or equivalently $k^0 \notin A_\infty$), then φ_A is locally Lipschitz on* $\operatorname{int}(\operatorname{dom} \varphi_A) = \mathbb{R}k^0 + \operatorname{int} A$.

Proof. Because A does not contain any line parallel with k^0, φ_A is proper (see Theorem 5.2.3 taking into account assumption $(A1)$). We know that $\operatorname{dom} \varphi_A = \mathbb{R}k^0 + A$, and so $\operatorname{int}(\operatorname{dom} \varphi_A) = \operatorname{int}(\mathbb{R}k^0 + A) = \mathbb{R}k^0 + \operatorname{int} A$ (see, e.g., [614, Exercise 1.4]). On the other hand it is clear that $A \subset \{y \in Y \mid \varphi_A(y) \leq 0\}$. Since $\operatorname{int} A \neq \emptyset$, we have that φ_A is bounded above on a neighbourhood of a point, and so φ_A is locally Lipschitz on $\operatorname{int}(\operatorname{dom} \varphi_A) = \mathbb{R}k^0 + \operatorname{int} A$ (see e.g. [614, Corollary 2.2.13]). □

We have seen in Theorem 5.2.12 that φ_A is Lipschitz even if A is not convex when $k^0 \in \operatorname{int} C$. So, in the sequel we are interested by the case in which A is not convex, $k^0 \notin \operatorname{int} C$ and A does not contain any line parallel with k^0.

Note that for A not convex and $y \in \operatorname{int}(\operatorname{dom} \varphi_A)$ we can have situations in which φ_A is not continuous at y or φ_A is continuous but not Lipschitz around y.

Example 5.2.19. Take $C := \mathbb{R}^2_+$, $k^0 := (1, 0)$ and

$$A_1 := (]-\infty, 0] \times]-\infty, 1]) \cup ([0, 1] \times]-\infty, 0])$$

$$A_2 := \{(a, b) \mid a \in]0, \infty[,\ b \leq -a^2\} \cup (]-\infty, 0] \times]-\infty, 1]).$$

Then

$$\varphi_{A_1,k^0}(u,v)=\begin{cases}+\infty & \text{if } v>1,\\ u & \text{if } 0<v\le 1,\\ u-1 & \text{if } v\le 0,\end{cases}\qquad \varphi_{A_2,k^0}(u,v)=\begin{cases}+\infty & \text{if } v>1,\\ u & \text{if } 0<v\le 1,\\ u-\sqrt{-v} & \text{if } v\le 0.\end{cases}$$

It is clear that $(0,0)\in \operatorname{int}(\operatorname{dom}\varphi_{A_1})$ but φ_{A_1} is not continuous at $(0,0)$, and $(0,0)\in \operatorname{int}(\operatorname{dom}\varphi_{A_2})$, φ_{A_2} is continuous at $(0,0)$ but φ_{A_2} is not Lipschitz at $(0,0)$.

In what concerns the Lipschitz continuity of φ_A around a point $y\in\operatorname{dom}\varphi_A$ in finite dimensional spaces this can be obtained using the notion of epi-Lipschitzianity of a set as introduced by Rockafellar [491] (see also [493]). We extend this notion in our context. We say that the set $A\subset Y$ is **epi-Lipschitz** at $\overline{y}\in A$ in the direction $v\in Y\setminus\{0\}$ if there exist $\varepsilon>0$ and a (closed convex symmetric) neighbourhood V_0 of 0 in Y such that

$$\forall y\in(\overline{y}+V_0)\cap A,\ \forall w\in v+V_0,\ \forall\lambda\in[0,\varepsilon]:\ y+\lambda w\in A. \tag{5.18}$$

Note that (5.18) holds for $v=0$ if and only if $\overline{y}\in\operatorname{int}A$. Moreover, if $\overline{y}\in\operatorname{int}A$ then A is epi-Lipschitz at $\overline{y}\in A$ in any direction.

Theorem 5.2.20. *Let A be a proper closed subset of Y and $k^0\in Y\setminus\{0\}$ be such that $A-\mathbb{R}_+k^0=A$. Assume that $y_0\in Y$ is such that $\varphi_A(y_0)\in\mathbb{R}$. Then φ_A is finite and Lipschitz on a neighbourhood of y_0 if and only if A is epi-Lipschitz at $\overline{y}:=y_0-\varphi_A(y_0)k^0$ in the direction $-k^0$.*

Proof. Using (5.7) we get $\varphi_A(\overline{y})=0$. Recall also that $A=\{y\in Y\mid \varphi_A(y)\le 0\}$ and the finite values of φ_A are attained (because A is closed).

Assume that there exist a closed convex symmetric neighbourhood V of 0 in Y and $p:Y\to\mathbb{R}$ a continuous seminorm such that φ_A is finite on y_0+V and $|\varphi_A(y)-\varphi_A(y')|\le p(y-y')$ for all $y,y'\in y_0+V$. Taking into account (5.7), we have that φ_A is finite on $\overline{y}+V$ and

$$\left|\varphi_A(y)-\varphi_A(y')\right|\le p(y-y')\quad \forall y,y'\in\overline{y}+V.$$

Take $V_0:=\{y\in\frac{1}{3}V\mid p(y)\le 1\}$ and $\varepsilon\in\,]0,1]$ such that $\varepsilon k^0\in V_0$. Let us show that (5.18) holds with v replaced by $-k^0$. For this take $y\in(\overline{y}+V_0)\cap A$, $w\in -k^0+V_0$ and $\lambda\in[0,\varepsilon]$. Then $y-\lambda k^0-\overline{y}\in V_0+V_0\subset V$ and $y+\lambda w-\overline{y}=y-\lambda k^0-\overline{y}+\lambda(w+k^0)\in V_0+V_0+V_0\subset V$, and so

$$\begin{aligned}\varphi_A(y+\lambda w)&\le\varphi_A(y-\lambda k^0)+p(\lambda(w+k^0))=\varphi_A(y)-\lambda+\lambda p(w+k^0)\\&\le\lambda(p(w+k^0)-1)\le 0.\end{aligned}$$

Hence $y+\lambda w\in A$.

Assume now that (5.18) holds with v replaced by $-k^0$. Let $r\in\,]0,\varepsilon]$ be such that $2r(1+p(k^0))<1$, where $p:=p_{V_0}$. Of course, $\{y\mid p(y)\le\lambda\}=\lambda V_0$ for every $\lambda>0$ and if $p(y)=0$ then $y\in\lambda V_0$ for every $\lambda>0$. Set

$$M := \{y \in \overline{y} + rV_0 \mid |\varphi_A(y)| \le p(y - \overline{y})\};$$

of course, $\overline{y} \in M$. We claim that $M = \overline{y} + rV_0$. Consider $y \in M$, $w \in V_0$ and $\lambda \in [0, r]$. Setting $y' := y - \varphi_A(y)k^0 \in A$, we have that $\varphi_A(y') = 0$ and

$$p(y' - \overline{y}) \le p(y - \overline{y}) + |\varphi_A(y)| \cdot p(k^0) \le r\left(1 + p(k^0)\right) < \tfrac{1}{2} \le 1, \tag{5.19}$$

and so, by (5.18), $y' + \lambda(w - k^0) \in A$; hence $\varphi_A(y' + \lambda w) \le \lambda$.

Take $v \in rV_0$. On one hand one has

$$\varphi_A(y' + v) = \varphi_A\left(y' + p(v) \cdot \frac{1}{p(v)} v\right) \le p(v)$$

if $p(v) > 0$, and $\varphi_A(y' + v) = \varphi_A(y' + \lambda(\lambda^{-1}v)) \le \lambda$ for every $\lambda \in]0, r]$, whence $\varphi_A(y' + v) \le 0 = p(v)$. Therefore, $\varphi_A(y' + v) \le p(v)$.

On the other hand, assume that $\varphi_A(y'+v) < -p(v)$. Because $2r(1+p(k^0)) < 1$, there exists $t > 0$ such that $r + (t + r)p(k^0) \le 1/2$ and $\varphi_A(y'+v) < -p(v) - t =: t' < 0$. It follows that $y' + v - t'k^0 \in A$. Moreover, taking into account (5.19),

$$p(y'+v-t'k^0-\overline{y}) \le p(y'-\overline{y})+p(v)+(t+p(v))p(k^0) \le 1/2+r+(t+r)p(k^0) \le 1,$$

and so $y' + v - t'k^0 \in (\overline{y} + V_0) \cap A$. Using (5.18), if $p(v) > 0$ then

$$y' + tk^0 = y' - \left(t' + p(v)\right)k^0 = y' + v - t'k^0 + p(v)\left(-k^0 - \frac{1}{p(v)}v\right) \in A,$$

while if $p(v) = 0$ then

$$y' + (1 - \gamma)tk^0 = y' + v - t'k^0 + \gamma t\left(-k^0 - (\gamma t)^{-1}v\right) \in A$$

for $\gamma := \min\{\frac{1}{2}, \varepsilon t^{-1}\}$. We get the contradiction $0 = \varphi_A(y') \le -t < 0$ in the first case and $0 = \varphi_A(y') \le -t(1 - \gamma) < 0$ in the second case. Hence $\varphi_A(y' + v) \in \mathbb{R}$ and $|\varphi_A(y' + v) - \varphi_A(y')| \le p(v)$ for every $v \in rV_0$, or equivalently,

$$\varphi_A(y + v) \in \mathbb{R}, \quad |\varphi_A(y + v) - \varphi_A(y)| \le p(v) \quad \forall v \in rV_0. \tag{5.20}$$

When $y := \overline{y} \in M$, from (5.20) we get $\overline{y} + rV_0 \subset M$, and so $M = \overline{y} + rV_0$ as claimed. Moreover, if $y, y' \in \overline{y} + \frac{1}{2}rV_0$, then $y \in M$ and $y' = y + v$ for some $v \in rV_0$; using again (5.20) we have that $|\varphi_A(y') - \varphi_A(y)| \le p(y' - y)$. The conclusion follows. □

The next result is similar to Corollary 5.2.11.

Corollary 5.2.21. *Let A be a proper closed subset of Y and $k^0 \in Y \setminus \{0\}$ be such that $A - \mathbb{R}_+k^0 = A$. Consider $\overline{y} \in \operatorname{bd} A$. If A is epi-Lipschitz at $\overline{y}$ in the direction $-k^0$ then $\varphi_A(\overline{y}) = 0$.*

Proof. Consider $\varepsilon \in]0,1[$ and V_0 provided by (5.18) with $v := -k^0$. Assume that $\varphi_A(\overline{y}) \neq 0$. Then there exists $t > 0$ such that $tp_{V_0}(k^0) \leq \varepsilon$ and $y := \overline{y} + tk^0 \in A$. Taking $\lambda := t$ in (5.18) we obtain that $y + t(-k^0 + V_0) = \overline{y} + tV_0 \subset A$, contradicting the fact that $\overline{y} \in \operatorname{bd} A$. □

Corollary 5.2.22. *Let A be a proper closed subset of Y and $k^0 \in Y \setminus \{0\}$ be such that $A - \mathbb{R}_+k^0 = A$. Assume that* $\dim Y < +\infty$ *and $\overline{y} \in \operatorname{bd} A$. Then φ_A is finite and Lipschitz on a neighbourhood of $\overline{y}$ if and only if $-k^0 \in \operatorname{int} C(A,\overline{y})$, where $C(A,\overline{y})$ is the Clarke's tangent cone of A at $\overline{y}$ (see Definition 4.2.7).*

Proof. By [493, Theorem 2], $-k^0 \in \operatorname{int} C(A,\overline{y})$ if and only if A is epi-Lipschitz at $\overline{y}$ in the direction $-k^0$. The conclusion follows from Corollary 5.2.11, Theorem 5.2.20 and Corollary 5.2.21. □

The fact that φ_A is Lipschitz on a neighbourhood of $\overline{y}$ under the condition $-k^0 \in \operatorname{int} C(A,\overline{y})$ is obtained in [55, Proposition 6] in the case $Y = \mathbb{R}^m$ (and $C = \mathbb{R}^m_+$).

Consider $y^* \in Y^*$ such that $\langle k^0, y^* \rangle \neq 0$, $H := \ker y^*$ and take

$$\varphi_0 : H \to \overline{\mathbb{R}}, \quad \varphi_0(z) := \varphi_A(z),$$

that is, $\varphi_0 = \varphi_A|_H$. Since φ_A is l.s.c., so is φ_0. Then any $y \in Y$ can be written uniquely as $z - tk^0$ with $z \in H$ and $t \in \mathbb{R}$. So, by (5.7), $\varphi_A(y) = \varphi_A(z - tk^0) = \varphi_0(z) - t$. Using (5.10) we obtain that $A = \{z - tk^0 \mid (z,t) \in \operatorname{epi} \varphi_0\}$. Conversely, if $g : H \to \overline{\mathbb{R}}$ is a l.s.c. function and $A := \{z - tk^0 \mid (z,t) \in \operatorname{epi} g\}$, then A is a closed set with $A - \mathbb{R}_+k^0 = A$ and $\varphi_0 = g$. Therefore, the closed set A with the property $A - \mathbb{R}_+k^0 = A$ is uniquely determined by a l.s.c. function $\varphi_0 : H \to \overline{\mathbb{R}}$. Moreover, for $\overline{y} = \overline{z} - \overline{t}k^0$ we have that φ_A is finite (resp. continuous) at $\overline{y}$ if and only if φ_0 is finite (resp. continuous) at $\overline{z}$. Moreover, because $Y = H + \mathbb{R}k^0$ and the sum is topological (that is, the projection onto H parallel to $\mathbb{R}k^0$ is continuous), we have that φ_A is finite and Lipschitz continuous on a neighbourhood of $\overline{y}$ if and only if φ_0 is finite and Lipschitz continuous on a neighbourhood of $\overline{z}$. Similarly, φ_A is finite and Lipschitz continuous if and only if φ_0 is finite and Lipschitz continuous.

Note that for Y a normed vector space in [217] one says that A is (globally) **epi-Lipschitz in the direction** $e \in Y \setminus \{0\}$ if there exist a closed linear subspace H of codimension 1 with $e \notin H$ and a Lipschitz function $g : H \to \mathbb{R}$ such that $A = \{y + \alpha e \mid y \in H,\ \alpha \in \mathbb{R},\ g(y) \leq \alpha\}$; A is epi-Lipschitz if there exists $e \in Y \setminus \{0\}$ such that A is epi-Lipschitz in the direction e. The main result of [217] asserts that the proper closed set $A \subset Y$ is epi-Lipschitz in the direction e if and only if $e \in \operatorname{int} A_\infty$, and so $A \subset Y$ is epi-Lipschitz if and only if $\operatorname{int} A_\infty \neq \emptyset$.

The discussion above shows that not only the main theorem of [217] can be obtained from Corollary 5.2.15, but this one extends the main theorem of [217] to locally convex spaces.

These properties of the functional φ_{A,k^0} are used for deriving minimal point theorems (cf. [214, 235]), optimality conditions (see [29]) and duality assertions(cf. [214]).

5.2.5 The Formula for the Conjugate and Subdifferential of φ_A for A Convex

The results of this section (except the second part of Corollary 5.2.24) were established in several papers; we give the proofs for reader's convenience. The formula for the conjugate of φ_A is derived by [488, Theorem 3] and [512, Theorem 2.2]. Results concerning the subdifferential of φ_A are given in [156, Theorem 2.2, Lemma 2.1]. In the statements below we use some usual notation from convex analysis. So, having X a separated locally convex space with topological dual X^* and a convex function $f : X \to \overline{\mathbb{R}}$, the conjugate of f is the function $f^* : X^* \to \overline{\mathbb{R}}$ defined by $f^*(x^*) := \sup\{\langle x, x^*\rangle - f(x) \mid x \in X\}$ and its subdifferential at $x \in X$ with $f(x) \in \mathbb{R}$ is the set $\partial f(x) := \{x^* \in X^* \mid \langle x' - x, x^*\rangle \leq f(x') - f(x)\ \forall x \in X\}$; $\partial f(x) := \emptyset$ if $f(x) \notin \mathbb{R}$. Recall that the **indicator function** of a set $A \subset X$ is the function $\iota_A : X \to \overline{\mathbb{R}}$ defined by $\iota_A(x) := 0$ for $x \in A$ and $\iota_A(x) := +\infty$ for $x \in X \setminus A$, while the support of A is the function $\sigma_A := (\iota_A)^*$. When A is nonempty, the domain of σ_A is a convex cone which is called the **barrier cone** of A and is denoted by bar A. Moreover, the normal cone of A at $a \in A$ is the set $N(A,a) := \partial\iota_A(a)$. Suppose that the assumptions of Sect. 5.2.2 are fulfilled. As before, we again use the notations

$$C^+ = \{y^* \in Y^* \mid \langle k, y^*\rangle \geq 0\ \forall k \in C\},$$

$$C^\# = \{y^* \in Y^* \mid \langle k, y^*\rangle > 0\ \forall k \in C \setminus \{0\}\}$$

for the positive dual cone of C and the quasi interior of C^+, respectively.

Proposition 5.2.23. *Assume that $A \subset Y$ is closed and convex with $A - C = A$ and $k^0 \notin A_\infty$. Then $\varphi_A \in \Gamma(Y)$, that is, φ_A is a proper l.s.c. convex function,*

$$\varphi_A^*(y^*) = \begin{cases} \sigma_A(y^*) & \text{if } y^* \in \operatorname{bar} A,\ \langle k^0, y^*\rangle = 1, \\ +\infty & \text{otherwise,} \end{cases} \tag{5.21}$$

and $\partial\varphi_A(y) \subset \{y^* \in \operatorname{bar} A \mid \langle k^0, y^*\rangle = 1\} \subset \{y^* \in C^+ \mid \langle k^0, y^*\rangle = 1\}$ for every $y \in Y$.

Proof. From Theorem 5.2.3 we have that $\varphi_A \in \Gamma(Y)$. Consider $y^* \in Y^*$. Then

$$\begin{aligned} \varphi_A^*(y^*) &= \sup\{\langle y, y^*\rangle - \varphi_A(y) \mid y \in Y\} \\ &= \sup\{\langle y, y^*\rangle - t \mid y \in Y,\ t \in \mathbb{R},\ y \in tk^0 + A\} \\ &= \sup\{\langle tk^0 + a, y^*\rangle - t \mid y \in Y,\ t \in \mathbb{R},\ a \in A\} \\ &= \sup\{\langle a, y^*\rangle \mid a \in A\} + \sup\{t(\langle k^0, y^*\rangle - 1) \mid t \in \mathbb{R}\}. \end{aligned}$$

Hence (5.21) holds.

Since $\partial f(y) \subset \operatorname{dom} f^*$ for every proper function $f : Y \to \overline{\mathbb{R}}$ and every $y \in Y$, the first estimate for $\partial\varphi_A(y)$ follows. Moreover, because $A = A - C$ we have $\sigma_A = \sigma_{A-C} = \sigma_A + \sigma_{-C} = \sigma_A + \iota_{C^+}$, and so $\operatorname{bar} A \subset C^+$. □

The estimate $\operatorname{bar} A \subset C^+$ becomes more precise when $C = -A_\infty$; in fact one has $(A_\infty)^+ = -\operatorname{cl}_{w^*}(\operatorname{bar} A)$. Indeed, from [614, Exercise 2.23] we have that $\iota_{A_\infty} = (\iota_A)_\infty = \sigma_{\operatorname{dom} \iota_A^*} = \sigma_{\operatorname{dom} \sigma_A}$, whence $\iota_{-(A_\infty)^+} = (\iota_{A_\infty})^* = \iota_{\operatorname{cl}_{w^*}(\operatorname{dom} \sigma_A)}$, and so $-(A_\infty)^+ = \operatorname{cl}_{w^*}(\operatorname{dom} \sigma_A)$.

Using Proposition 5.2.23 one deduces the expression of $\partial\varphi_A$ (see also [156, Theorem 2.2] for Y a normed vector space).

Corollary 5.2.24. *Assume that A is convex and $k^0 \notin A_\infty$. Then for all $\overline{y} \in Y$ one has*

$$\partial\varphi_A(\overline{y}) = \{y^* \in \operatorname{bar} A \mid \langle k^0, y^* \rangle = 1,\ \langle \overline{y}, y^* \rangle - \varphi_A(\overline{y}) \geq \langle y, y^* \rangle\ \forall y \in A\}. \tag{5.22}$$

Moreover, if $A - (C \setminus \{0\}) \subset \operatorname{int} A$ holds then $\partial\varphi_A(y) \subset C^{\#}$ for every $y \in Y$.

Proof. Fix $\overline{y} \in Y$. If $\overline{y} \notin \operatorname{dom} \varphi_A$ then both sets in (5.22) are empty. Let $\overline{y} \in \operatorname{dom} \varphi_A$. Then, of course, $\overline{y} - \varphi_A(\overline{y})k^0 \in A$. If $y^* \in \partial\varphi_A(\overline{y})$ then $\varphi_A(\overline{y}) + \varphi_A^*(y^*) = \langle \overline{y}, y^* \rangle$. Taking into account (5.21) we obtain that

$$y^* \in \operatorname{bar} A,\ \langle k^0, y^* \rangle = 1 \text{ and } \langle \overline{y}, y^* \rangle - \varphi_A(\overline{y}) \geq \langle y, y^* \rangle\ \forall y \in A, \tag{5.23}$$

that is, the inclusion $\subset$ holds in (5.22). Conversely, if $y^* \in Y^*$ is such that (5.23) holds, since $\overline{y} - \varphi_A(\overline{y})k^0 \in A$ and $\langle k^0, y^* \rangle = 1$, we obtain that $\langle \overline{y} - \varphi_A(\overline{y})k^0, y^* \rangle = \sigma_A(y^*)$, which shows that $\varphi_A(\overline{y}) + \varphi_A^*(y^*) = \langle \overline{y}, y^* \rangle$. Hence $y^* \in \partial\varphi_A(\overline{y})$. Therefore, (5.22) holds.

Assume now $A - (C \setminus \{0\}) \subset \operatorname{int} A$, and take $y^* \in \partial\varphi_A(\overline{y})$. Hence $\overline{y} \in \operatorname{dom} \varphi_A$. Consider $k \in C \setminus \{0\}$. Since $(\overline{y} - k) - \overline{y} = -k \in -(C \setminus \{0\})$, by Theorem 5.2.12 (iv), we have that $\langle -k, y^* \rangle \leq \varphi_A(\overline{y} - k) - \varphi_A(\overline{y}) < 0$, that is, $\langle k, y^* \rangle > 0$. Therefore, $y^* \in C^{\#}$. □

5.3 Scalarizing Functionals by Hiriart-Urruty and Zaffaroni

In this section, we characterize Q-minimal points of a subset A of a Banach space Y with respect to a proper pointed closed convex cone $C \subset Y$ (elements of $\operatorname{QMin}(A, C)$, see Definition 2.4.8) using a scalarization by means of Hiriart-Urruty's function and establish properties of subdifferentials of this function in some special cases.

In Sect. 12 we derive necessary and sufficient conditions for Q-minimal points/Q-minimizers (see Definitions 2.4.8, 2.6.3) in the form of being a solution for a scalar optimization problem. Here the scalar function is the oriented distance function introduced by Hiriart-Urruty [262] (see Example 5.1.3). In this section we suppose that Y is a Banach space and $C \subset Y$ a proper closed convex and pointed cone. Recall that for $\emptyset \neq A \subset Y$ and $Y \setminus A \neq \emptyset$, this function is defined by

$$\Delta_A(y) := d(y, A) - d(y, Y \setminus A).$$

Remark 5.3.1. Of course, if $\operatorname{int} A = \emptyset$, it holds $d(y, Y \setminus A) = 0$ and so for all $y \in Y$

$$\Delta_A(y) = d(y, A),$$

i.e., Hiriart-Urruty's function at $y \in Y$ coincides with the distance from y to A.

Furthermore, if $A(\neq Y)$ is convex and $\operatorname{int} A \neq \emptyset$ we have $\Delta_A(y) = \Delta_{\operatorname{cl} A}(y)$ for all $y \in Y$. Indeed, on the one hand side we have obviously $d(y, A) = d(y, \operatorname{cl} A)$ and $d(y, Y \setminus A) = d(y, \operatorname{cl}(Y \setminus A)) = d(y, Y \setminus \operatorname{int} A)$. On the other hand,

$$d(y, Y \setminus \operatorname{cl} A) = d(y, \operatorname{cl}(Y \setminus \operatorname{cl} A)) = d(y, Y \setminus \operatorname{int}(\operatorname{cl} A)) = d(y, Y \setminus \operatorname{int} A),$$

taking into account $\operatorname{int}(\operatorname{cl} A) = \operatorname{int} A$ since A is convex, and so $d(y, Y \setminus A) = d(y, Y \setminus \operatorname{cl} A)$.

It is well known that this function Δ_A has very good general properties (see Zaffaroni [607]) and we list below for the reader's convenience the properties we shall use in the sequel.

Proposition 5.3.2. *Assume that A is a nonempty subset of Y and $Y \setminus A \neq \emptyset$.*

(i) *Δ_A is Lipschitz of rank 1.*
(ii) $\Delta_{Y \setminus A} = -\Delta_A$.
(iii) *Δ_A is convex if A is convex and Δ_A is concave if A is reverse convex, i.e., $Y \setminus A$ is convex.*
(iv) *$\Delta_A(y) < 0$ iff $y \in \operatorname{int} A$, $\Delta_A(y) = 0$ iff $y \in \operatorname{bd} A$ and $\Delta_A(y) > 0$ iff $y \in Y \setminus \operatorname{int} A$.*
(v) *Suppose that A is convex and has a nonempty interior, and $x \in \operatorname{bd} A$. Then*

$$\partial \Delta_A(x) \subset N(x; A) \setminus \{0\}$$

where $N(x; A)$ is the normal cone in the sense of convex analysis of A at x.

In the following proposition we present a result by Ha [228, Proposition 21.10] that characterizes Q-minimal points introduced in Definition 2.4.8 via scalarization by means of the functional Δ_A.

Proposition 5.3.3. *Suppose that $A \subset Y$ is a nonempty set and $Q := \operatorname{int} D$ for a proper cone $D \subset Y$ with nonempty interior. Then, $\overline{a} \in \operatorname{QMin}(A, C)$ if and only if the function $\Delta_{-Q}(\cdot - \overline{a})$ attains its minimum at $\overline{a}$, that is*

$$\Delta_{-Q}(a - \overline{a}) \geq \Delta_{-Q}(0) = 0, \quad \forall a \in A. \tag{5.24}$$

Proof. Observe first that $0 \in \mathrm{bd}(-Q)$ because of Definition 2.4.8 and $\Delta_{-Q}(0) = 0$ taking into account Proposition 5.3.2 (iv). Now, let $\overline{a} \in \mathrm{QMin}(A, C)$. By Definition 2.4.8 we have $(A - \overline{a}) \subseteq Y \setminus (-Q)$. Because of Proposition 5.3.2 (iv), $\Delta_{-Q}(a - \overline{a}) \geq 0$ for all $a \in A$, i.e., (5.24) holds. Next, suppose that (5.24) holds. If $\overline{a} \notin \mathrm{QMin}(A, C)$ then there is $a' \in A$ such that $a' - \overline{a} \in -Q$. As the set $-Q$ is open it holds $\Delta_{-Q}(a' - \overline{a}) < 0$ since Proposition 5.3.2 (iv), a contradiction to (5.24). Thus, $\overline{a} \in \mathrm{QMin}(A, C)$. □

For deriving optimality conditions it is of interest to know more about the subdifferential of $\Delta_{-Q}(\cdot - \overline{a})$ at $\overline{a} \in \mathrm{QMin}(A, C)$. The results presented in the following proposition are shown by Ha [228, Proposition 21.11] and play an important role in formulating Fermat rules and Lagrange multiplier rules.

Proposition 5.3.4. (i) $\partial \Delta_{-\operatorname{int} C}(0) \subseteq C^+ \setminus \{0\}$.

(ii) *Let D be a proper cone with* $\operatorname{int} D \neq \emptyset$ *and* $Q = \operatorname{int} D$ *a convex dilation of* C. *Then*

$$\partial \Delta_{-Q}(0) \subseteq C^{\#}.$$

(iii) *Assume that C has a bounded base Θ, then it holds* $\partial \Delta_{-V_\eta}(0) \subseteq \operatorname{int} C^+$ *for* V_η *given by (2.41).*

(iv) *Suppose that A has a nonempty interior, then it holds*

$$\partial_A \Delta_{Y \setminus \mathrm{cl\,conv\,cone}[(A - \overline{a}) \cup C]}(0) \subseteq C^+ \setminus \{0\},$$

where ∂_A denotes the approximate subdifferential introduced in Definition 3.6.1.

Proof. (i) We get the result applying Proposition 5.3.2 (v) to $A = -\operatorname{int} C$ and take account of $N(0; -\operatorname{int} C) = C^+$.

(ii) Applying Proposition 5.3.2 (v) to $A = -C$ we get $\partial \Delta_{-Q}(0) \subset N(0; -Q) \setminus \{0\}$. Now take $y^* \in \partial \Delta_{-Q}(0)$ and $k \in C \setminus \{0\} \subset Q$. We have to show that $y^*(k) > 0$ for all $k \in C \setminus \{0\}$. As $y^* \neq 0$ because of Proposition 5.3.2 (v), there is $y \in Y$ such that $y^*(y) > 0$. On the other hand, since Q is open and $k \in Q$, there exists a scalar $t > 0$ such that $-k + ty \in -Q$. Hence, $y^*(-k + ty) \leq 0$ and $y^*(k) \geq ty^*(y) > 0$. Thus, $y^* \in C^{\#}$.

(iii) Suppose that Θ is bounded and denote

$$\overline{\delta} := \sup\{\|\theta\| \mid \theta \in \Theta\} < +\infty.$$

Let $y^* \in \partial \Delta_{-V_\eta}(0)$. It is known that $y^* \in \operatorname{int} C^+$ if and only if y^* is uniformly positive on C in the sense that there exists a scalar $\alpha > 0$ such that $y^*(k) \geq \alpha\|k\|$ for all $k \in C \setminus \{0\}$. Let $k \in C \setminus \{0\}$ be an arbitrary vector. As Θ is a base of C, there exist a scalar $t > 0$ and $\theta \in \Theta$ such that $k = t\theta$. Since $-\theta + \eta \overset{\circ}{B}_Y \subset -V_\eta$, it follows that

$$-k + t\eta \mathring{B}_Y = t(-\theta + \eta \mathring{B}_Y) \subset -V_\eta,$$

i.e., the open ball centered at $-k$ with the radius $t\eta$ is contained in $-V_\eta$. Therefore, we get

$$d(-k, Y \setminus (-V_\eta)) \geq t\eta.$$

On the other hand, $t\eta = (\|k\|/\|\theta\|)\eta \geq (\eta/\overline{\delta})\|k\|$. Hence,

$$d(-k, Y \setminus (-V_\eta)) \geq (\eta/\overline{\delta})\|k\|.$$

Of course, $\Delta_{-V_\eta}(0) = 0$. Because of $y^* \in \partial\Delta_{-V_\eta}(0)$, the definition of the convex subdifferential yields

$$\begin{aligned} y^*(-k) &\leq \Delta_{-V_\eta}(-k) - \Delta_{-V_\eta}(0) = d(-k, -V_\eta) - d(-k, Y \setminus (-V_\eta)) \\ &= -d(-k, Y \setminus (-V_\eta)) \leq -(\eta/\overline{\delta})\|k\| \end{aligned}$$

such that we have $\langle y^*, k\rangle \geq (\eta/\overline{\delta})\|k\|$. This means that y^* is uniformly positive on C, or $y^* \in \operatorname{int} C^+$ as it was to be shown.

(iv) By the relationship between approximate subdifferential and the Clarke subdifferential (see Proposition 3.6.3 (i)), it is sufficient to show that

$$\partial_C \Delta_{Y \setminus \text{cl conv cone}[(A-\overline{a})\cup C]}(0) \subseteq C^+ \setminus \{0\}.$$

For the simplicity, we denote $Q := Y \setminus -V$, where $V :=$ cl conv cone $[(A - \overline{a}) \cup C]$. Note that V is a closed convex cone with a nonempty interior and $C \subset V$. We have to show that

$$\partial_C \Delta_{-Q}(0) \subseteq C^+ \setminus \{0\}.$$

By Proposition 5.3.2 (ii), we have $\Delta_{-Q}(0) = \Delta_{Y\setminus V}(0) = -\Delta_V(0)$. The properties of the Clarke subdifferential and the subdifferential of convex analysis yield

$$\partial_C(-\Delta_V)(0) = -\partial_C \Delta_V(0) = -\partial\Delta_V(0).$$

Applying Proposition 5.3.2 (v) to the closed convex cone V which has a nonempty interior gives $-\partial\Delta_V(0) \subset -N(0; V) \setminus \{0\}$. Furthermore, since $C \subset V$ we get $-N(0; V) \subseteq C^+$. Therefore, we get $-\partial\Delta_V(0) \subseteq C^+ \setminus \{0\}$, which yields $\partial_C \Delta_{-Q}(0) \subseteq C^+ \setminus \{0\}$. □

Remark 5.3.5. Note that the properties described above for the oriented distance function are similar to those required for risk measures used in mathematical finance [256, 488, 497] (see Example 5.1.4).

5.4 Characterization of Solutions of Set-Valued Optimization Problems by Means of Nonlinear Scalarizing Functionals

5.4.1 An Extension of the Functional φ_A

In this section we assume that Y is a linear topological space ordered by a proper convex closed pointed cone $C \subset Y$ with nonempty interior, $A \subset Y$, $k^0 \in -\operatorname{int} C$. We consider the functional φ_{A,k^0} given by (5.5). Now, we take $d \in Y$ and $A = d + C$. Then, the functional φ_{A,k^0} has the form

$$\varphi_{d,C,k^0}(y) = \inf\{t \in \mathbb{R} \mid y \in tk^0 + d + C\}, \quad y \in Y. \tag{5.25}$$

As usual, the family of all nonempty subsets of Y is denoted by $\mathscr{P}(Y)$.

Replacing d in (5.25) by a set $D \in \mathscr{P}(Y)$, we get a functional $\varphi_{D,C,k^0} : Y \to \mathbb{R} \cup \{-\infty\}$ defined by

$$\varphi_{D,C,k^0}(y) := \inf\{t \in \mathbb{R} \mid y \in tk^0 + D + C\}, \quad y \in Y. \tag{5.26}$$

Setting $A = D + C$ the functional φ_{D,C,k^0} coincides with the functional φ_{A,k^0} given by (5.5). Obviously, we have for any $y \in Y$

$$\varphi_{D,C,k^0}(y) = \inf_{d \in D}\{\varphi_{d,C,k^0}(y)\}. \tag{5.27}$$

Definition 5.4.1. A set $D \in \mathscr{P}(Y)$ is called C**-proper**, if $D + C \neq Y$.

We denote the **family of all C-proper subsets of** Y by $\mathscr{P}_C(Y)$.

In this section we discuss an extension of the functional $\varphi_{D,C,k^0} : Y \to \mathbb{R} \cup \{-\infty\}$ defined by (5.26) to a function $\Phi_{k^0} : \mathscr{P}_C(Y)^2 \longrightarrow \mathbb{R} \cup \{+\infty\}$ (see Definition 5.4.5) that is very useful in set-valued optimization where the solution concept is based on set approach (compare Sect. 2.6.2).

The results in this section are derived by Hernández and Rodríguez-Marín [252]. Based on a corresponding scalarization technique by means of the functional Φ_{k^0} Hernández and Rodríguez-Marín have shown a characterization of solutions of set-valued optimization problems where the solution concept is given by the lower set less order relation $\preceq^l_C$ (compare Definition 2.6.9).

It is easy to check that the set-relation $\approx^l$ defined below in (5.28) is an equivalence relation on $\mathscr{P}(Y)$, such that we can consider an order on the **family of equivalence classes** which are denoted by $[\cdot]^l$:

$$A \approx^l B :\Longleftrightarrow A \preceq^l_C B \quad \text{and} \quad B \preceq^l_C A. \tag{5.28}$$

Obviously, $A \in [B]^l$ if and only if $A + C = B + C$ (compare Definition 2.6.9 and Remark 2.6.11).

We say that $A \in \mathscr{P}(Y)$ is **C-closed** if $A + C$ is a closed set and that A is **C-bounded** if for each neighborhood U of zero in Y there is some positive number t such that $A \subset tU + C$. Furthermore, we say that A is **C-compact** if any cover of A of the form $\{U_\alpha + C \mid U_\alpha \text{ are open}\}$ admits a finite subcover. Every C-compact set is C-closed and C-bounded (see Luc [402, Proposition 3.3]).

The following characterization of C-proper sets is given in [252, Lemma 2.16].

Lemma 5.4.2. *Let $D \in \mathscr{P}(Y)$. D is C-proper if and only if $\varphi_{D,C,k^0}(y) > -\infty$ for every $y \in Y$.*

From Lemma 5.4.2 we get that $\varphi_{D,C,k^0}(\cdot)$ is a real-valued function, if $D \in \mathscr{P}_C(Y)$.

From the properties (5.6), (5.12), (5.13) of the functional φ_{A,k^0} given by (5.5) in Theorems 5.2.3, 5.2.6 we get the corresponding assertions concerning φ_{D,C,k^0} in the following lemma:

Lemma 5.4.3. *Let $D \in \mathscr{P}_C(Y)$ and $t \in \mathbb{R}$. Then, for any $y \in Y$ it holds:*

(i) $\varphi_{D,C,k^0}(y) < t \iff y \in tk^0 + D + \operatorname{int} C$,
(ii) $\varphi_{D,C,k^0}(y) \leq t \iff y \in tk^0 + \operatorname{cl}(D + C)$,
(iii) $\varphi_{D,C,k^0}(y) \geq t \iff y \notin tk^0 + D + \operatorname{int} C$,
(iv) $\varphi_{D,C,k^0}(y) = t \iff y \in tk^0 + \operatorname{bd}(D + C)$,
(v) $\varphi_{D,C,k^0}(y) > t \iff y \notin tk^0 + \operatorname{cl}(D + C)$.

From the definition of the functional in (5.26) we get the following assertions:

Proposition 5.4.4. *Let $D, B \in \mathscr{P}_C(Y)$ and $y \in Y$. Then*

(i) *If D is C-closed, it holds $\varphi_{D,C,k^0}(y) = \min_{d \in D}\{\varphi_{d,C,k^0}(y)\}$.*
(ii) *If $D \preceq^l_C B$, then $\varphi_{D,C,k^0}(y) \leq \varphi_{B,C,k^0}(y)$. In particular, if $D \in [B]^l$ then $\varphi_{D,C,k^0}(y) = \varphi_{B,C,k^0}(y)$.*

Now, we study an extension of the functional (5.26) defined from Y to $\mathbb{R} \cup \{-\infty\}$ to a function defined from $\mathscr{P}_C(Y)^2$ to $\mathbb{R} \cup \{+\infty\}$.

Definition 5.4.5. Consider a fixed element $k^0 \in -\operatorname{int} C$. Let the function $\Phi_{k^0} : \mathscr{P}_C(Y)^2 \longrightarrow \mathbb{R} \cup \{+\infty\}$ defined by

$$\Phi_{k^0}(B, D) := \sup_{d \in D}\{\varphi_{B,C,k^0}(d)\} \quad \text{for} \quad (B, D) \in \mathscr{P}_C(Y)^2. \tag{5.29}$$

From Lemma 5.4.3, (ii), we get

$$\Phi_{k^0}(B, D) \leq t \iff D \subset tk^0 + B + C \tag{5.30}$$

if B is a C-closed set and $t \in \mathbb{R}$.

Using this assertion we get an equivalent formulation of (5.29) in the following proposition:

Proposition 5.4.6. *Let $B \in \mathscr{P}_C(Y)$ be a C-closed set and $D \in \mathscr{P}_C(Y)$. Assume $\Phi_{k^0}(B, D) < +\infty$. Then we get the following expression for Φ_{k^0}:*

$$\Phi_{k^0}(B, D) = \min\{t \in \mathbb{R} \mid D \subset tk^0 + B + C\}.$$

Proof. Consider the set

$$T := \{t \in \mathbb{R} \mid \Phi_{k^0}(B, D) \leq t\}.$$

Under the assumption $\Phi_{k^0}(B, D) < +\infty$ and taking into account (5.30) we get

$$\min_{t \in T} t = \Phi_{k^0}(B, D)$$

and the proof is completed. □

Lemma 5.4.7. *Consider $B \in \mathscr{P}_C(Y)$. Then B is C-bounded if and only if $\Phi_{k^0}(C, B) < +\infty$.*

Proof. ["⇒":] Assume that B is C-bounded, then for the neighborhood of zero

$$U := k^0 + \operatorname{int} C$$

there exists $\alpha > 0$ such that $B \subset \alpha(k^0 + \operatorname{int} C) + C \subset \alpha k^0 + C$. Thus, for each $b \in B$ it holds $\varphi_{C,C,k^0}(b) \leq \alpha$. Hence, $\Phi_{k^0}(C, B) \leq \alpha$.

["⇐":] Conversely, suppose that $\Phi_{k^0}(C, B) < +\infty$, then there exists $r \in \mathbb{R}$ such that $\Phi_{k^0}(C, B) < r$. Namely, for each $b \in B$ we have

$$\varphi_{C,C,k^0}(b) < r.$$

Taking into account Lemma 5.4.3 (i), the above inequality can be written as $b \in rk^0 + \operatorname{int} C$, that is

$$B \subset rk^0 + \operatorname{int} C. \tag{5.31}$$

Let U' be a neighborhood of zero. Chose $\lambda > 0$ such that $rk^0 \in \lambda U'$, then we get together with (5.31) that

$$B \subset \lambda U' + C,$$

which implies that B is C-bounded. □

Theorem 5.4.8. *Let B be a C-bounded set and $D \in \mathscr{P}_C(Y)$. Then D is C-bounded if and only if $\Phi_{k^0}(B, D) < +\infty$.*

Proof. ["⇒":] Suppose that D is C-bounded. Then, because of (5.27), for every $d \in D$ we have

$$\varphi_{B,C,k^0}(d) = \inf_{b \in B}\{\varphi_{b,C,k^0}(d)\}. \tag{5.32}$$

It is easy to prove that $\varphi_{b,C,k^0}(d) = \varphi_{0,C,k^0}(d-b)$ and $\varphi_{0,C,k^0}(d-b) = \varphi_{C,C,k^0}(d-b)$. Hence the equality (5.32) can be written as

$$\varphi_{B,C,k^0}(d) = \inf_{b \in B}\{\varphi_{C,C,k^0}(d - b)\}.$$

Then, for any fixed $b_0 \in B$ we obtain

$$\varphi_{B,C,k^0}(d) \leq \varphi_{C,C,k^0}(d - b_0) \quad \text{for all} \quad d \in D.$$

This yields

$$\Phi_{k^0}(B, D) \leq \sup_{d \in D}\{\varphi_{C,C,k^0}(d - b_0)\} = \Phi_{k^0}(C, D - b_0).$$

Because of the C-boundedness of D it follows $\Phi_{k^0}(C, D - b_0) < +\infty$ taking into account Lemma 5.4.7. Therefore, $\Phi_{k^0}(B, D) < +\infty$.

["$\Leftarrow$":] Conversely, we suppose $\Phi_{k^0}(B, D) = r$ with $r \in \mathbb{R}$, then for each $d \in D$ we get $\varphi_{B,C,k^0}(d) \leq r$. Let $\alpha > 0$, then $\varphi_{B,C,k^0}(d) < r + \alpha$ and applying Lemma 5.4.3 (i) it follows $d \in (r + \alpha)k^0 + B + C$ for all $d \in D$, thus

$$D \subset (r + \alpha)k^0 + B + C. \tag{5.33}$$

Let U be a neighbourhood of zero. Since B is C-bounded we get that the set $(r + \alpha)k^0 + B$ is also C-bounded. Then, there exists $\lambda > 0$ such that

$$(r + \alpha)k^0 + B \subset \lambda U + C.$$

Taking into account (5.33) this inclusion yields

$$D \subset \lambda U + C.$$

This means that D is C-bounded. □

In Sect. 5.4.2 we will give characterizations of solutions of set-valued optimization problems where the solution concept is given by the lower set less order relation $\preceq^l_C$ by means of the functional Φ_{k^0}. In order to prove such assertions we need certain monotonicity properties of the functional Φ_{k^0}.

Definition 5.4.9. Let $S \subset \mathcal{P}(Y)$. A given functional $\varphi : \mathcal{P}(Y) \longrightarrow \mathbb{R}$ is said to be **$\preceq^l_C$-increasing** on S if

$$A, B \in S, \ A \preceq^l_C B \Longrightarrow \varphi(A) \leq \varphi(B).$$

The following result follows directly from Proposition 5.4.4, (ii).

Theorem 5.4.10. *Let $D \in \mathscr{P}_C(Y)$, then $\Phi_{k^0}(\cdot, D)$ is $\preceq^l_C$-increasing on $\mathscr{P}_C(Y)$.*

Furthermore, the following assertion holds (see Hernández and Rodríguez-Marín [252, Theorem 3.10]).

Theorem 5.4.11. *Let $B \in \mathscr{P}_C(Y)$ be a C-closed set. Then the following statements are true:*

(i) $\Phi_{k^0}(B, B) = 0$,
(ii) *If $B \in [D]^l$, then $\Phi_{k^0}(B, D) = \Phi_{k^0}(D, B) = 0$,*
(iii) *$B \preceq^l_C D$ if and only if $\Phi_{k^0}(B, D) \leq 0$.*

5.4.2 Characterization of Solutions of Set-Valued Optimization Problems with Lower Set Less Order Relation $\preceq^l_C$ by Scalarization

In this section we assume that Y is a linear topological space partially ordered by a proper convex closed pointed cone $C \subset Y$ with nonempty interior. Furthermore, let S be a set and $F : S \rightrightarrows Y$ is a set-valued map with domain S. We consider a set-valued optimization problem where the solution concept is given with respect to the lower set less order relation $\preceq^l_C$ (see Definitions 2.6.9 and 2.6.19):

$$\preceq^l_C -\text{minimize } F(x) \quad \text{subject to} \quad x \in S. \qquad (SP - \preceq^l_C)$$

Let $\mathscr{M}(\mathscr{P}_C(Y), \mathbb{R})$ be the set of all functions from $\mathscr{P}_C(Y)$ to $\mathbb{R}$.

The next results are shown by Hernández and Rodríguez-Marín [252, Theorem 4.1 and Corollary 4.3]. First, we present certain kinds of separation theorems.

Theorem 5.4.12. *Consider the problem $(SP - \preceq^l_C)$. Assume that F is C-closed and C-bounded valued on S and $x^0 \in S$. Then, x^0 is a minimal solution of $(SP - \preceq^l_C)$ if and only if there exists a functional $\varphi \in \mathscr{M}(\mathscr{P}_C(Y), \mathbb{R})$ which is $\preceq^l_C$-increasing on $\mathscr{P}_C(Y)$ such that the following statements are true:*

(i) *If $x \in S$ and $F(x) \in [F(x^0)]^l$ then $\varphi(F(x)) = 0$.*
(ii) *If $x \in S$ and $F(x) \notin [F(x^0)]^l$ then $\varphi(F(x)) > 0$.*
(iii) *If $D \in \mathscr{P}_C(Y)$ and $D \preceq^l_C F(x^0)$ then $\varphi(D) \leq 0$.*

Proof. Assume that x^0 is a minimal solution of $(SP - \preceq^l_C)$. We fix $k^0 \in -\operatorname{int} C$. Taking into account Theorem 5.4.10, the functional

$$\varphi(\cdot) := \Phi_{k^0}(\cdot, F(x^0)) \in \mathscr{M}(\mathscr{P}_C(Y), \mathbb{R})$$

is $\preceq^l_C$-increasing on $\mathscr{P}_C(Y)$. So it is sufficient to show that φ satisfies the conditions (i), (ii) and (iii). The conditions (i) and (iii) are consequences of Theorem 5.4.11 (ii) and (iii), respectively.

Now, we will show that (ii) holds. x^0 is a minimal solution of $(SP - \preceq^l_C)$. Then, for each $x \in S$ with $F(x) \notin [F(x^0)]^l$ we have $F(x) \not\preceq^l_C F(x^0)$. Hence, by Theorem 5.4.11 (iii),

$$\varphi(F(x)) = \Phi_{k^0}(F(x), F(x^0)) > 0.$$

Conversely, we assume that (i), (ii) and (iii) are true for some $\varphi \in \mathscr{M}(\mathscr{P}_C(Y), \mathbb{R})$ which is $\preceq^l_C$-increasing on $\mathscr{P}_C(Y)$. We have to prove that x^0 is a minimal solution of $(SP - \preceq^l_C)$. Indeed, if we suppose that x^0 is not a minimal solution of $(SP - \preceq^l_C)$, there exists $x' \in S$ such that $F(x') \notin [F(x^0)]^l$ and

$$F(x') \preceq^l_C F(x^0).$$

Thus, by (ii), $\varphi(F(x')) > 0$. On the other hand, by (iii), it follows $\varphi(F(x')) \leq 0$ which is a contradiction. □

For the special case of a vector optimization problem we get from Theorem 5.4.12 well known characterizations of solutions by means of a sustainable scalarization method.

Consider the vector optimization problem

$$\mathrm{Min}(f(S), C), \tag{VOP}$$

where $f : S \longrightarrow Y$ is a single-valued map. We use the solution concept introduced in Definition 2.4.1.

In the assertion of the next corollary we use (strictly) increasing functionals $\varphi : Y \to \mathbb{R}$ in the following sense (compare $(A_{sf}1)$ and $(A_{sf}1')$): φ is called **increasing**, if $y, w \in Y$, $w - y \in C$ implies $\varphi(y) \leq \varphi(w)$. Furthermore, φ is called **strictly increasing**, if $y, w \in Y$, $w - y \in \operatorname{int} C$ implies $\varphi(y) < \varphi(w)$.

Corollary 5.4.13. *Consider the problem (VOP). Then,* $f(x^0) \in \mathrm{Min}(f(S), C)$ *if and only if there exists a continuous functional* $\varphi : Y \to \mathbb{R}$ *that is increasing and strictly increasing, such that:*

(i) $\varphi(f(x^0)) = 0$.
(ii) *If* $x \in S$ *and* $f(x) \neq f(x^0)$, *then* $\varphi(f(x)) > 0$.
(iii) *If* $y \in f(x^0) - C$, *then* $\varphi(y) \leq 0$.

Now, we derive sufficient conditions for minimal solutions of $(SP - \preceq^l_C)$ using the functional $\Phi_{k^0}(B, \cdot)$ (see (5.29)). In order to show these conditions we introduce the following notations.

Definition 5.4.14. A net $\{D_\alpha \mid \alpha \in I\}$ from $\mathscr{P}(Y)$ is said to be $\preceq^l_C$**-decreasing** (with respect to $\preceq^l_C$) and we write $\{D_\alpha\} \downarrow_l$ if $D_\alpha \preceq^l_C D_\beta$ for each $\alpha, \beta \in I$ and $\beta < \alpha$.

For a given $x^0 \in S$ we introduce

$$L(x^0) := \{x \in S \mid F(x) \preceq^l_C F(x^0)\}.$$

Definition 5.4.15. Let $x^0 \in X$ and $k^0 \in -\operatorname{int} C$. F is said to be Φ_{k^0}**-stable at** x^0 if for any $\{x_\alpha\} \subset L(x^0)$ and $t \in \mathbb{R}$ verifying $\{F(x_\alpha)\} \downarrow_l$ and $\Phi_{k^0}(F(x^0), F(x_\alpha)) = t$ for all α imply that there exists $\overline{x} \in S$ such that $F(\overline{x}) \preceq^l_C F(x^0) + tk^0$.

The following **existence result** for minimal solution of $(SP - \preceq^l_C)$ in the sense of Definition 2.6.19 is shown by Hernández and Rodríguez-Marín [252, Theorem 5.3].

Theorem 5.4.16. *Consider the problem* $(SP - \preceq^l_C)$*. Let* $x^0 \in S$ *and* $k^0 \in -\operatorname{int} C$*. Assume that* F *is* C*-closed valued,* $F(x^0)$ *is* C*-bounded,* F *is* Φ_{k^0}*-stable at* x^0 *and* $\cup_{x \in L(x_0)} F(x)$ *is* C*-compact. Then there exists* $x \in L(x^0)$ *and* x *is a minimal solution of* $(SP - \preceq^l_C)$*.*

Furthermore, Hernández and Rodríguez-Marín [252, Theorem 5.4] proved the following existence result using Cantor's intersection theorem.

Theorem 5.4.17. *Consider the problem* $(SP - \preceq^l_C)$*. Assume* (X, d) *is a complete metric space ,* $S \subset X$*,* F *is* C*-bounded valued and for each* $x \in S$ *the set* $L(x)$ *is closed. Assume that* $x^0 \in S$ *and that the following conditions are satisfied:*

(i) $\cup_{x \in L(x^0)} F(x)$ *is* C*-bounded,*
(ii) *there are an element* $k^0 \in -\operatorname{int} C$ *and a* C*-bounded set* $B \subset Y$ *such that*

$$x^1, x^2 \in S,\ F(x^1) \preceq^l_C F(x^2) \Longrightarrow d(x^1, x^2) \leq \Phi_{k^0}(B, F(x^1)) - \Phi_{k^0}(B, F(x^2)).$$

Then there exists $\overline{x} \in L(x^0)$ *with* $\overline{x}$ *a minimal solution of* $(SP - \preceq^l_C)$*.*

Proof. Suppose that x^0 is not a minimal solution of $(SP - \preceq^l_C)$. Taking into account Definition 2.6.19 there exists $x \in L(x^0)$ with

$$F(x) \preceq^l_C F(x^0) \quad \text{and} \quad F(x^0) \not\preceq^l_C F(x).$$

Because of assumption (ii) it follows

$$0 < \Phi_{k^0}(B, F(x)) - \Phi_{k^0}(B, F(x^0)).$$

On the other hand, since $\bigcup_{x \in L(x^0)} F(x)$ and $F(x^0)$ are C-bounded sets, by Theorem 5.4.8

$$0 < \sup_{x \in L(x_0)} \{\Phi_{k^0}(B, F(x))\} - \Phi_{k^0}(B, F(x^0)) < +\infty. \tag{5.34}$$

For $z \in S$ we introduce $v(z) := \sup_{x \in L(z)}\{\Phi_{k^0}(B, F(x))\}$. From (5.34) we get that there exists an element $x_1 \in L(x^0)$ with

$$v(x^0) - \Phi_{k^0}(B, F(x_1)) \leq 2^{-1}.$$

If $x_1 \in L(x^0)$ and x_1 is not a minimal solution of $(SP - \preceq^l_C)$, then there exists $x_2 \in L(x_1)$ such that

$$v(x_1) - \Phi_{k^0}(B, F(x_2)) \leq 2^{-2}.$$

In this way we obtain a sequence $\{x_n\}_{n\in\mathbb{N}}$ from $L(x^0)$ such that for each $n \in \mathbb{N}$ it holds $L(x_{n+1}) \subset L(x_n)$.

Moreover, since for each $n \in \mathbb{N}$ the set $L(x_n)$ is closed and

$$v(x_n) - \Phi_{k^0}(B, F(x_{n+1})) \leq 2^{-(n+1)}, \tag{5.35}$$

it follows $\{\operatorname{diam} L(x_n)\} \to 0$. Indeed, if that is not the case, then there exists $\delta > 0$ with $\operatorname{diam} L(x_n) > \delta$ for each $n \in \mathbb{N}$. By triangle inequality of $d(\cdot,\cdot)$, for fixed $n \in \mathbb{N}$, there exists $z' \in L(x_n)$ such that $d(z', x_n) > \frac{\delta}{2}$ and taking into account assumption (ii), we get

$$\frac{\delta}{2} < d(z', x_n) \leq \Phi_{k^0}(B, F(z')) - \Phi_{k^0}(B, F(x_n)). \tag{5.36}$$

On the other hand, we have

$$\Phi_{k^0}(B, F(z')) - \Phi_{k^0}(B, F(x_n)) \leq v(x_n) - \Phi_{k^0}(B, F(x_n))$$

and as $x_n \in L(x_{n-1})$, then $v(x_n) \leq v(x_{n-1})$. Therefore, because of (5.35) and (5.36) we get

$$\frac{\delta}{2} < d(z', x_n) \leq v(x_{n-1}) - \Phi_{k^0}(B, F(x_n)) \leq 2^{-n}$$

and for $n \to +\infty$ the above inequality implies $\delta < 0$, a contradiction.

So, because for every $n \in \mathbb{N}$ the set $L(x_n)$ is closed, (X, d) is a complete metric space, $L(x_{n+1}) \subset L(x_n)$ and $\operatorname{diam} L(x_n) \to 0$ we get from Cantor's intersection theorem

$$\bigcap_{n\in\mathbb{N}} L(x_n) = \{\overline{x}\}.$$

Consequently, $\overline{x} \in L(x^0)$ and $L(\overline{x}) = \{\overline{x}\}$. Hence, $\overline{x}$ is a minimal solution of $(SP - \preceq^l_C)$ and the proof is completed. □

The following corollary is a direct consequence of Theorem 5.4.17.

Corollary 5.4.18. *Consider the vector optimization problem*

$$\text{Min}(f(S), C), \tag{VOP}$$

where $f : S \to Y$, C is a proper pointed closed convex cone in the linear topological space Y. Assume (X, d) is a complete metric space , $S \subset X$ and for each $x \in S$ the set $L(x)$ is closed. Suppose that $x^0 \in S$ and the following conditions are satisfied:

(i) *$f(L(x^0))$ is C-bounded,*
(ii) *there are an element $k^0 \in -\text{int}\, C$ and a C-bounded set $B \subset Y$ such that*

$$x^1, x^2 \in S,\ f(x^1) \in f(x^2) - C \Longrightarrow d(x^1, x^2) \le \varphi_{B,C,k^0}(f(x^1)) - \varphi_{B,C,k^0}(f(x^2)).$$

Then there exists $\overline{x} \in L(x^0)$ with $f(\overline{x}) \in \text{Min}(f(S), C)$.

Remark 5.4.19. Kuroiwa [347, Theorems 4.1 and 4.2] obtained existence results for minimal solutions of $(SP - \preceq^l_C)$ under similar conditions to those supposed in Theorems 5.4.16 and 5.4.17. However, the assumptions in Theorems 5.4.16 and 5.4.17 are less restrictive because there are no continuity assumptions.

5.5 The Extremal Principle

In the book by Mordukhovich [430] a general concept of set extremality (see Mordukhovich [425, 427], Kruger, Mordukhovich [342], Kruger [343]) is presented. Its relationships to separation properties of nonconvex sets are discussed and corresponding applications for deriving optimality conditions in constrained optimization are developed in [425, 430]. In this section we will present these fundamental results, especially we will formulate several versions of the **extremal principle**. These extremal principles will be applied in Sect. 12.9 for deriving a subdifferential variational principle for set-valued mappings (Theorem 12.9.1) and in Sect. 12.11 in order to prove a first order necessary condition for fully localized minimizers of set-valued optimization problems in the sense of Definition 2.6.49.

First, we introduce local extremal points and extremal systems (see [430, Definition 2.1]).

Definition 5.5.1. Consider nonempty subsets $\Omega_1, \cdots, \Omega_n$ ($n \ge 2$) of a Banach space X. Let $\overline{x}$ be a common point of these sets. This point $\overline{x}$ is called a **local extremal point** of the set system $\{\Omega_1, \cdots, \Omega_n\}$ if there are sequences $\{a_{ik}\} \subset X$, $i = 1, \cdots, n$, and a neighborhood V of $\overline{x}$ such that $a_{ik} \overset{k\to\infty}{\to} 0$ and

$$\bigcap_{i=1}^{n}(\Omega_i - a_{ik}) \cap V = \emptyset \qquad \text{for all large } k \in \mathbb{N}.$$

In this case $\{\Omega_1, \cdots, \Omega_n, \overline{x}\}$ is said to be an **extremal system** in X.

Remark 5.5.2. For the case of two closed sets Ω_1 and Ω_2 in the Banach space X we say that a point $\bar{x} \in \Omega_1 \cap \Omega_2$ is *locally extremal* for the set system $\{\Omega_1, \Omega_2\}$ if there is a neighborhood V of $\bar{x}$ such that for any $\varepsilon > 0$ we can find an element $a \in \varepsilon B_X$ with

$$\Omega_1 \cap (\Omega_2 + a) \cap V = \emptyset. \tag{5.37}$$

It is important to mention that the condition $\Omega_1 \cap \Omega_2 = \{\overline{x}\}$ does not necessary imply that $\overline{x}$ is a local extremal point of $\{\Omega_1, \Omega_2\}$.

Mordukhovich [430, Definition 2.5] introduced the following three basic versions of the **extremal principle** in Banach spaces and explained that they can be considered as a kind of local separation of nonconvex sets around extremal points. In the following definition the set of ε-normals $\hat{N}_\varepsilon(x; \Omega)$ (see Definition 3.5.2), the Fréchet normal cone $\hat{N}(x; \Omega)$ (see (3.23)) and the limiting normal cone $N(x; \Omega)$ (see Definition 3.5.6) (with X Banach space, $\Omega \subset X$, $x \in \Omega$) are used.

Definition 5.5.3. Let $\{\Omega_1, \cdots, \Omega_n, \overline{x}\}$ be an extremal system in X. Then

(a) $\{\Omega_1, \cdots, \Omega_n, \overline{x}\}$ satisfies the **ε-extremal principle** if for every $\varepsilon > 0$ there are $x_j \in \Omega_j \cap (\overline{x} + \varepsilon B_X)$ and $x_j^* \in X^*$ such that

$$x_j^* \in \hat{N}_\varepsilon(x_j; \Omega_j) \quad j = 1, \cdots, n, \tag{5.38}$$

$$x_1^* + \cdots + x_n^* = 0, \qquad ||x_1^*||_* + \cdots + ||x_n^*||_* = 1. \tag{5.39}$$

(b) $\{\Omega_1, \cdots, \Omega_n, \overline{x}\}$ satisfies the **approximate extremal principle** if for every $\varepsilon >$ 0 there are $x_j \in \Omega_j \cap (\overline{x} + \varepsilon B_X)$ and $x_j^* \in X^*$ and

$$x_j^* \in \hat{N}(x_j; \Omega_j) + \varepsilon B_{X^*} \quad j = 1, \cdots, n, \tag{5.40}$$

such that (5.39) holds.

(c) $\{\Omega_1, \cdots, \Omega_n, \overline{x}\}$ satisfies the **exact extremal principle** if there are basic normals

$$x_j^* \in N(\overline{x}; \Omega_j), \quad j = 1, \cdots, n, \tag{5.41}$$

such that (5.39) holds.

The ε- (approximate, exact, respectively) extremal principle holds **in** the space X if it holds for every extremal system $\{\Omega_1, \cdots, \Omega_n, \overline{x}\}$ in X, where all the sets Ω_j $(j = 1, \cdots, n)$ are closed around $\overline{x}$.

Remark 5.5.4. Taking into account $\hat{N}(x; \Omega) + \varepsilon B_{X^*} \subset \hat{N}_\varepsilon(x; \Omega)$ it is clear that the ε-extremal principle follows from the approximate extremal principle for any extremal system $\{\Omega_1, \cdots, \Omega_n, \overline{x}\}$ in a Banach space X.

Remark 5.5.5. The assertions in the extremal principles provide necessary conditions for local extremal points of set systems and can be considered as generalized Euler equations in an abstract geometric setting. Furthermore, the extremal principles are related to local separation of nonconvex sets. Indeed, if we consider the exact extremal principle for two sets Ω_1 and Ω_2, then (5.39) and (5.41) reduce to the fact that there exists an element $x^* \in X^*$ with

$$0 \neq x^* \in N(\overline{x}; \Omega_1) \cap (-N(\overline{x}; \Omega_2)). \tag{5.42}$$

Such a kind of extremal principle we also get using the separation theorem for nonconvex sets (Theorem 5.2.9) where the nonlinear separating functional is given by (5.5). Applying Theorem 5.2.9 to an extremal system $\{\Omega_1, \Omega_2, \overline{x}\}$ and taking into account the properties of the subgradients of the nonlinear separating functional (see Lemma 12.8.1 (ii) and (iii)) we get a condition corresponding to (5.42) following the line of the proof of Theorem 12.8.2.

Remark 5.5.6. If we assume that Ω_1 and Ω_2 are convex, (5.42) is equivalent to

$$\forall\, x_1 \in \Omega_1,\ \forall x_2 \in \Omega_2 : \qquad \langle x^*, x_1 \rangle \leq \langle x^*, x_2 \rangle,$$

which is nothing else then the classical separation property for two convex sets.

Mordukhovich [430, Theorem 2.20] has shown the following fundamental assertion in an Asplund space X.

Theorem 5.5.7. *Suppose that X is an Asplund space. Then the approximate extremal principle holds in X.*

Remark 5.5.8. Taking into account Remark 5.5.4 we get from Theorem 5.5.7 that the ε-extremal principle holds in an Asplund space X (see [430, Theorem 2.20]).

We will use Theorem 5.5.7 for the case of two subsets Ω_1 and Ω_2 of an Asplund space X in Sects. 12.9 and 12.11. In the case of a system $\{\Omega_1, \Omega_2\}$, a locally extremal point $\bar{x} \in \Omega_1 \cap \Omega_2$ is given by (5.37). Then we get the following assertion from Theorem 5.5.7 (see [27]).

Theorem 5.5.9 (The Extremal Principle). *Let X be an Asplund space and $\bar{x}$ be a local extremal point of the set system $\{\Omega_1, \Omega_2\}$, where both Ω_1 and Ω_2 are closed around $\bar{x}$. Then for every $\varepsilon > 0$ there are*

$$x_j \in \Omega_j \cap (\bar{x} + \varepsilon B_X) \quad \textit{and} \quad x_j^* \in \hat{N}(x_j; \Omega_j) \quad j = 1, 2,$$

satisfying the relations

$$1-\varepsilon \le \|x_1^*\|_* + \|x_2^*\|_* \le 1+\varepsilon \qquad \textit{and} \qquad \|x_1^* + x_2^*\|_* \le \varepsilon.$$

As explained in Remark 5.5.5, the extremal principle can be considered as a nonconvex variational counterpart of the classical separation principle for convex sets, similar like the separation theorems for nonconvex sets (see Theorem 5.2.9 in Sect. 5.2). It plays in fact a fundamental role in variational analysis similar to that played by the convex separation and Bishop-Phelps theorems under convexity assumptions; see the books by Mordukhovich [430] and [431] for more details and numerous applications.

In order to derive necessary optimality conditions for minimizers of set-valued optimization problems with general geometric constraints by using suitable rules of subdifferential/coderivative calculus we need certain additional assumptions, especially we assume normal compactness properties of sets and mappings, which are automatically fulfilled in finite dimensional spaces while are indispensably needed in infinite-dimensional spaces due to the natural lack of compactness therein. Among the advantages of the underlying sequential normal compactness (SNC) properties recalled below one gets important SNC calculus results ensuring the preservation of these properties under various operations and describing broad classes of sets and mappings in infinite dimensions for which these properties are fulfilled; see Mordukhovich [430] for more details, discussions, and applications. These properties are used in the framework of Asplund spaces, and so the given definitions are specified to this setting; compare [430] for appropriate modifications in general Banach spaces.

Recall that a set $S \subset X$ is **sequentially normally compact (SNC) at** $\overline{x} \in S$ if for any sequences $x_k \overset{S}{\to} \overline{x}$ and $x_k^* \overset{w^*}{\to} 0$ with $x_k^* \in \hat{N}(x_k; S)$, $k \in \mathbb{N}$, we have $\|x_k^*\|_* \longrightarrow 0$ as $k \longrightarrow +\infty$. A set-valued mapping $F : X \rightrightarrows Y$ is *SNC* at $(\overline{x}, \overline{y}) \in \operatorname{graph} F$ if its graph is *SNC* at this point.

For deriving the necessary optimality conditions in Theorem 12.11.3 and corresponding applications in welfare economics (see Sect. 15.3) the following extension of Theorem 5.5.9 is important (see [430, Lemma 5.58]). The result is shown under certain partially sequentially normally compactness conditions. Recall that a set $\Omega \subset X$ in the product space $X = \prod_{i=1}^n X_i$ is **partially sequentially normally compact (PSNC)** at $\overline{x} \in \operatorname{cl}\Omega$ with respect to $\{X_i \mid i \in I\}$ as $I \subset \{1, \cdots, n\}$ (or with respect to the indices I) if for any sequences of elements $(x_k, x_k^*) \in X \times X^*$ with $x_k = (x_{1k}, \cdots, x_{nk})$ and $x_k^* = (x_{1k}^*, \cdots, x_{nk}^*)$ satisfying

$$x_k \overset{\Omega}{\to} \overline{x} \quad \text{and} \quad x_k^* \in \hat{N}(x_k; \Omega) \quad \text{for all} \quad k \in \mathbb{N} \tag{5.43}$$

the following implication holds

$$\left\{x_{ik}^* \overset{w^*}{\to} 0,\ i \in I, \text{ and } \|x_{ik}^*\|_* \to 0, i \in \{1, \cdots, n\} \setminus I\right\} \quad \Rightarrow \quad \|x_{ik}^*\|_* \to 0, i \in I.$$

The **strong PSNC property of Ω at $\overline{x}$** with respect to $\{X_i \mid i \in I\}$ with $I \subset \{1, \cdots, n\}$ says that for any sequences $\{(x_k, x_k^*)\}$ satisfying (5.43) we have the implication

$$\left\{ x_{ik}^* \overset{w^*}{\to} 0,\ i \in \{1, \cdots, n\} \right\} \quad \Rightarrow \quad ||x_{ik}^*||_* \to 0, i \in I.$$

Both PSNC and strong PSNC properties do not depend on the product structure and reduce to the **sequentially normally compact (SNC) property** of Ω at $\overline{x}$ in the extreme case of $I = \{1, \cdots, n\}$. Sufficient conditions ensuring the fulfillment of these properties and their preservation under various operations are given in Mordukhovich [430]. Especially, a set Ω is SNC at $\overline{x} \in \Omega$ if it is compactly epi-Lipschitzian (CEL) around this point. Both properties are fulfilled for a broad class of Banach spaces; see, e.g., [66, 278, 430] for more details and references.

The following formulation of the (extended) extremal principle is given in Bao and Mordukhovich [28, Section 2] (compare [431, Lemma 5.58]) and will be applied in Sect. 12.11 in order to show first order necessary conditions for localized minimizers in Theorem 12.11.3 as well as in Sect. 15.3 for deriving an extended welfare theorem (see Theorem 15.3.5).

Theorem 5.5.10 (Extended Extremal Principle).

Let $\overline{x}$ be a local extremal point of the set system $\{\Omega_1, \Omega_2\}$, where both Ω_1 and Ω_2 are closed around $\overline{x}$ in the product $\prod_{i=1}^n X_i$ of Asplund spaces X_i, $i = 1, \cdots, n$. Take two index sets $I, J \subset \{1, \cdots, n\}$ with $I \cup J = \{1, \cdots, n\}$ and $I \cap J = \emptyset$. Furthermore, suppose that one of the following PSNC conditions is satisfied for Ω_1 and Ω_2:

- *Ω_1 is PSNC at $\overline{x}$ with respect to I and Ω_2 is strongly PSNC at $\overline{x}$ with respect to J;*
- *Ω_1 is strongly PSNC at $\overline{x}$ with respect to I and Ω_2 is PSNC at $\overline{x}$ with respect to J.*

Then there exists an element $x^* \in X^*$ such that

$$0 \neq x^* \in N(\overline{x}; \Omega_1) \cap (-N(\overline{x}; \Omega_2)).$$

Proof. Consider the extremal system $\{\Omega_1, \Omega_2, \overline{x}\}$. Under the given assumptions we get with Theorem 5.5.7 that the approximate extremal principle holds in $(\prod_{i \in I} X_i) \times (\prod_{j \in J} X_j)$ with the corresponding index sets I and J therein satisfying the condition $I \cap J = \emptyset$. Following the line of the proof in [431, Lemma 5.58] we get the desired result. □

Chapter 6
Hahn-Banach Type Theorems

In this chapter we present a generalization of the Hahn–Banach–Kantorovich extension theorem to K-convex set-valued maps, as well as Yang's extension theorem. We also present classical separation theorems for convex sets, the core convex topology on a linear space, and a criterion for the convexity of the cone generated by a set.

In the sequel X, Y, Z are real linear spaces. Recall that the class of linear operators from X into Y is denoted by $L(X, Y)$; we set $X' := L(X, \mathbb{R})$. In the case in which X is a topological vector space, as usual, we denote by X^* its topological dual; of course, $X^* \subseteq X'$. Moreover, if X and Y are topological vector spaces, $\mathscr{L}(X, Y)$ denotes the set of continuous linear operators from X into Y. We consider $K \subseteq Z$ a proper (i.e. $\{0\} \neq K \neq Z$) convex cone. The cone K induces a quasi-order on Z, denoted $\leq_K$ or simply $\leq$ if there is no risk of confusion. So, for $z_1, z_2 \in Z$ one has $z_1 \leq z_2$ (or equivalently $z_2 \geq z_1$) if $z_2 - z_1 \in K$. We denote by core A the algebraic interior (or core) of $A \subseteq X$ and by icr A the relative algebraic interior (or intrinsic core) of A, that is icr A is the algebraic interior of A with respect to the affine hull aff A of A. It follows that core $A \neq \emptyset$ if and only if aff $A = X$ and icr $A \neq \emptyset$; in this case (that is aff $A \neq X$) icr $A =$ core A. Of course, if $A \subseteq X$ is an affine set (that is, aff $A = A$) then icr $A = A$. Recall that for A a convex set one has

$$a \in \operatorname{core} A \Leftrightarrow [\forall x \in X,\ \exists \lambda \in \mathbb{P} : a + \lambda x \in A], \tag{6.1}$$

and

$$a \in \operatorname{icr} A \Leftrightarrow [\forall x \in A,\ \exists \lambda \in \mathbb{P} : (1 + \lambda)a - \lambda x \in A], \tag{6.2}$$

where $\mathbb{P} :=]0, \infty[\subseteq \mathbb{R}$. When A is convex we have

$$a_0 \in \operatorname{icr} A,\ a \in A,\ \lambda \in]0, 1[\Rightarrow (1 - \lambda)a_0 + \lambda a \in \operatorname{icr} A; \tag{6.3}$$

A.A. Khan et al., *Set-valued Optimization*, Vector Optimization,
DOI 10.1007/978-3-642-54265-7_6

in particular, $\operatorname{icr} A$ is a convex set. Indeed, take $a_0 \in \operatorname{icr} A$, $a \in A$ and $\lambda \in]0,1[$, and set $\overline{a} := (1-\lambda)a_0 + \lambda a$. Consider $x \in A$. Because $a_0 \in \operatorname{icr} A$, by (6.2), there exists $\mu > 0$ such that $\overline{x} := (1+\mu)a_0 - \mu x \in A$. Then $\alpha := (1-\lambda)/(1+\mu\lambda) \in]0,1[$, $\gamma := \mu\alpha > 0$, and $(1+\gamma)\overline{a} - \gamma x = \alpha\overline{x} + (1-\alpha)a \in A$. Using again (6.2) we get $\overline{a} \in \operatorname{icr} A$.

When $K_0 := \operatorname{core} K \neq \emptyset$, we use $z_1 < z_2$ (or $z_2 > z_1$) when $z_2 - z_1 \in K_0$, and $z_1 \not< z_2$ (or $z_2 \not> z_1$) if $z_2 - z_1 \notin K_0$.

As in Sect. 2.1, we extend Z to $Z^\bullet := Z \cup \{-\infty, +\infty\}$, $\pm\infty \notin Z$, $-(\pm\infty) = \mp\infty$, and consider that $-\infty \leq z \leq +\infty$ (even $-\infty < z < +\infty$ if $K_0 \neq \emptyset$) for all $z \in Z$; moreover, $z + (\pm\infty) := \pm\infty$, $0 \cdot (+\infty) := +\infty$, $0 \cdot (-\infty) := 0$, $t \cdot (\pm\infty) := \pm\infty$ for all $z \in Z$ and $t \in \mathbb{P}$.

For $A, B \subseteq Z^\bullet$ and ρ a relation on $Z^\bullet$ (for example ρ can be $\leq$, $<$, $\not<$) we write $A\rho B$ whenever $a\rho b$ for all $a \in A$ and $b \in B$. If $a \in X$, we write $a\rho B$ if $\{a\}\rho B$.

Again, as in Sect. 2.1, for a function $f : X \to Z^\bullet$, $\operatorname{dom} f := \{x \in X \mid f(x) \neq +\infty\}$ and $\operatorname{epi} f := \{(x,z) \in X \times Z \mid f(x) \leq z\}$; f is proper if $\operatorname{dom} f \neq \emptyset$ and f does not take the value $-\infty$, while f is convex if $\operatorname{epi} f$ is convex. We say that $f : X \to Z^\bullet$ is sublinear if f is proper, $f(0) = 0$, $f(tx) = tf(x)$ and $f(x + x') \leq f(x) + f(x')$ for all $x, x' \in X$ and $t \in \mathbb{P}$; hence $\operatorname{epi} f$ is a convex cone containing the origin when f is sublinear.

In the sequel we identify a set-valued map $\Gamma : E \rightrightarrows F$ (that is, a function $\Gamma : E \to 2^F$) with its graph $\operatorname{graph} \Gamma := \{(x,y) \in E \times F \mid y \in \Gamma(x)\}$. So, several times in the sequel, having a subset A of $E \times F$ we interpret A as the graph of a set-valued map from E to F, and so $A(x) := \{y \in F \mid (x,y) \in A\}$.

6.1 The Hahn–Banach–Kantorovich Theorem

In this section the proper convex cone $K \subseteq Z$ is such that (Z, K) has the **least upper bound property** (l.u.b.p., for short), i.e. every nonempty and upper bounded set has a least upper bound; of course, in this situation every nonempty and lower bounded set has a greatest lower bound. If the nonempty set $B \subseteq Z$ is bounded from below (resp. above), we denote by $\inf B$ ($\sup B$) a greatest lower (resp. least upper) bound of B; if B is not bounded below (resp. above) we set $\inf B := -\infty$ (resp. $\sup B := +\infty$). We also use the notation $\inf\emptyset := +\infty$ and $\sup\emptyset := -\infty$.

We first give an extension theorem for linear mappings bounded above by convex set-valued maps. As we shall see below, the result is equivalent to the classical Hahn–Banach–Kantorovich theorem when the intrinsic core of the domain of the involved set-valued map is nonempty.

Theorem 6.1.1. *Let $\Gamma : X \rightrightarrows Z$ be a K-convex set-valued map, $X_0 \subseteq X$ a linear subspace and $T_0 \in L(X_0, Z)$. Suppose that $0 \in \operatorname{icr}(\operatorname{dom}\Gamma - X_0)$ and $T_0 x \leq \Gamma(x)$ for all $x \in X_0$ ($\cap \operatorname{dom}\Gamma$). Then there exists $T \in L(X, Z)$ such that $T|_{X_0} = T_0$ and $Tx \leq \Gamma(x)$ for all $x \in X$.*

Proof. Consider $\overline{X} := \operatorname{aff}(\operatorname{dom}\Gamma - X_0)$. Because $0 \in \operatorname{icr}(\operatorname{dom}\Gamma - X_0) \subseteq \operatorname{dom}\Gamma - X_0$, $\overline{X}$ is a linear space.

The case $\overline{X} = X$; hence $0 \in \operatorname{dom}(\operatorname{dom}\Gamma - X_0)$. Consider the set

$$\mathscr{F} := \big\{(E,T) \mid X_0 \subseteq E = \operatorname{span} E \subseteq X,\ T \in L(E,Z),\\ T|_{X_0} = T_0,\ Tx \leq \Gamma(x)\ \forall x \in E\big\};$$

clearly, $(X_0,T_0) \in \mathscr{F}$. For $(E_1,T_1),(E_2,T_2) \in \mathscr{F}$ we set $(E_1,T_1) \leq (E_2,T_2)$ if $E_1 \subseteq E_2$ and $T_2|_{E_1} = T_1$. Clearly, $\leq$ is a partial order on $\mathscr{F}$. Take $\mathscr{C} := \{(E_i,T_i) \mid i \in I\}$ a chain in $\mathscr{F}$ ($I \neq \emptyset$). Define $E := \cup_{i\in I} E_i$ and $T : E \to Z$ defined by $Tx := T_i x$ if $x \in E_i$. Because $\mathscr{C}$ is a chain, E is a linear subspace of X with $X_0 \subseteq E$. Moreover, T is a well defined linear operator and $Tx \leq z$ for $(x,z) \in \Gamma \cap (E \times Z)$; this is valid because x being in E, there exists $i \in I$ with $x \in E_i$, and so $(x,z) \in \Gamma \cap (E_i \times Z)$ and $Tx = T_i x$. It follows that $(E,T) \in \mathscr{F}$ and (E,T) is an upper bound for $\mathscr{C}$. Using Zorn's lemma, $(\mathscr{F},\leq)$ has maximal elements; let $(E,T) \in \mathscr{F}$ be maximal. If $E = X$ the linear operator T verifies the conditions in the conclusion of the theorem.

Assume that $E \neq X$. Then there exists $\overline{x} \in X \setminus E$. Because $X_0 \subseteq E$ we have that $\operatorname{dom}\Gamma - X_0 \subseteq \operatorname{dom}\Gamma - E$, and so $0 \in \operatorname{core}(\operatorname{dom}\Gamma - E)$. Set $E_1 := E + \mathbb{R}\overline{x}$. Clearly $E \subseteq E_1$, the inclusion being strict. Since $0 \in \operatorname{core}(\operatorname{dom}\Gamma - E)$, there exists $\lambda \in \mathbb{P}$ such that $\pm\lambda\overline{x} \in \operatorname{dom}\Gamma - E$, and so there exist $x_1 \in E$ and $z_1 \in Z$ such that $(x_1 + \lambda\overline{x}, z_1) \in \Gamma$ and $x_2 \in E$ and $z_2 \in Z$ such that $(x_2 - \lambda\overline{x}, z_2) \in \Gamma$. Set $A := \operatorname{epi}\Gamma$. It follows that the sets

$$B_1 := \left\{ \frac{z_1 - Tx_1}{\lambda_1} \;\middle|\; x_1 \in E,\ z_1 \in Z,\ \lambda_1 \in \mathbb{P} \;:\; (x_1 + \lambda_1\overline{x}, z_1) \in A \right\}\\ = \{z_1 - Tx_1 \mid x_1 \in E,\ z_1 \in Z \;:\; (x_1 + \overline{x}, z_1) \in \mathbb{P}A\},$$

and

$$B_2 := \left\{ \frac{Tx_2 - z_2}{\lambda_2} \;\middle|\; x_2 \in E,\ z_2 \in Z,\ \lambda_2 \in \mathbb{P} \;:\; (x_2 - \lambda_2\overline{x}, z_2) \in A \right\}\\ = \{Tx_2 - z_2 \mid x_2 \in E,\ z_2 \in Z \;:\; (x_2 - \overline{x}, z_2) \in \mathbb{P}A\},$$

are nonempty (and convex) sets. Moreover,

$$\forall\, b_1 \in B_1,\ \forall b_2 \in B_2 \;:\; b_1 \geq b_2.$$

Indeed, let $b_1 \in B_1$ and $b_2 \in B_2$; then $b_1 = z_1 - Tx_1$ and $b_2 = Tx_2 - z_2$ with $x_1, x_2 \in E$, $z_1, z_2 \in Z$ and $(x_1 + \overline{x}, z_1), (x_2 - \overline{x}, z_2) \in \mathbb{P}A$. It follows that $(x_1 + x_2, z_1 + z_2) \in \mathbb{P}A$ (because this set is a convex cone), and $x_1 + x_2 \in E$. From the hypothesis, we have that $T(x_1 + x_2) \leq z_1 + z_2$, which is equivalent to $b_2 \leq b_1$. It follows that B_2 is bounded from above and any element of B_1 is an upper bound for B_2. Take $\overline{z} \in Z$ a least upper bound of B_2; then $b_2 \leq \overline{z} \leq b_1$ for all $b_1 \in B_1$ and $b_2 \in B_2$. Let us define T_1 by $T_1(x+\lambda\overline{x}) := Tx+\lambda\overline{z}$ (for $x \in E$ and $\lambda \in \mathbb{R}$); we have that $T_1 : E_1 \to Z$ is linear, $T_1|_E = T$ and $T_1x \leq z$ for all $(x,z) \in \Gamma \cap (E_1 \times Z)$. Let us prove the last assertion. So, take $(x,z) \in \Gamma$ such that $x := x_0 + \lambda\overline{x}$ with $x_0 \in E$ and $\lambda \in \mathbb{R}$. If $\lambda = 0$ then $(x,z) = (x_0,z) \in \Gamma \cap (E \times Z)$, and so $T_1x = Tx_0 \leq z$. If $\lambda < 0$ then $\lambda_2 := -\lambda > 0$ and $T_1x = Tx_0 - \lambda_2\overline{z}$. Setting $b_2 := \lambda_2^{-1}(Tx_0 - z)$ we have that $b_2 \in B_2$. But

$$T_1x \leq z \iff Tx_0 - \lambda_2\overline{z} \leq z \iff \frac{Tx_0 - z}{\lambda_2} \leq \overline{z} \iff b_2 \leq \overline{z}.$$

Hence our assertion is true in this case by the choice of $\overline{z}$. The case $\lambda > 0$ is proven similarly. It follows that $(E_1, T_1) \in \mathscr{F}$ and $(E,T) \preceq (E_1, T_1)$. Since (E,T) is maximal we obtain that $(E,T) = (E_1,T_1)$, whence the contradiction $E = E_1$. Therefore our assumption that $E \neq X$ is false.

The case $\overline{X} \neq X$. Of course, $\operatorname{dom} \Gamma \subseteq \overline{X}$. Taking $x_0 \in X_0 \cap \operatorname{dom} \Gamma$, we have that $X_0 = x_0 - X_0 \subseteq \overline{X}$. Applying the first case we find $\overline{T} : \overline{X} \to Z$ a linear operator such that $\overline{T}|_{X_0} = T$ and $\overline{T}x \leq z$ for all $(x,z) \in \Gamma \subseteq \overline{X} \times Z$. Taking $\overline{Y}$ a linear subspace of X such that $X = \overline{X} \oplus \overline{Y}$ (that is $X = \overline{X} + \overline{Y}$ and $\overline{X} \cap \overline{Y} = \{0\}$) and $T : X \to Z$ defined by $T(\overline{x} + \overline{y}) := \overline{T}(\overline{x})$ for $\overline{x} \in \overline{X}$, $\overline{y} \in \overline{Y}$, T verifies the conclusion. The proof is complete. □

Remark 6.1.2. In the proof of Theorem 6.1.1 we used only the fact that $\mathbb{P} \cdot \operatorname{epi} \Gamma$ is convex (or, equivalently, $\mathbb{R}_+ \cdot \operatorname{epi} \Gamma$ is a convex cone) and not that $\operatorname{epi} \Gamma$ itself is convex.

Remark 6.1.3. The preceding result is obtained by Brumelle [81, Lemma 1] for $X_0 \neq \{0\}$ and Γ convex with $0 \in \Gamma(0)$ and $0 \in \operatorname{core}(\operatorname{dom} \Gamma)$, and by Nehse [439, Theorem 2]. Moreover, Chen and Wang [93, Theorem 1] obtained Theorem 6.1.1 for Γ convex-valued with $\operatorname{dom} \Gamma = X$ and $0 \in \Gamma(0)$. The proof here is from Zălinescu [615].

Recently, Du (see [147, Theorem 2.1]) obtained the following result (for the case in which $X_0 \cap \operatorname{core}(\operatorname{dom} \Gamma) \neq \emptyset$).

Theorem 6.1.4. *Let $\Gamma : X \rightrightarrows Z$ be a K-convex set-valued map, $X_0 \subseteq X$ a linear subspace and $\phi : X_0 \rightrightarrows Z$ a $(-K)$-convex set-valued map with $\operatorname{dom} \phi = X_0$. Suppose that $0 \in \operatorname{icr}(\operatorname{dom} \Gamma - X_0)$ and $\phi(x) \leq \Gamma(x)$ for all $x \in X_0$. Then there exists a $(-K)$-convex set-valued map $\Phi : X \rightrightarrows Z$ with $\operatorname{dom} \Phi = X$ such that $\Phi|_{X_0} = \phi$ and $\Phi(x) \leq \Gamma(x)$ for all $x \in X$.*

Proof. Consider the set-valued map $\tilde{\Gamma} : X \rightrightarrows Z$ whose graph is $\operatorname{graph} \tilde{\Gamma} := \operatorname{epi}_K \Gamma - \operatorname{epi}_{-K} \phi$. Clearly, $\operatorname{graph} \tilde{\Gamma} = \operatorname{epi}_K \tilde{\Gamma}$ is convex, and so $\tilde{\Gamma}$ is K-convex; moreover, $\operatorname{dom} \tilde{\Gamma} = \operatorname{dom} \Gamma - \operatorname{dom} \phi = \operatorname{dom} \Gamma - X_0$ and $\tilde{\Gamma}(0) = \cup_{x \in X_0}\big(\Gamma(x) - \phi(x) + K\big)$. It follows that $0 \in \operatorname{icr}(\operatorname{dom} \tilde{\Gamma})$ and $0 \leq \tilde{\Gamma}(0)$. Applying Theorem 6.1.1 for $\tilde{\Gamma}$ and $T_0 := 0 \in L(\{0\}, Z)$, we get $T \in L(X,Z)$ such that $Tx \leq \tilde{\Gamma}(x)$ for every $x \in X$. Hence

$$Tx - Tx' \le z - z' \quad \forall x \in \operatorname{dom} \Gamma,\ \forall z \in \Gamma(x),\ \forall x' \in X_0,\ \forall z' \in \phi(x'). \tag{6.4}$$

Let $X_1 \subseteq X$ be a linear space such that $X = X_0 \oplus X_1$ and define $\Phi : X \rightrightarrows Z$ by $\Phi(x_0 + x_1) := \phi(x_0) + Tx_1$ for $x_0 \in X_0$ and $x_1 \in X_1$. Clearly, $\operatorname{dom} \Phi = X$ and $\operatorname{epi}_{-K} \Phi = \operatorname{epi}_{-K} \phi + \operatorname{graph} T|_{X_1}$, and so Φ is $(-K)$-convex. Fix $x \in \operatorname{dom} \Gamma$, $z \in \Gamma(x)$ and $z' \in \Phi(x)$. Then $x = x_0 + x_1$ with $x_0 \in X_0$, $x_1 \in X_1$, and $z' = z_0 + Tx_1$ with $z_0 \in \phi(x_0)$. Using (6.4) we get $Tx_1 \le z - z_0$, or, equivalently, $z' \le z$. It follows that $\Phi(x) \le \Gamma(x)$ for every $x \in X$. □

Remark 6.1.5. Theorem 6.1.4 looks more general than Theorem 6.1.1. However it is not possible to obtain Theorem 6.1.1 from Theorem 6.1.4 because instead of a linear operator T with $T|_{X_0} = T_0$ we get a $(-K)$-convex set-valued map having as domain the entire space X. However, by the procedure used in the proof of Theorem 6.1.4 we can obtain Theorem 6.1.1 from its particular case in which $X_0 := \{0\}$.

Corollary 6.1.6. *Let $P \subseteq X$ be a convex cone and let $Y \subseteq X$ be a linear subspace. If $S_0 \in L(Y, Z)$ is monotone, that is $x_1, x_2 \in Y$ and $x_1 \le_P x_2$ imply $S_0x_1 \le S_0x_2$, and $Y - P$ is a linear subspace, then there exists $S \in L(X, Z)$ monotone with $S|_Y = S_0$.*

Proof. Consider $\Gamma : X \rightrightarrows Z$ with $\operatorname{graph} \Gamma := \operatorname{graph} S_0 - P \times \{0\}$; clearly, Γ is K-convex and $\operatorname{dom} \Gamma = Y - P$. Take $X_0 := \{0\}$ and $T_0 = 0 \in L(X_0, Z)$. Then $\operatorname{dom} \Gamma - X_0 = Y - P$, and so $0 \in \operatorname{icr}(\operatorname{dom} \Gamma - X_0)$. Moreover, if $(x, z) \in \operatorname{graph} \Gamma \cap (X_0 \times Z)$, then $x = 0 = y - v$ with $y \in Y$, $v \in P$ and $z = S_0 y$. Hence $y = v \ge_P 0$, whence $z = S_0 y \ge S_0 0 = 0$. Hence $0 = T_0 x \le z$. Applying Theorem 6.1.1 we get $T \in L(X, Z)$ such that $Tx \le z$ for all $(x, z) \in \operatorname{graph} S_0 - P \times \{0\}$, that is, $y \in Y$ and $v \in P$ imply $T(y - v) \le S_0 y$. In particular, $Ty - S_0 y \in K \cap (-K)$ for every $y \in Y$ and $Tv \ge S_0 0 = 0$ for every $v \in P$, that is T is monotone. Let us consider $Y' \subseteq X$ a linear space such that $X = Y \oplus Y'$ and define $T_0 : X \to Z$ by $T_0(y + y') := S_0 y - Ty$ for $y \in Y$ and $y' \in Y'$; then $T_0 x \in K \cap (-K)$ for every $x \in X$. Taking $S := T$ the conclusion follows. □

Corollary 6.1.7. *Let $f : X \to Z^\bullet$ be a proper convex operator, X_0 be a linear subspace of X and $T_0 \in L(X_0, Z)$. Suppose that $T_0 x \le f(x)$ for every $x \in X_0$. If $0 \in \operatorname{icr}(\operatorname{dom} f - X_0)$ then there exists $T \in L(X, Z)$ such that $T|_{X_0} = T_0$ and $Tx \le f(x)$ for every $x \in X$.*

Proof. In Theorem 6.1.1 take $\operatorname{graph} \Gamma := \operatorname{epi} f$; of course, the hypothesis of the theorem holds, so that there exists $T \in L(X, Z)$ such that $T|_{X_0} = T_0$ and $Tx \le \Gamma(x)$ for every $x \in X$. Since $f(x) \in \Gamma(x)$ for $x \in \operatorname{dom} f = \operatorname{dom} \Gamma$ and $f(x) = +\infty$ for $x \in X \setminus \operatorname{dom} f$, the conclusion follows. □

Remark 6.1.8. Corollary 6.1.6 is the implication (LUB)⇒(MP) in [127, Theorem VI.3.1]. For f sublinear with $\operatorname{dom} f = X$, Corollary 6.1.7 is the well-known Kantorovich' generalization of the Hahn–Banach extension theorem, and is the implication (LUB)⇒(HB) in [127, Theorem VI.3.1]; moreover, Corollary 6.1.7

for f sublinear is equivalent to [412, Theorem 3 (3)], which, at its turn, covers [412, Theorem 3 (1)]. Corollary 6.1.7 for f with $X_0 \cap \operatorname{core}(\operatorname{dom} f) \neq \emptyset$ is [629, Theorem 2.1].

Assume that f is sublinear; then condition $0 \in \operatorname{icr}(\operatorname{dom} f - X_0)$ in Corollary 6.1.7 is equivalent to (H) "$X_0 + \operatorname{dom} f$ is a linear subspace of X". We emphasize the importance of condition (H) in Corollary 6.1.7. Without this condition the conclusion of Corollary 6.1.7 could be false even for $Z = \mathbb{R}$ and $\dim X < \infty$; see [534] and [7] for interesting counterexamples.

Using Theorem 6.1.1 we get easily the following separation theorem obtained by Nehse in [439, Theorem 1].

Proposition 6.1.9. *Let $A, B \subseteq X \times Z$ be such that $\mathbb{R}_+(A - B)$ is convex. Assume that $0 \in \operatorname{icr}(\operatorname{Pr}_X(A - B))$ and $A(x) \leq B(x)$ for all $x \in \operatorname{Pr}_X(A) \cap \operatorname{Pr}_X(B)$. Then there exist $T \in L(X, Z)$ and $z_0 \in Z$ such that*

$$\forall (x_1, z_1) \in A,\ \forall (x_2, z_2) \in B : z_1 - Tx_1 \leq z_0 \leq z_2 - Tx_2. \tag{6.5}$$

Proof. Consider the set $C := B - A \subseteq X \times Z$. Then $\operatorname{Pr}_X(C) = \operatorname{Pr}_X(A) - \operatorname{Pr}_X(B)$ and if $(0, z) \in C$ then $z = z_1 - z_2$ with $z_1 \in A(x)$, $z_2 \in B(x)$ for some $x \in \operatorname{Pr}_X(A) \cap \operatorname{Pr}_X(B)$; hence $0 \leq z$. Taking $X_0 := \{0\}$ and $T(0) := 0$, we can apply Theorem 6.1.1 (see also Remark 6.1.2) for graph $\Gamma := C$ and T_0. Therefore, there exists $T \in L(X, Z)$ such that $Tx \leq z$ for every $(x, z) \in C$. It follows that

$$\forall (x_1, z_1) \in A,\ \forall (x_2, z_2) \in B : z_1 - Tx_1 \leq z_2 - Tx_2.$$

Taking $z_0 := \sup \{z_1 - Tx_1 \mid (x_1, z_1) \in A\}$, the conclusion follows. □

Before stating the next result recall that the subdifferential $\partial f(x_0)$ of the proper operator $f : X \to Z^\bullet$ at $x_0 \in \operatorname{dom} f$ is the set of those $T \in L(X, Z)$ such that

$$\forall x \in X : Tx - Tx_0 \leq f(x) - f(x_0).$$

(In the continuous case, $\partial f(x_0) \cap \mathscr{L}(X, Y)$ is denoted by $\partial^s f(x_0)$; see Sect. 7.1.)

Corollary 6.1.10. *Let $f : X \to Z^\bullet$ be a proper convex operator and consider $x_0 \in \operatorname{icr}(\operatorname{dom} f)$. Then $\partial f(x_0)$ is nonempty.*

Proof. Consider $g : X \to Z^\bullet$, $g(x) = f(x_0 + x) - f(x_0)$. Then g is a convex operator with $0 \in \operatorname{icr}(\operatorname{dom} g)$. Consider now graph $\Gamma = \operatorname{epi} g$, $X_0 = \{0\}$ and $T_0(0) := 0$; Γ is a convex set-valued map. As $\operatorname{dom} \Gamma = \operatorname{dom} g$, we have that $0 \in \operatorname{icr}(\operatorname{dom} \Gamma - X_0)$, and, of course, $(x, z) \in \Gamma \cap (X_0 \times Z)$ implies $T_0 x = 0 \leq z$. Applying Theorem 6.1.1 we get $T \in L(X, Z)$ such that $Tx \leq z$ for every $(x, z) \in \Gamma$. In particular, if $x \in \operatorname{dom} f$ then $(x - x_0, f(x) - f(x_0)) \in \operatorname{epi} g$, whence $T(x - x_0) \leq f(x) - f(x_0)$. □

Note that Corollary 6.1.10 can be viewed as a particular case of the next result; just take $g : X \to Z^\bullet$ defined by $g(x_0) := -f(x_0)$ and $g(x) := +\infty$ for $x \in X \setminus \{x_0\}$. The next result is the sandwich theorem proved by Zowe in [629, Theorem 3.1] (see also [610, Corollary 2.6]).

Corollary 6.1.11. *Let $f, g : X \to Z^\bullet$ be proper convex operators. Suppose that $0 \in \operatorname{icr}(\operatorname{dom} f - \operatorname{dom} g)$ and that $f(x) \geq -g(x)$ for all $x \in \operatorname{dom} f \cap \operatorname{dom} g$. Then there exists $T \in L(X, Z)$ and $z_0 \in Z$ such that*

$$\forall\, x \in X \ : \ -g(x) \leq Tx + z_0 \leq f(x).$$

Proof. Consider $\Gamma : X \rightrightarrows Z$ with

$$\operatorname{graph} \Gamma := \{(x, z) \in X \times Z \mid f(x) \leq z\} + \{(-x', z') \in X \times Z \mid g(x') \leq z'\}.$$

Then Γ is convex and $\operatorname{dom} \Gamma = \operatorname{dom} f - \operatorname{dom} g$. The hypothesis of Theorem 6.1.1 holds for $X_0 = \{0\}$ and $T_0(0) := 0$. Indeed, if $(0, z) \in \Gamma$ then $(0, z) = (x, z') + (-x, z'')$, with $f(x) \leq z'$ and $g(x) \leq z''$; it follows that $T_0(0) = 0 \leq f(x) + g(x) \leq z' + z'' = z$. Therefore, there exists $T \in L(X, Z)$ such that $Tx \leq z$ for $(x, z) \in \Gamma$. In particular, for $x_1 \in \operatorname{dom} f$ and $x_2 \in \operatorname{dom} g$ we get $T(x_1 - x_2) \leq f(x_1) + g(x_2)$, which yields $-g(x_2) - Tx_2 \leq f(x_1) - Tx_1$ for all $x_1 \in \operatorname{dom} f$ and $x_2 \in \operatorname{dom} g$. It follows that

$$\sup_{x_2 \in \operatorname{dom} g} (-g(x_2) - Tx_2) \leq \inf_{x_1 \in \operatorname{dom} f} (f(x_1) - Tx_1).$$

Taking z_0 between these two values, we get the desired conclusion. □

Note that Theorem 6.1.1 follows from Corollary 6.1.11 when $\operatorname{icr}(\operatorname{dom} \Gamma)$ is nonempty. To show this we need the following result (see [610, Proposition 1.5 (i)]).

Proposition 6.1.12. *Let $\Gamma : X \rightrightarrows Z$ be K-convex set-valued map. Assume that $\Gamma(x_0)$ is bounded below for some $x_0 \in \operatorname{icr}(\operatorname{dom} \Gamma)$. Then*

$$f_\Gamma : X \to Z^\bullet, \quad f_\Gamma(x) := \inf \Gamma(x), \tag{6.6}$$

is a proper convex operator with $\operatorname{dom} f_\Gamma = \operatorname{dom} \Gamma$.

Proof. From the definition of f_Γ we have that $\operatorname{dom} f_\Gamma = \operatorname{dom} \Gamma$. Let $x \in \operatorname{dom} \Gamma$. Because $\operatorname{dom} \Gamma = \operatorname{Pr}_X(\operatorname{epi} \Gamma)$ is convex and $x_0 \in \operatorname{icr}(\operatorname{dom} \Gamma)$, using (6.2), there exists $\lambda > 0$ such that $x' := (1 + \lambda)x_0 - \lambda x \in \operatorname{dom} \Gamma$. Fix $z' \in \Gamma(x')$ and take $z \in \Gamma(x)$ arbitrary. Then, by the convexity of $\operatorname{epi} \Gamma$, we have

$$\left(x_0, \frac{\lambda}{1 + \lambda} z + \frac{1}{1 + \lambda} z'\right) = \frac{\lambda}{1 + \lambda}(x, z) + \frac{1}{1 + \lambda}(x', z') \in \operatorname{epi} \Gamma,$$

and so $Z \ni f_\Gamma(x_0) \le \frac{\lambda}{1+\lambda}z + \frac{1}{1+\lambda}z'$. It follows that

$$(1+\lambda)\lambda^{-1} f_\Gamma(x_0) - z' \le z \quad \forall z \in \Gamma(x).$$

Hence $f_\Gamma(x) \in Z$, and so f_Γ is proper.

Take now $x_0, x_1 \in \operatorname{dom} f_\Gamma$ and $\lambda \in]0,1[$, and set $x_\lambda := (1-\lambda)x_0 + \lambda x_1$ ($\in \operatorname{dom}\Gamma = \operatorname{dom} f_\Gamma$). For $z_0 \in \Gamma(x_0)$ and $z_1 \in \Gamma(x_1)$ we have that

$$(x_\lambda, (1-\lambda)z_0 + \lambda z_1) = (1-\lambda)(x_0, z_0) + \lambda(x_1, z_1) \in \operatorname{epi}\Gamma,$$

whence $f_\Gamma(x_\lambda) \le (1-\lambda)z_0 + \lambda z_1$. Taking the infimum with respect to $z_0 \in \Gamma(x_0)$ and $z_1 \in \Gamma(x_1)$ we get $f_\Gamma(x_\lambda) \le (1-\lambda) f_\Gamma(x_0) + \lambda f_\Gamma(x_1)$. Hence f_Γ is convex. □

As mentioned above, Theorem 6.1.1 follows from Corollary 6.1.11 when $\operatorname{icr}(\operatorname{dom}\Gamma)$ is nonempty. Indeed, in this case there exists $x_0 \in X_0 \cap \operatorname{icr}(\operatorname{dom}\Gamma)$ [see Proposition 6.3.2 (iii)]. Taking f_Γ defined in (6.6), we have that $T_0(x_0) \le f_\Gamma(x_0) < +\infty$, and so $f_\Gamma(x_0) \in Z$. Using Proposition 6.1.12 we get $f_\Gamma(x) \in Z$ for every $x \in \operatorname{dom}\Gamma$. Taking $g(x) := -T_0 x$ for $x \in X_0$, $g(x) := +\infty$ for $x \in X \setminus X_0$ and applying Corollary 6.1.11 we get the conclusion of Theorem 6.1.1.

The preceding discussion shows that only the case when the intrinsic core of $\operatorname{dom}\Gamma$ is empty Theorem 6.1.1 is not covered by known results (however see [412, Theorem 3]). It is worth observing that under the hypotheses of Theorem 6.1.1 we have that $f_\Gamma(x) := \inf \Gamma(x) \in Z$ for every $x \in \operatorname{dom}\Gamma$ and $T_0 x \le f_\Gamma(x)$ for every $x \in X_0 \cap \operatorname{dom}\Gamma$. For this use the conclusion of Theorem 6.1.1. However a direct proof is not so obvious as that in the case $X_0 \cap \operatorname{icr}(\operatorname{dom}\Gamma) \ne \emptyset$.

Let us point out the following consequence of Theorem 6.1.1. Similar to [458, Definition 1.3], we say that $H : X \rightrightarrows Z$ is **affinelike** if there exist $T \in L(X,Z)$ and a nonempty convex set $M \subseteq Y$ such that $H(x) = T(x) + M$ for every $x \in X$.

Corollary 6.1.13. *Let $F : X \rightrightarrows Z$ be a convex set-valued map, $X_0 \subseteq X$ a linear subspace and $H_0 : X_0 \rightrightarrows Z$ an affinelike set-valued map. Suppose that $0 \in \operatorname{icr}(\operatorname{dom} F - X_0)$ and $F(x) - H_0(x) \subseteq K$ for every $x \in X_0 \cap \operatorname{dom} F$. Then there exists $H : X \rightrightarrows Z$ an affinelike set-valued map such that $H|_{X_0} = H_0$ and $F(x) - H(x) \subseteq K$ for every $x \in X \cap \operatorname{dom} F$.*

Proof. Let $T_0 \in L(X_0, Z)$ and $M \subseteq Y$ a nonempty convex set such that $H_0(x) = T_0(x) + M$ for every $x \in X_0$. Consider $\Gamma : X \rightrightarrows Z$ with

$$\operatorname{graph}\Gamma := \{(x, z - m + k) \mid z \in F(x),\ m \in M,\ k \in K\} = \operatorname{epi} F - \{0\} \times M.$$

One obtains immediately that Γ is convex, $\operatorname{dom}\Gamma = \operatorname{dom} F$ and $T_0(x) \le z$ for all $(x,z) \in \Gamma \cap (X_0 \times Z)$. Applying Theorem 6.1.1 we get $T \in L(X,Z)$ such that $T|_{X_0} = T_0$ and $T(x) \le z$ for all $(x,z) \in \Gamma$. Setting $H(x) := T(x) + M$, the conclusion follows. □

Concerning the affine and affinelike set-valued maps we mention the following result. The equivalence of (a), (b") and (c) (in an slightly different form) is mentioned in [572, 573].

Proposition 6.1.14. *Let $A \subseteq X \times Z$. Consider the following assertions.*

(a) *A is an affine manifold;*
(b) *$\lambda A(x) + (1-\lambda)A(x') \subseteq A(\lambda x + (1-\lambda)x')$ for all $x, x' \in X$ and $\lambda \in \mathbb{R}$;*
(b') *$\lambda A(x) + (1-\lambda)A(x') \subseteq A(\lambda x + (1-\lambda)x')$ for all $x, x' \in \mathrm{Pr}_X(A)$ and $\lambda \in \mathbb{R}$;*
(b") *$\lambda A(x) + (1-\lambda)A(x') = A(\lambda x + (1-\lambda)x')$ for all $x, x' \in \mathrm{Pr}_X(A)$ and $\lambda \in \mathbb{R}$;*
(c) *there exist linear subspaces $X_0 \subseteq X$, $Z_0 \subseteq Z$, a linear map $T_0 : X_0 \to Z$ and $(x_0, z_0) \in X \times Z$ such that $\mathrm{Pr}_X(A) = x_0 + X_0$ and $A(x) = T_0(x - x_0) + z_0 + Z_0$ for every $x \in \mathrm{Pr}_X(A)$;*
(d) *A is an affinelike set-valued map on a linear subspace $X_0 \subseteq X$.*

Then (a) $\Leftrightarrow$ (b) $\Leftrightarrow$ (b') $\Leftrightarrow$ (b") $\Leftrightarrow$ (c); *if $X_0 := \mathrm{Pr}_X(A)$ is a linear space then* (a) $\Rightarrow$ (d).

Proof. We may (and do) assume that $(0,0) \in A$; otherwise replace A by $A - (x_0, z_0)$ with $(x_0, z_0) \in A$.

The fact that (a) $\Leftrightarrow$ (b) $\Leftrightarrow$ (b') $\Leftarrow$ (b") is simple verification, (b) being a rewriting of (b') taking into account the fact that $B + \emptyset = \emptyset + B = \emptyset$ for $B \subseteq Z$.

(a) $\Rightarrow$ (c) Because A is a linear subspace, $X_0 := \mathrm{Pr}_X(A)$ and $Z_0 := A(0) = \mathrm{Pr}_Z\big(A \cap (\{0\} \times Z)\big)$ are linear spaces. Take $Z_1 \subseteq Z$ a linear subspace such that $Z = Z_0 \oplus Z_1$. Observe first that for $(x, z), (x, z') \in A$ we have $(0, z - z') \in A$, and so $z - z' \in Z_0$. Hence, taking $T_0 : X_0 \to Z$ defined by $T_0(x) := z_1$, where $z = z_0 + z_1 \in A(x)$ with $z_0 \in Z_0$, $z_1 \in Z_1$, T_0 is well defined. Since A is a linear subspace it follows immediately that T_0 is a linear operator and $A(x) = T_0(x) + Z_0$ for every $x \in X_0$.

(c) $\Rightarrow$ (b") follows by a simple verification.

(c) $\Rightarrow$ (d) is obvious when $X_0 := \mathrm{Pr}_X(A)$ is a linear space. □

Observe that (d) $\Rightarrow$ (a) if and only if the set M in the definition of an affinelike set-valued map is an affine set.

Note also that the spaces X_0 and Z_0 are uniquely determined by A; in fact X_0 is the parallel subspace of $\mathrm{Pr}_X(A)$ and Z_0 is the parallel subspace of $A(x_0)$ for some $x_0 \in \mathrm{Pr}_X(A)$. The equivalences of (a), (b") and (c) (the last one presented in the form there exist an affine mapping $T_0 : \mathrm{Pr}_X(A) \to Z$ and a linear subspace Z_0 of Z such that $A(x) = T_0(x) + Z_0$ for every $x \in \mathrm{Pr}_X(A)$) are provided in [573]. Note that when Z is ordered by the convex cone K, in [573] one says that the affine set-valued map $A \subseteq X \times Z$ is non-vertical when every two distinct elements from $A(x)$ are not comparable for each $x \in X$, which is equivalent, by [573, Lemma 3.1], to $Z_0 \cap K = \{0\}$ where Z_0 is provided by Proposition 6.1.14 (d).

We do not treat here the continuous versions of the preceding results, that is the case in which X, Z are topological linear space and the involved linear operators are taken to be continuous. This can be done as in Section 4 of [610].

6.2 Classical Separation Theorems for Convex Sets

We say that the nonempty subsets A, B of X are separated by the hyperplane $H_{x^*,\alpha} := \{x \in X \mid \langle x, x^*\rangle = \alpha\}$, where $x^* \in X' \setminus \{0\}$ and $\alpha \in \mathbb{R}$, if $\inf x^*(A) \geq \alpha \geq \sup x^*(B)$ [or $\inf x^*(B) \geq \alpha \geq \sup x^*(A)$]; the separation is proper or strict if, moreover, $A \cup B \not\subseteq H_{x^*,\alpha}$ or $\inf x^*(A) > \sup x^*(B)$ or $\inf x^*(B) > \sup x^*(A)$], respectively. Note that A and B can be (properly, strictly) separated (by some hyperplane) if and only if $A - B$ and $\{0\}$ can be (properly, strictly) separated. Moreover, note that the nonempty subsets A, B of X can be (properly, strictly) separated if and only if $\operatorname{conv} A$ and $\operatorname{conv} B$ can be (properly, strictly) separated.

For deriving separation theorems the Minkowski functional proves to be very useful.

Proposition 6.2.1. *Let $A \subseteq X$ be such that $0 \in \operatorname{core} A$ and $[0,1]A = A$. Then the Minkowski functional*

$$p_A : X \to \mathbb{R}, \quad p_A(x) := \inf\{\alpha > 0 \mid x \in \alpha A\},$$

is positively homogeneous and

$$\{x \in X \mid p_A(x) < 1\} \subseteq A \subseteq \{x \in X \mid p_A(x) \leq 1\}. \tag{6.7}$$

If, furthermore, A is convex, then A is sublinear and

$$\operatorname{core} A = \{x \in X \mid p_A(x) < 1\}. \tag{6.8}$$

Proof. For $x \in X$ set $\Lambda(x) := \{\alpha > 0 \mid x \in \alpha A\}$. Because $0 \in \operatorname{core} A$, we have that $\Lambda(x) \neq \emptyset$ for every $x \in X$ and $\Lambda(0) = \mathbb{P}$. Hence $p_A(0) = 0$ and $p_A(x) \in \mathbb{R}$ for $x \in X$. From the very definition of $\Lambda(x)$ we have that $\Lambda(\alpha x) = \alpha\Lambda(x)$ for all $x \in X$ and $\alpha \in \mathbb{P}$; hence p_A is positively homogeneous. Moreover, if $\alpha \in \Lambda(x)$ and $\alpha \leq \beta$ then $\beta \in \Lambda(x)$. Indeed, $\alpha\beta^{-1} \in [0,1]$, and so, because $A = [0,1]A$, we have that $\beta^{-1}x = \alpha\beta^{-1}\left(\alpha^{-1}x\right) \in A$; hence $\beta \in \Lambda(x)$. Hence, if $p_A(x) < 1$ then there exists $\alpha \in \Lambda(x)$ with $\alpha < 1$, and so $1 \in \Lambda(x)$, whence $x \in A$; if $x \in A$ then $1 \in \Lambda(x)$, and so $p_A(x) \leq 1$. Therefore, (6.7) holds.

Assume, moreover, that A is convex. Since $\Lambda(x) + \Lambda(x') \subseteq \Lambda(x + x')$ for all $x, x' \in X$, we obtain immediately that p_A is sub-additive, and so p_A is sublinear.

Let $x \in \operatorname{core} A$. By (6.1) there exists $\lambda > 0$ such that $x + \lambda x \in A$, and so $(1+\lambda)^{-1} \in \Lambda(x)$. Therefore, $p_A(x) \leq (1+\lambda)^{-1} < 1$. Conversely, assume that

$a \in X$ is such that $p_A(a) < 1$, and take $x \in X$. Then for $\lambda > 0$ sufficiently small, using the fact that p_A is sublinear, we have that $p_A(a+\lambda x) \leq p_A(a)+\lambda p_A(x) < 1$, whence $a + \lambda x \in A$. Therefore, $a \in \operatorname{core} A$, and so (6.8) holds. □

We provide now an algebraic separation theorem.

Theorem 6.2.2. *Let $A \subseteq X$ be a convex set with* $\operatorname{icr} A \neq \emptyset$ *and let $x_0 \in X \setminus \operatorname{icr} A$. Then A and $\{x_0\}$ can be properly separated, that is, there exists $x^* \in X'$ which is not constant on $A \cup \{x_0\}$ such that*

$$\forall x \in A : \langle x, x^* \rangle \leq \langle x_0, x^* \rangle .$$

Proof. W.l.o.g. we may assume that $0 \in \operatorname{icr} A$ (otherwise we do a translation).

Consider first the case $\operatorname{span} A = \operatorname{aff} A = X$. In this situation, A is a convex set with $0 \in \operatorname{core} A$. Using Proposition 6.2.1 we have that $p_A : X \to \mathbb{R}$ is sublinear and (6.7) holds. Consider $X_0 := \mathbb{R}x_0$ and $T_0 \in L(X_0, \mathbb{R})$ defined by $T_0(\lambda x_0) := \lambda p_A(x_0)$; clearly $T_0(x) \leq p_A(x)$ for every $x \in X_0$. By Corollary 6.1.7 we get $x^* \in L(X, \mathbb{R}) = X'$ such that $x^*|_{X_0} = T_0$ and $x^*(x) = \langle x, x^* \rangle \leq p_A(x)$ for every $x \in X$. Using (6.7), since $x_0 \notin \operatorname{core} A$ we have that $\langle x_0, x^* \rangle = T_0(x_0) = p_A(x_0) \geq 1$ and for $x \in A$ we get $\langle x, x^* \rangle \leq p_A(x) \leq 1$. Clearly, x^* is not constant on $A \cup \{x_0\}$ because $x^*(0) = 0 < 1 \leq x^*(x_0)$. The conclusion follows in this case.

Let now $X_0 := \operatorname{span} A = \operatorname{aff} A \neq X$. We have two possible situations: a) $x_0 \in X_0$, and b) $x_0 \notin X_0$. In case a), from the preceding situation, we get $x_0^* \in X_0'$ which is not constant on $A \cup \{x_0\}$ such that $x_0^*(x) \leq x_0^*(x_0)$ for every $x \in A$. Any $x^* \in X'$ with $x^*|_{X_0} = x_0^*$ does the job. In case b) there exists $x^* \in X'$ such that $x^*|_{X_0} = 0$ and $x^*(x_0) = 1$; this x^* does the job. □

We provide now other variants of the separation theorem.

Theorem 6.2.3. *Let $A, B \subseteq X$ be nonempty convex sets. The sets A and B can be properly separated in each of the following situations:*

(i) $\operatorname{core} A \neq \emptyset$ *and* $B \cap \operatorname{core} A = \emptyset$,
(ii) $\operatorname{icr} A \neq \emptyset$, $\operatorname{icr} B \neq \emptyset$ *and* $\operatorname{icr} A \cap \operatorname{icr} B = \emptyset$,
(iii) $0 \notin \operatorname{icr}(A - B) \neq \emptyset$.

Proof. Using Proposition 6.3.2 (iii) and (iv) we have that (ii) $\Rightarrow$ (iii) and (i) $\Rightarrow$ (iii), respectively. Assume that (iii) holds. By the Theorem 6.2.2, $\{0\}$ and $A - B$ can be separated properly, and so, as noticed above, A and B can be separated properly. □

In the case in which X is a topological linear space one prefers to have hyperplanes determined by continuous linear functionals when dealing with separation theorems; in this case we speak about separation by closed hyperplanes. This is motivated by the following result.

Proposition 6.2.4. *Let (X, τ) be a topological vector space and $x^* \in X' \setminus \{0\}$. Then the following statements are equivalent:*

(i) x^* *is continuous,*
(ii) $H_{x^*,\alpha}$ *is closed for some (for every)* $\alpha \in \mathbb{R}$,
(iii) *the interior of the half-space* $H^{\leq}_{x^*,\alpha} := \{x \in X \mid \langle x, x^*\rangle \leq \alpha\}$ *is nonempty for some (for every)* $\alpha \in \mathbb{R}$.

Proof. First observe that $H_{x^*,\alpha} = x_\alpha + H_{x^*,0}$ and $H^{\leq}_{x^*,\alpha} = x_\alpha + H^{\leq}_{x^*,0}$ for every $\alpha \in \mathbb{R}$ and $x_\alpha \in X$ with $\langle x_\alpha, x^*\rangle = \alpha$; such an x_α exists because $x^* \neq 0$. So, in (ii) and (iii) we can (and do) take $\alpha = 0$.

(i) $\Rightarrow$ (ii) is obvious.

(ii) $\Rightarrow$ (iii) Take $x_0 \in X$ with $\langle x_0, x^*\rangle < 0$. Clearly, $x_0 \notin H_{x^*,0}$ $(= \ker x^*)$, and so, because $\ker x^*$ is closed, there exists a balanced neighborhood U of 0 such that $(x_0+U)\cap\ker x^* = \emptyset$. Assume that $x_0+U \not\subseteq H^{\leq}_{x^*,0}$. Then there exists $u_0 \in U$ such that $x_0 + u_0 \notin H^{\leq}_{x^*,0}$, that is $\langle x_0 + u_0, x^*\rangle > 0$. Then $t := -\langle x_0, x^*\rangle / \langle u_0, x^*\rangle \in]0,1[$, and so $tu_0 \in U$. Because $\langle x_0 + tu_0, x^*\rangle = 0$, we get the contradiction $(x_0 + U) \cap \ker x^* \neq \emptyset$. Hence $x_0 \in \operatorname{int} H^{\leq}_{x^*,0}$, whence $\operatorname{int} H^{\leq}_{x^*,0} \neq \emptyset$.

(iii) $\Rightarrow$ (i) Let $x_0 \in \operatorname{int} H^{\leq}_{x^*,0}$. Then $\alpha := -\langle x_0, x^*\rangle > 0$; else x^* is non-positive on a neighborhood of the origin, and so $x^* = 0$. It follows that $0 \in \operatorname{int} H^{\leq}_{x^*,\alpha}$, and so there exists a balanced neighborhood U of 0 with $U \subseteq H^{\leq}_{x^*,\alpha}$. It follows that $|\langle u, x^*\rangle| \leq \alpha$ for every $u \in U$, whence $|\langle u, x^*\rangle| \leq \varepsilon\alpha$ for every $u \in \varepsilon U$, for every $\varepsilon > 0$. This proves the continuity of x^*. □

Using Theorem 6.2.3 and Proposition 6.2.4 we get the following result on separation by closed hyperplanes.

Theorem 6.2.5. *Let X be a separated locally convex space and let $A, B \subseteq X$ be nonempty convex sets.*

(i) *If* $\operatorname{int} A \neq \emptyset$ *and* $B \cap \operatorname{int} A = \emptyset$, *or more generally, if* $0 \notin \operatorname{int}(A - B) \neq \emptyset$, *then there exists* $x^* \in X^* \setminus \{0\}$ *such that* $\sup x^*(A) \leq \inf x^*(B)$.
(ii) *If A is compact, B is closed and* $A \cap B = \emptyset$, *or more generally, if* $0 \notin \operatorname{cl}(A-B)$, *then there exists* $x^* \in X^*$ *such that* $\sup x^*(A) < \inf x^*(B)$.

Proof. (i) If $\operatorname{int} A \neq \emptyset$ then $\operatorname{core}(A - B) = \operatorname{int}(A - B) = (\operatorname{int} A) - B \neq \emptyset$. So, let $0 \notin \operatorname{int}(A - B) \neq \emptyset$. Using Theorem 6.2.2 (iii) for $A - B$ and $x_0 := 0$, we get $x^* \in X' \setminus \{0\}$ such that $\sup x^*(A - B) \leq 0$, that is $A - B \subseteq H_{x^*,0}$. From Proposition 6.2.4 we obtain that $x^* \in X^*$. The conclusion follows.

(ii) Of course, if A is compact and B is closed, then $A - B$ is closed; moreover, if $A \cap B = \emptyset$ then $0 \notin A - B = \operatorname{cl}(A - B)$. So, assume that $0 \notin \operatorname{cl}(A - B)$. Then there exists a balanced convex neighborhood U of 0 such that $U \cap (B - A) = \emptyset$. Using (i) we get $x^* \in X^* \setminus \{0\}$ such that

$$\sup x^*(U) \leq \inf x^*(B - A) = \inf x^*(B) - \sup x^*(A).$$

Since U is a neighborhood of 0 and $x^* \neq 0$ we obtain that $\sup x^*(U) > 0$, and so $\sup x^*(A) < \inf x^*(B)$. The proof is complete. □

Remark 6.2.6. Observe that in the case in which X is a topological vector space and $A \subset X$ is a convex set such that $0 \in \operatorname{int} A$, then p_A is a continuous sublinear function and $\operatorname{int} A = \{x \in X \mid p_A(x) < 1\}$, $\operatorname{cl} A = \{x \in X \mid p_A(x) \leq 1\}$.

6.3 The Core Convex Topology

As seen in the previous sections, we have established several results in real linear spaces without topology. However, every real linear space X can be endowed with the strongest locally convex topology $\tau_c := \tau_{\mathscr{P}}$, where $\mathscr{P}$ is the family of all the semi-norms defined on X; τ_c is called the **core convex topology** in the sequel. In the next proposition we collect several results concerning this topology.

Proposition 6.3.1. *Let X be a real linear space. Then*

(i) *(X, τ_c) is a (separated) locally convex space.*
(ii) $X^* := (X, \tau_c)^* = X'$.
(iii) $\operatorname{int}_{\tau_c} C = \operatorname{core} C$ *for every convex set* $C \subseteq X$.
(iv) *Every convex function* $f : X \to \overline{\mathbb{R}}$ *is τ_c-continuous on* $\operatorname{core}(\operatorname{dom} f) = \operatorname{int}(\operatorname{dom} f)$.
(v) *Let* $f : X \to \overline{\mathbb{R}}$ *be a proper convex function. Then the restriction of f to* $\operatorname{aff}(\operatorname{dom} f)$ *is continuous on* $\operatorname{icr}(\operatorname{dom} f)$. *Moreover, f is subdifferentiable on* $\operatorname{icr}(\operatorname{dom} f)$.

Proof. (i) Note that $\{|\varphi| \mid \varphi \in X'\} \subseteq \mathscr{P}$. Because for every $x \in X \setminus \{0\}$ there exists $\varphi \in X'$ with $\varphi(x) \neq 0$, we obtain that $\mathscr{P}$ is sufficient, and so τ_c is separated.

(ii) We observed above that $X^* \subseteq X'$; let $\varphi \in X'$. Then $p := |\varphi| \in \mathscr{P}$. Since $|\varphi| \leq p$, φ is τ_c-continuous, and so $\varphi \in X^*$.

(iii) Let $C \subseteq X$ be a convex set. The inclusion $\operatorname{int} C \subseteq \operatorname{core} C$ is well known. Take $x \in \operatorname{core} C$. Then $A := C - x := \{c - x \mid c \in C\}$ is absorbing (and convex). Therefore, $A_0 := A \cap (-A)$ is an absorbing, symmetric convex set. It follows that the Minkowski functional p_{A_0} associated to A_0 is a semi-norm on X. Moreover, $\operatorname{core} A_0 = \{x \in X \mid p_{A_0}(x) < 1\} \subseteq A_0$. Because $p_{A_0} \in \mathscr{P}$, p_{A_0} is τ_c-continuous, and so $\{x \in X \mid p_{A_0}(x) < 1\}$ is τ_c-open and contains 0. Therefore, $0 \in \operatorname{int} A_0 \subseteq \operatorname{int}(C - x) = \operatorname{int} C - x$. Hence $x \in \operatorname{int} C$. Thus we have the inclusion $\operatorname{core} C \subseteq \operatorname{int} C$, too.

(iv) Let $f : X \to \overline{\mathbb{R}}$. From (iii) we have that $\operatorname{core}(\operatorname{dom} f) = \operatorname{int}(\operatorname{dom} f)$. It is well known that a convex function is $-\infty$ on $\operatorname{icr}(\operatorname{dom} f)$ if f takes the value $-\infty$ (see e.g. [614, Proposition 2.1.4]). Hence f is continuous on $\operatorname{core}(\operatorname{dom} f)$ if f takes the value $-\infty$. Thus assume that f is proper and take $x_0 \in \operatorname{core}(\operatorname{dom} f)$. Consider $A := \{x \in X \mid f(x) \leq f(x_0) + 1\}$. It is easy to see that $x_0 \in \operatorname{core} A$, that is, $x_0 \in \operatorname{int} A$. Hence A is a τ_c-neighborhood of x_0 and so f is bounded above on a neighborhood of x_0. Therefore, f is continuous at x_0 (see e.g. [614, Theorem 2.2.9]).

(v) Let $x_0 \in \operatorname{icr}(\operatorname{dom} f)$. Taking the usual translation ($g(x) := f(x_0 + x)$), we may assume that $x_0 = 0$. Then $X_0 := \operatorname{aff}(\operatorname{dom} f)$ is a linear subspace of X. Take $f_0 := f|_{X_0}$. Then $\operatorname{icr}(\operatorname{dom} f) = \operatorname{core}(\operatorname{dom} f_0)$, and so $0 \in \operatorname{core}(\operatorname{dom} f_0)$. By (iv) we have that f_0 is τ_c-continuous at 0. By [614, Theorem 2.4.12] $\partial f(x_0) = \partial f(0) \neq \emptyset$. □

Note that the subdifferential of a function on X when X is endowed with the core topology is called (generally) the algebraic subdifferential.

Related to operations with the intrinsic core we mention the following result which is well known in finite dimensional spaces.

Proposition 6.3.2. (i) *Let $A \subseteq X$ be a convex set and $T \in L(X, Y)$. If* $\operatorname{icr} A$ *is nonempty then* $\operatorname{icr}(T(A)) = T(\operatorname{icr} A)$.

(ii) *Let $A \subseteq X$ and $B \subseteq Y$ be nonempty sets. Then* $\operatorname{icr}(A \times B) = \operatorname{icr} A \times \operatorname{icr} B$.

(iii) *Let $A, C \subseteq X$ be convex sets such that* $\operatorname{icr} A$ *and* $\operatorname{icr} C$ *are nonempty. Then* $\operatorname{icr}(A - C) = \operatorname{icr} A - \operatorname{icr} C$.

(iv) *Let $A, C \subseteq X$ be convex sets such that* $\operatorname{core} A \neq \emptyset$. *Then* $\operatorname{core}(A - C) = \operatorname{core} A - C$.

Proof. (i) Let first $a \in \operatorname{icr} A$ and take $y \in T(A)$; then $y = Tx$ with $x \in A$. Because $a \in \operatorname{icr} A$, by (6.2) there exists $\lambda > 0$ such that $(1 + \lambda)a - \lambda x \in A$. It follows that $(1 + \lambda)Ta - \lambda y \in T(A)$. Since $T(A)$ is convex, using again (6.2), we get $Ta \in \operatorname{icr}(T(A))$. Conversely, fix $a \in \operatorname{icr} A$ and take $y \in \operatorname{icr}(T(A))$. By (6.2) we get $\lambda > 0$ such that $y' := (1 + \lambda)y - \lambda Ta \in T(A)$, and so $y' = Tx'$ with $x' \in A$. Then $y = Tx''$ with $x'' := \frac{1}{1+\lambda}x' + \frac{\lambda}{1+\lambda}a$. Since A is convex, $a \in \operatorname{icr} A$, $x' \in A$ and $1/(1 + \lambda) \in \,]0, 1[$ we have that $x'' \in \operatorname{icr} A$, and so $y \in T(\operatorname{icr} A)$.

(ii) It is clear that $\operatorname{aff}(A \times B) = \operatorname{aff} A \times \operatorname{aff} B$. Doing a translation we may assume that $0 \in A$ and $0 \in B$. In this way, replacing if necessary X by $\operatorname{aff} A$ and Y by $\operatorname{aff} B$, the conclusion reduces to $\operatorname{core}(A \times B) = \operatorname{core} A \times \operatorname{core} B$, which is immediate.

(iii) Consider $T \in L(X \times X, X)$ defined by $T(x, x') := x - x'$. Applying (ii) to A and C, then (i) to T and $A \times C$ we get the conclusion.

(iv) Let $a \in \operatorname{core} A$ and $c \in C$. Consider $x \in X$. Then there exists $\lambda > 0$ such that $a + \lambda x \in A$, and so $(a - c) + \lambda x \in A - C$. Hence $a - c \in \operatorname{core}(A - C)$.

Conversely, fix $a_0 \in \operatorname{core} A$, $c_0 \in C$, and take $u \in \operatorname{core}(A - C) = \operatorname{icr}(A - C)$. Then there exists $\lambda > 0$ such that $(1 + \lambda)u - \lambda(a_0 - c_0) = a - c$ for some $a \in A$ and $c \in C$. Because $\frac{\lambda}{1+\lambda} \in \,]0, 1[$, using (6.3), we get $u = \frac{\lambda}{1+\lambda}a_0 + \frac{1}{1+\lambda}a - \left(\frac{\lambda}{1+\lambda}c_0 + \frac{1}{1+\lambda}c\right) \in \operatorname{core} A - C$. Hence $\operatorname{core}(A - C) = \operatorname{core} A - C$. □

Note that assertion (iv) can be found in [562, Theorem 2.1], while assertion (iii) from the preceding result can be found in [581]. Note also that it is possible to have convex sets $A, B \subseteq X$ with $\operatorname{icr} A$ or $\operatorname{icr} B$ empty but $\operatorname{icr}(A - B)$ nonempty. For this take $A := B := \ell_2^+$ or $A := \{(x_n)_{n\geq 1} \in \ell_2 \mid x_1 = x_2\}$ and $B := \ell_2^+$. Since $\ell_2 = \ell_2^+ - \ell_2^+$ we have that $\operatorname{aff} \ell_2^+ = \ell_2$, and so $\operatorname{icr}(\ell_2^+) = \operatorname{core}(\ell_2^+)$. It is well known that $\operatorname{core}(\ell_2^+) = \emptyset$, and so $\operatorname{icr}(\ell_2^+) = \emptyset$. In the second case A is a linear subspace, and so $\operatorname{icr} A = A$. In both cases we have that $A - B = \ell_2$ and so $\operatorname{icr}(A - B) = \ell_2 \neq \emptyset$.

Another result related to the intrinsic core of convex sets is the following one; the assertion (ii) for X and Z finite dimensional linear spaces is just [489, Theorem 6.8].

Proposition 6.3.3. *Let $A \subseteq X \times Z$ be a convex set (set-valued map) and $(x_0, z_0) \in X \times Z$. Then:*

(i) *the following statements are equivalent:* a) $(x_0, z_0) \in \operatorname{core} A$, b) $x_0 \in \operatorname{core}(\operatorname{Pr}_X(A))$ *and* $z_0 \in \operatorname{core}(A(x_0))$, c) $x_0 \in \operatorname{core}\left(A^{-1}(z_0)\right)$ *and* $z_0 \in \operatorname{core}(A(x_0))$*;*

(ii) *the following statements are equivalent:* a) $(x_0, z_0) \in \operatorname{icr} A$, b) $x_0 \in \operatorname{icr}(\operatorname{Pr}_X(A))$ *and* $z_0 \in \operatorname{icr}(A(x_0))$.

Proof. First observe that, doing a translation, we may (and do) suppose that $(x_0, z_0) = (0, 0)$. Let us first prove (i).

a) $\Rightarrow$ c) Let $x \in X$; since $(0, 0) \in \operatorname{core} A$, there exists $\lambda > 0$ such that $\lambda(x, 0) \in A$, whence $\lambda x \in A^{-1}(0)$. Therefore $0 \in \operatorname{core}\left(A^{-1}(0)\right)$. Similarly, $0 \in \operatorname{core}(A(0))$.

c) $\Rightarrow$ b) is obvious (because $A^{-1}(0) \subseteq \operatorname{Pr}_X(A)$).

b) $\Rightarrow$ a) Let $(x, z) \in X \times Z$. Since $0 \in \operatorname{core}(\operatorname{Pr}_X(A))$, there exists $\lambda > 0$ such that $\lambda x \in \operatorname{Pr}_X(A)$, and so there exists $z' \in Z$ such that $(\lambda x, z') \in A$. Since $0 \in \operatorname{core}(A(0))$, there exists $\mu > 0$ such that $(0, \mu(\lambda z - z')) \in A$. Since A is convex, it follows that

$$\frac{\mu\lambda}{1+\mu}(x, z) = \frac{1}{1+\mu}\left(0, \mu(\lambda z - z')\right) + \frac{\mu}{1+\mu}\left(\lambda x, z'\right) \in A.$$

Therefore $(0, 0) \in \operatorname{core} A$.

(ii) a) $\Rightarrow$ b) Assume that $(0, 0) \in \operatorname{icr} A$. Consider first $x \in \operatorname{Pr}_X(A)$. Then $(x, z) \in A$ for some $z \in Z$. It follows that $(x', z') := -\lambda(x, z) \in A$ for some $\lambda > 0$, and so $x' = -\lambda x \in \operatorname{Pr}_X(A)$. Hence $0 \in \operatorname{icr}(\operatorname{Pr}_X(A))$. Let now $z \in A(0)$, that is, $(0, z) \in A$. As before, $-\lambda(0, z) \in A$ for some $\lambda > 0$, whence $-\lambda z \in A(0)$. Hence $0 \in \operatorname{icr}(A(0))$.

b) $\Rightarrow$ a) Let $(x, z) \in A$. Then $x \in \operatorname{Pr}_X(A)$. Since $0 \in \operatorname{icr}(\operatorname{Pr}_X(A))$, there exists $\lambda > 0$ such that $x' := -\lambda x \in \operatorname{Pr}_X(A)$ for some $\lambda > 0$, and so $(x', z') \in A$ for some $z' \in Z$. Then

$$\left(0, \frac{1}{1+\lambda}z' + \frac{\lambda}{1+\lambda}z\right) = \frac{1}{1+\lambda}(x', z') + \frac{\lambda}{1+\lambda}(x, z) \in A,$$

and so $z'' := \frac{1}{1+\lambda}z' + \frac{\lambda}{1+\lambda}z \in A(0)$. Because $0 \in \operatorname{icr}(A(0))$, there exists $\mu > 0$ such that $z''' := -\mu z'' \in A(0)$. Taking $\eta := \mu/(1 + \lambda + \mu) \in]0, 1[$ we obtain that $\eta x' = -\eta\lambda x$ and

$$\eta z' + (1 - \eta)z''' = \eta z' - \frac{\mu(1-\eta)}{1+\lambda}z' - \frac{\lambda\mu(1-\eta)}{1+\lambda}z = -\frac{\lambda\mu}{1+\lambda+\mu}z = -\eta\lambda z.$$

Hence $-\eta\lambda(x, z) = \eta(x', z') + (1 - \eta)(0, z''') \in A$, which proves that $(0, 0) \in \operatorname{icr} A$. □

Note that the implications a) $\Rightarrow$ c) and c) $\Rightarrow$ b) in (i) are valid for arbitrary sets A, but generally c) $\Rightarrow$ a) [and so b) $\Rightarrow$ a)] is not valid if A is not convex; take for example $A := ([-1,1] \times \{0\}) \cup (\{0\} \times [-1,1]) \subseteq \mathbb{R}^2$. Also note that even for A convex the fact that $0 \in \operatorname{icr}(A^{-1}(0))$ and $0 \in \operatorname{icr}(A(0))$ does not imply that $(0,0) \in \operatorname{icr} A$; for this take $A := \{(x,x) \mid x \in \mathbb{R}_+\} \subseteq \mathbb{R}^2$. Moreover, the assertion (ii) of the preceding proposition cannot be obtained from (i) because, even for $(0,0) \in A$, $\operatorname{span} A$ is not the product of two linear spaces.

In [572, Lemma 2.1] and [573] the implication b) $\Rightarrow$ a) of (i) is given in a weaker form: if $\operatorname{core}(\Pr_X(A)) \neq \emptyset$ and $\operatorname{core}(A(x)) \neq \emptyset$ for every $x \in \operatorname{core}(\Pr_X(A))$ then $\operatorname{core} A \neq \emptyset$.

6.4 Yang's Generalization of the Hahn–Banach Theorem

In this section the proper (that is, $\{0\} \neq K \neq Z$) convex cone K has nonempty algebraic interior $\operatorname{core} K$. Of course $0 \notin \operatorname{core} K$, $K + \operatorname{core} K = \operatorname{core} K + \operatorname{core} K = \operatorname{core} K$ and $\mathbb{P} \operatorname{core} K = \operatorname{core} K$.

Theorem 6.4.1. *Let $\Gamma : X \rightrightarrows Z$ be a K-convex set-valued map, $X_0 \subseteq X$ a linear subspace and $T_0 \in L(X_0, Z)$. Suppose that $0 \in \operatorname{icr}(\operatorname{dom} \Gamma - X_0)$ and $T_0 x \not> z$ for all $(x,z) \in \Gamma \cap (X_0 \times Z)$. Then there exists $T \in L(X,Z)$ such that $T|_{X_0} = T_0$ and $Tx \not> z$ for all $(x,z) \in \Gamma$.*

Proof. We give a proof similar to that of Theorem 6.1.1. Consider $\overline{X} := \operatorname{aff}(\operatorname{dom} \Gamma - X_0)$. Because $0 \in \operatorname{icr}(\operatorname{dom} \Gamma - X_0) \subseteq \operatorname{dom} \Gamma - X_0$, $\overline{X}$ is a linear space.

The case $\overline{X} = X$; hence $0 \in \operatorname{core}(\operatorname{dom} \Gamma - X_0)$. Consider the set

$$\mathscr{F} := \big\{(E,T) \mid X_0 \subseteq E = \operatorname{span} E \subseteq X,\ T \in L(E,Z),\\ T|_{X_0} = T_0,\ Tx \not> z\ \forall (x,z) \in \Gamma \cap (E \times Z)\big\};$$

clearly, $(X_0, T_0) \in \mathscr{F}$. For $(E_1,T_1),(E_2,T_2) \in \mathscr{F}$ we set $(E_1,T_1) \leq (E_2,T_2)$ if $E_1 \subseteq E_2$ and $T_2|_{E_1} = T_1$; $\leq$ is a partial order on $\mathscr{F}$. Exactly as in the proof of Theorem 6.1.1 we have that every chain $\mathscr{C}$ in $\mathscr{F}$ is bounded above. Using Zorn's lemma, $(\mathscr{F}, \leq)$ has maximal elements; let $(E,T) \in \mathscr{F}$ be maximal. If $E = X$ the linear operator T verifies the conditions in the conclusion of the theorem.

Assume that $E \neq X$, and take $\overline{x} \in X \setminus E$. As in the proof of Theorem 6.1.1 we obtain that the sets

$$B_1 := \{z_1 - Tx_1 \mid x_1 \in E,\ z_1 \in Z\ :\ (x_1 + \overline{x}, z_1) \in \mathbb{P}A\},$$
$$B_2 := \{Tx_2 - z_2 \mid x_2 \in E,\ z_2 \in Z\ :\ (x_2 - \overline{x}, z_2) \in \mathbb{P}A\}$$

are nonempty (and convex), where $A := \operatorname{epi} \Gamma$. We claim that

$$(B_1 + \operatorname{core} K) \cap (B_2 - \operatorname{core} K) = \emptyset.$$

In the contrary case there exist $k_1, k_2 \in \operatorname{core} K$, $x_1, x_2 \in E$, $z_1, z_2 \in Z$ such that $(x_1 + \overline{x}, z_1), (x_2 - \overline{x}, z_2) \in \mathbb{P}A$ and $k_1 + z_1 - Tx_1 = Tx_2 - z_2 - k_2$, whence $T(x_1 + x_2) > z_1 + z_2$. But $(x_1 + x_2, z_1 + z_2) \in \mathbb{P}A$ and $x_1 + x_2 \in E$, contradicting the hypothesis.

Assume that $(B_1 + \operatorname{core} K) \cup (B_2 - \operatorname{core} K) = Z$. Then $B_1 \subseteq B_1 + \operatorname{core} K$ and $B_2 \subseteq B_2 - \operatorname{core} K$. Indeed, if $b_1 \in B_1 \setminus (B_1 + \operatorname{core} K)$, then $b_1 \in B_2 - \operatorname{core} K$, whence $b_1 = b_2 - k$, with $b_2 \in B_2$ and $k \in \operatorname{core} K$. In this situation we get the contradiction $b_1 + k/2 = b_2 - k/2 \in (B_1 + \operatorname{core} K) \cap (B_2 - \operatorname{core} K) = \emptyset$. Fix now $b_1 \in B_1$ and $b_2 \in B_2$ and consider

$$\overline{\gamma} := \sup\{\gamma \in [0, 1] \mid \forall\, t \in [0, \gamma] : (1 - t)b_1 + tb_2 \in B_1 + \operatorname{core} K\}.$$

Set $\overline{b} := (1 - \overline{\gamma})b_1 + \overline{\gamma}b_2$ and suppose first that $\overline{b} \in B_1 + \operatorname{core} K$; of course, in this situation $\overline{\gamma} < 1$. Then $\overline{b} = \overline{b}_1 + \overline{k}$, with $\overline{b}_1 \in B_1$ and $\overline{k} \in \operatorname{core} K$. Since $\overline{k} \in \operatorname{core} K$, there exists $\delta \in \mathbb{P}$ such that $\overline{k} + \mu(b_2 - b_1) \in \operatorname{core} K$ for $\mu \in [-\delta, \delta]$. It follows that

$$\forall\, \mu \in [-\delta, \delta] : \overline{b} + \mu(b_2 - b_1) = \overline{b}_1 + \overline{k} + \mu(b_2 - b_1) \in B_1 + \operatorname{core} K.$$

Taking $\mu = \min\{\delta, 1 - \overline{\gamma}\}$, we get

$$\forall\, t \in [0, \mu] : [1 - (\overline{\gamma} + t)]\, b_1 + (\overline{\gamma} + t)b_2 \in B_1 + \operatorname{core} K,$$

contradicting the choice of $\overline{\gamma}$. Suppose now that $\overline{b} \in B_2 - \operatorname{core} K$; of course $\overline{\gamma} > 0$. In this situation $\overline{b} = \overline{b}_2 - \overline{k}$, with $\overline{b}_2 \in B_2$ and $\overline{k} \in \operatorname{core} K$. Since $\overline{k} \in \operatorname{core} K$, there exists $\delta \in \mathbb{P}$ such that $\overline{k} + \mu(b_2 - b_1) \in \operatorname{core} K$ for $\mu \in [-\delta, \delta]$. It follows that

$$\forall\, \mu \in [-\delta, \delta] : \overline{b} - \mu(b_2 - b_1) = \overline{b}_2 - \overline{k} - \mu(b_2 - b_1) \in B_2 - \operatorname{core} K.$$

Taking $\mu = \min\{\delta, \overline{\gamma}\}$, we get the contradiction

$$[1 - (\overline{\gamma} - \mu)]\, b_1 + (\overline{\gamma} - \mu)b_2 \in B_1 + \operatorname{core} K.$$

Therefore $(B_1 + \operatorname{core} K) \cup (B_2 - \operatorname{core} K) \neq Z$. Taking $\overline{z} \in Z \setminus \big[(B_1 + \operatorname{core} K) \cup (B_2 - \operatorname{core} K)\big]$ and $E_1 = E + \mathbb{R}\overline{x}$, then defining T_1 by $T_1(x + t\overline{x}) = Tx + t\overline{z}$, $T_1 : E_1 \to Z$ is a linear extension of T and $(x, z) \in \Gamma \cap (E_1 \times Z)$ implies $T_1 x \not> z$. Therefore, $(E_1, T_1) \in \mathscr{F}$ and $(E, T) \leq (E_1, T_1)$. Because (E, T) is maximal in $\mathscr{F}$ we obtain that $(E, T) = (E_1, T_1)$, whence the contradiction $E = E_1 \ni \overline{x}$. Hence $E = X$.

When $\operatorname{aff}(\operatorname{dom} \Gamma - X_0) \neq X$, proceeding as in the second case of the proof of Theorem 6.1.1 we get the desired conclusion. □

Remark 6.4.2. In the proof of Theorem 6.4.1 we used only the fact that $\mathbb{P} \cdot \operatorname{epi} \Gamma$ is convex (or, equivalently, $\mathbb{R}_+ \operatorname{epi} \Gamma$ is a convex cone) and not that $\operatorname{epi} \Gamma$ itself is convex.

The next result was obtained by Thierfelder in [573, Theorem 2.2].

Corollary 6.4.3. *Let $\Gamma_i : X \rightrightarrows Z$, $i \in \{1,2\}$, be two convex set-valued maps and $z^* \in Z' \setminus \{0\}$. Assume that $0 \in \operatorname{icr}(\operatorname{dom} \Gamma_1 - \operatorname{dom} \Gamma_2)$ and $\inf z^*(\Gamma_1(x)) \geq \sup z^*(\Gamma_2(x))$ for all $x \in \operatorname{dom} \Gamma_1 \cap \operatorname{dom} \Gamma_2$. Then there exists $T \in L(X,Z)$ such that $\inf z^*(\Gamma_1(x_1) - Tx_1) \geq \sup z^*(\Gamma_2(x_2) - Tx_2)$ for all $x_1 \in \operatorname{dom} \Gamma_1$ and $x_2 \in \operatorname{dom} \Gamma_2$.*

Proof. Consider $K := K_{z^*} := \{z \in Z \mid z^*(z) \geq 0\}$. Then $\operatorname{core} K = \{z \in Z \mid z^*(z) > 0\}$, and so $0 \not> z$ if and only if $z^*(z) \geq 0$. Consider also the set-valued map $\Gamma : X \rightrightarrows Z$ with $\operatorname{graph} \Gamma := \operatorname{graph} \Gamma_1 - \operatorname{graph} \Gamma_2$. Then Γ is convex, and so is K-convex, $\operatorname{dom} \Gamma = \operatorname{dom} \Gamma_1 - \operatorname{dom} \Gamma_2$ and

$$\Gamma(0) = \cup_{x \in X}(\Gamma_1(x) - \Gamma_2(x)) = \cup_{x \in \operatorname{dom} \Gamma_1 \cap \operatorname{dom} \Gamma_2}(\Gamma_1(x) - \Gamma_2(x)).$$

From our hypothesis we have that $0 \not> \Gamma(0)$. Taking $X_0 := \{0\}$ and $T_0 := 0 \in L(X_0, Z)$, the hypothesis of Theorem 6.4.1 holds. Therefore, there exists $T \in L(X,Z)$ such that $Tx \not> \Gamma(x)$ for every $x \in X$. This implies that for all $(x_1, z_1) \in \operatorname{graph} \Gamma_1$ and $(x_2, z_2) \in \operatorname{graph} \Gamma_2$ we have that $z^*(Tx_1 - Tx_2) \leq z^*(z_1) - z^*(z_2)$, or, equivalently, $z^*(z_1 - Tx_1) \geq z^*(z_2 - Tx_2)$. The conclusion follows. □

Corollary 6.4.4. *Let $p : X \to Z^\bullet$ be a sublinear operator, X_0 be a linear subspace of X and $T_0 \in L(X_0, Z)$. Suppose that $T_0 x \not> p(x)$ for every $x \in X_0$. If $X_0 + \operatorname{dom} p$ is a linear subspace then there exists $T \in L(X,Z)$ such that $T|_{X_0} = T_0$ and $Tx \not> p(x)$ for every $x \in X$.*

Proof. In Theorem 6.4.1 let us take $\Gamma : X \rightrightarrows Z$ with $\operatorname{graph} \Gamma = \operatorname{epi} p$; of course, the hypotheses of the theorem hold, so that there exists $T \in L(X,Z)$ such that $T|_{X_0} = T_0$ and $(x,z) \in \Gamma$ implies $Tx \not> z$. Taking $z = p(x)$, the conclusion follows. □

Before stating the next result recall that the weak subdifferential $\partial^w f(x_0)$ of the proper operator $f : X \to Z^\bullet$ at $x_0 \in \operatorname{dom} f$ is the set of those $T \in L(X,Z)$ such that

$$\forall\, x \in X \;:\; Tx - Tx_0 \not> f(x) - f(x_0).$$

Corollary 6.4.5. *Let $f : X \to Z^\bullet$ be a proper convex operator and consider $x_0 \in \operatorname{icr}(\operatorname{dom} f)$. Then $\partial^w f(x_0)$ is nonempty.*

Proof. Take $g(x) = f(x_0 + x) - f(x)$, $\operatorname{graph} \Gamma = \operatorname{epi} g$, $X_0 = \{0\}$ and $T_0 := 0 \in L(X_0, Z)$. Of course, if $(x,z) \in \Gamma \cap (X_0 \times Z)$, then $x = 0$ and $z \in K$, whence $T_0 x \not> z$. Using Theorem 6.4.1 we obtain $T \in L(X,Z)$ such that $Tx \not> z$

for $(x,z) \in \Gamma$. Taking $x \in \operatorname{dom} f$, we have that $(x - x_0, f(x) - f(x_0)) \in \Gamma$, so that $T(x - x_0) \not> f(x) - f(x_0)$. □

Corollary 6.4.6. *Let $f, g : X \to Z^\bullet$ be proper convex operators. Suppose that $0 \in \operatorname{icr}(\operatorname{dom} f - \operatorname{dom} g)$ and that $f(x) + g(x) \not< 0$ for all $x \in \operatorname{dom} f \cap \operatorname{dom} g$. Then there exists $T \in L(X, Z)$ such that*

$$\forall\, x_1 \in \operatorname{dom} f,\ \forall\, x_2 \in \operatorname{dom} g\ :\ f(x_1) - Tx_1 \not< -g(x_2) - Tx_2.$$

Proof. Consider $\Gamma : X \rightrightarrows Z$ with

$$\operatorname{graph} \Gamma := \{(x,z) \in X \times Z \mid f(x) \le z\} + \{(-x', z') \in X \times Z \mid g(x') \le z'\}.$$

Then Γ is convex and $\operatorname{dom} \Gamma = \operatorname{dom} f - \operatorname{dom} g$. Take $X_0 = \{0\}$ and $T_0 := 0 \in L(X_0, Z)$. The hypotheses of the preceding theorem hold. Indeed, if $(0,z) \in \Gamma$ then $(0,z) = (x, z') + (-x, z'')$, with $f(x) \le z'$ and $g(x) \le z''$; assuming that $0 = T_0(0) > z$, from $f(x) + g(x) \le z' + z'' = z$ and the known relation $K + \operatorname{core} K = \operatorname{core} K$ we get the contradiction $f(x) + g(x) < 0$. Hence $T_0(0) \not> z$. Therefore, there exists $T \in L(X, Z)$ such that $Tx \not> z$ for $(x,z) \in \Gamma$. In particular, for $x_1 \in \operatorname{dom} f$ and $x_2 \in \operatorname{dom} g$ we get $T(x_1 - x_2) \not> f(x_1) + g(x_2)$, which yields $-g(x_2) - Tx_2 \not> f(x_1) - Tx_1$ for all $x_1 \in \operatorname{dom} f$ and $x_2 \in \operatorname{dom} g$. The proof is complete. □

Note that applying Corollary 6.4.6 for $f = p$ and $g(x) = -T_0 x$ for $x \in X_0$, $g(x) = +\infty$ for $x \notin X_0$, where $p : X \to Z^\bullet$ is a sublinear operator and $T_0 \in L(X_0, Z)$ we get a weaker variant of Corollary 6.4.4.

As mentioned in [603, Lemma 1], Wang [588] obtained Theorem 6.4.1 for $X_0 = \{0\}$, $T_0 = 0$ and Γ a K-convex set-valued map with $\operatorname{core}(\operatorname{epi} \Gamma) \neq \emptyset$ and $0 \in \operatorname{core}\left(\Gamma^{-1}(z_0)\right)$ for some $z_0 \in Z$ (compare also with Proposition 6.3.3); based on results in [588], Yang [603, Theorem 1] obtained a weaker form of Theorem 6.4.1: the interiority hypothesis is stronger [more precisely, $\operatorname{core}(\operatorname{epi} \Gamma) \neq \emptyset$ and $X_0 \cap \operatorname{core}(\operatorname{dom} \Gamma) \neq \emptyset$ instead of $0 \in \operatorname{icr}(\operatorname{dom} \Gamma - X_0)$] and the conclusion is weaker [more precisely $0 \not< (T - T_0)(x) \not< 0$ for all $x \in X_0$ instead of $T|_{X_0} = T_0$].

As in the case $Z = \mathbb{R}$, the extension theorems can be used for separating convex sets in product spaces. For example, from Theorem 6.4.1 we can deduce the next separation result.

Proposition 6.4.7. *Let $A, B \subseteq X \times Z$ be such that $\mathbb{R}_+(A - B)$ is convex. Assume that $0 \in \operatorname{icr}(\operatorname{Pr}_X(A) - \operatorname{Pr}_X(B))$ and $A(x) \not< B(x)$ for all $x \in \operatorname{Pr}_X(A) \cap \operatorname{Pr}_X(B)$. Then there exists $T \in L(X, Z)$ such that*

$$\forall (x_1, z_1) \in A,\ \forall (x_2, z_2) \in B : z_1 - Tx_1 \not< z_2 - Tx_2. \tag{6.9}$$

Proof. Consider the set $C := A - B \subseteq X \times Z$. Then $\operatorname{Pr}_X(C) = \operatorname{Pr}_X(A) - \operatorname{Pr}_X(B)$ and if $(0,z) \in C$ then $z = z_1 - z_2$ with $z_1 \in A(x)$, $z_2 \in B(x)$ for some $x \in \operatorname{Pr}_X(A) \cap \operatorname{Pr}_X(B)$; hence $z \not< 0$. Taking $X_0 := \{0\}$ and $T_0(0) := 0$, we can apply

Theorem 6.4.1 (see also Remark 6.4.2) for graph $\Gamma := C$ and T_0. Therefore, there exists $T \in L(X,Z)$ such that $z \nless Tx$ for every $(x,z) \in C$. Hence (6.9) holds. □

Note that Thierfelder [572, 573] says that A and B are separable by an affine mapping if there exist $T \in L(X,Z)$ and $z_0 \in Z$ such that

$$\forall (x_1,z_1) \in A,\ \forall (x_2,z_2) \in B : z_1 - Tx_1 \nless z_0 \nless z_2 - Tx_2 \tag{6.10}$$

[compare with (6.5)].

One can ask which is the relationship between (6.9) and (6.10). Setting $A_0 := \{z_1 - Tx_1 \mid (x_1,z_1) \in A\}$, $B_0 := \{z_2 - Tx_2 \mid (x_2,z_2) \in B\}$, condition (6.9) becomes $B_0 \cap (A_0 + \operatorname{core} K) = \emptyset$, or equivalently, $(B_0 - \operatorname{core} K) \cap (A_0 + \operatorname{core} K) = \emptyset$, while condition (6.10) becomes $z_0 \notin (B_0 - \operatorname{core} K) \cup (A_0 + \operatorname{core} K)$. So, it is quite clear that (6.10) does not imply (6.9). (One can take $X := \mathbb{R}$, $Z := \mathbb{R}^2$, $K := \mathbb{R}^2_+$, $A := \{(0,(-1,-1))\}$, $B := \{(0,(1,1))\}$, $T := 0$ and $z_0 := (-2,2)$.)

However, when $A, B \subseteq X \times Z$ are convex and $\operatorname{core} K \neq \emptyset$ (which is the case in Proposition 6.4.7) we have that (6.9) implies (6.10). Indeed, taking A_0 and B_0 as above we have seen that condition (6.9) becomes $B_0 \cap (A_0 + \operatorname{core} K) = \emptyset$. Since A_0 and B_0 are convex sets and $A_0 + \operatorname{core} K$ is algebraically open, by the classic (algebraic) separation theorem we get $z^* \in Z'$, $z^* \neq 0$, and $\gamma \in \mathbb{R}$ such that $\langle b, z^* \rangle \leq \gamma \leq \langle a + k, z^* \rangle$ for all $a \in A_0$, $b \in B_0$, $k \in \operatorname{core} K$. It follows that

$$\forall\, k \in \operatorname{core} K : \langle k, z^* \rangle > 0, \quad \text{and}$$

$$\forall (x_1,z_1) \in A,\ \forall (x_2,z_2) \in B : \langle z_1 - Tx_1, z^* \rangle \geq \gamma \geq \langle z_2 - Tx_2, z^* \rangle .$$

Taking $z_0 \in Z$ such that $\langle z_0, z^* \rangle = \gamma$ we have that (6.10) holds.

The previous discussion shows that (6.9) is more adequate as separation property than (6.10). Another argument is the fact that taking $x_1 = x_2$ in (6.9) we recover a part of the hypothesis of Proposition 6.4.7, which is not the case for (6.10).

However, using directly the classic separation theorem we obtain a stronger conclusion than that of Theorem 6.4.1 under the same hypotheses.

Theorem 6.4.8. *Let $\Gamma : X \rightrightarrows Z$ be a K-convex set-valued map, $X_0 \subseteq X$ a linear subspace and $T_0 \in L(X_0, Z)$. Suppose that $0 \in \operatorname{icr}(\operatorname{dom} \Gamma - X_0)$ and $T_0 x \ngtr z$ for all $(x,z) \in \Gamma \cap (X_0 \times Z)$. Then there exists $z^* \in Z'$ and $T \in L(X,Z)$ such that*

$$T\,|_{X_0} = T, \ \ \forall\, z \in \operatorname{core} K : \langle z, z^* \rangle > 0, \quad \text{and} \quad \forall\, (x,z) \in \Gamma : \langle Tx, z^* \rangle \leq \langle z, z^* \rangle .$$

In particular $Tx \ngtr z$ for very $(x,z) \in \Gamma$. Moreover, T can be defined by $Tx = T_0 x_0 - \langle x_1, x^* \rangle \bar{z}$, where $\bar{z} \in Z$, $x^* \in X'$ and the linear subspace $X_1 \subseteq X$ with $X = X_0 \oplus X_1$ are fixed, and $x = x_0 + x_1$ with $x_0 \in X_0$, $x_1 \in X_1$.

Proof. Assume first that $0 \in \operatorname{core}(\operatorname{dom} \Gamma - X_0)$. Let $B = \operatorname{epi} \Gamma - \operatorname{graph} T_0$. It is obvious that $\operatorname{Pr}_X(B) = \operatorname{dom} \Gamma - X_0$, and so $0 \in \operatorname{core}(\operatorname{Pr}_X(B))$. Then $B(0) \neq \emptyset$ and $z \in B(0) \Rightarrow 0 \ngtr z$. Indeed, if $z \in B(0)$ then $(0,z) = (x,z') - (x_0, T_0 x_0) + (0,k')$

for some $(x,z') \in \Gamma$, $x_0 \in X_0$ and $k' \in K$. It follows that $x = x_0$ and $z = z' + k' - T_0x_0 \nless 0$. Since core $K \neq \emptyset$, it is clear that core $(B(0)) \neq \emptyset$. Therefore, by Proposition 6.3.3 (i), core $B \neq \emptyset$. Moreover, $(0,0) \notin$ core B. In the contrary case, again by Proposition 6.3.3 (i), $0 \in$ core$(B(0))$, contradicting the fact observed above that $z \nless 0$ for every $z \in B(0)$.

Using an algebraic separation theorem we get $(x^*, z^*) \in X' \times Z' \setminus \{(0,0)\}$ such that

$$\forall\,(x,z) \in \Gamma,\ \forall x_0 \in X_0,\ \forall k \in K : \langle x - x_0, x^* \rangle + \langle z - T_0x_0 + k, z^* \rangle \geq 0. \tag{6.11}$$

First notice that, because $0 \in$ core (dom $\Gamma - X_0$), if $z^* = 0$ then $x^* = 0$, a contradiction; therefore, $z^* \neq 0$. Next, from (6.11), it follows that $\langle k, z^* \rangle \geq 0$ for every $k \in K$,

$$\forall x_0 \in X_0 : \langle x_0, x^* \rangle + \langle T_0x_0, z^* \rangle = 0, \tag{6.12}$$

$$\forall (x,z) \in \Gamma : \langle x, x^* \rangle + \langle z, z^* \rangle \geq 0. \tag{6.13}$$

Of course, $\langle k, z^* \rangle > 0$ for every $k \in$ core K. Fix $\overline{z} \in Z$ such that $\langle \overline{z}, z^* \rangle = 1$, $X_1 \subseteq X$ a linear subspace such that $X_0 \oplus X_1 = X$ and $T : X \to Z$ defined by $Tx := T_0x_0 - \langle x_1, x^* \rangle \overline{z}$ for $x = x_0 + x_1$ with $x_0 \in X_0$ and $x_1 \in X_1$. Then, of course, $T\,|_{X_0} = T_0$ and for $(x,z) \in \Gamma$, taking $x_0 \in X_0$ and $x_1 \in X_1$ with $x = x_0 + x_1$, we have

$$\langle Tx, z^* \rangle = \langle T_0x_0 - \langle x_1, x^* \rangle \overline{z}, z^* \rangle = \langle T_0x_0, z^* \rangle - \langle x_1, x^* \rangle = -\langle x, x^* \rangle \leq \langle z, z^* \rangle .$$

If $\overline{X} :=$ aff(dom $\Gamma - X_0$) $=$ span(dom $\Gamma - X_0$) $\neq X$ we find first $x^* \in L(\overline{X}, \mathbb{R})$ and $z^* \in Z'$ as above and then extend x^* to an element of X'. The proof is complete. □

Notice that using the same hypotheses as in Theorem 6.4.1, the conclusion of Theorem 6.4.8 is stronger, not only because $\langle Tx, z^* \rangle \leq \langle z, z^* \rangle$ implies $Tx \ngtr z$, but also because we have a very special expression for T. This is quite surprising because in the proof of Theorem 6.4.8 we utilized the usual separation theorem from the scalar case which is equivalent to the classical Hahn–Banach theorem, while in the proof of Theorem 6.4.1 we utilized a similar technique to that used in the proof of Theorem 6.1.1, that is, for the proof of the Hahn–Banach–Kantorovich theorem. Furthermore, the technique utilized in the proof of Theorem 6.4.8 is more adequate for obtaining continuous versions of this theorem.

A continuous version of Theorem 6.4.8 is the next result.

Theorem 6.4.9. *Let X, Z be separated locally convex spaces, $\Gamma : X \rightrightarrows Z$ be a K-convex set-valued map, $X_0 \subseteq X$ a linear subspace and $T_0 \in \mathcal{L}(X_0, Z)$. Suppose that* int(epi Γ) $\neq \emptyset$*, $X_0 \cap$* int(dom Γ) $\neq \emptyset$*, and $T_0x \ngtr z$ for all $(x,z) \in \Gamma \cap (X_0 \times Z)$. If either* (a) *$X_0$ has a topological supplement, or* (b) $T_0x = \langle x, x_0^* \rangle z_0$

for every $x \in X_0$ *with fixed* $x_0^* \in X^*$ *and* $z_0 \in Z$*, then there exists* $z^* \in Z^*$ *and* $T \in \mathscr{L}(X,Z)$ *such that*

$$T\,|_{X_0} = T, \quad \forall\, z \in \operatorname{core} K : \langle z, z^* \rangle > 0, \quad \textit{and} \quad \forall\,(x,z) \in \Gamma : \langle Tx, z^* \rangle \le \langle z, z^* \rangle\,.$$

In particular $Tx \not> z$ *for very* $(x,z) \in \Gamma$*. Moreover, in case* (a) T *can be defined by* $Tx = T_0x_0 - \langle x_1, x^* \rangle \bar{z}$*, where* $\bar{z} \in Z$*,* $x^* \in X^*$ *are fixed and* $x = x_0 + x_1$ *with* $x_0 \in X_0$*,* $x_1 \in X_1$*, while in case* (b) T *can be defined by* $Tx := \langle x, x_0^* \rangle (z_0 - \langle z_0, z^* \rangle z_1) - \langle x, x^* \rangle z_1$ *for* $x \in X$ *with fixed* $x^* \in X^*$ *and* $z_1 \in Z$ *such that* $\langle z_1, z^* \rangle = 1$.

Proof. In the proof of Theorem 6.4.8, because $\operatorname{int}(\operatorname{epi}\Gamma) \neq \emptyset$, we have that $\operatorname{int} B$ ($= \operatorname{core} B$) is nonempty; it follows that $x^* \in X^*$ and $z^* \in Z^*$. In case (a), because the projections on X_0 and X_1 are continuous, taking into account the construction of T we obtain that T is continuous.

Assume now that $T_0x = \langle x, x_0^* \rangle z_0$ for every $x \in X_0$ with fixed $x_0^* \in X^*$ and $z_0 \in Z$. From (6.12) we obtain that $\langle x, x^* + \langle z_0, z^* \rangle x_0^* \rangle = 0$ for every $x \in X_0$. Taking $z_1 \in Z$ with $\langle z_1, z^* \rangle = 1$ and $Tx := \langle x, x_0^* \rangle (z_0 - \langle z_0, z^* \rangle z_1) - \langle x, x^* \rangle z_1$ for $x \in X$, we have that T is continuous and $Tx = T_0x$ for $x \in X_0$. Moreover, for $(x,z) \in \Gamma$ we have

$$\begin{aligned}
\langle Tx, z^* \rangle &= \langle \langle x, x_0^* \rangle z_0 - (\langle z_0, z^* \rangle \langle x, x_0^* \rangle - \langle x, x^* \rangle) z_1, z^* \rangle \\
&= \langle x, x_0^* \rangle \langle z_0, z^* \rangle - \langle z_0, z^* \rangle \langle x, x_0^* \rangle - \langle x, x^* \rangle \\
&= -\langle x, x^* \rangle \le \langle z, z^* \rangle\,,
\end{aligned}$$

the last inequality being obtained using (6.13). □

Thierfelder in [572] and [573, Theorem 2.5] obtains similar conclusions for Theorem 6.4.8 in the following two situations: 1) there exists z^* strictly positive on $K \setminus \{0\}$ such that $\langle z, z^* \rangle \ge \langle T_0x_0, z^* \rangle$ for $(x,z) \in \Gamma \cap (X_0 \times Z)$ and 2) there exists an algebraically open convex cone P such that $K \setminus \{0\} \subseteq P$ and $\Gamma(0) + P \subseteq \Gamma(0)$. Taking $P = \{z \in Z \mid \langle z, z^* \rangle > 0\}$ in case 1), in each situation we have that $\Gamma(0) \cap (-P) = \emptyset$, and so the conclusion is obtained applying Theorem 6.4.8 for K replaced by $P \cup \{0\}$.

Remark 6.4.10. Not that the condition $X_0 \cap \operatorname{int}(\operatorname{dom}\Gamma) \neq \emptyset$ in Theorem 6.4.9 is essential. To show this take $X := Z := \mathbb{R}$, $\Gamma := \{(x,z) \in X \times Z \mid x \ge 0,\ z \ge -\sqrt{x}\}$, $K := \mathbb{R}_+$, $X_0 := \{0\}$, $T_0 := 0$; in this case all the hypotheses of Theorem 6.4.9, but $X_0 \cap \operatorname{int}(\operatorname{dom}\Gamma) \neq \emptyset$, are verified (compare with [419, Theorem 2.1]).

Related to the non emptiness of the interior of $\operatorname{epi}\Gamma$ see Proposition 3.1.27 and Corollary 3.2.2.

6.5 A Sufficient Condition for the Convexity of $\mathbb{R}_+A$

We have seen that the convexity of $\operatorname{epi}\Gamma$ in Theorems 6.1.1 and 6.4.1 can be replaced by the convexity of $\mathbb{R}_+ \operatorname{epi}\Gamma$. In this section we consider $A \subseteq X \times Z$, where X, Z are real vector spaces. Let us denote by $\mathcal{N}_Z'$ a class of nonempty subsets of Z with the property that for every $W \in \mathcal{N}_Z'$ one has $[0,1]W = W$ and there exists $V \in \mathcal{N}_Z'$ such that $V + V \subseteq W$. Of course, $\mathcal{N}_Z'$ could contain only the set $\{0\}$, or could be the class of balanced neighborhoods of 0 with respect to a topology τ such that (Z, τ) is a topological vector space.

Theorem 6.5.1. *Let $A \subseteq X \times Z$ be a nonempty set. Suppose that the following conditions hold:*

(i) $\exists \alpha, \beta > 0,\ \forall u, v \in A : \alpha u + \beta v \in A$,
(ii) $\forall u \in A,\ \forall V \in \mathcal{N}_Z',\ \exists \delta \in\,]0,1[,\ \forall \lambda \in [\delta, 1] : \lambda u \in A + \{0\} \times V$.

Then

$$\tilde{A} := \cap_{V \in \mathcal{N}_Z'} (\mathbb{R}_+ A + \{0\} \times V) \tag{6.14}$$

is a convex cone.

Proof. Take $\alpha' := \alpha(\alpha+\beta)^{-1}$ and $\beta' := \beta(\alpha+\beta)^{-1}$ and define inductively the sequence of subsets I_n by

$$I_0 := \{0, 1\}, \quad I_{n+1} := \left\{\alpha' t + \beta' s \mid t, s \in I_n\right\} \text{ for } n \in \mathbb{N}.$$

We prove that $\tilde{A}$ is convex in several steps.

Step 1: $I_n \subseteq I_{n+1}$ for $n \in \mathbb{N}$ and $\operatorname{cl} I = [0,1]$, where $I := \cup_{n \in \mathbb{N}} I_n$. This is (almost) obvious.

Step 2: $t \in I_n,\ u, v \in A \Rightarrow tu + (1-t)v \in (\alpha+\beta)^{-n} A$. We prove this claim by induction. For $n = 0$ the claim is obvious. Suppose that it holds for $n \in \mathbb{N}$ and take $t \in I_{n+1}$ and $u, v \in A$. Then $t = \alpha' t_1 + \beta' t_2$ with $t_1, t_2 \in I_n$. So

$$tu + (1-t)v = \alpha' \left[t_1 u + (1-t_1)v\right] + \beta' \left[t_2 u + (1-t_2)v\right].$$

By our assumption, $t_i u + (1-t_i)v = (\alpha+\beta)^{-n} u_i$ with $u_i \in A$ for $i \in \{1,2\}$, and so

$$\begin{aligned} tu + (1-t)v &= (\alpha+\beta)^{-n} \left(\alpha' u_1 + \beta' u_2\right) \\ &= (\alpha+\beta)^{-n-1}(\alpha u_1 + \beta u_2) \in (\alpha+\beta)^{-n-1} A. \end{aligned}$$

Step 3: $t \in I,\ u, v \in A \Rightarrow tu + (1-t)v \in \mathbb{R}_+ A$. This is obvious by Step 1 and Step 2.

Step 4: $t \in [0,1]$, $u, v \in A \Rightarrow tu + (1-t)v \in \tilde{A}$. For $t \in \{0,1\}$ the assertion is obvious because $A \subseteq \tilde{A}$. Fix $t \in]0,1[$ and set $w := (1-t)u + tv$. For $\lambda \in]\frac{1}{2}, 1[$ consider $t_\lambda := 1 - \lambda t \in]0,1[$ and $u_\lambda := \lambda t_\lambda^{-1}(1-t)u$. We have that $\lambda w = t_\lambda u_\lambda + (1-t_\lambda)v$. Consider $V \in \mathscr{N}_Z'$ and take $W \in \mathscr{N}_Z'$ with $W + W \subseteq V$. As $u \in A$, by (ii), there exists $\delta \in]\frac{1}{2}, 1[$ such that $u_\lambda \in A + \{0\} \times W$ for every $\lambda \in [\delta, 1]$. Since $\operatorname{cl} I = [0,1]$, we have that $I \cap]1-t, 1-\delta t[\neq \emptyset$, and so there exists $\lambda \in]\delta, 1[$ such that $t_\lambda \in I \cap]1-t, 1-\delta t[$. Hence $u_\lambda = \tilde{u} + (0, z)$ with $\tilde{u} \in A$ and $z \in W$. Using Step 3, we obtain that

$$\lambda w = t_\lambda u_\lambda + (1-t_\lambda)v = t_\lambda \tilde{u} + (1-t_\lambda)v + t_\lambda(0, z),$$

and so $w \in \mathbb{R}_+ A + \{0\} \times 2W \subseteq \mathbb{R}_+ A + \{0\} \times V$ for every $V \in \mathscr{N}_Z'$. Hence $w \in \tilde{A}$.

Step 5: $\tilde{A}$ is a convex cone. Let $u \in \tilde{A}$ and $t > 0$. For $V \in \mathscr{N}_Z'$ there exists $W \in \mathscr{N}_Z'$ such that $tW \subseteq V$. By the definition of $\tilde{A}$, there exist $u' \in \mathbb{R}_+ A$ and $z \in W$ such that $u = u' + (0, z)$, and so $tu = tu' + (0, tz) \in \mathbb{R}_+ A + \{0\} \times V$. It follows that $tu \in \tilde{A}$.

Let now $u, v \in \tilde{A}$ and consider $V \in \mathscr{N}_Z'$; take $W \in \mathscr{N}_Z'$ such that $W + W + W \subseteq V$. Since $u, v \in \tilde{A}$, there exist $u', v' \in A$, $s, t \geq 0$ and $z, w \in W$ such that $u = su' + (0, z)$, $v = tv' + (0, w)$. If $s + t = 0$ then $u + v = 0 + (0, z + w) \in \mathbb{R}_+ A + \{0\} + V$. If $s + t > 0$ then

$$u + v = (s+t)\left[s(s+t)^{-1}u' + t(s+t)^{-1}v'\right] + (0, z + w).$$

Using Step 4 and the first part $\tilde{u} := (s+t)\left[s(s+t)^{-1}u' + t(s+t)^{-1}v'\right] \in \tilde{A}$. Therefore, there exists $\overline{u} \in \mathbb{R}_+ A$ and $\tilde{w} \in W$ such that $\tilde{u} = \overline{u} + (0, \tilde{w})$. It follows that $u + v = \overline{u} + (0, z + w + \tilde{w}) \in \mathbb{R}_+ A + \{0\} \times V$. Since $V \in \mathscr{N}_Z'$ is arbitrary we have that $u + v \in \tilde{A}$. Therefore, $\tilde{A}$ is a convex cone. □

Note that in the case $\{0\} \in \mathscr{N}_Z'$, if $A \subseteq X \times Z$ verifies conditions (i) and (ii) in the preceding theorem then $\mathbb{R}_+ A$ is convex. Observe that A satisfies condition (ii) if A is open (X, Z being t.v.s.) or $]0,1]A = A$.

Corollary 6.5.2. *Let Z be a topological vector space and let $K \subseteq Z$ be a proper convex cone. If $A \subseteq X \times Z$ is a nonempty set verifying conditions* (i) *and* (ii) *in Theorem 6.5.1 for $\mathscr{N}_Z'$ the class of balanced neighborhoods of* $0 \in Z$.

(j) *Assume that* $\operatorname{int} K \neq \emptyset$. *Then*

$$A \cap [\{0\} \times (-\operatorname{int} K)] = \emptyset \Rightarrow \operatorname{conv} A \cap [\{0\} \times (-\operatorname{int} K)] = \emptyset.$$

(jj) *Assume that K is closed. Then*

$$A(0) \subseteq K \Rightarrow (\operatorname{conv} A)(0) \subseteq K.$$

Proof. By Theorem 6.5.1 we have that $\operatorname{conv} A \subseteq \tilde{A}$.

(j) Assume that $\operatorname{int} K \neq \emptyset$ and $A \cap [\{0\} \times (-\operatorname{int} K)] = \emptyset$. It is sufficient to show that $\tilde{A} \cap [\{0\} \times (-\operatorname{int} K)] = \emptyset$. Suppose that $(0, -k) \in \tilde{A} \cap [\{0\} \times (-\operatorname{int} K)]$. Since $k \in \operatorname{int} K$, there exists $V \in \mathscr{N}_Z'$ such that $k + V \subseteq \operatorname{int} K$. On the other hand, from $(0, -k) \in \tilde{A}$ we get $u \in A$, $t \geq 0$ and $z \in V$ such that $(0, -k) = tu + (0, z)$, whence $tu = (0, -k - z) \in \{0\} \times (-\operatorname{int} K)$. Since K is proper, we have that $t > 0$, whence the contradiction $u \in A \cap [\{0\} \times (-\operatorname{int} K)] = \emptyset$.

(jj) Assume that K is closed and $A(0) \subseteq K$. It is sufficient to show that $\tilde{A}(0) \subseteq K$. Assume that $z \in \tilde{A}(0) \setminus K$. Because K is closed, there exists $V \in \mathscr{N}_Z'$ such that $W \subseteq V$ and $z \notin W$. Then there exist $u \in A$, $t \geq 0$ and $z' \in W$ such that $(0, z) = tu + (0, z')$. Hence $tu = (0, z - z')$. Because $z \notin W$ we have that $t > 0$. It follows that $t^{-1}(z - z') \in A(0)$, and so $z - z' \in K$. Hence $z \in K + W \subseteq K + V$. Since V is arbitrary we obtain the contradiction $z \in K$. Hence $\tilde{A}(0) \subseteq K$. □

Theorem 6.5.1 and its corollary can be found in [613].

and so $\{Tu \mid T \in \partial^s_\varsigma f(\overline{x})\}$ is relatively weakly compact for every $u \in U_0$. Because U_0 is absorbent we obtain that $\{Tu \mid T \in \partial^s_\varsigma f(\overline{x})\}$ is relatively weakly compact for every $u \in X$. The conclusion follows using [76, Proposition III.3.4].

(ii) By Lemma 7.1.2, we have that

$$\begin{aligned}\partial^s_\varsigma f(\overline{x}) &= \cap_{u\in X}\{T \in L(X,Y) \mid Tu \leq f'_\varsigma(\overline{x},u)\}\\ &= \cap_{u\in X}\{T \in L(X,Y) \mid Tu \in f'_\varsigma(\overline{x},u) - C\}.\end{aligned}$$

Since the mapping $\mathscr{L}_{\rm pw}(X,Y_\sigma) \ni T \mapsto Tu \in Y_\sigma$ is continuous for any $u \in X$ and C is (weakly) closed, $\partial^s_\varsigma f(\overline{x})$ is a closed subset of $\mathscr{L}_{\rm pw}(X,Y)$ (as intersection of a family of closed sets). Therefore, using (i), $\partial^s_\varsigma f(\overline{x})$ is compact in $\mathscr{L}_{\rm pw}(X,Y_\sigma)$.

(iii) Let us fix $y^* \in C^+$ and $\eta = 0$, and prove the inclusion $\subseteq$ in (7.2). Since the mapping $\mathscr{L}_{\rm pw}(X,Y_\sigma) \ni T \mapsto y^* \circ T \in (X^*,w^*)$ is continuous, where $w^* = \sigma(X^*,X)$, $y^* \circ \partial^s f(\overline{x})$ is w^*-compact. Hence $y^* \circ \partial^s f(\overline{x}) \subseteq X^*$ is nonempty, convex and w^*-compact (hence w^*-closed).
Set $g := y^* \circ f$. Since g is convex and continuous at $\overline{x}$, using Theorem 7.1.3 for Y replaced by $\mathbb{R}$ (or [614, Theorem 2.4.98]), we have that

$$\forall u \in X : g'(\overline{x},u) = g'_0(\overline{x},u) = \max\{\langle u,x^*\rangle \mid x^* \in \partial g(\overline{x})\}. \tag{7.3}$$

Take $\overline{x}^* \in X^* \setminus y^* \circ \partial^s f(\overline{x})$. Since, as seen above, $y^* \circ \partial^s f(\overline{x})$ is nonempty, convex and w^*-closed, there exists $\overline{u} \in X$ such that

$$\langle \overline{u},\overline{x}^*\rangle > \sup\{\langle \overline{u},x^*\rangle \mid x^* \in y^* \circ \partial^s f(\overline{x})\} = \sup\{\langle T\overline{u},y^*\rangle \mid T \in \partial^s f(\overline{x})\}. \tag{7.4}$$

By Theorem 7.1.3, there exists $T_0 \in \partial^s f(\overline{x})$ such that $T_0\overline{u} = f'(\overline{x},\overline{u})$. We have

$$\langle \overline{u}, y^* \circ T_0\rangle = \langle T_0\overline{u},y^*\rangle = \left\langle f'(\overline{x},\overline{u}),y^*\right\rangle = \left\langle \inf_{n\geq 1}\frac{f(\overline{x}+t_n\overline{u})-f(\overline{x})}{t_n},y^*\right\rangle,$$

where $(t_n)_{n\geq 1} \subseteq \mathbb{P}$ is a decreasing sequence with limit 0 such that $\overline{x} + t_n\overline{u} \in \operatorname{dom} f$ for every $n \geq 1$. By Lemma 7.1.7 (below), the sequence $\left(t_n^{-1}\left(f(\overline{x}+t_n\overline{u})-f(\overline{x})\right)\right)_{n\geq 1}$ weakly converges to its infimum. So, because y^* is continuous and positive,

$$\langle T_0\overline{u},y^*\rangle = \inf_{n\geq 1}\left\langle \frac{f(\overline{x}+t_n\overline{u})-f(\overline{x})}{t_n},y^*\right\rangle = g'(\overline{x},\overline{u}),$$

and so, by (7.4), $\langle \overline{u},\overline{x}^*\rangle > g'(\overline{x},\overline{u})$. Taking into account (7.3), we obtain that $\overline{x}^* \notin \partial g(\overline{x})$. Therefore, $y^* \circ \partial^s f(\overline{x}) \supseteq \partial(y^* \circ f)(\overline{x})$. The proof is complete. □

Lemma 7.1.7. *Assume that Y is a separated locally convex weakly sequentially complete space and $C \subseteq Y$ be normal and closed. If $(y_n)_{n\geq 1} \subseteq Y$ is a decreasing sequence which is bounded from below, then (y_n) converges weakly to $\inf\{y_n \mid n \geq 1\}$. Therefore, Y is weakly sequentially Daniell.*

Proof. Because C is normal, we have that C^+ is reproducing, that is $C^+ - C^+ = Y^*$.

For $y^* \in C^+$ we have that the sequence $(\langle y_n, y^*\rangle)_{n\geq 1} \subseteq \mathbb{R}$ is decreasing and bounded from below, and so $(\langle y_n, y^*\rangle)$ is convergent (in $\mathbb{R}$). Because C^+ is reproducing, we obtain that $(\langle y_n, y^*\rangle)$ is convergent for every $y^* \in Y^*$. In particular, the sequence $(y_n)_{n\geq 1}$ is a weakly Cauchy sequence. Because Y is weakly sequentially complete, there exists $y \in Y$ such that $(y_n)_{n\geq 1}$ weakly converges to y. Since $\langle y_n, y^*\rangle \searrow \langle y, y^*\rangle$ for $y^* \in C^+$, we obtain that $\langle y_n - y, y^*\rangle \geq 0$ for all $y^* \in C^+$ and all $n \geq 1$, whence $y_n - y \in C^{++} = \operatorname{cl} C = C$ for every $n \geq 1$. Hence y is a lower bound for $(y_n)_{n\geq 1}$. If $y' \in Y$ is another lower bound for $(y_n)_{n\geq 1}$, then $\langle y_n - y', y^*\rangle \geq 0$ for all $n \geq 1$ and $y^* \in C^+$, whence, taking the limit we get $\langle y - y', y^*\rangle \geq 0$ for every $y^* \in C^+$. Hence $y - y' \in C$, which shows that $y = \inf\{y_n \mid n \geq 1\}$. The proof is complete. □

Observe that assuming that Y is weakly complete instead of being weakly sequentially complete in Lemma 7.1.7 we obtain that Y is weakly Daniell.

Throughout the rest of this section we assume that C is a pointed proper normal convex cone such that (Y, C) has the l.u.b.p.

Definition 7.1.8. Let $f : X \to Y^\bullet$. The **(strong) conjugate** of f is

$$f_s^* : \mathscr{L}(X, Y) \to Y^\bullet, \quad f_s^*(T) := \sup\{Tx - f(x) \mid x \in X\}.$$

Of course, as in the scalar case, if $f(x) = +\infty$ for every $x \in X$ then $f_s^*(T) = -\infty$ for every $T \in \mathscr{L}(X, Y)$, that is $f_s^* = -\infty$; if there exists $x \in X$ such that $f(x) = -\infty$ then $f_s^* = +\infty$. (As in the scalar case, it is possible to have $f_s^* = +\infty$ even if f is C-convex and $f(x) \in Y$ for every $x \in X$.) So, it is natural to consider f a proper operator.

Setting $(-\infty) + (+\infty) := (+\infty) + (-\infty) := +\infty$, the Fenchel–Young inequality

$$\forall x \in X,\ \forall T \in \mathscr{L}(X, Y) : f(x) + f_s^*(T) \geq Tx,$$

holds.

Proposition 7.1.9. *Let $f : X \to Y^\bullet$ be proper. Then*

(i) *f_s^* is a convex operator.*
(ii) *Let $\varsigma \in C$ and $\overline{x} \in \operatorname{dom} f$; then*

$$T \in \partial_\varsigma^s f(\overline{x}) \Leftrightarrow f_s^*(T) + f(\overline{x}) \leq T\overline{x} + \varsigma, \quad T \in \partial f(\overline{x}) \Leftrightarrow f_s^*(T) + f(\overline{x}) = T\overline{x}.$$

(iii) *If X is ordered by a convex cone P and f is (P, C)-increasing then* $\operatorname{dom} f_s^* \subseteq \mathscr{L}_+(X, Y) := \{T \in \mathscr{L}(X, Y) \mid T(P) \subseteq C\}$.

Proof. The proofs for (i) and (ii) are immediate. For (iii) we proceed in the following way. Fix $z_0 \in \operatorname{dom} f$ and take $T \in \operatorname{dom} f_s^*$. Then for $z \in P$ we have that $T(z_0 - z) \leq f(z_0 - z) + f_s^*(T) \leq f(z_0) + f_s^*(T)$, whence $Y \ni y_0 := Tz_0 - f(z_0) - f_s^*(T) \leq Tz =: y$. Replacing z by αz with $\alpha \in \mathbb{R}_+$, we obtain that $y_0 \leq \alpha y$ for all $\alpha \in \mathbb{R}_+$. For $\alpha = 0$ we get $y_0 \leq 0$. Taking $\alpha := \beta^{-1}$ with $\beta \in \mathbb{P}$, we get $\beta y_0 \leq y$ for every $\beta \in \mathbb{P}$. Clearly, $\mathbb{P} \subseteq I := \{\beta \in \mathbb{R} \mid \beta y_0 \leq y\}$ and I is an interval. If $0 \in I$ then $y \in C$. Else $0 = \inf I$ (and $y_0 \neq 0$). Take $\overline{y}$ the least upper bound of $\{\beta y_0 \mid \beta \in \mathbb{P}\}$; hence $\overline{y} \leq y$. Since $0 \geq \beta y_0$ for $\beta \in \mathbb{P}$, we have that $\overline{y} \leq 0$. From $\beta y_0 \leq \overline{y}$ for every $\beta \in \mathbb{P}$, we get $\beta y_0 \leq 2\overline{y}$ for $\beta \in \mathbb{P}$, and so $\overline{y} \leq 2\overline{y}$, that is $0 \leq \overline{y}$. It follows that $\overline{y} = 0$, and so $y \in C$. Therefore, $T(P) \subseteq C$. □

The proof of (iii) uses an idea from the proof of [533, Theorem 1].

Let now Z be another topological vector space and $\Phi : X \times Z \to Y^\bullet$ be an operator. Consider the problem

$$\min \Phi(x, 0) \quad \text{s.t.} \quad x \in X, \tag{P}$$

called the **primal problem**. The perturbed problems for (P) are

$$\min \Phi(x, z) \quad \text{s.t.} \quad x \in X, \tag{P_z}$$

for $z \in Z$; of course, the initial problem (P) coincides with (P_0).

To the operator Φ (or problems (P_z)), we associate the **marginal** (or **value**) function

$$h : Z \to Y^\bullet, \quad h(z) := \inf\{\Phi(x, z) \mid x \in X\}; \tag{7.5}$$

hence, the value v_P of problem (P) is $h(0)$. Note that

$$\operatorname{dom} h = \operatorname{Pr}_Z(\operatorname{dom} \Phi) = \operatorname{Pr}_Z(\operatorname{epi} \Phi),$$

and so $0 \in \operatorname{dom} h$ iff $0 \in \operatorname{Pr}_Z(\operatorname{epi} \Phi)$ iff there exists $x \in X$ such that $\Phi(x, 0) < +\infty$.

It is worth observing that h is convex when Φ is convex. Furthermore, for every $T \in \mathscr{L}(Z, Y)$ one has

$$\begin{aligned} h_s^*(T) &= \sup\{Tz - h(z) \mid z \in Z\} = \sup\{Tz - \inf_{x \in X} \Phi(x, z) \mid z \in Z\} \\ &= \sup\{Tz - \Phi(x, z) \mid x \in X,\ z \in Z\} = \Phi_s^*(0, T). \end{aligned}$$

Because $h(0) \geq -h_s^*(T) = -\Phi_s^*(0, T)$ for every $T \in \mathscr{L}(Z, Y)$, it is natural to consider as **dual problem** of (P) the problem

$$\max[-\Phi_s^*(0, T)] \quad s.t. \quad T \in \mathscr{L}(Z, Y). \tag{D}$$

It follows that $v_P \geq v_D$, that is **weak duality** holds for problems (P) and (D). Of course, $v_P = v_D$ automatically when $v_P = -\infty$ or $v_D = +\infty$. A very important situation is when $v_P = v_D \in Y$ and (D) has optimal solutions, that is the problem (P) is **stable**. Taking into account Proposition 7.1.9, one has that (P) is stable if and only if $v_P \in Y$ and $\partial^s h(0) \neq \emptyset$; moreover, in this case, $\partial^s h(0)$ is the set of solutions of (D).

In the following result we provide quite general conditions in order to have $v_P = v_D$ and (D) have optimal solutions.

Theorem 7.1.10. *Let $\Phi \in \Lambda(X \times Z, Y)$ with $0 \in \mathrm{Pr}_Z(\mathrm{dom}\,\Phi)$. Assume that one of the following conditions holds:*

(i) *there exists $x_0 \in X$ such that $\Phi(x_0, \cdot)$ is finite and continuous at 0;*
(ii) *X is a locally convex space and there exists $x_0 \in X$ such that $\Phi(x_0, \cdot)|_{Z_0}$ is finite and continuous at 0, where $Z_0 := \mathrm{span}(\mathrm{Pr}_Z(\mathrm{dom}\,\Phi))$;*
(iii) *X, Y are Fréchet spaces, Z_0 is a barreled space (verified if Z is a Fréchet space and Z_0 is closed), $\mathrm{epi}\,\Phi$ is closed and $0 \in \mathrm{icr}(\mathrm{Pr}_Z(\mathrm{dom}\,\Phi))$.*

Then

$$\inf_{x \in X} \Phi(x, 0) = \max_{T \in \mathscr{L}(Z,Y)} \left[-\Phi_s^*(0, T)\right]. \tag{7.6}$$

Moreover, $\overline{x}$ is an optimal solution for (P) if and only if there exists $T \in \mathscr{L}(Z, Y)$ such that $(0, T) \in \partial^s \Phi(\overline{x}, 0)$.

Proof. Consider the marginal function h defined by (7.5); because Φ is convex, h is convex, too. Since $0 \in \mathrm{Pr}_Z(\mathrm{dom}\,\Phi) = \mathrm{dom}\,h$, we have that $h(0) < +\infty$. If $h(0) = -\infty$ then the conclusion clearly holds because $\Phi_s^*(0, T) = h_s^*(T) = +\infty$ for every $T \in \mathscr{L}(Z, Y)$.

Assume that $h(0) \in Y$. Because in anyone of the conditions (i)–(iii) $0 \in \mathrm{icr}(\mathrm{Pr}_Z(D))$, by Remark 7.1.1, $h \in \Lambda(Z, Y)$.

(i) It is obvious that $h \leq \Phi(x_0, \cdot)$; applying Proposition 3.2.11 we obtain that h is continuous at 0. Using now Theorem 7.1.3 we have that $\partial^s h(0) \neq \emptyset$. Taking $T_0 \in \partial^s h(0)$ we have that

$$v_D \leq v_P = h(0) = -h_s^*(T_0) = -\Phi_s^*(0, T_0) \leq v_D.$$

It follows that (7.6) holds.

(ii) As in (i), $h|_{Z_0} \leq \Phi(x_0, \cdot)|_{Z_0}$, and so $h|_{Z_0}$ is continuous at $0 \in \mathrm{dom}\,h|_{Z_0}$. Using now Corollary 7.1.5 we get $T_0 \in \partial^s h(0)$. As in (i), the conclusion holds.

(iii) Let $\Phi_0 := \Phi|_{X \times Z_0}$. Then $\Phi_0 \in \Lambda(X \times Z_0)$ has closed epigraph (because $\mathrm{epi}\,\Phi = \mathrm{epi}\,\Phi_0 \subseteq X \times Z_0 \times Y$). Consider

$$F : X \times Y \rightrightarrows Z_0, \quad \mathrm{graph}\,F := \{(x, y, z) \in X \times Y \times Z_0 \mid (x, z, y) \in \mathrm{epi}\,\Phi_0\}.$$

It is obvious that $\mathrm{graph}\,F$ is a closed convex set and $\mathrm{Im}\,F = \mathrm{Pr}_Z(\mathrm{dom}\,\Phi) = \mathrm{Pr}_{Z_0}(\mathrm{dom}\,\Phi) \subseteq Z_0$, and so $0 \in \mathrm{core}(\mathrm{Im}\,F)$. Take $(x_0, y_0) \in X \times Y$ such that $0 \in F(x_0, y_0)$ ($\Leftrightarrow \Phi_0(x_0, 0) \leq y_0$). Using [576, Theorem] (see also [614, Theorem 1.3.7]), we obtain that

$$\forall W \in \mathcal{N}_Y,\ U \in \mathcal{N}_X,\ \exists V \in \mathcal{N}_{Z_0} : V \subseteq F\left((x_0, y_0) + U \times W\right).$$

Therefore,

$$\forall W \in \mathcal{N}_Y,\ U \in \mathcal{N}_X,\ \exists V \in \mathcal{N}_{Z_0},\ \forall z \in V,\ \exists x \in x_0 + U,\ y \in y_0 + W : z \in F(x, y),$$

or, equivalently

$$\forall W \in \mathcal{N}_Y,\ U \in \mathcal{N}_X,\ \exists V \in \mathcal{N}_{Z_0},\ \forall z \in V,\ \exists x \in x_0 + U : \Phi_0(x, z) \in y_0 + W - C.$$

In conclusion,

$$\exists x_0 \in X,\ y_0 \in Y,\ \forall W \in \mathcal{N}_Y,\ U \in \mathcal{N}_X,\ \exists V \in \mathcal{N}_{Z_0},$$
$$\forall z \in V,\ \exists x \in x_0 + U : \Phi_0(x, z) \in y_0 + W - C. \tag{7.7}$$

From the previous relation we obtain that

$$\exists y_0 \in Y,\ \forall W \in \mathcal{N}_Y,\ \exists V \in \mathcal{N}_{Z_0},\ \forall z \in V : h|_{Z_0}(z) \in y_0 + W - C.$$

Applying Proposition 3.2.11, we obtain that h_0 is continuous at 0. Using (ii) we get the conclusion.

Assume that $\overline{x}$ is an optimal solution for (P). Then $\Phi(\overline{x}, 0) = \inf_{x \in X} \Phi(x, 0)$, and so, by (7.6), there exists $T \in \mathcal{L}(Z, Y)$ such that $\Phi(\overline{x}, 0) = -\Phi_s^*(0, T)$. It follows that $\Phi(\overline{x}, 0) + \Phi_s^*(0, T) = 0 = 0\overline{x} + T0$, and so $(0, T) \in \partial^s \Phi(\overline{x}, 0)$. The converse implication is immediate. □

As in the scalar case, the preceding result gives the possibility to compute conjugates and subdifferentials for functions obtained by operations that preserve convexity.

Theorem 7.1.11. *Let $f \in \Lambda(X, Y)$, $g \in \Lambda(Z, Y)$ and $S \in \mathcal{L}(X, Z)$. Assume that one of the following conditions holds:*

(i) *there exists $x_0 \in \operatorname{dom} f \cap S^{-1}(\operatorname{dom} g)$ such that g is continuous at Sx_0;*
(ii) *X is a locally convex space and there exists $x_0 \in \operatorname{dom} f \cap S^{-1}(\operatorname{dom} g)$ such that $g|_{\operatorname{aff}(\operatorname{dom} g)}$ is continuous at Sx_0;*
(iii) *X, Y are Fréchet spaces, C is closed, $0 \in \operatorname{icr}(\operatorname{dom} g - S(\operatorname{dom} f))$ and $\operatorname{span}(\operatorname{dom} g - S(\operatorname{dom} f))$ is a barreled space, f and g are C-lower continuous.*

Then

$$(f + g \circ S)_s^*(T) = \min\left\{f_s^*(T - U \circ S) + g_s^*(U) \mid U \in \mathcal{L}(Z, Y)\right\} \tag{7.8}$$

for all $T \in \mathscr{L}(X,Y)$, *and*

$$\partial^s_\varsigma (f + g \circ S)(x) = \bigcup \left\{ \partial^s_{\varsigma_1} f(x) + \left(\partial^s_{\varsigma_2} g(Sx)\right) \circ S \mid \varsigma_1, \varsigma_2 \in C,\ \varsigma_1 + \varsigma_2 = \varsigma \right\} \tag{7.9}$$

for all $x \in X$ *and* $\varsigma \in C$.

Proof. Fix $T \in \mathscr{L}(X,Y)$ and $U \in \mathscr{L}(Z,Y)$. Then

$$(T - U \circ S)x \leq f(x) + f_s^*(T - U \circ S), \quad Uz \leq g(z) + g_s^*(U) \quad \forall x \in X,\ z \in Z.$$

Taking $z := Sx$ and summing up both inequalities we get

$$Tx \leq f(x) + g(Sx) + f_s^*(T - U \circ S) + g_s^*(U) \quad \forall x \in X.$$

This shows that

$$(f + g \circ S)_s^*(T) \leq f_s^*(T - U \circ S) + g_s^*(U).$$

Hence

$$(f + g \circ S)_s^*(T) \leq \inf\{ f_s^*(T - U \circ S) + g_s^*(U) \mid U \in \mathscr{L}(Z,Y)\} \tag{7.10}$$

(for arbitrary f, g and $S \in \mathscr{L}(X,Z)$). It follows that (7.8) holds if $T \notin \operatorname{dom}(f + g \circ S)_s^*$.

Take $T \in \operatorname{dom}(f + g \circ S)_s^*$ and consider $\Phi : X \times Z \to Y^\bullet$ defined by $\Phi(x,z) := f(x) + g(Sx + z) - Tx$. Clearly, Φ is convex and $0 \in \operatorname{Pr}_Z(\operatorname{dom} \Phi) = \operatorname{dom} g - S(\operatorname{dom} f)$; hence $\Phi \in \Lambda(X \times Z, Y)$; moreover $Z_0 := \operatorname{span}(\operatorname{Pr}_Z(\operatorname{dom} \Phi)) = \operatorname{span}(\operatorname{dom} g - S(\operatorname{dom} f))$.

Assume that (i) holds. Then the mapping $z \mapsto \Phi(x_0, z) = f(x_0) + g(Sx_0 + z) - Tx_0$ is clearly continuous at 0 (because g is continuous at Sx_0). Hence condition (i) of Theorem 7.1.10 holds. In a similar way, condition (ii) of Theorem 7.1.10 holds when (ii) is verified.

Assume that (iii) holds. Set $\tilde{f}(x,z) := f(x) - Tx = (f - T)(\operatorname{Pr}_X(x,z))$ and $\tilde{g}(x,z) := g(Sx + z) = g((S, \operatorname{Id}_Z)(x,z))$, where $(S, \operatorname{Id}_Z)(x,z) := Sx + z$. Since T is continuous and f is C-lower continuous we have that $f - T$ is C-lower continuous by Proposition 3.1.31 (ii); since $\operatorname{Pr}_X \in \mathscr{L}(X \times Z, X)$, $\tilde{f}$ is C-lower continuous by Proposition 3.1.31 (iii). Similarly, $\tilde{g}$ is C-lower continuous because g is C-lower continuous and $(S, \operatorname{Id}_Z) \in \mathscr{L}(X \times Z, Z)$. Since $\Phi = \tilde{f} + \tilde{g}$, Φ is C-lower continuous by Proposition 3.1.31 (ii). Using now Proposition 3.1.31 (i) we have that epi Φ is closed. Hence condition (iii) in Theorem 7.1.10 holds. Therefore, (7.6) holds.

Clearly, $\inf_{x \in X} \Phi(x,0) = -(f + g \circ S)_s^*(T) \in Y$. Using (7.6), there exists $U \in \mathscr{L}(Z,Y)$ such that $(f + g \circ S)_s^*(T) = \Phi_s^*(0, U)$. But

$$\begin{aligned}
\Phi_s^*(0,U) &= \sup\{Uz - f(x) - g(Sx+z) + Tx \mid x \in X,\ z \in Z\} \\
&= \sup\{U(z' - Sx) - f(x) - g(z') + Tx \mid x \in X,\ z' \in Z\} \\
&= \sup\{Uz - g(z) \mid x \in X,\ z \in Z\} + \sup\{(T - U \circ S)x - f(x) \mid x \in X\} \\
&= g_s^*(U) + f_s^*(T - U \circ S).
\end{aligned}$$

Taking into account (7.10) we obtain that (7.8) holds.

Take now $T \in \partial_{\varsigma_1}^s f(x)$ and $U \in \partial_{\varsigma_2}^s g(Sx)$, where $\varsigma_1, \varsigma_2 \in C$. Then

$$f(x) + f_s^*(T) \leq Tx + \varsigma_1, \quad g(Sx) + g_s^*(U) \leq U(Sx) + \varsigma_2,$$

and so

$$(f + g \circ S)(x) + g_s^*(U) \leq (T + U \circ S)x + \varsigma_1 + \varsigma_2,$$

which proves that $T + U \circ S \in \partial_{\varsigma_1+\varsigma_2}^s (f + g \circ S)(x)$. Hence the inclusion $\supseteq$ in (7.9) holds (for arbitrary f, g and $S \in \mathscr{L}(X,Z)$).

Conversely, take $T \in \partial_\varsigma^s (f + g \circ S)(x)$. By Proposition 7.1.9 one has that $(f + g \circ S)(x) + (f + g \circ S)_s^*(T) \leq Tx + \varsigma$. From (7.8) we get $U \in \mathscr{L}(Z,Y)$ such that $(f + g \circ S)_s^*(T) = f_s^*(T - U \circ S) + g_s^*(U)$, and so

$$f(x) + g(Sx) + f_s^*(T - U \circ S) + g_s^*(U) \leq Tx + \varsigma = (T - U \circ S)x + U(Sx) + \varsigma,$$

or, equivalently,

$$\left[f(x) + f_s^*(T - U \circ S) - (T - U \circ S)x\right] + \left[g(Sx) + g_s^*(U) - U(Sx)\right] \leq \varsigma.$$

Taking $\varsigma_1 := \left[f(x) + f_s^*(T - U \circ S) - (T - U \circ S)x\right] \in C$ and $\varsigma_2 := \varsigma - \varsigma_1 \in C$, we obtain that $T - U \circ S \in \partial_{\varsigma_1}^s f(x)$, $U \in \partial_{\varsigma_2}^s g(Sx)$. Hence, the inclusion $\subseteq$ in (7.9) holds, too. □

There are two important particular cases of the preceding result: a) $X = Z$, $S = \mathrm{Id}_X$, and b) $f = 0$.

Corollary 7.1.12. *Let $f, g \in \Lambda(X,Y)$. Assume that one of the following conditions holds:*

(i) *there exists $x_0 \in \operatorname{dom} f \cap \operatorname{dom} g$ such that g is continuous at x_0;*
(ii) *X is a locally convex space and there exists $x_0 \in \operatorname{dom} f \cap \operatorname{dom} g$ such that $g|_{\operatorname{aff}(\operatorname{dom} g)}$ is continuous at x_0;*
(iii) *X, Y are Fréchet spaces, C is closed, $0 \in \operatorname{icr}(\operatorname{dom} f - \operatorname{dom} g)$ and $\operatorname{span}(\operatorname{dom} f - \operatorname{dom} g)$ is a barreled space, and f, g are C-lower continuous.*

Then

$$\forall T \in \mathscr{L}(X,Y) : (f + g)_s^*(T) = \min\{f_s^*(T - S) + g_s^*(S) \mid S \in \mathscr{L}(X,Y)\},$$

$$\forall x \in X,\ \forall \varsigma \in C : \partial_\varsigma^s (f + g)(x) = \bigcup\{\partial_{\varsigma_1}^s f(x) + \partial_{\varsigma_2}^s g(x) \mid \varsigma_1, \varsigma_2 \in C,\ \varsigma_1 + \varsigma_2 = \varsigma\}.$$

Corollary 7.1.13. *Let $g \in \Lambda(Z,Y)$ and $S \in \mathscr{L}(X,Z)$. Assume that one of the following conditions holds:*

(i) *there exists $x_0 \in S^{-1}(\operatorname{dom} g)$ such that g is continuous at Sx_0;*
(ii) *X is a locally convex space and there exists $x_0 \in S^{-1}(\operatorname{dom} g)$ such that $g|_{\operatorname{aff}(\operatorname{dom} g)}$ is continuous at Sx_0;*
(iii) *X, Y are Fréchet spaces, C is closed, $0 \in \operatorname{icr}(\operatorname{dom} g - \operatorname{Im} S)$ and $\operatorname{span}(\operatorname{dom} g - \operatorname{Im} S)$ is a barreled space, and g is C-lower continuous.*

Then

$$\forall T \in \mathscr{L}(X,Y) : (g \circ S)_s^*(T) = \min\{g_s^*(U) \mid U \in \mathscr{L}(Z,Y),\ U \circ S = T\},$$

$$\forall x \in X,\ \forall \varsigma \in C : \partial_\varsigma^s(g \circ S)(x) = \left(\partial_\varsigma^s g(Sx)\right) \circ S.$$

Theorem 7.1.14. *Let Z be ordered by the convex cone P, $f \in \Lambda(Z,Y)$ be (P,C)-increasing, $g \in \Lambda(X,Y)$, and $\operatorname{dom} f \cap g(\operatorname{dom} g) \neq \emptyset$. Assume that one of the following conditions holds:*

(i) *there exists $x_0 \in g^{-1}(\operatorname{dom} f)$ ($\subseteq \operatorname{dom} g$) such that f is continuous at $g(x_0)$;*
(ii) *X is a locally convex space and there exists $x_0 \in g^{-1}(\operatorname{dom} f)$ such that $f|_{\operatorname{aff}(\operatorname{dom} f)}$ is continuous at $g(x_0)$;*
(iii) *X, Y are Fréchet spaces, $0 \in \operatorname{icr}(\operatorname{dom} f - g(\operatorname{dom} g))$ and $\operatorname{span}(\operatorname{dom} f - g(\operatorname{dom} g))$ is a barreled space, C is closed, f, g are C-lower continuous and for every $y \in Y$ there exists $z \in Z$ with $f(z) \geq y$.*

Then, setting $f(+\infty) := +\infty$, one has

$$\forall T \in \mathscr{L}(X,Y) : (f \circ g)_s^*(T) = \min\{f_s^*(U) + (U \circ g)_s^*(T) \mid U \in \mathscr{L}_+(Z,Y)\}, \tag{7.11}$$

where $U(+\infty) := +\infty$, and

$$\partial_\varsigma^s(f \circ g)(x) = \bigcup \left\{\partial_{\varsigma_1}^s(U \circ g)(x) \mid U \in \partial_{\varsigma_2}^s f\left(g(x)\right),\ \varsigma_1, \varsigma_2 \in C,\ \varsigma_1 + \varsigma_2 = \varsigma\right\} \tag{7.12}$$

for all $x \in X$ and $\varsigma \in C$.

Proof. Using Proposition 7.1.9 (iii) we have that $\operatorname{dom} f_s^* \subseteq \mathscr{L}_+(Z,Y)$. Taking $T \in \mathscr{L}(X,Y)$ and $U \in \mathscr{L}_+(Z,Y)$, we have that

$$\forall x \in X,\ \forall z \in Z : f(z) + f_s^*(U) \geq Uz, \quad (U \circ g)(x) + (U \circ g)_s^*(T) \geq Tx.$$

Adding side by side both inequalities for $z = g(x)$ with $x \in \operatorname{dom} g$, we get

$$f(g(x)) + f_s^*(U) + (U \circ g)_s^*(T) \geq Tx.$$

It follows that $f_s^*(U) + (U \circ g)_s^*(T) \geq (f \circ g)_s^*(T)$, and so

$$\forall T \in \mathscr{L}(X,Y) : (f \circ g)_s^*(T) \le \inf\{f_s^*(U) + (U \circ g)_s^*(T) \mid U \in \mathscr{L}_+(Z,Y)\}. \tag{7.13}$$

A similar computation shows that the inclusion $\supseteq$ in (7.12) holds.

In order to prove (7.11) we use Theorem 7.1.10. For this (it is sufficient to) take $T \in \operatorname{dom}(f \circ g)_s^*$. Consider $\Phi : X \times Z \to Y$, $\Phi(x,z) = f(g(x)+z) - Tx$. Since f and g are convex and f is increasing, Φ is convex. Moreover, $\operatorname{Pr}_Z(\operatorname{dom}\Phi) = \operatorname{dom} f - g(\operatorname{dom} g)$, and so $0 \in \operatorname{Pr}_Z(\operatorname{dom}\Phi)$.

In case (i) we have that $\Phi(x_0,\cdot)$ is continuous at 0 because f is continuous at $g(x_0) \in \operatorname{dom} f$. Similarly, in case (ii), $\Phi(x_0,\cdot)|_{Z_0}$ is continuous at 0, where $Z_0 := \operatorname{span}(\operatorname{Pr}_Z(\operatorname{dom}\Phi)) = \operatorname{span}(\operatorname{dom} f - g(\operatorname{dom} g))$. In case (iii) we have that $0 \in \operatorname{icr}(\operatorname{Pr}_Z(\operatorname{dom}\Phi))$ and Z_0 is barreled. We have to prove that $\operatorname{epi}\Phi$ is closed. For this, by Proposition 3.1.31 (i), it is sufficient to show that Φ is C-lower continuous. But $\tilde{g} : X \times Z \to Y^\bullet$ defined by $\tilde{g}(x,z) := g(x) + z$ is proper and C-lower continuous. Using Proposition 3.1.31 (iv), we have that Φ is C-lower continuous. Therefore condition (i), (ii) or (iii) of Theorem 7.1.10 is verified if (i), (ii) or (iii) holds, respectively. Using Theorem 7.1.10 we get $U \in \mathscr{L}(Z,Y)$ such that $\inf_{x\in X}\Phi(x,0) = -\Phi_s^*(0,U)$, or, equivalently, $(f \circ g)_s^*(T) = \Phi_s^*(0,U)$. But

$$\begin{aligned}
\Phi_s^*(0,U) &= \sup\{Uz - f(g(x)+z) + Tx \mid x \in X,\ z \in Z\} \\
&= \sup\left\{U(z' - g(x)) - f(z') + Tx \mid x \in \operatorname{dom} g,\ z' \in Z\right\} \\
&= \sup\{Tx - U(g(x)) \mid x \in \operatorname{dom} g\} + \sup\{Uz - f(z) \mid z \in Z\} \\
&= (U \circ g)_s^*(T) + f_s^*(U).
\end{aligned}$$

It follows that $U \in \operatorname{dom} f_s^* \subseteq \mathscr{L}_+(Z,Y)$. Using (7.13) we have that (7.11) holds.

Take now $T \in \partial_\varsigma^s(f \circ g)(x)$ with $x \in X$ and $\varsigma \in C$. Using Proposition 7.1.9 (ii) we have that $(f \circ g)(x) + (f \circ g)_\varsigma^s(T) \le Tx + \varsigma$; hence $x \in g^{-1}(\operatorname{dom} f) \subseteq \operatorname{dom} g$. From (7.11) we get $U \in \mathscr{L}_+(Z,Y)$ such that $(f \circ g)_\varsigma^s(T) = f_s^*(U) + (U \circ g)_s^*(T)$, and so

$$(f \circ g)(x) + f_s^*(U) + (U \circ g)_s^*(T) \le Tx + \varsigma,$$

whence

$$\left[f(g(x)) + f_s^*(U) - U(g(x))\right] + \left[(U \circ g)(x) + (U \circ g)_s^*(T) - Tx\right] \le \varsigma.$$

Taking $\varsigma_1 := (U \circ g)(x) + (U \circ g)_s^*(T) - Tx \in C$ and $\varsigma_2 := \varsigma - \varsigma_1 \in C$, we have that $U \in \partial_{\varsigma_2}^s f(g(x))$ and $T \in \partial_{\varsigma_1}^s(U \circ g)(x)$. Hence (7.12) holds. □

The strong conjugate and subdifferential of vector valued functions were introduced by Raffin [485] and studied by many authors; see the book [358] by Kusraev and Kutateladze for a detailed list of references. Theorem 7.1.6 is stated by Valadier in [580, Corollary 7]; its proof and its preliminaries follow the presentation in [580]. The other results in this section can be found, essentially, in [610] (see also [609, 611, 616]).

7.2 The Weak Subdifferential

Throughout this section we assume that $C \subseteq Y$ is a proper pointed convex cone with nonempty interior; moreover, we use the notation

$$E^{w,C}(A) := E^w(A) := \mathrm{WMin}\, A := \mathrm{WMin}(A, C).$$

Moreover, for $\varsigma \in Y$, we set

$$E_\varsigma^{w,C}(A) := E_\varsigma^w(A) := \{a \in A \mid A \cap (a - \varsigma - \operatorname{int} C) = \emptyset\} = A \setminus (\varsigma + A + \operatorname{int} C). \tag{7.14}$$

Of course, $E^w(A) = E_0^w(A)$. Clearly,

$$E_\varsigma^w(A) \neq \emptyset \Rightarrow \varsigma \in D^w := Y \setminus (-\operatorname{int} C).$$

The next result represents an extension of Proposition 2.4.5 (iii).

Lemma 7.2.1. *Let $\emptyset \neq A \subseteq Y$ be closely C-convex and $\overline{y} \in Y$. Then $A \cap (\overline{y} - \operatorname{int} C) = \emptyset$ iff*

$$\exists y^* \in C^+ \setminus \{0\},\ \forall y \in A : \langle \overline{y}, y^* \rangle \leq \langle y, y^* \rangle . \tag{7.15}$$

In particular for $\varsigma \in Y$, $\overline{y} \in E_\varsigma^w(A)$ iff $\overline{y} \in A$ and there exists $y^* \in C^+ \setminus \{0\}$ such that $\langle \overline{y} - \varsigma, y^* \rangle \leq \langle y, y^* \rangle$ for every $y \in A$.

Proof. Since $C + \operatorname{int} C = \operatorname{int} C$, we have that

$$A \cap (\overline{y} - \operatorname{int} C) = \emptyset \Leftrightarrow \overline{y} \notin A + \operatorname{int} C \Leftrightarrow (A + C) \cap (\overline{y} - \operatorname{int} C) = \emptyset.$$

So, assuming that $A \cap (\overline{y} - \operatorname{int} C) = \emptyset$, we have that $(A + C) \cap (\overline{y} - \operatorname{int} C) = \emptyset$. Using a separation theorem, there exists $y^* \in Y^* \setminus \{0\}$ such that $\langle y + v, y^* \rangle \geq \langle \overline{y} - v', y^* \rangle$ for all $y \in A, v \in C, v' \in \operatorname{int} C$. It follows that $y^* \in C^+$ and $\langle \overline{y}, y^* \rangle \leq \langle y, y^* \rangle$ for every $y \in A$. Conversely, let y^* provided by (7.15). Assuming that $y \in A \cap (\overline{y} - \operatorname{int} C)$, we get the contradiction $0 \geq \langle \overline{y} - y, y^* \rangle > 0$ because $\overline{y} - y \in \operatorname{int} C$ and $y^* \in C^+ \setminus \{0\}$. □

Let us recall now the notion of (approximate) weak subdifferential of set-valued maps. First note that $F + T$ is C-convex (resp. C-closed) for $T \in \mathcal{L}(X, Y)$ provided $F : X \rightrightarrows Y$ is C-convex (resp. C-closed). Let $F : X \rightrightarrows Y$; the **weak subdifferential of** F **at** $(\overline{x}, \overline{y}) \in \operatorname{graph} F$ is the set

$$\partial^w F(\overline{x}, \overline{y}) := \partial^{w,C} F(\overline{x}, \overline{y}) := \{T \in \mathcal{L}(X, Y) \mid \overline{y} - T\overline{x} \in E^w(\operatorname{Im}(F - T))\};$$

moreover, if $\overline{x} \in \operatorname{dom} F$, the **weak subdifferential of** F **at** $\overline{x}$ is the set

$$\partial^w F(\overline{x}) := \bigcup_{y \in F(\overline{x})} \partial^w F(\overline{x}, y)$$

(see [566, Definition 4.1]). Similarly, if $\varsigma \in Y$, the ς-**weak subdifferential of** F **at** $(\overline{x}, \overline{y}) \in \operatorname{graph} F$ is the set

$$\partial_\varsigma^w F(\overline{x}, \overline{y}) := \partial_\varsigma^{w,C} F(\overline{x}, \overline{y}) := \{T \in \mathscr{L}(X, Y) \mid \overline{y} - T\overline{x} \in E_\varsigma^w(\operatorname{Im}(F - T))\};$$

of course, $\partial^w F(\overline{x}, \overline{y}) = \partial_0^w F(\overline{x}, \overline{y})$.

Chen and Jahn [92] have taken as the weak subdifferential of F at $\overline{x} \in \operatorname{dom} F$ the set $\bigcap\{\partial^w F(\overline{x}, \overline{y}) \mid \overline{y} \in F(\overline{x})\}$.

Note that $\overline{y} \in E_\varsigma^w(F(\overline{x}))$ if $\partial_\varsigma^w F(\overline{x}, \overline{y}) \neq \emptyset$; consequently, $F(\overline{x}) = E^w(F(\overline{x}))$ if $\bigcap\{\partial^w F(\overline{x}, \overline{y}) \mid \overline{y} \in F(\overline{x})\} \neq \emptyset$.

The next result establishes a characterization of the ς-weak sub-gradients of a convex set-valued map at a point in its graph.

Proposition 7.2.2. *Let $F : X \rightrightarrows Y$ be C-convex, $(\overline{x}, \overline{y}) \in \operatorname{graph} F$, $T \in \mathscr{L}(X, Y)$ and $\varsigma \in Y$. Then $T \in \partial_\varsigma^w F(\overline{x}, \overline{y})$ iff there exists $y^* \in C^+ \setminus \{0\}$ such that*

$$\forall (x, y) \in \operatorname{graph} F : \langle y - \overline{y} - T(x - \overline{x}) + \varsigma, y^*\rangle \geq 0. \tag{7.16}$$

Proof. By definition, $T \in \partial_\varsigma^w F(\overline{x}, \overline{y})$ iff $\overline{y} - T\overline{x} \in E_\varsigma^w(\operatorname{Im}(F - T))$. Since $F - T$ is C-convex we have that $\operatorname{Im}(F - T)$ is C-convex. Using Lemma 7.2.1 we obtain that $T \in \partial_\varsigma^w F(\overline{x}, \overline{y})$ iff there exists $y^* \in C^+ \setminus \{0\}$ such that $\langle y, y^*\rangle \geq \langle \overline{y} - T\overline{x} - \varsigma, y^*\rangle$ for all $y \in \operatorname{Im}(F - T)$. The conclusion holds. □

A simple existence result for the ς-weak subdifferential of F at $(\overline{x}, \overline{y}) \in \operatorname{graph} F$ is the following.

Proposition 7.2.3. *Let $F : X \rightrightarrows Y$ be C-convex, $(\overline{x}, \overline{y}) \in \operatorname{graph} F$ and $\varsigma \in Y$. Assume that $\operatorname{int}(\operatorname{epi} F) \neq \emptyset$. If $\partial_\varsigma^w F(\overline{x}, \overline{y}) \neq \emptyset$ then $(\overline{x}, \overline{y} - \varsigma) \notin \operatorname{int}(\operatorname{epi} F)$. Conversely, if $(\overline{x}, \overline{y} - \varsigma) \notin \operatorname{int}(\operatorname{epi} F)$ and $\overline{x} \in \operatorname{int}(\operatorname{dom} F)$ then $\partial_\varsigma^w F(\overline{x}, \overline{y}) \neq \emptyset$.*

Proof. Take $T \in \partial_\varsigma^w F(\overline{x}, \overline{y})$. By Proposition 7.2.2 there exists $y^* \in C^+ \setminus \{0\}$ such that (7.16) holds. Then $\langle x - \overline{x}, -T^* y^*\rangle + \langle y - (\overline{y} - \varsigma), y^*\rangle \geq 0$ for every $(x, y) \in \operatorname{epi} F$. Since $(-T^* y^*, y^*) \neq (0, 0)$, it follows that $(\overline{x}, \overline{y} - \varsigma) \notin \operatorname{int}(\operatorname{epi} F)$.

Conversely, assume that $(\overline{x}, \overline{y} - \varsigma) \notin \operatorname{int}(\operatorname{epi} F)$ and $\overline{x} \in \operatorname{int}(\operatorname{dom} F)$. Since $\operatorname{epi} F$ is convex, by a separation theorem, there exists $(x^*, y^*) \in X^* \times Y^* \setminus \{(0, 0)\}$ such that

$$\forall (x, y) \in \operatorname{epi} F : \langle x - \overline{x}, x^*\rangle + \langle y - (\overline{y} - \varsigma), y^*\rangle \geq 0. \tag{7.17}$$

It follows that $y^* \in C^+$. Assuming that $y^* = 0$, we obtain that $\langle x - \overline{x}, x^*\rangle \geq 0$ for every $x \in \operatorname{dom} F$. Since $\overline{x} \in \operatorname{int}(\operatorname{dom} F)$ we get $x^* = 0$, and so we get the contradiction $(x^*, y^*) = (0, 0)$. Therefore, $y^* \in C^+ \setminus \{0\}$. There exists $k^0 \in \operatorname{int} C$ such that $\langle k^0, y^*\rangle = 1$. Taking $T : X \to Y$ defined by $Tx := -\langle x, x^*\rangle k^0$, from (7.17) we obtain that (7.16) holds, and so $T \in \partial_\varsigma^w F(\overline{x}, \overline{y})$. □

Before establishing other existence results and some rules for the ς-weak subdifferential let us recall the scalarization function associated to $\operatorname{cl} C$ and $k^0 \in \operatorname{int} C$:

$$\varphi = \varphi_{k^0} : Y \to \mathbb{R}, \quad \varphi(y) := \inf\{t \in \mathbb{R} \mid y \in tk^0 - \operatorname{cl} C\}. \tag{7.18}$$

As usual, we use the conventions

$$\inf \emptyset := +\infty, \quad \sup \emptyset := -\infty, \quad (+\infty) + (-\infty) := (-\infty) + (+\infty) := +\infty.$$

It is known (see Corollary 5.2.8 for C replaced by cl$C$) that φ is a C-increasing continuous sublinear function such that

$$\forall \lambda \in \mathbb{R} : [\varphi \le \lambda] = \lambda k^0 - \operatorname{cl} C, \quad [\varphi < \lambda] = \lambda k^0 - \operatorname{int} C, \tag{7.19}$$

and

$$\forall y \in Y, \ \forall t \in \mathbb{R} : \varphi(y + tk^0) = \varphi(y) + t.$$

Moreover, the Fenchel conjugate of φ is given by the following formula (see Proposition 5.2.23):

$$\varphi^*(y^*) = \begin{cases} 0 & \text{if } y^* \in C^+, \ \langle k^0, y^* \rangle = 1, \\ +\infty & \text{otherwise.} \end{cases} \tag{7.20}$$

The next result is immediate from (7.19).

Lemma 7.2.4. *Let $A \subseteq Y$ and $\overline{y} \in Y$. Then $A \cap (\overline{y} - \operatorname{int} C) = \emptyset$ iff $\varphi(y - \overline{y}) \ge 0$ for every $y \in A$. In particular for $\varsigma \in Y$, one has $\overline{y} \in E^w_\varsigma(A)$ iff $\overline{y} \in A$ and $\varphi(y - \overline{y} + \varsigma) \ge 0$ for every $y \in A$.*

Lemma 7.2.5. *If $A, B \subseteq Y$ then $\inf \varphi(A+B) \le \inf \varphi(A) + \inf \varphi(B)$. In particular, if $\inf \varphi(A + B) \in \mathbb{R}$ then $\inf \varphi(A), \inf \varphi(B) \in \mathbb{R}$.*

Proof. The assertions are obvious. □

Lemma 7.2.6. *Let $F : X \rightrightarrows Y$ and take*

$$\varphi F : X \to \overline{\mathbb{R}}, \quad (\varphi F)(x) := \inf \varphi\,(F(x))\,.$$

Then (i) $\operatorname{dom} \varphi F = \operatorname{dom} F$, (ii) *$\varphi F$ is convex provided F is C-convex,* (iii) *φF is lsc provided F is C-closed.*

Proof. We may (and do) assume that graph $F \ne \emptyset$. Let us set $\psi := \varphi F$.

(i) The assertion is obvious.

(ii) Assume that F is C-convex. Since $\psi(x) = \inf_{y \in Y} \Phi(x, y)$ for $x \in X$, where $\Phi(x, y) := \varphi(y) + \iota_{\operatorname{epi} F}(x, y)$, we have that ψ is convex (ψ being the marginal function associated to the convex function Φ).

(iii) Let us set

$$A^r := \{x \in X \mid rk^0 \in F(x) + C\} = \{x \in X \mid (x, rk^0) \in \operatorname{epi} F\}.$$

Since epi F is closed, A^r is closed for every $r \in \mathbb{R}$. Clearly, $A^r \subseteq A^s$ if $r < s$. Moreover, if $x \in A^r$ then there exists $y \in F(x)$ such that $rk^0 \in y + C$, that is $y \in rk^0 - C$; hence $\psi(x) \leq \varphi(y) \leq r$. Therefore, $x \in [\psi \leq r]$. On the other hand, if $r < s$ and $x \in [\psi \leq r]$ then there exists $y \in F(x)$ such that $\varphi(y) < s$, and so $y \in sk^0 - C$; this implies that $x \in A^s$. Therefore, $A^r \subseteq [\psi \leq r] \subseteq A^s$ for all $r, s \in \mathbb{R}$ with $r < s$. Then

$$A^r \subseteq [\psi \leq r] \subseteq \cap_{s>r} A^s \subseteq \cap_{s>r} [\psi \leq s] = [\psi \leq r],$$

and so $[\psi \leq r] = \cap_{s>r} A^s$ is closed. Hence ψ is lsc. □

Using functions of the type φF we derive subdifferentiability criteria for C-convex set-valued maps. First let us introduce the **strong intrinsic core** of $A \subseteq X$ as being

$$\text{sic}\, A := {}^{ic}A := \begin{cases} \text{icr}\, A & \text{if aff}\, A \text{ is closed,} \\ \emptyset & \text{otherwise.} \end{cases}$$

Theorem 7.2.7. *Let $F : X \rightrightarrows Y$ be C-convex, $\overline{x} \in \text{dom}\, F$ and $\overline{y} \in E^w_\varsigma(F(\overline{x}))$ for some $\varsigma \in \{0\} \cup \text{int}\, C$. Assume that one of the following conditions holds:*

(i) $\text{int}(\text{epi}\, F) \neq \emptyset$ *and* $\overline{x} \in \text{int}(\text{dom}\, F)$,
(ii) *X is Fréchet, F is C-closed and* $\overline{x} \in \text{sic}(\text{dom}\, F)$.

Then $\partial^w_\varsigma F(\overline{x}, \overline{y}) \neq \emptyset$; more precisely, there exists $x^* \in X^*$ such that the operator $T \in \mathcal{L}(X, Y)$ defined by $T(x) := \langle x, x^* \rangle\, k^0$ for $x \in X$ is in $\partial^w_\varsigma F(\overline{x}, \overline{y})$, where $k^0 := \varsigma$ if $\varsigma \in \text{int}\, C$ and $k^0 \in \text{int}\, C$ if $\varsigma = 0$.

Proof. Let φ be defined by (7.18) with $k^0 := \varsigma$ if $\varsigma \in \text{int}\, C$ and $k^0 \in \text{int}\, C$ if $\varsigma = 0$; set $\eta := 1$ if $\varsigma \neq 0$ and $\eta := 0$ if $\varsigma = 0$. Because $\overline{y} \in E^w_{\eta k^0}(F(\overline{x}))$ (hence $\left(F(\overline{x}) - \overline{y} + \eta k^0\right) \cap (-\text{int}\, C) = \emptyset$), we have that $-\eta \leq -\delta := \inf \varphi(F(\overline{x}) - \overline{y}) \leq 0$. Consider $\psi : X \to \overline{\mathbb{R}}$ be defined by $\psi(x) := \inf \varphi(F(x) - \overline{y} + \delta k^0)$. Clearly, $\psi(\overline{x}) = 0$. Because F is C-convex, so is the set-valued map $x \mapsto F(x) - \overline{y} + \delta k^0$, and so, by Lemma 7.2.6 (ii), ψ is a convex function with $\text{dom}\, \psi = \text{dom}\, F$.

(i) Take $(x_0, y_0) \in \text{int}(\text{epi}\, F)$. Then there exists $U \in \mathcal{N}_X(x_0)$ such that $y_0 \in F(x) + C$ for every $x \in U$. Since φ is C-increasing, we obtain that $\psi(x) \leq \varphi(y_0 - \overline{y} + \delta k^0)$ for all $x \in U$. Since ψ is convex and finite at $\overline{x} \in \text{int}(\text{dom}\, F) = \text{int}(\text{dom}\, \psi)$, we obtain that ψ is finite-valued on $\text{dom}\, \psi$ and continuous on $\text{int}(\text{dom}\, \psi)$ (see e.g. [614, Theorem 2.2.9]). In particular, $\partial \psi(\overline{x}) \neq \emptyset$ (see e.g. [614, Theorem 2.4.9]).
(ii) By Lemma 7.2.6 (iii) we have that ψ is lsc, and so $\psi \in \Gamma(X)$ because $\psi(\overline{x}) \in \mathbb{R}$. Since $\overline{x} \in \text{sic}(\text{dom}\, \psi)$, $\partial \psi(\overline{x}) \neq \emptyset$ (which is a consequence of [614, Theorem 2.2.12]).

In both cases take $x^* \in \partial\psi(\overline{x})$. Then $\langle x - \overline{x}, x^*\rangle \le \psi(x) - \psi(\overline{x}) = \psi(x)$ for every $x \in X$. Therefore, $\langle x - \overline{x}, x^*\rangle \le \varphi(y - \overline{y} + \delta k^0)$ for all $(x, y) \in \operatorname{graph} F$. Taking $Tx := \langle x, x^*\rangle k^0$, we obtain that $\varphi(y - (\overline{y} - T\overline{x}) + \delta k^0) \ge 0$ for every $y \in (F - T)(X)$. Using Lemma 7.2.4 we obtain that $\overline{y} - T\overline{x} \in E^w_{\delta k^0}(\operatorname{Im}(F - T))$, whence $T \in \partial^w_{\eta k^0} F(\overline{x}, \overline{y})$ because $\delta k^0 \le_C \eta k^0$. □

As mentioned in [535, Theorem 2.1] (in an equivalent form), $\operatorname{int}(\operatorname{epi} F) \neq \emptyset$ iff there exists $y_0 \in Y$ such that $F^{-1}(y_0 - C)$ has nonempty interior. Other (not necessarily independent) sufficient conditions for $\operatorname{int}(\operatorname{epi} F) \neq \emptyset$ are provided in the next result.

Proposition 7.2.8. *Let $F : X \rightrightarrows Y$ and $x_0 \in \operatorname{int}(\operatorname{dom} F)$. Assume that one of the following conditions holds:*

(a) *there exists $y \in Y$ such that $F^{-1}(y - C) \in \mathcal{N}_X(x_0)$,*
(b) *there exists a (topologically) bounded set $E \subseteq Y$ such that $F^{-1}(E - C) \in \mathcal{N}_X(x_0)$,*
(c) *there exist $U \in \mathcal{N}_X(x_0)$ and a function $f : U \to Y$ such that $f(x) \in F(x)$ for every $x \in U$ and f is continuous at x_0,*
(d) *there exist $U \in \mathcal{N}_X(x_0)$ and a function $f : U \to Y$ such that $f(x) \in F(x)$ for every $x \in U$ and $f(U)$ is (topologically) bounded,*
(e) *there exists $y \in F(x_0)$ such that F is lower semicontinuous at (x_0, y).*

Then there exists $y_0 \in Y$ such that $(x_0, y_0) \in \operatorname{int}(\operatorname{epi} F)$.

Proof. It is clear that $F^{-1}(y_0 - C) \in \mathcal{N}_X(x_0)$ if $(x_0, y_0) \in \operatorname{int}(\operatorname{epi} F)$. Assume that (a) holds, that is $U := F^{-1}(y - C) \in \mathcal{N}_X(x_0)$, and take $k^0 \in \operatorname{int} C$. Then there exists $V \in \mathcal{N}_Y$ such that $k^0 + V \subseteq C$. Take $x \in U$ and $y' \in y_0 + V$, where $y_0 := y + k^0$. Since $(x, y) \in \operatorname{epi} F$ and $y \le y'$, we have that $(x, y') \in \operatorname{epi} F$; hence $U \times (y_0 + V) \subseteq \operatorname{epi} F$, which proves that $(x_0, y_0) \in \operatorname{int}(\operatorname{epi} F)$.

Even if $\operatorname{int} C = \emptyset$, the implications (a) ⇒ (b) and (d) ⇒ (b) are obvious.

(b) ⇒ (a) Consider $y' \in \operatorname{int} C$. Because $y' - C \in \mathcal{N}_Y(0)$, there exists $\gamma > 0$ such that $E \subseteq \gamma(y' - C) = y - C$, where $y := \gamma y'$. It follows that $F^{-1}(E - C) \subseteq F^{-1}(y - C)$.

(c) ⇒ (a) Take $y \in f(x_0) + \operatorname{int} C$; from the continuity of f at x_0, $U_0 := f^{-1}(y - C) \in \mathcal{N}_X(x_0)$. Because $f^{-1}(y - C) \subseteq F^{-1}(y - C)$, (a) holds.

(e) ⇒ (a) Take $y' \in y + \operatorname{int} C$. Because $y' - C \in \mathcal{N}_Y(y)$ and F is lsc at x_0, there exists $U \in \mathcal{N}(x_0)$ such that $F(x) \cap (y' - C) \neq \emptyset$ for every $x \in U$; hence (a) holds. □

Recall that F is **connected at** $x_0 \in X$ (see [569, Definition 3.4], [389, Definition 4]) if there exist $U \in \mathcal{N}_X(x_0)$ and a continuous function $f : U \to Y$ such that $f(x) \in F(x)$ for every $x \in U$.

Corollary 7.2.9. *Let $F : X \rightrightarrows Y$ be C-convex and $\overline{x} \in \operatorname{int}(\operatorname{dom} F)$ be such that $F(\overline{x}) = E^w(F(\overline{x}))$ and $F(\overline{x}) - C$ is convex. Assume that either* (i) $\operatorname{int}(\operatorname{epi} F) \neq \emptyset$ *or* (ii) *X is Fréchet, F is C-closed and $F(\overline{x})$ is compact. Then there exists $T \in \bigcap_{\overline{y} \in F(\overline{x})} \partial^w F(\overline{x}, \overline{y})$ such that $\operatorname{Im} T \subseteq \mathbb{R} \operatorname{int} C$. In fact, T can be taken in the form $Tx = \langle x, x^*\rangle k^0$ with $x^* \in X^*$ and $k^0 \in \operatorname{int} C$.*

Proof. Let us define $H : X \rightrightarrows Y$ by $H(x) := F(x) - F(\overline{x})$. Clearly, $\operatorname{dom} H = \operatorname{dom} F$. The set-valued map H is C-convex. Indeed, for $x, x' \in \operatorname{dom} H$ and $\lambda \in]0,1[$ we have that

$$\begin{aligned}\lambda H(x) + (1-\lambda)H(x') &= \lambda F(x) + (1-\lambda)F(x') - [\lambda F(\overline{x}) + (1-\lambda)F(\overline{x})]\\ &\subseteq F(\lambda x + (1-\lambda)x') + C - [F(\overline{x}) - C]\\ &= H(\lambda x + (1-\lambda)x') + C.\end{aligned}$$

Because $F(\overline{x}) = E^w(F(\overline{x}))$ we have that $H(\overline{x}) \cap (-\operatorname{int} C) = [F(\overline{x}) - F(\overline{x})] \cap (-\operatorname{int} C) = \emptyset$, and so $0 \in E^w(H(\overline{x}))$.

(i) Fixing $y_0 \in F(\overline{x})$ and taking $H_0(x) := F(x) - y_0$, we have that $\operatorname{epi} F - (0, y_0) = \operatorname{epi} H_0 \subseteq \operatorname{epi} H$, and so $\operatorname{int}(\operatorname{epi} H) \neq \emptyset$.
(ii) Because $F(\overline{x})$ is compact, H is C-closed.
Applying Theorem 7.2.7 we get $x^* \in X^*$ and $k^0 \in \operatorname{int} C$ such that the operator $T \in \mathcal{L}(X, Y)$ defined by $Tx = \langle x, x^* \rangle k^0$ is in $\partial^w H(\overline{x}, 0)$. Therefore, $y' - T(x - \overline{x}) \notin (-\operatorname{int} C)$ for all $(x, y') \in \operatorname{graph} H$, that is $y - \overline{y} - T(x - \overline{x}) \notin (-\operatorname{int} C)$ for all $(x, y) \in \operatorname{graph} F$ and $\overline{y} \in F(\overline{x})$. This shows that $T \in \partial^w F(\overline{x}, \overline{y})$ for all $\overline{y} \in F(\overline{x})$. □

The next result will be used for deriving estimates of the weak subdifferential of the sum of two set-valued maps.

Lemma 7.2.10. *Let $\Phi : X \times Z \rightrightarrows Y$ be a C-convex set-valued map with $0 \in \operatorname{Pr}_Z(\operatorname{dom} \Phi)$ and $\overline{y} \in Y$ be such that $(x, 0, y) \in \operatorname{graph} \Phi \Rightarrow y \notin \overline{y} - \operatorname{int} C$. Set $Z_0 := \operatorname{span}(\operatorname{Pr}_Z(\operatorname{dom} \Phi))$. Assume that one of the following conditions holds:*

(i) *for some $k^0 \in \operatorname{int} C$ one has:*

$$\exists \gamma \in \mathbb{R},\ \exists V \in \mathcal{N}_Z(0),\ \forall z \in V \cap Z_0,\ \exists x \in X : \Phi(x, z) \cap (\overline{y} + \gamma k^0 - C) \neq \emptyset; \tag{7.21}$$

(ii) *X, Y, Z are Fréchet spaces, Φ is C-closed and $0 \in \operatorname{sic}(\operatorname{Pr}_Z(\operatorname{dom} \Phi))$.*

Then there exists $z^* \in Z^*$ such that $(x, z, y) \in \operatorname{graph} \Phi \Rightarrow y - \langle z, z^* \rangle k^0 \notin \overline{y} - \operatorname{int} C$.

Proof. Consider the function $\psi : X \times Z \times Y \to \overline{\mathbb{R}}$ defined by $\psi(x, z, y) := \varphi(y - \overline{y}) + \iota_{\operatorname{epi} \Phi}(x, z, y)$, where φ is defined in (7.18). Clearly, ψ is a proper convex function and $\operatorname{dom} \psi = \operatorname{epi} \Phi$. Moreover, $\operatorname{Pr}_Z(\operatorname{dom} \psi) = \operatorname{Pr}_Z(\operatorname{epi} \Phi) = \operatorname{Pr}_Z(\operatorname{dom} \Phi)$.

In case (i) we have that $V \cap Z_0 \subseteq \{z \in Z \mid \exists (x, y) \in X \times Y : \psi(x, z, y) \leq \gamma\}$, while in case (ii) we have that ψ is lsc. Using [614, Theorem 2.7.1 (i)] in case (i) and [614, Theorem 2.7.1 (vii)] in case (ii) we obtain that

$$\infty > \overline{\alpha} := \inf_{(x,y) \in X \times Y} \psi(x, 0, y) = -\max_{z^* \in Z^*} \psi^*(0, z^*, 0). \tag{7.22}$$

Take $(x, y) \in X \times Y$. If $(x, 0, y) \notin \operatorname{epi} \Phi$ then $\psi(x, 0, y) = \infty > 0$. Let $(x, 0, y) \in \operatorname{epi} \Phi$; then there exists $y' \leq_C y$ such that $(x, 0, y') \in \operatorname{graph} \Phi$. It follows that $y' \notin \overline{y} - \operatorname{int} C$, and so $y \notin \overline{y} - \operatorname{int} C$. From the properties of φ we get $\varphi(y - \overline{y}) \geq 0$, whence $\psi(x, 0, y) \geq 0$. Hence $\overline{\alpha} \geq 0$. By (7.22) there exists $z^* \in Z^*$ such that $\psi^*(0, z^*, 0) = -\overline{\alpha} \leq 0$. Because φ is continuous and using e.g. [614, Theorem 2.8.7] together with (7.20), there exists $y^* \in C^+$ with $\langle k^0, y^* \rangle = 1$ such that $0 \geq \psi^*(0, z^*, 0) = \langle \overline{y}, y^* \rangle + \iota^*_{\operatorname{epi} \Phi}(0, z^*, -y^*)$. It follows that for $(x, z, y) \in \operatorname{graph} \Phi$ we have that

$$-\langle \overline{y}, y^* \rangle \geq \iota^*_{\operatorname{epi} \Phi}(0, z^*, -y^*) \geq \langle z, z^* \rangle - \langle y, y^* \rangle .$$

Therefore,

$$\forall (x, z, y) \in \operatorname{graph} \Phi : \left\langle \overline{y} - y + \langle z, z^* \rangle k^0, y^* \right\rangle = \langle z, z^* \rangle - \langle y - \overline{y}, y^* \rangle \leq 0,$$

which proves that $y - \langle z, z^* \rangle k^0 \notin \overline{y} - \operatorname{int} C$ if $(x, z, y) \in \operatorname{graph} \Phi$. □

Theorem 7.2.11. *Let $F_1 : X \rightrightarrows Y$, $F_2 : Z \rightrightarrows Y$ be two C-convex set-valued maps, and $S \in L(X, Z)$. Assume that there exist $x_0 \in \operatorname{dom} F_1 \cap S^{-1}(\operatorname{dom} F_2)$ and $y_0 \in Y$ such that $(Sx_0, y_0) \in \operatorname{int}(\operatorname{epi} F_2)$. Consider $(\overline{x}, \overline{y}_1) \in \operatorname{graph} F_1$, $(S\overline{x}, \overline{y}_2) \in \operatorname{graph} F_2$, $\varsigma \in C$ and $T \in \partial^w_\varsigma (F_1 + F_2 \circ S)(\overline{x}, \overline{y}_1 + \overline{y}_2)$. Then there exists $z^* \in Z^*$ such that $T_2 \in \partial^w_\varsigma F_2(S\overline{x}, \overline{y}_2)$, where $T_2(z) := \langle z, z^* \rangle k^0$ $(z \in Z)$ for some given $k^0 \in \operatorname{int} C$, and $T_1 := T - T_2 \circ S \in \partial^w_\varsigma F_1(\overline{x}, \overline{y}_1)$. Consequently,*

$$\partial^w_\varsigma (F_1 + F_2 \circ S)(\overline{x}, \overline{y}_1 + \overline{y}_2) \subseteq \partial^w_\varsigma F_1(\overline{x}, \overline{y}_1) + \left(\partial^w_\varsigma F_2(S\overline{x}, \overline{y}_2)\right) \circ S. \tag{7.23}$$

Proof. Let us fix $k^0 \in \operatorname{int} C$. Consider

$$\Phi : X \times Z \rightrightarrows Y, \quad \Phi(x, z) := F_1(x) + F_2(Sx + z) - Tx. \tag{7.24}$$

It follows easily that Φ is C-convex. Let us set $\overline{y} := \overline{y}_1 + \overline{y}_2 - T\overline{x} - \varsigma$. Clearly, $(\overline{x}, 0) \in \operatorname{dom} \Phi$; moreover, because $T \in \partial^w_\varsigma (F_1 + F_2 \circ S)(\overline{x}, \overline{y}_1 + \overline{y}_2)$, we have that $(x, 0, y) \in \operatorname{graph} \Phi \Rightarrow y \notin \overline{y} - \operatorname{int} C$. We wish to show that (7.21) holds. First observe that

$$\operatorname{epi} \Phi = \{(x, z, y_1 + y_2) \mid (x, y_1) \in \operatorname{epi}(F_1 - T),\ (Sx + z, y_2) \in \operatorname{epi} F_2)\} .$$

Fix $y_0' \in F_1(x_0) - Tx_0$ and take $\gamma := 1 + \varphi(y_0 + y_0' - \overline{y})$. Because $(Sx_0, y_0) \in \operatorname{int}(\operatorname{epi} F_2)$, there exist $V \in \mathscr{N}_Z(0)$ such that $(Sx_0 + V) \times \{y_0\} \subseteq \operatorname{epi} F_2$. For $z \in V$ we have that $Sx_0 + z \in Sx_0 + V$, and so there exist $y_z \in F_2(Sx_0 + z) \cap (y_0 - C)$. Then $y_0' + y_z \in F_1(x_0) + F_2(Sx_0 + z) - Tx_0$ and $y_0' + y_z - \overline{y} \in y_0 + y_0' - \overline{y} - C \subseteq \gamma k^0 - C$, whence $y_0' + y_z \in \Phi(x_0, z) \cap (\overline{y} + \gamma k^0 - C)$. Therefore, condition (7.21) holds. Using Lemma 7.2.10 (i) we get $z^* \in Z^*$ such that $(x, z, y) \in \operatorname{graph} \Phi \Rightarrow y - \langle z, z^* \rangle k^0 \notin \overline{y} - \operatorname{int} C$.

Let $T_2 : Z \to Y$ be defined by $T_2 z := \langle z, z^* \rangle k^0$; clearly $T_2 \in L(Z, Y)$. Let $(z, y) \in \operatorname{graph} F_2$. Then $\overline{y}_1 + y - T\overline{x} \in F_1(\overline{x}) + F_2(z) - T\overline{x} = \Phi(\overline{x}, z - S\overline{x})$, and so $\overline{y}_1 + y - T\overline{x} - \langle z - S\overline{x}, z^* \rangle k^0 \notin \overline{y} - \operatorname{int} C$. It follows that $y - T_2 z \notin (\overline{y}_2 - T_2(S\overline{x})) - \varsigma - \operatorname{int} C$ for all $(z, y) \in \operatorname{graph} F_2$, and so $T_2 \in \partial^w_\varsigma F_2(S\overline{x}, \overline{y}_2)$.

Let us take $(x, y) \in \operatorname{graph} F_1$. Then $y + \overline{y}_2 - Tx \in F_1(x) + F_2(S\overline{x}) - Tx = \Phi(x, S(\overline{x} - x))$, and so $y + \overline{y}_2 - Tx - \langle S(\overline{x} - x), z^* \rangle k^0 \notin \overline{y} - \operatorname{int} C$. It follows that $y - T_1 x \notin (\overline{y}_1 - T_1\overline{x}) - \varsigma - \operatorname{int} C$ for all $(x, y) \in \operatorname{graph} F_1$, and so $T_1 \in \partial^w_\varsigma F_1(\overline{x}, \overline{y}_1)$.

The inclusion (7.23) is now obvious. The proof is complete. □

The set-valued map Φ defined in (7.24) is not necessarily C-closed if F_1 and F_2 are C-closed. So, we can not apply Lemma 7.2.10 (ii) for getting the corresponding results in the case $0 \in \operatorname{sic}(S(\operatorname{dom} F_1) - \operatorname{dom} F_2)$. In the next result we obtain a similar conclusion as that of Theorem 7.2.11 in the case in which $\varsigma \in \{0\} \cup \operatorname{int} C$.

Theorem 7.2.12. *Let X, Y be Fréchet spaces, $F_1 : X \rightrightarrows Y$, $F_2 : Z \rightrightarrows Y$ be two C-convex and C-closed set-valued maps, and $S \in L(X, Z)$ be such that* $0 \in \operatorname{sic}(S(\operatorname{dom} F_1) - \operatorname{dom} F_2)$. *Assume that $(\overline{x}, \overline{y}_1) \in \operatorname{graph} F_1$, $(S\overline{x}, \overline{y}_2) \in \operatorname{graph} F_2$ and $T \in \partial^w_\varsigma (F_1 + F_2 \circ S)(\overline{x}, \overline{y}_1 + \overline{y}_2)$ for some $\varsigma \in \{0\} \cup \operatorname{int} C$. Then the conclusion of Theorem 7.2.11 holds.*

Proof. Take $k^0 := \varsigma$ if $\varsigma \neq 0$ and $k^0 \in \operatorname{int} C$ if $\varsigma = 0$, and let φ be defined by (7.18). Because $T \in \partial^w_\varsigma (F_1 + F_2 \circ S)(\overline{x}, \overline{y}_1 + \overline{y}_2)$ we have that $\overline{y} := \overline{y}_1 + \overline{y}_2 - T\overline{x} \in E^w_\varsigma [\operatorname{Im}(F_1 + F_2 \circ S - T)]$. Using Lemma 7.2.4 we obtain that

$$\varphi(y - \overline{y} + \varsigma) \geq 0 \quad \forall y \in (F_1 + F_2 \circ S - T)(X). \tag{7.25}$$

Setting $-\delta_1 := \inf(F_1(\overline{x}) - \overline{y}_1) \leq 0$, $-\delta_2 := \inf(F_2(S\overline{x}) - \overline{y}_2) \leq 0$, from (7.25) and Lemma 7.2.5 we get

$$-\eta \leq \inf \varphi\left((F_1(\overline{x}) - \overline{y}_1) + (F_2(\overline{x}) - \overline{y}_2)\right) \leq (-\delta_1) + (-\delta_2),$$

where $\eta = 1$ if $\varsigma \neq 0$ and $\eta := 0$ if $\varsigma = 0$. Hence $\delta_1, \delta_2 \in \mathbb{R}_+$ and $\delta_1 + \delta \leq \eta$.

Consider $\psi_1 : X \to \overline{\mathbb{R}}$, $\psi_2 : Z \to \overline{\mathbb{R}}$ be defined by $\psi_1(x) := \inf \varphi(F_1(x) - Tx - (\overline{y}_1 - T\overline{x}) + \delta_1 k^0)$ and $\psi_2(z) := \inf \varphi(F_2(z) - \overline{y}_2 + \delta_2 k^0)$. Clearly, $\psi_1(\overline{x}) = 0$ and $\psi_2(S\overline{x}) = 0$. Because F_1, F_2 are C-convex and C-closed, so are the set-valued maps $x \mapsto F_1(x) - Tx - (\overline{y}_1 - T\overline{x}) + \delta_1 k^0$, $z \mapsto F_2(z) - \overline{y}_2 + \delta_2 k^0$, and so, by Lemma 7.2.6, ψ_1 and ψ_2 are lsc convex functions. Because $\psi_1(\overline{x}) = 0$ and $\psi_2(S\overline{x}) = 0$ it follows that ψ_1 and ψ_2 are also proper. Clearly, $\operatorname{dom} \psi_1 = \operatorname{dom} F_1$ and $\operatorname{dom} \psi_2 = \operatorname{dom} F_2$. Let $x \in \operatorname{dom} \psi_1 \cap S^{-1}(\operatorname{dom} \psi_2) = \operatorname{dom} F_1 \cap S^{-1}(\operatorname{dom} F_2)$. Because

$$(F_1 + F_2 \circ S - T)(x) - \overline{y} = (F_1(x) - Tx - (\overline{y}_1 - T\overline{x})) + (F_2(Sx) - \overline{y}_2),$$

from (7.25) and Lemma 7.2.5 we get $(\delta_1 + \delta_2) - \eta \leq \psi_1(x) + \psi_2(Sx) < \infty$. Hence $\eta' := \eta - (\delta_1 + \delta_2) \geq 0$ and

$$\forall x \in X : \psi_1(\overline{x}) + \psi_2(S\overline{x}) \leq \psi_1(x) + \psi_2(x) + \eta'.$$

It follows that $0 \in \partial_{\eta'}(\psi_1 + \psi_2 \circ S)(\overline{x})$. Because $\psi_1, \psi_2 \in \Gamma(X)$ and $0 \in \operatorname{sic}(S(\operatorname{dom}\psi_1) - \operatorname{dom}\psi_2)$, using [614, Theorem 2.8.3 (vii)] we have that there exist $\eta_1', \eta_2' \geq 0$ with $\eta_1' + \eta_2' = \eta'$ and $z^* \in \partial_{\eta_2'}\psi_2(S\overline{x})$ such that $-S^*z^* \in \partial_{\eta_1'}\psi_1(\overline{x})$. Because $z^* \in \partial_{\eta_2'}\psi_2(S\overline{x})$ we have that $\langle z - S\overline{x}, z^*\rangle \leq \psi_2(z) - \psi_2(S\overline{x}) + \eta_2' = \psi_2(z) + \eta_2'$ for $z \in Z$, whence

$$\forall (z, y) \in \operatorname{graph} F_2 : \langle z - S\overline{x}, z^*\rangle \leq \varphi(y - \overline{y}_2 + \delta_2 k^0) + \eta_2'. \tag{7.26}$$

Let us set $T_2 z := \langle z, z^*\rangle\, k^0$ and $\eta_2 := \delta_2 + \eta_2' \geq 0$. From (7.26) we obtain that $\varphi(y - T_2 z - (\overline{y}_2 - T_2(S\overline{x})) + \eta_2 k^0) \geq 0$ for all $(z, y) \in \operatorname{graph} F_2$, which shows that $\big((F_2 - T_2)(Z) - (\overline{y}_2 - T_2(S\overline{x})) + \eta_2 k^0\big) \cap (-\operatorname{int} C) = \emptyset$. This proves that $T_2 \in \partial_{\eta_2 k^0 - w} F_2(S\overline{x}, \overline{y}_2) \subseteq \partial^w_\eta F_2(S\overline{x}, \overline{y}_2)$. In a similar way we obtain that $T_1 := T - T_2 \circ S \in \partial_{\eta_1 k^0 - w} F_1(\overline{x}, \overline{y}_1) \subseteq \partial^w_\eta F_1(\overline{x}, \overline{y}_1)$, where $\eta_1 := \delta_1 + \eta_1' \geq 0$; clearly $\eta = \eta_1 + \eta_2$. □

The weak subdifferential of a vector function was introduced by Tanino and Sawaragi in [569]. Then the notion was studied by many other authors (see [91, 92, 389, 535, 549–551, 566]). The results of this section are mainly extensions of results of Taa from [549–551]. Theorem 7.2.2 is stated in [550, Theorem 4.1] in Banach spaces, Lemma 7.2.6 is practically contained in the proof of [551, Theorem 3.1], Theorem 7.2.7 (i) for $\varsigma = 0$ is stated in [566, Proposition 4.3] and completed in [535, Proposition 2.4] (see also [537, Proposition 2.3]), Corollary 7.2.9 was obtained in [92, Theorem 7] under more stringent conditions. In the case $X = Z$, $S = \operatorname{Id}_X$ and $\varsigma = 0$ Theorem 7.2.11 was obtained by Lin in [389, Theorem 3.1] for F_2 connected at some $x_0 \in \operatorname{int}(\operatorname{dom} F_1) \cap \operatorname{int}(\operatorname{dom} F_2)$ and by Taa in [549, Theorem 3.2] for X and Y Banach spaces, with different proofs; the case $\varsigma \neq 0$ is considered in [550, Theorem 4.2], but the estimate for $\partial^w_\varsigma(F_1 + F_2)(\overline{x}, \overline{z}_1 + \overline{z}_2)$ is less precise. Theorem 7.2.12 for $Z = X$, $S := \operatorname{Id}_X$ and $\varsigma = 0$ is practically [551, Theorem 3.1].

7.3 Subdifferentials Corresponding to Henig Proper Efficiency

Throughout this section Y is a locally convex space, $C \subseteq Y$ is a proper pointed convex cone with $C^\# \neq \emptyset$ and Θ is a base of C. As in (2.25), we set $\mathcal{N}_Y^\Theta := \{V \in \mathcal{N}_Y^c \mid V \cap \Theta = \emptyset\}$, where $\mathcal{N}_Y^c := \{V \in \mathcal{N}_Y \mid V \text{ convex}\}$. As already observed, for $V \in \mathcal{N}_Y^\Theta$, $P_V^\Theta := \operatorname{cone}(\Theta + V)$ is a proper convex cone with $\operatorname{int} P_V = \mathbb{P}(\Theta + \operatorname{int} V)$. Note that $P_{V_1}^\Theta \subseteq P_{V_2}^\Theta$ (and so $\operatorname{WMin}(A; P_{V_1}) \supseteq \operatorname{WMin}(A; P_{V_2})$) for $V_1, V_2 \in \mathcal{N}_Y^\Theta$ with $V_1 \subseteq V_2$.

Similarly to ς-weak efficiency, for $\emptyset \neq A \subseteq Y$ and $\varsigma \in Y$, we define

$$E_\varsigma^{\sigma,C}(A) := E_\varsigma^\sigma(A) := \bigcup\{E_\varsigma^{w,D}(A) \mid D \in \mathcal{D}^\sigma\} \quad (\sigma \in \{h, t\}), \tag{7.27}$$

where

$$\mathscr{D}^{\sigma} := \begin{cases} \mathscr{D}_C \text{ if } \sigma = h, \\ \mathscr{D}_\Theta \text{ if } \sigma = t, \end{cases} \tag{7.28}$$

$\mathscr{D}_C$ and $\mathscr{D}_\Theta$ being defined in (2.31) and (2.34), respectively; $\sigma = h$ corresponds to global Henig efficiency, while $\sigma = t$ corresponds to Henig efficiency w.r.t. the base Θ.

Correspondingly, for $\sigma \in \{h, t\}$ we define the following ς-subdifferentials ($\varsigma \in Y$) of the set-valued map $F : X \rightrightarrows Y$ at $(\overline{x}, \overline{y}) \in \operatorname{graph} F$:

$$\partial_\varsigma^{\sigma,C} F(\overline{x},\overline{y}) := \partial_\varsigma^\sigma F(\overline{x},\overline{y}) := \bigcup\{\partial_\varsigma^{w,D} F(\overline{x},\overline{y}) \mid D \in \mathscr{D}^\sigma\}, \quad \partial^\sigma F(\overline{x},\overline{y}) := \partial_0^\sigma F(\overline{x},\overline{y}).$$

Applying the results from the previous section we get easily the following ones.

Proposition 7.3.1. *Let $F : X \rightrightarrows Y$ be C-convex, $(\overline{x}, \overline{y}) \in \operatorname{graph} F$, $\varsigma \in Y$ and $T \in \mathscr{L}(X, Y)$. Then for $\sigma \in \{t, h\}$, $T \in \partial_\varsigma^\sigma F(\overline{x}, \overline{y})$ iff there exists $y^* \in C^\sigma$ such that*

$$\forall (x, y) \in \operatorname{graph} F : \langle y - \overline{y} - T(x - \overline{x}) + \varsigma, y^* \rangle \geq 0, \tag{7.29}$$

where $C^h := C^\#$ and $C^t := \{y^* \in Y^* \mid \inf y^*(\Theta) > 0\}$.

Proof. Let $\sigma = t$ (the proof for $\sigma = h$ being similar). Take $T \in \partial_\varsigma^t F(\overline{x}, \overline{y})$; hence there exists $D = P_V^\Theta \in \mathscr{D}_\Theta$ ($V \in \mathscr{N}_Y^\Theta$) such that $T \in \partial_\varsigma^{w,D} F(\overline{x}, \overline{y})$. Applying Proposition 7.2.2 for C replaced by D, there exists $y^* \in D^+ \setminus \{0\}$ such that (7.29) holds. It follows that $\inf_\Theta y^* > 0$, and so $y^* \in C^t$. Conversely, assume that (7.29) holds for $y^* \in C^t$. Consider $V := \{y \in Y \mid |\langle y, y^* \rangle| \leq \frac{1}{2} \inf_\Theta y^*\}$. Then clearly $V \in \mathscr{N}_Y^\Theta$ and $y^* \in \left(P_V^\Theta\right)^+ \setminus \{0\}$. Applying again Proposition 7.2.2 for C replaced by $D := P_V^\Theta$ we obtain that $T \in \partial_\varsigma^{w,D} F(\overline{x}, \overline{y})$, and so $T \in \partial_\varsigma^t F(\overline{x}, \overline{y})$. □

Theorem 7.3.2. *Let $F : X \rightrightarrows Y$ be C-convex and $\overline{x} \in \operatorname{int}(\operatorname{dom} F)$. Assume that for some $x_0 \in X$ one of the following conditions holds:*

(a) *there exists a (topologically) bounded set $E \subseteq Y$ such that $F^{-1}(E - C) \in \mathscr{N}_X(x_0)$,*
(b) *there exist $U \in \mathscr{N}_X(x_0)$ and a function $f : U \to Y$ such that $f(x) \in F(x)$ for every $x \in U$ and f is continuous at x_0,*
(c) *there exist $U \in \mathscr{N}_X(x_0)$ and a function $f : U \to Y$ such that $f(x) \in F(x)$ for every $x \in U$ and $f(U)$ is (topologically) bounded,*
(d) *there exists $y_0 \in F(x_0)$ such that F is lower semicontinuous at (x_0, y_0).*

Then for $\sigma \in \{h, t\}$ and $\varsigma \in C$ we have that $\partial_\varsigma^\sigma F(\overline{x}, \overline{y}) \neq \emptyset$ provided $\overline{y} \in E_\varsigma^\sigma(F(\overline{x}))$.

Proof. Consider now $\sigma = h$ and $\varsigma \in C$ (the proof for $\sigma = t$ being similar). Because $\overline{y} \in E_\varsigma^\sigma(F(\overline{x}))$, there exists $D \in \mathscr{D}_C$ such that $\overline{y} \in E_\varsigma^{w,D}(F(\overline{x}))$; hence

$C\setminus\{0\} \subseteq \operatorname{int} D$. In case (a), because $F^{-1}(E-C) \subseteq F^{-1}(E-D)$ and $\operatorname{int} D \neq \emptyset$, we obtain that condition (b) in Proposition 7.2.8 holds. If (b), or (c), or (d) holds, clearly the condition (c), or (d), or (e) in Proposition 7.2.8 holds, respectively. Therefore, there exists $y_0 \in Y$ such that $(x_0, y_0) \in \operatorname{int}(\operatorname{epi}_D F) \neq \emptyset$. Using now Theorem 7.2.7 for the present C replaced by D, $k^0 = \varsigma$ ($\in \operatorname{int} D$) and $\eta = 1$ if $\varsigma \neq 0$, respectively $k^0 \in \operatorname{int} D$ and $\eta = 0$ if $\varsigma = 0$, we have that $\partial_\varsigma^{w,D} F(\overline{x}, \overline{y}) \neq \emptyset$. □

Of course, another sufficient condition instead of conditions (a)–(d) in Theorem 7.3.2 is the fact that $\operatorname{int}(\operatorname{epi}_C F) \neq \emptyset$.

Theorem 7.3.3. *Let $F_1 : X \rightrightarrows Y$, $F_2 : Z \rightrightarrows Y$ be two C-convex set-valued maps, and $S \in L(X, Z)$. Assume that there exists $x_0 \in \operatorname{dom} F_1 \cap S^{-1}(\operatorname{dom} F_2)$ such that one of the following conditions holds:*

(a) *there exists a (topologically) bounded set $E \subseteq Y$ such that $F_2^{-1}(E - C) \in \mathcal{N}_Z(Sx_0)$,*
(b) *there exist $V \in \mathcal{N}_Z(Sx_0)$ and a function $f_2 : V \to Y$ such that $f_2(z) \in F_2(z)$ for every $z \in V$ and f_2 is continuous at Sx_0,*
(c) *there exist $V \in \mathcal{N}_Z(Sx_0)$ and a function $f_2 : V \to Y$ such that $f_2(z) \in F_2(z)$ for every $z \in V$ and $f_2(V)$ is (topologically) bounded,*
(d) *there exists $y_0 \in F_2(Sx_0)$ such that F_2 is lower semicontinuous at (Sx_0, y_0).*

Consider $(\overline{x}, \overline{y}_1) \in \operatorname{graph} F_1$, $(S\overline{x}, \overline{y}_2) \in \operatorname{graph} F_2$ and $\varsigma \in C$. Then

$$\forall \sigma \in \{h, t\} : \partial_\varsigma^\sigma (F_1 + F_2 \circ S)(\overline{x}, \overline{y}_1 + \overline{y}_2) \subseteq \partial_\varsigma^\sigma F_1(\overline{x}, \overline{y}_1) + \left(\partial_\varsigma^\sigma F_2(S\overline{x}, \overline{y}_2)\right) \circ S. \tag{7.30}$$

Proof. Take $\sigma = t$. Let $T \in \partial_\varsigma^t (F_1 + F_2 \circ S)(\overline{x}, \overline{y}_1 + \overline{y}_2)$. Then there exists $V \in \mathcal{N}_Y^\Theta$ such that $T \in \partial_\varsigma^{w,D} (F_1 + F_2 \circ S)(\overline{x}, \overline{y}_1 + \overline{y}_2)$, where $D := P_V^\Theta$. As in the proof of Theorem 7.3.2, we get $y_0 \in Y$ such that $(Sx_0, y_0) \in \operatorname{int}(\operatorname{epi}_D F_2)$. Applying Theorem 7.2.11 for C replaced by D we obtain that $T \in \partial_\varsigma^{w,D} F_1(\overline{x}, \overline{y}_1) + \left(\partial_\varsigma^{w,D} F_2(S\overline{x}, \overline{y}_2)\right) \circ S$. The conclusion follows. □

In the case $\sigma = t$, Proposition 7.3.1 and Theorems 7.3.2, 7.3.3 were obtained in [386] for $\varsigma = 0$ and in [623] for $\operatorname{int} C \neq \emptyset$ under slightly less general conditions.

7.4 Exact Formulas for the Subdifferential of the Sum and the Composition

We assume that X, Y, Z are topological vector spaces, Y being ordered by the proper pointed convex cone C. Besides the notations in (7.14) and (7.27), for $A \subseteq Y$ and $\varsigma \in Y$ we set

$$E_\varsigma^{s,C}(A) := E_\varsigma^s(A) := \left\{y \in A \mid y \leqq y' + \varsigma \ \forall y \in A\right\}.$$

In the sequel, when dealing with $\sigma \in \{s, w, h, t\}$ we assume:

- *C is closed and Y is a locally convex space if $\sigma = s$,*
- *$\operatorname{int} C \neq \emptyset$ if $\sigma = w$,*
- *$C^{\#} \neq \emptyset$ if $\sigma = h$,*
- *Θ is a base of C and Y is a locally convex space if $\sigma = t$.*

Remark 7.4.1. Note that for a nonempty set $A \subseteq Y$, $\varsigma \in Y$ and $\sigma \in \{s, w, h, t\}$ one has

$$E^{\sigma}_{\varsigma}(A) \neq \emptyset \Rightarrow \varsigma \in D^{\sigma},$$

where

$$D^{\sigma} := \begin{cases} C & \text{if } \sigma = s, \\ Y \setminus (-\operatorname{int} C) & \text{if } \sigma = w, \\ Y \setminus (\bigcap \{\operatorname{int} D \mid D \in \mathscr{D}^{\sigma}\}) & \text{if } \sigma \in \{h, t\}, \end{cases} \tag{7.31}$$

Moreover, for $\varsigma, \varsigma' \in Y$ and $\sigma \in \{s, w, h, t\}$ one has

$$\varsigma \leqq \varsigma' \Rightarrow E^{\sigma}_{\varsigma}(A) \subseteq E^{\sigma}_{\varsigma'}(A). \tag{7.32}$$

Besides C^h and C^t used in Proposition 7.3.1, we set $C^s := C^w := C^+ \setminus \{0\}$. So

$$C^{\sigma} := \begin{cases} C^+ \setminus \{0\} & \text{if } \sigma \in \{w, s\} \\ C^{\#} & \text{if } \sigma = h, \\ \{y^* \in Y^* \mid \inf y^*(\Theta) > 0\} & \text{if } \sigma = t. \end{cases}$$

Lemma 7.4.2. *Let $A \subseteq Y$ be a nonempty set and $\varsigma \in Y$.*

(i) *One has*

$$E^{s}_{\varsigma}(A) = \bigcap_{y^* \in C^s} \langle \varsigma, y^* \rangle\text{-}\operatorname{arg\,min}_A y^*. \tag{7.33}$$

(ii) *For $\sigma \in \{w, h, t\}$ one has*

$$\bigcup_{y^* \in C^{\sigma}} \langle \varsigma, y^* \rangle\text{-}\operatorname{arg\,min}_A y^* \subseteq E^{\sigma}_{\varsigma}(A); \tag{7.34}$$

moreover, if A is closely C-convex then equality holds in (7.34).

Proof. The proof of (i) is easy.

(ii) Take $\sigma = w$; hence $\operatorname{int} C \neq \emptyset$. Take $y^* \in C^w = C^+ \setminus \{0\}$ and $\overline{y} \in \langle \varsigma, y^* \rangle$-$\operatorname{arg\,min}_A y^*$. Then $\langle \overline{y} - \varsigma, y^* \rangle \leq \langle y, y^* \rangle$ for all $y \in A$. By Lemma 7.2.1 we obtain that $A \cap (\overline{y} - \varsigma - \operatorname{int} C) = \emptyset$, and so $\overline{y} \in E^{w}_{\varsigma}(A)$.

Conversely, assume that A is closely C-convex and take $\overline{y} \in E^w_\varsigma(A)$. Then $\overline{y} \in A$ and $A \cap (\overline{y} - \varsigma - \operatorname{int} C) = \emptyset$. Using again Lemma 7.2.1, there exists $y^* \in C^+ \setminus \{0\} = C^w$ such that $\langle \overline{y} - \varsigma, y^* \rangle \leq \langle y, y^* \rangle$ for all $y \in A$. It follows that $\overline{y} \in \langle \varsigma, y^* \rangle$-$\arg\min_A y^*$.

The proof for the case $\sigma \in \{h, t\}$ is similar. One uses the fact that A is closely D-convex for $D \in \mathscr{D}^\sigma$. □

From the very definition of $\partial^\sigma_\varsigma f(\overline{x})$, for every proper function $f : X \to Y^\bullet$ and every $x \in X$ one has

$$\forall T \in L(X,Y),\ \forall \sigma \in \{s,w,h,t\},\ \forall \varsigma \in Y : \partial^\sigma_\varsigma (f+T)(x) = \partial^\sigma_\varsigma f(x) + T; \tag{7.35}$$

moreover, if $S \in L(Z,X)$ is an isomorphism (of t.v.s.), then

$$\forall z \in Z,\ \forall \sigma \in \{s,w,h,t\},\ \forall \varsigma \in Y : \partial^\sigma_\varsigma (f \circ S)(z) = \partial^\sigma_\varsigma f(Sz) \circ S. \tag{7.36}$$

Proposition 7.4.3. *Let $f : X \to Y^\bullet$ be a proper positively homogeneous operator (that is $f(0) = 0$ and $f(tx) = tf(x)$ for $t > 0$ and $x \in X$). Then*

$$\forall \sigma \in \{s,w,h,t\} : \partial^\sigma_\varsigma f(0) = \begin{cases} \partial^\sigma f(0) & \textit{if } \varsigma \in D^\sigma, \\ \emptyset & \textit{otherwise} \end{cases}$$

where D^σ is defined by (7.31).

Proof. Consider first $\sigma = s$ (hence C is closed). Take $T \in \partial^s_\varsigma f(0)$. Then $(f - T)(x) + \varsigma \in C$ for every $x \in \operatorname{dom} f$. Taking $x = 0$ we get $\varsigma \in C = D^s$. For $x \in \operatorname{dom} f$ and $n \in \mathbb{N}^*$ we have that $(f - T)(nx) + \varsigma \in C$, and so $(f - T)(x) + n^{-1}\varsigma \in C$ for $n \geq 1$. Because C is closed we get $(f - T)(x) \in C$. It follows that $T \in \partial^s f(0)$. Conversely, take $\varsigma \in C$ and $T \in \partial^s f(0)$. Then $(f - T)(x) + \varsigma \in \varsigma + C \subseteq C$, and so $T \in \partial^s_\varsigma f(0)$.

Consider $\sigma = w$ (hence $\operatorname{int} C \neq \emptyset$). Take $T \in \partial^s_\varsigma f(0)$. Then $(f - T)(x) + \varsigma \in Y \setminus (-\operatorname{int} C)$ for every $x \in \operatorname{dom} f$. Taking $x = 0$ we get $\varsigma \in Y \setminus (-\operatorname{int} C) = D^w$. For $x \in \operatorname{dom} f$ and $n \in \mathbb{N}^*$ we have that $(f - T)(nx) + \varsigma \in Y \setminus (-\operatorname{int} C)$, and so $(f - T)(x) + n^{-1}\varsigma \in Y \setminus (-\operatorname{int} C)$ for $n \geq 1$. Taking the limit for $n \to \infty$ we get $(f - T)(x) \in Y \setminus (-\operatorname{int} C)$. It follows that $T \in \partial^w f(0)$. Conversely, take $\varsigma \in Y \setminus (-\operatorname{int} C)$ and $T \in \partial^w f(0)$. Then $(f - T)(x) + \varsigma \in \varsigma + [Y \setminus (-\operatorname{int} C)] \subseteq Y \setminus (-\operatorname{int} C)$, and so $T \in \partial^w_\varsigma f(0)$.

The proof of the case $\sigma \in \{h, t\}$ is similar to the case $\sigma = w$ (just use some $D \in \mathscr{D}^\sigma$ instead of C). □

Of course, if $f : X \to Y^\bullet$ is proper and $g : X \to Y^\bullet$ is defined by $g(x) := f(x - x_0) + y_0$ for some $x_0 \in X$ and $y_0 \in Y$, then

$$\forall x \in X,\ \forall \sigma \in \{s,w,h,t\} : \partial^\sigma_\varsigma g(x) = \partial^\sigma_\varsigma f(x - x_0). \tag{7.37}$$

Let us set

$$\vartheta_\sigma^C(X,Y) := \vartheta_\sigma(X,Y) := \partial^\sigma\theta(0) \quad (\sigma \in \{s,w,h,t\}), \tag{7.38}$$

where $\theta : X \to Y$, $\theta(x) := 0$ for every $x \in X$.

Proposition 7.4.4. *Let $\vartheta_\sigma(X,Y)$ be defined by (7.38). Then*

$$\vartheta_\sigma(X,Y) = \begin{cases} \{0\} & \text{if } \sigma = s, \\ \{T \in L(X,Y) \mid \ker T^* \cap C^\sigma \neq \emptyset\} & \text{if } \sigma = \{w,h,t\}. \end{cases} \tag{7.39}$$

Proof. The assertion is obvious for $\sigma = s$.

Take $\sigma = w$; hence $\operatorname{int} C \neq \emptyset$. Let $T \in \partial^w\theta(0)$. Then $0 \in E_w(-T(X))$, and so $T(X) \cap \operatorname{int} C = \emptyset$. By a separation there exists $y^* \in Y^* \setminus \{0\}$ such that $\langle Tx, y^*\rangle \leq \langle y, y^*\rangle$ for all $x \in X$ and $y \in C$. It follows that $\langle x, T^*y^*\rangle = \langle Tx, y^*\rangle = 0$ for every $x \in X$, and so $T^*y^* = 0$, and $0 \leq \langle y, y^*\rangle$ for every $y \in C$, and so $y^* \in C^+$. Therefore, the inclusion $\subseteq$ holds in (7.39) holds for $\sigma = w$. Conversely, assume that $T \in L(X,Y)$ is such that $\ker T^* \cap C^w \neq \emptyset$ and take $y^* \in \ker T^* \cap (C^+ \setminus \{0\})$. Then $\langle Tx, y^*\rangle = 0 \leq \langle y, y^*\rangle$ for all $x \in X$ and $y \in C$. Because $y^* \neq 0$, we have that $\langle y, y^*\rangle > 0$ for $y \in \operatorname{int} C$, and so $T(X) \cap (-\operatorname{int} C) = \emptyset$. Hence $T \in \partial^w\theta(0)$. Therefore, (7.39) holds for $\sigma = w$.

Take $\sigma \in \{h,t\}$; hence $C^\sigma \neq \emptyset$. Then

$$\begin{aligned} \vartheta_\sigma(X,Y) &= \bigcup\{\vartheta_w^D(X,Y) \mid D \in \mathscr{D}^\sigma\} = \bigcup_{D\in\mathscr{D}^\sigma}\{T \in L(X,Y) \mid \ker T^* \cap (D^+ \setminus \{0\}) \neq \emptyset\} \\ &= \left\{T \in L(X,Y) \mid \ker T^* \cap \left(\bigcup\{D^+ \setminus \{0\} \mid D \in \mathscr{D}^\sigma\}\right) \neq \emptyset\right\} \\ &= \{T \in L(X,Y) \mid \ker T^* \cap C^\sigma \neq \emptyset\}. \end{aligned}$$

The proof is complete. □

Corollary 7.4.5. *Let $T \in L(X,Y)$. Then for every $x \in X$ we have that*

$$\forall x \in X,\ \forall \sigma \in \{s,w,h,t\} : \partial_\varsigma^\sigma T(x) = \begin{cases} T + \vartheta_\sigma(X,Y) & \text{if } \varsigma \in D^\sigma, \\ \emptyset & \text{otherwise.} \end{cases}$$

Proof. From (7.37) we get $\partial_\varsigma^\sigma T(x) = \partial_\varsigma^\sigma T(0)$, while from Proposition 7.4.3 we get $\partial_\varsigma^\sigma T(0) = \partial^\sigma T(0)$ for $\varsigma \in D^\sigma$ and $\partial_\varsigma^\sigma T(0) = \emptyset$ if $\varsigma \in Y \setminus D^\sigma$. Using (7.35) we have that $\partial^\sigma T(0) = \partial^\sigma\theta(0) + T = T + \vartheta_\sigma(X,Y)$. The conclusion follows. □

In the next result one provides formulas for the approximate subdifferentials using scalar functions. As usual, for $y^* \in C^+$ we set $y^*(+\infty) = +\infty$.

Theorem 7.4.6. *Let $f : X \to Y^\bullet$ be proper and $\varsigma \in Y$. Then:*

(i) *for every $x \in X$ one has*

$$\partial_\varsigma^s f(x) = \bigcap_{y^*\in C^s}\left\{T \in L(X,Y) \mid y^* \circ T \in \partial_{\langle\varsigma,y^*\rangle}\left(y^* \circ f\right)(x)\right\}; \tag{7.40}$$

(ii) *for* $\sigma \in \{w,h,t\}$ *and* $x \in X$ *one has*

$$\partial_\varsigma^\sigma f(x) \supseteq \bigcup_{y^*\in C^\sigma} \left\{T \in L(X,Y) \mid y^* \circ T \in \partial_{\langle\varsigma,y^*\rangle}\left(y^*\circ f\right)(x)\right\}, \tag{7.41}$$

with equality if f *is* C*-convex.*

Proof. The result is a direct consequence of Lemma 7.4.2 and the definition of $\partial_\varsigma^\sigma f(x)$. □

Corollary 7.4.7. *Let* $f : X \to Y^\bullet$ *be a proper* C*-convex function,* $\overline{x} \in \operatorname{dom} f$ *and* $\sigma \in \{w,h,t\}$.

(i) *If* $\partial^\sigma f(\overline{x}) \neq \emptyset$ *then there exists* $y^* \in C^\sigma$ *such that* $y^* \circ f$ *is lsc at* $\overline{x}$.
(ii) *Conversely, assume that* X *is a Fréchet space and* $\overline{x} \in \operatorname{sic}(\operatorname{dom} f)$*. If* $y^* \circ f$ *is lsc at* $\overline{x}$ *for some* $y^* \in C^\sigma$*, then* $\partial^\sigma f(\overline{x}) \neq \emptyset$.

Proof. (i) Because f is C-convex, we have equality in (7.41). Taking $T \in \partial^\sigma f(\overline{x})$, there exists $y^* \in C^\sigma$ such that $y^* \circ T \in \partial\left(y^* \circ f\right)(\overline{x})$, and so $y^* \circ f$ is lsc at $\overline{x}$.

(ii) Because $\operatorname{dom} y^* \circ f = \operatorname{dom} f$ we have that $\overline{x} \in \operatorname{sic}(\operatorname{dom} y^* \circ f)$. The function $y^* \circ f$ being a proper convex function, and lsc at $\overline{x}$, applying [359, Corollary 3.2] we obtain that $\partial(y^* \circ f)(\overline{x}) \neq \emptyset$. Take $x^* \in \partial(y^* \circ f)(\overline{x})$ and $T_0 \in L(X,Y)$ defined by $T_0x := \langle x, x^*\rangle\, y_0$, where $y_0 \in C$ is such that $\langle y_0, y^*\rangle = 1$ (the existence of y_0 is ensured by the fact that $y^* \in C^\sigma$). Then $y^* \circ T = x^*$. From (7.41) we obtain that $T_0 \in \partial^\sigma f(\overline{x}) \neq \emptyset$. □

The η-regularity ($\eta \in \mathbb{R}_+$) of the proper function $f : X \to Y^\bullet$ at $\overline{x} \in \operatorname{dom} f$ was defined in Sect. 7.1:

$$\forall y^* \in C^+ \setminus \{0\} : \partial_\eta(y^*\circ f)(\overline{x}) = y^* \circ \left(\bigcup\{\partial_\varsigma^s f(\overline{x}) \mid \varsigma \in \eta C,\ \langle\varsigma, y^*\rangle = \eta\}\right), \tag{7.42}$$

Theorem 7.1.6 provides a sufficient condition for the 0-regularity of $f \in \Lambda(X,Y)$ at $\overline{x} \in \operatorname{int}(\operatorname{dom} f)$. For example, if $Y = \mathbb{R}^m$ and $C \subseteq \mathbb{R}^m$ is a pointed closed convex cone, any proper C-convex function $f : X \to Y^\bullet$ is regular at $\overline{x} \in \operatorname{int}(\operatorname{dom} f)$ provided f is continuous at $\overline{x}$.

Remark 7.4.8. Let $y^* \in C^+ \setminus \{0\}$ be such that there exists $k \in C$ with $\langle k, y^*\rangle > 0$ (or, equivalently, $\exists k \in C : \langle k, y^*\rangle = 1$). Consider $\emptyset \neq A \subseteq Y$ and $I_A : X \to Y^\bullet$, $I_A(x) := 0$ for $x \in A$, $I_A(x) := +\infty$ for $x \in X \setminus A$. Then $f := I_A$ satisfies (7.2) at $\overline{x} \in A$ for every $\eta \in \mathbb{R}_+$. In particular, if for every $y^* \in C^+ \setminus \{0\}$ there exists $k \in C$ with $\langle k, y^*\rangle > 0$ then I_A is η-regular at $\overline{x} \in A$ for every $\eta \geq 0$.

Indeed, fix $\overline{x} \in A$ and $\eta \in \mathbb{R}_+$. As already observed, the inclusion $\supseteq$ in (7.42) holds always. Clearly, $y^* \circ I_A = \iota_A$, where ι_A is the indicator function of A. Take $x^* \in \partial_\eta \iota_A(\overline{x})$, that is $\langle x - \overline{x}, x^*\rangle \leq \eta$ for every $x \in A$. Consider $T \in L(X,Y)$ defined by $Tx := \langle x, x^*\rangle k$, where $k \in C$ is such that $\langle k, y^*\rangle = 1$. Then

$Tx - T\overline{x} = \langle x - \overline{x}, x^* \rangle k \leq \varsigma := \eta k$ for every $x \in A$. Therefore, $T \in \partial^s_\varsigma I_A(\overline{x})$. Since $y^* \circ T = x^*$, we have that $x^* \in y^* \circ \partial^s_\varsigma I_A(\overline{x})$. Hence (7.42) holds.

For $\sigma \in \{w, h, t\}$ and $\eta \in \mathbb{R}_+$ one says that f is *σ-η-regular* at $\overline{x} \in \operatorname{dom} f$ if (7.42) holds for every $y^* \in C^\sigma$; f is *σ-regular* at $\overline{x} \in \operatorname{dom} f$ if f is *σ-0-regular* at $\overline{x}$.

Of course, if f is η-regular at $\overline{x} \in \operatorname{dom} f$ then f is σ-η-regular at $\overline{x}$ for $\sigma \in \{w, h, t\}$.

Let us observe that the inclusion $\partial^\sigma(f + g)(x) \supseteq \partial^\sigma f(x) + \partial^\sigma g(x)$ for $\overline{x} \in \operatorname{dom} f \cap \operatorname{dom} g$, well known in the scalar case, is not true in general for $\sigma \in \{w, h, t\}$. This can be seen taking $f = g = \theta$. The above inclusion becomes $\vartheta_\sigma(X, Y) + \vartheta_\sigma(X, Y) \subseteq \vartheta_\sigma(X, Y)$. Since $\vartheta_\sigma(X, Y)$ is a cone, the relation above would imply that $\vartheta_\sigma(X, Y)$ is a convex cone. Taking $X := \mathbb{R}$, $Y := \mathbb{R}^2$ and $C := \mathbb{R}^2_+$, we have that $T, S \in \vartheta_w(X, Y)$ for $Tx := (2x, -x)$, $Sx := (-x, 2x)$, but $T + S \notin \vartheta_w(X, Y)$.

The above example shows that we can not expect to obtain equality in Theorem 7.2.11 (for example), even for $\varsigma = 0$ and for (multi) functions with very good properties if $\dim Y > 1$. For this reason the conclusion of the following result is important.

Theorem 7.4.9. *Let $f, g : X \to Y^\bullet$ be proper operators, $\sigma \in \{w, h, t\}$, and $\varsigma \in Y$. Then*

$$\forall x \in X : \partial^\sigma_\varsigma(f + g)(x) \supseteq \bigcup \{\partial^\sigma_{\varsigma_1} f(x) + \partial^s_{\varsigma_2} g(x) \mid \varsigma_1 \in Y,\ \varsigma_2 \in C,\ \varsigma_1 + \varsigma_2 = \varsigma\}. \tag{7.43}$$

Moreover, assume that $f, g \in \Lambda(X, Y)$ and either (i) *X is a l.c.s., f or g is continuous at some point in* $\operatorname{dom} f \cap \operatorname{dom} g$, *or* (ii) *$f, g$ are C-lower continuous, X is a Fréchet space and* $0 \in \operatorname{sic}(\operatorname{dom} f - \operatorname{dom} g)$.

If g is σ-regular at $\overline{x} \in \operatorname{dom} f \cap \operatorname{dom} g$ then

$$\partial^\sigma(f + g)(\overline{x}) = \partial^\sigma f(\overline{x}) + \partial^s g(\overline{x}). \tag{7.44}$$

If $\varsigma \neq 0$ and g is σ-η-regular at $\overline{x} \in \operatorname{dom} f \cap \operatorname{dom} g$ for every $\eta \in \mathbb{R}_+$ then

$$\partial^\sigma_\varsigma(f + g)(\overline{x}) = \bigcup \{\partial^\sigma_{\varsigma_1} f(\overline{x}) + \partial^s_{\varsigma_2} g(\overline{x}) \mid \varsigma_1 \in Y,\ \varsigma_2 \in C,\ \varsigma_1 + \varsigma_2 = \varsigma\}. \tag{7.45}$$

Proof. Take $\varsigma_1 \in Y$, $\varsigma_2 \in C$ with $\varsigma_1 + \varsigma_2 = \varsigma$ and $T \in \partial^w_{\varsigma_1} f(x)$, $S \in \partial^s_{\varsigma_2} g(x)$. Assume that $T + S \notin \partial^w_\varsigma(f + g)(x)$, and so there exists $x' \in \operatorname{dom} f \cap \operatorname{dom} g$ such that $(f(x') + g(x') - Tx' - Sx') - (f(x) + g(x) - Tx - Sx) \in (-\varsigma - \operatorname{int} C)$. Because $S \in \partial^s_{\varsigma_2} g(x)$ we have that $g(x) - Sx - g(x') + Sx' \in (\varsigma_2 - C)$, and so we get $\big(f(x') - Tx'\big) - \big(f(x) - Tx\big) \in \varsigma_2 - \varsigma + (-\operatorname{int} C) + (-C) = (-\varsigma_1 - \operatorname{int} C)$, contradicting the fact that $T \in \partial^w_{\varsigma_1} f(x)$. Hence (7.43) holds.

A similar argument (just replacing C by some $D \in \mathscr{D}^\sigma$) shows that the inclusion (7.43) holds in the case $\sigma \in \{h, t\}$, too.

Assume now that (i) or (ii) holds. Let $\varsigma \neq 0$ and g be σ-η-regular for every $\eta \geq 0$ at $\overline{x} \in \operatorname{dom} f \cap \operatorname{dom} g$, and take $T \in \partial^\sigma_\varsigma(f + g)(\overline{x})$. Since $f + g$ is convex, by Theorem 7.4.6 there exists $y^* \in C^\sigma$ such that $y^* \circ T \in \partial_{\langle \varsigma, y^* \rangle}(y^* \circ (f + g))(\overline{x})$.

Clearly, $y^* \circ (f + g) = y^* \circ f + y^* \circ g$. In case (i) we have that one of the convex functions $y^* \circ f$, $y^* \circ g$ is continuous at some point in $\operatorname{dom} f \cap \operatorname{dom} g = \operatorname{dom}(y^* \circ f) \cap \operatorname{dom}(y^* \circ g)$. Using Theorem 7.1.11 (i) for $Y = \mathbb{R}$, we get

$$\partial_{\langle \varsigma, y^* \rangle} \left(y^* \circ f + y^* \circ g\right)(\overline{x}) = \bigcup_{\eta_1, \eta_2 \geq 0,\ \eta_1 + \eta_2 = \langle \varsigma, y^* \rangle} \left[\partial_{\eta_1}(y^* \circ f)(\overline{x}) + \partial_{\eta_2}(y^* \circ g)(\overline{x})\right]. \tag{7.46}$$

In case (ii), by Proposition 3.1.31 (iv), $y^* \circ f$ and $y^* \circ g$ are lsc. Applying Theorem 7.1.11 (iii) for $Y = \mathbb{R}$, we have that (7.46) holds again.

Hence there exist $\eta_1, \eta_2 \geq 0$ with $\eta_1 + \eta_2 = \langle \varsigma, y^* \rangle$ and $u^* \in \partial_{\eta_1}(y^* \circ f)(\overline{x})$, $v^* \in \partial_{\eta_2}(y^* \circ g)(\overline{x})$ such that $y^* \circ T = u^* + v^*$. Because g is σ-η_2-regular at $\overline{x}$, there exist $\varsigma_2 \in \eta_2 C \subseteq C$ with $\langle \varsigma_2, y^* \rangle = \eta_2$ and $S \in \partial^s_{\varsigma_2} g(\overline{x})$ such that $v^* = y^* \circ S$. Then $u^* = y^* \circ T - y^* \circ S = y^* \circ (T - S) \in \partial_{\langle \varsigma_1, y^* \rangle}(y^* \circ f)(\overline{x})$, where $\varsigma_1 := \varsigma - \varsigma_2$. Using again Theorem 7.4.6 we have that $u^* \in \partial^\sigma_{\varsigma_1} f(\overline{x})$. Therefore, (7.45) holds.

For $\varsigma = 0$ and g σ-regular at $\overline{x}$, the proof is similar (but simpler, because $\eta_1 = \eta_2 = 0$, $\varsigma_1 = \varsigma_2 = 0$). □

The next result relates $\partial^\sigma f(\overline{x})$ and $\partial^s f(\overline{x})$ for $\sigma \in \{w, h, t\}$.

Corollary 7.4.10. *Let $f \in \Lambda(X, Y)$ and $\sigma \in \{w, h, t\}$. If f is σ-regular at $\overline{x} \in \operatorname{dom} f$ then*

$$\partial^\sigma f(\overline{x}) = \partial^s f(\overline{x}) + \vartheta_\sigma(X, Y).$$

If f is σ-η-regular at $\overline{x} \in \operatorname{dom} f$ for every $\eta \in \mathbb{R}_+$ then

$$\forall \varsigma \in Y \setminus \{0\} : \partial^\sigma_\varsigma f(\overline{x}) = \vartheta_\sigma(X, Y) + \bigcup \{\partial^s_{\varsigma'} f(\overline{x}) \mid \varsigma' \in C \cap (\varsigma - D^\sigma)\}.$$

Proof. Apply Theorem 7.4.9 for θ and f, taking into account Corollary 7.4.5. □

The next result deals with the composition of convex functions. We assume that Z is ordered by a pointed convex cone K.

Theorem 7.4.11. *Let $g : X \to Z^\bullet$, $f : Z \to Y^\bullet$ be proper operators with $\operatorname{dom} f \circ g$ $\left(= g^{-1}(\operatorname{dom} f)\right) \neq \emptyset$, $\varsigma \in Y$ and $\sigma \in \{w, h, t\}$. Then*

$$\partial^\sigma_\varsigma (f \circ g)(x) \supseteq \bigcup \left\{\partial^\sigma_{\varsigma_1}(T \circ g)(x) \mid \varsigma_1 \in Y,\ \varsigma_2 \in C,\ \varsigma_1 + \varsigma_2 = \varsigma,\ T \in \partial^s_{\varsigma_2} f(g(x))\right\} \tag{7.47}$$

for every $x \in X$, where $f(+\infty) := +\infty$, $T(+\infty) := +\infty$.

Moreover, assume that $f \in \Lambda(Z, Y)$ is (K, C)-increasing, $g \in \Lambda(X, Z)$ and either (i) *X and Z are l.c.s., f is continuous at some point in* $\operatorname{dom} f \cap g(\operatorname{dom} g)$, *or* (ii) *$f, g$ are C-lower continuous,* $\operatorname{dom} g$ *is closed, X and Z are Fréchet spaces, and* $0 \in \operatorname{sic}(\operatorname{dom} f - g(\operatorname{dom} g))$. *Consider $\overline{x} \in \operatorname{dom} f \circ g$.*

If f is σ-regular at $g(\overline{x})$ then

$$\partial^{\sigma}(f \circ g)(\overline{x}) = \bigcup \{\partial^{\sigma}(T \circ g)(\overline{x}) \mid T \in \partial^{s} f(g(\overline{x}))\}. \tag{7.48}$$

If $\varsigma \neq 0$ and f is σ-η-regular at $g(\overline{x})$ for every $\eta \geq 0$ then

$$\partial^{\sigma}_{\varsigma}(f \circ g)(\overline{x}) = \bigcup \left\{\partial^{\sigma}_{\varsigma_1}(T \circ g)(\overline{x}) \mid \varsigma_1 \in Y,\ \varsigma_2 \in C,\ \varsigma_1 + \varsigma_2 = \varsigma,\ T \in \partial^{s}_{\varsigma_2} f(g(\overline{x}))\right\} \tag{7.49}$$

Proof. For $x \in X \setminus \operatorname{dom} f \circ g$ (7.47) holds clearly. Let $x \in \operatorname{dom} f \circ g$, $\varsigma_1 \in Y, \varsigma_2 \in C$ with $\varsigma_1 + \varsigma_2 = \varsigma$ and $T \in \partial^{s}_{\varsigma_2} f(g(x))$, $S \in \partial^{w}_{\varsigma_1}(T \circ g)(x)$. Assume that $S \notin \partial^{w}_{\varsigma}(f \circ g)(x)$. Then there exists $x' \in \operatorname{dom} f \circ g$ such that $\left[f(g(x')) - Sx'\right] - [f(g(x)) - Sx] \in (-\varsigma - \operatorname{int} C)$. Since $T \in \partial^{s}_{\varsigma_1} f(g(x))$ we have that $T(g(x') - g(x)) \in f(g(x')) - f(g(x)) + \varsigma_2 - C$. Summing up both relations we get

$$\left[T(g(x')) - Sx'\right] - [T(g(x)) - Sx] \in [-\varsigma_1 + (-C) + (-\operatorname{int} C)] = -\varsigma_1 - \operatorname{int} C,$$

a contradiction because $S \in \partial^{w}_{\varsigma_1}(T \circ g)(x)$.

The proof for $\sigma \in \{h, t\}$ is similar (in fact reduces to the preceding case taking some $D \in \mathscr{D}^{\sigma}$).

Assume now $f \in \Lambda(Z, Y)$ is (K, C)-increasing, $g \in \Lambda(X, Z)$ and (i) or (ii) holds. Let $\varsigma \neq 0$ and f be σ-η-regular for every $\eta \geq 0$ at $\overline{x} \in \operatorname{dom} f \circ g$, and take $T \in \partial^{\sigma}_{\varsigma}(f \circ g)(\overline{x})$. Since $f \circ g$ is convex, by Theorem 7.4.6 there exists $y^* \in C^{\sigma}$ such that $y^* \circ S \in \partial_{\eta}\left(y^* \circ (f \circ g)\right)(\overline{x}) = \partial_{\eta}\left((y^* \circ f) \circ g\right)(\overline{x})$, where $\eta := \langle \varsigma, y^* \rangle$. Since $y^* \in C^{\sigma} \subseteq C^{+} \setminus \{0\}$, we have that $\tilde{f} := y^* \circ f$ is convex and increasing. In case (i) $\tilde{f}$ is continuous at some point in $\operatorname{dom} \tilde{f} \cap g(\operatorname{dom} g)$, while in case (ii) $\tilde{f}$ is convex and lsc. Using [102, Proposition 4.11] we get

$$\begin{aligned} \partial_{\eta}(\tilde{f} \circ g)(\overline{x}) &= \partial_{\eta}\left((y^* \circ f) \circ g\right)(\overline{x}) \\ &= \bigcup \left\{\partial_{\eta_1}(z^* \circ g)(\overline{x}) \mid z^* \in \partial_{\eta_2} \tilde{f}(g(\overline{x})),\ \eta_1, \eta_2 \geq 0,\ \eta_1 + \eta_2 = \eta\right\}. \end{aligned}$$

Hence there exist $\eta_1, \eta_2 \geq 0$ with $\eta_1 + \eta_2 = \eta$ and $z^* \in \partial_{\eta_2} \tilde{f}(g(\overline{x})) = \partial_{\eta_2}(y^* \circ f)(g(\overline{x}))$ such that $y^* \circ S \in \partial_{\eta_1}(z^* \circ g)(\overline{x})$. Because f is σ-η_2-regular at $g(\overline{x})$, there exist $\varsigma_2 \in \eta_2 C \subseteq C$ with $\langle \varsigma_2, y^* \rangle = \eta_2$ and $T \in \partial^{s}_{\varsigma_2} f(g(\overline{x}))$ such that $z^* = y^* \circ T$. Taking $\varsigma_1 := \varsigma - \varsigma_2$, we have that $\langle \varsigma_1, y^* \rangle = \eta_1$. Hence $y^* \circ S \in \partial_{\eta_1}(z^* \circ g)(\overline{x}) = \partial_{\eta_1}(y^* \circ (T \circ g))(\overline{x})$, and so, using again Theorem 7.4.6, we obtain that $S \in \partial^{\sigma}_{\varsigma_1}(T \circ g)(\overline{x})$. Hence (7.49) holds.

For $\kappa = 0$ and f σ-regular at $g(\overline{x})$ the proof is similar (but simpler, because $\eta_1 = \eta_2 = 0$, $\varsigma_1 = \varsigma_2 = 0$). □

Corollary 7.4.12. *Let Y be a l.c.s., $T \in L(X, Z)$, $f \in \Lambda(Z, Y)$, $\overline{x} \in T^{-1}(\operatorname{dom} f)$ and $\sigma \in \{w, h, t\}$. Assume that either* (a) *X, Z are l.c.s., f is continuous at Tx for some $x \in T^{-1}(\operatorname{dom} f)$, or* (b) *$X$, Z are Fréchet spaces, f C-lower continuous and $0 \in \operatorname{sic}(\operatorname{dom} f - \operatorname{Im} T)$.*

If f is σ-regular at $T\overline{x}$ then

$$\partial^{\sigma}(f \circ T)(\overline{x}) = \partial^{s} f(T\overline{x}) \circ T + \vartheta_{\sigma}(X,Y). \tag{7.50}$$

If f is σ-η-regular at $T\overline{x}$ for every $\eta \geq 0$ then

$$\forall \varsigma \in Y \setminus \{0\} : \partial^{\sigma}_{\varsigma}(f \circ T)(\overline{x}) = \vartheta_{\sigma}(X,Y) + \bigcup \{\partial^{s}_{\varsigma'} f(T\overline{x}) \circ T \mid \varsigma' \in C \cap (\varsigma - D_{\sigma})\}. \tag{7.51}$$

Proof. We prove only the case in which $\varsigma \in Y \setminus \{0\}$ and f is σ-η-regular at $T\overline{x}$ for every $\eta \geq 0$ (the proof in the other case being similar). Using Theorem 7.4.11 and Corollary 7.4.5 we obtain that

$$\begin{aligned}
\partial^{\sigma}_{\varsigma}(f \circ T)(\overline{x}) &= \bigcup \left\{ \partial^{\sigma}_{\varsigma_1}(S \circ T)(\overline{x}) \mid \varsigma_1 \in Y,\ \varsigma_2 \in C,\ \varsigma_1 + \varsigma_2 = \varsigma,\ S \in \partial^{s}_{\varsigma_2} f(T\overline{x}) \right\} \\
&= \bigcup \left\{ S \circ T + Y_{\sigma}(X,Y) \mid \varsigma_1 \in D_{\sigma},\ \varsigma_2 \in C,\ \varsigma_1 + \varsigma_2 = \varsigma,\ S \in \partial^{s}_{\varsigma_2} f(T\overline{x}) \right\} \\
&= \vartheta_{\sigma}(X,Y) + \bigcup \left\{ \partial^{s}_{\varsigma_2} f(T\overline{x}) \circ T \mid \varsigma_1 \in D_{\sigma},\ \varsigma_2 \in C,\ \varsigma_1 + \varsigma_2 = \varsigma \right\}.
\end{aligned}$$

Hence (7.51) holds. □

The results presented in this section are mainly obtained by El Maghri and Laghdir in [175] (for $\varsigma = 0$) and [174] (for general ς), where the cases $\sigma = w$ and $\sigma = h$ (denoted p) were considered. For simplifying the presentation we have taken the cone C to be pointed.

Chapter 8
Duality

In this chapter we present duality assertions for set-valued optimization problems in infinite dimensional spaces where the solution concept is based on vector approach, on set approach as well as on lattice approach. For set-valued optimization problems where the solution concept is based on vector approach we present conjugate duality statements. The notions of conjugate maps, subdifferential and a perturbation approach used for deriving these duality assertions are given. Furthermore, Lagrange duality for set-valued problems based on vector approach is shown. Moreover, we consider set-valued optimization problems where the solution concept is given by a set order relation introduced by Kuroiwa and derive corresponding saddle point assertions. For set-valued problems where the solution concept is based on lattice structure, we present duality theorems that are based on an consequent usage of infimum and supremum (in the sense greatest lower and least upper bounds with respect to a partial ordering). We derive conjugate duality assertions as well as Lagrange duality statements.

A comparison of different approaches to duality in set-valued optimization is given at the end of this chapter.

It is an old idea to study additionally to a given optimization problem ($p(x) \to$ inf with infimal value I) a corresponding dual problem ($d(u) \to$ sup with supremal value $S, S \leq I$), remember the dual variational principles of Dirichlet and Thompson or simply the pair of dual programs in linear optimization. The reasons for the introduction of a useful dual problem are the following:

- The dual problem has (under additional conditions) the same optimal value as the given "primal" optimization problem, but solving the dual problem could be done with other methods of analysis or numerical mathematics.
- An approximate solution of the given minimization problem gives an estimation of the infimal value I from above, whereas an approximate solution of the dual problem is an estimation of I from below, so that one gets intervals containing I.

A.A. Khan et al., *Set-valued Optimization*, Vector Optimization,
DOI 10.1007/978-3-642-54265-7_8

- Recalling Lagrange method, saddle points, equilibrium points of two person games, shadow prices in economics, perturbation methods or dual variational principles, it becomes clear, that optimal dual variables often have a special meaning for the given problem.

Of course, the just listed advantages require a skilfully chosen dual program. Nevertheless, the mentioned points are motivation enough, to look for dual problems in set-valued optimization with corresponding properties too. There are a lot of papers, which are dedicated to that aim, also a lot of survey papers (see the books in [73, 292, 293, 402] and references therein).

8.1 Duality Assertions for Set-Valued Problems Based on Vector Approach

8.1.1 Conjugate Duality for Set-Valued Problems Based on Vector Approach

A comprehensive and detailed theory for set-valued conjugate duality in infinite dimensional spaces is presented in the book by Boţ, Grad and Wanka [73]. In [73] the notions of conjugate maps, subdifferential and a perturbation approach introduced by Kawasaki [315], Tanino and Sawaragi [569] are extended and applied.

Assume that $F : X \rightrightarrows Y^\bullet = Y \cup \{-\infty\} \cup \{+\infty\}$, X and Y are linear topological spaces, X^* and Y^* the corresponding topological dual spaces. The domain of F is given by $\operatorname{Dom} F := \{x \in X \mid F(x) \neq \emptyset \text{ and } F(x) \neq \{+\infty\}\}$ and the graph of F by $\operatorname{graph} F := \{(x, y) \in X \times Y^\bullet \mid y \in F(x)\}$. Suppose that $C \subset Y$ is a proper pointed closed convex cone. However, it is possible to show the results presented in this section without the closedness assumption concerning C (see [73]). We denote the set of linear continuous mappings from X to Y by $\mathscr{L}(X, Y)$. In this section we consider the set-valued optimization problem

$$\text{minimize} \quad F(x) \quad \text{subject to} \quad x \in X, \tag{SP}$$

where we use the notations from Definition 2.4.1 and the solution concept introduced in Definition 2.6.1 adapted for set-valued functions $F : X \rightrightarrows Y^\bullet$. Consider $A \subseteq Y^\bullet$. In the case that $A = \emptyset$, take by convention $\operatorname{Min}(A, C) = \{+\infty\}$ and $\operatorname{Max}(A, C) = \{-\infty\}$. The operations with sets in $Y^\bullet$ are taken like for sets in Y (see Sect. 2.6.3). It is important to mention that if $-\infty \in \operatorname{Max}(A, C)$, then $A = \{-\infty\}$ or $A = \emptyset$. Furthermore, if $+\infty \in \operatorname{Min}(A, C)$, one has $A = \{+\infty\}$ or $A = \emptyset$.

In order to construct a dual problem to (SP) using the concept of conjugation we introduce in the following definition the conjugate map of F, the biconjugate map of F and the subgradient of F (compare also Sects. 7.1 and 7.2).

Definition 8.1.1. Consider a set-valued map $F : X \rightrightarrows Y^\bullet$.

(a) The set-valued map $F^* : \mathscr{L}(X, Y) \rightrightarrows Y^\bullet$ defined by

$$F^*(T) := \text{Max}(\cup_{x \in X}[T(x) - F(x)], C)$$

is called the **conjugate map** of F.

(b) The set-valued map $F^{**} : X \rightrightarrows Y^\bullet$ defined by

$$F^{**}(x) := \text{Max}(\cup_{T \in \mathscr{L}(X,Y)}[T(x) - F^*(T)], C)$$

is called the **biconjugate map** of F.

(c) The operator $T \in \mathscr{L}(X, Y)$ is said to be a **subgradient** of F at $(x, y) \in \text{graph}\, F \cap (X \times Y)$ if

$$Tx - y \in \text{Max}(\cup_{u \in X}[Tu - F(u)], C).$$

The set of all subgradients of F at $(x, y) \in \text{graph}\, F \cap (X \times Y)$ is called the **subdifferential** of F at $(x, y) \in \text{graph}\, F \cap (X \times Y)$ and it is denoted by $\partial F(x; y)$. Moreover, for all $x \in X$ denote

$$\partial F(x) := \cup_{y \in F(x) \cap Y} \partial F(x; y).$$

If for all $y \in F(x) \cap Y$ it holds $\partial F(x; y) \neq \emptyset$ then F is said to be **subdifferentiable** at x.

Concerning the existence of subgradients in the sense of Definition 8.1.1 (c) the following assertion is given by Boţ, Grad, Wanka [73, Proposition 7.1.4].

Proposition 8.1.2. *Let $F : X \rightrightarrows Y^\bullet$ be a set-valued map and $(x, y) \in \text{graph}\, F \cap (X \times Y)$. Then $\partial F(x; y) \neq \emptyset$ if and only if $y \in F^{**}(x)$. Consequently, F is subdifferentiable at x if and only if $F(x) \cap Y \subseteq F^{**}(x)$.*

In Definition 8.1.1 (a) the conjugate map F^* is defined for linear continuous operators T and in Definition 8.1.1 (c) the subgradients of F are linear continuous operators. An other possibility would be to replace $T \in \mathscr{L}(X, Y)$ by linear continuous functionals $x^* \in X^*$ (see Tanino and Sawaragi [569] for the finite dimensional case, Boţ, Grad and Wanka [73, Definition 7.1.9] for the infinite dimensional case and Definition 8.1.3 below). The relationship between subgradients in the sense of Definition 8.1.1 and the following Definition 8.1.3 is given for $T \in \mathscr{L}(X, Y)$, $x^* \in X^*$ and $c \in Y \setminus \{0\}$ by

$$Tx = \langle x^*, x \rangle c \quad \text{for} \quad x \in X.$$

The following definitions of conjugates and subgradients with respect to an element $c \in Y \setminus \{0\}$ are introduced in [73, Definition 7.1.9].

Definition 8.1.3. Consider a set-valued map $F : X \rightrightarrows Y^\bullet$ and $c \in Y \setminus \{0\}$.

(a) The set-valued map $F_c^* : X^* \rightrightarrows Y^\bullet$ defined by

$$F_c^*(x^*) := \mathrm{Max}(\cup_{x\in X}[\langle x^*, x\rangle c - F(x)], C),$$

is called the **c-conjugate map** of F.

(b) The set-valued map $F_c^{**} : X \rightrightarrows Y^\bullet$ defined by

$$F_c^{**}(x) := \mathrm{Max}(\cup_{x^*\in X^*}[\langle x^*, x\rangle c - F_c^*(x^*)], C),$$

is called the **c-biconjugate map** of F.

(c) The vector $x^* \in X^*$ is called **c-subgradient** of F at $(x, y) \in \operatorname{graph} F \cap (X \times Y)$ if

$$\langle x^*, x\rangle c - y \in \mathrm{Max}(\cup_{u\in X}[\langle x^*, u\rangle c - F(u)], C).$$

The set of all c-subgradients of F at $(x, y) \in \operatorname{graph} F \cap (X \times Y)$ is called the **c-subdifferential** of F at $(x, y) \in \operatorname{graph} F \cap (X \times Y)$ and it is denoted by $\partial_c F(x; y)$. Moreover, for all $x \in X$ denote

$$\partial_c F(x) := \cup_{y\in F(x)\cap Y}\, \partial_c F(x; y).$$

If for all $y \in F(x) \cap Y$ it holds $\partial_c F(x; y) \neq \emptyset$ then F is said to be **c-subdifferentiable** at x.

Remark 8.1.4. The definition of c-subgradients given above is closed to Definition 8.4.5 where we use infimal sets instead of the Pareto maximum like in Definition 8.1.3 (c).

Remark 8.1.5. In Boţ, Grad and Wanka [73] the definition of c-subgradients is used in order to introduce the property that the primal set-valued optimization problem is c-stable: The problem (SP) is called **c-stable** with respect to a certain set-valued perturbation map if the corresponding minimal value map is c-subdifferentiable at 0.

In the following we present weak and strong conjugate duality statements for set-valued optimization problems where the solution concept is based on vector approach. For deriving duality assertions for set-valued optimization problems via conjugation the following generalization of the **Young-Fenchel inequality** given by Boţ, Grad, Wanka [73, Proposition 7.1.1] is important.

Proposition 8.1.6. *Let $F : X \rightrightarrows Y^\bullet$ be a set-valued map, $x \in X$ and $T \in \mathscr{L}(X, Y)$. Then for all $y \in F(x)$ and all $y^* \in F^*(T)$ it holds*

$$y + y^* \notin Tx - (C \setminus \{0\}). \tag{8.1}$$

Consider now $F : X \rightrightarrows Y \cup \{+\infty\}$. Let us introduce a **set-valued perturbation map**

$$\Phi : X \times V \rightrightarrows Y \cup \{+\infty\} \quad \text{such that} \quad \Phi(x, 0) = F(x) \text{ for all } x \in X,$$

where V is a linear topological space, the so called **perturbation space**, its topological dual space is denoted by V^*. Then the set-valued problem (SP) is embedded into a family of perturbed problems

$$\text{Min}(\Phi(X, v), C), \tag{SP_v}$$

where $v \in V$ is the perturbation variable.

Using the conjugate of the perturbation map

$$\Phi^* : L(X,Y) \times L(V,Y) \rightrightarrows Y^\bullet, \Phi^*(T,Q) = \text{Max}(\cup_{x\in X, v\in V}[Tx + Qv - \Phi(x,v)], C)$$

we can attach a set-valued dual problem

$$\text{maximize} \quad -\Phi^*(0, Q) \quad \text{subject to} \quad Q \in \mathcal{L}(V, Y) \tag{DSP}$$

to the primal problem (SP). The following weak duality statement is shown in Boţ, Grad, Wanka [73, Theorem 7.1.11].

Theorem 8.1.7 (Weak Duality for (SP) and (DSP)). *For all $x \in X$ and all $Q \in \mathcal{L}(V, Y)$ it holds*

$$\Phi(x, 0) \cap \{-\Phi^*(0, Q) - (C \setminus \{0\})\} = \emptyset.$$

Proof. Let $x \in X$ and $Q \in \mathcal{L}(V, Y)$ be arbitrarily chosen. Suppose that $\Phi(x, 0) \cap \{-\Phi^*(0, Q) - (C \setminus \{0\})\} \neq \emptyset$. Then there exists $y \in \Phi(x, 0) \cap \{-\Phi^*(0, Q) - (C \setminus \{0\})\}$. This means that there exist $y \in F(x)$ and $y^* \in \Phi^*(0, Q)$ with $y + y^* \in -C \setminus \{0\}$. So we get a contradiction to (8.1) in Proposition 8.1.6. □

Boţ, Grad, Wanka [73, Corollary 7.1.12] gave the following equivalent formulation of the weak duality assertion in Theorem 8.1.7:

Corollary 8.1.8. *For all $x \in X$ and all $Q \in \mathcal{L}(V, Y)$, it holds $y \notin y^* - (C \setminus \{0\})$ whenever $y \in F(x)$ and $y^* \in -\Phi^*(0, Q)$.*

Furthermore, from the weak duality in Theorem 8.1.7 we get:

Corollary 8.1.9. *Let $\overline{y} \in F(\overline{x}) \cap \{-\Phi^*(0, \overline{Q})\}$ for $\overline{x} \in X$ and $\overline{Q} \in \mathcal{L}(V, Y)$. Then $(\overline{x}, \overline{y})$ is a minimizer of (SP) and $(\overline{Q}, \overline{y})$ is a maximizer of (DSP).*

Proof. Suppose that $(\overline{x}, \overline{y})$ is not a minimizer of (SP). Then there exist $x \in X$ and $y \in F(x) = \Phi(x, 0)$ such that $y \in \overline{y} - (C \setminus \{0\})$, i.e., $y \in \Phi(x, 0) \cap (-\Phi^*(0, \overline{Q}) - (C \setminus \{0\}))$ in contradiction to Theorem 8.1.7. Assuming $(\overline{Q}, \overline{y})$ not being a maximizer of (DSP) we get analogously a contradiction to Theorem 8.1.7. □

In order to give a reformulation of the dual problem (DSP) and to formulate a certain regularity assumption for deriving strong duality statements we introduce the **minimal value map** $H : V \rightrightarrows Y \cup \{+\infty\}$ defined by

$$H(v) := \mathrm{Min}(\cup_{x \in X} \Phi(x, v), C).$$

Definition 8.1.10. Let $A \subseteq Y^\bullet$ be a given set.

(a) The set $\mathrm{Min}(A,C)$ is said to be **externally stable** if $A \backslash \{+\infty\} \subseteq \mathrm{Min}(A, C) + C$.
(b) The set $\mathrm{Max}(A,C)$ is said to be **externally stable** if $A \backslash \{-\infty\} \subseteq \mathrm{Max}(A, C) - C$.

This property is also called **domination property** (see Luc [402, 404]).

For some arbitrary $v \in V$ one says that $H(v)$ is externally stable if $\mathrm{Min}(\cup_{x \in X} \Phi(x, v), C)$ is externally stable in the sense of Definition 8.1.10.

Subsequentially we suppose that $H(v)$ is externally stable for all $v \in V$. Then it holds $H^*(Q) = \Phi^*(0, Q)$ and the dual problem (DSP) can be equivalently written in the form

$$\mathrm{Max}(\cup_{Q \in \mathscr{L}(V,Y)}[-H^*(Q)], C). \tag{DSP}$$

Furthermore, the following assertion is shown by Boţ, Grad, Wanka [73, Lemma 7.1.14].

Lemma 8.1.11. *It holds*

$$\mathrm{Max}(\cup_{Q \in \mathscr{L}(V,Y)}[-\Phi^*(0, Q)], C) = H^{**}(0).$$

For deriving strong duality assertions we need a certain kind of regularity assumption:

Definition 8.1.12. The problem (SP) is called **stable with respect to the perturbation map** Φ if the minimal value map H is subdifferentiable at 0 in the sense of Definition 8.1.1 (c).

It is important to mention that $\pm\infty \notin H(0)$ under our assumptions.

The following theorem shows the existence of a common element $\overline{y}$ in the objective values of the primal and dual set-valued problems. So it can be considered as a strong duality assertion (see [73, Theorem 7.1.15]).

Theorem 8.1.13 (Strong Duality). *If the set-valued problem (SP) is stable in the sense of Definition 8.1.12, then for each minimizer $(\overline{x}, \overline{y})$ of the problem (SP) there exists a maximizer $(\overline{Q}, \overline{y})$ of the problem (DSP).*

Proof. Suppose that (SP) is stable. Then H is subdifferentiable at 0 because of Definition 8.1.12. This is equivalent to $H(0) \subseteq H^{**}(0)$ taking into account

Proposition 8.1.2, as $\pm\infty \notin H(0)$. Take a minimizer $(\overline{x}, \overline{y})$ of the problem (SP) such that $\overline{y} \in H(0)$. From Lemma 8.1.11 we get the existence of $\overline{Q} \in L(V, Y)$ such that $\overline{y} \in -\Phi^*(0, \overline{Q})$. Moreover, Corollary 8.1.9 yields that $(\overline{Q}, \overline{y})$ is a maximizer of the dual problem (DSP). □

Remark 8.1.14. A comparison of different approaches to duality in set-valued optimization is given in Sect. 8.4. Using the concept of supremum/infimum by Tanino and Sawaragi (cf. [569]) of a set, defined in terms of the closure of the set, Lalitha and Arora [360] introduce the notions of conjugate and biconjugate maps as well as that of subgradients of a set-valued map. Conjugate duality results are established for a set-valued optimization problem based on vector approach in [360].

In the paper by Li, Chen and Wu [379] two conjugate dual problems are proposed by considering the different perturbations for a set-valued vector optimization problem with explicit constraints. Weak duality assertions, inclusion relations between the image sets of dual problems, strong duality and stability criteria are investigated.

8.1.2 Lagrange Duality for Set-Valued Optimization Problems Based on Vector Approach

In this section we derive duality statements for solutions of set-valued optimization problems in the sense of Definition 2.6.6. Let X, Y, Z be real locally convex Hausdorff spaces, Y, Z be ordered by proper pointed closed convex cones C, K, respectively, $F : X \rightrightarrows Y$ and $G : X \rightrightarrows Z$. Under these assumptions we consider the set-valued optimization problem (SP) with restrictions given in (2.49) (compare Tasset [570]):

$$\text{minimize} \quad F(x) \quad \text{subject to} \quad x \in S, \tag{SP}$$

where $M \subseteq X$ is a convex set with $M \subseteq \operatorname{dom} F \cap \operatorname{dom} G$ and

$$S := \{x \in M \mid G(x) \cap (-K) \neq \emptyset\}.$$

Here the "minimization" is to be understood in the sense of Definition 2.6.6, i.e., we are looking for *quasi-weak minimizers* of (SP). In this section we will show duality assertions for the problem (SP) and a corresponding dual problem based on the solution concept introduced in Definition 2.6.6.

Recall that the multifunction $F : X \rightrightarrows Y$ is C–**convex** if and only if epi F is convex (compare Sect. 2.3).

We introduce

$$\begin{aligned}\mathscr{E} := \mathscr{E}_C &:= (F,G)(M) + (C \times K)\\ &= \{(y+u, z+v) \mid x \in M,\ y \in F(x),\ z \in G(x),\ u \in C,\ v \in K\}.\end{aligned}$$

If $x_0 \in S$ and $y_0 \in F(x_0)$ then $(y_0, 0) \in \mathscr{E}$.

Lemma 8.1.15. *If F is C–convex and G is K–convex, then $\mathscr{E}$ is convex .*

Proof. Under the given assumptions $(F,G) : X \rightrightarrows Y \times Z$ defined by $(F,G)(x) := F(x) \times G(x)$ is $C \times K$–convex. Since $\mathscr{E} = \operatorname{epi}_{C\times K}(F,G)$, we have that $\mathscr{E}$ is convex. □

Theorem 8.1.16. *Let F be C–convex and G be K–convex. Assume that*

(a) $x_0 \in S$ *and* $y_0 \in F(x_0)$,
(b) $(y_0, 0) \notin \operatorname{qri} \mathscr{E}$,
(c) $0 \in \operatorname{qi}(G(M) + K)$.

Then there exists $y^* \in C^+ \setminus \{0\}$, $z^* \in K^+$ such that

$$\forall (y,z) \in (F,G)(M) : \langle y, y^*\rangle + \langle z, z^*\rangle \geq \langle y_0, y^*\rangle .$$

Moreover, $z^*\,(G(x_0) \cap (-K)) = \{0\}$.

Proof. Because $(y_0, 0) \in \mathscr{E} \setminus \operatorname{qri}\mathscr{E}$, by (2.52), there exists $(y^*, z^*) \in Y^* \times Z^* \setminus \{(0,0)\}$ such that

$$\langle y+u, y^*\rangle + \langle z+v, z^*\rangle \geq \langle y_0, y^*\rangle \quad \forall (y,z) \in (F,G)(M),\ \forall (u,v) \in C \times K.$$

It follows that $y^* \in C^+$ and $z^* \in K^+$. Assume that $y^* = 0$. Then $\langle z, z^*\rangle \geq 0$ for every $z \in G(M) + K$, and so $\langle z, z^*\rangle \geq 0$ for every $z \in \operatorname{cl}(\mathbb{R}_+ (G(M)+K))$. Since $0 \in \operatorname{qi}(G(M)+K)$ we obtain that $z^* = 0$. This contradicts the fact that $(y^*, z^*) \neq (0,0)$. Hence $y^* \neq 0$.

Let now $z_0 \in G(x_0) \cap (-K)$. Then $0 \geq \langle z_0, z^*\rangle = \langle y_0 - y_0, y^*\rangle + \langle z_0, z^*\rangle \geq 0$, and so $\langle z_0, z^*\rangle = 0$. The proof is complete. □

Proposition 8.1.17. *Assume that $\operatorname{qi} C \neq \emptyset$, $x_0 \in S$ and $y_0 \in F(x_0)$. Then the assertion of Theorem 8.1.16 imply that $y_0 \in \operatorname{Min}(F(S), \operatorname{qi} C)$. In particular, under the conditions of Theorem 8.1.16 one has $y_0 \in \operatorname{Min}(F(S), \operatorname{qi} C)$.*

Proof. Assume that $y \in (F(S) - y_0) \cap (-\operatorname{qi} C)$. Then there exists $\overline{x} \in S$ and $\overline{y} \in F(\overline{x})$ such that $y = \overline{y} - y_0 \in (-\operatorname{qi} C)$. By (2.53) we obtain that $\langle \overline{y} - y_0, y^*\rangle < 0$. Since $\overline{x} \in S$, there exists $\overline{z} \in G(\overline{x}) \cap (-K)$. So we get the contradiction $0 > \langle \overline{y} - y_0, y^*\rangle + \langle \overline{z}, z^*\rangle \geq 0$. Hence $y_0 \in \operatorname{Min}(F(S), \operatorname{qi} C)$. □

In the next theorem we show an assertion concerning the existence of linear continuous operators related to our duality statements in Theorem 8.1.21.

Theorem 8.1.18. *Let F be C–convex, G be K–convex, and* $\operatorname{qi} C \neq \emptyset$. *Assume that*

(a) $x_0 \in S$ *and* $y_0 \in F(x_0)$,
(b) $(y_0, 0) \notin \operatorname{qri} \mathscr{E}$,
(c) $0 \in \operatorname{qi}(G(M) + K)$.

Then there exists $T \in \mathscr{L}_+ := \mathscr{L}_+(Z, Y)$ *such that* $y_0 \in (F + TG)(x_0)$,

$$y_0 \in \operatorname{Min}((F + TG)(M), \operatorname{qi} C)$$

and

$$T(G(x_0) \cap (-K)) = \{0\}.$$

Proof. Taking into account Theorem 8.1.16 we get the existence of $y^* \in C^+ \setminus \{0\}$ and $z^* \in K^+$ such that $\langle y, y^* \rangle + \langle z, z^* \rangle \geq \langle y_0, y^* \rangle$ for all $(y, z) \in (F \times G)(M)$. Since $\operatorname{qi} C \neq \emptyset$, we have that $C \subset \operatorname{cl}(\operatorname{qi} C)$, and so there exists $\overline{y} \in \operatorname{qi} C$ such that $\langle \overline{y}, y^* \rangle = 1$. Let $T : Z \to Y$ be defined by $T(z) := \langle z, z^* \rangle \, \overline{y}$. Clearly $T \in \mathscr{L}_+$. From Theorem 8.1.16 we have that $T(G(x_0) \cap (-K)) = \{0\}$, and so $y_0 \in (F + TG)(x_0)$ because $G(x_0) \cap (-K) \neq \emptyset$. Moreover, for $x \in M$, $y \in F(x)$, $z \in G(x)$ we have

$$\langle y + Tz, y^* \rangle = \langle y, y^* \rangle + \langle z, z^* \rangle \langle \overline{y}, y^* \rangle = \langle y, y^* \rangle + \langle z, z^* \rangle \geq \langle y_0, y^* \rangle .$$

Since $y^* \in C^+ \setminus \{0\}$, this shows that $y + Tz - y_0 \notin (-\operatorname{qi} C)$, and so $((F + TG)(M) - y_0) \cap (-\operatorname{qi} C) = \emptyset$. Hence $y_0 \in \operatorname{Min}((F + TG)(M), \operatorname{qi} C)$. □

Remark 8.1.19. In comparison with the corresponding result by Tasset [570, Theorem 3.7] we do not assume $y_0 \in \operatorname{Min}(F(S), \operatorname{qi} C)$ in Theorem 8.1.18 (compare with Proposition 8.1.17).

In order to derive Lagrange duality assertions we introduce the set-valued Lagrangian $L : M \times \mathscr{L}_+ \rightrightarrows Y$ for (SP) with restrictions given in (2.49) by

$$L(x, T) := F(x) + TG(x).$$

Furthermore, we introduce the set-valued map $\psi : \mathscr{L}_+ \rightrightarrows Y$ by

$$\psi(T) := \operatorname{Min}(L(M, T), \operatorname{qi} C) = \{y \in L(M, T) \mid (L(M, T) - y) \cap (-\operatorname{qi} C) = \emptyset\}.$$

Using these notions we introduce a set-valued dual problem to (SP) given by

$$\text{maximize} \quad \psi(T) \quad \text{subject to} \quad T \in \mathscr{L}_+. \tag{SD}$$

In (SD) the "maximization" is to understand in the following sense: We are looking for a **quasi-weak maximizer**, i.e., for $(T_0, y_0) \in \operatorname{graph} \psi$ with

$$(\psi(\mathscr{L}_+) - y_0) \cap \operatorname{qi} C = \emptyset. \tag{8.2}$$

Now, we formulate a weak duality assertion.

Theorem 8.1.20 (Weak Duality). *For every* $\overline{y} \in \psi(\mathscr{L}_+)$ *it holds*

$$(F(S) - \overline{y}) \cap (-\operatorname{qi} C) = \emptyset$$

or equivalently, $(F(S) - \psi(\mathscr{L}_+)) \cap (-\operatorname{qi} C) = \emptyset$.

Proof. Take $x_0 \in S$ and $\overline{y} \in \psi(T) = \operatorname{Min}((F + TG)(M), \operatorname{qi} C)$ with $T \in \mathscr{L}_+$. Because $G(x_0) \cap (-K) \neq \emptyset$ and $T \in \mathscr{L}_+$, we have that

$$F(x_0) - \overline{y} \subset F(x_0) + TG(x_0) + C - \overline{y} \subset (F + TG)(M) + C - \overline{y}.$$

Since $C + \operatorname{qi} C = \operatorname{qi} C$ and $((F + TG)(M) - \overline{y}) \cap (-\operatorname{qi} C) = \emptyset$, we obtain that $(F(x_0) - \overline{y}) \cap (-\operatorname{qi} C) = \emptyset$. The conclusion follows. □

Theorem 8.1.21 (Duality Result). *Assume that* $\operatorname{qi} C \neq \emptyset$ *and the hypotheses of Theorem 8.1.16 hold. Then*

$$y_0 \in \operatorname{Min}(F(S), \operatorname{qi} C) \iff y_0 \in \operatorname{Max}(\psi(\mathscr{L}_+), \operatorname{qi} C).$$

Proof. We have seen in Proposition 8.1.17 that under the present conditions $y_0 \in \operatorname{Min}(F(S), \operatorname{qi} C)$. We have to show that $y_0 \in \operatorname{Max}(\psi(\mathscr{L}_+), \operatorname{qi} C)$. We know already that $y_0 \in \psi(\mathscr{L}_+)$. Moreover, from Theorem 8.1.20 we have that $(\psi(\mathscr{L}_+) - F(S)) \cap \operatorname{qi} C = \emptyset$. In particular, $(\psi(\mathscr{L}_+) - y_0) \cap \operatorname{qi} C = \emptyset$, and so $y_0 \in \operatorname{Max}(\psi(\mathscr{L}_+), \operatorname{qi} C)$. □

Furthermore, we will show a Lagrange multiplier rule for Henig global proper minimizer of (SP) in the sense of Definition 2.6.3. First, we recall some notations. Let X be a separated locally convex space and $K \subset X$ a convex cone. The convex set $B \subset K$ is a base of K if $0 \notin \operatorname{cl} B$ and $K = \mathbb{R}_+ B$. Then there exists a closed convex and symmetric neighborhood U of 0 such that $(2U) \cap B = \emptyset$. Let $C_U(B) := \mathbb{R}_+(U + B)$. Clearly, $C_U(B)$ is a pointed convex cone with non empty interior; moreover, $K \setminus \{0\} \subset \operatorname{int} C_U(B)$. Indeed, let $x \in K \setminus \{0\}$. Then $x = tb$ for some $t > 0$ and $b \in B$. It follows that $x \in t(B + \operatorname{int} U) \subset C_U(B)$. The claim follows because $t(B + \operatorname{int} U)$ is open.

Set

$$C^{\Delta}(B) := \{x^* \in X^* \mid \inf x^*(B) > 0\}.$$

The following result is shown by Gong [213].

Proposition 8.1.22. *Let* K *be a convex cone with a base* B.

(a) *For any open balanced convex neighborhood* U *of zero we have* $(C_U(B))^+ \setminus \{0\} \subset C^{\Delta}(B)$.
(b) *For any* $x^* \in C^{\Delta}(B)$, *there exists an open convex balanced neighborhood* U *of zero such that* $x^* \in (C_U(B))^+$.

Proof. (a) is immediate. For (b) take $x^* \in C^{\Delta}(B)$; then $r := \inf x^*(B) > 0$. Consider $U := \{x \in X \mid |\langle x, x^*\rangle| \leq r/2\}$. Then for $x \in C_U(B)$ we have that $x = t(b+u)$ with $t \geq 0$, $b \in B$, $u \in U$, and so $\langle x, x^*\rangle = t(\langle b, x^*\rangle + \langle u, x^*\rangle) \geq t(r - r/2) = tr/2 \geq 0$. Hence $x^* \in (C_U(B))^+$. □

Theorem 8.1.23. *Let F be C–convex and G be K–convex. If*

(a) $x_0 \in S$ *and* $y_0 \in F(x_0)$,
(b) *there exists a symmetric convex neighborhood U of 0 in Y such that* $(2U) \cap B = \emptyset$ *and* $(y_0, 0) \notin \operatorname{qri} \mathscr{E}'_U$, *where $\mathscr{E}'_U$ denotes the set* $(F,G)(M) + (C_U(B), K)$,
(c) $0 \in \operatorname{qi}(G(M) + K)$,

then (x_0, y_0) is a Henig minimizer of (SP) with restrictions given by (2.49) and

(d) *there exist* $y^* \in C^{\Delta}(B)$, $z^* \in K^+$ *such that* $z^*(G(x_0) \cap (-K)) = \{0\}$ *and*

$$\langle y - y_0, y^*\rangle + \langle z, z^*\rangle \geq 0 \quad \forall (y, z) \in (F, G)(M). \tag{8.3}$$

Conversely, if (a) and (d) hold then (x_0, y_0) is a Henig global proper minimizer of (SP) in the sense of Definition 2.6.3 and (b) holds.

Proof. Assume that (a), (b) and (c) hold. Applying Theorem 8.1.16 for C replaced by $C_U(B)$, there exist $y^* \in (C_U(B))^* \setminus \{0\}$, $z^* \in K^+$ such that $z^*(G(x_0) \cap (-K)) = \{0\}$ and (8.3) holds. Moreover, (x_0, y_0) is a quasi-weak minimal solution of (SP) w.r.t. $C_U(B)$, and so it is a Henig global proper minimizer of (SP) in the sense of Definition 2.6.3. By Proposition 8.1.22 we have that $y^* \in C^{\Delta}(B)$.

Conversely, assume that (a) and (d) hold. By Proposition 8.1.22 there exists a symmetric convex neighborhood U of 0 in Y such that $(2U) \cap B = \emptyset$ and $y^* \in (C_U(B))^* \setminus \{0\}$. It follows that $(y_0, 0) \notin \operatorname{qri} \mathscr{E}'_U$, and so (b) holds. Using Proposition 8.1.17 we have that (x_0, y_0) is a quasi-weak minimizer of (SP) w.r.t. $C_U(B)$, and so it is a Henig global proper minimizer of (SP) in the sense of Definition 2.6.3. □

8.2 Duality Assertions for Set-Valued Problems Based on Set Approach

In this section we present Lagrange duality statements going back to Kuroiwa [351]. We consider set-valued optimization problems, where the solution concept is given by the relation $\preceq^l_C$ (see Definition 2.6.9):

$$A \preceq^l_C B \iff A + C \supseteq B \iff \forall b \in B, \exists a \in A \text{ such that } a \leq_C b$$

for a proper closed convex and pointed cone $C \subset Y$ and arbitrarily chosen sets $A, B \subset Y$.

Let X be a linear topological space, $(Y, \leq_C)$, $(Z, \leq_K)$ linear topological spaces with the proper pointed closed convex ordering cones C and K, respectively, $F : X \rightrightarrows Y$, and $G : X \rightrightarrows Z$, then the primal set-valued optimization problem is given by

$$\preceq^l_C -\text{minimize} \quad F(x), \quad \text{subject to} \quad G(x) \preceq^l_K 0. \tag{$SP- \preceq^l_C$}$$

Here the minimization is to understand in the sense of Definition 2.6.19 with respect to the preorder $\preceq^l_C$:

- An element $x_0 \in X$ is said to be an $\preceq^l_K$-type feasible point of $(SP- \preceq^l_C)$ if $G(x_0) \preceq^l_K 0$.
- An element $x_0 \in X$ is said to be an $\preceq^l_C$-type minimal solution of $(SP- \preceq^l_C)$ if it is $\preceq^l_K$-type feasible and if

$$F(x) \preceq^l_C F(x_0) \qquad \Longrightarrow \qquad F(x_0) \preceq^l_C F(x)$$

for each $\preceq^l_K$-type feasible x.

It is easy to see that $\{x \in X \mid G(x) \preceq^l_K 0\} \subset \operatorname{dom} G$ and this set coincides with $\{x \in X \mid G(x) \cap (-K) \neq \emptyset\}$, which is the usual inequality restriction in set-valued optimization. So the problem $(SP - \preceq^l_C)$ can be written as

$$\preceq^l_C -\text{minimize} \quad F(x), \quad \text{subject to} \quad x \in S, \tag{$SP - \preceq^l_C$}$$

where the feasible set is given by

$$S := \{x \in X \mid G(x) \cap (-K) \neq \emptyset\}.$$

Consider $\mathscr{L}(Z, Y) = \{T : Z \to Y \mid T \text{ is linear and continuous}\}$, $\mathscr{L}_+(Z, Y) = \{T \in \mathscr{L}(Z, Y) \mid T(K) \subset C\}$, $\operatorname{graph} G = \{(x, z) \in X \times Z \mid z \in G(x)\}$. In order to construct a dual problem to $(SP - \preceq^l_C)$, we introduce $\Phi : \mathscr{L}(Z, Y) \rightrightarrows Y$ defined by

$$\Phi(T) := \preceq^l_C -\operatorname{Min}\{F(x) + T(z) \mid (x, z) \in \operatorname{graph} G\},$$

where "$\preceq^l_C -\operatorname{Min}\{F(x) + T(z) \mid (x, z) \in \operatorname{graph} G\}$" denotes the set of objective function values $F(x) + T(z)$ for minimal solutions $(x, z) \in \operatorname{graph} G$ with respect to $\preceq^l_C$ in the sense of Definition 2.6.19.

Now, we introduce a generalized Lagrange function corresponding to $(SP - \preceq^l_C)$.

Definition 8.2.1. For $x \in X$ and $z \in Z$ and $T \in \mathscr{L}(Z, Y)$, let

$$L(x, z, T) := F(x) + T(z).$$

This function L is said to be the **Lagrange function** corresponding to $(SP - \preceq^l_C)$.

We will study saddle points of the Lagrange function L in the following sense:

Definition 8.2.2 (Saddle Points). A triple $(x_0, z_0, T_0) \in \operatorname{graph} G \times \mathcal{L}_+(Z, Y)$ is said to by an $\preceq^l_C$-type **saddle point** of L if the following two conditions (a) and (b) are satisfied:

(a) $L(x, z, T_0) \preceq^l_C L(x_0, z_0, T_0)$, $(x, z) \in \operatorname{graph} G \Longrightarrow L(x_0, z_0, T_0) \preceq^l_C L(x, z, T_0)$,
(b) $L(x_0, z_0, T_0) \preceq^l_C L(x_0, z_0, T)$, $T \in \mathcal{L}_+(Z, Y) \Longrightarrow L(x_0, z_0, T) \preceq^l_C L(x_0, z_0, T_0)$.

Remark 8.2.3. The triple (x_0, z_0, T_0) satisfies condition (a) in Definition 8.2.2 if and only if $L(x_0, z_0, T_0) \in \Phi(T_0)$.

The dual problem to $(SP - \preceq^l_C)$ is the following set-valued optimization problem

$$\preceq^l_C -\text{maximize} \quad \Phi(T), \quad \text{subject to} \quad T \in \mathcal{L}_+(Z, Y). \qquad (SD - \preceq^l_C)$$

For this dual problem we use the following solution concept:

- An element $T_0 \in \mathcal{L}(Z, Y)$ is said to be an $\preceq^l_C$-type feasible point of $(SD- \preceq^l_C)$ if $T_0 \in \mathcal{L}_+(Z, Y)$ and $\Phi(T_0) \neq \emptyset$.
- An element $T_0 \in \mathcal{L}(Z, Y)$ is said to be an $\preceq^l_C$-type maximal solution of $(SD- \preceq^l_C)$ if it is $\preceq^l_C$-type feasible and if

$$\Phi(T_0) \preceq^l_C \Phi(T) \qquad \Longrightarrow \qquad \Phi(T) \preceq^l_C \Phi(T_0)$$

for each $\preceq^l_C$-type feasible T.

Theorem 8.2.4 (Weak Duality Between $(SP - \preceq^l_C)$ **and** $(SD - \preceq^l_C)$**).** *Let* x_0 *be an* $\preceq^l_K$*-type feasible element of* $(SP - \preceq^l_C)$*,* T_1 *an* $\preceq^l_C$*-type feasible element of* $(SD - \preceq^l_C)$*, and* (x_1, z_1) *an element of* $\operatorname{graph} G$ *with* $F(x_1) + T_1(z_1) \in \Phi(T_1)$*. Then,*

$$F(x_0) \preceq^l_C F(x_1) + T_1(z_1) \quad \Rightarrow \quad F(x_1) + T_1(z_1) \preceq^l_C F(x_0).$$

Proof. Suppose that

$$F(x_0) \preceq^l_C F(x_1) + T_1(z_1), \tag{8.4}$$

$G(x_0) \preceq^l_K 0$, $(x_1, z_1) \in \operatorname{graph} G$ and $T_1 \in \mathcal{L}_+(Z, Y)$. Because of $G(x_0) \preceq^l_K 0$ we can find $z_0 \in G(x_0)$ such that $z_0 \in -K$. So we get from (8.4)

$$F(x_0) + T_1(z_0) \preceq^l_C F(x_1) + T_1(z_1).$$

Then, taking into account $F(x_1) + T_1(z_1) \in \Phi(T_1)$ and Definition 2.6.19 we get

$$F(x_1) + T_1(z_1) \preceq^l_C F(x_0) + T_1(z_0) \preceq^l_C F(x_0).$$

□

Theorem 8.2.5 (Saddle Point Theorem). *Suppose that $(Y, \leq_C)$, $(Z, \leq_K)$ are linear topological spaces, C and K are proper pointed closed convex cones. Additionally, assume that $\operatorname{int} C \neq \emptyset$. If $(x_0, z_0, T_0) \in \operatorname{graph} G \times \mathscr{L}_+(Z, Y)$ is an $\preceq^l_C$-type saddle point of L, then it holds*

(i) $z_0 \in -K$ *and* $T_0(z_0) = 0$,
(ii) x_0 *is an* $\preceq^l_C$*-minimal solution of* $(SP - \preceq^l_C)$,
(iii) T_0 *is an* $\preceq^l_C$*-maximal solution of* $(SD - \preceq^l_C)$.

Proof. (i): In order to show $z_0 \in -K$ and $T_0(z_0) = 0$ we first suppose $z_0 \notin -K$. By a separation argument there exists $z_0^* \in K^+$ such that $\langle z_0^*, z_0\rangle > 0$. Using an element $c \in \operatorname{int} C$ we define $T_n \in \mathscr{L}_+(Z, Y)$ by

$$T_n(z) := n\langle z_0^*, z\rangle c, \quad n = 1, 2, \cdots$$

for each $z \in Z$. Then, we can choose an element $n_0 \in N$ such that $T_0(z_0) \in T_n(z_0) - C$ for each $n \geq n_0$.

Taking into account Definition 8.2.2 (b), from $L(x_0, z_0, T_0) \preceq^l_C L(x_0, z_0, T_n)$ we have

$$L(x_0, z_0, T_n) \preceq^l_C L(x_0, z_0, T_0),$$

hence $T_n(z_0) \in T_0(z_0) - C$ for each $n \geq n_0$. This implies $n\langle z_0^*, z_0\rangle c = 0$ for all $n \geq n_0$, a contradiction to $\langle z_0^*, z_0\rangle > 0$ and $c \in \operatorname{int} C$. So we get $z_0 \in -K$ which implies

$$T_0(z_0) \in -C. \tag{8.5}$$

For the zero element $\mathscr{O}$ of $\mathscr{L}_+(Z, Y)$ it holds

$$F(x_0) + T_0(z_0) \preceq^l_C F(x_0) + \mathscr{O}(z_0) \tag{8.6}$$

because of (8.5). From (8.6) and Definition 8.2.2 (b) we get

$$F(x_0) + \mathscr{O}(z_0) \preceq^l_C F(x_0) + T_0(z_0).$$

This yields $T_0(z_0) \in C$ such that we have $T_0(z_0) = 0$ because of (8.5) and C is pointed.

(ii): Consider an element $x \in X$ such that $G(x) \preceq^l_K 0$ and

$$F(x) \preceq^l_C F(x_0). \tag{8.7}$$

For an $\preceq^l_K$-type feasible point z of $(SP\!- \preceq^l_C)$ it holds $z \in G(x) \cap (-K)$ such that we get

$$L(x, z, T_0) = F(x) + T_0(z) \preceq^l_C F(x_0) + T_0(z_0) = L(x_0, z_0, T_0).$$

Since (x_0, y_0, T_0) is a saddle point, the converse inequality holds. Since $z \in -K$ and $T_0(z_0) = 0$ we get

$$F(x_0) \preceq^l_C F(x) + T_0(z) \preceq^l_C F(x).$$

This implies together with (8.7) that x_0 is an $\preceq^l_C$-minimal solution of $(SP - \preceq^l_C)$.

(iii): Taking into account the definition of saddle points, we have $L(x_0, z_0, T_0) \in \Phi(T_0)$. Consider $T_1 \in \mathscr{L}_+(Z, Y)$, $(x_1, z_1) \in \operatorname{graph} G$ with $L(x_1, y_1, T_1) \in \Phi(T_1)$ such that

$$L(x_0, z_0, T_0) \preceq^l_C L(x_1, z_1, T_1). \tag{8.8}$$

From (i) we know $T_0(z_0) = 0$, such that we get $F(x_0) \preceq^l_C F(x_1) + T_1(z_1)$. Furthermore, we have $T_1(z_0) \in -C$ since $z_0 \in -K$ and $T_1 \in \mathscr{L}_+(Z, Y)$. This yields

$$L(x_0, z_0, T_1) = F(x_0) + T_1(z_0) \preceq^l_C F(x_1) + T_1(z_1) = L(x_1, z_1, T_1).$$

The converse relation is satisfied since $L(x_1, z_1, T_1) \in \Phi(T_1)$ such that we can conclude

$$L(x_1, z_1, T_1) = F(x_1) + T_1(z_1) \preceq^l_C F(x_0) + T_1(z_0) \preceq^l_C F(x_0) = L(x_0, z_0, T_0).$$

This yields together with (8.8) that T_0 is an $\preceq^l_C$-maximal solution of $(SD - \preceq^l_C)$. □

Corollary 8.2.6. *Suppose that the assumptions of Theorem 8.2.5 are fulfilled. Then,* $(x_0, z_0, T_0) \in \operatorname{graph} G \times \mathscr{L}_+(Z, Y)$ *is an* $\preceq^l_C$*-type saddle point of* L *if and only if the following conditions hold:*

(i) $L(x, z, T_0) \preceq^l_C L(x_0, z_0, T_0)$, $(x, z) \in \operatorname{graph} G \implies L(x_0, z_0, T_0) \preceq^l_C L(x, z, T_0)$,

(ii) $z_0 \in -K$ *and* $T_0(z_0) = 0$.

Proof. First, suppose that (x_0, z_0, T_0) is an $\preceq^l_C$-type saddle point of L, then (i) and (ii) hold because of Theorem 8.2.5.

Furthermore, suppose that (x_0, z_0, T_0) satisfies the conditions (i) and (ii). Because of (i) it is sufficient to show that (b) of Definition 8.2.2 is fulfilled. If $T \in \mathscr{L}_+(Z, Y)$ such that

$$L(x_0, z_0, T_0) \preceq^l_C L(x_0, z_0, T),$$

then it follows that $0 = T_0(z_0) \in T(z_0) - C$ since $T_0(z_0) = 0$ in (ii) and $T(z_0) = 0$ because of $z_0 \in -K$ and C is pointed, therefore $L(x_0, z_0, T) = F(x_0) = L(x_0, z_0, T_0)$. This yields that (x_0, z_0, T_0) is an $\preceq^l_C$-type saddle point of L. □

Further results concerning Lagrange duality for set-valued optimization problems based on the set approach where the solution concept is given by Definition 2.6.9 are derived in the paper by Hernández, Rodríguez-Marín [251], compare also Hernández, Rodríguez-Marín [249].

8.3 Duality Assertions for Set-Valued Problems Based on Lattice Structure

In this section, duality theorems will be presented that are based on an consequent usage of infimum and supremum (in the sense greatest lower and least upper bounds with respect to a partial ordering) (see Sect. 2.6.3).

We consider set-valued optimization problems where the objective map takes its values in $\mathscr{F}$ (see Definition 2.6.36) as well as in $\mathscr{I}$ (see Definition 2.6.37). In the book by Rockafellar and Wets [499] set-valued maps $F : \mathbb{R}^n \rightrightarrows \mathbb{R}^q$ are considered but we formulate conjugate duality results only for $\mathscr{F}$-valued maps, i.e., we assume the upper closedness of the values. However, this is no loss of generality taking into account that semi-continuity properties are not influenced by the closure operation to the values of the map F (compare [499, Proposition 4.4, Definition 5.4]).

By the fact $\mathscr{F}$ and $\mathscr{I}$ being isomorphic and isotone (see Proposition 2.6.39) it is clear that it is sufficient to prove the duality assertions just for one case, either for the $\mathscr{F}$-valued case or for the $\mathscr{I}$-valued case. The advantage of the $\mathscr{F}$-valued case is that the operations, the ordering and the infimum and supremum have an easier structure, which is beneficial for proofs. In this case we speak about **set inclusion problems** (see Löhne [393, 394]). In Sect. 8.3.1 we show Fenchel duality for $\mathscr{F}$-valued optimization problems, especially that under the usual assumptions weak as well as strong duality assertions can be obtained. Furthermore, in Sect. 8.3.2 we derive Lagrange duality assertions for the $\mathscr{I}$-valued case. The advantage of the $\mathscr{I}$-valued case is that it is closely related to vector optimization problems (see Sect. 15.1).

Let Y be a separated locally convex space ordered by a proper pointed closed convex cone $C \subset Y$ such that $\emptyset \neq \operatorname{int} C \neq Y$ and $Y^\bullet = Y \cup \{-\infty\} \cup \{+\infty\}$. Let Y^* be the topological dual space and let X be a linear space.

In the subsequent considerations let $\mathscr{F} := \mathscr{F}_C(Y)$ and $\mathscr{I} := \mathscr{I}_C(Y^\bullet)$.

The main tool for the proofs of the following duality assertions is a scalarization of the $\mathscr{F}$-valued and $\mathscr{I}$-valued problems depending on a parameter $y^* \in Y^* \setminus \{0\}$.

Let us start with some statements concerning a suitable scalarization technique.

The **support function** $\sigma_A : Y^* \longrightarrow \overline{\mathbb{R}}$ with respect to $A \subseteq Y$ is defined as usual by

$$\sigma_A(y^*) := \sigma(y^* | A) := \sup_{y \in A} \langle y^*, y \rangle\,, \tag{8.9}$$

where $\overline{\mathbb{R}}$ is equipped with the sup-addition, that is, $-\infty+(+\infty) = +\infty+(-\infty) = -\infty$. This ensures that the expression

$$\forall y^* \in Y^* : \qquad \sigma(y^* | A + B) = \sigma(y^* | A) + \sigma(y^* | B)$$

is valid for all (not necessarily nonempty) sets $A, B \subseteq Y$.

8.3.1 *Conjugate Duality for $\mathscr{F}$-Valued Problems*

We will show conjugate duality assertions for optimization problems with a $\mathscr{F}$-valued objective function introduced in Sect. 2.6.3. Löhne [394, 395] has shown a Fenchel duality theorem for optimization problems with a $\mathscr{F}$-valued objective function and being based on the order relation "set inclusion" and infimal / supremal sets introduced by Tanino [566], Nieuwenhuis [442]) (compare Sect. 2.6.3). A more general result for closed (but not necessarily upper closed) sets was recently obtained in [393, 394].

The notions and solution concepts that we will use in this section are introduced in Sect. 2.6.3. We consider the space $(\mathscr{F}, \supseteq)$ (see Definition 2.6.36), where the order relation is given by

$$A \preccurlyeq_{\mathscr{F}} B :\Longleftrightarrow A \supseteq B$$

for subsets A, B of $\mathscr{F}$ (compare Sect. 2.6.3). Furthermore, for deriving the duality assertions we use the property that $(\mathscr{F}, \supseteq)$ is a complete lattice (Proposition 2.6.40).

In order to formulate the $\mathscr{F}$-valued optimization problem we study an objective map $F : X \to \mathscr{F}$, i.e., the objective function values of F are subsets of the hyperspace of upper closed sets introduced in Definition 2.6.36.

For simplicity we assume in this section $X = X^* = \mathbb{R}^n$, $Z = Z^* = \mathbb{R}^m$ and $Y = Y^* = \mathbb{R}^q$. Suppose that $C \subset \mathbb{R}^q$ is a proper pointed closed convex cone.

Remark 8.3.1. The following duality assertions can be extended to more general spaces than $X = \mathbb{R}^n$, $Z = \mathbb{R}^m$ and $Y = \mathbb{R}^q$ as long as the corresponding scalar result, which is used in the proof, is valid in these spaces. Then, one usually has to modify the constraint qualification.

For deriving a Fenchel dual problem associated to the primal $\mathscr{F}$-valued problem we introduce the conjugate of a $\mathscr{F}$-valued map with respect to $c \in Y$. In difference to Definition 8.1.3, where a solution concept given by the vector approach is applied in the definition of a conjugate, we use in the following definition the solution concept in the complete lattice $\mathscr{F}$.

Definition 8.3.2. Let $c \in Y$. The function $F_c^* : X^* \to \mathscr{F}^\diamond$ (where $\mathscr{F}^\diamond$ is the space of lower closed subsets of Y), defined by

$$F_c^*(x^*) := -\inf_{x \in X} \{F(x) - \langle x^*, x\rangle \cdot \{c\}\},$$

is said to be the **conjugate of** F with respect to $c \in Y$.

Remark 8.3.3. In order to avoid calculations in the space $\mathscr{F}^\diamond$ of lower closed subsets of $\mathbb{R}^q$ we will prefer to use the term $-F_c^*(x^*) \in \mathscr{F}$ rather than $F_c^*(x^*) \in \mathscr{F}^\diamond$ in the following. Nevertheless a calculus in $\mathscr{F}^\diamond$ is possible in the same way as in $\mathscr{F}$, if we replace the inf-addition in $\overline{\mathbb{R}}^q$ by the sup-addition in $\overline{\mathbb{R}}^q$, C by $-C$, and $\subseteq$ by $\supseteq$ in the definition of the order relation. By a consequent usage of the notions in $\mathscr{F}^\diamond$ we could express the conjugate as $F_c^*(x^*) = \sup_{x \in X} \{\langle x^*, x\rangle \cdot \{c\} - F(x)\}$, where "sup" now means the supremum in $\mathscr{F}^\diamond$. For more details on this kind of duality, using the concept of oriented sets by Rockafellar [489], see [393, 394]. Corresponding conjugate duality statements for $\mathscr{I}$-valued functions are shown by Löhne [395, Theorem 3.5] in a similar way. This is quite natural because $\mathscr{F}$-valued and $\mathscr{I}$-valued problems are isomorphic and isotone (compare Proposition 2.6.39).

For given $\mathscr{F}$-valued maps $F : X \to \mathscr{F}$ and $G : Z \to \mathscr{F}$, a linear map $A : X \to Z$ and a vector $c \in Y$, let

$$P : X \to \mathscr{F} \quad \text{and} \quad D_c : Z^* \to \mathscr{F}$$

be defined, respectively, by

$$P(x) = F(x) \oplus G(Ax) \quad \text{and} \quad D_c(u^*) = -\big(F_c^*(A^T u^*) \oplus G_c^*(-u^*)\big).$$

We consider the following $\mathscr{F}$-valued optimization problems (compare Proposition 2.6.40), the **primal problem**

$$\bar{P} := \inf_{x \in X} P(x) = \operatorname{cl} \bigcup_{x \in X} P(x), \qquad (SP- \preceq_{\mathscr{F}})$$

and the **dual problem** associated to $(SP- \preceq_{\mathscr{F}})$ for $c \in Y$

$$\bar{D}_c := \sup_{u^* \in Z^*} D_c(u^*) = \bigcap_{u^* \in Z^*} D_c(u^*). \qquad (SD_F- \preceq_{\mathscr{F}})$$

Note that the **convexity** of a function $F : X \to \mathscr{F}$ is equivalent to the convexity of the graph of the corresponding set-valued map (compare Sect. 2.3). In particular, if F and G are convex in problem $(SP- \preceq_{\mathscr{F}})$, the values $\bar{P}$ and $\bar{D}_c$ are convex sets.

We define **weak** and **strong duality** in the usual way, i.e., by $\bar{D}_c \supseteq \bar{P}$ (that is $\bar{D}_c \preceq_{\mathscr{F}} \bar{P}$) and $\bar{D}_c = \bar{P}$, respectively.

In the following theorem (see Löhne, Tammer [397]) we will show weak duality for the problems $(SP- \preceq_{\mathscr{F}})$ and $(SD_F- \preceq_{\mathscr{F}})$ as well as strong duality under convexity and certain regularity assumptions. The next theorem will be used in the proof of Theorem 15.1.4 in Sect. 15.1 for deriving Fenchel duality for vector optimization problems and it is a special case of Löhne [394, Theorem 4.3]. .

Theorem 8.3.4 (Weak and Strong Duality for $(SP\!-\preceq_{\mathscr{F}})$ and $(SD_F\!-\preceq_{\mathscr{F}})$). *The problems $(SP\!-\preceq_{\mathscr{F}})$ and $(SD_F\!-\preceq_{\mathscr{F}})$ (with arbitrary $c \in Y$) satisfy the weak duality inequality, i.e., $\bar{D}_c \supseteq \bar{P}$.*

Furthermore, let F and G be convex, let $c \in \operatorname{int} C$ and let the following regularity assumption be fulfilled:

$$0 \in \operatorname{ri}(\operatorname{dom} G - A \operatorname{dom} F).$$

Then we have strong duality, i.e., $\bar{D}_c = \bar{P}$.

Proof. The weak duality is obvious from the definition. In order to prove the strong duality assertion we use a scalarization method by the **support function** $\sigma_A : \mathbb{R}^q \to \overline{\mathbb{R}}$ with respect to $A \subseteq Y$ given by (8.9). Note that for (not necessarily nonempty) subsets $A, B \subseteq \mathscr{F}$ we have $\sigma_A + \sigma_B = \sigma_{A \oplus B}$ and $\alpha\sigma_A = \sigma_{\alpha A}$ for $\alpha > 0$. Moreover, for all (not necessarily nonempty) sets $\mathscr{A} \subseteq \mathscr{F}$ we have $-\sigma_{\inf \mathscr{A}} = \inf_{A \in \mathscr{A}} -\sigma_A$ (see e.g. [489, Corollary 16.5.1]).

It holds $-\sigma_{\bar{P}} = \inf_{x \in X} -\sigma_{P(x)}$. By the extended real–valued functions $\bar{F}_{y^*} : X \rightrightarrows \overline{\mathbb{R}}$ and $\bar{G}_{y^*} : Z \rightrightarrows \overline{\mathbb{R}}$ being defined, respectively, by $\bar{F}_{y^*}(x) := -\sigma\big(y^* |\, F(x)\big)$ and $\bar{G}_{y^*}(u) := -\sigma\big(y^* |\, G(u)\big)$ this can be rewritten as a collection of scalar optimization problems

$$\forall y^* \in Y : \; -\sigma(y^* |\, \bar{P}) = \inf_{x \in X} \big\{\bar{F}_{y^*}(x) + \bar{G}_{y^*}(Ax)\big\}. \tag{8.10}$$

The convexity of F and G implies the convexity of $\bar{F}_{y^*}$ and $\bar{G}_{y^*}$, respectively. Clearly, we have $\operatorname{dom} F = \operatorname{dom} \bar{F}_{y^*}$ and $\operatorname{dom} G = \operatorname{dom} \bar{G}_{y^*}$, whence $0 \in \operatorname{ri}(\operatorname{dom} \bar{G}_{y^*} - A \operatorname{dom} \bar{F}_{y^*})$. A scalar duality theorem, for instance [69, Theorem 3.3.5], yields that

$$\forall y^* \in Y : \; -\sigma(y^* |\, \bar{P}) = \sup_{u^* \in Z^*} \left\{-\bar{F}^*_{y^*}(A^T u^*) - \bar{G}^*_{y^*}(-u^*)\right\},$$

where the supremum is attained whenever $-\sigma(y^* |\, \bar{P})$ is finite, i.e.,

$$\forall y^* \in \operatorname{dom} \sigma(\cdot \,|\, \bar{P}),\, \exists \bar{u}^* \in Z^* : \; -\sigma(y^* |\, \bar{P}) = -\bar{F}^*_{y^*}(A^T \bar{u}^*) - \bar{G}^*_{y^*}(-\bar{u}^*). \tag{8.11}$$

Let $y^* \in \operatorname{dom} \sigma(\cdot \,|\, \bar{P})$ be arbitrarily given (hence $\bar{p} \neq Y$). Since $\emptyset \neq \bar{P} \neq Y$ and $\bar{P} = \operatorname{Cl}_+ \bar{P}$, we have $\operatorname{dom} \sigma(\cdot \,|\, \bar{P}) \subset C^-$. By the choice $c \in \operatorname{int} C$, it follows that $\langle y^*, c\rangle < 0$. Hence, there exists $\alpha_{y^*} > 0$ such that $\big\langle \alpha_{y^*} y^*, c\big\rangle = -1$. This can be rewritten as

$$\forall t \in \mathbb{R} : \;\; -\sigma\big(\alpha_{y^*} y^* |\, \{t \cdot c\}\big) = -\big\langle \alpha_{y^*} y^*, t \cdot c\big\rangle = t. \tag{8.12}$$

For $\alpha := \alpha_{y^*} > 0$ we have

$$\begin{aligned}
\alpha \cdot \left(-\sigma\left(y^* |\, \bar{P}\right)\right) &= -\sigma\left(\alpha y^* |\, \bar{P}\right) \overset{(8.11)}{=} -\bar{F}^*_{\alpha y^*}(A^T \bar{u}^*) - \bar{G}^*_{\alpha y^*}(-\bar{u}^*) \\
&= \inf_{x \in X} \left\{-\left\langle A^T \bar{u}^*, x\right\rangle + \bar{F}_{\alpha y^*}(x)\right\} + \inf_{u \in Z} \left\{\langle \bar{u}^*, u\rangle + \bar{G}_{\alpha y^*}(u)\right\} \\
&\overset{(8.12)}{=} \inf_{x \in X} \left\{-\sigma\left(\alpha y^* \middle| -\left\langle A^T \bar{u}^*, x\right\rangle \cdot \{c\}\right) - \sigma\left(\alpha y^* \middle| F(x)\right)\right\} \\
&+ \inf_{u \in Z} \left\{-\sigma\left(\alpha y^* \middle| \langle \bar{u}^*, u\rangle \cdot \{c\}\right) - \sigma\left(\alpha y^* \middle| G(u)\right)\right\} \\
&= -\sigma\left(\alpha y^* \middle| \inf_{x \in X} \left\{-\left\langle A^T \bar{u}^*, x\right\rangle \{c\} + F(x)\right\} \oplus \inf_{u \in Z} \left\{\langle \bar{u}^*, u\rangle \{c\} + G(u)\right\}\right) \\
&= -\sigma\left(\alpha y^* \middle| - F^*_c(A^T \bar{u}^*) \oplus -G^*_c(-\bar{u}^*)\right) = -\sigma\left(\alpha y^* | D_c(\bar{u}^*)\right) \\
&= \alpha \cdot \left(-\sigma\left(y^* | D_c(\bar{u}^*)\right)\right).
\end{aligned}$$

We deduce that

$$\forall y^* \in \operatorname{dom} \sigma(\cdot \,|\, \bar{P}\,),\ \exists \bar{u}^* \in Z^* : \quad \sigma(y^* | D_c(\bar{u}^*)) = \sigma(y^* |\, \bar{P}\,). \tag{8.13}$$

It follows that $\sigma(y^* |\, \bar{D}_c\,) \leq \sigma(y^* |\, \bar{P}\,)$ for all $y^* \in Y$, hence ($\bar{D}_c$ and $\bar{P}$ being closed and convex) $\bar{D}_c \subseteq \bar{P}$. By the weak duality inequality we obtain $\bar{D}_c = \bar{P}$.□

Remark 8.3.5. Note that we have not the usual dual attainment assertion, i.e., the dual value $\bar{D}_c$ is not attained by a single element $\bar{u}^* \in Z^*$. Instead we have condition (8.13), which describes the present situation.

Remark 8.3.6. In Theorem 8.3.4 (and its conclusion Theorem 15.1.4) we suppose the constraint qualification $0 \in \operatorname{ri}(\operatorname{dom} G - A \operatorname{dom} F)$. In the proof we use this condition in order to obtain the corresponding condition for the family of scalarized problems in (8.10). If all these problems are polyhedral, the constraint qualification can be replaced by $\operatorname{dom} G \cap A \operatorname{dom} F \neq \emptyset$, compare e.g. [69, Corollary 5.1.9].

An comprehensive discussion of conjugate duality theory for problems where the solution concept is based on lattice approach is given in the book by Löhne [395].

8.3.2 Lagrange Duality for $\mathscr{I}$-Valued Problems

The results in this section are derived by Löhne and Tammer [398], Hernández, Löhne, Rodríguez-Marín, Tammer [247] and Löhne [395].

First, we introduce the $\mathscr{I}$-valued primal problem for that we will derive Lagrange duality statements.

Throughout this section let $Y^\bullet$ be an extended locally convex space ordered by a proper pointed closed convex cone $C \subset Y$ such that $\emptyset \neq \operatorname{int} C \neq Y$. Let X be a linear space.

We put $\mathcal{I} := \mathcal{I}_C(Y^\bullet)$ and consider the space $(\mathcal{I}, \preccurlyeq_{\mathcal{I}})$ (see Definition 2.6.37), where the order relation is given by

$$A \preccurlyeq_{\mathcal{I}} B :\Longleftrightarrow A \preccurlyeq B :\Longleftrightarrow \mathrm{Cl}_+ A \supseteq \mathrm{Cl}_+ B$$

for subsets A, B of $\mathcal{I}$ (compare Sect. 2.6.3). The property that $(\mathcal{I}, \preccurlyeq_{\mathcal{I}})$ is a complete lattice (Proposition 2.6.40) will be used intensively for deriving duality assertions.

In this section we derive Lagrange duality statements for $\mathcal{I}$-valued problems using the notations and results given in Sect. 2.6.3, especially the order relation $\preccurlyeq := \preccurlyeq_{\mathcal{I}}$.

Given $S \subseteq X$ a nonempty set and an $\mathcal{I}$-valued map $F : S \to \mathcal{I}$ we consider the following $\mathcal{I}$-valued problem

$$P := \inf_{x \in S} F(x). \qquad (SP- \preceq_{\mathcal{I}})$$

The domain of the $\mathcal{I}$-valued map F is given by $\operatorname{Dom} F = \{x \in S \mid F(x) \neq \emptyset, F(x) \neq \{+\infty\}\}$. Note that we have

$$P = \inf_{x \in S} F(x) = \operatorname{Inf} \bigcup_{x \in S} F(x),$$

where $\operatorname{Inf} \bigcup_{x \in S} F(x)$ denotes the infimal set of $\bigcup_{x \in S} F(x)$ (see Theorem 2.6.41).

We emphasize that problem $(SP- \preceq_{\mathcal{I}})$ is related with several well-known problems as we see in Sect. 15.1. Indeed, if $Y = \mathbb{R}$ and $C = \{0\}$ we obtain the real case ($\preccurlyeq := \leq$).

In this section we establish several Lagrangian duality problems in the framework of $\mathcal{I}$-valued problems and give weak and strong duality theorems. To do that we need to define an $\mathcal{I}$-valued Lagrange dual problem associated to $(SP- \preceq_{\mathcal{I}})$.

In a natural way we assign to $(SP- \preceq_{\mathcal{I}})$ a dual problem

$$D := \sup_{u \in U} \phi(u) \qquad (SD_L- \preceq_{\mathcal{I}})$$

where U is a nonempty set and $\phi : U \to \mathcal{I}$ is the dual objective function that is constructed in the following by a Lagrangian technique.

We say that $(SP- \preceq_{\mathcal{I}})$ and $(SD_L- \preceq_{\mathcal{I}})$ satisfy **weak duality** if

$$D \preccurlyeq_{\mathcal{I}} P.$$

Furthermore we speak about **strong duality** if

$$D = P.$$

Taking into account that $\mathcal{I}$ is a complete lattice and Lemma 2.6.38 we can rewrite the weak duality property as follows:

Lemma 8.3.7. *Consider problems (SP–* $\preceq_{\mathcal{I}}$*) and (SD*$_L$*–* $\preceq_{\mathcal{I}}$*). The following conditions are equivalent*

(i) $D \preccurlyeq_{\mathcal{I}} P$
(ii) *For any* $x \in S$ *and* $u \in U$,

$$\phi(u) \preccurlyeq_{\mathcal{I}} F(x).$$

If, in addition, $F(x) \neq -\infty$ *and* $\phi(x) \neq -\infty$ *for all* $x \in S$ *and* $u \in U$, *the above conditions are equivalent to*

(iii) *For any* $x \in S$ *and* $u \in U$,

$$(F(x) + \operatorname{int} C) \cap \phi(u) = \emptyset.$$

To establish the strong duality theorem we need the notion of convex $\mathcal{I}$-valued objective function $F : X \to \mathcal{I}$.

Definition 8.3.8. Let X be real a linear space. The function $F : X \to \mathcal{I}$ is **convex** if for all $t \in [0, 1]$ and all $x^1, x^2 \in X$ it holds

$$F\big(tx^1 + (1-t)x^2\big) \preccurlyeq t \odot F(x^1) \oplus (1-t) \odot F(x^2).$$

We next recall the notion of cone-convexity (compare Sect. 2.3) which is used for the set-valued constraints: Let X, Z be real linear spaces and let $K \subset Z$ be a proper closed convex cone. A set–valued map $G : X \rightrightarrows Z$ is said to be K**–convex** if

$$\forall x^1, x^2 \in X,\ \forall t \in [0, 1] : \qquad G\big(tx^1 + (1-t)x^2\big) + K \supseteq tG(x^1) + (1-t)G(x^2).$$

Given a set-valued map $G\colon X \rightrightarrows Z$ we denote by $G_{\mathcal{I}}\colon X \to \mathcal{I}$ the corresponding $\mathcal{I}$-valued map associated to G as follows: $G_{\mathcal{I}}(x) = \operatorname{Inf} G(x)$ for $x \in X$.

Lemma 8.3.9. *Let* $G\colon X \rightrightarrows Z$. *If* G *is* K*-convex then* $G_{\mathcal{I}}$ *is convex.*

Proof. Let $x^1, x^2 \in X$ and $t \in [0, 1]$. The conclusion follows from

$$G_{\mathcal{I}}\big(tx^1 + (1-t)x^2\big) \preccurlyeq t \odot G_{\mathcal{I}}(x^1) \oplus (1-t) \odot G_{\mathcal{I}}(x^2)$$

is equivalent to

$$\operatorname{Cl}_+(t \cdot (G(x^1)) + (1-t) \cdot (G(x^2))) \subset \operatorname{Cl}_+ G(tx^1 + (1-t)x^2).$$

□

In the following we will derive duality assertions for optimization problems with $\mathscr{I}$-valued objective function and set-valued constraints. First we formulate the corresponding assumptions and notations.

Let X be a linear space and let Z be a separated locally convex space. Moreover, let $\langle Z, Z^* \rangle$ be a dual pair. Let $F : X \to \mathscr{I}$, let $G : X \rightrightarrows Z$ be a set-valued map with $\operatorname{dom} G = X$ and let $K \subseteq Z$ be a proper closed convex cone. We denote by $\mathscr{L}(Z, Y)$ the set of all linear continuous operators from Z to Y and by $\mathscr{L}_+(Z, Y) = \{T \in \mathscr{L}(Z, Y) \mid T(K) \subseteq C\}$ the subset of all positive operators.

From now on, we suppose that the restrictions in $(SP- \preceq_{\mathscr{I}})$ are given in the form

$$S := \{x \in X \mid G(x) \cap -K \neq \emptyset\}. \tag{8.14}$$

Constraints of this type have been considered by many authors, such as Borwein [56]; Corley [112]; Jahn [290]; Luc [402]; Götz and Jahn [218]; Jahn [293]; Crespi, Ginchev and Rocca [116]; Boţ and Wanka [52]; Hernández, Löhne, Rodríguez-Marín, Tammer [248].

Then the $\mathscr{I}$-valued **primal problem** $(SP- \preceq_{\mathscr{I}})$ is rewritten as follows:

$$\bar{P} := \inf_{x \in S} F(x), \qquad S := \{x \in X \mid G(x) \cap -K \neq \emptyset\}. \tag{$SP- \preceq_{\mathscr{I}}$}$$

We first recall a scalar Lagrange duality result with this type of set-valued constraints. The scalar result is later used in the proof of the strong duality for the general problem. We consider a scalar problem with an objective $f : X \to \overline{\mathbb{R}}$:

$$\hat{p} := \inf_{x \in S} f(x), \qquad S := \{x \in X \mid G(x) \cap -K \neq \emptyset\}. \tag{$\hat{\mathrm{P}}$}$$

The Lagrangian map associated to $(\hat{\mathrm{P}})$ is defined as

$$L : X \times Z^* \to \overline{\mathbb{R}}, \qquad L(x, u^*) = f(x) + \inf_{u \in G(x) + K} \langle u^*, u \rangle. \tag{8.15}$$

The dual objective is

$$\phi : Z^* \to \overline{\mathbb{R}}, \qquad \phi(u^*) := \inf_{x \in X} L(x, u^*)$$

and the dual problem is

$$\hat{d} := \sup_{u^* \in Z^*} \phi(u^*). \tag{$\hat{\mathrm{D}}$}$$

Note that we have

$$\inf_{x \in X} L(x, u^*) = \begin{cases} \inf_{x \in X} \{f(x) + \inf_{u \in G(x)} \langle u^*, u \rangle\} & \text{if} \quad u^* \in K^+ \\ -\infty & \text{else}. \end{cases}$$

Thus we obtain the following version of the dual problem

$$\hat{d} := \sup_{u^* \in K^+} \phi(u^*). \tag{$\hat{\mathrm{D}}'$}$$

Of course it holds weak duality, i.e., $\hat{d} \leq \hat{p}$. Under certain convexity assumptions and some constraint qualification we get the following strong duality assertion, which we prove in the same way as [69, Proposition 4.3.5].

Theorem 8.3.10 (Strong Duality for the Scalar Problems ($\hat{\mathrm{P}}$) and ($\hat{\mathrm{D}}$)). *Let $f : X \to \overline{\mathbb{R}}$ be convex, let $G : X \rightrightarrows Z$ be K-convex and let*

$$G(\operatorname{dom} f) \cap -\operatorname{int} K \neq \emptyset. \tag{8.16}$$

Then, we have strong duality, i.e., $\hat{d} = \hat{p}$.

Proof. The value function is defined by

$$v : Z \to \overline{\mathbb{R}} : \qquad v(u) := \inf\{f(x) |\ G(x) \cap (\{u\} - K) \neq \emptyset\}$$

It is easily shown to be convex and $v(0) = \hat{p}$. For the conjugate $v^* : Z^* \to \overline{\mathbb{R}}$ of v it holds

$$\begin{aligned}
v^*(-u^*) &= \sup\{-\langle u^*, u\rangle - v(u) |\ u \in Z\} \\
&= \sup\{-\langle u^*, u\rangle - f(x) |\ G(x) \cap (\{u\} - K) \neq \emptyset,\ u \in Z,\ x \in \operatorname{dom} f\} \\
&= \sup\{-\langle u^*, u\rangle - f(x) |\ u \in G(x) + K,\ x \in \operatorname{dom} f\} \\
&= -\inf_{x \in X} L(x, u^*) = -\phi(u^*).
\end{aligned}$$

It follows $v^{**}(0) = \hat{d}$. We next show that v is lower semi-continuous at 0 (even continuous). Indeed, by (8.16) there is some $\bar{x} \in \operatorname{dom} f$ and some $\bar{z} \in -\operatorname{int} K$ such that $\bar{z} \in G(\bar{x})$. There exists some neighborhood V of 0 such that $\{\bar{z}\} - V \subset -\operatorname{int} K$. Now it follows that $f(\bar{x})$ is an upper bound of v on V. This implies that v is continuous at 0, see e.g. [172, Lemma 2.1].

If $\hat{p} = -\infty$, we get $\hat{p} = \hat{d} = -\infty$ from the weak duality. So we can assume that $\hat{p} = v(0)$ is finite. Thus we have $v(0) = \operatorname{cl} v(0)$, compare [490, Theorem 4]. By the bipolar theorem, e.g. [172, Proposition 4.1], we have $\operatorname{cl} v = v^{**}$. This yields $\hat{p} = v(0) = v^{**}(0) = \hat{d}$. □

Our main aim is to derive duality results for $\mathscr{I}$-valued optimization problems using the ideas of Sect. 2.6.3. To this end we develop an $\mathscr{I}$-valued version of Theorem 8.3.10.

We will use the following scalarization method of the $\mathscr{I}$-valued problem depending on a parameter $y^* \in C^- \setminus \{0\}$. For $\sigma : Y^* \longrightarrow \overline{\mathbb{R}}$ given by (8.9) and $A \in \mathscr{I}$, we set

$$\varphi_A(y^*) := \varphi(y^* | A) := -\sigma(y^* | \mathrm{Cl}_+ A). \tag{8.17}$$

For fixed y^* we get by (8.17) a functional from $\mathcal{I}$ to $\overline{\mathbb{R}}$. The addition, the multiplication by positive reals, the infimum and the supremum for extended real-valued functions are defined pointwise. Here we use the inf-addition, i.e., $-\infty + \infty = +\infty - \infty = +\infty$.

For fixed $A \in \mathcal{I}$, we consider φ_A to be a function from $C^- \setminus \{0\}$ into $\overline{\mathbb{R}}$, that is,

$$\varphi_A : C^- \setminus \{0\} \to \overline{\mathbb{R}}.$$

For some $\gamma \in \overline{\mathbb{R}}$ we write $\varphi_A \equiv \gamma$ whenever $\varphi_A(y^*) = \gamma$ for all $y^* \in C^- \setminus \{0\}$.

In the next theorem we collect several properties of the scalarization by means of the functional (8.17) that are useful for the proof of the duality statements (compare [395, Theorem 1.65]).

Theorem 8.3.11. *For any $A, B \in \mathcal{I}$ and $\alpha > 0$ it holds*

(i) $[\mathrm{Cl}_+ A = \mathrm{conv}\,\mathrm{Cl}_+ A \ \wedge\ \varphi_A \equiv -\infty] \iff A = \{-\infty\}$,
(ii) $\varphi_A \equiv +\infty \iff [\exists y^* \in C^- \setminus \{0\} : \varphi_A(y^*) = +\infty] \iff A = \{+\infty\}$,
(iii) $A \preccurlyeq B \implies \varphi_A \leq \varphi_B$,
(iv) $[\mathrm{Cl}_+ A = \mathrm{conv}\,\mathrm{Cl}_+ A \ \wedge\ \varphi_A \leq \varphi_B] \implies A \preccurlyeq B$,
(v) $\varphi_{A \oplus B} = \varphi_A + \varphi_B$,
(vi) $\alpha \cdot \varphi_A = \varphi_{\alpha \odot A}$.

For a nonempty subset $\mathcal{A} \subseteq \mathcal{I}$ we have

(vii) $\varphi_{\inf \mathcal{A}} = \inf_{A \in \mathcal{A}} \varphi_A$
(viii) $\varphi_{\sup \mathcal{A}} \geq \sup_{A \in \mathcal{A}} \varphi_A$

Proof. (i) As $\mathrm{Cl}_+ \{-\infty\} = Y$, we get $\varphi_{\{-\infty\}} = -\sigma(y^* \mid Y) = -\infty$ for all $y^* \in C^- \setminus \{0\}$. On the other hand, $\varphi_A \equiv -\infty$ implies that

$$+\infty = \sigma(y^* \mid \mathrm{Cl}_+ A) + \sigma(y^* \mid C) = \sigma(y^* \mid \mathrm{Cl}_+ A + C) = \sigma(y^* \mid \mathrm{Cl}_+ A)$$

for all $y^* \in Y^* \setminus \{0\}$. It follows $\mathrm{Cl}_+ A = \mathrm{cl}\,\mathrm{conv}\,\mathrm{Cl}_+ A = Y$ and hence $A = \{-\infty\}$.

(ii) It holds $\mathrm{Cl}_+ \{+\infty\} = \emptyset$ and so $A = \{+\infty\}$ implies $\varphi_A(y^*) = -\sigma(y^* | \emptyset) = +\infty$ for all $y^* \in C^- \setminus \{0\}$. If $\sigma(y^* | \mathrm{Cl}_+ A) = -\infty$ for some $y^* \in C^- \setminus \{0\}$, then $\mathrm{Cl}_+ A = \emptyset$. This implies $A = \{+\infty\}$.

(iii) Let $A \preccurlyeq B$. If $A = \{-\infty\}$ or $B = \{+\infty\}$, from (i) and (ii) we get $\varphi_A \leq \varphi_B$. Otherwise we get $\mathrm{Cl}_+ A \supseteq \mathrm{Cl}_+ B$ and hence $\varphi_A(y^*) = -\sigma(y^* | \mathrm{Cl}_+ A) \leq -\sigma(y^* | \mathrm{Cl}_+ B) = \varphi_B(y^*)$ for all $y^* \in Y$.

(iv) Let $\varphi_A \leq \varphi_B$, i.e., for all $y^* \in Y$ it holds $-\sigma(y^* | \mathrm{Cl}_+ A) \leq -\sigma(y^* | \mathrm{Cl}_+ B)$. As $\mathrm{Cl}_+ A$ is convex and closed we get $\mathrm{Cl}_+ A \supseteq \mathrm{cl}\,\mathrm{conv}\,\mathrm{Cl}_+ B \supseteq \mathrm{Cl}_+ B$ and so $A \preccurlyeq B$.

In order to prove the statements (v) to (vii), let $y^* \in C^- \setminus \{0\}$ be arbitrarily given.

(v) If A or B equals $\{+\infty\}$ then $A \oplus B = \{+\infty\}$ and the statement follows as $\mathrm{Cl}_+\{+\infty\} = \emptyset$. If A and B are not $\{+\infty\}$ but one of them or both equal $\{-\infty\}$ then the result follows from the fact $\mathrm{Cl}_+\{-\infty\} = X$. So we can now assume $A, B \subseteq Y$ and in this case we have $\mathrm{Cl}_+ A = \mathrm{cl}\,(A+C)$. It holds $\varphi_{A+B}(y^*) = -\sigma(y^* | \mathrm{cl}\,(A+B+C)) = -\sigma(y^* | \mathrm{cl}\,(A+C)) - \sigma(y^* | \mathrm{cl}\,(B+C)) = \varphi_A(y^*) + \varphi_B(y^*)$.

(vi) If $A = \{+\infty\}$, then $\alpha A = \{+\infty\}$ and so $\alpha\varphi_A(y^*) = \varphi_{\alpha\odot A}(y^*) = +\infty$. If $A = \{-\infty\}$, then $\alpha A = \{-\infty\}$ and so $\alpha\varphi_A(y^*) = \varphi_{\alpha\odot A}(y^*) = -\infty$. If $A \subseteq Y$, then we have $\alpha \cdot \varphi_A(y^*) = -\alpha\sigma(y^* | \mathrm{cl}\,(A+C)) = -\sigma(y^* | \mathrm{cl}\,(\alpha A + C)) = \varphi_{\alpha\odot A}(y^*)$.

(vii) It remains to show the statement for the case $\{+\infty\} \notin \mathscr{A}$, because omitting $\{+\infty\}$ does not change anything. If $\{-\infty\} \in \mathscr{A}$ the equality can be easily shown. So let $A \subseteq Y$ for all $A \in \mathscr{A}$. It holds

$$\varphi_{\inf \mathscr{A}}(y^*) = -\sigma\left(y^* \,\middle|\, \mathrm{Cl}_+ \bigcup_{A\in\mathscr{A}} A\right) = -\sigma\left(y^* \,\middle|\, \mathrm{cl} \bigcup_{A\in\mathscr{A}} (A+C)\right)$$
$$= \inf_{A\in\mathscr{A}} -\sigma\left(y^* | \mathrm{cl}\,(A+C)\right) = \inf_{A\in\mathscr{A}} \varphi_A(y^*).$$

(viii) We have $\sup \mathscr{A} \succcurlyeq A$ and hence $\varphi_{\sup \mathscr{A}}(y^*) \geq \varphi_A(y^*)$ for all $A \in \mathscr{A}$. Taking the supremum the statement follows. □

The following example shows that statement (viii) in Theorem 8.3.11 does not hold with equality:

Example 8.3.12 (See [393, Example 1.3.6]). Let $C := \mathbb{R}^2_+$, $A_1 := \{(0,1)^T\} + \mathrm{bd}\,C$, $A_2 = \{(1,0)^T\} + \mathrm{bd}\,C$, then $\sup\{A_1, A_2\} = \{(1,1)^T\} + \mathrm{bd}\,C$. For $y^* = (-1,-1)^T$ we get $\varphi_{A_1}(y^*) = \varphi_{A_2}(y^*) = 1$ but $\varphi_{\sup\{A_1,A_2\}}(y^*) = 2$.

Coming back to the $\mathscr{I}$-valued optimization problem (SP– $\preceq_{\mathscr{I}}$) we introduce a corresponding Lagrangian map. For a given fixed vector $c \in \mathrm{int}\,C$, the **Lagrangian map of problem** (SP– $\preceq_{\mathscr{I}}$) is defined by

$$L_c : X \times Z^* \to \mathscr{I}, \qquad L_c(x,u^*) := F(x) \oplus \inf_{u\in G(x)+K} \left\{\langle u^*, u\rangle \{c\} + \mathrm{bd}\,C\right\}. \tag{8.18}$$

In the special case $q = 1$, $C = \mathbb{R}_+$, $c = 1$, the Lagrangian coincides with the Lagrangian of the scalar problem $\hat{\mathrm{P}}$ in (8.15). For every choice of $c \in \mathrm{int}\,C$ we have a different Lagrangian map and a different corresponding dual problem, but we show that weak duality and strong duality hold for all these problems.

The scalar counterpart of the following result (compare [398], [395, Proposition 3.23]) is well known.

Proposition 8.3.13. *For all $x \in S$ and all $c \in \mathrm{int}\,C$ it holds*

$$\sup_{u^*\in Z^*} L_c(x,u^*) = F(x).$$

Proof. For arbitrary $c \in \operatorname{int} C$ it holds

$$\begin{aligned}
\sup_{u^* \in Z^*} L_c(x,u^*) &\preccurlyeq F(x) \oplus \sup_{u^* \in Z^*} \inf_{u \in G(x)+K} \left\{ \langle u^*, u \rangle \{c\} + \operatorname{bd} C \right\} \\
&= F(x) \oplus \sup_{u^* \in Z^*} \left(\inf_{z \in G(x)} \left\{ \langle u^*, z \rangle \{c\} + \operatorname{bd} C \right\} \right. \\
&\qquad \left. + \inf_{d \in K} \left\{ \langle u^*, d \rangle \{c\} + \operatorname{bd} C \right\} \right)
\end{aligned}$$

Taking into account

$$\inf_{d \in K} \left\{ \langle u^*, d \rangle \{c\} + \operatorname{bd} C \right\} = \begin{cases} \operatorname{bd} C & \text{if} \quad u^* \in K^+ \\ \{-\infty\} & \text{otherwise} \end{cases}$$

we get

$$\begin{aligned}
\sup_{u^* \in Z^*} L_c(x,u^*) &\preccurlyeq F(x) \oplus \sup_{u^* \in K^+} \inf_{z \in G(x)} \left\{ \langle u^*, z \rangle \{c\} + \operatorname{bd} C \right\} \\
&\preccurlyeq F(x) \oplus \sup_{u^* \in K^+} \inf_{z \in G(x) \cap -K} \left\{ \langle u^*, z \rangle \{c\} + \operatorname{bd} C \right\} \\
&\preccurlyeq F(x) \oplus \inf_{z \in G(x) \cap -K} \sup_{u^* \in K^+} \left\{ \langle u^*, z \rangle \{c\} + \operatorname{bd} C \right\} \\
&\preccurlyeq F(x).
\end{aligned}$$

Since $L(x,0) = F(x)$, it follows that $\sup_{u^* \in Z^*} L_c(x,u^*) = F(x)$. □

We next define the $\mathscr{I}$-valued Lagrange dual problem associated to $(SP- \preceq_{\mathscr{I}})$. The dual objective function (with respect to $c \in \operatorname{int} C$) is defined by

$$\phi_c : Z^* \to \mathscr{I}, \qquad \phi_c(u^*) := \inf_{x \in X} L_c(x,u^*)$$

and the $\mathscr{I}$-valued dual problem (with respect to $c \in \operatorname{int} C$) associated to $(SP- \preceq_{\mathscr{I}})$ is defined by

$$\bar{D}_c := \sup_{u^* \in Z^*} \phi_c(u^*). \qquad (SD_L^c - \preceq_{\mathscr{I}})$$

Similarly to the real case we have

$$\phi_c(u^*) = \inf_{x \in X} L_c(x,u^*) = \begin{cases} \inf_{x \in X} \{ F(x) + \inf_{u \in G(x)} \langle u^*, u \rangle \{c\} + \operatorname{bd} C \} & \text{if } u^* \in K^+ \\ \{-\infty\} & \text{otherwise.} \end{cases}$$

Therefore we obtain the following equivalent formulation of the dual problem $(SD_L^c - \preceq_{\mathscr{I}})$

$$\bar{D}_c := \sup_{u^* \in K+} \phi_c(u^*). \qquad (SD_L'^c - \preceq_{\mathscr{I}})$$

Theorem 8.3.14 (Weak Duality for $(SP- \preceq_{\mathscr{I}})$ and $(SD_L^c - \preceq_{\mathscr{I}})$). *For every $c \in \operatorname{int} C$, the problems $(SP- \preceq_{\mathscr{I}})$ and $(SD_L^c - \preceq_{\mathscr{I}})$ satisfy the weak duality inequality, i.e., $\bar{D}_c \preccurlyeq \bar{P}$.*

Proof. Since $\mathscr{I}$ is a complete lattice, we immediately have

$$\sup_{u^* \in Z^*} \inf_{x \in X} L_c(x, u^*) \preccurlyeq \inf_{x \in X} \sup_{u^* \in Z^*} L_c(x, u^*), \tag{8.19}$$

(even if L_c would be replaced by an arbitrary function from $X \times Z^*$ into $\mathscr{I}$). By Proposition 8.3.13 we know that $\inf_{x \in X} \sup_{u^* \in Z^*} L_c(x, u^*) \preccurlyeq \bar{P}$. □

It follows a strong duality assertion under a certain constraint qualification (compare [398] and [395, Theorem 3.26]).

Theorem 8.3.15 (Strong Duality for $(SP- \preceq_{\mathscr{I}})$ and $(SD_L^c - \preceq_{\mathscr{I}})$). *Let F be convex and let G be K-convex, let*

$$G(\operatorname{dom} F) \cap -\operatorname{int} K \neq \emptyset, \tag{8.20}$$

and let $c \in \operatorname{int} C$. Then, we have strong duality, i.e., $\bar{D}_c = \bar{P}$.

Proof. Of course, we have $\bar{P} \neq \{+\infty\}$ (by (8.20)). If $\bar{P} = \{-\infty\}$ we obtain $\bar{P} = \bar{D}_c$ from the weak duality. So we can assume $\bar{P} \subseteq Y$. As $\bar{P}$ is self-infimal, it is nonempty.

Let $y^* \in C^-$ such that $\langle c, y^* \rangle = -1$. It holds

$$\begin{aligned} \varphi\left(y^* \,\middle|\, \inf_{u \in G(x)+K} \langle u^*, u \rangle \{c\} + \operatorname{bd} C\right) &\overset{\text{Th. 8.3.11 (vii)}}{=} \inf_{u \in G(x)+K} \varphi\big(y^* \big| \ \langle u^*, u \rangle \{c\} + \operatorname{bd} C\big) \\ &= \inf_{u \in G(x)+K} -\sigma\big(y^* \big| \ \langle u^*, u \rangle \{c\} + C\big) \\ &= \inf_{u \in G(x)+K} \langle u^*, u \rangle . \end{aligned}$$

It follows that

$$\begin{aligned} \varphi(y^* | \bar{P}) &= \varphi\left(y^* \,\middle|\, \inf_{G(x) \cap -K \neq \emptyset} F(x)\right) \\ &\overset{\text{Th. 8.3.11 (vii)}}{=} \inf_{G(x) \cap -K \neq \emptyset} \varphi\big(y^* \big| \ F(x)\big) \end{aligned}$$

$$\overset{\text{Th. 8.3.10}}{=} \sup_{u^* \in Z^*} \inf_{x \in X} \left\{ \varphi\big(y^* \big|\, F(x)\big) + \inf_{u \in G(x)+K} \langle u^*, u \rangle \right\}$$

$$= \sup_{u^* \in Z^*} \inf_{x \in X} \left\{ \varphi\big(y^* \big|\, F(x)\big) + \varphi\left(y^* \middle| \inf_{u \in G(x)+K} \langle u^*, u \rangle \{c\} + \operatorname{bd} C \right) \right\}$$

$$\overset{\text{Th. 8.3.11 (v, vii)}}{=} \sup_{u^* \in Z^*} \varphi\big(y^* \big|\, \phi_c(u^*)\big)$$

$$\overset{\text{Th. 8.3.11 (viii)}}{\leq} \varphi\big(y^* \big|\, \sup_{u^* \in Z^*} \phi_c(u^*)\big) = \varphi\big(y^* \big|\, \bar{D}_c\big).$$

The same is true for arbitrary $y^* \in C^-$. If $y^* \notin C^-$, we have

$$\varphi(y^* |\, \bar{P}) = -\sigma(y^* |\, \mathrm{Cl}_+ \bar{P}) = -\sigma(y^* |\, \bar{P} + C) = -\sigma(y^* |\, \bar{P}) - \sigma(y^* |\, C) = -\infty.$$

Therefore we have $\varphi\left(y^* |\, \bar{P}\right) \leq \varphi\left(y^* |\, \bar{D}_c\right)$ for all $y^* \in Y$. As $\mathrm{Cl}_+ \bar{P}$ is convex, Theorem 8.3.11 (iv) yields $\bar{P} \preccurlyeq \bar{D}_c$. From the weak duality we get equality. □

Note that strong duality (i.e., $\bar{P} = \bar{D}_c$) implies that (8.19) is satisfied with equality.

Proposition 8.3.13 clarifies the relation between the problem ($SP- \preceq_{\mathscr{I}}$) and the Lagrangian only for feasible points $x \in S$. It remains the question what happens if x is not feasible. From special cases in scalar optimization, we expect that $\sup_{u^* \in Z^*} L_c(x, u^*) = \{+\infty\}$. This can be shown under additional assumptions to the constraints. Note that the following proposition is not used in the proof of the duality theorem. A similar result was shown in [393]. The next result goes back to [398] and [395, Proposition 3.24].

Proposition 8.3.16. *Let $F : X \to \mathscr{I}$ be a proper function. Assume that the set $G(x) + K$ is convex and closed for all $x \in X$. Then for all $c \in \operatorname{int} C$ it holds*

$$\sup_{u^* \in Z^*} L_c(x, u^*) = \begin{cases} F(x) & \text{if} \quad x \in S \\ \{+\infty\} & \text{else.} \end{cases}$$

Proof. The case $x \in S$ follows from Proposition 8.3.13. Let $x \notin S$, i.e., $G(x) \cap -K = \emptyset$. Thus, $0 \notin G(x) + K$. Applying the well-known separation result [32, Theorem 1.15], yields the existence of some $\bar{u}^* \in K^+ \setminus \{0\}$ such that

$$\inf_{u \in G(x)+K} \langle \bar{u}^*, u \rangle = \inf_{u \in G(x)} \langle \bar{u}^*, u \rangle - \sup_{u \in -K} \langle \bar{u}^*, u \rangle > 0. \tag{8.21}$$

Choose some $y^* \in C^-$ such that $c^T y^* = -1$. It holds

$$\begin{aligned}
\varphi\left(y^* \,\middle|\, \sup_{u^* \in Z^*} L_c(x,u^*)\right) &\overset{\text{Th. 8.3.11 (x)}}{\geq} \sup_{u^* \in Z^*} \varphi\left(y^* \,|\, L_c(x,u^*)\right) \\
&\geq \sup_{t>0} \varphi\left(y^* \,|\, L_c(x,t\bar{u}^*)\right) \\
&= \sup_{t>0} \left(\varphi\left(y^* \,|\, F(x)\right) + t \inf_{u \in G(x)+K} \langle \bar{u}^*, u \rangle\right) = +\infty
\end{aligned}$$

Theorem 8.3.11 (ii) yields that $\sup_{u^* \in Z^*} L_c(x,u^*) = \{+\infty\}$. □

The next example shows that the assumption $G(x) + K$ is closed in the preceding proposition cannot be omitted, not even in the scalar case. However, this assumption is fulfilled in many important special cases, such as for vector–valued or compact–valued functions G.

Example 8.3.17. Let $F : \mathbb{R}^2 \to \mathscr{I}$, $F(x) \equiv \operatorname{Inf}\{0\} = \operatorname{bd} C$. We set $A := \left\{y \in \mathbb{R}^2_+ \mid y_2 \geq \frac{1}{y_1}\right\}$, $G(x) := \{x\} + A$ and $K := \left\{u \in \mathbb{R}^2 \mid u_1 \geq 0\right\}$. Then $K^- \setminus \{0\} = \left\{u^* \in \mathbb{R}^2 \mid u_1^* < 0, u_2^* = 0\right\}$. An easy computation shows

$$L_c(0,u^*) := \begin{cases} F(0) & \text{if } u^* \in K^+ \\ \{-\infty\} & \text{else.} \end{cases}.$$

It follows that $\sup_{u^* \in Z^*} L_c(0,u^*) = F(0) \neq \{+\infty\}$, $G(0) + K$ is not closed but $G(0) \cap -K = \emptyset$, i.e., $0 \notin S$.

Now we study another dual problem associated to (SP– $\preceq_{\mathscr{I}}$) where the **dual variables are operators** instead of vectors as above. The following results are given by Hernández, Löhne, Rodríguez-Marín, Tammer [247]. Exactly, we consider the formulation of the Lagrangian map more used in the Literature (for instance, see [112, 387, 402]) but in the framework of $\mathscr{I}$-valued problems. We obtain the weak duality from the completeness of the lattice and prove the strong duality applying Theorem 8.3.15.

Given $T \in \mathscr{L}$ and $\emptyset \neq A \subseteq Z$ we write $T(A) = \bigcup_{a \in A} T(a)$.

Now the Lagrangian map $L : X \times \mathscr{L} \to \mathscr{I}$ is defined by

$$L(x,T) := F(x) \oplus \inf_{u \in G(x)+K} \left(T(u) + \operatorname{bd} C\right), \tag{8.22}$$

the dual objective function $\phi : \mathscr{L} \to \mathscr{I}$ is

$$\phi(T) := \inf_{x \in X} L(x,T), \tag{8.23}$$

and the dual problem is defined by

$$\tilde{D} := \sup_{T \in \mathscr{L}_+} \phi(T). \qquad (SD_L^T - \preceq_{\mathscr{I}})$$

Again we can similarly prove the following result like in Proposition 8.3.13 with respect to the above Lagrangian.

Proposition 8.3.18. *For all $x \in S$ it holds*

$$\sup_{T \in \mathscr{L}_+} L(x,T) = F(x).$$

Theorem 8.3.19 (Weak Duality for $(SP- \preceq_{\mathscr{I}})$ and $(SD_L^T- \preceq_{\mathscr{I}})$). *Problems $(SP- \preceq_{\mathscr{I}})$ and $(SD_L^T- \preceq_{\mathscr{I}})$ satisfy the weak duality inequality, that is, $\tilde{D} \preccurlyeq \bar{P}$.*

Proof. Since $\mathscr{I}$ is a complete lattice (compare Proposition 2.6.40), we immediately have

$$\sup_{T \in \mathscr{L}_+} \inf_{x \in X} L(x,T) \preccurlyeq \inf_{x \in X} \sup_{T \in \mathscr{L}_+} L(x,T), \tag{8.24}$$

By Proposition 8.3.18 we know that $\inf_{x \in X} \sup_{T \in \mathscr{L}_+} L(x,T) \preccurlyeq \bar{P}$. □

To prove the strong duality from Theorem 8.3.15 we establish another formulation for $\bar{D}_c$ (see $(SD_L'^c - \preceq_{\mathscr{I}})$, $c \in \operatorname{int} C$) in terms of linear operators as follows. Let $\mathscr{L}_c$ be a subset of $\mathscr{L}$ defined by

$$\mathscr{L}_c := \left\{T \in \mathscr{L} \mid T = \langle u^*, \cdot \rangle\, c \text{ for some } u^* \in K^+\right\}.$$

Is is easy to check that $\mathscr{L}_c$ is a subset of $\mathscr{L}_+$, i.e.,

$$\mathscr{L}_c \subseteq \mathscr{L}_+, \tag{8.25}$$

and is isomorphic to $K^+ \subseteq Z^*$. The dual problem $(SD_L'^c - \preceq_{\mathscr{I}})$ can be equivalently expressed by

$$\bar{D}_c := \sup_{T \in \mathscr{L}_c} \phi(T), \tag{8.26}$$

where the duality function is defined in (8.23).

Remark 8.3.20. It is standard in the literature to consider the larger class $\mathscr{L}_+$ instead of $\mathscr{L}_c$. Nevertheless, in [402] $\mathscr{L}_c$ has also been considered to establish strong duality by using properly efficient solutions. On the other hand, we point out that, in general, the Lagrangian multipliers are always given via elements $T \in \mathscr{L}_c$

for any $c \in \operatorname{int} C$ see, for instance, [112, Theorem 4.1], [387, Theorem 3], [402]. Therefore, according to the obtained results the dual problem defined in such papers could define by using $\mathscr{L}_c$ instead $\mathscr{L}_+$. So, we can consider vectors as dual variables instead of operators by virtue of (8.26).

Strong duality between $(SP- \preceq_{\mathscr{I}})$ and $(SD_L^T- \preceq_{\mathscr{I}})$ can be easily derived from previous results.

Theorem 8.3.21 (Strong Duality Between $(SP- \preceq_{\mathscr{I}})$ and $(SD_L^T- \preceq_{\mathscr{I}})$). *Let F be convex and G be K-convex. Suppose that*

$$G(\operatorname{dom} F) \cap -\operatorname{int} K \neq \emptyset,$$

then we have strong duality, i.e., $\tilde{D} = \bar{P}$.

Proof. By Theorems 8.3.19 and 8.3.15 we obtain strong duality for $(SP- \preceq_{\mathscr{I}})$ and $(SD_L^T- \preceq_{\mathscr{I}})$ taking into account (8.25) and (8.26). □

The proof of the following result is analogous to that in Proposition 8.3.16 taking into account that given $u^* \in K^+\backslash\{0\}$ and $y^* \in C^-$ there always exists $T \in \mathscr{L}_+$ such that $T^*y^* = -u^*$.

Proposition 8.3.22. *If for some $x \in \operatorname{dom} F \cap \operatorname{dom} G$ the set $G(x) + K$ is closed and convex, then*

$$\sup_{T \in \mathscr{L}_+} L(x,T) = \begin{cases} F(x) & \text{if} \quad x \in S \\ \{+\infty\} & \text{else.} \end{cases}$$

8.4 Comparison of Different Approaches to Duality in Set-Valued Optimization

In order to derive duality assertions in set-valued optimization one has the possibilities to use an approach via conjugates, via Lagrangian technique or an axiomatic approach. Conjugate duality statements, based on different types of perturbation of the original problem, have been derived by Tanino and Sawaragi [569], Sawaragi, Nakayama and Tanino [521], Löhne [394, 395], Boţ, Grad and Wanka [73], and others. Lagrange duality for set-valued problems has been studied, for instance, by Luc [402], Ha [226], Hernández and Rodríguez-Marín [249, 251], Li, Chen and Wu [379], Löhne [395] and Hamel and Löhne [236]. An axiomatic approach was given by Luc [402]. Furthermore, duality assertions can be developed for different solution concepts, this means for the vector approach (cf. Luc [402], Boţ, Grad and Wanka [73], Li, Chen and Wu [379]), for the set-approach (cf. Kuroiwa [351], Hernández and Rodríguez-Marín [251]) and by using supremal and infimal sets (see Nieuwenhuis [442], Tanino [563,566]) and/or infimum and supremum in a complete lattice for the lattice approach (cf. Tanino [566], Song [535,536], Löhne [393–395], Lalitha and Arora [360]).

Another important difference in constructing the dual problems concerns the type of dual variables, which can be vectors like in the scalar case (cf. Löhne [394, 395], Boţ, Grad and Wanka [73], Section 7.1.3), extended vectors (cf. Hamel and Löhne [236]) or operators (cf. Corley [112], Luc [402], Tanino [566], Boţ, Grad and Wanka [73], Section 7.1.2, Hernández and Rodríguez-Marín [251], Lalitha and Arora [360], Li, Chen and Wu [379]).

In the mentioned papers one can observe that it is very easy to derive weak duality statements without additional assumptions. In order to get strong duality one assumes additionally convexity and certain regularity assumptions. An important approach for the formulation of regularity assumptions is the stability of the primal set-valued problem. In the literature, stability of the primal set-valued problem is formulated using the subdifferential of the minimal value map (cf. Tanino and Sawaragi [569], Boţ, Grad and Wanka [73]). In this section we will introduce a subdifferential notion based on infimal sets where subgradients are vectors. A corresponding stability notion is used to prove strong duality statements. Furthermore, we will discuss the space of dual variables and explain the relations between some other approaches from the literature. Subdifferentials notions for vector and set-valued maps have been investigated by many authors, see e.g., by Tanino [566], Jahn [293], Boţ, Grad and Wanka [73].

Lagrange duality theory is an important tool in optimization and there are many approaches to a corresponding theory for vector optimization problems, see e.g. Corley [111, 112, 509], Boţ and Wanka [52], Li and Chen [387], Jahn [293], Göpfert, Tammer, Riahi and Zălinescu [214], Boţ, Grad and Wanka [73] and references therein.

The following results are given in the paper by Hernández, Löhne, Rodríguez-Marín, Tammer [248].

In order to compare different approaches to duality in set-valued optimization we discuss Lagrange duality assertions from Sect. 8.3.2 for optimization problems with $\mathscr{I}$-valued objective function and set-valued constraints using a notation adapted to [566].

Assume that $(Y, \leq)$ is a partially ordered linear topological space, where the order is induced by a proper pointed closed convex cone C satisfying $\emptyset \neq \operatorname{int} C \neq Y$. Let X be a linear space and let Z be a separated locally convex space. Moreover, let $\langle Z, Z^* \rangle$ be a dual pair. Let $F : X \rightrightarrows Y^\bullet$ and $G : X \rightrightarrows Z$ be set-valued maps. We set $\operatorname{Dom} F := \{x \in X | F(x) \neq \emptyset,\ F(x) \neq \{+\infty\}\}$ and $\operatorname{dom} G := \{x \in X \mid G(x) \neq \emptyset\}$. Let $K \subset Z$ be a proper closed convex cone with nonempty interior. We denote by K^- the negative polar cone of K.

8.4.1 Lagrange Duality

In this section we use the notation introduced in Sect. 2.6.3, especially the infimal set (Definition 2.6.31) and the hyperspace of self-infimal sets $(\mathscr{I}, \oplus_{\mathscr{I}}, \odot_{\mathscr{I}}, \preccurlyeq_{\mathscr{I}})$

(Definition 2.6.37). We now study a reformulation of problem ($SP- \preceq_{\mathscr{I}}$) from Sect. 8.3.2 taking into account Theorem 2.6.41.

$$\bar{p} := \operatorname{Inf} \bigcup_{x \in S} F(x), \qquad S := \{x \in X \mid G(x) \cap -K \neq \emptyset\}. \tag{SP}$$

Restrictions of this type have been considered in many papers (see Borwein [56]; Corley [112]; Jahn [290]; Luc [402]; Götz and Jahn [218]; Crespi, Ginchev and Rocca [116]; Boţ and Wanka [52]; Hernández, Löhne, Rodríguez-Marín, Tammer [248]).

For simplicity we put $\preceq := \preceq_{\mathscr{I}}$.

In the sequel we consider the primal problem (SP). We assume throughout that a fixed vector

$$c \in \operatorname{int} C \tag{8.27}$$

is given. Several concepts, for instance, the Lagrangian, the dual objective function and subgradients will depend on the choice of this vector c. Note that we do not mention this dependance explicitly.

The Lagrangian map of problem (SP) is defined by (compare the $\mathscr{I}$-valued formulation in (8.18))

$$L : X \times Z^* \rightrightarrows Y^\bullet, \qquad L(x, u^*) = F(x) + \operatorname{Inf} \bigcup_{u \in G(x)+K} \langle u^*, u \rangle \{c\}. \tag{8.28}$$

In the special case $q = 1$, $C = \mathbb{R}_+$, $c = 1$, the well-known Lagrangian coincides with the Lagrangian of the scalar problem. For every choice of $c \in \operatorname{int} C$ we have a different Lagrangian map and a different corresponding dual problem, but we show that weak duality and strong duality hold for all these problems.

With the notations of this section we get an equivalent formulation to Proposition 8.3.13).

Proposition 8.4.1. *For all $x \in S$ it holds*

$$\operatorname{Sup} \bigcup_{u^* \in Z^*} L(x, u^*) = F(x).$$

We next define the dual problem. The dual objective function is defined by

$$\phi : Z^* \rightrightarrows Y^\bullet, \qquad \phi(u^*) := \operatorname{Inf} \bigcup_{x \in X} L(x, u^*)$$

and the dual problem (with respect to $c \in \operatorname{int} C$) associated to (SP) is defined by

$$\bar{d} := \operatorname{Sup} \bigcup_{u^* \in Z^*} \phi(u^*). \tag{SD_L}$$

As shown in Theorem 8.3.14, we have weak duality. Taking into account Theorem 2.6.41, we get the following formulation.

Theorem 8.4.2 (Weak Duality Between (SP) and (SD_L)). *The problems (SP) and (SD_L) satisfy the weak duality inequality, i.e.,*

$$\operatorname{Sup} \bigcup_{u^* \in Z^*} \phi(u^*) \preccurlyeq \operatorname{Inf} \bigcup_{x \in S} F(x).$$

The following strong duality theorem has been proven in Theorem 8.3.15 and is reformulated now using Theorem 2.6.41.

Theorem 8.4.3 (Strong Duality Between (SP) and (SD_L)). *Let F be C-convex, let G be K-convex, and let*

$$G(\operatorname{Dom} F) \cap -\operatorname{int} K \neq \emptyset. \tag{8.29}$$

Then, strong duality holds, that is,

$$\operatorname{Sup} \bigcup_{u^* \in Z^*} \phi(u^*) = \operatorname{Inf} \bigcup_{x \in S} F(x).$$

By strong duality and Theorem 2.6.41 we get

$$\operatorname{Sup} \bigcup_{u^* \in Z^*} \operatorname{Inf} \bigcup_{x \in X} L(x, u^*) = \operatorname{Inf} \bigcup_{x \in X} \operatorname{Sup} \bigcup_{u^* \in Z^*} L(x, u^*).$$

The next statement extends Proposition 8.4.1. Note that the additional assumption of $G(x) + K$ being a closed convex set originates from the set-valued constraints and could not be omitted if the objective function would be scalar-valued, see [395, Example 3.21] for further details.

Proposition 8.4.4. *Let $F : X \rightrightarrows Y^\bullet$ a set valued map with $F(x) = \operatorname{Inf} F(x) \neq \{-\infty\}$ for all $x \in X$, $\operatorname{Dom} F \neq \emptyset$ and let the set $G(x) + K$ be closed and convex for every $x \in X$, then*

$$\operatorname{Sup} \bigcup_{u^* \in Z^*} L(x, u^*) = \begin{cases} F(x) & \text{if} \quad x \in S \\ \{+\infty\} & \text{else.} \end{cases}$$

8.4.2 Subdifferentials and Stability

In this section we will use the notations introduced in Sect. 8.4.1 and study subdifferential notions and corresponding duality statements.

We start by introducing the notion of subgradient and stability for our framework. These concepts are mainly motivated by Tanino [566] (see also Boţ, Grad and Wanka [73] for numerous related results).

Let φ be a set-valued map from $X \times Z$ to Y defined by

$$\varphi(x,u) := \begin{cases} F(x) & \text{if } G(x) \cap (-K-u) \neq \emptyset \\ \emptyset & \text{else.} \end{cases}$$

Denote by $W : Z \rightrightarrows Y^\bullet$ the **perturbation map** defined by

$$W(u) := \operatorname{Inf} \bigcup_{x \in X} \varphi(x,u).$$

Clearly, we have

$$W(0) = \operatorname{Inf} \bigcup_{x \in S} F(x) = \bar{p}.$$

This leads to the following definition of a subgradient.

Definition 8.4.5. A point $u^* \in K^+$ is a called **positive subgradient** of W at $(\bar{u}, \bar{y}) \in Z \times Y$ with $\bar{y} \in W(\bar{x})$, written $u^* \in \partial^+ W(\bar{u}, \bar{y})$ for short, if

$$\bar{y} - \langle u^*, \bar{u} \rangle\, c \in \operatorname{Inf} \bigcup_{u \in Z} (W(u) - \langle u^*, u \rangle\, c).$$

Remark 8.4.6. Note that the subgradient in [379, Definition 2.5] (see Sect. 15.1.2, especially Definition 15.1.10) is a stronger notion than the one in Definition 8.4.5. In particular, in [379, Definition 2.5], subgradients are operators whereas in Definition 8.4.5 subgradients are vectors. Furthermore, the definition of subgradients in Definition 8.4.5 is closely related to that given by Boţ, Grad and Wanka [73, Definition 7.1.9 (c)] (see Definition 8.1.3, (c)), where the subgradients are defined using the vector approach (Pareto maximum) instead of the supremal set like in Definition 8.4.5. However, in [73] (Definition 8.1.3, (c)) the so called c-subgradients are vectors too.

Lemma 8.4.7. *Consider the set-valued maps $\partial^+ W(0,\cdot) : Y \rightrightarrows Z^*$ and $\phi : Z^* \rightrightarrows Y^\bullet$. For $u^* \in Z^*$ with $\phi(u^*) \subseteq Y$, one has*

$$u^* \in \partial^+ W(0,y) \iff y \in \phi(u^*).$$

Proof. By definition, $u^* \in \partial^+ W(0,y)$ means

$$y \in \operatorname{Inf} \bigcup_{u \in Z} (W(u) - \langle u^*, u \rangle\, c).$$

We have

$$\begin{aligned}
\operatorname{Inf} \bigcup_{u\in Z} (W(u) - \langle u^*, u\rangle\, c) &= \operatorname{Inf} \bigcup_{u\in Z} \operatorname{Inf} \bigcup_{x\in X} (\varphi(x,u) - \langle u^*, u\rangle\, c) \\
&= \operatorname{Inf} \bigcup_{-u\in G(x)+K,\, x\in X} (F(x) - \langle u^*, u\rangle\, c) \\
&= \operatorname{Inf} \bigcup_{u\in G(x)+K,\, x\in X} (F(x) + \langle u^*, u\rangle\, c) \\
&= \operatorname{Inf} \bigcup_{x\in X} \Big(F(x) + \operatorname{Inf} \bigcup_{u\in G(x)+K} \langle u^*, u\rangle\, c \Big) \\
&= \operatorname{Inf} \bigcup_{x\in X} L(x, u^*) \;=\; \phi(u^*),
\end{aligned}$$

which proves the claim. □

Definition 8.4.8. Problem (SP) is called **stable** if $W(0) \neq \{+\infty\}$, $W(0) \neq \{-\infty\}$ and $\partial^+ W(0, y) \neq \emptyset$ for all $y \in W(0)$.

Remark 8.4.9. In Boţ, Grad and Wanka [73] the definition of c-subgradients (see Definition 8.1.3) is used in order to introduce the property that the primal set-valued optimization problem is c-stable, compare Remark 8.1.5.

Theorem 8.4.10. *If (SP) is stable, then strong duality holds for (SP) and (SD_L), that is,*

$$\operatorname{Sup} \bigcup_{u^*\in Z^*} \phi(u^*) = \operatorname{Inf} \bigcup_{x\in S} F(x).$$

Proof. We set $\bar{d} = \operatorname{Sup} \bigcup_{u^*\in Z^*} \phi(u^*)$ and $\bar{p} = W(0) = \operatorname{Inf} \bigcup_{x\in S} F(x)$. By assumption, we have $W(0) \neq \{+\infty\}$ and $W(0) \neq \{-\infty\}$, which implies $\emptyset \subsetneq \mathrm{Cl}_+ W(0) \subsetneq Y$. Take some $y \in W(0)$. Since (SP) is stable, there is some $u^* \in Z^*$ with $y \in \phi(u^*)$ (by Lemma 8.4.7). Using weak duality we get $\phi(u^*) \preccurlyeq \bar{d} \preccurlyeq \bar{p}$. Thus $\bar{d} \neq \{-\infty\}$ and $\bar{d} \neq \{+\infty\}$, which implies $\emptyset \subsetneq \mathrm{Cl}_+ \bar{d} \subsetneq Y$.

By weak duality, it remains to prove $\bar{p} \preccurlyeq \bar{d}$. Taking into account Lemma 2.6.38, we have to prove that $(\bar{p} - \operatorname{int} C) \cap \bar{d} = \emptyset$. On the contrary, suppose that there is $y \in Y$ with $y \in (\bar{p} - \operatorname{int} C) \cap \bar{d}$. Then there exists $z \in \bar{p} = W(0)$ and $c \in \operatorname{int} C$ such that $y = z - c$. On the other hand, there exists $u^* \in \partial^+ W(0, z)$. By Lemma 8.4.7, this means $z \in \phi(u^*)$. Hence $y \in (\phi(u^*) - \operatorname{int} C) \cap \bar{d}$. By Lemma 2.6.38 this contradicts $\phi(u^*) \preccurlyeq \bar{d}$. □

Theorem 8.4.11. *If F is C-convex, G is K-convex,*

$$G(\operatorname{Dom} F) \cap (-\operatorname{int} K) \neq \emptyset, \tag{8.30}$$

and $W(0) \neq \{-\infty\}$, then (SP) is stable.

Proof. From (8.30), we get $W(0) \neq \{+\infty\}$.

Let $\bar{y} \in W(0)$. By Lemma 8.4.7 we have to show that there exists $u^* \in Z^*$ with $\bar{y} \in \phi(u^*)$. The map $Q\colon X \rightrightarrows Y \times Z$ defined by $Q(x) := (F(x), G(x))$ is $C \times K$-convex. Thus, $Q(X) + C \times K$ is a convex set. We next show that

$$(Q(S) + C \times K) \cap \operatorname{int}(B \times (-K)) = \emptyset \tag{8.31}$$

where $B = \{\bar{y}\} - C$. Indeed, if there exist $x' \in S$ and (y, u) such that

$$(y, u) \in \big((F(x'), G(x')) + C \times K\big) \cap \operatorname{int}(B \times (-K)),$$

then $y \in (F(x') + C) \cap (\{\bar{y}\} - \operatorname{int} C)$ and $u \in (G(x') + K) \cap -\operatorname{int} K$. Thus, $y' = \bar{y} - c'$ where $y' \in F(x')$ and $c' \in \operatorname{int} C$ and $(G(x') + K) \cap (-K) \neq \emptyset$ (that is, $x' \in S$) which contradicts $\bar{y} \in W(0) = \operatorname{Inf} \bigcup_{x \in S} F(x)$.

By (8.31), applying a separation theorem there exists a pair $(y^*, u^*) \in Y^* \times Z^* \setminus \{(0, 0)\}$ such that

$$\langle y^*, y\rangle + \langle u^*, u\rangle \leq \langle y^*, b\rangle + \langle u^*, -d\rangle \tag{8.32}$$

for all $(y, u) \in Q(S) + C \times K$, $b \in B$ and $d \in K$. We deduce that $(y^*, u^*) \in (C^- \times K^-) \setminus \{(0, 0)\}$. This implies

$$\forall (y, u) \in Q(S) + C \times K : \; \langle y^*, y\rangle + \langle u^*, u\rangle \leq \langle y^*, \bar{y}\rangle .$$

Since, by (2.62), $\operatorname{Cl}_+ W(0) = \operatorname{Cl}_+ F(S) = \operatorname{cl}(F(S) + C)$, we get

$$\forall y \in \operatorname{Cl}_+ F(S), \; \forall u \in G(S) + K : \; \langle y^*, y\rangle + \langle u^*, u\rangle \leq \langle y^*, \bar{y}\rangle . \tag{8.33}$$

We show that $y^* \neq 0$. Assuming the contrary, we get $u^* \neq 0$ and, by (8.33), we have

$$\forall u \in G(S) + K : \; \langle u^*, u\rangle \leq 0.$$

On the other hand, by (8.30), there exists $x \in S$ and $u' \in G(x) \cap (-\operatorname{int} K)$, i.e., $\langle u^*, u'\rangle > 0$ as $u^* \in K^- \setminus \{0\}$. Since $u' \in G(x) \subseteq G(x) + K$ this is a contradiction.

Since $y^* \in C^- \setminus \{0\}$, for the fixed vector $c \in \operatorname{int} C$ according to (8.27), we have $\langle y^*, c\rangle < 0$. Without loss of generality we can assume $\langle y^*, c\rangle = -1$.

Since $\bar{y} \in W(0) \subseteq \operatorname{Cl}_+ F(S)$, by (8.33), we have $\langle u^*, u\rangle \leq 0$ for all $u \in G(\bar{x}) \subseteq G(\bar{x}) + K$. Since $u^* \in K^-$, we have $\langle u^*, \bar{u}\rangle = 0$ for all $u \in G(\bar{x}) \cap -K$. Thus, (8.33) can be written as

$$\forall y \in \operatorname{Cl}_+ F(S), \; \forall u \in G(S) + K : \; \langle y^*, y - \langle u^*, u\rangle \{c\}\rangle \leq \langle y^*, \bar{y}\rangle .$$

From weak duality, we know that $\bar{y} \in \mathrm{Cl}_{+}\phi(u^*)$. Assuming that $\bar{y} \in \phi(u^*) + \mathrm{int}\, C$, we obtain a contradiction to the latter inequality. Hence, by Proposition 2.6.33, we have $\bar{y} \in \phi(u^*)$. □

8.4.3 Duality Statements with Operators as Dual Variables

We establish in this section another type of dual problem where the dual variables are operators instead of vectors as in problem $(SD_L^c - \preceq_{\mathscr{I}})$ (compare also problem $(SD_L^T - \preceq_{\mathscr{I}})$, where the dual variables are operators too). The usage of operators is more common in the literature (see, for instance [112, 251, 351, 387, 402]). We will see, however, that a duality theory based on operators as dual variables is an easy consequence of the above results.

Denote by $\mathscr{L}$ the set of all linear continuous operators from Z to Y and by $\mathscr{L}_+$ the subset of all positive operators, that is, $\mathscr{L}_+ := \{T \in \mathscr{L} : T(K) \subseteq C\}$. Given $T \in \mathscr{L}$ and $A \subseteq Z$ we write $T(A) = \{T(a) \mid a \in A\}$. Let $\mathscr{L}_c$ be a subset of $\mathscr{L}$ defined by $\mathscr{L}_c := \{T \in \mathscr{L} \mid T = \langle u^*, \cdot\rangle\, c$ for some $u^* \in K^+\}$. Obviously, we have

$$\mathscr{L}_c \subseteq \mathscr{L}_+. \tag{8.34}$$

Moreover, $\mathscr{L}_c$ is isomorphic to $K^+ \subseteq Z^*$.

The Lagrangian map $L : X \times \mathscr{L} \to Y^\bullet$ is defined by

$$L(x,T) := F(x) + T(G(x)). \tag{8.35}$$

The dual objective function $\Phi : \mathscr{L} \to Y^\bullet$ is defined by

$$\Phi(T) := \operatorname{Inf} \bigcup_{x \in X} L(x,T),$$

and the associated dual problem is defined by

$$\tilde{d} := \operatorname{Sup} \bigcup_{T \in \mathscr{L}_+} \Phi(T). \tag{SD_L^T}$$

Comparing the two dual problems (SD_L) (with vectors as dual variables) and (SD_L^T) (with operators as dual variables), we observe that Lagrangian (8.28) for problem (SD_L) involves the cone K but Lagrangian (8.35) for problem (SD_L^T) does not. On the other hand, the supremum in (SD_L) is taken over the whole linear space Z^* whereas in (SD_L^T) only the subspace $\mathscr{L}_+$ of the linear space $\mathscr{L}$ is considered. A reformulation of problem (SD_L) clarifies the connection. Consider, instead of (8.28), the Lagrangian

$$\hat{L} : X \times Z^* \rightrightarrows Y^\bullet, \qquad \hat{L}(x, u^*) = F(x) + \operatorname{Inf} \bigcup_{u \in G(x)} \langle u^*, u \rangle \{c\}. \tag{8.36}$$

and the corresponding dual objective function

$$\hat{\phi} : Z^* \rightrightarrows Y^\bullet, \qquad \hat{\phi}(u^*) := \operatorname{Inf} \bigcup_{x \in X} \hat{L}(x, u^*).$$

Lemma 8.4.12. *The dual objective function of problem (SD_L) can be expressed as*

$$\phi(u^*) = \begin{cases} \hat{\phi}(u^*) & \text{if} \quad u^* \in K^+ \\ \{-\infty\} & \text{otherwise.} \end{cases} \tag{8.37}$$

Proof. Since $c \in \operatorname{int} C$, we have

$$\operatorname{Inf} \bigcup_{d \in K} \langle d, u^* \rangle\, c = \begin{cases} \{0\} & \text{if } u^* \in K^+ \\ \{-\infty\} & \text{otherwise.} \end{cases}$$

It follows

$$\begin{aligned}
\phi(u^*) &= \operatorname{Inf} \bigcup_{x \in X} L(x, u^*) \\
&= \operatorname{Inf} \bigcup_{x \in X} \Big(F(x) + \operatorname{Inf} \bigcup_{u \in G(x)+K} \langle u, u^* \rangle\, c \Big) \\
&= \operatorname{Inf} \bigcup_{x \in X} \Big(F(x) + \operatorname{Inf} \bigcup_{u \in G(x), k \in K} \big(\langle u, u^* \rangle\, c + \langle k, u^* \rangle\, c \big) \Big) \\
&= \operatorname{Inf} \bigcup_{x \in X} \Big(F(x) + \operatorname{Inf} \bigcup_{u \in G(x)} \langle u, u^* \rangle \{c\} + \operatorname{Inf} \bigcup_{k \in K} \langle k, u^* \rangle \{c\} \Big) \\
&= \operatorname{Inf} \bigcup_{x \in X} \hat{L}(x, u^*) + \operatorname{Inf} \bigcup_{k \in K} \langle k, u^* \rangle \{c\} \\
&= \hat{\phi}(u^*) + \operatorname{Inf} \bigcup_{k \in K} \langle k, u^* \rangle \{c\}.
\end{aligned}$$

Combining the two equations we obtain the result. □

As a consequence we can define a dual problem

$$\hat{d} := \operatorname{Sup} \bigcup_{u^* \in K^+} \hat{\phi}(u^*), \tag{$\hat{D}$}$$

where we obviously have

$$\bar{d} = \operatorname{Sup} \bigcup_{u^* \in Z^*} \phi(u^*) = \operatorname{Sup} \bigcup_{u^* \in K^+} \hat{\phi}(u^*) = \hat{d}. \tag{8.38}$$

Since K^+ is isomorphic to $\mathscr{L}_c$ and $\mathscr{L}_c \subseteq \mathscr{L}_+$, we get (using Theorem 2.6.41)

$$\bar{d} = \operatorname{Sup} \bigcup_{u^* \in Z^*} \phi(u^*) \preccurlyeq \operatorname{Sup} \bigcup_{T \in \mathscr{L}_+} \Phi(T) = \tilde{d}. \tag{8.39}$$

We next prove weak duality.

Theorem 8.4.13 (Weak Duality Between (SP) and (SD_L^T)). *The problems (SP) and (SD_L^T) satisfy the weak duality inequality, i.e.,*

$$\operatorname{Sup} \bigcup_{T \in \mathscr{L}_+} \Phi(T) \preccurlyeq \operatorname{Inf} \bigcup_{x \in S} F(x).$$

Proof. By Theorem 2.6.41, we have

$$\operatorname{Sup} \bigcup_{T \in \mathscr{L}_+} \operatorname{Inf} \bigcup_{x \in X} L(x, T) \preccurlyeq \operatorname{Inf} \bigcup_{x \in X} \operatorname{Sup} \bigcup_{T \in \mathscr{L}_+} L(x, T).$$

Since $\Phi(T) = \operatorname{Inf} \bigcup_{x \in X} L(x, T)$, it remains to show

$$\operatorname{Inf} \bigcup_{x \in X} \operatorname{Sup} \bigcup_{T \in \mathscr{L}_+} L(x, T) \preccurlyeq \operatorname{Inf} \bigcup_{x \in S} F(x).$$

But this follows from Proposition 8.4.1 and

$$\operatorname{Inf} \bigcup_{x \in X} \operatorname{Sup} \bigcup_{T \in \mathscr{L}_+} L(x, T) \preccurlyeq \operatorname{Inf} \bigcup_{x \in S} \operatorname{Sup} \bigcup_{T \in \mathscr{L}_+} L(x, T),$$

which is a consequence of Theorem 2.6.41. □

Finally we obtain strong duality as a conclusion of the Lagrange duality theorem with vectors as variables.

Theorem 8.4.14 (Strong Duality Between (SP) and (SD_L^T)). *Let F be C-convex, let G be K-convex, and let*

$$G(\operatorname{Dom} F) \cap (-\operatorname{int} K) \neq \emptyset.$$

Then strong duality holds, that is,

$$\operatorname{Sup} \bigcup_{T \in \mathscr{L}_+} \Phi(T) = \operatorname{Inf} \bigcup_{x \in S} F(x).$$

Proof. From Theorem 8.4.3, inequality (8.39), and Theorem 8.4.13, we get

$$\operatorname{Inf} \bigcup_{x \in S} F(x) = \operatorname{Sup} \bigcup_{u^* \in Z^*} \phi(u^*) \preccurlyeq \operatorname{Sup} \bigcup_{T \in \mathscr{L}_+} \Phi(T) \preccurlyeq \operatorname{Inf} \bigcup_{x \in S} F(x),$$

which yields the desired equation. □

Remark 8.4.15. Further relationships between approaches to duality are discussed in Sect. 15.1.2, where the notations well-known in vector optimization are used (compare Li, Chen and Wu [379]).

Chapter 9
Existence Results for Minimal Points

In this chapter we establish several existence results for minimal points with respect to transitive relations; then we apply them in topological vector spaces for quasi-orders generated by convex cones. We continue with the presentation of several types of convex cones and compactness notions with respect to cones. We end the chapter with existence results for vector and set optimization problems.

The presentation in Sects. 9.1, 9.2, 9.3 and 9.4 follows that in [214] or [539], where one can find the proofs for the results which are only stated.

9.1 Preliminary Notions and Results Concerning Transitive Relations

In the sequel Y is a nonempty set and $t \subseteq Y \times Y$, that is, t is a relation on Y. If $\emptyset \neq A \subseteq Y$ and $t \subseteq Y \times Y$, the restriction of t to A is denoted by t^A; i.e., $t^A := t \cap (A \times A)$.

With the relation t on Y we associate the following relations:

$$t_R := t \cup \Delta_Y, \qquad t_N := t \setminus t^{-1} = t \setminus (t \cap t^{-1}), \qquad t_{NR} = (t_N)_R \,.$$

Hazen and Morin [240] call t_N the asymmetric part of t. Some properties of these relations are given in the following proposition; the first three properties mentioned in (ii) are stated by Dolecki and Malivert in [144].

Proposition 9.1.1. *Let t be a transitive relation on Y and $\emptyset \neq A \subseteq Y$.*

(i) *t_R is reflexive and transitive; $t_N \cap \Delta_Y = \emptyset$;*
(ii) *$t \circ t_N \subseteq t_N$, $t_N \circ t \subseteq t_N$, $t_N \circ t_N \subseteq t_N$, $(t_R)_N = t_N$, $t_R \circ t_N = t_N \circ t_R = t_N$; $(t_N)_N = t_N$;*

A.A. Khan et al., *Set-valued Optimization*, Vector Optimization,
DOI 10.1007/978-3-642-54265-7_9

(iii) *t_{NR} is reflexive, antisymmetric, and transitive;* $(t_N)_N = (t_{NR})_N = t_N$*;*
(iv) $\left(t^A\right)_R = (t_R)^A$, $\left(t^A\right)_N = (t_N)^A$.

Taking into account the above proposition, we denote by t_R^A, t_N^A, and t_{NR}^A the relations $\left(t^A\right)_R$, $\left(t^A\right)_N$, and $\left(t^A\right)_{NR}$, respectively.

The preceding proposition shows that with every transitive relation one can associate a partial order. It is useful to know whether they determine the same maximal and minimal points. As noted in Sect. 2.1, $\mathrm{Max}(A;t) = \mathrm{Min}(A, t^{-1})$ for $\emptyset \neq A \subseteq Y$; so it is sufficient to study only the problems related to minimal points.

Corollary 9.1.2. *Let t be a transitive relation on Y and $\emptyset \neq A \subseteq Y$; then*

$$\mathrm{Min}(A;t) = \mathrm{Min}(A;t_R) = \mathrm{Min}(A;t_N) = \mathrm{Min}(A;t_{NR}).$$

The above corollary shows that the problem of existence (for example) of minimal points w.r.t. to a transitive relation t reduces, theoretically, to the same problem for the partial order t_{NR}. Another way to reduce this problem to one for a partial order is given by the following known result.

Proposition 9.1.3. *Let t be a transitive relation on Y and take $\rho := t_R \cap (t_R)^{-1}$. Then ρ is an equivalence relation, and $\hat{t} = \{(\hat{x}, \hat{y}) \mid (x, y) \in t\}$ is a partial order on $\hat{Y} := Y/\rho$, where $\hat{x}$ is the class of $x \in Y$ with respect to ρ. Moreover, if $x \in Y$, $x \in \mathrm{Min}(Y;t)$ if and only if $\hat{x} \in \mathrm{Min}(\hat{Y};\hat{t})$.*

In the sequel we shall also use the notation $A_t^+(x)$ and $A_t^-(x)$ for the **upper** and **lower sections** of $A \subseteq Y$ with respect to t and $x \in Y$. So

$$A_t^+(x) := \{y \in A \mid (x, y) \in t_R\}, \quad A_t^-(x) := \{y \in A \mid (y, x) \in t_R\},$$

respectively; the most common case is for $x \in A$. Similarly, for $E \subseteq Y$, we consider

$$A_t^+(E) := \{y \in A \mid \exists z \in E : (z, y) \in t_R\} = \bigcup_{z \in E} A_t^+(z),$$

$$A_t^-(E) := \{y \in A \mid \exists z \in E : (y, z) \in t_R\} = \bigcup_{z \in E} A_t^-(z);$$

clearly, $E \cap A \subseteq A_t^-(E) \cap A_t^+(E)$.

In the rest of this section we assume that t is a transitive relation on Y.

Note that $A_t^-\left(A_t^-(E)\right) = A_t^-(E)$ (so one may suppose that $E \subseteq A$) and

$$\mathrm{Min}(A_t^-(E);t) = A_t^-(E) \cap \mathrm{Min}(A;t), \quad \mathrm{Max}(A_t^+(E);t) = A_t^+(E) \cap \mathrm{Max}(A;t); \tag{9.1}$$

moreover,

$$\mathrm{Min}(A;t) = A \cap \mathrm{Min}(Y_t^+(A);t), \quad \mathrm{Max}(A;t) = A \cap \mathrm{Max}(Y_t^-(A);t), \tag{9.2}$$

and

$$\operatorname{Min}(A;t) = \operatorname{Min}(Y_t^+(A);t), \quad \operatorname{Max}(A;t) = \operatorname{Max}(Y_t^-(A);t) \tag{9.3}$$

provided that $Y_t^-(y) \cap Y_t^+(y) = \{y\}$ for every $y \in A$ (for example if t_R is antisymmetric).

Indeed, let us prove the first relation in (9.2). Take $\overline{y} \in \operatorname{Min}(A;t)$. Then $\overline{y} \in A \subseteq Y_t^+(A)$. Take $z \in Y_t^+(A)$ with $(z,\overline{y}) \in t_R$; then there exists $y \in A$ such that $(y,z) \in t_R$, and so $(y,\overline{y}) \in t_R$. It follows that $(\overline{y},y) \in t_R$, whence $(\overline{y},z) \in t_R$, and so the inclusion $\subseteq$ holds. Conversely, consider $\overline{y} \in A \cap \operatorname{Min}(Y_t^+(A);t)$ and take $y \in A$ $(\subseteq Y_t^+(A))$ with $(y,\overline{y}) \in t_R$. Then $(\overline{y},y) \in t_R$, and so $\overline{y} \in \operatorname{Min}(A;t)$.

Assume now that $Y_t^-(y) \cap Y_t^+(y) = \{y\}$ for every $y \in A$ and take $\overline{y} \in \operatorname{Min}(Y_t^+(A);t)$ $(\subseteq Y_t^+(A))$; there exists $\tilde{y} \in A$ $(\subseteq Y_t^+(A))$ such that $(\tilde{y},\overline{y}) \in t_R$, and so $(\overline{y},\tilde{y}) \in t_R$. Thus $\overline{y} \in Y_t^-() \cap Y_t^+(\tilde{y}) = \{\tilde{y}\}$, whence $\overline{y} = \tilde{y} \in A$. Therefore, $\overline{y} \in \operatorname{Min}(A,y)$ by (9.2), and so (9.3) holds.

Note that (9.2) makes more precise [189, Proposition 2.1 (a)].

Taking E the singleton $\{x\}$ in (9.1), we get

$$\operatorname{Min}(A_t^-(x);t) \subseteq \operatorname{Min}(A;t), \quad \operatorname{Max}(A_t^+(x);t) \subseteq \operatorname{Max}(A;t). \tag{9.4}$$

Relation (9.4) shows that in order to have that $\operatorname{Min}(A;t)$ is nonempty it is sufficient to get the nonemptiness of $\operatorname{Min}(A_t^-(x);t)$ for some $x \in A$.

We say that $\emptyset \neq A \subseteq Y$ has the **domination property** (DP) (w.r.t. t) if $\operatorname{Min}(A_t^-(y);t) \neq \emptyset$, for every $y \in A$ (i.e., every element of A is dominated by a minimal element of A).

A quite important problem is how to extend other notions related to partially ordered sets, like chain or increasing net, to sets endowed with transitive relations. Related to this problem we have the next result.

Proposition 9.1.4. *Let $A \subseteq Y$ be nonempty.*

(i) *If t_R^A is a partial order on A, then $t_R^A \subseteq t_{NR}$.*
(ii) *If μ is a total order on A such that $\mu \subseteq t_{NR}$, then $\mu = t_{NR}^A$.*

Note that in (ii) we cannot replace t_{NR} by t or t_R (take $Y = \mathbb{R}$ and $t = \mathbb{R} \times \mathbb{R}$).

Let $(x_i)_{i\in I} \subseteq Y$ be a net. We say that (x_i) is **t-increasing** [**t-decreasing**] if $(x_i,x_j) \in t$ $[(x_j,x_i) \in t]$ for all $i,j \in I$ with $i \preceq j$ and $i \neq j$; (x_i) is **strictly t-increasing** [**strictly t-decreasing**] if (x_i) is t_N-increasing [t_N-decreasing].

In the sequel we say that $\emptyset \neq A \subseteq (Y,t)$ is a **chain** (w.r.t. to t) if t_R^A is a total order on A, while A is **well-ordered** (w.r.t. to t) if A is well-ordered by t_R^A.

The following result is due to Gajek and Zagrodny (see [193]).

Proposition 9.1.5. *Let t be a transitive relation on Y. Then there exists a nonempty well-ordered subset W of Y such that for every $x \in Y \setminus W$ there exists $w \in W$ with $(x,w) \in t$ or $(w,x) \notin t$.*

9.2 Existence of Minimal Elements with Respect to Transitive Relations

We begin with the following result.

Proposition 9.2.1. *Let $A \subseteq Y$ be nonempty. Suppose that one of the following conditions holds:*

(i) *every nonempty set $E \subseteq A$ with $E \times E \subseteq t \cup t^{-1} \cup \Delta_Y$ is lower bounded in A;*
(ii) *every chain in A is lower bounded in A;*
(iii) *every well-ordered subset of A w.r.t. t^{-1} is t-lower bounded in A.*

Then A has the domination property. In particular, $\operatorname{Min}(A;t) \neq \emptyset$.

Corollary 9.2.2. *Let $A \subseteq Y$ be nonempty. Suppose that one of the following conditions holds:*

(i) *every t-decreasing net of A is lower bounded in A;*
(ii) *every strictly t-decreasing net of A is lower bounded in A;*
(iii) *every t-decreasing net $(y_i)_{i\in I} \subseteq A$ with I totally ordered is lower bounded in A;*
(iv) *every strictly t-decreasing net $(y_i)_{i\in I} \subseteq A$ with I totally ordered is lower bounded in A;*
(v) *every t-decreasing net $(y_i)_{i\in I} \subseteq A$ with I well-ordered is lower bounded in A;*
(vi) *every strictly t-decreasing net $(y_i)_{i\in I} \subseteq A$ with I well-ordered is lower bounded in A.*

Then A has the domination property. In particular, $\operatorname{Min}(A;t) \neq \emptyset$.

Note that conditions (i), (ii), and (iii) of Corollary 9.2.2 are equivalent to

$$\forall\, (y_i)_{i\in I} \subseteq A,\ (y_i)\ t\text{–decreasing} \ :\ A \cap \left(\bigcap_{i\in I} Y_t^-(y_i) \right) \neq \emptyset, \tag{9.5}$$

$$\forall\, (y_i)_{i\in I} \subseteq A,\ (y_i)\ \text{strictly}\ t\text{–decreasing} \ :\ A \cap \left(\bigcap_{i\in I} Y_{t_{NR}}^-(y_i) \right) \neq \emptyset, \tag{9.6}$$

$$\forall\, D \subseteq A,\ D\ \text{totally ordered} \ :\ A \cap \left(\bigcap_{y\in D} Y_t^-(y) \right) \neq \emptyset, \tag{9.7}$$

respectively.

Corollary 9.2.3. *Let $A \subseteq Y$ be nonempty. Suppose that*

$$\left[\forall\, E \subseteq A,\ A \subseteq \bigcup_{a\in E} (Y \setminus Y_t^-(a)) \right],\ \exists\, E_0 \in \mathscr{F}(E) : A \subseteq \bigcup_{a\in E_0} (Y \setminus Y_t^-(a)), \tag{9.8}$$

where $\mathscr{F}(E)$ is the class of finite subsets of E. Then A has the domination property. In particular, $\operatorname{Min}(A;t) \neq \emptyset$.

Proof. Let us observe that (9.8) implies (9.5). Indeed, take $(y_i)_{i\in I} \subseteq A$ a t-decreasing net and assume that $A \cap \left(\bigcap_{i\in I} Y_t^-(y_i)\right) = \emptyset$. Then $E := \{y_i \mid i \in I\} \subseteq A$ and $A \subseteq \bigcup_{a\in E}\left(Y \setminus Y_t^-(a)\right)$. Using our hypothesis, there exists $E_0 \in \mathscr{F}(E)$ with $A \subseteq \bigcup_{a\in E_0}\left(Y \setminus Y_t^-(a)\right)$. Therefore, there exists $I_0 \in \mathscr{F}(I)$ such that $E_0 = \{y_i \mid i \in I_0\}$, and so $A \cap \left(\bigcap_{i\in I_0} Y_t^-(y_i)\right) = \emptyset$. Because I is directed, there exists $i_0 \in I$ such that $i_0 \geq i$ for every $i \in I_0$. It follows that (y_{i_0}, y_i) (and so $y_{i_0} \in Y_t^-(y_i)$) for every $i \in I_0$, getting so a contradiction. The conclusion follows using Corollary 9.2.2 (i). □

An even stronger condition than (9.8) is

$$\left[\forall\, E \subseteq Y,\ A \subseteq \bigcup_{a\in E}\left(Y \setminus Y_t^-(a)\right)\right],\ \exists\, E_0 \in \mathscr{F}(E) : A \subseteq \bigcup_{a\in E_0}\left(Y \setminus Y_t^-(a)\right). \tag{9.9}$$

The following result is due, essentially, to Gajek and Zagrodny [193].

Proposition 9.2.4. *Let $A \subseteq Y$ be nonempty. Suppose that the following two conditions hold:*

(i) *every nonempty well-ordered subset W of A w.r.t. t^{-1} is at most countable;*
(ii) *every strictly t-decreasing sequence of A is lower bounded in A.*

Then A has the domination property. In particular, $\operatorname{Min}(A;t) \neq \emptyset$.

Observe that the set

$$\left\{E \mid E \subseteq A,\ A \subseteq \bigcup_{a\in E}\left(Y \setminus Y_t^-(a)\right)\right\}$$

is empty if and only if there exists $\overline{y} \in A$ such that $(\overline{y}, y) \in t$ for every $y \in A$, that is, A has a smallest element w.r.t. t; similarly, the set

$$\left\{E \mid E \subseteq Y,\ A \subseteq \bigcup_{a\in E}\left(Y \setminus Y_t^-(a)\right)\right\}$$

is empty if and only if there exists $\overline{y} \in A$ such that $(\overline{y}, y) \in t$ for every $y \in Y$, that is Y has a smallest element w.r.t. t which belongs to A. Of course, in these situations conditions (9.8) and (9.9) are verified, respectively.

If τ is a topology on Y and the lower sections of Y are closed (for example, if t is closed in $Y \times Y$ and τ is Hausdorff), then the sets $Y \setminus Y_t^+(a)$ are open; in this situation condition (9.8) is a kind of compactness of A. This is the reason for which a set $A \subseteq Y$ verifying condition (9.8) is called order-semicompact in [189, Definition 2.4 (a)]. Note also that condition (9.7) is equivalent to the fact that A is order-totally-complete in the sense of [189, Definition 2.2 (a)]; so, [189, Theorem 3.1] is practically Corollary 9.2.2 (iii).

The following result is related to this kind of conditions.

Proposition 9.2.5. *Let $A \subseteq Y$ be nonempty. Assume that there exists a relation s on Y such that*

$$s \circ t_N \subseteq t_R \tag{9.10}$$

and one of the following conditions holds:

$$\forall\, (y_i)_{i\in I} \subseteq A,\ (y_i) \text{ strictly } t\text{-decreasing} \ :\ A \cap \Big(\bigcap_{i\in I} Y_s^-(y_i)\Big) \neq \emptyset, \tag{9.11}$$

$$\forall\, D \subseteq A,\ D \text{ totally order w.r.t. } t \ :\ A \cap \Big(\bigcap_{y\in D} Y_s^-(y)\Big) \neq \emptyset. \tag{9.12}$$

Then A has the domination property. In particular, $\mathrm{Min}(A;t) \neq \emptyset$.

In the sequel, in this section, we suppose that (Y,τ) is a topological space. In this situation one can formulate other conditions for the existence of minimal points. Having a net $(x_i)_{i\in I}$, by $(x_{\varphi(j)})_{j\in J}$, or simply $(x_{\varphi(j)})$, we denote a subnet of (x_i); this means that J is directed and $\varphi : J \to I$ has the property that for every $i \in I$ there exists $j_i \in J$ such that $\varphi(j) \geq i$ for all $j \in J$, $j \geq j_i$.

Corollary 9.2.6. *Let Y be endowed with the topology τ and $\emptyset \neq A \subseteq Y$. Assume that one of the following conditions holds:*

$$\forall\, (y_i)_{i\in I} \subseteq A \ t\text{–decreasing},\ \exists\, (y_{\varphi(j)})_{j\in J} \to y \in A,\ \forall\, j \in J \ :\ (y, y_{\varphi(j)}) \in t_R, \tag{9.13}$$

$$\forall\, (y_i)_{i\in I} \subseteq A \text{ strictly } t\text{–decreasing},\ \exists\, (y_{\varphi(j)})_{j\in J} \to y \in A, \forall\, j \in J \ :\ (y, y_{\varphi(j)}) \in t_R. \tag{9.14}$$

Then A has the domination property. In particular, $\mathrm{Min}(A;t) \neq \emptyset$.

When the upper or lower sections of Y are closed, we may consider other conditions, too.

Proposition 9.2.7. *Let Y be endowed with the topology τ and $\emptyset \neq A \subseteq Y$. Consider the following conditions:*

$$\forall\, (y_i)_{i\in I} \subseteq A \ t\text{–decreasing} \ :\ \exists\, (y_{\varphi(j)})_{j\in J} \to y \in A, \tag{9.15}$$

$$\forall\, (y_i)_{i\in I} \subseteq A \text{ strictly } t\text{–decreasing} \ :\ \exists\, (y_{\varphi(j)})_{j\in J} \to y \in A, \tag{9.16}$$

$$\forall\, (y_i)_{i\in I} \subseteq A \ t\text{–decreasing},\ \exists\, y \in A \ :\ y_i \to y, \tag{9.17}$$

and

$$\forall\,(y_i)_{i\in I} \subseteq A \text{ strictly } t\text{–decreasing},\ \exists\, y \in A\ :\ y_i \to y. \tag{9.18}$$

If the lower sections of Y are closed, then (9.15) ⇔ (9.13) *and* (9.16) ⇔ (9.14), *while if t is a partial order and the upper and lower sections of Y (w.r.t. t) are closed, then* (9.15) ⇔ (9.17) *and* (9.16) ⇔ (9.18).

Corollary 9.2.8. *Let Y be endowed with the topology τ and $\emptyset \neq A \subseteq Y$. Assume that the lower sections of Y are closed and A is compact. Then A has the domination property. In particular,* $\mathrm{Min}(A;t) \neq \emptyset$.

Hazen and Morin proved Corollary 9.2.8 in [240, Corollary 2.8]. The next result is [189, Theorem 3.2].

Corollary 9.2.9. *Let Y be endowed with the topology τ and $\emptyset \neq A \subseteq Y$. Suppose that*

$$\forall x,y,z \in Y : \left[x \in Y_t^-(y) \setminus Y_t^+(y),\ z \in \mathrm{cl}\, Y_t^-(y)\right] \Rightarrow z \in Y_t^-(x), \tag{9.19}$$

and

$$\forall\, D \subseteq A,\ D \text{ totally ordered w.r.t. } t\ :\ A \cap \left(\bigcap_{y\in D} Y_s^-(y)\right) \neq \emptyset \tag{9.20}$$

hold, where

$$s := \{(y,z) \in Y \times Y \mid y \in \mathrm{cl}\, Y_t^-(z)\}. \tag{9.21}$$

Then A has the domination property; in particular, $\mathrm{Min}(A;t) \neq \emptyset$.

Proof. With our notation, condition (9.19) can be written as $s \circ t_N \subseteq t_R$, that is (9.10), while condition (9.20) is just (9.12). Applying Proposition 9.2.5 we get the conclusion. □

9.3 Existence of Minimal Points with Respect to Cones

To begin with, let Y be a real vector space and $\emptyset \neq C \subseteq Y$ a convex cone. As usual, with C we associate the reflexive and transitive relation

$$\leq_C := t := \{(x,y) \in Y \times Y \mid y - x \in C\}. \tag{9.22}$$

Taking

$$L := C \cap (-C),$$

the **lineality space** of C, the equivalence relation $\rho := t \cap t^{-1}$ is $\{(x, y) \in Y \times Y \mid y - x \in L\}$. So we get

$$t_N = \{(x, y) \in Y \times Y \mid y - x \in C \setminus L\}. \tag{9.23}$$

Using Proposition 9.1.1 we obtain that

$$C + (C \setminus L) = C \setminus L, \qquad (C \setminus L) + (C \setminus L) \subseteq C \setminus L;$$

the above formulae were obtained by Luc in [401, 402]. So,

$$t_{NR} = \{(x, y) \in Y \times Y \mid y - x \in (C \setminus L) \cup \{0\}\}.$$

It follows that $(C \setminus L) \cup \{0\}$ is a pointed convex cone. Note that for $x \in Y$ and $A \subseteq Y$ one has $A_C^+(x) := A_{\leq_C}^+(x) = A \cap (x + C)$ and $A_C^-(x) = A \cap (x - C)$. Therefore the upper and lower sections of Y (w.r.t. C) are closed if and only if C is closed. Similarly, for $E \subseteq Y$, $A_C^+(E) := A \cap (E + C)$ and $A_C^-(E) := A \cap (E - C)$.

In accordance with the notions introduced in Sect. 9.1, the net $(x_i)_{i \in I} \subseteq Y$ is **[strictly] C-increasing** if $x_j - x_i \in C$ $[x_j - x_i \in (C \setminus L)]$ for all $i, j \in I, i \preceq j$, $i \neq j$; $(x_i)_{i \in I}$ is **[strictly] C-decreasing** if $(-x_i)_{i \in I}$ is [strictly] C-increasing. Of course, the set $A \subseteq Y$ is **C-upper (lower) bounded** if $A \subseteq x_0 - C$ $(A \subseteq x_0 + C)$ for some $x_0 \in Y$. Moreover, the set $\mathrm{Min}(A, \leq_C)$ will be denoted by $\mathrm{Min}(A; C)$. An element $y \in \mathrm{Min}(A; C)$ is also called an **efficient point** of A (w.r.t. C).

It is obvious that for $\emptyset \neq A \subseteq Y$ and $C = Y$ or $C = \{0\}$ we have that $\mathrm{Min}(A; C) = A$. Taking into account this fact, in the sequel we shall suppose that C is proper, i.e., $\{0\} \neq C \neq Y$.

To Proposition 9.2.1 and Corollary 9.2.2 corresponds the following result.

Proposition 9.3.1. *Let $\emptyset \neq A \subseteq Y$. Suppose that one of the following conditions holds:*

(i) *every nonempty set $E \subseteq A$ such that $E - E \subseteq C \cup (-C)$ is C-lower bounded in A;*
(ii) *every chain in A w.r.t. $\leq_C$ is C-lower bounded in A;*
(iii) *every well-ordered subset of A w.r.t. $\leq_{(-C)}$ is C-lower bounded in A;*
(iv) *every C–decreasing net of A is C-lower bounded in A.*

Then A has the domination property. In particular, $\mathrm{Min}(A; C) \neq \emptyset$.

In the following proposition we gather some properties of efficient sets.

Proposition 9.3.2. *Let $C, K \subseteq Y$ be convex cones such that $C \subseteq K$, $x \in Y$, and $\emptyset \neq A, E \subseteq Y$.*

(i) $\mathrm{Min}(A_K^-(E); C) = A_K^-(E) \cap \mathrm{Min}(A; C)$. *In particular, we have that* $\mathrm{Min}(A_C^-(x); C) = A_C^-(x) \cap \mathrm{Min}(A; C)$.

(ii) *If* $A \subseteq E \subseteq A + K$ *and*

$$K \cap (-C) \subseteq C, \tag{9.24}$$

then

$$\operatorname{Min}(E; K) \subseteq \operatorname{Min}(A; C) + (K \cap (-K)).$$

In particular, if $\operatorname{Min}(E; K) \neq \emptyset$, *then* $\operatorname{Min}(A; C) \neq \emptyset$.

(iii) *If (9.24) holds, then* $\operatorname{Min}(A; K) \subseteq \operatorname{Min}(A; C)$.

(iv) *If* $A \subseteq E \subseteq A + C$, *then* $\operatorname{Min}(A; K) \subseteq \operatorname{Min}(E; K)$; *moreover, if* $K \cap (-C) = \{0\}$, *then* $\operatorname{Min}(A; K) = \operatorname{Min}(E; K)$.

(v) *Suppose that (9.24) and*

$$K + (C \setminus L) \subseteq C \tag{9.25}$$

hold. If $A \subseteq \operatorname{Min}(A; K) + K$, *then* $A \subseteq \operatorname{Min}(A; C) + C$ *(i.e., if* A *has (DP) w.r.t.* K, *then* A *has (DP) with respect to* C*).*

(vi) *Suppose that* $K \cap (-C) = \{0\}$ *and* $A \subseteq \operatorname{Min}(A; K) + C$. *Then* $\operatorname{Min}(A; C) = \operatorname{Min}(A; K)$.

In the sequel we suppose that Y is a real Hausdorff topological vector space (H.t.v.s. for short), its topology being denoted by τ.

Applying the results from Sect. 9.2 we obtain several existence theorems for efficient points w.r.t. cones. Before stating them, let us recall or introduce some notions or, more exactly, some possible properties of the cone C.

So, C is **(sequentially Daniell) Daniell** if every C-upper bounded and C-increasing (sequence) net in Y has a least upper bound (or supremum) and converges to it. Because we suppose that Y is Hausdorff, every (sequentially) Daniel cone is pointed.

Other similar conditions are:

(P1) Every C-increasing and C-upper bounded net $(x_i)_{i \in I} \subseteq C$ is convergent to an element of C.

(P2) Every C-increasing and C-upper bounded net $(x_i)_{i \in I} \subseteq Y$ is Cauchy.

(P3) Every C-increasing and τ-bounded net$(x_i)_{i \in I} \subseteq C$ is convergent to an element of C.

(P4) Every C-increasing and τ-bounded net $(x_i)_{i \in I} \subseteq Y$ is Cauchy.

(P5) Every C-increasing and τ-bounded net $(x_i)_{i \in I} \subseteq Y$ which is contained in a complete set is convergent.

The sequential variants of (P1)–(P4) are:

(SP1) Every C-increasing and C-upper bounded sequence $(x_n)_{n \in \mathbb{N}} \subseteq C$ is convergent to an element of C.

(SP2) Every C-increasing and C-upper bounded sequence $(x_n)_{n \in \mathbb{N}} \subseteq Y$ is Cauchy.

(SP3) Every C-increasing and τ-bounded sequence $(x_n)_{n\in\mathbb{N}} \subseteq C$ is convergent to an element of C.

(SP4) Every C-increasing and τ-bounded sequence $(x_n)_{n\in\mathbb{N}} \subseteq Y$ is Cauchy.

If one of the above conditions holds, then, necessarily, C is pointed.

Note that an equivalent formulation for (P1) [and similarly for (P3), (SP1), and (SP3)] is:

(P1′) Every C-increasing and C-upper bounded net $(x_i)_{i\in I} \subseteq Y$ is convergent to an element of Y which is a C-upper bound for $\{x_i \mid i \in I\}$.

Of course, (P1′) $\Rightarrow$ (P1). For the converse implication consider a C-increasing and C-upper bounded net $(x_i)_{i\in I} \subseteq Y$. Fix $k \in I$. Then the net $(y_i)_{k\preceq i}$, with $y_i := x_i - x_k \in C$, is C-increasing and C-upper bounded. By (P1) there exists $y \in C$ such that $y_i \to y$. It follows that $x_i \to x = x_k + y$, and so $x_k \leq_C x$. Since $k \in I$ is arbitrary, (P1′) holds.

Note that when Y is a normed space, Krasnosel'skij [339] says that a convex cone C satisfying (SP2) is *regular*, while a convex cone C satisfying (SP4) is *completely regular*. Isac [280] says that C is *completely regular* if every C-increasing and τ-bounded net of C is convergent. Of course, if Y is **quasi-complete** (i.e. every closed and bounded subset of Y is complete), then C is completely regular in Isac's sense if and only if C has property (P4). Németh [440] says that a convex cone C satisfying (P1) [(SP1)] is *regular [sequentially regular]*; moreover, he says that C is *fully regular* if C satisfies (P3). Condition (P5) was introduced by Ha [224] under the name of property (**).

We have the following scheme of implications, where (D) means Daniell, while (sD) means sequentially Daniell.

$$\begin{array}{ccccccccccc}
(D) & \overset{\leftharpoonup}{\Rightarrow} & (P1) & \overset{\leftarrow}{\Rightarrow} & (P2) & & (P3) & \overset{\leftarrow}{\Rightarrow} & (P4) & \Rightarrow & (P5) \\
\Downarrow\uparrow & & \Downarrow\uparrow & & \Updownarrow & & \Downarrow\uparrow & & \Updownarrow & & \downarrow \\
(sD) & \overset{\leftharpoonup}{\Rightarrow} & (SP1) & \overset{\leftarrow}{\Rightarrow} & (SP2) & & (SP3) & \overset{\leftarrow}{\Rightarrow} & (SP4) & & (P3).
\end{array}$$

The implications marked by $\Rightarrow$, $\Downarrow$, and $\Updownarrow$ always hold; those marked by $\uparrow$ are valid when C is complete or when $0 \in Y$ has a countable basis of neighborhoods; the implications $\downarrow$ and $\leftarrow$ are valid if C is complete; those marked $\leftharpoonup$ are valid for C closed. The nontrivial implications are stated in the next proposition.

Moreover, the implications $\rightarrow$ below hold if C is normal, while the implications $\leftarrow$ hold if C is **boundedly order complete**, i.e., every C-increasing and τ-bounded net of C has a supremum:

$$(P3) \leftrightarrows (P1),\ (P4) \leftrightarrows (P2),\ \ (SP3) \leftrightarrows (SP1) \text{ and } (SP4) \leftrightarrows (SP2). \tag{9.26}$$

The proof of the following result uses an idea from [440]. Note that Borwein (see Proposition 2.7 (ii) of [61]) showed that C is Daniell iff C is sequentially Daniell when "C admits a strict monotone functional."

Proposition 9.3.3. *Let $C \subseteq Y$ be a convex cone. Then*

(i) $(P2) \Leftrightarrow (SP2)$ *and* $(P4) \Leftrightarrow (SP4)$.
(ii) *If C is complete, or $0 \in Y$ has a countable basis of neighborhoods, then* $(P1) \Leftrightarrow (SP1)$, $(P3) \Leftrightarrow (SP3)$, *and C is Daniell if and only if C is sequentially Daniell.*

Before stating the next result we introduce other two notions. We say that the subset A of Y is **C-complete** (resp. **sequentially C-complete**) if every Cauchy C-increasing net (resp. Cauchy C-increasing sequence) is convergent to an element of C.

Proposition 9.3.4. *Let $A \subseteq Y$ be a nonempty set. Assume that one of the following conditions holds:*

(i) *C satisfies (P1), while A is closed and C-lower bounded;*
(ii) *C is closed and satisfies (SP2), while A is C-complete and C-lower bounded;*
(iii) *C satisfies (P3), while A is closed and τ-bounded;*
(iv) *C is closed and satisfies (SP4), while A is C-complete and τ-bounded.*

Then A has the domination property. In particular, $\mathrm{Min}(A; C) \neq \emptyset$.

When every well-ordered subset W of Y w.r.t. $\leq_{(-C)}$ (see page 351) is at most countable (i.e., $(Y, \leq_{(-C)})$ is countable orderable in the sense of Gajek and Zagrodny [193]) the closedness and completeness in the preceding result can be taken in the weaker sequential sense.

Proposition 9.3.5. *Let $A \subseteq Y$ be a nonempty set. Assume that every well-ordered subset W of Y w.r.t. $\leq_{(-C)}$ is at most countable and one of the following conditions holds:*

(i) *C satisfies (SP1), while A is sequentially closed and C-lower bounded;*
(ii) *C is closed and satisfies (SP2), while A is sequentially C-complete and C-lower bounded;*
(iii) *C satisfies (SP3), while A is sequentially closed and τ-bounded;*
(iv) *C is closed and satisfies (SP4), while A is sequentially C-complete and τ-bounded.*

Then A has the domination property. In particular, $\mathrm{Min}(A; C) \neq \emptyset$.

The next result is due to Ha [225].

Proposition 9.3.6. *Assume that* $\mathrm{cl}\, C$ *satisfies (P5) and $A \subseteq Y$ is nonempty, complete, and τ-bounded. Then* $\mathrm{Min}(A; C) \neq \emptyset$.

Proposition 9.3.7. *Assume that C is closed and $A \subseteq Y$ is nonempty and compact. Then A has the domination property. In particular,* $\mathrm{Min}(A; C) \neq \emptyset$.

Corollary 9.3.8. *Let $\emptyset \neq A \subseteq Y$ and assume that*

$$\mathrm{cl}\, C \cap (-C) \subseteq C. \tag{9.27}$$

If there exists $E \subseteq Y$ such that $A^-_C(E)$ or $A^-_{\mathrm{cl}\,C}(E)$ is nonempty and compact (even weakly compact if Y is locally convex), then $\mathrm{Min}(A; C) \neq \emptyset$.

Proposition 9.3.9. *Let $C, K \subseteq Y$ be convex cones and $\emptyset \neq A \subseteq Y$. Assume that $K + (C \setminus L) \subseteq C$ and*

$$\forall\, (y_i)_{i\in I} \subseteq A \text{ strictly } C\text{-decreasing} \;:\; A \cap \Big(\bigcap_{i\in I}(y_i - K)\Big) \neq \emptyset \tag{9.28}$$

holds. Then A has the domination property. In particular, $\mathrm{Min}(A; C) \neq \emptyset$.

Note that in locally convex spaces, every result stated above has a weak version when the topology τ is replaced by the weak topology on Y.

9.4 Types of Convex Cones and Compactness with Respect to Cones

The following result is stated in Jameson [303, Theorem 3.8.7].

Theorem 9.4.1. *Let Y be an H.t.v.s. and C be a nontrivial well-based convex cone. Then condition (P4) holds.*

Proposition 9.4.2. *Let $(Y, \mathscr{P})$ be an H.l.c.s. If C is supernormal, then C satisfies condition (P4), and therefore (P2).*

Another type of cone is the (π)-cone. Suppose that Y is a l.c.s.; the cone C is a (π)-**cone** if there exists $x^* \in C^+$ such that $\{x \in C \mid \langle x, x^*\rangle \leq 1\}$ is relatively weakly compact (see [542, Definition 2.1] and [87, Definition 4.1]). Sterna-Karwat showed in [542, Proposition 2.1] that

$$C \text{ is } (\pi)\text{-cone} \Leftrightarrow \mathrm{cl}\, C \text{ has a weakly compact base.}$$

If the cone C has a bounded base Θ, then C is supernormal; furthermore, if Θ is complete, it follows easily that C is complete, too. This discussion shows that a cone with compact base is Daniell, supernormal, and complete. In particular, a closed (π)-cone is Daniell and complete w.r.t. the weak topology.

Luc in [401, Definition 2.3] and [402, Definition I.1.1] says that C is **correct** if (9.25) holds for $K = \mathrm{cl}\, C$. Every domination cone in the sense of Henig [241, p. 112] is correct.

Concerning cone compactness, recall the following notions.

Hartley [239, p. 214] says that A is C**-compact** if $A^-_{\mathrm{cl}\,C}(y)$ is compact for every $y \in A$.

Corley in Definition 2.5 of [110] says that $A \subseteq Y$ is C**-semicompact** if (9.8) holds for t equal to $\leq_{\mathrm{cl}\,C}$. As a generalization of this notion, Luc ([401,

Definition 2.1] and [402, Definition II.3.2]) says that A is *C-complete* 'if (9.28) holds for $K = \operatorname{cl} C$.

Postolică [479] and Isac [281] say that A is *C-bounded* if there exists a τ-bounded set $A_0 \subseteq A$ such that $A \subseteq A_0 + C$, A is *C-closed* if $A + C$ is closed, and A is *C-semicompact* if A is C-bounded and C-closed; in fact, these notions are also used in [280, Definition 3], but with A_0 a singleton.

Dedieu [129] says that A is **asymptotically compact** (a.c. for short) if there exist $\gamma > 0$ and U a neighborhood of $0 \in Y$ such that $([0, \gamma]A) \cap U$ is relatively compact; note that A is a.c. iff cl A is a.c. (see [612, Proposition 2.2(i)]). Of course, every subset of $\mathbb{R}^m$ is asymptotically compact. Several properties of a.c. sets can be found in [612, Proposition 2.2].

The **asymptotic cone** of the nonempty set $A \subseteq Y$ is

$$A_\infty := \{x \in Y \mid \exists\, (t_i)_{i \in I} \subseteq (0, \infty),\ t_i \to 0,\ \exists\, (a_i)_{i \in I} \subseteq A\ :\ t_i a_i \to x\};$$

if Y is a normed space, in particular, for $Y = \mathbb{R}^m$, one can use sequences instead of nets. Note that $A_\infty = (\operatorname{cl} A)_\infty$. If A is closed and convex, then A_∞ is given by the known formula from convex analysis $A_\infty = \bigcap_{t>0} t(A - a)$ for some fixed $a \in A$. The importance of this notion in our context is shown by the following result.

Proposition 9.4.3. *Let $C \subseteq Y$ be a closed convex cone and $\emptyset \neq A \subseteq Y$ closed.*

(i) *If A is asymptotically compact and $A_\infty \cap (-C) = \{0\}$, then $A \cap (x - C)$ is compact for every $x \in Y$ (hence A is C-compact).*
(ii) *Suppose that there exists a compact set $Q \subseteq Y$ such that $0 \notin Q$, $C = [0, \infty)\cdot Q$ and $A_\infty \cap (-C) = \{0\}$. Then $A \cap (x - C)$ is compact for every $x \in Y$. Moreover, if C is pointed, then $A + C$ is closed and $(A + C) \cap (x - C)$ is compact for every $x \in Y$.*
Conversely, if A is convex and $A \cap (x - C)$ is nonempty and compact for some $x \in Y$, then $A_\infty \cap (-C) = \{0\}$.

Corollary 9.4.4. *Let $C \subseteq Y$ be a closed convex cone and A a nonempty closed subset of Y. Assume that $A_\infty \cap (-C) = \{0\}$. If A is asymptotically compact or $C = [0, \infty)\cdot Q$ for some compact set $Q \subseteq Y$ with $0 \notin Q$, then A has the domination property.*

Using the results in the previous section and the present one, one can deduce the most part of existence theorems spread in the literature.

For example, using Proposition 9.3.1 one gets the results obtained in Corley [110, Theorem 3.1] and Chew [95]; using Propositions 9.3.4 and 9.3.5 one gets the results obtained in Jameson [303, Corollary 3.8.10], Penot [460, Theorem 3.3], Cesari and Suryanarayana [87, Lemma 4.1], Borwein [68, Theorem 1], Isac [280, Theorem 2], Postolică [480, Theorem 3.2, Corollary 3.2.1], Attouch and Riahi [13, Theorem 2.5], Ha [224, 225], Ng and Zheng [441, Theorem 3.1]; using Proposition 9.3.7 one gets the results obtained in Bitran and Magnanti [51, Proposition 3.1], Nieuwenhuis [442, Theorem I-14], Henig [241, Theorem 2.1], Borwein [68, Theorem 1], Penot and Sterna-Karwat [472, Remark 3.3], Jahn [291, Theorem 2.3(b), Theorem 2.6(b)], Tanaka [561, Lemma 2.4]; using Proposition 9.3.9 one gets the results obtained in Luc [401, Theorem 2.6, Corollary 2.12], [402, Theorem II.3.3], Malivert [411, Theorem 3.5].

Note that there are other results concerning existence of efficient points w.r.t. cones that can not be deduced from the existence results in Sect. 9.2.

9.5 Existence of Optimal Solutions for Vector and Set Optimization Problems

In this section (X, τ) is a Hausdorff topological space and t is a reflexive transitive relation on Y. Consider first the problem

$$\text{Min}\, f(x) \quad \text{s.t.} \quad x \in X, \tag{P}$$

where $f : X \to Y$ is a function. A solution of problem (P) is an element $x_0 \in X$ such that $f(x_0) \in \text{Min}(f(X), t)$.

We say that f is t-**monotonically lower semicontinuous** (m.l.s.c. for short) at $x_0 \in X$ if for every net $(x_i)_{i \in I} \subseteq X$ such that $x_i \to x_0$ and $(f(x_i))_{i \in I}$ is t-decreasing one has that $(f(x_0), f(x_i)) \in t$ for every $i \in I$; f is t-**monotonically lower semicontinuous** if f is t-m.l.s.c. at any $x \in X$.

The next theorem subsumes the existence results established in Theorems 5.1, 5.2 and 5.8 of Flores-Bazán et al. [189].

Theorem 9.5.1. *Let X be compact. Assume that one of the following conditions holds:*

(i) *f is t-m.l.s.c.;*
(ii) *$f^{-1}\left(Y_t^-(y)\right)$ is closed for every $y \in f(X)$;*
(iii) *Y is a topological space, $f^{-1}\left(\text{cl}\, Y_t^-(y)\right)$ is closed for every $y \in f(X)$ and (9.19) holds.*

Then $f(X)$ has the domination property. In particular, problem (P) has optimal solutions.

Proof. Set $A := f(X) \subseteq Y$.

(i) Let us apply Corollary 9.2.2 (i). For this take $(y_i)_{i \in I} \subseteq A$ be t-decreasing. Hence, there exists $(x_i)_{i \in I} \subseteq X$ such that $y_i = f(x_i)$ for every $i \in I$. Because X is compact, taking into account [417, Definition 2.1.26, Proposition 2.1.37], there exists J a directed set and $\psi : J \to I$ an increasing function with $\psi(J)$ cofinal such that $(x_{\psi(j)})_{j \in J}$ converges to $x \in X$. Since $\left(f(x_{\psi(j)})\right)_{j \in J}$ is t-decreasing and f is t-m.l.s.c., it follows that $(f(x), y_{\psi(j)}) \in t$ for every $j \in J$. Since $\psi(J)$ is cofinal and $(y_i)_{i \in I}$ is t-decreasing, we get $(f(x), y_i) \in t$ for every $i \in I$. Using Corollary 9.2.2 (i) we get the conclusion.

(ii) $\Rightarrow$ (i) (However, a direct proof of (ii), using Corollary 9.2.3, is possible.) Indeed, take $(x_i)_{i \in I} \subseteq X$ such that $x_i \to x$ and $(f(x_i))_{i \in I}$ is t-decreasing. Setting $y_i := f(x_i) \in f(X)$, we have that $x_j \in f^{-1}\left(Y_t^-(y_i)\right)$ for all $i, j \in I$ with $j \geq i$.

Since $f^{-1}\left(Y_t^-(y_i)\right) \supseteq (x_j)_{j\geq i} \to x$, it follows that $x \in \operatorname{cl} f^{-1}\left(Y_t^-(y_i)\right) = f^{-1}\left(Y_t^-(y_i)\right)$, and so $f(x) \in Y_t^-(y_i)$ for every $i \in I$.

(iii) In order to apply Corollary 9.2.9, let us consider $D \subseteq A$ a totally ordered set and assume that $A \cap \left(\bigcap_{y\in D} Y_s^-(y)\right) = \emptyset$, where s is defined by (9.21). Then $A \subseteq \bigcup_{y\in D}\left(Y \setminus Y_s^-(y)\right)$. It follows that

$$X \subseteq f^{-1}\left(\bigcup_{y\in D}\left(Y \setminus Y_s^-(y)\right)\right) = \bigcup_{y\in D} f^{-1}(Y \setminus Y_s^-(y)) \ (\subseteq X).$$

Because X is compact and $f^{-1}(Y \setminus Y_s^-(y)) = X \setminus f^{-1}\left(Y_s^-(y)\right)$ is open, there exists $D_0 \in \mathscr{F}(D)$ such that $X = \bigcup_{y\in D_0} f^{-1}(Y \setminus Y_s^-(y))$, and so

$$A = f(X) = \bigcup_{y\in D_0} f\left(f^{-1}(Y \setminus Y_s^-(y))\right) \subseteq \bigcup_{y\in D_0}\left(Y \setminus Y_s^-(y)\right).$$

The set D_0 being finite and totally ordered w.r.t. t, there exists $y_0 \in D_0$ $(\subseteq A)$ such that $Y \setminus Y_s^-(y_0) = \bigcup_{y\in D_0}\left(Y \setminus Y_s^-(y)\right)$, and so $A \subseteq Y \setminus Y_s^-(y_0)$. It follows that $y_0 \in A \cap \operatorname{cl} Y_t^-(y_0) = \emptyset$, a contradiction. Hence, by Corollary 9.2.9, $f(X)$ has the domination property. □

When Y is a topological vector space ordered by the proper convex cone C, taking $t :=\leq_C$, as seen in Sect. 9.3, $Y_t^-(y) = y - C$. In such a situation the fact that $f^{-1}\left(Y_t^-(y)\right)$ is closed for every $y \in Y$ means that f is C-lower semicontinuous. So, in this special case Theorem 9.5.1 becomes the following result.

Corollary 9.5.2. *Let X be compact. Assume that Y is a topological vector space ordered by the proper closed convex cone C. If either* (a) *f is C-lower semicontinuous or* (b) *f is $\operatorname{cl} C$-lower semicontinuous and $\operatorname{cl} C + (C \setminus (-C)) \subseteq C$, then $f(X)$ has the domination property. In particular, problem (P) has optimal solutions.*

Consider now the set optimization problem

$$\operatorname{Min} F(x) \quad \text{s.t.} \quad x \in X, \tag{VP}$$

where $F : X \rightrightarrows Y$ is a set-valued map; one says that $x_0 \in X$ is a solution of problem (*VP*) if $F(x_0) \cap \operatorname{Min}(F(X), t) \neq \emptyset$.

The next result is obtained in Flores-Bazán et al. [189, Theorem 7.2].

Theorem 9.5.3. *Assume that Y is a topological space and there exists $y_0 \in F(X)$ such that $F^{-1}(Y_t^-(y))$ is compact for every $y \in Y_t^-(y_0)$. If $F(x) \cap Y_t^-(y_0)$ verifies condition (9.9) for all $x \in X$, then the problem (VP) has optimal solutions.*

Proof. It is sufficient to show that $\operatorname{Min}(A, t)$ is nonempty, where $A := F(X) \cap Y_t^-(y_0) = (F(X))_t^-(y_0) \ni y_0$. For this let us prove that condition (9.5) holds. So, consider $(y_i)_{i\in I} \subseteq A$ a t-decreasing net and assume that $A \cap \left(\bigcap_{i\in I} Y_t^-(y_i)\right) = \emptyset$. For $i \in I$ set

$$U_i := \{x \in X \mid F(x) \cap Y_t^-(y_i) = \emptyset\} = X \setminus F^{-1}\left(Y_t^-(y_i)\right).$$

Clearly, $F^{-1}(Y_t^-(y_i)) \neq \emptyset$ and $U_i \supseteq U_{i'}$ for all $i, i' \in I$ with $i \geq i'$. Because $y_i \in Y_t^-(y_0)$, we have that U_i is open for every $i \in I$. Moreover,

$$F^{-1}(Y_t^-(y_0)) \subseteq \bigcup_{i\in I} U_i. \tag{9.29}$$

Indeed, if $x \in F^{-1}(Y_t^-(y_0))$ then $F(x) \cap Y_t^-(y_0) \neq \emptyset$. Because $F(x) \cap Y_t^-(y_0) \subseteq F(X) \cap Y_t^-(y_0) = A$ we obtain that $F(x) \cap Y_t^-(y_0) \subseteq \bigcup_{i\in I}(Y \setminus Y_t^-(y_i))$. Using condition (9.9) we get $I_0 \in \mathscr{F}(I)$ such that $F(x) \cap Y_t^-(y_0) \subseteq \bigcup_{i\in I_0}(Y \setminus Y_t^-(y_i))$. The subset I_0 of I being finite, there exists $i_x \in I$ such that $i_x \geq i$ for every $i \in I_0$. It follows that $Y \setminus Y_t^-(y_i) \supseteq Y \setminus Y_t^-(y_{i_x})$ for $i \in I_0$, and so $F(x) \cap Y_t^-(y_0) \subseteq Y \setminus Y_t^-(y_{i_x})$. Moreover, taking into account that $Y \setminus Y_t^-(y_0) \subseteq Y \setminus Y_t^-(y_{i_x})$, we get

$$F(x) = \left[F(x) \cap Y_t^-(y_0)\right] \cup \left[F(x) \cap (Y \setminus Y_t^-(y_0))\right] \subseteq Y \setminus Y_t^-(y_{i_x}),$$

whence $x \in U_{i_x}$. Therefore, (9.29) holds. Because $F^{-1}(Y_t^-(y_0))$ is compact, there exists $I_0 \in \mathscr{F}(I)$ such that $F^{-1}(Y_t^-(y_0)) \subseteq \bigcup_{i\in I_0} U_i$. Taking $i_0 \in I$ such that $i_0 \geq i$ for every $i \in I_0$ we obtain that $F^{-1}(Y_t^-(y_0)) \subseteq U_{i_0}$, and so $\emptyset = F^{-1}(Y_t^-(y_0)) \cap F^{-1}(Y_t^-(y_{i_0})) = F^{-1}(Y_t^-(y_{i_0}))$, a contradiction. Applying Corollary 9.2.2 we obtain that $\operatorname{Min}(A, t) \neq \emptyset$. □

An immediate consequence of the preceding result is the following (see [189, Corollary 7.4]).

Corollary 9.5.4. *Assume that X is compact, $F^{-1}(Y_t^-(y))$ is closed for every $y \in Y$ and $F(x) \cap Y_t^-(y)$ verifies condition (9.9) for all $x \in X$ and $y \in F(X)$. Then $F(X)$ has the domination property.*

In the case in which Y is a topological vector space ordered by a proper convex cone C we have the following result.

Theorem 9.5.5. *Let* dom F *be compact and Y be a topological space. Assume that F is upper continuous and $F(x)$ verifies condition (9.9) with t replaced by s defined by (9.21) for all $x \in$* dom F. *If (9.19) holds, then $F(X)$ has the domination property.*

Proof. W.l.o.g. we suppose that $X = \operatorname{dom} F$. In order to apply Proposition 9.2.5 let us show that $A := F(X)$ verifies condition (9.9) with t replaced by s defined by (9.21). As observed in the proof of Corollary 9.2.9, condition (9.19) is (9.10) and condition (9.20) is (9.12) in this case. Let $E \subseteq Y$ with $A \subseteq \bigcup_{y\in E}(Y \setminus Y_s^-(y))$. Take $x \in X$. Because $F(x) \subseteq A$ and $F(x)$ verifies condition (9.9), there exists $E_x \in \mathscr{F}(E)$ such that $F(x) \subseteq \bigcup_{y\in E_x}(Y \setminus Y_s^-(y)) = \bigcup_{y\in E_x}(Y \setminus \operatorname{cl} Y_t^-(y)) =: D_x$. Since D_x is open and F is upper continuous at x, there exists an open neighborhood U_x of x such that $F(U_x) \subseteq D_x$. Since X is compact and $X = \cup_{x\in X} U_x$, there exists $X_0 \in \mathscr{F}(X)$ such that $X = \cup_{x\in X_0} U_x$. Taking $E_0 := \cup_{x\in X_0} E_x \in \mathscr{F}(E)$ we have that

$$A = F(X) = F(\cup_{x\in X_0} U_x) = \cup_{x\in X_0} F(U_x) \subseteq \cup_{x\in X_0} D_x = \cup_{y\in E_0}\left(Y \setminus Y_s^-(y)\right).$$

Because (9.9) with s instead of t implies (9.12), the conclusion follows. □

Note that in the case in which Y is a topological vector space ordered by a (proper) convex cone C, condition (9.19) for $t = \leq_C$ is equivalent to $\operatorname{cl} C + (C \setminus (-C)) \subseteq C$, and $Y_s^-(y) = \operatorname{cl} Y_t^-(y) = y - \operatorname{cl} C$ for every $y \in Y$; in this case, instead of asking F to be upper continuous it is sufficient to ask that F is C-upper continuous, getting so [189, Theorem 7.5, Corollary 7.6].

In the rest of this section we consider Y a topological vector space ordered by a (proper) convex cone C. On the family $\mathscr{P}_0(Y)$ of nonempty sets of Y we consider the relation $\leq^l := \leq^l_C$ and $\leq^u := \leq^u_C$ introduced by Kuroiwa [350] (see also Definition 2.6.9) in the following way:

$$A \leq^l B \text{ if } B \subseteq A + C, \quad A \leq^u B \text{ if } A \subseteq B - C.$$

Note that

$$A \leq^u_C B \Leftrightarrow (-B) \leq^l_C (-A) \Leftrightarrow B \leq^l_{-C} A. \tag{9.30}$$

It follows that for $A, B \in \mathscr{P}_0(Y)$ one has

$$A \in (\mathscr{P}_0(Y))^-_{\leq^l}(B) \Leftrightarrow [\forall b \in B : A \cap (b - C) \neq \emptyset], \tag{9.31}$$

$$A \in \mathscr{P}_0(Y) \setminus (\mathscr{P}_0(Y))^-_{\leq^l}(B) \Leftrightarrow [\exists b \in B : A \subseteq Y \setminus (b - C)] \tag{9.32}$$

and

$$A \in \mathscr{P}_0(Y) \setminus (\mathscr{P}_0(Y))^-_{\leq^u}(B) \Leftrightarrow A \cap [Y \setminus (B - C)] \neq \emptyset. \tag{9.33}$$

It is clear that $\leq^l$ and $\leq^u$ are reflexive and transitive. So, all the results stated in Sects. 9.1, 9.2 and the present one for a transitive relation t can be adapted for $\leq^l$ and $\leq^u$.

In order to translate Corollary 9.2.2 (i) to the present context for $t = \leq^l$ let us introduce some more notation. Let us fix $\mathscr{A} \subseteq \mathscr{P}_0(Y)$ and take $\mathscr{E} \subseteq \mathscr{A}$; we set

$$C_{\mathscr{E}} := C^{\mathscr{A}}_{\mathscr{E}} := \{H \subseteq H_{\mathscr{E}} \mid \forall A \in \mathscr{A}, \exists b \in H : A \subseteq Y \setminus (b - C)\},$$

where

$$H_{\mathscr{E}} := \bigcup \{E \mid E \in \mathscr{E}\}. \tag{9.34}$$

Clearly,

$$\mathscr{E}' \subseteq \mathscr{E} \subseteq \mathscr{A} \Rightarrow H_{\mathscr{E}'} \subseteq H_{\mathscr{E}} \subseteq H_{\mathscr{A}}, \quad C_{\mathscr{E}} \ni H \subseteq H' \subseteq H_{\mathscr{E}} \Rightarrow H' \in C_{\mathscr{E}}.$$

Consequently, if $\mathscr{E} \subseteq \mathscr{A}$, from (9.32) we have that

$$\mathscr{A} \subseteq \bigcup_{B \in \mathscr{E}} \left[\mathscr{P}_0(Y) \setminus (\mathscr{P}_0(Y))^-_{\leq^l}(B)\right] \Leftrightarrow H_{\mathscr{E}} \in C_{\mathscr{E}} \Leftrightarrow C_{\mathscr{E}} \neq \emptyset. \tag{9.35}$$

We have the following result stated in [250, Theorem 4.1].

Corollary 9.5.6. *Assume that* $\emptyset \neq \mathscr{A} \subseteq \mathscr{P}_0(Y)$ *verifies the condition*

$$\forall (A_i)_{i\in I} \subseteq \mathscr{A} : \ (A_i)_{i\in I} \ \leq^l \text{-decreasing} \ \Rightarrow C_{\mathscr{E}_I} = \emptyset, \tag{9.36}$$

where $\mathscr{E}_I := \{A_i \mid i \in I\} \subseteq \mathscr{A}$*. Then* $\mathscr{A}$ *has the domination property with respect to* $\leq^l$*; in particular* $\operatorname{Min}(\mathscr{A}, \leq^l) \neq \emptyset$.

Proof. Let us show that (9.5) holds for $t :=\leq^l$, that is

$$\left[\forall (A_i)_{i\in I} \subseteq \mathscr{A},\ (A_i)_{i\in I} \ \leq^l \text{-decreasing}\right],\ \exists A \in \mathscr{A},\ \forall i \in I : A \leq^l A_i. \tag{9.37}$$

For this, take a $\leq^l$-decreasing net $(A_i)_{i\in I} \subseteq \mathscr{A}$. By (9.35) we have that $H_{\mathscr{E}_I} \notin C_{\mathscr{E}_I}$, that is there exists $A \in \mathscr{A}$ such that for all $i \in I$ and $b \in A_i$ we have that $A \cap (b - C) \neq \emptyset$. This can be rewritten as $A_i \subseteq A + C$, that is $A \leq^l A_i$, for every $i \in I$. Hence (9.37) holds. Using Corollary 9.2.2 (i) we obtain that $\mathscr{A}$ has the domination property w.r.t. $\leq^l$. □

It is easy to prove that (9.37) ⇒ (9.36), too.

The transcription of (9.8) for $\mathscr{A} \subseteq \mathscr{P}_0(Y)$ using (9.35) is

$$[\forall \mathscr{E} \subseteq \mathscr{A},\ C_{\mathscr{E}} \neq \emptyset],\ \exists \mathscr{E}' \in \mathscr{F}(\mathscr{E}) : C_{\mathscr{E}'} \neq \emptyset. \tag{9.38}$$

A stronger condition is

$$\forall \mathscr{E} \subseteq \mathscr{A} : C_{\mathscr{E}} \neq \emptyset \Rightarrow C_{\mathscr{E}} \cap \mathscr{F}(H_{\mathscr{E}}) \neq \emptyset, \tag{9.39}$$

and an even stronger condition is

$$\forall H \in C_{\mathscr{A}} : C_{\mathscr{A}} \cap \mathscr{F}(H) \neq \emptyset. \tag{9.40}$$

Of course, using Corollary 9.2.3, if $\mathscr{A} \subseteq \mathscr{P}_0(Y)$ verifies (9.40) then $\mathscr{A}$ has the domination property w.r.t. $\leq^l$.

In the present context we consider the set optimization problem

$$\operatorname{Min} F(x) \quad \text{s.t.} \quad x \in X, \tag{SP}$$

where $F : X \rightrightarrows Y$ is a set-valued map with nonempty domain; one says that $x_0 \in \operatorname{dom} F$ is a $\leq^l$ (resp. $\leq^u$) solution of problem (*SP*) if $F(x_0) \in \operatorname{Min}(\mathscr{A}_F, \leq^l)$ (resp. $F(x_0) \in \operatorname{Min}(\mathscr{A}_F, \leq^u)$), where

$$\mathscr{A}_F := \{F(x) \mid x \in \operatorname{dom} F\}. \tag{9.41}$$

The next theorem is the first result established in the literature concerning the existence of solutions for problem (*SP*) w.r.t. $\preceq^l$, and is due to Kuroiwa [352, Theorem 4.1] (see also [250, Theorem 5.8]). Before stating the result let us introduce the limit superior of a net $(E_i)_{i\in I}$ of subsets of the topological space (X, τ) as being the set of those $x \in X$ such that there exist $X \supseteq (x_j)_{j\in J} \to x$ and $\psi : J \to I$ increasing and cofinal (that is $\psi(J)$ is cofinal in I) such that $x_j \in E_{\psi(j)}$ for every $j \in J$. Let us observe that

$$\limsup_{j\in J} E_{\psi(j)} \subseteq \limsup_{i\in I} E_i \tag{9.42}$$

for each directed set J and each increasing cofinal mapping $\psi : J \to I$; moreover, one has equality in (9.42) if $(E_i)_{i\in I}$ is increasing. Observe also that $E_{i_0} \subseteq \limsup_{i\in I} E_i$ for each $i_0 \in I$ when $(E_i)_{i\in I}$ is increasing.

Theorem 9.5.7. *Let $X = \operatorname{dom} F$ be compact. Assume that $F(x) \preceq^l \limsup\big(F(x_i) + C\big)$ whenever $(x_i)_{i\in I} \subseteq X$ is such that $x_i \to x \in X$ and $(F(x_i))_{i\in I}$ is $\preceq^l$-decreasing. Then $\mathscr{A}_F := \{F(x) \mid x \in \operatorname{dom} F\}$ has the domination property with respect to $\preceq^l$. In particular, (SP) has optimal solutions with respect to $\preceq^l$.*

Proof. Let us consider $\mathscr{A} := \{F(x) + C \mid x \in X\}$. Observe that $\mathscr{A}$ and $\mathscr{A}_F$ have the same minimal points w.r.t. $\preceq^l$. We apply Corollary 9.5.6 to $\mathscr{A}$. For this take a $\preceq^l$-decreasing net $(A_i)_{i\in I} \subseteq \mathscr{A}$, that is $A_i \subseteq A_{i'}$ for all $i, i' \in I$ with $i' \geq i$. Then $A_i = F(x_i) + C$ for some $(x_i)_{i\in I} \subseteq X$. It follows that $(F(x_i))_{i\in I}$ is $\preceq^l$-decreasing. Because X is compact, taking into account [417, Definition 2.1.26, Proposition 2.1.37], there exists J a directed set and an increasing cofinal mapping $\psi : J \to I$ such that $(x_{\psi(j)})_{j\in J}$ converges to $x \in X$. Since ψ is increasing, $(F(x_{\psi(j)}))_{j\in J}$ is $\preceq^l$-decreasing. Using our assumption, $F(x) \preceq^l \limsup_{j\in J}\big(F(x_{\psi(j)})+C\big) = \limsup_{j\in J} A_{\psi(j)}$. Since $(A_{\psi(j)})_{j\in J}$ is increasing, we have that $A_{\psi(j)} \subseteq \limsup_{j\in J} A_{\psi(j)} \subseteq A := F(x) + C$ for every $j \in J$. Because (A_i) is increasing and ψ is cofinal we obtain that $A_i \subseteq A$, that is $A \preceq^l A_i$, for every $i \in I$. Therefore, (9.37) holds, and so (9.36) holds, too. Applying Corollary 9.5.6 we obtain that $\mathscr{A}$ has the domination property. □

Using the considerations above we get the next result established essentially by Alonso and Rodríguez-Marín in [3] (see also [189, Theorem 6.4]).

Theorem 9.5.8. *Assume that $X = \operatorname{dom} F$ is compact and F is C-upper continuous. If C is closed, then $\mathscr{A}_F := \{F(x) \mid x \in \operatorname{dom} F\}$ has the domination property with respect to $\preceq^l$. In particular, (SP) has optimal solutions with respect to $\preceq^l$.*

Proof. Let us show that condition (9.40) holds. Take $H \in C_{\mathscr{A}_F}$. For $x \in X$ there exists $b_x \in H$ such that $F(x) \subseteq Y \setminus (b_x - C)$. Because C is closed and F is C-u.c. at x, there exists an open neighborhood U_x of $x \in X$ such that $F(U_x) \subseteq [Y \setminus (b_x - C)] + C = Y \setminus (b_x - C)$. Because $X = \cup_{x\in X} U_x$, there exists $X_0 \in \mathscr{F}(X)$ such that $X = \cup_{x\in X_0} U_x$. Set $H' := \{b_x \mid x \in X_0\} \in \mathscr{F}(H)$. For $x' \in X$ we have that $x' \in U_x$ for some $x \in X$, and so $F(x') \subseteq Y \setminus (b_x - C)$; hence $H' \in C_{\mathscr{A}_F}$. Therefore, condition (9.40) holds, and so (9.38) holds, too. The conclusion follows using Corollary 9.2.3. □

We present now an existence result for problem (*SP*) w.r.t. $\leq^u$. First observe that using (9.33) we have that

$$\mathscr{A} \subseteq \bigcup_{B\in\mathscr{E}}\big[\mathscr{P}_0(Y) \setminus (\mathscr{P}_0(Y))^-_{\leq^u}(B)\big] \Leftrightarrow \forall A \in \mathscr{A},\ \exists B \in \mathscr{E} : A \cap [Y \setminus (B - C)] \neq \emptyset. \tag{9.43}$$

The next result is practically established in Alonso–Rodríguez-Marín [3, Proposition 22].

Theorem 9.5.9. *Assume that* $X = \operatorname{dom} F$ *is compact and* F *is* C*-lower continuous with* $F(x) - C$ *closed for every* $x \in X$. *Then* $\mathscr{A}_F$ *has the domination property with respect to* $\leq^u$. *In particular, (SP) has optimal solutions with respect to* $\leq^u$.

Proof. Let $\mathscr{E} \subseteq \mathscr{A}_F$ be such that $\mathscr{A} \subseteq \bigcup_{B\in\mathscr{E}}\big[\mathscr{P}_0(Y) \setminus (\mathscr{P}_0(Y))^-_{\leq^u}(B)\big]$; hence $\mathscr{E} = \{F(x) \mid x \in E\}$ for some $E \subseteq X$. Take $x \in X$; using (9.43), there exists $u_x \in E$ such that $F(x) \cap [Y \setminus (F(u_x) - C)] \neq \emptyset$. Because $F(u_x) - C$ is closed and F is C-l.c., there exists an open neighborhood U_x of x such that

$$F(x') \cap \big[\big(Y \setminus (F(u_x) - C)\big) + C\big] = F(x') \cap [Y \setminus (F(u_x) - C)] \neq \emptyset \quad \forall x' \in U_x.$$

Since X is compact and $X = \cup_{x\in X} U_x$, there exists $X_0 \in \mathscr{F}(X)$ such that $X = \cup_{x\in X_0} U_x$. Taking $E_0 := \{u_x \mid x \in X_0\}$ $(\in \mathscr{F}(E))$ and $\mathscr{E}_0 := \{F(x) \mid x \in E_0\}$ $(\in \mathscr{F}(\mathscr{E}))$, and using again (9.43), we have that $\mathscr{A} \subseteq \bigcup_{B\in\mathscr{E}_0}\big[\mathscr{P}_0(Y) \setminus (\mathscr{P}_0(Y))^-_{\leq^u}(B)\big]$. The conclusion follows using Corollary 9.2.3. □

Chapter 10
Ekeland Variational Principle

In this chapter we present existence results for minimal points of subsets of the Cartesian product of a complete metric space and a topological vector space with respect to order relations determined by generalized set-valued metrics; such results are useful for deriving EVP type results for vector and scalar functions. Then we derive EVP results of Ha's type as well as an EVP result for bi-set-valued maps. We end the chapter with an application to error bounds for set-valued optimization problems.

The celebrated Ekeland variational principle (see [171]) has many equivalent formulations and generalizations. Quite rapidly after the publication of the Ekeland variational principle (EVP) in 1974 there were formulated extensions to functions $f : (X, d) \to Y$, where Y is a real (topological) vector space. A systematization of such results was done in [215] (see also [214]), where instead of a function f it was considered a subset of $X \times Y$; said differently, it was considered a set-valued map from X to Y. The common feature of these results is the presence of a certain term $d(x, x')k^0$ in the perturbed objective function, where K is the convex ordering cone in Y and $k^0 \in K \setminus \{0\}$. Very recently $d(x, x')k^0$ was replaced by $d(x, x')H$ with H a bounded convex subset of K (see [40]) or by $F(x, x') \subseteq K$, F being a so called K-metric (see [220]); in both papers one deals with functions $f : X \to Y$. Our approach is that of [558], being similar to that in [215]. The results without bibliographical references can be found in [558] or are new.

10.1 Preliminary Notions and Results

In the sequel (X, d) is a complete metric space, Y is a real topological vector space, Y^* is its topological dual, and $K \subseteq Y$ is a proper convex cone; as usual, K^+ is the positive dual cone of K and $K^{\#}$ is the quasi interior of K^+:

A.A. Khan et al., *Set-valued Optimization*, Vector Optimization,
DOI 10.1007/978-3-642-54265-7_10

$$K^+ = \{y^* \in Y^* \mid y^*(y) \geq 0 \ \forall\, y \in K\},$$
$$K^\# = \{y^* \in Y^* \mid y^*(y) > 0 \ \forall\, y \in K \setminus \{0\}\}.$$

If Y is just a real linear space we (can) endow it with the finest locally convex topology, that is, the core convex topology (see Sect. 6.3).

We say that $E \subseteq Y$ is **K^+-bounded** if $y^*(E)$ is bounded from below for every $y^* \in K^+$. It is clear that any quasi (K-lower) bounded set is K^+-bounded.

Let $F : X \times X \rightrightarrows K$ satisfy the conditions:

(F1) $0 \in F(x,x)$ for all $x \in X$,
(F2) $F(x_1,x_2) + F(x_2,x_3) \subseteq F(x_1,x_3) + K$ for all $x_1,x_2,x_3 \in X$.

Using F we introduce a quasi-order on $X \times Y$, denoted by $\preceq_F$, in the following manner:

$$(x_1,y_1) \preceq_F (x_2,y_2) \iff y_2 \in y_1 + F(x_1,x_2) + K. \tag{10.1}$$

Indeed, $\preceq_F$ is reflexive by (F1). If $(x_1,y_1) \preceq_F (x_2,y_2)$ and $(x_2,y_2) \preceq_F (x_3,y_3)$, then

$$y_2 = y_1 + v_1 + k_1, \quad y_3 = y_2 + v_2 + k_2 \tag{10.2}$$

with $v_1 \in F(x_1,x_2)$, $v_2 \in F(x_2,x_3)$ and $k_1,k_2 \in K$. By (F2) we have that $v_1 + v_2 = v_3 + k_3$ for some $v_3 \in F(x_1,x_3)$ and $k_3 \in K$, and so

$$y_3 = y_1 + v_1 + k_1 + v_2 + k_2 = y_1 + v_3 + k_1 + k_2 + k_3 \in y_1 + F(x_1,x_3) + K;$$

hence $(x_1,y_1) \preceq_F (x_3,y_3)$, and so $\preceq_F$ is transitive. Of course,

$$(x_1,y_1) \preceq_F (x_2,y_2) \Rightarrow y_1 \leq_K y_2; \tag{10.3}$$

moreover, by (F1), we have that

$$(x,y_1) \preceq_F (x,y_2) \iff y_2 \in y_1 + K \iff y_1 \leq_K y_2. \tag{10.4}$$

Besides conditions (F1) and (F2) we shall consider also the condition

(F3) there exists $z^* \in K^+$ such that

$$\forall \delta > 0 : \eta(\delta) := \inf\left\{z^*(v) \mid v \in \cup_{d(x,x') \geq \delta} F(x,x')\right\} > 0. \tag{10.5}$$

Let us set

$$F_\delta := \cup_{d(x,x') \geq \delta} F(x,x') \tag{10.6}$$

for $\delta \geq 0$; clearly, $F_\delta \subseteq F_{\delta'}$ if $0 \leq \delta' < \delta$. So, condition (F3) can be rewritten as

$$\exists z^* \in K^+, \ \forall \delta > 0 \ : \ \inf z^*(F_\delta) > 0. \tag{10.7}$$

A weaker condition is

$$\forall \delta > 0, \ \exists z^* \in K^+ \ : \ \inf z^*(F_\delta) > 0. \tag{10.8}$$

An even weaker condition is the following

$$\forall \delta > 0, \ \forall (z_n) \subseteq F_\delta, \ \exists z^* \in K^+ \ : \ \limsup z^*(z_n) > 0. \tag{10.9}$$

Clearly, if (F3) holds then $0 \notin \overline{\operatorname{conv}} F(x, x')$ for $x \neq x'$.
Condition (F3) holds also if one has

$$\forall x, x' \in X : \inf_{z \in F(x,x')} z^*(z) \geq d(x, x'). \tag{10.10}$$

In the next lemma we give an important example of set-valued map F satisfying conditions (F1) and (F2).

Lemma 10.1.1. *Let $\emptyset \neq H \subseteq K$ be a K-convex set, that is $H + K$ is convex. Consider*

$$F_H : X \times X \rightrightarrows K, \quad F_H(x, x') := d(x, x')H. \tag{10.11}$$

(i) *F_H verifies (F1) and (F2).*
(ii) *F_H verifies condition (F3) iff F_H verifies condition (10.8) iff there exists $z^* \in K^+$ such that $\inf z^*(H) > 0$; if Y is a separated locally convex space, then F_H verifies condition (F3) iff $0 \notin \operatorname{cl}(H + K)$. Moreover, F_H verifies condition (10.9) iff*

$$\forall (h_n) \subseteq H, \ \exists z^* \in K^+ \ : \ \limsup z^*(h_n) > 0. \tag{10.12}$$

Proof. (i) Clearly $F_H(x, x) = \{0\}$ for every $x \in X$. Take $x_1, x_2, x_3 \in X$ and $v_1 \in F_H(x_1, x_2)$, $v_2 \in F_H(x_2, x_3)$; hence $d(x_1, x_2) + d(x_2, x_3) = d(x_1, x_3) + \alpha$ for some $\alpha > 0$ and $v_1 = d(x_1, x_2)h_1$, $v_2 = d(x_2, x_3)h_2$ with $h_1, h_2 \in H$. Using the convexity of $H + K$ we get

$$\begin{aligned} v_1 + v_2 &\in d(x_1, x_2)H + d(x_2, x_3)H \subseteq d(x_1, x_2)(H + K) + d(x_2, x_3)(H + K) \\ &= [d(x_1, x_2) + d(x_2, x_3)]\,(H + K) = [d(x_1, x_3) + \alpha]\,(H + K) \\ &\subseteq d(x_1, x_3)H + K = F_H(x_1, x_3) + K. \end{aligned}$$

Hence (F2) is verified, too.
(ii) is obvious. □

If (10.10) holds then

$$(x_1, y_1) \preceq_F (x_2, y_2) \Rightarrow d(x_1, x_2) \leq z^*(y_2) - z^*(y_1). \tag{10.13}$$

Indeed, since $F(x_1, x_2) \subseteq K$, from (10.3) we get first that $y_1 \leq_K y_2$; then from (10.2)

$$z^*(y_2) = z^*(y_1) + z^*(v_1) + z^*(k_1) \geq z^*(y_1) + \inf_{v \in F(x_1, x_2)} z^*(v) \geq z^*(y_1) + d(x_1, x_2),$$

and so (10.13) holds. Using (10.13) we obtain that

$$[(x_1, y_1) \preceq_F (x_2, y_2),\ (x_2, y_2) \preceq_F (x_1, y_1)] \Rightarrow \left[x_1 = x_2,\ z^*(y_1) = z^*(y_2)\right]. \tag{10.14}$$

In fact (10.14) holds if F verifies conditions (F1)–(F3). Indeed for $\delta := d(x_1, x_2)$, from (10.2) we get

$$z^*(y_1) \geq z^*(y_2) + \inf z^*(F_\delta), \quad z^*(y_2) \geq z^*(y_1) + \inf z^*(F_\delta),$$

and so $-|z^*(y_1) - z^*(y_2)| \geq \inf z^*(F_\delta) \geq 0$.

Equation (10.14) shows that $\preceq_F$ is anti-symmetric (and so $\preceq_F$ is a partial order) when F verifies conditions (F1)–(F3) with $z^* \in K^\#$.

For F satisfying conditions (F1) and (F2), and $z^* \in K^+$, we introduce the partial order $\preceq_{F,z^*}$ on $X \times Y$ by

$$(x_1, y_1) \preceq_{F,z^*} (x_2, y_2) \iff \begin{cases} (x_1, y_1) = (x_2, y_2) \text{ or} \\ (x_1, y_1) \preceq_F (x_2, y_2) \text{ and } z^*(y_1) < z^*(y_2). \end{cases}$$

It is easy to verify that $\preceq_{F,z^*}$ is reflexive, transitive, and antisymmetric. Of course, for $z^* = 0$ the relation $\preceq_{F,z^*}$ coincides with the equality relation on $X \times Y$, while for $z^* \in K^\#$ and $0 \notin F(x, x')$ for $x \neq x'$ one has that $\preceq_{F,z^*}$ and $\preceq_F$ coincide.

The following result is established in [484].

Lemma 10.1.2. *Let Y be a separated locally convex space. One has* (10.8) $\Rightarrow$ (10.7) *[that is,* (10.8) $\Rightarrow$ (F3)*] in each of the following situations:*

(i) *for every sequence $(z_n^*)_{n \geq 1} \subseteq K^+$ there exists a sequence $(\mu_n)_{n \geq 1} \subseteq \mathbb{P}$ such that $\overline{\operatorname{conv}}^{w^*}\{\mu_n z_n^* \mid n \geq 1\}$ is w^*-compact;*

(ii) *K^+ has a weakly-star compact base (e.g. if* $\operatorname{int} K \neq \emptyset$*);*

(iii) *for every sequence $(z_n^*)_{n \geq 1} \subseteq Z^*$ there exists a sequence $(\mu_n)_{n \geq 1} \subseteq \mathbb{P}$ such that $\overline{\operatorname{conv}}^{w^*}\{\mu_n z_n^* \mid n \geq 1\}$ is w^*-compact;*

(iv) *Y is a normed vector space.*

Proof. Note first that (iv) $\Rightarrow$ (iii) $\Rightarrow$ (i). Indeed, the implication (iii) $\Rightarrow$ (i) is obvious. For (iv) $\Rightarrow$ (iii) just note that having $(z_n^*)_{n\geq 1} \subseteq Y^*$, there exists $(\mu_n)_{n\geq 1} \subseteq \mathbb{P}$ such that $\mu_n z_n^* \in U_{Y^*} := \{z^* \mid \|z^*\| \leq 1\}$ for every $n \geq 1$, and so $\overline{\text{conv}}^{w^*} \{\mu_n z_n^* \mid n \geq 1\}$ is w^*-compact.

(ii) $\Rightarrow$ (i) Let B be a weakly-star compact base of K^+. It follows that $B_0 := [0,1]B$ is a w^*-compact convex subset of K^+ such that $K^+ = \mathbb{R}_+ B_0$. This shows that for every $z^* \in K^+$ there exists $\mu > 0$ such that $\mu z^* \in B_0$. So, for $(z_n^*)_{n\geq 1} \subseteq K^+$, there exists $(\mu_n)_{n\geq 1} \subseteq \mathbb{P}$ such that $\{\mu_n z_n^* \mid n \geq 1\} \subseteq B_0$, whence $\overline{\text{conv}}^{w^*} \{\mu_n z_n^* \mid n \geq 1\}$ is w^*-compact.

(i) Assume that (10.8) holds. Then for every $n \in \mathbb{N}^*$ there exists $z_n^* \in K^+$ such that $\gamma_n := \inf z_n^*(F_{1/n}) > 0$. By hypothesis, there exists a sequence $(\mu_n)_{n\geq 1} \subseteq \mathbb{P}$ such that $B := \overline{\text{conv}}^{w^*} \{\mu_n z_n^* \mid n \geq 1\}$ is w^*-compact, and so

$$B_0 := [0,1]B = \overline{\text{conv}}^{w^*} \left(\{0\} \cup \{\mu_n z_n^* \mid n \geq 1\}\right)$$

is w^*-compact, too. Then $z^* := \sum_{n\geq 1} 2^{-n} \mu_n z_n^*$ exists and belongs to $B_0 \subseteq K^+$. Indeed, because B_0 is w^*-bounded, the sequence $(z_n'^*)_{n\geq 1}$ with $z_n'^* := \sum_{k=1}^{n} 2^{-k} \mu_k z_k^* \in B_0$ is w^*-Cauchy. Since $(z_n'^*)_{n\geq 1}$ has a subnet w^*-converging to some $z^* \in B_0$, we obtain that $z_n'^* \to^{w^*} z^*$. It follows that $z^* = \sum_{n\geq 1} 2^{-n} \mu_n z_n^*$. Clearly, for $z \in F_{1/n}$ we have that $z^*(z) \geq 2^{-n} \mu_n z_n^*(z)$, and so $\inf z^*(F_{1/n}) \geq 2^{-n} \mu_n \gamma_n > 0$. Since for every $\delta > 0$ we have that $\delta \geq 1/n$ for some $n \in \mathbb{N}^*$, we obtain that (10.5) holds. □

10.2 Minimal Points in Product Spaces

We take X, Y, K, F as above, that is, F verifies conditions (F1) and (F2).

Consider a nonempty set $A \subseteq X \times Y$. In the sequel we shall use the following condition on A:

(H1) for every $\preceq_F$–decreasing sequence $((x_n, y_n)) \subseteq A$ with $x_n \to x \in X$ there exists $y \in Y$ such that $(x, y) \in A$ and $(x, y) \preceq_F (x_n, y_n)$ for every $n \in \mathbb{N}$.

The next theorem is the main result of this section.

Theorem 10.2.1. *Let (X, d) be a complete metric space, Y a real topological vector space and $K \subseteq Y$ a proper convex cone. Assume that:*

(i) *$F : X \times X \rightrightarrows Y$ verifies conditions (F1), (F2) and (10.9), and*
(ii) *$A \subseteq X \times Y$ verifies (H1).*

Then for every $(x_0, y_0) \in A$ and every $z^* \in K^+$ such that $z^*(\mathrm{Pr}_Y(A))$ is bounded from below, there exists a minimal element $(\bar{x}, \bar{y})$ of A with respect to $\preceq_{F,z^*}$ such that $(\bar{x}, \bar{y}) \preceq_{F,z^*} (x_0, y_0)$. Moreover, if $z^*(z) > 0$ for every $z \in \cup_{\delta>0} F_\delta$, then $A \ni (x, y) \preceq_F (\bar{x}, \bar{y})$ implies $x = \bar{x}$.

Proof. Let us fix $(x_0, y_0) \in A$ and $z^* \in K^+$ such that $z^*(\mathrm{Pr}_Y(A))$ is bounded from below. Then

$$\alpha := \inf\{z^*(y) \mid \exists x \in X : (x, y) \in A,\ (x, y) \preceq_{F,z^*} (x_0, y_0)\} \in \mathbb{R}.$$

Let us set

$$A_{F,z^*}(x, y) := \left\{(x', y') \in A \mid (x', y') \preceq_{F,z^*} (x, y)\right\}.$$

We construct a sequence $((x_n, y_n))_{n\geq 0} \subseteq A$ as follows: Having $(x_n, y_n) \in A$, we take $(x_{n+1}, y_{n+1}) \in A_{F,z^*}(x_n, y_n)$ such that

$$z^*(y_{n+1}) \leq \inf\{z^*(y) \mid (x, y) \in A_{F,z^*}(x_n, y_n)\} + 1/(n + 1).$$

Of course, $((x_n, y_n))$ is $\preceq_{F,z^*}$–decreasing. It follows that $(y_n)_{n\geq 0} \subseteq \mathrm{Pr}_Y(A)$ is $\leq_K$–decreasing, and so the sequence $(z^*(y_n))_{n\geq 0}$ is non increasing and bounded from below; hence $\gamma := \lim z^*(y_n) \in \mathbb{R}$.

If $A_{F,z^*}(x_{n_0}, y_{n_0})$ is a singleton (that is, $A_{F,z^*}(x_{n_0}, y_{n_0}) = \{(x_{n_0}, y_{n_0})\}$) for some $n_0 \in \mathbb{N}$, then clearly $(\overline{x}, \overline{y}) := (x_{n_0}, y_{n_0})$ is the desired element. In the contrary case the sequence $(z^*(y_n))$ is (strictly) decreasing; moreover, $\gamma < z^*(y_n)$ for every $n \in \mathbb{N}$.

Assume that (x_n) is not a Cauchy sequence. Then there exist $\delta > 0$ and the sequences (n_l), (p_l) from $\mathbb{N}^*$ such that $n_l \to \infty$ and $d(x_{n_l}, x_{n_l+p_l}) \geq \delta$ for every l. Since $(x_{n_l+p_l}, y_{n_l+p_l}) \preceq_{F,z^*} (x_{n_l}, y_{n_l})$ we obtain that $y_{n_l} - y_{n_l+p_l} = v_l + k_l'$ with $v_l \in F(x_{n_l+p_l}, x_{n_l}) \subseteq F_\delta$ and $k_l' \in K$. By (10.9), there exists $v^* \in K^+$ such that $v^*(v_l) \not\to 0$. On the other hand, because $(y_n)_{n\geq 0} \subseteq \mathrm{Pr}_Y(A)$ and $v^* \in K^+$, $(v^*(y_n))_{n\geq 0}$ is bounded from below. The sequence $(y_n)_{n\geq 0}$ being K–decreasing we obtain that $\lim v^*(y_n)$ exists in $\mathbb{R}$. Since $v^*(y_{n_l}) - v^*(y_{n_l+p_l}) \geq z^*(v_l) \geq 0$ for every $l \in \mathbb{N}$, we get the contradiction $v^*(v_l) \to 0$. Therefore, (x_n) is a Cauchy sequence in the complete metric space (X, d), and so (x_n) converges to some $\bar{x} \in X$.

Since $((x_n, y_n))$ is $\preceq_F$–decreasing, by (H1) there exists some $\bar{y} \in Y$ such that $(\bar{x}, \bar{y}) \in A$ and $(\bar{x}, \bar{y}) \preceq_F (x_n, y_n)$ for every $n \in \mathbb{N}$. It follows that $z^*(\bar{y}) \leq \lim z^*(y_n)$, and so $z^*(\bar{y}) < z^*(y_n)$ for every $n \in \mathbb{N}$. Therefore $(\bar{x}, \bar{y}) \preceq_{F,z^*} (x_n, y_n)$ for every $n \in \mathbb{N}$. Let $(x', y') \in A$ be such that $(x', y') \preceq_{F,z^*} (\bar{x}, \bar{y})$. Since $(\bar{x}, \bar{y}) \preceq_{F,z^*} (x_n, y_n)$, we have that $(x', y') \preceq_{F,z^*} (x_n, y_n)$ for every $n \in \mathbb{N}$. It follows that

$$\forall\, n \geq 1 : 0 \leq z^*(\bar{y}) - z^*(y') \leq z^*(y_n) - z^*(y') \leq 1/n,$$

whence $z^*(y') = z^*(\bar{y})$. By the definition of $\preceq_{F,z^*}$ we obtain that $(x', y') = (\bar{x}, \bar{y})$.

Assume now that $z^*(z) > 0$ for every $z \in \cup_{\delta>0} F_\delta$. Consider $A \ni (x, y) \preceq_F (\bar{x}, \bar{y})$ and assume that $x \neq \overline{x}$. Then $\overline{y} = y + v + k$ with $v \in F(x, \overline{x})$ and $k \in K$. By our assumption we have that $z^*(v) > 0$, and so $z^*(y) < z^*(\overline{y})$. It follows that $(x, y) \preceq_{F,z^*} (\bar{x}, \bar{y})$, contradicting the minimality of $(\bar{x}, \bar{y})$ with respect to $\preceq_{F,z^*}$. Hence $x = \overline{x}$. □

Remark 10.2.2. As seen from the proof, to have the conclusion of Theorem 10.2.1 for a fixed $(x_0, y_0) \in A$, in its hypothesis, it is sufficient that $A_{F,z^*}(x_0, y_0)$ verify (H1) and z^* $(\mathrm{Pr}_Y\,(A_{F,z^*}(x_0, y_0)))$ be bounded from below. For this just replace A by $A_{F,z^*}(x_0, y_0)$ and apply the preceding theorem.

Theorem 10.2.3. *Assume that (X, d) is a complete metric space, Y is a real topological vector space and $K \subseteq Y$ is a proper convex cone. Let $F : X \times X \rightrightarrows Y$ satisfy conditions (F1)–(F3) and $A \subseteq X \times Y$ satisfy (H1). Furthermore, suppose that z^* (from (F3)) is bounded from below on $\mathrm{Pr}_Y(A)$. Then for every $(x_0, y_0) \in A$ there exists a minimal element $(\bar{x}, \bar{y})$ of A with respect to $\preceq_{F,z^*}$ such that $(\bar{x}, \bar{y}) \preceq_{F,z^*} (x_0, y_0)$. Moreover, $(\bar{x}, \bar{y}) \preceq_F (x_0, y_0)$ and $A \ni (x, y) \preceq_F (\bar{x}, \bar{y})$ implies $x = \bar{x}$.*

Proof. The proof is the same as that of Theorem 10.2.1. One must observe that for every $\delta > 0$ we can take as z^* in (10.9) the element z^* provided by (F3). □

Remark 10.2.4. As in Remark 10.2.2, to have the conclusion of Theorem 10.2.3 for a fixed $(x_0, y_0) \in A$, in its hypothesis, it is sufficient that $A_{F,z^*}(x_0, y_0)$ verify (H1) and z^* be bounded from below on the set $A_{F,z^*}(x_0, y_0)$.

As seen above, a possible choice for F is F_H given in (10.11) with $\emptyset \neq H \subseteq K$ a K-convex set; if $z^* \in K^+$ is such that $\inf z^*(H) > 0$, then conditions (F1)–(F3) are verified (see Lemma 10.1.1). In the case $H := \{k^0\}$ condition (H1) becomes condition (H1) in [214, p. 199]. So Theorem 10.2.3 extends [214, Theorem 3.10.7] to this framework, using practically the same proof.

For H as above, that is, $H \subseteq K \setminus \{0\}$ is a nonempty set such that $H + K$ is convex, and F_H defined in (10.11), we set $\preceq_H := \preceq_{F_H}$; moreover, for $z^* \in K^+$ we set $\preceq_{H,z^*} := \preceq_{F_H,z^*}$.

An immediate consequence of the preceding theorem is the next result.

Corollary 10.2.5. *Assume that (X, d) is a complete metric space, Y is a real topological vector space, $K \subseteq Y$ is a proper convex cone and $\emptyset \neq H \subseteq K$ is a nonempty K-convex set. Let $A \subseteq X \times Y$ satisfy (H1) with respect to $\preceq_H$. If there exists $z^* \in K^+$ such that $\inf z^*(H) > 0$ and $\inf z^*\,(\mathrm{Pr}_Y(A)) > -\infty$, then for every $(x_0, y_0) \in A$ there exists $(\bar{x}, \bar{y}) \in A$ a minimal element of A with respect to $\preceq_{H,z^*}$ such that $(\bar{x}, \bar{y}) \preceq_{H,z^*} (x_0, y_0)$. Moreover, $(\bar{x}, \bar{y}) \preceq_H (x_0, y_0)$ and $A \ni (x, y) \preceq_H (\bar{x}, \bar{y})$ implies $x = \bar{x}$.*

Of course, a variant of Corollary 10.2.5 can be formulated corresponding to condition (10.12).

A condition related to (H1) is the next one.

(H2) for every sequence $\big((x_n, y_n)\big) \subseteq A$ with $x_n \to x \in X$ and (y_n) $\preceq_K$–decreasing there exists $y \in Y$ such that $(x, y) \in A$ and $y \preceq_K y_n$ for every $n \in \mathbb{N}$.

Remark 10.2.6. Note that (H2) holds if A is closed with $\mathrm{Pr}_Y(A) \subseteq y_0 + K$ for some $y_0 \in Y$ and every $\preceq_K$–decreasing sequence in K is convergent (i.e., K is a sequentially Daniell cone). In fact, instead of asking that A is closed we may assume that

$$\forall\, ((x_n, y_n))_{n\geq 1} \subseteq A : [x_n \to x,\ y_n \to y,\ (y_n) \text{ is } \leq_K\text{–decreasing} \Rightarrow (x, y) \in A].$$

Remark 10.2.7. Note that (H1) is verified whenever A satisfies (H2) and

$$\forall u \in X,\ \forall X \supseteq (x_n) \to x \in X\ :\ \cap_{n\in\mathbb{N}} (F(x_n, u) + K) \subseteq F(x, u) + K.$$

Indeed, let $\big((x_n, y_n)\big) \subseteq A$ be $\preceq_F$–decreasing with $x_n \to x$. It is obvious that (y_n) is $\leq_K$–decreasing. By (H2), there exists $y \in Y$ such that $(x, y) \in A$ and $y \leq_K y_n$ for every $n \in \mathbb{N}$. It follows that

$$\forall\, n, p \in \mathbb{N} : y_n \in y_{n+p} + F(x_{n+p}, x_n) + K \subseteq y + F(x_{n+p}, x_n) + K.$$

Fix n; then $y_n - y \in F(x_{n+p}, x_n) + K$ for every $p \in \mathbb{N}$, and so, by our hypothesis, $y_n - y \in F(x, x_n) + K$ because $\lim_{p\to\infty} x_{n+p} = x$. Therefore, $(x, y) \preceq_F (x_n, y_n)$.

Remark 10.2.8. When d generates the discrete topology on X, both conditions (H1) and (H2) are equivalent to:

(H12) for every $x \in \Pr_X(A)$ and every $\leq_K$–decreasing sequence $(y_n)_{n\geq 1} \subseteq A(x) := \{y \in Y \mid (x, y) \in A\}$ there exists $y \in A(x)$ such that $y \leq_K y_n$ for every $n \geq 1$.

Indeed, it is clear that (H1) $\wedge$ (H2) $\Rightarrow$ (H12) (for any metric d). Assume that (H12) holds and take $((x_n, y_n))_{n\geq 1} \subseteq A$ a $\preccurlyeq_F$-decreasing sequence with $x_n \to x$; in particular, (y_n) is $\leq_K$–decreasing. Since $x_n \to x$, there exists $n_0 \geq 1$ such that $x_n = x$ for $n \geq n_0$. It follows that $(y_n)_{n\geq n_0} \subseteq A(x)$. By (H12) we have that $y \in A(x)$, and so $(x, y) \preccurlyeq_F (x, y_n) = (x_n, y_n) \preccurlyeq_F (x_{n_0}, y_{n_0}) \preccurlyeq_F (x_m, y_m)$ for $n \leq n_0 \leq m$. Hence (H1) holds. The proof of (H12) $\Rightarrow$ (H2) is similar.

Remark 10.2.9. In the case $F = F_H$ with $\emptyset \neq H \subseteq K$ a K-convex set, (H1) is verified whenever A satisfies (H2) and $H + K$ is closed.

Indeed, let $\big((x_n, y_n)\big) \subseteq A$ be a $\preceq_H$–decreasing sequence with $x_n \to x$. It is obvious that (y_n) is $\leq_K$–decreasing. By (H2), there exists $y \in Y$ such that $(x, y) \in A$ and $y \leq_K y_n$ for every $n \in \mathbb{N}$.

Fix n. If $x_n = x$ then clearly $(x, y) = (x_n, y) \preceq_H (x_n, y_n)$. Else, because $d(x_{n+p}, x_n) \to d(x, x_n) > 0$ for $p \to \infty$, we get $d(x_{n+p}, x_n) > 0$ for sufficiently large p, and so

$$y_n \in y_{n+p} + d(x_{n+p}, x_n)H + K \subseteq y + d(x_{n+p}, x_n)H + K = y + d(x_{n+p}, x_n)(H + K)$$

for sufficiently large p. Since $H + K$ is closed we obtain that

$$y_n \in y + d(x_n, x)(H + K) = y + d(x_n, x)H + K,$$

that is, $(x, y) \preceq_H (x_n, y_n)$.

Another condition to be added to (H2) in order to have (H1) was suggested by the hypotheses of [40, Theorem 4.1]. Recall that a set $E \subseteq Y$ is **cs-complete** (see [614, p. 9]) if for all sequences $(\lambda_n)_{n\geq 1} \subseteq \mathbb{R}_+$, $(y_n)_{n\geq 1} \subseteq E$ such that $\sum_{n\geq 1} \lambda_n = 1$ and the sequence $\left(\sum_{m=1}^{n} \lambda_m y_m\right)_{n\geq 1}$ is Cauchy, the series $\sum_{n\geq 1} \lambda_n y_n$ is convergent and its sum belongs to E. One says that $E \subseteq Y$ is **cs-closed** if the sum of the series $\sum_{n\geq 1} \lambda_n y_n$ belongs to E whenever $\sum_{n\geq 1} \lambda_n y_n$ is convergent and $(y_n) \subseteq E$, $(\lambda_n)_{n\geq 1} \subseteq \mathbb{R}_+$ and $\sum_{n\geq 1} \lambda_n = 1$. Of course, any cs-complete set is cs-closed; if Y is complete then the converse is true. Moreover, notice that any cs-closed set is convex.

It is worth observing that the closed convex subsets as well as the open convex subsets of a separated locally convex space are cs-closed; moreover, all the convex subsets of finite dimensional normed spaces are cs-closed (hence cs-complete).

Note that the sequence $\left(\sum_{m=1}^{n} \lambda_m y_m\right)_{n\geq 1}$ is Cauchy whenever $(\lambda_n)_{n\geq 1} \subseteq \mathbb{R}_+$ is such that the series $\sum_{n\geq 1} \lambda_n$ is convergent and $(y_n)_{n\geq 1} \subseteq Y$ is such that $\operatorname{conv}\{y_n \mid n \geq 1\}$ is bounded; of course, if Y is a locally convex space then $B \subseteq Y$ is bounded iff $\operatorname{conv} B$ is bounded. Indeed, let $(\lambda_n)_{n\geq 1} \subseteq \mathbb{R}_+$ with $\sum_{n\geq 1} \lambda_n$ convergent and $(y_n)_{n\geq 1} \subseteq Y$ with $B := \operatorname{conv}\{y_n \mid n \geq 1\}$ bounded. Fix $V \subseteq Y$ a balanced neighborhood of 0. Because B is bounded, there exists $\alpha > 0$ such that $B \subseteq \alpha V$. Since the series $\sum_{n\geq 1} \lambda_n$ is convergent there exists $n_0 \geq 1$ such that $\sum_{k=n}^{n+p} \lambda_k \leq \alpha^{-1}$ for all $n, p \in \mathbb{N}$ with $n \geq n_0$. Then for such n, p and some $b_{n,p} \in B$ we have

$$\sum_{k=n}^{n+p} \lambda_k y_k = \left(\sum_{k=n}^{n+p} \lambda_k\right) b_{n,p} \in [0, \alpha^{-1}] B \subseteq [0, \alpha^{-1}] \alpha V = V.$$

Proposition 10.2.10. *Assume that (X, d) is a complete metric space, Y is a real topological vector space and $K \subseteq Y$ is a proper closed convex cone. Furthermore, suppose that $H \subseteq K$ is a nonempty cs-complete bounded set with $0 \notin \operatorname{cl}(H + K)$. If A satisfies (H2) then A satisfies (H1), too.*

Proof. Let $\left((x_n, y_n)\right)_{n\geq 1} \subseteq A$ be a $\preceq_H$–decreasing sequence with $x_n \to x$. It follows that (y_n) is $\leq_K$–decreasing. By (H2), there exists $y \in Y$ such that $(x, y) \in A$ and $y \leq_K y_n$ for every $n \in \mathbb{N}$.

Because $\left((x_n, y_n)\right)_{n\geq 1}$ is $\preceq_H$–decreasing we have that

$$y_n = y_{n+1} + d(x_n, x_{n+1}) h_n + k_n \tag{10.15}$$

with $h_n \in H$ and $k_n \in K$ for $n \geq 1$. If $x_n = x_{\overline{n}}$ for $n \geq \overline{n} \geq 1$ we take $x := x_{\overline{n}}$; then $(x, y) \preceq_H (x_n, y_n)$ for every $n \in \mathbb{N}$. Indeed, for $n \leq \overline{n}$ we have that $(x_{\overline{n}}, y_{\overline{n}}) \preceq_H (x_n, y_n)$; because $y \leq_K y_{\overline{n}}$, by (10.4) we get $(x, y) \preceq_H (x, y_{\overline{n}}) = (x_{\overline{n}}, y_{\overline{n}})$, and so $(x, y) \preceq (x_n, y_n)$. If $n > \overline{n}$, using again (10.4), we have $(x, y) = (x_n, y) \preceq_H (x_n, y_n)$.

Assume that (x_n) is not constant for large n. Fix $n \geq 1$. From (10.15), for $p \geq 0$, we have

$$y_n = y_{n+p+1} + \sum_{l=n}^{n+p} d(x_l, x_{l+1})h_l + \sum_{l=n}^{n+p} k_l = y_{n+p+1} + \left(\sum_{l=n}^{n+p} d(x_l, x_{l+1})\right)h_{n,p}$$
$$+ \sum_{l=n}^{n+p} k_l = y + k'_{n,p} + \left(\sum_{l=n}^{n+p} d(x_l, x_{l+1})\right)h_{n,p} \tag{10.16}$$

for some $h_{n,p} \in H$ and $k'_{n,p} \in K$. Assuming that $\sum_{l \geq n} d(x_l, x_{l+1}) = \infty$, from

$$\left(\sum_{l=n}^{n+p} d(x_l, x_{l+1})\right)^{-1} (y_n - y) = h_{n,p} + \left(\sum_{l=n}^{n+p} d(x_l, x_{l+1})\right)^{-1} k'_{n,p} \in H + K,$$

we get the contradiction $0 \in \operatorname{cl}(H + K)$ taking the limit for $p \to \infty$. Hence $0 < \mu := \sum_{l \geq n} d(x_l, x_{l+1}) < \infty$. Set $\lambda_l := \mu^{-1} d(x_l, x_{l+1})$ for $l \geq n$. Since H is cs-complete and $\operatorname{conv}\{h_l \mid l \geq n\}$ ($\subseteq H$) is bounded we obtain that the series $\sum_{l \geq n} \lambda_l h_l$ is convergent and its sum $\overline{h}_n$ belongs to H. It follows that $\sum_{l \geq n} d(x_l, x_{l+1})h_l = \mu \overline{h}_n$, and so

$$\overline{k}_n := \lim_{p \to \infty} k'_p = y_n - y - \mu \overline{h}_n \in K$$

because K is closed. Since $d(x_n, x_{n+p}) \leq \sum_{l=n}^{n+p-1} d(x_l, x_{l+1})$, we obtain that $d(x_n, x) \leq \mu$, and so

$$y_n = y + d(x_n, x)\overline{h}_n + \overline{k}_n + (\mu - d(x_n, x))\, \overline{h}_n \in y + d(x_n, x)H + K.$$

Hence $(x, y) \preceq_H (x_n, y_n)$ for every $n \in \mathbb{N}$. □

The most part of vector EVP type results are established for Y a separated locally convex space. However, there are topological vector spaces Y whose topological dual reduce to $\{0\}$. In such a case it is not possible to find z^* satisfying the conditions of Corollary 10.2.5.

Theorem 10.2.11. *Assume that (X, d) is a complete metric space, Y is a real topological vector space. Let $K \subseteq Y$ be a proper closed convex cone and $H \subseteq K$ be a nonempty cs-complete bounded set with $0 \notin \operatorname{cl}(H + K)$. Suppose that $A \subseteq X \times Y$ satisfies (H2) and that $\operatorname{Pr}_Y(A)$ is quasi bounded. Then for every $(x_0, y_0) \in A$ there exists $(\bar{x}, \bar{y}) \in A$ such that $(\bar{x}, \bar{y}) \preceq_H (x_0, y_0)$ and $(x, y) \in A$, $(x, y) \preceq_H (\bar{x}, \bar{y})$ imply $x = \bar{x}$.*

Proof. First observe that A satisfies condition (H1) by Proposition 10.2.10. Moreover, because $\operatorname{Pr}_Y(A)$ is quasi bounded, there exists a bounded set $B \subseteq Y$ such that $\operatorname{Pr}_Y(A) \subseteq B + K$.

Note that for $(x, y) \in A$ the set $\mathrm{Pr}_X(A_H(x, y))$ is bounded, where

$$A_H(x, y) := \{(x', y') \in A \mid (x', y') \preceq_H (x, y)\}.$$

In the contrary case there exists a sequence $((x_n, y_n))_{n\geq 1} \subseteq A_H(x, y)$ with $d(x_n, x) \to \infty$. Hence $y = y_n + d(x_n, x)h_n + k_n = b_n + d(x_n, x)h_n + k'_n$ with $h_n \in H$, $b_n \in B$, $k_n, k'_n \in K$. It follows that $d(x_n, x)^{-1}(y - b_n) \in H + K$, whence the contradiction $0 \in \mathrm{cl}(H + K)$.

Let us construct a sequence $((x_n, y_n))_{n\geq 0} \subseteq A$ in the following way: Having $(x_n, y_n) \in A$, where $n \in \mathbb{N}$, because $D_n := \mathrm{Pr}_X(A_H(x_n, y_n))$ is bounded, there exists $(x_{n+1}, y_{n+1}) \in A_H(x_n, y_n)$ such that

$$d(x_{n+1}, x_n) \geq \tfrac{1}{2} \sup\{d(x, x_n) \mid x \in D_n\} \geq \tfrac{1}{4} \operatorname{diam} D_n.$$

We obtain in this way the sequence $((x_n, y_n))_{n\geq 0} \subseteq A$, which is $\preceq_H$–decreasing. Since $A_H(x_{n+1}, y_{n+1}) \subseteq A_H(x_n, y_n)$, we have that $D_{n+1} \subseteq D_n$ for every $n \in \mathbb{N}$. Of course, $x_n \in D_n$. Let us show that $\operatorname{diam} D_n \to 0$. In the contrary case there exists $\delta > 0$ such that $\operatorname{diam} D_n \geq 4\delta$, and so $d(x_{n+1}, x_n) \geq \delta$ for every $n \in \mathbb{N}$. As in the proof of Proposition 10.2.10, for every $p \in \mathbb{N}$, we obtain that

$$\begin{aligned} y_0 &= y_{p+1} + \Big(\sum_{l=0}^{p} d(x_l, x_{l+1})\Big)h_p + \sum_{l=0}^{p} k_l = b_p + \Big(\sum_{l=0}^{p} d(x_l, x_{l+1})\Big)h_p + k'_p \\ &= b_p + (p+1)\delta h_p + k''_p, \end{aligned}$$

where $h_p \in H$, $b_p \in B$, $k_l, k'_p, k''_p \in K$. It follows that $[(p+1)\delta]^{-1}(y_0 - b_p) \in H + K$ for every $p \in \mathbb{N}$. Since (b_p) is bounded we obtain the contradiction $0 \in \mathrm{cl}(H + K)$. Thus we have that the sequence $(\mathrm{cl}\, D_n)$ is a decreasing sequence of nonempty closed subsets of the complete metric space (X, d), whose diameters tend to 0. By Cantor's theorem, $\bigcap_{n\in\mathbb{N}} \mathrm{cl}\, D_n = \{\bar{x}\}$ for some $\bar{x} \in X$. Of course, $x_n \to \bar{x}$. Since $((x_n, y_n)) \subseteq A$ is a $\preceq_H$–decreasing sequence, from (H1) we get an $\bar{y} \in Y$ such that $(\bar{x}, \bar{y}) \preceq_H (x_n, y_n)$ for every $n \in \mathbb{N}$; $(\bar{x}, \bar{y})$ is the desired element. Indeed, $(\bar{x}, \bar{y}) \preceq_H (x_0, y_0)$. Let $(x', y') \in A_H(\bar{x}, \bar{y})$. It follows that $(x', y') \in A_H(x_n, y_n)$, and so $x' \in D_n \subseteq \mathrm{cl}\, D_n$ for every n. Thus $x' = \bar{x}$. □

If Y is a separated locally convex space, the preceding result follows immediately from Corollary 10.2.5.

Of course, the set $A \subseteq X \times Y$ can be viewed as the graph of a set-valued map $\Gamma : X \rightrightarrows Y$; then $\mathrm{Pr}_X(A) = \operatorname{dom} \Gamma$ and $\mathrm{Pr}_Y(A) = \operatorname{Im} \Gamma$.

For the nonempty set $E \subseteq Y$ let us set

$$\operatorname{BMMin} E := \{\overline{y} \in E \mid E \cap (\overline{y} - K) = \{\overline{y}\}\}$$

(see [26, (1.2)]); note that this set is different of the usual set

$$\operatorname{Min} E := \{\overline{y} \in E \mid E \cap (\overline{y} - K) \subseteq \overline{y} + K\},$$

but they coincide if K is pointed, that is, $K \cap (-K) = \{0\}$. As in [26, Definition 3.2], we say that $\Gamma : X \rightrightarrows Y$ satisfies the **limiting monotonicity condition** at $\overline{x} \in \operatorname{dom} \Gamma$ if for every sequence $((x_n, y_n))_{n\geq 1} \subseteq \operatorname{graph} \Gamma$ with $x_n \to \overline{x}$ and (y_n) $\leq_K$–decreasing, there exists $\overline{y} \in \operatorname{BMMin} \Gamma(\overline{x})$ such that $\overline{y} \leq y_n$ for every $n \geq 1$. As observed in [26], if Γ satisfies the limiting monotonicity condition at $\overline{x} \in \operatorname{dom} \Gamma$ then $\Gamma(\bar{x}) \subseteq \operatorname{BMMin} \Gamma(\overline{x}) + K$, that is, $\Gamma(\bar{x})$ satisfies the domination property.

In [26, Proposition 3.3], in the case Y a Banach space, there are mentioned sufficient conditions in order that Γ satisfy the limiting monotonicity condition at $\overline{x} \in \operatorname{dom} \Gamma$.

When X and Y are Banach spaces and H is a singleton the next result is practically [26, Theorem 3.5].

Corollary 10.2.12. *Assume that (X, d) is a complete metric space, Y is a real topological vector space. Let $K \subseteq Y$ be a proper closed convex cone and $H \subseteq K$ be a nonempty cs-complete bounded set with $0 \notin \operatorname{cl}(H + K)$. Suppose that $\Gamma : X \rightrightarrows Y$ is K-l.s.c., satisfies the limiting monotonicity condition on* $\operatorname{dom} \Gamma$ *and* $\operatorname{Im} \Gamma$ *is quasi-bounded. Then for every $(x_0, y_0) \in \operatorname{graph} \Gamma$ there exist $\bar{x} \in \operatorname{dom} \Gamma$ and $\bar{y} \in \operatorname{BMMin} \Gamma(\bar{x})$ such that $(\bar{x}, \bar{y}) \preceq_H (x_0, y_0)$ and $(x, y) \in \operatorname{graph} \Gamma$, $(x, y) \preceq_H (\bar{x}, \bar{y})$ imply $x = \bar{x}$.*

Proof. In order to apply Theorem 10.2.11 for $A := \operatorname{graph} \Gamma$ we have only to show that A verifies condition (H2). For this consider the sequence $((x_n, y_n))_{n\geq 1} \subseteq A$ such that (y_n) is $\leq_K$–decreasing and $x_n \to \overline{x}$. Clearly, $x_n \in \operatorname{lev}_\Gamma(y_1)$ for every n; since Γ is K-l.s.c., we have that $\overline{x} \in \operatorname{lev}_\Gamma(y_1) \subseteq \operatorname{dom} \Gamma$. Since Γ satisfies the limiting monotonicity condition at $\overline{x}$, we find $\overline{y} \in \operatorname{BMMin} \Gamma(\overline{x}) \subseteq \Gamma(\overline{x})$ such that $\overline{y} \leq y_n$ for every n. Hence (H2) holds. By Theorem 10.2.11 there exists $(x, y) \in A$ such that $(x, y) \preceq_H (x_0, y_0)$ and $(x', y') \in \operatorname{graph} \Gamma$, $(x', y') \preceq_H (x, y)$ imply $x' = x$. Set $\bar{x} := x$ and take $\bar{y} \in \operatorname{BMMin} \Gamma(\bar{x})$ such that $\bar{y} \leq_K y$. By (10.4) we have that $(\bar{x}, \bar{y}) \preceq_H (x_0, y_0)$. Let now $(x', y') \in \operatorname{graph} \Gamma = A$ with $(x', y') \preceq_H (\bar{x}, \bar{y})$. Since $(\bar{x}, \bar{y}) = (x, \bar{y}) \preceq_H (x, y)$, we have that $(x', y') \preceq_H (x, y)$, and so $x' = x = \bar{x}$. The proof is complete. □

In the case when H is a singleton the next result is practically [25, Theorem 1] under the supplementary hypothesis that $\operatorname{Min} \Gamma(x)$ is compact for every $x \in X$.

Corollary 10.2.13. *Assume that (X, d) is a complete metric space, Y is a real topological vector space. Let $K \subseteq Y$ be a proper closed convex cone and $H \subseteq K$ be a nonempty cs-complete bounded set with $0 \notin \operatorname{cl}(H + K)$. Suppose that $\Gamma : X \rightrightarrows Y$ is K-l.s.c.,* $\operatorname{Min} \Gamma(x)$ *is compact and $\Gamma(x) \subseteq K +$* $\operatorname{Min} \Gamma(x)$ *for every $x \in$* $\operatorname{dom} \Gamma$, *and* $\operatorname{Im} \Gamma$ *is quasi-bounded. Then for every $(x_0, y_0) \in \operatorname{graph} \Gamma$ there exist $\bar{x} \in \operatorname{dom} \Gamma$ and $\bar{y} \in \operatorname{Min} \Gamma(\bar{x})$ such that $(\bar{x}, \bar{y}) \preceq_H (x_0, y_0)$ and $(x, y) \in \operatorname{graph} \Gamma$, $(x, y) \preceq_H (\bar{x}, \bar{y})$ imply $x = \bar{x}$.*

Proof. In order to apply Theorem 10.2.11 for $A := \operatorname{graph} \Gamma$ we have only to show that A verifies condition (H2). For this consider the sequence $((x_n, y_n))_{n\geq 1} \subseteq A$ such that (y_n) is $\leq_K$–decreasing and $x_n \to \overline{x}$. As in the proof of the preceding corollary, $\overline{x} \in \operatorname{lev}_\Gamma(y_n)$ for every $n \in \mathbb{N}$. Because $\Gamma(\overline{x}) \subseteq K + \operatorname{Min} \Gamma(\overline{x})$, for every $n \in \mathbb{N}$ there exists $y'_n \in \operatorname{Min} \Gamma(\overline{x})$ such that $y'_n \leq y_n$. Because $\operatorname{Min} \Gamma(\overline{x})$ is compact, (y'_n) has a subnet $(y'_{\psi(i)})_{i \in I}$ converging to some $\overline{y} \in \operatorname{Min} \Gamma(\overline{x})$; here $\psi : (I, \succeq) \to \mathbb{N}$ is such that for every n there exists $i_n \in I$ with $\psi(i) \geq n$ for $i \succeq i_n$. Hence $y'_{\psi(i)} \leq y_{\psi(i)} \leq y_n$ for $i \succeq i_n$, whence $\overline{y} \leq y_n$ because K is closed. Therefore, (H2) holds. By Theorem 10.2.11, for $(x_0, y_0) \in \operatorname{graph} \Gamma$, there exists $(x, y) \in A$ such that $(x, y) \preceq_H (x_0, y_0)$ and $(x', y') \in \operatorname{graph} \Gamma$, $(x', y') \preceq_H (x, y)$ imply $x' = x$. Set $\bar{x} := x$ and take $\bar{y} \in \operatorname{Min} \Gamma(\bar{x})$ such that $\bar{y} \leq_K y$. As in the proof of Corollary 10.2.12 we find that $(\bar{x}, \bar{y})$ is the desired element. The proof is complete. □

10.3 Minimal Points in Product Spaces of Isac–Tammer's Type

Besides $F : X \times X \rightrightarrows K$ considered in the preceding section we consider also $F' : Y \times Y \rightrightarrows K$ satisfying conditions (F1) and F(2), that is, $0 \in F'(y, y)$ for all $y \in Y$ and $F'(y_1, y_2) + F'(y_2, y_3) \subseteq F'(y_1, y_3) + K$ for all $y_1, y_2, y_3 \in Y$. Then $\Phi : Z \times Z \rightrightarrows K$ with $Z := X \times Y$, defined by $\Phi((x_1, y_1), (x_2, y_2)) := F(x_1, x_2) + F'(y_1, y_2)$, satisfies conditions (F1) and (F2), too. As in Sect. 10.2 we obtain that the relation $\preccurlyeq_{F,F'}$ defined by

$$(x_1, y_1) \preccurlyeq_{F,F'} (x_2, y_2) \iff y_2 \in y_1 + F(x_1, x_2) + F'(y_1, y_2) + K$$

is reflexive and transitive. Moreover, for $x, x_1, x_2 \in X$ and $y_1, y_2 \in Y$ we have

$$\begin{aligned} (x_1, y_1) \preccurlyeq_{F,F'} (x_2, y_2) &\Longrightarrow (x_1, y_1) \preccurlyeq_F (x_2, y_2) \Longrightarrow y_1 \leq_K y_2, \\ (x, y_1) \preccurlyeq_{F,F'} (x, y_2) &\iff y_2 \in y_1 + F'(y_1, y_2) + K \Rightarrow y_1 \leq_K y_2. \end{aligned}$$

As in the preceding section, for F satisfying (F1)–(F3), F' satisfying (F1), (F2) and z^* from (F3) we define the partial order $\preceq_{F,F',z^*}$ by

$$(x_1, y_1) \preceq_{F,F',z^*} (x_2, y_2) \iff \begin{cases} (x_1, y_1) = (x_2, y_2) \text{ or} \\ (x_1, y_1) \preceq_{F,F'} (x_2, y_2) \text{ and } z^*(y_1) < z^*(y_2). \end{cases}$$

Theorem 10.3.1. *Assume that (X, d) is a complete metric space, Y is a real topological vector space and $K \subseteq Y$ is a proper convex cone. Let $F : X \times X \rightrightarrows K$ satisfy conditions (F1)–(F3), let $F' : Y \times Y \rightrightarrows K$ satisfy (F1) and (F2), and let $A \subseteq X \times Y$ satisfy the condition*

(H1b) *for every $\preccurlyeq_{F,F'}$–decreasing sequence $((x_n, y_n)) \subseteq A$ with $x_n \to x \in X$ there exists $y \in Y$ such that $(x, y) \in A$ and $(x, y) \preccurlyeq_{F,F'} (x_n, y_n)$ for every $n \in \mathbb{N}$.*

Suppose that z^* (from (F3)) is bounded from below on $\mathrm{Pr}_Y(A)$. *Then for every $(x_0, y_0) \in A$ there exists a minimal element $(\bar{x}, \bar{y})$ of A with respect to $\preceq_{F,F',z^*}$ such that $(\bar{x}, \bar{y}) \preceq_{F,F',z^*} (x_0, y_0)$.*

Proof. It is easy to verify that $\preceq_{F,F',z^*}$ is reflexive, transitive and antisymmetric. To get the conclusion one follows the lines of the proof of Theorem 10.2.3. □

Clearly, taking $F' = 0$ in Theorem 10.3.1 we get Theorem 10.2.3. As mentioned after the proof of Theorem 10.2.3, this extends significantly [214, Theorem 3.10.7], keeping practically the same proof. One can ask if [214, Theorem 3.10.15] could be extended to this framework, taking into account that the boundedness condition on A in [214, Theorem 3.10.15] is much less restrictive. In [214, Theorem 3.10.15] it is used a functional φ_{C,k^0} (defined in (10.19) below) in order to prove the minimal point theorem. Because an element k^0 does not impose itself naturally, and we need a stronger condition on the functional φ_{C,k^0} even if $k^0 \in K \subseteq \operatorname{int} C$, we consider an abstract K–monotone functional φ to which we impose some conditions φ_{C,k^0} has already.

Theorem 10.3.2. *Assume that (X, d) is a complete metric space, Y is a real topological vector space and $K \subseteq Y$ is a proper convex cone. Let $F : X \times X \rightrightarrows K$ satisfy conditions (F1)–(F3), let $F' : Y \times Y \rightrightarrows K$ satisfy (F1) and (F2), and let $A \subseteq X \times Y$ satisfy the condition (H1b). Assume that there exists a functional $\varphi : Y \to \overline{\mathbb{R}}$ such that*

(F4) $(x_1, y_1) \preceq_{F,F'} (x_2, y_2) \Longrightarrow \varphi(y_1) + d(x_1, x_2) \le \varphi(y_2).$

If φ is bounded below on $\mathrm{Pr}_Y(A)$ *then for every point $(x_0, y_0) \in A$ with $\varphi(y_0) \in \mathbb{R}$, there exists $(\bar{x}, \bar{y}) \in A$ such that $(\bar{x}, \bar{y}) \preceq_{F,F'} (x_0, y_0)$, and $(x', y') \in A$, $(x', y') \preceq_{F,F'} (\bar{x}, \bar{y})$ imply $x' = \bar{x}$. Moreover, if φ is strictly K–monotone on* $\mathrm{Pr}_Y(A)$, *that is, $y_1, y_2 \in$* $\mathrm{Pr}_Y(A)$, *$y_2 - y_1 \in K \setminus \{0\} \Longrightarrow \varphi(y_1) < \varphi(y_2)$, then $(\bar{x}, \bar{y})$ is a minimal point of A with respect to $\preceq_{F,F'}$.*

Proof. Let us construct a sequence $((x_n, y_n))_{n\ge 0} \subseteq A$ as follows: Having $(x_n, y_n) \in A$, we take $(x_{n+1}, y_{n+1}) \in A$, $(x_{n+1}, y_{n+1}) \preceq_{F,F'} (x_n, y_n)$, such that

$$\varphi(y_{n+1}) \le \inf\{\varphi(y) \mid (x, y) \in A,\ (x, y) \preceq_{F,F'} (x_n, y_n)\} + 1/(n+1). \tag{10.17}$$

Of course, the sequence $((x_n, y_n))$ is $\preceq_{F,F'}$–decreasing. From (F4) we obtain that $0 \le d(x_{n+1}, x_n) \le \varphi(y_n) - \varphi(y_{n+1})$, and so the sequence $(\varphi(y_n))$ is non-increasing. Because $(y_n) \subseteq \mathrm{Pr}_Y(A)$, $(\varphi(y_n))$ is also bounded (from below), and so $(\varphi(y_n))$ is convergent in $\mathbb{R}$. Because $(x_{n+p}, y_{n+p}) \preceq_{F,F'} (x_n, y_n) \preceq_{F,F'} (x_{n-1}, y_{n-1})$, using again (F4) and (10.17) we get

$$\forall\, n, p \in \mathbb{N}^* : d(x_{n+p}, x_n) \le \varphi(y_n) - \varphi(y_{n+p}) \le 1/n.$$

It follows that (x_n) is a Cauchy sequence in the complete metric space (X, d), and so (x_n) is convergent to some $\bar{x} \in X$.

By (H1b) there exists $\overline{y} \in Y$ such that $(\overline{x}, \overline{y}) \in A$ and $(\bar{x}, \bar{y}) \preceq_{F,F'} (x_n, y_n)$ for every $n \in \mathbb{N}$, and so

$$\forall n \geq 1 : \varphi(\overline{y}) \leq \varphi(\overline{y}) + d(\overline{x}, x_n) \leq \varphi(y_n) \leq \varphi(\overline{y}) + 1/n, \tag{10.18}$$

whence $\varphi(y_n) \to \varphi(\overline{y})$. Let us show that $(\overline{x}, \overline{y})$ is the desired element. Indeed, $(\bar{x}, \bar{y}) \preceq_{F,F'} (x_0, y_0)$. Suppose that $(x', y') \in A$ is such that $(x', y') \preceq_{F,F'} (\bar{x}, \bar{y})$ $(\preceq_{F,F'} (x_n, y_n)$ for every $n \in \mathbb{N})$. Thus $\varphi(y') + d(x', \overline{x}) \leq \varphi(\overline{y})$ by (F4); moreover, (10.18) holds with $\overline{y}$ replaced by y', and so $\varphi(\overline{y}) = \varphi(y') = \lim \varphi(y_n)$. Hence $d(x', \bar{x}) = \varphi(\bar{y}) - \varphi(y') = 0$, whence $x' = \overline{x}$.

Assuming that φ is strictly K–monotone, because $y' \leq_K \overline{y}$ and $\varphi(\bar{y}) - \varphi(y') = 0$, we have necessarily $y' = \overline{y}$. Hence $(\bar{x}, \bar{y})$ is a minimal point with respect to $\preceq_{F,F'}$. □

Note that if $C \subseteq Y$ is a proper convex cone such that $K \setminus \{0\} \subseteq \operatorname{int} C$ and $k^0 \in K \setminus \{0\}$, the functional $\varphi_{C,k^0} : Y \to \mathbb{R}$ defined by

$$\varphi_{C,k^0}(y) := \inf \left\{ t \in \mathbb{R} \mid y \in tk^0 - C \right\} \tag{10.19}$$

is a strictly K–monotone continuous sublinear functional (see Corollary 5.2.8). Moreover, if the condition

(B) $\operatorname{Pr}_Y(A) \cap (\tilde{y} - \operatorname{int} C) = \emptyset$ for some $\tilde{y} \in Y$

holds, then $\varphi := \varphi_{C,k^0}$ is bounded from below on $\operatorname{Pr}_Y(A)$. Indeed, by Corollary 5.2.8, we have that $\varphi(y) + \varphi(-\tilde{y}) \geq \varphi(y - \tilde{y}) \geq 0$ for $y \in \operatorname{Pr}_Y(A)$, whence $\varphi(y) \geq -\varphi(-\tilde{y})$ for $y \in \operatorname{Pr}_Y(A)$.

Another example for a function φ is that defined by

$$\varphi(y) := \varphi_{K,k^0}(y - \hat{y}), \tag{10.20}$$

where K is a proper convex cone, $k^0 \in K \setminus \{0\}$, and $\hat{y} \in Y$ is such that

(B1) $y_0 - \hat{y} \in \mathbb{R}k^0 - K, \quad \operatorname{Pr}_Y(A) \cap (\hat{y} - K) = \emptyset.$

Then φ is K–monotone, $\varphi(y_0) < \infty$ and $\varphi(y) \geq 0$ for every $y \in \operatorname{Pr}_Y(A)$.

For both of these functions in (10.19) and (10.20) we have to impose condition (F4) in order to be used in Theorem 10.3.2.

Remark 10.3.3. Using the function $\varphi = \varphi_{K,k^0}(\cdot - \hat{y})$ (defined by (10.20)) in Theorem 10.3.2 we can derive [237, Theorem 4.2] taking $F(x_1, x_2) := \{d(x_1, x_2)k^0\}$ and $F'(y_1, y_2) := \{\varepsilon \|y_1 - y_2\| k^0\}$ when Y is a Banach space; note that, at its turn, [237, Theorem 4.2] extends [285, Theorem 8].

10.4 Ekeland's Variational Principles of Ha's Type

The previous EVP type results correspond to Pareto optimality. Ha [227] established an EVP type result which corresponds to Kuroiwa optimality. Theorem 10.4.4 below is an extension of this type of result. For its proof we need the following result established by Hamel and Tammer in [237, Theorem 2.2] which generalizes [575, Theorem 3.1] (where $S(x)$ is assumed to be closed for every $x \in X$).

Theorem 10.4.1. *Let (X, d) be a metric space, quasi-ordered by $\preccurlyeq$ such that (X, d) is $\preccurlyeq$–complete, that is any $\preccurlyeq$–increasing Cauchy sequence $(x_n) \subseteq X$ is convergent. Assume that*

(i) *$S(x) := \{x' \in X \mid x \preccurlyeq x'\}$ is $\preccurlyeq$–upper closed for every $x \in X$, that is for every $\preccurlyeq$–increasing sequence $(x_n) \subseteq S(x)$ with $x_n \to u$ one has $u \in S(x)$, and*
(ii) *any $\preccurlyeq$–increasing sequence $(x_n) \subseteq X$ is asymptotic, that is $d(x_{n+1}, x_n) \to 0$.*

Then for every $x \in X$ there exists $\overline{x} \in X$ such that $x \preccurlyeq \overline{x}$ and $S(\overline{x}) = \{\overline{x}\}$.

Proof. Let $x \in X$ and set $x_0 := x$; fix also $\alpha \in]0, 1/2[$. There exists $x_1 \in S(x_0)$ such that $d(x_1, x_0) \geq \min\{1, \alpha \operatorname{diam} S(x_0)\}$. Then there exists $x_2 \in S(x_1)$ such that $d(x_2, x_1) \geq \min\{1, \alpha \operatorname{diam} S(x_1)\}$. Continuing in this way we get a sequence $(x_n)_{n\geq 0} \subseteq X$ such that $x_{n+1} \in S(x_n)$ and $d(x_{n+1}, x_n) \geq \min\{1, \alpha \operatorname{diam} S(x_n)\}$ for every n. By (ii) we have that $d(x_{n+1}, x_n) \to 0$, whence $\operatorname{diam} S(x_n) \to 0$. For $\varepsilon > 0$ there exists $n_0 \in \mathbb{N}$ such that $\operatorname{diam} S(x_n) < \varepsilon$ for $n \geq n_0$. Hence for $n, m \geq n_0$ we have that $x_n, x_m \in S(x_{n_0})$, whence $d(x_n, x_m) < \varepsilon$. It follows that (x_n) is an increasing Cauchy sequence, and so (x_n) is convergent to some $\overline{x} \in X$. Since $(x_n)_{n\geq k} \subseteq S(x_k)$ is a $\preccurlyeq$–increasing sequence and $S(x_k)$ is $\preccurlyeq$–upper closed, we obtain that $\overline{x} \in S(x_k)$ [and so $S(\overline{x}) \subseteq S(x_k)$] for every $k \geq 0$. It follows that $0 \leq \operatorname{diam} S(\overline{x}) \leq \operatorname{diam} S(x_k) \to 0$, which shows that $\operatorname{diam} S(\overline{x}) = 0$. Since $\overline{x} \in S(\overline{x})$ we get $S(\overline{x}) = \{\overline{x}\}$. As $\overline{x} \in S(x) = S(x_0)$, the conclusion follows. □

Note that in Theorem 10.4.1 $\preccurlyeq$ is in fact anti-symmetric. Indeed, take $x, x' \in X$ with $x \preccurlyeq x'$ and $x' \preccurlyeq x$. Then the sequence $(x_n)_{n\geq 0}$ defined by $x_{2n} := x$ and $x_{2n+1} := x'$ is $\preccurlyeq$-increasing; by (ii) we get $d(x, x') = d(x_n, x_{n+1}) \to 0$, and so $x = x'$.

Remark 10.4.2. Conditions (i) and (ii) in Theorem 10.4.1 are equivalent to (i') and (ii'), respectively, where

(i') for any $\preccurlyeq$–increasing sequence $(x_n) \subseteq X$ with $x_n \to x \in X$ one has $x_n \preccurlyeq x$ for every $n \geq 1$,
(ii') any $\preccurlyeq$–increasing sequence $(x_n) \subseteq X$ is Cauchy.

Indeed, assume that (i) in Theorem 10.4.1 holds and take $(x_n) \subseteq X$ a $\preccurlyeq$–increasing sequence with $x_n \to x \in X$. Since $(x_n)_{n\geq k} \subseteq S(x_k)$ and $S(x_k)$ is $\preccurlyeq$–upper closed, we have that $x \in S(x_k)$, and so $x_k \preccurlyeq x$ for every k. Conversely, assume that (i') holds and take $(x_n) \subseteq S(x)$ a $\preccurlyeq$–increasing sequence with $x_n \to u$. Then $x \preccurlyeq x_1 \preccurlyeq u$, whence $u \in S(x)$.

Clearly, (ii') $\Rightarrow$ (ii). Conversely, assume that (ii) holds, but the increasing sequence $(x_n)_{n\geq 1} \subseteq X$ is not Cauchy. Then there exists $\varepsilon_0 > 0$ such that for every $p \geq 1$ there exists $n > p$ such that $d(x_n, x_p) \geq \varepsilon_0$. So, for $p = 1 =: n_1$, there exists $n_2 > n_1$ such that $d(x_{n_2}, x_{n_1}) \geq \varepsilon_0$. For $p := n_2$, there exists $n_3 > n_2$ such that $d(x_{n_3}, x_{n_2}) \geq \varepsilon_0$. Continuing in this way we get the (strictly) increasing sequence $(n_k)_{k\geq 1} \subseteq \mathbb{N}^*$ such that $d(x_{n_{k+1}}, x_{n_k}) \geq \varepsilon_0$ for all $k \geq 1$. Setting $x'_k := x_{n_k}$ for $k \geq 1$ we have that $(x'_k)_{k\geq 1}$ is $\preccurlyeq$–increasing, and so (x'_k) is asymptotic. This contradicts the fact that $d(x'_{k+1}, x'_k) \geq \varepsilon_0$ for every $k \geq 1$.

Remark 10.4.2 shows that Theorem 10.4.1 can be reformulated as in the following result (see [391, Lemma 2.2]).

Theorem 10.4.3. *Let (X, d) be a metric space, quasi-ordered by $\preccurlyeq$, such that any $\preccurlyeq$–increasing sequence $(x_n) \subseteq X$ converges to some $x \in X$ with $x_n \preccurlyeq x$ for every n. Then for every $x \in X$ there exists $\overline{x} \in X$ such that $x \preccurlyeq \overline{x}$ and $S(\overline{x}) = \{\overline{x}\}$.*

Note also that Theorem 10.4.1 is slightly more general than the Dancs–Hegedüs–Medvegyev Theorem (see [120]), in which (X, d) is assumed to be complete instead of being $\preccurlyeq$–complete.

Theorem 10.4.4. *Assume that (X, d) is a complete metric space, Y is a real topological vector space and $K \subseteq Y$ is a proper convex cone. Let $F : X \times X \rightrightarrows K$ satisfy conditions (F1)–(F3) and $\Gamma : X \rightrightarrows Y$ be such that z^* (from (F3)) is bounded below on $\Gamma(X)$. If $\{x \in X \mid \Gamma(u) \subseteq \Gamma(x) + F(x, u) + K\}$ is closed for every $u \in X$, then for every $x_0 \in \operatorname{dom} \Gamma$ there exists $\overline{x} \in X$ such that $\Gamma(x_0) \subseteq \Gamma(\overline{x}) + F(\overline{x}, x_0) + K$, and $\Gamma(\overline{x}) \subseteq \Gamma(x) + F(x, \overline{x}) + K$ implies $x = \overline{x}$.*

Proof. Let us consider the relation $\preccurlyeq$ on X defined by

$$x \preccurlyeq x' \iff \Gamma(x) \subseteq \Gamma(x') + F(x', x) + K. \tag{10.21}$$

By our hypotheses we have that $S(x) := \{x' \in X \mid x \preccurlyeq x'\}$ is closed for every $x \in X$. Note that for $x \in X \setminus \operatorname{dom} \Gamma$ we have that $S(x) = X$, while for $x \in \operatorname{dom} \Gamma$ we have that $S(x) \subseteq \operatorname{dom} \Gamma$. The relation $\preccurlyeq$ is reflexive and transitive. The reflexivity of $\preccurlyeq$ is obvious. Let $x \preccurlyeq x'$ and $x' \preccurlyeq x''$. Then $\Gamma(x) \subseteq \Gamma(x') + F(x', x) + K$ and $\Gamma(x') \subseteq \Gamma(x'') + F(x'', x') + K$. Using (F2) we get

$$\Gamma(x) \subseteq \Gamma(x'') + F(x'', x') + K + F(x', x) + K \subseteq \Gamma(x'') + F(x'', x) + K,$$

that is, $x \preccurlyeq x''$. Consider

$$\varphi : X \to \overline{\mathbb{R}}, \quad \varphi(x) := \inf z^* \left(\Gamma(x)\right), \tag{10.22}$$

with the usual convention $\inf \emptyset := +\infty$. Clearly, $\varphi(x) \geq m := \inf z^*(\Gamma(X)) > -\infty$. Moreover, if $\operatorname{dom} \Gamma \ni x \preccurlyeq x'$ then $z^*(\Gamma(x)) \subseteq z^*(\Gamma(x')) + z^*(F(x', x)) + z^*(K)$, whence $\varphi(x) \geq \varphi(x') + \inf z^*(F(x', x)) \geq \varphi(x')$.

Fix $x_0 \in \operatorname{dom} \Gamma$. The conclusion of the theorem asserts that there exists $\overline{x} \in X$ such that $\overline{x} \in S(x_0)$ and $S(\overline{x}) = \{\overline{x}\}$. To get this conclusion we apply Theorem 10.4.1. Because (X, d) is complete and $S(x)$ is closed for every $x \in X$, we may (and do) assume that $\operatorname{dom} \Gamma = X$ (otherwise we replace X by $S(x_0)$).

In order to apply Theorem 10.4.1 let $(x_n)_{n\geq 1} \subseteq X$ be $\preccurlyeq$–increasing and let us show that (x_n) is a Cauchy sequence. In the contrary case there exist $\delta > 0$ and $(n_p)_{p\geq 1} \subseteq \mathbb{N}^*$ an increasing sequence such that $d(x_{n_p}, x_{n_{p+1}}) \geq \delta$ for every $p \geq 1$. Then, as seen above, $\varphi(x_{n_p}) \geq \varphi(x_{n_{p+1}}) + \inf z^*\big(F(x_{n_{p+1}}, x_{n_p})\big)$, and so

$$\varphi(x_{n_1}) \geq \varphi\big(x_{n_{p+1}}\big) + \sum_{l=1}^{p} \inf z^*\big(F(x_{n_{l+1}}, x_{n_l})\big) \geq m + p \cdot \eta(\delta)$$

with $\eta(\delta) > 0$ from (F3). Letting $p \to \infty$ we get a contradiction. Hence (x_n) is Cauchy, and so asymptotic. Applying Theorem 10.4.1 the conclusion follows. □

Note that instead of assuming $S(u)$ to be closed for every $u \in X$ it is sufficient to have that $S(u)$ is $\preccurlyeq$–upper closed, where $\preccurlyeq$ is defined by (10.21).

Remark 10.4.5. Taking Y to be a separated locally convex space, $K \subseteq Y$ a pointed closed convex cone and $F(x, x') := \{d(x, x')k^0\}$ with $k^0 \in K \setminus \{0\}$, we can deduce [227, Theorem 3.1]. For this assume that $\Gamma(X)$ is quasi bounded, $\Gamma(x) + K$ is closed for every $x \in X$ and Γ is K-l.s.c. Since clearly z^* is bounded from below on $\operatorname{Im} \Gamma$, in order to apply the preceding theorem we need to have that $S(u)$ is closed for every $u \in X$; this is done in [227, Lemma 3.2]. Below we provide another proof for the closedness of $S(u)$.

First, if $x \notin \operatorname{lev}_\Gamma(b)$ then there exists $\delta > 0$ such that $B(x, \delta) \cap \operatorname{lev}_\Gamma(b + \delta k^0) = \emptyset$. Indeed, because $x \notin \operatorname{lev}_\Gamma(b)$ we have that $b \notin \Gamma(x) + K$, and so $b + \delta' k^0 \notin \Gamma(x) + K$, that is, $x \notin \operatorname{lev}_\Gamma(b + \delta' k^0)$, for some $\delta' > 0$ (since $\Gamma(x) + K$ is closed). Because $\operatorname{lev}_\Gamma(b + \delta' k^0)$ is closed, there exists $\delta \in \,]0, \delta']$ such that $B(x, \delta) \cap \operatorname{lev}_\Gamma(b + \delta' k^0) = \emptyset$, and so $B(x, \delta) \cap \operatorname{lev}_\Gamma(b + \delta k^0) = \emptyset$.

Fix $u \in X$ and take $x \in X \setminus S(u)$, that is, $\Gamma(u) \not\subseteq \Gamma(x) + d(x, u)k^0 + K$. Then there exists $y \in \Gamma(u)$ with $b := y - d(x, u)k^0 \notin \Gamma(x) + K$. By the argument above there exists $\delta' > 0$ such that $B(x, \delta') \cap \operatorname{lev}_\Gamma(b + \delta' k^0) = \emptyset$, that is, $y - d(x, u)k^0 + \delta' k^0 \notin \Gamma(x') + K$ for every $x' \in B(x, \delta')$. Taking $\delta \in \,]0, \delta']$ sufficiently small we have that $d(x', u) \geq d(x, u) - \delta'$ for $x' \in B(x, \delta)$, and so $y \notin \Gamma(x') + d(x', u)k^0 + K$ for every $x' \in B(x, \delta)$, that is, $B(x, \delta) \cap S(u) = \emptyset$.

If we assume that $\Gamma(x_0) \not\subseteq \Gamma(x) + k^0 + K$ for every $x \in X$, then $\overline{x}$ provided by the preceding theorem satisfies $d(\overline{x}, x_0) < 1$. Indeed, in the contrary case, because $\Gamma(x_0) \subseteq \Gamma(\overline{x}) + d(\overline{x}, x_0)k^0 + K$ and $d(\overline{x}, x_0)k^0 + K \subseteq k^0 + K$, we get the contradiction $\Gamma(x_0) \subseteq \Gamma(\overline{x}) + k^0 + K$. Replacing k^0 by εk^0 and d by $\lambda^{-1} d$ for some $\varepsilon, \lambda > 0$ we obtain exactly the statement of [227, Theorem 3.1].

With a similar proof we get the next result; for Y a separated locally convex space, $K \subseteq Y$ a closed convex cone, $k^0 \in K \setminus \{0\}$, $F(x, x') := d(x, x')k^0$ for $x, x' \in X$, and $\Gamma(X)$ quasi-bounded, one gets [390, Corollary 3.1].

Theorem 10.4.6. *Assume that (X, d) is a complete metric space, Y is a real topological vector space and $K \subseteq Y$ is a proper convex cone. Let $F : X \times X \rightrightarrows K$ satisfy conditions (F1) and (F3), and $\Gamma : X \rightrightarrows Y$ be such that z^* (from (F3)) is bounded below on $\Gamma(X)$. Assume that* (a) $\{x \in X \mid \Gamma(u) \subseteq \Gamma(x) + F(x, u)\}$ *is closed for every $u \in X$, and* (b) $\Gamma(y) \subseteq \Gamma(x) + F(x, y)$, $\Gamma(z) \subseteq \Gamma(y) + F(y, z)$ *imply $\Gamma(z) \subseteq \Gamma(x) + F(x, z)$ for all $x, y, z \in X$. Then for every $x_0 \in \operatorname{dom} \Gamma$ there exists $\overline{x} \in X$ such that $\Gamma(x_0) \subseteq \Gamma(\overline{x}) + F(\overline{x}, x_0)$, and $\Gamma(\overline{x}) \subseteq \Gamma(x) + F(x, \overline{x}) + K$ implies $x = \overline{x}$.*

Proof. The proof is practically the same as that of Theorem 10.4.4. This time the relation $\preccurlyeq$ on X is defined by $x \preccurlyeq x'$ if $\Gamma(x) \subseteq \Gamma(x') + F(x', x)$. By (a) we have that $S(x) := \{x' \in X \mid x \preccurlyeq x'\}$ is closed for every $x \in X$. Note that for $x \in X \setminus \operatorname{dom} \Gamma$ we have that $S(x) = X$, while for $x \in \operatorname{dom} \Gamma$ we have that $S(x) \subseteq \operatorname{dom} \Gamma$. The reflexivity of $\preccurlyeq$ is obvious, while its transitivity is given by (b). Consider φ given by (10.22). The rest of the proof is the same as that of Theorem 10.4.4. □

In the case in which Y is just a topological vector space we have the following version of Theorem 10.4.4.

Theorem 10.4.7. *Assume that (X, d) is a complete metric space, Y is a real topological vector space and $K \subseteq Y$ is a proper closed convex cone. Let $H \subseteq K$ be nonempty convex set with $0 \notin \operatorname{cl}(H + K)$, and $\Gamma : X \rightrightarrows Y$. If $\{x \in X \mid \Gamma(u) \subseteq \Gamma(x) + d(x, u)H + K\}$ is closed for every $u \in X$ and $\Gamma(X)$ is quasi bounded, then for every $x_0 \in \operatorname{dom} \Gamma$ there exists $\overline{x} \in X$ such that $\Gamma(x_0) \subseteq \Gamma(\overline{x}) + d(\overline{x}, x_0)H + K$ and $\Gamma(\overline{x}) \subseteq \Gamma(x) + d(x, \overline{x})H + K$ implies $x = \overline{x}$.*

Proof. Let $B \subseteq Y$ be a bounded set such that $\Gamma(X) \subseteq B + K$.

Consider $F := F_H$ with F_H defined in (10.11). As seen before, F satisfies conditions (F1) and (F2), and so the relation $\preccurlyeq$ defined in the proof of Theorem 10.4.4 is a quasi-order; moreover, by our hypotheses, $S(x) := \{x' \in X \mid x \preccurlyeq x'\}$ is closed for every $x \in X$. As in the proof of Theorem 10.4.4 we may (and do) assume that $X = \operatorname{dom} \Gamma$ and it is sufficient to show that every $\preccurlyeq$–increasing sequence $(x_n)_{n \geq 1} \subseteq X$ is Cauchy. In the contrary case there exist $\delta > 0$ and $(n_p)_{p \geq 1} \subseteq \mathbb{N}^*$ an increasing sequence such that $d(x_{n_p}, x_{n_{p+1}}) \geq \delta$ for every $p \geq 1$.

Fixing $y_1 \in \Gamma(x_1)$, inductively we find the sequences $(y_n)_{n \geq 0} \subseteq Y$, $(h_n)_{n \geq 0} \subseteq H$ and $(k_n)_{n \geq 0} \subseteq K$ such that $y_n = y_{n+1} + d(x_n, x_{n+1})h_n + k_n$ for every $n \geq 1$. Using the convexity of H, and the facts that $H \subseteq K$ and $\Gamma(X) \subseteq B + K$, for $p \in \mathbb{N}$ we get $h'_p \in H$, $b_p \in B$ and $k'_p, k''_p \in K$ such that

$$\begin{aligned} y_1 &= y_{n_{p+1}} + \sum_{l=1}^{n_p} d(x_l, x_{l+1})h_l + \sum_{l=0}^{p} k_l = b_p + \delta(h_{n_1} + \ldots + h_{n_p}) + k'_p \\ &= b_p + p\delta h'_p + k''_p. \end{aligned}$$

It follows that $(p\delta)^{-1}(y_1 - b_p) \in H + K$ for every $p \geq 1$. Since (b_p) is bounded we obtain the contradiction $0 \in \operatorname{cl}(H + K)$. The conclusion follows. □

Remark 10.4.8. In Theorem 10.4.7, instead of assuming that $S(u)$ is closed for every $u \in X$, it is sufficient to assume that $S(u)$ is $\preccurlyeq$–upper closed for $u \in X$, where $\preccurlyeq$ is defined by (10.21) for $F(x,x') = d(x,x')H$.

Theorem 10.4.7 covers [227, Theorem 3.1], too.

Theorem 10.4.9. *Assume that (X,d) is a complete metric space, Y is a real topological vector space, $K \subseteq Y$ is a proper closed convex cone, and $H \subseteq K$ is a nonempty convex set. Let $\Gamma : X \rightrightarrows Y$ be such that* epi Γ *:= $\{(x,y) \in X \times Y \mid y \in \Gamma(x) + K\}$ is closed and $\Gamma(X)$ is quasi bounded. Assume that one of the following conditions holds:*

(i) *H is bounded, cs-complete, and $0 \notin \operatorname{cl}(H+K)$;*
(ii) *Y is a n.v.s., and H is cs-complete with $0 \notin \operatorname{cl} H$ and satisfying*

$$\exists \overline{y}^* \in K^+ \setminus \{0\},\ \forall h \in H : \langle h, \overline{y}^* \rangle \geq \|h\|; \tag{10.23}$$

(iii) *Y is a reflexive Banach space, and H is closed with $0 \notin H$ and satisfying*

$$\exists \gamma > 0,\ \forall h \in H,\ \forall k \in K : \|h+k\| \geq \gamma \|h\|. \tag{10.24}$$

Then for every $x_0 \in \operatorname{dom} \Gamma$ there exists $\overline{x} \in X$ such that $\Gamma(x_0) \subseteq \Gamma(\overline{x}) + d(\overline{x},x_0)H + K$ and $\Gamma(\overline{x}) \subseteq \Gamma(x) + d(x,\overline{x})H + K$ implies $x = \overline{x}$.

Proof. We apply Theorem 10.4.7, taking into account Remark 10.4.8. For this first observe that $0 \notin \operatorname{cl}(H+K)$ in each case. Indeed, (10.23) implies (10.24) with $\gamma := \|\overline{y}^*\|^{-1} > 0$, and (10.24) together $0 \notin \operatorname{cl} H$ imply $\|y\| \geq \gamma d(0,H) > 0$ for every $y \in H + K$.

We have to show that $S(u)$ is $\preccurlyeq$-upper closed for every $u \in X$. Let $B \subseteq Y$ be a bounded set such that $\Gamma(X) \subseteq B + K$.

Consider $(x_n)_{n\geq 1} \subseteq S(u)$ a $\preccurlyeq$–increasing sequence with $x_n \to x \in X$ and $y \in \Gamma(u)$. We have to show that $y \in \Gamma(x) + d(x,u)H + K$. Setting $x_0 := u$ and $y_0 := y$ we have that

$$y_n = y_{n+1} + d(x_n,x_{n+1})h_n + k_n \geq_K y_{n+1} \quad \forall n \geq 0, \tag{10.25}$$

where $y_n \in \Gamma(x_n)$, $h_n \in H$ and $k_n \in K$ for all $n \geq 0$. It follows that

$$\begin{aligned} y &= y_{n+1} + \sum_{i=0}^{n} d(x_i,x_{i+1})h_i + k_n' = b_n + \sum_{i=0}^{n} d(x_i,x_{i+1})h_i + k_n'' \\ &= y_{n+1} + \mu_n h_n' + k_n' = b_n + \mu_n h_n' + k_n'' \end{aligned} \tag{10.26}$$

for every $n \geq 0$, where $k_n' := \sum_{i=0}^{n} k_n \in K$, $k_n'' \in K$, $\mu_n := \sum_{i=0}^{n} d(x_i,x_{i+1})$, $h_n' \in H$ and $b_n \in B$; we have taken into account the convexity of H and that $\Gamma(X) \subseteq B + K$.

Set $\mu := \sum_{n\geq 0} d(x_n, x_{n+1}) = \lim \mu_n \in [0, \infty]$. Clearly, $d(x_{n+1}, u) \leq \mu_n \leq \mu_{n+1} \leq \mu$ for every $n \geq 0$. If $\mu = 0$ then $x = u$, and so $y \in \Gamma(x) + d(x,u)H + K$. Assume that $\mu = \infty$. From the fourth expression of y in (10.26) we have that $\mu_n^{-1}(y - b_n) \in H + K$ for n sufficiently large. Since (b_n) is bounded we obtain the contradiction $0 \in \operatorname{cl}(H + K)$.

Let $\mu \in]0, \infty[$. For $n \geq 0$ let us set

$$y_n' := \mu_n h_n' = \sum_{i=0}^{n} d(x_i, x_{i+1}) h_i.$$

Assume first that (i) holds. Because H is bounded and convex we obtain that the sequence $(y_n')_{n\geq 1}$ is Cauchy. Assume now that (ii) holds. From the second expression of y in (10.26) and (10.23) we have that $\langle y, \overline{y}^* \rangle \geq \langle b_n, \overline{y}^* \rangle + \sum_{i=0}^{n} d(x_i, x_{i+1}) \|h_i\|$. Because B is bounded we obtain that the series $\sum_{n\geq 0} d(x_n, x_{n+1}) h_n$ is absolutely convergent, and so the sequence $(y_n')_{n\geq 1}$ is Cauchy. In both cases, because H is cs-complete, we have that $y_n' \to \mu h$ for some $h \in H$. From the first expression of y in (10.26) we have that $y - y_n' \in \Gamma(x_{n+1}) + K$, and so $(x_{n+1}, y - y_n') \in \operatorname{epi} \Gamma$. Hence $(x_{n+1}, y - y_n') \to (x, y - \mu h) \in \operatorname{cl}(\operatorname{epi} \Gamma) = \operatorname{epi} \Gamma$. It follows that $y \in \Gamma(x) + \mu H + K \subseteq \Gamma(x) + d(x,u)H + K$ (because $d(x,u) \leq \mu$), and so $x \in S(u)$.

Assume now that (iii) holds. Using again the fourth expression of y in (10.26), from (10.24) we obtain that $\|y - b_n\| \geq \gamma \mu_n \|h_n'\|$, and so (h_n') is bounded. Because Y is reflexive, $(h_n')_{n\geq 1}$ has a subsequence converging weakly to $h \in H$, and so $h \in H_p := \overline{\operatorname{conv}}\{h_n' \mid n \geq p\}$ for every $p \geq 1$. Fix some $n_1 \geq 1$ such that $d(x_n, x) < 1$ for every $n \geq n_1$. Because $h \in H_{n_1}$, there exists $n_2 > n_1$ and $(\lambda_n^1)_{n_1 \leq n < n_2} \subseteq \mathbb{R}_+$ such that $\sum_{n=n_1}^{n_2-1} \lambda_n^1 = 1$ and $\left\|h^1 - h\right\| < 1$, where $h^1 := \sum_{n=n_1}^{n_2-1} \lambda_n^1 h_n' \in H$. Increasing if necessary n_2, we may (and do) assume that $d(x_n, x) < 1/2$ for every $n \geq n_2$. Continuing in this way, we find an increasing sequence $(n_k)_{k\geq 1} \subseteq \mathbb{N}^*$ such that for $k \geq 1$ we have $d(x_n, x) < 1/k$ for every $n \geq n_k$ and there exists $(\lambda_n^k)_{n_k \leq n < n_{k+1}} \subseteq \mathbb{R}_+$ such that $\sum_{n=n_k}^{n_{k+1}-1} \lambda_n^k = 1$ and $\left\|h^k - h\right\| < 1/k$, where $h^k := \sum_{n=n_k}^{n_{k+1}-1} \lambda_n^k h_n' \in H$. Because $\mu_n = \sum_{n\geq 0} d(x_n, x_{n+1}) \geq d(x,u) - d(x_{n+1}, x)$, we have that $\mu_n \geq d(x,u) - 1/k$ for every $n \geq n_k$. Taking into account the third expression of y in (10.26), then (10.24), it follows that

$$y \geq_K y_{n+1} + \mu_n h_n' \geq y_{n+1} + (d(x,u) - 1/k) h_n' \geq_K y_{n_{k+1}} + (d(x,u) - 1/k) h_n'$$

for every $n \in \overline{n_k, n_{k+1} - 1}$, whence $y \geq y_{n_{k+1}} + (d(x,u) - 1/k) h^k$. Hence $(x_{n_{k+1}}, y - (d(x,u) - 1/k) h^k) \in \operatorname{epi} \Gamma$ for every $k \geq 1$. Since $h^k \to h$ for $k \to \infty$, from the closedness of $\operatorname{epi} \Gamma$, we get $(x, y - d(x,u)h) \in \operatorname{epi} \Gamma$, and so $y \in \Gamma(x) + d(x,u)H + K$. The proof is complete. □

Remark 10.4.10. Theorem 10.4.9 (iii) is exactly [391, Theorem 3.5 (iii)], Theorem 10.4.9 (ii) is slightly more general than [391, Theorem 3.5 (i)] (in which, moreover, Y is a Banach space, and H is closed and convex), and Theorem 10.4.9

(i) is much more general than [391, Theorem 3.5 (ii)] (in which, moreover, Y is a Banach space, H is closed and convex, and (10.24) holds).

10.5 Ekeland's Variational Principle for Bi-Set-Valued Maps

In [48] Bianchi, Kassay and Pini obtained an EVP type result for vector functions of two variables; previously such results were obtained by Isac [282] and Li *et al.* [384]. The next result extends [48, Theorem 1] in two directions: $d(\cdot,\cdot)e$ is replaced by F satisfying (F1)–(F3) and instead of a single-valued function $f : X \times X \to Y$ we take a multi-valued one.

Theorem 10.5.1. *Assume that (X,d) is a complete metric space, Y is a real topological vector space and $K \subseteq Y$ is a proper convex cone. Let $F : X \times X \rightrightarrows K$ satisfy conditions (F1)–(F3). Assume that $G : X \times X \rightrightarrows Y$ has the properties:*

(i) *$0 \in G(x,x)$ for every $x \in X$,*
(ii) *$G(x_1,x_2) + G(x_2,x_3) \subseteq G(x_1,x_3) + K$ for all $x_1,x_2,x_3 \in X$,*
(iii) *z^* (from (F3)) is bounded below on the set* $\operatorname{Im} G(x,\cdot)$ *for every $x \in X$.*

If

(iv) *$\{x' \in X \mid [G(x,x') + F(x,x')] \cap (-K) \neq \emptyset\}$ is closed for every $x \in X$,*

then for every $x_0 \in X$ there exists $\overline{x} \in X$ such that $[G(x_0,\overline{x}) + F(x_0,\overline{x})] \cap (-K) \neq \emptyset$ and $[G(\overline{x},x) + F(\overline{x},x)] \cap (-K) \neq \emptyset$ implies $x = \overline{x}$.

Proof. Let us consider the relation $\preccurlyeq$ on X defined by

$$x \preccurlyeq x' \iff \left[G(x,x') + F(x,x')\right] \cap (-K) \neq \emptyset.$$

Then $\preccurlyeq$ is reflexive and transitive. The reflexivity is immediate from (i) and (F1). Assume that $x \preccurlyeq x'$ and $x' \preccurlyeq x''$. Then $-k \in G(x,x') + F(x,x')$ and $-k' \in G(x',x'') + F(x',x'')$ with $k,k' \in K$. Hence, by (ii) and (F2),

$$-k-k' \in G(x,x')+F(x,x')+G(x',x'')+F(x',x'') \subseteq G(x,x'')+F(x,x'')+K,$$

whence $[G(x,x'') + F(x,x'')] \cap (-K) \neq \emptyset$, that is, $x \preccurlyeq x''$.

Setting $S(x) := \{x' \in X \mid x \preccurlyeq x'\}$, by (iv) we have that $S(x)$ is closed for every $x \in X$. We have to show that for $(x_n)_{n\geq 1} \subseteq X$ a $\preccurlyeq$–increasing sequence one has $d(x_n,x_{n+1}) \to 0$. In the contrary case there exist an increasing sequence $(n_l)_{l\geq 1} \subseteq \mathbb{N}$ and $\delta > 0$ such that $d(x_{n_l},x_{n_l+1}) \geq \delta$ for every $l \geq 1$. Because (x_n) is $\preccurlyeq$–increasing, we have that $-k_n \in G(x_n,x_{n+1}) + F(x_n,x_{n+1})$ for some $k_n \in K$ and every $n \geq 1$. Then

$$-k_1 - \ldots - k_n \in G(x_1,x_{n+1}) + F(x_1,x_2) + \ldots + F(x_n,x_{n+1}) + K,$$

and so

$$\forall n \geq 1 : \inf z^* (\operatorname{Im} G(x_1, \cdot)) + \inf z^* (F(x_1, x_2)) + \ldots + \inf z^* (F(x_n, x_{n+1})) \leq 0.$$

Since $\inf z^* (F(x_n, x_{n+1})) \geq 0$ for every $n \geq 1$ and $\inf z^* \left(F(x_{n_l}, x_{n_l+1})\right) \geq \eta(\delta) > 0$ for every $l \geq 1$, taking $n := n_p$ with $p \geq 1$, we obtain that $p\eta(\delta) \leq -\inf z^* (\operatorname{Im} G(x_1, \cdot))$ for every $p \geq 1$. This yields the contradiction $\eta(\delta) \leq 0$. Hence $d(x_n, x_{n+1}) \to 0$. Applying Theorem 10.4.1 we get some $\overline{x} \in S(x_0)$ with $S(\overline{x}) = \{\overline{x}\}$, that is, our conclusion holds. □

Remark 10.5.2. If we need the conclusion only for a fixed (given) point $x_0 \in X$, we may replace condition (iii) by the fact that z^* (from (F3)) is bounded below on the set $\operatorname{Im} G(x_0, \cdot)$.

Indeed, $X_0 := S(x_0)$ is closed by (iv), and so (X_0, d) is complete. If $x \in X_0$ then $-k \in G(x_0, x) + F(x_0, x) \subseteq G(x_0, x) + K$ for some $k \in K$, and so $-k' \in G(x_0, x)$ for some $k' \in K$. It follows that $-k' + G(x, u) \subseteq G(x_0, x) + G(x, u) \subseteq G(x_0, u) + K$, whence $G(x, u) \subseteq G(x_0, u) + K$ for every $u \in X$. Hence condition (iii) is verified on X_0, and so the conclusion of the theorem holds for x_0.

Remark 10.5.3. For $F(x, x') := \{d(x, x')k^0\}$ with $k^0 \in K \setminus \{0\}$ and G single-valued, using Theorem 10.5.1 and the preceding remark one obtains [282, Theorem 8] and [384, Theorem 3]; in [282] K is normal and closed, while in [384] $k^0 \in \operatorname{int} K$.

Note that condition (iv) in the preceding theorem holds when G is compact-valued, $G(u, \cdot)$ is K-l.s.c., K is closed and $F(x, x') := \{d(x, x')k^0\}$ for some $k^0 \in K$. Indeed, assume that $-k_n \in G(u, x_n) + d(x_n, u)k^0$ for every $n \geq 1$, where $k_n \in K$. Take $\varepsilon > 0$. Then there exists $n_\varepsilon \geq 1$ such that $d(x_n, u) \geq d(x, u) - \varepsilon =: \gamma_\varepsilon$ for every $n \geq n_\varepsilon$. Then for such n we have that $G(u, x_n) \cap \left(-\gamma_\varepsilon k^0 - K\right) \neq \emptyset$, whence $G(u, x) \cap \left(-\gamma_\varepsilon k^0 - K\right) \neq \emptyset$. Hence there exists $y_\varepsilon \in G(u, x)$ such that $y_\varepsilon + \gamma_\varepsilon k^0 \in -K$. Since $G(u, x)$ is compact, $(y_\varepsilon)_{\varepsilon > 0}$ has a subnet converging to $y \in G(x, u)$. Since $\lim_{\varepsilon \to 0} \gamma_\varepsilon = d(x, u)$ and K is closed, we obtain that $y + d(x, u)k^0 \in -K$.

If Y is a separated locally convex space then we may assume that G is weakly compact-valued instead of being compact-valued.

When G is single-valued and $F(x, x') := \{d(x, x')k^0\}$ with $k^0 \in K$, where K is closed and $z^*(k^0) = 1$, the preceding theorem reduces to [48, Theorem 1].

10.6 EVP Type Results

The framework is the same as in the previous sections. We want to apply the preceding results to obtain vectorial EVPs. To envisage functions defined on subsets of X we add to Y the distinct elements $-\infty$, ∞ not belonging to the space Y, obtaining thus the space $Y^\bullet := Y \cup \{-\infty, \infty\}$. We consider that $-\infty \leq_K y \leq_K \infty$

for all $y \in Y$. Consider now the function $f : X \to Y^\bullet$. As usual, the domain of f is $\operatorname{dom} f = \{x \in X \mid f(x) \neq \infty\}$; the epigraph of f is $\operatorname{epi} f = \{(x,y) \in X \times Y \mid f(x) \leq_K y\}$; f is proper if $\operatorname{dom} f \neq \emptyset$ and $f(x) \neq -\infty$ for every $x \in X$; the graph of the proper function f is $\operatorname{graph} f = \{(x, f(x)) \mid x \in \operatorname{dom} f\}$. For $y^* \in K^+$ we set $(y^* \circ f)(x) := +\infty$ for $x \in X \setminus \operatorname{dom} f$.

Theorem 10.6.1. *Assume that (X,d) is a complete metric space, Y is a real topological vector space and $K \subseteq Y$ is a proper convex cone. Let $F : X \times X \rightrightarrows K$ satisfy the conditions (F1), (F2) and (10.9), and let $f : X \to Y^\bullet$ be a proper function such that the next condition holds:*

(H3) *for every sequence $(x_n) \subseteq \operatorname{dom} f$ with $x_n \to x \in X$ and $f(x_n) \in f(x_{n+1}) + F(x_{n+1}, x_n) + K$ for every $n \in \mathbb{N}$ one has $f(x_n) \in f(x) + F(x, x_n) + K$ for every $n \in \mathbb{N}$.*

Then for every $x_0 \in \operatorname{dom} f$ and every $z^* \in K^+$ such that $z^* \circ f$ is bounded from below on the set $\{x \in \operatorname{dom} f \mid f(x_0) \in f(x) + F(x, x_0) + K\}$, there exists $\overline{x} \in \operatorname{dom} f$ such that

$$f(x_0) \in f(\overline{x}) + F(\overline{x}, x_0) + K$$

and

$$\forall x \in \operatorname{dom} f \; : \; f(\overline{x}) \in f(x) + F(\overline{x}, x) + K \Rightarrow x = \overline{x}.$$

Proof. Consider $A := \operatorname{graph} f := \{(x, f(x)) \mid x \in \operatorname{dom} f\}$. Condition (H3) says nothing else than that (H1) is verified. Applying Theorem 10.2.1 (taking into account Remark 10.2.2) we get the conclusion. □

Note that the above result is equivalent to [484, Theorem 3.6], which, at its turn, extends the main result of [220]. Moreover, the result above can be used when F satisfies also (F3) and $z^* \circ f$ is bounded from below, where $z^* \in K^+$ is the element appearing in (F3).

In the case in which $F(x, x') = d(x, x')H$ for some $H \subseteq K$ the condition (H3) becomes

(H4) for every sequence $(x_n) \subseteq \operatorname{dom} f$ with $x_n \to x \in X$ and $f(x_n) \in f(x_{n+1}) + d(x_{n+1}, x_n)H + K$ for every $n \in \mathbb{N}$ one has $f(x_n) \in f(x) + d(x, x_n)H + K$ for every $n \in \mathbb{N}$.

In the case $H := \{k^0\}$ condition (H4) is nothing else than condition (E1) in [237].

Using Theorem 10.2.11 we have the following variant of the preceding result.

Theorem 10.6.2. *Assume that (X,d) is a complete metric space, Y is a real topological vector space and $K \subseteq Y$ is a proper closed convex cone. Let $f : X \to Y^\bullet$ be a proper function and $H \subseteq K$ be a nonempty cs-complete bounded set with $0 \notin \operatorname{cl}(H + K)$. Assume that $f(\operatorname{dom} f)$ is quasi bounded. If*

(H5) *for every sequence* $(x_n) \subseteq \operatorname{dom} f$ *such that* $x_n \to x \in X$ *and* $(f(x_n))$ *is* $\leq_K$*–decreasing one has* $f(x) \leq_K f(x_n)$ *for every* $n \in \mathbb{N}$

holds, then for every $x_0 \in \operatorname{dom} f$ *there exists* $\overline{x} \in \operatorname{dom} f$ *such that*

$$(f(x_0) - K) \cap (f(\overline{x}) + d(\overline{x}, x_0)H) \neq \emptyset$$

and

$$\forall x \in \operatorname{dom} f \setminus \{\overline{x}\} : (f(\overline{x}) - K) \cap (f(x) + d(\overline{x}, x)H) = \emptyset.$$

Proof. Let us apply Theorem 10.2.11 for $A := \operatorname{graph} f$. To have the conclusion of the theorem it is sufficient to observe that (H2) is verified. Indeed, let $((x_n, y_n)) \subseteq \operatorname{graph} f$ be such that $x_n \to x \in X$ and let (y_n) be $\leq_K$–decreasing. Hence $y_n = f(x_n)$ for every n. By (H5) we have that $y := f(x) \leq_K f(x_n) = y_n$ for every $n \in \mathbb{N}$ and, of course, $(x, y) \in \operatorname{graph} f$. The proof is complete. □

Remark 10.6.3. According to L. Gajek and D. Zagrodny in [194], condition (H5) is saying that f is monotonically semicontinuous with respect to K at any $x \in X$.

Remark 10.6.4. Taking H to be complete, convex and bounded, then H is cs-complete. In this case, Theorem 10.6.2 reduces to the main result in [40], that is, [40, Theorem 4.1].

Remark 10.6.5. Taking $H := \{k^0\}$ in the preceding theorem one obtains [214, Corollary 3.10.6]; there K is assumed to be closed in the direction k^0, the present condition (H5) being condition (H4) in [214, Corollary 3.10.6].

Remark 10.6.6. Similar results can be stated using Theorems 10.3.1 and 10.3.2. When specializing to $F(x_1, x_2) = \{d(x_1, x_2)k^0\}$ and $F'(y_1, y_2) = \{\varepsilon \|y_1 - y_2\| k^0\}$ one recovers [237, Corollary 3.1] and [237, Theorem 4.2].

In the conclusions of [483, Lemma 2.2, Theorem 2.1] $\psi \circ d$ appears instead of d, as in the usual EVP type results, where Ψ is the class of those subadditive and non decreasing functions $\psi : \mathbb{R}_+ \to \mathbb{R}_+$ with $\psi(t) = 0$ only for $t = 0$. Based on the next simple result we can show that Lemma 2.2 and Theorem 2.1 of [483] can be deduced easily from the classical (scalar) EVP and [214, Corollary 3.10.14], respectively.

Let us denote by Ψ_0 the set $\{\psi \in \Psi \mid \psi \text{ is continuous at } 0\}$; in [288, Theorem 6] one observed that ψ is continuous if $\psi \in \Psi_0$. Clearly, for $\psi \in \Psi$ (resp. $\psi \in \Psi_0$) and $(t_n) \subseteq \mathbb{R}_+$ one has: $\psi(t_n) \to 0 \Rightarrow t_n \to 0$ (resp. $\psi(t_n) \to 0 \Leftrightarrow t_n \to 0$).

The case $\psi \in \Psi_0$ of the next result is practically stated in the proof of [288, Theorem 6].

Proposition 10.6.7. *Let* $\psi \in \Psi$ *and let* (M, d) *be a metric space. Then* $d' := \psi \circ d$ *is a metric on* M*. Moreover, if* $\psi \in \Psi_0$ *then* d *and* d' *determine the same topology and the same Cauchy sequences; in particular* (M, d) *is complete if and only if* (M, d') *is complete. If* $\psi \in \Psi \setminus \Psi_0$ *then the topology determined by* d' *is the* *discrete topology, and so, if* $(x_n) \subseteq M$ *is asymptotic, that is* $d'(x_n, x_{n+1}) \to 0$*, then* (x_n) *is constant for large* n.

Proof. The fact that d' is a metric is well known. Let $\psi \in \Psi_0$, $(x_n)_{n\geq 1} \subseteq M$ and $x \in M$. As seen above, for $(t_n) \subseteq \mathbb{R}_+$ one has $\psi(t_n) \to 0 \Leftrightarrow t_n \to 0$, we get:

$$\lim_{n\to\infty} d'(x_n, x) = 0 \Leftrightarrow \lim_{n\to\infty} d(x_n, x) = 0,$$

$$\lim_{n,m\to\infty} d'(x_n, x_m) = 0 \Leftrightarrow \lim_{n,m\to\infty} d(x_n, x_m) = 0.$$

The first equivalence implies that the closed sets with respect to d and d' are the same, while the second equivalence shows that the Cauchy sequences with respect to d and d' are the same.

Let $\psi \in \Psi \setminus \Psi_0$. Then necessarily $\delta := \inf\{\psi(t) \mid t > 0\} > 0$. It follows that $B_{d'}(x, r) = \{x\}$ for every $r \in]0, \delta[$, and so $\{x\}$ is open for every $x \in X$. Hence the topology determined by d' is the discrete topology. Let $(x_n) \subseteq M$ be such that $d'(x_n, x_{n+1}) \to 0$, that is $\psi\,(d(x_n, x_{n+1})) \to 0$. Then there exists $n_0 \geq 1$ such that $\psi\,(d(x_n, x_{n+1})) < \delta$, and so $d(x_n, x_{n+1}) = 0$, for $n \geq n_0$. □

Using the preceding proposition, if $\psi \in \Psi$, then f from [483, Lemma 2.2] is $\psi \circ d$-lsc; the conclusion of [483, Lemma 2.2] follows from [614, Theorem 1.4.1] just taking $\varepsilon = 1$ and replacing d by $\psi \circ d$. A similar argument shows that [483, Theorem 2.1] follows from [214, Corollary 3.10.14] by replacing d by $\psi \circ d$.

10.7 Error Bounds

Let us consider $\Gamma : X \rightrightarrows Y$ with graph $\Gamma \neq \emptyset$ ($\Leftrightarrow$ dom $\Gamma \neq \emptyset$). We have the following result of Hamel type.

Theorem 10.7.1. *Let* (X, d) *be a complete metric space,* Y *a real topological vector space,* $K \subseteq Y$ *a proper closed convex cone,* $H \subseteq K$ *a nonempty convex set with* $0 \notin \operatorname{cl}(H + K)$*, and* $\Gamma : X \rightrightarrows Y$ *a set-valued map with* $\operatorname{dom}\Gamma \neq \emptyset$*. Assume that* $\{x \in X \mid \Gamma(u) \subseteq \Gamma(x) + d(x, u)H + K\}$ *is closed for every* $u \in X$*, and* $\Gamma(X)$ *is quasi bounded. If* $S \subseteq X$ *is such that*

$$\forall x \in \operatorname{dom}\Gamma \setminus S,\ \exists u \in X \setminus \{x\} : \Gamma(x) \subseteq \Gamma(u) + d(x, u)H + K, \tag{10.27}$$

then $S \cap \operatorname{dom}\Gamma \neq \emptyset$ *and*

$$\forall x \in X : \Gamma(x) \subseteq \Gamma(S) + d(x, S \cap \operatorname{dom}\Gamma)H + K. \tag{10.28}$$

Proof. Let $x \in \operatorname{dom}\Gamma$. By Theorem 10.4.7, there exists $x_1 \in X$ such that

$$\Gamma(x) \subseteq \Gamma(x_1) + d(x_1, x)H + K \tag{10.29}$$

and

$$\Gamma(x_1) \subseteq \Gamma(x) + d(x, x_1)H + K \Rightarrow x = x_1; \tag{10.30}$$

clearly, $x_1 \in \operatorname{dom} \Gamma$. Assume that $x_1 \notin S$. Using (10.27), there exists $u \in X \setminus \{x_1\}$ such that $\Gamma(x_1) \subseteq \Gamma(u) + d(x_1, u)H + K$. From (10.30) we get the contradiction $u = x_1$. In particular, because $\operatorname{dom} \Gamma \neq \emptyset$, we have that $S' := S \cap \operatorname{dom} \Gamma \neq \emptyset$, too. From (10.29) we obtain that

$$\Gamma(x) \subseteq \Gamma(x_1) + d(x_1, x)H + K \subseteq \Gamma(S) + d(x, S')H + K, \tag{10.31}$$

because $d(x_1, x)H = d(x, S')H + [d(x_1, x) - d(x, S')]\, H \subseteq d(x, S')H + K$. Since (10.31) is obvious for $x \notin \operatorname{dom} \Gamma$, the conclusion holds. □

A slightly more general result is the following.

Theorem 10.7.2. *Let (X, d) be a complete metric space, Y a real topological vector space, $K \subseteq Y$ a proper closed convex cone, $H \subseteq K$ a nonempty convex set with $0 \notin \operatorname{cl}(H + K)$, and $\Gamma : X \rightrightarrows Y$ a set-valued map with* $\operatorname{dom} \Gamma \neq \emptyset$. *Assume that $\{x \in X \mid \Gamma(u) \subseteq \Gamma(x) + d(x, u)H + K\}$ is closed for every $u \in X$, and $\Gamma(X)$ is quasi bounded. Furthermore, assume that $S, W \subseteq \operatorname{dom} \Gamma$ with $S \neq \emptyset$ are such that for every $x \in \operatorname{dom} \Gamma \setminus (S \cup W)$ there exists $u \in X \setminus \{x\}$ with $\Gamma(x) \subseteq \Gamma(u) + d(x, u)H + K$, and*

$$\forall x \in W : \Gamma(x) \subseteq \Gamma(S) + \eta d(x, S)H + K \tag{10.32}$$

for some $\eta > 0$. Then, for $\mu := \min(\eta, 1)$,

$$\forall x \in X : \Gamma(x) \subseteq \Gamma(S) + \mu d(x, S)H + K. \tag{10.33}$$

Proof. As in the proof of Theorem 10.7.1, consider $x \in \operatorname{dom} \Gamma$. As there, there exists $x_1 \in \operatorname{dom} \Gamma$ such that (10.29) and (10.30) hold. If $x_1 \in W$ then, by (10.32), $\Gamma(x) \subseteq \Gamma(S) + \eta d(x, S)H + K$, and so $\Gamma(x) \subseteq \Gamma(S) + \mu d(x, S)H + K$ because $0 < \mu \leq \eta$. If $x_1 \in S$ then (10.31) holds, and so again $\Gamma(x) \subseteq \Gamma(S) + \mu d(x, S)H + K$ because $\mu \leq 1$. In the contrary case, $x_1 \in X \setminus (S \cup W)$, and so, by our hypothesis, there exists $u \in X \setminus \{x_1\}$ such that $\Gamma(x_1) \subseteq \Gamma(u) + d(x_1, u)H + K$. From (10.30) we get the contradiction $u = x_1$. □

Remark 10.7.3. The hypothesis "$\{x \in X \mid \Gamma(u) \subseteq \Gamma(x) + d(x, u)H + K\}$ is closed for every $u \in X$" in Theorems 10.7.1 and 10.7.2 can be replaced by "epi Γ is closed and one of the conditions (i), (ii), (iii) of Theorem 10.4.9 holds" for getting the same conclusions of Theorem 10.7.1 and Theorem 10.7.2.

A natural candidate for S in Theorems 10.7.1 and 10.7.2 is the set of solutions (in a certain sense) of the set minimization problem

$$\min\nolimits_K \Gamma(x), \quad \text{s.t.} \quad x \in X. \tag{P}$$

For example, $\overline{x}$ is a solution of (P) in the sense of Kuroiwa if $\Gamma(\overline{x}) \subseteq \Gamma(x) + K$ implies $\Gamma(x) \subseteq \Gamma(\overline{x}) + K$; the choice of solutions in the sense of Kuroiwa is natural in the case of Theorem 10.7.1 because the conclusion of Theorem 10.4.7 is saying that $\overline{x}$ is a strict solution (or strict minimizer, [227]) of (P) in the sense of Kuroiwa for Γ replaced by Γ', where $\Gamma'(x) := \Gamma(x) + d(x, \overline{x})H$.

A particular case of Theorem 10.7.1 is the following corollary which subsumes results of Takahashi [555] and Hamel [233].

Corollary 10.7.4. *Let (X, d) be a complete metric space and $f : X \to \overline{\mathbb{R}}$ be a proper lsc lower bounded function. If for a fixed $\gamma \in]0, \infty[$ and every $x \in X$ with $f(x) > \inf f$ there exists $u \in X \setminus \{x\}$ such that $f(u) + \gamma d(x, u) \leq f(x)$, then $S := \arg\min f$ is nonempty and*

$$\forall x \in X : \gamma d(x, S) \leq f(x) - \inf f. \tag{10.34}$$

Proof. Consider $Y := \mathbb{R}$, $K := \mathbb{R}_+$, $H := \{\gamma\} \subseteq K$, and $\Gamma : X \rightrightarrows \mathbb{R}$ defined by $\Gamma(x) := \{f(x)\}$ for $x \in \operatorname{dom} f$ and $\Gamma(x) := \emptyset$ elsewhere. Clearly, Γ and S satisfy the conditions of Theorem 10.7.1. Hence $S \neq \emptyset$ and $\Gamma(S) = \{\inf f\}$, and so the conclusion (10.28) of Theorem 10.7.1 becomes (10.34). □

Corollary 10.7.5. *Let (X, d) be a complete metric space and $f : X \to \overline{\mathbb{R}}$ be a proper lsc function. If for a fixed $\gamma \in]0, \infty[$ and every $x \in X$ there exists $u \in X \setminus \{x\}$ such that $f(u) + \gamma d(x, u) \leq f(x)$, then $\inf f = -\infty$.*

Proof. Assume that $\inf f > -\infty$ and take $S = \emptyset$. Applying Corollary 10.7.4 we obtain the contradiction $S \neq \emptyset$. □

Applying Corollary 10.7.4 for $f_+ := \max\{f, 0\}$ we observe that [391, Lemma 2.3] is valid without assuming that $[f \leq 0] := \{x \in X \mid f(x) \leq 0\}$ is nonempty.

A similar result to Corollary 10.7.4, corresponding to Theorem 10.7.2, is the following.

Corollary 10.7.6. *Let (X, d) be a complete metric space and $f : X \to \overline{\mathbb{R}}$ be a proper lsc lower bounded function. Assume that $S := \arg\min f \neq \emptyset$, $W \subseteq \operatorname{dom} f$, and $\gamma, \eta \in]0, \infty[$ are such that for every $x \in \operatorname{dom} f \setminus (S \cup W)$ there exists $u \in X \setminus \{x\}$ such that $f(u) + \gamma d(x, u) \leq f(x)$, and $\eta d(x, S) \leq f(x) - \inf f$ for every $x \in W$. Then, for $\mu := \min(\gamma, \eta)$,*

$$\forall x \in X : \mu d(x, S) \leq f(x) - \inf f.$$

Let $f_i : X \to \overline{\mathbb{R}}$ be proper for every $i \in \overline{1, n}$ and assume that $S := \cap_{i=1}^{n} [f_i \leq 0] \neq \emptyset$ (that is, the solution set of the system $f_i(x) \leq 0$, $i \in \overline{1, n}$, is nonempty). As in [391], for $x \in X$ set $I_>(x) := \{i \in \overline{1, n} \mid f_i(x) > 0\}$. For $\gamma > 0$, Liu and Ng in [391] say that $x \in X \setminus S$ has the γ-descent property if there exists $\hat{x} \in X \setminus \{x\}$ such that: (a) $I_>(\hat{x}) \subseteq I_>(x)$, (b) $f_i(\hat{x}) \leq f_i(x)$ for every $i \in I_>(x)$, and (c) $\gamma d(x, \hat{x}) \leq \sum_{i \in I_>(x)} [(f_i)_+(x) - (f_i)_+(\hat{x})]$. In fact, taking into account (a), for

$i \in \overline{1,n} \setminus I_>(x)$ we have that $i \in \overline{1,n} \setminus I_>(\hat{x})$, and so $(f_i)_+(x) = (f_i)_+(\hat{x}) = 0$. This shows that if $x \in X \setminus S$ has the γ-descent property then there exists $\hat{x} \in X \setminus \{x\}$ such that $f(\hat{x}) + \gamma d(x, \hat{x}) \le f(x)$, where $f := \sum_{i=1}^n (f_i)_+$. It is obvious that $S = [f \le 0] = \arg\min f$. Using the preceding corollary we get the next result.

Corollary 10.7.7. *Let (X, d) be a complete metric space and $f_i : X \to \overline{\mathbb{R}}$, $i \in \overline{1,n}$, be proper lsc functions with $S := \cap_{i=1}^n [f_i \le 0]$ nonempty. Assume that there exist $W \subseteq D := \cap_{i=1}^n \operatorname{dom} f_i$ and $\gamma, \eta \in \,]0, \infty[$ satisfying the following conditions:*

(i) *for each $x \in D \setminus (S \cup W)$ there exists $\hat{x} \in X \setminus \{x\}$ such that $I_>(\hat{x}) \subseteq I_>(x)$ and $\gamma d(x, \hat{x}) \le \sum_{i \in I_>(x)} [(f_i)_+(x) - (f_i)_+(\hat{x})]$, and*
(ii) *$\eta d(x, S) \le \sum_{i \in I_>(x)} (f_i)_+(x)$ for every $x \in W$.*

Then, $\mu d(x, S) \le \sum_{i=1}^n (f_i)_+(x)$ for every $x \in X$, where $\mu := \min(\gamma, \eta)$.

Proof. Consider $f := \sum_{i=1}^n (f_i)_+$. Clearly, $D = \operatorname{dom} f$, $\inf f = 0$, and $S = \arg\min f \neq \emptyset$. Let $x \in \operatorname{dom} f \setminus (S \cup W)$. By (i), there exists $\hat{x} \in X \setminus \{x\}$ such that conditions (a) and (c) (above) are verified. The argument provided before the statement of the present corollary shows that $f(\hat{x}) + \gamma d(x, \hat{x}) \le f(x)$. Since $\eta d(x, S) \le f(x)$ $(= f(x) - \inf f)$ for every $x \in W$, the conclusion follows using Corollary 10.7.6. □

Remark 10.7.8. Corollary 10.7.7 provides sufficient conditions of error bounds for the system $f_i(x) \le 0$, $i \in \overline{1,n}$. This result is essentially [391, Theorem 4.1]; there X is a Banach space and instead of (i) one asks that any $x \in X \setminus (S \cup W)$ has the γ-descend property. For the proof of [391, Theorem 4.1] one uses [391, Theorem 3.5] (see also Remark 10.4.10).

Chapter 11
Derivatives and Epiderivatives of Set-Valued Maps

In set-valued optimization, derivatives, epiderivatives, and coderivatives of set-valued maps play the most fundamental role. We give optimality conditions by using derivatives, epiderivatives, and coderivatives. Sensitivity analysis is another important area where these objects are the building blocks. Numerical solutions of set-valued optimization problems are also computed by expressing optimality in terms of derivatives, epiderivatives, and coderivatives.

This chapter presents a detailed study of derivatives and epiderivatives of set-valued maps. In the literature, several approaches have been proposed to define derivatives of set-valued maps. To show the wide spectrum of ideas available to define the derivatives of set-valued maps, we refer the reader to the interesting articles by Banks and Jacobs [23], Daniilidis and Pang [124], Dentcheva [132], De Blasi [128], Gorokhovik and Zabreiko [216], Minchenko and Volosevich [421], Nurminskiĭ [444], Ovcharova and Gwinner [447, 448], Plotnikov [477], Rolewicz [507], Sach [514], Schinas and Boudourides [523], Xia, Wang, and Zhang [600], and the cited references therein. See also Chen, Nashed, and Qi [94] for an elegant treatment of nonsmoothness.

Our focus in this chapter is on the derivatives of set-valued maps which are motivated by the geometric interpretation of the classical notion of derivative for single-valued maps as a local approximation of its graph. We study the contingent derivative, the contingent epiderivative, and several analogues of these notions such as the adjacent derivative, the Clarke derivative, the lower Dini derivative, and the radial derivative. We also investigate their second-order analogues. In particular, we study the second-order contingent derivative, the second-order adjacent derivative, the second-order lower Dini derivative, and their epiderivatives counterparts. We give several calculus rules for the first-order derivatives and epiderivatives and their second-order extensions. The existence of contingent epiderivatives is an important issue in set-valued optimization. We present a comprehensive study of the various issues pertaining to the existence of contingent epiderivatives, their generalized analogues, and their various characterizations. We also study important issues

A.A. Khan et al., *Set-valued Optimization*, Vector Optimization,
DOI 10.1007/978-3-642-54265-7_11

such as proto-differentiability, semi-differentiability, and several related notions. Numerous other concepts which play an important role in differentiability of set-valued maps are given.

There are excellent monographs that address the notion of derivatives of set-valued maps and related issues such as by Aubin and Ekeland [16], Aubin and Frankowska [18], Borwein and Zou [66], Clarke [100], Hu and Papageorgiou [267], Jahn [293], Mordukhovich [430, 431], Penot [470], Rockafellar and Wets [499], and Schirotzek [524], among others. Our treatment of the subject has benefited particularly from [16, 18, 267, 499].

11.1 Contingent Derivatives of Set-Valued Maps

We begin by introducing the notion of the contingent derivative which plays a fundamental role in various aspects of set-valued optimization.

Definition 11.1.1. Let X and Y be normed spaces, let $F : X \rightrightarrows Y$ be a set-valued map, and let $(\bar{x}, \bar{y}) \in \operatorname{graph}(F)$. A set-valued map $DF(\bar{x}, \bar{y}) : X \rightrightarrows Y$ is called the **contingent derivative** of F at $(\bar{x}, \bar{y})$, if

$$\operatorname{graph}(DF(\bar{x}, \bar{y})) = T(\operatorname{graph}(F), (\bar{x}, \bar{y})),$$

where $T(\operatorname{graph}(F), (\bar{x}, \bar{y}))$ is the contingent cone of $\operatorname{graph}(F)$ at $(\bar{x}, \bar{y})$.

A geometric interpretation of the contingent derivative is given in Fig. 11.1. The graph of the contingent derivative is obtained by locally approximating the graph of the underlying set-valued map by means of the contingent cone.

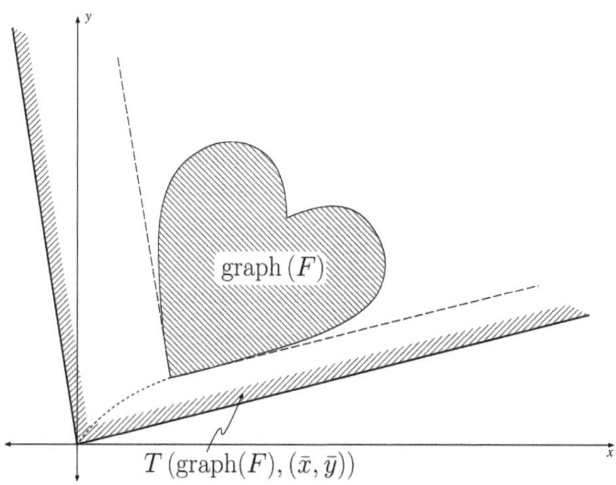

Fig. 11.1 Contingent derivative of the set-valued map F

In the subsequent development, for a single-valued map $f : X \to Y$, the contingent derivative of f at $(\bar{x}, f(\bar{x})) \in X \times Y$ will be denoted by $Df(\bar{x})$. That is, $Df(\bar{x}) := Df(\bar{x}, f(\bar{x}))$.

Remark 11.1.2. The above notion of the contingent derivative was introduced by Aubin [14] and has been used extensively in variational analysis, nonsmooth analysis, optimization, viability theory, finance, among others. It has now become a standard derivative notion for set-valued maps and its properties can be found in most of the recent books on these subjects. In our treatment of the subject, besides following the classical paper [14], we used the excellent books by Aubin and Ekeland [16], Aubin and Frankowska [18], and Ha and Papageorgiou [267].

We recall that given normed spaces X and Y, a function $f : X \to Y$ is said to be Hadamard directionally differentiable at $\bar{x} \in X$ in the direction $x \in X$, if the Hadamard directional derivative

$$f'(\bar{x}, x) = \lim_{(t,z)\to(0_+,x)} \frac{f(\bar{x} + tz) - f(\bar{x})}{t},$$

exists as an element of Y.

We have the following insightful result which is quite easy to prove (see Ward [596]):

Proposition 11.1.3. *Let X and Y be normed spaces and let $f : X \to Y$ be Hadamard directionally differentiable at $\bar{x} \in X$ in the direction $x \in X$. Then*

$$Df(\bar{x}, f(\bar{x}))(x) = \{f'(\bar{x}, x)\}.$$

The following simple examples illustrate the notion of the contingent derivative (see [563]):

Example 11.1.4. Let $F : \mathbb{R} \rightrightarrows \mathbb{R}$ be a set-valued map given by

$$F(x) := \begin{cases} \{y \in \mathbb{R} |\ y \geq 0\}, & \text{if } x = 0, \\ \{y \in \mathbb{R} |\ y > |x|\}, & \text{if } x \neq 0. \end{cases}$$

Then, for every $x \in \mathbb{R}$, the contingent derivatives of the set-valued maps F and $(F + \mathbb{R}_+)$ at $(0, 0)$ are given by $D(F + \mathbb{R}_+)(0, 0)(x) = DF(0, 0)(x) = \{y |\ y \geq |x|\}$. □

Example 11.1.5. Let $F : \mathbb{R} \rightrightarrows \mathbb{R}^2$ be a set-valued map given by

$$F(x) := \begin{cases} \{(0, 0)\}, & \text{if } x \leq 0, \\ \{(y_1, y_2) \in \mathbb{R}^2 |\ y_2 = -y_1^2,\ 0 \leq y_1 \leq x\}, & \text{if } x > 0. \end{cases}$$

Then,

$$DF(0,(0,0))(x) = \begin{cases} \{(0,0)\}, & \text{if } x \le 0, \\ \{(y_1, y_2) \in \mathbb{R}^2 |\ y_2 = 0,\ 0 \le y_1 \le x\}, & \text{if } x > 0, \end{cases}$$

whereas

$$D(F + \mathbb{R}^2_+)(0,(0,0))(x) = \{(y_1, y_2) |\ y_1 \ge 0,\ y_2 \ge 0\},$$

for every $x \in \mathbb{R}$. □

Example 11.1.6. Let $F : \mathbb{R} \rightrightarrows \mathbb{R}$ be a set-valued map given by

$$F(x) := \begin{cases} \{0\}, & \text{if } x \le 0, \\ \{0, -\sqrt{x}\}, & \text{if } x > 0. \end{cases}$$

Then,

$$DF(0,0)(x) = \begin{cases} \{0\}, & \text{if } x \ne 0, \\ \{y |\ y \le 0\}, & \text{if } x = 0, \end{cases}$$

$$D(F + \mathbb{R})(0,0)(x) = \begin{cases} \{y |\ y \ge 0\}, & \text{if } x < 0, \\ \mathbb{R}, & \text{if } x \ge 0, \end{cases}$$

showing the difference in the derivative of a set-valued map and its profile map. □

Example 11.1.7. Let $F : \mathbb{R} \rightrightarrows \mathbb{R}$ be a set-valued map given by

$$F(x) := \begin{cases} \{0\}, & \text{if } x \le 0, \\ \{-\sqrt{x}\}, & \text{if } x > 0. \end{cases}$$

Then,

$$DF(0,0)(x) = \begin{cases} \{0\}, & \text{if } x < 0, \\ \{y |\ y \le 0\}, & \text{if } x = 0, \\ \{\emptyset\}, & \text{if } x > 0, \end{cases}$$

depicting that the domain of the contingent derivative is not necessarily the whole space. □

In the following, we give various characterizations of the contingent derivative. For this, we first recall that given normed spaces X and Y, a set-valued map $F : X \rightrightarrows Y$ is called **Lipschitz around** $\bar{x} \in X$, if there exist a positive constant L and a neighborhood $U(\bar{x}) \subset \operatorname{dom}(F)$ such that for every $x, z \in U(\bar{x})$, we have $F(x) \subseteq F(z) + L\|x - z\|B$.

Proposition 11.1.8. *Let X and Y be normed spaces, let $F : X \rightrightarrows Y$ be a set-valued map, and let $(\bar{x}, \bar{y}) \in \operatorname{graph}(F)$. Then:*

1. *A pair $(x, y) \in \operatorname{graph}(DF(\bar{x}, \bar{y}))$ if and only if there are sequences $\{t_n\} \subset \mathbb{P}$ and $\{(x_n, y_n)\} \subset X \times Y$ with $t_n \downarrow 0$ and $(x_n, y_n) \to (x, y)$ such that $\bar{y} + t_n y_n \in F(\bar{x} + t_n x_n)$ for every $n \in \mathbb{N}$.*
2. *A pair $(x, y) \in \operatorname{graph}(DF(\bar{x}, \bar{y}))$ if and only if*

$$\liminf_{(\hat{x},t)\to(x,0+)} d\left(\frac{F(\bar{x}+t\hat{x})-\bar{y}}{t}, y\right) = 0.$$

3. *The contingent derivative of F at $(\bar{x}, \bar{y})$ is the set-valued map $DF(\bar{x}, \bar{y}) : X \rightrightarrows Y$ given by*

$$DF(\bar{x}, \bar{y})(x) = \limsup_{(\hat{x},t)\to(x,0+)} \frac{F(\bar{x}+t\hat{x})-\bar{y}}{t}.$$

4. *If $\bar{x} \in \operatorname{int}(\operatorname{dom}(F))$ and F is Lipschitz around $\bar{x}$, then $(x, y) \in \operatorname{graph}(DF(\bar{x}, \bar{y}))$ if and only if*

$$\liminf_{t\downarrow 0} d\left(\frac{F(\bar{x}+tx)-\bar{y}}{t}, y\right) = 0.$$

Proof. The proof follows from the characterizations of the contingent cone (see [18, 267]). □

It is at once evident from the first characterization given above that the contingent derivative of the inverse of a set-valued map is the inverse of the contingent derivative. That is,

$$DF^{-1}(\bar{y}, \bar{x}) = DF(\bar{x}, \bar{y})^{-1}.$$

It is also clear that the domain of the contingent derivative does not need to be the whole space X. The following result sheds some light on this issue. We recall that a set-valued map F is **pseudo-convex** at $(\bar{x}, \bar{y}) \in \operatorname{graph}(F)$, if its graph is a pseudo-convex set at $(\bar{x}, \bar{y})$. That is,

$$\operatorname{graph}(F) - (\bar{x}, \bar{y}) \subseteq T(\operatorname{graph}(F), (\bar{x}, \bar{y})),$$

or equivalently, for every $(x, y) \in \operatorname{graph}(F)$, we have

$$F(x) - \bar{y} \subseteq DF(\bar{x}, \bar{y})(x - \bar{x}).$$

Proposition 11.1.9. *Let X and Y be normed spaces, let $F : X \rightrightarrows Y$ be a set-valued map, and let $(\bar{x}, \bar{y}) \in \operatorname{graph}(F)$. Then:*

1. *The following inclusion holds*

$$\mathrm{cl}\,(\mathrm{dom}\,(DF(\bar{x},\bar{y}))) \subseteq T(\mathrm{dom}\,(F),\bar{x}).$$

2. *If the map F is pseudo-convex at $(\bar{x},\bar{y})$, then*

$$\mathrm{cl}\,(\mathrm{dom}\,(DF(\bar{x},\bar{y}))) = T(\mathrm{dom}\,(F),\bar{x}).$$

3. *If Y is finite dimensional, $\bar{x} \in \mathrm{int}(\mathrm{dom}\,(F))$, and F is Lipschitz around $\bar{x}$, then* $\mathrm{dom}\,(DF(\bar{x},\bar{y})) = X$. *Moreover, the contingent derivative $DF(\bar{x},\bar{y})$ is Lipschitz (with the same modulus).*

Proof. The proof of the first statement being obvious, let us verify the second one. By using the properties of the projection P_X of $X \times Y$ into X and the fact that F is pseudo-convex, we have

$$\begin{aligned}
\mathrm{cl}\,(\mathrm{dom}\,(DF(\bar{x},\bar{y}))) &= \mathrm{cl}\,(P_X(\mathrm{graph}\,(DF(\bar{x},\bar{y})))),\\
&= \mathrm{cl}\,(P_X(T(\mathrm{graph}\,(F),(\bar{x},\bar{y})))),\\
&= T(P_X(\mathrm{graph}\,(F)),\bar{x}),\\
&= T(\mathrm{dom}\,(F),\bar{x}).
\end{aligned}$$

For the final part, let $x \in X$ be arbitrary. Then for every sufficiently small $t > 0$, we have

$$\bar{y} \in F(\bar{x}) \subseteq F(\bar{x}+tx) + Lt\|x\|B_Y,$$

implying that there exists $\bar{y}_t \in F(\bar{x}+tx)$ such that

$$y_t = \frac{\bar{y}_t - \bar{y}}{t} \in L\|x\|B_Y,$$

which, due to the compactness of the ball B_Y in the finite-dimensional space Y, ensures that a subsequence $\{\bar{y}_t\}$ converges to some y. It follows at once that $y \in DF(\bar{x},\bar{y})(x)$, proving that $\mathrm{dom}\,(DF(\bar{x},\bar{y})) = X$.

Finally, for the Lipschitz continuity of $DF(\bar{x},\bar{y})$, let $\hat{x},\tilde{x} \in X$ and let $\hat{y} \in DF(\bar{x},\bar{y})(\hat{x})$. We claim that there exists $\tilde{y} \in DF(\bar{x},\bar{y})(\tilde{x})$ such that $\|\hat{y}-\tilde{y}\| \leq L\|\hat{x}-\tilde{x}\|$. Let $\{t_n\} \subset \mathbb{P}$ and $\{\hat{y}_n\} \subset Y$ be such that $t_n \to 0$, $\hat{y}_n \to \hat{y}$ and $\bar{y} + t_n\hat{y}_n \in F(\bar{x}+t_n\hat{x}_n)$. Then the Lipschitz continuity of F ensures that we can find $z_n \in F(\bar{x}+t_n\tilde{x}_n)$ such that $\|z_n - \bar{y} - t_n\hat{y}_n\| \leq Lt_n\|\hat{x}-\tilde{x}\|$.

By choosing a subsequence, if necessary, we may assume that $\dfrac{z_n - \bar{y}}{t_n} \to \tilde{y} \in DF(\bar{x},\bar{y})(\tilde{x})$. Evidently, $\tilde{y}$ satisfy the estimate $\|\hat{y}-\tilde{y}\| \leq L\|\hat{x}-\tilde{x}\|$. The proof is complete. □

The following simple result sheds some light on the kernel of the contingent derivative:

Proposition 11.1.10. *Let X and Y be normed spaces, let $F : X \rightrightarrows Y$ be a set-valued map, and let $(\bar{x}, \bar{y}) \in \operatorname{graph}(F)$. Then:*

$$T(F^{-1}(\bar{y}), \bar{x}) \subset \operatorname{Ker}(DF(\bar{x}, \bar{y})) := DF(\bar{x}, \bar{y})^{-1}(0). \tag{11.1}$$

Furthermore, the equality holds in the above formula, if F^{-1} is Lipschitz-like around $(\bar{y}, \bar{x})$.

Proof. The proof of (11.1) follows from the definition of the contingent cone and the converse inclusion is a consequence of the Lipschitz-like property as depicted in many other results. □

For our next result, we recall the notion of calmness which is weaker than the Aubin property (see [243]). Given normed spaces X and Y, a set-valued map $F : X \rightrightarrows Y$ is said to be **calm** at $(\bar{x}, \bar{y}) \in \operatorname{graph}(F)$, if there exist $c > 0$ and neighborhoods $U(\bar{x})$ and $V(\bar{y})$ such that for all $x \in U(\bar{x})$ and for all $y \in F(x) \cap V(\bar{y})$, we have

$$d(y, F(\bar{x})) \leq c\, d(x, \bar{x}). \tag{11.2}$$

We now give a useful property of the contingent derivatives of calm maps (see [380]):

Proposition 11.1.11. *Let X and Y be normed spaces, and let $F : X \rightrightarrows Y$ be a set-valued map that is calm at $(\bar{x}, \bar{y}) \in \operatorname{graph}(F)$. Then*

$$DF(\bar{x}, \bar{y})(0) = T(F(\bar{x}), \bar{y}).$$

Proof. Let $y \in DF(\bar{x}, \bar{y})(0)$. Then there exist sequences $\{t_n\} \subset \mathbb{P}$, $\{(x_n, y_n)\} \subset X \times Y$ such that $t_n \downarrow 0$, $(x_n, y_n) \to (0, y)$ and $\bar{y} + t_n y_n \in F(\bar{x} + t_n x_n)$, for every $n \in \mathbb{N}$. Let $U(\bar{x})$ and $V(\bar{y})$ be the neighborhoods appearing in the definition of calmness. Then for sufficiently large n, we have $\bar{x} + t_n x_n \in U(\bar{x})$ and $\bar{y} + t_n y_n \in V(\bar{y})$ such that there exists $\{\bar{y}_n\} \subset F(\bar{x})$ and $\|\bar{y} + t_n y_n - \bar{y}_n\| \leq c t_n \|x_n\| + t_n^2$. Then, for $z_n := t_n^{-1}(\bar{y}_n - \bar{y})$, we have $\|z_n - y_n\| \leq c\|x_n\| + t_n$, which confirms that z_n converges to y. Consequently, $y \in T(F(\bar{x}), \bar{y})$. The proof is complete. □

The following result gives useful information about the contingent derivative.

Theorem 11.1.12. *Let X and Y be normed spaces, let L and S be nonempty sets in X, where S is assumed to be open, let $G : X \rightrightarrows Y$ be a set-valued map, and let $f : S \to Y$ be a single-valued map. We define a set-valued map $F : X \rightrightarrows Y$ by*

$$F(x) = \begin{cases} f(x) - G(x) & \text{if } x \in L, \\ \emptyset & \text{if } x \notin L. \end{cases}$$

If the map f is Fréchet differentiable at $\bar{x} \in S \cap \mathrm{dom}\,(F)$, then for every $(\bar{x}, \bar{y}) \in \mathrm{graph}\,(F)$, we have

$$DF(\bar{x}, \bar{y})(x) \subseteq \begin{cases} f'(\bar{x})(x) - DG(\bar{x}, f(\bar{x}) - \bar{y})(x) & \text{if } x \in T(L, \bar{x}), \\ \emptyset & \text{if } x \notin T(L, \bar{x}). \end{cases}$$

Furthermore, the converse of the above inclusion holds if the set L is convex and the map G is Lipschitz at $\bar{x}$. In particular, if $G(x) = G$, for every $x \in \mathrm{dom}\,(G)$, then

$$DF(\bar{x}, \bar{y})(x) = f'(\bar{x})(x) - T(G, f(\bar{x}) - \bar{y}) \quad \text{for every } x \in T(L, \bar{x}).$$

Proof. Let $(x, y) \in \mathrm{graph}\,(DF(\bar{x}, \bar{y}))$ be arbitrary. Then there are sequences $\{(x_n, y_n)\} \subset X \times Y$ and $\{t_n\} \subset \mathbb{P}$ such that $(x_n, y_n) \to (x, y)$, $t_n \downarrow 0$ and $\bar{y} + t_n y_n \in F(\bar{x} + t_n x_n)$, for every $n \in \mathbb{N}$, implying that

$$\begin{aligned} &\bar{x} + t_n x_n \in L, \\ &\bar{y} + t_n y_n \in f(\bar{x} + t_n x_n) - G(\bar{x} + t_n x_n). \end{aligned}$$

Clearly, the first containment, in view of the properties of $\{t_n\}$ and $\{x_n\}$, at once confirms that $x \in T(L, \bar{x})$. Moreover, since the map f is Fréchet differentiable at $\bar{x}$, we have

$$f(\bar{x} + t_n x_n) = f(\bar{x}) + t_n f'(\bar{x})(x) + t_n h(t_n),$$

with $h(t_n) \to 0$ as $t_n \downarrow 0$. Therefore,

$$f(\bar{x}) - \bar{y} + t_n \left(f'(\bar{x})(x) + h(t_n) - y_n\right) \in G(\bar{x} + t_n x_n),$$

which implies that

$$y \in f'(\bar{x})(x) - DG(\bar{x}, f(\bar{x}) - \bar{y})(x).$$

For the converse, let $(x, y) \in X \times Y$ be such that

$$\begin{aligned} x &\in T(L, \bar{x}), \\ f'(\bar{x})(x) - y &\in DG(\bar{x}, f(\bar{x}) - \bar{y})(x). \end{aligned}$$

Since the set L is convex, we have $T(L, \bar{x}) = A(L, \bar{x})$, and therefore there are sequences $\{t_n\} \subset \mathbb{P}$ with $t_n \downarrow 0$ and $\{x_n\}$ and $\{\bar{x}_n\}$ in X with $x_n \to x$, $\bar{x}_n \to x$ and $\{y_n\} \subset Y$ with $y_n \to (f'(\bar{x})x - y)$ such that for every $n \in \mathbb{N}$, we have

$$\begin{aligned} \bar{x} + t_n x_n &\in L, \\ f(\bar{x}) - \bar{y} + t_n y_n &\in G(\bar{x} + t_n x_n). \end{aligned}$$

Since the map G is locally Lipschitz at $\bar{x}$, we can find $\{\bar{y}_n\} \subset Y$ with $\bar{y}_n \to f'(\bar{x})(x) - y$ such that

$$f(\bar{x}) - \bar{y} + t_n \bar{y}_n \in G(\bar{x} + t_n x_n) \quad \text{for every } n \in \mathbb{N}.$$

Using the Fréchet differentiability of the map f at $\bar{x}$, we have

$$f(\bar{x} + t_n x_n) = f(\bar{x}) + t_n f'(\bar{x})(x) + t_n h(t_n),$$

with $h(t_n) \to 0$ as $t_n \downarrow 0$. Consequently, for every $n \in \mathbb{N}$, we have

$$f(\bar{x} + t_n x_n) - t_n(f'(\bar{x})(x) + h(t_n)) - \bar{y} + t_n \bar{y}_n \in G(\bar{x} + t_n x_n),$$

implying

$$\bar{y} + t_n(f'(\bar{x})(x) + h(t_n) - \bar{y}_n) \in f(\bar{x} + t_n x_n) - G(\bar{x} + t_n x_n),$$

where

$$f'(\bar{x})(x) + h(t_n) - \bar{y}_n \to \big(f'(\bar{x})(x) - f'(\bar{x})(x) + y\big) = y,$$

which confirms that $y \in DG(\bar{x}, \bar{y})(x)$.

Finally, if G is constant, then $DG(\bar{x}, f(\bar{x}) - \bar{y}) = T(G, f(\bar{x}) - \bar{y})$. The proof is complete. □

11.1.1 Miscellaneous Graphical Derivatives of Set-valued Maps

The following definition introduces some other notions of derivatives of set-valued maps.

Definition 11.1.13. Let X and Y be normed spaces, let $F : X \rightrightarrows Y$ be a set-valued map, and let $(\bar{x}, \bar{y}) \in \text{graph}\,(F)$. Then:

1. A set-valued map $D_A F(\bar{x}, \bar{y}) : X \rightrightarrows Y$ is called the **adjacent derivative** of F at $(\bar{x}, \bar{y})$, if

$$\text{graph}\,(D_A F(\bar{x}, \bar{y})) = A(\text{graph}\,(F), (\bar{x}, \bar{y})),$$

   where $A(\text{graph}\,(F), (\bar{x}, \bar{y}))$ is the adjacent cone of graph (F) at $(\bar{x}, \bar{y})$.
2. A set-valued map $D_C F(\bar{x}, \bar{y}) : X \rightrightarrows Y$ is called the **Clarke derivative** of F at $(\bar{x}, \bar{y})$, if

$$\text{graph}\,(D_C F(\bar{x}, \bar{y})) = C(\text{graph}\,(F), (\bar{x}, \bar{y})),$$

where $C(\text{graph}\,(F), (\bar{x}, \bar{y}))$ is the Clarke tangent cone of graph (F) at $(\bar{x}, \bar{y})$.

3. A set-valued map $D_R F : X \rightrightarrows Y$ is called the **radial derivative** of F at $(\bar{x}, \bar{y})$, if

$$\text{graph}\,(D_R F(\bar{x}, \bar{y})) = \bar{R}(\text{graph}\,(F), (\bar{x}, \bar{y})),$$

where $\bar{R}(\text{graph}\,(F), (\bar{x}, \bar{y}))$ is the closed radial cone of graph (F) at $(\bar{x}, \bar{y})$.

4. A set-valued map $D_L F : X \rightrightarrows Y$ is called the **lower Dini derivative** of F at $(\bar{x}, \bar{y})$, if for every sequence $\{t_n\} \subset \mathbb{P}$ and $\{x_n\} \subset$ with $t_n \downarrow 0$ and $x_n \to x$, there exist a sequence $\{y_n\} \subset Y$ and an index $m \in \mathbb{N}$ such that $\bar{y} + t_n y_n \in F(\bar{x} + t_n x_n)$ for every $n \geq m$.
5. A set-valued map $D_Q F : X \rightrightarrows Y$ is called the **quasi-interiorly Clarke derivative** of F at $(\bar{x}, \bar{y})$, if for every sequence $\{t_n\} \subset \mathbb{P}$ and every sequence $\{(x_n, y_n)\} \subset \text{graph}\,(F)$ with $t_n \downarrow 0$ and $(\bar{x}_n, \bar{y}_n) \to (\bar{x}, \bar{y})$, and for every $\{x_n\} \to x$, there exist a sequence $\{y_n\} \subset Y$ with $y_n \to y$ and an index $m \in \mathbb{N}$ such that $\bar{y}_n + t_n y_n \in F(\bar{x}_n + t_n x_n)$ for every $n \geq m$.

For a single-valued map $f : X \to Y$, the adjacent derivative, the Clarke derivative, the radial derivative, the lower Dini derivative, and the quasi-interiorly Clarke derivative are denoted by $D_A f(\bar{x})$, $D_C f(\bar{x})$, $D_L f(\bar{x})$, $D_R f(\bar{x})$, and $D_Q f(\bar{x})$, respectively.

Remark 11.1.14. The lower Dini derivative was proposed in the celebrated paper by Penot [462], where one also finds the definition of the contingent derivative termed as the upper Dini derivative. This notion turns out to be quite useful in giving optimality conditions and sensitivity analysis in set-valued optimization. The notions of the adjacent derivative and the Clarke derivative are more akin to the idea of the contingent derivative and have also been used very often in set-valued optimization. The notion of the quasi-interiorly Clarke derivative was introduced by Thibault [571]. However, the above sequential definition of this derivative was used by Amahroq and Thibault [6]. The notion of the radial derivative was proposed by Taa [548].

The following result gives characterizations of the adjacent derivative, the Clarke derivative, and the lower Dini derivative.

Proposition 11.1.15. *Let X and Y be normed spaces, let $F : X \rightrightarrows Y$ be a set-valued map, and let $(\bar{x}, \bar{y}) \in \text{graph}\,(F)$. Then:*

1. *A pair $(x, y) \in \text{graph}\,(D_A F(\bar{x}, \bar{y}))$ if and only if for every $\{t_n\} \subset \mathbb{P}$ with $t_n \downarrow 0$ there exists $\{(x_n, y_n)\} \subset X \times Y$ with $(x_n, y_n) \to (x, y)$ such that $\bar{y} + t_n y_n \in F(\bar{x} + t_n x_n)$ for every $n \in \mathbb{N}$.*
2. *A pair $(x, y) \in \text{graph}\,(D_C F(\bar{x}, \bar{y}))$ if and only if for every $\{t_n\} \subset \mathbb{P}$ with $t_n \downarrow 0$ and for every $\{\bar{x}_n, \bar{y}_n)\} \subset \text{graph}\,(F)$ with $(\bar{x}_n, \bar{y}_n) \to (\bar{x}, \bar{y})$, there exists $\{(x_n, y_n)\} \subset X \times Y$ with $(x_n, y_n) \to (x, y)$ such that $\bar{y}_n + t_n y_n \in F(\bar{x}_n + t_n x_n)$ for every $n \in \mathbb{N}$.*

3. *An element* $(x, y) \in \operatorname{graph}(D_A F(\bar{x}, \bar{y}))$ *if and only if*

$$\limsup_{t\to 0+} \inf_{\hat{x}\to x} d\left(\frac{F(\bar{x}+t\hat{x})-\bar{y}}{t}, y\right) = 0.$$

4. *If* $\bar{x} \in \operatorname{int}(\operatorname{dom}(F))$ *and* F *is Lipschitz around* $\bar{x}$, *then* $(x, y) \in \operatorname{graph}(D_A F(\bar{x}, \bar{y}))$ *if and only if*

$$\lim_{t\downarrow 0} d\left(\frac{F(\bar{x}+tx)-\bar{y}}{t}, y\right) = 0.$$

5. *An element* $(x, y) \in \operatorname{graph}(D_C F(\bar{x}, \bar{y}))$ *if and only if*

$$\limsup_{t\to 0+,\, (\hat{x},\hat{y}) \xrightarrow{\operatorname{graph}(F)} (\bar{x},\bar{y})} \inf_{\tilde{x}\to x} d\left(\frac{F(\hat{x}+t\tilde{x})-\hat{y}}{t}, y\right) = 0.$$

6. *If* $\bar{x} \in \operatorname{int}(\operatorname{dom}(F))$ *and* F *is Lipschitz around* $\bar{x}$, *then* $(x, y) \in \operatorname{graph}(D_C F(\bar{x}, \bar{y}))$ *if and only if*

$$\lim_{t\to 0+,\, (\hat{x},\hat{y}) \xrightarrow{\operatorname{graph}(F)} (\bar{x},\bar{y})} d\left(\frac{F(\hat{x}+tx)-\hat{y}}{t}, y\right) = 0.$$

7. *An element* $(x, y) \in \operatorname{graph}(D_L F(\bar{x}, \bar{y}))$ *if and only if*

$$\lim_{(\hat{x},t)\to(x,0+)} d\left(\frac{F(\bar{x}+t\hat{x})-\bar{y}}{t}, y\right) = 0.$$

8. *The lower Dini derivative of* F *at* $(\bar{x}, \bar{y})$ *is a set-valued map* $D_L F(\bar{x}, \bar{y}) : X \rightrightarrows Y$ *given by*

$$D_L F(\bar{x}, \bar{y})(x) = \liminf_{(\hat{x},t)\to(x,0+)} \frac{F(\bar{x}+t\hat{x})-\bar{y}}{t}.$$

9. *An element* $(x, y) \in \operatorname{graph}(D_Q F(\bar{x}, \bar{y}))$ *if and only if for each neighborhood* $U(y)$, *there exist a neighborhood* $U(\bar{x})$, *a neighborhood* $U(\bar{y})$, *a real number* $s > 0$, *and a neighborhood* $U(x)$ *such that for every* $(u, v) \in (U(\bar{x}) \times U(\bar{y})) \cap \operatorname{graph}(F)$, $t \in (0, s)$, *and* $w \in U(x)$, *we have*

$$[(u, v) + t(\{w\} \times U(y))] \cap \operatorname{graph}(F) \neq \emptyset.$$

The above result shows that for certain Lipschitz maps, the contingent derivative, the Clarke derivative, and the adjacent derivative have a simpler structure. As it is evident from the properties of the tangent cones studied earlier, when the graph of the involved set-valued map is a convex set, all these derivatives coincide with the radial derivative and they share the individual properties of these derivatives listed above. Nonetheless, we have the following interesting property of the adjacent derivative for set-valued maps taking convex values (see [267]):

Proposition 11.1.16. *Let X and Y be normed spaces, let $F : X \rightrightarrows Y$ be a convex valued map, that is, for every $x \in \operatorname{dom}(F)$, the set $F(x)$ is convex. If F is Lipschitz around $\bar{x}$, then for every $(\bar{x}, \bar{y}) \in \operatorname{graph}(F)$, the adjacent derivative $D_A F(\bar{x}, \bar{y})$ is a convex valued map.*

Proof. For $x \in \operatorname{dom}(D_A F(\bar{x}, \bar{y}))$, let $y, z \in D_A F(\bar{x}, \bar{y})(x)$ be arbitrary. Then for any $\{t_n\} \subset \mathbb{P}$ with $t_n \downarrow 0$, there are sequences $\{(x_n, y_n)\}$ and $\{(\tilde{x}_n, \tilde{z}_n)\}$ with $(x_n, y_n) \to (x, y)$ and $(\tilde{x}_n, \tilde{z}_n) \to (x, z)$ such that $\bar{y} + t_n y_n \in F(\bar{x} + t_n, x_n)$ and $\bar{y} + t_n \tilde{z}_n \in F(\bar{x} + t_n, \tilde{x}_n)$, for every $n \in \mathbb{N}$. By using the fact that F is Lipschitz around $\bar{x}$, we can show that there exists a sequence $\{z_n\}$ such that $z_n \to z$ and $\bar{y} + t_n z_n \in F(\bar{x} + t_n x_n)$. Since F is convex valued, for every $s \in [0, 1]$, we have

$$\bar{y} + t_n(s y_n + (1 - s) z_n) \in F(\bar{x} + t_n x_n),$$

and since $(s y_n + (1 - s) z_n) \to (sy - (1 - s)z)$, we deduce that $(sy + (1 - s)z) \in D_A(\bar{x}, \bar{y})(x)$, proving the convexity of the set $D_A F(\bar{x}, \bar{y})(x)$. The proof is complete. □

The following is an analogue of Theorem 11.1.12 for the adjacent and the Clarke derivatives.

Theorem 11.1.17. *Let X and Y be normed spaces, let L and S be nonempty sets in X, where S is assumed to be open, let $G : X \rightrightarrows Y$ be a set-valued map and let $f : S \to Y$ be a single-valued map. We define a set-valued map $F : X \rightrightarrows Y$ by*

$$F(x) = \begin{cases} f(x) - G(x) & \text{if} \quad x \in L, \\ \emptyset & \text{if} \quad x \notin L. \end{cases}$$

Then:

1. *If f is Fréchet differentiable at $\bar{x} \in S \cap \operatorname{dom}(F)$, then for every $(\bar{x}, \bar{y}) \in \operatorname{graph}(F)$, we have*

$$D_A F(\bar{x}, \bar{y})(x) = \begin{cases} f'(\bar{x})(x) - D_A G(\bar{x}, f(\bar{x}) - \bar{y}) & \text{if} \quad x \in A(L, \bar{x}), \\ \emptyset & \text{if} \quad x \notin A(L, \bar{x}). \end{cases}$$

2. *If f is continuously differentiable at $\bar{x} \in S \cap \operatorname{dom}(F)$, then for every $(\bar{x}, \bar{y}) \in \operatorname{graph}(F)$, we have*

$$D_C F(\bar{x}, \bar{y})(x) = \begin{cases} f'(\bar{x})(x) - D_C G(\bar{x}, f(\bar{x}) - \bar{y}) & \text{if} \quad x \in C(L, \bar{x}), \\ \emptyset & \text{if} \quad x \notin C(L, \bar{x}). \end{cases}$$

As a direct consequence of the above result, we obtain the following useful corollary. Here, by $f_{|L}$, we denote the restriction of the map f to the set L.

Corollary 11.1.18. *Let X and Y be normed spaces, let $L \subset X$ be nonempty, and let $f : X \to Y$ be a single-valued map.*

1. *If f is Fréchet differentiable at $\bar{x} \in L$, then*

$$D f_{|L}(\bar{x}) = f'(x)_{|T(L,\bar{x})}$$
$$D_{A} f_{|L}(\bar{x}) = f'(x)_{|A(L,\bar{x})}.$$

2. *If f is continuously differentiable at $\bar{x} \in L$, then*

$$D_{C} f_{|L}(\bar{x}) = f'(x)_{|C(L,\bar{x})}.$$

In the following, we give a result concerning the computation of the contingent derivative of a composition map. Let X, Y, and U be normed spaces, let $F : U \rightrightarrows X$ be a set-valued map, and let $f : X \times U \to Y$ be a single-valued map. With these we associate the set-valued map $fF : U \rightrightarrows Y$ defined by

$$fF(u) := f\left(F(u) \times \{u\}\right).$$

Our objective is to evaluate $DfF(\bar{u}, \bar{y})$ for some $(\bar{u}, \bar{y}) \in \operatorname{graph}(fF)$. Recall that a set-valued map $G : X \rightrightarrows Y$ is **upper Lipschitz at** $\bar{x} \in \operatorname{dom}(G)$, if there exist $L, \rho > 0$ such that $G(x) \subset G(\bar{x}) + L\,\|x - \bar{x}\|\, B_Y$, for every $x \in B(\bar{x}, \rho)$.

We have the following:

Theorem 11.1.19. *Suppose that $\bar{x} \in F(\bar{u})$, $\bar{y} = f(\bar{x}, \bar{u})$ and that f is Fréchet differentiable at $(\bar{x}, \bar{u})$. Then the following inclusions hold:*

$$\nabla_x f(\bar{x}, \bar{u})\,(DF(\bar{u}, \bar{x})(u)) + \nabla_u f(\bar{x}, \bar{u})(u) \subset DfF(\bar{u}, \bar{y})(u) \quad \textit{for every } u \in U, \tag{11.3}$$

$$\nabla_x f(\bar{x}, \bar{u})\,(D_{L} F(\bar{u}, \bar{x})(u)) + \nabla_u f(\bar{x}, \bar{u})(u) \subset D_{L} fF(\bar{u}, \bar{y})(u) \quad \textit{for every } u \in U. \tag{11.4}$$

Furthermore, define a set-valued map $\tilde{F} : U \times Y \rightrightarrows X$ by

$$\tilde{F}(u, y) := \{x \in F(u) \mid f(x, u) = y\},$$

and assume that X is finite-dimensional and one of the following two conditions holds:

(a) *$F(\bar{u}) = \{\bar{x}\}$ and F is upper Lipschitz at $\bar{u}$.*
(b) *$\tilde{F}(\bar{u}, \bar{y}) = \{\bar{x}\}$ and $\tilde{F}$ is upper Lipschitz at $(\bar{u}, \bar{y})$.*

Then

$$\nabla_x f(\bar{x}, \bar{u})\,(DF(\bar{u}, \bar{x})(u)) + \nabla_u f(\bar{x}, \bar{u})(u) = DfF(\bar{u}, \bar{y})(u) \quad \textit{for every } u \in U. \tag{11.5}$$

Proof. The proof is adopted from [214]. Let $x \in DF(\bar{u},\bar{x})(u)$ and take $y = \nabla_x f(\bar{x},\bar{u})(x) + \nabla_u f(\bar{x},\bar{u})(u)$. Then there exist $\{t_n\} \to 0_+$, $\{(u_n,x_n)\} \to (u,x)$ such that $\bar{x} + t_n x_n \in F(\bar{u} + t_n u_n)$, for every $n \in \mathbb{N}$. Since f is differentiable at $(\bar{x},\bar{u})$, we have that for some $Y \supset \{v_n\} \to 0$,

$$f(\bar{x} + t_n x_n, \bar{u} + t_n u_n) = \bar{y} + t_n \left(\nabla_x f(\bar{x},\bar{u})(x_n) + \nabla_u f(\bar{x},\bar{u})(u_n) + v_n\right),$$

for every $n \in \mathbb{N}$. Taking

$$y_n := \nabla_x f(\bar{x},\bar{u})(x_n) + \nabla_u f(\bar{x},\bar{u})(u_n) + v_n,$$

since $\nabla f(\bar{x},\bar{u})$ is continuous, we have that $\{y_n\} \to y$. Since $\bar{y} + t_n y_n \in fF(\bar{u} + t_n u_n)$, for every $n \in \mathbb{N}$, $y \in DfF(\bar{u},\bar{y})(u)$. Hence (11.3) holds.

Consider $x \in D_L F(\bar{u},\bar{x})(u)$ and y as above. Let $\{(t_n,u_n)\} \to (0_+,u)$ with $\bar{u} + t_n u_n \in \operatorname{dom} fF$ $(= \operatorname{dom} F)$, for every n. Then there exists $\{x_n\} \to x$ such that $\bar{x} + t_n x_n \in F(\bar{u} + t_n u_n)$ for $n \geq n_0$. Continuing as above, we get $\{y_n\} \to y$ with $\bar{y} + t_n y_n \in fF(\bar{u} + t_n u_n)$ for $n \geq n_0$. Therefore $y \in D_L fF(\bar{u},\bar{y})(u)$, whence (11.4).

Suppose now that $\dim X < \infty$ and (a) or (b) holds. Consider $y \in DfF(\bar{u},\bar{y})(u)$. Then there exist $\{t_n\} \to 0_+$, $\{(u_n,y_n)\} \to (u,y)$ such that $\bar{y} + t_n y_n \in fF(\bar{u} + t_n u_n)$ for every $n \in \mathbb{N}$. Let $\tilde{x}_n \in F(\bar{u} + t_n u_n)$ be such that $\bar{y} + t_n y_n = f(\tilde{x}_n, \bar{u} + t_n u_n)$, that is, $\tilde{x}_n \in \tilde{F}(\bar{u} + t_n u_n, \bar{y} + t_n y_n)$, for every n. Since F is upper Lipschitz at $\bar{u}$ and $F(\bar{u}) = \{\bar{x}\}$ in case (a) and $\tilde{F}$ is upper Lipschitz at $(\bar{u},\bar{y})$ and $\tilde{F}(\bar{u},\bar{y}) = \{\bar{x}\}$ in case (b), there exists a bounded sequence $\{x_n\} \subset X$ such that $\tilde{x}_n = \bar{x} + t_n x_n$, for every $n \in \mathbb{N}$. Since $\dim X < \infty$, there exists a subsequence $\{x_{n_p}\}_{p\in\mathbb{N}}$ converging to some $x \in X$. It follows that $x \in DF(\bar{u},\bar{x})(u)$, and so $y \in \nabla_x f(\bar{x},\bar{u})\,(DF(\bar{u},\bar{x})(u)) + \nabla_u f(\bar{x},\bar{u})(u)$. Taking into account (11.3), we obtain that (11.5) holds. The proof is complete. □

In the following example, which is taken from [214], one of the conditions in (a) and (b) of Theorem 11.1.19 is not satisfied, and (11.5) does not hold.

Example 11.1.20. Let $U = X = Y = \mathbb{R}$, $F, F' : U \rightrightarrows X$ and $f : X \times U \to Y$ be defined by

$$F(u) := [0, \max(1, 1+u)], \quad F'(u) := [0, \max(1, 1+u)[, \quad f(x,u) := x^2 - x,$$

$\bar{u} = 0$, $\bar{x} = 0$, $\bar{y} = f(0,0) = 0$. Then F, F' are upper Lipschitz at $\bar{u}$, but $F(\bar{u}) \neq \{\bar{x}\} \neq F'(\bar{u})$ (so (a) does not hold for F and F') and

$$fF(u) = \begin{cases} [-\frac{1}{4}, 0] & \text{if } u \leq 0, \\ [-\frac{1}{4}, u(1+u)] & \text{if } u > 0, \end{cases} \qquad f'F(u) = \begin{cases} [-\frac{1}{4}, 0] & \text{if } u \leq 0, \\ [-\frac{1}{4}, u(1+u)[& \text{if } u > 0. \end{cases}$$

Let $\{x_1^y, x_2^y\} = \{x \in \mathbb{R} \mid x^2 - x = y\}$, with $x_1^y \leq x_2^y$, for $y \geq -\frac{1}{4}$. Then

$$\tilde{F}(u,y) = \begin{cases} \{x_1^y, x_2^y\} & \text{if } u \in \mathbb{R},\ y \in [-\frac{1}{4}, 0], \\ \{x_2^y\} & \text{if } u > 0,\ y \in]0, u(1+u)], \\ \emptyset & \text{otherwise,} \end{cases}$$

$$\widetilde{F'}(u,y) = \begin{cases} \{x_1^y, x_2^y\} & \text{if } u \in \mathbb{R},\ y \in [-\frac{1}{4}, 0[, \\ \{0\} & \text{if } u \in \mathbb{R},\ y = 0, \\ \{x_2^y\} & \text{if } u > 0,\ y \in]0, u(1+u)[, \\ \emptyset & \text{otherwise.} \end{cases}$$

Now, $\tilde{F}$ is upper Lipschitz at $(\bar{u}, \bar{y})$ but $\tilde{F}(\bar{u}, \bar{y}) = \{0, 1\} \neq \{\bar{x}\}$, while $\widetilde{F'}$ is not upper Lipschitz at $(\bar{u}, \bar{y})$ (so (b) does not hold for F and F'). One obtains that

$$DF(\bar{u}, \bar{x})(u) = DF'(\bar{u}, \bar{x})(u) = \mathbb{R} \times \mathbb{R}_+, \text{for every } u \in \mathbb{R},$$

$$DfF(\bar{u}, \bar{y})(u) = DfF'(\bar{u}, \bar{y})(u) = \{(u, y) \in \mathbb{R}^2 \mid y \leq \max(0, u)\} \quad \text{for every } u \in \mathbb{R}.$$

Since $\nabla_x f(\bar{x}, \bar{u}) = -1$ and $\nabla_u f(\bar{x}, \bar{u}) = 0$,

$$\nabla_x f(\bar{x}, \bar{u})\,(DF(\bar{u}, \bar{x})(u)) + \nabla_u f(\bar{x}, \bar{u})(u) =]-\infty, 0] \neq]-\infty, u] = DfF(\bar{u}, \bar{x})(u),$$

for every $u > 0$, and similarly for F'.

Remark 11.1.21. Formulae (11.3) and (11.5) are stated by Tanino [561] under the assumption that U, X, and Y are finite-dimensional and f is of class $\mathscr{C}^1$ and by Klose [330, Theorem 4.1].

The following result highlights a useful feature of the radial derivative (see Taa [548]):

Theorem 11.1.22. *Let X and Y be normed spaces, let F be a set-valued map, and let $(\bar{x}, \bar{y}) \in \operatorname{graph}(F)$. Then, for every $x \in X$, we have*

$$F(x) - \bar{y} \subseteq D_R F(\bar{x}, \bar{y})(x - \bar{x}). \tag{11.6}$$

Furthermore, if $\operatorname{graph}(F)$ *is a convex set, then*

$$F(x) - \bar{y} \subseteq DF(\bar{x}, \bar{y})(x - \bar{x}). \tag{11.7}$$

Proof. For any $y \in F(x) - \bar{y}$, we have $\bar{y} + y \in F(x)$. This, by setting $t_n := 1$, $y_n := y$, and $x_n := x - \bar{x}$, can be expressed as $\bar{y} + t_n y_n \in F(\bar{x} + t_n x_n)$, for every $n \in \mathbb{N}$. Since $(x_n, y_n) \to (x - \bar{x}, y)$, we deduce that $(x - \bar{x}, y) \in \bar{R}(\operatorname{graph}(F), (\bar{x}, \bar{y})) = \operatorname{graph}(D_R F(\bar{x}, \bar{y}))$. Consequently, $y \in D_R F(\bar{x}, \bar{y})(x - \bar{x})$ and the proof of (11.6) is complete. Inclusion (11.7) follows from the fact that, for a convex set, the closed radial tangent cone and the contingent cone coincide. □

We have the following useful corollary of the above result.

Corollary 11.1.23. *Let X and Y be normed spaces, let $C \subset Y$ be a pointed, closed and convex cone, let F be a set-valued map, and let $(\bar{x}, \bar{y}) \in \operatorname{graph}(F)$. If $\operatorname{epi}(F)$ is convex, then for every $x \in X$, we have*

$$F(x) - \bar{y} \subseteq D(F + C)(\bar{x}, \bar{y})(x - \bar{x}). \tag{11.8}$$

Proof. This follows from using the fact that, for a convex set, the closed radial tangent cone and the contingent cone coincide and the chain of inclusions

$$F(x) - \bar{y} \subseteq F(x) - \bar{y} + C \subseteq D(F + C)(\bar{x}, \bar{y})(x - \bar{x}),$$

where the first inclusion is obvious and the second inclusion is (11.7) written for $F + C$. □

11.1.2 Convexity Characterization Using Contingent Derivatives

Having shown a useful property of the convex set-valued maps, we take a pause to give a characterization of a certain convexity notion for set-valued maps. Recall that given a normed space X and a continuously differentiable map $f : X \to \mathbb{R}$, the following two characterizations of convexity are well-known:

1. The map f is convex if and only if its Fréchet derivative ∇f is monotone. That is,

$$\langle \nabla f(x) - \nabla f(z), x - z \rangle \geq 0 \quad \text{for every } x, z \in X. \tag{11.9}$$

2. The map f is convex if and only if

$$f(z) - f(x) \geq \langle \nabla f(x), z - x \rangle \quad \text{for every } x, z \in X. \tag{11.10}$$

The above two characterizations of convexity have been extended in several directions. For instance, generalized notions of convexity have been characterized by using generalized notions of differentiability. See Clarke, Stern, and Wolenski [101], Correa, Jofré, and Thibault [114, 115], Daniilidis and Hadjisavvas [123], Ellaia and Hassouni [176], Luc and Swaminathan [406], Penot and Sach [471], Sach [515, 516], Sach and Yen [518], Zagrodny [608] and the cited references therein. The primary objective of this short section is to give an analogue of the first statement given above by giving a characterization of certain convexity notion for set-valued maps in terms of the contingent derivative. Here our intention is only to briefly discuss the involvement of set-valued maps and contingent derivatives in this interesting area. The interested reader can find more elaborate results in the literature.

Let X and Y be normed spaces, let $C \subset Y$ be a pointed, closed, and convex cone, and let $F : X \rightrightarrows Y$ be a set-valued map. We recall that the map F is called almost C-convex, if for all $x, z \in X$ and every $t \in [0, 1]$, we have

$$tF(x) + (1-t)F(z) \subset \text{cl}\,[F(tx + (1-t)z) + C]. \tag{11.11}$$

We have the following characterization of the almost C-convexity:

Theorem 11.1.24. *Let X and Y be normed spaces, let $C \subset Y$ be a pointed, closed, and convex cone, and let $F : X \rightrightarrows Y$ be a set-valued map. We formulate the following statements:*

(a) *The map F is almost C-convex.*
(b) *For every $(x_1, y_1), (x_2, y_2) \in \text{graph}\,(F)$, we have*

$$y_2 - y_1 \in DF(x_1, y_1)(x_2 - x_1). \tag{11.12}$$

(c) *The contingent derivative DF is monotone on* graph (F) *in the following sense: for every $(x_1, y_1), (x_2, y_2) \in \text{graph}\,(F)$, we have*

$$0 \in DF(x_1, y_1)(x_2 - x_1) + DF(x_2, y_2)(x_1 - x_2). \tag{11.13}$$

Then: (a) $\Longrightarrow$ (b) $\Longrightarrow$ (c).

Proof. We begin with a proof of the first implication, namely, (a) $\Longrightarrow$ (b). Let F be almost C-convex and let $(x_1, y_1), (x_2, y_2) \in \text{graph}\,(F)$ be arbitrary. We claim that since the map F is almost C-convex, the cl (epi (F)) is a convex set. Indeed, by the almost C-convexity of F, and by setting $\bar{F}(x) := \text{cl}\,(F(x) + C)$, for every $x_1, x_2 \in X$ and every $t \in (0, 1]$, we have

$$t(F + C)(x_1) + (1-t)(F + C)(x_2) \subset \text{cl}\,((F + C)(tx_1 + (1-t)x_2)) + C,$$

which implies that

$$t\bar{F}(x_1) + (1-t)\bar{F}(x_2) \subset \text{cl}\,\left[\bar{F}(tx_1 + (1-t)x_2) + C\right] = \bar{F}(tx_1 + (1-t)x_2),$$

which shows that graph $(\bar{F})$ is a convex set. Note that the identity cl (graph $(\bar{F})$) = cl (graph $((F + C))$) also holds. Then, for every $(\bar{x}, \bar{y}) \in \text{graph}\,((F + C))$, we have

$$\begin{aligned} C(\text{graph}\,((F + C)), (\bar{x}, \bar{y})) &= T(\text{graph}\,((F + C)), (\bar{x}, \bar{y})) \\ &= \text{cl}\,\text{cone}\,(\text{graph}\,((F + C)) - (\bar{x}, \bar{y})), \end{aligned}$$

and, by setting $(\bar{x}, \bar{y}) := (x_1, y_1)$, we deduce $(x_2 - x_1, y_2 - y_1) \in T(\text{graph}\,((F + C)), (x_1, y_1))$, which confirms that (b) holds.

For the second implication, namely, (b) $\Longrightarrow$ (c), we note that (11.12) implies that

$$y_2 - y_1 \in DF(x_1, y_1)(x_2 - x_1),$$
$$y_1 - y_2 \in DF(x_2, y_2)(x_1 - x_2),$$

and by summing up the above inclusions, we get (11.13). The proof is complete. □

The above result is based on the work by Sach and Yen [518] where more elaborate relationships among the statements of Theorem 11.1.24, under various settings, can be found.

11.1.3 Proto-Differentiability, Semi-Differentiability, and Related Concepts

In the previous sections, we witnessed various derivatives of set-valued maps. It turns out that, in the setting when two or more derivative notions coincide, the resulting derivative benefits from their individual features and is typically equipped with nicer features. In his celebrated paper, Rockafellar [495] introduced the notion of proto-differentiable set-valued maps to recapture the combined benefits of the contingent and the adjacent derivative of set-valued maps and investigated related concepts. He also studied the proto-differentiability of certain constraint set-valued maps appearing in parametric optimization. In this section, we briefly discuss some of these situations to which many authors have contributed. This section is partly adopted from Amahroq and Thibault [6]. See also [19,495].

We begin with the following definition:

Definition 11.1.25. Let X and Y be normed spaces, let $F : X \rightrightarrows Y$ be a set-valued map, and let $(\bar{x}, \bar{y}) \in \text{graph}\,(F)$.

1. The map F is said to be **proto-differentiable** or **derivable** at $(\bar{x}, \bar{y})$, if the contingent derivative and the adjacent derivative of F at $(\bar{x}, \bar{y})$ coincide. That is,

$$T(\text{graph}\,(F), (\bar{x}, \bar{y})) = A(\text{graph}\,(F), (\bar{x}, \bar{y})).$$

2. The map F is said to be **strictly proto-differentiable** at $(\bar{x}, \bar{y})$, if the contingent derivative and the Clarke derivative of F at $(\bar{x}, \bar{y})$ coincide. That is,

$$T(\text{graph}\,(F), (\bar{x}, \bar{y})) = C(\text{graph}\,(F), (\bar{x}, \bar{y})).$$

3. The map F is said to be **semi-differentiable** at $(\bar{x}, \bar{y})$, if the contingent derivative $DF(\bar{x}, \bar{y})$ and the lower Dini derivative $D_{\text{L}}F(\bar{x}, \bar{y})$ coincide. That is, if for every $x \in X$, we have

$$\limsup_{(\hat{x},t)\to(x,0+)} \frac{F(\bar{x}+t\hat{x})-\bar{y}}{t} = \liminf_{(\hat{x},t)\to(x,0+)} \frac{F(\bar{x}+t\hat{x})-\bar{y}}{t}. \tag{11.14}$$

4. The map F is said to be **strictly semi-differentiable** at $(\bar{x}, \bar{y})$, if the contingent derivative $DF(\bar{x}, \bar{y})$ and the quasi-interiorly Clarke derivative $D_{Q}F(\bar{x}, \bar{y})$ coincide. That is, if for every $x \in X$,

$$\limsup_{(t,z)\to(0+,x)} \frac{F(\bar{x}+tz)-\bar{y}}{t} = \liminf_{(t,z)\to(0+,x),(x,y)\overset{\text{graph}(F)}{\longrightarrow}(\bar{x},\bar{y})} \frac{F(x+tz)-y}{t}. \tag{11.15}$$

Remark 11.1.26. Clearly any set-valued map with a convex graph is strictly proto-differentiable and hence derivable. However, the converse in not true in general. For example, consider a set-valued map $F : \mathbb{R} \rightrightarrows \mathbb{R}$ defined by $F(x) := \{|x|, -|x|\}$, for all $x \in \mathbb{R}$. Then F is not strictly proto-differentiable at the point $(0, 0) \in \text{graph}(F)$, but only derivable at this point (see [18]).

We have the following result which was originally proved by Amahroq and Thibault [6]:

Theorem 11.1.27. *Let X and Y be normed spaces, let $F : X \to Y$ be a set-valued map which is Lipschitz-like around $(\bar{x}, \bar{y}) \in \text{graph}(F)$. Then F is strictly semi-differentiable at $(\bar{x}, \bar{y})$ if and only if F is strictly proto-differentiable at $(\bar{x}, \bar{y})$.*

Proof. In view of the definitions of the strictly semi-differentiability and the strictly proto-differentiability, it is enough to show that for any $(v, w) \in C(\text{graph}(F), (\bar{x}, \bar{y}))$, we have

$$w \in \liminf_{(t,z)\to(0+,v),(x,y)\overset{\text{graph}(F)}{\longrightarrow}(\bar{x},\bar{y})} \frac{F(x+tz)-y}{t}. \tag{11.16}$$

Let $\{t_n\} \subset \mathbb{P}$, $\{(x_n, y_n)\} \subset \text{graph}(F)$, and $\{v_n\} \subset Y$ be sequences such that $(t_n, x_n, y_n) \to (0, \bar{x}, \bar{y})$ and $v_n \to v$. Then there exists $(\bar{v}_n, \bar{w}_n) \to (v, w)$ such that $y_n + t_n\bar{w}_n \in F(x_n + t_n\bar{v}_n)$ for every $n \in \mathbb{N}$.

Since F is Lipschitz-like around $(\bar{x}, \bar{y})$, for sufficiently large $n \in \mathbb{N}$, we have

$$y_n + t_n\bar{w}_n \in W \cap F(x_n + t_n\bar{v}_n) \subseteq F(x_n + t_n v_n) + L\, t_n\|\bar{v}_n - v_n\| B_Y,$$

where L is a positive constant and W is a neighborhood of $\bar{y}$. Hence there exists $b_n \in B_Y$ and $w_n := (\bar{w}_n + L\,\|\bar{v}_n - v_n\| b_n) \to w$ such that for sufficiently large $n \in \mathbb{N}$, we have $y_n + t_n w_n \in G(x_n + t_n v_n)$ and hence (11.16) holds. The proof is complete. □

We have the following analogue of the above result.

Theorem 11.1.28. *Let X and Y be normed spaces, let $F : X \to Y$ be a set-valued map which is Lipschitz-like around $(\bar{x}, \bar{y}) \in \operatorname{graph}(F)$. Then F is derivable at $(\bar{x}, \bar{y})$ if and only if F is semi-differentiable at $(\bar{x}, \bar{y})$.*

Proof. Evidently, if F is semi-differentiable at $(\bar{x}, \bar{y})$, then it is derivable at $(\bar{x}, \bar{y})$. For the converse, let F is derivable at $(\bar{x}, \bar{y})$ and let $(x, y) \in \operatorname{graph}(DF(\bar{x}, \bar{y}))$ be arbitrary. Then, for every $\{t_n\} \subset \mathbb{P}$, there exists $\{(z_n, w_n)\} \subset X \times Y$ such that $t_n \downarrow 0$, $(z_n, w_n) \to (x, y)$ and $\bar{y} + t_n w_n \in F(\bar{x} + t_n z_n)$. Let $\{x_n\} \subset X$ be an arbitrary sequence such that $x_n \to x$. Using the fact that F is Lipschitz-like at $(\bar{x}, \bar{y})$, we can show that there exist a sequence $\{y_n\}$ and an index $m \in \mathbb{N}$ such that $(\bar{x} + t_n x_n, \bar{y} + t_n y_n) \in \operatorname{graph}(F)$, for every $n \geq m$. Consequently, $y \in D_{\mathrm{L}} F(\bar{x}, \bar{y})(x)$. The proof is complete. □

We will now investigate the proto-differentiability of a set-valued map associated to a specific constraint set. For this we first give the following two results which are of interest (see [6]):

Theorem 11.1.29. *Let X, Y and Z be Banach spaces, let $G : X \times Y \rightrightarrows Z$ be a set-valued map, let $S \subset Y$ and $Q \subset Z$ be nonempty closed sets. Assume that the following metric regularity condition holds: For $(\bar{x}, \bar{y}, \bar{z}) \in X \times S \times Q \cap \operatorname{graph}(G)$, there exist $k \geq 0$ and neighborhoods $U(\bar{x})$, $V(\bar{y})$, and $W(\bar{z})$ such that for every $x \in U(\bar{x})$, $y \in V(\bar{y}) \cap S$, and $z \in G(x, y) \cap W(\bar{z})$, we have*

$$d((x, y), (X \times S) \cap G^{-1}(Q)) \leq k\, d(z, Q). \tag{11.17}$$

Then for the set-valued map $F : X \rightrightarrows Y$ given by

$$F(x) = \{y \in S \mid G(x, y) \cap Q \neq \emptyset\}, \tag{11.18}$$

the following inclusion holds

$$[X \times A(S, \bar{y})] \cap D_L G(\bar{x}, \bar{y}, \bar{z}))^{-1}[A(Q, \bar{z})] \subseteq A(\operatorname{graph}(F), (\bar{x}, \bar{y})). \tag{11.19}$$

Moreover, if $G(\cdot, \cdot) \cap Q$ is lower semicontinuous at $(\bar{x}, \bar{y}, \bar{z})$ relative to its domain, then

$$[X \times C(S, \bar{y})] \cap D_Q G(\bar{x}, \bar{y}, \bar{z}))^{-1}[C(Q, \bar{z})] \subseteq C(\operatorname{graph}(F), (\bar{x}, \bar{y})). \tag{11.20}$$

Proof. We will only prove (11.20) as (11.19) can be proved by similar arguments. Let

$$(v, w) \in [X \times C(S, \bar{y})] \cap D_Q G(\bar{x}, \bar{y}, \bar{z}))^{-1}[C(Q, \bar{z})]$$

be arbitrary. Let $\{t_n\} \subset \mathbb{P}$ and $\{(x_n, y_n)\} \subset \operatorname{graph}(F)$ be arbitrary sequences such that $t_n \downarrow 0$, and $(x_n, y_n) \to (\bar{x}, \bar{y})$. To prove that $(v, w) \in C(\operatorname{graph}(F), (\bar{x}, \bar{y}))$, we have to show that there exists $\{(v_n, w_n)\}$ with $(v_n, w_n) \to (v, w)$ such that $(x_n + t_n v_n, y_n + t_n w_n) \in \text{graph}\,(F)$.

In view of the containment $w \in C(S, \bar{y})$, there exists a sequence $\{\hat{w}_n\}$ with $\hat{w}_n \to w$ such that $y_n + t_n \hat{w}_n \in S$, for every $n \in \mathbb{N}$. Furthermore, due to $(v, w) \in D_Q G(\bar{x}, \bar{y}, \bar{z}))^{-1}[C(Q, \bar{z})]$, there exists $u \in C(Q, \bar{z})$ such that $(v, w, u) \in \text{graph}\,(D_Q G(\bar{x}, \bar{y}, \bar{z}))$.

Moreover, since $G(\cdot, \cdot) \cap Q$ is lower semicontinuous relative to its domain, there is a sequence $z_n \in G(x_n, y_n) \cap Q$ such that $(t_n, x_n, y_n, z_n) \to (0, \bar{x}, \bar{y}, \bar{z})$. Also, from the containment $u \in C(Q, \bar{z})$, there exists a sequence $\{u_n\}$ with $u_n \to u$ such that $z_n + t_n u_n \in Q$ for every $n \in \mathbb{N}$. By the definition of $D_Q G(\bar{x}, \bar{y}, \bar{z})$ and the containment $(v, w, u) \in \text{graph}\,(D_Q G(\bar{x}, \bar{y}, \bar{z}))$, there are a sequence $\{\hat{u}_n\}$ with $\hat{u}_n \to u$ and an index $m \in \mathbb{N}$ such that for every $n \geq m$, we have $z_n + t_n \hat{u}_n \in G(x_n + t_n v, y_n + t_n \hat{w}_n)$.

Since

$$d\,(z_n + t_n \hat{u}_n, Q) \leq \|z_n + t_n \hat{u}_n - z_n - t_n u_n\| = t_n \|u_n - \hat{u}_n\|,$$

we obtain from the metric regularity hypothesis that for sufficiently large $n \in \mathbb{N}$, we have

$$d\,((x_n + t_n v, y_n + t_n \hat{w}_n), \text{graph}\,(F)) \leq k\, t_n \|u_n - \hat{u}_n\| < t_n^2 + k\, t_n \|u_n - \hat{u}_n\|,$$

and hence for large n there exists $\{(v_n, w_n)\} \subset X \times Y$ with $(x_n + t_n v_n, y_n + t_n w_n) \in \text{graph}\,(F)$, and

$$\|(x_n + t_n v, y_n + t_n \hat{w}_n) - (x_n + t_n v_n, y_n + t_n w_n)\| < t_n^2 + k\, t_n \|u_n - \hat{u}_n\|,$$

implying that

$$\|(v, \hat{w}_n) - (v_n, w_n)\| < t_n + k\, \|u_n - \hat{u}_n\|,$$

and consequently $(v_n, w_n) \to (v, w)$ as $\hat{w}_n \to w$. The proof is complete. □

For our next result, we recall the notion of compactly approximable set-valued maps.

Definition 11.1.30. Let X and Y be normed spaces, let $F : X \rightrightarrows Y$ be a set-valued map, and let $(\bar{x}, \bar{y}) \in \text{graph}\,(F)$. The map $F : X \rightrightarrows Y$ is called **compactly approximable** at $(\bar{x}, \bar{y})$ in a direction $z \in X$, if there are a compact set $L(z)$ in Y, a number $\alpha > 0$, and a function $r :]0, \infty[\times X \to \mathbb{R}$ such that

1. $\lim_{(t,x) \to (0,z)} r(t, x) = 0.$
2. For each $x \in U_\alpha(z)$ and $t \in]0, \alpha]$ with $F(\bar{x} + tx) \neq \emptyset$, we have

$$F(\bar{x} + tx) \cap (\bar{y} + t(L(z) + r(t, x) B_Y)) \neq \emptyset.$$

Remark 11.1.31. The above notion of compactly approximable sets was originally proposed by Amahroq and Thibault [6]. Numerous illustrative examples and useful properties of the above class of set-valued maps can be found in [6]. See also Taa [548].

The following result makes use of the notion of compactly approximiable set-valued maps.

Theorem 11.1.32. *Let X, Y and Z be Banach spaces, let $G : X \times Y \rightrightarrows Z$ be a set-valued map, let $S \subset Y$ and $Q \subset Z$ be closed. If for $(\bar{x}, \bar{y}, \bar{z}) \in X \times S \times Q \cap \operatorname{graph}(G)$, the map $G(\cdot,\cdot) \cap Q$ is compactly approximiable at $(\bar{x}, \bar{y}, \bar{z})$ in all directions $(v, w) \in T(\operatorname{graph}(F), (\bar{x}, \bar{y}))$, then for F given in (11.18),*

$$T(\operatorname{graph}(F), (\bar{x}, \bar{y})) \subseteq [X \times T(S, \bar{y})] \cap DG(\bar{x}, \bar{y}, \bar{z}))^{-1}[T(Q, \bar{z})]. \tag{11.21}$$

Proof. Let $(v, w) \in T(\operatorname{graph}(F), (\bar{x}, \bar{y}))$ be arbitrary. Then there exist sequences $\{t_n\} \subset \mathbb{P}$ and $\{(v_n, w_n)\}$ with $t_n \downarrow 0$ and $(v_n, w_n) \to (v, w)$ such that $(\bar{x} + t_n v_n, \bar{y} + t_n w_n) \in \operatorname{graph}(F)$, which in view of (11.18) implies that for every $n \in \mathbb{N}$, we have $\bar{y} + t_n w_n \in S$ and $G(\bar{x} + t_n v_n, \bar{y} + t_n w_n) \cap Q \neq \emptyset$.

By the definition of compactly approximability, for sufficiently large $n \in \mathbb{N}$, we have

$$G(\bar{x} + t_n v_n, \bar{y} + t_n w_n) \cap S \cap [\bar{z} + t_n L(v, w) + t_n r_n B_Z] \neq \emptyset,$$

where $r_n \to 0$. Therefore, for sufficiently large $n \in \mathbb{N}$, there exists $\ell_n \in L(v, w)$ and $b_n \in B_Z$ such that

$$\bar{z} + t_n(\ell_n + r_n b_n) \in G(\bar{x} + t_n v_n, \bar{y} + t_n w_n) \cap Q. \tag{11.22}$$

By the compactness of $L(v, w)$, there exists a subsequence $\{\ell_n\}_{n \in J}$, where J is a subset of $\mathbb{N}$, converging to some $\ell \in L(v, w) \subset Z$. Therefore, using (11.22), we have $\ell \in T(Q, \bar{z})$ and $(v, w, \ell) \in T(\operatorname{graph}(G), (\bar{x}, \bar{y}, \bar{z}))$ which means that $(v, w) \in DG(\bar{x}, \bar{y}, \bar{z}))^{-1}[T(Q, \bar{z})$. However, we have also shown that $w \in T(S, \bar{y})$. The proof is complete. □

With the above preparation,we have the following main result (see [6]):

Theorem 11.1.33. *Let X, Y and Z be Banach spaces, let $G : X \times Y \rightrightarrows Z$ be a set-valued map, let $S \subset Y$ and $Q \subset Z$ be closed. If G is Lipschitz-like and derivable at $(\bar{x}, \bar{y}, \bar{z})$, the map $G(\cdot,\cdot) \cap Q$ is compactly approximable at $(\bar{x}, \bar{y}, \bar{z})$ in all directions $(v, w) \in T(\operatorname{graph} F, (\bar{x}, \bar{y}))$ where F is given by (11.18), (11.17) holds at $(\bar{x}, \bar{y}, \bar{z})$, and S and Q are derivable at $\bar{y}$ and $\bar{z}$, then F is derivable at $(\bar{x}, \bar{y})$ with*

$$DF(\bar{x}, \bar{y})(x) = \{y \in A(S, \bar{y}) |\ DG(\bar{x}, \bar{y}, \bar{z})(x, y) \cap A(Q, \bar{z}) \neq \emptyset\},$$

where G is derivable.

Moreover, if, in addition, the map $G(\cdot,\cdot) \cap Q$ is lower-semicontinuous at $(\bar{x},\bar{y},\bar{z})$ relative to its domain, G is strictly proto-differentiable at $(\bar{x},\bar{y},\bar{z})$, and S and Q are tangentially regular at $\bar{y}$ and $\bar{z}$, then F is strictly proto-differentiable at $(\bar{x},\bar{y})$ with

$$DF(\bar{x},\bar{y})(x) = \{y \in C(S,\bar{y}) \mid DG(\bar{x},\bar{y},\bar{z})(x,y) \cap C(Q,\bar{z}) \neq \emptyset\},$$

where G is strictly proto-differentiable.

Proof. Since the set-valued map G is Lipschitz-like at $(\bar{x},\bar{y},\bar{z}))$, we have $D_A G(\bar{x},\bar{y},\bar{z})) = D_L G(\bar{x},\bar{y},\bar{z}))$, and $D_C G(\bar{x},\bar{y},\bar{z})) = D_Q G(\bar{x},\bar{y},\bar{z}))$. The proof now follows from Theorems 11.1.29 and 11.1.32. □

In the remaining part of this section, we give sufficient conditions for the semidifferentiability of convex set-valued maps. For the proof of this theorem, we need the following technical lemmas whose proofs can be found in [214]. We recall that by $\mathrm{icr}A$ we mean the **algebraic relative interior** of A.

Lemma 11.1.34. *Let $\emptyset \neq A \subset X$ be a convex set and $T : X \to Y$ be a linear operator.*

(i) *If* $\mathrm{int}A \neq \emptyset$ *and T is open, then* $\mathrm{int}(]0,\infty[\cdot A) =]0,\infty[\cdot \mathrm{int}A$ *and* $T(\mathrm{int}A) = \mathrm{int}T(A)$.
(ii) *If* $\mathrm{icr}A \neq \emptyset$ *(e.g., if X is finite-dimensional), then* $\mathrm{icr}(]0,\infty[\cdot A) =]0,\infty[\cdot \mathrm{icr}A$ *and* $T(\mathrm{icr}A) = \mathrm{icr}T(A)$.

Lemma 11.1.35. *Let $A \subset X \times Y$ be a nonempty convex set. Suppose that either $(x,y) \in \mathrm{int}A$ or $(x,y) \in \mathrm{icr}A$ and $\dim X < \infty$. Then for every sequence $\{x_n\} \subset \mathrm{Pr}_X(A)$ with $x_n \to x$ there exists $\{y_n\} \to y$ such that $(x_n,y_n) \in A$ for every $n \in \mathbb{N}$.*

The following result, originally due to Penot [462], is adopted from [214].

Theorem 11.1.36. *Let $F : X \rightrightarrows Y$ be a convex set-valued map. Suppose that one of the following conditions holds:*

(i) $\mathrm{int}(\mathrm{graph}\, F) \neq \emptyset$, $x \in \mathrm{int}(\mathrm{dom}\,(F))$ *and* $y \in F(x)$;
(ii) *X and Y are finite-dimensional,* $x \in \mathrm{icr}(\mathrm{dom}\,(F))$, *and* $y \in F(x)$.

Then F is semidifferentiable at (x,y).

Proof. We give the proof of (ii), that of (i) being similar. We already know that for every $u \in X$, we have

$$D_L F(x,y)(u) \subset DF(x,y)(u) = \{v \in Y \mid (u,v) \in T(\mathrm{graph}\, F,(x,y))\}.$$

Let us show that $v \in D_L F(x,y)(u)$, if $(u,v) \in \mathrm{icr}T(\mathrm{graph}\, F,(x,y))$, even without requiring that $x \in \mathrm{icr}(\mathrm{dom}\,(F))$.

Without loss of generality, we consider that $(x,y) = (0,0)$. Take

$$(u,v) \in \mathrm{icr}T(\mathrm{graph}\, F,(0,0)) = \mathrm{icr}(]0,\infty[\cdot \mathrm{graph}\, F) =]0,\infty[\cdot \mathrm{icr}(\mathrm{graph}\, F),$$

where we used the relation $\mathrm{icr}A = \mathrm{icr}(\mathrm{cl}A)$, valid if A is a convex subset of a finite-dimensional space, for the first equality and Lemma 11.1.34 for the last one. Then there exist $\lambda' > 0$ and $(u', v') \in \mathrm{icr}(\mathrm{graph}\, F)$ such that $(u, v) = \lambda'(u', v')$. Let $\{t_n\} \to 0_+$ and $\{u_n\} \to u$ such that $t_n u_n = x + t_n u_n \in \mathrm{dom}\,(F)$, for every $n \in \mathbb{N}$. It follows that $\{u_n\} \subset \mathrm{aff}(\mathrm{dom}\,(F)) = \mathrm{span}(\mathrm{dom}\,(F))$ and $(u_n/\lambda') \to u' \in \mathrm{icr}(\mathrm{dom}\,(F))$. So there exists $n_0 \in \mathbb{N}$ such that $u_n/\lambda' \in \mathrm{dom}\,(F)$ for $n \geq n_0$. Therefore, by using Lemma 11.1.35, there exists a sequence $(v_n') \subset Y$ such that $\{v_n'\} \to v'$ and $(u_n/\lambda', v_n') \in \mathrm{graph}\,(F)$, for $n \geq n_0$. Let $v_n := \lambda' v_n'$ for $n \geq n_0$ and be arbitrary for $n < n_0$. So $v_n \to v$. Since $t_n \to 0_+$, there exists $n_1 \geq n_0$ such that $t_n \leq 1/\lambda'$ for $n \geq n_1$. Since $(0, 0) \in \mathrm{graph}\, F$, we obtain that $(x, y) + t_n(u_n, v_n) = t_n\lambda'(u_n/\lambda', v_n') \in \mathrm{graph}\, F$ for $n \geq n_1$. Therefore $v \in D_L F(x, y)(u)$.

Consider now $u \in \mathrm{icr}T(\mathrm{dom}\,(F); 0) =]0, \infty[\cdot \mathrm{icr}(\mathrm{dom}\,(F))$ and $v \in DF(x, y)(u)$. Therefore, we have $u \in \mathrm{icr}\,(]0, \infty[\cdot \mathrm{dom}\,(F))$ and $(u, v) \in \mathrm{cl}\,(]0, \infty[\cdot \mathrm{graph}\, F)$. By Lemma 11.1.34, there exists $v_0 \in Y$ such that $(u, v_0) \in \mathrm{icr}\,(]0, \infty[\cdot \mathrm{graph}\, F)$. It follows that for every $\lambda \in]0, 1[$,

$$(u, (1 - \lambda)v + \lambda v_0) = (1 - \lambda)(u, v) + \lambda(u, v_0) \in \mathrm{icr}\,(]0, \infty[\cdot \mathrm{graph}\, F)\,.$$

By what precedes, we have that $(1 - \lambda)v + \lambda v_0 \in D_L F(x, y)(u)$ for every $\lambda \in]0, 1[$. Taking into account that $D_L F(x, y)(u)$ is a closed set, we obtain that $v \in D_L F(x, y)(u)$ for $\lambda \to 0$. Therefore for every $u \in \mathrm{icr}T(\mathrm{dom}\,(F), x)$,

$$D_L F(x, y)(u) = \{v \in Y \mid (u, v) \in T(\mathrm{graph}\, F, (x, y))\} = DF(x, y)(u),$$

that is, F is semidifferentiable at every $u \in \mathrm{icr}T(\mathrm{dom}\,(F), x)$. Under our initial condition that $x \in \mathrm{icr}(\mathrm{dom}\,(F))$, we have $T(\mathrm{dom}\,(F), x) = \mathrm{span}(\mathrm{dom}\,(F) - x) = \mathrm{icr}T(\mathrm{dom}\,(F), x)$, and consequently F is semidifferentiable. The proof is complete. □

11.1.4 Weak Contingent Derivatives of Set-Valued Maps

For optimization problems posed in infinite-dimensional spaces, a fruitful approach is to modify the definition of tangent cones by replacing the strong convergence of sequences by the weak-convergence. The usefulness of this approach was depicted in a thorough study of this idea by Borwein [59]. On the other hand, while using tangent cones in product spaces, for instance to define graphical or epigraphical derivatives of set-valued maps, one possibility is to use a combination of the strong and the weak-convergence. This idea was explored by Rodríguez-Marín and Sama [504] on which the forgoing discussion is based.

In the following, the notion $(u_n, v_n) \underset{s,w}{\to} (u, v)$ indicates that $\{u_n\}$ converges to u strongly and $\{v_n\}$ converges to v weakly. Equipped with this notion, we propose the following:

Definition 11.1.37. Let X and Y be normed spaces, let $A \subset X \times Y$, and let $(\bar{x}, \bar{y}) \in A$. The **quasi-weak contingent cone** $T_{\mathrm{QW}}(A, (\bar{x}, \bar{y}))$ of A at $(\bar{x}, \bar{y})$ consists of all $(x, y) \in X \times Y$ such that there are sequences $\{t_n\} \subset \mathbb{P}$ and $\{(x_n, y_n)\} \subset A$ with $(x_n, y_n) \underset{s,w}{\to} (\bar{x}, \bar{y})$ and $t_n(x_n - \bar{x}, y_n - \bar{y}) \underset{s,w}{\to} (x, y)$.

The quasi-weak contingent cone $T_{\mathrm{QW}}(A, (\bar{x}, \bar{y}))$ is the sequential weak-closure of a subset of sequences,

$$\begin{aligned} T_{QW}(A, (\bar{x}, \bar{y})) &= (\tau^w \times \|\cdot\|) - \mathrm{cl} \left\{ (t_n ((x_n, y_n) \right. \\ &\qquad \left. - (\bar{x}, \bar{y}))) \mid \{t_n\} \subset \mathbb{P},\ A \supset ((x_n, y_n)) \underset{s,w}{\to} (\bar{x}, \bar{y}) \right\}, \end{aligned}$$

with respect to the product topology defined by the norm topology in the first space and the weak topology in the second space. Consequently, $T_{\mathrm{W}}(A, (\bar{x}, \bar{y}))$ is sequentially closed with respect to this product topology. That is, for every sequence $\{(u_n, v_n)\} \subset T_{QW}(A, (\bar{x}, \bar{y}))$ such that $(u_n, v_n) \underset{s,w}{\to} (u, v)$ we have $(u, v) \in T_{QW}(A, (\bar{x}, \bar{y}))$.

We equip the notion of the quasi-weak contingent cone to propose the following:

Definition 11.1.38. A set-valued map $D_w F(\bar{x}, \bar{y}) : X \rightrightarrows Y$ is called the **weak contingent derivative** of F at $(\bar{x}, \bar{y})$, if

$$\mathrm{graph}\,(D_w F(\bar{x}, \bar{y})) = T_{QW}(\mathrm{graph}\,(F), (\bar{x}, \bar{y})),$$

where $T_{QW}(\mathrm{graph}\,(F), (\bar{x}, \bar{y}))$ is the quasi-weak contingent cone of $\mathrm{graph}\,(F)$ at $(\bar{x}, \bar{y})$.

To establish some connection between the contingent and weak contingent derivatives, we introduce the following notions.

Definition 11.1.39. Let X and Y be normed spaces, let $F : X \rightrightarrows Y$ be a set-valued map, and let $(\bar{x}, \bar{y}) \in \mathrm{graph}\,(F)$.

1. The map F is said to be **stable** at $(\bar{x}, \bar{y})$, if there exist constants $\varepsilon > 0$ and $k > 0$ such that

$$F(x) \subseteq \{\bar{y}\} + k\ \|x - \bar{x}\|\ B_Y \quad \text{for every } x \in U_\varepsilon(\bar{x}) \cap \mathrm{dom}\,(F).$$

2. The map F is said to be **directionally compact** at $(\bar{x}, \bar{y})$ in the direction $x \in T(\mathrm{dom}\,(F), \bar{x})$, if for every sequence $\{t_n\} \subset \mathbb{P}$ with $t_n \downarrow 0$ and for every sequence $\{x_n\} \subset X$ with $x_n \to x$, any sequence $\{y_n\} \subset Y$ satisfying $\bar{y} + t_n y_n \in F(\bar{x} + t_n x_n)$, for each $n \in \mathbb{N}$, contains a convergent subsequence. If F is directionally compact at $(\bar{x}, \bar{y})$ for every direction $x \in T(\mathrm{dom}\,(F), \bar{x})$, then F is said to be directionally compact at $(\bar{x}, \bar{y})$.

Remark 11.1.40. Clearly, if F is stable then $F(\bar{x}) = \{\bar{y}\}$. Moreover, if Y is finite-dimensional, then as a consequence of the compactness of the closed unit ball, it follows that, if the map F is stable at $(\bar{x}, \bar{y})$, then F is directionally compact at $(\bar{x}, \bar{y})$.

The following result gives some useful implications of the directional compactness.

Theorem 11.1.41. *Let X and Y be normed spaces, let $F : X \rightrightarrows Y$ be a set-valued map, and let $(\bar{x}, \bar{y}) \in \operatorname{graph}(F)$. Assume that the map F is directionally compact at $(\bar{x}, \bar{y})$. Then:*

1. *For every $\{(x_n, y_n)\} \subset \operatorname{graph}(F)$ with $x_n \to \bar{x}$, we have $y_n \to \bar{y}$.*
2. *The map F is stable at $(\bar{x}, \bar{y})$ if and only if $DF(\bar{x}, \bar{y})(0) = \{0\}$.*

Proof. For the first part, we begin by defining

$$t_n = \|x_n - \bar{x}\|^{1/2},$$
$$\bar{x}_n = \frac{(x_n - \bar{x})}{t_n}.$$

Clearly $t_n \downarrow 0$ and $\bar{x}_n \to 0$. For $\bar{y}_n := (y_n - \bar{y})/h_n$, by the directional compactness of F at $(\bar{x}, \bar{y})$, there is a subsequence $\{\bar{y}_{n_k}\}$ converging to an element y. Therefore,

$$y_{n_k} - \bar{y} = t_{n_k} \bar{y}_{n_k} \underset{k}{\to} 0.$$

Indeed, following this reasoning, every subsequence of $\{y_n\}$ has a convergent subsequence converging to $\bar{y}$ and this implies that $y_n \to \bar{y}$.

Next we show that the stability of the map F implies that $DF(\bar{x}, \bar{y})(0) = \{0\}$. Let $y \in DF(\bar{x}, \bar{y})(0)$ be arbitrary. Then there are sequences $\{t_n\} \subset \mathbb{P}$ and $\{(x_n, y_n)\} \subset \operatorname{graph}(F)$ such that $(x_n, y_n) \to (\bar{x}, \bar{y})$, and $t_n(x_n - \bar{x}, y_n - \bar{y}) \to (0, y)$.

Since the map F is stable, we deduce that for sufficiently large n, we have

$$\|t_n (y_n - \bar{y})\| \leq k \ \|t_n(x_n - \bar{x})\|,$$

which, by passing to the limit, yields that $\|y\| \leq 0$, and hence $y = 0$.

For the converse, assume that the map F is directionally compact but not stable at $(\bar{x}, \bar{y})$. In this case, we can find a sequence $\{(x_n, y_n)\} \subset \operatorname{graph}(F)$ with $x_n \neq 0$ such that $x_n \to \bar{x}$ and

$$\frac{\|y_n - \bar{y}\|}{\|x_n - \bar{x}\|} \to \infty. \tag{11.23}$$

We define

$$t_n = \|y_n - \bar{y}\|,$$
$$\bar{x}_n = \frac{(x_n - \bar{x})}{h_n},$$
$$\bar{y}_n = \frac{(y_n - \bar{y})}{h_n}.$$

Then $t_n \downarrow 0$ and hence $\bar{x}_n \to 0$ by (11.23). By the directional compactness of the map F at $(\bar{x}, \bar{y})$, there exists a subsequence $\{\bar{y}_{n_k}\}$ of $\{\bar{y}_n\}$ converging to an element $y \in Y$. Since $\|\bar{y}_n\| = 1$, we get a contradiction to the hypothesis that $DF(\bar{x}, \bar{y})(0) = \{0\}$. The proof is complete. □

With the above preparation, we give the following result:

Theorem 11.1.42. *Let X and Y be normed spaces, let $C \subset Y$ be a pointed, closed, and convex cone, let $F : X \rightrightarrows Y$ be a set-valued map, and let $(\bar{x}, \bar{y}) \in \operatorname{graph}(F)$. Let F be directionally compact at $(\bar{x}, \bar{y})$. Then the following equivalences hold:*

1. $D(F + C)(\bar{x}, \bar{y})(\cdot) = D_w(F + C)(\bar{x}, \bar{y})(\cdot)$.
2. $DF(\bar{x}, \bar{y})(\cdot) = D_w F(\bar{x}, \bar{y})(\cdot)$.

Proof. Since $T_{QW}(\operatorname{epi}(F), (\bar{x}, \bar{y})) \supseteq T(\operatorname{epi}(F), (\bar{x}))$ is always true, it suffices to show that

$$T_{QW}(\operatorname{epi}(F), (\bar{x}, \bar{y})) \subseteq T(\operatorname{epi}(F), (\bar{x}, \bar{y})).$$

Let $(x, y) \in T_{QW}(\operatorname{epi}(F), (\bar{x}, \bar{y}))$ be arbitrary. Then there are sequences $\{t_n\} \subset \mathbb{P}$, $\{c_n\} \subset C$ and $\{(x_n, y_n)\} \subset \operatorname{graph}(F)$ such that

$$(x_n, y_n + c_n) \underset{s,w}{\to} (\bar{x}, \bar{y})$$
$$t_n(x_n - \bar{x}, y_n + c_n - \bar{y}) \underset{s,w}{\to} (x, y).$$

From the directional compactness of the map F at $(\bar{x}, \bar{y})$, Theorem 11.1.41 ensures that $y_n \to \bar{y}$. Furthermore, by the same result, we can assume, taking subsequences if necessary, that the sequence $(t_n(y_n - \bar{y}))$ converges to an element $y_1 \in Y$. Consequently, by definition $y_1 \in DF(\bar{x}, \bar{y})(x)$. In particular, we have

$$(t_n(y_n - \bar{y})) \underset{w}{\to} y_1,$$

which, due to the convergence properties of the sequences, confirms that the sequence $\{t_n c_n\}$ converges weakly to an element $c \in C$. Therefore $y = y_1 + c$ and we deduce that

$$y = y_1 + c \in DF(\bar{x}, \bar{y})(x) + C \subset D(F + C)(\bar{x}, \bar{y})(x).$$

Since (x, y) was arbitrary, we have shown that $T_{QW}(\text{epi}(F), (\bar{x}, \bar{y})) \subset T(\text{epi}(F), (\bar{x}, \bar{y}))$. The proof of the second part can be obtained by similar arguments. □

11.1.5 A Lyusternik-Type Theorem Using Contingent Derivatives

In this section, our objective is to give an analogue of the well-known Lusternik theorem for set-valued maps by employing a weaker notion of metric regularity recently proposed by Pühl and Schirotzek [481]. We first recall the notion of metric regularity and related notions:

Definition 11.1.43. Let X and Y be Banach spaces, let $F : X \rightrightarrows Y$ be a closed set-valued map, and let $(\bar{x}, \bar{y}) \in \text{graph}(F)$.

1. The map F is said to be an **open covering** with linear rate r at $(\bar{x}, \bar{y})$, if there exist neighborhoods of $U(\bar{x})$ and $V(\bar{y})$ such that for any $x \in U(\bar{x})$ and t sufficiently small, we have

$$F(x) \cap V(\bar{y}) + t\, r\, B_Y \subset F(\bar{x} + tB_X). \tag{11.24}$$

2. The map F is said to be **metrically regular** at $(\bar{x}, \bar{y})$ with modulus r, if there exist neighborhoods $U(\bar{x})$ and $V(\bar{y})$ such that for any $x \in U(\bar{x})$ and $y \in V(\bar{y})$, we have

$$d(x, F^{-1}(y)) \leq r\, d(y, F(x)). \tag{11.25}$$

We have the following result whose proof can be found in Penot [461] or Borwein and Zhu [66]:

Theorem 11.1.44. *Let X and Y be Banach spaces and let $F : X \rightrightarrows Y$ be a closed set-valued map. Then the following statements are equivalent:*

1. *F is an open covering with linear rate r at $(\bar{x}, \bar{y}) \in \text{graph}(F)$.*
2. *F is metrically regular with modulus $1/r$ at $(\bar{x}, \bar{y}) \in \text{graph}(F)$.*
3. *F^{-1} is pseudo-Lipschitz with modulus $1/r$ at $(\bar{y}, \bar{x}) \in \text{graph}(F^{-1})$.*

We now give an extension of the notions used in the above classical result:

Definition 11.1.45. Let X and Y be Banach spaces, let $F : X \rightrightarrows Y$ be a set-valued map, and let $(\bar{x}, \bar{y}) \in \text{graph}(F)$.

1. The map F is said to be **linearly semi-open** around $(\bar{x}, \bar{y})$, if there are numbers $c > 0$ and $s > 0$ such that for every $(x, y) \in \text{graph}(F)$ with $x \in U_s(\bar{x})$ and $y \in U_s(\bar{y})$, and for every $t \in [0, s]$, we have

$$y + c\,t\|x - \bar{x}\|B \subset F(x + t\|x - \bar{x}\|B). \tag{11.26}$$

2. The map F is said to be **metrically semi-regular** around $(\bar{x}, \bar{y})$, if there are numbers $c > 0$ and $s > 0$ such that for every (x, y) with $x \in U_s(\bar{x}),\ y \in U_s(\bar{y})$ and $d(y, F(x)) \leq s\|x - \bar{x}\|$, we have

$$d(x, F^{-1}(y)) \leq c\, d(y, F(x)). \tag{11.27}$$

3. The map F is said to be **semi-pseudo-Lipschitz** around $(\bar{x}, \bar{y})$, if there are numbers $c > 0$ and $s > 0$ such that for every $(x, y) \in \operatorname{graph}(F)$ with $x \in U_s(\bar{x})$ and $y \in U_s(\bar{y})$, there is a neighborhood $V(y)$ such that for every $\hat{x} \in x + s\|y - \bar{y}\|B$, we have

$$F(x) \cap V(y) \subset F(\hat{x}) + c\,\|x - \hat{x}\|B. \tag{11.28}$$

We have the following result connecting the above three concepts:

Theorem 11.1.46. *Let X and Y be Banach spaces, let $F : X \rightrightarrows Y$ be a set-valued map, and let $(\bar{x}, \bar{y}) \in \operatorname{graph}(F)$. Let the map F be closed-valued. Then the following properties are equivalent:*

1. *The map F is linearly semi-open around $(\bar{x}, \bar{y})$.*
2. *The map F is metrically semi-regular around $(\bar{x}, \bar{y})$.*
3. *The map F^{-1} is semi-pseudo-Lipschitz around $(\bar{y}, \bar{x})$.*

Proof. See Pühl and Schirotzek [481]. □

With the above preparation, we give the following Lyusternik-type theorem for set-valued maps. This result was originally proved by Pühl and Schirotzek [481].

Theorem 11.1.47. *Let X and Y be Banach spaces, let $F : X \rightrightarrows Y$ be a set-valued map, and let $(\bar{x}, \bar{y}) \in \operatorname{graph}(F)$. Let the map F be metrically semi-regular around $(\bar{x}, 0)$. Then:*

$$T(\operatorname{Ker}(F), \bar{x}) = \operatorname{Ker}(DF(\bar{x}, 0)). \tag{11.29}$$

Proof. Let $x \in T(\operatorname{Ker}(F), \bar{x})$ be arbitrary. Then there are sequence $\{t_n\} \subset \mathbb{P}$ and $\{x_n\} \subset X$ with $t_n \downarrow 0$ and $x_n \to x$ such that $\bar{x} + t_n x_n \in \operatorname{Ker}(F)$, for every $n \in \mathbb{N}$. Therefore, $0 \in F(\bar{x} + t_n x_n)$ for every $n \in \mathbb{N}$, and consequently,

$$0 \in \limsup_{n\to\infty} \frac{F(\bar{x} + t_n x_n)}{t_n} \subset DF(\bar{x}, 0)(x),$$

which implies that $x \in \operatorname{Ker}(DF(\bar{x}, 0))$, proving the inclusion $T(\operatorname{Ker}(F), \bar{x}) \subset \operatorname{Ker}(DF(\bar{x}, 0))$.

For the converse, let $x \in \operatorname{Ker}(DF(\bar{x}, 0))$ be arbitrary. If $x = 0$, then $x \in T(\operatorname{Ker}(F), \bar{x})$. Therefore, we assume that $x \neq 0$. Let k and s be the semi-regularity parameters. By the definition of $DF(\bar{x}, 0)$, there are sequences $\{t_n\} \subset \mathbb{P}$ and $\{x_n\}$ such that $t_n \downarrow 0,\ x_n \to x$, and $y_n \in F(\bar{x} + t_n v_n)$ such that

$$\lim_{n\to\infty} \frac{y_n}{t_n} = 0. \tag{11.30}$$

Furthermore, for sufficiently large n, we have $\bar{x} + t_n x_n \in U_s(\bar{x})$. Since $x \neq 0$, for sufficiently large n, we have

$$\frac{\|y_n\|}{t_n} \leq \frac{1}{2} s \|x\| \leq s \|x_n\|,$$

which implies that $y_n \in s\, t_n \|x_n\| B$ and that $y_n \in sB$. The metric semi-regularity of F then applied to $(\bar{x} + t_n x_n, 0)$ gives

$$d(\bar{x} + t_n x_n, \mathrm{Ker}(F)) \leq k \cdot d(F(\bar{x} + t_n x_n), 0) \leq k\, \|y_n\|.$$

Therefore, there is a $w_n \in \mathrm{Ker}(F)$ such that

$$\|\bar{x} + t_n x_n - w_n\| \leq 2k\, \|y_n\|. \tag{11.31}$$

We set $u_n = t_n^{-1}(w_n - \bar{x})$, and hence $\bar{x} + t_n u_n \in \mathrm{Ker}(F)$. We claim that $u_n \to x$. For this, we note that

$$\begin{aligned}
\|u_n - x\| &\leq \|u_n - x_n\| + \|x_n - x\| \\
&= \left\| \frac{w_n - \bar{x}}{t_n} - x_n \right\| + \|x_n - x\| \\
&= \frac{\|w_n - (\bar{x} + t_n x_n)\,\|}{t_n} + \|x_n - x\| \\
&\leq 2k\, \frac{y_n}{t_n} + \|x_n - x\| \to 0,
\end{aligned}$$

and the proof is complete. □

Several particular cases of the above result, including the classical Lusternik theorem, can be found in [481, 524].

11.2 Calculus Rules for Derivatives of Set-Valued Maps

In this section, we give calculus rules for graphical derivatives of set-valued maps using two techniques. Namely, using a direct approach and by the calculus of the involved tangent cones.

11.2.1 Calculus Rules by a Direct Approach

We begin with a scalar multiplication rule for the contingent derivative.

Theorem 11.2.1. *Let X and Y be normed spaces and let $F : X \rightrightarrows Y$ be a set-valued map. For every $\lambda \in \mathbb{R}_+$ and for every $(\bar{x}, \bar{y}) \in \operatorname{graph}(F)$, we have*

$$D(\lambda F)(\bar{x}, \lambda \bar{y})(x) = \lambda DF(\bar{x}, \bar{y})(x). \tag{11.32}$$

Proof. For $(x, y) \in \operatorname{graph}(\lambda DF(\bar{x}, \bar{y}))$, there are sequences $\{(x_n, y_n)\} \subset X \times Y$ and $\{t_n\} \subset \mathbb{P}$ such that $t_n \downarrow 0,\ x_n \to x,\ y_n \to \frac{y}{\lambda}$ and $\bar{y} + t_n y_n \in F(\bar{x} + t_n x_n)$, which implies that $\lambda \bar{y} + \lambda t_n y_n \in \lambda F(\bar{x} + t_n x_n)$, and hence $(x, y) \in \operatorname{graph}(D(\lambda F)(\bar{x}, \lambda \bar{y}))$. We have shown that $\lambda DF(\bar{x}, \bar{y})(x) \subseteq D(\lambda F(\bar{x}, t\bar{y})(x)$. The converse inclusion can also be proved by similar arguments. □

The next three results, whose proofs follow from the definitions of the involved objects, give calculus rules which either involve the lower Dini derivative or a combination of the lower Dini derivative and the contingent derivative (see [462]).

Theorem 11.2.2. *Let X, Y and Z be normed spaces, let $F : X \rightrightarrows Y$ and $G : Y \rightrightarrows Z$ be set-valued maps, let $(\bar{x}, \bar{y}) \in \operatorname{graph}(F)$, and let $(\bar{y}, \bar{z}) \in \operatorname{graph}(G)$. Then the following chain rules hold:*

$$D_L G(\bar{y}, \bar{z}) \circ D_L F(\bar{x}, \bar{y}) \subseteq D_L (G \circ F)(\bar{x}, \bar{z}),$$
$$D_L G(\bar{y}, \bar{z}) \circ DF(\bar{x}, \bar{y}) \subseteq D(G \circ F)(\bar{x}, \bar{z}).$$

Theorem 11.2.3. *Let X and Y be normed spaces, let $F_1, F_2 : X \rightrightarrows Y$ be set-valued maps, let $(\bar{x}, \bar{y}_1) \in \operatorname{graph}(F_1)$, and let $(\bar{x}, \bar{y}_2) \in \operatorname{graph}(F_2)$. Then the following sum formulae hold:*

$$D_L F_1(\bar{x}, \bar{y}_1) + D_L F_2(\bar{x}, \bar{y}_2) \subseteq D_L (F_1 + F_2)(\bar{x}, \bar{y}_1 + \bar{y}_2),$$
$$DF_1(\bar{x}, \bar{y}_1) + D_L F_2(\bar{x}, \bar{y}_2) \subseteq D(F_1 + F_2)(\bar{x}, \bar{y}_1 + \bar{y}_2).$$

Theorem 11.2.4. *Let X, Y_1, and Y_2 be normed spaces, let $F_1 : X \rightrightarrows Y_1$ and $F_2 : X \to Y_2$ be set-valued maps, let $(\bar{x}, \bar{y}_1) \in \operatorname{graph}(F_1)$, and let $(\bar{x}, \bar{y}_2) \in \operatorname{graph}(F_2)$. Then:*

$$D_L (F_1 \times F_2)(\bar{x}, \bar{y}_1, \bar{y}_2) = D_L F_1(\bar{x}, \bar{y}_1) \times D_L F_2(\bar{x}, \bar{y}_2),$$
$$D(F_1 \times F_2)(\bar{x}, \bar{y}_1, \bar{y}_2) \supset DF_1(\bar{x}, \bar{y}_1) \times D_L F_2(\bar{x}, \bar{y}_2),$$
$$D(F_1 \times F_2)(\bar{x}, \bar{y}_1, \bar{y}_2) \subseteq DF_1(\bar{x}, \bar{y}_1) \times DF_2(\bar{x}, \bar{y}_2).$$

In the following, we give a strengthening of some of the above results by giving calculus rules for the contingent derivatives by an approach proposed by Luc [400]. We need the following:

Definition 11.2.5. Let X and Y be Banach spaces and let $F : X \rightrightarrows Y$ be a set-valued map.

1. F is called **lower semidifferentiable** at $(\bar{x}, \bar{y}) \in \operatorname{graph}(F)$, if for any sequence $\{x_n\} \subset \operatorname{dom}(F)$ converging to $\bar{x}$ and for any sequence $\{t_n\} \subset \mathbb{P}$ with $\{t_n(x_n - \bar{x})\}$ converging to some $x \in X$, there exists a subsequence $\{y_{n_m}\}$ such that the sequence $\{t_{n_m}(y_{n_m} - \bar{y})\}$ is convergent.
2. F is called **upper semidifferentiable** at $(\bar{x}, \bar{y}) \in \operatorname{graph}(F)$, if for any sequence $\{(x_n, y_n)\} \subset \operatorname{graph}(F)$, not identical to $(\bar{x}, \bar{y})$, and converging to $(\bar{x}, \bar{y})$, there exists a subsequence $\{t_{n_m}\} \subset \mathbb{P}$ such that $\{t_{n_m}(x_{n_m} - \bar{x}, y_{n_m} - \bar{y})\}$ converges to some nonzero $(x, y) \in X \times Y$.
3. F is called **compact** at $\bar{x} \in \operatorname{dom}(F)$, if any sequence $\{(x_n, y_n)\} \subset \operatorname{graph}(F)$ possesses a convergent subsequence provided that $\{x_n\}$ converges to $\bar{x}$.

Many useful results concerning lower semidifferentiable and upper semidifferentiable set-valued maps can be found in [400].

The following result gives a chain rule for the contingent derivatives.

Theorem 11.2.6. *Let X, Y, and Z be Banach spaces and let $F : X \rightrightarrows Y$ and $G : Y \rightrightarrows Z$ be set-valued maps. For $(\bar{x}, \bar{y}) \in \operatorname{graph}(F)$ and $(\bar{y}, \bar{z}) \in \operatorname{graph}(G)$, let F be compact at $\bar{x}$ and upper semidifferentiable at $(\bar{x}, \bar{y})$, let $DF(\bar{x}, \bar{y})(0) = \{0\}$, and let G be closed on $F(\bar{x})$. Then:*

$$D(G \circ F)(\bar{x}, \bar{z}) \subseteq DG(\bar{y}, \bar{z}) \circ DF(\bar{x}, \bar{y}). \tag{11.33}$$

Proof. Let $(x, z) \in \operatorname{graph}(D(G \circ F)(\bar{x}, \bar{z}))$ be arbitrary. Then, $(x, z) \in T(\operatorname{graph}(G \circ F), (\bar{x}, \bar{z}))$ and consequently there are sequences $\{t_n\} \subset \mathbb{P}$ and $\{(x_n, z_n)\} \subset \operatorname{graph}(G \circ F)$ such that $(x_n, z_n) \to (\bar{x}, \bar{z})$ and $t_n(x_n - \bar{x}, z_n - \bar{z}) \to (x, z)$. Since $\{(x_n, z_n)\} \subset \operatorname{graph}(G \circ F)$, we can find $\{y_n\} \subset Y$ such that $y_n \in F(x_n)$ and $z_n \in G(y_n)$, for every $n \in \mathbb{N}$. Since the map F is compact at $\bar{x}$, we can assume that $\{y_n\}$ converges to some $\bar{y} \in F(\bar{x})$. Furthermore, since the map G is closed, we deduce that $\bar{z} \in G(\bar{y})$.

At this juncture, we use the upper semidifferentiability of the map F to ensure that there exists a sequence $\{s_n\} \subset \mathbb{P}$ such that a subsequence of $\{s_n(x_n - \bar{x}, y_n - \bar{y})\}$ converges to a nonzero (u, v). We retain the same notations for subsequences as well. We consider the sequences $\left\{\frac{t_n}{s_n}\right\}$ and $\left\{\frac{s_n}{t_n}\right\}$, and by taking subsequences if necessary, we can assume that one of the two sequences converges to t. We first assume that $\frac{t_n}{s_n}$ converges to t. Then $t_n(y_n - \bar{y}, z_n - \bar{z})$ converges to (tv, z) and $t_n(x_n - \bar{x}, y_n - \bar{y})$ converges to $(tu, tv) = (x, tv)$. Consequently, $z \in DG(\bar{y}, \bar{z})(tv)$, where $tv \in DF(\bar{x}, \bar{y})(x)$, confirming (11.33).

Next, we assume that $\frac{s_n}{t_n}$ converges to t. Then $s_n(y_n - \bar{y}, z_n - \bar{z})$ converges to (v, tz) whereas $s_n(x_n - \bar{x}, y_n - \bar{y})$ converges to $(tx, v) = (u, v)$. Assume that $t \neq 0$. Then $z \in DG(\bar{y}, \bar{z})(v/t)$ where $v/t \in DF(\bar{x}, \bar{y})(x)$ confirming (11.33). It remains to show that $t \neq 0$. Note that if $t = 0$, then $u = 0$, and due to the assumption that $DF(\bar{x}, \bar{y})(0) = \{0\}$, we obtain $v = 0$. This, however, is a contraction. The proof is complete. □

We give another result recently obtained by Ward [596] who noted the usefulness of the notion of directional compactness for deriving calculus rules.

Theorem 11.2.7. *Let X, Y, and Z be normed spaces, let $F : X \rightrightarrows Y$ and $G : Y \rightrightarrows Z$ be set-valued maps, let $(\bar{x}, \bar{y}) \in \operatorname{graph}(F)$, and let $(\bar{y}, \bar{z}) \in \operatorname{graph}(G)$. Define $H : X \rightrightarrows Z$ be*

$$H(x, z) := \{y \in Y \mid y \in F(x),\ z \in G(y)\}.$$

Let H be directionally compact at $((\bar{x}, \bar{z}), \bar{y})$ in the direction $(x, z) \in \operatorname{graph}(D(G \circ F)(\bar{x}, \bar{z}))$. Then

$$(x, z) \in \operatorname{graph}(DG(\bar{y}, \bar{z}) \circ DF(\bar{x}, \bar{y})).$$

Proof. Let $(x, z) \in \operatorname{graph}(D(G \circ F)(\bar{x}, \bar{z}))$ with H directionally compact at $((\bar{x}, \bar{z}), \bar{y})$ in the direction (x, z). Then there exist sequences $\{t_n\} \subset \mathbb{P}$ and $\{(x_n, z_n)\}$ with $t_n \downarrow 0$, $(x_n, z_n) \to (x, z)$ such that $(\bar{z}+t_n z_n) \in (G \circ F)(\bar{x}+t_n x_n)$, for every $n \in \mathbb{N}$. Therefore, there exists $\{w_n\}$ such that $w_n \in F(\bar{x} + t_n x_n)$ and $\bar{z} + t_n z_n \in G(w_n)$. By setting $y_n := t_n^{-1}(w_n - \bar{y})$, we have $\bar{y} + t_n y_n \in H(\bar{x} + t_n x_n, \bar{z} + t_n z_n)$. Now, by invoking the directional compactness, we may assume, by taking subsequence if necessary, that $\{y_n\} \to y$ for some $y \in Y$. Then $(x, y) \in T(\operatorname{graph}(F), (\bar{x}, \bar{y}))$ and $(y, z) \in T(\operatorname{graph}(G), (\bar{y}, \bar{z}))$ implying that $(x, z) \in \operatorname{graph}(DG(\bar{y}, \bar{z}) \circ DF(\bar{x}, \bar{y}))$. The proof is complete. □

The following result gives a sum formula for the contingent derivatives.

Theorem 11.2.8. *Let X and Y be Banach spaces and let $F_1 : X \rightrightarrows Y$ and $F_2 : X \rightrightarrows Y$ be set-valued maps. Let F_1 and F_2 be closed at $\bar{x} \in X$, let either F_1 or F_2 be compact at $\bar{x}$, let for each $\bar{y}_1 \in F_1(\bar{x})$, F_1 be upper semidifferentiable at $(\bar{x}, \bar{y}_1)$ with $DF_1(\bar{x}, \bar{y}_1)(0) = \{0\}$, and let $(\bar{x}, \bar{y}_2) \in \operatorname{graph}(F_2)$. Then:*

$$D(F_1 + F_2)(\bar{x}, \bar{y}_1 + \bar{y}_2) \subseteq DF_1(\bar{x}, \bar{y}_1) + DF_2(\bar{x}, \bar{y}_2). \tag{11.34}$$

Proof. The proof follows the same line of arguments as in Theorem 11.2.6 and is omitted. □

Our objective now is to give some calculus rules for the Clarke derivative and the quasi-interiorly Clarke derivative. For this, we recall the following notion introduced by Thibault [571]:

Definition 11.2.9. Let X and Y be normed spaces, let $F_1, F_2 : X \rightrightarrows Y$ be set-valued maps, let $(\bar{x}, \bar{y}_1) \in \operatorname{graph}(F_1)$, and let $(\bar{x}, \bar{y}_2) \in \operatorname{graph}(F_2)$. The maps F_1 and F_2 are **additively separate** at $(\bar{x}, \bar{y}_1, \bar{y}_2)$, if for every neighborhood $U(\bar{y}_1)$ and every neighborhood $U(\bar{y}_2)$, there exist a neighborhood $U(\bar{y}_1 + \bar{y}_2)$ and a neighborhood $U(\bar{x})$ such that for every $x \in U(\bar{x})$, we have

$$(F_1(x) + F_2(x)) \cap U(\bar{y}_1 + \bar{y}_2) \subset U(\bar{y}_1) \cap F_1(x) + U(\bar{y}_2) \cap F_2(x).$$

With the above notion, we are now in a position to announce the following sum formula:

Theorem 11.2.10. *Let X and Y be normed spaces, let $F_1, F_2 : X \rightrightarrows Y$ be set-valued maps, let $(\bar{x}, \bar{y}_1) \in \text{graph}(F_1)$, and let $(\bar{x}, \bar{y}_2) \in \text{graph}(F_2)$. Let F_1 and F_2 be additively separate at $(\bar{x}, \bar{y}_1, \bar{y}_2)$. Then:*

$$D_Q F_1(\bar{x}, \bar{y}_1) + D_C F_2(\bar{x}, \bar{y}_2) \subset D_C(F_1 + F_2)(\bar{x}, \bar{y}_1 + \bar{y}_2)$$
$$D_Q F_1(\bar{x}, \bar{y}_1) + D_Q F_2(\bar{x}, \bar{y}_2) \subset D_Q(F_1 + F_2)(\bar{x}, \bar{y}_1 + \bar{y}_2).$$

Proof. See Thibault [571]. □

We now recall the following notion of lower semicontinuity:

Definition 11.2.11. Let X and Y be normed spaces and let $F : X \rightrightarrows Y$ be a set-valued map. The map F is said to be lower-semicontinuous at $(\bar{x}, \bar{y}) \in \text{graph}(F)$ relative to $S \subset X$, if for each neighborhood $U(\bar{y})$, there is a neighborhood $U(\bar{x})$ such that $U(\bar{y}) \cap F(x) \neq \emptyset$, for every $x \in S \cap U(\bar{x})$.

We are now in a position to give the following formula for a composition:

Theorem 11.2.12. *Let X, Y and Z be normed spaces, let $F : X \rightrightarrows Y$ and $G : Y \rightrightarrows Z$ be two set-valued maps, let $(\bar{x}, \bar{y}) \in \text{graph}(F)$ and let $(\bar{y}, \bar{z}) \in \text{graph}(G)$. Let the set-valued map defined from $X \times Z$ to Y by $(x,z) \to G(x) \cap F^{-1}(z)$ be lower-semicontinuous at $((\bar{x}, \bar{z}), \bar{y})$ relative to $\text{graph}(F \circ G)$. Then:*

$$D_Q G(\bar{y}, \bar{z}) \circ D_Q F(\bar{x}, \bar{y}) \subset D_Q(G \circ F)(\bar{x}, \bar{z}),$$
$$D_Q G(\bar{y}, \bar{z}) \circ D_C F(\bar{x}, \bar{y}) \subset D_C(G \circ F)(\bar{x}, \bar{z}).$$

Furthermore, if G is Lipschitzian, then

$$D_C G(\bar{y}, \bar{z}) \circ D_C F(\bar{x}, \bar{y}) \subset D_C(G \circ F)(\bar{x}, \bar{z}).$$

Proof. See Thibault [571]. □

Recently in an interesting work, Li, Meng, and Penot [380] derived numerous calculus rules for derivatives of set-valued maps and gave numerous examples and counterexamples.

11.2.2 Derivative Rules by Using Calculus of Tangent Cones

In set-valued optimization, an efficient approach for obtaining various calculus rules can be devised by using certain properties of the tangent cones which are being used to define the derivatives of the involved set-valued maps. This approach, which seems to be devised independently, by Ward and Borwein [597] and Aubin and Frankowska [17], has been extensively pursued ever since [17, 597]. Given the generality of this approach and the fact that many derivatives of set-valued maps defined via suitable tangent cones can benefit from this strategy, we would give the calculus rules for abstract tangent cone approximations, relying only on certain features of the underlying cones. Our approach is based on Aubin and Frankowska [17, 18] and Ward [590, 596, 597].

In the following, we deal with abstract tangent cones in the sense of Sect. 4.5. Furthermore, given an abstract tangent cone, the associated graphical derivative is defined in a way analogous to the contingent derivative, that is, by taking the tangent cone to the graph of the set-valued map.

We begin with the following general scheme for deriving an abstract chain rule:

Theorem 11.2.13. *Let X, Y, and Z be normed spaces, let $F : X \rightrightarrows Y$ and $G : Y \rightrightarrows Z$ be set-valued maps, let $(\bar{x}, \bar{y}) \in \operatorname{graph}(F)$, and let $(\bar{y}, \bar{z}) \in \operatorname{graph}(G)$. Assume that P and Q are closed tangent cones such that for the maps $\ell : X \times Y \times Y \times Z \to X \times Z$ and $g : X \times Y \times Y \times Z \to Y$, defined by*

$$\ell(a, b, c, d) := (a, d), \tag{11.35}$$

$$g(a, b, c, d) := b - c, \tag{11.36}$$

and for $S_1 := \operatorname{graph}(F)$, $S_2 := \operatorname{graph}(G)$, $S := S_1 \times S_2$, and $x_0 := (\bar{x}, \bar{y}, \bar{y}, \bar{z})$, we have

$$\ell(P(S, x_0)) \subseteq P(\ell(S), \ell(x_0)), \tag{11.37}$$

$$P(S_1, (\bar{x}, \bar{y})) \times Q(S_2, (\bar{y}, \bar{z})) \subseteq P(S_1 \times S_2, (\bar{x}, \bar{y}, \bar{y}, \bar{z})), \tag{11.38}$$

$$P(S, x_0) \cap \nabla g(x_0)^{-1}(0) \subseteq P(S \cap g^{-1}(0), x_0). \tag{11.39}$$

Then, for the graphical derivatives $D_P G(\bar{y}, \bar{z})$ and $D_Q F(\bar{y}, \bar{z})$ defined via P and Q, we have

$$D_Q G(\bar{y}, \bar{z}) \circ D_P F(\bar{x}, \bar{y}) \subseteq D_P (G \circ F)(\bar{x}, \bar{z}). \tag{11.40}$$

Proof. Following the reasoning used by Aubin and Frankowska [18] and by Ward [590], we have

$$\begin{aligned}
\operatorname{graph}(D_P (G \circ F)(\bar{x}, \bar{z})) &= P(\operatorname{graph}(G \circ F), (\bar{x}, \bar{z})) \\
&= P(\ell\left([\operatorname{graph}(F) \times \operatorname{graph}(G)] \cap g^{-1}(0)\right), (\bar{x}, \bar{z})) \\
&\supset \ell\left(P\left([\operatorname{graph}(F) \times \operatorname{graph}(G)] \cap g^{-1}(0), (\bar{x}, \bar{y}, \bar{y}, \bar{z})\right)\right) \\
&\supset \ell\Big(P(\operatorname{graph}(F) \times \operatorname{graph}(G), (\bar{x}, \bar{y}, \bar{y}, \bar{z})) \\
&\cap \nabla g(\bar{x}, \bar{y}, \bar{y}, \bar{z})^{-1}(0)\Big)
\end{aligned}$$

$$\supset \ell\big(P\,(\mathrm{graph}\,(F), (\bar{x}, \bar{y})) \times Q\,(\mathrm{graph}(G), (\bar{y}, \bar{z}))$$
$$\cap\, g^{-1}(0)\big)$$
$$= \ell\,\big(\mathrm{graph}\,(D_{\mathrm{P}}F(\bar{x}, \bar{y})) \times \mathrm{graph}(D_{\mathrm{Q}}\,G(\bar{y}, \bar{z})) \cap g^{-1}(0)\big)$$
$$= \mathrm{graph}\;(D_{\mathrm{Q}}\,G(\bar{y}, \bar{z})) \circ D_{\mathrm{P}}\,F(\bar{x}, \bar{y}))\,,$$

and (11.40) is proved. The proof is complete. □

The above result conveys that to give a chain rule formula like (11.40), a combination of cones can be used that satisfy (11.37)–(11.39). It has been shown by Ward [590,593] that general calculus rules for epiderivative of nonsmooth functions can be derived by the strategy depicted in the above result. We remark that, in set-valued optimization, many cones discussed in Chap. 4 have not been employed yet. However, based on their properties studied in Chap. 4, new derivatives can be defined which will enjoy a rich calculus.

Remark 11.2.14. In Chap. 4, several inversion theorems have been given which ensure (11.39). Inclusions (11.37) and (11.38) hold for a various combination of cones. Note that (11.38) holds as an equality for the contingent cone for derivable sets.

In the following, we collect a few results and identify the conditions where (11.37)–(11.39) are met. We begin with the following result (see [18, Theorem 5.3.3]):

Theorem 11.2.15. *Let X, Y, and Z be Banach spaces where Y is finite-dimensional, let $F : X \rightrightarrows Y$ and $G : Y \rightrightarrows Z$ be set-valued maps, let* $(\bar{x}, \bar{y}) \in \mathrm{graph}\,(F)$, *and let* $(\bar{y}, \bar{z}) \in \mathrm{graph}\,(G)$. *Assume that F and G are closed and the following condition that involves the Clarke's derivative, holds:*

$$\mathrm{Im}(D_C F(\bar{x}, \bar{y})) - \mathrm{dom}\,(D_C G(\bar{y}, \bar{z})) = Y. \tag{11.41}$$

Then the following inclusions hold:

$$D_A G(\bar{y}, \bar{z}) \circ DF(\bar{x}, \bar{y}) \subseteq D(G \circ F)(\bar{x}, \bar{z}), \tag{11.42}$$
$$D_A G(\bar{y}, \bar{z}) \circ D_A F(\bar{x}, \bar{y}) \subseteq D_A(G \circ F)(\bar{x}, \bar{z}), \tag{11.43}$$
$$D_C G(\bar{y}, \bar{z}) \circ D_C F(\bar{x}, \bar{y}) \subseteq D_C(G \circ F)(\bar{x}, \bar{z}). \tag{11.44}$$

Proof. The proof follows the strategy of Theorem 11.2.13, and using Theorem 4.7.7, which holds due to (11.41), ensures (11.37)–(11.39) for ℓ and g and for T, C, and A. □

A useful extension, in finite dimensional spaces, is the following result by Ward [596]:

Theorem 11.2.16. *Let $F : \mathbb{R}^m \rightrightarrows \mathbb{R}^n$ and $G : \mathbb{R}^n \rightrightarrows \mathbb{R}^p$ be set-valued maps, let* graph (F) *be locally closed at* $(\bar{x}, \bar{y}) \in \mathrm{graph}\,(F)$, *and let* graph (G) *locally closed* *at* $(\bar{y}, \bar{z}) \in \operatorname{graph}(G)$. *If*

$$\operatorname{Ker}(D^* F(\bar{x}, \bar{y})) \cap D^* G(\bar{y}, \bar{z})(0) = \{0\}, \tag{11.45}$$

where $\operatorname{Ker}(D^* F(\bar{x}, \bar{y})) := \{y^* \in \mathbb{R}^n \mid 0 \in D^* F(\bar{x}, \bar{y})(y^*)\}$, *then*

$$DG(\bar{y}, \bar{z}) \circ D_A F(\bar{x}, \bar{y}) \subseteq D(G \circ F)(\bar{x}, \bar{z}), \tag{11.46}$$

$$D_A G(\bar{y}, \bar{z}) \circ D_A F(\bar{x}, \bar{y}) \subseteq D_A(G \circ F)(\bar{x}, \bar{z}). \tag{11.47}$$

Proof. For the maps $\ell : \mathbb{R}^m \times \mathbb{R}^n \times \mathbb{R}^n \times \mathbb{R}^p \to \mathbb{R}^n \times \mathbb{R}^p$ and $g : \mathbb{R}^m \times \mathbb{R}^n \times \mathbb{R}^n \times \mathbb{R}^p \to \mathbb{R}^n$, defined by (11.35) and (11.36), we assume that

$$0 \in y^* \nabla g(\bar{x}, \bar{y}, \bar{y}, \bar{z}) + N(\bar{x}, \bar{y}, \bar{y}, \bar{z}, S).$$

Since the normal cone is product preserving, we deduce that $(0, -y^*) \in N(\operatorname{graph}(F), (\bar{x}, \bar{y}))$ and $(y^*, 0) \in N(\operatorname{graph}(G), (\bar{y}, \bar{z}))$. Condition (11.45) then implies that $y^* = 0$, which allows us to apply Theorem 4.7.10. Therefore, the conditions (11.35) and (11.36) are satisfied and the arguments used in the proof of Theorem 11.2.13 are applicable. □

We now give a general scheme for deriving an abstract sum formula.

Theorem 11.2.17. *Let* X *and* Y *be normed spaces, let* $F_1, F_2 : X \rightrightarrows Y$ *be set-valued maps, let* $f : X \to X$ *be a single-valued map, let* $(\bar{x}, \bar{y}_1) \in \operatorname{graph}(F_1)$, *and let* $(\bar{x}, \bar{y}_2) \in \operatorname{graph}(F_2 \circ f)$. *Assume that* P *and* Q *are closed tangent cones such that for* $\ell : (X \times Y)^2 \to X \times Y$ *and* $g : (X \times Y)^2 \to X$ *given by*

$$\ell(a, b, c, d) := (a, c + d), \tag{11.48}$$

$$g(a, b, c, d) := f(a) - c, \tag{11.49}$$

and for $S_1 := \operatorname{graph}(F_1)$, $S_2 := \operatorname{graph}(F_2)$, $S := S_1 \times S_2$, *and* $x_0 := (\bar{x}, \bar{y}_1, f(\bar{x}), \bar{y}_2)$, *we have*

$$\ell(P(S, x_0)) \subseteq P(\ell(S), \ell(x_0)),$$

$$P(S_1, (\bar{x}, \bar{y}_1)) \times Q(S_2, (f(\bar{x}), \bar{y}_2)) \subseteq P(S_1 \times S_2, (\bar{x}, \bar{y}_1, f(\bar{x}), \bar{y}_2)),$$

$$P(S, x_0) \cap \nabla g(x_0)^{-1}(0) \subseteq P(S \cap g^{-1}(0), x_0).$$

Then the following sum formula, for the graphical derivatives given by P *and* Q, *holds:*

$$D_P F_1(\bar{x}, \bar{y}_1)(\cdot) + D_Q F_2(f(\bar{x}), \bar{y}_2)(\nabla f(\bar{x})(\cdot)) \subseteq D_P(F_1 + F_2 \circ f)(\bar{x}, \bar{y}_1 + \bar{y}_2)(\cdot). \tag{11.50}$$

Proof. Proceeding as in the proof of Theorem 11.2.13, we have

$$\begin{aligned}
&\text{graph}\,(D_P\,(F_1 + F_2 \circ f)\,(\bar{x}, \bar{y}_1 + \bar{y}_2)) \\
&\quad = P\,(\text{graph}\,(F_1 + F_2 \circ f), (\bar{x}, \bar{y}_1 + \bar{y}_2)) \\
&\quad = P(\ell\left([\text{graph}\,(F_1) \times \text{graph}\,(F_2)] \cap g^{-1}(0)\right), \ell(\bar{x}, \bar{y}_1, f(\bar{x}), \bar{y}_2)) \\
&\quad \supset \ell\left(P([\text{graph}\,(F_1) \times \text{graph}\,(F_2)] \cap g^{-1}(0), (\bar{x}, \bar{y}_1, f(\bar{x}), \bar{y}_2)\right) \\
&\quad \supset \ell\,(P\,(\text{graph}\,(F_1) \\
&\times \text{graph}\,(F_2), (\bar{x}, \bar{y}_1, f(\bar{x}), \bar{y}_2)) \bigcap \nabla g(\bar{x}, \bar{y}_1, f(\bar{x}), \bar{y}_2)^{-1}(0)\Big) \\
&\supset \ell\,(P\,(\text{graph}\,(F_1), (\bar{x}, \bar{y}_1)) \\
&\times Q\,(\text{graph}\,(F_2), (f(\bar{x}), \bar{y}_2)) \bigcap \nabla g(\bar{x}, \bar{y}_1, f(\bar{x}), \bar{y}_2)^{-1}(0)\Big) \\
&= \ell\,(\text{graph}\,(D_P\,F_1(\bar{x}, \bar{y}_1)) \\
&\times \text{graph}\,(D_Q\,F_2(f(\bar{x}), \bar{y}_2)) \bigcap \nabla g(\bar{x}, \bar{y}_1, f(\bar{x}), \bar{y}_2)^{-1}(0)\Big) \\
&= \{(a, b + d) |\ (a, b) \in \text{graph}\,(D_P\,F_1(\bar{x}, \bar{y}_1)),\ (c, d) \\
&\in \text{graph}\,(D_Q\,F_2(f(\bar{x}), \bar{y}_2),\ \nabla f(\bar{x})(a) = c\} \\
&= \text{graph}\ (D_P\,F_1(\bar{x}, \bar{y}_1) + D_Q\,F_2(f(\bar{x}), \bar{y}_2)(\nabla f(\bar{x}))),
\end{aligned}$$

and the inclusion (11.50) is established. The proof is complete. □

Remark 11.2.18. In Theorem 11.2.17, f should be sufficiently smooth to ensure the smoothness of g. The above sum-chain type calculus rule for nonsmooth maps was first devised by Ward [590].

We recall that given Banach spaces X and Y, a nonempty set $S \subset X$, a map $f : X \to Y$ is called metrically subregular at $(\bar{x}, f(\bar{x}))$, where $\bar{x} \in S$, with respect to S if there exist real numbers $k > 0$ and $\tau > 0$ such that for every $x \in U_\tau(\bar{x}) \cap S$, we have

$$d(x, f^{-1}(f(\bar{x}) \cap S) \leq k\|f(\bar{x}) - f(x)\|.$$

Using the above notion, recently Durea and Strugariu [155] gave the following:

Theorem 11.2.19. *Let X and Y be Banach spaces, let $F_1, F_2 : X \rightrightarrows Y$ be closed set-valued maps, let $f : X \to X$ be a continuously differentiable map, let $(\bar{x}, \bar{y}_1) \in$ graph (F_1), and let $(\bar{x}, \bar{y}_2) \in$ graph $(F_2 \circ f)$. Assume that the map $g : X \times Y \times X \times Y \to X$ given by $g(a, b, c, d) = f(a) - c$ is metrically subregular at $(\bar{x}, \bar{y}_1, f(\bar{x}), \bar{y}_2, 0)$ with respect to* graph $(F_1) \times$ graph (F_2). *Assume that either* graph (F_1) *is derivable at $(\bar{x}, \bar{y}_1)$ or* graph (F_2) *is derivable at $(f(\bar{x}), \bar{y}_2)$. Then:*

$$D_P\,F_1(\bar{x}, \bar{y}_1)(\cdot) + DF_2(f(\bar{x}), \bar{y}_2)(\nabla f(\bar{x})(\cdot)) \subseteq D(F_1 + F_2 \circ f)(\bar{x}, \bar{y}_1 + \bar{y}_2)(\cdot).$$

Proof. The proof follows the scheme of Theorem 11.2.17 and employes Theorem 4.7.6. □

We conclude this section by giving a chain rule for convex set-valued maps acting between finite-dimensional spaces. In this case, it is natural to give corresponding results for the associated coderivative. Recall that for a convex set-valued map $F : X \rightrightarrows Y$, the coderivative $D^*F(\bar{x}, \bar{y}) : Y \rightrightarrows X$ is a set-valued map given by

$$D^*F(\bar{x}, \bar{y})(y^*) = \{x^* | (x^*, -y^*) \in N(\text{graph}(F), (\bar{x}, \bar{y}))\}.$$

The following result is taken from [459]:

Theorem 11.2.20. *Let X, Y, and Z be finite-dimensional spaces, let $F : X \rightrightarrows Y$ and $G : Y \rightrightarrows Z$ be convex set-valued maps, let $(\bar{x}, \bar{z}) \in \text{graph}(G \circ F)$, and let $\bar{y} \in F(\bar{x}) \cap G^{-1}(\bar{z})$. Then,*

$$D(G \circ F)(\bar{x}, \bar{z}) \subseteq \text{cl}(DG(\bar{y}, \bar{z}) \circ DF(\bar{x}, \bar{y})).$$
$$D^*(G \circ F)(\bar{x}, \bar{z}) \supset D^*G(\bar{y}, \bar{z}) \circ D^*F(\bar{x}, \bar{y}),$$

and if ri range$(F) \cap$ ri dom$(F) \neq \emptyset$, *then equalities hold.*

Proof. The proof follows by the arguments used in Theorem 11.2.13 and makes use of the intersection formula (4.31): □

Numerous particular cases of the above result can be found in [459]. Stronger conditions can be imposed for obtaining analogues of the formulas given in the above result for convex maps acting between general Banach spaces. Details can be found in [16].

11.3 Contingently C-Absorbing Maps

As we have noted earlier in this chapter, under some nice properties of the graph of a set-valued map, its contingent derivative enjoys useful properties. For instance, when a set-valued map has a convex graph, then, under mild additional assumptions, the contingent derivative of such a map is defined on the whole space. However, the convexity of the graph is a very stringent assumption whereas the convexity of its epigraph is a relatively milder assumption. Therefore, the question arises as to whether the contingent derivative of a set-valued map can provide some insight in looking at the contingent derivative of the profile map of the underlying set-valued map. It turns out that the following notion plays an important role in studying this relationship:

Definition 11.3.1. Let X and Y be normed spaces and let $C \subset Y$ be a suitable cone. A set-valued $F : X \rightrightarrows Y$ is called **contingently C-absorbing** at $(\bar{x}, \bar{y}) \in \text{graph}(F)$, if the following identity holds:

$$DF(\bar{x},\bar{y})(x)+C = D(F+C)(\bar{x},\bar{y})(x) \quad \text{for every } x \in \text{dom}\,(D(F+C)(\bar{x},\bar{y})). \tag{11.51}$$

The above notion plays an important role in several issues such as the existence of contingent epiderivatives, generalized epiderivatives, and sensitivity analysis, among others. In the following, we will give examples to show that (11.51) is in general false. We will also give several conditions on the map F and on the cone C which ensure that (11.51) remains valid.

Before any advancement, we recall the following notion proposed by Shi [530].

Definition 11.3.2. Let X and Y be normed spaces, let $F : X \rightrightarrows Y$ be a set-valued map and let $(\bar{x},\bar{y}) \in \text{graph}\,(F)$. A set-valued map $D_s F : X \rightrightarrows Y$ is called the **S-derivative** of F at $(\bar{x},\bar{y})$ if there are sequences $\{t_n\} \subset \mathbb{P}$ and $\{(x_n,y_n)\} \subset X \times Y$ with $(x_n,y_n) \to (x,y)$ and $t_n x_n \to 0$ such that $\bar{y}+t_n y_n \in F(\bar{x}+t_n x_n)$ for every $n \in \mathbb{N}$.

The following result, whose proof is adopted from [214, Theorem 3.6.1] will be used shortly.

Theorem 11.3.3. *Let X and Y be normed spaces, let $F : X \rightrightarrows Y$ be a set-valued map, and let $(\bar{x},\bar{y}) \in \text{graph}\,(F)$. Let $C \subseteq Y$ be a pointed, closed and convex cone. Then*

$$D_s F(\bar{x},\bar{y})(0) \cap (-C) = \{0\}, \tag{11.52}$$

holds, provided that any of the following conditions is verified:

(i) $\bar{y} \in \text{Bo-PMin}\,(F(\bar{x}),C)$*, F is upper Lipschitz at $\bar{x}$, that is, there exist $L,\rho > 0$ such that*

$$F(x) \subset F(\bar{x}) + L\,\|x-\bar{x}\|\,B_Y \quad \textit{for all } x \in B(\bar{x},\rho);$$

(ii) $\bar{y} \in \text{Bo-PMin}\,(F(\bar{x});C)$*, X, Y are finite-dimensional,* $\bar{x} \in \text{icr(dom}\,(F))$*, and F is C-convex;*

(iii) $\bar{y} \in \text{GHe-PMin}\,(F(\bar{x});C)$*, X, Y are Banach spaces,* $\bar{x} \in \text{int(dom}\,(F))$*, and F is C-convex and C-closed.*

Proof. (i) Let $y \in D_s F(\bar{x},\bar{y})(0)\cap(-C)$. Then there exist $\{t_n\} \subset \mathbb{P}$, $\{(x_n,y_n)\} \subset \text{graph}\,(F)$ such that $x_n \to 0$ and $t_n((x_n,y_n)-(\bar{x},\bar{y})) \to (0,y)$. We may assume that $x_n \in B(\bar{x},\rho)$, for every $n \in \mathbb{N}$. It follows that for every n there exist $\bar{y}_n \in F(\bar{x})$ and $v_n \in B_Y$ such that $y_n = \bar{y}_n + L\,\|x_n-\bar{x}\|\cdot v_n$. Hence

$$\|t_n(\bar{y}_n-\bar{y}) - t_n(y_n-\bar{y})\| \le L t_n\,\|x_n-\bar{x}\|\,,$$

for every $n \in \mathbb{N}$, whence $t_n(\bar{y}_n-\bar{y}) \to y$. Therefore $y \in \text{cl cone}(F(\bar{x})-\bar{y}) \cap (-C) = \{0\}$, which is a contradiction.

(ii) Taking $(\bar{x},\bar{y}) = (0,0)$ (replacing F by $F-(\bar{x},\bar{y})$) and replacing X by span(dom$\,(F)$), if necessary, we may suppose that $\bar{x} \in \text{int(dom}\,(F))$.

Since $\dim Y < \infty$ and $\bar{y} \in$ Bo-PMin $(F(\bar{x}); C)$, we have that $\bar{y} \in$ GHe-PMin $(F(\bar{x}), C)$. Consequently, there exists a convex cone K such that $C \setminus \{0\} \subset \text{int}K \neq Y$ and $F(\bar{x}) \cap (\bar{y} - K) \subset \bar{y} + K$. Since

$$\text{graph}\,(F + K) = \text{graph}\,(F) + \{0\} \times C + \{0\} \times K,$$

the set graph $(F + K)$ is convex. We have that $(\bar{x}, \bar{y}) \notin \text{icr}(\text{graph}\,(F + K))$. In the contrary case, for $c \in C \setminus \{0\}$, there exists some $\lambda > 0$ such that $(\bar{x}, \bar{y}) - \lambda(0, c) = (\bar{x}, \bar{y} - \lambda c) \in \text{graph}\,(F) + \{0\} \times K$; i.e., $\bar{y} - \lambda c - k' \in F(\bar{x})$ for some $k' \in K$. This contradicts the hypothesis since $\lambda c + k' \in C \setminus \{0\} + K \subset \text{int}K$. Using a separation theorem, there exists $(u^*, y^*) \in X^* \times Y^* \setminus \{(0,0)\}$ such that

$$\langle \bar{x}, u^* \rangle + \langle \bar{y}, y^* \rangle \leq \langle x, u^* \rangle + \langle y + k, y^* \rangle \quad \forall\, (x, y) \in \text{graph}\,(F),\ \forall\, k \in K. \tag{11.53}$$

If $y^* = 0$, then $\langle \bar{x}, u^* \rangle \leq \langle x, u^* \rangle$ for every $x \in \text{dom}\,(F)$, and so $u^* = 0$. Therefore $y^* \neq 0$. From (11.53) we obtain $\langle k, y^* \rangle \geq 0$ for $k \in K$, whence $\langle k, y^* \rangle > 0$ for every $k \in \text{int}K \supset C \setminus \{0\}$. Now assume that there exists $y \in D_s F(\bar{x}, \bar{y})(0) \cap (-C) \setminus \{0\}$. Then there are $\{t_n\} \subset (0, \infty)$ and $\{(x_n, y_n)\} \subset \text{graph}\,(F)$ such that $x_n \to 0$ and $t_n\,((x_n, y_n) - (\bar{x}, \bar{y})) \to (0, y)$. From (11.53) we obtain that

$$\langle t_n(x_n - \bar{x}), u^* \rangle + \langle t_n(y_n - \bar{y}), y^* \rangle \geq 0,$$

for every n, whence $\langle -y, y^* \rangle \leq 0$, a contradiction.

(iii) Let K be a convex cone such that $C \setminus \{0\} \subset \text{int}K \neq Y$ and $\bar{y} \in \text{Min}(F(\bar{x}), K)$. If we show that int(graph $(F + K) \neq \emptyset$, arguments like those in the proof of (ii) show that (11.52) holds. For this aim, consider the relation

$$\mathscr{R} := \{(y, x) \mid (x, y) \in \text{graph}\,(F + K)\}.$$

Clearly, $\mathscr{R}$ is closed, convex, and $0 \in \text{int}(\text{Im}\mathscr{R})$. Applying the Robinson–Ursescu theorem, we have that $\mathscr{R}(V)$ is a neighborhood of $\bar{x}$ for every neighborhood V of $\bar{y}$. Let $k_0 \in \text{int}K$. There exists $\rho > 0$ such that $B(k_0, \rho) \subset K$. It follows that for some $\delta > 0$ we have $B(\bar{x}, \delta) \subset \mathscr{R}\,(B(\bar{y}, \rho/2))$. Let us show that $B(\bar{x}, \delta) \times B(k_0 + \bar{y}, \rho/2) \subset \text{graph}\,(F + K)$. Indeed, let (u, y) belong to the first set. There exists $y' \in \mathscr{R}^{-1}(u)$ such that $\|y' - \bar{y}\| \leq \rho/2$. It follows that $y = y' + (y - y')$ and

$$\left\|k_0 - (y - y')\right\| \leq \|(k_0 + \bar{y} - y)\| + \left\|y' - \bar{y}\right\| \leq \rho/2 + \rho/2 = \rho,$$

which shows that $y - y' \in B(k_0, \rho) \subset K$. Therefore $(x, y) \in \mathscr{R}^{-1} + \{0\} \times K = \text{graph}\,(F + K)$. The proof is complete. □

The following example, taken from [564], shows that the condition that F is upper Lipschitz at $\bar{x}$ in Theorem 11.3.3 (i) cannot be dropped.

Example 11.3.4. Let $X = Y = \mathbb{R}$, $C = \mathbb{R}_+$, and let $F : \mathbb{R} \rightrightarrows \mathbb{R}$ be defined by

$$F(x) := \begin{cases} \{0\} & \text{if } x \le 0, \\ \{-\sqrt{x}, 0\} & \text{if } x > 0. \end{cases}$$

Then,

$$(F + C)(x) = \begin{cases} [0, \infty[& \text{if } x \le 0, \\ [-\sqrt{x}, \infty[& \text{if } x > 0. \end{cases}$$

Taking $\bar{x} = 0$ and $\bar{y} = 0$, we have

$$D_s F(\bar{x}, \bar{y})(x) = DF(\bar{x}, \bar{y})(x) = \begin{cases} \{0\} & \text{if } x \ne 0, \\ -C & \text{if } x = 0, \end{cases}$$

$$D(F + C)(\bar{x}, \bar{y})(x) = \begin{cases} C & \text{if } x < 0, \\ \mathbb{R} & \text{if } x \ge 0, \end{cases}$$

showing that (11.52) is violated. □

We remark that the condition $\bar{y} \in \text{GHe-PMin}\,(F(\bar{x}), C)$ in the above theorem cannot be replaced by $\bar{y} \in \text{Min}\,(F(\bar{x}); C)$. In the next example F is upper Lipschitz at $\bar{x} \in \text{int}(\text{dom}\,(F))$, C-convex, and C-closed.

Example 11.3.5. Let $X = \mathbb{R}$, $Y = \mathbb{R}^2$, $C = \mathbb{R}^2_+$, and $F : X \rightrightarrows Y$ be defined by

$$F(x) := \left\{(y_1, y_2) \mid y_2 \ge y_1^2\right\}.$$

We take $\bar{x} = 0$ and $\bar{y} = 0$. Note that $\bar{y} \in \text{Min}(F(\bar{x}), C) \setminus \text{BMin}(F(\bar{x}), C)$. Then

$$D_s F(\bar{x}, \bar{y})(x) = DF(\bar{x}, \bar{y})(x) = \mathbb{R} \times \mathbb{R}_+,$$

and consequently,

$$D_s F(\bar{x}, \bar{y})(x) \cap (-C) =]-\infty, 0] \times \{0\} \ne \{0\}.$$

The following example shows that Theorem 11.3.3 (ii) does not hold when the space X is infinite dimensional.

Example 11.3.6. Consider X an infinite-dimensional normed vector space, $Y := \mathbb{R}$, and $\phi : X \to \mathbb{R}$ a noncontinuous linear functional. Let $F : X \rightrightarrows Y$ be defined by $F(x) := \{\phi(x)\}$ and $C := \mathbb{R}_+$. Then it is easy to verify that

$$D_s F(\bar{x}, \bar{y})(x) = DF(\bar{x}, \bar{y})(x) = \mathbb{R},$$

for all $(\bar{x}, \bar{y}) \in \text{graph}\,(F)$ and all $x \in X$. □

The following result gives conditions for set-valued maps to be contingently C-absorbing. The original proofs of the second and the third parts were given by Tanino [564, Proposition 2.1, Theorem 2.1] and Shi [530, Proposition 3.1], however, our presentation is taken from [214, Theorem 3.6.6].

Theorem 11.3.7. *Let X and Y be normed spaces, let $F : X \rightrightarrows Y$ be a set-valued map and let $(\bar{x}, \bar{y}) \in \operatorname{graph}(F)$. Let $C \subseteq Y$ be a pointed, closed and convex cone. Then:*

(i) *For every $x \in X$, we have*

$$DF(\bar{x}, \bar{y})(x) + C \subset D(F + C)(\bar{x}, \bar{y})(x).$$

(ii) *If the set $\{y \in C \mid \|y\| = 1\}$ is compact, then for every $u \in X$,*

$$\operatorname{Min}(D(F + C)(\bar{x}, \bar{y})(x), C) \subset \operatorname{Min}(DF(\bar{x}, \bar{y})(x); C) \subset DF(\bar{x}, \bar{y})(x). \tag{11.54}$$

(iii) *If the set $\{y \in C \mid \|y\| = 1\}$ is compact and (11.52) holds, then F is contingently C-absorbing, that is,*

$$DF(\bar{x}, \bar{y})(x) + C = D(F + C)(\bar{x}, \bar{y})(x) \quad \text{for every } x \in X. \tag{11.55}$$

Proof. (i) Let $y \in DF(\bar{x}, \bar{y})(x)$ and $c \in C$. Then there exist $\{t_n\} \to 0_+$ and $\{(x_n, y_n)\} \to (x, y)$ such that $(\bar{x}, \bar{y}) + t_n(x_n, y_n) \in \operatorname{graph}(F)$ for every $n \in \mathbb{N}$. It follows that $(\bar{x}, \bar{y}) + t_n(x_n, y_n + c) \in \operatorname{graph}(F + C)$ and $(x_n, y_n + c) \to (x, y + c)$. Therefore $y + c \in D(F + C)(0, 0)(x)$.

In the rest of the proof we suppose that $Q := \{y \in C \mid \|y\| = 1\}$ is compact. Of course, $C = \operatorname{cone} Q$.

(ii) Let $y \in \operatorname{Min}(D(F + C)(\bar{x}, \bar{y})(x), C)$. Then there exists $\{t_n\} \to 0_+$, $\{\lambda_n\} \subset [0, \infty)$, $\{q_n\} \subset Q$, $\{(x_n, y_n)\} \subset X \times Y$ such that $(x_n, y_n + \lambda_n q_n) \to (x, y)$ and $(\bar{x}, \bar{y}) + t_n(x_n, y_n) \in \operatorname{graph}(F)$ for every $n \in \mathbb{N}$. Since Q is compact, we may suppose that $q_n \to q \in Q$. Assume that there are a subsequence $\{\lambda_{n_k}\}$ and $\gamma > 0$ such that $\lambda_{n_k} \geq \gamma$ for every $k \in \mathbb{N}$. Then $(\bar{x}, \bar{y}) + t_{n_k}(x_{n_k}, y_{n_k} + (\lambda_{n_k} - \gamma)q_{n_k}) \in \operatorname{graph}(F + C)$ and $(x_{n_k}, y_{n_k} + (\lambda_{n_k} - \gamma)q_{n_k}) \to (u, y - \lambda q)$. It follows that $y - \gamma q \in D(F + C)(\bar{x}, \bar{y})(x) \cap (y - C) \setminus \{y\}$, a contradiction. Therefore $\lambda_n \to 0$, and so $(x_n, y_n) \to (u, y)$. It follows that $y \in DF(\bar{x}, \bar{y})(x)$. Using also (i), we have that for every $x \in X$,

$$\operatorname{Min}(D(F + C)(\bar{x}, \bar{y})(x); C) \subset DF(\bar{x}, \bar{y})(x) \subset D(F + C)(\bar{x}, \bar{y})(x).$$

From the above inclusions we obtain immediately that (11.54) holds.

(iii) The inclusion $\supset$ being proven in (i), let us prove the converse one. Of course, we assume that (11.52) holds. Let $y \in D(F + C)(\bar{x}, \bar{y})(x)$; there exist $\{t_n\} \to 0_+$, $\{\lambda_n\} \subset [0, \infty)$, $\{q_n\} \subset Q$, $\{(x_n, y_n)\} \subset X \times Y$ such that $(x_n, y_n + \lambda_n q_n) \to (x, y)$ and $(\bar{x}, \bar{y}) + t_n(x_n, y_n) \in \operatorname{graph}(F)$ for every $n \in \mathbb{N}$. We may suppose that $q_n \to q \in Q$. Taking also a subsequence if necessary, we may assume that $\lambda_n \to \lambda \in [0, \infty]$. Suppose that $\lambda = \infty$. Since $y_n + \lambda_n q_n \to y$, we obtain that $\lambda_n^{-1} y_n \to -q$. Of course, $\bar{x} + t_n x_n \to \bar{x}$ and

$$\frac{1}{\lambda_n t_n}\left((\bar{x} + t_n x_n, \bar{y} + t_n y_n) - (\bar{x}, \bar{y})\right) = \left(\frac{u_n}{\lambda_n}, \frac{y_n}{\lambda_n}\right) \to (0, -q).$$

It follows that $-q \in D_s F(\bar{x}, \bar{y})(0) \cap (-C) \setminus \{0\}$, a contradiction. Therefore $\lambda < \infty$, whence $(x_n, y_n) \to (u, y - \lambda q)$. Hence $y \in DF(\bar{x}, \bar{y})(x) + C$. The proof is complete. □

The following example, originally constructed in [214, Example 3.6.8], shows that the hypothesis "$\{y \in C \mid \|y\| = 1\}$ is compact" in the above theorem is essential for having (11.54) and (11.55).

Example 11.3.8. Let $X = \mathbb{R}$, $Y = \ell^2$, $C = \ell^2_+ = \{(x_n)_{n\in\mathbb{N}} \in \ell^2 \mid x_n \geq 0 \ \forall n \in \mathbb{N}\}$, and $F : X \rightrightarrows Y$ be given by

$$F(x) := \begin{cases} \{0\} & \text{if } x \in]-\infty, 0] \cup [1, \infty[, \\ \frac{1}{n}[-a - e_n, a] & \text{if } x \in [\frac{1}{n+1}, \frac{1}{n}[, \ n \in \mathbb{N}^*, \end{cases}$$

where $a \in \ell^2_+ \setminus \{0\}$ and $(e_n)_{n\in\mathbb{N}}$ is the canonical base of ℓ^2. Then

$$(F + C)(x) := \begin{cases} C & \text{if } x \in]-\infty, 0] \cup [1, \infty[, \\ -\frac{1}{n}(a + e_n) + C & \text{if } x \in [\frac{1}{n+1}, \frac{1}{n}[, \ n \in \mathbb{N}^*. \end{cases}$$

Take $\bar{x} = 0$ and $\bar{y} = 0$. Then

$$DF(\bar{x}, \bar{y})(u) = \begin{cases} \{0\} & \text{if } u \leq 0, \\ \{ua\} & \text{if } u > 0, \end{cases}$$

$$D(F + C)(\bar{x}, \bar{y})(u) = \begin{cases} C & \text{if } u \leq 0, \\ -ua + C & \text{if } u > 0. \end{cases}$$

So (11.54) and (11.55) do not hold for $u > 0$.

Let us prove the formulae for $DF(\bar{x}, \bar{y})(u)$ and $D(F + C)(\bar{x}, \bar{y})(u)$ for $u \geq 0$, those for $u < 0$ being obvious. So, let $u \geq 0$ and $v \in DF(\bar{x}, \bar{y})(u)$. Then there exist $\{t_n\} \subset]0, \infty[$ and graph $F \supset \big((x_n, y_n)\big) \to (0, 0)$ such that $(t_n x_n) \to u$ and $(t_n y_n) \to v$. If $P := \{n \in \mathbb{N} \mid x_n \leq 0\}$ is infinite, then $y_n = 0$ for every $n \in P$, and so $(u, v) = (0, 0)$. Suppose that P is finite (this is the case when $u > 0$). Since $(x_n) \to 0$, there exists $n_0 \in \mathbb{N}^*$ such that $x_n \in]0, 1[$ for $n \geq n_0$. For every $n \geq n_0$ there exists $p_n \in \mathbb{N}^*$ such that $x_n \in [\frac{1}{p_n+1}, \frac{1}{p_n}[$; of course, $(p_n) \to \infty$. Then for every $n \geq n_0$ there exists $\lambda_n \in [0, 1]$ such that $y_n = \frac{1}{p_n}\big((1 - 2\lambda_n)a - \lambda_n e_n\big)$. It follows that $\big(\frac{t_n}{p_n}\big) \to u$ and $\big(\frac{t_n}{p_n}((1 - 2\lambda_n)a - \lambda_n e_{p_n})\big) \to v$. If $u = 0$, it is obvious that $v = 0$. Suppose that $u > 0$. Taking eventually a subsequence, we may assume that $(\lambda_n) \to \lambda \in [0, 1]$. We obtain that $(\lambda_n e_{p_n}) \to (1 - 2\lambda)a - u^{-1}v$. If $\lambda \neq 0$, we get the contradiction that (e_n) contains a norm-convergent subsequence. Therefore $(\lambda_n) \to 0$, whence $v = ua$. It is easy to get that $ua \in DF(\bar{x}, \bar{y})(u)$ for $u \geq 0$ (in fact, similar to the proof in the next paragraph). So the formula for $DF(\bar{x}, \bar{y})(u)$ holds.

Consider first $u > 0$ and $v = -ua + k$ with $k \in C$. Taking $x_n = \frac{1}{n+1}$, $y_n = \frac{1}{n}(-a + u^{-1}k) = \frac{1}{n}(-a - e_n + u^{-1}k + e_n)$ $(\in (F + C)(x_n))$, and $t_n = (n+1)u > 0$ for $n \in \mathbb{N}^*$, we have that $\big((x_n, y_n)\big) \to (0,0)$ and $\big(t_n(x_n, y_n)\big) \to (u, v)$; if $u = 0$ and $v = k \in k$ just take $x_n = 0$, $y_n = \frac{1}{n}k$, and $t_n = n$. Therefore $v \in DF_C(\bar{x}, \bar{y})(u)$.

Let $u \geq 0$ and $v \in D(F + C)(\bar{x}, \bar{y})(u)$. Then there exist $(t_n) \subset]0, \infty[$ and $\operatorname{graph}(F + C) \supset \big((x_n, y_n)\big) \to (0,0)$ such that $\big(t_n(x_n, y_n)\big) \to (u, v)$. As in the first part we may suppose that $x_n \in]0, 1[$ for $n \geq n_0$. So there exists $\mathbb{N}^* \supset (p_n) \to \infty$ with $x_n \in [\frac{1}{p_n+1}, \frac{1}{p_n}[$ for $n \geq n_0$. Of course, $y_n = -\frac{1}{p_n}(a + e_{p_n}) + k_n$ with $k_n \in C$. It follows that $\big(\frac{t_n}{p_n}\big) \to u$ and $\big(-\frac{t_n}{p_n}(a + e_{p_n}) + t_n k_n\big) \to v$. Since $0 = w\text{-}\lim e_n$, it follows that $v + ua = w\text{-}\lim(t_n k_n)$. Because C is weakly closed, $v + ua \in C$, whence $v \in -ua + C$. □

In Example 11.3.4, (11.52) is not satisfied and $F(x)$ is not convex for $x > 0$ and (11.55) do not hold for $u > 0$.

For various purposes, a number of compactness assumptions have been introduced in optimization. In the following, we recall two notions of compactness and employ them to show that a given map set-valued map is contingently C-absorbing. We recall that, given normed spaces X and Y, a set-valued map $F : X \rightrightarrows Y$ is called directionally compact at $(\bar{x}, \bar{y}) \in \operatorname{graph}(F)$ in a direction $x \in X$, if for every sequence $\{t_n\} \subset \mathbb{P}$ with $t_n \downarrow 0$ and every sequence $\{x_n\} \subset X$ with $x_n \to x$, any sequence $\{y_n\} \subset Y$ contains a convergent subsequence provided that $\bar{y} + t_n y_n \in F(\bar{x} + t_n x_n)$, for every $n \in \mathbb{N}$.

We have the following result which was originally given by Bednarczuk and Song [39].

Theorem 11.3.9. *Let X and Y be normed spaces and let $C \subset Y$ be a pointed, closed, and convex cone. Let $F : X \rightrightarrows Y$ be a set-valued map and let $(\bar{x}, \bar{y}) \in \operatorname{graph}(F)$. Assume that F is directionally compact at $(\bar{x}, \bar{y})$ in any direction $x \in X$. Then, the map F is contingently C-absorbing at $(\bar{x}, \bar{y})$, that is, (11.90) holds.*

Proof. In view of (11.88), it suffices to show that

$$D(F + C)(\bar{x}, \bar{y})(x) \subset DF(\bar{x}, \bar{y})(x) + C.$$

Let $y \in D(F + C)(\bar{x}, \bar{y})(x)$. Then there exist sequences $\{t_n\} \subset \mathbb{P}$, $\{(x_n, y_n)\} \subset X \times Y$, and $\{c_n\} \subset C$ such that $t_n \downarrow 0$, $(x_n, y_n) \to (x, y)$, and

$$\bar{y} + t_n(y_n - c_n/t_n) = \bar{y} + t_n y_n - c_n \in F(\bar{x} + t_n x_n).$$

Since the map F is directionally compact, we may assume that $y_n - c_n/t_n$ converges to some $\tilde{y} \in Y$. Hence $\tilde{y} \in DF(\bar{x}, \bar{y})(x)$ and $c_n/t_n \to (y - \tilde{y}) \in C$, confirming that

$$y \in DF(\bar{x}, \bar{y})(x) + C,$$

and the proof is complete. □

The following result which was originally given by Taa [548, Proposition 2.2], is obtained by using the concept of compactly approximable maps.

Theorem 11.3.10. *Let X and Y be normed spaces, and let $C \subset Y$ be a pointed, closed, and convex cone. Let $F : X \rightrightarrows Y$ be a set-valued map and let $(\bar{x}, \bar{y}) \in$ graph (F). If the map F is compactly approximable at $(\bar{x}, \bar{y})$, then the map F is contingently C-absorbing at $(\bar{x}, \bar{y})$, that is (11.90) holds.*

Proof. For any $x \in \text{dom}(D(F+C)(\bar{x}, \bar{y}))$, let $y \in D(F+C)(\bar{x}, \bar{y})(x)$ be arbitrary. Hence there exist sequences $\{t_n\} \subset \mathbb{P}$ and $\{(x_n, y_n)\} \subset X \times Y$ such that $t_n \downarrow 0$, $(x_n, y_n) \to (x, y)$ and

$$\bar{y} + t_n y_n \in F(\bar{x} + t_n x_n) + C.$$

Then,

$$\bar{y} + t_n y_n = z_n + c_n \quad \text{with } z_n \in F(\bar{x} + t_n x_n) \text{ and } c_n \in C.$$

Since F is compactly approximable, there exists $\{k_n\} \subset R(x)$, $r_n := r(t_n, x_n)$, $b_n \in B_Y$ and $n_0 \in \mathbb{N}$ such that $r_n \to 0$, and

$$z_n = \bar{y} + t_n(k_n + r_n b_n) \quad \text{for all } n > n_0,$$

which implies

$$\bar{y} + t_n(k_n + r_n b_n) \in F(\bar{x} + t_n x_n),$$

for all $n > n_0$. In view of the compactness of $R(x)$, we may assume that $k_n \to k \in R(x)$. Because of the fact that $(k_n + r_n b_n) \to k$, we deduce that $k \in DF(\bar{x}, \bar{y})(x)$. Since

$$\bar{y} + t_n y_n = \bar{y} + t_n(k_n + r_n b_n) + c_n,$$

for sufficiently large n, we have $y_n - k_n - r_n b_n = t_n^{-1} c_n \in C$. By passing to the limit, we have $y - k \in C$ and hence

$$y \in k + C \subset DF(\bar{x}, \bar{y})(x) + C,$$

and the proof is complete. □

11.4 Epiderivatives of Set-Valued Maps

In this section, we introduce the notion of the contingent epiderivative and some related epiderivatives which play a fundamental role in set-valued optimization. The main motivation behind considering the epiderivatives is the fact that the epigraph of a set-valued map has considerably better structure than the graph of the set-valued map.

We begin with formulating the following notion of the contingent epiderivative which was originally proposed by Jahn and Rauh [301].

Definition 11.4.1. Let X and Y be normed spaces and let $C \subset Y$ be a pointed, closed, and convex cone. Let $F : X \rightrightarrows Y$ be a set-valued map and let $(\bar{x}, \bar{y}) \in \text{graph}\,(F)$. A single-valued map $D_E F(\bar{x}, \bar{y}) : X \to Y$ is called the **contingent epiderivative** of F at $(\bar{x}, \bar{y})$, if the following identity holds

$$\text{epi}\,(D_E F(\bar{x}, \bar{y})) = T(\text{epi}\,(F), (\bar{x}, \bar{y})),$$

where $T(\text{epi}\,(F), (\bar{x}, \bar{y}))$ is the contingent cone of $\text{epi}\,(F)$ at $(\bar{x}, \bar{y})$. The map F is called **contingently epidifferentiable**, if the effective domain of $D_E F(\bar{x}, \bar{y})$ is given by

$$L := \text{dom}\,(D(F + C)(\bar{x}, \bar{y})) = \{x \in X \,|\, (x, y) \in T(\text{epi}\,(F), (\bar{x}, \bar{y}))\}.$$

A geometric interpretation of the contingent epiderivative is given in Fig. 11.2. The epigraph of the single-valued contingent epiderivative is obtained by locally approximating the epigraph of the underlying set-valued map by means of the contingent cone.

In the following, for a single-valued map $f : X \to Y$, we denote the contingent epiderivative of f at $(\bar{x}, f(\bar{x})) \in \text{graph}\,(f)$ by $D_E f(\bar{x})$. That is, $D_E f(\bar{x}) := D_E f(\bar{x}, f(\bar{x}))$.

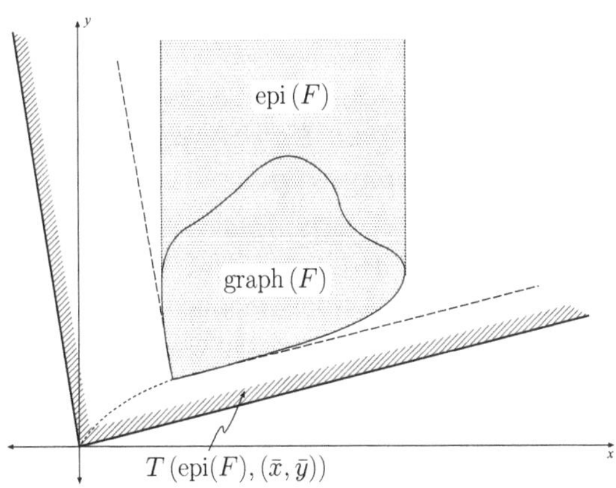

Fig. 11.2 Contingent epiderivative of the set-valued map F

Remark 11.4.2. If $(\bar{x}, \bar{y})$ belongs to the interior of epi (F), then $T(\text{epi}\,(F), (\bar{x}, \bar{y}))$ coincides with the product space $X \times Y$, and, in this case, the contingent epiderivative $D_E F(\bar{x}, \bar{y})$ does not exist. It is clear that the domain of the contingent epiderivative does not need to be the whole space X.

Although our focus will primarily be on the contingent epiderivative, the following definition collects some of the related concepts which are based on the same idea but take a different tangent cone for the local approximation.

Definition 11.4.3. Let X and Y be normed spaces and let $C \subset Y$ be a pointed, closed, and convex cone. Let $F : X \rightrightarrows Y$ be a set-valued map and let $(\bar{x}, \bar{y}) \in$ graph (F).

1. A single-valued map $\hat{D}_E F(\bar{x}, \bar{y}) : X \to Y$ is called the **adjacent epiderivative** of F at $(\bar{x}, \bar{y})$, if the following identity holds

$$\text{epi}\,(\hat{D}_E F(\bar{x}, \bar{y})) = A(\text{epi}\,(F), (\bar{x}, \bar{y})),$$

   where $A(\text{epi}\,(F), (\bar{x}, \bar{y}))$ is the adjacent cone of epi (F) at $(\bar{x}, \bar{y})$. The map F is called **adjacently epidifferentiable**, if the effective domain of $\hat{D}_E F(\bar{x}, \bar{y})$ is given by

$$\hat{L} := \text{dom}\,(D_A(F + C)(\bar{x}, \bar{y})) = \{x \in X \mid (x, y) \in A(\text{epi}\,(F), (\bar{x}, \bar{y}))\}.$$

2. A single-valued map $\tilde{D}_E F(\bar{x}, \bar{y}) : X \to Y$ is called the **Clarke epiderivative** of F at $(\bar{x}, \bar{y})$, if the following identity holds

$$\text{epi}\,(\tilde{D}_E F(\bar{x}, \bar{y})) = C(\text{epi}\,(F), (\bar{x}, \bar{y})),$$

   where $C(\text{epi}\,(F), (\bar{x}, \bar{y}))$ is the Clarke cone of epi (F) at $(\bar{x}, \bar{y})$. The map F is called **Clarke epidifferentiable**, if the effective domain of $\tilde{D}_E F(\bar{x}, \bar{y})$ is given by

$$\tilde{L} := \text{dom}\,(\tilde{D}(F + C)(\bar{x}, \bar{y})) = \{x \in X \mid (x, y) \in C(\text{epi}\,(F), (\bar{x}, \bar{y}))\}.$$

11.4.1 Contingent Epiderivatives of Set-Valued Maps with Images in $\mathbb{R}$

To illustrate the underlying idea, we investigate the contingent epiderivative of set-valued maps defined from a normed space to the real line and give some simple illustrative examples. Contingent epiderivatives of set-valued maps acting between general spaces will be studied in a later part of this section. Our exposition of this subsection is based on [298].

Let X be a normed space and let $F : X \rightrightarrows \mathbb{R}$ be a given set-valued map. To the map F, we associate another set-valued map $F_+ : X \rightrightarrows \mathbb{R}$ defined by

$$F_+(x) := \mathrm{cl}\,(F(x) + \mathbb{R}_+).$$

We define a single-valued map $f : X \to \mathbb{R}$ by

$$f(x) = \{y \in F_+(x) |\ y \leq z, \quad \text{for every } z \in F_+(x)\}. \tag{11.56}$$

We begin with the following intuitive result for the existence of contingent epiderivatives.

Theorem 11.4.4. *Let X be a normed space and let $F : X \rightrightarrows \mathbb{R}$ be a set-valued map. Let the single-valued function f defined by* (11.56) *be convex and lower-semicontinuous. Let $\bar{x} \in \mathrm{int}(\mathrm{dom}\,(F))$ and $(\bar{x}, f(\bar{x})) \in \mathrm{graph}\,(F)$. We set $\bar{y} = f(\bar{x})$. Then the contingent epiderivative $D_E F(\bar{x}, \bar{y})$ of F at $(\bar{x}, \bar{y})$ exists and is given by*

$$D_E F(\bar{x}, \bar{y})(x) = \lim_{\lambda \downarrow 0} \frac{f(\bar{x} + \lambda x) - f(\bar{x})}{\lambda}, \quad \textit{for every } x \in X.$$

Proof. The proof is based on straightforward arguments. It is evident from (11.56) that the epigraph of F_+ is characterized by f. By using the lower-semicontinuity of f, we obtain

$$\mathrm{epi}\,(f) = \mathrm{cl}\,(\mathrm{epi}\,(f)) = \mathrm{cl}\,(\mathrm{epi}\,(F)).$$

Due to the properties of the contingent cone, we have

$$T(\mathrm{epi}\,(f), (\bar{x}, \bar{y})) = T(\mathrm{cl}\,(\mathrm{epi}\,(F)), (\bar{x}, \bar{y})) = T(\mathrm{epi}\,(F), (\bar{x}, \bar{y})).$$

Since the function f is convex and lower-semicontinuous and $\bar{x} \in \mathrm{int}(\mathrm{dom}\,(f))$, it follows that the contingent epiderivative $D_E f(\bar{x}, \bar{y})(x)$ of the single-valued functional f exists for all $x \in X$ (see [16, pp.198–199]). This amounts to say that

$$Df(\bar{x}, \bar{y})(x) = \inf\{y |\ (x, y) \in T((f), (\bar{x}, \bar{y}))\}$$

is finite for all $x \in X$. Since the contingent cone is closed, the above identity implies that

$$\mathrm{epi}\,(D_E f(\bar{x}, \bar{y})) = T(\mathrm{epi}\,(f), (\bar{x}, \bar{y})).$$

By virtue of the definition of the contingent epiderivative $D_E F(\bar{x}, \bar{y})$, we have

$$\operatorname{epi}(D_E F(\bar{x},\bar{y})) = T(\operatorname{epi}(F),(\bar{x},\bar{y})) = T(\operatorname{epi}(f),(\bar{x},\bar{y})) = \operatorname{epi}(D_E f(\bar{x},\bar{y})),$$

and hence the existence of $D_E F(\bar{x},\bar{y})(x)$ follows from the existence of $D_E f(\bar{x},\bar{y})(x)$. The given characterization then follows from [16, pp.198–199]. The proof is complete. □

Next we slightly improve the above result by relaxing the assumption of convexity to *local convexity*. To achieve this goal, the following result will be useful:

Theorem 11.4.5. *Let X be a normed space, let $F : X \rightrightarrows \mathbb{R}$ be a set-valued map, and let $(\bar{x},\bar{y}) \in \operatorname{graph}(F)$. Assume that there are functionals $f_1, f_2 : \operatorname{dom}(f_1) = \operatorname{dom}(f_2) = X \to \mathbb{R}$ such that*

$$\operatorname{epi}(f_1) \supseteq T(\operatorname{epi}(F),(\bar{x},\bar{y})) \supseteq \operatorname{epi}(f_2).$$

Then the contingent epiderivative $D_E F(\bar{x},\bar{y})$ is given by:

$$D_E F(\bar{x},\bar{y})(x) = \min\{y \in \mathbb{R} \mid (x,y) \in T(\operatorname{epi}(F),(\bar{x},\bar{y}))\}, \quad \textit{for every } x \in X.$$

It is clear that the above theorem assures that $\operatorname{dom}(D_E F(\bar{x},\bar{y})) = X$. In fact, it is a consequence of the assumptions that $T(\operatorname{epi}(F),(\bar{x},\bar{y})) \supseteq \operatorname{epi}(f_2)$ and $\operatorname{dom}(f_2) = X$. Here the domain of F is of less relevance. The following simple example shows that if $\operatorname{dom}(f_2)$ is a proper subset of X, then $D_E F(\bar{x},\bar{y})$ might not be well defined in some directions.

Example 11.4.6. Let $F : \mathbb{R} \rightrightarrows \mathbb{R}$ be a set-valued map defined as follows (see Fig. 11.3):

$$F(x) = \begin{cases} \left]\sqrt{|x|}, \infty\right[& \text{if } x \neq 0, \\ [0, \infty[& \text{if } x = 0. \end{cases}$$

We choose $(\bar{x},\bar{y}) = (0,0)$. Then,

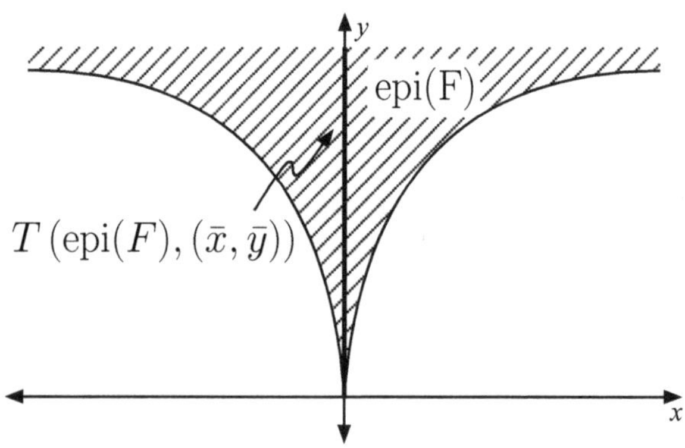

Fig. 11.3 A set-valued map and its contingent epiderivative

$$T(\text{epi}\,(F), (\bar{x}, \bar{y})) = \left\{(0, y) \in \mathbb{R}^2 | \; y \geq 0\right\}.$$

Now let us choose a function $p_1 : \mathbb{R} \to \mathbb{R}$ defined by $p_1(x) = |x|$, for every $x \in \mathbb{R}$, and another function $p_2 : \{0\} \to \mathbb{R}$ defined by $p_2(0) = 0$. It is easy to check that if the contingent epiderivative of p_i at $(\bar{x}, \bar{y})$ is denoted by f_i, where $i = 1, 2$, then the required condition

$$\text{epi}\,(f_1) \supseteq T(\text{epi}\,(F), (\bar{x}, \bar{y})) \supseteq \text{epi}\,(f_2),$$

holds. However, the contingent epiderivative $D_E F(\bar{x}, \bar{y})$, whose effective domain consists of the singleton $\{0\}$, is given by $D_E F(\bar{x}, \bar{y})(0) = 0$. □

Let Z be a normed space and let $\omega \in \Omega \subset Z$ be arbitrary. We reserve the notation $U(\Omega, \omega)$ to represent the set $V(\omega) \cap \Omega$, where $V(\omega)$ is an arbitrary neighborhood of ω.

Our next result for the existence of contingent epiderivative is as follows:

Theorem 11.4.7. *Let X be a normed space and let $F : X \rightrightarrows \mathbb{R}$ be a set-valued map. Assume that $\bar{x} \in \text{int}(\text{dom}\,(F))$ and $(\bar{x}, f(\bar{x})) \in \text{graph}\,(F)$, where the functional f is defined via (11.56). We set $\bar{y} = f(\bar{x})$. Assume that there exist functionals $f_1, f_2 : X \to \mathbb{R}$, satisfying the following conditions:*

1. $(\bar{x}, \bar{y}) \in \text{graph}\,(f_1) \cap \text{graph}\,(f_2)$ *and* $\bar{x} \in \text{int}(\text{dom}\,(f_1)) \cap \text{int}(\text{dom}\,(f_2))$.
2. $U(\text{epi}\,(f_1), (\bar{x}, \bar{y})) \supseteq U(\text{epi}\,(f), (\bar{x}, \bar{y})) \supseteq U(\text{epi}\,(f_2), (\bar{x}, \bar{y}))$.
3. *The functionals f_1 and f_2 are convex and lower-semicontinuous.*

Then the contingent epiderivative $D_E F(\bar{x}, \bar{y})(x)$ of F at $(\bar{x}, \bar{y})$ exists for every $x \in X$.

Proof. In view of the first two assumptions and the isotony property of the contingent cone, we have

$$\begin{aligned} T(U(\text{epi}\,(f_1), (\bar{x}, \bar{y})), (\bar{x}, \bar{y})) &\supseteq T(U(\text{epi}\,(f), (\bar{x}, \bar{y})), (\bar{x}, \bar{y})) \\ &\supseteq T(U(\text{epi}\,(f_2), (\bar{x}, \bar{y})), (\bar{x}, \bar{y})). \end{aligned}$$

Moreover, from the local approximation nature of the contingent cone, it follows that

$$\begin{aligned} T(U(\text{epi}\,(f_1), (\bar{x}, \bar{y})), (\bar{x}, \bar{y})) &= T(\text{epi}\,(f_1), (\bar{x}, \bar{y})), \\ T(U(\text{epi}\,(f_2), (\bar{x}, \bar{y})), (\bar{x}, \bar{y})) &= T(\text{epi}\,(f_2), (\bar{x}, \bar{y})). \end{aligned}$$

By combining this with the preceding chain of inclusions, we obtain

$$T(\text{epi}\,(f_1), (\bar{x}, \bar{y})) \supseteq T(\text{epi}\,(f), (\bar{x}, \bar{y})) \supseteq T(\text{epi}\,(f_2), (\bar{x}, \bar{y})). \tag{11.57}$$

Since f_i is convex and lower-semicontinuous, we have the existence of function $p_i : X \to \mathbb{R}$, $i = 1, 2$, such that

$$\text{epi}\,(p_i) = T(\text{epi}\,(f_i), (\bar{x}, \bar{y})), \quad i = 1, 2.$$

The above identity, when combined with (11.57)), yields

$$\text{epi}\,(p_1) \supseteq T(\text{epi}\,(f), (\bar{x}, \bar{y})) \supseteq \text{epi}\,(p_2).$$

The existence of $D_E F(\bar{x}, \bar{y})(x)$, for every $x \in X$, now follows from Theorem 11.4.5. □

Remark 11.4.8. In the above result, if we use the so-called lower-semicontinuous regularization (see Rockafellar and Wets [499]) of f_i to generate p_i, then we can dispense with the assumption that f_i is lower-semicontinuous. Of course, it is also possible to impose other conditions on f_i which assure the existence of p_i.

The following example highlights the use of the above result.

Example 11.4.9. Consider a set-valued map $F : [-1, 1] \rightrightarrows \mathbb{R}$ given by (cf. Fig. 11.4 (i)):

$$F(x) = \begin{cases} \frac{21}{8}x - \frac{5}{8} & \text{if } \frac{1}{3} < x \leq 1, \\ \left[0, \frac{1}{4}\right] & \text{if } 0 \leq x \leq \frac{1}{3}, \\ 2x & \text{if } -\frac{1}{3} < x < 0, \\ -\frac{3}{2}x - \frac{7}{6} & \text{if } -\frac{2}{3} < x \leq -\frac{1}{3}, \\ \left[-\frac{x}{2} - \frac{1}{2}, 0\right] & \text{if } -1 \leq x \leq -\frac{2}{3}. \end{cases}$$

In the present case, the single-valued selection function f (cf. 11.56) of F is given by:

$$f(x) = \begin{cases} \frac{21}{8}x - \frac{5}{8} & \text{if } \frac{1}{3} < x \leq 1, \\ 0 & \text{if } 0 \leq x \leq \frac{1}{3}, \\ 2x & \text{if } -\frac{1}{3} < x < 0, \\ -\frac{3}{2}x - \frac{7}{6} & \text{if } -\frac{2}{3} < x \leq -\frac{1}{3}, \\ -\frac{x}{2} - \frac{1}{2} & \text{if } -1 \leq x \leq -\frac{2}{3}. \end{cases}$$

The map f is discontinuous and $\text{dom}\,(f) = [-1, 1]$ (see Fig. 11.4 (ii)). Let us consider the point $(-1/3, -2/3) \in \text{graph}\,(f)$. We define two functions $f_1, f_2 : [-1, 1] \to \mathbb{R}$ as by

$$f_1(x) = \frac{3}{2}x^2 + x - \frac{1}{2}, \quad \text{for every } x \in [-1, 1],$$

and

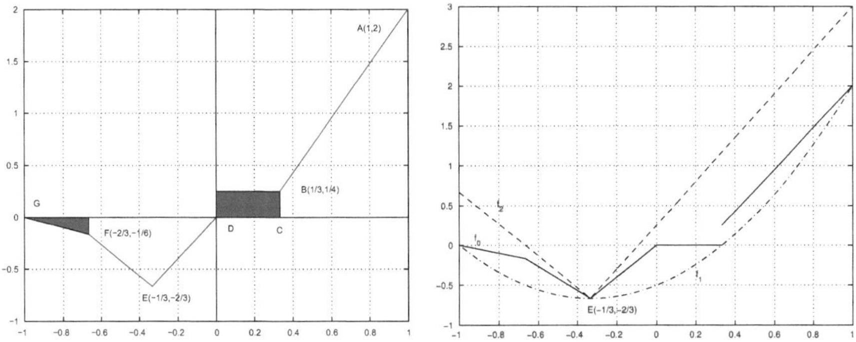

Fig. 11.4 (*i*) The set-valued map F. (*ii*) Functions f_i, $i = 0, 1, 2$

$$f_2(x) = \begin{cases} \frac{11}{4}x + \frac{1}{4} & \text{if } -\frac{1}{3} < x \leq 1, \\ -2x - \frac{4}{3} & \text{if } \quad -1 \leq x \leq -\frac{1}{3}. \end{cases}$$

The functions f_1 and f_2 are convex and satisfy the following conditions (see Fig. 11.4 (ii)):

I. $U(\text{epi}\,(f_1), (-1/3, -2/3)) \supset U(\text{epi}\,(f), (-1/3, -2/3)) \supset U(\text{epi}\,(f_2), (-1/3, -2/3))$.
II. $(-1/3, -2/3) \in \text{graph}\,(f_1) \cap \text{graph}\,(f_2)$ and $-1/3 \in \text{int}(\text{dom}\,(f_1)) \cap \text{int}(\text{dom}\,(f_2))$.

Therefore, all the hypotheses of Theorem 11.4.7 are satisfied and hence the existence of the contingent epiderivative is assured at $(-\frac{1}{3}, -\frac{2}{3})$. Note that the epigraph of the set-valued map f is not convex, but only convex around $(-\frac{1}{3}, -\frac{2}{3})$.

Note that Theorem 11.4.7 does not provide any information for the existence of the contingent epiderivative at $(-\frac{2}{3}, -\frac{1}{6})$ as there does not exist any convex map satisfying the hypotheses on the function f_1. Though the contingent epiderivative exists at $(-\frac{2}{3}, -\frac{1}{6})$.

In order to illustrate the importance of the assumption $\bar{x} \in \text{int}(\text{dom}\,(f_1)) \cap \text{int}(\text{dom}\,(f_2))$, let us consider the point $(\frac{1}{3}, 0) \in \text{graph}\,(f)$. Since

$$T(\text{epi}\,(F), (1/3, 0)) = \{(x, y) \in \mathbb{R}^2 | \quad 0 \geq x, \quad y \geq 0\},$$

the contingent epiderivative of F at $(\frac{1}{3}, 0)$ with restricted domain is given by

$$D_E F(1/3, 0)(x) = 0 \quad \text{for every } x \leq 0,$$

and consequently,

$$\text{dom}\,(D_E F(1/3, 0)) = \{x \in \mathbb{R} | \quad x \leq 0\}. \qquad \square$$

Remark 11.4.10. In the above discussion, our objective was to show the relevance of the single-valued map marginal function (11.56) and, for clarity, we have imposed the convexity assumption. Clearly, all the results can be modified to dispense with the convexity, and analogous results can be given by assuming that the marginal functional is differentiable in the Hadamard sense.

11.4.2 Contingent Epiderivatives in General Spaces

Having studied contingent epiderivatives of set-valued maps with images in $\mathbb{R}$, we now move to investigate this notion for maps acting between general spaces. For the treatment of this subject in the remaining part of this section, besides relying on the contribution by Jahn and Rauh [301], we use extensively the results based on a thorough investigation of the subject by Rodríguez-Marín and Sama [500, 502] and Sama [520]. See also [210, 246, 501, 503, 504, 519].

We begin with the following result that gives some insight on the existence of contingent epiderivatives.

Theorem 11.4.11. *Let X and Y be normed spaces and let $C \subset Y$ be a convex cone. Let $F : X \rightrightarrows Y$ be a set-valued map and let $(\bar{x}, \bar{y}) \in \operatorname{graph}(F)$. If the contingent epiderivative $D_E F(\bar{x}, \bar{y}) : X \rightrightarrows Y$ of F at $(\bar{x}, \bar{y})$ exists, then for every $x \in L := \operatorname{dom}(D(F + C)(\bar{x}, \bar{y}))$, we have*

$$D_E F(\bar{x}, \bar{y})(x) \in \operatorname{SMin}(D(F + C)(\bar{x}, \bar{y})(x), C),$$

where $\operatorname{SMin}(D(F + C)(\bar{x}, \bar{y})(x), C)$ is the set of all strongly minimal points of $D(F + C)(\bar{x}, \bar{y})(x)$ with respect to the cone C.

Proof. By the definition of the contingent epiderivative, we have

$$\operatorname{epi}(D_E F(\bar{x}, \bar{y})) = T(\operatorname{epi}(F), (\bar{x}, \bar{y})),$$

which implies that for every $x \in (D(F + C)(\bar{x}, \bar{y}))$, the following identity holds

$$D_E F(\bar{x}, \bar{y})(x) + C = D(F + C)(\bar{x}, \bar{y})(x),$$

and consequently $D_E F(\bar{x}, \bar{y})(x) \in \operatorname{SMin}(D(F + C)(\bar{x}, \bar{y})(x), C)$. The proof is complete. □

Remark 11.4.12. In the above result, we did not explicitly imposed the condition that the cone C is closed. However, the existence of the contingent epiderivative ensures that C is closed. In fact, if the contingent derivative exists, then we have

$$D_E F(\bar{x}, \bar{y})(x) + C = D(F + C)(\bar{x}, \bar{y})(x) \quad \text{for any } x \in \operatorname{dom}(D(F + C)(\bar{x}, \bar{y})),$$

which ensures that $C = D(F + C)(\bar{x}, \bar{y})(x) - D_E F(\bar{x}, \bar{y})(x)$. Because the set $D(F + C)(\bar{x}, \bar{y})(x)$ is closed, we deduce the closedness of C from the fact that it is a difference of a closed and a compact set.

The following result addresses the uniqueness issue for the contingent epiderivative.

Theorem 11.4.13. *Let X and Y be normed spaces and let $C \subset Y$ be a pointed, closed, and convex cone. Let $F : X \rightrightarrows Y$ be a set-valued map and let $(\bar{x}, \bar{y}) \in \operatorname{graph}(F)$. If the set-valued map F is contingent epidifferentiable at $(\bar{x}, \bar{y})$, then the contingent epiderivative $D_E F(\bar{x}, \bar{y})$ is unique.*

Proof. Suppose that $\breve{D}_E F(\bar{x}, \bar{y})$ and $D_E F(\bar{x}, \bar{y})$ are two contingent epiderivatives such that

$$\breve{D}_E F(\bar{x}, \bar{y})(x) \neq D_E F(\bar{x}, \bar{y})(x) \quad \text{for any } x \in \operatorname{dom}(D(F + C)(\bar{x}, \bar{y})).$$

Then, for such $x \in \operatorname{dom}(D(F + C)(\bar{x}, \bar{y}))$, we have

$$\breve{D}_E F(\bar{x}, \bar{y})(x) \in \operatorname{SMin}(D(F + C)(\bar{x}, \bar{y})(x), C),$$
$$D_E F(\bar{x}, \bar{y})(x) \in \operatorname{SMin}(D(F + C)(\bar{x}, \bar{y})(x), C),$$

implying that

$$\breve{D}_E F(\bar{x}, \bar{y})(x) \in D_E F(\bar{x}, \bar{y})(x) + C,$$
$$D_E F(\bar{x}, \bar{y})(x) \in \breve{D}_E F(\bar{x}, \bar{y})(x) + C,$$

and hence

$$\breve{D}_E F(\bar{x}, \bar{y})(x) - D_E F(\bar{x}, \bar{y})(x) \in C \cap -C,$$

which, by taking into account the pointedness of C, ensures that $\breve{D}_E F(\bar{x}, \bar{y})(x) = D_E F(\bar{x}, \bar{y})(x)$. □

The fact that the pointedness of the ordering cone C plays a vital role in the uniqueness of the contingent epiderivative was singled out in [500]. It turns out that if the ordering cone C is non-pointed, then there exists a family of maps such that each member of the family satisfies the definition of the contingent epiderivative. The following example illustrates this situation:

Example 11.4.14. We define a set-valued map $F : \mathbb{R} \rightrightarrows \mathbb{R}^2$ by

$$F(x) = \{(0, 0), (|x|, |x|)\}, \quad \text{for every } x \in \mathbb{R}.$$

Then, for $C = \mathbb{R} \times \mathbb{R}_+$ and $(\bar{x}, \bar{y}) = (0, (0, 0))$, we have

$$T(\operatorname{epi}(F), (\bar{x}, \bar{y})) = \mathbb{R} \times \mathbb{R} \times \mathbb{R}_+$$

which means that for every $x \in \operatorname{dom}(D(F+C)(\bar{x},\bar{y})) = \mathbb{R}$, we have

$$D(F+C)(\bar{x},\bar{y})(x) = \mathbb{R} \times \mathbb{R}_+.$$

Clearly, a map $\varphi : \mathbb{R} \to \mathbb{R}^2$ is a contingent epiderivative of F at $(\bar{x},\bar{y})$, if for any $x \in \mathbb{R}$, we have

$$\varphi(x) + \mathbb{R} \times \mathbb{R}_+ = D(F+C)(\bar{x},\bar{y})(x).$$

Evidently, the maps

$$\varphi_1(x) = (x,0),$$
$$\varphi_2(x) = (-x,0)$$

are both contingent epiderivatives of F at $(\bar{x},\bar{y})$, although $\varphi_1 \neq \varphi_2$. In fact, there exists a family of epiderivatives given by $\Gamma = \{\varphi : \mathbb{R} \to \mathbb{R}^2 | \ \varphi(x) = (h(x),0)$ where $h : \mathbb{R} \to \mathbb{R}$ is any map$\}$. □

We have the following useful property of the contingent epiderivative (see [301]).

Theorem 11.4.15. *Let X and Y be normed spaces and let $C \subset Y$ be a pointed, closed, and convex cone. Let $F : X \rightrightarrows Y$ be a set-valued map and let $(\bar{x},\bar{y}) \in$ graph (F). Assume that the map F is C-convex. Then the contingent epiderivative $D_E F(\bar{x},\bar{y})$ is a sublinear map, that is, for every $\lambda \in \mathbb{R}_+$ and all $x, y \in$ dom $(D(F+C)(\bar{x},\bar{y}))$, we have*

$$D_E F(\bar{x},\bar{y})(\lambda x) = \lambda D_E F(\bar{x},\bar{y})(x)),$$
$$D_E F(\bar{x},\bar{y})(x+y) \leq D_E F(\bar{x},\bar{y})(x) + D_E F(\bar{x},\bar{y})(y).$$

Proof. The proof can be extracted from the proof of Theorem 11.4.16 given below. □

Note that if the ordering cone C is not pointed, then not necessarily every member of the family of contingent epiderivatives is sublinear. In fact, this is evident from Example 11.4.14. However, if epi (F) is locally convex around $(\bar{x},\bar{y})$, then there exists a sublinear map in the family of contingent epiderivatives of F at $(\bar{x},\bar{y})$, which in the following, is denoted by $E_F(\bar{x},\bar{y})$. This is shown in the following result.

Theorem 11.4.16. *Let X and Y be normed spaces and let $C \subset Y$ be a closed, and convex cone. Let $F : X \rightrightarrows Y$ be a set-valued map and let $(\bar{x},\bar{y}) \in$ graph (F). Assume that $E_F(\bar{x},\bar{y}) \neq \emptyset$. Then:*

1. *If the map F is C-convex, then there exists a sublinear map $\Phi \in E_F(\bar{x},\bar{y})$.*
2. *If $F(\bar{x})$ convex, then $\bar{y}$ is a strong minimal point of $F(\bar{x})$.*

Proof. By the proof Theorem 11.4.11, it follows that $E_F(\bar{x},\bar{y}) \neq \emptyset$ implies

$$\operatorname{SMin}((D(F+C)(\bar{x},\bar{y})(x), C) \neq \emptyset \quad \text{for any } x \in L = \operatorname{dom}(D(F+C)(\bar{x},\bar{y})).$$

Consider the set $B = \{e \in L \mid \|e\| = 1\}$. For each $e \in B$, we assign a fixed element $y_e \in \text{SMin}(D(F + C)(\bar{x}, \bar{y})(e))$. For the unique $\alpha > 0$ such that $x = \alpha e$, we define $\Phi : L \to Y$ by

$$\Phi(0) = 0$$
$$\Phi(e) = y_e$$
$$\Phi(x) = \alpha y_e.$$

We claim that $\Phi \in E_F(\bar{x}, \bar{y})$. In fact, if $x \neq 0$, then

$$\begin{aligned}\Phi(x) &= \alpha y_e \in \text{SMin}(\alpha D(F + C)(\bar{x}, \bar{y})(e), C) \\ &= \text{SMin}(D(F + C)(\bar{x}, \bar{y})(\alpha e), C) = \text{SMin}(D(F + C)(\bar{x}, \bar{y})(x), C).\end{aligned}$$

If $x = 0$, then we claim that

$$\Phi(0) = 0 \in \text{SMin}(D(F + C)(\bar{x}, \bar{y})(0)).$$

Indeed, if the above containment does not hold, then

$$0 \notin \text{SMin}(D(F + C)(\bar{x}, \bar{y})(0), C). \tag{11.58}$$

Therefore, by Theorem 11.4.11, there exists $w \neq 0$ with $w \in \text{SMin}(D(F + C)(\bar{x}, \bar{y})(0), C)$. Since $0 \in D(F + C)(\bar{x}, \bar{y})(0)$, by the minimality of w, we have $w \in -C$. Furthermore, by (11.58), $w \notin C$. Since the contingent derivative is strictly positive homogeneous,

$$\lambda w \in D(F + C)(\bar{x}, \bar{y})(0) \quad \text{for any } \lambda > 0.$$

This, in view of the minimality of w, confirms that $w \leq \lambda w$. Therefore $w \in C \cap -C$, but this contradicts $w \notin C$.

By applying Theorem 11.4.11, we conclude $\Phi \in E_F(\bar{x}, \bar{y})$. It is easily seen that Φ is positive homogeneous because $\Phi(0) = 0$ and for each $x \neq 0$, we have

$$\Phi(\lambda x) = \Phi(\lambda(\alpha e)) = \Phi((\lambda\alpha)e) = \lambda\alpha y_e = \lambda\Phi(x).$$

We not proceed to show that Φ is subadditive. Since $T(\text{epi}\,(F), (\bar{x}, \bar{y}))$ is convex and because $\Phi \in E_F(\bar{x}, \bar{y})$, we deduce that the cone $\text{epi}\,(\Phi) = T(\text{epi}\,(F), (\bar{x}, \bar{y}))$ is convex as well. Therefore,

$$\left(\tfrac{1}{2}x + \tfrac{1}{2}y, \tfrac{1}{2}\Phi(x) + \tfrac{1}{2}\Phi(y)\right) \in \text{epi}\,(\Phi), \quad \text{for any } x, y \in L,$$

implying that

$$\tfrac{1}{2}\Phi(x) + \tfrac{1}{2}\Phi(y) \in D(F + C)(\bar{x}, \bar{y})\left(\tfrac{1}{2}x + \tfrac{1}{2}y\right) \subset \Phi\left(\tfrac{1}{2}x + \tfrac{1}{2}y\right) + C,$$

which, when multiplied by 2 on both sides, yields

$$\Phi(x) + \Phi(y) \subset \Phi(x + y) + C.$$

For the second part, we suppose that the result is false. Then, there exists an element $y \in F(\bar{x})$ such that $y - \bar{y} \notin C$. Since $F(\bar{x})$ is a convex set, the sequence $\{y_n\}$, where $y_n := \bar{y} + \dfrac{y - \bar{y}}{n}$, belongs to $F(\bar{x})$ and satisfies that

$$\frac{y_n - \bar{y}}{\|y_n - \bar{y}\|} = \frac{y - \bar{y}}{\|y - \bar{y}\|} \to v.$$

Obviously, $v \notin C$ and $v \in D(F + C)(\bar{x}, \bar{y})(0)$. From the first part, we note that there is an element $\varphi \in E_F(\bar{x}, \bar{y})$ such that $\varphi(0) = 0$. By the definition of φ, we have $v \geq 0 = \varphi(0)$ which contradicts that $v \notin C$ and the proof is complete. □

As noted above, the notion of the contingent epiderivative is connected to $\mathrm{SMin}(D(F + C)(\bar{x}, \bar{y}), C)$. The following result connects $\mathrm{SMin}(DF(\bar{x}, \bar{y})(\cdot), C)$ and $\mathrm{SMin}(D(F + C)(\bar{x}, \bar{y})(\cdot), C)$.

Theorem 11.4.17. *Assume that F is a contingently C-absorbing map at $(\bar{x}, \bar{y})$, that is,* (11.51) *holds. Then, for every $x \in \mathrm{dom}\,(D(F + C)(\bar{x}, \bar{y}))$, the set $\mathrm{SMin}(DF(\bar{x}, \bar{y})(x), C)$ is nonempty, if and only if, $D_E F(\bar{x}, \bar{y})(x)$ exists, and, in that case, the two sets coincide.*

Proof. Let any $x \in \mathrm{dom}\,(D(F + C)(\bar{x}, \bar{y}))$, be arbitrary. Using (11.51), we have

$$\begin{aligned} \mathrm{SMin}\,(D(F + C)(\bar{x}, \bar{y})(x), C) &= \mathrm{SMin}\,(DF(\bar{x}, \bar{y})(x) + C), C) \\ &= \mathrm{SMin}(DF(\bar{x}, \bar{y})(x), C). \end{aligned}$$

By the proof of Theorem 11.4.11, if $D_E F(\bar{x}, \bar{y})$ exists then $\mathrm{SMin}\,(DF(\bar{x}, \bar{y})(x) + C), C) \neq \emptyset$. The conclusion then follows from the above identity. The converse implication is obvious. □

Clearly, the above result is a consequence of (11.51). Several examples have been given in [500, 502] which show that the above result does not hold without (11.51), and, in general, $\mathrm{SMin}(DF(\bar{x}, \bar{y})(x), C)$ is different from $D_E F(\bar{x}, \bar{y})(x)$.

11.4.3 Existence Theorems for Contingent Epiderivatives

Our objective in this section is to give some existence results for contingent epiderivatives. Throughout this section, we assume that the ordering cone C is strongly minihedral. That is, every subset which is C-bounded from below has a supremum.

The following result gives some insight into the existence of contingent epiderivatives.

Theorem 11.4.18. *Let X and Y be normed spaces and let $C \subset Y$ be a pointed, closed, and convex cone. Let $F : X \rightrightarrows Y$ be a set-valued map and let $(\bar{x}, \bar{y}) \in \text{graph}\,(F)$. Assume that for every $x \in L := \text{dom}\,(D(F+C)(\bar{x}, \bar{y}))$, the set $D(F+C)(\bar{x}, \bar{y})(x)$ is C-lower bounded. Then for the map $M : L \to Y$ defined by*

$$M(x) = \text{SMax}(\{y \in Y \mid D(F+C)(\bar{x}, \bar{y})(x) \subset y + C\}, C), \tag{11.59}$$

the following properties hold:

1. *$M(\lambda x) = \lambda M(x)$, for any $x \in L$ and $\lambda \geq 0$.*
2. *If $D_E F(\bar{x}, \bar{y})$ exists, then it coincides with M.*
3. *$D_E F(\bar{x}, \bar{y})$ exists, if and only if,*

$$M(x) \in D(F+C)(\bar{x}, \bar{y})(x) \quad \textit{for every } x \in L.$$

Proof. We begin by proving the first part. By the strong minihedrality of the cone, the map M is well-defined. To prove the first part, we fix $x \in L$ and $\lambda > 0$. Since $D(F+C)(\bar{x}, \bar{y})(x)$ is C-lower bounded and the map $D(F+C)(\bar{x}, \bar{y})(\cdot)$ is positive homogenous, we have

$$D(F+C)(\bar{x}, \bar{y})(\lambda x) = \lambda D(F+C)(\bar{x}, \bar{y})(x) \geq \lambda M(x),$$

and consequently, by the definition of M, we get

$$M(\lambda x) \geq \lambda M(x).$$

Analogously, because

$$\lambda D(F+C)(\bar{x}, \bar{y})(x) = D(F+C)(\bar{x}, \bar{y})(\lambda x) \geq M(\lambda x),$$

we have

$$D(F+C)(\bar{x}, \bar{y})(x) \geq \frac{1}{\lambda} M(\lambda x), \tag{11.60}$$

implying

$$M(x) \geq \frac{1}{\lambda} M(\lambda x),$$

and it follows that $\lambda M(x) = M(\lambda x)$.

If $x = 0$, then, using (11.60), we have

$$M(0) \geq \lambda M(0) \quad \text{for every } \lambda \in \mathbb{R}_+,$$

and hence $M(0) = 0$.

The second part follows from the fact that for a C-lower bounded subset of X, it holds that

$$\text{SMin}(A, C) = \text{SMax}(B, C),$$

where $B = \{y \in Y \mid A \subset y + C\}$, provided that $\text{SMin}(A, C) \neq \emptyset$.

We now proceed to prove the third part. Let $x \in L$ be arbitrary. If $D_E F(\bar{x}, \bar{y})$ exists, then by Theorem 11.4.11, we have $\text{SMin}(D(F + C)(x, y)(x), C) \neq \emptyset$. Therefore, by Theorem 11.4.11 and the preceding arguments, we deduce that $M(x) = \text{SMin}(D(F + C)(x, y)(x), C)$.

Conversely, if $M(x) \in D(F + C)(x, y)(x)$, then by $D(F + C)(x, y)(x) \subset M(x) + C$, we have

$$M(x) = \text{SMin}(D(F + C)(x, y)(x), C),$$

and hence Theorem 11.4.11 ensures that $D_E F(\bar{x}, \bar{y})$ exists. The proof is complete.□

In the above result, the assumption that $D(F + C)(\bar{x}, \bar{y})(x)$ remains C-lower bounded for every $x \in L$ has played an important role. The following result gives a necessary criteria to ensure this condition.

Theorem 11.4.19. *Let X and Y be normed spaces and let $C \subset Y$ be a pointed, closed, and convex cone with* $\text{int}(C) \neq \emptyset$. *Let $F : X \rightrightarrows Y$ be a set-valued map and let $(\bar{x}, \bar{y}) \in$* graph (F). *If F is upper locally Lipschitz at $\bar{x}$ and $\bar{y} \in$* $\text{SMin}(F(\bar{x}) + C, C)$, *then the set $D(F + C)(\bar{x}, \bar{y})(x)$ is C-lower bounded for every $x \in L$.*

Proof. Let$(x, y) \in T(\text{epi}\,(F), (\bar{x}, \bar{y}))$ be arbitrary. Then it follows from the definition of the contingent cone that there exist sequences $\{t_n\} \subset \mathbb{P}$, $\{(x_n, y_n + c_n)\} \subset \text{epi}\,(F)$, such that $(x_n, y_n + c_n) \to (\bar{x}, \bar{y})$ and $t_n(x_n - \bar{x}, y_n + c_n - \bar{y}) \to (x, y)$. Since F is upper locally Lipschitz at $\bar{x}$, there exists a sequence $\{z_n\} \subset F(\bar{x})$ such that

$$y_n - z_n \in l\,\|x_n - \bar{x}\|\, B_Y \quad \text{for every } n \in \mathbb{N},$$

implying that $y_n - z_n \to 0$, and therefore $(z_n + c_n - \bar{y}) \to 0$. Since $\text{int}(C) \neq \emptyset$, there exists a C-lower bound of B_Y, $z \in Y$ such that

$$t_n(y_n - z_n) \geq t_n l\,\|x_n - \bar{x}\|\, z.$$

Since $\bar{y}$ is a local strongly minimal point of $(F(\bar{x})+C)$, we can assume without loss of generality that $(z_n + c_n - \bar{y}) \in C$, and, consequently

$$t_n(y_n + c_n - \bar{y}) = t_n(y_n - z_n) + t_n(z_n + c_n - \bar{y}) \geq l \, \|t_n(x_n - \bar{x})\| \, z,$$

which, when passed to the limit as $n \to \infty$, confirms that $y \geq l \, \|u\| \, z$. □

For a preparation for our next result, we need the following auxiliary result:

Proposition 11.4.20. *Let X and Y be normed space and let $C \subset Y$ be a pointed, closed, and convex cone. Let $S \subset X$ be nonempty, let $F : S \rightrightarrows Y$ be a set-valued map, and let $(0, \bar{y}) \in \operatorname{graph}(F)$. Assume $D(F + C)(0, \bar{y})(x)$ is C-lower bounded for every x in its domain. If $S \subset L = \operatorname{dom}(M)$ and M, defined in (11.59), is continuous, then*

$$D(F + M)(0, \bar{y}) = M + DF(0, \bar{y}).$$

Proof. For any $(x, y) \in \operatorname{graph}(D(F + M)(0, \bar{y}))$, we have $(x, y) \in T(\operatorname{graph}(F + M), (0, \bar{y}))$. Then there exist sequences $\{t_n\} \subset \mathbb{P}$, $\{(x_n, y_n)\} \subset X \times Y$ with $t_n \to 0$, $(x_n, y_n) \to (x, y)$ such that $t_n x_n \in S$ and

$$\bar{y} + t_n y_n \in F(t_n x_n) + M(t_n y_n),$$

implying

$$\bar{y} + t_n(y_n - M(x_n)) \in F(t_n x_n).$$

Since M is continuous, we have $y_n - M(x_n) \to y - M(x)$ and hence $(x, y - M(x)) \in \operatorname{graph}(DF(0, \bar{y}))$.

For the converse, let $(x, y) \in \operatorname{graph}((M + DF(0, \bar{y}))$ and $y - M(x) \in DF(0, \bar{y})(x)$, then there exist sequences $t_n \downarrow 0$, $x_n \to x$, $w_n \to y - M(x)$ with $t_n x_n \in S$ and $\bar{y} + t_n w_n \in F(t_n x_n)$ or equivalently,

$$\bar{y} + t_n(w_n + M(x_n)) \in F(t_n u_n) + M(t_n x_n).$$

Since $y_n + M(x_n) \to y$, we obtain $(x, y) \in \operatorname{graph}(D(F + M)(0, \bar{y}))$. The proof is complete. □

We have the following result regarding the existence of the contingent epiderivative.

Theorem 11.4.21. *Let X and Y be normed spaces and let $C \subset Y$ be a pointed, closed, and convex cone. Let $S \subset X$ be nonempty, let $F : S \rightrightarrows Y$ be a set-valued map, and let $(\bar{x}, \bar{y}) \in \operatorname{graph}(F)$. Assume $D(F + C)(\bar{x}, \bar{y})(x)$ is C-lower bounded for every $x \in L = \operatorname{dom}(D(F + C)(\bar{x}, \bar{y}))$, the map M, defined in (11.59), is continuous and $S \subset \bar{x} + \operatorname{dom}(M)$. Then the contingent epiderivative $D_E F(\bar{x}, \bar{y})$ exists, if and only if,*

$$0 \in D(F(\cdot + \bar{x}) - M + C)(0, \bar{y})(x) \quad \textit{for any } x \in L.$$

Proof. Let $x \in L$ and $H(x) = F(x + \bar{x})$, then $(0, \bar{y}) \in \text{graph}(H)$. It is easy to check that $D(H + C)(0, \bar{y})$ coincides with $D(F + C)(\bar{x}, \bar{y})$. By Theorem 11.4.18, a necessary and sufficient condition for the existence of $D_E F(\bar{x}, \bar{y})$ is

$$M(x) \in D(H + C)(0, \bar{y})(x) \quad \text{for every } \; x \in L. \tag{11.61}$$

Since $\text{dom}(H + C) = S - \{\bar{x}\} \subset \text{dom}(M)$ and M is continuous, by Proposition 11.4.20, we get

$$D(H + C - M)(0, \bar{y})(x) = D(F + C)(\bar{x}, \bar{y})(x) - M(x) \quad \text{for any } u \in S - \{\bar{x}\}.$$

Therefore (11.61) is equivalent to

$$0 \in D(F(\cdot + \bar{x}) - M + C)(0, \bar{y})(x),$$

and the proof is complete. □

In the following, we give another characterization for the existence of $D_E F(\bar{x}, \bar{y})$:

Theorem 11.4.22. *Let X and Y be normed spaces and let $C \subset Y$ be a pointed, closed, and convex cone. Let $S \subset X$, let $\bar{x} \in \text{int}(S)$, let $F : S \rightrightarrows Y$ be a set-valued map, and let $\bar{y} \in \text{SMin}(F(\bar{x}), C)$. Assume that $L := \text{dom}\,(D(F + C)(\bar{x}, \bar{y})) = X$ and there exists a neighborhood $U(0)$ with $\bar{x} + U(0) \subset S$ satisfying that, for every $x \in U(0)$, the following conditions hold:*

1. $D(F + C)(\bar{x}, \bar{y})(x) \subset F(x + \bar{x}) - \bar{y} + C$.
2. $\text{SMin}(F(x + \bar{x}), C) \neq \emptyset$.

Then, $D_E F(\bar{x}, \bar{y})$ exists provided that the function M is continuous.

Proof. We begin by claiming that $D(F+C)(\bar{x}, \bar{y})(x)$ is C-lower bounded for every $x \in L$. In fact, by the imposed conditions, the set $D(F + C)(\bar{x}, \bar{y})(0)$ is C-lower bounded and $D(F + C)(\bar{x}, \bar{y})(0) \subset F(\bar{x}) - \bar{y} + C \subset C$.

For any $0 \neq x \in L$, there exists $\alpha \in \mathbb{R}_+ \backslash \{0\}$ such that $\alpha^{-1} x \in U(0)$. Therefore, by the imposed conditions, we have

$$D(F + C)(\bar{x}, \bar{y})(\alpha^{-1} x) \subset \text{SMin}(F(\alpha^{-1} x + \bar{x}), C) - \bar{y} + C.$$

By the positive homogeneity of the contingent derivative, we deduce

$$D(F+C)(\bar{x}, \bar{y})(x) = \alpha D(F+C)(\bar{x}, \bar{y})(\alpha^{-1} x) \subset \alpha(\text{SMin}(F(\alpha^{-1} x+\bar{x}), C)-\bar{y})+C,$$

ensuring that the element $\alpha(\text{SMin}(F(\alpha^{-1} x + \bar{x}), C) - \bar{y})$ is a C-lower bound of $D(F + C)(\bar{x}, \bar{y})(x)$.

For $x \in \text{dom}\,(D(F(\cdot+\bar{x})-M+C)(0,\bar{y}))$, there exist $\{x_n\} \subset \text{dom}\,(F(\cdot+\bar{x})-M+C)\cap U(0)$, $\{t_n\} \subset \mathbb{P}$ such that $x_n \to 0$ and $t_n x_n \to x$.

Using the imposed conditions once again, we have $D(F+C)(\bar{x},\bar{y})(x_n) \subset z_n - \bar{y} + C$, with $z_n = \text{SMin}(F(x_n+\bar{x}), C)$, and hence

$$M(x_n) \geq z_n - \bar{y}.$$

By choosing $c_n = -(z_n - M(x_n) - \bar{y}) \in C$, we get $t_n(x_n, z_n - M(x_n) - \bar{y} + c_n) \to (x, 0)$ which confirms that

$$0 \in D(F(\cdot+\bar{x}) - M + C)(0,\bar{y})(x),$$

and the proof finishes by Theorem 11.4.21. □

In the remaining part of this section, we focus on the set-valued maps taking values in $\mathbb{R}^n$.

We recall that, given a normed space X, if a map $f : X \to \mathbb{R}$ is upper locally Lipschitz map at $\bar{x} \in X$, then the contingent epiderivative $D_E f(\bar{x})$ exists and is given by

$$D_E f(\bar{x})(x) = \inf\{y \in \mathbb{R} |\; y \in Df(\bar{x}, f(\bar{x}))(x)\}.$$

Based on this observation, we have the following:

Lemma 11.4.23. *Let X be a normed space, let $S \subset X$ be nonempty, and let $f : X \to \mathbb{R}^n$ be a given map. Assume that $f = (f_1, f_2, \ldots f_n)$ is upper locally Lipschitz at $\bar{x} \in S$. For any $i \in \{1, .., n\}$ and for given $y \in Df_i(\bar{x}, f_i(\bar{x}))(x)$ with $x \in L$, there exists $(w_j)_{j \neq i} \in \mathbb{R}^{n-1}$ such that*

$$(w_1, w_2, .., y, \ldots w_n) \in Df(\bar{x}, f(\bar{x}))(x).$$

Proof. For $x \in L$, let $y \in Df_i(\bar{x}, f_i(\bar{x}))(x)$ be arbitrary. Then there exist sequences $\{t_m\} \subset \mathbb{R}_+$ and $\{x_m\} \subset S$ such that $x_m \to \bar{x}$, $t_m(f_i(x_m) - f_i(\bar{x})) \to y$. By the fact that the component functions $\{f_i\}_{i \in I}$ are Lipschitz, it is possible to extract subsequences $\{t_{m_k}\} \subset \{t_m\}$ and $\{x_{m_k}\} \subset \{x_m\}$ such that there exists $(w_j)_{j \neq i} \in \mathbb{R}^{n-1}$ which verifies that $t_{m_k}(f_i(x_{m_k}) - f_i(\bar{x})) \to w_j$ for any $j \in \{1, .., n\} \backslash \{i\}$. Consequently, $(w_1, w_2, .., y, \ldots w_n) \in Df(\bar{x}, f(\bar{x}))(x)$. □

In the following, by M_f we denote the map M associated with f (see (11.59)).

Theorem 11.4.24. *Let X be a normed space, let $S \subset X$ be nonempty, and let $\mathbb{R}^n$ be partially ordered by $C = \mathbb{R}^n_+$. If $f = (f_1, f_2, \ldots f_n)$ is upper locally Lipschitz at $\bar{x} \in S$, then*

$$M_f = (D_E f_1(\bar{x}), D_E f_2(\bar{x}), \ldots, D_E f_n(\bar{x})).$$

Proof. By Theorem 11.4.19, we have

$$M_f(x) = \text{SMax}\,(y \in Y : D(f + C)(\bar{x}, \bar{y})(x) \subset y + C, C),$$

and since the upper locally Lipschitz maps are contingently C-absorbing, it is equivalent to

$$M_f(x) = \text{SMax}\,(\{y \in Y : D(f)(\bar{x}, \bar{y})(x) \subset y + C\}, C).$$

Since

$$D_E f_i(\bar{x})(x) = \inf\{y \mid y \in Df_i(\bar{x}, \bar{y})(x)\} \quad \text{for any } i = 1, 2, .., n,$$

we have

$$Df(\bar{x}, \bar{y})(x) \subset (D_E f_1(\bar{x}), D_E f_2(\bar{x}), .., D_E f_n(\bar{x})) + C,$$

which, by the definition of M_f, yields that

$$M_f(x) \geq (D_E f_1(\bar{x}), D_E f_2(\bar{x}), .., D_E f_n(\bar{x})).$$

Suppose that $M_f(x) \neq (D_E f_1(\bar{x}), D_E f_2(\bar{x}), .., D_E f_n(\bar{x}))$. Therefore, there exists an index i such that

$$M_{f_i}(x) > D_E f_i(\bar{x})(x).$$

Since $D_E f_i(\bar{x})(x) \in Df_i(\bar{x}, \bar{y})(x)$, it follows by Lemma 11.4.23 that there exists $(w_j)_{j \neq i} \in \mathbb{R}^{n-1}$ such that

$$(w_1, w_2, \ldots, D_E f_i(\bar{x})(x), \ldots, w_n) \in Df(\bar{x}, \bar{y})(x) \subset M_f(x) + C,$$

which contradicts $M_{f_i}(x) > D_E f_i(\bar{x})(x)$. The proof is complete. □

We intend to close this discussion by giving an existence result for the contingent epiderivative for convex set-valued maps. For this we will first need the following result.

Lemma 11.4.25. *Let X be a normed space, let $S \subset X$ be nonempty, and let $\mathbb{R}^n$ be partially ordered by $C := \mathbb{R}^n_+$. If $f : S \to \mathbb{R}^n$ is C-convex and continuous at $\bar{x} \in \text{int}(S)$, then the contingent epiderivative $D_E f(\bar{x}, f(\bar{x}))$ exists and is given by*

$$D_E f(\bar{x}, f(\bar{x})) = (D_E f_1(\bar{x}), D_E f_2(\bar{x}), \ldots, D_E f_n(\bar{x})).$$

Proof. Because f is C-convex, each component f_i is convex and locally lipschitz at $\bar{x} \in \text{int}(S)$. Therefore, it follows that (see [18]),

$$D_E f_i(\bar{x})(x) = \lim_{t \to 0+} \frac{f_i(\bar{x} + tx) - f_i(\bar{x})}{t}, \quad \text{for any } x \in X,\ i \in \{1, .., n\}.$$

A direct computation shows that

$$(x, D_E f_1(\bar{x})(x), .., D_E f_n(\bar{x})(x)) \in T(\text{epi}\,(f), (\bar{x}, f(\bar{x}))).$$

According to Theorem 11.4.24, we have

$$M_f(x) = (D_E f_1(\bar{x})(x), .., D_E f_n(\bar{x})(x)),$$

asserting

$$(x, M_f(x)) \in T(\text{epi}\,(f), (\bar{x}, f(\bar{x}))),$$

and, as a consequence, we get

$$M_f(x) \in D(f + C)(\bar{x}, f(\bar{x}))(x),$$

and the existence of the contingent epiderivative $Df(\bar{x}, f(\bar{x})) = M_f$ follows from Theorem 11.4.18. The proof is complete. □

We have the following existence result for the contingent epiderivative of set-valued map convex maps.

Theorem 11.4.26. *Let X be a normed space, let $S \subset X$ be convex, let $\bar{x} \in \text{int}(S)$, and let $\mathbb{R}^n$ be partially ordered by $C := \mathbb{R}^n_+$. Let $F : S \to 2^{\mathbb{R}^n}$ be a C-convex set-valued map such that* SMin$(F(x), C) \neq \emptyset$*, for any $x \in S$. If $h(x) =$* SMin$(F(x), C)$ *is continuous at $\bar{x}$, then the contingent epiderivatives $D_E h(\bar{x})$ and $D_E F(\bar{x}, h(\bar{x}))$ exists. Furthermore, we have*

$$D_E F(x, h(\bar{x})) = D_E h(x, h(\bar{x})) = (D_E h_1(\bar{x}), D_E h_2(\bar{x}), \ldots, D_E h_n(\bar{x})).$$

Proof. Since epi$\,(h) =$ epi$\,(F)$, the existence of $D_E F(\bar{x}, h(\bar{x}))$ is equivalent to the existence of $D_E h(\bar{x}, h(\bar{x}))$. Since F is C-convex, for any $x_1, x_2 \in S$ and $\lambda \in [0, 1]$, we have

$$h(\lambda x_1 + (1 - \lambda)x_2) \in F(\lambda x_1 + (1 - \lambda)x_2) \subset \lambda F(x_1) + (1 - \lambda)F(x_2) + C,$$

and by using the condition $h(x) = \text{SMin}(F(x), C)$, we deduce that

$$h(\lambda x_1 + (1 - \lambda)x_2) \leq \lambda h(x_1) + (1 - \lambda)h(x_2)$$

which confirms that the map h is C-convex. It follows from Lemma 11.4.25 that the contingent epiderivative $D_E h(\bar{x})$ exists and is given by

$$Dh(x, h(\bar{x})) = (D_E h_1(\bar{x}), D_E h_2(\bar{x}), .., D_E h_n(\bar{x})),$$

and the proof is complete. □

11.4.4 Variational Characterization of the Contingent Epiderivatives

In this section, our objective is to give a variational characterization for the contingent epiderivative by associating it to a variational system. This is done in two ways. In the first approach, the variational system is obtained by scalarizing the contingent derivative of the involved set-valued map, whereas in the second approach, we employ the contingent epiderivative of the scalarization of the set-valued map. Our exposition of this subsection is based on the investigations by Rodríguez-Marín and Sama [502] and Sama [520].

We begin with the first approach by considering the following family of variational systems: For a given $x \in L := \text{dom}\,(D(F + C)(\bar{x}, \bar{y}))$, find $y \in Y$ such that

$$\ell(y) = \inf\{\ell(y) | \; y \in D(F + C)(\bar{x}, \bar{y})(x)\} \quad \text{for any } \ell \in C^+. \tag{11.62}$$

The following is the promised variational characterization of the contingent epiderivative.

Theorem 11.4.27. *Let X and Y be normed space and let $C \subset Y$ be a pointed, closed, and convex cone. Let $F : X \rightrightarrows Y$ be a set-valued map and let $(\bar{x}, \bar{y}) \in$ graph (F). Assume that for any $x \in L$, there exists a weakly compact subset $B_x \subset D(F + C)(\bar{x}, \bar{y})(x)$ such that $D(F + C)(\bar{x}, \bar{y})(x) \subset B_x + C$. Then the contingent epiderivative $D_E F(\bar{x}, \bar{y})$ exists, if and only if, for each $x \in L$, the system* (11.62) *is solvable. Moreover, if $D_E F(\bar{x}, \bar{y})$ exists, then $D_E F(\bar{x}, \bar{y})(x)$ is the unique solution of* (11.62) *for $x \in L$.*

Proof. We recall that the contingent epiderivative $D_E F(\bar{x}, \bar{y})$ exists, if and only if, for every $x \in L$, we have $\text{SMin}(D(F + C)(\bar{x}, \bar{y})(x), C) \neq \emptyset$. Furthermore, if $D_E F(\bar{x}, \bar{y})$ exists, then we have

$$D_E F(\bar{x}, \bar{y})(x) = \text{SMin}(D(F + C)(\bar{x}, \bar{y})(x), C).$$

We will finish the proof in the following three steps:

1. For any weakly compact set $A \subset Y$, if $\text{SMin}(A, C)$ is nonempty, then it solves the variational system: Find $y \in Y$ such that

$$\ell(y) = \inf_{a \in A} \ell(a), \quad \text{for any } \ell \in C^+. \tag{11.63}$$

2. If the variational system (11.63) is solvable, then the solution is unique and is given by $\mathrm{SMin}(A, C)$.
3. If $B \subset Y$ be such that $A \subset B$, $B \subset A + C$, then $\mathrm{SMin}(B, C) \neq \emptyset$ provided that (11.63) is solvable.

Assuming that A is weakly compact, we will first show that if (11.63) has solution then it coincides with $\mathrm{SMin}(A, C)$. Suppose that $y_0 \in Y$ is such that

$$\ell(y_0) = \inf_{a \in A} \ell(a), \text{ for every } \ell \in C^+.$$

We define the set

$$S_\ell := \{a \in A |\ \ell(a) \leq \ell(\tilde{a}) \quad \text{for every } \tilde{a} \in A\},$$

which, due to the weak-compactness of A, is nonempty and weakly compact. We claim that

$$\cap_{\ell \in C^+} S_\ell \neq \emptyset.$$

In view of the weak-compactness of A, it suffices to show that any finite subfamily $(S_{\ell_i})_{i \in \overline{1,n}}$ of $(S_\ell)_{\ell \in C^+}$ has a nonempty intersection. Note that,

$$S_{\ell_1 + \ldots + \ell_n} \subset \cap_{i \in \overline{1,n}} S_{\ell_i}, \text{ for all } \ell_1, \ldots, \ell_n \in C^+.$$

In fact, by setting $\ell := \ell_1 + \ldots + \ell_n$ and taking $a \in S_\ell$, we have

$$\ell_i(y_0) \leq \ell_i(a), \text{ for every } i \in \overline{1,n},$$

and by summing these inequalities term by term , we get

$$\ell(y_0) \leq \ell(a) = \ell(y_0),$$

and consequently

$$\ell_i(y_0) = \ell_i(a) = \min \ell_i(A), \text{ for every } i \in \overline{1,n}.$$

It follows that $a \in \cap_{i \in \overline{1,n}} S_{\ell_i}$ which confirms the claim that $\cap_{\ell \in C^+} S_\ell \neq \emptyset$. For $\bar{a} \in \cap_{\ell \in C^+} S_\ell \neq \emptyset$, we have

$$\ell(a - \bar{a}) \geq 0, \text{ for every } \ell \in C^+, \quad \text{for every } a \in A,$$

confirming that $a - \bar{a} \in C^{++} = C$, for every $a \in A$. Therefore, $\bar{a}$ is the strong minimum point of A. The last two parts follow immediately. □

We now proceed to give a variational characterization by taking the contingent epiderivative of the scalarization of the involved set-valued map. For the sake of a clear presentation, we only consider the case when $Y = \mathbb{R}^n$ and $C = \mathbb{R}^n_+$. A treatment of this issue in general spaces can be found in the useful contribution by Sama [520].

Given a normed space X and a set-valued map $F : X \rightrightarrows \mathbb{R}^n$, we define the following variational system associated with the scalar set-valued maps $\ell \circ F$ for $\ell \in C^+$: Find $y \in Y$ such that

$$\ell(y) = D(\ell \circ F)(\bar{x}, \ell(\bar{y}))(x) \quad \text{for any } \ell \in C^+. \tag{11.64}$$

It follows that, for contingently C-absorbing maps, we have

Lemma 11.4.28. *Let $\ell \in C^+$, let $\operatorname{SMin}(F(\bar{x}), C) = \{\bar{y}\}$, and let F be stable at $(\bar{x}, \bar{y})$. Then:*

$$\inf\{\ell(y) \mid y \in D(F + C)(\bar{x}, \bar{y})(x)\} = \inf\{\ell(y) \mid y \in DF(\bar{x}, \bar{y})(x)\} \textit{ for any}$$
$$x \in L := \operatorname{dom}(D(F + C)(\bar{x}, \bar{y})).$$

We begin with exploring some properties of set-valued maps which get transferred to their scalarized analogues.

Lemma 11.4.29. *Given a normed space X, let $F : X \rightrightarrows Y$, where $Y = \mathbb{R}^n$, be a set-valued map. Let $\ell \in Y^*$. Assume that $F(\bar{x}) = \{\bar{y}\}$ and that F is stable at $(\bar{x}, \bar{y})$. Then the following assertions hold:*

1. *If F is stable at $(\bar{x}, \bar{y})$, then $\ell \circ F$ is stable at $(\bar{x}, \ell(\bar{y}))$.*
2. *$T(\operatorname{dom}(F), \bar{x}) = \operatorname{dom}(DF(\bar{x}, \bar{y})) = \operatorname{dom}(D(\ell \circ F)(\bar{x}, \ell(\bar{y})))$.*
3. *$\ell \circ DF(\bar{x}, \bar{y})(x)) = D(\ell \circ F)(\bar{x}, \ell(\bar{y}))(x)$, for any $x \in T(\operatorname{dom}(F), \bar{x})$.*
4. *For every $x \in T(\operatorname{dom}(F), \bar{x})$, the set $DF(\bar{x}, \bar{y})(x)$ is compact.*
5. *For any $x \in T(\operatorname{dom}(F), \bar{x})$, we have*

$$\inf\{\ell(y) \mid y \in DF(\bar{x}, \bar{y})(x)\} = \inf\{y \mid y \in D(\ell \circ F)(\bar{x}, \ell(\bar{y}))(x)\} \in \mathbb{R}.$$

6. *For any $x \in L$, there exists a (weakly) compact subset $B_x \subset D(F + C)(\bar{x}, \bar{y})(x)$ such that*

$$D(F + C)(\bar{x}, \bar{y})(x) \subset B_x + C.$$

Proof. We begin with a proof of the first claim. Since the map F is stable at $(\bar{x}, \bar{y})$, there exists a neighborhood $U(\bar{x})$ and a constant $c > 0$ such that

$$y - \bar{y} \in B(0, c\,\|x - \bar{x}\|) \quad \text{for any } y \in F(x),\ x \in U \backslash \{\bar{x}\}.$$

The containment $\ell \in Y^*$ ensures that there exists a constant $k > 0$ such that

$$|\ell(y - \bar{y})| \leq k\,\|x - \bar{x}\|,$$

implying that

$$|\ell(y-\bar{y})| \leq ck \ \|x-\bar{x}\|,$$

which completes the proof of the first part.

For the second part, we set $\text{dom}\,(F) = S$ and note that the inclusion $\text{dom}\,(DF(\bar{x},\bar{y})) \subset T(S,\bar{x})$ is always true. Therefore, it suffices to show that $T(S,\bar{x}) \subset \text{dom}\,(DF(\bar{x},\bar{y}))$. Let $x \in T(S,\bar{x})$ be arbitrary. Then there are sequences $\{x_n\} \subset S$, $\{t_n\} \subset \mathbb{P}$ such that $t_n \to \infty$, $x_n \to \bar{x}$ and $t_n(x_n - \bar{x}) \to x$. Since the map F is stable at $(\bar{x},\bar{y})$ and $F(\bar{x}) = \{\bar{y}\}$, there exists $\{y_n\} \subset Y$ with $y_n \in F(x_n)$ such that $y_n \to \bar{y}$ and for sufficiently large $n \in \mathbb{N}$, we have

$$t_n(y_n - \bar{y}) \in B(0, ct_n \|x_n - \bar{x}\|). \tag{11.65}$$

Since Y is finite-dimensional, we can assume that there exists $y \in Y$ such that $t_n(y_n - \bar{y}) \to y$. Consequently, $(x,y) \in T(\text{graph}\,(F),(\bar{x},\bar{y}))$ implying that $y \in DF(\bar{x},\bar{y})(x)$ and $x \in \text{dom}\,(DF(\bar{x},\bar{y}))$. This proves that $T(\text{dom}\,(F),\bar{x}) = \text{dom}\,(DF(\bar{x},\bar{y}))$. For the second identity, it suffices to note that from the first part, the map $\ell \circ F$ is stable at $(\bar{x},\ell(\bar{y}))$.

We now proceed to prove the third part. We claim that for any $x \in T(S,\bar{x})$, we have

$$D(\ell \circ F)(\bar{x},\ell(\bar{y}))(x) \subset \ell(DF(\bar{x},\bar{y})(x)).$$

Let $y \in D(\ell \circ F)(\bar{x},\ell(\bar{y}))(y)$ be arbitrary. Then there exist $\{t_n\} \subset \mathbb{P}$ with $t_n \to \infty$, $\{(x_n,y_n)\} \subset \text{graph}\,(F)$ such that $(x_n,\ell(y_n)) \to (\bar{x},\ell(\bar{y}))$ and $t_n(\ell(y_n) - \ell(\bar{y})) \to y$. Since the map F is stable at $(\bar{x},\bar{y})$ and $\text{SMin}(F(\bar{x}),C) = \{\bar{y}\}$, we have $y_n \to \bar{y}$ and there exists $z \in Y$ such that a subsequence of $\{t_n(y_n - \bar{y})\}$ converges to z. This implies that $z \in DF(\bar{x},\bar{y})(x)$. It is clear that $y = \ell(z)$. For the converse, let $y \in DF(\bar{x},\bar{y})(x)$ be arbitrary. By the linearity and continuity of ℓ, we can show that $\ell(y) \in D(\ell \circ F)(\bar{x},\ell(\bar{y}))(x)$.

The fourth part follows from the observation that, using (11.65), we get $y \in B(0,c\,\|x\|)$ which means that $DF(\bar{x},\bar{y})(x) \subset B(0,c\,\|u\|)$.

For the fourth part, we begin by choosing $x \in T(S,\bar{x})$ arbitrarily. Since ℓ is continuous, we deduce that the set of reals $\ell(DF(\bar{x},\bar{y})(x))$ is compact and therefore $\inf\{\ell(y)|\ y \in DF(\bar{x},\bar{y})(x)\}$ is finite. Then, from the third part, we have

$$\inf\{\ell(y)|\ y \in DF(\bar{x},\bar{y})(x)\} = \inf\{y|\ y \in D(\ell \circ F)(\bar{x},\ell(\bar{y}))(x)\}.$$

Finally, we proceed to prove the last part. For $x \in T(S,\bar{x})$, we deduce from Lemma 11.4.28 that

$$D(F+C)(\bar{x},\bar{y})(u) = DF(\bar{x},\bar{y})(x) + C,$$

and hence, by taking $B_x = DF(\bar{x},\bar{y})(x)$, we have that $B_x \subset D(F+C)(\bar{x},\bar{y})(x)$ and $D(F+C)(\bar{x},\bar{y})(x) \subset B_x + C$. The proof is complete. □

We remark that the above results have been extended to Banach spaces by Sama [520]).

The following is the variational characterization by using scalarization of the involved map.

Theorem 11.4.30. *Given a normed space X, let $F : X \rightrightarrows Y = \mathbb{R}^n$ be a set-valued map. Let $\ell \in Y'$. Assume that $F(\bar{x}) = \{\bar{y}\}$ and the map F is stable at $(\bar{x}, \bar{y})$. Then $D_E F(\bar{x}, \bar{y})$ exists if and only if each system* (11.64) *has a solution for every $x \in T(\mathrm{dom}\,(F), \bar{x})$. Furthermore if it exists, $D_E F(\bar{x}, \bar{y})$ is given by the unique solution of* (11.64) *for any $x \in T(\mathrm{dom}\,(F), \bar{x})$.*

Proof. Let $\ell \in C^+$. we set $S = \mathrm{dom}\,(F)$ and define a set-valued map $\hat{F} : X \rightrightarrows Y$ by setting $\hat{F}(\bar{x}) = \{\bar{y}\}$ and $\hat{F}(x) = F(x)$ for any $x \in S\backslash\{\bar{x}\}$. Clearly, $L = \mathrm{dom}\,(D(\hat{F} + C)(\bar{x}, \bar{y})) = T(S, \bar{x})$. In fact, the inclusion $\mathrm{dom}\,(D\hat{F}(\bar{x}, \bar{y})) \subset \mathrm{dom}\,(D(\hat{F} + C)(\bar{x}, \bar{y})) \subset T(S, \bar{x})$, is always true and it follows from the above lemma that $L = T(S, \bar{x})$.

It is also evident that $(F + C) = (\hat{F} + C)$ and $(\ell \circ F + \mathbb{R}_+) = (\ell \circ \hat{F} + \mathbb{R}_+)$. Therefore, $D(\ell \circ \hat{F} + \mathbb{R}_+)(\bar{x}, \ell(\bar{y}))$ coincides with $D_E(\ell \circ F + \mathbb{R}_+)(\bar{x}, \ell(\bar{y}))$ and $D(\hat{F} + C)(\bar{x}, \bar{y})$ coincides with $D(F + C)(\bar{x}, \bar{y})$.

Note that it suffices to prove that $D_E(\ell \circ \hat{F})(\bar{x}, \ell(\bar{y}))$ exists and

$$D_E(\ell \circ \hat{F})(\bar{x}, \ell(\bar{y}))(\cdot) = \inf\{\ell(v) |\ v \in D(\hat{F} + C)(\bar{x}, \bar{y})(\cdot)\}, \tag{11.66}$$

which would imply that $D_E(\ell \circ F)(\bar{x}, \ell(\bar{y}))$ exists and

$$D_E(\ell \circ F)(\bar{x}, \ell(\bar{y}))(\cdot) = \inf\{\ell(y) |\ y \in D(F + C)(\bar{x}, \bar{y})(\cdot)\}.$$

As a consequence the family of variational systems (11.64) is equivalent to (11.63).

From the last part of Lemma 11.4.29, the map $\hat{F}$ verifies the compactness condition of Theorem 11.4.27 and hence the existence of the contingent epiderivative ensues.

Let us now prove (11.66). For $x \in L$, it follows from Lemma 11.4.29 that the map $\ell \circ \hat{F}$ is stable at $(\bar{x}, \ell(\bar{y}))$, and hence by the fifth part of Lemma 11.4.29 and by Lemma 11.4.28, we deduce that

$$\begin{aligned} \inf\{y |\ \in D(\ell \circ \hat{F})(\bar{x}, \ell(\bar{y}))(x)\} &= \inf\{\ell(y) |\ D\hat{F}(\bar{x}, \bar{y})(x)\} \\ &= \inf\{\ell(y) |\ y \in D(\hat{F} + C)(\bar{x}, \bar{y})(x)\}. \end{aligned}$$

From the previous equalities, we conclude that $D_E(\ell \circ \hat{F})(\bar{x}, \ell(\bar{x}))$ exists and verifies

$$D_E(\ell \circ \hat{F})(\bar{x}, \ell(\bar{x}))(x) = \inf\{\ell(y) |\ y \in D(\hat{F} + C)(\bar{x}, \bar{y})(u)\}, \text{ for any } x \in L,$$

and the proof is complete. □

The following example illustrates the above approach for a single-valued map.

Example 11.4.31. Let $X = \mathbb{R}$, $Y = \mathbb{R}^2$, and

$$C = \{(x, y) \in \mathbb{R}^2 | \, x - y \geq 0, \; -x \geq 0\}.$$

We set $\bar{x} = 0$ and define $f : \mathbb{R} \to \mathbb{R}^2$ by

$$f(x) = \begin{cases} (e^x \sin x, e^x \sin x) & \text{if } x \in \left\{\dfrac{1}{n}, \; n \in \mathbb{N}\right\} \\ (x^2, x^2) & \text{if } x \notin \left\{\dfrac{1}{n}, \; n \in \mathbb{N}\right\}. \end{cases}$$

The map f is stable at $\bar{x} = 0$, and the dual cone C^+ is characterized by

$$C^+ = \{(\alpha - \beta, -\alpha) | \, \alpha, \beta \in \mathbb{R}_+\}.$$

For $\ell = (\alpha - \beta, -\alpha) \in C^+$, we have

$$\ell \circ f(x) = \begin{cases} -\beta e^x \sin x & \text{if } x \in \left\{\dfrac{1}{n}, \; n \in \mathbb{N}\right\} \\ -\beta x^2 & \text{if } x \notin \left\{\dfrac{1}{n}, \; n \in \mathbb{N}\right\}, \end{cases}$$

and hence

$$D_E(\ell \circ f)(0, \ell(0))(x) = \begin{cases} -\beta x & \text{if } x > 0, \\ 0 & \text{if } x \leq 0. \end{cases}$$

Therefore, for any $x > 0$, if the contingent epiderivative $Df(\bar{x}, f(\bar{x}))(x)$ exists, then it is the solution of the following variational system: Find $(x, y) \in \mathbb{R}^2$ such that

$$(\alpha - \beta)x - \alpha y = -\beta x \quad \text{for any } \alpha, \beta \in \mathbb{R}_+.$$

It can easily be verified that $(x, y) = (u, u)$ is a solution.

Analogously, for any $x < 0$, if the contingent epiderivative $Df(\bar{x}, f(\bar{x}))(x)$ exists, then it is the solution of the following variational system: Find $(x_1, x_2) \in \mathbb{R}^2$ such that

$$(\alpha - \beta)x_1 - \alpha x_2 = 0, \; \text{for any } \alpha, \beta \in \mathbb{R}_+.$$

It is easily seen that $(x, y) = (0, 0)$ is a solution. Therefore, $Df(\bar{x}, f(\bar{x}))$ exists and is given by

$$Df(\bar{x}, f(\bar{x}))(x) = \begin{cases} (x, x) & \text{if } x > 0, \\ (0, 0) & \text{if } x \leq 0. \end{cases}$$

11.5 Generalized Contingent Epiderivatives of Set-Valued Maps

The existence theory for the contingent epiderivative of set-valued maps acting between general spaces requires strong conditions on the data. This motivates the introduction of the generalized contingent epiderivatives for which the existence theorems can be given under more general conditions. This section is devoted to the generalized contingent epiderivative and two other notions of epiderivatives which are very closely related to it. Besides some simple examples, we give existence theorems and characterizations of the generalized contingent epiderivative and its variants. Our treatment of the subject is based on [297], however, we have also benefited from the contributions by Chen and Jahn [92] and Bednarczuk and Song [39], where the notion of the generalized contingent epiderivative was originally introduced.

We begin with the following:

Definition 11.5.1. Let X and Y be normed spaces, let $C \subset Y$ be a pointed, closed and convex cone, let $F : X \rightrightarrows Y$ be a set-valued map, and let $(\bar{x}, \bar{y}) \in \text{graph}\,(F)$. Define a set-valued map $G : X \rightrightarrows Y$ by

$$G(x) := D(F + C)(\bar{x}, \bar{y})(x) \quad \text{for every } x \in L := \text{dom}\,(D(F + C)(\bar{x}, \bar{y})). \tag{11.67}$$

1. The **generalized contingent epiderivative** of F at $(\bar{x}, \bar{y})$ is the set-valued map $D_G F(\bar{x}, \bar{y}) : X \rightrightarrows Y$ defined by

$$D_G F(\bar{x}, \bar{y})(x) = \text{Min}(G(x), C).$$

2. The **generalized proper contingent epiderivative** of F at $(\bar{x}, \bar{y})$ is the set-valued map $D_P F(\bar{x}, \bar{y}) : X \rightrightarrows Y$ defined by

$$D_P F(\bar{x}, \bar{y})(x) = \text{GHe-PMin}(G(x), C),$$

where $\text{GHe-PMin}(G(x), C)$ represents the set of all Henig global properly minimal points.

3. The **generalized weak contingent epiderivative** of F at $(\bar{x}, \bar{y})$ is the set-valued map $D_W F(\bar{x}, \bar{y}) : X \rightrightarrows Y$ defined by

$$D_W F(\bar{x}, \bar{y})(x) = \text{WMin}(G(x), C),$$

provided that the ordering cone C is solid.

In view of the relationship among $\text{GHe-PMin}(\cdot, \cdot)$, $\text{Min}(\cdot, \cdot)$ and $\text{WMin}(\cdot, \cdot)$, the following inclusions are immediate:

$$D_P F(\bar{x}, \bar{y})(\cdot) \subseteq D_G F(\bar{x}, \bar{y})(\cdot) \subseteq D_W F(\bar{x}, \bar{y})(\cdot). \tag{11.68}$$

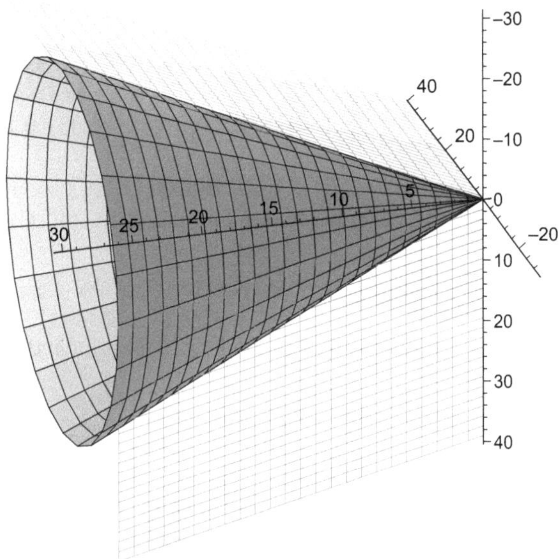

Fig. 11.5 (*i*) A view of epi (F). (*ii*) A contradiction to $(-1,0) \in$ GHe-PMin$(G(1), \mathbb{R}^2_+)$

In order to show that the above inclusion is strict in general and to illustrate the above notions, we give some simple examples. We begin with an example showing that $D_P F(\bar{x}, \bar{y})(\cdot)$ is a proper subset of $D_G F(\bar{x}, \bar{y})(\cdot)$.

Example 11.5.2. Consider a set-valued map $F : \mathbb{R} \rightrightarrows \mathbb{R}^2$ defined as follows:

$$F(x) := \begin{cases} \emptyset & \text{if } x < 0, \\ \{(y_1, y_2) \in \mathbb{R}^2 |\ y_1^2 + y_2^2 \leq x^2\} & \text{if } x \geq 0. \end{cases}$$

Let $\bar{x} = 0$ and let $\bar{y} = (0,0)$. We note that (see Fig. 11.5 (i))

$$T(\text{epi}\,(F), (\bar{x}, \bar{y})) = \text{epi}\,(F),$$

and hence the map $G : \mathbb{R}_+ \rightrightarrows \mathbb{R}^2$ defined in (11.67) is given by:

$$G(x) = \{(y_1, y_2) \in \mathbb{R}^2 |\ (y_1, y_2) \in F(x) + \mathbb{R}^2_+\}.$$

Consequently, the generalized contingent epiderivative $D_G F(\bar{x}, \bar{y}) : \mathbb{R}_+ \rightrightarrows \mathbb{R}^2$ is defined by:

$$D_G F(\bar{x}, \bar{y})(x) = \left\{(y_1, y_2) \in \mathbb{R}^2 |\ y_1^2 + y_2^2 = x^2,\ y_1 \leq 0,\ y_2 \leq 0\right\}.$$

Let $x = 1$. Then $(-1,0) \in D_G F(\bar{x}, \bar{y})(1)$. We claim that $(-1,0) \notin D_P F(\bar{x}, \bar{y})(1)$. We have (see Fig. 11.5 (ii))

$$T(G(1) + \mathbb{R}^2_+, (-1, 0)) = \{(y_1, y_2) \in \mathbb{R}^2 |\ y_1 \geq 0,\ y_2 \in \mathbb{R}\},$$

and therefore

$$T(G(1) + \mathbb{R}^2_+, (-1, 0)) \cap (-\mathbb{R}^2_+) = \{(y_1, y_2) \in \mathbb{R}^2 |\ y_1 = 0,\ y_2 \leq 0\} \neq \{0\},$$

confirming that $(-1, 0) \notin$ Bo-PMin$(G(1), \mathbb{R}^2_+) =$ GHe-PMin$(G(1), \mathbb{R}^2_+)$.

In fact, the generalized proper contingent epiderivative at $x = 1$ is given by:

$$D_P F(\bar{x}, \bar{y})(1) = \left\{(y_1, y_2) \in \mathbb{R}^2 | \quad y_1^2 + y_2^2 = 1,\ y_1 < 0,\ y_2 < 0\right\}$$

which is a proper subset of $D_G F(\bar{x}, \bar{y})(1)$. Indeed this is true for all $x \in \mathbb{R}_+/\{0\}$. However

$$D_P F(\bar{x}, \bar{y})(0) = D_G F(\bar{x}, \bar{y})(0) = \{0\}. \qquad \square$$

Next we give an example showing that $D_G F(\bar{x}, \bar{y})(\cdot)$ is a proper subset of $D_W F(\bar{x}, \bar{y})(\cdot)$. This example also makes it clear that the set $D_W F(\bar{x}, \bar{y})(\cdot) \setminus D_G F(\bar{x}, \bar{y})(\cdot)$ may be very large.

Example 11.5.3. Consider a set-valued map $F : \mathbb{R} \rightrightarrows \mathbb{R}^2$ defined as follows:

$$F(x) = \begin{cases} \emptyset & \text{if } x < 0 \\ \left\{(y_1, y_2) \in \mathbb{R}^2 |\ y_1,\ y_2 \in [0, x]\right\} & \text{if } x \geq 0. \end{cases}$$

Let $\bar{x} = 0$ and let $\bar{y} = (0, 0)$. We note that

$$T(\text{epi}\,(F), (\bar{x}, \bar{y})) = \{(x, (y_1, y_2)) \in \mathbb{R}^3 |\, x \geq 0,\ y_1 \geq 0,\ \ y_2 \geq 0\},$$

and hence the map $G : \mathbb{R}_+ \rightrightarrows \mathbb{R}^2$ defined in (11.67) is given by:

$$G(x) = \left\{(y_1, y_2) \in \mathbb{R}^2 |\ y_1 \geq 0,\ y_2 \geq 0\right\}.$$

Consequently, the generalized contingent epiderivative $D_G F(\bar{x}, \bar{y}) : \mathbb{R}_+ \rightrightarrows \mathbb{R}^2$ is given by

$$D_G F(\bar{x}, \bar{y})(x) := \{(0, 0)\},$$

and the weak contingent epiderivative $D_W F(\bar{x}, \bar{y}) : \mathbb{R}_+ \rightrightarrows \mathbb{R}^2$ is given by (see Fig. 11.6 (i)):

$$D_W F(\bar{x}, \bar{y})(x) = \left\{(0, y_2) \in \mathbb{R}^2 |\ \ y_2 \geq 0\right\} \bigcup \left\{(y_1, 0) \in \mathbb{R}^2 |\ \ y_1 \geq 0\right\},$$

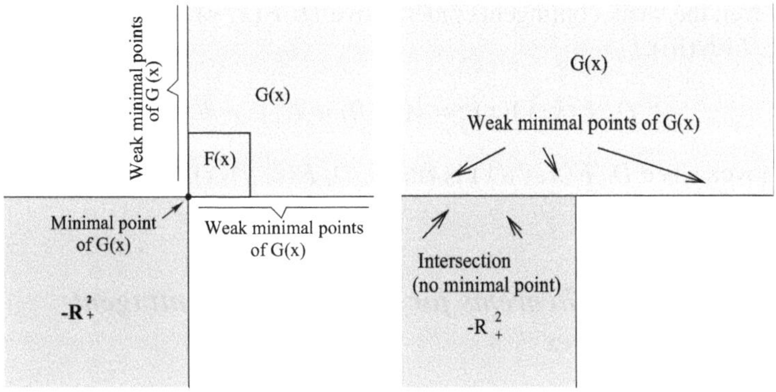

Fig. 11.6 (*i*) Example 11.5.3: Minimal points. (*ii*) Example 11.5.4: Only weakly minimal points

depicting the fact that the generalized weak contingent epiderivative $D_W F(\bar{x}, \bar{y})(x)$ is much larger than $D_G F(\bar{x}, \bar{y})(x)$ for all $x \in \operatorname{dom}(D_G F(\bar{x}, \bar{y})) = \operatorname{dom}(D_W F(\bar{x}, \bar{y})) = \mathbb{R}_+$. □

By virtue of the connections among properly minimal, minimal and weakly minimal points, it is clear that when $D_P F(\bar{x}, \bar{y})(\cdot)$ or $D_G F(\bar{x}, \bar{y})(\cdot)$ exists, then $D_W F(\bar{x}, \bar{y})(\cdot)$ also exists, provided that the ordering cone C has a nonempty interior. However, it is possible that, out of these three epiderivatives, only $D_W F(\bar{x}, \bar{y})(x)$ exists. Such a situation is furnished by the following example.

Example 11.5.4. Consider a set-valued map $F : \mathbb{R} \rightrightarrows \mathbb{R}^2$ defined as follows:

$$F(x) = \begin{cases} \emptyset & \text{if } x < 0 \\ \{(y_1, -y_1^2) \in \mathbb{R}^2 |\ 0 \le y_1 \le x\} \bigcup \{(y_1, y_2) \in \mathbb{R}^2 |\ y_1 \le 0,\ y_2 \ge 0\} & \text{if } x \ge 0. \end{cases}$$

Let $\bar{x} = 0$ and let $\bar{y} = (0, 0)$. We note that

$$T(\operatorname{epi}(F), (\bar{x}, \bar{y})) = \{(x, (y_1, y_2)) \in \mathbb{R}^3 |\ x \ge 0,\ \ y_2 \ge 0,\ \ y_1 \in \mathbb{R}\},$$

and hence the map $G : \mathbb{R}_+ \rightrightarrows \mathbb{R}^2$ defined in (11.67) is given by:

$$G(x) := \{(y_1, y_2) \in \mathbb{R}^2 |\ y_2 \ge 0,\ \ y_1 \in \mathbb{R}\}.$$

Consequently, the generalized contingent epiderivative $D_G F(\bar{x}, \bar{y}) : \mathbb{R}_+ \rightrightarrows \mathbb{R}^2$ is not defined as

$$D_G F(\bar{x}, \bar{y})(x) = \{\emptyset\}.$$

However, the weak contingent epiderivative $D_w F(\bar{x}, \bar{y}) : \mathbb{R}_+ \rightrightarrows \mathbb{R}^2$ is given by (see Fig. 11.6 (ii)):

$$D_w F(\bar{x}, \bar{y})(x) := \left\{(y_1, 0) \in \mathbb{R}^2_+ \mid \; y_1 \in \mathbb{R}\right\}.$$

Moreover, since $D_G F(\bar{x}, \bar{y})(x)$ is empty, $D_P F(\bar{x}, \bar{y})(x)$ is empty as well. □

11.5.1 Existence Theorems for Generalized Contingent Epiderivatives

To give existence results for the generalized contingent epiderivatives, we need to recall some notions and results (see Chap. 9). Let Y be a real Hausdorff topological vector space. We formulate the following condition:

$(P1)$ Every C-increasing and C-upper bounded net $(x_i)_{i \in I} \subseteq C$ is convergent to an element of C.

Furthermore, following Németh [440], a convex cone C is called *regular* if it satisfies $(P1)$.

The following result is needed for our next result.

Proposition 11.5.5. *Let $D \subset Y$ be nonempty. Assume that D is closed and C-lower bounded, C satisfies the condition $(P1)$. Then* $\mathrm{Min}(D, C) \neq \emptyset$ *and the domination property holds, that is,* $D \subseteq \mathrm{Min}(D, C) + C$.

A proof of the above result as well as many other results concerning the existence of minimal points, which could be useful in proving the existence of the generalized contingent epiderivatives, can be found in Chap. 9.

We now give an existence theorem for the generalized contingent epiderivatives and the generalized weak contingent epiderivatives.

Theorem 11.5.6. *Let X and Y be normed spaces and let $C \subset Y$ be a closed, convex, and regular cone. Let $F : X \rightrightarrows Y$ be a set-valued map and let $(\bar{x}, \bar{y}) \in$* graph(F)*. Let, for every $x \in$* dom(G)*, the set-valued map G given by* (11.67)*, be C-lower bounded. Then, for every $x \in$* dom(G)*, the generalized contingent epiderivative $D_G F(\bar{x}, \bar{y})(x)$ exists. Moreover, the following identity holds:*

$$\mathrm{epi}\,(D_G F(\bar{x}, \bar{y})) = T(\mathrm{epi}\,(F), (\bar{x}, \bar{y})). \tag{11.69}$$

If additionally $\mathrm{int}(C) \neq \emptyset$*, then for every $x \in$* dom(G)*, the generalized weak contingent epiderivative $D_W F(\bar{x}, \bar{y})(x)$ also exists and the following identity holds:*

$$\mathrm{epi}\,(D_W F(\bar{x}, \bar{y})) = T(\mathrm{epi}\,(F), (\bar{x}, \bar{y})). \tag{11.70}$$

Proof. Since the contingent cone is always closed, the set $G(x)$ is closed for every $x \in \operatorname{dom}(G)$. It has a C-lower bound by the assumption. Therefore, the existence of $D_G F(\bar{x}, \bar{y})(x)$ now follows from Proposition 11.5.5 which ensures the existence of minimal points for $G(x)$ with respect to the convex cone C. Moreover, it follows from the same result that $G(x)$ has the domination property, that is

$$G(x) \subseteq D_G F(\bar{x}, \bar{y})(x) + C. \tag{11.71}$$

We claim that equality holds in the above inclusion. For this, let $y \in D_G F(\bar{x}, \bar{y})(x) + C$ be arbitrary. Then $y \in G(x) + C$. Now making use of the fact that $G(x) = G(x) + C$, we obtain $y \in G(x)$. This confirms the inclusion

$$G(x) \supseteq D_G F(\bar{x}, \bar{y})(x) + C.$$

Therefore

$$D_G F(\bar{x}, \bar{y})(x) + C = G(x) = D(F + C)(\bar{x}, \bar{y})(x),$$

where $D(F + C)(\bar{x}, \bar{y})(x)$ is the contingent derivative of the profile map $(F + C)$. Furthermore, the above identity and the definition of the contingent derivative implies

$$\begin{aligned} \operatorname{epi}(D_G F(\bar{x}, \bar{y})) &= \operatorname{graph}(D(F + C)(\bar{x}, \bar{y})) \\ &= T(\operatorname{graph}(F + C), (\bar{x}, \bar{y})) \\ &= T(\operatorname{epi}(F), (\bar{x}, \bar{y})), \end{aligned}$$

which confirms the claim.

The existence of $D_W F(\bar{x}, \bar{y})(x)$ follows from the existence of $D_G F(\bar{x}, \bar{y})(x)$ and the inclusion $D_G F(\bar{x}, \bar{y})(x) \subseteq D_W F(\bar{x}, \bar{y})(x)$ (see ((11.68)). It remains to prove that the identity (11.70) holds. In fact, in view of the inclusions (11.68) and (11.70), it follows that

$$G(x) \subseteq D_W F(\bar{x}, \bar{y})(x) + C.$$

Moreover, the reverse inclusion follows from arguments similar to those in the proof of (11.69). The proof is complete. □

Remark 11.5.7. Note that if the generalized continent epiderivative $D_G F(\bar{x}, \bar{y})$ of F at $(\bar{x}, \bar{y})$, then (11.69) holds if and only if the domination property holds for the set $D(F+C)(\bar{x}, \bar{y})(x)$ for any $x \in \operatorname{dom}(G)$. In fact, note that if $\operatorname{epi}(D_G F(\bar{x}, \bar{y})) \neq T(\operatorname{epi}(F), (\bar{x}, \bar{y}))$ then there exists (x, y) such that $(x, y) \in T(\operatorname{epi}(F), (\bar{x}, \bar{y}))$ and $(x, y) \notin \operatorname{epi}(D_G F(\bar{x}, \bar{y}))$. This means that $y \in D(F + C)(\bar{x}, \bar{y})(x)$ but $y \notin D_G F(\bar{x}, \bar{y})(x) + C$, contradicting domination property holds for $D(F + C)(\bar{x}, \bar{y})(x)$. See [500].

The relations (11.69) and (11.70) play a fundamental role in set-valued optimization. The main reason for this is the fact that these identities enable the use of several properties of the contingent cones to obtain various conclusions about the corresponding epiderivatives. It should also be clear that in deriving the identities (11.69) and (11.70), the domination property (11.70) has played a decisive role.

We note that if the weak contingent epiderivative exists, then we can prove (11.70) by assuming the following **weak-domination property**:

$$G(x) \subseteq \mathrm{WMin}(G(x), C) + C.$$

It is easy to check that the weak-domination property is much weaker than the domination property.

The following existence result can be useful when less information is available about the ordering cone C. Recall that given a set $A \subset Y$ and a point $s \in Y$, the C-lower section of A at s, denoted by A_s, is given by the set $A \cap (s - C)$.

Theorem 11.5.8. *Let X and Y be normed spaces and let $C \subset Y$ be a pointed, closed, and convex cone. Let $F : X \rightrightarrows Y$ be a set-valued map and let $(\bar{x}, \bar{y}) \in$ graph (F). Assume that for each $x \in$ dom (G), there exists $s \in Y$ such that the C-lower section $G_s(x)$ is nonempty and compact. Then for every $x \in$ dom (G), the generalized contingent epiderivative $D_G F(\bar{x}, \bar{y})(x)$ exists. Moreover, if* $\mathrm{int}(C) \neq \emptyset$ *then the generalized weak contingent epiderivative $D_W F(\bar{x}, \bar{y})(x)$ also exists.*

Proof. The existence of $D_G F(\bar{x}, \bar{y})(x)$ (and hence of $D_W F(\bar{x}, \bar{y})(x)$) follows from [57, Theorem 1], which assures the existence of minimal points for $G(x)$ with respect to the cone C. □

The following existence theorem in a finite-dimensional setting is also of importance.

Theorem 11.5.9. *Let X be a normed space and let $C \subset \mathbb{R}^n$ be a pointed, closed, and convex cone. Let $F : X \rightrightarrows \mathbb{R}^n$ be a set-valued map and let $(\bar{x}, \bar{y}) \in$ graph (F). Let for each $x \in$ dom (G), the set $G(x)$, given by* (11.67)*, be convex and let $s \in \mathbb{R}^n$ be such that the C-lower section $G_s(x)$ is nonempty and compact. Then, for every $x \in$ dom (G), the epiderivatives $D_P F(\bar{x}, \bar{y})(x)$, $D_G F(\bar{x}, \bar{y})(x)$ and $D_W F(\bar{x}, \bar{y})(x)$ exist. Moreover,*(11.69) *and* (11.70) *hold.*

Proof. The existence of $D_G F(\bar{x}, \bar{y})(x)$ and $D_W F(\bar{x}, \bar{y})(x)$ follow from Theorem 11.5.8 and the existence of $D_P F(\bar{x}, \bar{y})(x)$ follow from Luc [403, Corollary 3.16]. Recall that in $\mathbb{R}^n$ every closed convex and pointed cone has a compact base (see, for example, [126]). It follows from Lemma 11.5.5 (2) that the hypotheses imposed on $G(x)$ together with its closedness assures the domination property. The rest of the proof is similar to the proof of Theorem 11.5.6. □

In the following, we give some important properties of the generalized contingent epiderivatives. In these results, the equations (11.69) and (11.70) play an important role.

Theorem 11.5.10. *If, besides the assumptions of Theorem 11.5.6 or Theorem 11.5.9, we suppose that the contingent cone $T(\text{epi}\,(F), (\bar{x}, \bar{y}))$ is convex, then the generalized contingent epiderivative $D_G F(\bar{x}, \bar{y})(\cdot)$ is a subadditive set-valued map. That is*

$$D_G F(\bar{x}, \bar{y})(x) + D_G F(\bar{x}, \bar{y})(y) \subset D_G F(\bar{x}, \bar{y})(x + y) + C \quad \text{for all}$$
$$x, y \in \text{dom}\,(D_G F(\bar{x}, \bar{y})).$$

Proof. It follows immediately from (11.69) and the convexity of $T(\text{epi}\,(F), (\bar{x}, \bar{y}))$. □

Since

$$D_G F(\bar{x}, \bar{y})(\lambda x) = \lambda D_G F(\bar{x}, \bar{y})(x)$$

for every $x \in \text{dom}\,(D_G F(\bar{x}, \bar{y}))$ and for every $\lambda > 0$, the above result states that the generalized contingent epiderivative $D_G F(\bar{x}, \bar{y})(\cdot)$ is a strictly sublinear set-valued map. That is, for all $x, y \in \text{dom}\,(D_G F(\bar{x}, \bar{y}))$, we have

$$\lambda D_G F(\bar{x}, \bar{y})(x) + \beta D_G F(\bar{x}, \bar{y})(y) \subset D_G F(\bar{x}, \bar{y})(\lambda x + \beta y) + C \quad \lambda > 0,\ \beta > 0.$$

We have the following:

Corollary 11.5.11. *Under the assumptions of Theorem 11.5.6 or Theorem 11.5.9, the following holds:*

$$\text{epi}\,(D_W F(\bar{x}, \bar{y})) = \text{epi}\,(D_G F(\bar{x}, \bar{y})).$$

Proof. This is a direct consequence of (11.69) and (11.70). □

It is well-known that single-valued functionals are identical if and only if their epigraphs are identical, provided that the ordering cone is pointed. However, contrary to this, two set-valued maps may not be identical even if their epigraphs are identical. Clearly, Corollary 11.5.11 does not imply that $D_G F(\bar{x}, \bar{y}) \equiv D_W F(\bar{x}, \bar{y})$. Indeed the difference between $D_G F(\bar{x}, \bar{y})$ and $D_W F(\bar{x}, \bar{y})$ can be very large (see Example 11.5.4).

Corollary 11.5.12. *Under the assumptions of Theorem 11.5.6 or Theorem 11.5.9, the following holds:*

$$\text{dom}\,(D_W F(\bar{x}, \bar{y})) \subset T(\text{dom}\,(F), \bar{x}),$$
$$\text{dom}\,(D_G F(\bar{x}, \bar{y})) \subset T(\text{dom}\,(F), \bar{x}).$$

Proof. We only prove the second inclusion because the first follows directly from Theroem 11.5.11. Let $(u, v) \in \text{epi}\,(D_G F(\bar{x}, \bar{y}))$. Therefore $(u, v) \in T(\text{epi}\,(F), (\bar{x}, \bar{y}))$ and hence there are sequences $\{t_n\} \subset \mathbb{P}$ and $\{(u_n, v_n)\}$ with $t_n \downarrow 0$ and $(u_n, v_n) \to (u, v)$ such that

$$\bar{y} + t_n v_n \in F(\bar{x} + t_n u_n) + C.$$

Therefore $\bar{x} + t_n u_n \in \mathrm{dom}\,(F)$ implying that $(u, v) \in T(\mathrm{dom}\,(F), \bar{x}) \times Y$. The proof is complete. □

The following result connects the contingent derivative and the generalized contingent derivative.

Theorem 11.5.13. *Let X and Y be normed spaces and let $C \subset Y$ be a pointed, closed, and convex cone. Let $F : X \rightrightarrows Y$ be a set-valued map and let $(\bar{x}, \bar{y}) \in$ graph (F). Let*

$$\mathrm{SMin}(G(x), C) \neq \emptyset \quad \text{for every } x \in \mathrm{dom}\,(G).$$

Then the contingent epiderivative $D_{\mathrm{E}} F(\bar{x}, \bar{y})(\cdot)$ and the generalized contingent epiderivative $D_{\mathrm{G}} F(\bar{x}, \bar{y})(\cdot)$ coincide.

Proof. Since the set $\mathrm{SMin}(G(x), C)$ is nonempty, it coincides with $\mathrm{Min}(G(x), C)$. By the assumption, the convex cone C is pointed and therefore $\mathrm{Min}(G(x), C)$ is a singleton set. It is clear from the definition of a strongly minimal point that it satisfies the domination property and hence we deduce that (11.69) holds. Consequently, $D_G F(\bar{x}, \bar{y})(\cdot)$ is a single-valued map satisfying (11.69). However, (11.69) is also satisfied by the contingent epiderivative $D_E F(\bar{x}, \bar{y})(\cdot)$, which under the imposed conditions is unique. This completes the proof. □

In view of the above theorems, we make a brief discussion of Example 11.4.6–11.5.4. It is easy to verify that, for $G(x)$ given in Example 11.4.6, all the hypotheses of Theorem 11.5.6-11.5.9 are satisfied. Hence the existence of $D_G F(\bar{x}, \bar{y})(x)$ is assured by these theorems. The same is the case for Example 11.5.3. However, for $G(x)$ defined in Example 11.5.4, there is no $\mathbb{R}^2_+$-lower bound. Neither does it have any compact section. Hence, the above theorems do not assure the existence of $D_G F(\bar{x}, \bar{y})(\cdot)$. Indeed, we have seen that $D_G F(\bar{x}, \bar{y})(\cdot)$ does not exist.

11.5.2 Characterizations of Generalized Contingent Epiderivatives

We begin with the following simple result.

Theorem 11.5.14. *Let X and Y be normed spaces, let $C \subset Y$ be pointed, closed, and convex cone, let $F : X \rightrightarrows Y$ be a set-valued map, and let $(\bar{x}, \bar{y}) \in$ graph (F). If for some $x \in X$, the proper contingent epiderivative $D_{\mathrm{P}} F(\bar{x}, \bar{y})(x)$ exists, then $D_{\mathrm{P}} F(\bar{x}, \bar{y})(x)$ and $D_{\mathrm{G}} F(\bar{x}, \bar{y})(x)$ belong to the boundary of the contingent derivative $D(F + C)(\bar{x}, \bar{y})(x)$.*

Proof. Assume that for some $x \in X$ the proper contingent epiderivative $D_P F(\bar{x}, \bar{y})(x)$ exists. Since $D(F + C)(\bar{x}, \bar{y})(x)$ is closed, it suffices to show

$$\{D_P F(\bar{x}, \bar{y})(x), D_G F(\bar{x}, \bar{y})(x)\} \not\subseteq \operatorname{int}(D(F + C)(\bar{x}, \bar{y})(x)).$$

Because $D_P F(\bar{x}, \bar{y})(x) \subseteq D_G F(\bar{x}, \bar{y})(x)$, we will prove the statement for $D_G F(\bar{x}, \bar{y})(x)$ only. Let $y \in D_G F(\bar{x}, \bar{y})(x)$ be arbitrary. Clearly $y \in D(F + C)(\bar{x}, \bar{y})(x)$. Let $y \in \operatorname{int}(D(F + C)(\bar{x}, \bar{y})(x))$. Then there exists $\varepsilon > 0$ so that $B_\varepsilon(y)(= y + B_\varepsilon(0)) \subset D(F + C)(\bar{x}, \bar{y})(x)$, where $B_\varepsilon(y)$ is an open ball of radius ε at y. $B_\varepsilon(0)$ is absorbing, that is, for all $z \in Y$ there exists some $\ell_0 > 0$ such that $|\ell| \leq \ell_0$ implies $\ell z \in B_\varepsilon(0)$. For an arbitrary $c \in C\backslash\{0\}$, there exists μ such that $\mu c \in B_\varepsilon(0)$ and $y - \mu c \in D(F + C)(\bar{x}, \bar{y})(x)$, implying $y \notin D_G F(\bar{x}, \bar{y})(x)$. This contradiction finishes the proof. □

For the next result, we need to recall the following (see Borwein [57]):

Theorem 11.5.15. *Let Z be a locally convex space, let P and Q be cones in Z such that $P \cap Q = \{0\}$. Assume that one of the following conditions hold:*

1. *P is weakly-closed and Q has a weakly-compact base.*
2. *P is closed and Q has a compact base.*

Then there is a pointed, convex cone R such that $Q\backslash\{0\} \subset \operatorname{int}(R)$ and $P \cap R = \{0\}$.

We begin with the following auxiliary result.

Theorem 11.5.16. *Let Y be a normed space, let $C \subset Y$ be pointed, closed, and convex cone, and let $D \subset Y$ be arbitrary. Then* GHe-PMin$(D, C) \subseteq$ Bo-PMin(D, C). *Furthermore, if either* $\operatorname{cl}(D + C)$ *is convex and C has a weakly-compact base or C has a compact base, then the converse holds.*

Proof. Let $\bar{y} \in$ GHe-PMin(D, C), that is, there exists a pointed convex cone $C_1 \neq Y$ such that $C\backslash\{0\} \subseteq \operatorname{int}(C_1)$ and $\bar{y} \in \operatorname{Min}(D, C_1)$. Let $\bar{y} \notin$ Bo-PMin(D, C) and let $y \in T(D + C, \bar{y}) \cap (-C\backslash\{0\})$. There are $\{\lambda_n\} \downarrow 0$ and $\{y_n\} \to y$ such that $\bar{y} + \lambda_n y_n \in D + C$. Since $y \in -C\backslash\{0\} \subseteq -\operatorname{int}(C_1)$, there exists $N \in \mathbb{N}$ such that $y_n \in -\operatorname{int}(C_1)$, $\forall n \geq N$. Since $\lambda_n > 0$ we deduce that $\lambda_n y_n \in -\operatorname{int}(C_1)$, for every $n \geq N$. Let $w_n \in D$ be such that $\bar{y} + \lambda_n y_n \in w_n + C$ which implies

$$w_n \in \bar{y} + \lambda_n y_n - C \subset \bar{y} - \operatorname{int}(C_1) - C \subset \bar{y} - \operatorname{int}(C_1) - C_1\backslash\{0\} - C_1 \subset \bar{y} - C_1\backslash\{0\}.$$

Therefore $\bar{y} \in w_n + C_1\backslash\{0\}$, which contradicts that $\bar{y} \in \operatorname{Min}(D, C_1)$.

Now, under the additional hypotheses, we show the converse. Let $\bar{y} \in$ Bo-PMin(D, C), that is, $T(D + C, \bar{y}) \cap (-C) = \{0\}$. If $cl(D + C)$ is convex, $T(D + C, \bar{y}) = T(\operatorname{cl}(D + C), \bar{y})$ is convex as well, and hence weakly closed. Therefore, by virtue of Proposition 11.5.15, there is a pointed convex cone C_1 such that $-C\backslash\{0\} \subset -\operatorname{int}(C_1)$ and $T(D+C, \bar{y}) \cap (-C_1) = \{0\}$. Now taking into account the convexity of the set $\mathrm{cl}\,(D + C)$, we conclude that $(D - \bar{y}) \cap (-C_1\setminus\{0\}) = \emptyset$. This implies that $\bar{y} \in \mathrm{Min}(D, C_1)$ and consequently $\bar{y} \in$ GHe-PMin(D, C). □

Recall that for a convex cone $C \subset Y$, the quasi-interior $C^{\#}$ of the dual C^+ of C is given by

$$C^{\#} = \{\ell \in Y^* |\ \ell(x) > 0 \ \ \forall x \in C\setminus\{0\}\}.$$

In the following, we give a characterization of the proper contingent epiderivative.

Theorem 11.5.17. *Let X and Y be normed spaces, let $C \subset Y$ be pointed, closed, and convex cone with a compact base, let $F : X \rightrightarrows Y$ be a set-valued map, and let $(\bar{x}, \bar{y}) \in \mathrm{graph}\,(F)$. Let $T(\mathrm{epi}\,(F), (\bar{x}, \bar{y}))$ be convex. Assume that for some $x \in X$, the proper contingent epiderivative $D_{\mathrm{P}} F(\bar{x}, \bar{y})(x)$ exists. Then*

$$D_{\mathrm{P}} F(\bar{x}, \bar{y})(x) = \bigcup_{\ell \in C^{\#}} \ell(G(x))^-, \tag{11.72}$$

where the map G is given by (11.67) *and* $\ell(G(x))^- := \{y \in Y \,|\, \ell(y) \leq \ell(G(x))\}$.

Proof. Let $y^* \in \cup_{\ell \in C^{\#}} \ell(G(x))^-$. Then there exists some $\ell_0 \in C^{\#}$ such that

$$\ell_0(y^*) \leq \ell_0(y), \quad \forall\, y \in G(x). \tag{11.73}$$

Let $y^* \notin D_P F(\bar{x}, \bar{y})(x)$. In view of Proposition 11.5.16 we have that $y^* \notin$ Bo-PMin$(G(x), C)$, where $G(x)$ is defined by (11.5.1). Let $w \in T(G(x)+C, y^*) \cap (-C\setminus\{0\})$ be arbitrary. Since $G(x) + C = G(x)$, we have $w \in T(G(x), y^*)$. Therefore, there are a sequences $\{\lambda_n\} \subset \mathbb{P}$ and $\{w_n\} \subset G(x)$ such that $\lambda_n(w_n - y^*) \to w$ and $w_n \to y^*$. In view of (11.73), we have $\ell_0(y^*) \leq \ell_0(w_n)$ for every $n \in \mathbb{N}$ and accordingly $\lim_{n\to\infty} \lambda_n(\ell_0(w_n) - \ell_0(y^*)) \geq 0$. Consequently $\ell_0(w) \geq 0$ for each $w \in T(G(x), y^*) \cap (-C\setminus\{0\})$. This contradiction to $\ell_0(w) < 0$ confirms that $y^* \in D_P F(\bar{x}, \bar{y})(x)$. Hence $\cup_{\ell \in C^{\#}} \ell(G(x) \subseteq D_P F(\bar{x}, \bar{y})(x)$.

For the converse, let $y^* \in D_P F(\bar{x}, \bar{y})(x)$. Since $T(\mathrm{epi}\,(F), (\bar{x}, \bar{y}))$ is convex, $G(x)$ is convex for all $x \in \mathrm{dom}\,(G)$ and consequently $T(G(x) + C, y^*)$ is convex. We have

$$T(G(x) + C, y^*) = T(G(x), y^*) \supseteq G(x) - y^*.$$

From the definition of y^* and Proposition 11.5.16 we have

$$T(G(x), y^*) \cap (-C\setminus\{0\}) = \emptyset.$$

Therefore, by a cone separation theorem (see [426] [293]), there exists $\ell_0 \in C^{\#}$ such that $\ell_0 \in [T(G(x), y^*)]^+$, or, $\ell_0(G(x) - y^*) \geq 0$, which ensures $\ell_0(y^*) \leq \ell_0(G(x))$. The proof is complete. □

In the following we characterize the weak contingent epiderivatives.

Theorem 11.5.18. *Let X, Y, F, $\ell(G(x))^-$ and $(\bar{x},\bar{y})$ be as in Theorem 11.5.17. Let* $\text{int}(C) \neq \emptyset$. *Assume that for some $x \in X$, the weak contingent epiderivative $D_{\text{w}} F(\bar{x},\bar{y})(x)$ exists. Then:*

$$D_{\text{w}} F(\bar{x},\bar{y})(x) \supseteq \bigcup_{\ell \in C^+ \setminus \{0\}} \ell(G(x))^-. \tag{11.74}$$

Moreover, if $T(\text{epi}\,(F),(\bar{x},\bar{y}))$ *is convex, then equality holds in* (11.74).

Proof. Since $D_{\text{w}} F(\bar{x},\bar{y})(x) = \text{WMin}(G(x),C)$, we have $G(x) \cap (y - \text{int}(C)) = \emptyset \quad \forall y \in D_{\text{w}} F(\bar{x},\bar{y})(x)$. Hence the standard separation theorems can be applied. The arguments are quite similar to that of the previous theorems and hence omitted. □

We now proceed to give some connections between the epiderivatives and the contingent derivatives. For this we the recall the following result (see also Jahn [293]).

Lemma 11.5.19. *Let Y be a normed space, let $C \subset Y$ be pointed, closed, and convex cone, and let A and B be nonempty subsets of Y. Assume that $A + C = B$. Then the following relationships hold:*

(a) $\text{GHe-PMin}(A,C) = \text{GHe-PMin}(B,C)$.
(b) $\text{Min}(A,C) = \text{Min}(B,C)$.
(c) $\text{WMin}(A,C) \subseteq \text{WMin}(B,C)$.

We have the following result:

Theorem 11.5.20. *Let X and Y be normed spaces and let $C \subset Y$ be pointed, closed, and convex cone. Let $F : X \rightrightarrows Y$ be a set-valued map and let $(\bar{x},\bar{y}) \in$* graph (F). *Assume that for $x \in X$ the epiderivatives $D_{\text{P}} F(\bar{x},\bar{y})(x)$, $D_{\text{G}} F(\bar{x},\bar{y})(x)$ and $D_{\text{w}} F(\bar{x},\bar{y})(x)$ exist. If the map F is contingently C-absorbing at $(\bar{x},\bar{y})$, then the following relationships hold:*

(a) $D_{\text{P}} F(\bar{x},\bar{y})(x) = \text{GHe-PMin}(DF(\bar{x},\bar{y})(x),C)$.
(b) $D_{\text{G}} F(\bar{x},\bar{y})(x) = \text{Min}(DF(\bar{x},\bar{y})(x),C)$.
(c) $D_{\text{w}} F(\bar{x},\bar{y})(x) \supset \text{WMin}(DF(\bar{x},\bar{y})(x),C)$.

Proof. (a) By virtue of

$$DF(\bar{x},\bar{y})(x) + C = DF(\bar{x},\bar{y})(x) \quad \text{for all } x \in L,$$

we have

$$\text{GHe-PMin}(D(F + C)(\bar{x},\bar{y})(x),C) = \text{GHe-PMin}(DF(\bar{x},\bar{y})(x) + C,C),$$

which, in view of Lemma 11.5.19 (a), implies that $D_P F(\bar{x}, \bar{y})(x) = \text{GHe-PMin}(DF(\bar{x}, \bar{y})(x), C)$.

The proof of (b) and (c) follows from the last two parts of Lemma 11.5.19. □

Finally, we relate the epiderivatives with a contingent cone in the image space.

Theorem 11.5.21. *Let X and Y be normed spaces, let $C \subset Y$ be pointed, closed, convex, and solid cone, and let $A \subset X$ be nonempty. Let $F : A : X \rightrightarrows Y$ be a set-valued map and let $(\bar{x}, \bar{y}) \in \text{graph}(F)$. Let $S := \text{dom}(D_W F(\bar{x}, \bar{y}))$. Then:*

$$\bigcup_{x \in S} D_W F(\bar{x}, \bar{y})(x) \subseteq T(F(A) + C, \bar{y}). \tag{11.75}$$

Proof. Let $T(F(A) + C, \bar{y}) \neq Y$ otherwise there is nothing to prove. Let $y \in D_W F(\bar{x}, \bar{y})(S)$ be arbitrary and let $x \in S$ be the corresponding element such that $y \in D_W F(\bar{x}, \bar{y})(x)$. Therefore

$$y \in \text{WMin}(G(x), C) \subseteq G(x),$$

where G is defined in (11.5.1). Hence $(x, y) \in \text{graph}(G) = T(\text{epi}(F), (\bar{x}, \bar{y}))$. By the definition of the contingent cone, there are sequences $\{\lambda_n\} \subset \mathbb{P}$, $\{x_n\} \subset X$ and $\{y_n\} \subset Y$ such that $\lambda_n \to 0$, $x_n \to x$, and $y_n \to y$ with $\bar{y} + \lambda_n y_n \in F(\bar{x}+\lambda_n x_n)+C$. This implies $\bar{y}+\lambda_n y_n \in F(\bar{x}+\lambda_n x_n)+C \subseteq F(A)+C$. Therefore, in view of the fact that $\lambda_n \to 0$ and $y_n \to y$, we deduce that $y \in T(F(A) + C, \bar{y})$. Since y is chosen arbitrarily, we have $D_W F(\bar{x}, \bar{y})(S) \subset T(F(A) + C, \bar{y})$, and the proof is complete. □

11.6 Calculus Rules for Contingent Epiderivatives

In this section we present some calculus rules for the contingent epiderivatives. A formula for scalar multiplication, sum formulae, and chain rules are given. This section is based on [299] and is strongly influenced by the general approach of deriving calculus rules devised by Ward [590, 593, 597]). This section is an extension of results given in Sect. 11.2.2 for the contingent derivatives to contingent epiderivatives.

We begin with exploring the contingent epiderivative for the product of a set-valued map and a strictly positive scalar. Let X and Y be normed spaces and let $C \subset Y$ be a pointed, closed, and convex cone. Given a set-valued map $F : X \rightrightarrows Y$ and a scalar $\lambda > 0$, we first define a set-valued map $\lambda \otimes F : X \rightrightarrows Y$ with

$$\lambda \otimes F(x) = \lambda F(\lambda^{-1}x), \quad \text{for every } x \in X.$$

We show that $\lambda \otimes F$ is a set-valued analogue of the epi-multiple function (see [499]).

Lemma 11.6.1. *Let X and Y be normed spaces, let $C \subset Y$ be a pointed, closed, and convex cone C, let $F : X \rightrightarrows Y$ be a set-valued map, and let $\lambda > 0$. Then:*

$$\text{epi}\,(\lambda \otimes F) = \lambda\, \text{epi}\,(F).$$

Proof. Clearly,

$$\begin{aligned} \text{epi}\,(\lambda \otimes F) &= \{(x, y) \in X \times Y \mid \; y \in \lambda \otimes F(x) + C\} \\ &= \{(x, y) \in X \times Y \mid \; y \in \lambda F(\tfrac{1}{\lambda}\, x) + C\} \\ &= \{(\lambda x, \lambda y) \in X \times Y \mid \; y \in F(x) + C\} \\ &= \lambda\, \text{epi}\,(F), \end{aligned}$$

which justifies the claim. The proof is complete. □

Proposition 11.6.2. *Let X and Y be normed spaces, let $C \subset Y$ be a pointed, closed, and convex cone C, let $F : X \rightrightarrows Y$ be a set-valued map, and let $(\bar{x}, \bar{y}) \in$ graph (F). If the contingent epiderivative $D_E F(\bar{x}, \bar{y})$ exists, then for $\lambda > 0$, the contingent epiderivative $D_E(\lambda \otimes F)(\lambda\bar{x}, \lambda\bar{y})$ exists, and,*

$$D_E(\lambda \otimes F)(\lambda\, \bar{x}, \lambda \bar{y})(\cdot) = D_E F(\bar{x}, \bar{y})(\cdot).$$

Proof. From Lemma 11.6.1 and from the definition of the contingent epiderivative, we have

$$\begin{aligned} \text{epi}\,(D_E(\lambda \otimes F)(\lambda\bar{x}, \lambda\bar{y})) &= T(\text{epi}\,(\lambda \otimes F), (\lambda\bar{x}, \lambda\bar{y})) \\ &= T(\lambda\, \text{epi}\,(F), \lambda(\bar{x}, \bar{y})) \\ &= \lambda T(\text{epi}\,(F), (\bar{x}, \bar{y})) \\ &= T(\text{epi}\,(F), (\bar{x}, \bar{y})) \\ &= \text{epi}\,\,(D_E F(\bar{x}, \bar{y})) \end{aligned}$$

which implies the assertion. The proof is complete. □

We now present a formula for scalar multiplication.

Theorem 11.6.3. *Let X and Y be normed spaces, let $C \subset Y$ be a pointed, closed, and convex cone C, let $F : X \rightrightarrows Y$ be a set-valued map, and let $(\bar{x}, \bar{y}) \in$ graph (F). If the contingent epiderivative $D_E F(\bar{x}, \bar{y})$ exists, then for $\lambda > 0$, the contingent epiderivative $D_E(\lambda \otimes F)(\lambda\bar{x}, \lambda\bar{y})$ exists, and,*

$$D_E(\lambda F)(\bar{x}, \lambda\bar{y})(\cdot) = \lambda D_E F(\bar{x}, \bar{y})(\cdot).$$

Proof. For a homomorphism $\Upsilon_\lambda : X \times Y \to X \times Y$ defined by $\Upsilon_\lambda(\bar{x}, \bar{y}) := (\bar{x}, \lambda\bar{y})$, it is obvious that

$$\begin{aligned}\text{epi}\,(\lambda F) &= \{(\bar{x}, \bar{y}) \in X \times Y, \ : \ \bar{y} \in (\lambda F)(\bar{x}) + C\} \\ &= \{(\bar{x}, \lambda\bar{y}) \in X \times Y, \ : \ \bar{y} \in F(\bar{x}) + C\} \\ &= \Upsilon_\lambda(\text{epi}\,(F)).\end{aligned}$$

Then the definition of the contingent epiderivative implies that

$$\begin{aligned}\text{epi}\,(D_E(\lambda F)(\bar{x}, \lambda\bar{y})) &= T(\text{epi}\,(\lambda F), (\bar{x}, \lambda\bar{y})) \\ &= T(\Upsilon_\lambda(\text{epi}\,(F)), \Upsilon_\lambda(\bar{x}, \bar{y})) \\ &= \Upsilon_\lambda(T(\text{epi}\,(F), (\bar{x}, \bar{y}))) \\ &= \Upsilon_\lambda(\text{epi}\,(D_E F(\bar{x}, \bar{y}))),\end{aligned}$$

which confirms that $D_E(\lambda F)(\bar{x}, \lambda\bar{y})(\cdot) = \lambda D_E F(\bar{x}, \bar{y})(\cdot)$. The proof is complete. □

Our next result of this section is the following sum formula:

Theorem 11.6.4. *Let X, Y, and Z be real Banach spaces, and let $C \subset Z$ be a pointed, closed, and convex cone. Let $F_1 : X \rightrightarrows Z$ and $F_2 : Y \rightrightarrows Z$ be set-valued maps, and let $f : X \to Y$ be a single valued map being strictly differentiable on $B_\delta(x)$ for $\delta > 0$. Let $(x, y_1) \in \text{graph}\,(F_1)$ and $(f(x), y_2) \in \text{graph}\,(F_2)$. Let the contingent epiderivatives $D_E(F_1 + F_2 \circ f)(x, y_1 + y_2)$, $D_E F_1(x, y_1)$ and $D_E F_2(f(x), y_2)$ exist. Let $\text{epi}\,(F_1)$ and $\text{epi}\,(F_2)$ be epi-Lipschitz-like and derivable at (x, y_1) and $(f(x), y_2)$, respectively. Let*

$$\nabla\Psi(x, y_1, f(x), y_2)C(\text{epi}\,(F_1) \times \text{epi}\,(F_2), (x, y_1, f(x), y_2)) = Y, \qquad (11.76)$$

where $\Psi : X \times Z \times Y \times Z \to Y$ is defined by $\Psi(\alpha, \beta, \gamma, \delta) := f(\alpha) - \gamma$. Then

$$D_E(F_1 + F_2 \circ f)(x, y_1 + y_2)(\cdot) \leq D_E F_1(x, y_1)(\cdot) + D_E F_2(f(x), y_2)(\nabla f(x)(\cdot)).$$

Proof. Let us define a map $\Phi : X \times Z \times Y \times Z \to X \times Z$ by

$$\Phi(\alpha, \beta, \gamma, \delta) = (\alpha, \beta + \delta).$$

Clearly Φ is linear and continuous. For simplicity we set $F := F_1 + F_2 \circ f$, $\Omega := \text{epi}\,(F_1) \times \text{epi}\,(F_2)$ and $K := \Omega \cap \Psi^{-1}(0)$. From the definitions of the maps Ψ and Φ and of Ω, it is evident

$$\begin{aligned}\Phi(K) &= \Phi((\text{epi}\,(F_1) \times \text{epi}\,(F_2)) \cap \{(x, y, w, z) \ : \ f(x) = w\}) \\ &= \Phi(\{(x, y, w, z) \in X \times Z \times Y \times Z \ : \ y \in F_1(x) + C,\end{aligned}$$

$$
\begin{aligned}
& z \in F_2(w) + C,\, f(x) = w\}) \\
&= \{(x, y + z) \in X \times Z \,:\, y \in F_1(x) + C,\, z \in F_2 \circ f(x) + C\} \\
&= \operatorname{epi}(F). \qquad (11.77)
\end{aligned}
$$

It follows from the definition of the contingent epiderivative, (11.77), and well-known properties of the contingent cone that

$$
\begin{aligned}
\operatorname{epi}(D_E F(x, y_1 + y_2)) &= T(\operatorname{epi}(F), (x, y_1 + y_2)) \\
&= T(\Phi(K), \Phi(x_0)) \text{ with } x_0 := (x, y_1, f(x), y_2) \\
&\supseteq \Phi(T(K, x_0)). \qquad (11.78)
\end{aligned}
$$

Since (11.76) holds and the set Ω being a product of epi-Lipschitz-like sets is epi-Lipschitz-like (see Definition 4.7.1), it follows that Theorem 4.7.5 is applicable and consequently

$$
T(K, x_0) = T(\Omega \cap \Psi^{-1}(0), x_0) \supseteq T(\Omega, x_0) \cap \nabla\Psi(x_0)^{-1}(0),
$$

which, in view of (11.78), leads to

$$
\begin{aligned}
\operatorname{epi}(D_E F(x, y_1 + y_2)) &\supseteq \Phi(T(\Omega, x_0) \cap \nabla\Psi(x_0)^{-1}(0)) \\
&= \Phi(T(\operatorname{epi}(F_1) \times \operatorname{epi}(F_2), x_0) \cap \nabla\Psi(x_0)^{-1}(0)). \qquad (11.79)
\end{aligned}
$$

Observe that derivability of $\operatorname{epi}(F_1)$ at (x, y_1) and of $\operatorname{epi}(F_2)$ at $(f(x), y_2)$ implies derivability of Ω at x_0. Consequently,

$$
\begin{aligned}
& T(\operatorname{epi}(F_1), (x, y_1)) \times T(\operatorname{epi}(F_2), (f(x), y_2)) \\
&\quad = T(\operatorname{epi}(F_1) \times \operatorname{epi}(F_2), (x, y_1, f(x), y_2)). \qquad (11.80)
\end{aligned}
$$

Therefore, with (11.79) and (11.80), we obtain

$$
\begin{aligned}
\operatorname{epi}(D_E F(x, y_1 + y_2)) &\supseteq \Phi(T(\operatorname{epi}(F_1), (x, y_1)) \\
&\qquad \times T(\operatorname{epi}(F_2), (f(x), y_2)) \cap \nabla\Psi(x_0)^{-1}(0)) \\
&= \Phi((\operatorname{epi}(D_E F_1(x, y_1)) \\
&\qquad \times \operatorname{epi}(D_E F_2(f(x), y_2))) \cap \nabla\Psi(x_0)^{-1}(0)) \\
&= \Phi(\{u, v, w, r) \in X \times Z \times Y \times Z \;: \\
&\qquad (u, v) \in \operatorname{epi}(D_E F_1(x, y_1)), \\
&\qquad (w, r) \in \operatorname{epi}(D_E F_2(f(x), y_2)), \nabla f(x)(x) = w\})
\end{aligned}
$$

$$= \{(u, v + r) \in X \times Z \ : (u, v) \in \text{epi}\,(D_E F_1(x, y_1),$$
$$(\nabla f(x)(x), r) \in \text{epi}\,(D_E F_2(f(x), y_2)\}$$
$$= \text{epi}\,(D_E F_1(x, y_1)(\cdot) + D_E F_2(f(x), y_2)(\nabla f(x)(\cdot))),$$

which confirms that

$$\text{epi}\,(D_E F(x, y_1 + y_2)(\cdot)) \supseteq \text{epi}\,(D_E F_1(x, y_1)(\cdot) + D_E F_2(f(x), y_2)(\nabla f(x)(\cdot))),$$

implying

$$D_E(F_1 + F_2 \circ f)(x, y_1 + y_2)(\cdot) \leq D_E F_1(x, y_1)(\cdot) + D_E F_2(f(x), y_2)(\nabla f(x)(\cdot)),$$

and this completes the proof. □

By setting $X = Y$ in the above theorem, we get the following result.

Corollary 11.6.5. *Let the assumptions of Theorem 11.6.4 hold with f as the identity map. Then:*

$$D_E(F_1 + F_2)(x, y_1 + y_2)(\cdot) \leq D_E F_1(x, y_1)(\cdot) + D_E F_2(x, y_2)(\cdot).$$

We now investigate chain rules for the contingent epiderivatives. Throughout the remaining part of this section, let X, Y and Z be Banach spaces. The space Z is partially ordered by a pointed, closed, and convex cone C. Let $F : X \rightrightarrows Y$ and $G : Y \rightrightarrows Z$ be given set-valued maps. In the following by $P_{X \times Z}$ we shall represent the canonical projection from $X \times Y \times Z$ onto $X \times Z$.

The following lemma will be used shortly (see Pennanen [468]):

Lemma 11.6.6. *For any $F : X \rightrightarrows Y$ and $G : Y \rightrightarrows Z$ the following holds:*

(*a*) $\text{graph}\,(G \circ F) = P_{X \times Z}[(\text{graph}\,(F) \times Z) \cap (X \times \text{graph}\,(G))]$,
(*b*) $\text{epi}\,(G \circ F) = P_{X \times Z}[(\text{graph}\,(F) \times Z) \cap (X \times \text{epi}\,(G))]$.

Proof. We only prove part (b). Since

$$\text{epi}\,(G \circ F) = \{(x, z) \in X \times Z \ : \exists\, y \in F(x), \ \ z \in G(y) + C\}$$
$$= \{(x, z) \in X \times Z : \exists\, y : (x, y, z) \in (\text{graph}\,(F) \times Z) \cap (X \times \text{epi}\,(G))\},$$

the proof is complete. □

We are now ready to give the following chain rule:

Theorem 11.6.7. *Let* graph (F) *and* epi (G) *be epi-Lipschitz-like and derivable at* $(x, y) \in$ graph (F) *and* $(y, z) \in$ graph (G), *respectively. Assume that the contingent epiderivative $D_E G(y, z)$ exists. Moreover, assume that*

$$\nabla \Upsilon(x, y, y, z) C(\text{graph}\,(F) \times \text{epi}\,(G), (x, y, y, z)) = Y, \tag{11.81}$$

where $\Upsilon : X \times Y \times Y \times Z \to Y$ *is defined by* $\Upsilon(\alpha, \beta, \gamma, \delta) := \beta - \gamma$. *Then*:

$$\operatorname{epi}(D_E(G \circ F)(x, z)) \supseteq \operatorname{epi}(D_E G(y, z) \circ DF(x, y)). \tag{11.82}$$

Proof. Clearly $\Upsilon(\cdot)$, as defined above, is linear and continuous. Set $\Sigma := (\operatorname{graph}(F) \times \operatorname{epi}(G)) \cap \Upsilon^{-1}(0)$ and $\omega := (x, y, y, z)$. In view of Lemma 11.6.6, we can express $\operatorname{epi}(G \circ F)$ as follows

$$\operatorname{epi}(G \circ F) = P_{X \times Z}((\operatorname{graph}(F) \times \operatorname{epi}(G)) \cap \Upsilon^{-1}(0)),$$

where $P_{X \times Z}$ is the canonical projection from $X \times Y \times Y \times Z$ onto $X \times Z$.

From the definition of the contingent epiderivative, we obtain

$$\operatorname{epi}(D_E(G \circ F)(x, z)) = T(\operatorname{epi}(G \circ F), (x, z)) = T(P_{X \times Z}(\Sigma), P_{X \times Z}(\omega)),$$

and by using the fact that $P_{X \times Z}$ is linear and continuous, we also have

$$\operatorname{epi}(D_E(G \circ F)(x, z)) \supseteq P_{X \times Z} T(\Sigma, \omega). \tag{11.83}$$

Since (11.81) holds and $\operatorname{graph}(F) \times \operatorname{epi}(G)$ is epi-Lipschitz-like, we get

$$\begin{aligned} T(\Sigma, \omega) &= T((\operatorname{graph}(F) \times \operatorname{epi}(G)) \cap \Upsilon^{-1}(0), \omega) \\ &\supseteq T(\operatorname{graph}(F) \times \operatorname{epi}(G), \omega) \cap \nabla \Upsilon(\omega)^{-1}(0), \end{aligned} \tag{11.84}$$

which, in view of the derivability assumptions, yields

$$\begin{aligned} T(\Sigma, \omega) &\supseteq T(\operatorname{graph}(F), (x, y)) \times T(\operatorname{epi}(G), (y, z)) \cap \nabla \Upsilon(\omega)^{-1}(0), \\ &= [\operatorname{graph}(DF(x, y)) \times \operatorname{epi}(D_E G(y, z))] \cap \nabla \Upsilon(\omega)^{-1}(0). \end{aligned}$$

Combining the above inclusion with (11.83) gives

$$\begin{aligned} &\operatorname{epi}(D_E(G \circ F)(x, z)) \\ &\quad \supseteq P_{X \times Z}(\operatorname{graph}(DF(x, y)) \times \operatorname{epi}(D_E G(y, z)) \cap \nabla \Upsilon(\omega)^{-1}(0)) \\ &\quad = \operatorname{epi}(D_E G(y, z) \circ DF(x, y)), \end{aligned}$$

which completes the proof. □

Recall that $f : X \to Y$ is said to be isotone, if $f(y) \in f(x) + C$, provided that $y \in x + C$.

We now give a chain rule for the contingent epiderivatives.

Theorem 11.6.8. *Besides the hypotheses of Theorem 11.6.7, assume that* F *is the profile map of a set-valued map* H, *the contingent epiderivative* $D_E G(y, z)$ *exists, and the contingent epiderivative* $D_E G(x, y)$ *is isotone. Then*

$$D_E(G \circ F)(x,z) \leq D_E G(y,z) \circ D_E H(x,y).$$

Proof. By virtue of the isotony property of $D_E G(x,y)$, the following identity holds:

$$\text{epi}\,(D_E G(y,z) \circ D_E H(x,y)) = P_{X\times Z}(\text{epi}\,(D_E H(x,y)) \times \text{epi}\,(D_E G(y,z)) \cap \Upsilon^{-1}(0)).$$

Hence the above equation in view of (11.82) and the identity graph $(DF(x,y)) =$ epi $(D_E H(x,y))$ gives

$$\text{epi}\,(D_E(G \circ F)(x,z)) \supseteq \text{epi}\,(D_E G(y,z) \circ D_E H(x,y)),$$

which implies that $D_E(G \circ F)(x,z) \leq D_E G(y,z) \circ D_E H(x,y)$. The proof is complete. □

11.7 Second-Order Derivatives of Set-Valued Maps

In this section, we study second-order extensions of the graphical derivatives and epiderivatives of set-valued maps studied earlier in this chapter. These notions are defined by using second-order tangent cones and second-order tangent sets studied in Chap. 4.

The following definition introduces some second-order derivatives of set-valued maps.

Definition 11.7.1. Let X and Y be normed spaces, let $F : X \rightrightarrows Y$ be a set-valued map, let $(\bar{x}, \bar{y}) \in$ graph (F), and let $(\bar{u}, \bar{v}) \in X \times Y$. Then:

1. A set-valued map $D^2 F(\bar{x}, \bar{y}, \bar{u}, \bar{v}) : X \rightrightarrows Y$ is called the **second-order contingent derivative** of F at $(\bar{x}, \bar{y})$ in the direction $(\bar{u}, \bar{v})$, if

$$\text{graph}\,(D^2 F(\bar{x}, \bar{y}, \bar{u}, \bar{v})) = T^2(\text{graph}\,(F), (\bar{x}, \bar{y}), (\bar{u}, \bar{v})),$$

   where $T^2(\text{graph}\,(F), (\bar{x}, \bar{y}), (\bar{u}, \bar{v}))$ is the second-order contingent set.
2. A set-valued map $\mathscr{D}^2 F(\bar{x}, \bar{y}, \bar{u}, \bar{v}) : X \rightrightarrows Y$ is called the **second-order compound contingent derivative** of F at $(\bar{x}, \bar{y})$ in the direction $(\bar{u}, \bar{v})$, if

$$\text{graph}\,(\mathscr{D}^2 F(\bar{x}, \bar{y}, \bar{u}, \bar{v})) = T(T(\text{graph}\,(F), (\bar{x}, \bar{y})), (\bar{u}, \bar{v})),$$

   where $T(T(\text{graph}\,(F), (\bar{x}, \bar{y})), (\bar{u}, \bar{v}))$ is the contingent cone of $T(\text{graph}\,(F), (\bar{x}, \bar{y}))$ at $(\bar{u}, \bar{v})$.
3. A set-valued map $D_A^2 F(\bar{x}, \bar{y}, \bar{u}, \bar{v}) : X \rightrightarrows Y$ is called the **second-order adjacent derivative** of F at $(\bar{x}, \bar{y})$ in the direction $(\bar{u}, \bar{v})$, if

$$\text{graph}\,(D_A^2 F(\bar{x}, \bar{y}, \bar{u}, \bar{v})) = A^2(\text{graph}\,(F), (\bar{x}, \bar{y}), (\bar{u}, \bar{v})),$$

   where $A^2(\text{graph}\,(F), (\bar{x}, \bar{y}), (\bar{u}, \bar{v}))$ is the second-order adjacent set.

4. A set-valued map $\mathscr{D}^2_A F(\bar{x}, \bar{y}, \bar{u}, \bar{v}) : X \rightrightarrows Y$ is called the **second-order compound adjacent derivative** of F at $(\bar{x}, \bar{y})$ in the direction $(\bar{u}, \bar{v})$, if

$$\text{graph}\,(\mathscr{D}^2_A F(\bar{x}, \bar{y}, \bar{u}, \bar{v})) = A(A(\text{graph}\,(F), (\bar{x}, \bar{y})), (\bar{u}, \bar{v})),$$

where $A(A(\text{graph}\,(F), (\bar{x}, \bar{y})), (\bar{u}, \bar{v}))$ is the adjacent cone of $A(\text{graph}\,(F), (\bar{x}, \bar{y}))$ at $(\bar{u}, \bar{v})$.
5. A set-valued map $\mathscr{D}^2_C F(\bar{x}, \bar{y}, \bar{u}, \bar{v}) : X \rightrightarrows Y$ is called the **second-order compound Clarke derivative** of F at $(\bar{x}, \bar{y})$ in the direction $(\bar{u}, \bar{v})$, if

$$\text{graph}\,(\mathscr{D}^2_C F(\bar{x}, \bar{y}, \bar{u}, \bar{v})) = C(C(\text{graph}\,(F), (\bar{x}, \bar{y})), (\bar{u}, \bar{v})),$$

where $C(C(\text{graph}\,(F), (\bar{x}, \bar{y})), (\bar{u}, \bar{v}))$ is the adjacent cone of $A(\text{graph}\,(F), (\bar{x}, \bar{y}))$ at $(\bar{u}, \bar{v})$.
6. A set-valued map $\mathbb{D}^2 F(\bar{x}, \bar{y}, \bar{u}, \bar{v}) : X \rightrightarrows Y$ is called the **second-order asymptotic contingent derivative** of F at $(\bar{x}, \bar{y})$ in the direction $(\bar{u}, \bar{v})$, if

$$\text{graph}\,(\mathbb{D}^2 F(\bar{x}, \bar{y}, \bar{u}, \bar{v})) = \mathbb{T}^2(\text{graph}\,(F), (\bar{x}, \bar{y}), (\bar{u}, \bar{v})),$$

where $\mathbb{T}^2(\text{graph}\,(F), (\bar{x}, \bar{y}), (\bar{u}, \bar{v}))$ is the second-order asymptotic contingent cone.
7. A set-valued map $\mathbb{D}^2_A F(\bar{x}, \bar{y}, \bar{u}, \bar{v}) : X \rightrightarrows Y$ is called the **second-order asymptotic adjacent derivative** of F at $(\bar{x}, \bar{y})$ in the direction $(\bar{u}, \bar{v})$, if

$$\text{graph}\,(\mathbb{D}^2_A F(\bar{x}, \bar{y}, \bar{u}, \bar{v})) = \mathbb{A}^2(\text{graph}\,(F), (\bar{x}, \bar{y}), (\bar{u}, \bar{v})),$$

where $\mathbb{A}^2(\text{graph}\,(F), (\bar{x}, \bar{y}), (\bar{u}, \bar{v}))$ is the second-order asymptotic adjacent cone.
8. A set-valued map $D^2_L F(\bar{x}, \bar{y}, \bar{u}, \bar{v}) : X \rightrightarrows Y$ is called the **second-order lower Dini derivative** of F at $(\bar{x}, \bar{y})$ in the direction $(\bar{u}, \bar{v})$ provided that $y \in D^2_L F(\bar{x}, \bar{y}, \bar{u}, \bar{v})(x)$, if and only if, for every $(t_n) \subset P$ and for every $\{x_n\} \subset X$ with $t_n \downarrow 0$ and $x_n \to x$, there are a sequence $\{y_n\} \subset Y$ with $y_n \to y$ and an index $m \in N$ such that $\bar{y} + t_n \bar{v} + t_n^2 y_n \in F(\bar{x} + t_n \bar{u} + t_n^2 x_n)$, for every $n \geq m$.

Remark 11.7.2. It is clear that if $(\bar{u}, \bar{v}) = (0, 0)$ in the above definition, then the second-order contingent derivative, the second-order compound contingent derivative, and the second-order asymptotic contingent derivative recover the contingent derivative $DF(\bar{x}, \bar{y})$ of F at $(\bar{x}, \bar{y})$, and the second-order adjacent derivative, the second-order compound adjacent derivative, and second-order asymptotic adjacent derivative recover the adjacent derivative $D_A F(\bar{x}, \bar{y})$ of F at $(\bar{x}, \bar{y})$. The notion of second-order lower Dini derivative is inspired by the lower Dini derivative introduced by Penot [462]. In fact, if $(\bar{u}, \bar{v}) = (0, 0)$, then the above notion recovers the lower Dini derivative of [462]. The second-order analogue given above was first introduced and used in [313] to give some necessary optimality conditions in set-valued optimization.

It is natural to have a second-order version of derivability and related notions:

Definition 11.7.3. Let X and Y be normed spaces, let $F : X \rightrightarrows Y$ be a set-valued map, let $(\bar{x}, \bar{y}) \in \operatorname{graph}(F)$, and let $(\bar{u}, \bar{v}) \in X \times Y$.

1. The map F is said to be **second-order proto-differentiable** or **second-order derivable** at $(\bar{x}, \bar{y})$ in the direction $(\bar{u}, \bar{v}) \in X \times Y$, if the second-order contingent derivative and the second-order adjacent derivative of F at $(\bar{x}, \bar{y})$ in direction $(\bar{u}, \bar{v})$ coincide, that is,
$$T^2(\operatorname{graph}(F), (\bar{x}, \bar{y}), (\bar{u}, \bar{v})) = A^2(\operatorname{graph}(F), (\bar{x}, \bar{y}), (\bar{u}, \bar{v})).$$
2. The map F is said to be **second-order compound proto-differentiable** or **second-order compound derivable** at $(\bar{x}, \bar{y})$ in the direction $(\bar{u}, \bar{v}) \in X \times Y$, if the second-order compound contingent derivative and the second-order compound adjacent derivative of F at $(\bar{x}, \bar{y})$ in direction $(\bar{u}, \bar{v})$ coincide, that is,
$$T(T(\operatorname{graph}(F), (\bar{x}, \bar{y})), (\bar{u}, \bar{v})) = A(A(\operatorname{graph}(F), (\bar{x}, \bar{y})), (\bar{u}, \bar{v})).$$
3. The map F is said to be **second-order semi-differentiable** at $(\bar{x}, \bar{y})$ in the direction $(\bar{u}, \bar{v}) \in X \times Y$, if the second-order contingent derivative and the second-order lower Dini derivative of F at $(\bar{x}, \bar{y})$ in the direction $(\bar{u}, \bar{v})$ coincide.

Note that if a set-valued map is strict proto-differentiable, then it is second-order compound proto-differentiable.

The following is a second-order analogue of Theorem 11.1.28:

Theorem 11.7.4. *Let X and Y be normed spaces, let $F : X \to Y$ be a set-valued map which is Lipschitz-like around $(\bar{x}, \bar{y}) \in \operatorname{graph}(F)$. Then F is second-order derivable at $(\bar{x}, \bar{y})$ in any direction $(\bar{u}, \bar{v}) \in X \times Y$ if and only if F is second-order semi-differentiable at $(\bar{x}, \bar{y})$ direction $(\bar{u}, \bar{v}) \in X \times Y$.*

Proof. In view of the chain of inclusions
$$D^2_L F(\bar{x}, \bar{y}, \bar{u}, \bar{v}) \subseteq D^2_A F(\bar{x}, \bar{y}, \bar{u}, \bar{v}) \subseteq D^2 F(\bar{x}, \bar{y}, \bar{u}, \bar{v}),$$
it is evident that if F is second-order semi-differentiable at $(\bar{x}, \bar{y})$ in any direction $(\bar{u}, \bar{v}) \in X \times Y$ then it is second-order derivable at $(\bar{x}, \bar{y})$ in that direction. For the converse, let F be second-order derivable at $(\bar{x}, \bar{y})$ in the direction $(\bar{u}, \bar{v}) \in X \times Y$ and let $(x, y) \in \operatorname{graph}(D^2 F(\bar{x}, \bar{y}, \bar{u}, \bar{v}))$ be arbitrary. Then, for every $\{t_n\} \subset \mathbb{P}$, there exists $\{(z_n, w_n)\} \subset X \times Y$ such that $t_n \downarrow 0$, $(z_n, w_n) \to (x, y)$ and $\bar{y} + t_n \bar{v} + t_n^2 w_n \in F(\bar{x} + t_n \bar{u} + t_n^2 z_n)$. Let $\{x_n\} \subset X$ be an arbitrary sequence such that $x_n \to x$. Using the fact that F is Lipschitz-like at $(\bar{x}, \bar{y})$, we can show that there exist a sequence $\{y_n\}$ and an index $m \in \mathbb{N}$ such that $(\bar{x} + t_n \bar{u} + t_n^2 x_n, \bar{y} + t_n \bar{v} + t_n^2 y_n) \in \operatorname{graph}(F)$ for every $n \geq m$. Consequently, $y \in D^2_L F(\bar{x}, \bar{y}, \bar{u}, \bar{v})(x)$. The proof is complete. □

The following result connects the second-order contingent and the second-order adjacent derivatives to the second-order Fréchet derivative of a single-valued smooth map (see [594]).

Theorem 11.7.5. *Let X and Y be normed spaces, and let $F : X \to Y$ be twice Fréchet differentiable at $\bar{x} \in X$ with derivative $\nabla F(\bar{x})$ and second-order derivative $\nabla^2 F(\bar{x})$. Then for every $\bar{u} \in X$,*

$$T^2(\text{graph}\,(F), (\bar{x}, F(\bar{x})), (\bar{u}, \nabla F(\bar{x})(\bar{u}))) = \{(\bar{v}, \nabla F(\bar{x})\bar{v} + \nabla^2 F(\bar{x})(\bar{u}, \bar{u}))|\ \bar{v} \in X\},$$
$$A^2(\text{graph}\,(F), (\bar{x}, F(\bar{x})), (\bar{v}, \nabla F(\bar{x})(\bar{v}))) = \{(\bar{v}, \nabla F(\bar{x})\bar{v} + \nabla^2 F(\bar{x})(\bar{u}, \bar{u}))|\ \bar{v} \in X\}.$$

Proof. Let $t > 0$, and let $\bar{u}, x \in X$ be arbitrary. By Taylor's theorem, we have

$$F(\bar{x}+t\bar{u}+t^2x/2) = F(\bar{x})+t\nabla F(\bar{x})(\bar{u})+\frac{t^2}{2}(\nabla F(\bar{x})(x)+\nabla^2 F(\bar{x})(\bar{u}, \bar{u}))+o(t^2),$$

and the claim follows at once from the above equality. □

Remark 11.7.6. In fact, if $F : X \to Y$ is a single valued map which is twice continuously Fréchet differentiable around $\bar{x} \in K \subset X$, then the second-order contingent derivative of the restriction F_K of F to K at $\bar{x}$ in a direction $\bar{u}$ is given by the formula (see [18, p. 215]):

$$D^2 F_K(\bar{x}, F(\bar{x}), \bar{u}, \nabla F(\bar{x})(\bar{u}))(x) = \begin{cases} \nabla F(\bar{x})(x) + \nabla^2 F(\bar{x})(\bar{u}, \bar{u}) & \text{if} \quad x \in T^2(K, \bar{x}, \bar{u}), \\ \emptyset & \text{if} \quad x \notin T^2(K, \bar{x}, \bar{u}). \end{cases} \tag{11.85}$$

The following result is a second-order analogue of Theorem 11.1.12 given for the first-order contingent derivative (see [18, Theorem 5.6.2]).

Theorem 11.7.7. *Let X and Y be normed spaces, let L and S be nonempty subsets of X where S is assumed to be open, let $G : X \rightrightarrows Y$ be a set-valued map and let $f : S \to Y$ be a twice continuously differentiable single-valued map. We define a set-valued map $F : X \rightrightarrows Y$ by*

$$F(x) := \begin{cases} f(x) - G(x), & \text{if} \ \ x \in L, \\ \emptyset, & \text{if} \ \ x \notin L. \end{cases}$$

Then for any $\bar{x} \in S \cap \text{dom}\,(F)$, $(\bar{x}, \bar{y}) \in \text{graph}\,(F)$, and $(\bar{u}, \bar{v}) \in \text{graph}\,(DF(\bar{x}, \bar{y}))$, we have

$$\begin{aligned} D^2 F(\bar{x}, \bar{y}, \bar{u}, \bar{v})(x) \subseteq\ & \nabla f(\bar{x})(x) \\ & + \frac{1}{2}\nabla^2 f(\bar{x})(\bar{u}, \bar{u}) \\ & - D^2 G(\bar{x}, f(\bar{x}) - \bar{y}, \bar{u}, \nabla f(\bar{x})(\bar{u}) - \bar{v})(x), \end{aligned} \tag{11.86}$$

for every $x \in T^2(L, \bar{x}, \bar{u})$. Furthermore, $D^2 F(\bar{x}, \bar{y}, \bar{u}, \bar{v})(x)$ is empty whenever $x \notin T^2(L, \bar{x}, \bar{u})$. Equality holds in (11.86) if either L is twice derivable or G is twice derivable and G is Lipschitz.

Proof. Let $(x, y) \in \operatorname{graph}(D^2F(\bar{x}, \bar{y}, \bar{u}, \bar{v}))$ be arbitrary. Then there are sequences $\{(x_n, y_n)\} \subset X \times Y$ and $\{t_n\} \subset \mathbb{P}$ such that $(x_n, y_n) \to (x, y)$, $t_n \downarrow 0$ and $\bar{y} + t_n\bar{v} + t_n^2 y_n \in F(\bar{x} + t_n\bar{u} + t_n^2 x_n)$, implying that

$$\begin{aligned}
&\bar{x} + t_n\bar{u} + t_n^2 x_n \in L, \\
&\bar{y} + t_n\bar{v} + t_n^2 y_n \in f(\bar{x} + t_n\bar{u} + t_n^2 x_n) - G(\bar{x} + t_n\bar{u} + t_n^2 x_n),
\end{aligned}$$

where the first containment confirms that $x \in T^2(L, \bar{x}, \bar{u})$. Moreover, since the map f is twice continuously differentiable at $\bar{x}$, we have

$$f(\bar{x}+t_n\bar{u}+t_n^2x_n) = f(\bar{x})+t_n\nabla f(\bar{x})(\bar{u})+t_n^2\left(\nabla f(\bar{x})(x) + \frac{1}{2}\nabla^2 f(\bar{x})(\bar{u}, \bar{u}) + h(t_n)\right)$$

with $h(t_n) \to 0$ as $t_n \downarrow 0$. Therefore,

$$\begin{aligned}
&f(\bar{x}) - \bar{y} + t_n\left(\nabla f(\bar{x})(\bar{u}) - \bar{v}\right) \\
&\quad + t_n^2\left(\nabla f(\bar{x})(x) + \frac{1}{2}\nabla^2 f(\bar{x})(\bar{u}, \bar{u}) - y + h(t_n)\right) \in G(\bar{x} + t_n\bar{u} + t_n^2 x_n)
\end{aligned}$$

which implies that

$$\nabla f(\bar{x})(x) + \frac{1}{2}\nabla^2 f(\bar{x})(\bar{u}, \bar{u}) - y \in D^2G(\bar{x}, f(\bar{x}) - \bar{y}, \bar{u}, \nabla f(\bar{x})(\bar{u}) - \bar{v})(x), \tag{11.87}$$

and the inclusion (11.86) follows at once.

For the converse, we work under the additional assumption that G is second-order derivable and Lipschitz, $x \in T^2(L, \bar{x}, \bar{u})$, and that (11.87) holds. Then there are sequences $\{t_n\} \subset \mathbb{P}$ with $t_n \downarrow 0$ and $\{(x_n, y_n)\} \subset X \times Y$ such that $t_n \to 0$, $x_n \to x$ and $y_n \to \left(\nabla f(\bar{x})(x) + \nabla^2 f(\bar{x})(\bar{u}, \bar{u}) - y\right)$ such that for every $n \in \mathbb{N}$, we have $\bar{x} + t_n\bar{u} + t_n^2 x_n \in L$ and

$$f(\bar{x}) - \bar{y} + t_n\left(\nabla f(\bar{x})(\bar{u}) - \bar{v}\right) + t_n^2 y_n \in G(\bar{x} + t_n\bar{u} + t_n^2 x_n),$$

where we used the Lipschitz continuity of F. Then, for some $h(t_n) \to 0$, the sequence

$$z_n := \nabla f(\bar{x})(x) + \frac{1}{2}\nabla^2 f(\bar{x})(\bar{u}, \bar{u}) + h(t_n) - y_n$$

converges to y and satisfies $\bar{y} + t_n\bar{v} + t_n^2 z_n \in f(\bar{x} + t_n\bar{u} + t_n^2 x_n) - G(\bar{x} + t_n\bar{u} + t_n^2 x_n)$, and this completes the proof. □

In the following, we collect some of the useful features of second-order derivatives of set-valued maps. We give these results in terms of the contingent derivatives but most of the results can easily be extended to other variants.

We begin with the following:

Proposition 11.7.8. *Let X and Y be normed spaces, and let $C \subset Y$ be a pointed, closed, and convex cone. Let $F : X \rightrightarrows Y$ be a set-valued map, let $(\bar{x}, \bar{y}) \in \operatorname{graph}(F)$, and let $(\bar{u}, \bar{v}) \in X \times Y$. Then for every $x \in \operatorname{dom}(D^2 F(\bar{x}, \bar{y}, \bar{u}, \bar{v}))$, the following inclusion holds:*

$$D^2 F(\bar{x}, \bar{y}, \bar{u}, \bar{v})(x) + C \subset D^2(F + C)(\bar{x}, \bar{y}, \bar{u}, \bar{v})(x). \tag{11.88}$$

Proof. For an arbitrary $x \in \operatorname{dom}(D^2 F(\bar{x}, \bar{y}, \bar{u}, \bar{v}))$, let $y \in D^2 F(\bar{x}, \bar{y}, \bar{u}, \bar{v})(x)$ be arbitrary. We also choose an arbitrary element $c \in C$. Then there exist sequences $\{t_n\} \subset \mathbb{P}$ and $\{(x_n, y_n)\} \subset X \times Y$ such that $t_n \downarrow 0$, $(x_n, y_n) \to (x, y)$, and $\bar{y} + t_n \bar{v} + t_n^2 y_n \in F(\bar{x} + t_n \bar{u} + t_n^2 x_n)$. Setting $\bar{y}_n := y_n + c$, we note that $\bar{y} + t_n \bar{v} + t_n^2 \bar{y}_n \in F(\bar{x} + t_n \bar{u} + t_n^2 x_n) + C$. Since the sequence $\{\bar{y}_n\}$ converges to $y + c$, we conclude that $y + c \in D^2(F + C)(\bar{x}, \bar{y}, \bar{u}, \bar{v})(x)$. This establishes (11.88) and the proof is complete. □

The following variant of the above result is useful when working with the second-order contingent epiderivatives which will be defined shortly.

Proposition 11.7.9. *Let X and Y be normed spaces, and let $C \subset Y$ be a pointed, closed, and convex cone. Let $F : X \rightrightarrows Y$ be a set-valued map, let $(\bar{x}, \bar{y}) \in \operatorname{graph}(F)$, and let $(\bar{u}, \bar{v}) \in X \times Y$. Then for every $x \in \operatorname{dom}(D^2(F + C)(\bar{x}, \bar{y}, \bar{u}, \bar{v}))$, the following identity holds:*

$$D^2(F + C)(\bar{x}, \bar{y}, \bar{u}, \bar{v})(x) + C = D^2(F + C)(\bar{x}, \bar{y}, \bar{u}, \bar{v})(x). \tag{11.89}$$

Proof. The inclusion

$$D^2(F + C)(\bar{x}, \bar{y}, \bar{u}, \bar{v})(x) + C \supseteq D^2(F + C)(\bar{x}, \bar{y}, \bar{u}, \bar{v})(x)$$

holds trivially whereas the inclusion

$$D^2(F + C)(\bar{x}, \bar{y}, \bar{u}, \bar{v})(x) + C \subseteq D^2(F + C)(\bar{x}, \bar{y}, \bar{u}, \bar{v})(x)$$

follows from Proposition 11.7.8 by using the identity $(F + C)(\cdot) + C = (F + C)(\cdot)$. □

The converse inclusion of (11.88) does not hold in general. Moreover, the effective domains of the two derivatives can be very different. We illustrate this in the context of the first-order derivative (the case where $(\bar{u}, \bar{v}) = (0, 0)$).

Example 11.7.10. We set $X = Y = \mathbb{R}$ and $C = \mathbb{R}_+$. Define a set-valued map $F : \mathbb{R} \rightrightarrows \mathbb{R}$ as follows:

$$F(x) = \begin{cases} 0 & \text{if } x \neq 0 \\ [1,2] & \text{if } x = 0. \end{cases}$$

Then $\operatorname{dom}(DF(0,1)) = \{0\}$ and

$$DF(0,1)(0) = \mathbb{R}_+,$$
$$D(F + \mathbb{R}_+)(0,1)(x) = \mathbb{R} \quad \text{for every } x \in \mathbb{R}.$$

Consequently,

$$D(F + \mathbb{R}_+)(0,1)(0) \not\subset DF(0,1)(0) + \mathbb{R}_+.$$

It is worth noting that for every $x \in \mathbb{R}$, we have

$$\operatorname{Min}(D(F + \mathbb{R}_+)(0,1)(x), \mathbb{R}_+) = \emptyset,$$

whereas

$$\operatorname{Min}(DF(0,1)(0), \mathbb{R}_+) = \{0\}.$$

Therefore,

$$\operatorname{Min}(D(F + \mathbb{R}_+)(0,1)(0), \mathbb{R}_+) \neq \operatorname{Min}(DF(0,1)(0), \mathbb{R}_+).$$

We remark that the set-valued map F is convex-valued. □

Given the importance of the converse of the inclusion given in (11.88), we introduce the following concept:

Definition 11.7.11. Let X and Y be normed spaces and let $C \subset Y$ be a suitable ordering cone. A set-valued map $F : X \rightrightarrows Y$ is called **second-order contingently C-absorbing** at $(\bar{x}, \bar{y}) \in \operatorname{graph}(F)$ in direction $(\bar{u}, \bar{v}) \in X \times Y$ if for every $x \in \operatorname{dom}(D^2(F + C)(\bar{x}, \bar{y}, \bar{u}, \bar{v}))$, we have

$$D^2 F(\bar{x}, \bar{y}, \bar{u}, \bar{v})(x) + C = D^2(F + C)(\bar{x}, \bar{y}, \bar{u}, \bar{v})(x). \tag{11.90}$$

To prove identity (11.90), we need to impose certain restrictions on the map F and/or on the ordering cone C. In the following, we will give several such conditions. We begin by proposing a second-order extension of the S-derivative that has an important role to play.

Definition 11.7.12. Let X and Y be normed spaces, and let $C \subset Y$ be a pointed, closed, and convex cone. Let $F : X \rightrightarrows Y$ be a set-valued map, let $(\bar{x}, \bar{y}) \in \operatorname{graph}(F)$, and let $(\bar{u}, \bar{v}) \in X \times Y$. The second-order S-derivative of F at $(\bar{x}, \bar{y})$ in direction $(\bar{u}, \bar{v})$ is the set-valued map $D_S^2 F(\bar{x}, \bar{y}, \bar{u}, \bar{v}) : X \rightrightarrows Y$ such that

$y \in D_s^2 F(\bar{x}, \bar{y}, \bar{u}, \bar{v})(x)$ if and only if there are sequences $\{\alpha_n\} \subset \mathbb{P}$, $\{\beta_n\} \subset \mathbb{P}$, and $\{(x_n, y_n)\} \subset \operatorname{graph}(F)$ satisfying that

$$x_n \to \bar{x},$$
$$\alpha_n[(x_n, y_n) - (\bar{x}, \bar{y})] \to (\bar{u}, \bar{v}),$$
$$\beta_n[\alpha_n[(x_n, y_n) - (\bar{x}, \bar{y})] - (\bar{u}, \bar{v})] \to (x, y).$$

We are now ready to give a result that ensures that a set-valued map is second-order contingently C-absorbing.

Theorem 11.7.13. *Let X and Y be normed spaces, and let $C \subset Y$ be a pointed, closed, and convex cone. Let $F : X \rightrightarrows Y$ be a set-valued map, let $(\bar{x}, \bar{y}) \in \operatorname{graph}(F)$, and let $(\bar{u}, \bar{v}) \in X \times Y$. Assume that the set $B := \{z \in C \mid \|z\| = 1\}$ is compact. Assume that*

$$D_s^2 F(\bar{x}, \bar{y}, \bar{u}, \bar{v})(0) \cap (-C/\{0\}) = \emptyset. \tag{11.91}$$

Then the map F is second-order contingently C-absorbing, that is, for every $x \in \operatorname{dom}(D^2(F + C)(\bar{x}, \bar{y}, \bar{u}, \bar{v}))$, the following identity holds

$$D^2(F + C)(\bar{x}, \bar{y}, \bar{u}, \bar{v})(x) = D^2 F(\bar{x}, \bar{y}, \bar{u}, \bar{v})(x) + C.$$

Proof. In view of (11.85), it suffices to show that

$$D^2(F + C)(\bar{x}, \bar{y}, \bar{u}, \bar{v})(x) \subset D^2 F(\bar{x}, \bar{y}, \bar{u}, \bar{v})(x) + C.$$

For an arbitrary $x \in \operatorname{dom}(D^2(F + C)(\bar{x}, \bar{y}, \bar{u}, \bar{v}))$, let $y \in D^2(F + C)(\bar{x}, \bar{y}, \bar{u}, \bar{v})(x)$ be arbitrary. Therefore, there are sequences $\{t_n\} \subset \mathbb{P}$, $\{c_n\} \subset C$, and $\{(x_n, y_n)\} \subset X \times Y$ with $t_n \downarrow 0$, $(x_n, y_n + c_n) \to (x, y)$ such that $\bar{y} + t_n\bar{v} + t_n^2 y_n \in F(\bar{x} + t_n\bar{u} + t_n^2 x_n)$. Since the set B is compact and $C = \operatorname{cone}(B)$, for some $r_n > 0$ and for some $b_n \in B$, we have $c_n = r_n b_n$. Without any loss of generality, we assume that $r_n \to r \in [0, \infty]$ and $b_n \to b$, for some $b \in B$.

We claim that $r < \infty$. For the sake of argument, we assume that $r_n \to \infty$. This, in view of the convergence $(y_n + r_n b_n) \to y$, implies that $r_n^{-1} y_n \to -b$, and hence

$$\frac{1}{r_n t_n}\left[\frac{1}{t_n}\left[(\bar{x} + t_n\bar{u} + t_n^2 x_n, \bar{y} + t_n\bar{v} + t_n^2 y_n) - (\bar{x}, \bar{y})\right] - (\bar{u}, \bar{v})\right]$$
$$= \left(\frac{x_n}{r_n}, \frac{y_n}{r_n}\right) \to (0, -b).$$

Therefore $-b \in D_s^2 F(\bar{x}, \bar{y}, \bar{u}, \bar{v})(0) \cap (-C/\{0\})$ which is a contradiction to (11.91). Hence $r < \infty$. This ensures that $y \in D^2 F(\bar{x}, \bar{y}, \bar{u}, \bar{v})(x) + C$. The proof is complete. □

The following result is then immediate.

Corollary 11.7.14. *Let X and Y be normed spaces where the space Y is of finite dimension, and let $C \subset Y$ be a pointed, closed, and convex cone. Let $F : X \rightrightarrows Y$ be a set-valued map, let $(\bar{x}, \bar{y}) \in \operatorname{graph}(F)$, and let $(\bar{u}, \bar{v}) \in X \times Y$. Assume that* (11.91) *remains valid. Then* (11.90) *holds.*

Proof. The proof follows by noticing that, for finite-dimensional Y, the set $B = \{z \in C \mid \|z\| = 1\}$ is a compact base for the cone C. □

By setting $(\bar{u}, \bar{v}) = (0, 0)$ in the above result, we deduce the following corollary:

Corollary 11.7.15. *Let X and Y be normed spaces, let $C \subset Y$ be a pointed, closed, and convex cone. Let $F : X \rightrightarrows Y$ be a set-valued map and let $(\bar{x}, \bar{y}) \in \operatorname{graph}(F)$. Assume that the set $B = \{z \in C \mid \|z\| = 1\}$ is compact. Assume that*

$$D_s F(\bar{x}, \bar{y})(0) \cap (-C/\{0\}) = \emptyset.$$

Then for every $x \in \operatorname{dom}(D(F + C)(\bar{x}, \bar{y}))$, the following identity holds:

$$D(F + C)(\bar{x}, \bar{y})(x) = DF(\bar{x}, \bar{y})(x) + C.$$

Remark 11.7.16. We note that in Example 11.7.10, we have $-y \in D_S F(0, 1)(0)$ for all $y > 0$, consistent with the above corollary.

In the following we extend two notions of compactness and employ them to prove (11.90).

We begin with the following:

Definition 11.7.17. Let X and Y be normed spaces, let $F : X \rightrightarrows Y$ be a set-valued map, let $(\bar{x}, \bar{y}) \in \operatorname{graph}(F)$, and let $(\bar{u}, \bar{v}) \in X \times Y$. The map F is called **second-order directionally compact** at $(\bar{x}, \bar{y})$ with respect to $(\bar{u}, \bar{v})$ in a direction $x \in X$, if for every sequence $\{t_n\} \subset \mathbb{P}$ with $t_n \downarrow 0$ and every sequence $\{x_n\} \subset X$ with $x_n \to x$, any sequence $\{y_n\} \subset Y$ contains a convergent subsequence provided that $\bar{y} + t_n\bar{v} + t_n^2 y_n \in F(\bar{x} + t_n\bar{u} + t_n^2 x_n)$.

Remark 11.7.18. If $(\bar{u}, \bar{v}) = (0, 0)$, then the second-order directionally compact map F at $(\bar{x}, \bar{y})$ with respect to $(\bar{u}, \bar{v})$ in a direction $x \in X$, reduces to the notion of directional compactness at $(\bar{x}, \bar{y})$ in the direction x. The notion of directional compactness, introduced in [39] is very important in set-valued optimization. The above notion of second-order directional compactness was introduced in [323]. Second-order directional compactness holds, in particular, in the case where F is single-valued, the Hadamard directional derivative

$$\bar{v} := F'(\bar{x}; \bar{u}) := \lim_{t \downarrow 0, u \to \bar{u}} \frac{F(\bar{x} + tu) - F(\bar{x})}{t}$$

exists, and the parabolic second-order directional derivative

$$F''(\bar{x},\bar{u};x) := \lim_{t\downarrow 0, w\to x} \frac{F(\bar{x}+t\bar{u}+t^2w)-F(\bar{x})-tF'(\bar{x};\bar{u})}{t^2}$$

exists. To see this, suppose that $t_n \downarrow 0$, $x_n \to x$, and $\bar{y}+t_n\bar{v}+t_n^2y_n \in F(\bar{x}+t_n\bar{u}+t_n^2x_n)$. Then

$$y_n = \frac{F(\bar{x}+t_n\bar{u}+t_n^2x_n)-F(\bar{x})-t_nF'(\bar{x};\bar{u})}{t_n^2},$$

so that $y_n \to F''(\bar{x},\bar{u};x)$. □

With the concept of second-order directional compactness, we give another proof of (11.90).

Theorem 11.7.19. *Let X and Y be normed spaces, and let $C \subset Y$ be a pointed, closed, and convex cone. Let $F : X \rightrightarrows Y$ be a set-valued map, let $(\bar{x},\bar{y}) \in$ graph (F), and let $(\bar{u},\bar{v}) \in X\times Y$. Assume that F is second-order directionally compact at $(\bar{x},\bar{y})$ with respect to $(\bar{u},\bar{v})$ in any direction $x\in X$. Then (11.90) holds.*

Proof. In view of (11.88), it suffices to show that

$$D^2(F+C)(\bar{x},\bar{y},\bar{u},\bar{v})(x) \subset D^2F(\bar{x},\bar{y},\bar{u},\bar{v})(x)+C.$$

Let $y \in D^2(F+C)(\bar{x},\bar{y},\bar{u},\bar{v})(x)$. Then there exist sequences $(t_n)\subset \mathbb{P}$, $(x_n,y_n)\subset X\times Y$, and $c_n\in C$ such that $t_n\downarrow 0$, $(x_n,y_n)\to(x,y)$, and

$$\bar{y}+t_n\bar{v}+t_n^2(y_n-c_n/t_n^2) = \bar{y}+t_n\bar{v}+t_n^2y_n-c_n \in F(\bar{x}+t_n\bar{u}+t_n^2x_n).$$

Since F is second-order directionally compact, we may assume that $y_n - c_n/t_n^2$ converges to some $\tilde{y}\in Y$. Hence $\tilde{y}\in D^2F(\bar{x},\bar{y},\bar{u},\bar{v})(x)$ and $c_n/t_n^2 \to (y-\tilde{y})\in C$, confirming that $y\in D^2F(\bar{x},\bar{y},\bar{u},\bar{v})(x)+C$. The proof is complete. □

In the following definition, we give a notion of second-order compactly approximable set-valued maps.

Definition 11.7.20. Let X and Y be normed spaces, and let $C\subset Y$ be a pointed, closed, and convex cone. Let $F : X \rightrightarrows Y$ be a set-valued map, let $(\bar{x},\bar{y})\in$ graph (F), and let $(\bar{u},\bar{v})\in X\times Y$. The map $F : X\rightrightarrows Y$ is called **second-order compactly approximable** at $(\bar{x},\bar{y})$ with respect to $(\bar{u},\bar{v})$, if for each $x'\in X$, there exists a set-valued map R from X into the set of all nonempty compact subsets of Y, a neighborhood $\mathscr{N}(\bar{x})$ of $\bar{x}$ in X, and a function $r:]0,1]\times X \to]0,\infty]$ such that

1. $\lim_{(t,x)\to(0,x')} r(t,x) = 0.$
2. For each $x\in X$ and $t\in]0,1]$, we have

$$F(\bar{x}+t\bar{u}+t^2x) \subset \bar{y}+t\bar{v}+t^2(R(x')+r(t,x)B_Y).$$

Remark 11.7.21. If $(\bar{u}, \bar{v}) = (0, 0)$, then the second-order compactly approximable map F at $(\bar{x}, \bar{y})$ with respect to $(\bar{u}, \bar{v})$ in a direction $x \in X$, reduces to the notion of the compactly approximable map at $(\bar{x}, \bar{y})$ in the direction $x \in X$ used in Taa [548]. We remark that set-valued compactly approximable map were originally introduced by Amahroq and Thibault [6].

The following result is obtained by the above concept of second-order compactly approximable maps.

Theorem 11.7.22. *Let X and Y be normed spaces, and let $C \subset Y$ be a pointed, closed, and convex cone. Let $F : X \rightrightarrows Y$ be a set-valued map, let $(\bar{x}, \bar{y}) \in \operatorname{graph}(F)$, and let $(\bar{u}, \bar{v}) \in X \times Y$. If the map F is second-order compactly approximable at $(\bar{x}, \bar{y})$ with respect to $(\bar{u}, \bar{v})$, then (11.90) holds.*

Proof. For $x \in \operatorname{dom}(D^2(F + C)(\bar{x}, \bar{y}, \bar{u}, \bar{v}))$, let $y \in D^2(F + C)(\bar{x}, \bar{y}, \bar{u}, \bar{v})(x)$. Hence there exist sequences $\{t_n\} \subset \mathbb{P}$ and $\{(x_n, y_n)\} \subset X \times Y$ such that $t_n \downarrow 0$, $(x_n, y_n) \to (x, y)$ and $\bar{y} + t_n\bar{v} + t_n^2 y_n \in F(\bar{x} + t_n\bar{u} + t_n^2 x_n) + C$.

Then $\bar{y} + t_n\bar{v} + t_n^2 y_n = z_n + c_n$ with $z_n \in F(\bar{x} + t_n\bar{u} + t_n^2 x_n)$ and $c_n \in C$. Since F is second-order compactly approximable, there exists $\{k_n\} \subset R(x)$, $r_n := r(t_n, x_n)$, $b_n \in B_Y$ and $n_0 \in \mathbb{N}$ such that $r_n \to 0$, and $z_n = \bar{y} + t_n\bar{v} + t_n^2(k_n + r_n b_n)$ for all $n > n_0$. This implies $\bar{y} + t_n\bar{v} + t_n^2(k_n + r_n b_n) \in F(\bar{x} + t_n\bar{u} + t_n^2 x_n)$ for all $n > n_0$.

In view of the compactness of $R(x)$, we may assume that $k_n \to k \in R(x)$. Because of the fact that $(k_n + r_n b_n) \to k$, we deduce that $k \in D^2 F(\bar{x}, \bar{y}, \bar{u}, \bar{v})(x)$. Since $\bar{y} + t_n\bar{v} + t_n^2 y_n = \bar{y} + t_n\bar{v} + t_n^2(k_n + r_n b_n) + c_n$ for sufficiently large n, we have $y_n - k_n - r_n b_n = t_n^{-2} c_n \in C$. By passing to the limit, we have $y - k \in C$ and hence $y \in k + C \subset D^2 F(\bar{x}, \bar{y}, \bar{u}, \bar{v})(x) + C$. The proof is complete. □

In view of Lemma 11.5.19, we obtain the following result.

Theorem 11.7.23. *Let X and Y be normed spaces, and let $C \subset Y$ be a pointed, closed, and convex cone. Let $F : X \rightrightarrows Y$ be a set-valued map, let $(\bar{x}, \bar{y}) \in \operatorname{graph}(F)$, and let $(\bar{u}, \bar{v}) \in X \times Y$. Assume that F is second-order contingently absorbing at $(\bar{x}, \bar{y})$ in direction $(\bar{u}, \bar{v})$. Then for every $x \in \operatorname{dom}(D^2 F(\bar{x}, \bar{y}, \bar{u}, \bar{v}))$, we have*

1. $\operatorname{GHe-PMin}(D^2(F + C)(\bar{x}, \bar{y}, \bar{u}, \bar{v})(x), C) = \operatorname{GHe-PMin}(D^2 F(\bar{x}, \bar{y}, \bar{u}, \bar{v})(x), C)$.
2. $\operatorname{Min}(D^2(F + C)(\bar{x}, \bar{y}, \bar{u}, \bar{v})(x), C) = \operatorname{Min}(D^2 F(\bar{x}, \bar{y}, \bar{u}, \bar{v})(x), C)$.
3. $\operatorname{WMin}(D^2(F + C)(\bar{x}, \bar{y}, \bar{u}, \bar{v})(x), C) \supseteq \operatorname{WMin}(D^2 F(\bar{x}, \bar{y}, \bar{u}, \bar{v})(x), C)$.

Proof. Since for every $x \in \operatorname{dom}(D^2 F(\bar{x}, \bar{y}, \bar{u}, \bar{v}))$, we have

$$D^2 F(\bar{x}, \bar{y}, \bar{u}, \bar{v})(x) + C = D^2(F + C)(\bar{x}, \bar{y}, \bar{u}, \bar{v})(x),$$

the proof follows directly from Lemma 11.5.19 by setting $A = D^2 F(\bar{x}, \bar{y}, \bar{u}, \bar{v})(x)$, and $B = D^2(F + C)(\bar{x}, \bar{y}, \bar{u}, \bar{v})(x)$. □

The following result is an analogue of Theorem 11.7.23 without (11.90):

Theorem 11.7.24. *Let X and Y be normed spaces, and let $C \subset Y$ be a pointed, closed, and convex cone. Let $F : X \rightrightarrows Y$ be a set-valued map, let $(\bar{x}, \bar{y}) \in \operatorname{graph}(F)$, and let $(\bar{u}, \bar{v}) \in X \times Y$. Assume that the set $B = \{z \in C \mid \|z\| = 1\}$ is compact. Then for every $x \in \operatorname{dom}(D^2(F + C)(\bar{x}, \bar{y}, \bar{u}, \bar{v}))$, the following inclusion holds*

$$\operatorname{Min}(D^2(F + C)(\bar{x}, \bar{y}, \bar{u}, \bar{v})(x), C) \subset D^2 F(\bar{x}, \bar{y}, \bar{u}, \bar{v})(x). \tag{11.92}$$

Proof. For $x \in \operatorname{dom}(D^2(F + C)(\bar{x}, \bar{y}, \bar{u}, \bar{v}))$, let $y \in \operatorname{Min}(D^2(F + C)(\bar{x}, \bar{y}, \bar{u}, \bar{v})(x), C)$ be arbitrary. Then $y \in D^2(F + C)(\bar{x}, \bar{y}, \bar{u}, \bar{v})(x)$, and hence there exist sequences $(t_n) \subset \mathbb{P}$, $(x_n, y_n) \subset X \times Y$, and $c_n \in C$ such that $t_n \downarrow 0$, $(x_n, y_n) \to (x, y)$, and $\bar{y} + t_n\bar{v} + t_n^2(y_n - c_n) \in F(\bar{x} + t_n\bar{u} + t_n^2 x_n)$.

Since $C = \operatorname{cone}(B)$, we have $c_n = r_n b_n$ for some $r_n > 0$ and $b_n \in B$. Moreover, because B is compact, we may assume that $b_n \to b \in B$. We will show that $r_n \to 0$. If not, we may assume, taking a subsequence if necessary, that there exists $\varepsilon > 0$ such that $r_n > \varepsilon$. We set $\bar{c}_n = \varepsilon c_n / r_n$. Notice that $c_n - \bar{c}_n \in C$, from which it follows that $\bar{y} + t_n\bar{v} + t_n^2(y_n - \bar{c}_n) \in F(\bar{x} + t_n\bar{u} + t_n^2 x_n) + C$.

Since $\bar{c}_n \to \varepsilon b$, we have that $y - \varepsilon b \in D^2(F + C)(\bar{x}, \bar{y}, \bar{u}, \bar{v})(x)$ contradicting the C-minimality of y. Therefore $r_n \to 0$, and this implies that $(y_n - c_n) \to y$. Consequently, $y \in D^2 F(\bar{x}, \bar{y}, \bar{u}, \bar{v})(x)$, and the proof is complete. □

By setting $(\bar{u}, \bar{v}) = (0, 0)$ in the above result, we recover the following result.

Corollary 11.7.25. *Let X and Y be normed spaces, let $C \subset Y$ be a pointed, closed, and convex cone, let $F : X \rightrightarrows Y$ be a set-valued map, and let $(\bar{x}, \bar{y}) \in \operatorname{graph}(F)$ be arbitrary. Assume that the set $B = \{z \in C \mid \|z\| = 1\}$ is compact. Then for every $x \in \operatorname{dom}(D(F + C)(\bar{x}, \bar{y}))$, the following inclusion holds*

$$\operatorname{Min}(D(F + C)(\bar{x}, \bar{y})(x), C) \subset DF(\bar{x}, \bar{y})(x).$$

Example 11.7.26. As an illustration of Theorem 11.7.24 and Corollary 11.7.25, let $F : \mathbb{R} \rightrightarrows \mathbb{R}^2$ be defined by $F(x) := \{(y_1, y_2) \in \mathbb{R}^2_+ \mid y_1 y_2 = x\}$, and let $C = \mathbb{R}^2_+$, $\bar{x} = 1$, and $\bar{y} = (1, 1)$. Then for all $x \in \mathbb{R}$,

$$\begin{aligned} DF(\bar{x}, \bar{y})(x) &= \{(y_1, y_2) \mid x = y_1 + y_2\}, \\ D(F + C)(\bar{x}, \bar{y})(x) &= \{(y_1, y_2) \mid x \leq y_1 + y_2\}, \end{aligned}$$

so that $\operatorname{Min}_C D(F + C)(\bar{x}, \bar{y})(x) = DF(\bar{x}, \bar{y})(x)$, consistent with Corollary 11.7.25.

For $\bar{v} = (v_1, v_2)$ and $\bar{u} = v_1 + v_2$, one obtains (e.g., by Proposition 2.3 of [594])

$$\begin{aligned} D^2 F(\bar{x}, \bar{y}, \bar{u}, \bar{v})(x) &= \{(y_1, y_2) \mid x = y_1 + y_2 + v_1 v_2\}, \\ D^2(F + C)(\bar{x}, \bar{y}, \bar{u}, \bar{v})(x) &= \{(y_1, y_2) \mid x \leq y_1 + y_2 + v_1 v_2\}, \end{aligned}$$

so that equality holds in (11.92). □

11.8 Calculus Rules for Second-Order Contingent Derivatives

We have seen in Sect. 11.2 of this chapter that a quite elegant approach for various calculus rules for derivatives of set-valued maps can be developed by exploiting certain properties of the involved tangent cones. In this section, we propose a second-order extension of this approach and obtain calculus rules for the second-order graphical derivatives. Following the scheme of Sect. 11.2, we present abstract chain rules and sum formulas and then derive some particular cases of interest. This section is influenced by Aubin and Frankowska [18] and Ward [594, 596].

We begin with the following general scheme for an abstract chain rule:

Theorem 11.8.1. *Let X, Y, and Z be normed spaces, let $F : X \rightrightarrows Y$ and $G : Y \rightrightarrows Z$ be set-valued maps, let $(\bar{x}, \bar{y}) \in \operatorname{graph}(F)$, let $(\bar{y}, \bar{z}) \in \operatorname{graph}(G)$, and let $(\bar{u}, \bar{v}, \bar{w}) \in X \times Y \times Z$. Assume that P^2 and Q^2 are second-order closed tangent sets such that for the maps $\ell : X \times Y \times Y \times Z \to X \times Z$ and $g : X \times Y \times Y \times Z \to Y$, given by*

$$\ell(a, b, c, d) := (a, d), \tag{11.93}$$

$$g(a, b, c, d) := b - c, \tag{11.94}$$

and for $S_1 := \operatorname{graph}(F)$, $S_2 := \operatorname{graph}(G)$, $S := S_1 \times S_2$, $x_0 := (\bar{x}, \bar{y}, \bar{y}, \bar{z})$, and $u_0 := (\bar{u}, \bar{v}, \bar{v}, \bar{w})$, we have

$$\ell(P^2(S, x_0, u_0)) \subseteq P^2(\ell(S), \ell(x_0), \ell(u_0)), \tag{11.95}$$

$$P^2(S_1, (\bar{x}, \bar{y}), (\bar{u}, \bar{v})) \times Q^2(S_2, (\bar{y}, \bar{z}), (\bar{v}, \bar{w})) \subseteq P^2(S_1 \times S_2, (x_0, u_0)), \tag{11.96}$$

$$P^2(S, x_0, u_0) \cap \nabla g(x_0)^{-1}(0) \subseteq P^2(S \cap g^{-1}(0), x_0, u_0). \tag{11.97}$$

Then, for $D^2_P G(\bar{y}, \bar{z}, \bar{v}, \bar{w})$ and $D^2_Q F(\bar{x}, \bar{y}, \bar{u}, \bar{v})$, we have

$$D^2_Q G(\bar{y}, \bar{z}, \bar{v}, \bar{w}) \circ D^2_P F(\bar{x}, \bar{y}, \bar{u}, \bar{v}) \subseteq D^2_P\,(G \circ F)\,(\bar{x}, \bar{z}, \bar{u}, \bar{w}). \tag{11.98}$$

Proof. Following the reasoning used in Theorem 11.2.13, we have

$$\begin{aligned}
&\operatorname{graph}\left(D^2_P\,(G \circ F)\,(\bar{x}, \bar{z}, \bar{u}, \bar{w})\right)\\
&\quad = P^2\,(\operatorname{graph}(G \circ F), (\bar{x}, \bar{z}), (\bar{u}, \bar{w}))\\
&\quad = P^2(\ell\left([\operatorname{graph}(F) \times \operatorname{graph}(G)] \cap g^{-1}(0)\right), (\bar{x}, \bar{z}), (\bar{u}, \bar{w}))\\
&\quad \supset \ell\left(P^2([\operatorname{graph}(F) \times \operatorname{graph}(G)] \cap g^{-1}(0), (\bar{x}, \bar{y}, \bar{y}, \bar{z}), (\bar{u}, \bar{v}, \bar{v}, \bar{w})\right)\\
&\quad \supset \ell\left(P^2\,(\operatorname{graph}(F) \times \operatorname{graph}(G), (\bar{x}, \bar{y}, \bar{y}, \bar{z}), (\bar{u}, \bar{v}, \bar{v}, \bar{w})\right)
\end{aligned}$$

$$\begin{aligned}
&\cap \nabla g(\bar{x},\bar{y},\bar{y},\bar{z})^{-1}(0)\big) \\
&\supset \ell\left(P^2\,(\mathrm{graph}\,(F),(\bar{x},\bar{y}),(\bar{u},\bar{v}))\right. \\
&\quad \left.\times Q^2\,(\mathrm{graph}\,(G),(\bar{y},\bar{z}),(\bar{v},\bar{w}))\cap g^{-1}(0)\right) \\
&= \ell\left(\mathrm{graph}\,(D_{\mathrm{P}}^2 F(\bar{x},\bar{y},\bar{u},\bar{v}))\times \mathrm{graph}\,(D_{\mathrm{Q}} G(\bar{y},\bar{z},\bar{v},\bar{w}))\cap g^{-1}(0)\right) \\
&= \mathrm{graph}\,\left(D_{\mathrm{Q}}^2 G(\bar{y},\bar{z},\bar{v},\bar{w})\circ D_{\mathrm{P}}^2 F(\bar{x},\bar{y},\bar{u},\bar{v})\right),
\end{aligned}$$

and the proof is complete. □

The above result depicts that a chain rule can be given by using a combination of second-order tangent sets satisfying (11.95)–(11.97). In the following, we collect a few results and identify the conditions where these requirements are met. The following result was given by Ward [594]:

Theorem 11.8.2. *Let X, Y, and Z be finite dimensional, let $F : X \rightrightarrows Y$ and $G : Y \rightrightarrows Z$ be set-valued maps, let $(\bar{x},\bar{y}) \in \mathrm{graph}\,(F)$, let $(\bar{y},\bar{z}) \in \mathrm{graph}\,(G)$, and let $(\bar{u},\bar{v},\bar{w}) \in X\times Y\times Z$. Assume that F is closed near $(\bar{x},\bar{y})$, G is closed near $(\bar{y},\bar{z})$ and the following condition holds:*

$$\mathrm{range}(D_c F(\bar{x},\bar{y})) - \mathrm{dom}\,(D_c G(\bar{y},\bar{z})) = Y. \tag{11.99}$$

Then the following inclusions hold:

$$D_A^2 G(\bar{y},\bar{z},\bar{v},\bar{w})\circ D^2 F(\bar{x},\bar{y},\bar{u},\bar{v}) \subseteq D^2(G\circ F)(\bar{x},\bar{z},\bar{u},\bar{w}), \tag{11.100}$$

$$D_A^2 G(\bar{y},\bar{z},\bar{v},\bar{w})\circ D_A^2 F(\bar{x},\bar{y},\bar{u},\bar{v}) \subseteq D_A^2(G\circ F)(\bar{x},\bar{z},\bar{u},\bar{w}), \tag{11.101}$$

$$D^2 G(\bar{y},\bar{z},\bar{v},\bar{w})\circ D_A^2 F(\bar{x},\bar{y},\bar{u},\bar{v}) \subseteq D^2(G\circ F)(\bar{x},\bar{z},\bar{u},\bar{w}). \tag{11.102}$$

Proof. The proof follows by applying Theorem 4.7.7 to $\{0\}$ and $\mathrm{graph}\,(F)\times \mathrm{graph}\,(G)$. We have

$$\begin{aligned}
&\nabla g(\bar{x},\bar{y},\bar{y},\bar{z}) C(\mathrm{graph}\,(F)\times \mathrm{graph}\,(G),(\bar{x},\bar{y},\bar{y},\bar{z})) - C(0,0) \\
&\quad = \mathrm{range}(D_c F(\bar{x},\bar{y})) - \mathrm{dom}\,(D_c G(\bar{y},\bar{z})),
\end{aligned}$$

and hence the required properties hold. □

We now give a general scheme for an abstract sum formula.

Theorem 11.8.3. *Let X and Y be normed spaces, let $F_1, F_2 : X \rightrightarrows Y$ be set-valued maps, let $f : X \to X$ be a linear map, let $(\bar{x},\bar{y}_1) \in \mathrm{graph}\,(F_1)$, and let $(\bar{x},\bar{y}_2) \in \mathrm{graph}\,(F_2\circ f)$. Assume that P^2 and Q^2 are closed second-order tangent sets such that for the maps $\ell : (X\times Y)^2 \to X\times Y$ and $g : (X\times Y)^2 \to X$ defined by*

$$\ell(a,b,c,d) := (a, c+d), \tag{11.103}$$

$$g(a,b,c,d) := f(a) - c, \tag{11.104}$$

and for $S_1 :=$ graph (F_1), $S_2 :=$ graph (F_2), $S := S_1 \times S_2$, $x_0 := (\bar{x}, \bar{y}_1, f(\bar{x}), \bar{y}_2)$, *and* $u_0 := (\bar{u}, \bar{v}_1, f(\bar{u}), \bar{v}_2)$, *we have*

$$\ell(P^2(S, x_0, u_0)) \subset P^2(\ell(S), \ell(x_0), \ell(u_0)),$$

$$P^2(S_1, (\bar{x}, \bar{y}_1), (\bar{u}, \bar{v}_1)) \times Q^2(S_2, (f(\bar{x}), \bar{y}_2), (f(\bar{u}), \bar{v}_2)) \subset P^2(S_1 \times S_2, (x_0, u_0)),$$

$$P^2(S, x_0, \bar{u}_0) \cap \nabla g(x_0)^{-1}(0) \subset P^2(S \cap g^{-1}(0), x_0, u_0).$$

Then the following inclusion holds:

$$\begin{aligned} &D^2_P F_1(\bar{x}, \bar{y}_1, \bar{u}, \bar{v}_1)(\cdot) + D^2_Q F_2(f(\bar{x}), \bar{y}_2, f(\bar{u}), \bar{v}_2)(f(\cdot)) \\ &\subseteq D^2_P (F_1 + F_2 \circ f)(\bar{x}, \bar{y}_1 + \bar{y}_2, f(\bar{u}), \bar{v}_1 + \bar{v}_2)(\cdot). \end{aligned}$$

Proof. Proceeding as in the proof of Theorem 11.8.1, we have

$$\begin{aligned} &\text{graph}\left(D^2_P (F_1 + F_2 \circ f)(\bar{x}, \bar{y}_1 + \bar{y}_2, \bar{u}, \bar{v}_1 + \bar{v}_2)\right) \\ &\quad = P^2(\text{graph}(F_1 + F_2 \circ f), (\bar{x}, \bar{y}_1 + \bar{y}_2), (\bar{u}, \bar{v}_1 + \bar{v}_2)) \\ &\quad = P^2(\ell([\text{graph}(F_1) \\ &\qquad \times \text{graph}(F_2)] \cap g^{-1}(0)), \ell(\bar{x}, \bar{y}_1, f(\bar{x}), \bar{y}_2), \ell(\bar{u}, \bar{v}_1, f(\bar{u}), \bar{v}_2)) \\ &\quad \supset \ell\left(P^2([\text{graph}(F_1) \\ &\qquad \times \text{graph}(F_2)] \cap g^{-1}(0), (\bar{x}, \bar{y}_1, f(\bar{x}), \bar{y}_2), (\bar{u}, \bar{v}_1, f(\bar{u}), \bar{v}_2)\right) \\ &\quad \supset \ell\left(P^2(\text{graph}(F_1) \\ &\qquad \times \text{graph}(F_2), (\bar{x}, \bar{y}_1, f(\bar{x}), \bar{y}_2), (\bar{u}, \bar{v}_1, f(\bar{u}), \bar{v}_2)) \bigcap \nabla g(\bar{x}, \bar{y}_1, f(\bar{x}), \bar{y}_2)^{-1}(0)\right) \\ &\quad \supset \ell\left(P^2(\text{graph}(F_1), (\bar{x}, \bar{y}_1), (\bar{u}, \bar{v}_1)) \\ &\qquad \times Q^2(\text{graph}(F_2), (f(\bar{x}), \bar{y}_2), (f(\bar{u}), \bar{v}_2)) \bigcap \nabla g(\bar{x}, \bar{y}_1, f(\bar{x}), \bar{y}_2)^{-1}(0)\right) \\ &\quad = \ell\left(\text{graph}(D^2_P F_1(\bar{x}, \bar{y}_1, \bar{u}, \bar{v}_1)) \\ &\qquad \times \text{graph}(D^2_Q F_2(f(\bar{x}), \bar{y}_2, f(\bar{u}), \bar{v}_2)) \bigcap \nabla g(\bar{x}, \bar{y}_1, f(\bar{x}), \bar{y}_2)^{-1}(0)\right) \\ &\quad = \ell\{(a,b,c,d) | (a,b) \in \text{graph}(D^2_P F_1(\bar{x}, \bar{y}_1, \bar{u}, \bar{v}_1)) \\ &\qquad (c,d) \in \text{graph}(D^2_Q F_2(f(\bar{x}), \bar{y}_2, f(\bar{u}), \bar{v}_2)),\ \nabla f(\bar{x})(a) = c\}, \\ &\quad = \{(a, b+d) | (a,b) \in \text{graph}(D^2_P F_1(\bar{x}, \bar{y}_1), (\bar{u}, \bar{v}_1)), \end{aligned}$$

$$(c,d) \in \text{graph}\,(D^2_P F_2(f(\bar{x}),\bar{y}_1),(f(\bar{u}),\bar{v}_1),\ f(a)=c\},$$
$$= \text{graph}\,\left(D^2_P F_1(\bar{x},\bar{y}_1,\bar{u},\bar{v}_1)) + D^2_Q F_2(f(\bar{x}),\bar{y}_2,f(\bar{u}),\bar{v}_2)(f(\cdot))\right),$$

where we used the fact that $\nabla f(\bar{x})(x) = f(x)$. The proof is complete. □

The following is a second-order sum formula for the second-order contingent derivatives. This is an analogue of Theorem 11.2.19 and was recently given in [155].

Theorem 11.8.4. *Let X and Y be Banach spaces, let $F_1, F_2 : X \rightrightarrows Y$ be closed set-valued maps, let $f : X \to X$ be a linear, let $(\bar{x},\bar{y}_1) \in$ graph (F_1), and let $(\bar{x},\bar{y}_2) \in$ graph $(F_2 \circ f)$. Assume that the map $g : X \times Y \times X \times Y \to X$ given by $g(a,b,c,d) = f(a)-c$ is metrically subregular at $(\bar{x},\bar{y}_1,f(\bar{x}),\bar{y}_2,0)$ with respect to* graph $(F_1) \times$ graph (F_2). *Assume that either* graph (F_1) *is second-order derivable at $(\bar{x},\bar{y}_1)$ in any direction or* graph (F_2) *is second-order derivable at $(f(\bar{x}),\bar{y}_2)$ in any direction.*

Then the following sum formula, for the second-order contingent derivatives, holds:

$$D^2 F_1(\bar{x},\bar{y}_1,\bar{u},\bar{v}_1)(\cdot) + D^2 F_2(f(\bar{x}),\bar{y}_2,f(\bar{u}),\bar{v}_2)(f(\cdot))$$
$$\subseteq D^2\,(F_1 + F_2 \circ f)\,(\bar{x},\bar{y}_1+\bar{y}_2,f(\bar{u}),\bar{v}_1+\bar{v}_2)(\cdot).$$

We conclude this section by a second-order analogue of Theorem 11.2.7 given by Ward [596]. This results employs the notion of the second-order directional compact:

Theorem 11.8.5. *Let X, Y, and Z be normed spaces, let $F : X \rightrightarrows Y$ and $G : Y \rightrightarrows Z$ be set-valued maps, let $(\bar{x},\bar{y}) \in$ graph (F), let $(\bar{y},\bar{z}) \in$ graph (G), and let $(\bar{u},\bar{v},\bar{w}) \in X \times Y \times Z$. Define $H : X \rightrightarrows Z$ be*

$$H(x,z) := \{y \in Y \mid y \in F(x),\ z \in G(y)\}.$$

Let H be second-order directionally compact at $((\bar{x},\bar{z}),\bar{y})$ with respect to $((\bar{u},\bar{w}),\bar{v})$ in the direction $(x,z) \in T^2(\text{graph}\,(G \circ F)),(\bar{x},\bar{z}),(\bar{u},\bar{w}))$. Then

$$(x,z) \in \text{graph}\,(D^2G(\bar{y},\bar{z},\bar{v},\bar{w}) \circ D^2F(\bar{x},\bar{y},\bar{u},\bar{v})).$$

Proof. Let $(x,z) \in$ graph $(D^2(G\circ F)(\bar{x},\bar{z},\bar{u},\bar{w}))$ with H second-order directionally compact at $((\bar{x},\bar{z}),\bar{y})$ with respect to $((\bar{u},\bar{w}),\bar{v})$ in the direction (x,z). Then there exist sequences $\{t_n\} \subset \mathbb{P}$ and $\{(x_n,z_n)\}$ with $t_n \downarrow 0$, $(x_n,z_n) \to (x,z)$ such that $(\bar{z}+t_n\bar{w}+t_n^2 z_n^2) \in (G \circ F)(\bar{x}+t_n\bar{u}+t_n^2 x_n)$ for every $n \in \mathbb{N}$. Therefore, there exists $\{w_n\}$ such that $w_n \in F(\bar{x}+t_n\bar{u}+t_n^2 x_n)$ and $\bar{z}+t_n\bar{w}+t_n^2 z_n^2 \in G(w_n)$. By setting $y_n := t_n^{-1}(w_n - \bar{y} - t_n\bar{v})$, we have $\bar{y}+t_n\bar{v}+t_n^2 y_n \in H(\bar{x}+t_n\bar{u}+t_n^2 x_n, \bar{z}+t_n\bar{w}+t_n^2 z_n^2)$. Now, by invoking the second-order directional compactness, we may assume, by taking subsequence if necessary, that $\{y_n\} \to y$ for some $y \in Y$. Then $(x,y) \in$

$T^2(\text{graph}(F), (\bar{x}, \bar{y}), (\bar{u}, \bar{v}))$ and $(y, z) \in T^2(\text{graph}(G), (\bar{y}, \bar{z}), (\bar{v}, \bar{z}))$ implying that $(x, z) \in \text{graph}(D^2G(\bar{y}, \bar{z}, \bar{v}, \bar{w}) \circ D^2F(\bar{x}, \bar{y}, \bar{u}, \bar{v}))$. The proof is complete. □

11.9 Second-Order Epiderivatives of Set-Valued Maps

In this section, we study second-order epiderivatives of set-valued maps. We begin with the following:

Definition 11.9.1. Let X and Y be normed spaces and let $C \subset Y$ be a pointed, closed and convex cone. Let $F : X \rightrightarrows Y$ be a set-valued map, let $(\bar{x}, \bar{y}) \in \text{graph}(F)$, and let $(\bar{u}, \bar{v}) \in X \times Y$.

1. A single-valued map $D^2_E F(\bar{x}, \bar{y}, \bar{u}, \bar{v}) : X \to Y$ is called the **second-order contingent epiderivative** of F at $(\bar{x}, \bar{y})$ in the direction $(\bar{u}, \bar{v})$, if
$$\text{epi}(D^2_E F(\bar{x}, \bar{y}, \bar{u}, \bar{v})) = T^2(\text{epi}(F), (\bar{x}, \bar{y}), (\bar{u}, \bar{v})),$$
where $T^2(\text{epi}(F), (\bar{x}, \bar{y}), (\bar{u}, \bar{v}))$ is the second-order contingent set.
2. A set-valued map $D^2_G F(\bar{x}, \bar{y}, \bar{u}, \bar{v}) : X \rightrightarrows Y$ is called the **second-order generalized contingent epiderivative** of F at $(\bar{x}, \bar{y})$ in direction $(\bar{u}, \bar{v})$, if
$$D^2_G F(\bar{x}, \bar{y}, \bar{u}, \bar{v})(x) = \text{Min}(D^2(F + C)(\bar{x}, \bar{y}, \bar{u}, \bar{v})(x), C)$$
$$x \in \text{dom}(D^2(F + C)(\bar{x}, \bar{y}, \bar{u}, \bar{v})),$$
where $D^2(F + C)(\bar{x}, \bar{y}, \bar{u}, \bar{v}))$ is the second-order contingent derivative.
3. A single-valued map $\mathbb{D}^2_E F(\bar{x}, \bar{y}, \bar{u}, \bar{v}) : X \to Y$ is called the **second-order asymptotic contingent epiderivative** of F at $(\bar{x}, \bar{y})$ in the direction $(\bar{u}, \bar{v})$, if
$$\text{epi}(\mathbb{D}^2_E F(\bar{x}, \bar{y}, \bar{u}, \bar{v})) = \mathbb{T}^2(\text{epi}(F), (\bar{x}, \bar{y}), (\bar{u}, \bar{v})),$$
where $\mathbb{T}^2(\text{epi}(F), (\bar{x}, \bar{y}), (\bar{u}, \bar{v}))$ is the second-order asymptotic contingent cone.
4. A set-valued map $\mathbb{D}^2_G F(\bar{x}, \bar{y}, \bar{u}, \bar{v}) : X \rightrightarrows Y$ is called the **second-order generalized asymptotic contingent epiderivative** of F at $(\bar{x}, \bar{y})$ in direction $(\bar{u}, \bar{v})$, if
$$\mathbb{D}^2_G F(\bar{x}, \bar{y}, \bar{u}, \bar{v})(x) = \text{Min}(\mathbb{D}^2(F + C)(\bar{x}, \bar{y}, \bar{u}, \bar{v})(x), C)$$
$$x \in \text{dom}(\mathbb{D}^2(F + C)(\bar{x}, \bar{y}, \bar{u}, \bar{v})),$$
where $\mathbb{D}^2(F + C)(\bar{x}, \bar{y}, \bar{u}, \bar{v}))$ is the second-order asymptotic contingent derivative.
5. A single-valued map $\mathscr{D}^2_E F(\bar{x}, \bar{y}, \bar{u}, \bar{v}) : X \to Y$ is called the **second-order compound contingent epiderivative** of F at $(\bar{x}, \bar{y})$ in the direction $(\bar{u}, \bar{v})$ if

$$\text{epi}\,(\mathscr{D}^2_E F(\bar{x},\bar{y},\bar{u},\bar{v})) = T(T((\text{epi}\,(F),(\bar{x},\bar{y})),(\bar{u},\bar{v})),$$

where $T(T((\text{epi}\,(F),(\bar{x},\bar{y})),(\bar{u},\bar{v}))$ is the contingent cone of $T(\text{epi}\,(F),(\bar{x},\bar{y}))$ at $(\bar{u},\bar{v})$.

6. A set-valued map $\mathscr{D}^2_G F(\bar{x},\bar{y},\bar{u},\bar{v}) : X \rightrightarrows Y$ is called the **second-order compound generalized contingent epiderivative** of F at $(\bar{x},\bar{y})$ in direction $(\bar{u},\bar{v})$ if

$$\mathscr{D}^2_G F(\bar{x},\bar{y},\bar{u},\bar{v})(x) = \text{Min}(\mathscr{D}^2(F+C)(\bar{x},\bar{y},\bar{u},\bar{v})(x), C)$$
$$x \in \text{dom}\,(\mathscr{D}^2(F+C)(\bar{x},\bar{y},\bar{u},\bar{v})),$$

where $\mathscr{D}^2(F+C)(\bar{x},\bar{y},\bar{u},\bar{v}))$ is the second-order compound contingent derivative.

Remark 11.9.2. If, in the above $(\bar{u},\bar{v}) = (0,0)$, then we recover the contingent epiderivative $D_E F(\bar{x},\bar{y})$ and the generalized contingent epiderivative $D_G F(\bar{x},\bar{y})$ of F at $(\bar{x},\bar{y})$. Clearly, several analogues of the above second-order epiderivatives can be defined by using other notions of second-order tangent cones/sets.

The following result clarifies the relationship between the second-order contingent epiderivative, the second-order compound contingent epiderivative, and the second-order asymptotic contingent epiderivative.

Proposition 11.9.3. *Let X and Y be normed spaces and let $C \subset Y$ be a pointed, closed and convex cone. Let $F : X \rightrightarrows Y$ be a set-valued map, let $(\bar{x},\bar{y}) \in \text{graph}\,(F)$, and let $(\bar{u},\bar{v}) \in X \times Y$. Assume that the map F is C-convex. Then the second-order compound epiderivative $\mathscr{D}^2_E(\bar{x},\bar{y},\bar{u},\bar{v})$ coincides with the second-order contingent derivative $D^2F(\bar{x},\bar{y},\bar{u},\bar{v})$ of $(F+C)$ provided that $0 \in D^2(F+C)(\bar{x},\bar{y},\bar{u},\bar{v})(0)$.*

Proof. The proof follows from the fact that, for a convex subset S of a normed space, the equality $T^2(S,x,y) = T(T(S,x),y)$ holds provided that $0 \in T^2(S,x,y)$. □

The following example illustrates the above notions.

Example 11.9.4. Let $F : \mathbb{R} \rightrightarrows \mathbb{R}$ be a set-valued map given by

$$F(x) := \{y \in \mathbb{R} \mid y \geq x^4\} \quad \text{for all } x \in \mathbb{R}.$$

Let $(\bar{x},\bar{y}) = (0,0)$ and let $(\bar{u},\bar{v}) = (1,0)$. Then:

$$T^2(\text{epi}\,(F),(0,0),(1,0)) = \left\{ z \in \mathbb{R}^2 \mid \exists\,(z_n) \to z,\ (\lambda_n) \downarrow 0 : \lambda_n(1,0) + \frac{\lambda_n^2}{2} z_n \in \text{epi}\,(F) \right\}.$$

The condition $\lambda_n(1,0) + \frac{\lambda_n^2}{2} z_n \in \operatorname{epi}(F)$ is equivalent to

$$\frac{\lambda_n^2}{2} z_{n_2} \geq (\lambda_n + \frac{\lambda_n^2}{2} z_{n_1})^4 = \lambda_n^4 (1 + \frac{\lambda_n}{2} z_{n_1})^4,$$

and hence

$$z_{n_2} \geq 2\lambda_n^2 (1 + \frac{\lambda_n}{2} z_{n_1})^4.$$

Since $z_{n_1} \to z_1,\ z_{n_2} \to z_2$ and $\lambda_n \downarrow 0$, we obtain that

$$T^2(\operatorname{epi}(F), (0,0), (1,0) = \mathbb{R} \times \mathbb{R}_+,$$

implying that

$$D^2 F_+(0,0,1,0)(x) = \mathbb{R}_+ \quad \text{for every } x \in \mathbb{R}.$$

On the other hand,

$$D^2 F(0,0,1,0)(x) = 0 \quad \text{for every } \ x \in \mathbb{R},$$
$$D_G^2 F(0,0,1,0)(x) = \operatorname{Min}(D^2 F_+(0,0,1,0)(x), \mathbb{R}_+) = \{0\} \quad \text{for every } \ x \in \mathbb{R}.$$

The following examples will further clarify the differences and the similarities between two of the above notions.

Example 11.9.5. Let $F : \mathbb{R} \rightrightarrows \mathbb{R}$ be a set-valued map given by:

$$F(x) := \{y \in \mathbb{R} \mid y \geq x^{3/2}\} \quad \text{for all } x \in \mathbb{R}.$$

Let $(\bar{x}, \bar{y}) = (0,0)$ and let $(\bar{u}, \bar{v}) = (1,0)$. Define $R_+ = \{r \in R \mid r \geq 0\}$. Then

$$\mathscr{D}^2(F + \mathbb{R}_+)(0,0,1,0)(x) = \mathbb{R}_+ \quad \text{for every } \ x \in \mathbb{R},$$

whereas the second-order contingent derivative is not defined.

The following is an existence theorem for the second-order asymptotic contingent epiderivatives. Analogous results for other derivatives can be given by repeating the arguments.

Theorem 11.9.6. *Let X and Y be normed spaces and let $C \subset Y$ be a proper, closed, convex and regular cone. Let $F : X \rightrightarrows Y$ be a set-valued map, let $(\bar{x}, \bar{y}) \in \operatorname{graph}(F)$ and let $(\bar{u}, \bar{v}) \in X \times Y$. Let for every $x \in A := \operatorname{dom}(\mathbb{D}^2(F + C)(\bar{x}, \bar{y}, \bar{u}, \bar{v}))$, the set $\mathbb{D}^2(F + C)(\bar{x}, \bar{y}, \bar{u}, \bar{v})(x)$ have a C-lower bound. Then*

$\mathbb{D}^2_G F(\bar{x}, \bar{y}, \bar{u}, \bar{v})(x) \neq \emptyset$, *for every* $x \in A$. *Moreover the following relationships hold:*

$$\mathbb{D}^2(F + C)(\bar{x}, \bar{y}, \bar{u}, \bar{v})(x) \subseteq \mathbb{D}^2_G F(\bar{x}, \bar{y}, \bar{u}, \bar{v})(x) + C \quad \text{for every } x \in A. \tag{11.105}$$

$$\operatorname{epi}(\mathbb{D}^2_G F(\bar{x}, \bar{y}, \bar{u}, \bar{v})) = \mathbb{T}^2(\operatorname{epi}(F), (\bar{x}, \bar{y}), (\bar{u}, \bar{v})). \tag{11.106}$$

Proof. For the existence of the generalized asymptotic epiderivative $\mathbb{D}^2_G F(\bar{x}, \bar{y}, \bar{u}, \bar{v})(x)$ and for (11.105) we used arguments analogous to those used for the first-order case. For this, we define a set-valued map $H : X \rightrightarrows Y$ by

$$H(x) := \mathbb{D}^2(F + C)(\bar{x}, \bar{y}, \bar{u}, \bar{v})(x) \quad \text{for every } x \in A.$$

Since the second-order asymptotic contingent cone is always closed, the set $H(x)$ is closed for every $x \in \operatorname{dom}(H) = A$. It has a C-lower bound by the assumption. Therefore, the existence of $\mathbb{D}^2_G F(\bar{x}, \bar{y}, \bar{u}, \bar{v})(x)$ now follows from Proposition 11.5.5, which ensures the existence of minimal points for $H(x)$ with respect to the convex cone C. Moreover, it follows from the same result that $H(x)$ has the domination property, that is

$$H(x) \subseteq \mathbb{D}^2_G F(\bar{x}, \bar{y}, \bar{u}, \bar{v})(x) + C,$$

and hence (11.105) is proved.

We claim that equality holds in the above inclusion. For this, let $y \in \mathbb{D}^2_G F(\bar{x}, \bar{y}, \bar{u}, \bar{v})(x) + C$ be arbitrary. Then $y \in H(x) + C$. Now making use of the fact that $H(x) = H(x) + C$, we obtain $y \in H(x)$. This confirms the inclusion

$$H(x) \supseteq \mathbb{D}^2_G F(\bar{x}, \bar{y}, \bar{u}, \bar{v})(x) + C.$$

Therefore

$$\mathbb{D}^2_G F(\bar{x}, \bar{y}, \bar{u}, \bar{v})(x) + C = H(x) = \mathbb{D}^2_G F(\bar{x}, \bar{y}, \bar{u}, \bar{v})(x).$$

Furthermore, the above identity and the definition of the second-order asymptotic contingent derivative implies

$$\begin{aligned}
\operatorname{epi}(\mathbb{D}^2_G F(\bar{x}, \bar{y}, \bar{u}, \bar{v})) &= \operatorname{graph}(\mathbb{D}^2(F + C)(\bar{x}, \bar{y}, \bar{u}, \bar{v})) \\
&= \mathbb{T}^2(\operatorname{graph}(F + C), (\bar{x}, \bar{y}), (\bar{u}, \bar{v})) \\
&= \mathbb{T}^2(\operatorname{epi}(F), (\bar{x}, \bar{y}), (\bar{u}, \bar{v})),
\end{aligned}$$

which confirms the claim. □

Remark 11.9.7. Numerous results given for the first-order contingent epiderivatives in this chapter have their second-order analogues which can be proved by employing similar arguments. Some of such results are available in [300].

Chapter 12
Optimality Conditions in Set-Valued Optimization

Let X and Y be normed spaces, let $S \subseteq X$ be a nonempty set, let $C \subset Y$ be a cone inducing a partial ordering in Y, and let $F : X \rightrightarrows Y$ be a set-valued map. We begin this chapter with optimization problems that can be written in the following form

$$\text{Minimize}\, F(x) \quad \text{subject to} \quad x \in S, \tag{12.1}$$

where the minimum is taken in the vector optimization sense. For instance, one option is to seek an element $(\bar{x}, \bar{y}) \in \text{graph}\,(F)$ such that $F(S) \cap (\{\bar{y}\} - C) = \{\bar{y}\}$, where $F(S) := \{F(x) | \; x \in S\}$. Clearly, such a minimizer of (12.1) is a point $(\bar{x}, \bar{y}) \in X \times Y$ that satisfies $\bar{y} \in F(\bar{x}) \cap \text{Min}(F(S), C)$.

Note that if the map F is single-valued, then (12.1) reduces to a vector optimization problem. Additionally, if $Y := \mathbb{R}$ and $C := \mathbb{R}_+$, then we are in the framework of classical optimization.

Numerous other possibilities for choosing a minimizer of (12.1) have been thoroughly discussed in Chap. 2 and will be considered in this chapter. This includes the proper minimizer, the weak-minimizer, and the strong-minimizer, among others. In this chapter, unless stated otherwise, the proper minimizer refers to the Henig global proper minimizer.

The problems of type (12.1) belong to the realm of set-valued optimization. This chapter is devoted to necessary and sufficient optimality conditions of first-order and second-order for set-valued optimization problems. We note that primarily there are two approaches for giving optimality conditions in set-valued optimization. The first approach uses suitable derivatives of the involved set-valued maps whereas the second approach yields optimality conditions by means of alternative-type theorems. Although interesting results have been obtained by means of alternative-type theorems, the focus of this chapter is mainly on obtaining optimality conditions using some suitable derivatives and epiderivatives of the involved set-valued maps.

Before describing the approach adopted in this chapter and giving the main results, let us first recall some of the major developments in the context of

A.A. Khan et al., *Set-valued Optimization*, Vector Optimization,
DOI 10.1007/978-3-642-54265-7_12

derivative-based optimality conditions in set-valued optimization. We recall that set-valued optimization problems explicitly appear in the interesting paper by Corley [109], where the contingent derivative and the circatangent derivative were employed to give general optimality conditions. However, it should be pointed out that already in 1982, W. Oettli in [445] studied optimization problems with set-valued data. Moreover, S. Tagawa in [554] also investigated optimization problems with set-valued data, and, in fact, introduced derivatives of set-valued maps. Tagawa's work was based on his Ph. D. thesis supervised by W. Oettli (see [553]). Due to the fact that the theory of variational analysis has been enriched by various notions of tangent cones, many extensions of Corley's results have been given using other approximating notions. For instance, Taa [548] replaced the contingent cone by the closed radial cone (see also Flores-Bazan [188] and Kasimbeyli [314]).

As another direction of generalization and strengthening of the known results in set-valued optimization, some authors considered more general optimization problems, but still used the contingent derivative as the derivative notion to give optimality conditions (see Luc [400]). Starting from Corley's work, many subsequent contributions revolved around the graphs of the involved set-valued maps. Although this idea gives suitable necessary optimality conditions, the sufficient optimality conditions demand that graph of the involved set-valued map be convex. This, however, is a quite stringent assumption. Of course, a quick remedy is to work with the profile map of the involved set-valued maps. In fact, Corley [109] himself used this idea. However, the profile map does not seem to be a natural candidate for studying optimality conditions, particularly when the interest lies also in analyzing particular cases of single-valued maps. Motivated by this unsatisfactory situation, an attractive alternative approach was proposed by Jahn and Rauh [301] where the optimality conditions were given using contingent epiderivatives. Here the main idea is to use a certain part of the boundary of the epigraphs of the involved set-valued maps (see also Ward [594]). Since [301], the epigraph-based approach has been rigorously pursued by many authors (see, for instance, Chen and Jahn [92], Sama [519] and [296, 297]). Some of the recent developments in set-valued optimization can be found in [8, 88, 116, 117, 318, 319, 325, 326, 361, 399], and the cited references therein. See also [450–452].

During the last several decades, the so-called Dubovitskii-Milyutin approach has been used to study various optimal control and optimization problems with ordinary or partial differential equations as constraints. Several authors have also worked on extending this approach to nonsmooth optimization problems. More recently, in [284], an extension of the Dubovitskii-Milyutin approach to set-valued optimization problems was proposed, where first-order necessary optimality conditions for the notion of the proper-minimality, the weak-minimality, and the strong-minimality were given.

Another fruitful approach in set-valued optimization is based on various notions of coderivatives and has attracted a great deal of attention in recent years. Abdouni and Thibault [2] were the ones who initiated this study by considering a set-valued constrained optimization problem. They established optimality conditions with Lagrange-Kuhn-Tucker and Lagrange-Fritz-John multipliers in terms of the coderivatives of the involved set-valued mapping. Since the appearance of [2], numerous authors have employed coderivatives to optimality conditions for various set-valued optimization problems (see [27, 151, 620, 621]). In this chapter, such optimality conditions will be discussed thoroughly.

All of the aforementioned results deal with first-order optimality conditions. Although many new refinements and interesting ideas related to the first-order optimality conditions in set-valued optimization are still in the making, recent developments in non-smooth scalar and vector optimization have shown an acute interest in the development of second-order or higher-order optimality conditions. This remains particularly evident for vector optimization problems (see the interesting work [304] and the cited references therein). Motivated by these developments, in [300], the second-order contingent epiderivatives were introduced and employed to give new second-order optimality conditions in set-valued optimization. These results were further refined in [313], where the second-order asymptotic epiderivatives were used. Some extensions of these and related results for higher-order optimality conditions are given in [586, 587]. An extension of the first-order Dubovitskii-Milyutin approach (see [148]) to set-valued optimization with an emphasis on the case of multi-equality constraints, and new first-order and second-order optimality conditions for several notions of optimality have been recently given in [321, 322].

This chapter covers a wide variety of optimality conditions in set-valued optimization. To be specific, the following directions of deriving optimality conditions are investigated in sufficient details:

- Necessary first-order and second-order optimality conditions in set-valued optimization by a direct approach using graphical derivatives and epiderivatives.
- Necessary first-order and second-order optimality conditions in set-valued optimization by the Dubovitskii-Milyutin approach using graphical derivatives and epiderivatives.
- Sufficient optimality conditions using graphical derivatives and epiderivatives.
- Optimality conditions for Q-minimizers using graphical derivatives and coderivatives.
- Lagrange multiplier rules based on limiting subdifferentials.
- Necessary optimality conditions for the notions of approximate solutions for set-valued optimization problems.
- Necessary and sufficient optimality conditions in set-valued optimization for solution concepts based on the set approach.
- Necessary optimality conditions for solution concepts with respect to a general preference relation.
- Set-valued optimization problems from the view of KKT-Points and corresponding stability results.

12.1 First-Order Optimality Conditions by the Direct Approach

We begin with various necessary and sufficient optimality conditions of first-order. In set-valued optimization, a common approach is to give necessary optimality conditions by showing a disjunction of certain sets in the image space. We refer to this approach as the direct approach. As a derivative notion, we choose the contingent derivative of the profile maps of the involved set-valued maps. However, we emphasize that the used arguments easily yield optimality conditions for several other notions of derivatives and epiderivatives of set-valued maps.

We now formulate the set-valued optimization problems we intend to explore. Let W, X, Y, and Z be normed spaces, and let $C \subset Y$ and $D \subset Z$ be pointed, closed, and convex cones. Let $S \subseteq X$ be a nonempty set and let $h \in W$ be a given element. Let $F : X \rightrightarrows Y$, $G : X \rightrightarrows Z$, and $H : X \rightrightarrows W$ be given set-valued maps.

We are interested in studying the following three set-valued optimization problems:

(P_0) Minimize $F(x)$ subject to $x \in S$.

(P_1) Minimize $F(x)$ subject to $x \in S_1 := \{x \in S \mid G(x) \cap -D \neq \emptyset\}$.

(P_2) Minimize $F(x)$ subject to $x \in S_2 := \{x \in S \mid G(x) \cap -D \neq \emptyset,\ h \in H(x)\}$.

Recall that $(\bar{x}, \bar{y}) \in \operatorname{graph}(F)$ is a minimizer of set-valued optimization problem (P_2), if and only if, $F(S_2) \cap (\bar{y} - C) = \{\bar{y}\}$. Analogous relationships are valid for other kinds of minimizers and have been defined earlier in Chap. 2. Observe that (P_2) reduces to (P_0), if $G(x) = 0$ and $H(x) = h$ uniformly on S. In this case, the set S defines abstract constraints. If, additionally, we have $S = X$, then (P_2) is an unconstrained set-valued optimization problem. Note that the optimality notions given above are the global ones, that is, the whole set $F(S_2)$ has been taken into account. We recall that their local versions are naturally defined as follows: The point $(\bar{x}, \bar{y}) \in \operatorname{graph}(F)$ is said to be a local minimizer of (P_2) if there exists a neighborhood U of $\bar{x}$ such that $\bar{y} \in \operatorname{Min}(F(S_2 \cap U), C)$. Local analogues of other minimizers have been defined in Chap. 2.

We have the following necessary optimality condition for local proper optimality for (P_1).

Theorem 12.1.1. *Let X, Y, and Z be normed spaces, let $C \subset Y$ and $D \subset Z$ be pointed, closed, and convex cones, where D is also solid, and let $S \subseteq X$. Let $F : S \rightrightarrows Y$ and $G : S \rightrightarrows Z$ be set-valued maps. For $\bar{x} \in S$, let $(\bar{x}, \bar{y}) \in \operatorname{graph}(F)$ be a local proper minimizer of (P_1) and let $\bar{z} \in G(\bar{x}) \cap (-D)$. Then, for every $x \in \Omega := \operatorname{dom}(D(F + C, G + D)(\bar{x}, \bar{y}, \bar{z}))$, the following disjunction holds:*

$$D(F + C, G + D)(\bar{x}, \bar{y}, \bar{z})(x) \bigcap [(-C \backslash \{0\}) \times (-\operatorname{int} D - \bar{z})] = \emptyset. \tag{12.2}$$

Proof. Since the element $(\bar{x}, \bar{y})$ is a local proper minimizer of (P_1), there are a neighborhood U of $\bar{x}$ and a proper, pointed, closed, convex, and solid cone K with $C\setminus\{0\} \subset \operatorname{int} K$ such that $\bar{y} \in \operatorname{Min}(F(S_1 \cap U), K)$. We will show that for every $x \in \Omega$, we have

$$[D(F + C, G + D)(\bar{x}, \bar{y}, \bar{z})(x) + (0, \bar{z})] \bigcap [(-\operatorname{int} K) \times (-\operatorname{int} D)] = \emptyset, \qquad (12.3)$$

which, due the fact that $C\setminus\{0\} \subset \operatorname{int} K$, at once ensures (12.2).

We will establish (12.3) by a contradiction. For this, we assume that (12.3) does not hold, and hence, for some $x \in \Omega$, we have

$$[D(F + C, G + D)(\bar{x}, \bar{y}, \bar{z})(x) + (0, \bar{z})] \bigcap [(-\operatorname{int} K) \times (-\operatorname{int} D)] \neq \emptyset,$$

which implies that there exists $(y, z) \in D(F + C, G + D)(\bar{x}, \bar{y}, \bar{z})(x)$ such that

$$(y, z + \bar{z}) \in [D(F + C, G + D)(\bar{x}, \bar{y}, \bar{z})(x) + (0, \bar{z})] \bigcap [(-\operatorname{int} K) \times (-\operatorname{int} D)], \qquad (12.4)$$

confirming that

$$(x, y, z) \in \operatorname{graph}(D(F+C, G+D)(\bar{x}, \bar{y}, \bar{z})) = T(\operatorname{epi}(F, G), (\bar{x}, \bar{y}, \bar{z})). \qquad (12.5)$$

Therefore, we may use the definition of the contingent cone $T(\operatorname{epi}(F, G), (\bar{x}, \bar{y}, \bar{z}))$ which ensures that there are sequences $\{(x_n, y_n, z_n)\} \subset \operatorname{epi}(F, G)$ and $\{t_n\} \subset \mathbb{P}$ such that $(x_n, y_n, z_n) \to (\bar{x}, \bar{y}, \bar{z})$ and $t_n(x_n - \bar{x}, y_n - \bar{y}, z_n - \bar{z}) \to (x, y, z)$.

Since $x_n \to \bar{x}$ and $\{x_n\} \subset \operatorname{dom}(F) = S$, there exists $n_1 \in \mathbb{N}$ such that

$$x_n \in S \cap U \quad \text{for every } n \geq n_1, \qquad (12.6)$$

where U is the neighborhood of $\bar{x}$, tied to the notion of local proper minimality.

Moreover, since $y \in -\operatorname{int} K$ (see (12.4)) and $t_n(y_n - \bar{y}) \to y$, there exists $n_2 \in \mathbb{N}$ satisfying that $t_n(y_n - \bar{y}) \in -\operatorname{int} K$, for every $n \geq n_2$. Now, using the fact that $t_n > 0$, we have

$$y_n \in \bar{y} - \operatorname{int} K \quad \text{for every } n \geq n_2. \qquad (12.7)$$

Furthermore, because $(x_n, y_n) \in \operatorname{epi}(F)$, there exists a sequence $\{w_n\}$ with $w_n \in F(x_n)$ such that $y_n \in w_n + C$, for every $n \in \mathbb{N}$. Therefore, $w_n \in y_n - C$, and by (12.7), we have

$$w_n \in \bar{y} - \operatorname{int} K - C \subset \bar{y} - \operatorname{int} K - K = \bar{y} - \operatorname{int} K,$$

which means that $w_n \in \bar{y} - \operatorname{int} K$, for every $n \geq n_2$.

We claim that $x_n \in S_1$, that is, $G(x_n) \cap (-D) \neq \emptyset$, for sufficiently large n. For this, we note that since $z + \bar{z} \in -\text{int}D$ (see (12.4)) and $t_n(z_n - \bar{z}) \to z$, there exists $n_3 \in \mathbb{N}$ such that

$$t_n(z_n - \bar{z}) + \bar{z} \in -\text{int}D \quad \text{for every } n \geq n_3,$$

and accordingly,

$$z_n - \left(1 - \frac{1}{t_n}\right)\bar{z} \in -\text{int}D \quad \text{for every } n \geq n_3.$$

To make a good use of the above containment, we note that $z_n \to \bar{z}$ and $t_n(z_n - \bar{z}) \to z$ imply that there exists $n_4 \in \mathbb{N}$ such that $t_n > 1$, for every $n \geq n_4$. Therefore, since we have $\bar{z} \in -D$, we also have that $(1 - \frac{1}{t_n})\bar{z} \in -D$, for every $n \geq \max\{n_3, n_4\}$. Consequently, we deduce that

$$z_n \in \left(1 - \frac{1}{t_n}\right)\bar{z} - \text{int}D \subset -D - \text{int}D = -\text{int}D \quad \text{for every } n \geq \max\{n_3, n_4\}.$$

Since $\{(x_n, z_n)\} \subset \text{epi}\,(G)$, there exists $u_n \in G(x_n)$ such that $z_n \in u_n + D$, for every $n \in \mathbb{N}$. This implies that

$$u_n \in -D - \text{int}D = -\text{int}D \quad \text{for every } n \geq \max\{n_3, n_4\},$$

confirming that

$$G(x_n) \cap (-D) \neq \emptyset \quad \text{for every } n \geq \max\{n_3, n_4\}.$$

Summarising the above observations, we have shown that, for sufficiently large $n \in \mathbb{N}$ (indeed for $n \geq \{n_1, n_2, n_3, n_4\}$), there exists $x_n \in S_1 \cap U$ such that

$$F(x_n) \cap (\bar{y} - \text{int}K) \neq \emptyset,$$

contradicting the assumption that $(\bar{x}, \bar{y})$ is a local proper minimizer. The proof is complete. □

From the above result, we recover the following necessary optimality condition for (P_0).

Theorem 12.1.2. *Let X and Y be normed spaces, let $C \subset Y$ be a pointed, closed, and convex cone, let $S \subseteq X$, and let $F : S \rightrightarrows Y$ be a set-valued map. If $(\bar{x}, \bar{y}) \in$ graph (F) is a local proper minimizer of (P_0), then for every $x \in$ dom $(D(F + C)(\bar{x}, \bar{y}))$, the following disjunction holds:*

$$D(F + C)(\bar{x}, \bar{y})(x) \bigcap (-C \backslash \{0\}) = \emptyset. \tag{12.8}$$

Remark 12.1.3. Disjunctions (12.2) and (12.8) remain valid as first-order necessary optimality conditions for local proper optimality for (P_1) and (P_0) if the involved contingent derivatives are replaced by the adjacent derivative, the lower Dini derivative, the circatangent derivative, the radial derivative, the contingent epiderivative, the adjacent epiderivative, the radial epiderivative, the circatangent epiderivative, or by their generalized and weakly generalized counterparts. Note that for epiderivatives, identities analogous to (12.5) will play an important role. We emphasize that the disjunctions for the radial derivative and the radial epiderivative do not directly follow from disjunctions (12.2) and (12.8) but can be proved using similar arguments.

We have the following example to illustrate the above results:

Example 12.1.4. Consider a set-valued map $F : [0,1] \rightrightarrows \mathbb{R}^2$ defined by

$$F(x) = \left\{(y_1, y_2) \in \mathbb{R}^2 |\ y_1^2 + y_2^2 \leq x^2\right\},$$

and let $C = \{(x,x) \in \mathbb{R}^2 |\ x \in \mathbb{R}_+\}$. It is easy to verify that $(1,(0,-1)) \in \operatorname{graph}(F)$ is a proper minimizer. For $x \in \operatorname{dom}(D(F+C)(\bar{x},\bar{y})) = \mathbb{R}_-$, we have

$$D(F+C)(\bar{x},\bar{y})(x) = \{(y_1,y_2) \in \mathbb{R}^2 |\ y_1 \in \mathbb{R},\ y_2 \geq |x|\},$$

and consequently,

$$D(F+C)(\bar{x},\bar{y})(x) \cap (-C \backslash \{0\}) = \emptyset \quad \text{for every } x \in \operatorname{dom}(D(F+C)(\bar{x},\bar{y})),$$

as claimed by the result. □

The arguments used in Theorem 12.1.1 readily yield the following necessary optimality conditions for local weak-optimality.

Theorem 12.1.5. *Let X, Y, and Z be normed spaces, let $C \subset Y$ and $D \subset Z$ be pointed, closed, convex and solid cones, and let $S \subseteq X$. Let $F : S \rightrightarrows Y$ and $G : S \rightrightarrows Z$ be set-valued maps. For $\bar{x} \in S$, let $(\bar{x},\bar{y}) \in \operatorname{graph}(F)$ be a local weak minimizer of (P_1) and let $\bar{z} \in G(\bar{x}) \cap (-D)$. Then, for every $x \in \operatorname{dom}(D(F+C,G+D)(\bar{x},\bar{y},\bar{z}))$, the following disjunction holds:*

$$D(F+C,G+D)(\bar{x},\bar{y},\bar{z})(x) \bigcap [(-\operatorname{int}C) \times (-\operatorname{int}D - \bar{z})] = \emptyset. \tag{12.9}$$

Theorem 12.1.6. *Let X and Y be normed spaces, let $C \subset Y$ be a pointed, closed, convex, and solid cone, let $S \subseteq X$, and let $F : S \rightrightarrows Y$ be a set-valued map. If $(\bar{x},\bar{y}) \in \operatorname{graph}(F)$ is a local weak minimizer of (P_0), then, for every $x \in \operatorname{dom}(D(F+C)(\bar{x},\bar{y}))$, the following disjunction holds:*

$$D(F+C)(\bar{x},\bar{y})(x) \bigcap (-\operatorname{int}C) = \emptyset. \tag{12.10}$$

In the following result, we extend the well-known Lagrange multiplier rule for the local proper optimality concept.

Theorem 12.1.7. *Let X, Y, and Z be normed spaces, let $C \subset Y$ and $D \subset Z$ be pointed, closed, and convex cones, where D is also solid, and let $S \subseteq X$. Let $F : S \rightrightarrows Y$ and $G : S \rightrightarrows Z$ be set-valued maps. For $\bar{x} \in S$, let $(\bar{x}, \bar{y}) \in \mathrm{graph}\,(F)$ be a local proper minimizer of (P_1), and let $\bar{z} \in G(\bar{x}) \cap (-D)$. Assume that $D(F + C, G + D)(\bar{x}, \bar{y}, \bar{z})(\Omega)$ is convex, where $\Omega := \mathrm{dom}\,(D(F + C, G + D)(\bar{x}, \bar{y}, \bar{z}))$. Then, there exists $(\xi, \zeta) \in Y^* \times Z^* \backslash \{(0,0)\}$ with $\xi \in C^+$ and $\zeta \in D^+$ such that $\zeta(\bar{z}) = 0$ and*

$$\xi(y) + \zeta(z) \geq 0 \quad \text{for every } (y,z) \in D(F + C, G + D)(\bar{x}, \bar{y}, \bar{z})(\Omega). \tag{12.11}$$

Proof. We set $M := D(F + C, G + D)(\bar{x}, \bar{y}, \bar{z})(\Omega)$. Then Theorem 12.1.1 ensures that the convex set $[M + (0, \bar{z})]$ satisfies the following disjunction:

$$[M + (0, \bar{z})] \bigcap [-(\mathrm{int}K \times \mathrm{int}D)] = \emptyset.$$

By the Eidelheit separation theorem, there are $(\xi, \zeta) \in Y^* \times Z^* \backslash \{(0,0)\}$ and $\gamma \in \mathbb{R}$ such that

$$\xi(y) + \zeta(z + \bar{z}) \geq \gamma \quad \text{for every } (y,z) \in M, \tag{12.12}$$

$$\xi(c) + \zeta(d) > \gamma \quad \text{for every } (c,d) \in \mathrm{int}K \times \mathrm{int}D. \tag{12.13}$$

It follows from the above inequalities that we can set $\gamma = 0$. Furthermore, in view of the continuity of (ξ, ζ), by taking $c \in \mathrm{int}K$ arbitrary close to zero, we obtain that $\zeta \in D^+$ and by taking $d \in D$ arbitrarily close to zero, we obtain $\xi \in C^+$.

Since $(0,0) \in M$, we get from (12.12) that $\zeta(\bar{z}) \geq 0$. However, the inequality $\zeta(\bar{z}) \leq 0$ is also true from $\zeta \in D^+$ and $\bar{z} \in -D$. Finally, (12.11) follows from (12.12). The proof is complete. □

The following result sheds some light on the convexity assumption used in the above result:

Proposition 12.1.8. *Under the setting of Theorem 12.1.7, the set $D(F + C, G + D)(\bar{x}, \bar{y}, \bar{z})(\Omega)$ is convex if the contingent cone $T(\mathrm{graph}\,(F + C, G + D), (\bar{x}, \bar{y}, \bar{z}))$ is convex.*

Proof. Let $T(\mathrm{graph}\,(F + C, G + D), (\bar{x}, \bar{y}, \bar{z}))$ be convex. We set $M := D(F + C, G + D)(\bar{x}, \bar{y}, \bar{z})(\Omega)$ and choose $(y_1, z_1), (y_2, z_2) \in M$ arbitrarily. Then, there exist $x_1, x_2 \in \Omega$ such that $(y_1, z_1) \in D(F + C, G + D)(\bar{x}, \bar{y}, \bar{z})(x_1)$ and $(y_2, z_2) \in D(F + C, G + D)(\bar{x}, \bar{y}, \bar{z})(x_2)$. Therefore, $(x_1, y_1, z_1), (x_2, y_2, z_2) \in T(\mathrm{graph}\,(F + C, G + D), (\bar{x}, \bar{y}, \bar{z}))$. Since $T(\mathrm{graph}\,(F + C, G + D), (\bar{x}, \bar{y}, \bar{z}))$ is convex, for every $t \in [0,1]$, we have

$$t(x_1, y_1, z_1) + (1 - t)(x_2, y_2, z_2) \in T(\mathrm{graph}\,(F + C, G + D), (\bar{x}, \bar{y}, \bar{z})),$$

confirming that $(tx_1 + (1-t)x_2) \in \Omega$ and $(t(y_1, z_1) + (1-t)(y_2, z_2)) \in M$. The proof is complete. □

Remark 12.1.9. The above result shows that the convexity assumption can be dropped by replacing contingent derivatives by Clarke derivatives.

Note that Theorem 12.1.7 does not exclude the possibility of the Lagrange multiplier ξ being zero, and if ξ happens to be zero, then (12.11) does not provide much information. Therefore, it is of interest to find conditions ensuring that $\xi \neq 0$. It is known from scalar and vector optimization theory that such conditions include a derivative of the map defining the constraints and are referred to as constraint qualifications. However, the existence of the Lagrange multipliers in Theorem 12.1.7 is proved in a setting where the contingent derivative $D(F + C, G + D)(\bar{x}, \bar{y}, \bar{z})$ of the composite map $(F + C, G + D)$ is being used. This, in general, is different from the composition $(D(F + C)(\bar{x}, \bar{y}), D(G + D)(\bar{x}, \bar{z}))$ of the derivatives. Therefore, first an analogue of Theorem 12.1.1 should be given that takes into account the derivatives $D(F + C)(\bar{x}, \bar{y})$ and $D(G + D)(\bar{x}, \bar{z})$.

Theorem 12.1.10. *Let X, Y, and Z be normed spaces, let $C \subset Y$ and $D \subset Z$ be pointed, closed, convex, and solid cones, and let $S \subset X$. Let $F : S \rightrightarrows Y$ and $G : S \rightrightarrows Z$ be set-valued maps. For $\bar{x} \in S$, let $(\bar{x}, \bar{y}) \in \operatorname{graph}(F)$ be a local weak minimizer of (P_1), and let $\bar{z} \in G(\bar{x}) \cap (-D)$. Assume that either $(F + C)$ is Lipschitz-like around $(\bar{x}, \bar{y})$ or $(G + D)$ is Lipschitz-like around $(\bar{x}, \bar{z})$. Assume that either* epi (F) *is derivable at $(\bar{x}, \bar{y})$ or* epi (G) *is derivable at $(\bar{x}, \bar{z})$. Then, for every $x \in \operatorname{dom}(D(F + C)(\bar{x}, \bar{y})) \cap \operatorname{dom}(D(G + D)(\bar{x}, \bar{z}))$, the following disjunction holds:*

$$(D(F + C)(\bar{x}, \bar{y}), D(G + D)(\bar{x}, \bar{z}))\,(x) \bigcap [(-\operatorname{int} C) \times (-\operatorname{int} D - \bar{z})] = \emptyset. \tag{12.14}$$

Proof. The proof follows by combining the arguments used above and in Theorem 12.2.1. □

We have the following Lagrange multiplier rule along with a constraint qualification.

Theorem 12.1.11. *Besides the hypotheses of Theorem 12.1.10, assume that $(D(F + C)(\bar{x}, \bar{y}), D(G + D)(\bar{x}, \bar{z}))(\Omega)$ is convex, where $\Omega := \operatorname{dom}(D(F + C)(\bar{x}, \bar{y})) \cap \operatorname{dom}(D(G + D)(\bar{x}, \bar{z}))$. Then there exists $(\xi, \zeta) \in Y^* \times Z^* \setminus \{(0, 0)\}$ with $\xi \in C^+$ and $\zeta \in D^+$ such that $\zeta(\bar{z}) = 0$ and*

$$\xi(y) + \zeta(z) \geq 0 \quad \text{for every } (y, z) \in (D(F + C)(\bar{x}, \bar{y}), D(G + D)(\bar{x}, \bar{z}))\,(\Omega). \tag{12.15}$$

Furthermore, $\xi \neq 0$ if additionally we assume that

$$D(G + D)(\bar{x}, \bar{y})(\Omega) + \operatorname{cone}(D + \bar{z}) = Z. \tag{12.16}$$

Proof. The proof of (12.15) is identical to the proof of (12.11) and hence has been omitted. It remains to show that $\xi \neq 0$. For this we assume that $\xi = 0$ and choose $\hat{z} \in Z$ arbitrarily. Then, for $s > 0$ and for some $z \in D(G + D)(\bar{x}, \bar{z})(\Omega)$, we have $\hat{z} = z + s(D + \bar{z})$, which implies that, for every $d \in D$, we get $\zeta(\hat{z}) = \zeta(z) + s\,(\zeta(d) + \zeta(\bar{z})) \geq 0$, confirming that $\zeta = 0$. This, however, is a contradiction to the fact that ξ and ζ are not both simultaneously zero. The proof is complete. □

Note that the imposed convexity in Theorem 12.1.11 allowed the use of classical separation arguments which ensured that the multipliers are linear and continuous. It turns out that within the absence of convexity, it will still be possible to prove the existence of multipliers, however, the linearity of the multipliers will be lost in this procedure. We explain this by giving an optimality condition for local weak minimality for (P_0).

We recall that a function $z : Y \to \mathbb{R}$ is called strictly C-monotone, where Y is a normed space and $C \subset Y$ is a cone, if $y_1 \in y_2 + C\backslash\{0\}$ implies that $z(y_1) > z(y_2)$. We also need to recall the following result (see [198]), where the notion of a C-monotone function is being used.

Lemma 12.1.12. *Let W be a linear topological space, let $C \subset W$ be a convex and solid cone, and let $A \subset W$ be nonempty. Then the following two statements are equivalent:*

1. $A \cap (-\text{int}C) = \emptyset$.
2. *There exists a continuous sublinear functional $s : W \to \mathbb{R}$ which is strictly* int$C$*-monotone with the range $(-\infty, +\infty)$ and possesses the following properties:*

$$s(A) \geq 0, \;\; s(\text{int}(A)) > 0, \;\; s(-C) \leq 0, \;\; s(-b(C)) = 0,$$
$$s(-\text{int}C) < 0, \;\; s(C) \geq 0, \;\; s(\text{int}C) > 0.$$

The following result is a direct application of the above result and Theorem 12.1.6:

Theorem 12.1.13. *Let X and Y be normed spaces, let $C \subset Y$ be a pointed, closed, convex, and solid cone, let $S \subseteq X$, and let $F : S \rightrightarrows Y$ be a set-valued map. Let $(\bar{x}, \bar{y}) \in$* graph(F) *be a local weak minimizer of (P_0) and let $\Omega :=$* dom$(D(F + C)(\bar{x}, \bar{y}))$*. Then, there exists a continuous, sublinear, strictly* int$C$*-monotone functional $s : W \to \mathbb{R} \cup \{\pm\infty\}$ possessing the following properties:*

$$s(D(F + C)(\bar{x}, \bar{y})(\Omega)) \geq 0, \;\; s(\text{int}(D(F + C)(\bar{x}, \bar{y})(\Omega))) > 0, \;\; s(-C) \leq 0,$$
$$s(-b(C)) = 0, \;\; s(-\text{int}C) < 0, \;\; s(C) \geq 0, \;\; s(\text{int}C) > 0.$$

In optimization theory, theorems of alternative-type play a crucial role (see Gwinner and Oettli [222]). In the following, we give another such result:

Theorem 12.1.14. *Let X and Y be normed spaces, let $C \subset Y$ be a pointed, closed, solid, and convex cone $C \subset Y$, let $S \subseteq X$, and let $F : S \rightrightarrows Y$ be a set-valued map. For $\Omega := \operatorname{dom}(D(F+C)(\bar{x},\bar{y}))$, let $D(F+C)(\bar{x},\bar{y})(\Omega)$ be convex. Then, only one of the following statements hold:*

1. *There exists $x \in \Omega$ such that*

$$D(F+C)(\bar{x},\bar{y})(x) \cap (-\text{int}C) \neq \emptyset.$$

2. *There exists $\ell \in C^+\backslash\{0\}$ such that*

$$\ell(y) \geq 0 \quad \textit{for every } y \in D(F+C)(\bar{x},\bar{y})(\Omega). \tag{12.17}$$

Consequently (12.17) *is a necessary condition for $(\bar{x},\bar{y})$ to be a local weak-minimizer of (P_0).*

Proof. The proof follows from standard separation arguments used above. □

For the next optimality condition, we recall the following result due to Kurcyusz [346]:

Lemma 12.1.15. *Let Y be a normed space and let K_1 and K_2 be two cones in Y with $\text{int}(K_2) \neq \emptyset$. If $K_1 \cap \text{int}(K_2) = \emptyset$ then $(K_1 - K_2)$ cannot be dense in Y.*

Proof. We will show that $(K_1 - K_2) \cap \text{int}(K_2) = \emptyset$ so $(K_1 - K_2)$ cannot be dense. Indeed, if the disjunction does not hold, then $(k_1 - k_2) = k$, for some $k_1 \in K_1$, $k_2 \in K_2$ and $k \in \text{int}(K_2)$. Therefore, for $k_1 \in K_1$, we have $k_1 = k_2 + k \in K_2 + \text{int}(K_2) = \text{int}(K_2)$. This, however, contradicts the assumption that $K_1 \cap \text{int}(K_2) = \emptyset$. The proof is complete. □

We now give a necessary optimality condition as a non-denseness of certain sets:

Theorem 12.1.16. *Let X and Y be normed spaces, let $C \subset Y$ be a pointed, closed, solid and convex cone, let $S \subseteq X$, and let $F : S \rightrightarrows Y$ be a set-valued map. If $(\bar{x},\bar{y})$ is a local weak minimizer of (P_0), then, $(D(F+C)(\bar{x},\bar{y})(\Omega) + C)$, where $\Omega := \operatorname{dom}(D(F+C)(\bar{x},\bar{y}))$, cannot be dense in Y.*

Proof. Since $D(F+C)(\bar{x},\bar{y})(\Omega)$ is a cone, and since $D(F+C)(\bar{x},\bar{y})(\Omega) \cap (-\text{int}C) = \emptyset$, the claim follows directly from Lemma 12.1.15. The proof is complete. □

In the following, we state a result for (P_2) when the map H is single-valued, that is,

$$\text{Minimize } F(x) \quad \text{subject to} \quad x \in S_2 := \{x \in S \mid G(x) \cap -D \neq \emptyset,\ H(x) = 0\}. \tag{12.18}$$

We have the following result that employs the lower Dini derivative:

Theorem 12.1.17. *Let* $(\bar{x}, \bar{y}, \bar{z})$ *be local weak-minimizer of* (12.18) *and let* $\operatorname{int} D$ *be nonempty. For any* $x \in \operatorname{dom}(D_L(F,G)(\bar{x}, \bar{y}, \bar{z}))$ *with* $H'(\bar{x})(x) = 0$, *if the set* $(D_L(F,G)(\bar{x}, \bar{y}, \bar{z})(x) + C \times D)$ *is convex, then there exists nonzero vector* $(\xi, \zeta) \in (C \times D)^+$ *such that*

$$\xi(y) + \zeta(z + \bar{z}) \geq 0 \text{ for every } (y,z) \in D_L(F,G)(\bar{x}, \bar{y}, \bar{z})(x). \tag{12.19}$$

Moreover, if the set $(D_L(F + C, G + D)(\bar{x}, \bar{y}, \bar{z})(X) + (C \times D))$ *is convex, then there exists nonzero vector* $(\xi, \zeta, \nu) \in (C \times D \times \{0\})^+$ *such that*

$$\xi(y) + \zeta(z) + \nu(w) \geq 0 \text{ for every } (y,z,w) \in D_L(F + C, G + D, H)(X).$$

The element ξ *is nonzero if the following condition holds:*

$$D_L(F + C, G + D, H)(\bar{x}, \bar{y}, \bar{z}, 0)(X) \bigcap (Y \times \{-\operatorname{int} D\} \times \{0\}) \neq \emptyset. \tag{12.20}$$

Proof. The proof is based on arguments similar to those used above and can be found in [405]. □

Finally, we give some necessary optimality conditions for the notion of strong minimality.

Theorem 12.1.18. *Let* X, Y, *and* Z *be normed spaces, let* $C \subset Y$ *and* $D \subset Z$ *be pointed, closed, and convex cones, where* D *is also solid, and let* $S \subseteq X$. *Let* $F : S \rightrightarrows Y$ *and* $G : S \rightrightarrows Z$ *be set-valued maps. Let* $(\bar{x}, \bar{y}) \in \operatorname{graph}(F)$ *be a local strong minimizer of* (P_1), *and let* $\bar{z} \in G(\bar{x}) \cap -D$. *Then:*

1. *For every* $x \in \Omega := \operatorname{dom}(D(F + C, G + D)(\bar{x}, \bar{y}, \bar{z}))$, *we have*

$$[D(F + C, G + D)(\bar{x}, \bar{y}, \bar{z})(x) + (0, \bar{z})] \cap [(Y \backslash C) \times (-\operatorname{int} D)] = \emptyset. \tag{12.21}$$

2. *Assume that either* $(F + C)$ *is Lipschitz-like around* $(\bar{x}, \bar{y})$ *or* $(G + D)$ *is Lipschitz-like around* $(\bar{x}, \bar{z})$. *Assume that either* $\operatorname{epi}(F)$ *is derivable at* $(\bar{x}, \bar{y})$ *or* $\operatorname{epi}(G)$ *is derivable at* $(\bar{x}, \bar{z})$. *Then for every* $x \in \Omega := \operatorname{dom}(D(F + C)(\bar{x}, \bar{y}) \cap D(G + D)(\bar{x}, \bar{z}))$, *we have*

$$(D(F + C)(\bar{x}, \bar{y}), D(G + D)(\bar{x}, \bar{z}))\,(x) \cap [(Y \backslash C) \times (-\operatorname{int} D - \bar{z})] = \emptyset. \tag{12.22}$$

Proof. We will prove (12.21) by contraposition. Assume that there exists $x \in \Omega$ such that

$$(y, z + \bar{z}) \in [D(F + C, G + D)(\bar{x}, \bar{y}, \bar{z})(x) + (0, \bar{z})] \cap [(Y \backslash C) \times (-\operatorname{int} D)].$$

Therefore, $(x,y,z) \in T(\text{epi}\,(F,G),(\bar{x},\bar{y},\bar{z}))$, and consequently, there are a sequence $\{t_n\} \subset \mathbb{P}$ and a sequence $\{(x_n,y_n,z_n)\} \subset X \times Y \times Z$ such that $t_n \downarrow 0$, $(x_n,y_n,z_n) \to (x,y,z)$ and for every $n \in \mathbb{N}$, we have $\bar{y} + t_n y_n \in F(\bar{x}+t_n x_n) + C$ and $\bar{z} + t_n z_n \in G(\bar{x}+t_n x_n) + D$. Since $u_n := (\bar{x}+t_n x_n) \to \bar{x}$, there exists $n_1 \in \mathbb{N}$ such that $u_n \in S \cap U$ for every $n \geq n_1$ and for some neighborhood U of $\bar{x}$. Because $y \notin C$, $y_n \to y$, and the cone C is closed, there exists $n_2 \in \mathbb{N}$ such that $y_n \notin C$ for every $n \geq n_2$. Since $t_n > 0$, we have $t_n y_n \notin C$. Let $w_n \in F(u_n)$ be such that for every $n \in \mathbb{N}$ and for some $c_n \in C$, we have $\bar{y} + t_n y_n = w_n + c_n$ and hence $c_n \notin \bar{y} + C$ for $n \geq \max\{n_1,n_2\}$. Since $C + C = C$, we have $w_n \notin \bar{y} + C$ for $n \geq \max\{n_1,n_2\}$. As in the arguments in the previous results, we can show that there exists $n_3 \in \mathbb{N}$ such that $G(u_n) \cap -D \neq \emptyset$ for every $n \geq n_3$.

Therefore, we have shown that for sufficiently large $n \in \mathbb{N}$, there are $w_n \in F(S_1 \cap U)$ such that $w_n \notin \bar{y} + C$. This, however, is a contradiction to the assumption that $(\bar{x},\bar{y})$ is a local strong minimizer. Hence (12.21) must hold. The proof of (12.22) is analogous. □

As a direct consequence of the above result, we get the following:

Theorem 12.1.19. *Let X and Y be normed spaces, let $C \subset Y$ be a pointed, closed, and convex cone, and let $S \subseteq X$. Let $F : S \rightrightarrows Y$ be a set-valued map, and let $(\bar{x},\bar{y}) \in \text{graph}\,(F)$ be a local strong minimizer of (P_0). Then, for every $x \in \Omega :=$ $\text{dom}\,(D(F+C)(\bar{x},\bar{y}))$, we have*

$$D(F+C)(\bar{x},\bar{y})(\Omega) \subseteq C. \tag{12.23}$$

The following example illustrates the above results:

Example 12.1.20. Consider the set-valued map $F : [0,1] \rightrightarrows \mathbb{R}$ given by:

$$F(x) := \left\{(y_1,y_2) \in \mathbb{R}^2 |\; y_1^2 + y_2^2 \leq x^2\right\}.$$

Let $C = \mathbb{R}^2_+$ be the ordering cone. We see that the point $(1,(0,-1)) \in \text{graph}\,(F)$ is a weak-minimizer. We employ the weak contingent epiderivative of F at $(1,(0,-1))$, which, for every $x \in \mathbb{R}_-$, is given by

$$D_{\text{w}}F(1,(0,-1))(x) = \left\{(y_1,y_2) \in \mathbb{R}^2 |\; y_2 \geq |x| \quad y_1 \in \mathbb{R}\right\}.$$

The necessary optimality for weak-minimality holds because

$$D_{\text{w}}F(1,(0,-1))(\mathbb{R}_-) \cap \left(-\text{int}\left(\mathbb{R}^2_+\right)\right) = \emptyset.$$

On the other hand, the necessary optimality condition for proper minimality is violated as

$$D_{\text{w}}F(1,(0,-1))(0) \cap \left(-\mathbb{R}^2_+ \backslash \{(0,0)\}\right) = \left\{(y_1,0) \in \mathbb{R}^2 |\; y_1 < 0\right\}.$$

Furthermore, the point $\left(1, \left(-\frac{1}{\sqrt{2}}, -\frac{1}{\sqrt{2}}\right)\right)$ is a proper minimizer of (P_0), but it is not a strong minimizer. The generalized contingent epiderivative is

$$D_{\mathrm{G}}F\left(1, \left(-\frac{1}{\sqrt{2}}, -\frac{1}{\sqrt{2}}\right)\right)(x) = \left\{(y_1, y_2) \in \mathbb{R}^2 \mid y_1 + y_2 \geq -\sqrt{2}x\right\}.$$

The necessary optimality condition for proper optimality holds because

$$D_{\mathrm{G}}F\left(1, \left(-\frac{1}{\sqrt{2}}, -\frac{1}{\sqrt{2}}\right)\right)(\mathbb{R}_-) \cap \left(-\mathbb{R}^2_+ \backslash \{0\}\right) = \emptyset.$$

Finally, since

$$D_{\mathrm{G}}F\left(1, \left(-\frac{1}{\sqrt{2}}, -\frac{1}{\sqrt{2}}\right)\right)(\mathbb{R}_-) \not\subseteq \mathbb{R}^2_+,$$

the necessary optimality condition for strong minimality is violated. □

12.2 First-Order Optimality Conditions by the Dubovitskii-Milyutin Approach

Let Ξ be a normed space and let Σ_i, $i = 1, \ldots, n+1$, be convex cones with Σ_i, $i = 1, \ldots, n$, open. The Dubovitskii-Milyutin-lemma asserts an equivalence between the following two statements:

(α) $\bigcap_{i=1}^{n+1} \Sigma_i = \emptyset$.

(β) There exist $l_i \in \Sigma_i^*$, $i = 1, \ldots, n+1$, not all simultaneously equal to zero, such that

$$l_1 + l_2 + \cdots + l_{n+1} = 0.$$

In optimization theory, the relevance of the so-called Dubovitskii-Milyutin-lemma (see Theorem 12.2.21) comes from the fact that the optimality of a point can be conveniently expressed as the disjunction of certain sets. Now, if these sets are locally approximated by using suitable tangent cones so that the disjunction is maintained, then, under suitable convexity assumptions, the functionals appearing in (β) give rise to the Lagrange multipliers. In applications of the Dubovitskii-Milyutin-lemma, the cone Σ_{n+1} is often an approximation of the equality constraints and the cones Σ_i, $i = 1, \ldots, n$, contain some information about the behavior of the derivatives of the objective map and the inequality constraints. The approach based on the Dubovitskii-Milyutin-lemma is very flexible for many applications and has proven to be of great use in dealing with scalar and vector optimization problems. In this chapter, we present optimality conditions by presenting an extension of the Dubovitskii-Milyutin approach to set-valued optimization.

Some of the works on the Dubovitskii-Milyutin approach are by Bazaraa and Goode [35], Das [125], Ioffe and Tihomirov [279], Halkin [231, 232], Kotarski [332–334], Kowalewski and Kotarski [337], Ledzewicz-Kowalewska [372–374], Ledzewicz and Schättler [365–369], Ledzewicz and Walczak [370, 371], Leung [375], Rigby [487], Studniarski [546], Virsan [583], and Watkins [598], among others.

12.2.1 Necessary Optimality Conditions by the Dubovitskii-Milyutin Approach

We begin with necessary optimality conditions for the following set-valued optimization problem (P_1), which was introduced earlier in this chapter:

$$(P_1) \quad \text{Minimize } F(x) \quad \text{subject to} \quad x \in S_1 = \{x \in S \mid G(x) \cap -D \neq \emptyset\}.$$

Before giving the main result, we recall that, given normed spaces X and Y and a set-valued map $F : X \rightrightarrows Y$, the **weak-inverse image** $F[A]^{-1}$ of F with respect to a set $A \subseteq Y$ is defined by:

$$F[A]^{-1} := \{x \in X \mid F(x) \cap A \neq \emptyset\}.$$

The following is a necessary optimality condition for the notion of weak minimality:

Theorem 12.2.1. *Let X, Y, and Z be normed spaces, $C \subset Y$ and $D \subset Z$, be pointed, closed, convex, and solid cones, and let $S \subseteq X$. Let $F : X \rightrightarrows Y$ and $G : X \rightrightarrows Z$ be set-valued maps. For $\bar{x} \in S$, let $(\bar{x}, \bar{y}) \in$ graph (F) be a local weak minimizer of (P_1). Assume that $(F + C)$ is Lipschitz-like around $(\bar{x}, \bar{y})$. Assume that either* epi (F) *is derivable at $(\bar{x}, \bar{y})$ or $G[-D]^{-1}$ is derivable at $\bar{x}$. Then the following disjunction holds:*

$$D(F + C)(\bar{x}, \bar{y})[-\text{int}C]^{-1} \bigcap IT(S, \bar{x}) \bigcap T(G[-D]^{-1}, \bar{x}) = \emptyset. \tag{12.24}$$

Proof. We will show that if (12.24) fails, then a feasible x can be obtained in a vicinity of $\bar{x}$ with $F(x) \cap (\bar{y} - \text{int}C) \neq \emptyset$, hence violating the local weak minimality of $(\bar{x}, \bar{y})$. Assume that

$$x \in D(F + C)(\bar{x}, \bar{y})[-\text{int}C]^{-1} \bigcap IT(S, \bar{x}) \bigcap T(G[-D]^{-1}, \bar{x}).$$

From $x \in T(G[-D]^{-1}, \bar{x})$, we deduce that there are sequences $\{t_n\} \subset \mathbb{P}$ and $\{z_n\} \subset X$ such that $t_n \downarrow 0$, $z_n \to x$ and $\bar{x} + t_n z_n \in G[-D]^{-1}$ for every $n \in \mathbb{N}$. Therefore, $G(\bar{x} + t_n z_n) \cap (-D) \neq \emptyset$ for all $n \in \mathbb{N}$.

Since $x \in D(F+C)(\bar{x}, \bar{y})[-\text{int}C]^{-1}$, there exists $y \in -\text{int}C$ such that $(x, y) \in \text{graph}\,(D(F+C)(\bar{x}, \bar{y}))$. This confirms that there are sequences $\{s_n\} \subset \mathbb{P}$ and $\{(x_n, y_n)\} \subset X \times Y$ such that $s_n \downarrow 0$, $(x_n, y_n) \to (x, y)$ and $\bar{y} + s_n y_n \in F(\bar{x} + s_n x_n) + C$, for every $n \in \mathbb{N}$. By the derivability hypothesis, we can set $t_n = s_n$, and hence, $\bar{y} + t_n y_n \in F(\bar{x} + t_n x_n) + C$ for every $n \in \mathbb{N}$. At this juncture, we intend to use the Lipschitz-like property at $(\bar{x}, \bar{y})$. Because $u_n := (\bar{x} + t_n x_n) \to \bar{x}$ and $v_n := (\bar{x} + t_n z_n) \to \bar{x}$, there exists $n_1 \in \mathbb{N}$ such that $u_n, v_n \in U := U_1 \cap U_2$ for $n \geq n_1$, where U_1 is the neighborhood tied to the notion of local weak minimality and U_2 is a neighborhood of $\bar{x}$ which exists, along with a neighborhood V of $\bar{y}$, as a consequence of the Lipschitz-like property. Moreover, since $(\bar{y} + t_n y_n) \to \bar{y}$, there exists $n_2 \in \mathbb{N}$ such that $\bar{y} + t_n y_n \in V$ for all $n \geq n_2$. Since $(F+C)$ is Lipschiz-like at $(\bar{x}, \bar{y})$, we get

$$\begin{aligned}\bar{y} + t_n y_n &\in [(F+C)(\bar{x} + t_n x_n)] \cap V \quad (\text{for } n \geq n_2) \subseteq (F+C)(\bar{x} + t_n z_n)\\ &+L \parallel (\bar{x} + t_n x_n) - (\bar{x} + t_n z_n) \parallel B_Y \quad (\text{for } n \geq \max\{n_1, n_2\}) = F(\bar{x} + t_n z_n)\\ &+C + L\, t_n \parallel x_n - z_n \parallel B_Y \quad (\text{for } n \geq \max\{n_1, n_2\}),\end{aligned}$$

and hence we can choose $b_n \in B_Y$ such that, for $n \geq \max\{n_1, n_2\}$, we have $\bar{y} + t_n w_n \in F(\bar{x} + t_n z_n) + C$ where $w_n = (y_n - Lb_n \parallel x_n - z_n \parallel) \to y$.

Since $y \in -\text{int}C$, $t_n > 0$, and $w_n \to y$, there exists $n_3 \in \mathbb{N}$ such that $t_n w_n \in -\text{int}C$. We choose $\tilde{w}_n \in F(v_n)$ such that for every $n \geq n_3$, we have $\bar{y} + t_n w_n \in \tilde{w}_n + C$ for every $n \in \mathbb{N}$. We have

$$\tilde{w}_n \in \bar{y} + t_n w_n - C \subseteq \bar{y} - \text{int}C - C \subset \bar{y} - \text{int}C$$

and consequently

$$\tilde{w}_n \in F(v_n) \cap (\bar{y} - \text{int}C) \text{ for } n \geq \max\{n_1, n_2, n_3\}.$$

Finally, we note that because $t_n \downarrow 0$ and $z_n \to x$, the containment $x \in IT(S, \bar{x})$ implies that there exists $n_4 \in \mathbb{N}$ such that $v_n := \bar{x} + t_n z_n \in S$ for every $n \geq n_4$.

Summarizing, we have shown that for every $n \geq \max\{n_1, n_2, n_3, n_4\}$, there are $v_n \in S_1 \cap U_1$, such that $F(v_n) \cap (\bar{y} - \text{int}C) \neq \emptyset$, which, however, contradicts the local weak optimality of $(\bar{x}, \bar{y})$. The proof is complete. □

In view of the proof of Theorem 12.2.1, the following results are immediate.

Theorem 12.2.2. *Let X, Y, and Z be normed spaces, let $C \subset Y$ and $D \subset Z$ be pointed, closed, convex, and solid cones, and let $S \subseteq X$. Let $F : X \rightrightarrows Y$ and $G : X \rightrightarrows Z$ be set-valued maps. Let $(\bar{x}, \bar{y}) \in \text{graph}\,(F)$ be a local weak minimizer to (P_1). Let either* $\text{dom}\,(F) = \text{dom}\,(G) = S$ *or* $\bar{x} \in \text{int}S$, *and let* $(F + C)$ *is*

Lipschitz-like around $(\bar{x}, \bar{y})$. *Assume that either* epi (F) *is derivable at* $(\bar{x}, \bar{y})$ *or* $G[-D]^{-1}$ *is derivable at* $\bar{x}$. *Then:*

$$D(F + C)(\bar{x}, \bar{y})[-\text{int}C]^{-1} \bigcap T(G[-D]^{-1}, \bar{x}) = \emptyset.$$

Theorem 12.2.3. *Let* X *and* Y *be normed spaces, let* $C \subset Y$ *be a pointed, closed, convex, and solid cone, let* $F : X \rightrightarrows Y$ *be a set-valued map, and* $(\bar{x}, \bar{y}) \in$ graph (F) *be a local weak minimizer to* (P_0). *Then:*

$$IT(S, \bar{x}) \bigcap D(F + C)(\bar{x}, \bar{y})[-\text{int}C]^{-1} = \emptyset.$$

Moreover, if additionally either dom $(F) = S$ *or* $\bar{x} \in \text{int}S$. *Then:*

$$D(F + C)(\bar{x}, \bar{y})[-\text{int}C]^{-1} = \emptyset.$$

Remark 12.2.4. Note that all the above results remain valid if, instead of $D(F + C)(\bar{x}, \bar{y})$, we take either the adjacent derivative, the lower Dini derivative, the circatangent derivative, the radial derivative, the contingent epiderivative, the adjacent epiderivative, the radial epiderivative, the circatangent epiderivative, or their generalized and weakly generalized counterparts. We remark that all the above results, in fact, will characterize the proper minimality if we replace int$C$ by int$K$, where K is a proper, pointed, closed, convex, and solid cone such that $C \backslash \{0\} \subset \text{int}K$. Also notice that for weak minimality, Theorem 12.2.3 states that $D(F + C)(\bar{x}, \bar{y})(x) \subset Y \backslash - \text{int}C$ for every $x \in \text{dom}\,(D(F + C)(\bar{x}, \bar{y}))$. In fact, in most known results, this necessary optimality conditions is proved by direct arguments, which, primarily because of the definition of the derivative, demands that dom $(F) = S$.

Next we give an example to illustrate the above remark:

Example 12.2.5. Consider the set-valued map $F : [0, 2] \rightrightarrows \mathbb{R}^2$ defined by:

$$F(x) := \left\{(y_1, y_2) \in \mathbb{R}^2 \mid \quad y_1^2 + y_2^2 \leq x^2\right\}.$$

Let $C := \left\{(x, x) \in \mathbb{R}^2 \mid \quad x \in \mathbb{R}_+\right\}$ be the ordering cone and let $S := [0, 1]$ be the set of implicit constraints. It is easy to verify that the point $(1, (0, -1)) \in \text{graph}\,(F)$ is a (global) proper minimizer to (P_0). For $x \in \text{dom}\,(D_G F(1, (0, -1))) = \mathbb{R}$, we have

$$D_G F(1, (0, -1))(x) = \left\{(y_1, y_2) \in \mathbb{R}^2 \mid \quad y_1 \in \mathbb{R}, \;\; y_2 = -x\right\},$$

and hence $D_G F(1, (0, -1))(x) \cap (-C \backslash \{0\}) \neq \emptyset$ for every $x \in \mathbb{P}$. However, since $IT(S, \bar{x}) = -\mathbb{P}$ and

$$D_G F(1, (0, -1))[-C \backslash \{0\}]^{-1} = \mathbb{P},$$

we have

$$D_{\mathrm{G}}F(1,(0,-1))[-C\setminus\{0\}]^{-1}\bigcap IT(S,\bar{x})=\emptyset,$$

ensuring that the necessary optimality condition holds. □

Another approach to prove (12.24), in fact more akin to the classical Dubovitskii-Milyutin approach, is to express the optimality as a disjunction of certain sets and then use some separation theorems. In the present setting, the following two results generalize the Dubovitskii-Milyutin approach and give an alternative proof of Theorem 12.2.1.

Proposition 12.2.6. *Let X, Y, and Z be normed spaces, let $C\subset Y$ and $D\subset Z$ be pointed, closed, convex, and solid cones, and let $S\subseteq X$. Let $F:X\rightrightarrows Y$ and $G:X\rightrightarrows Z$ be set-valued maps. For $\bar{x}\in S$, let $(\bar{x},\bar{y})\in \operatorname{graph}(F)$ be a local weak minimizer to (P_1). Then:*

$$IT(S,\bar{x})\bigcap IT(F[\bar{y}-\operatorname{int}C]^{-1},\bar{x})\bigcap T(G[-D]^{-1},\bar{x})=\emptyset. \tag{12.25}$$

Proof. We claim that

$$U\bigcap S\bigcap F[\bar{y}-\operatorname{int}C]^{-1}\bigcap G[-D]^{-1}=\emptyset,$$

where U is a neighborhood of $\bar{x}$ corresponding to the local weak minimality. In fact, if there exists

$$x\in U\bigcap S\bigcap F[\bar{y}-\operatorname{int}C]^{-1}\bigcap G[-D]^{-1},$$

then from $x\in U\bigcap S\bigcap G[-D]^{-1}$ we notice that x is feasible and from $x\in F[\bar{y}-\operatorname{int}C]^{-1}$ we obtain

$$F(x)\cap(\bar{y}-\operatorname{int}C)\neq\emptyset$$

which is a contradiction to the optimality of $(\bar{x},\bar{y})$. The assertion now follows from the properties of the interiorly contingent cones and the contingent cones. The proof is complete. □

Remark 12.2.7. We emphasize that in (12.25) and in all such results in this work, it is implicitly assumed that all the involved sets are nonempty. For instance the use of $IT(S,\bar{x})$ demands that the set S has a nonempty interior.

Besides giving an alternative proof for (12.24), the following result has its own significance.

Proposition 12.2.8. *Let X and Y be normed spaces, let $F : X \rightrightarrows Y$ be a set valued map, let $(F + C)$ is Lipschitz-like around $(\bar{x}, \bar{y}) \in \text{graph}\,(F)$ and derivable at this point. Let C be a pointed, closed, convex, and solid cone. Then:*

$$D(F + C)(\bar{x}, \bar{y})[-\text{int}C]^{-1} \subseteq IT(F[\bar{y} - \text{int}C]^{-1}, \bar{x}).$$

Proof. The proof is based on the arguments given in the proof of Theorem 12.2.1. □

Let $\text{dom}\,(F) = \text{dom}\,(G) = S$, let $(\bar{x}, \bar{y}) \in \text{graph}\,(F)$ and let $\bar{z} \in G(\bar{x}) \cap (-D)$. Then a necessary optimality condition for $(\bar{x}, \bar{y})$ to be a weak minimizer to (P_1), by the direct approach, reads as follows

$$D(F, G)(\bar{x}, \bar{y}, \bar{z})(\Omega) \cap ((-\text{int}C) \times (-\text{int}D - \bar{z})) = \emptyset,$$

where $D(F, G)(\bar{x}, \bar{y}, \bar{z})$ is the contingent derivative of the map $(F, G) := F \times G$ at $(\bar{x}, \bar{y}, \bar{z})$ and Ω is the domain of this derivative. It has been shown in the previous section that, under suitable convexity assumptions, these disjoint sets given in the image space can be separated and a multiplier rule can be obtained. However the optimality condition (12.24) is of different nature. Firstly, the disjunction in (12.24) is taking place in the objective space. Secondly, in (12.24) we have not taken into account any derivative of G. To get a multiplier rule from Theorem 12.2.1, we need to employ the Dubovitskii-Milyutin lemma to (12.24). For this, among other things, we need to gather some information about the dual of $D(F + C)(\bar{x}, \bar{y})[-\text{int}C]^{-1}$ and a suitable regularity condition relating the cone $T(G[-D]^{-1}, \bar{x})$ with some derivative of G at a feasible point. All this is done in the next section.

12.2.2 Inverse Images and Subgradients of Set-Valued Maps

Recall that given a normed space X with X^* as its dual, the polar M^- of $M \subset X$ is defined by

$$M^- = \{\ell \in X^* \,|\, \ell(x) \leq 0 \text{ for every } x \in M\},$$

and the positive dual M^+ is defined by $M^+ = -M^-$. It is known that $M_1 \subseteq M_2$ implies $M_2^- \subseteq M_1^-$.

Given a set $A \subset \mathbb{R}$ and $b \in \mathbb{R}$, by the inequality $A \geq b$ we understand that $a \geq b$ for every $a \in A$.

We formulate the following:

Definition 12.2.9. Let X and Y be normed spaces, let $C \subset Y$ be a pointed, closed and convex cone, and let $F : X \rightrightarrows Y$ be set-valued.

1. Given $A \subset Y^*$, the **A-scalarized subgradient** $\partial_A F(\bar{x}, \bar{y})$ of F at $(\bar{x}, \bar{y}) \in \text{graph}\,(F)$ is defined by

$$\partial_A F(\bar{x}, \bar{y}) := \{L \in X^* \,|\, \exists\, y^* \in A \text{ such that } L(x) \leq (y^* \circ D(F + C)(\bar{x}, \bar{y}))(x)$$
$$\forall\, x \in \text{dom}\,(D(F + C)(\bar{x}, \bar{y}))\}.$$

2. Given $B \subset Y$, the **B-subgradient** $\partial_B F(\bar{x}, \bar{y})$ of F at $(\bar{x}, \bar{y}) \in \text{graph}\,(F)$ is defined by

$$\partial_B F(\bar{x}, \bar{y}) := \{L \in \mathscr{L}(X, Y) |\ (D(F + C)(\bar{x}, \bar{y})(x) - L(x)) \cap B = \emptyset$$
$$\forall\, x \in \text{dom}\,(D(F + C)(\bar{x}, \bar{y}))\}.$$

Remark 12.2.10. We shall use the terms, **generalized subgradient**, **proper subgradient** and **weak subgradient**, if $B = -C\backslash\{0\}$, $-\text{int}K$, where K is a closed, convex, pointed and solid cone with $C\backslash\{0\} \subset \text{int}K$, and $-\text{int}C$, and denote these variants by $\partial_G F(\bar{x}, \bar{y})$, $\partial_P F(\bar{x}, \bar{y})$ and $\partial_W F(\bar{x}, \bar{y})$, respectively. If, in the above definition of the B-subgradient, we replaced $D(F + C)(\bar{x}, \bar{y})$ by $D_G F(\bar{x}, \bar{y})$, then we shall replace the term subgradient by subdifferential. In the following, the **generalized subdifferential**, **generalized proper subdifferential**, **generalized weak subdifferential** of F at $(\bar{x}, \bar{y})$ will be denoted by $\Gamma_G F(\bar{x}, \bar{y})$, $\Gamma_P F(\bar{x}, \bar{y})$ and $\Gamma_W F(\bar{x}, \bar{y})$, respectively. Sticking to our terminology, the notion of generalized weak subdifferential was introduced and analyzed by Song [538]. All the other notions have first appeared in [284].

The above notions are similar to a notion due to Bair-Jahn [22]. Recall that, given a set-valued map $F : X \rightrightarrows Y$ and $(\bar{x}, \bar{y}) \in \text{graph}\,(F)$, the contingent epiderivative $D_E F(\bar{x}, \bar{y}) : X \to Y$ is a *single-valued map* such that $\text{epi}\,(D_E F(\bar{x}, \bar{y})) = T(\text{epi}\,(F), (\bar{x}, \bar{y}))$. By employing this epiderivative, a notion of subgradients for set-valued maps, given in [22], is as follows:

$$\hat{\partial} F(\bar{x}, \bar{y}) := \{L \in \mathscr{L}(X, Y) |\ L(x) \leq D_E F(\bar{x}, \bar{y})(x) \quad \forall\, x \in X\}.$$

Though our primary goal is to state some constraint qualifications in terms of the above notions, we pause to give some of their basic properties. The following result focuses on the generalized weak subdifferential $\Gamma_W F(\bar{x}, \bar{y})$, which can also be written as:

$$\Gamma_W F(\bar{x}, \bar{y}) = \{T \in \mathscr{L}(X, Y) |\ T(x) \notin D_G F(\bar{x}, \bar{y})(\bar{x}, \bar{y})(x) + \text{int}C, \ \text{ for every } x \in X\}.$$

The next two results are based on Song [538]. In these results, it is assumed that the generalized contingent epiderivative exists.

Theorem 12.2.11. *Let X and Y be normed spaces, let $C \subset Y$ be a proper, pointed, closed, and convex cone, let $F : X \rightrightarrows Y$ be a set-valued map, and let $(\bar{x}, \bar{y}) \in$* graph (F). *Then:*

(a) *The set* $\Gamma_W F(\bar{x}, \bar{y})$ *is closed in* $\mathscr{L}(X, Y)$.
(b) *For any* $\lambda > 0$, *the following formula holds:*

$$\lambda \Gamma_W F(\bar{x}, \bar{y}) = \Gamma_W(\lambda F)(\bar{x}, \lambda \bar{y}).$$

(c) *If* $f : X \to Y$ *is Fréchet differentiable at* $\bar{x} \in S \cap \mathrm{dom}(F)$, *where* $S \subset X$ *is open, then*

$$\Gamma_W(f + F)(\bar{x}, f(\bar{x}) + \bar{y}) = f'(\bar{x}) + \Gamma_W F(\bar{x}, \bar{y}).$$

Proof. We begin with a proof of (a). Let $\{T_n\} \subset \Gamma_W F(\bar{x}, \bar{y})$ be an arbitrary sequence such that $T_n \to T$. We will prove the closedness of $\Gamma_W F(\bar{x}, \bar{y})$ by contradiction. Assume that $T \notin \Gamma_W F(\bar{x}, \bar{y})$. Then there exists $x \in X$ such that $Tx \in D_G F(\bar{x}, \bar{y})(x) + \mathrm{int} C$. From the convergence $T_n \to T$, we deduce that $T_n x \to Tx$. Therefore, there exists an index $m \in \mathbb{N}$ such that $T_n x \in D_G F(\bar{x}, \bar{y})(x) + \mathrm{int} C$ for every $n \geq m$. This contradicts the fact that $\{T_n\} \subset \Gamma_W F(\bar{x}, \bar{y})$ and hence $\Gamma_W F(\bar{x}, \bar{y})$ must be closed.

For (b), it suffices to note that, for every $\lambda > 0$, we have $\lambda D_G F(\bar{x}, \bar{y})(x) = D_G(\lambda F)(\bar{x}, \lambda \bar{y})(x)$.

For (c), we note that, from the identity (which is easy to prove),

$$D(f + F + C)(\bar{x}, f(\bar{x}) + \bar{y})(x) = f'(\bar{x})(x) + D(F + C)(\bar{x}, \bar{y})(x),$$

it follows that $D_G(f + F)(\bar{x}, f(\bar{x}) + \bar{y})(x) = f'(\bar{x})(x) + D_G F(\bar{x}, \bar{y})(x)$.

For $T \in \Gamma_W(f + F)(\bar{x}, f(\bar{x}) + \bar{y})$, we have $Tx \notin D_G F(\bar{x}, \bar{y})(x) + f'(\bar{x})(x) + \mathrm{int} C$ for any $x \in X$ and hence $Tx - f'(\bar{x})(x) \notin D_G F(\bar{x}, \bar{y})(x) + \mathrm{int} C$, which confirms that $T - f'(\bar{x}) \in \Gamma_W F(\bar{x}, \bar{y})(x)$. This shows $\Gamma_W(f + F)(\bar{x}, f(\bar{x}) + \bar{y}) \subseteq f'(\bar{x}) + \Gamma_W F(\bar{x}, \bar{y})$. The converse can be proved by similar arguments. □

Recall that given normed spaces X and Y, a solid cone $C \subset Y$, a set-valued map $F : X \rightrightarrows Y$, and $(\bar{x}, \bar{y}) \in \mathrm{graph}\,(F)$, the weakly-subdifferential $\partial F(\bar{x}, \bar{y})$, proposed by Tanino [567], reads:

$$\partial F(\bar{x}, \bar{y}) := \{T \in \mathscr{L}(X, Y) |\ \bar{y} - T(\bar{x}) \in \mathrm{WMin}\,(\cup_{x \in X}[F(x) - T(x)], C)\}.$$

We recall the following result that gives some information on $\partial F(\bar{x}, \bar{y}) \neq \emptyset$ (see Song [535, 538]):

Lemma 12.2.12. *Let* X *and* Y *be normed spaces, and let* $C \subset Y$ *be a pointed, closed, convex, and solid cone. Assume that* $F : X \rightrightarrows Y$ *is a* C*-convex set-valued map such that* $\mathrm{int}(\mathrm{epi}\,(F)) \neq \emptyset$. *If* $\bar{x} \in \mathrm{int}(\mathrm{dom}\,(F))$ *and* $\bar{y} \in \mathrm{WMin}(F(\bar{x}), C)$, *then* $\partial F(\bar{x}, \bar{y}) \neq \emptyset$.

The following result relates the weakly subdifferential and the weak generalized gradient.

Theorem 12.2.13. *Let X and Y be normed spaces, let $C \subset Y$ be a proper, pointed, closed, convex, and solid cone, let $F : X \rightrightarrows Y$ be a set-valued map, and let $(\bar{x}, \bar{y}) \in$* graph (F). *Then*

$$\partial F(\bar{x}, \bar{y}) \subseteq \Gamma_W F(\bar{x}, \bar{y}),$$

and the converse holds if F is C-convex and epi $(D_G F(\bar{x}, \bar{y})) = T(\text{epi}\,(F), (\bar{x}, \bar{y}))$.

Proof. Let $T \in \partial F(\bar{x}, \bar{y})$ be arbitrary. Assume that $T \notin \Gamma_W F(\bar{x}, \bar{y})$. Then for some $x \in X$, we have $D_G F(\bar{x}, \bar{y})(x) \cap (Tx - \text{int}C) \neq \emptyset$. Let $y \in D_G F(\bar{x}, \bar{y})(x) \cap (Tx - \text{int}C)$ be arbitrary. Consequently, there are sequences $\{(x_n, y_n)\} \subset X \times Y$ and $\{t_n\} \subset \mathbb{P}$ such that $t_n \downarrow 0$, $x_n \to x$, $y_n \to y$, and $\bar{y} + t_n y_n \in F(\bar{x} + t_n x_n) + C$. Therefore, there exists a sequence $\{k_n\} \subset C$ such that $\bar{y}_n := \bar{y} - T(\bar{x}) + t_n(y_n - T(x_n)) - k_n \in F(\bar{x} + t_n x_n) - T(\bar{x} + t_n x_n)$.

Since $y - T(x) \in -\text{int}C$, we have $t_n(y_n - T(x_n)) \in -\text{int}C$, for sufficiently large $n \in \mathbb{N}$, confirming that $y_n - (\bar{y} - T(\bar{x})) \in -\text{int}C$ for such $n \in \mathbb{N}$. However, this contradicts the assumption that $T \in \partial F(\bar{x}, \bar{y})$ and hence $T \in \Gamma_W F(\bar{x}, \bar{y})$.

For the converse inclusion, let $T \in \Gamma_W F(\bar{x}, \bar{y})$ be arbitrary. Then $Tx \notin D_G F(\bar{x}, \bar{y})(x) + \text{int}C$ for every $x \in X$. Since F is C-convex, we have $F(x) - \bar{y} \subseteq D_G F(\bar{x}, \bar{y})(x - \bar{x}) + C$, and hence, for every $x \in X$ we have $(F(x) - T(x) - (\bar{y} - T(\bar{x}))) \cap (-\text{int}C) = \emptyset$ which implies $\bar{y} - T(\bar{x}) \in \text{WMin}(\bigcup_{x \in X}(Fx - Tx), C)$. Therefore $T \in \partial F(\bar{x}, \bar{y})$. The proof is complete. □

We next connect the scalarized subgradients and the generalized subgradients.

Proposition 12.2.14. *Let X and Y be normed spaces and let $F : X \rightrightarrows Y$ be set-valued which is assumed to be locally C-convex at $(\bar{x}, \bar{y}) \in$* graph (F). *Then $\partial_P F(\bar{x}, \bar{y}) \neq \emptyset$ implies that $\partial_{C^+\setminus\{0\}} F(\bar{x}, \bar{y}) \neq \emptyset$.*

Proof. Let $\partial_P F(\bar{x}, \bar{y}) \neq \emptyset$ and let $L \in \partial_P F(\bar{x}, \bar{y})$ be arbitrary. By the definition of the proper subgradient, there is a closed, convex, pointed and solid cone K with $C \setminus \{0\} \subset \text{int}K$, such that

$$(D(F + C)(\bar{x}, \bar{y}) - L)(x) \cap (-\text{int}K) = \emptyset$$
$$\text{for every } x \in \Omega := \text{dom}\,(D(F + C)(\bar{x}, \bar{y})). \tag{12.26}$$

We claim that $(D(F + C)(\bar{x}, \bar{y}) - L)(\Omega)$ is convex. For $i \in \{1, 2\}$, we choose $y_i \in (D(F + C)(\bar{x}, \bar{y}) - L)(\Omega)$, ensuring that there are $x_i \in \Omega$ such that $y_i \in D(F + C)(\bar{x}, \bar{y})(x_i) - L(x_i)$ and hence $y_i + L(x_i) \in D(F + C)(\bar{x}, \bar{y})(x_i)$. This implies that, for some $\lambda \in (0, 1]$, we have $\lambda y_1 + (1 - \lambda)y_2 \in (D(F + C)(\bar{x}, \bar{y}) - L)(\lambda x_1 + (1 - \lambda)x_2)$, and by using the convexity of the set Ω, we conclude that $\lambda y_1 + (1 - \lambda)y_2 \in (D(F + C)(\bar{x}, \bar{y}) - L)(\Omega)$. Now, by employing a separation theorem to the disjunction given in (12.26), we assure the existence of $y^* \in K^+ \setminus \{0\} \subseteq C^+ \setminus \{0\}$ such that

$$(y^* \circ D(F + C)(\bar{x}, \bar{y}))(x) \geq (y^* \circ L)(x) \quad \text{for every } x \in \Omega.$$

Since $L \in \mathscr{L}(X,Y)$ and $y^* \in Y^*$, we have $y^* \circ L \in X^*$ and hence $y^* \circ L \in \partial_{C^+\setminus\{0\}} F(\bar{x},\bar{y})$. □

The following technical result will play an important role in the subsequent development of this chapter. This result, which is taken from [284], is an extension of a result by Rigby [487, Proposition 5] given for single-valued maps.

Theorem 12.2.15. *Let X and Y be normed spaces, let $M \subseteq X$ be convex set, and let $A \subset Y$ be a closed, convex, and solid cone. Let $T : M \rightrightarrows Y$ be a A-convex set-valued map. Then:*

1. *If $T[-\text{int}(A)]^{-1} \neq \emptyset$, then for every $\ell \in L^-$, where $L := T[-A]^{-1}$, there exists $t \in A^+$ such that*

$$t \circ T(x) \geq \ell(x) \quad \textit{for every } x \in M.$$

2. *If $T[-\text{int}(A)]^{-1} = \emptyset$, then there exists $t \in A^+\setminus\{0\}$ such that*

$$t \circ T(x) \geq 0 \quad \textit{for every } x \in M.$$

Proof. Let us begin with the case when the set $T[-\text{int}(A)]^{-1}$ is nonempty. Then the (negative) dual L^- of $L := T[-A]^{-1}$ is nonempty as well. We choose $\ell \in L^-$ arbitrarily and define a set

$$E := \{(y, \ell(x)) \in Y \times \mathbb{R} \mid y \in T(x) + A,\ x \in M\}.$$

Using the facts that M is convex, T is A-convex and $\ell \in Y^*$, we deduce that E is a convex set. Indeed, let $(y_1, z_1), (y_2, z_2) \in E$ be arbitrary. Then by the definition of E, for $i = 1, 2$, there exists $x_i \in X$ with $z_i = \ell(x_i)$ and $y_i \in T(x_i) + A$. For $\lambda \in (0, 1]$, we have $\lambda z_1 + (1 - \lambda)z_2 = \ell(\lambda x_1 + (1 - \lambda)x_2)$, which, in view of the A-convexity of T, ensures that

$$\lambda y_1 + (1 - \lambda)y_2 \in \lambda T(x_1) + (1 - \lambda)T(x_2) + A \subseteq T(\lambda x_1 + (1 - \lambda)x_2) + A.$$

Since the set M is convex, we at once obtain that $\lambda(y_1, z_1) + (1 - \lambda)(y_2, z_2) \in E$. Having established the convexity of the set E, we claim that

$$E \cap (-\text{int}(A) \times \mathbb{P}) = \emptyset. \tag{12.27}$$

In fact, if this is not the case, then there exists $(x, y) \in X \times Y$ such that $y \in (T(x) + A) \cap (-\text{int}(A))$ and $\ell(x) > 0$. Let $w \in T(x)$ be such that $y \in w + A$. Then $w \in y - A \subset -\text{int}(A) - A = -\text{int}(A)$, which contradicts that $\ell \in L^-$.

Therefore (12.27) holds, and hence by a separation theorem, we get the existence of $(f, g) \in Y^* \times \mathbb{R}\setminus\{0, 0\}$ and a real number α such that we have

$$f(u) + g(v) \geq \alpha \quad \text{for every } (u, v) \in E \tag{12.28a}$$

$$f(c) + g(d) < \alpha \quad \text{for every } (c, d) \in -\text{int}(A) \times \mathbb{P}. \tag{12.28b}$$

Since A is a cone, we can set $\alpha = 0$ in (12.28a) and (12.28b). By taking $d \in \mathbb{P}$ arbitrarily close to 0 and $c \in -\text{int}(A)$ arbitrarily close to 0, we obtain $f \in A^+$ and $g \leq 0$, respectively. We claim that $g < 0$. Indeed, if $g = 0$, we get

$$f(c) < 0 \quad \text{for every } c \in -\text{int}(A),$$
$$f(u) \geq 0 \quad \text{for every } u \in T(M) + A,$$

which, however, is impossible because $(T(M) + A) \cap (-\text{int}(A)) \neq \emptyset$ and hence $g < 0$.

Moreover, from (12.28a), for every $x \in M$, we have $f \circ (T + A)(x) \geq -(g \circ \ell)(x)$. Therefore, by setting $t = (-f/g) \in A^*$ and noticing that $0 \in A$, we finish the proof of the first part.

For the second part, we notice that if $T[-\text{int}(A)]^{-1} = \emptyset$, we have $T(M) \cap -\text{int}(A) = \emptyset$, and hence using arguments similar to those given above, we can prove the existence of $t \in A^* \backslash \{0\}$ such that $t \circ T(x) \geq 0$ for every $x \in M$. The proof is complete. □

The following particular case is quite useful. Henceforth we set $(T(\cdot,\cdot))^- = N(\cdot,\cdot)$.

Theorem 12.2.16. *If in Theorem 12.2.15, we assume that* $A = T(B, -\bar{w})$, *where* $B \subset Y$ *is a pointed, closed, convex, and solid cone with* $\bar{w} \in -B$ *and* T *is* B*-convex, then* $t \in B^+$ *and* $t(\bar{w}) = 0$.

Proof. To apply Theorem 12.2.15, it suffices to show that the set-valued map T is $T(B, -\bar{w})$-convex. To show that $B \subseteq T(B, -\bar{w})$, it suffices to notice that $B + \bar{w} \supset B$ is equivalent to $B \supset B + (-\bar{w})$. For this, we notice that, due to $\bar{w} \in -B$, we have $B + (-\bar{w}) \subset B + B = B$. Now, the $T(B, -\bar{w})$-convexity of the map T follows from its B-convexity.

Furthermore, by invoking Theorem 12.2.15, for every $\ell \in \left(T[-A]^{-1}\right)^-$, with $A = T(B, -\bar{w})$, there exists a functional $t \in (T(B, -\bar{w}))^+$ which satisfies the assertion. Since B is convex, we have $T(B, -\bar{w}) \supseteq B + \bar{w}$ and consequently

$$t(b + \bar{w}) \geq 0 \quad \text{for every } b \in B. \tag{12.29}$$

By setting $b = 0$ in the above inequality, we obtain $t(\bar{w}) \geq 0$. Because B is a cone and $\bar{w} \in -B$, we can substitute $b = -2\bar{w} \in B$ in (12.29) to obtain $t(\bar{w}) \leq 0$. Therefore, combining the preceding two inequalities, we have $t(\bar{w}) = 0$ and this, in view of (12.29), yields $t \in B^+$. □

We have the following result which will play an important role in deriving optimality conditions in set-valued optimization.

Theorem 12.2.17. *Let* X *and* Y *be normed spaces and let* $F : X \rightrightarrows Y$ *be a set-valued map which we assume to be locally* C*-convex at* $(\bar{x}, \bar{y}) \in \text{graph}\,(F)$. *That is,*

$D(F + C)(\bar{x}, \bar{y})$ is a convex set-valued map. Let $\mathbb{K}$ be a solid, closed, and convex cone such that $C \subseteq \mathbb{K}$. Then:

1. *If*

$$D(F + C)(\bar{x}, \bar{y})[-\mathrm{int}(\mathbb{K})]^{-1} \neq \emptyset,$$

then

$$(D(F + C)(\bar{x}, \bar{y})[-\mathbb{K}]^{-1})^{-} \subset (D(F + C)(\bar{x}, \bar{y})[-\mathrm{int}(\mathbb{K})]^{-1})^{-} \subset \partial_{C^+} F(\bar{x}, \bar{y}). \tag{12.30}$$

2. *If*

$$D(F + C)(\bar{x}, \bar{y})[-\mathrm{int}(\mathbb{K})]^{-1} = \emptyset,$$

then

$$0 \in \partial_{C^+\backslash\{0\}} F(\bar{x}, \bar{y}).$$

Proof. The first inclusion is a consequence of

$$D(F + C)(\bar{x}, \bar{y})[-\mathrm{int}(\mathbb{K})]^{-1} \subset D(F + C)(\bar{x}, \bar{y})[-\mathbb{K}]^{-1},$$

and the second inclusion is essentially a restatement of Theorem 12.2.15 when we notice that the map $D(F+C)(\bar{x}, \bar{y})$ is $\mathbb{K}$-convex and set $T \equiv D(F+C)(\bar{x}, \bar{y})$. □

We conclude this section by using the scalarized subgradient to express a necessary optimality condition for the local proper minimality in set-valued optimization.

Theorem 12.2.18. *Let $(\bar{x}, \bar{y}) \in \mathrm{graph}\,(F)$ be a local proper minimizer to (P_0). Let the map $F : X \rightrightarrows Y$ be locally C-convex at $(\bar{x}, \bar{y})$ with $\mathrm{dom}\,(F) = S$. Then,*

$$0 \in \partial_{C^+\backslash\{0\}} F(\bar{x}, \bar{y}).$$

Proof. Since $(\bar{x}, \bar{y})$ is a local proper minimizer, by the similar arguments as in Theorem 12.2.1 we can show that

$$D(F + C)(\bar{x}, \bar{y})[-\mathrm{int} K]^{-1} = \emptyset$$

for some pointed, closed, convex, and solid cone K satisfying $C\backslash\{0\} \subset \mathrm{int} K$. (In fact this link has lead to the proof of Theorem 12.2.2.) Now, by applying Theorem 12.2.17, we obtain that $0 \in \partial_{C^+\backslash\{0\}} F(\bar{x}, \bar{y})$. The proof is complete. □

12.2.3 Separation Theorems and the Dubovitskii-Milyutin Lemma

In this section, we give a separation theorem and from that we deduce the so-called Dubovitskii-Milyutin lemma. This is a classical result and can be found in many books. Our proofs are based on the excellent paper by Ben-Tal and Zowe [45].

We recall that, for a subset S of a normed space X, its **support functional** $\sigma(\cdot, S)$, defined on the topological dual X^* of X with values in the extended real line $\mathbb{R} \cup \{\pm\infty\}$, is given by:

$$\sigma(x^*, S) = \sup_{x \in S} \langle x^*, x\rangle \quad \text{for } x^* \in X^*. \tag{12.31}$$

The effective domain of $\sigma(\cdot, S)$ is denoted and defined as $\Lambda(S) := \{x^* \in X^* \mid \sigma(x^*, S) < \infty\}$. It is easy to verify that $\sigma(\cdot, S)$ is a positively homogeneous, closed, and convex function, and that $\Lambda(S)$ is a convex cone. Furthermore, whenever S is a cone, we have

$$\Lambda(S) = -S^+ \tag{12.32}$$

$$\sigma(x^*, S) = \begin{cases} 0 & \text{if } x^* \in \Lambda(S), \\ \infty & \text{otherwise.} \end{cases} \tag{12.33}$$

We begin with the first main result of this section:

Theorem 12.2.19. *Let X be a normed space, let $S_1, \ldots, S_n$ be convex subsets of X, and let $x^* \in \Lambda\left(\bigcap_{i=1}^n S_i\right)$. If*

$$\left(\bigcap_{i=1}^{n-1} \operatorname{int}(S_i)\right) \cap S_n \neq \emptyset, \tag{12.34}$$

then

$$\sigma\left(x^*, \bigcap_{i=1}^n S_i\right) = \min\left\{\sum_{i=1}^n \sigma(x_i^* | S_i) \mid x^* = x_1^* + \cdots x_n^*,\ x_i^* \in \Lambda(S_i)\right\}.$$

Proof. The proof is by induction over n. For $n = 1$, there is nothing to prove. Now assume that the assertion holds for $n - 1$ and proceed to n sets. We set

$$K_1 := \bigcap_{i=1}^{n-1} S_i,$$

$$K_2 := S_n.$$

With the given $x^* \in \Lambda\left(\bigcap_{i=1}^n S_i\right) = \Lambda\left(K_1 \bigcap K_2\right)$, we define the following two disjoint sets in $X \times \mathbb{R}$:

$$A := \left\{ \begin{pmatrix} x \\ \langle x^*, x\rangle - \varepsilon \end{pmatrix} \text{ such that } x \in K_1,\ \varepsilon > 0 \right\},$$

$$B := \left\{ \begin{pmatrix} x \\ \sigma(x^*, K_1 \cap K_2) \end{pmatrix} \mid x \in K_2 \right\}.$$

The sets A and B are nonempty, convex, and because of (12.34), A has a nonempty interior. Hence, there exists a closed hyperplane in $X \times \mathbb{R}$ separating A and B. That is, we can choose $y^* \in X^*$, $\lambda \in \mathbb{R}$, not both simultaneously zero, and $\beta \in \mathbb{R}$ such that, for every $x \in K_1$ and every $z \in K_2$, we have

$$\langle y^*, x\rangle + \lambda\left(\langle x^*, x\rangle - \varepsilon\right) \geq \beta \geq \langle y^*, z\rangle + \lambda\sigma\left(x^*, K_1 \cap K_2\right), \tag{12.35}$$

from which it follows that $\lambda \leq 0$. Without any loss of generality, we set $\lambda = -1$. Indeed, if $\lambda = 0$, then $\langle y^*, x - z\rangle \geq 0$ for every $x - z \in K_1 - K_2$, and because $\operatorname{int}(K_1) \cap K_2 \neq \emptyset$, we deduce that $y^* = 0$, which is a contradiction. Therefore, we get from (12.35) that

$$z^* := x^* - y^* \in \Lambda(K_1) \text{ and } \sigma\left(z^*, K_1\right) \leq \varepsilon - \beta,$$
$$y^* \in \Lambda(K_2) \text{ and } \sigma\left(y^*, K_2\right) \leq \beta + \sigma(x^*, K_1 \cap K_2),$$

and hence $x^* = z^* + y^*$ with $z^* \in \Lambda(K_1)$ and $y^* \in \Lambda\left(K_2\right)$ with

$$\sigma(z^*, K_1) + \sigma(y^*, K_2) \leq \varepsilon + \sigma(x^*, K_1 \cap K_2). \tag{12.36}$$

However, the following inequality also holds trivially,

$$\sigma(x^*, K_1 \cap K_2) \leq \varepsilon + \sigma(z^*, K_1 \cap K_2) + \sigma(y^*, K_1 \cap K_2) \leq \varepsilon + \sigma(z^*, K_1) + \sigma(y^*, K_2),$$

and consequently, we have $\sigma(z^*, K_1) + \sigma(y^*, K_2) = \sigma(x^*, K_1 \cap K_2)$. This forces $\varepsilon = 0$, and this proves the assertion for n. The proof is complete. □

Our second result is the following separation theorem:

Theorem 12.2.20. *Let X be a normed space, let $S_1, \ldots, S_n, S_{n+1}$, be nonempty convex subsets of X, where $S_1, \ldots, S_n$ are open. Then*

$$\bigcap_{i=1}^{n+1} S_i = \emptyset \tag{12.37}$$

if and only if there are $x_i^* \in \Lambda(S_i)$, $i = 1, \ldots, n+1$, *such that*

$$x_1^* + x_2^* + \cdots + x_{n+1}^* = 0, \tag{12.38}$$

$$\sigma(x_1^*, S_1) + \sigma(x_2^*, S_2) + \cdots + \sigma(x_{n+1}^*, S_{n+1}) \leq 0. \tag{12.39}$$

Proof. Assume that (12.38) and (12.39) hold but (12.37) fails. That is, there exists $\bar{x} \in \bigcap_{i=1}^{n+1} S_i$. We set $K_i := S_i - \bar{x}$, for $i = 1, 2, \ldots, n+1$. Then $0 \in K_i$ and hence $\sigma(x_i^*, K_i) \geq 0$, for every i. Note that $x_j^* \neq 0$ for at least one $j \leq n$, as otherwise $-x_{n+1}^* = x_1^* + \ldots + x_n^* = 0$, implying that $x_1^* = \cdots = x_{n+1}^* = 0$. However, since $0 \in \operatorname{int}(K_j)$, we have $\sigma(x_j^*, K_j) > 0$ for this special j. Therefore,

$$\sum_{i=1}^{n+1} \sigma(x_i^*, S_i) = \sum_{i=1}^{n+1} \sigma(x_i^*, K_i) + \sum_{i=1}^{n+1} \langle x_i^*, \bar{x} \rangle = \sum_{i=1}^{n+1} \sigma(x_i^*, K_i) > 0,$$

which is a contradiction to (12.39).

For the converse, we set $S := \bigcap_{i=1}^{n} S_i$ and assume that $K \neq \emptyset$. Then S and S_{n+1} are nonempty convex sets, S is open, and by (12.37), $S \bigcap S_{n+1} = \emptyset$. Therefore, by a standard separation theorem, we have a suitable $x^* \in X^*$, $x^* \neq 0$ such that

$$\langle x^*, x \rangle \leq \langle x^*, z \rangle \quad \text{for all } x \in S,\ z \in S_{n+1},$$

implying that $x^* \in \Lambda(S)$, $-x^* \in \Lambda(S_{n+1})$, and

$$\sigma\left(x^*, \bigcap_{i=1}^{n} S_i\right) + \sigma\left(-x^*, S_{n+1}\right) \leq 0.$$

We now apply Theorem 12.2.19 to ensure that, with suitable $x_i^* \in \Lambda(S_i)$, $i = 1, 2, \ldots, n$ and

$$x^* = x_1^* + x_2^* + \cdots x_n^*,$$

$$\sigma\left(x^*, \bigcap_{i=1}^{n} S_i\right) = \sum_{i=1}^{n} \sigma\left(x_i^*, S_i\right),$$

equations (12.38) and (12.39) follow if we put $x_{n+1} := -x^*$. Note that, if $S = \bigcap_{i=1}^{n} S_i = \emptyset$, then there exists $1 \leq m < n$ such that $\bigcap_{i=1}^{m} S_i \neq \emptyset$ and $\bigcap_{i=1}^{m+1} S_i = \emptyset$. We apply the result just proved above with n replaced by m and set $x_{m+2}^* = \cdots = x_{n+1}^* = 0$. The proof is complete. □

We have the following well-known separation theorem for convex cones, also known as the Dubovitskii-Milyutin lemma. An alternative proof of this is given in Girsanov [211, pp 37].

Theorem 12.2.21. *Let X be a normed space, let $S_1, \dots, S_n, S_{n+1}$ be nonempty convex cones in X, and let $S_1, \dots, S_n$ be open. Then,*

$$\bigcap_{i=1}^{n+1} S_i = \emptyset$$

if and only if there are $x_i^* \in S_i^+$, $i = 1, 2, \dots, n+1$, not all zero, such that

$$x_1^* + \cdots + x_{n+1}^* = 0.$$

Proof. The proof follows from Theorem 12.2.20 and the observation that (12.39) becomes the trivial inequality $0 \le 0$. □

12.2.4 Lagrange Multiplier Rules by the Dubovitskii-Milyutin Approach

In our next result we give the promised Lagrange multiplier rule.

Theorem 12.2.22. *Let X, Y, and Z be normed spaces, let $C \subset Y$, and $D \subset Z$ be pointed, closed, and convex cones, where C is also solid, and let $F : X \rightrightarrows Y$ and $G : X \rightrightarrows Z$ be set-valued maps. For $\bar{x} \in S \subset X$, let $(\bar{x}, \bar{y}) \in \operatorname{graph}(F)$ and let $\bar{z} \in G(\bar{x}) \cap (-D)$. Let $(\bar{x}, \bar{y})$ be a local weak minimizer of (P_1). Assume that the following conditions hold:*

1. *The map $(F+C)$ is Lipschitz-like around $(\bar{x}, \bar{y})$ and locally convex and derivable at this point.*
2. *There exists an open convex cone $M \subset IT(S, \bar{x})$.*
3. *Either $IT(F[\bar{y} - \operatorname{int}C]^{-1}, \bar{x})$ is convex or $D(F + C)(\bar{x}, \bar{y})[-\operatorname{int}C]^{-1}$ is open.*
4. $\operatorname{dom}(D(F + C)(\bar{x}, \bar{y})) = \operatorname{dom}(D(G + D)(\bar{x}, \bar{z})) = X$, *and the following regularity condition $(\mathscr{RC})$ holds: If*

$$D(G + D)(\bar{x}, \bar{z})[-T(D, -\bar{z}) \setminus \{0\}]^{-1} = \emptyset,$$

then $0 \in \partial_{-N(D,-\bar{z})\setminus\{0\}} G(\bar{x}, \bar{z})$, otherwise $T(G[-D]^{-1}, \bar{x})$ is convex and we have:

$$N(G[-D]^{-1}, \bar{x}) \subseteq \partial_{-N(D,-\bar{z})} G(\bar{x}, \bar{z}). \tag{12.40}$$

Then there exists $(s, t, u) \in X^* \times Y^* \times Z^*$, not all zero, such that $s \in M^+$, $t \in C^+$ and $u \in D^+$. Moreover the complementary slackness condition $u(\bar{z}) = 0$ and the following inequality hold:

$$t \circ D(F + C)(\bar{x}, \bar{y})(x) + u \circ D(G + D)(\bar{x}, \bar{z})(x) \ge s(x) \quad \text{for every } x \in X. \tag{12.41}$$

If either $\bar{x} \in \text{int}S$ *or* $\text{dom}\,(F) = \text{dom}\,(G) = S$ *then* $s \in Z^*$ *can be set to zero. In this particular case, we have* $t \neq 0$ *if the following regularity condition holds:*

$$D(G + D)(\bar{x}, \bar{z})(X) + \text{cone}(D + \bar{z}) = Z. \tag{12.42}$$

Proof. We begin by showing that the assertions would hold trivially if either of the following disjunctions hold:

$$D(F + C)(\bar{x}, \bar{y})[-\text{int}C]^{-1} = \emptyset,$$
$$D(G + D)(\bar{x}, \bar{z})[-T(D, -\bar{z})\backslash\{0\}]^{-1} = \emptyset.$$

Indeed, if $D(F + C)(\bar{x}, \bar{y})[-\text{int}C]^{-1} = \emptyset$, then, in view of Theorem 12.2.17, there exists $t \in C^+\backslash\{0\}$ such that for every $x \in X$, we have

$$t \circ D(F + C)(\bar{x}, \bar{y})(x) \geq 0,$$

and hence by choosing $s = 0$ and $u = 0$, we get the desired result.

On the other hand, if $D(G + D)(\bar{x}, \bar{z})[-T(D, -\bar{z})\backslash\{0\}]^{-1} = \emptyset$, then, in view of the condition

$$0 \in \partial_{-N(D,-\bar{z})\backslash\{0\}} G(\bar{x}, \bar{z}),$$

we assure the existence of

$$u \in -N(D, -\bar{z})\backslash\{0\} = -T(D, -\bar{z})^{-}\backslash\{0\} = T(D, -\bar{z})^{+}\backslash\{0\},$$

such that for every $x \in X$, we have

$$u \circ D(G + D)(\bar{x}, \bar{z})(x) \geq 0,$$

and by choosing $s = 0$ and $t = 0$, we obtain (12.41). The proof for $u \in D^+$ and the complementary slackness condition follows from Theorem 12.2.16.

Therefore, without any loss of generality, we can assume that

$$D(F + C)(\bar{x}, \bar{y})[-\text{int}C]^{-1} \neq \emptyset,$$
$$D(G + D)(\bar{x}, \bar{z})[-T(D, -\bar{z})\backslash\{0\}]^{-1} \neq \emptyset.$$

We first consider the case when $IT(F[\bar{y} - \text{int}C]^{-1}, \bar{x})$ is convex. Now, by using Proposition 12.2.6 and the containment $M \subset IT(S, \bar{x})$, we have

$$IT(F[\bar{y} - \text{int}C]^{-1}, \bar{x}) \cap M \cap T(G[-D]^{-1}, \bar{x}) = \emptyset.$$

The above disjunction, in view of the Dubovitskii-Milyutin lemma, ensures the existence of functionals

$$l_1 \in M^-,$$
$$l_2 \in IT(F[\bar{y} - \text{int}C]^{-1}, \bar{x})^-,$$
$$l_3 \in N(G[-D]^{-1}, \bar{x}),$$

such that

$$l_1 + l_2 + l_3 = 0. \tag{12.43}$$

At this juncture, we recall the following inclusion (see Proposition 12.2.8),

$$D(F + C)(\bar{x}, \bar{y})[-\text{int}C]^{-1} \subset IT(F[\bar{y} - \text{int}C]^{-1}, \bar{x}),$$

and consequently, we have

$$l_1 \in M^-,$$
$$l_2 \in (D(F + C)(\bar{x}, \bar{y})[-\text{int}C]^{-1})^-,$$
$$l_3 \in N(G[-D]^{-1}, \bar{x}).$$

Note that, if $D(F + C)(\bar{x}, \bar{y})[-\text{int}C]^{-1}$ is open then since it is a convex cone, we can directly obtain the above conclusion. Now, because $D(G + D)(\bar{x}, \bar{z})[-T(D, -\bar{z})\backslash\{0\}]^{-1} \neq \emptyset$, we have

$$N(G[-D]^{-1}, \bar{x}) \subseteq \partial_{-N(D,-\bar{z})} G(\bar{x}, \bar{z}),$$

and consequently $l_3 \in \partial_{-N(D,-\bar{z})} G(\bar{x}, \bar{z})$.

This observation and Theorem 12.2.17, with $A = C$, ensure the existence of functionals $t \in C^+$ and $u \in -N(D, -\bar{z}) = T(D, -\bar{z})^+$ such that

$$t \circ D(F + C)(\bar{x}, \bar{y})(x) \geq l_2(x) \quad \text{for every } x \in X$$
$$u \circ D(G + D)(\bar{x}, \bar{z})(x) \geq l_3(x) \quad \text{for every } x \in X.$$

Combining the above two inequalities with (12.43) yields

$$t \circ D(F + C)(\bar{x}, \bar{y})(x) + u \circ D(G + D)(\bar{x}, \bar{z})(x) \geq -l_1(x) \quad \text{for every} \quad x \in X.$$

By setting $s = -l_1 \in M^+$ we finish the proof of (12.40). The remaining proof for $u \in D^+$ and for the complementary slackness condition $u(\bar{z}) = 0$ is same as in Theorem 12.2.16.

Finally, if either $\bar{x} \in \operatorname{int} S$ (that is, $IT(S,\bar{x}) = X$) or $\operatorname{dom}(F) = \operatorname{dom}(G) = S$, then we can apply the above arguments to the following disjunction given in Theorem 12.2.2:

$$IT(F[\bar{y} - \operatorname{int} C]^{-1}, \bar{x}) \bigcap T(G[-D]^{-1}, \bar{x}) = \emptyset.$$

It remains to show that under the regularity condition (12.42), we have $t \neq 0$. For this, assume that $t = 0$ and choose $\tilde{z} \in Z$ arbitrarily. Since

$$\tilde{z} = z + \beta(d + \bar{z}) \quad \text{where} \quad z \in D(G + D)(\bar{x}, \bar{z})(X) \quad d \in D, \;\; \beta > 0,$$

we deduce that for an arbitrary $\tilde{z} \in Z$, we have

$$u(\tilde{z}) = u(z) + \beta(u(d) + u(\bar{z})) \geq 0.$$

This, however, implies that $u = 0$, and we have a contradiction to the fact that $(t,u) \neq (0,0)$. The proof is complete. □

As evident from the above proof, the assumption ($\mathscr{RC}$) is vital to ensure the existence of the multipliers. A detailed discussion of this condition, which is partly motivated by the Guignard constraint qualification, can be found in [284]. Furthermore, this condition can be bypassed when the cone D is solid. This can be seen in Sect. 12.5 where an extension of the Dubovitskii-Milyutin is presented. We also remark that a treatment of (P_2) for optimality condition by the Dubovitskii-Milyutin approach is given in [284]. To avoid unnecessary repetition, we treat an analogue of (P_2) in Sect. 12.5.

We now focus on the strong minimizers. Recall that a point $(\bar{x}, \bar{y}) \in \operatorname{graph}(F)$ is called a local strong minimizer of (P_2) if there exists a neighborhood U of $\bar{x}$ such that $F(S_2 \cap U) \subseteq \bar{y} + C$. In the following result, we give necessary optimality conditions for the local strong minimality. Let the cone C be closed and convex and the cone D be solid.

Theorem 12.2.23. *Let X and Y be normed spaces, let $F : X \rightrightarrows Y$ be set-valued and let $(\bar{x}, \bar{y}) \in \operatorname{graph}(F)$. Let the map $(F + C)$ possess the Aubin property around $(\bar{x}, \bar{y})$.*

(a) If $(\bar{x}, \bar{y})$ is a strong minimizer to (P_1) then the following holds:

$$D(F + C)(\bar{x}, \bar{y})[Y \backslash C]^{-1} \bigcap IT(S, \bar{x}) \bigcap T(G[-D]^{-1}, \bar{x}) = \emptyset.$$

(b) If $(\bar{x}, \bar{y})$ is a strong minimizer to (P_1) then for $\bar{z} \in G(\bar{x}) \cap -D$, the following holds:

$$D(F + C)(\bar{x}, \bar{y})[Y \backslash C]^{-1} \bigcap IT(S, \bar{x}) \bigcap \operatorname{cl}(D(G + D)(\bar{x}, \bar{z})[IT(-D, \bar{z})]^{-1}) = \emptyset.$$

(*c*) *If* $(\bar{x}, \bar{y})$ *is a strong minimizer to* (P_2) *then the following holds:*

$$D(F + C)(\bar{x}, \bar{y})[Y \backslash C]^{-1} \bigcap IT(S, \bar{x}) \bigcap IT(G[-D]^{-1}, \bar{x}) \bigcap T(H[h]^{-1}, \bar{x}) = \emptyset.$$

(*d*) *Let the map* $(G + D)$ *possess the Aubin property around* $(\bar{x}, \bar{z})$ *where* $\bar{z} \in G(\bar{x}) \cap -D$. *If* $(\bar{x}, \bar{y})$ *is a strong minimizer to* (P_2) *then the following holds:*

$$D(F + C)(\bar{x}, \bar{y})[Y \backslash C]^{-1} \bigcap IT(S, \bar{x}) \bigcap D(G + D)[IT(-D), \bar{z})]^{-1}$$
$$\bigcap T(H[h]^{-1}, \bar{x}) = \emptyset.$$

Proof. (a) Assume that there exists an element $x \in X$ such that

$$x \in D(F + C)(\bar{x}, \bar{y})[Y \backslash C]^{-1} \bigcap IT(S, \bar{x}) \bigcap T(G[-D]^{-1}, \bar{x}).$$

In view of the containment $x \in D(F + C)(\bar{x}, \bar{y})[Y \backslash C]^{-1}$, there exists $y \in Y \backslash C$ such that $(x, y) \in \text{graph}\,(D(F + C)(\bar{x}, \bar{y}))$. This implies that there are sequences $(\lambda_n) \subset \mathbb{P}$ and $((x_n, y_n)) \subset X \times Y$ such that $\lambda_n \to 0$, $(x_n, y_n) \to (x, y)$ and $\bar{y} + \lambda_n y_n \in F(\bar{x} + \lambda_n x_n) + C$ for all $n \in \mathbb{N}$. Also, since $x \in IT(S, \bar{x})$, there exists $n_1 \in \mathbb{N}$ such that $\bar{x} + \lambda_n x_n \in S$ for $n \geq n_1$. Since $y \notin C$, $y_n \to y$ and the cone C is closed, there exists $n_2 \in \mathbb{N}$ such that $y_n \notin C$ for all $n \geq n_2$. Because $\lambda_n > 0$, we have $\lambda_n y_n \notin C$. For $u_n := \bar{x} + \lambda_n x_n$, let $w_n \in F(u_n)$ be such that for $n \in \mathbb{N}$, we have $\bar{y} + \lambda_n y_n = w_n + c_n$, where $c_n \in C$. Hence $w_n + c_n \notin \bar{y} + C$ for $n \geq n_2$. Since $C + C = C$, we have $w_n \notin \bar{y} + C$, for every $n \geq n_2$. As in the proof of Theorem 12.2.1, we can show that there exists $n_3 \in \mathbb{N}$ such that $G(u_n) \cap (-D) \neq \emptyset$ for all $n \geq n_3$. Therefore, we have shown that, for sufficiently large $n \in \mathbb{N}$, there are $w_n \in F(S_1)$ such that $w_n \notin \bar{y} + C$, which, however, is a contradiction to the assumption that $(\bar{x}, \bar{y})$ is a strong minimizer. Hence part (a) is true. The parts (b)–(d) can now be shown by unifying the above arguments with those given in the preceding section. □

We conclude this section by giving a simple example which shows that the various optimality conditions given above are not, in general, sufficient optimality conditions:

Example 12.2.24. Consider a set-valued map $F : \mathbb{R} \rightrightarrows \mathbb{R}$ given by:

$$F(x) = [-x^2, -\infty) \quad \text{for every } x \in \mathbb{R}.$$

Let $\mathbb{R}_+$ be the ordering cone. We choose a point $(\bar{x}, \bar{y}) = (0, 0) \in \text{graph}\,(F)$. Then the generalized contingent epiderivative $D_G F(\bar{x}, \bar{y})$ is given by:

$$D_G F(\bar{x}, \bar{y})(x) = 0 \quad \text{for every } x \in \text{dom}\,(D_G F(\bar{x}, \bar{y})) = \mathbb{R}.$$

Since $D_{\mathrm{G}}F(\bar{x},\bar{y})(\mathbb{R}) \subset \mathbb{R}_+$, the optimality condition for strong minimality holds, however, the point $(\bar{x},\bar{y})$ is not a strong minimizer. □

12.3 Sufficient Optimality Conditions in Set-Valued Optimization

It is evident from scalar and vector optimization that, for sufficient optimality conditions, some additional assumptions must be imposed on the involved data. Most naturally, this remains valid in set-valued optimization as well (see Example 12.2.24). In this section, our objective is to give some sufficient optimality conditions for set-valued optimization problems.

12.3.1 Sufficient Optimality Conditions Under Convexity and Quasi-Convexity

Before giving more general sufficient optimality conditions for constrained set-valued optimization problems, we give sufficient optimality conditions for the unconstrained problem (P_0). The following result illustrates an important aspect of the behaviour of the involved derivatives that is instrumental in deriving sufficient optimality conditions.

Theorem 12.3.1. *Let X and Y be normed spaces, let $S \subset X$ be convex, let $C \subset Y$ be a pointed, closed, and convex cone, and let $F : S \rightrightarrows Y$ be a C-convex set-valued map.*

1. *An element $(\bar{x},\bar{y}) \in \mathrm{graph}\,(F)$ is a weak-minimizer, if for every $x \in S$, we have*

$$D(F+C)(\bar{x},\bar{y})(x-\bar{x}) \cap (-\mathrm{int}C) = \emptyset. \tag{12.44}$$

2. *An element $(\bar{x},\bar{y}) \in \mathrm{graph}\,(F)$ is a minimizer, if for every $x \in S$, we have*

$$D(F+C)(\bar{x},\bar{y})(x-\bar{x}) \cap (-C\backslash\{0\}) = \emptyset. \tag{12.45}$$

Proof. Since the map F is C-convex, for every $x \in S$, we have

$$F(x) - \bar{y} \subseteq (F(x) + C - \bar{y}) \subseteq D(F+C)(\bar{x},\bar{y})(x-\bar{x}), \tag{12.46}$$

and consequently, it follows from (12.44) that $(F(S) - \bar{y}) \cap -\mathrm{int}C = \emptyset$, which confirms that $(\bar{x},\bar{y})$ is a weak minimizer of (P_0). The proof of the second part follows from similar arguments. □

Remark 12.3.2. The key ingredient in the above proof is the inclusion (12.46). Since this inclusion always holds for the radial derivatives, the convexity hypothesis is not needed when using the radial derivatives. Although we have used the contingent derivative of the profile map, the same arguments are used when employing contingent epiderivatives or their generalized analogues.

We now formulate a notion of convexity taken from [297]. This notion was inspired by a related notion proposed by Jahn and Sachs [302] for single-valued maps. See also [218].

Definition 12.3.3. Let X and Y be normed spaces, let $S \subset X$, and let $\Gamma \subset Y$. Let $F : S \subseteq X \rightrightarrows Y$ be a set-valued map, and let $(\bar{x}, \bar{y}) \in \operatorname{graph}(F)$. The map F is called Γ**-contingently-quasi-convex** at $(\bar{x}, \bar{y})$ if for every $x \in S$, the condition $(F(x) - \bar{y}) \cap \Gamma \neq \emptyset$ ensures that $(DF(\bar{x}, \bar{y})(x - \bar{x})) \cap \Gamma \neq \emptyset$.

Recall that given a convex cone in a normed space Y, the quasi-interior $C^{\#}$ of C^{+} is given by

$$C^{\#} = \{\ell \in Y^{*} \mid \ \ell(x) > 0 \ \ \forall x \in C \backslash \{0\}\}.$$

In the following result, we give a sufficient optimality condition for minimality.

Theorem 12.3.4. *Let X, Y, and Z be normed spaces, let $C \subset Y$ and $D \subset Z$ be pointed, closed, and convex cones, let $S \subset X$ be convex, let $F : S \rightrightarrows Y$ and $G : S \rightrightarrows Z$ be set-valued maps, let $(\bar{x}, \bar{y}, \bar{z}) \in \operatorname{graph}(F, G)$, and let $S - \bar{x} \subseteq \operatorname{dom}(D(F, G)(\bar{x}, \bar{y}, \bar{z}))$. Assume that there are $t \in C^{\#}$ and $u \in D^{+}$ such that $u(\bar{z}) = 0$ and*

$$t(y) + u(z) \geq 0 \quad \text{for every } (y, z) \in D(F, G)(\bar{x}, \bar{y}, \bar{z})(x - \bar{x}) \quad \text{for all } \in S. \tag{12.47}$$

Then the element $(\bar{x}, \bar{y})$ is a minimizer of F on

$$\hat{S} := \{x \in S \mid G(x) \cap (-D + \operatorname{cone}(\bar{z}) - \operatorname{cone}(\bar{z})) \neq \emptyset\} \tag{12.48}$$

if and only if $(F + C, G + D) : \hat{S} \rightrightarrows Y \times Z$ is Γ-contingently quasi convex at $(\bar{x}, \bar{y}, \bar{z})$ with $\Gamma = (-C \backslash \{0\}) \times (-D + \operatorname{cone}(\bar{z}) - \operatorname{cone}(\bar{z}))$.

Proof. We claim that for every $x \in \hat{S}$, the following disjunction holds:

$$[D(F + C, G + D)(\bar{x}, \bar{y}, \bar{z})(x - \bar{x})] \cap \Gamma = \emptyset. \tag{12.49}$$

Indeed, if (12.49) is violated, then there exists $\tilde{x} \in \hat{S}$ such that $(\tilde{y}, \tilde{z}) \in D(F + C, G + D)(\tilde{x} - x)$, $\tilde{y} \in -C \backslash \{0\}$, and $\tilde{z} \in (-D + \operatorname{cone}(\bar{z}) - \operatorname{cone}(\bar{z}))$. Since $t \in C^{\#}$ and $u(\bar{z}) = 0$, we have $t(\tilde{y}) + u(\tilde{z}) < 0$ which is a contradiction to (12.47).

Therefore, (12.49) holds. Since $(F+C, G+D)$ is Γ-contingently-quasi convex at $(\bar{x}, \bar{y}, \bar{z})$, (12.49) ensures that there is no $x \in \hat{S}$ with

$$(F(x)+C-\bar{y}) \times (G(x)+D-\bar{z}) \cap \Gamma \neq \emptyset,$$

or, there is no $x \in \hat{S}$ with $(F(x)+C-\bar{y}) \cap (-C\backslash\{0\}) \neq \emptyset$ and $(G(x)+D-\bar{z}) \cap (-D + \text{cone}(\bar{z}) - \text{cone}(\bar{z})) \neq \emptyset$. This means that there is no $x \in \hat{S}$ with $(F(x)-\bar{y}) \cap (-C\backslash\{0\}) \neq \emptyset$ and $G(x) \cap (-D + \text{cone}(\bar{z}) - \text{cone}(\bar{z})) \neq \emptyset$. Therefore, there is no feasible point x such that $(F(x)-\bar{y}) \cap (-C\backslash\{0\}) \neq \emptyset$. Consequently, $(\bar{x}, \bar{y})$ is a minimizer of F on $\hat{S}$.

For the converse, assume that $(\bar{x}, \bar{y})$ is a minimizer of F on $\hat{S}$. Therefore, there is no $x \in \hat{S}$ with $(F(x)-\bar{y}) \cap (-C\backslash\{0\}) \neq \emptyset$ and $(G(x)) \cap (-D + \text{cone}(\bar{z}) - \text{cone}(\bar{z})) \neq \emptyset$. Consequently, for every $x \in \hat{S}$,

$$((F(x)+C-\bar{y}) \times (G(x)+D-\{\bar{z}\}) \cap \Gamma = \emptyset,$$

which, using (12.49), confirms that $(F+C, G+D)$ is Γ-contingently-quasi convex at $(\bar{x}, \bar{y}, \bar{z})$. □

We have the following particular case of interest:

Corollary 12.3.5. *Under the setting of Theorem 12.3.4, if the map* $(F+C, G+D): \hat{S} \rightrightarrows Y \times Z$ *is* Γ*-contingently quasi convex at a minimizer of* (P_1).

Proof. The proof follows from the fact that $S_1 \subset \hat{S}$ and hence $(\bar{x}, \bar{y})$ is a minimizer of (P_1). □

We have the following analogue of Theorem 12.3.4 for the weak-minimality.

Theorem 12.3.6. *Let* X, Y, *and* Z *be normed spaces, let* $C \subset Y$ *and* $D \subset Z$ *be pointed, closed, and convex cones where* C *is also solid, let* $S \subset X$ *be convex, let* $F: S \rightrightarrows Y$ *and* $G: S \rightrightarrows Z$ *be set-valued maps, let* $(\bar{x}, \bar{y}, \bar{z}) \in \text{graph}(F, G)$, *and let* $S - \bar{x} \in \text{dom}(D(F,G)(\bar{x}, \bar{y}, \bar{z}))$. *Assume that there are* $t \in C^+\backslash\{0\}$ *and* $u \in D^+$ *with* $u(\bar{z}) = 0$ *satisfying* (12.47)*.Then* $(\bar{x}, \bar{y})$ *is a weak-minimizer of* F *on* $\hat{S}$ *given in* (12.48) *if and only if* $(F+C, G+D): \hat{S} \rightrightarrows Y \times Z$ *is* Γ*-contingently quasi convex at* $(\bar{x}, \bar{y}, \bar{z})$ *with* $\Gamma = (-\text{int}C) \times (-D + \text{cone}(\bar{z}) - \text{cone}(\bar{z}))$.

Proof. The proof is analogous to that of Theorem 12.3.4 and hence omitted. □

The following result is immediate:

Corollary 12.3.7. *Under the setting of Theorem 12.3.6, if the map* $(F+C, G+D): \hat{S} \rightrightarrows Y \times Z$ *is* Γ*-contingently quasi convex at a weak-minimizer of* (P_1).

Proof. The proof follows from the fact that $S_1 \subset \hat{S}$ and hence $(\bar{x}, \bar{y})$ is a weak-minimizer of (P_1). □

Some results related to this subsection can be found in [218, 297].

12.3.2 Sufficient Optimality Conditions Under Paraconvexity

In the following, we explore the notion of γ-paraconvex and compactly γ-paraconvex set-valued maps and use these notions to give some first-order sufficient optimality conditions for set-valued optimization problems. The notion of paraconvex functions was proposed by S. Rolewicz (see [505, 506]) and used by many authors (see [270, 311, 552] and the cited references therein). Our treatment of this section is based on Amahroq and Taa [5].

We begin with the following:

Definition 12.3.8. Let X and Y be Banach spaces and let $F : X \rightrightarrows Y$ be a set-valued map.

1. The map F is called γ-**paraconvex**, where $\gamma > 0$, if there exists a constant $c > 0$ such that for all $x, z \in X$ and $\alpha \in (0,1)$, the following inclusion holds:

$$\alpha F(x) + (1-\alpha)F(z) \subseteq F(\alpha x + (1-\alpha)z) + c\|x-z\|^{\gamma} B_Y. \tag{12.50}$$

2. The map F is called **compactly** γ-**paraconvex** with $\gamma > 1$, if there exists a convex and compact subset Π of Y with $0 \in \Pi$ such that

$$\alpha F(x)+(1-\alpha)F(z) \subseteq F(\alpha x+(1-\alpha)z)+\min(\alpha, 1-\alpha)\|x-z\|^{\gamma}\Pi. \tag{12.51}$$

Evidently, any convex set-valued map is compactly γ-paraconvex and hence γ-paraconvex but the converse may be false. The following example, given by Jourani [310], shows that γ-paraconvex set-valued maps need not to be convex:

Example 12.3.9. Let $f : \mathbb{R} \to \mathbb{R}$ be given by $f(x) = ||x|-1|$. Then the set-valued map $F : \mathbb{R} \rightrightarrows \mathbb{R}$ defined by $F(x) = f(x) + \mathbb{R}_+$ is 1-paraconvex but it fails to be convex.

The following result is a useful implication of the notion of compact γ-paraconvexity.

Proposition 12.3.10. *Let X and Y be Banach spaces, let $F : X \rightrightarrows Y$ be a set-valued map, let $(\bar{x}, \bar{y}) \in \operatorname{graph}(F)$, and let $\gamma > 1$. If F is compactly γ-paraconvex, then for every $x \in X$ we have*

$$F(x) - \bar{y} \subseteq DF(\bar{x}, \bar{y})(x-\bar{x}) + \|x-\bar{x}\|^{\gamma}\Pi, \tag{12.52}$$

where Π is the convex and compact set appearing in (12.51).

Proof. Let $\alpha \in (0,1)$. Since the map F is compactly γ-paraconvex, we have

$$\alpha F(x) + (1-\alpha)\bar{y} \subseteq \alpha F(x) + (1-\alpha)F(\bar{x}) \subseteq F(\alpha x + (1-\alpha)\bar{x}) + \alpha\|x-\bar{x}\|^{\gamma}\Pi,$$

which can be rearranged as

$$F(x) - \bar{y} \subseteq \alpha^{-1}\left[F(\bar{x} + \alpha(x - \bar{x})) - \bar{y}\right] + \|x - \bar{x}\|^{\gamma} \Pi,$$

and hence

$$F(x) - \bar{y} \subseteq \limsup \alpha^{-1}\left[F(\bar{x} + \alpha(x - \bar{x})) - \bar{y}\right] + \|x - \bar{x}\|^{\gamma} \Pi.$$

Let $y \in \limsup \alpha^{-1}\left[F(\bar{x} + \alpha(x - \bar{x})) - \bar{y}\right] + \|x - \bar{x}\|^{\gamma} \Pi$ be arbitrary. Then there exist sequences $\{(\alpha_n, y_n)\} \subset \mathbb{P} \times Y$ with $(\alpha_n, y_n) \to (0_+, y)$, $\{p_n\} \subset \Pi$ and $\bar{y} + \alpha_n(y_n - \|x - \bar{x}\|^{\gamma} p_n) \in F(\bar{x} + \alpha_n(x - \bar{x}))$.

Since Π is compact, there exists an infinite subset $I \subset \mathbb{N}$ and $p \in \Pi$ with $(y_n - \|x - \bar{x}\|^{\gamma} p_n)_{n \in I} \to (y - \|x - \bar{x}\|^{\gamma} p)$, confirming that $y \in DF(\bar{x}, \bar{y})(x - \bar{x}) + \|x - \bar{x}\|^{\gamma} \Pi$. The proof is complete. □

We have the following special case of the above result:

Corollary 12.3.11. *Let X and Y be Banach spaces, where Y is finite-dimensional, let $F : X \rightrightarrows Y$ be a set-valued map, and let $(\bar{x}, \bar{y}) \in \operatorname{graph}(F)$. If F is γ-paraconvex with $\gamma > 1$, then for any $x \in X$,*

$$F(x) - \bar{y} \subseteq DF(\bar{x}, \bar{y})(x - \bar{x}) + c\|x - \bar{x}\|^{\gamma} B_Y.$$

In the following, we summarize some other useful features of paraconvex set-valued maps:

Proposition 12.3.12. *Let X, Y and Z be Banach spaces, let $F : X \rightrightarrows Y$ and $G : X \rightrightarrows Z$ be set-valued maps, and let $(\bar{x}, \bar{y}) \in \operatorname{graph}(F)$ and $(\bar{x}, \bar{z}) \in \operatorname{graph}(G)$.*

1. *If F and G are γ-paraconvex, then (F, G) is γ-paraconvex.*
2. *If F and G are compactly γ-paraconvex, then (F, G) is compactly γ-paraconvex.*
3. *Let $\gamma > 1$. If F and G are γ-paraconvex, then for every $x \in X$,*

$$(F, G)(x) \subset (\bar{y}, \bar{z}) + (DF(\bar{x}, \bar{y}), DG(\bar{x}, \bar{z}))(x - \bar{x}) + \|x - \bar{x}\|^{\gamma} (\Pi_1 \times \Pi_2),$$

where Π_1 and Π_2 are compact, convex subsets of Y and Z, respectively.

Proof. The proof follows from Proposition 12.3.10 and the involved definitions. □

We are now ready to give the promised sufficient optimality conditions for (P_1) (with $S = X$).

Theorem 12.3.13. *Let X, Y, and Z be Banach spaces with X finite-dimensional, let $C \subset Y$ and $D \subset Z$ be pointed, closed, and convex cones where C is also solid and $C \times D$ compactly based, let $F : X \rightrightarrows Y$ and $G : X \rightrightarrows Z$ be compactly*

γ-paraconvex set-valued maps with $\gamma > 1$, let $(\bar{x}, \bar{y}) \in \operatorname{graph}(F)$ and let $\bar{z} \in G(\bar{x}) \cap (-D)$. Assume that the following conditions hold:

1. $(DF(\bar{x}, \bar{y}), DG(\bar{x}, \bar{z}))\,(0) \cap (-C, -T(D, -\bar{z})) = \{(0,0)\}$.
2. $(DF(\bar{x}, \bar{y}), DG(\bar{x}, \bar{z}))\,(x) \cap (-C, -T(D, -\bar{z})) = \emptyset$ *for every* $0 \neq x \in \operatorname{dom}(DF(\bar{x}, \bar{y})) \cap \operatorname{dom}(DG(\bar{x}, \bar{z}))$.

Then $(\bar{x}, \bar{y})$ is a local minimizer of (P_1) (with $S = X$).

Proof. Assume that $(\bar{x}, \bar{y})$ is not a local minimizer of (P_1). There there are sequences $\{x_n\} \subset X$, $\{y_n\} \subset Y$, and $\{z_n\} \subset Z$ such that $x_n \to \bar{x}$ and for every $n \in \mathbb{N}$, we have $y_n \in F(x_n)$, $z_n \in G(x_n)$ and

$$(\bar{y} - y_n, \bar{z} - z_n) \in (C \times (D + \bar{z})) \setminus \{(0,0)\}. \tag{12.53}$$

For $n \in \mathbb{N}$, we set $\alpha_n(p_n, q_n) := (\bar{y} - y_n, -z_n)$, with

$$\begin{aligned} \alpha_n &= \|(\bar{y} - y_n, -z_n)\|, \\ p_n &= \alpha_n^{-1}(\bar{y} - y_n), \\ q_n &= \alpha_n^{-1}(-z_n). \end{aligned}$$

Since D is convex and since $-\bar{z} \in D$, we have $D \subset -\bar{z} + T(D, -\bar{z})$. Moreover, it follows from (12.53) that for every $n \in \mathbb{N}$, we have

$$\bar{z} + \alpha_n q_n \in T(D, -\bar{z}). \tag{12.54}$$

By using the assumption that the maps F and G are compactly γ-paraconvex, it follows from Proposition 12.3.12 that there are sequences $\{a_n\} \subset \Pi_1$ and $\{b_n\} \subset \Pi_2$ so that for any $n \in \mathbb{N}$, we get

$$-\alpha_n p_n - \|x_n - \bar{x}\|^{\gamma} a_n \in DF(\bar{x}, \bar{y})(x_n - \bar{x}), \tag{12.55}$$

$$-\bar{z} - \alpha_n q_n - \|x_n - \bar{x}\|^{\gamma} b_n \in DG(\bar{x}, \bar{z})(x_n - \bar{x}). \tag{12.56}$$

Since $C \times D$ has a compact base, by extracting subsequences if necessary, we can assume that there are $p \in C$ and $q \in D$ with $\|(p,q)\| = 1$ such that $p_n \to p$ and $q_n \to q$.

We shall analyze various possibilities. First assume that $\{\alpha_n\}$ has no convergent subsequence. Then $\{\alpha_n\} \to \infty$. It follows from (12.55) and (12.56) that

$$-(p,q) \in (DF(\bar{x}, \bar{y}), DG(\bar{x}, \bar{z}))\,(0), \tag{12.57}$$

$$-(p,q) \in (-C, -T(D, -\bar{z})) \setminus \{(0,0)\}, \tag{12.58}$$

which is a contradiction to the hypothesis.

Next we assume that $\{\alpha_n\}$ has a convergent subsequence. By using the same notation for the subsequences as well, we assume that $\{\alpha_n\}$ converges to some $\alpha \in [0,\infty)$. We will analyze the cases when $\alpha = 0$ and $\alpha \neq 0$ separately. If $\alpha \neq 0$, then by dividing (12.55) and (12.56) by α_n and taking the limit as $n \to \infty$, we obtain

$$(-p, -\alpha^{-1}\bar{z} - q) \in (DF(\bar{x},\bar{y}), DG(\bar{x},\bar{z}))\,(0),$$
$$(-p, -\alpha^{-1}\bar{z} - q) \in (-C, -T(D,-\bar{z}))\setminus\{(0,0)\},$$

and hence the first hypothesis is once again contradicted.

For the case when $\alpha = 0$, we need to analyze the two cases, namely, $\bar{z} = 0$ and $\bar{z} \neq 0$. First assume that $\bar{z} \neq 0$. Then by taking the limit $n \to \infty$ in (12.56), we obtain $-z \in DG(\bar{x},\bar{z}) \cap -T(D,-\bar{z})$ which is a contradiction. Let's now focus on the case when $\bar{z} = 0$. In this setting, we first analyze the case when $\alpha_n^{-1}(x_n - \bar{x})$ has a subsequence converging to some $x_0 \in X$. Dividing (12.55) and (12.56) by α_n, we deduce that

$$-p_n - \|x_n - \bar{x}\|^{\gamma-1}\alpha_n^{-1}\|x_n - \bar{x}\|a_n \in DF(\bar{x},\bar{y})(\alpha_n^{-1}(x_n - \bar{x}))\,,$$
$$-q_n - \|x_n - \bar{x}\|^{\gamma-1}\alpha_n^{-1}\|x_n - \bar{x}\|b_n \in DG(\bar{x},\bar{z})(\alpha_n^{-1}(x_n - \bar{x}))\,,$$

and by taking the limit $n \to \infty$, we obtain

$$-(p,q) \in (DF(\bar{x},\bar{y}), DG(\bar{x},\bar{z}))\,(x_0)$$
$$-(p,q) \in (-C, -T(D,-\bar{z}))\,,$$

which is a contradiction to the hypothesis for both cases $x_0 = 0$ and $x_0 \neq 0$.

Finally, we analyze the case when $\alpha_n^{-1}(x_n - \bar{x})$ has no convergent sequences. In this case, we have $\alpha_n(\|x_n - \bar{x}\|^{-1}) \to 0$. Since X is finite-dimensional, by extracting subsequences if necessary, we can assume that $[(x_n - \bar{x})\|x_n - \bar{x}\|^{-1}] \to y$ with $\|y\| = 1$. Then by dividing (12.55) by $\|x_n - \bar{x}\|$ and taking the limit $n \to \infty$, we deduce that $0 \in DF(\bar{x},\bar{y})(y)$, which is again a contradiction. The proof is complete. □

From the above result, we obtain the following sufficient optimality condition for (P_0) :

Theorem 12.3.14. *Let X and Y be Banach spaces where X finite-dimensional, let $C \subset Y$ be a pointed, closed, convex and compactly based, let $F : X \rightrightarrows Y$ be compactly γ-paraconvex set-valued map with $\gamma > 1$, and let $(\bar{x},\bar{y}) \in$ graph (F). Assume that the following conditions hold:*

1. $DF(\bar{x},\bar{y})(0) \cap (-C) = \{0\}$.
2. $DF(\bar{x},\bar{y})(x) \cap (-C) = \emptyset$, *for every* $0 \neq x \in$ dom $(DF(\bar{x},\bar{y}))$.

Then $(\bar{x},\bar{y})$ is a local minimizer of (P_0) (with $S = X$).

12.3.3 Sufficient Optimality Conditions Under Semidifferentiability

In this section, to show yet another interesting direction of deriving sufficient optimality conditions in set-valued optimization, we present a brief discussion of results obtained by Luc [400]. In the following by using the notion of upper semi-differentiability, we state sufficient optimality conditions for (P_0) (with $S = X$) and (P_1).

Theorem 12.3.15. *Let X and Y be normed spaces, let $C \subset Y$ be pointed, closed, and convex cone, let $F : X \rightrightarrows Y$ be set-valued, let $\bar{x} \in \operatorname{dom}(F)$, and let $\bar{y} \in \operatorname{Min}(F(\bar{x}), C)$. Assume that:*

1. $DF(\bar{x}, \bar{y})(x) \bigcap (-C) = \emptyset$ *for every* $0 \neq x \in \operatorname{dom}(DF(\bar{x}, \bar{y}))$.
2. $DF(\bar{x}, \bar{y})(0) \bigcap (-C) = \{0\}$.
3. *F is compact at $\bar{x}$ and upper semidifferentiable at $(\bar{x}, \bar{y})$.*

Then $(\bar{x}, \bar{y})$ is a local minimizer of (P_0) (with S=X).

Proof. See Luc [400]. □

Theorem 12.3.16. *Let X, Y, and Z be normed spaces, let $C \subset Y$ and $D \subset Z$ be pointed, closed, and convex cones, let $F : X \rightrightarrows Y$ and $G : X \rightrightarrows Z$ be set-valued maps, let $\bar{x} \in \operatorname{dom}(F)$ with $G(\bar{x}) \cap -D \neq \emptyset$, and let $\bar{y} \in \operatorname{Min}(F(\bar{x}), C)$. Assume that the following conditions hold:*

1. *F and G are compact at $\bar{x}$, upper semidifferentiable at $(\bar{x}, \bar{y})$ and $(\bar{x}, \bar{z})$ for every $\bar{z} \in G(\bar{x}) \cap (-D)$.*
2. $(DF(\bar{x}, \bar{y})(x), DG(\bar{x}, \bar{z})(x)) \cap (-C, -T(D, -\bar{z})) = \emptyset$ *for every* $0 \neq x \in \operatorname{dom}(DF(\bar{x}, \bar{y})) \cap \operatorname{dom}(DG(\bar{x}, \bar{z}))$ *and* $\bar{z} \in G(\bar{x}) \cap (-D)$.
3. $(DF(\bar{x}, \bar{y})(0), DG(\bar{x}, \bar{z})(0)) \cap (-C, -T(D, -\bar{z})) = \{0\}$ *for every* $\bar{z} \in G(\bar{x}) \cap (-D)$.

Then $(\bar{x}, \bar{y}, \bar{z})$ is a local minimizer of (P_1) for every $\bar{z} \in G(\bar{x}) \cap (-D)$.

Proof. See Luc [400]. □

12.4 Second-Order Optimality Conditions in Set-Valued Optimization

In this section, we employ various second-order derivatives of set-valued maps to give second-order optimality conditions in set-valued optimization. We note that given a set-valued map F, we shall use the second-order contingent derivative $D^2F(\bar{x}, \bar{y}, \bar{u}, \bar{v})$, the second-order lower Dini derivative $D_{\mathrm{L}}^2 F(\bar{x}, \bar{y}, \bar{u}, \bar{v})$, the second-order asymptotic derivative $\mathbb{D}^2 F(\bar{x}, \bar{y}, \bar{u}, \bar{v})$, and the second-order compound derivative $\mathscr{D}^2 F(\bar{x}, \bar{y}, \bar{u}, \bar{v})$. Here $(\bar{u}, \bar{v})$ is the direction in which the second-order derivative is being computed. In view of the properties of the second-order tangent sets and the second-order tangent cones, it is evident that the second-order objects can only be nonempty if the directions are included in the first-order tangent cones. This, and the first-order optimality conditions, impose restriction on $(\bar{u}, \bar{v})$. Such conditions are always evident from the context. Therefore, for notational simplicity, throughout this work we will work under the assumption that the directions are suitably chosen and the second-order objects are non-trivial. This section is based on [283, 300, 313, 321, 322]. See also [381, 385, 626]

12.4.1 Second-Order Optimality Conditions by the Dubovitskii-Milyutin Approach

In this subsection, we present a second-order extension of the Dubovitski-Milyutin approach explored in Sect. 12.2. We begin with the following necessary optimality condition for (P_1):

Theorem 12.4.1. *Let X, Y, and Z be normed spaces and let $C \subset Y$ and $D \subset Z$ be pointed, closed, convex, and solid cones. Let $F : X \rightrightarrows Y$ and $G : X \rightrightarrows Z$ be set-valued maps. For $\bar{x} \in S \subseteq X$, let $(\bar{x}, \bar{y}) \in \text{graph}(F)$ be a local weak-minimizer of (P_1). For $(\bar{u}, \bar{v}) \in X \times Y$, let $(F + C)$ be second-order asymptotic derivable at $(\bar{x}, \bar{y}, \bar{u}, \bar{v})$ and Lipschitz-like around $(\bar{x}, \bar{y})$. Then*

$$\mathbb{D}^2(F + C)(\bar{x}, \bar{y}, \bar{u}, \bar{v})[IT(-C, \bar{v})]^{-1} \cap \mathbb{IT}^2(G[-D]^{-1}, \bar{x}, \bar{u}) \cap \mathbb{T}^2(S, \bar{x}, \bar{u}) = \emptyset. \tag{12.59}$$

Proof. Since $(\bar{x}, \bar{y})$ is a local weak-minimizer of (P_1), there exists a neighborhood U_1 of $\bar{x}$ such that $\bar{y} \in \text{WMin}(F(S_1 \cap U_1), C)$. Therefore, $F(S_1 \cap U_1) \cap (\{\bar{y}\} - \text{int}C) = \emptyset$. We will show that if (12.59) fails, then a feasible u can be constructed in a suitable vicinity of $\bar{x}$ such that $F(u) \cap (\bar{y} - \text{int}C) \neq \emptyset$, and hence violating the local weak minimality of $(\bar{x}, \bar{y})$.

For the sake of argument, we assume that there exists an $x \in X$ that violates (12.59), that is

$$x \in \mathbb{D}^2(F + C)(\bar{x}, \bar{y}, \bar{u}, \bar{v})[IT(-C, \bar{v})]^{-1} \cap \mathbb{IT}^2(G[-D]^{-1}, \bar{x}, \bar{u}) \cap \mathbb{T}^2(S, \bar{x}, \bar{u}). \tag{12.60}$$

Since $x \in \mathbb{T}^2(S, \bar{x}, \bar{u})$ from (12.60), the definition of the second-order asymptotic contingent cone ensures that there are a sequence $\{x_n\} \subset X$ with $x_n \to x$ and a sequence $\{(s_n, t_n)\} \subset \mathbb{P} \times \mathbb{P}$ with $(s_n, t_n) \downarrow (0, 0)$ and $s_n/t_n \to 0$ such that $\bar{x} + s_n\bar{u} + s_n t_n x_n \in S$ for every $n \in \mathbb{N}$.

From $x \in \mathbb{IT}^2(G[-D]^{-1}, \bar{x}, \bar{u})$ and the facts that $x_n \to x$, $(s_n, t_n) \downarrow (0,0)$ and $s_n/t_n \to 0$, there exists $n_1 \in \mathbb{N}$ such that $\bar{x} + s_n\bar{u} + s_n t_n x_n \in G[-D]^{-1}$ for every $n \geq n_1$, or $G(\bar{x}+s_n\bar{u}+s_n t_n x_n) \cap -D \neq \emptyset$ for every $n \geq n_1$. Due to the containment $x \in \mathbb{D}^2(F + C)(\bar{x}, \bar{y}, \bar{u}, \bar{v})[IT(-C, \bar{v})]^{-1}$, there exists $y \in IT(-C, \bar{v})$ with $(x, y) \in \mathbb{T}^2(\text{epi}\,(F), (\bar{x}, \bar{y}), (\bar{u}, \bar{v}))$. Hence, there are a sequence $\{(\hat{x}_n, \hat{y}_n)\} \subset X \times Y$ with $(\hat{x}_n, \hat{y}_n) \to (x, y)$ and a sequence $\{(p_n, q_n)\} \subset \mathbb{P} \times \mathbb{P}$ with $(p_n, q_n) \downarrow (0,0)$ and $p_n/q_n \to 0$ such that $\bar{y} + p_n\bar{v} + p_n q_n \hat{y}_n \in F(\bar{x} + p_n\bar{u} + p_n q_n \hat{x}_n) + C$, for every $n \in \mathbb{N}$.

By using the second-order asymptotic derivability of epi (F), we can set $p_n = s_n$ and $q_n = t_n$. Next, we define $u_n := (\bar{x}+s_n\bar{u}+s_n t_n x_n)$, and $\hat{u}_n := (\bar{x}+s_n\bar{u}+s_n t_n \hat{x}_n)$. Since both $\{u_n\}$ and $\{\hat{u}_n\}$ converge to $\bar{x}$, there exists $n_2 \in \mathbb{N}$ such that $u_n, \hat{u}_n \in U := U_1 \cap U_2$ for $n \geq n_2$ where U_1 is the neighborhood of $\bar{x}$ described above and U_2 is a neighborhood of $\bar{x}$ which exists, along with a neighborhood V of $\bar{y}$, as a consequence of the Lipschitz-like property. Moreover, since $(\bar{y}+s_n\bar{v}+s_n t_n \hat{y}_n) \to \bar{y}$, there exists $n_3 \in \mathbb{N}$ such that $\bar{y} + s_n\bar{v} + s_n t_n \hat{y}_n \in V$ for all $n \geq n_3$.

Since $(F + C)$ is Lipschitz-like at $(\bar{x}, \bar{y})$, we get

$$\begin{aligned}
\bar{y} + s_n\bar{v} + s_n t_n \hat{y}_n &\in [F(\bar{x} + s_n\bar{u} + s_n t_n \hat{x}_n) + C] \cap V \quad (\text{for } n \geq n_3) \\
&\subseteq F(\bar{x} + s_n\bar{u} + s_n t_n x_n) + C \\
&\quad + L\, s_n t_n \|x_n - \hat{x}_n\| B_Y \quad (\text{for } n \geq \max\{n_1, n_2, n_3\}).
\end{aligned}$$

In view of the above inclusion, there exists $b_n \in B_Y$ such that for $n \geq \max\{n_1, n_2\}$, we have $\bar{y} + s_n\bar{v} + s_n t_n y_n \in F(\bar{x} + s_n\bar{u} + s_n t_n x_n) + C$, where $y_n = (\hat{y}_n - L b_n \|x_n - \hat{x}_n\|) \to y$.

Since $y \in IT(-\text{int}C, \bar{v})$, $t_n \downarrow 0$ and $y_n \to y$, it follows from the definition of the interiorly contingent cone that there exists $n_4 \in \mathbb{N}$ such that for $n \geq n_4$, we have $\bar{v}+t_n y_n \in -\text{int}C$. Using the fact that $s_n > 0$, we obtain $\bar{y}+s_n\bar{v}+s_n t_n y_n \in \bar{y}-\text{int}C$. Next we choose $w_n \in F(u_n)$ such that $\bar{y} + s_n\bar{v} + s_n t_n y_n \in w_n + C$ for every $n \in \mathbb{N}$. We have $w_n \in \bar{y}+s_n\bar{v}+s_n t_n y_n - C \subseteq \bar{y}-\text{int}C - C \subset \bar{y} - \text{int}C$, and consequently $w_n \in F(u_n) \cap (\bar{y} - \text{int}C)$, for $n \geq \max\{n_1, n_2, n_3, n_4\}$.

Therefore, we have shown that for every $n \geq \max\{n_1, n_2, n_3, n_4\}$ there are $u_n \in S_1 \cap U_1$ such that $F(u_n) \cap (\bar{y} - \text{int}C) \neq \emptyset$, contradicting the weak optimality of $(\bar{x}, \bar{y})$. The proof is complete. □

Let us now deduce the following necessary optimality condition for (P_0):

Theorem 12.4.2. *Let X and Y be normed spaces, let $C \subset Y$ be a pointed, closed, convex, and solid cone, and let $S \subset X$. Let $F : X \rightrightarrows Y$ be a set-valued map and let $(\bar{x}, \bar{y}) \in$ graph (F) be a local weak minimizer of (P_0). For $(\bar{u}, \bar{v}) \in X \times Y$, let $(F + C)$ be second-order asymptotic derivable at $(\bar{x}, \bar{y}, \bar{u}, \bar{v})$ and Lipschitz-like around $(\bar{x}, \bar{y})$. Then*

$$\mathbb{D}^2(F + C)(\bar{x}, \bar{y}, \bar{u}, \bar{v})[IT(-C, \bar{v})]^{-1} \cap \mathbb{T}^2(S, \bar{x}, \bar{u}) = \emptyset. \tag{12.61}$$

Clearly, Theorem 12.4.1 and Theorem 12.4.2 remain valid if the asymptotic contingent derivatives and cones are replaced by second-order contingent derivatives. Note that Theorem 12.4.1 does not involve any derivative of the constraint map G. Therefore, in order to give multiplier rules, it becomes necessary to either impose some constraint qualifications that connect the asymptotic contingent derivative of the map G to the cone $\mathbb{IT}^2(G[-D]^{-1}, \bar{x}, \bar{u})$. Another possibility that serves this purpose is furnished in the following:

Theorem 12.4.3. *Let X and Z be normed spaces and let $D \subset Z$ be a pointed, closed, convex, and solid cone. Let $G : X \rightrightarrows Z$ be a set-valued map, let $(\bar{x}, \bar{z}) \in$ graph (G), and let $(\bar{u}, \bar{w}) \in X \times Z$. Assume that* epi (G) *is second-order asymptotic derivable at $(\bar{x}, \bar{z}, \bar{u}, \bar{w})$ and $(G + D)$ is Lipschitz-like around $(\bar{x}, \bar{z})$. Then the following inclusion holds:*

$$\mathbb{D}^2(G + D)(\bar{x}, \bar{z}, \bar{u}, \bar{w})[\mathbb{IT}^2(-D, \bar{z}, \bar{w})]^{-1} \subset \mathbb{IT}^2(G[-D]^{-1}, \bar{x}, \bar{u}) \tag{12.62}$$

Proof. Let $x \in \mathbb{D}^2(G + D)(\bar{x}, \bar{z}, \bar{u}, \bar{w})[\mathbb{IT}^2(-D, \bar{z}, \bar{w})]^{-1}$ be arbitrary. Then there exists an element $z \in Z$ such that $z \in \mathbb{D}^2(G + D)(\bar{x}, \bar{y}, \bar{u}, \bar{w})(x) \cap \mathbb{IT}^2(-D, \bar{z}, \bar{u})$. Since $(x, z) \in \mathbb{T}^2(\text{epi}\,(G), (\bar{x}, \bar{z}), (\bar{u}, \bar{w}))$, there are a sequence $\{(\hat{x}_n, \hat{z}_n)\} \subset X \times Z$ with $(\hat{x}_n, \hat{z}_n) \to (x, z)$ and a sequence $\{(p_n, q_n)\} \subset \mathbb{P} \times \mathbb{P}$ with $(p_n, q_n) \downarrow (0, 0)$ and $p_n/q_n \to 0$ so that for every $n \in \mathbb{N}$, we have $\bar{z} + p_n\bar{w} + p_nq_n\hat{z}_n \in G(\bar{x} + p_n\bar{u} + p_nq_n\hat{x}_n) + D$.

Choose arbitrary sequences $\{x_n\} \subset X$ with $x_n \to x$ and $\{(s_n, t_n)\} \subset \mathbb{P} \times \mathbb{P}$ with $(s_n, t_n) \downarrow (0, 0)$ and $s_n/t_n \to 0$. To show that $x \in \mathbb{IT}^2(G[-D]^{-1}, \bar{x}, \bar{u})$, it suffices to show that there exists $m \in \mathbb{N}$ such that $\bar{x} + s_n\bar{u} + s_nt_nx_n \in G[-D]^{-1}$ for every $n \geq m$, or $G(\bar{x} + s_n\bar{u} + s_nt_nx_n) \cap -D \neq \emptyset$ for every $n \geq m$.

In view of the derivability, we can set $p_n = s_n$ and $q_n = t_n$. Furthermore, by following the same lines of argument as in the proof of Theorem 12.4.1, by using the Lipschitz-like property, we can show that there exists $n_1 \in \mathbb{N}$ such that for every $n \geq n_1$, we have $\bar{z} + s_n\bar{w} + s_nt_nz_n \in G(\bar{x} + s_n\bar{u} + s_nt_nx_n) + D$, where for some $b_n \in B_Y$, we have $z_n = (\hat{z}_n - Lb_n\|x_n - \hat{x}_n\|) \to z$. Because $z \in \mathbb{IT}^2(-D, \bar{z}, \bar{w})$, we also have $\bar{z} + s_n\bar{w} + s_nt_nz_n \in -D$ for sufficiently large $n \in \mathbb{N}$.

We now choose $w_n \in G(\bar{x} + s_n\bar{u} + s_nt_nx_n)$ so that $\bar{z} + s_n\bar{w} + s_nt_nz_n \in w_n + D$ implying that $w_n \in (\bar{z} + s_n\bar{w} + s_nt_nz_n) - D$ and hence for sufficiently large $n \in \mathbb{N}$, we have $w_n \in -D$. Summarizing, for sufficiently large n, we have $\bar{x} + s_n\bar{u} + s_nt_nx_n \in G[-D]^{-1}$. This, however, confirms that $x \in \mathbb{IT}^2(G[-D]^{-1}, \bar{x}, \bar{u})$. The proof is complete. □

A direct consequence of the above result is the following analogue of Theorem 12.4.1:

Theorem 12.4.4. *Let X, Y, and Z be normed spaces, let $C \subset Y$ and $D \subset Z$ be pointed, closed, convex, and solid cones, and let $S \subseteq X$. Let $F : X \rightrightarrows Y$ and $G : X \rightrightarrows Z$ be set-valued maps. Let $(\bar{x}, \bar{y}) \in$ graph (F) be a local weak minimizer of (P_1) and let $\bar{z} \in G(\bar{x}) \cap (-D)$. For $(\bar{u}, \bar{v}, \bar{w}) \in X \times Y \times Z$, let $(F + C)$ be second-order asymptotic derivable at $(\bar{x}, \bar{y}, \bar{u}, \bar{v})$ and Lipschitz-like around $(\bar{x}, \bar{y})$.*

Let $(G + D)$ be second-order asymptotic derivable at $(\bar{x}, \bar{z}, \bar{u}, \bar{w})$ and Lipschitz-like around $(\bar{x}, \bar{z})$. Then

$$\mathbb{D}^2(F + C)(\bar{x}, \bar{y}, \bar{u}, \bar{v})[IT(-C, \bar{v})]^{-1} \cap \mathbb{D}^2(G + D)(\bar{x}, \bar{y}, \bar{u}, \bar{w})$$
$$[\mathbb{IT}^2(-D, \bar{z}, \bar{w})]^{-1} \cap \mathbb{T}^2(S, \bar{x}, \bar{u}) = \emptyset. \tag{12.63}$$

For notational simplicity, for the following Lagrange multiplier rule, we set

$$L := IT(C, -\bar{v});$$
$$N := \mathbb{IT}^2(D, -\bar{z}, -\bar{w});$$
$$P := \mathbb{D}^2(F + C)(\bar{x}, \bar{y}, \bar{u}, \bar{v})[IT(-C, \bar{v})]^{-1};$$
$$Q := \mathbb{D}^2(G + D)(\bar{x}, \bar{y}, \bar{u}, \bar{w})[\mathbb{IT}^2(-D, \bar{z}, \bar{w})]^{-1}.$$

Theorem 12.4.5. *Let X, Y, and Z be normed spaces, let $C \subset Y$ and $D \subset Z$ be pointed, closed, convex, and solid cones, and let $S \subseteq X$. Let $(\bar{x}, \bar{y}) \in \text{graph}(F)$ be a local weak minimizer of (P_1) and let $\bar{z} \in G(\bar{x}) \cap (-D)$. For $(\bar{u}, \bar{v}, \bar{w}) \in X \times Y \times Z$, let $(F + C)$ be second-order asymptotic derivable at $(\bar{x}, \bar{y}, \bar{u}, \bar{v})$ and Lipschitz-like around $(\bar{x}, \bar{y})$. Let $(G + D)$ be second-order asymptotic derivable at $(\bar{x}, \bar{z}, \bar{u}, \bar{w})$ and Lipschitz-like around $(\bar{x}, \bar{z})$. Assume that P and Q are open convex cones and assume that there exists a nonempty, closed, and convex cone $M \subseteq \mathbb{T}^2(Q, \bar{x}, \bar{u})$. Let $\mathbb{D}^2(F + C)(\bar{x}, \bar{y}, \bar{u}, \bar{v}) : X \rightrightarrows Y$ be L-convex and $\mathbb{D}^2(G + D)(\bar{x}, \bar{y}, \bar{u}, \bar{w}) : X \rightrightarrows Z$ be N-convex. Then there exist functionals $p \in M^+$, $t \in L^+$, $s \in N^+$, not all zero, such that for every $x \in X$, the following inequality holds*

$$t \circ \mathbb{D}^2(F + C)(\bar{x}, \bar{y}, \bar{u}, \bar{v})(x) + s \circ \mathbb{D}^2(G + D)(\bar{x}, \bar{z}, \bar{u}, \bar{w})(x) \geq p(x). \tag{12.64}$$

Proof. We will prove the theorem by analyzing the following three possibilities:

(i) $P = \emptyset$;
(ii) $Q = \emptyset$;
(iii) $P \neq \emptyset$ and $Q \neq \emptyset$.

Let us begin with case (i) where $P = \emptyset$, we have

$$\mathbb{D}^2(F + C)(\bar{x}, \bar{y}, \bar{u}, \bar{v})[IT(-C, \bar{v})]^{-1} = \emptyset.$$

Then it follows from Theorem 12.2.15 that there exists $t \in L^+ \backslash \{0\}$ such that

$$t \circ \mathbb{D}^2(F + C)(\bar{x}, \bar{y}, \bar{u}, \bar{v}) \geq 0 \quad \text{for every } x \in X.$$

We choose $s = 0$ and $p = 0$ and obtain the desired result.
For case (ii) where $Q = \emptyset$, we have

$$\mathbb{D}^2(G + D)(\bar{x}, \bar{y}, \bar{u}, \bar{w})[\mathbb{IT}^2(-D, \bar{z}, \bar{w})]^{-1} = \emptyset.$$

By invoking Theorem 12.2.15 once again, we obtain $s \in N^+ \backslash \{0\}$ such that

$$s \circ \mathbb{D}^2(G + D)(\bar{x}, \bar{y}, \bar{u}, \bar{w})(x) \geq 0 \quad \text{for every } x \in X.$$

By setting $t = 0$, and $p = 0$, we obtain the desired result.

We now consider case (iii). Since $(\bar{x}, \bar{y})$ is a weak minimizer of (P_1), Theorem 12.4.4 yields:

$$P \cap Q \cap M = \emptyset.$$

Since P, Q, and M are all nonempty, we can apply the Dubovitskii-Milyutin lemma to assure the existence of

$$\begin{aligned} l &\in M^-, \\ l_0 &\in \left(\mathbb{D}^2(F + C)(\bar{x}, \bar{y}, \bar{u}, \bar{v})[IT(-C, \bar{v})]^{-1}\right)^- \\ l_1 &\in \left(\mathbb{D}^2 G(\bar{x}, \bar{y}, \bar{u}, \bar{w})[\mathbb{IT}^2(-D, \bar{z}, \bar{w})]^{-1}\right)^-, \end{aligned}$$

such that

$$l + l_0 + l_1 = 0. \tag{12.65}$$

Now, in view of Theorem 12.2.15, we get the existence of functionals $t \in L^+$ and $s \in N^+$ such that for all $x \in X$, the following inequalities hold

$$\begin{aligned} t \circ \mathbb{D}^2(F + C)(\bar{x}, \bar{y}, \bar{u}, \bar{v})(x) &\geq l_0(x); \\ s \circ \mathbb{D}^2(G + D)(\bar{x}, \bar{y}, \bar{u}, \bar{w})(x) &\geq l_1(x). \end{aligned}$$

Combining the above inequalities with (12.65) and setting $p = -l$ yields (12.64). □

12.4.2 Second-Order Optimality Conditions by the Direct Approach

In this section, we give some second-order optimality conditions by using the direct approach. The main objective here is that we want to shed some light on the approach which is more commonly used in set-valued optimization and contrast it with the Dubovitskii-Milyutin approach. We begin with the following:

Theorem 12.4.6. *Let X, Y, and Z be normed spaces, let $C \subset Y$ and $D \subset Z$ be pointed, closed, convex, and solid cones, and let $S \subseteq X$. Let $F : S \rightrightarrows Y$ and*

$G : S \rightrightarrows Z$ be set-valued maps, let $(\bar{x}, \bar{y}) \in \text{graph}(F)$ be a local weak minimizer of (P_1), and let $\bar{z} \in G(\bar{x}) \cap (-D)$. For $(\bar{u}, \bar{v}, \bar{w}) \in X \times Y \times Z$, let $(F + C)$ be second-order asymptotic derivable at $(\bar{x}, \bar{y}, \bar{u}, \bar{v})$ and Lipschitz-like around $(\bar{x}, \bar{y})$. Then for every $x \in \text{dom}((\mathbb{D}^2 F(\bar{x}, \bar{y}, \bar{u}, \bar{v}), \mathbb{D}^2 G(\bar{x}, \bar{z}, \bar{u}, \bar{w}))$, we have

$$(\mathbb{D}^2 F(\bar{x}, \bar{y}, \bar{u}, \bar{v}), \mathbb{D}^2 G(\bar{x}, \bar{z}, \bar{u}, \bar{w})(x) \cap IT(-C, \bar{v}) \times \mathbb{IT}^2(-D, \bar{z}, \bar{w}) = \emptyset. \quad (12.66)$$

Proof. Assume that (12.66) is false. Then there exists $x \in \text{dom}((\mathbb{D}^2 F(\bar{x}, \bar{y}, \bar{u}, \bar{v}), \mathbb{D}^2 G(\bar{x}, \bar{z}, \bar{u}, \bar{w}))$ with

$$(y, z) \in (\mathbb{D}^2 F(\bar{x}, \bar{y}, \bar{u}, \bar{v}), \mathbb{D}^2 G(\bar{x}, \bar{z}, \bar{u}, \bar{w}))(x) \cap IT(-C, \bar{v}) \times \mathbb{IT}^2(-D, \bar{z}, \bar{w}).$$

Consequently, we have $(x, y) \in \mathbb{T}^2(\text{graph}(F), (\bar{x}, \bar{y}), (\bar{u}, \bar{v}))$ and $(x, z) \in \mathbb{T}^2(\text{graph}(G), (\bar{x}, \bar{z}), (\bar{u}, \bar{w}))$. Since $(x, y) \in \mathbb{T}^2(\text{graph}(F), (\bar{x}, \bar{y}), (\bar{u}, \bar{v}))$, there are a sequence $\{(\hat{x}_n, \hat{y}_n)\} \subset X \times Y$ with $(\hat{x}_n, \hat{y}_n) \to (x, y)$ and a sequence $\{(p_n, q_n)\} \subset \mathbb{P} \times \mathbb{P}$ with $(p_n, q_n) \downarrow (0, 0)$ and $p_n/q_n \to 0$ so that for every $n \in \mathbb{N}$, we have $\bar{y} + p_n\bar{v} + p_n q_n \hat{y}_n \in F(\bar{x} + p_n\bar{u} + p_n q_n \hat{x}_n)$.

From $(x, z) \in \mathbb{T}^2(\text{graph}(G), (\bar{x}, \bar{z}), (\bar{u}, \bar{w}))$, we deduce that there are a sequence $\{(\bar{x}_n, \bar{z}_n)\} \subset X \times Y$ with $(\bar{x}_n, \bar{z}_n) \to (x, z)$ and a sequence $\{(s_n, t_n)\} \subset P \times P$ with $(s_n, t_n) \downarrow (0, 0)$ and $s_n/t_n \to 0$ so that for every $n \in \mathbb{N}$, we have $\bar{z} + s_n\bar{w} + s_n t_n \bar{z}_n \in G(\bar{x} + s_n\bar{u} + s_n t_n \bar{x}_n)$. By using the second-order asymptotic derivability of graph (G), we can set $s_n = p_n$ and $t_n = q_n$. Furthermore, by the arguments given in the proof of Theorem 12.4.1, we can show that for sufficiently large n there exists a feasible sequence $\{a_n\}$ that violates the weak minimality of $(\bar{x}, \bar{y})$. □

The following second-order necessary optimality for (P_0), which employs the notion of the contingent derivative, can be proved by arguments used above.

Theorem 12.4.7. *Let X and Y be normed spaces, let $C \subset Y$ be a pointed, closed, convex, and solid cone, let $S \subseteq X$, and let $F : S \rightrightarrows Y$ be a set-valued map. Let $(\bar{x}, \bar{y}) \in \text{graph}(F)$ be a local weak minimizer of (P_0). Then for every $\bar{u} \in \Omega_0 := \text{dom}(D(F + C)(\bar{x}, \bar{y}))$, for every $\bar{v} \in D(F + C)(\bar{x}, \bar{y})(\bar{u}) \bigcap (-\partial C)$ (∂ denotes the boundary) and every $x \in \Omega_1 := \text{dom}(D^2(F + C)(\bar{x}, \bar{y}, \bar{u}, \bar{v}))$, we have*

$$D^2(F + C)(\bar{x}, \bar{y}, \bar{u}, \bar{v})(x) \bigcap (-\text{int}C - \{\bar{v}\}) = \emptyset. \quad (12.67)$$

Remark 12.4.8. Note that by setting $(\bar{u}, \bar{v}) = (0, 0)$ in (12.67), we obtain the following first order optimality condition: For every $x \in \text{dom}(D(F + C)(\bar{x}, \bar{y}))$, we have $D(F + C)(\bar{x}, \bar{y})(x) \subset Y \backslash \{-\text{int}C)\}$.

Theorem 12.2.1 extends the second-order theory known from nonlinear programming to the general set-valued optimization (and nonsmooth vector optimization) case. For instance, the following result, which generalizes [86, Theorem 5.3], is a direct consequence of Theorem 12.2.1.

Corollary 12.4.9. *Let $Y = \mathbb{R}$ and let $C = \mathbb{R}_+ := \{t \in \mathbb{R} | \ t \geq 0\}$ in Theorem 12.2.1. Then:*

$$D(F+C)(\bar{x},\bar{y})(x) \subseteq \mathbb{R}_+ \quad \text{for every } x \in \Omega_0 := \text{dom}\,(D(F+C)(\bar{x},\bar{y})).$$

Furthermore, for $\bar{u} \in \Omega_0$ with $0 \in D(F+C)(\bar{x},\bar{y})(\bar{u})$, we have

$$D^2(F+C)(\bar{x},\bar{y},\bar{u},0)(x), \subseteq \mathbb{R}_+ \quad \text{for every } \ x \in \text{dom}\,(D^2(F+C)(\bar{x},\bar{y},\bar{u},0)).$$

The following example illustrates the above result.

Example 12.4.10. Consider the following set optimization problem

$$\min_{x \in \mathbb{R}} F(x)$$

with $F : \mathbb{R} \rightrightarrows \mathbb{R}$ given by $F(x) := \{y \in \mathbb{R} \mid y \geq x^2\}$ for every$x \in \mathbb{R}$.

Clearly, $(\bar{x},\bar{y}) := (0,0)$ is a local weak minimizer. For every $x \in \mathbb{R}$, we have

$$D(F+C)(0,0)(x) = \{y \in \mathbb{R} \mid (x,y) \in T(\text{epi}\,(F),(0,0))\} = \mathbb{R}_+,$$

and for every $\bar{u} \in \mathbb{R}$ and for every $x \in \mathbb{R}$, we have

$$D^2(F+C)(\bar{x},\bar{y},\bar{u},0)(x) = D^2(F+C)(0,0,\bar{u},0)(x) = \left\{y \in \mathbb{R} \mid y \geq 2\bar{u}^2\right\}.$$

Summarizing, for the local minimizer $(0,0)$, for $\bar{u} \in \mathbb{R}$, and for every $x \in \mathbb{R}$, we have

$$D(F+C)(0,0)(x) = \mathbb{R}_+,$$
$$D^2(F+C)(0,0,\bar{u},0)(x) \subseteq \mathbb{R}_+,$$

that is, the necessary optimality conditions of Corollary 12.4.9 are fulfilled. □

We next give optimality conditions by using the second-order compound contingent derivative defined by taking the contingent cone of the contingent cone. Since this approach allows to combine two different tangent cone of first-order, it seems to have some potential (see [283]).

The following is yet another necessary optimality conditions for (P_1).

Theorem 12.4.11. *Let X, Y, and Z be normed spaces, let $C \subset Y$ and $D \subset Z$ be pointed, closed, convex, and solid cones, and let $S \subseteq X$. Let $F : S \rightrightarrows Y$ and $G : S \rightrightarrows Z$ be set-valued maps, let $(\bar{x},\bar{y}) \in \text{graph}\,(F)$ be a local weak minimizer of (P_1), and let $\bar{z} \in G(\bar{x}) \cap (-D)$. Let $(\bar{u},\bar{v},\bar{w}) \in X \times Y \times Z$. Then, for every $x \in \text{dom}\,(\mathscr{D}^2(FG)(\bar{x},\bar{y},\bar{z},\bar{u},\bar{v},\bar{w}))$, we have*

$$\mathscr{D}^2(FG)(\bar{x},\bar{y},\bar{z},\bar{u},\bar{v},\bar{w})(x) \cap IT(-C,\bar{v}) \times IT(IT(-D,\bar{z}),\bar{w}) = \emptyset. \qquad (12.68)$$

Proof. Assume that (12.68) is violated. Then there exists $x \in \operatorname{dom}(\mathscr{D}^2(FG)(\bar{x}, \bar{y}, \bar{z}, \bar{u}, \bar{v}, \bar{w}))$ such that

$$(y, z) \in \mathscr{D}^2(F, G)(\bar{x}, \bar{y}, \bar{z}, \bar{u}, \bar{v}, \bar{w})(x) \cap IT(-C, \bar{v}) \times IT(IT(-\operatorname{int}D, \bar{z}), \bar{w}),$$

which implies that $(x, y, z) \in T(T(\operatorname{graph}(F, G), (\bar{x}, \bar{y}, \bar{z})), (\bar{u}, \bar{v}, \bar{w}))$. Therefore, there are exists $\{t_n\} \subset \mathbb{P}$, $\{(x_n, y_n, z_n)\} \subset X \times Y \times Z$ such that $t_n \downarrow 0$, $(x_n, y_n, z_n) \to (x, y, z)$ and $(\bar{u} + t_n x_n, \bar{v} + t_n y_n, \bar{w} + t_n z_n) \in T(\operatorname{graph}(F, G), (\bar{x}, \bar{y}, \bar{z}))$. Since $y \in IT(-\operatorname{int}C, \bar{v})$, there exists $n_1 \in \mathbb{N}$ such that $\bar{v} + t_n y_n \in -\operatorname{int}C$ for $n > n_1$. Analogously, since $z \in IT(IT(-\operatorname{int}D, \bar{z}), \bar{w})$, there exists $n_2 \in \mathbb{N}$ such that $\bar{w} + t_n z_n \in IT(-\operatorname{int}D, \bar{z})$ for $n > n_2$. For $n \geq \max\{n_1, n_2\}$, we fix elements $u_n := \bar{u} + t_n x_n$, $v_n := \bar{v} + t_n y_n$, and $w_n := \bar{w} + t_n z_n$. Note that we have $(u_n, v_n, w_n) \in T(\operatorname{graph}(FG), (\bar{x}, \bar{y}, \bar{z}))$ and $(v_n, w_n) \in -\operatorname{int}C \times IT(-\operatorname{int}D, \bar{z})$. In view of the definition of the contingent cone, there are sequences $\{t_m\} \subset \mathbb{P}$, $\{(x_m, y_m, z_m)\} \subset X \times Y \times Z$, such that $t_m \downarrow 0$, $(x_m, y_m, z_m) \to (u_n, v_n, w_n)$ and $(\bar{y} + t_m y_m, \bar{z} + t_m z_m) \in (F, G)(\bar{x} + t_m x_m)$ which means that $\bar{y} + t_m y_m \in F(\bar{x} + t_m x_m) + C$ and $\bar{z} + t_m z_m \in G(\bar{x} + t_m x_m) + D$. Since $y_m \to v_n$ and $v_n \in -\operatorname{int}C$, there exists $m_1 > 0$ such that $t_m y_m \in -\operatorname{int}C$ for every $m > m_1$. Let $a_m \in F(\bar{x} + t_m x_m)$ be such that $\bar{y} - \operatorname{int}C \in a_m + C$, and consequently we have $a_m \in \bar{y} - \operatorname{int}C$. Similarly, because $w_n \in IT(-\operatorname{int}D, \bar{z})$, there exists $m_2 > 0$ such that $\bar{z} + t_m z_m \in -\operatorname{int}D$ for every $m > m_4$. Let $b_m \in G(\bar{x} + t_m x_m)$ be such that $\bar{z} + t_m z_m \in b_m + D$, and consequently we have $b_m \in -\operatorname{int}D$.

Therefore, we have shown that for sufficiently large m, we have $c_m := \bar{x} + t_m x_m \in U(\bar{x})$, $G(c_m) \cap -D \neq \emptyset$ and $F(c_m) \cap (\bar{y} - \operatorname{int}C) \neq \emptyset$. This contradicts the optimality of $(\bar{x}, \bar{y})$. □

The following necessary optimality conditions for (P_0) following at once.

Theorem 12.4.12. *Let X and Y be normed spaces, let $C \subset Y$ be pointed, closed, convex, and solid cone, and let $S \subseteq X$. Let $F : S \rightrightarrows Y$ be a set-valued map and let $(\bar{x}, \bar{y}) \in \operatorname{graph}(F)$ be a local weak minimizer of (P_0). Then, for every $x \in \operatorname{dom}(\mathscr{D}^2(F)(\bar{x}, \bar{y}, \bar{u}, \bar{v}))$, we have*

$$\mathscr{D}^2 F(\bar{x}, \bar{y}, \bar{u}, \bar{v})(x) \cap IT(-\operatorname{int}(C), \bar{v}) = \emptyset. \tag{12.69}$$

Remark 12.4.13. Lagrange multiplier rules can be obtained by combining the contingent cone and the Clarke tangent cone to define new convex derivatives.

12.5 Generalized Dubovitskii-Milyutin Approach in Set-Valued Optimization

In this section, we consider a more general set-valued optimization problem and give necessary optimality conditions by giving an extension of the Dubovitskii-Milyutin approach.

Let $I := \{1,2,\ldots,m\}$ and $J := \{1,2,\ldots,n\}$ be two index sets. Let X and Y be Banach spaces, let $C \subset Y$ be a pointed, closed, convex, and solid cone, let $S \subset X$, and let $F : X \rightrightarrows Y$ be a given set-valued map. For $i \in I$, let Z_i be a normed space, let $C_i \subset Z_i$ be a pointed, closed, convex, and solid cone, and let $G_i : X \rightrightarrows Z_i$ be a given set-valued map. Analogously, for $j \in J$, let W_j be a Banach space, let $D_j \subset W_j$ be closed and convex set, and let $H_j : X \rightrightarrows W_j$ be a given set-valued map.

In this section, we focus on the following set-valued optimization problem (GP_1):

$$\text{Minimize } F(x) \quad \text{subject to } x \in S_1 := \{x \in S \mid G_i(x) \cap -C_i \neq \emptyset,$$
$$\forall\, i \in I,\ H_j(x) \cap -D_j \neq \emptyset,\ \forall\, j \in J\}.$$

Here we only study the local weak-optimality, however, other optimality notions can be treated analogously. We will also focus on the following particular case:

$$\text{Minimize } F(x) \quad \text{subject to } x \in S_1 := \{x \in S \mid G_i(x) \cap -C_i \neq \emptyset,\ \forall\, i \in I\}.$$

By choosing the maps G_i and H_j to be single valued, from (P_1) we recover the following set-valued optimization problem with single-valued constraints:

$$\text{Minimize } F(x) \quad \text{subject to } x \in \{x \in S \mid G_i(x) \geq 0,\ \forall\, i \in I,\ H_j(x) = 0, \forall\, j \in J\}.$$

We remark that the above optimization problem with multi-equality constraints can not be tackled by the classical Dubovitskii-Milyutin approach studied in the previous sections. This is mainly due to the fact that the separation arguments used in the classical Dubovitskii-Milyutin approach are applicable to an intersection of cones in which at most one cone can be closed. On the other hand, a proper formulation of multi-equality constraints leads to an intersection with several closed cones. To handle optimization problems with multi-equality constraints, a generalized Dubovitskii-Milyutin theory has been developed for smooth optimization problems (see [334]). This theory is enriched by the notion of cones of the same sense and cones of the opposite sense introduced by Walczak [585]. Although many interesting contributions have appeared in this direction, the generalized Dubovitskii-Milyutin theory so far has only been used to study vector optimization problems with differentiable data. The primary objective of this section is to present an extension of the generalized Dubovitskii-Milyutin theory to set-valued optimization problems. In this process, we also present new extensions of this theory to nonsmooth optimization and to more general vector optimization problems.

12.5.1 *A Separation Theorem for Multiple Closed and Open Cones*

This section is devoted to some specific properties of cones and their use in giving a separation theorem for a disjunction involving multiple closed and multiple open cones.

We begin with the following notion (see Walczak [585]):

Definition 12.5.1. Let X be a normed space, let $\{K_i\}_{i=1}^k$ be a system of cones in X. Then:

1. The cones $\{K_i\}_{i=1}^k$ are of the **same sense** , if there exists $\varepsilon_1, \ldots, \varepsilon_k > 0$ such that for any $x \in B_\varepsilon \cap (K_1 + \cdots + K_k)$ where $x = x_1 + \cdots + x_k$ with $x_i \in K_i$, we have
$$x_i \in B_{\varepsilon_i} \cap K_i \quad \text{for } i = 1, \ldots, k.$$
2. The cones $\{K_i\}_{i=1}^k$ are of the **opposite sense**, if there exists nontrivial vectors $\{x_i\}_i^n$, that is, $(x_1, \ldots, x_k) \neq (0, \ldots, 0)$ with $x_i \in K_i$, such that $x_1 + x_2 + \cdots + x_k = 0$.

Remark 12.5.2. 1. It follows from the definition that the cones $\{K_i\}_{i=1}^k$ are of the same sense if for any $\varepsilon > 0$ there exists $\varepsilon_1, \ldots, \varepsilon_k > 0$ such that for each $x = x_1 + x_2 + \ldots + x_k$, with $x_i \in K_i$, the inequality $\|x\| \leq \varepsilon$ implies the inequalities $\|x_i\| \leq \varepsilon_i \quad i = 1, 2, \ldots .n$.
2. We also note that the system of same sense is disjoint with the system of opposite sense. Furthermore, any subsystem of a system of the same sense is of the same sense. If any subsystem is of opposite sense, then the whole system is of the opposite sense.
3. If two cones K_1 and K_2 are subspaces and if $K_1 \cap K_2$ contains a nonzero element then these cones are of the opposite sense.

We following result can be found in Walczak [585, Theorem 3.5].

Theorem 12.5.3. *Let X be a Banach space and let X^* be the topological dual of X. Then the algebraic sum of a finite number of weak* closed convex cones of the same sense is weak* closed.*

With the above preparation, we are ready to prove the following result given by Kotarski [333]:

Theorem 12.5.4. *Let X be a Banach space and let $\{K_1, K_2, \ldots, K_p, \ldots, K_n\}$ be a collection of cones in X. Assume that the following conditions hold:*

1. *The cones $K_1, K_2, \ldots, K_p$ are open and convex with vertices at* 0.
2. *The cones $K_{p+1}, \ldots, K_n$ are closed and convex with vertices at* 0.
3. *The dual cones $K_{p+1}^+, \ldots, K_n^+$ are either of the same sense or the opposite sense.*

Then the following two statements are equivalent:

I. $\bigcap_{i=1}^{n} K_i = \emptyset$.

II. There exist functions $\ell_i \in K_i^+$, $i = 1, \ldots, n$, *not all simultaneously zero, such that*

$$\ell_1 + \ell_2 + \cdots + \ell_n = 0.$$

Proof. Let us consider the case when the cones $K_{p+1}^+, \ldots, K_n^+$ are of the same sense because the case when the cones are of the opposite sense follows trivially. We set

$$K := \bigcap_{i=1}^{p} K_i,$$

$$C := \bigcap_{i=p+1}^{n} K_i.$$

There are the following four possibilities:

1. $K \neq \emptyset \quad C \neq \emptyset$,
2. $K = \emptyset \quad C \neq \emptyset$,
3. $K \neq \emptyset \quad C = \emptyset$,
4. $K = \emptyset \quad C = \emptyset$.

We begin by analyzing the first scenario. Clearly K is an open convex cone with vertex at the origin and C is a convex cone with vertex at the origin. Therefore, by using the classical separation theorem, there exists a nonzero linear continuous functional ℓ such that

$$\ell \in \left(\bigcap_{i=1}^{p} K_i \right)^+$$

$$-\ell \in \left(\bigcap_{i=p+1}^{n} K_i \right)^+.$$

It is known that ℓ has the following representation:

$$\ell = \sum_{i=1}^{p} \ell_i \quad \text{where } \ell_i \in K_i^+ \quad i = 1, \ldots, p. \tag{12.70}$$

Furthermore, from Theorem 12.5.3, we have

$$\left(\bigcap_{i=p+1}^{n} K_i\right)^+ = w\text{-cl conv}\left(\sum_{i=p+1}^{n} K_i^+\right) = w\text{-cl}\left(\sum_{i=p+1}^{n} K_i^+\right) = \sum_{i=p+1}^{n} K_i^+,$$

where w-cl denotes the closure in the weak-star topology. Consequently, $-\ell$ admits the representation

$$-\ell = \sum_{i=p+1}^{n} \ell_i \quad \text{where } \ell_i \in K_i^+ \quad i = p+1, \ldots, n. \tag{12.71}$$

Using (12.70) and (12.71), we get $\sum_{i=1}^{n} \ell_i = 0$ and the proof is complete for this case.

For the second case, we have $\bigcap_{i=1}^{p} K_i = \emptyset$, which implies that there exists $1 \leq s < p$ such that

$$\bigcap_{i=1}^{s} K_i \neq \emptyset,$$

$$\bigcap_{i=1}^{s+1} K_i = \emptyset,$$

and the conclusion follows by using a separation theorem and setting $\ell_{s+2} = \cdots = \ell_n = 0$.

The remaining two cases can be proved in an analogous manner.

For the converse, we assume that

$$\sum_{i=1}^{n} \ell_i = 0,$$

and, if possible, assume that

$$\bigcap_{i=1}^{n} K_i \neq \emptyset,$$

Let $x \in K_i$ for $i = 1, \ldots, n$ be arbitrary. For a certain $1 \leq j \leq p$, we have $\ell_j \neq 0$. However, since K_j is an open cone, we also have $\ell_j(x) > 0$, implying

$$0 = \sum_{i=1}^{n} \ell_i(x) \geq \ell_j(x) > 0$$

which is impossible. The proof is complete. □

We conclude this section by formulating the following notion:

Definition 12.5.5. A system of sets $\{\Omega_1, \Omega_2, \ldots, \Omega_n\}$ is called **optimally positioned**, if for any $z \in \Omega_1 \cap \Omega_2 \cap \cdots \cap \Omega_n$, we have

$$T(\Omega_1 \cap \Omega_2 \cap \cdots \cap \Omega_n, z) = T(\Omega_1, z) \cap T(\Omega_2, z) \cap \cdots \cap T(\Omega_n, z). \tag{12.72}$$

Numerous conditions are given in Chap. 4 which ensure (12.72).

12.5.2 First-Order Generalized Dubovitskii-Milyutin Approach

In the following, we give some technical results to be used later.

Proposition 12.5.6. *Let* $(\bar{x}, \bar{y}) \in \operatorname{graph}(F)$ *be a local weak-minimizer to (GP). Then:*

$$U \bigcap S \bigcap F[\bar{y} - \operatorname{int}(C)]^{-1} \bigcap_{i=1}^{m} G_i[-C_i]^{-1} \bigcap_{j=1}^{n} H_j[-D_j]^{-1} = \emptyset, \tag{12.73}$$

where U *is a neighborhood of* $\bar{x}$ *corresponding to the definition of the local weak-minimality.*

Proof. Assume that there exists

$$x \in U \bigcap S \bigcap F[\bar{y} - \operatorname{int}(C)\}]^{-1} \bigcap_{i=1}^{m} G_i[-C_i]^{-1} \bigcap_{j=1}^{n} H_j[-D_j]^{-1}.$$

From the containment

$$x \in S \bigcap_{i=1}^{m} G_i[-C_i]^{-1} \bigcap_{j=1}^{n} H_j[-D_j]^{-1},$$

we notice that $x \in S$ and for all $i \in I$ and all $j \in J$, we have

$$G_i(x) \cap -C_i \neq \emptyset,$$
$$H_j(x) \cap -D_j \neq \emptyset,$$

ensuring that x is a feasible point. From $x \in U \cap F[\bar{y} - \operatorname{int}(C)\}]^{-1}$, we obtain that there exists $x \in U$, a neighborhood of $\bar{x}$, such that $F(x) \cap (\bar{y} - \operatorname{int}(C)) \neq \emptyset$. This, however, in view of the feasibility of x, contradicts the local weak-minimality of $(\bar{x}, \bar{y})$. □

The above result then leads to the following new optimality condition:

Proposition 12.5.7. *Let* $(\bar{x}, \bar{y}) \in \text{graph}(F)$ *be a local weak minimizer to* (P). *Assume that the sets* $H_j[-D_j]^{-1}$ *for* $j \in J$, *are optimally positioned. Then:*

$$IT(S, \bar{x}) \bigcap IT(F[\bar{y} - \text{int}(C)]^{-1}, \bar{x}) \bigcap_{i=1}^{m} IT(G_i[-C_i]^{-1}, \bar{x}) \bigcap_{j=1}^{n} T(H_j[-D_j]^{-1}, \bar{x}) = \emptyset. \tag{12.74}$$

Proof. In view of Proposition 12.5.6 and the properties of the interiorly contingent cones and the contingent cones, we obtain

$$IT(S, \bar{x}) \bigcap IT(F[\bar{y} - \text{int}(C)]^{-1}, \bar{x}) \bigcap_{i=1}^{m} IT(G_i[-C_i]^{-1}, \bar{x})$$

$$\bigcap T(\bigcap_{j=1}^{n} H_j[-D_j]^{-1}, \bar{x})) = \emptyset.$$

The assertion then follows by the assumption that the sets $H_j[-D_j]^{-1}$ for $j \in J$ are optimally positioned and hence

$$T(\bigcap_{j=1}^{n} H_j[-D_j]^{-1}, \bar{x})) = \bigcap_{j=1}^{n} T(H_j[-D_j]^{-1}, \bar{x}),$$

and the proof is complete. □

We will also need the following result where the lower Dini derivative is being used:

Proposition 12.5.8. *Let* X *and* Y *be normed spaces, let* $F : X \rightrightarrows Y$ *be a set valued map and let* $(\bar{x}, \bar{y}) \in \text{graph}(F)$. *Let* C *be a proper, closed, convex, and solid cone. Then:*

$$D_L F(\bar{x}, \bar{y})[-\text{int}C]^{-1} \subseteq IT(F[\bar{y} - \text{int}C]^{-1}, \bar{x}).$$

Proof. Let $x \in D_L F(\bar{x}, \bar{y})[-\text{int}C]^{-1}$ be arbitrary. Therefore, there exists $y \in D_L F(\bar{x}, \bar{y})(x) \cap -\text{int}C$. Let $\{x_n\} \subset X$ and $\{\lambda_n\} \subset \mathbb{P}$ be arbitrary sequences such that $x_n \to x$ and $\lambda_n \downarrow 0$. It suffices to show that there exists $m \in \mathbb{N}$ such that $\bar{x} + \lambda_n x_n \in F[\bar{y} - \text{int}C]^{-1}$ for every $n \geq m$. By the definition of $D_L F(\bar{x}, \bar{y})(\cdot)$, there exist $(y_n) \subset Y$ with $y_n \to y$ and $n_1 \in \mathbb{N}$ such that $\bar{y} + \lambda_n y_n \in F(\bar{x} + \lambda_n x_n)$ for every $n \geq n_1$. Since $y \in -\text{int}C$ and $y_n \to y$, there exists $n_2 \in \mathbb{N}$ such that $\lambda_n y_n \in -\text{int}C$ for every $n \geq n_2$. This implies that $\bar{y} + \lambda_n y_n \in F(\bar{x} + \lambda_n x_n) \cap (\bar{y} - \text{int}C)$ for $n \geq m := \max\{n_1, n_2\}$. Hence for the sequences (x_n) and (λ_n), we have $\bar{x} + \lambda_n x_n \in F[\bar{y} - \text{int}C]^{-1}$ for $n \geq m$. This is equivalent to saying that $x \in IT(F[\bar{y} - \text{int}C]^{-1}, \bar{x})$. The proof is complete. □

Proposition 12.5.9. *Let X and Z be normed spaces, let $G : X \rightrightarrows Z$ be a set valued map and let $(\bar{x}, \bar{z}) \in \operatorname{graph}(G)$. Let $A \subset Z$ with $\operatorname{int}(A) \neq \emptyset$. Then the following holds:*

$$D_L G(\bar{x}, \bar{z})[IT(-A, \bar{z})]^{-1} \subseteq IT(G[-A]^{-1}, \bar{x}).$$

Proof. Let $u \in D_L G(\bar{x}, \bar{z})[IT(-A, \bar{z})]^{-1}$ be arbitrary. Let $\{u_n\} \subset X$ and $\{\lambda_n\} \subset \mathbb{P}$ be arbitrary sequences such that $u_n \to u$ and $\lambda_n \downarrow 0$. It suffices to show that there exists $m \in \mathbb{N}$ such that $\bar{x} + \lambda_n u_n \in G[-A]^{-1}$ for every $n \geq m$. Since $u \in D_L G(\bar{x}, \bar{z})[IT(-A, \bar{z})]^{-1}$, there exists $v \in D_L G(\bar{x}, \bar{z})(u) \cap IT(-A, \bar{z})$. Therefore, there are a sequence $\{v_n\} \subset Z$ and an integer $n_1 \in \mathbb{N}$ such that $v_n \to v$ and $\bar{z} + \lambda_n v_n \in G(\bar{x} + \lambda_n u_n)$ for every $n \geq n_1$. Because of the containment $v \in IT(-A, \bar{z})$, there exists $n_2 \in \mathbb{N}$ such that $\bar{z} + \lambda_n v_n \in -A$ for every $n \geq n_2$. Therefore we have $\bar{z} + \lambda_n v_n \in G(\bar{x} + \lambda_n u_n) \cap (-A)$ for every $n \geq m := \max\{n_1, n_2\}$. Consequently, $u \in IT(G[-A]^{-1}, \bar{x})$. □

We recall the following notion.

Definition 12.5.10. Let X and Y be normed spaces and let $R : X \rightrightarrows Y$ be a set-valued map. The map R is called **Dini locally convex** at $(\bar{x}, \bar{y}) \in \operatorname{graph}(R)$, if the lower Dini derivative $D_L R(\bar{x}, \bar{y})$ of R at $(\bar{x}, \bar{y})$ is a convex set-valued map. The map R is called **Dini locally regular** at $(\bar{x}, \bar{y}) \in \operatorname{graph}(R)$, if additionally $\operatorname{dom}(D_L R(\bar{x}, \bar{y})) = X$ and the weak-inverse image $D_L R(\bar{x}, \bar{y})$ with respect to an open cone is an open cone.

Given $A \subset \mathbb{R}$ and $b \in \mathbb{R}$, by the inequality $A \geq b$ we understand that $a \geq b$ for every $a \in A$. With this convention in mind we are ready to give the promised multiplier rule.

Theorem 12.5.11. *Let $(\bar{x}, \bar{y}) \in \operatorname{graph}(F)$ be a local weak-minimizer of (GP_1) and let $\bar{z}_i \in G_i(\bar{x}) \cap (-C_i)$ where $i \in I := \{1, 2, \ldots, m\}$. Let there exist an open convex cone $L \subseteq IT(S, \bar{x})$. Assume that the sets $\{H_j[-D_j]^{-1}\}_{j=1}^n$ are optimally positioned. Assume that $M_j := T(H_j[-D_j]^{-1}, \bar{x})$ for $j \in J$ are nontrivial and convex and $\{M_j^+\}_{j=1}^n$ are either of the same sense or the opposite sense. Let F be Dini locally regular at $(\bar{x}, \bar{y})$ and let G_i be Dini locally regular at $(\bar{x}, \bar{z}_i)$ for $i \in I$. Then there exist functionals $s \in L^+$, $t \in C^+$, $u_i \in (T(C_i, -\bar{z}_i))^+$, $v_j \in M_j^+$, not all zero, such that $u_i(\bar{z}_i) = 0$. Moreover, the following inequality holds for every $x \in X$:*

$$\begin{aligned} &t \circ D_L F(\bar{x}, \bar{y})(x) + u_1 \circ D_L G_1(\bar{x}, \bar{z}_1)(x) + \cdots + u_m \circ D_L G_m(\bar{x}, \bar{z}_m)(x) \\ &\quad \geq s(x) + v_1(x) + \cdots + v_n(x). \end{aligned} \tag{12.75}$$

Proof. Note that the identity

$$IT(S, \bar{x}) \bigcap IT(F[\bar{y} - \operatorname{int}(C)]^{-1}, \bar{x}) \bigcap_{i=1}^{m} IT(G_i[-C_i]^{-1}, \bar{x}) \bigcap_{j=1}^{n} T(H_j[-D_j]^{-1}, \bar{x}) = \emptyset,$$

and the following inclusions

$$D_L F(\bar{x}, \bar{y})[-\text{int}C]^{-1} \subseteq IT(F[\bar{y} - \text{int}C]^{-1}, \bar{x}),$$
$$D_L G_i(\bar{x}, \bar{z})[IT(-C_i, \bar{z}_i)]^{-1} \subseteq IT(G[-C_i]^{-1}, \bar{x}),$$

which follow from Propositions 12.5.8 and 12.5.9 (by taking $A = C_i$ and $G = G_i$) ensures that

$$IT(S, \bar{x}) \bigcap D_L F(\bar{x}, \bar{y})[-\text{int}C]^{-1} \bigcap_{i=1}^{m} D_L G_i(\bar{x}, \bar{z}_i)[IT(-C_i, \bar{z}_i)]^{-1}$$
$$\bigcap_{j=1}^{n} T(H_j[-D_j]^{-1}, \bar{x}) = \emptyset.$$

We define

$$\Phi := D_L F(\bar{x}, \bar{y})[-\text{int}C]^{-1}$$
$$\Psi_i := D_L G_i(\bar{x}, \bar{z}_i)[IT(-C_i, \bar{z}_i)]^{-1} \quad (i \in I).$$

We shall prove the theorem by analyzing the three possibilities, namely:

(i) $\Phi = \emptyset$;
(ii) $\Psi_i = \emptyset$ for some $i \in I$;
(iii) $\Phi \neq \emptyset$ and $\Psi_i \neq \emptyset$ for every $i \in I$.

We begin with case (i). Since $\Phi = \emptyset$, we can apply Theorem 12.2.15 for

$$T := D_L F(\bar{x}, \bar{y}) : X \rightrightarrows Y, \ M := X, \ A := C$$

such that there exists $t \in C^+ \backslash \{0\}$ with

$$t \circ D_L F(\bar{x}, \bar{y})(x) \geq 0 \quad \text{for every } x \in X.$$

Setting $s = 0$, $u_i = 0$ for every $i \in I$, and $v_j = 0$ for every $j \in J$, we get the desired result.

For case (ii), let there exist $i \in I$ such that $\Psi_i = \emptyset$. Then again by invoking Theorem 12.2.15 with

$$T := D_L G_i(\bar{x}, \bar{z}_i) : X \rightrightarrows Z_i, \ D := X, \ A := T(C_i, -\bar{z}_i), \ (i \in I)$$

we obtain $u_i \in (T(C_i, -\bar{z}_i))^+ \backslash \{0\}$ such that

$$(u_i \circ DG_i(\bar{x}, \bar{z}_i))(x) \geq 0 \quad \text{for every } x \in X.$$

By setting $s = 0$, $v_j = 0_{W^*}$ for every $j \in J$, and $u_j = 0_{Z_j^*}$, $i \neq j \in I$, we obtain (12.75). For $u_i(C_i + \bar{z}_i) \geq 0$, it suffices to notice that, in view of the convexity of C_i, we have $T(C_i, -\bar{z}_i) \supseteq C_i + \bar{z}_i$ and hence $u_i(z + \bar{z}_i) \geq 0$ for every $z \in C_i$.

Finally, we consider the case (iii). Since $(\bar{x}, \bar{y})$ is a local-minimizer of (P), it follows from Proposition 12.5.7 and the imposed conditions that we have

$$L \bigcap \Phi \bigcap_{i=1}^{m} \Psi_i \bigcap_{j=1}^{n} M_j = \emptyset.$$

Since L, Φ, Ψ_i, $(i \in I)$ and M_j, $(j \in J)$ are all nonempty, we can apply Theorem 12.5.4 to assure the existence of

$$\begin{aligned} \ell &\in L^+ \\ \ell_0 &\in (D_L F(\bar{x}, \bar{y})[-\mathrm{int} C]^{-1})^+ \\ \ell_i &\in (D_L G_i(\bar{x}, \bar{z}_i)[IT(-C_i, \bar{z}_i)]^{-1})^+, \quad i \in I \\ \ell_{m+j} &\in M_j^+ \quad j \in J \end{aligned}$$

such that

$$-\ell - \ell_0 - \ell_1 - \ell_2 - \cdots - \ell_m - \ell_{m+1} - \cdots - \ell_{m+n} = 0. \tag{12.76}$$

Now, in the case $\Phi \neq \emptyset$, in view of Theorem 12.2.15 for

$$T := D_L F(\bar{x}, \bar{y}) : X \rightrightarrows Y, \; M := X, \; A := C,$$

we get the existence of functionals $t \in C^+$ such that for all $x \in X$ and $-\ell_0 \in (D_L F(\bar{x}, \bar{y})[-C]^{-1})^-$ the following inequality holds

$$(t \circ D_L F(\bar{x}, \bar{y}))(x) \geq -\ell_0(x). \tag{12.77}$$

Analogously, in the case $\Psi \neq \emptyset$ with Theorem 12.2.15 for

$$T := D_L G_i(\bar{x}, \bar{y}) : X \rightrightarrows Z_i, \; D := X, \; A := T(C_i, -\bar{z}_i), \; (i \in I)$$

we get the existence of functionals $u_i \in (T(C_i, -\bar{z}_i))^+$ such that for all $x \in X$ and $-\ell_i \in (D_L G_i(\bar{x}, \bar{z}_i)[T(-C_i, \bar{z}_i)]^{-1})^-$ the following inequality holds

$$(u_i \circ D_L G_i(\bar{x}, \bar{z}_i))(x) \geq -\ell_i(x) \quad i \in I. \tag{12.78}$$

Combining the above inequalities (12.77) and (12.78) with (12.65) and setting $s = \ell$ and $v_j = \ell_{m+j}$ with $j \in J$ yield (12.75). This completes the proof. □

12.5.3 Second-Order Generalized Dubovitskii-Milyutin Approach

We begin with the following necessary optimality condition for Problem (GP_1):

Theorem 12.5.12. *Let $(\bar{x}, \bar{y}) \in \text{graph}\,(F)$ be a local weak minimizer of (GP_1). Let $(\bar{u}, \bar{v}) \in X \times Y$. Then:*

$$\mathbb{IT}^2(S, \bar{x}, \bar{u}) \bigcap \mathbb{D}_L^2 F(\bar{x}, \bar{y}, \bar{u}, \bar{v})[IT(-C, \bar{v})]^{-1} \bigcap_{i=1}^{m} \mathbb{IT}^2(G_i[-C_i]^{-1}, \bar{x}, \bar{u})$$

$$\bigcap \mathbb{T}^2(\bigcap_{j=1}^{n} H_j[-D_j]^{-1}, \bar{x}, \bar{u}) = \emptyset. \tag{12.79}$$

Proof. Since $(\bar{x}, \bar{y})$ is a local weak-minimizer of (P), there exists a neighborhood U of $\bar{x}$ such that $F(S \cap U) \cap (\{\bar{y}\} - \text{int}C) = \emptyset$. We will show that if (12.79) fails then this criteria for the weak-minimality will be violated.

For the sake of argument, we assume that there exists an $x \in X$ such that

$$x \in \mathbb{IT}^2(S, \bar{x}, \bar{u}) \bigcap \mathbb{D}_L^2 F(\bar{x}, \bar{y}, \bar{u}, \bar{v})[IT(-C, \bar{v})]^{-1} \bigcap_{i=1}^{m} \mathbb{IT}^2(G_i[-C_i]^{-1}, \bar{x}, \bar{u})$$

$$\bigcap \mathbb{T}^2(\bigcap_{j=1}^{n} H_j[-D_j]^{-1}, \bar{x}, \bar{u}).$$

In view of the containment $x \in \mathbb{T}^2(\bigcap_{j=1}^{n} H_i[-D_i]^{-1}, \bar{x}, \bar{u})$ there are a sequence $(x_n) \subset X$ with $x_n \to x$ and a sequence $(s_n, t_n) \subset \mathbb{P} \times \mathbb{P}$ with $(s_n, t_n) \downarrow (0, 0)$ and $s_n/t_n \to 0$ so that for every $n \in \mathbb{N}$, we have $\bar{x} + s_n\bar{u} + s_nt_nx_n \in \bigcap_{j=1}^{n} H_j[-D_j]^{-1}$, implying that $H_j(\bar{x} + s_n\bar{u} + s_nt_nx_n) \cap -D_j \neq \emptyset$ for every $j \in J$.

Furthermore, due to the containment $x \in \mathbb{IT}^2(G_i[C_i]^{-1}, \bar{x}, \bar{u})$, and the facts that $x_n \to x$, $(s_n, t_n) \downarrow (0, 0)$ and $s_n/t_n \to 0$, we ensure the existence of $n_1 \in \mathbb{N}$ such that $\bar{x} + s_n\bar{u} + s_nt_nx_n \in G_i[-C_i]^{-1}$ for every $n \geq n_1$ and for every$i \in I$, or equivalently $G_i(\bar{x} + s_n\bar{u} + s_nt_nx_n) \cap -C_i \neq \emptyset$ for every $n \geq n_1$ and every $i \in I$. Moreover, since $x \in \mathbb{D}_L^2 F(\bar{x}, \bar{y}, \bar{u}, \bar{v})[IT(-C, \bar{v})]^{-1}$, there exists $y \in IT(-C, \bar{v})$ such that $(x, y) \in \text{graph}(\mathbb{D}_L^2 F(\bar{x}, \bar{y}, \bar{u}, \bar{v}))$. This, in view of the definition of $\mathbb{D}_L^2 F(\bar{x}, \bar{y}, \bar{u}, \bar{v})$ and the sequences (s_n, t_n), ensure that there are a sequence $(y_n) \subset Y$ and an index $n_2 \in \mathbb{N}$ such that $\bar{y} + s_n\bar{v} + s_nt_ny_n \in F(\bar{x} + s_n\bar{u} + s_nt_nx_n)$. Since $y \in IT(-\text{int}C, \bar{v})$, $t_n \downarrow 0$ and $y_n \to y$, it follows from the definition of the interior tangent cones that there exists $n_3 \in \mathbb{N}$ such that for $n \geq n_3$, we have $\bar{v} + t_ny_n \in -\text{int}C$. Using the fact that $s_n > 0$, we obtain that $\bar{y} + s_n\bar{v} + s_nt_ny_n \in \bar{y} - \text{int}C$.

Finally, from the containment $x \in \mathbb{IT}^2(S, \bar{x}, \bar{u})$ and the definition of sequences (x_n) and (s_n, t_n), we infer that there exists $n_4 \in \mathbb{N}$ such that for every $n > n_4$, we have $\bar{x} + s_n\bar{u} + s_nt_nx_n \in U \cap S$.

Therefore, we have shown that for every $n \geq \max\{n_1, n_2, n_3, n_4\}$ there are $u_n := \bar{x}+s_n\bar{u}+s_nt_nx_n \in S\cap U$, satisfying the constraints such that $F(u_n)\cap(\bar{y}-\mathrm{int}C) \neq \emptyset$. This, however, contradicts the weak optimality of $(\bar{x}, \bar{y})$. The proof is complete. □

To obtain a variant of the above optimality condition that involves the derivative of the maps G_i for $i \in I$ as well, we need the following:

Theorem 12.5.13. *Let X and Z be normed spaces, let $G : X \to Z$ be a set-valued map, let $(\bar{x}, \bar{z}) \in \mathrm{graph}\,(G)$ and let $(\bar{u}, \bar{w}) \in \mathrm{graph}\,(D_L G(\bar{x}, \bar{z}))$. Let $A \subset Z$ with $\mathrm{int}(A) \neq \emptyset$. Then,*

$$\mathbb{D}^2_L G(\bar{x}, \bar{z}, \bar{u}, \bar{w})[\mathbb{IT}^2(-A, \bar{z}, \bar{w})]^{-1} \subset \mathbb{IT}^2(G[-A]^{-1}, \bar{x}, \bar{u}) \tag{12.80}$$

Proof. Let $x \in \mathbb{D}^2_L G(\bar{x}, \bar{z}, \bar{u}, \bar{w})[\mathbb{IT}^2(-A, \bar{z}, \bar{w})]^{-1}$ be arbitrary. Choose arbitrary sequences $(x_n) \subset X$ with $x_n \to x$, and $(s_n, t_n) \subset P \times P$ with $(s_n, t_n) \downarrow (0, 0)$ and $s_n/t_n \to 0$. It suffices to show that there exists $m \in \mathbb{N}$ such that $\bar{x} + s_n\bar{u} + s_nt_nx_n \in G[-A]^{-1}$ for every $n \geq m$. Since $x \in \mathbb{D}^2_L G(\bar{x}, \bar{z}, \bar{u}, \bar{w})[\mathbb{IT}^2(-A, \bar{z}, \bar{w})]^{-1}$, there exists $z \in \mathbb{D}^2_L G(\bar{x}, \bar{z}, \bar{u}, \bar{w})(x) \cap \mathbb{IT}^2(-A, \bar{z}, \bar{w})$. Therefore, there are a sequence $\{z_n\} \subset Z$ and an integer $n_1 \in \mathbb{N}$ such that $z_n \to z$ and $\bar{z}+s_n\bar{w}+s_nt_nz_n \in G(\bar{x}+s_n\bar{u}+s_nt_nx_n)$ for every $n \geq n_1$. Because of the containment $z \in \mathbb{IT}^2(-A, \bar{z}, \bar{w})$ there exists $n_2 \in \mathbb{N}$ such that $\bar{z} + s_n\bar{w} + s_nt_nz_n \in -A$ for every $n \geq n_2$. Therefore we have $\bar{z} + s_n\bar{w} + s_nt_nz_n \in G(\bar{x} + s_n\bar{u} + s_nt_nx_n) \cap (-A)$ for every $n \geq m := \max\{n_1, n_2\}$. Consequently $x \in \mathbb{IT}^2(G[-A]^{-1}, \bar{x}, \bar{u})$. □

Combining the above two results, we obtain the following optimality condition:

Theorem 12.5.14. *Let $(\bar{x}, \bar{y}) \in \mathrm{graph}\,(F)$ be a local weak minimizer of (P). Let $(\bar{u}, \bar{v}) \in \mathrm{graph}\,(D_L F(\bar{x}, \bar{y}))$. Let $(\bar{x}, \bar{z}_i) \in \mathrm{graph}\,(G_i)$, let $\bar{u} \in \cap_{i=1}^m \mathrm{dom}\,(D_L G_i(\bar{x}, \bar{z}_i))$, and let $\bar{w}_i \in D_L G_i(\bar{x}, \bar{z}_i)(\bar{u})$. Then*

$$\mathbb{IT}^2(S, \bar{x}, \bar{u}) \bigcap \mathbb{D}^2_L F(\bar{x}, \bar{y}, \bar{u}, \bar{v})[IT(-C, \bar{v})]^{-1} \bigcap_{i=1}^{m} \mathbb{D}^2_L G_i(\bar{x}, \bar{z}_i, \bar{u}, \bar{w}_i)$$

$$[\mathbb{IT}^2(-C_i, \bar{z}, \bar{w}_i)]^{-1} \bigcap \mathbb{T}^2(\bigcap_{j=1}^{n} H_j[-D_j]^{-1}, \bar{x}, \bar{u}) = \emptyset.$$

12.6 Set-Valued Optimization Problems with a Variable Order Structure

As has already been discussed in Chap. 2, in recent years, optimality conditions for vector optimization and set-valued optimization problems with a variable order structure have attracted a great deal of attention. The primary objective of this section is to give first-order and second-order necessary and sufficient optimality conditions for a weak v-minimizer of set-valued optimization problems by using the lower Dini derivative and the contingent derivative.

Let X and Y be normed spaces, $\emptyset \neq S \subset X$, let $C : X \rightrightarrows Y$ be a set-valued map such that for each $x \in X$, $C(x)$ is a pointed, closed, convex, and solid cone, and let $F : X \rightrightarrows Y$ be a set-valued map. We consider the following set-valued optimization problem with a variable order structure

$$(SP)_v \qquad \text{v-minimize } F(x) \quad \text{subject to} \quad x \in S,$$

where "v-minimize" stands for problems with variable ordering structure with respect to a cone-valued mapping $C : X \rightrightarrows Y$. Here our interest is in the **weak v-minimizer**, that is, we seek a pair $(\bar{x}, \bar{y}) \in \text{graph}\,(F)$ such that

$$(F(S) - \bar{y}) \cap (-\text{int}C(\bar{x})) = \emptyset.$$

We begin with the following result using the lower Dini derivatives:

Theorem 12.6.1. *Let X and Y be normed spaces and let $C : X \rightrightarrows Y$ be a set-valued map such that for each $x \in X$, $C(x)$ is a pointed, closed, convex, and solid cone. For $S \subset X$, let $F : S \rightrightarrows Y$ be a set-valued map and let $(\bar{x}, \bar{y}) \in \text{graph}\,(F)$ be a local weak v-minimizer of $(SP)_v$.*

1. *For every $x \in T(S, \bar{x})$, we have*

$$D_L F(\bar{x}, \bar{y})(x) \cap (-\text{int}(C(\bar{x}))) = \emptyset. \tag{12.81}$$

2. *Let $\bar{u} \in \text{dom}\,(D_L F(\bar{x}, \bar{y}))$ and $\bar{v} \in D_L F(\bar{x}, \bar{y})(\bar{u}) \bigcap (-\partial C(\bar{x}))$, where ∂ denotes the boundary, be arbitrary. Then, for every $x \in T^2(S, \bar{x}, \bar{u})$, we have*

$$D_L^2 F(\bar{x}, \bar{y}, \bar{u}, \bar{v})(x) \bigcap IT(-C(\bar{x}), \bar{v}) = \emptyset. \tag{12.82}$$

Proof. Since (12.81) follows from (12.82), we will prove (12.82). Let $x \in T^2(S, \bar{x}, \bar{u})$ be such that

$$D_L^2 F(\bar{x}, \bar{y}, \bar{u}, \bar{v})(x) \bigcap IT(-C(\bar{x}), \bar{v}) \neq \emptyset.$$

Then, there exists $y \in D_L^2 F(\bar{x}, \bar{y}, \bar{u}, \bar{v})(x)$ such that $y \in IT(-C(\bar{x}), \bar{v})$. Since $x \in T^2(S, \bar{x}, \bar{u})$, there are sequences $\{t_n\} \subset \mathbb{P}$ and $\{x_n\} \subset X$ with $t_n \downarrow 0$, $x_n \to x$, and $\bar{x} + t_n\bar{u} + t_n^2 x_n \in S$. Therefore, it follows from the definition of the second-order lower Dini derivative that there are a sequence $\{y_n\} \subset Y$ and an index $n \in \mathbb{N}$ such that $\bar{y} + t_n\bar{v} + t_n^2 y_n \in F(\bar{x} + t_n\bar{u} + t_n^2 x_n)$ for every $n \geq n_1$. Since $y \in IT(-C(\bar{x}), \bar{v}) = IT(-\text{int}(C(\bar{x})), \bar{v})$, we have $\bar{v} + t_n y_n \in -\text{int}(C(\bar{x}))$ and hence $\bar{y} + t_n\bar{v} + t_n^2 y_n \in \bar{y} - \text{int}(C(\bar{x}))$ for sufficiently large n. Therefore, for such $n \in \mathbb{N}$, $u_n := \bar{x} + t_n\bar{u} + t_n^2 x_n \in S$ satisfies that $F(u_n) \cap (\bar{y} - \text{int}(C(\bar{x}))) \neq \emptyset$. This, however, contradicts that $(\bar{x}, \bar{y}) \in \text{graph}\,(F)$ is a weak v-minimizer of $(SP)_v$. The proof is complete. □

Remark 12.6.2. If in the above result we use the first-order and the second-order contingent derivatives, then (12.81) and (12.82) would only hold on the effective domains of the derivatives.

In any case, the following analogue remains valid for the contingent derivative:

Theorem 12.6.3. *Let X and Y be normed spaces and let $C : X \rightrightarrows Y$ be a cone-valued map such that* $\operatorname{int} C(x) \neq \emptyset$ *for all $x \in X$. For $S \subset X$, let $F : S \rightrightarrows Y$ be a set-valued map and let* $(\bar{x}, \bar{y}) \in \operatorname{graph}(F)$ *be a weak v-minimizer of $(SP)_v$. Then, for every $x \in S$, we have*

$$DF(\bar{x}, \bar{y})(x - \bar{x}) \cap (-\operatorname{int}(C(\bar{x}))) = \emptyset. \tag{12.83}$$

Proof. We show the result by contradiction. Let $(\bar{x}, \bar{y}) \in S \times F(\bar{x})$ be a weak v-minimizer of $(SP)_v$. Suppose that there is an element $x_0 \in S$ such that $DF(\bar{x}, \bar{y})(x_0 - \bar{x}) \not\subset Y \backslash \{-\operatorname{int}(C(\bar{x}))\}$. Therefore, there exists an element z such that $z \in DF(\bar{x}, \bar{y})(x_0 - \bar{x})$ and $z \in -\operatorname{int}(C(\bar{x}))$. By the definition of the contingent derivative, we get $(x_0 - x, z) \in T(\operatorname{graph}(F), (\bar{x}, \bar{y}))$. This means that there are sequences $\{(x_k, y_k)\} \subset \operatorname{graph}(F)$ and $\{\lambda_k\} \subset \mathbb{P}$ with $(x_k, y_k) \to (\bar{x}, \bar{y})$ and $\lambda_k(x_k - \bar{x}, y_k - \bar{y}) \to (x_0 - \bar{x}, z)$. We assert that there is a positive integer μ such that for every $k \geq \mu$ we have $y_k \in \bar{y} - \operatorname{int} C(\bar{x})$. Since $y_k \in F(x_k) \subset F(S)$, it follows that $(\bar{x}, \bar{y})$ cannot be a weak v-minimizer of $(SP)_v$, a contradiction. □

For sufficient optimality conditions for a weak v-minimizer of $(SP)_v$, we need certain convexity assumptions which are proposed in the following:

Definition 12.6.4. Let X and Y be normed spaces, let $C : X \rightrightarrows Y$ be a cone-valued map, and let $F : X \rightrightarrows Y$ be a set-valued map. The map F is said to be **$C(\bar{x})$-pseudo-convex** at $(\bar{x}, \bar{y}) \in \operatorname{graph}(F)$ if and only if for every $x \in \operatorname{dom}(F)$, we have

$$F(x) - \bar{y} \subset DF(\bar{x}, \bar{y})(x - \bar{x}) + C(\bar{x}).$$

Definition 12.6.5. Let X and Y be normed spaces, let $C : X \rightrightarrows Y$ be a cone-valued map and let $F : X \rightrightarrows Y$ be a set-valued map. For a convex set $S \subset X$, let $\bar{x} \in S$ and let $\bar{y} \in F(S)$. The map F is said to be $C(\bar{x})$-convex at $(\bar{x}, \bar{y}) \in \operatorname{graph}(F)$, if for any $x, z \in S$ and $\lambda \in (0, 1)$,

$$\lambda F(x) + (1 - \lambda) F(z) \subset F(\lambda x + (1 - \lambda) z) + C(\bar{x}).$$

Remark 12.6.6. The $C(\bar{x})$-pseudo-convexity reduces to the usual C-convexity of F.

The next proposition is important for the proof of the sufficient optimality condition in Theorem 12.6.8. We show the sufficient optimality condition under the assumption that F is compactly approximable at $(\bar{x}, \bar{y}) \in \operatorname{graph} F$ (see Definition 11.1.30).

Proposition 12.6.7. *Let X and Y be normed spaces, let $C : X \rightrightarrows Y$ be a cone-valued map, and let $F : X \rightrightarrows Y$ be a set-valued map with a convex domain. Let the map F be compactly approximable and $C(\bar{x})$-convex at $(\bar{x}, \bar{y}) \in \text{graph}(F)$. Then, F is $C(\bar{x})$-pseudo-convex at $(\bar{x}, \bar{y})$.*

Proof. Consider an element $(x, y) \in \text{graph}(F)$ and define a sequence $\{(x_k, y_k)\}$ in $X \times Y$ by

$$x_k := \bar{x} + \frac{1}{k}(x - \bar{x}), \quad y_k := \bar{y} + \frac{1}{k}(y - \bar{y}).$$

Since $\text{dom}(F)$ is convex and since F is $C(\bar{x})$-convex at $(\bar{x}, \bar{y})$, it follows that for all $k \in N$,

$$x_k = \left(1 - \frac{1}{k}\right)\bar{x} + \frac{1}{k}x \in \text{dom}\, F$$

and

$$y_k = \left(1 - \frac{1}{k}\right)\bar{y} + \frac{1}{k}y \in F\left(\left(1 - \frac{1}{k}\right)\bar{x} + \frac{1}{k}x\right) + C(\bar{x}) = F(x_k) + C(\bar{x}).$$

Thus, $(x_k, y_k) \in \text{graph}(F + C(\bar{x}))$ for every $k \in N$ and $(x_k, y_k) \to (\bar{x}, \bar{y})$ for $k \to \infty$. Moreover, we have $k(x_k - \bar{x}, y_k - \bar{y}) \to (x - \bar{x}, y - \bar{y})$. This implies $(x - \bar{x}, y - \bar{y}) \in T(\text{graph}(F + C(\bar{x})), (\bar{x}, \bar{y})$. So we get, for every $x \in \text{dom}\,(F)$ and for every $y \in F(x)$ that $y - \bar{y} \in D(F + C(\bar{x}))(\bar{x}, \bar{y})(x - \bar{x})$. Since F is supposed to be compactly approximable at $(\bar{x}, \bar{y})$, we get (see Theorem 11.3.10 or [91, Proposition 2.72])

$$D(F + C(\bar{x}))(\bar{x}, \bar{y})(x - \bar{x}) = DF(\bar{x}, \bar{y})(x - \bar{x}) + C(\bar{x}).$$

Thus, for every $x \in \text{dom}\,(F)$, we have $y - \bar{y} \in DF(\bar{x}, \bar{y})(x - \bar{x}) + C(\bar{x})$. This means that F is $C(\bar{x})$-pseudo-convex at $(\bar{x}, \bar{y})$. The proof is complete. □

Now, we will show the announced sufficient condition for a weak v-minimizer of $(SP)_v$ under certain convexity assumptions (see [91, Theorem 2.73]).

Theorem 12.6.8. *Let X and Y be normed spaces, let $C : X \rightrightarrows Y$ be a cone-valued function such that $\text{int}(C(x)) \neq \emptyset$ for all $x \in X$, let $S \subset X$ be nonempty convex set, let $F : X \rightrightarrows Y$ be compactly approximable and $C(\bar{x})$-convex at $(\bar{x}, \bar{y}) \in \text{graph}\,(F)$, and let $C(\bar{x})$ be proper, pointed, closed, and convex cone. If for every $x \in S$, we have*

$$DF(\bar{x}, \bar{y})(x - \bar{x}) \cap (-\,\text{int}(C(\bar{x}))) = \emptyset,$$

then $(\bar{x}, \bar{y})$ is a weak v-minimizer of $(SP)_v$.

Proof. Because of our assumptions, we have

$$DF(\bar{x}, \bar{y})(x - \bar{x}) \cap (-\operatorname{int} C(\bar{x})) = \emptyset.$$

Furthermore, because $C(\bar{x})$ is pointed and convex, for every $x \in S$, we have

$$(DF(\bar{x}, \bar{y})(x - \bar{x}) + C(\bar{x})) \cap (-\operatorname{int} C(\bar{x})) = \emptyset.$$

Since F is $C(\bar{x})$-pseudo-convex at $(\bar{x}, \bar{y})$ and because of Proposition 12.6.7, we get

$$(F(x) - \bar{y}) \cap (-\operatorname{int} C(\bar{x})) = \emptyset,$$

which means that $(\bar{x}, \bar{y})$ is a weak v-minimizer of $(SP)_v$. □

12.7 Optimality Conditions for Q-Minimizers in Set-Valued Optimization

We have noted that the optimality conditions for the proper minimality and the weak minimality are quite similar and largely depend on using the interior of the cones involved. In fact, as seen in Chap. 2, the notion of Q minimizers also builds on exploiting the fact that numerous other notions of minimality in set-valued optimization can be stated by using an open cone Q. The objective of this section is to give some optimality conditions for the abstract notion of a Q-minimizer and then derive optimality conditions for many other optimality notions. This section is based on [223].

12.7.1 Optimality Conditions for Q-Minimizers Using Radial Derivatives

In the following, we consider the following **unconstrained** set-valued optimization problem:

$$\text{minimize} \quad F(x) \quad \text{subject to} \quad x \in X, \tag{P}$$

where X and Y are normed or Banach spaces and $F : X \rightrightarrows Y$ is a set-valued map.

In the following result, we use the radial derivative (see Definition 11.1.13).

Proposition 12.7.1. *Let X and Y be normed spaces, let $Q \subset Y$ be an open cone, let $F : X \rightrightarrows Y$ be a set-valued map, and let $(\bar{x}, \bar{y}) \in \operatorname{graph}(F)$. Then $(\bar{x}, \bar{y})$ is a Q-minimizer of (P) if and only if*

$$D_R F(\bar{x}, \bar{y})(x) \cap (-Q) = \emptyset \quad \text{for every } x \in X. \tag{12.84}$$

Proof. We recall that $(\bar{x}, \bar{y})$ is a Q-minimizer of (SP) if and only if

$$(F(X) - \bar{y}) \cap (-Q) = \emptyset. \tag{12.85}$$

Assume that $(\bar{x}, \bar{y})$ is a Q-minimizer of (P) but (12.84) does not hold. Then, there exists $x \in X$ such that $y \in -Q$ and $y \in D_R F(\bar{x}, \bar{y})(x)$. By the definition of the radial derivative, there exist sequences $\{t_n\} \subset \mathbb{P}$ and $\{(x_n, y_n)\}$ such that $x_n \to x$ and $y_n \to y$ and $\bar{y} + t_n y_n \in F(\bar{x} + t_n x_n)$ for every $n \in \mathbb{N}$. Since $y_n \to y \in -Q$ and since Q is open, we have $t_n y_n \in -Q$, for sufficiently large n. Therefore, $t_n y_n \in (F(\bar{x} + t_n x_n) - \bar{y}) \cap (-Q)$ for sufficiently large n. This, however, is a contradiction to (12.85).

For the converse, assume that (12.84) holds. Then, for any $x \in X$, we have

$$(F(x) - \bar{y}) \cap (-Q) \subset D_R F(\bar{x}, \bar{y})(x - \bar{x}) \cap (-Q) = \emptyset.$$

Therefore, (12.85) holds, confirming that $(\bar{x}, \bar{y})$ is a Q-minimizer of (P). The proof is complete. □

The following result shows the usefulness of the abstract notion of Q-minimality:

Theorem 12.7.2. *Let X and Y be Banach spaces, let $Q \subset Y$ be an open cone, let $F : X \rightrightarrows Y$ be a set-valued map, and let $(\bar{x}, \bar{y}) \in$ graph(F).*

1. *The element $(\bar{x}, \bar{y})$ is a weak minimizer of (P) if and only if (12.84) holds with $Q = -\text{int}C$.*
2. *The element $(\bar{x}, \bar{y})$ is a S-proper minimizer of (P) if and only if (12.84) holds with $Q = \{y \in Y \mid \varphi(y) < 0\}$ for some $\varphi \in C^{\#}$.*
3. *The element $(\bar{x}, \bar{y})$ is a Hartley proper minimizer of (P) if and only if (12.84) hold with $Q = C(\varepsilon)$ for some scalar $\varepsilon > 0$.*
4. *The element $(\bar{x}, \bar{y})$ is a Borwein proper minimizer of (P) if and only if (12.84) holds with Q being some dilation of C.*
5. *The element $(\bar{x}, \bar{y})$ is a Henig global proper minimizer of (P) if and only if (12.84) holds with $Q = \text{int}(K)$ where K is a proper, pointed, convex, and solid cone in Y such that $C\{0\} \subset \text{int}K$.*
6. *The element $(\bar{x}, \bar{y})$ is a Henig proper minimizer of (P) if and only if (12.84) holds with $Q = V_\eta$ for some scalar η satisfying $0 < \eta < \delta$.*
7. *Assuming that C has a bounded base, the element $(\bar{x}, \bar{y})$ is a super proper minimizer of (P) if and only if (12.84) holds with $Q = V_\eta$ for some scalar η satisfying $0 < \eta < \delta$, provided that C has a compact base.*

Proof. The proof is a direct consequence of the fact that a certain notion of minimality satisfies (12.85) for the associated value of the cone Q (see Definition 2.6.3 and Theorem 2.4.11). □

Remark 12.7.3. Note that in Theorem 12.7.2, we only included cases where either Q is expressed directly in terms of C, or Q is a dilation of C.

12.7.2 Optimality Conditions for Q-Minimizers Using Coderivatives

In the following, our objective is to give optimality conditions for the notion of Q-minimality by using the Ioffe approximate coderivative (introduced in Definition 3.6.2 and the Clarke coderivative (see Definition 3.4.5). These optimality conditions are given by using a scalarization by means of the functional Δ_{-Q} introduced and discussed in Sect. 5.3.

Proposition 12.7.4. *Let X and Y be Banach spaces, let $Q \subset Y$ be an open cone, let $F : X \rightrightarrows Y$ be a set-valued map with a closed graph, and let $(\bar{x}, \bar{y}) \in \operatorname{graph}(F)$. If $(\bar{x}, \bar{y})$ is a Q-minimizer of (P) then there exists $y^* \in \partial_A \Delta_{-Q}(0)$ such that the following inclusions hold:*

$$0 \in D_A^* F(\bar{x}, \bar{y})(y^*), \tag{12.86}$$

$$0 \in D_C^* F(\bar{x}, \bar{y})(y^*). \tag{12.87}$$

Proof. By using Proposition 5.3.3, we note that $(\bar{x}, \bar{y})$ is a solution of (P) if an only if $(\bar{x}, \bar{y})$ is a solution of the following problem

$$\text{Minimize } \Delta_{-Q}(y - \bar{y}) + \iota((x, y); \operatorname{graph}(F)).$$

Therefore, by Proposition 3.6.3, we have

$$\begin{aligned}(0,0) &\in \partial_A(\Delta_{-Q}(0) + \iota_{\operatorname{graph}(F)}(\bar{x}, \bar{y}))\\ &\subseteq \{0\} \times \partial_A \Delta_{-Q}(0) + \partial_A \iota_{\operatorname{graph}(F)}(\bar{x}, \bar{y})\\ &= \{0\} \times \partial_A \Delta_{-Q}(0) + N_A((\bar{x}, \bar{y}); \operatorname{graph}(F)),\end{aligned}$$

and hence there exists $y^* \in \partial_A \Delta_{-Q}(0)$ with $(0, -y^*) \in N_A((\bar{x}, \bar{y}); \operatorname{graph}(F))$, which further implies that $0 \in D_A^* F(\bar{x}, \bar{y})(y^*)$, showing that (12.86) holds. Finally, (12.87) is a direct consequence of (12.86) and Proposition 3.6.3 (iv). The proof is complete. □

The following is an analogue of the above result for other notions of minimality.

Theorem 12.7.5. *Let X and Y be Banach spaces, let $Q \subset Y$ be an open cone, let $F : X \rightrightarrows Y$ be a set-valued map with a closed graph, and let $(\bar{x}, \bar{y}) \in \operatorname{graph}(F)$.*

1. *If* $(\bar{x}, \bar{y})$ *is a weak minimizer of* (P) *or* $(\bar{x}, \bar{y})$ *is a Hurwicz proper minimizer of* (P) *and the image* $F(X)$ *has a nonempty interior, then there exists* $y^* \in C^+ \setminus \{0\}$ *such that (12.86)–(12.87) hold.*
2. *If* $(\bar{x}, \bar{y})$ *is a global Henig proper minimizer, (in particular, if* $(\bar{x}, \bar{y})$ *is either a positively proper minimizer, or a Henig proper minimizer, or a super minimizer) of* (P)*, then there exists* $y^* \in C^\#$ *such that (12.86)–(12.87) hold.*
3. *If* $(\bar{x}, \bar{y})$ *is a super minimizer of* (P) *and* C *has a bounded base, then there exists* $y^* \in \text{int}(C^+)$ *such that (12.86)–(12.87) hold.*

Proof. In view of Definition 2.6.3 and Theorem 2.4.11, the element $(\bar{x}, \bar{y})$ is a minimizer of (P) in some sense if an only if it is a Q-minimizer of (P) with Q being some open set which is chosen in correspondence with Theorem 2.4.11. The proof then follows from Propositions 5.3.4 and 12.7.4. □

In the following, we establish the Lagrange multiplier rule for the **constrained set-valued optimization problem** (SP). We consider a set-valued optimization problem

$$\text{minimize} \quad F(x) \quad \text{subject to} \quad x \in S_1, \tag{SP}$$

where X, Y and Z are Banach spaces and $F : X \rightrightarrows Y$ and $G : X \rightrightarrows Z$ are set-valued maps, $S \subset X$, $D \subset Z$ and the feasible set S_1 of (SP) is given by

$$S_1 := \{x \in S \mid G(x) \cap D \neq \emptyset\}. \tag{12.88}$$

Remark 12.7.6. Note that the only difference between (SP) and (P_1) is that in the former we have more general constraint set. To emphasize this difference, we have preferred renaming this set-valued optimization problem.

In order to prove the following assertion we assume that the map F is Lipschitz-like around $(\bar{x}, \bar{y}) \in \text{graph}\,(F)$ (see Definition 3.3.9) and the map G is metrically regular around $(\bar{x}, \bar{z}) \in \text{graph}\,(G)$ relative to $S \times D$, (see Remark 12.8.10 and [2])). That is, there exist scalars $r > 0$ and $\gamma > 0$ such that, for all $(x, z) \in [(\bar{x} + rB_X) \times (\bar{z} + rB_Z)] \cap (S \times D)$,

$$d((x, z), S \times D) \cap \text{graph } G) \leq \gamma d(z, G(x)).$$

We give the following Lagrange multiplier rule for the Q-minimizers of (SP).

Proposition 12.7.7. *Let* X*,* Y*, and* Z *be Banach spaces, let* $F : X \rightrightarrows Y$ *and* $G : X \rightrightarrows Z$ *be set-valued maps, and let* $S \subset X$ *and* $D \subset Z$ *be closed. Let* F *and* G *be closed and Lipschitz-like around* $(\bar{x}, \bar{y}) \in \text{graph}\,(F)$ *and* $(\bar{x}, \bar{z}) \in \text{graph}\,(G)$ *respectively, where* $\bar{z} \in G(\bar{x}) \cap D$*, and let* G *be metrically regular around* $(\bar{x}, \bar{z})$ *relative to* $S \times D$*. If* $(\bar{x}, \bar{y})$ *is a* Q*-minimizer of* (SP)*, then there exists* $y^* \in \partial_A \Delta_{-Q}(0)$ *such that*

$$0 \in D_A^* F(\bar{x}, \bar{y})(y^*) + D_A^* G(\bar{x}, \bar{z})(z^*) + N_A(\bar{x}; S) \tag{12.89}$$

and

$$0 \in D_C^* F(\bar{x}, \bar{y})(y^*) + D_C^* G(\bar{x}, \bar{z})(z^*) + N_C(\bar{x}; S), \tag{12.90}$$

where $z^* \in N_A(\bar{z}; D)$ *and* $z^* \in N_C(\bar{z}; D)$*, respectively.*

Proof. The proof is similar to that of [2, Theorem 3.7]. By Definition 2.6.3, $(\bar{x}, \bar{y})$ is a Q-minimizer of (SP) if an only if

$$(F(S_1) - \bar{y}) \cap (-Q) = \emptyset.$$

Proposition 5.3.3 yields that $(\bar{x}, \bar{y}, \bar{z})$ is a minimal solution of the scalar problem

$$\text{minimize } q(x, y, z) \text{ subject to } (x, y) \in \text{graph } F \text{ and } (x, z) \in (S \times D) \cap \text{graph } G,$$

where $q(x, y, z) := \Delta_{-Q}(y - \bar{y})$. Then by Clarke's penalization (see [100, Proposition 2.4.3]), the metric regularity assumption and [2, Proposition 2.6], for some integer $l > 0$ large enough, $(\bar{x}, \bar{y}, \bar{z})$ is an unconstrained local minimizer of

$$(x, y, z) \to q(x, y, z) + ld((x, y), \text{graph } F) + ld((x, z), \text{graph } G) + ld((x, z), S \times D).$$

Taking into account the assertions (ii) and (iv) of Proposition 3.6.3, zero is in the sum of the subdifferentials. That is, there exist

$$\begin{aligned}
y_1^* &\in \partial_A q(\bar{x}, \bar{y}, \bar{z}) = \partial_A \Delta_{-Q}(0) \subset \partial_C \Delta_{-Q}(0), \\
(x_2^*, y_2^*) &\in l \partial_A d((\bar{x}, \bar{y}), \text{graph } F) \subset N_A((\bar{x}, \bar{y}); \text{graph } F), \\
(x_3^*, z_3^*) &\in l \partial_A d((\bar{x}, \bar{z}), \text{graph } G) \subset N_A((\bar{x}, \bar{z}); \text{graph } G), \\
(x_4^*, -z_4^*) &\in l \partial_A d((\bar{x}, \bar{z}), S \times D) \subset N_A((\bar{x}, \bar{z}); S \times D),
\end{aligned}$$

such that

$$0 = x_2^* + x_3^* + x_4^*, \quad 0 = y_1^* + y_2^*, \quad \text{and} \quad 0 = z_3^* + z_4^*.$$

Putting $y^* = y_1^* = -y_2^*$ and $z^* = z_4^* = -z_3^*$, we obtain

$$0 \in D_A^* F(\bar{x}, \bar{y})(y^*) + D_A^* G(\bar{x}, \bar{z})(z^*) + N_A(\bar{x}; S)$$

and (12.89) holds. Finally, (12.90) follows from (12.89) and Proposition 3.6.3 (iv). □

The following Lagrange multiplier rule for several types of minimizers of (SP) are derived in [228, Theorem 21.20].

Theorem 12.7.8. *Let X, Y, and Z be Banach spaces, let $F : X \rightrightarrows Y$ and $G : X \rightrightarrows Z$ be set-valued maps, and let $S \subset X$ and $D \subset Z$ be closed. Let F and G be closed and Lipschitz-like around $(\bar{x}, \bar{y}) \in \operatorname{graph}(F)$ and $(\bar{x}, \bar{z}) \in \operatorname{graph}(G)$ respectively, where $\bar{z} \in G(\bar{x}) \cap D$, and let G be metrically regular around $(\bar{x}, \bar{z})$ relative to $S \times D$.*

1. *If $(\bar{x}, \bar{y})$ is a weak minimizer of (SP) or $(\bar{x}, \bar{y})$ is a Hurwicz proper minimizer of (SP) and the image $F(S_1)$ has a nonempty interior, then there exists $y^* \in C^+ \setminus \{0\}$ such that (12.89)–(12.90) hold.*
2. *If $(\bar{x}, \bar{y})$ is a global Henig proper minimizer (in particular, a positive proper minimizer, or a Henig proper minimizer) of (SP), then there exists $y^* \in C^\#$ such that (12.89)–(12.90) hold.*
3. *If $(\bar{x}, \bar{y})$ is a super minimizer of (SP) and C has a bounded base, then there exists $y^* \in \operatorname{int} C^+$ such that (12.89)–(12.90) hold.*

Proof. The proof can devised in the same way as Theorem 12.7.5 (but applying Proposition 12.7.7 instead of Proposition 12.7.4) and hence is omitted. □

Remark 12.7.9. The above version of the Lagrange multiplier rule for weak minimizers were derived in [2], for strong minimizers or positive proper minimizers in [228], and are given in [228] for the case of Hurwicz proper minimizer, global Henig proper minimizer, Henig proper minimizer and super minimizers. We note that for weak minimizers to (SP), there have also been obtained Lagrange-Kuhn-Tucker multipliers in terms of the subdifferential of set-valued maps introduced by Sawaragi and Tanino [568, 569]. Recently, the Lagrange multiplier rule for super minimizers to a constrained set-valued optimization problem has been obtained by Bao and Mordukhovich in Asplund spaces [26] and by Huang in Banach spaces [269]. Bao and Mordukhovich used the *extremal principle* of variational analysis in their proof while Huang's technique is based on exploiting a property of the normal cone due to Clarke. In contrast to Ha's work, they considered super minimizers of the set-valued optimization problem with the geometric constraint $x \in S_1$.

Using a characterization of a Benson proper efficient point in a separable or reflexive Banach space obtained in [482], it is possible to derive the Lagrange claim, the Fermat rule and the Lagrange multiplier rule for Benson proper minimizers from Theorems 12.7.2, 12.7.5 and 12.7.8.

Theorem 12.7.10. *Assume that Y is a separable Banach space, or that Y is a reflexive Banach space and C has a base.Consider the problem (SP) where the feasible set is given by (12.88).*

1. *Let $\bar{x} \in X$ and $(\bar{x}, \bar{y}) \in \operatorname{graph} F$. Assume that F has a closed graph and $F - \bar{y}$ is nearly C-subconvexlike on X, that is, $\operatorname{cl}\operatorname{cone}(F(X) - \bar{y} + C)$ is convex. Then $(\bar{x}, \bar{y})$ is a Benson proper minimizer of (SP) if and only if there exists $y^* \in C^\#$ such that* (12.84) *hold with $Q = \{y : \varphi(y) < 0\}$ and only if there exists $y^* \in C^\#$ such that (12.86)–(12.87) hold.*

2. *Let* $\bar{x} \in S_1$ *and* $(\bar{x}, \bar{y}) \in \operatorname{graph} F$. *let* $S \subset X$ *and* $D \subset Z$ *be closed. Let* F *and* G *be closed and Lipschitz-like around* $(\bar{x}, \bar{y}) \in \operatorname{graph}(F)$ *and* $(\bar{x}, \bar{z}) \in \operatorname{graph}(G)$ *respectively, where* $\bar{z} \in G(\bar{x}) \cap D$, *and let* G *be metrically regular around* $(\bar{x}, \bar{z})$ *relative to* $S \times D$. *Assume in addition that* $F - \bar{y}$ *is nearly* C*-subconvexlike on* S_1, *that is,* $\operatorname{cl}\operatorname{cone}(F(S_1) - \bar{y} + C)$ *is convex. If* $(\bar{x}, \bar{y})$ *is a Benson minimizer of* (SP), *then there exists* $y^* \in C^\#$ *such that (12.89)–(12.90) hold.*

Proof. It has been established (see [482, Corollaries 4.2 and 4.5]) that if Y is a separable Banach space, or Y is a reflexive Banach space and C has a base and if $F - \bar{y}$ is nearly C-subconvexlike on X (on S_1, respectively), then $(\bar{x}, \bar{y})$ being a Benson proper minimizer of the unconstrained problem (SP) (of the constraint problem (SP) with the feasible set S_1, respectively) is also a positive proper minimizer. The assertion follows from this fact and Theorems 12.7.2, 12.7.5 and 12.7.8. □

Remark 12.7.11. Recently, Benson proper minimizers to set-valued optimization problems have received more attention, and several optimality conditions for them were obtained under assumptions on generalized convexity of set-valued data such as convexlikeness, subconvexlikeness, near convexlikeness and near subconvexlikeness (see [482] for references). Near subconvexlikeness, first presented in [602] and also used in [513] is the weakest convexity among the above four kinds of generalized convexity. Under the assumptions on the near subconvexlikeness of the objective and constraints, it has been established in [513] that a Benson proper minimizer of (SP) can be expressed in terms of saddle points defined in a suitable sense.

12.8 Lagrange Multiplier Rules Based on Limiting Subdifferential

The aim of this section is to show necessary optimality conditions using the limiting subdifferential by Mordukhovich ([342, 426, 430], see Sect. 3.5) for the set-valued optimization problem (SP) introduced in Sect. 2.6. The corresponding calculus rules are working in Asplund spaces such that we assume in this section that X and Y are Asplund spaces. We will use the solution concept based in vector approach (see Sect. 2.6.1) introduced in Definitions 2.6.1. This means that we will derive our results for minimizers of the following set-valued optimization problem (SP):

$$\text{minimize} \quad F(x) \quad \text{subject to} \quad x \in S, \tag{12.91}$$

where the cost mapping $F : X \rightrightarrows Y$ is a set-valued mapping, Y and X are Asplund spaces, $C \subset Y$ is a proper pointed convex closed cone, S is a subset of X, and we are looking for minimizers of (SP) in the sense of Definition 2.6.1.

In many papers Lagrange multiplier rules are shown for weak minimizers of the set-valued optimization problem (SP) (see Definition 2.6.2 and Sect. 12.7.2). The most important characteristic of the weakly minimal points of a subset A of Y with respect to C (cf. Definition 2.4.2 and the corresponding solution concept for the set-valued problem (SP) given in Definition 2.6.3) is the possibility to study its optimality conditions via numerous powerful scalarization techniques under the additional condition

$$\operatorname{int} C \neq \emptyset. \tag{12.92}$$

However, the condition $\operatorname{int} C \neq \emptyset$ is a very strong restriction in infinite dimensional spaces since "the class of ordered topological vector spaces possessing cones with nonempty interiors is not very broad" ([475, page 183]); in particular, the natural ordering cones in the Lebesgue spaces l^p and L^p for $1 \leq p < +\infty$ have empty interior.

So the aim of this section is to present optimality conditions without assuming (12.92) derived by Bao and Tammer [29].

Obviously, every Pareto minimal point of a vector optimization problem is *weakly minimal* provided that the nonempty interior condition (12.92) holds; otherwise, the weak concept is no longer defined. In the latter case, the *Henig global proper minimality* (see Definition 2.4.4 (f)) takes the center and plays the role of the weak optimality since it is defined via the Pareto optimality with respect to a bigger convex cone with a nonempty interior (compare Definition 2.4.4 (f)).

Recently, a few works have challenged the lack of the nonempty interior condition (12.92) imposed on ordering cones. In [151] some necessary conditions were established for the so-called approximate Pareto minimizers via a suitable scalarization scheme. In [149] some Lagrange multiplier rules for Pareto minimal solutions were obtained by using the separation theorem for nonconvex sets (see Lemma 12.8.1) provided that the ordering cone enjoys the **asymptotical compactness** (AC) condition, i.e., the intersection of it and the closed unit ball is compact. This requirement is strict in infinite-dimensional settings. Moreover, the necessary results therein are in fact applied to properly minimal points since $\operatorname{Min}(A, C) = \text{GHe-PMin}(A, C)$ under the AC condition and other standard assumptions and since the first set is bigger than the second in general. In difference to this scalarization approach, the variational approach is mainly based on the *extremal principle* (Theorem 5.5.9), a variational counterpart of local separation for nonconvex sets. It is initiated by Kruger and Mordukhovich in [342, 426], and recently applied to multiobjective optimization in [24, 25, 28, 430]. The advantage of the latter approach is the possibility of deriving necessary conditions to Pareto optimal points of a multiobjective optimization problem directly without converting it to a scalar problem in both finite- and infinite-dimensional settings. However, to implement it, the *sequential normal compactness* (SNC) condition (see Sect. 5.5) of ordering cones is unavoidable due to the natural lack of the compactness property in infinite dimensional spaces which is automatic in finite dimensions. It is known from [431, Theorem 1.21] that a convex cone fulfills the SNC condition if it has a nonempty relative interior r-int $C \neq \emptyset$ and $\mathrm{cl}\,(C - C)$ has a finite-codimensions. In summary, both approaches can not be applied to the positive ordering cones of the Lebesgue spaces l^p and L^p for $1 < p < +\infty$ since they are neither AC nor the SNC at the origin.

The main aim of this section is to establish *subdifferential* necessary optimality conditions for Pareto minimal points of sets and minimizers of constrained set-valued optimization problems where the ordering cone of a infinite-dimensional Banach space does not necessarily have an empty interior (even relative interior), or a finite-dimensional span following the paper by Bao and Tammer [29]. Our approach combines the scalarization scheme in [149, 151] and several advanced tools of variational analysis used in [24, 25, 28]. Precisely, we proceed as follows: Given a Pareto minimal point $\overline{y}$ to a set $A \subset Y$ with respect to an ordering cone C with $\mathrm{int}\, C = \emptyset$. In contrast to [149] in which the existence of a bigger cone $D \subset Y$ satisfying the condition $C \setminus \{0\} \subset \mathrm{int}\, D$ is required (so that, $\overline{y}$ is proper efficient,) in this section we will choose an appropriate nonempty interior cone that might not contain in the given cone or be contained in it. Such a cone automatically enjoys the SNC condition, but it might not have the AC property. Then utilizing the strategy in [24, 28] which deals with epigraphical multifunctions, we derive subdifferential necessary conditions thanks to the full calculus for generalized differentiation in [430, Chapter 3] (see Sect. 3.5).

In order to derive multiplier rules we use the fundamentals from the book by Mordukhovich [431] (see Sect. 3.5), especially subdifferential constructions for set-valued mappings with values in partially ordered spaces first introduced in [25]. We derive necessary optimality conditions for Pareto minimal points of sets, and minimizers of set-valued optimization problems without the nonempty interior requirement imposed on ordering cones.

We use the standard notation of variational analysis presented in Sect. 3.5, especially the limiting normal cone (see Definition 3.5.6) and the limiting subdifferential (see Definition 3.5.11); cf. the books [431,499], and assume that all the spaces under consideration are Asplund (see Definition 3.5.3) unless otherwise stated.

In the proofs of our assertions we apply the results concerning the scalarizing functional $\varphi_A := \varphi_{A,k^0} : Y \longrightarrow \overline{\mathbb{R}}$ given in (5.5) with $A \subset Y$ and $k^0(\neq 0) \in Y$. The most important properties (from Theorem 5.2.3, Proposition 5.2.23 and Corollary 5.2.24) for deriving multiplier rules are collected in the following lemma:

Lemma 12.8.1 (Scalarization Functions for Nonconvex Sets). *Let $D \subset Y$ be a proper, closed and convex cone with a nonempty interior in a Banach space Y. Given $k^0 \in \mathrm{int}\, D$, the functional $\varphi_{D,k^0} : Y \to \mathbb{R}$ defined by*

$$\varphi_{D,k^0}(y) = \inf\{\lambda \in \mathbb{R} \mid \lambda \cdot k^0 \in y + D\} \tag{12.93}$$

*is continuous, sublinear, strictly-*int D*-monotone. Moreover, the following relations hold:*

(i) $\varphi_{D,k^0}(y + tk^0) = \varphi_{D,k^0}(y) + t \quad \forall y \in Y,\ \forall t \in \mathbb{R}$.
(ii) $\partial\varphi_{D,k^0}(0) = \{y^* \in D^* \mid \langle y^*, k^0\rangle = 1\}$ *with* $D^* := \{y^* \in Y^* \mid \langle y^*, y\rangle \geq 0,\ \forall y \in D\} = -N(0; D)$.
(iii) $\partial\varphi_{D,k^0}(y) = \{y^* \in D^* \mid \langle y^*, k^0\rangle = 1, \langle y^*, y\rangle = \varphi_{D,k^0}(y)\}$ *for any* $y \in Y$.
(iv) *Given a nonempty set* $A \subset Y$, *if* $\overline{y} \in A$ *is a weakly minimal point of* A *with respect to* D *(see Definition 2.4.2), then one has*

$$\varphi_{D,k^0}(y - \overline{y}) \geq 0 \quad \textit{for all} \quad y \in A.$$

This real-valued function plays the role of an utility function. In this section we present necessary conditions for Pareto minimal points of sets, for Pareto minimizers of vector-valued optimization problems and for Pareto solutions of set-valued optimization problems in the Asplund setting due to the full calculus for (Mordukhovich) generalized differentiation. It is worth to stressing that our results can be extended to the general Banach setting by using the Ioffe approximate differentiation by imposing stronger assumptions on the given data such that the corresponding calculus rules are applied (compare Sect. 12.7.2).

Let Y be an Asplund space, and let $C \subset Y$ be a proper, convex and pointed cone which generates a partial order "$\leq_C$" defined by

$$y_1 \leq_C y_2 \quad \text{if and only if} \quad y_2 - y_1 \in C \quad \text{for all } y_1, y_2 \in Y. \tag{12.94}$$

Given a nonempty subset A in Y and $\overline{y} \in \text{Min}(A, C)$ (see Definition 2.4.1). For each $k^0 \in C \setminus \{0\}$ and for each $\varepsilon \in (0, \|k^0\|)$ we consider the following cone

$$C_{k^0,\varepsilon} := \text{cone}\left(B(k^0, \varepsilon)\right) = \left\{t \cdot z \mid t \geq 0,\ z \in B(k^0, \varepsilon)\right\}, \tag{12.95}$$

where $B(k^0, \varepsilon)$ is a closed ball with the center k^0 and the radius ε. Obviously, it is a proper, closed, convex and pointed cone and its interior is nonempty since $k^0 \in \text{int}\, C_{k^0,\varepsilon}$. This cone will take the center in our procedure in formulating necessary conditions for Pareto points with respect to an empty interior ordering cone (Fig. 12.1).

It is worth emphasizing that the approach in [149] involved the Henig global proper minimality (Definition 2.4.4 (f)). Recall that Henig minimality is defined in the same way of proper minimality by specifying the proper cone $D \subset Y$ with $C \setminus \{0\} \subset \text{int}\, D$ (see Theorem 2.4.11) in form of a dilation of C (see Definition 2.4.9) given by

$$C_{H,\varepsilon} := \bigcup_{k^0 \in B} C_{k^0,\varepsilon},$$

where B is a base of the cone C with B is a convex set such that $0 \notin \text{cl}\, B$ and $C = \text{cone}\, B$, i.e., C is well-based (see Definition 2.2.6). In our approach, we just use one component of the Henig cone.

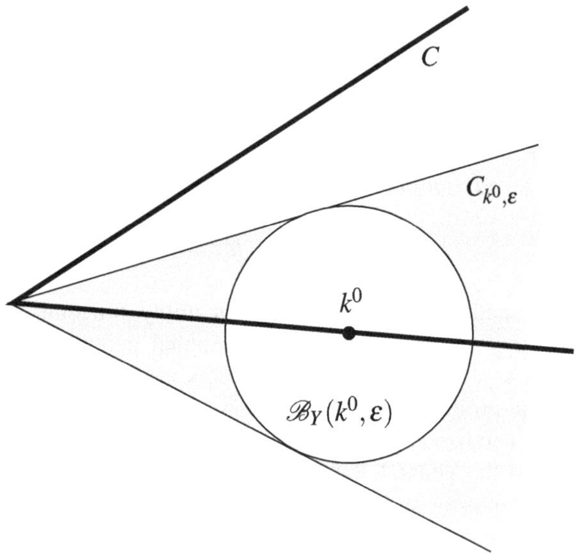

Fig. 12.1 The cone C with $\operatorname{int} C = \emptyset$ and the corresponding cone $C_{k^0,\varepsilon}$ with $\operatorname{int} C_{k^0,\varepsilon} \neq \emptyset$

Theorem 12.8.2 (Necessary Conditions for Pareto Minimal Points of Sets). *Let $\overline{y} \in \operatorname{Min}(A, C)$. Assume that the epigraphical set of A with respect to C and $A + C$ are locally closed at $\overline{y}$. Then for every $k^0 \in C \setminus \{0\}$ satisfying*

$$-k^0 \notin \operatorname{cl}\operatorname{cone}(A + C - \overline{y}), \tag{12.96}$$

there are a positive number $\varepsilon > 0$ and $y^* \in Y^*$ such that

$$-y^* \in N(\overline{y}; A + C), \quad -y^* \in N(0; C) \cap N(0; C_{k^0,\varepsilon}) \quad \text{and} \quad \langle y^*, k^0 \rangle = 1, \tag{12.97}$$

where $C_{k^0,\varepsilon}$ is given in (12.95).

Proof. First employing [293, Lemma 4.7 (a)], we have the implication

$$\overline{y} \in \operatorname{Min}(A, C) \quad \Longrightarrow \quad \overline{y} \in \operatorname{Min}(A + C, C) \tag{12.98}$$

for any proper, convex and pointed ordering cone C. Taking into account Definition 2.4.1, we get from (2.19) that

$$(A + C - \overline{y}) \cap (-C) = \{0\}, \text{ and thus } \operatorname{cone}(A + C - \overline{y}) \cap (-C) = \{0\}. \tag{12.99}$$

Since $-k^0 \notin \operatorname{cl}\operatorname{cone}(A+C-\overline{y})$ by (12.96), there is a positive number $\varepsilon \in (0, \|k^0\|)$ such that

$$\operatorname{cl}\operatorname{cone}(A+C-\overline{y}) \cap \big(-B(k^0,\varepsilon)\big) = \emptyset, \text{ and thus } \operatorname{cl}\operatorname{cone}(A+C-\overline{y}) \cap (-C_{k^0,\varepsilon}) = \{0\},$$

which justifies $\overline{y} \in \operatorname{Min}(A+C, C_{k^0,\varepsilon})$, where $C_{k^0,\varepsilon}$ is given in (12.95).

Now we can apply Lemma 12.8.1 for $D := C_{k^0,\varepsilon}$ and the element $k^0 \in \operatorname{int} C_{k^0,\varepsilon}$ to get that $\overline{y}$ is a minimum of the scalarization function φ_{D,k^0} defined by (12.93) over $A+C$:

$$\text{minimize} \quad \varphi_{D,k^0}(y-\overline{y}) \quad \text{subject to} \quad y \in A+C.$$

Since the functional $\varphi_{D,k^0}(\cdot)$ is Lipschitz continuous (in fact, sublinear and continuous) by Lemma 12.8.1, and since the constraint set $A+C$ is assumed to be locally closed at $\overline{y}$, all the assumptions of the lower-subdifferential necessary condition for local minima in [430, Proposition 5.3] are satisfied. Employing it to the minimum $\overline{y}$ of the above problem, we have

$$0 \in \partial\varphi_{D,k^0}(\cdot-\overline{y})(\overline{y}) + N(\overline{y}; A+C) = \partial\varphi_{D,k^0}(0) + N(\overline{y}; A+C). \qquad (12.100)$$

Taking into account the description of the subdifferential of φ_{D,k^0} at the origin in Lemma 12.8.1 (ii)

$$\partial\varphi_{D,k^0}(0) = \{y^* \in C^+_{k^0,\varepsilon} \mid \langle y^*, k^0\rangle = 1\} \quad \text{with} \quad C^+_{e,\varepsilon} = -N(0; C_{k^0,\varepsilon}),$$

we find from the inclusion (12.100) a subgradient $y^* \in \partial\varphi_{D,k^0}(0)$ with $-y^* \in N(\overline{y}; A+C)$ and $y^* \in -N(0; C_{k^0,\varepsilon})$, which justifies the necessary conditions in (12.97) provided that $-y^* \in N(0; C)$. To show this, we further elaborate the definition of the normal cone (3.25) to the set $A+C$ at $\overline{y}$. We get from $-y^* \in N(\overline{y}; A+C)$ a sequence $(y_k, y_k^*) \in Y \times Y^*$ satisfying

$$y_k \overset{A+C}{\longrightarrow} \overline{y}, \quad -y_k^* \overset{w^*}{\to} -y^* \quad \text{with} \quad -y_k^* \in \hat{N}(y_k; A+C).$$

The last inclusion gives

$$\begin{aligned}
\limsup_{y \overset{C}{\to} 0} \frac{\langle -y_k^*, y\rangle}{\|y\|} &= \limsup_{y-y_k \overset{C}{\to} 0} \frac{\langle -y_k^*, y-y_k\rangle}{\|y-y_k\|} \\
&= \limsup_{y \overset{C+y_k}{\longrightarrow} y_k} \frac{\langle -y_k^*, y-y_k\rangle}{\|y-y_k\|} \\
&\le \limsup_{y \overset{A+C}{\longrightarrow} y_k} \frac{\langle -y_k^*, y-y_k\rangle}{\|y-y_k\|} \quad \le \quad 0,
\end{aligned}$$

which verifies $-y_k^* \in \hat{N}(0; C)$, and thus $-y^* \in N(0; C)$ as $k \to +\infty$. The proof is completed. □

Taking into account the definition of the normal cones $N(0; C)$, $N(0; C_{k^0,\varepsilon})$ and condition (12.97) in Theorem 12.8.2 we get the following corollary:

Corollary 12.8.3 (Necessary Conditions for Pareto Minimal Points of Sets). *Let $\overline{y} \in \mathrm{Min}(A, C)$. Assume that the epigraphical set of A with respect to C and $A + C$ are locally closed at $\overline{y}$. Then for every $k^0 \in C \setminus \{0\}$ satisfying (12.96) there are a positive number $\varepsilon > 0$ and $y^* \in C^+$ such that*

$$-y^* \in N(\overline{y}; A + C), \quad \langle y^*, k^0 \rangle \geq \varepsilon ||y^*||_* \quad \textit{and} \quad \langle y^*, k^0 \rangle = 1. \tag{12.101}$$

Proof. Because of $-y^* \in N(0; C) \cap N(0; C_{k^0,\varepsilon})$ (with $C_{k^0,\varepsilon}$ is given by (12.95)) in (12.97) in Theorem 12.8.2 and the properties of the included normal cones we get $y^* \in -N(0; C) = C^+$ (cf. Lemma 12.8.1 (ii)). Furthermore, $-y^* \in N(0; C_{k^0,\varepsilon})$ implies $y^* \in C^+_{k^0,\varepsilon}$ and so we have

$$\forall \lambda > 0, \ \forall y \in Y \text{ with } ||y|| = 1 : \qquad \langle y^*, \lambda(k^0 + \varepsilon y) \rangle \geq 0.$$

This yields for $y^* \neq 0$ and for all $y \in Y$ with $||y|| = 1$

$$\langle \frac{y^*}{||y^*||_*}, k^0 \rangle \geq -\varepsilon \langle \frac{y^*}{||y^*||_*}, y \rangle,$$

and so

$$\langle \frac{y^*}{||y^*||_*}, k^0 \rangle \geq \sup_{||y||=1} -\varepsilon \langle \frac{y^*}{||y^*||_*}, y \rangle = \varepsilon,$$

i.e., $\langle y^*, k^0 \rangle \geq \varepsilon ||y^*||_*$. □

Remark 12.8.4. Note that the condition (12.96) is fulfilled if the cone cone $(A + C - \overline{y})$ happens to be closed. Furthermore, (12.96) is satisfied for any $k^0 \in \mathrm{int}\, C$ provided that $\mathrm{int}\, C \neq \emptyset$.

Note that the condition (12.96) is essential since we do not assume that the ordering cone has a nonempty interior. Analyzing the proof of Theorem 12.8.2, the existence of a vector k^0 satisfying (12.96) is ensured provided that

$$-C \cap \mathrm{bd}\, \mathrm{cone}\,(A + C - \overline{y}) \neq \emptyset.$$

Remark 12.8.5. It is worth emphasizing that if the cone cone $(A + C - \overline{y})$ is closed at the origin, then every Pareto minimal point $\overline{y}$ of A with respect to C is a **Benson minimal point** (see Definition 2.4.4 (d)) in the sense that

$$\operatorname{cl}\operatorname{cone}(A + C - \overline{y}) \cap (-C) = \{0\}. \tag{12.102}$$

In the book chapter by Ha [229] necessary conditions for Benson proper minimality are (indirectly) established under the condition that the cone $D := Y \setminus -\operatorname{cl}\operatorname{cone}(A + C - \overline{y})$ has a nonempty interior. Ha [229] used the concept of Benson proper minimality to obtain sufficient optimality conditions.

Remark 12.8.6. Theorem 12.8.2 can be applied to Pareto minimality and not Benson proper minimality. Take A be a subset of $\mathbb{R}^3$ defined by

$$A := \{(x, y, z) \mid x^2 + y^2 + (z - 1)^2 = 1\},$$

where $\mathbb{R}^3$ is partially ordered by the non-solid cone $C := \{0\} \times \mathbb{R}^2_+$, i.e., $\operatorname{int} C = \emptyset$. Obviously, the origin is a Pareto point of A with respect to C, but it is not a Benson properly minimal point since $\operatorname{cl}\operatorname{cone}(A + C - 0) = \{(x, y, z) \mid z \geq 0\}$ is a closed half-space of $\mathbb{R}^3$ and $\{(0, 0)\} \times \mathbb{R}_- = \operatorname{cl}\operatorname{cone}(A + C - 0) \cap (-C)$. Moreover, every vector $k^0 = (0, y, z)$ with $z > 0$ satisfies the condition (12.96). Therefore, the necessary conditions in Theorem 12.8.2 can be applied to this example.

The necessary conditions in Theorem 12.8.2 do not hold true without the condition (12.96). Consider a closed set in $\mathbb{R}^2$ given by $A := \operatorname{epi}(-\sqrt{x})$, where $\mathbb{R}^2$ is partially ordered by the cone $C = \{0\} \times \mathbb{R}^+$. Obviously, the origin is a Pareto minimal point but not a Benson properly minimal point. It is easy to check that

$$N(0; A + C) = N(0; A) = \mathbb{R}_- \times 0.$$

There is only the unique choice of $y^* = (-1, 0) \in N(0; A + C)$ and $k^0 = (0, 1) \in C$. We can easily check that $-y^* \notin N(0; C_{k^0,\varepsilon})$ for every $\varepsilon > 0$, and thus the necessary conditions in Theorem 12.8.2 is not applicable to this example.

Next, we will establish a revised version of Theorem 12.8.2

Theorem 12.8.7 (Refined Necessary Conditions for Pareto Minimal Points of Sets). *Let* $\overline{y} \in \operatorname{Min}(A, C)$. *Assume that* $A + C$ *is locally closed at* $\overline{y}$ *and* $\operatorname{cone}(A + C - \overline{y})$ *is closed. Then for every* $k^0 \in C \setminus \{0\}$, *there are a positive number* $\varepsilon > 0$ *and* $y^* \in Y^*$ *such that*

$$-y^* \in N(\overline{y}; A + C), \quad -y^* \in N(0; C) \cap N(0; C_{k^0,\varepsilon}) \quad \textit{and} \quad \langle y^*, k^0 \rangle = 1, \tag{12.103}$$

where $C_{k^0,\varepsilon}$ *is given in (12.95).*

Proof. The assertion follows by the proof of Theorem 12.8.2. Since the cone $\operatorname{cone}(A + C - \overline{y})$ is closed, we get from (12.99) that for every $k^0 \in C \setminus \{0\}$, the condition (12.96) is fulfilled, and thus the necessary condition is proved. □

Note that the third condition in (12.103) implies the nontriviality condition $y^* \neq 0$ in the Lagrange-type necessary conditions while the second provides the range of multipliers. Note also that the necessary conditions in Theorem 12.8.7 are new when the ordering cone is not SNC at the origin; otherwise, they are weaker than the existing necessary results due to the additional closedness assumption imposed on cone $(A + C - \overline{y})$.

Remark 12.8.8 (Alternative Necessary Conditions).

(1) It is important to note that the closedness of the cone cone $(A + C - \overline{y})$ and the closedness of the set itself are different conditions. Indeed, in $\mathbb{R}^2$ equipped with the usual Pareto order $C = \mathbb{R}^2_+$, the set $A := \text{graph}(x^2)$ has its epigraphical set $(A + C - \mathbf{0})$ which is closed, but its cone is not closed at the origin, while the set $A := \{(x, y) \mid x > (y-1)^2 - 1 \quad \text{and} \quad y \geq x\} \cup \{\mathbf{0}\}$ has the epigraphical set which is not closed, but its cone is closed at the origin.

(2) Observe that the closedness assumption of the set $(A+C-\overline{y})$ in Theorem 12.8.7 can be dropped. In such a case, i.e., $A + C$ is not locally closed at $\overline{y}$, the first condition in (12.103) will be formulated by

$$-y^* \in N\big(\overline{y}; \text{cl}\,(A + C)\big).$$

Observe also that the closedness assumption of the cone cone $(A + C - \overline{y})$ is essential. Indeed, if A is the graph of the cardioid $r(\alpha) := 1 + \sin(\alpha)$ in $\mathbb{R}^2$, $C := \{(0,t) \in \mathbb{R}^2 \mid t \geq 0\}$, and $\overline{y} = 0$, then 0 is a Pareto minimal point to A. Since cone $(A + C - \overline{y}) = \mathbb{R}^2 \setminus (-C)$, and since the only vector $k^0 \in C$ with $\|k^0\| = 1$ is $k^0 = (0, 1)$, there is no $\varepsilon > 0$ such that $0 \in \text{Min}(A, C_{k^0,\varepsilon})$. Therefore, the arguments in the proof are no longer valid.

(3) The first condition in (12.103) formulated with the epigraphical set $(A + C)$ is much better than the conventional condition with the set itself

$$-y^* \in N(\overline{y}; A)$$

provided that $A + C$ is *order continuous* at $\overline{y}$ since we have the following inclusion

$$N(\overline{y}; A + C) \subset N(\overline{y}; A). \tag{12.104}$$

To justify this, let us first recall that A is *order continuous* at $\overline{y}$ if for any sequence $\{y_k\} \subset A + C$ converging to $\overline{y}$, there is a sequence $\{v_k\} \subset A$ with $v_k \leq_C y_k$ and $v_k \to \overline{y}$; cf. [26, Definition 4.1] for the order continuity of set-valued mappings. In fact, a set A is order continuous if the constant mapping $F : \mathbb{R} \rightrightarrows Y$ with $F(x) \equiv A$ is order continuous at $(0, \overline{y})$. The reader is referred to [26] for several efficient conditions ensuring this property.

To justify inclusion (12.104), we mimic the proof of [26, Proposition 4.3]. Take $y^* \in N(\overline{y}; A+C)$, then find a sequence $(y_k, y_k^*) \in Y \times Y^*$ such that

$$y_k \xrightarrow{A+C} \overline{y} \quad y_k^* \xrightarrow{w^*} y^* \quad \text{with} \quad y_k^* \in \hat{N}(y_k; A+C).$$

Fix k and write y_k in the form $y_k = v_k + t_k$ with $v_k \in A$ and $t_k \in C$. For every $\gamma > 0$ we get from the definition of Fréchet normal $x_k^* \in \hat{N}(y_k; A+C)$ that

$$\langle y_k^*, y - y_k \rangle \leq \gamma \|y - y_k\|$$

for all $y \in (A+C) \cap (y_k + \eta B_Y)$ for sufficiently small $\eta > 0$. Taking now a neighborhood of v_k in form $v_k + \eta B_Y$, for every $\tilde{y} \in A \cap (v_k + \eta B_Y)$ we have $y = \tilde{y} + t_k \in (A+C) \cap (y_k + \eta B_Y)$, and thus

$$\langle y_k^*, \tilde{y} - v_k \rangle = \langle y_k^*, (y + t_k) - (v_k + t_k) \rangle = \langle y_k^*, y - y_k \rangle \leq \gamma \|y - y_k\| = \gamma \|\tilde{y} - v_k\|,$$

which justifies that $y_k^* \in \hat{N}(v_k; A)$. This together with the convergent sequence $v_k \to \overline{y}$ as $k \to +\infty$ derived from the order continuity assumption implies $y^* \in N(\overline{y}; A)$. We have proved the inclusion (12.104) under the order continuity assumption.

Note that the normal cones to a set and its epigraphical set are incomparable in general. To illustrate this, consider the set

$$A := \{(x, y) \in \mathbb{R}^2 \mid x^2 + y^2 = 2 \quad \text{and} \quad x + y \geq 0\}$$

in $\mathbb{R}^2$ equipped with the usual ordering cone $C := \mathbb{R}^2_+$, then it is easy to check that

$$N((1,1); A) := \text{span}\,\{(1,1)\} \quad \text{and} \quad N((1,1); A+C) := \text{bd}\,\mathbb{R}^2_-.$$

Now we consider a set-valued optimization problem with a general geometric constraint:

$$\text{minimize} \quad F(x) \quad \text{subject to} \quad x \in S, \tag{SP}$$

where the cost mapping $F : X \rightrightarrows Y$ is a set-valued mapping, Y and X are Asplund spaces, S is a subset of X, and "minimization" is understood in the sense of Definition 2.6.1. In the following theorem we assume that F is *epigraphically Lipschitz-like* (ELL) around $(\overline{x}, \overline{y})$ (see Definition 3.3.19) and we use the limiting subdifferential introduced in Definition 3.5.11.

Theorem 12.8.9 (Necessary Conditions for Minimizers in Set-Valued Optimization). *Let* $(\overline{x}, \overline{y}) \in \operatorname{graph} F$ *with* $\overline{x} \in S$ *be a minimizer of the problem* (SP). *Assume that* F *is ELL around* $(\overline{x}, \overline{y})$, $\operatorname{epi} F$ *is closed around this point, and* S *is* *closed around* $\overline{x}$. *Assume furthermore that* $\operatorname{cone}(F(S)+C-\overline{y})$ *is closed. Then for every* $k^0 \in C \setminus \{0\}$, *there are a positive number* $\varepsilon > 0$ *and a dual element* $y^* \in Y^*$ *such that*

$$0 \in \partial F(\overline{x},\overline{y})(y^*) + N(\overline{x};S), \quad -y^* \in N(0;C) \cap N(0;C_{k^0,\varepsilon}) \quad and \quad \langle y^*, k^0\rangle = 1. \tag{12.105}$$

Proof. Since $\operatorname{cone}(F(S)+C-\overline{y})$ is assumed to be closed and $\operatorname{epi} F$ is closed around $(\overline{x},\overline{y})$ (this yields that $(F(S)+C-\overline{y})$ is locally closed at the origin), we can show that for every $k^0 \in C\setminus\{0\}$, there exists a positive number $\varepsilon \in (0, \|k^0\|)$ such that $\overline{y} \in \operatorname{Min}(F(S)+C, C_{k^0,\varepsilon})$. Then, we get from Lemma 12.8.1 (iv) with $D = C_{k^0,\varepsilon}$ that $(\overline{x},\overline{y})$ is a minimum of the problem

$$\text{minimize} \quad \varphi(x,y) + \iota_A\big((x,y)\big),$$

where the cost $\varphi : X \times Y \to \mathbb{R}$ is given by

$$\varphi(x,y) := \varphi_{D,k^0}(y-\overline{y}),$$

and the indicator function $\iota_A(\,\cdot\,)$ has the value 0 for all elements of A and the value $+\infty$ for all elements of $X \times Y$ not in A, and $A := \operatorname{epi} F_S$. Recall that F_S is the restriction of the mapping F on S given by

$$F_S(x) := F(x) + \Delta(x;S) \quad \text{with} \quad \Delta(x;S) := \begin{cases} \{0\}(\subset Y) & \text{if} \quad x \in S, \\ \emptyset & \text{otherwise.} \end{cases}$$

Since $A = \operatorname{epi} F \cap (S \times Y)$ is locally closed at $(\overline{x},\overline{y})$, the indicator function $\iota_A(\cdot)$ is lower semicontinuous at that point. Since the function φ_{D,k^0} is Lipschitz continuous, so is the cost function φ. Employing first the Fermat rule to this auxiliary scalar optimization problem and then the sum rule to the sum of one Lipschitz continuous function and one lower semicontinuous one, we have

$$\begin{aligned} 0 &\in \partial\big(\varphi + N(\cdot;A)\big)(\overline{x},\overline{y}) \\ &\subset \partial\varphi(\overline{x},\overline{y}) + N\big((\overline{x},\overline{y});A\big) \\ &= \{0\} \times \partial\varphi_{D,k^0}(0) + N\big((\overline{x},\overline{y}); \operatorname{epi} F \times (S \times Y)\big), \end{aligned}$$

where the equality holds due to the definitions of the functions.

Taking into account now the subdifferential of φ_{D,k^0} (see Lemma 12.8.1 (ii)) and the fact that $A = \operatorname{epi} F_S = \operatorname{graph} \mathscr{E}_{F_S}$ we have the existence of $y^* \in -N(0;C_{k^0,\varepsilon})$ with $\langle y^*, k^0\rangle = 1$ such that

$$(0,-y^*) \in N\big((\overline{x},\overline{y});A\big) = N\big((\overline{x},\overline{y}); \operatorname{epi} F_S\big),$$

which is equivalent (see Definition 3.5.8) to the following inclusion

$$0 \in D^*\mathscr{E}_{F_S}(\overline{x},\overline{y})(y^*) = D^*\big(\mathscr{E}_F + \Delta(\cdot;S)\big)(\overline{x},\overline{y})(y^*). \qquad (12.106)$$

Note that the graphs of both set-valued mappings $\mathscr{E}_F$ and $\Delta(\cdot;S)$ are locally closed at $(\overline{x},\overline{y})$ under the closedness assumptions imposed on F and S. Note also that the imposed ELL property of F at $(\overline{x},\overline{y})$ is nothing but the Lipschitz-like property of the epigraphical multifunction $\mathscr{E}_F$ at the same point. Hence, the pair of mappings $\mathscr{E}_F$ and $\Delta(\cdot;S)$ meet all the assumptions of the coderivative sum rule from [431, Proposition 3.12], and thus we get from this rule that

$$D^*\big(\mathscr{E}_F + \Delta(\cdot;S)\big)(\overline{x},\overline{y})(y^*) \subset D^*\mathscr{E}_F(\overline{x},\overline{y})(y^*) + N(\overline{x};S). \qquad (12.107)$$

Substituting (12.107) into (12.106) and taking into account the definition of subdifferential (3.29) and the implication $\big[\ D^*\mathscr{E}_F(\overline{x},\overline{y})(y^*) \neq \emptyset \implies -y^* \in N(0;C)\ \big]$ in (3.30), we arrive at the necessary condition (12.105)

$$0 \in \partial F(\overline{x},\overline{y})(y^*) + N(\overline{x};S)$$

and complete the proof. □

Note that if the cost mapping $F = f : X \to Y$ in Theorem 12.8.9 happens to be single-valued, the necessary conditions in Theorem 12.8.9 reduce to those in [29, Theorem 3.8] while the reduced assumption that epi f is closed around $(\overline{x}, f(\overline{x}))$ is abundant since it is automatic from the Lipschitz continuity of f. However, the ELL property of a set-valued mapping does not imply the closedness of the epigraph of F in general. Consider the set-valued mapping $F : \mathbb{R} \rightrightarrows \mathbb{R}$ defined by

$$F(x) := \begin{cases} (-x,+\infty) & \text{if} \quad x < 0, \\ [0,+\infty) & \text{if} \quad x = 0, \\ (x,+\infty) & \text{if} \quad x > 0, \end{cases}$$

where the usual order on real numbers is used, i.e., $C := [0,+\infty)$. It is easy to check that F is ELL at $(0,0)$, but its epigraph epi F = graph F is not locally closed at this point.

In comparison with necessary conditions in [24, 25, 27, 149, 151, 157, 431, 620, 621] and the references therein, the results in Theorem 12.8.7 and Theorem 12.8.9 hold even when the ordering cone has an empty interior, and enjoys neither AC nor SNC conditions.

Remark 12.8.10 (Necessary Conditions Under Regularity Assumptions). Mimicking the penalization scheme used in [149, Theorem 4.4], we are able to obtain a refined necessary condition for the problem (SP) under the **metric regularity** assumption, i.e., there is $\gamma > 0$ and a neighborhood U of $(\overline{x}, \overline{y}) \in \operatorname{epi} F \cap (S \times Y)$ such that

$$d\big((x,y), \operatorname{epi} F \cap (S \times Y)\big) \le \gamma\big(d(x,S) + d((x,y), \operatorname{epi} F)\big), \quad \text{for all} \quad (x,y) \in U. \tag{12.108}$$

Necessary conditions with regularity properties. Let $(\overline{x}, \overline{y}) \in \operatorname{graph} F$ with $\overline{x} \in S$ be a Pareto minimizer of the problem (SP), where the image space Y is partially ordered by a proper closed, convex and pointed cone C whose interior might be empty. Assume that S is closed around $\overline{x}$, $\big(F(S)+C-\overline{y}\big)$ and cone $\big(F(S)+C-\overline{y}\big)$ are closed around the origin. Assume also that the regularity condition (12.108) is fulfilled. Then for every $k^0 \in C \setminus \{0\}$, there is a positive number $\varepsilon > 0$ and $y^* \in Y^*$ with $-y^* \in N(0; C) \cap N(0; C_{k^0,\varepsilon})$ and $\langle y^*, k^0 \rangle = 1$ satisfying

$$(-\partial d(\overline{x}, S) \times \{-y^*\}) \cap \partial d\big((\overline{x}, \overline{y}), \operatorname{epi} F\big) \neq \emptyset, \tag{12.109}$$

which surely implies the relationships in (12.105).

Remark 12.8.11 (Variants of Necessary Optimality Conditions in Banach Spaces). Although the results in this section were established in Asplund spaces, they are easily to extend to general Banach spaces by using other types of generalized differentiation which have the calculus rules used in the proofs of Theorem 12.8.7 and Theorem 12.8.9; for example, Clarke's generalized differentiations in [100], and the Ioffe's approximate constructions in [275–277]. In Sect. 12.7.2 we presented corresponding results in general Banach spaces for Q-minimizers of set-valued optimization problems (see Definition 2.6.3). In this section we used the concept of generalized differentiability by Kruger/Mordukhovich (see Sect. 3.5) since it has a full calculus for normal cones to sets and coderivatives for set-valued mappings under the SNC conditions and the Mordukhovich qualification conditions; they are satisfied under the Lipschitz-like properties, and since it is smaller or at least not bigger than the other generalized differentiation constructions.

If we want to have necessary conditions in terms of Ioffe's normal cones and coderivatives, we need the corresponding conditions for the fulfillment of the calculus rules such as the sum rule [277, Theorem 5.6], the chain rule [277, Theorem 6.2], the chain rule for coderivatives [312, Theorem 3.1], and the sum rule for coderivatives [312, Theorem 3.4]. Loosely speaking, instead of the Lipchitz-like assumptions needed for the Mordukhovich's calculus, the Ioffe's calculus requires the strongly compactness Lipschitz continuity property whose definition is quite complicated even for vector-valued functions (compare Definition 3.3.20). We were dealing with necessary conditions based on Ioffes's subdifferential calculus in Sect. 12.7.

12.9 Necessary Conditions for Approximate Solutions of Set-Valued Optimization Problems

In order to show necessary conditions for approximate solutions of a set-valued optimization problem we will use the *extremal principle of variational analysis* [430, Theorem 2.20] for the case of two closed sets Ω_1 and Ω_2 in the Asplund space X (see Sect. 5.5).

In this section we present a subdifferential variational principle for set-valued mappings established in [27, Theorem 3.9] as an extension of the corresponding scalar result by Mordukhovich and Wang. This version of the subdifferential variational principle is derived under the *limiting monotonicity condition* (compare Sect. 10.2) imposed on the set-valued mapping $F : X \rightrightarrows Y$ with no pointedness assumption supposed on the proper closed convex ordering cone $C \subset Y$. Furthermore, we assume that F is **epiclosed**, i.e., its epigraph is closed in $X \times Y$. It is easy to see that every epiclosed mapping is level-closed, but the opposite may not be true in the case of set-valued mappings. Similarly to [25, 430, 434], the proof of the subdifferential variational principle given below is based on the extremal principle (Theorem 5.5.9) and the variational principle given in Corollary 10.2.12 (see [27, Theorem 3.5]), and hence it requires the Asplund property of the Banach spaces in question. The necessary condition in the following theorem is formulated for approximate solutions of set-valued optimization problems in the sense of Definition 2.6.61 using the Fréchet subdifferential of F at $(\overline{x}, \overline{y}) \in \operatorname{graph} F$ introduced in (3.32). The following result is shown by Bao and Mordukhovich [27, Theorem 3.9].

Theorem 12.9.1 (Subdifferential Variational Principle for Set-Valued Mappings). *Let X and Y be Asplund spaces, $C \subset Y$ a proper closed convex cone and $F : X \rightrightarrows Y$. Suppose that the assumptions of Corollary 10.2.12 are fulfilled. Furthermore, assume additionally that F is epiclosed with respect to the ordering cone $C \subset Y$.*

Then for any $\varepsilon > 0$, $\lambda > 0$, $k^0 \in C \setminus (-C)$ with $||k^0|| = 1$, and a strict approximate εk^0-minimizer $(x^0, y^0) \in \operatorname{graph} F$ for the mapping F, there exists $(\overline{x}, \overline{y}) \in \operatorname{graph} F$ such that $||\overline{x} - x^0|| \leq \lambda$ and

$$\hat{\partial} F(\overline{x}, \overline{y}) \cap \frac{\varepsilon}{\lambda} B_{X^*} \neq \emptyset. \tag{12.110}$$

Proof. We consider the strict approximate εk^0-minimizer $(x^0, y^0) \in \operatorname{graph} F$ for the mapping F. Taking into account Definition 2.6.61 there is a positive number $\tilde{\varepsilon} < \varepsilon$ such that (x^0, y^0) is an approximate $\tilde{\varepsilon} k^0$-minimizer for this mapping. Set

$$\tilde{\lambda} := \frac{\varepsilon + \tilde{\varepsilon}}{2\varepsilon} \lambda \quad \text{with} \quad 0 < \tilde{\lambda} < \lambda \tag{12.111}$$

and apply the variational principle in Corollary 10.2.12 to the mapping F and its approximate $\tilde{\varepsilon}k^0$-minimizer (x^0, y^0) with the chosen parameters $\tilde{\varepsilon}$ and $\tilde{\lambda}$.

The variational principle in Corollary 10.2.12 yields the existence of $(\overline{u}, \overline{v}) \in \operatorname{graph} F$ with

$$\overline{v} \in \operatorname{BMMin} F(\overline{u}), \quad ||x^0 - \overline{u}|| \leq \tilde{\lambda}, \tag{12.112}$$

and

$$y - \overline{v} + \frac{\tilde{\varepsilon}}{\tilde{\lambda}}||x - \overline{u}||k^0 \notin -C \quad \text{for all} \quad (x, y) \in \operatorname{graph} F \quad \text{with} \quad (x, y) \neq (\overline{u}, \overline{v}). \tag{12.113}$$

We introduce a single-valued Lipschitz continuous mapping $g : X \to Y$ defined by

$$g(x) := \overline{v} - \frac{\tilde{\varepsilon}}{\tilde{\lambda}}||x - \overline{u}||k^0 \tag{12.114}$$

and consider the following two closed subsets of the Asplund product space $X \times Y$ given by

$$\Omega_1 := \operatorname{epi} F \quad \text{and} \quad \Omega_2 := \operatorname{graph} g. \tag{12.115}$$

We will show that $(\overline{u}, \overline{v})$ is an extremal point of the set system $\{\Omega_1, \Omega_2\}$ from (12.115) in the sense of Definition 5.5.1. Indeed, the condition $(\overline{u}, \overline{v}) \in \Omega_1 \cap \Omega_2$ is obvious. Furthermore, by (5.37) we have to check the existence of a sequence $\{a_i\} \subset X \times Y$ such that $a_i \to 0$ as $i \to +\infty$ and

$$\Omega_1 \cap (\Omega_2 + a_i) = \emptyset \quad \text{for all} \quad i \in \mathbb{N}.$$

In order to do this, we show

$$\Omega_1 \cap (\Omega_2 + (0, -i^{-1}k^0)) = \emptyset \quad \text{for all} \quad i \in \mathbb{N}^*, \tag{12.116}$$

i.e., (5.37) holds with $a_i = (0, -i^{-1}k^0)$. By the contrary, suppose that (12.116) does not hold for some fixed $i \in \mathbb{N}^*$. Then by the constructions of Ω_1 and Ω_2 in (12.115) there is (x, v) satisfying the conditions

$$v = g(x) - i^{-1}k^0 \quad \text{and} \quad (x, v) \in \operatorname{epi} F. \tag{12.117}$$

Taking into account the definition of the epigraph of the set-valued map $F : X \rightrightarrows Y$, we find $y \in F(x)$ and $c \in C$ with $v = y + c$. We get with $v = y + c$ and (12.117)

$$y = v - c = g(x) - i^{-1}k^0 - c \in g(x) - C,$$

because of $-k^0 \in -C$ and $-c \in -C$. This, together with condition (12.113) and the definition of g in (12.114), implies that $(x, y) = (\overline{u}, \overline{v})$. From $(x, y) = (\overline{u}, \overline{v})$, (12.114), (12.117), and $v = y + c$ we get that

$$y = \overline{v} = g(\overline{u}) = g(x) = v + i^{-1}k^0 = y + c + i^{-1}k^0,$$

and so $c + i^{-1}k^0 = 0$. This yields $k^0 = -ic \in -C$ in contradiction to the choice of $k^0 \in C \setminus (-C)$. This contradiction implies that (12.116) holds and thus $\{\Omega_1, \Omega_2, (\overline{u}, \overline{v})\}$ is an extremal system in the sense of Definition 5.5.1.

Now, we can apply the *extremal principle* (Theorem 5.5.9) to the system $\{\Omega_1, \Omega_2, (\overline{u}, \overline{v})\}$ imposing the sum norm $||(x, y)|| := ||x|| + ||y||$ on $X \times Y$ that generates the dual norm

$$||(x^*, y^*)||_* = \max\{||x^*||_*, ||y^*||_*\} \quad \text{for} \quad (x^*, y^*) \in X^* \times Y^*$$

on $X^* \times Y^*$. Proceeding in this way for any $\rho > 0$, we find elements $(x_j, y_j, x_j^*, y_j^*) \in X \times Y \times X^* \times Y^*$ with $j = 1, 2$ satisfying the relations

$$\begin{aligned} &(x_j, y_j) \in \Omega_j, \quad ||x_j - \overline{u}|| + ||y_j - \overline{v}|| \le \rho, \\ &(x_j^*, -y_j^*) \in \hat{N}((x_j, y_j); \Omega_j), \quad j = 1, 2, \end{aligned} \tag{12.118}$$

$$\begin{aligned} &\frac{1}{2} - \rho \le \max\{||x_j^*||_*, ||y_j^*||_*\} \le \frac{1}{2} + \rho, \; j = 1, 2, \\ &\max\{||x_1^* + x_2^*||_*, ||y_1^* + y_2^*||_*\} \le \rho. \end{aligned} \tag{12.119}$$

From the Lipschitz continuity of the mapping g in (12.114) with constant $L = \tilde{\varepsilon}/\tilde{\lambda}$ and the coderivative estimate from [430, Theorem 1.43] it follows that

$$||x_2^*||_* \le \frac{\tilde{\varepsilon}}{\tilde{\lambda}}||y_2^*||_* \quad \text{and hence} \quad y_2^* \ne 0$$

by (12.118), (12.119) with $\rho > 0$ sufficiently small. The latter yields by Theorem 5.5.9 that

$$||y_1^*||_* \ne 0 \quad \text{and} \quad \frac{||x_1^*||_*}{||y_1^*||_*} < \frac{\varepsilon}{\lambda}, \tag{12.120}$$

see more details in the similar setting of [25, Theorem 3.5]. Moreover, from the third condition in (12.118) with $j = 1$ we find $\tilde{y}_1 \in F(x_1)$ with

$$(x_1, \tilde{y}_1) \in \operatorname{graph} F, \quad (x_1^*, -y_1^*) \in \hat{N}((x_1, \tilde{y}_1); \operatorname{epi} F), \quad \text{and} - y_1^* \in \hat{N}(0; C). \tag{12.121}$$

Finally, denoting $(\overline{x}, \overline{y}) := (x_1, \tilde{y}_1)$, $x^* := x_1^*/||y_1^*||_*$, and $y^* := y_1^*/||y_1^*||_*$ and taking into account the definition of the Fréchet subdifferential (see (3.32)), we get the desired subdifferential condition (12.110) from the relations in (12.120) and (12.121). The condition $||\overline{x} - x^0|| < \lambda$ in the assertions of the theorem follows from the second condition in (12.112), condition (12.118) for $j = 1$, and the choice of $\tilde{\lambda}$ in (12.111). The proof is completed. □

Remark 12.9.2. In Theorem 12.9.1 it is supposed that $||k^0|| = 1$. If we don't assume $||k^0|| = 1$, then condition (12.110) can be replaced by the modified subdifferential condition

$$\hat{\partial} F(\overline{x}, \overline{y}) \cap \frac{\varepsilon}{\lambda} ||k^0|| B_{X^*} \neq \emptyset,$$

if $||k^0||$ is selected arbitrarily from $C \setminus (-C)$, with no change in the proof.

Remark 12.9.3. The proof of this theorem, based on the extremal principle (Theorem 5.5.9) and the variational principle in [27, Theorem 3.5] (see Corollary 10.2.12) is similar to that one given in [25, Theorem 2]. In [25] the reader can find more details.

12.10 Necessary and Sufficient Conditions for Solution Concepts Based on Set Approach

In this section we derive necessary and sufficient conditions for solutions described by the set approach using a directional derivative of the set-valued map. We assume that Y is a n.v.s., C is a proper closed convex pointed cone in Y with $\operatorname{int} C \neq \emptyset$, $\operatorname{int} C^+ \neq \emptyset$ and we use the notations introduced in Sect. 2.6.4. Especially, let $\mathcal{G}$ be the family of all nonempty C-convex and C^+-bounded subsets of Y, $\mathcal{V} = \{[A, B] | (A, B) \in \mathcal{G}^2\}$.

Moreover, we assume that S is a convex subset of a normed space $(X, \| \cdot \|)$. Consider $c \in \operatorname{int} C$ and a weak* compact base $W := \{y^* \in C^+ \mid \langle y^*, c\rangle = 1\}$ of C^+, $\mathcal{V} = \mathcal{V}(W)$, that is $|[A, B]| < +\infty$ when $[A, B] \in \mathcal{V}$, and $F : S \rightrightarrows \mathcal{G}$.

In this section we are dealing with weakly minimal solutions defined using the following binary relation $\prec_C^l$ on $\mathcal{G}$: For $A, B \in \mathcal{G}$,

$$A \prec_C^l B \quad :\Longleftrightarrow \quad \exists V \subset Y \text{ such that } A + C \supseteq B + V, \tag{12.122}$$

where V is a neighborhood of 0. Furthermore, we discuss minimal solutions based on the set relation introduced in (2.64)

$$A \preceq_C^{cl} B \quad :\Longleftrightarrow \quad \operatorname{cl}(A + C) \supset B, \tag{12.123}$$

where $A, B \in \mathcal{G}$.

We consider the set-valued problem

$$\preceq_C \text{-minimize} \quad F(x), \quad \text{subject to} \quad x \in S, \qquad (SP - \preceq_C)$$

where the set order relation $\preceq_C$ is given by (12.122) or by (12.123).
The results presented in the following are shown by Kuroiwa [354]. An important and interesting discussion of this approach is given in a recent paper by J. Jahn: Directional derivatives in set optimization with the set less order relation (2014), page 8.

The derivatives introduced in the following definition are used for deriving the optimality conditions for solutions of $(SP - \preceq_C)$.

Definition 12.10.1. Let $x \in S$ and $d \in X$.

$$CF(x,d) := \{[A, B] \in \mathscr{V} \mid \exists\{\lambda_k\} \longrightarrow +0 \text{ s.t. } \{\frac{1}{\lambda_k}[F(x + \lambda_k d), F(x)]\} \longrightarrow [A, B]\}$$

is said to be **$\mathscr{V}$-directional derivative cluster of F at x in the direction d** . If $CF(x,d)$ is a singleton, then the element is written by $DF(x,d)$ and called **$\mathscr{V}$-directional derivative** of F at x in the direction d , and F is said to be **$\mathscr{V}$-directional differentiable** at x in the direction d .

Example 12.10.2. Let $F : \mathbb{R} \rightrightarrows \mathbb{R}^2$ be a set-valued map defined for all $x \in \mathbb{R}$ by

$$F(x) = \operatorname{conv}\{(|x|, -|x| + 1), (-|x| + 1, |x|)\},$$

and let $C = \mathbb{R}^2_+ = \{(x_1, x_2) | x_1, x_2 \geq 0\}$. Then F is $\mathscr{V}$-directional differentiable in any direction $d \in \mathbb{R}$, for example, if $x_0 = 0$:

$$DF(x_0, d) = [\{(0,0)\},\ |d| \operatorname{conv}\{(1,-1),(-1,1)\}] \quad (d \in \mathbb{R}),$$

if $0 < x_0 < \frac{1}{2}$:

$$DF(x_0, d) = \begin{cases} [\{(0,0)\}, |d| \operatorname{conv}\{(1,-1),(-1,1)\}] & \text{if } d \geq 0, \\ [|d| \operatorname{conv}\{(1,-1),(-1,1)\}, \{(0,0)\}] & \text{if } d < 0, \end{cases}$$

and if $x_0 = \frac{1}{2}$:

$$DF(x_0, d) = [|d| \operatorname{conv}\{(1,-1),(-1,1)\}, \{(0,0)\}] \quad (d \in \mathbb{R}).$$

Example 12.10.3. Consider a set-valued map $F : S \rightrightarrows Y$ defined by

$$F(x) := g(x) + \sum_{i \in I} r_i(x) A_i, \quad x \in S,$$

where g is a function from S to Y which is directional differentiable at $x_0 \in S$, I is a nonempty finite set, and for each $i \in I$, $r_i : S \to (0, +\infty)$ is a function which is directional differentiable at $x_0 \in S$, and $A_i \in \mathscr{G}$. Then F is $\mathscr{V}$-directional differentiable at x_0 for each direction $d \in X$, and it holds

$$DF(x_0, d) = \left[g'(x_0, d), \{0\}\right] + \sum_{i \in I} r_i'(x_0, d)\,[A_i, \{0\}]$$

$$= \left[g'(x_0, d) + \sum_{i \in I_+(d)} r_i'(x_0, d)A_i,\ -\sum_{i \in I_-(d)} r_i'(x_0, d)A_i\right]$$

where $I_+(d) = \{i \in I \mid r_i'(x_0, d) > 0\}$ and $I_-(d) = \{i \in I \mid r_i'(x_0, d) < 0\}$.

We present necessary and sufficient optimality conditions for solutions of $(SP - \preceq_C)$ using the derivatives introduced in Definition 12.10.1 and the cone $\mu(C) = \{[A, B] \in \mathscr{V} | B \preceq_C^{cl} A\}$ (see (2.65)), where $\preceq_C^{cl}$ is defined by (12.123).

The following implication is given in [354, Proposition 1].

Proposition 12.10.4. *For any $A, B \in \mathscr{G}$ it holds*

$$[A, B] \in \operatorname{Int} \mu(C) \Longrightarrow B \prec_C^l A,$$

where $\operatorname{Int} \mu(C)$ *is the set of all interior points w.r.t.* $(\mathscr{V}, |\cdot|)$.

Definition 12.10.5. An element $x_0 \in S$ is said to be a **weakly minimal solution** of $(SP - \preceq_C)$ if

$$\nexists x \in S \quad \text{s.t.} \quad F(x) \prec_C^l F(x_0).$$

Furthermore, we define local solutions of $(SP - \preceq_C)$ using the order relation $\preceq_C^{cl}$ given by (12.123).

Definition 12.10.6. An element $x_0 \in S$ is said to be a **local minimal solution** of $(SP - \preceq_C)$ if there exists a neighborhood U of x_0 such that

$$x \in U \cap S, \quad F(x) \preceq_C^{cl} F(x_0) \Longrightarrow F(x_0) \preceq_C^{cl} F(x);$$

and $x_0 \in S$ is called a **local weakly minimal solution of** $(SP - \preceq_C)$ if there exists a neighborhood U of x_0 such that

$$\nexists x \in U \cap S \quad \text{s.t.} \quad F(x) \prec_C^l F(x_0).$$

The following necessary condition of local weakly minimal solutions of $(SP - \preceq_C)$ is given in Kuroiwa [354, Theorem 1].

Theorem 12.10.7. *If x_0 is a local weakly minimal solution of $(SP - \preceq_C)$, then for all $x \in S$ it holds*

$$CF(x_0, x - x_0) \cap (-\operatorname{Int} \mu(C)) = \emptyset.$$

Furthermore, the following sufficient condition for local minimality concerning $(SP - \preceq_C)$ is presented in Kuroiwa [354, Theorem 2].

Theorem 12.10.8. *Assume that X is a finite dimensional space, and F is $\mathscr{V}$-directional differentiable at $x_0 \in S$ in each direction $d \in X$. Moreover, suppose that $DF(x_0, d) = \lim_{\lambda \to +0} \frac{1}{\lambda}[F(x_0 \in S + \lambda d), F(x_0)]$ converges uniformly and continuous with respect to $d \in X$ on the unit ball. If for all $d \in T(S, x_0) \setminus \{0\}$*

$$DF(x_0, d) \notin -\mu(C),$$

then x_0 is a local minimal solution of $(SP - \preceq_C)$.

Example 12.10.9. We discuss Example 12.10.2, in view of the optimality conditions given above.

At first, it is easy to check that there are no (global) minimal, and no (global) weakly minimal solutions of $(SP - \preceq_C)$, 0 is the only local minimal solution of $(SP - \preceq_C)$, and for each $x \in \mathbb{R}$, x is a local weakly minimal solution of $(SP - \preceq_C)$.

If $x_0 = 0$, $DF(x_0, d) \in -\mu(C)$ holds if and only if

$$\mathbb{R}^2_+ \supset |d| \operatorname{conv}\{(1, -1), (-1, 1)\},$$

but it does not hold when $d \neq 0$. From Theorem 12.10.8, we get that x_0 is a local minimal solution of $(SP - \preceq_C)$.

Furthermore, if $0 < x_0 < \frac{1}{2}$, $DF(x_0, d) \in -\mu(C)$ holds if and only if

$$\begin{cases} \mathbb{R}^2_+ \supset |d| \operatorname{conv}\{(1, -1), (-1, 1)\} & \text{if } \ d \geq 0, \\ \mathbb{R}^2_+ + |d| \operatorname{conv}\{(1, -1), (-1, 1)\} \ni (0, 0) & \text{if } \ d < 0 \end{cases}$$

and

$$\mathbb{R}^2_+ + |d| \operatorname{conv}\{(1, -1), (-1, 1)\} \ni (0, 0)$$

is always true when $d < 0$. This is consistent with Theorem 12.10.8, since x_0 is not a local minimal solution.

Moreover, $DF(x_0, d) \in -\operatorname{Int}\mu(C)$ holds if and only if

$$\begin{cases} (0, +\infty)^2 \supset |d| \operatorname{conv}\{(1, -1), (-1, 1)\} & \text{if } \ d \geq 0 \\ (0, +\infty)^2 + |d| \operatorname{conv}\{(1, -1), (-1, 1)\} \ni (0, 0) & \text{if } \ d < 0 \end{cases}$$

does not hold for each $d \in \mathbb{R}$. This is also consistent with Theorem 12.10.7, since x_0 is a local weakly minimal solution of $(SP - \preceq_C)$.

12.11 Necessary Conditions for Solution Concepts with Respect to a General Preference Relation

In this section we present necessary conditions for solutions introduced in Sect. 2.6.5 that are related to applications in welfare economics (see Sect. 15.3). The main tool for the proof of the necessary conditions is the extended extremal principle in Theorem 5.5.10.

We derive the necessary optimality conditions for set-valued optimization problems with geometric constraints

$$\text{minimize} \quad F(x) \quad \text{subject to} \quad x \in S, \tag{SP}$$

where X and Y are Banach spaces, S is a subset of X and the cost mapping $F : S \rightrightarrows Y$ is a set-valued mapping.

The "minimization" in (SP) is understood with respect to some preference relation on Y (see Sect. 2.6.5). We consider a general (abstract) preference relation $\prec$ on Y given by a nonempty subset $\mathscr{R} \subset Y \times Y$ (y^1 is preferred to y^2 ($y^1 \prec y^2$) if $(y^1, y^2) \in \mathscr{R}$) and the **level-set mapping** $L : Y \rightrightarrows Y$ associated with the preference relation $\prec$:

$$L(y) := \{u \in Y \mid u \prec y\}. \tag{12.124}$$

To derive necessary conditions for fully localized minimizers of the set-valued optimization problem (SP) (Definition 2.6.49), Bao and Mordukhovich [28] developed a variational approach based on the extremal principle (Theorem 5.5.10) and generalized differential calculus (see Sect. 3.5, compare [430]).

In order to prove first order necessary conditions for localized minimizers in Theorem 12.11.3 the following asymptotic closedness property is supposed concerning the level-set mapping L.

Definition 12.11.1 (Asymptotic Closedness Property). Let $A \subset Y$ be a subset in a Banach space Y, and let $\overline{y} \in \operatorname{cl} A$. We say that A is **asymptotically closed** at $\overline{y}$ if there is a neighborhood V of $\overline{y}$ and for any $\varepsilon > 0$ there is $c \in \varepsilon B_Y$ such that

$$(\operatorname{cl} A + c) \cap V \subset A \setminus \{\overline{y}\}.$$

This asymptotic closedness property is not very restrictive and fulfilled for many models in welfare economics (compare [28, Section 4]). This is illustrated by the following examples given by Bao and Mordukhovich [28].

Example 12.11.2.

- Each proper closed convex pointed cone $C \subset Y$ with $C \setminus (-C) \neq \emptyset$ has the asymptotic closedness property at the origin.
- The asymptotic closedness property at the origin is satisfied for each proper convex cone $C \subset Y$ with nonempty interior and for its nonconvex complement $Y \setminus C$.

- The epigraph of a functional $\varphi : X \to \mathbb{R} \cup \{+\infty\}$ that is lower semicontinuous around $\overline{x} \in X$ has the asymptotic closedness property at $(\overline{x}, \varphi(\overline{x}))$.

The following necessary optimality condition for all the types of fully localized minimizers of the set-valued optimization problem (SP) in the sense of Definition 2.6.49 is shown in [28, Theorem 3.4] using the extended extremal principle in Theorem 5.5.10 and will be applied in Sect. 15.3 to set-valued problems in welfare economics. The *SNC assumptions* in the following theorem are introduced in Sect. 5.5.

Theorem 12.11.3 (First Order Necessary Conditions for Localized Minimizers of (SP)**).** *Let X and Y be Asplund spaces and $F : X \rightrightarrows Y$ be a set-valued mapping. Assume that* graph F *is closed around some point* $(\overline{x}, \overline{y}) \in \operatorname{graph} F$, *let $S \subset X$ be closed around $\overline{x}$, and let $L : Y \rightrightarrows Y$ be a preference mapping on Y locally satiated at $\overline{y}$. Furthermore, suppose that one of the following* ***SNC assumptions*** *on the initial data (F, S, L) is fulfilled:*

(a) graph F *is SNC at* $(\overline{x}, \overline{y})$;
(b) *S is SNC at $\overline{x}$ and* $\operatorname{cl} L(\overline{y})$ *is SNC at* $\overline{y}$;
(c) *F is PSNC at $(\overline{x}, \overline{y})$ and* $\operatorname{cl} L(\overline{y})$ *is SNC at* $\overline{y}$;
(d) *S is SNC at $\overline{x}$ and F^{-1} is PSNC at $(\overline{y}, \overline{x})$.*

Then there exists a pair $(0, 0) \neq (x^*, y^*) \in X^* \times Y^*$ satisfying the necessary optimality conditions

$$x^* \in D^* F(\overline{x}, \overline{y})(y^*) \cap (-N(\overline{x}; S)) \quad \textit{and} \quad y^* \in N(\overline{y}; \operatorname{cl} L(\overline{y})) \tag{12.125}$$

in each of the following cases:

- *$(\overline{x}, \overline{y})$ is a fully localized weak minimizer for (SP) provided that the set $L(\overline{y})$ is asymptotically closed at $\overline{y} \in \operatorname{cl} L(\overline{y})$;*
- *$(\overline{x}, \overline{y})$ is a fully localized minimizer for (SP) provided that the set* $\operatorname{cl} L(\overline{y})$ *is asymptotically closed at $\overline{y}$;*
- *$(\overline{x}, \overline{y})$ is a fully localized strong minimizer for (SP) provided that either* $\operatorname{cl} L(\overline{y})$ *or S is asymptotically closed at $\overline{y}$ or $\overline{x}$, respectively.*

Proof. The main tool for the proof is the application of the extended extremal principle in Theorem 5.5.10. Therefore, it is to check that a fully localized minimizer $(\overline{x}, \overline{y})$ is a local extremal point of the set system $\{\Omega_1, \Omega_2\}$ (in the Asplund space $X \times Y$) with

$$\Omega_1 := \operatorname{graph} F \quad \text{and} \quad \Omega_2 := S \times \operatorname{cl} L(\overline{y})$$

(see [28, Theorem 3.4]). □

12.12 KKT-Points and Corresponding Stability Results

The aim of this section is to study some set-valued optimization problems from the point of view of stability of *KKT* points. All of the results presented in this section are derived by Durea, Dutta, Tammer in [152].

Let us consider the set-valued map $F : X \rightrightarrows Y$, $G : X \rightrightarrows Z$, where X, Y and Z are finite dimensional normed vector spaces. Furthermore, $C \subset Y$ and $K \subset Z$ are proper closed convex pointed cones.

In this section, we study the following set-valued optimization problems where we use the vector approach as solution concept, this means that the "minimization" is to understand in the sense of Definition 2.6.1. Consider the problem

$$\text{minimize } F(x), \text{ subject to } \quad x \in S \tag{SP}$$

where the feasible set S is given by

$$S := \{x \in X \mid 0 \in G(x) + K\}. \tag{12.126}$$

In fact, we shall also consider a simplified version of problem (SP) where G is not a set-valued map but a single-valued one denoted by g. Then the problem above is represented as

$$\text{minimize } F(x), \text{ subject to } x \in S', \tag{SP'}$$

where the feasible set is given by

$$S' := \{x \in X \mid g(x) \in -K\}. \tag{12.127}$$

The main tools we use in this section are the generalized differentiability concepts introduced by Mordukhovich (see [430, Chapters 1, 3] and Definitions 3.5.6, 3.5.8). It is well known that these objects enjoy rich calculus in Asplund spaces. However, as we already said, we work here on finite dimensional vector spaces.

A set-valued mapping H acting between two finite dimensional normed vector spaces X and Y is said to be inner-semicompact at $\overline{x} \in X$ if for every sequence $x_k \to \overline{x}$ there is a sequence $y_k \in H(x_k)$ which has a convergent subsequence. For example if H is uniformly bounded at $\overline{x}$ then H is inner-semicompact at $\overline{x}$.

We start by considering the problem (SP) assuming that F is Lipschitz-like at $(\overline{x}, \overline{y})$ in the sense of Definition 3.3.9. The following theorem is shown in [150, Theorem 4.5].

Theorem 12.12.1. *Consider the problem* (SP) *and let* $(\overline{x},\overline{y}) \in \operatorname{graph} F$ *be a minimizer of* (SP) *in the sense of Definition 2.6.1. Assume that* F *is Lipschitz-like around* $(\overline{x},\overline{y})$ *and* $G(\overline{x})$ *is compact. Further assume that the mapping* $x \mapsto G(x) \cap (-K)$ *is inner-semicompact at* $\overline{x}$*. Moreover, we assume to be true the following qualification condition: For any* $z \in G(\overline{x}) \cap (-K)$ *one has*

$$N(z;-K) \cap \ker D^*G(\overline{x},z) = \{0\}. \tag{12.128}$$

Then there exist $y^* \in C^+$ *with* $||y^*||_* = 1$ *such that the set*

$$L^\circ_{y^*} = \{z^* \in Z^* \mid \exists \overline{z} \in G(\overline{x}) \cap (-K), \quad \text{with} \quad z^* \in N(\overline{z};-K) \quad \text{such that}$$
$$0 \in D^*F(\overline{x},\overline{y})(y^*) + D^*G(\overline{x},\overline{z})(z^*)\}. \tag{12.129}$$

is nonempty and bounded.

Now, we introduce a notion of approximate *KKT* points in this set-valued setting. We emphasize the fact that, as always in the case of generalized differentiation objects, the main motivation for their study is related to the nice theoretical properties. For example, in many cases the set-valued maps can actually be single-valued in the neighborhood of a point where it is single-valued. This is indeed true for the subdifferential of a cone-convex vector-valued function. So, from the computational point of view, in the minimizing sequence of $\varepsilon - KKT$ points, we may just get points where the subdifferential is actually single-valued.

Definition 12.12.2. Let $\varepsilon \geq 0$. A feasible point $(\overline{x},\overline{y}) \in \operatorname{graph} F$ of problem (SP) is said to be an $\varepsilon-$KKT point for (SP) if there exist $y^* \in C^+$, $||y^*||_* = 1$, $z^* \in Z$, $\overline{z} \in G(\overline{x}) \cap -K$ with $z^* \in N(\overline{z},-K)$ and $x^* \in X^*$, $\|x^*\|_* \leq \varepsilon$ s.t.

$$x^* \in D^*F(\overline{x},\overline{y})(y^*) + D^*G(\overline{x},\overline{z})(z^*).$$

Denote by T° the set-valued map which assigns to every $\varepsilon \in \mathbb{R}_+$ the subset of $X \times Y$ of all $\varepsilon - KKT$ points of (SP). In this nonsmooth set-valued case the stability result we have is the following. When $\varepsilon = 0$ in Definition 12.12.2, $(\overline{x},\overline{y})$ is called a *KKT* point for (SP).

Theorem 12.12.3. *Suppose that* F *is Lipschitz-like around* $(\overline{x},\overline{y}) \in \operatorname{graph} F$*,* G *has closed graph, there exists* $r > 0$ *such that* $G(B(\overline{x},r))$ *is bounded and for any* $z \in G(\overline{x}) \cap -K$

$$N(z;-K) \cap \ker D^*G(\overline{x},z) = \{0\}.$$

If $\varepsilon \geq 0$ *and* $(\overline{x},\overline{y}) \in \operatorname{Limsup}_{\mu\to\varepsilon} T^\circ(\mu)$*, then* $(\overline{x},\overline{y}) \in T^\circ(\varepsilon)$*.*

Proof. Let $\varepsilon \in \mathbb{R}_+$ and take $(\overline{x}, \overline{y}) \in \operatorname{Limsup}_{\mu \to \varepsilon} T^\circ(\mu)$. Thus, there exist some sequences $\{\mu_n\} \to \varepsilon$, $\{(x_n, y_n)\} \subset \operatorname{graph} F$, $\{(x_n, y_n)\} \to (\overline{x}, \overline{y})$, $\{y_n^*\} \subset C^+$, $\|y_n^*\|_* = 1$, $\{z_n^*\} \subset Z^*$, $z_n \in G(x_n) \cap -K$ with $z_n^* \in N(z_n; -K)$, $\{x_n^*\} \subset X^*$, $\|x_n^*\|_* \leq \mu_n$ s.t.

$$x_n^* \in D^* F(x_n, y_n)(y_n^*) + D^* G(x_n, z_n)(z_n^*). \tag{12.130}$$

First, we show that $\{z_n^*\}$ is bounded. Suppose that this is not the case. There exists a subsequence (denoted $\{z_n^*\}$ as well) s.t. $\|z_n^*\|_* \to +\infty$. It is clear that $z_n^* \neq 0$ for n large enough, whence, by division with $\|z_n^*\|$, one gets

$$\|z_n^*\|_*^{-1} x_n^* \in D^* F(x_n, y_n)(\|z_n^*\|_*^{-1} y_n^*) + D^* G(x_n, z_n)(\|z_n^*\|_*^{-1} z_n^*),$$

i.e., there exist $\{u_n^*\} \subset X^*$ s.t.

$$u_n^* \in D^* F(x_n, y_n)(\|z_n^*\|_*^{-1} y_n^*)$$
$$-u_n^* + \|z_n^*\|_*^{-1} x_n^* \in D^* G(x_n, z_n)(\|z_n^*\|_*^{-1} z_n^*).$$

Note that in finite dimensions the graph of the set-valued mapping $N(\cdot; -K)$ is closed (see [499]). Now we employ the hypotheses in order to deduce, by moving to subsequences if needed, that $\|z_n^*\|_*^{-1} y_n^* \to 0$, $\|z_n^*\|_*^{-1} x_n^* \to 0$, $u_n^* \to u^* \in X^*$ (since F is Lipschitz-like around $(\overline{x}, \overline{y})$, $\{u_n^*\}$ is bounded, see [430, Theorem 1.43]) and $z_n \to z \in G(\overline{x}) \cap -K$ (since $x_n \subset B(\overline{x}, r)$ for n large enough), $\|z_n^*\|_*^{-1} z_n^* \to z^* \in N(z; -K) \setminus \{0\}$, and

$$u^* \in D^* F(\overline{x}, \overline{y})(0)$$
$$-u^* \in D^* G(\overline{x}, z)(z^*).$$

Therefore,

$$0 \in D^* F(\overline{x}, \overline{y})(0) + D^* G(\overline{x}, z)(z^*).$$

One uses again the fact that F is Lipschitz-like around $(\overline{x}, \overline{y})$ to observe that $D^* F(\overline{x}, \overline{y})(0) = \{0\}$, whence

$$0 \in D^* G(\overline{x}, z)(z^*)$$

in contradiction with the qualification condition we have imposed. Therefore, for the second step of the proof, we know that in relation (12.130) the sequence $\{z_n^*\}$ is bounded so it is convergent (on a subsequence, if necessary) towards an element $z^* \in N(z; -K)$ for an $z \in G(\overline{x}) \cap -K$. Similarly, $\{y_n^*\}$ is convergent towards an element y^* of norm 1 from C^+. Moreover $\{x_n^*\}$ is bounded too, whence it is convergent to some $x^* \in X^*$ with $\|x^*\|_* \leq \varepsilon$. Passing to the limit (as above) in (12.130) we get

$$x^* \in D^*F(\overline{x}, \overline{y})(y^*) + D^*G(\overline{x}, z)(z^*)$$

i.e., $(\overline{x}, \overline{y}) \in T^\circ(\varepsilon)$. The proof is complete. □

We have the following consequence.

Corollary 12.12.4. *Suppose that* $\{(x_n, y_n)\} \subset \operatorname{graph} F$ *are* $\varepsilon_n - KKT$ *points for* (SP) *where* $\{\varepsilon_n\} \to 0$ *s.t.* $\{\varepsilon_n\} \subset \mathbb{R}_+$ *and* $(\overline{x}, \overline{y}) \in \operatorname{graph} F$ *s.t.* $\{(x_n, y_n)\} \to (\overline{x}, \overline{y})$. *Suppose that* F *is Lipschitz-like around* $(\overline{x}, \overline{y}) \in \operatorname{graph} F$, G *has closed graph,* $\{G(x_n)\}$ *is uniformly bounded and for any* $z \in G(\overline{x}) \cap -K$

$$N(z; -K) \cap \ker D^*G(\overline{x}, z) = \{0\}.$$

Then $(\overline{x}, \overline{y})$ *is a KKT point for* (SP).

We turn now our attention to the problem (SP'). If we adapt [150, Theorem 4.4] to the case of finite dimensional spaces we get the following result.

Theorem 12.12.5. *Let* $(\overline{x}, \overline{y})$ *be a minimizer of* (SP') *in the sense of Definition 2.6.1. Suppose that* F *is Lipschitz-like around* $(\overline{x}, \overline{y})$ *and* g *is locally Lipschitz at* $\overline{x}$. *If*

$$z^* \in K^+, 0 \in D^*g(\overline{x})(z^*) \Rightarrow z^* = 0, \tag{12.131}$$

then there exists $y^* \in C^+$, $\|y^*\|_* = 1$ *s. t.*

$$L_{y^*}^{\circ\prime} := \{z^* \in K^+ \mid 0 \in D^*F(\overline{x}, \overline{y})(y^*) + D^*g(\overline{x})(z^*)\}. \tag{12.132}$$

is non-empty and bounded.

Definition 12.12.6. Let $\varepsilon \geq 0$. A feasible point $(\overline{x}, \overline{y}) \in \operatorname{graph} F$ of problem (SP') is said to be an $\varepsilon - KKT$ point for (SP') if there exists $y^* \in C^+, ||y^*||_* = 1$, $z^* \in K^+$ and $x^* \in X^*$, $\|x^*\|_* \leq \varepsilon$ s.t.

$$x^* \in D^*F(\overline{x}, \overline{y})(y^*) + D^*g(\overline{x})(z^*).$$

Denote by $T^{\circ\prime}$ the set-valued map which assigns to every $\varepsilon \in \mathbb{R}_+$ the subset of $X \times Y$ of all $\varepsilon - KKT$ points of (SP'). If $\varepsilon = 0$ in Definition 12.12.6, then $(\overline{x}, \overline{y})$ is called a *KKT* point for (SP'). Our stability results for this problem uses the same hypotheses as above. The proof is also on similar lines and therefore we omit it.

Theorem 12.12.7. *Suppose that F is Lipschitz-like around $(\overline{x}, \overline{y}) \in \operatorname{graph} F$, and g is locally Lipschitz at $\overline{x}$ and the following implication holds:*

$$z^* \in K^+,\ 0 \in D^*g(\overline{x})(z^*) \Rightarrow z^* = 0.$$

If $\varepsilon \geq 0$ and $(\overline{x}, \overline{y}) \in \operatorname{Limsup}_{\mu \to \varepsilon} T^{\circ\prime}(\mu)$, then $(\overline{x}, \overline{y}) \in T^{\circ\prime}(\varepsilon)$.

Again, we record a corollary.

Corollary 12.12.8. *Suppose that $(x_n, y_n) \subset \operatorname{graph} F$ are ε_n-KKT points for (SP') where $(\varepsilon_n) \to 0$ s.t. $(\varepsilon_n) \subset \mathbb{R}_+$ and $(\overline{x}, \overline{y}) \in \operatorname{graph} F$ s.t. $(x_n, y_n) \to (\overline{x}, \overline{y})$. Suppose that F is Lipschitz-like around $(\overline{x}, \overline{y}) \in \operatorname{graph} F$, and g is locally Lipschitz at $\overline{x}$ and the following implication holds:*

$$z^* \in K^+,\ 0 \in D^*g(\overline{x})(z^*) \Rightarrow z^* = 0.$$

Then $(\overline{x}, \overline{y})$ is a KKT point for (SP').

Chapter 13
Sensitivity Analysis in Set-Valued Optimization and Vector Variational Inequalities

Sensitivity analysis, the quantitative analysis of the perturbation map, is of paramount interest in optimization theory and has applications in several branches of pure and applied mathematics. During the last five decades, substantial progress has been made in sensitivity analysis for optimization problems with scalar objectives. On the other hand, the differentiability issues of the perturbation map for vector optimization problems and set optimization problems are rather involved and they require modern tools from variational analysis. The main difficulty here stems from the fact that the perturbation map for such problems is, in general, set-valued.

To fix these ideas, we begin by describing the general framework for such problems. Let X, Y, and Z be normed spaces, and let $C \subset Y$ be a nonempty cone in Y. Let $H : X \rightrightarrows Z$ and $G : X \times Z \rightrightarrows Y$ be set-valued maps. We consider the following parametrized family of set-valued optimization problems defined by

$$P(x) = \text{Min}(F(x), C), \tag{13.1}$$

where

$$F(x) := \{y \in G(x, z) |\, z \in H(x)\}. \tag{13.2}$$

Here we interpret X as the parameter space, Z as the decision space, and Y as the objective space, with the cone C inducing a partial ordering on objective values in Y. For each value of the parameter x, the set $H(x)$ defines a feasible set in Z, whereas the set $G(x, z)$ gives a set of objectives in the space Y. The set $F(x)$ comprises the graph of the objective values over the entire feasible region of the optimization problem. In this setting, $P(x)$ singles out those objective values that cannot be improved relative to the ordering induced by C. Evidently, when G is single-valued, the above model then recovers the setting of parametric vector optimization. If additionally, $Y + \mathbb{R}$ and $C = \mathbb{R}_+$, then we are in the framework of scalar parametric optimization. The objective of this chapter is to employ derivatives

A.A. Khan et al., *Set-valued Optimization*, Vector Optimization,
DOI 10.1007/978-3-642-54265-7_13

of set-valued maps to estimate how sets $F(x)$ and $P(x)$ vary with changes in the parameter x. For this purpose, we will use graphical derivatives as well as the coderivatives of these set-valued maps. For this study, a crucial role is played by the derivatives of F, $F+C$, P, $P+C$ and their minimal points. In this chapter, we will also study differentiability properties of set-valued gap functions associated to vector variational inequalities and sensitivity analysis for vector variational inequalities. For set-valued optimization and for vector variational inequalities, both the first-order as well as the second-order results are given.

13.1 First Order Sensitivity Analysis in Set-Valued Optimization

Many authors have focused on sensitivity analysis using first-order contingent derivatives. For instance, Tanino [564, 565] derived many interesting relationships among the first-order contingent derivative of the perturbation maps associated with some set-valued maps. Since then, several authors have investigated the differentiability of set-valued perturbation maps (see [99, 344, 345, 530, 531, 547] and the cited references therein). Interesting results on sensitivity analysis for non-smooth optimization problems are available in [462, 594], among others. A very comprehensive treatment of sensitivity analysis based on graphical derivatives can be found in [214, Section 3.6] and which is the main source for our presentation in this section.

Throughout this section, X and Y are normed spaces, and $\{0\} \neq C \subset Y$ is a pointed, closed, and convex cone. When dealing with weakly minimal points of a set, we also assume that $\operatorname{int}(C) \neq \emptyset$. Let $F : X \rightrightarrows Y$ be a given set-valued map. We are interested in the assessment of the differentiability properties of the set-valued perturbation maps P, Q, and R, defined for $x \in X$, by

$$P(x) = \text{GHe-PMin}(F(x), C), \tag{13.3a}$$

$$Q(x) = \text{Min}(F(x), C), \tag{13.3b}$$

$$R(x) = \text{WMin}(F(x), C), \tag{13.3c}$$

where $\text{GHe-PMin}(F(x), C)$ is the set of global Henig-proper minimal points of $F(x)$ with respect to C, $\text{Min}(F(x), C)$ is the set of minimal points of $F(x)$ with respect to C, and $\text{WMin}(F(x), C)$ is the set of weakly minimal points of $F(x)$ with respect to C.

The perturbation maps P, Q, and R will be termed as the proper perturbation map, the perturbation map, and the weak perturbation map, respectively. Our goal here is to investigate the relationships among the first-order contingent derivatives of the perturbation maps P, Q, and R, the first-order contingent derivative of the map F, and various minimal points of these derivatives.

We begin by recalling the following result, which is a restatement of Theorem 11.5.20 and which was proved under the assumption that the map F is contingently C-absorbing. In the following, the imposed conditions ensure that F is contingently C-absorbing.

Theorem 13.1.1. *Assume that* $\{y \in C \mid \|y\| = 1\}$ *is compact and*

$$D_s F(\bar{x}, \bar{y})(0) \cap (-C) = \{0\}, \tag{13.4}$$

where $D_s F(\bar{x}, \bar{y})$ *is the* S*-derivative of* F *at* $(\bar{x}, \bar{y})$*. Then, for every* $x \in X$*, we have*

$$\text{Min}\,(DF(\bar{x}, \bar{y})(x), C) = \text{Min}\,(D(F + C)(\bar{x}, \bar{y})(x), C)\,. \tag{13.5}$$

$$\text{GHe-PMin}\,(DF(\bar{x}, \bar{y})(x), C) = \text{GHe-PMin}\,(D(F + C)(\bar{x}, \bar{y})(x), C)\,. \tag{13.6}$$

Moreover, if $\text{int}C \neq \emptyset$*, then for every* $x \in X$*, we have*

$$\text{WMin}\,(DF(\bar{x}, \bar{y})(x), C) \subset \text{WMin}\,(D(F + C)(\bar{x}, \bar{y})(x), C)\,, \tag{13.7}$$

$$\text{WMin}\,(DF(\bar{x}, \bar{y})(x), C) = \text{WMin}\,\big(D(F + \tilde{C})(\bar{x}, \bar{y})(x), C\big)\,, \tag{13.8}$$

where $\tilde{C}$ *is a closed convex cone such that* $\tilde{C} \subset \{0\} \cup \text{int}(C)$.

Proof. See [214]. □

Remark 13.1.2. When X, Y are finite-dimensional, formulae (13.5), (13.6), and (13.8) are established in [345, Theorem 2.1], while formulae (13.6) and (13.8) are established in [344, Theorem 2.1] for $\bar{x} \in \text{int}(\text{dom}F)$ and F locally C-convex near $\bar{x}$.

Remark 13.1.3. It is easy to verify that in Example 11.3.4, (13.4) is not satisfied and (13.6), (13.7), (13.8) (for $\tilde{C} = C$) do not hold for $x > 0$.

We have the following result concerning the differentiability of P and Q.

Theorem 13.1.4. *Let* $(\bar{x}, \bar{y}) \in \text{graph}\,F$*. Assume that* $\{y \in C \mid \|y\| = 1\}$ *is compact and* F *is* C*-dominated by* Q *near* $\bar{x}$*. That is,* $F(x) \subset Q(x) + C$ *for every* $x \in V$ *for some neighborhood* V *of* $\bar{x}$*.*

(i) *If* $\bar{y} \in Q(\bar{x})$*, then for every* $x \in X$*,*

$$\text{Min}\,(D(F + C)(\bar{x}, \bar{y})(x), C) \subset \text{Min}\,(DQ(\bar{x}, \bar{y})(x), C) \subset DQ(\bar{x}, \bar{y})(x). \tag{13.9}$$

Moreover, if $F(x)$ *is convex for* $x \in V$*, with* V *a neighborhood of* $\bar{x}$*, then for every* $x \in X$*,*

$$\text{Min}\,(DF(\bar{x}, \bar{y})(x), C) \subset \text{Min}\,(DQ(\bar{x}, \bar{y})(x), C)\,. \tag{13.10}$$

(ii) *If* (13.4) *holds, then* $\bar{y} \in Q(\bar{x})$ *and for every* $x \in X$,

$$\text{Min}\,(DF(\bar{x},\bar{y})(x),C) = \text{Min}\,(DQ(\bar{x},\bar{y})(x),C) \subset DQ(\bar{x},\bar{y})(x), \tag{13.11}$$

$$\text{GHe-PMin}\,(DF(\bar{x},\bar{y})(x),C) = \text{GHe-PMin}\,(DQ(\bar{x},\bar{y})(x),C) \subset DQ(\bar{x},\bar{y})(x). \tag{13.12}$$

Proof. See [214]. □

Remark 13.1.5. Formula (13.9) of Theorem 13.1.4 is proved by Tanino [561, Theorem 3.1] for finite-dimensional Y and by Klose [330, Theorem 3.4] for arbitrary Y; for finite-dimensional spaces X and Y, formula (13.10) is proved by Shi [531, Theorem 4.1], Theorem 13.1.4 (ii) is proved by Shi [530, Theorem 4.1].

In the next example, which is taken from [345], $F(x)$ is not C-dominated for $x < 0$, but the other conditions of Theorem 13.1.4 hold. As we show below, (13.10), (13.11), and (13.12) do not hold for $x < 0$.

Example 13.1.6. Let $X = \mathbb{R}$, $Y = \mathbb{R}^2$, $C = \mathbb{R}^2_+$, and $F : \mathbb{R} \rightrightarrows \mathbb{R}^2$ be defined by

$$F(x) := \begin{cases} \{(y_1,y_2) \mid y_1 \geq 0,\ y_2 \geq y_1^2\} & \text{if } x \geq 0, \\ \{(y_1,y_2) \mid y_1 > 0,\ y_2 \geq y_1^2\} & \text{if } x < 0. \end{cases}$$

Then

$$Q(x) := \begin{cases} \{(0,0)\} & \text{if } x \geq 0, \\ \emptyset & \text{if } x < 0. \end{cases}$$

Take $\bar{x} = 0$, $\bar{y} = (0,0)$. Then

$$DF(\bar{x},\bar{y})(x) = D(F+C)(\bar{x},\bar{y})(x) = C,$$
$$DQ(\bar{x},\bar{y})(x) = Q(x),$$

for every $x \in X$. □

In the following corollary Bo-PMin stands for the Borwein minimal points.

Corollary 13.1.7. *Suppose that* $\bar{y} \in \text{Bo-PMin}\,(F(\bar{x}),C)$ *and* F *is* C*-dominated by* Q *near* $\bar{x}$*. Then (13.11) and (13.12) hold if one of the following conditions is satisfied:*

(i) F *is upper Lipschitz at* $\bar{x}$ *and* $\{y \in C \mid \|y\| = 1\}$ *is compact;*
(ii) X, Y *are finite-dimensional,* $\bar{x} \in \text{int}(\text{dom} F)$*, and* F *is* C*-convex near* $\bar{x}$.

Remark 13.1.8. Corollary 13.1.7 (i) was obtained by Tanino [561, Theorem 3.2] for Y finite-dimensional, Shi [530, Corollary 4.1], and Klose [330, Theorem 3.5], while part (ii) by Shi [531, Theorem 4.2].

The following is an analogue of Theorem 13.1.4 for the weak perturbation map:

Theorem 13.1.9. *Let* $(\bar{x}, \bar{y}) \in \operatorname{graph}(F)$. *Assume that (13.4) holds,* $\dim Y < \infty$, $\operatorname{int} C \neq \emptyset$, *and* F *is* $\tilde{C}$*-dominated by* R *near* $\bar{x}$ *for some closed convex cone* $\tilde{C}$ *with* $\tilde{C} \subset \{0\} \cup \operatorname{int} C$. *Then for every* $x \in X$,

$$\operatorname{WMin}(DF(\bar{x}, \bar{y})(x), C) = \operatorname{WMin}(DR(\bar{x}, \bar{y})(x), C) \subset DR(\bar{x}, \bar{y})(x). \qquad (13.13)$$

Proof. Since (13.4) holds, we have that $\bar{y} \in Q(\bar{x}) \subset R(\bar{x})$. Since the conditions of Corollary 13.1.1 hold, equality (13.8) holds, too. Since $\operatorname{graph}(R) \subset \operatorname{graph}(F)$, condition (13.4) is satisfied with F replaced by R, hence (13.8) holds for the same substitution. Because F is $\tilde{C}$-dominated by R near $\bar{x}$, we have that $(F + \tilde{C})(x) = (R + \tilde{C})(x)$ for every x in a neighborhood of $\bar{x}$, hence

$$D(F + \tilde{C})(\bar{x}, \bar{y}) = D(R + \tilde{C})(\bar{x}, \bar{y}).$$

The conclusion is obtained immediately by using the properties of minimal points. □

Remark 13.1.10. Theorem 13.1.9 was proved by Kuk, Tanino, and Tanaka in [345, Theorem 3.1] in the same conditions.

Remark 13.1.11. In Examples 11.3.4 and 13.1.6 relation (13.13) does not hold. Note that in Example 11.3.4, (13.4) is violated and in Example 13.1.6, $F(x)$ is not $\tilde{C}$-dominated by $R(x)$ for $x < 0$ for some closed convex cone $\tilde{C}$ with $\tilde{C} \subset \{0\} \cup \operatorname{int} C$.

The following result is immediate:

Corollary 13.1.12. *Suppose that* X, Y *are finite-dimensional,* $\bar{x} \in \operatorname{int}(\operatorname{dom} F)$, $\bar{y} \in \operatorname{Bo-PMin}(F(\bar{x}), C)$, *and* F *is* C*-convex near* $\bar{x}$. *If* F *is* $\tilde{C}$*-dominated by* R *near* $\bar{x}$ *for some closed convex cone* $\tilde{C}$ *with* $\tilde{C} \subset \{0\} \cup \operatorname{int} C$, *then (13.13) holds.*

In the following example, which is taken from [531], F is not C-convex (near $\bar{x}$) and relation (13.13) does not hold for every $x \in X$.

Example 13.1.13. Let $X = Y = \mathbb{R}$, $C = \mathbb{R}_+$, and let $F : \mathbb{R} \rightrightarrows \mathbb{R}$ be defined by

$$F(x) := \begin{cases} [-|x|, \infty[\cup \{-\sqrt{|x|}\} & \text{if } |x| \leq 1, \\ [-\sqrt{|x|}, \infty[& \text{if } |x| > 1. \end{cases}$$

Then, for every $x \in X$,

$$(F + C)(x) = [-\sqrt{|x|}, \infty[,$$
$$Q(x) = \{-\sqrt{|x|}\}.$$

Take $\bar{x} = 0$, $\bar{y} = 0$. Then,

$$DF(\bar{x}, \bar{y})(x) = [-|x|, \infty[, \quad D(F+C)(\bar{x}, \bar{y})(x) = X, \quad \forall\, x \in X,$$

$$DQ(\bar{x}, \bar{y})(x) = \begin{cases}]-\infty, 0] & \text{if } x = 0, \\ \emptyset & \text{if } x \neq 0. \end{cases}$$

Corollary 13.1.14. *Suppose that Y is finite-dimensional, $\bar{y} \in$ Bo-PMin $(F(\bar{x}), C)$, $F(x)$ is C-closed for all x in a neighborhood of $\bar{x}$, and one of the following two conditions is satisfied:*

(i) *(13.4) holds and $F(x)$ is C-bounded (i.e., $(F(x))_\infty \cap (-C) = \{0\}$) for all x in a neighborhood of $\bar{x}$;*
(ii) *X is finite-dimensional, $\bar{x} \in$ int(dom$F$), and F is C-convex and C-dominated by Q near $\bar{x}$.*

Then for every $x \in X$,

$$\text{Min}\,(DF(\bar{x}, \bar{y})(x), C) = \text{Min}\,(DP(\bar{x}, \bar{y})(x), C) \subset DP(\bar{x}, \bar{y})(x).$$

Remark 13.1.15. In Example 13.1.13, $D_S F(\bar{x}, \bar{y})(0) = X$. So (13.4) does not hold and F is not C-convex near $\bar{x}$; the conclusion of the preceding corollary does not hold.

We have the following analogue for the weak perturbation map:

Theorem 13.1.16. *Suppose that* int$C \neq \emptyset$ *and consider $\bar{y} \in R(\bar{x})$. If one of the following conditions holds,*

(i) *F is semidifferentiable at $(\bar{x}, \bar{y})$;*
(ii) *F is C-convex and $\bar{x} \in$ int(dom$F$);*
(iii) *X, Y are finite-dimensional, F is C-convex and $\bar{x} \in$ raint(dom$F$),*

then for every $x \in X$, we have

$$DR(\bar{x}, \bar{y})(x) \subset \text{WMin}\,(DF(\bar{x}, \bar{y})(x), C). \tag{13.14}$$

Suppose that either (a) *(13.4) and* (i) *hold or* (b) *$\bar{y} \in$ Bo-PMin $(F(\bar{x}), C)$ and* (iii) *hold. If $\{y \in C \mid \|y\| = 1\}$ is compact and F is $\tilde{C}$-dominated by R or Q near $\bar{x}$, where $\tilde{C} \subset \{0\} \cup$ int$C$ is a closed convex cone, then for every $x \in X$,*

$$DR(\bar{x}, \bar{y})(x) = \text{WMin}\,(DF(\bar{x}, \bar{y})(x), C), \tag{13.15}$$

or

$$DQ(\bar{x}, \bar{y})(x) = \text{WMin}\,(DF(\bar{x}, \bar{y})(x), C), \tag{13.16}$$

respectively.

Proof. Assume that condition (i) holds and consider $y \in DR(\bar{x}, \bar{y})(x)$. If y does not belong to WMin $(DF(\bar{x}, \bar{y})(x), C)$, then there exists $k \in \text{int}(C)$ such that $y - k \in DR(\bar{x}, \bar{y})(x)$. Therefore there are $\{t_n\} \to 0_+$ and $\{(x_n, y_n)\} \to (x, y - k)$ such that $(\bar{x}, \bar{y}) + t_n(x_n, y_n) \in \text{graph}\,(R) \subset \text{graph}\,(F)$. Since F is semidifferentiable at $(\bar{x}, \bar{y})$ and $\{t_n\} \to 0_+$, $x_n \to u$, there exists $\{z_n\} \to y - k$ such that $(\bar{x}, \bar{y}) + t_n(x_n, z_n) \in \text{graph}\, F$ for every $n \geq n_0$. Therefore $\bar{y} + t_n z_n \in F(\bar{x} + t_n x_n)$ and $\bar{y} + t_n y_n \in R(\bar{x} + t_n x_n) \subset F(\bar{x} + t_n x_n)$. Since

$$\big((\bar{y} + t_n z_n) - (\bar{y} + t_n y_n)\big)/t_n = z_n - y_n \to y - k - y = -k \in -\text{int}(C),$$

there exists some $n_1 \geq n_0$ such that $(\bar{y} + t_n z_n) - (\bar{y} + t_n y_n) \in -\text{int}(C)$ for every $n \geq n_1$, contradicting the fact that $\bar{y} + t_n y_n \in R(\bar{x} + t_n x_n)$.

Assume now that (ii) holds; since $\text{int}C \neq \emptyset$, $(\bar{x}, \bar{y}, \bar{k}) \in \text{int}(\text{graph}\,(F + C))$ for $\bar{k} \in \text{int}C$. By Theorem 11.1.36 (i) $(F + C)$ is semidifferentiable at $(\bar{x}, \bar{y})$. Taking $y \in DR(\bar{x}, \bar{y})(x)$, with the same argument as in the proof of (i), we obtain the same sequences, the sole difference being that $\bar{y} + t_n z_n \in F(\bar{x} + t_n x_n) + C$ for $n \geq n_0$. The same contradiction is obtained.

If (iii) holds, by Theorem 11.1.36 (ii), $(F + C)$ is semidifferentiable at $(\bar{x}, \bar{y})$. The proof is the same as for (ii).

(b) $\Rightarrow$ (a). Indeed, if (iii) holds, using Theorem 11.3.3 (ii) we get that (13.4) is satisfied.

Assume that (a) is satisfied and $F(x)$ is $\tilde{C}$-dominated by $R(x)$ for all $x \in V$, for some neighborhood V of $\bar{x}$. It follows that $F(x) + \tilde{C} = R(x) + \tilde{C}$ for $x \in V$. So, $D(F + \tilde{C})(\bar{x}, \bar{y}) = D(R + \tilde{C})(\bar{x}, \bar{y})$. Applying Theorem 11.3.7 for $(F, \tilde{C})$ and $(R, \tilde{C})$, we obtain that $DF(\bar{x}, \bar{y})(x) + \tilde{C} = DR(\bar{x}, \bar{y})(u) + \tilde{C}$ for every $x \in X$. Applying now Proposition p12 (iv) for the pair $(\tilde{C}, C_0)$, we get for every $x \in X$,

$$\text{WMin}\,(DF(\bar{x}, \bar{y})(x), C) = \text{WMin}\,(DR(\bar{x}, \bar{y})(x),) \subset DR(\bar{x}, \bar{y})(x).$$

From (13.14) and the above relation (13.15) follows.

If F is $\tilde{C}$-dominated by Q near $\bar{x}$, all we said above is true if we replace R by Q. Therefore (13.16) holds, too. Note that $\bar{y} \in Q(\bar{x})$ because (13.4) holds. □

Remark 13.1.17. In Example 11.3.4 F is neither semidifferentiable at $\bar{x}$, nor C-convex; relation (13.14) does not hold for $x = 0$. In Example 13.1.6 (a) and (b) are satisfied but F is not $\tilde{C}$-dominated by R or Q near $\bar{x}$, for some closed convex cone $\tilde{C} \subset \{0\} \cup \text{int}C$; relations (13.15) and (13.16) do not hold for $x < 0$.

In the next example, which is taken from [561], neither (a) nor (b) is satisfied; relations (13.15) and (13.16) do not hold for $x > 0$.

Example 13.1.18. Let $X = \mathbb{R}$, $Y = \mathbb{R}^2$, $C = \mathbb{R}^2_+$, and let $F : \mathbb{R} \rightrightarrows \mathbb{R}^2$ be defined by

$$F(x) := \{(y_1, -y_1) \mid y_1 \leq x\} \cup \{(y_1, -1 - y_1) \mid y_1 > 0\}.$$

Then

$$Q(x) = R(x) = \{(y_1, -y_1) \mid y_1 \leq \min(0, x)\} \cup \{(y_1, -1 - y_1) \mid y_1 > 0\}.$$

Let $\bar{x} = 0$ and $\bar{y} = (0, 0)$. Then

$$DF(\bar{x}, \bar{y})(x) = \text{Min}\,(DF(\bar{x}, \bar{y})(x), C) = \{(v_1, -v_1) \mid v_1 \leq x\},$$
$$DQ(\bar{x}, \bar{y})(x) = \text{Min}\,(DF(\bar{x}, \bar{y})(x), C) = \{(v_1, -v_1) \mid v_1 \leq \min(0, x)\}.$$

Remark 13.1.19. For R replaced by Q and X, Y finite-dimensional, (13.14) was obtained by Shi [531, Theorem 5.1] under condition (iii) and by Kuk, Tanino, and Tanaka [345, Theorem 3.3] under (i). Kuk, Tanino, and Tanaka obtained (13.14) under condition (iii) and (13.15) under condition (b) in [344, Theorem 3.2] and (13.16) under condition (b) in [344, Theorem 3.3].

We now consider another normed vector space U, a set-valued map $G : U \rightrightarrows X$, a single-valued map $f : X \times U \to Y$, and define a set-valued $F : U \rightrightarrows Y$ by $F(x) := f(G(x) \times \{x\})$.

With this preparation, we have the following:

Theorem 13.1.20. *Suppose that the following conditions hold:*

(i) *G is upper Lipschitz at $\bar{u} \in \text{dom}(G)$ and $G(u)$ is compact for $u \in U$, with U a neighborhood of $\bar{u}$;*
(ii) *$\bar{x} \in G(\bar{u})$ and $\bar{y} = f(\bar{x}, \bar{u}) \in \text{BMin}(F(\bar{u}); C)$;*
(iii) *X is finite-dimensional, C has a compact base, f is Lipschitz on bounded sets and Fréchet differentiable at $(\bar{x}, \bar{u})$;*
(iv) *$G(\bar{u}) = \{\bar{x}\}$ or the set-valued map*

$$\tilde{G} : U \times Y \rightrightarrows X, \quad \tilde{G}(u, y) := \{x \in G(u) \mid f(x, u) = y\},$$

is upper Lipschitz at $(\bar{u}, \bar{y})$ and $\tilde{G}(\bar{u}, \bar{y}) = \{\bar{x}\}$.

Then for every $u \in X$,

$$\text{Min}\,(\nabla_x f(\bar{x}, \bar{u})\,(DG(\bar{u}, \bar{x})(u)) + \nabla_u f(\bar{x}, \bar{u})(u), C) = \text{Min}\,(DQ(\bar{u}, \bar{y})(u), C)\,.$$

In order to prove this result we need the following lemma.

Lemma 13.1.21. *Let G, f, F be as above and $\bar{u} \in \text{dom}G$. If f is Lipschitz on bounded sets, $G(\bar{u})$ is bounded, and G is upper Lipschitz at $\bar{u}$, then F is upper Lipschitz at $\bar{u}$.*

Proof. See [214, Lemma 3.6.23]. □

Remark 13.1.22. Evidently, a sufficient condition for f to be Lipschitz on bounded sets is that f be of class $\mathscr{C}^1$ and U and X have finite dimension.

Proof of Theorem 13.1.20. From (i) and (iii), using Lemma 13.1.21, we obtain that F is upper Lipschitz at $\bar{u}$. Since $\bar{y} \in \text{Bo-PMin}(F(\bar{u}), C)$, from Theorem 11.3.3 (i), we obtain that (13.4) holds for F at $(\bar{u}, \bar{y})$. Since $G(u)$ is compact for $u \in U$ and f is continuous, $F(u)$ is compact for $u \in U$. Therefore $F(u) \subset Q(u) + C$ for $u \in U$. Now using Theorem 13.1.4(ii), for every $u \in U$, we have

$$\text{Min}\,(DF(\bar{u}, \bar{y})(u); C) = \text{Min}\,(DQ(\bar{u}, \bar{y})(u); C)\,.$$

However, from the second part of Theorem 11.1.19, for every $u \in U$, we have

$$DF(\bar{u}, \bar{y})(u) = \nabla_x f(\bar{x}, \bar{u})\,(DG(\bar{u}, \bar{x})(u)) + \nabla_u f(\bar{x}, \bar{u})(u).$$

The conclusion follows. □

Remark 13.1.23. The above result is taken from [214, Theorem 3.6.22]. A slightly weaker version of this result was originally given by Tanino in [561, Theorem 4.1] for U, X, Y finite-dimensional and f of class $\mathscr{C}^1$ and by Klose [330, Theorem 4.4]. Variants of Lemma 13.1.21 are stated and proved in [561, Lemma 4.1] and [330, Lemmas 4.2, 4.3].

13.2 Second Order Sensitivity Analysis in Set-Valued Optimization

This section is devoted to the second-order sensitivity analysis for set-valued perturbation maps. To obtain second-order analogues of the known first-order sensitivity analysis results presented in the previous section, we introduce and employ several concepts. This includes a second-order analogue of the S-derivative and second-order directional compactness, among others.

Let X and Y be normed spaces, and let $C \subset Y$ be a pointed, closed, and convex cone. When dealing with weak minimality, we will assume that $\text{int}(C)$ is non-empty. Let $F : X \rightrightarrows Y$ be a given set-valued map. We are interested in the assessment of differentiability properties of the set-valued perturbation maps P, Q, and R, defined in (13.3). That is,

$$\begin{aligned} P(x) &= \text{GHe-PMin}(F(x), C), \\ Q(x) &= \text{Min}(F(x), C), \\ R(x) &= \text{WMin}(F(x), C). \end{aligned}$$

Our goal here is to investigate the relationships among the second-order contingent derivatives of the perturbation maps P, Q, and R, the second-order contingent derivative of the map F, and various minimal points of these derivatives.

We begin with the following result.

Theorem 13.2.1. *Let X and Y be normed spaces, let $C \subset Y$ be a proper, pointed, closed, and convex cone, and let $F : X \rightrightarrows Y$ be a set-valued map. Let $(\bar{x}, \bar{y}) \in \operatorname{graph}(F)$ and let $(\bar{u}, \bar{v}) \in X \times Y$ be arbitrary. Assume that the set $B = \{z \in C \mid \|z\| = 1\}$ is compact. Assume that F is C-dominated by Q near $\bar{x}$. Then for every $x \in U(\bar{x})$, where $U(\bar{x})$ is a neighbourhood of $\bar{x}$, we have*

$$\operatorname{Min}(D^2(F + C)(\bar{x}, \bar{y}, \bar{u}, \bar{v})(x), C) \subset D^2 Q(\bar{x}, \bar{y}, \bar{u}, \bar{v})(x). \tag{13.17}$$

Proof. Since $Q(x) \subset F(x)$ and F is C-dominated by Q near $\bar{x}$, there exists a neighbourhood $U(\bar{x})$ of $\bar{x}$ such that $F(x) + C = Q(x) + C$ for every $x \in U(\bar{x})$. Consequently

$$D^2(F + C)(\bar{x}, \bar{y}, \bar{u}, \bar{v})(x) = D^2(Q + C)(\bar{x}, \bar{y}, \bar{u}, \bar{v})(x), \quad \text{for every } x \in U(\bar{x}),$$

which implies that

$$\begin{aligned} &\operatorname{Min}(D^2(F + C)(\bar{x}, \bar{y}, \bar{u}, \bar{v})(x), C) \\ &\quad = \operatorname{Min}(D^2(Q + C)(\bar{x}, \bar{y}, \bar{u}, \bar{v})(x), C), \quad \text{for every } x \in U(\bar{x}). \end{aligned} \tag{13.18}$$

Because B is compact, we have

$$\operatorname{Min}(D^2(Q + C)(\bar{x}, \bar{y}, \bar{u}, \bar{v})(x), C) \subset D^2 Q(\bar{x}, \bar{y}, \bar{u}, \bar{v})(x), \quad \text{for every } x \in U(\bar{x}),$$

and hence (13.17) follows from the above inclusion and (13.18). The proof is complete. □

Example 13.2.2. Consider the map F defined in Example 11.7.26. Let $C = \mathbb{R}^2_+$, $\bar{x} = 1$, and $\bar{y} = (1, 1)$. Since $Q(x) = F(x)$ for $x > 0$, F is C-dominated by Q near $\bar{x}$. As in Example 11.7.26, for $\bar{v} = (v_1, v_2)$ and $\bar{u} = v_1 + v_2$, we have

$$\begin{aligned} D^2 Q(\bar{x}, \bar{y}, \bar{u}, \bar{v})(x) &= \{(y_1, y_2) \mid x = y_1 + y_2 + v_1 v_2\}, \\ D^2(F + C)(\bar{x}, \bar{y}, \bar{u}, \bar{v})(x) &= \{(y_1, y_2) \mid x \le y_1 + y_2 + v_1 v_2\}, \end{aligned}$$

so that both sides of (13.17) are equal. □

In our next result we derive analogues of (13.17). We recall that $D^2_S F(\bar{x}, \bar{y}, \bar{u}, \bar{v})$ stands for the second-order S derivative of F.

Theorem 13.2.3. *Let X and Y be normed spaces, let $C \subset Y$ be a proper, pointed, closed, and convex cone, and let $F : X \rightrightarrows Y$ be a set-valued map. Let $(\bar{x}, \bar{y}) \in \operatorname{graph}(F)$ and let $(\bar{u}, \bar{v}) \in X \times Y$ be arbitrary. Assume that one of the following conditions holds.*

(a) $D^2_S F(\bar{x}, \bar{y}, \bar{u}, \bar{v})(0) \cap (-C/\{0\}) = \emptyset$.

(b) *Q is second-order directionally compact at $(\bar{x}, \bar{y})$ with respect to $(\bar{u}, \bar{v})$ in any direction $x \in X$.*
(c) *Q is second-order compactly approximable at $(\bar{x}, \bar{y})$ with respect to $(\bar{u}, \bar{v})$.*

Then the following are valid:

(1) *If F is dominated by Q near $\bar{x}$, then there exists a neighborhood $U(\bar{x})$ of $\bar{x}$ such that for every $x \in U(\bar{x})$, we have*

$$\text{Min}(D^2F(\bar{x}, \bar{y}, \bar{u}, \bar{v})(x), C) \subset D^2Q(\bar{x}, \bar{y}, \bar{u}, \bar{v})(x). \tag{13.19}$$

(2) *If F is $\tilde{C}$-dominated by R near $\bar{x}$, then there exists a neighborhood $U(\bar{x})$ of $\bar{x}$ such that for every $x \in U(\bar{x})$, we have*

$$\text{WMin}(D^2F(\bar{x}, \bar{y}, \bar{u}, \bar{v})(x), C) \subset D^2R(\bar{x}, \bar{y}, \bar{u}, \bar{v})(x). \tag{13.20}$$

(3) *If F is C-bounded and C-closed, Y is finite dimensional, and F is dominated by Q near $\bar{x}$, then there exists a neighborhood $U(\bar{x})$ of $\bar{x}$ such that for every $x \in U(\bar{x})$, we have*

$$\text{GHe-PMin}(D^2F(\bar{x}, \bar{y}, \bar{u}, \bar{v})(x), C) \subset D^2P(\bar{x}, \bar{y}, \bar{u}, \bar{v})(x). \tag{13.21}$$

Proof. We begin by deducing some simple implications of (a)-(c). Due to the chain of inclusions

$$D_S^2Q(\bar{x}, \bar{y}, \bar{u}, \bar{v})(0) \subset D_S^2R(\bar{x}, \bar{y}, \bar{u}, \bar{v})(0) \subset D_S^2F(\bar{x}, \bar{y}, \bar{u}, \bar{v})(0),$$

we obtain

$$\begin{aligned} D_S^2F(\bar{x}, \bar{y}, \bar{u}, \bar{v})(0) \cap (-C/\{0\}) = \emptyset \ &\Rightarrow\ D_S^2R(\bar{x}, \bar{y}, \bar{u}, \bar{v})(0) \cap (-C/\{0\}) = \emptyset \\ &\Rightarrow\ D_S^2Q(\bar{x}, \bar{y}, \bar{u}, \bar{v})(0) \cap (-C/\{0\}) = \emptyset. \end{aligned}$$

Therefore, the following formula holds for $\Gamma = F, Q, R$:

$$\begin{aligned} &D^2(\Gamma + C)(\bar{x}, \bar{y}, \bar{u}, \bar{v})(x) \\ &\quad = D^2\Gamma(\bar{x}, \bar{y}, \bar{u}, \bar{v})(x) + C, \quad \text{for all } x \in \text{dom}(D^2(\Gamma + C)(\bar{x}, \bar{y}, \bar{u}, \bar{v})). \end{aligned}$$

In fact, the above formula also remains valid under (b) and (c), because when Q is either second-order directionally compact or second-order compactly approximable, then so are the maps Q and R.

Since F is C-dominated by Q near $\bar{x}$, for each $x \in U(\bar{x})$, we have

$$\text{Min}_C\, D^2F(\bar{x}, \bar{y}, \bar{u}, \bar{v})(x) = \text{Min}(D^2(F + C)(\bar{x}, \bar{y}, \bar{u}, \bar{v})(x), C)$$

$$
\begin{aligned}
&= \mathrm{Min}(D^2(Q + C)(\bar{x}, \bar{y}, \bar{u}, \bar{v})(x), C) \\
&= \mathrm{Min}(D^2Q(\bar{x}, \bar{y}, \bar{u}, \bar{v})(x), C) \\
&\subset D^2Q(\bar{x}, \bar{y}, \bar{u}, \bar{v})(x),
\end{aligned}
$$

establishing (13.19).

Analogously, since F is $\tilde{C}$-dominated by R near $\bar{x}$, for each $x \in U(\bar{x})$, we have

$$
\begin{aligned}
\mathrm{WMin}(D^2F(\bar{x}, \bar{y}, \bar{u}, \bar{v})(x), C) &= \mathrm{WMin}(D^2(F + \tilde{C})(\bar{x}, \bar{y}, \bar{u}, \bar{v})(x), C) \\
&= \mathrm{WMin}(D^2(R + \tilde{C})(\bar{x}, \bar{y}, \bar{u}, \bar{v})(x), C) \\
&= \mathrm{WMin}(D^2R(\bar{x}, \bar{y}, \bar{u}, \bar{v})(x), C) \\
&\subset D^2R(\bar{x}, \bar{y}, \bar{u}, \bar{v})(x),
\end{aligned}
$$

proving (13.20).

Finally, for (13.21), we begin by noticing that $D^2P(\bar{x}, \bar{y}, \bar{u}, \bar{v})(x) \subset D^2Q(\bar{x}, \bar{y}, \bar{u}, \bar{v})(x)$. Moreover, due to a known relationship between minimal and properly minimal points we also have that

$$
Q(x) \subset \mathrm{cl}(P(x)), \quad \text{for every } x \in U(\bar{x}),
$$

which implies that $D^2Q(\bar{x}, \bar{y}, \bar{u}, \bar{v})(x) \subset D^2P(\bar{x}, \bar{y}, \bar{u}, \bar{v})(x)$, or, $D^2Q(\bar{x}, \bar{y}, \bar{u}, \bar{v})(x) = D^2P(\bar{x}, \bar{y}, \bar{u}, \bar{v})(x)$.

Consequently

$$
\begin{aligned}
\mathrm{PMin}_C\, D^2F(\bar{x}, \bar{y}, \bar{u}, \bar{v})(x) &= \text{GHe-PMin}(D^2(F + C)(\bar{x}, \bar{y}, \bar{u}, \bar{v})(x), C) \\
&= \text{GHe-PMin}(D^2(Q + C)(\bar{x}, \bar{y}, \bar{u}, \bar{v})(x), C) \\
&= \text{GHe-PMin}(D^2Q(\bar{x}, \bar{y}, \bar{u}, \bar{v})(x), C) \\
&= \text{GHe-PMin}(D^2P(\bar{x}, \bar{y}, \bar{u}, \bar{v})(x), C) \\
&\subset D^2P(\bar{x}, \bar{y}, \bar{u}, \bar{v})(x),
\end{aligned}
$$

confirming (13.21). The proof is complete. □

The following notion of normally minimal points will be used in the sequel.

Definition 13.2.4. Let Y be a normed space, let $C \subset Y$ be a proper, closed, and convex cone, and let S be a nonempty subset of Y such that $S + C$ is convex. A point $z \in \mathrm{Min}(S, C)$ is called a **normally C-minimal point** of S if

$$
T(S + C, z)^{-} \subset int(C^{-}) \cup \{0\},
$$

where C^{-} is the negative dual of C and $T(S + C, z)^{-}$ is the negative dual of $T(S + C, z)$.

For our next result, we also need to recall that, given a normed space Z, a set $S \subset Z$ is second-order derivable at $\bar{z}$ in direction $w \in Z$, if $A^2(S, \bar{z}, w) = T^2(S, \bar{z}, w)$.

Example 13.2.5. To illustrate these notions, let

$$S = \{(x_1, x_2) \in \mathbb{R}^2_+ \mid x_1 x_2 = 1\}.$$

Then, for $\bar{z} = (1, 1)$, we have

$$T(S, \bar{z}) = \{(z_1, z_2) \in \mathbb{R}^2 \mid z_2 = -z_1\},$$

and for $w = (w_1, -w_1)$,

$$T^2(S, \bar{z}, w) = A^2(S, \bar{z}, w) = \{(z_1, z_2) \in \mathbb{R}^2 \mid z_2 = -z_1 + {w_1}^2\},$$

which means that S is second-order derivable at $\bar{z}$ in direction w. For $C = \mathbb{R}^2_+$,

$$T(S + C, \bar{z}) = \{(z_1, z_2) \in \mathbb{R}^2 \mid z_2 \geq -z_1\},$$

so that

$$T(S + C, \bar{z})^- = \{(z_1, z_1) \mid z_1 \leq 0\},$$

and hence $\bar{z}$ is a normally C-minimal point of S. □

Theorem 13.2.6. *Let X and Y be finite-dimensional normed spaces, let $C \subset Y$ be a proper, pointed, closed, and convex cone, and let $F : X \rightrightarrows Y$ be a set-valued map. Let $(\bar{x}, \bar{y}) \in$ graph (F) and let $(\bar{u}, \bar{v}) \in X \times Y$ be arbitrary. We make the following assumptions:*

1. $\bar{x} \in \operatorname{int}(\operatorname{dom}(F))$ *and* graph $(F + C)$ *is convex.*
2. graph (F) *is second-order derivable at $(\bar{x}, \bar{y})$ in direction $(\bar{u}, \bar{v})$.*
3. *$\bar{y}$ is a normally C-minimal point of $F(\bar{x})$.*

Then for every $x \in$ dom $(D^2Q(\bar{x}, \bar{y}, \bar{u}, \bar{v}))$, the following inclusion holds:

$$D^2Q(\bar{x}, \bar{y}, \bar{u}, \bar{v})(x) \subset \operatorname{Min}(D^2F(\bar{x}, \bar{y}, \bar{u}, \bar{v})(x), C). \tag{13.22}$$

Proof. Assume that $y \in D^2Q(\bar{x}, \bar{y}, \bar{u}, \bar{v})(x)$ which implies that $y \in D^2F(\bar{x}, \bar{y}, \bar{u}, \bar{v})$ (x). Now suppose that $y \notin \operatorname{Min}(D^2F(\bar{x}, \bar{y}, \bar{u}, \bar{v})(x), C)$, and consequently there exists $\tilde{y} \in D^2F(\bar{x}, \bar{y}, \bar{u}, \bar{v})(x)$ such that $y - \tilde{y} \in C\backslash\{0\}$. Since $\tilde{y} \in D^2F(\bar{x}, \bar{y}, \bar{u}, \bar{v})(x)$, there are sequences $\{s_n\} \subset \mathbb{P}$ and $\{(\tilde{x}_n, \tilde{y}_n)\} \subset X \times Y$ such that $s_n \downarrow 0$, $(\tilde{x}_n, \tilde{y}_n) \to (x, \tilde{y})$, and

$$\bar{y} + s_n\bar{v} + s_n^2\tilde{y}_n \in F(\bar{x} + s_n\bar{u} + s_n^2\tilde{x}_n). \tag{13.23}$$

Analogously, because $y \in D^2Q(\bar{x}, \bar{y}, \bar{u}, \bar{v})(x)$, there are sequences $\{t_n\} \subset \mathbb{P}$ and $\{(x_n, y_n)\} \subset X \times Y$ such that $t_n \downarrow 0$, $(x_n, y_n) \to (x, y)$, and

$$\bar{y} + t_n\bar{v} + t_n^2 y_n \in Q(\bar{x} + t_n\bar{u} + t_n^2 x_n) = \text{Min}(F(\bar{x} + t_n\bar{u} + t_n^2 x_n), C).$$

Therefore, $(\bar{x} + t_n\bar{u} + t_n^2 x_n, \bar{y} + t_n\bar{v} + t_n^2 y_n)$ is a boundary point of the convex set graph $(F + C)$. By a standard separation argument, there are nonzero vectors $(\lambda_n, \mu_n) \in X \times Y$ such that

$$\langle \lambda_n, \bar{x} + t_n\bar{u} + t_n^2 x_n\rangle + \langle \mu_n, \bar{y} + t_n\bar{v} + t_n^2 y_n\rangle$$
$$\geq \langle \lambda_n, \hat{x}\rangle + \langle \mu_n, \hat{y}\rangle \quad \text{for all } (\hat{x}, \hat{y}) \in \text{graph}\,(F + C). \tag{13.24}$$

We normalize these vectors so that $\|(\lambda_n, \mu_n)\| = 1$, and assume that $(\lambda_n, \mu_n) \to (\lambda, \mu)$. Passing to the limit as $n \to \infty$ in (13.24), we obtain

$$\langle \lambda, \bar{x}\rangle + \langle \mu, \bar{y}\rangle \geq \langle \lambda, \hat{x}\rangle + \langle \mu, \hat{y}\rangle \quad \text{for all } (\hat{x}, \hat{y}) \in \text{graph}\,(F + C). \tag{13.25}$$

Let $z \in F(\bar{x}) + C$ be arbitrary. The map $F + C$, being convex, is lower-semicontinuous at $\bar{x}$ (see [565]). Therefore, there exists a sequence $\{z_n\} \subset Y$ such that $z_n \to z$ and $z_n \in F(\bar{x}+t_n\bar{u}+t_n^2 x_n)+C$. By setting $(\hat{x}, \hat{y}) = (\bar{x}+t_n\bar{u}+t_n^2 x_n, z_n)$ in (13.24), we obtain

$$\langle \lambda_n, \bar{x} + t_n\bar{u} + t_n^2 x_n\rangle + \langle \mu_n, \bar{y} + t_n\bar{v} + t_n^2 y_n\rangle \geq \langle \lambda_n, \bar{x} + t_n\bar{u} + t_n^2 x_n\rangle + \langle \mu_n, z_n\rangle.$$

Passing to the limit as $n \to \infty$ in the above inequality, we obtain $\langle \mu, \bar{y}\rangle \geq \langle \mu, z\rangle$, and because $z \in F(\bar{x}) + C$ was chosen arbitrarily, the above inequality confirms that $\mu \in T(F(\bar{x}) + C, \bar{y})^-$. Since $\bar{y}$ is a normally C-minimal point, we deduce that $\mu \in (\text{int}(C^-) \cup \{0\})$. However, due to the assumption that $\bar{x} \in \text{int}(\text{dom}(F))$, we have $\mu \neq 0$, ensuring that $\mu \in \text{int}(C^-)$. This, in view of the fact that $y - \tilde{y} \in C\backslash\{0\}$, confirms that

$$\langle \mu, y\rangle < \langle \mu, \tilde{y}\rangle. \tag{13.26}$$

Due to the assumption that F is derivable, we can set $t_n = s_n$. Using (13.23) in (13.24), we have

$$\langle \lambda_n, \bar{x}+t_n\bar{u}+t_n^2 x_n\rangle + \langle \mu_n, \bar{y}+t_n\bar{v}+t_n^2 y_n\rangle \geq \langle \lambda_n, \bar{x}+t_n\bar{u}+t_n^2 \tilde{x}_n\rangle + \langle \mu_n, \bar{y}+t_n\bar{v}+t_n^2 \tilde{y}_n\rangle,$$

which after simplifying yields

$$\langle \lambda_n, x_n\rangle + \langle \mu_n, y_n\rangle \geq \langle \lambda_n, \tilde{x}_n\rangle + \langle \mu_n, \tilde{y}_n\rangle.$$

Passing to the limit as $n \to \infty$ in the above inequality, we obtain $\langle \mu, y \rangle \geq \langle \mu, \tilde{y} \rangle$, which, however, is a contradiction to (13.26). Consequently $y \in \mathrm{Min}(D^2F(\bar{x}, \bar{y}, \bar{u}, \bar{v})(x), C)$. The proof is complete. □

By setting $(\bar{u}, \bar{v}) = (0, 0)$, we obtain the following result.

Corollary 13.2.7. *Let X and Y be finite-dimensional normed spaces, let $C \subset Y$ be a proper, pointed, closed, and convex cone, let $F : X \rightrightarrows Y$ be a set-valued map, and let $(\bar{x}, \bar{y}) \in$ graph (F). Assume that $\bar{x} \in$ int(dom(F)) and graph $(F + C)$ is convex. If $\bar{y}$ is a normally C-minimal point of $F(\bar{x})$, then for every $x \in$ dom$(DQ(\bar{x}, \bar{y}))$,*

$$DQ(\bar{x}, \bar{y})(x) \subset \mathrm{Min}_{CDF}(\bar{x}, \bar{y})(x).$$

The following results show that, when dealing with weak-perturbation maps, we may dispense with the requirement that $\bar{y}$ is a normally C-minimal point of $F(\bar{x})$.

Theorem 13.2.8. *Let X and Y be finite-dimensional normed spaces, let $C \subset Y$ be a proper, pointed, closed, and convex cone, and let $F : X \rightrightarrows Y$ be a set-valued map. Let $(\bar{x}, \bar{y}) \in$ graph (F) and let $(\bar{u}, \bar{v}) \in X \times Y$ be arbitrary. We make the following assumptions:*

1. $\bar{x} \in$ int(dom (F)) *and* graph $(F + C)$ *is convex.*
2. graph (F) *is derivable at* $(\bar{x}, \bar{y})$ *in direction* $(\bar{u}, \bar{v})$.
3. $\bar{y}$ *is a normally* C*-minimal point of* $F(\bar{x})$.

Then

$$D^2R(\bar{x}, \bar{y}, \bar{u}, \bar{v})(x) \subset \mathrm{WMin}(D^2F(\bar{x}, \bar{y}, \bar{u}, \bar{v})(x), C). \tag{13.27}$$

Proof. Assume that $y \in D^2R(\bar{x}, \bar{y}, \bar{u}, \bar{v})(x)$ but $y \notin \mathrm{WMin}(D^2F(\bar{x}, \bar{y}, \bar{u}, \bar{v})(x), C)$. Therefore, there exists $\tilde{y} \in D^2F(\bar{x}, \bar{y}, \bar{u}, \bar{v})(x)$ such that $y - \tilde{y} \in \mathrm{int}(C)$. As in the proof of (13.25), we prove that

$$\langle \lambda, \bar{x} \rangle + \langle \mu, \bar{y} \rangle \geq \langle \lambda, \hat{x} \rangle + \langle \mu, \hat{y} \rangle \quad \text{for all } (\hat{x}, \hat{y}) \in \text{graph}\,(F + C),$$

which, as above, will confirm that $\mu \in C^{-}\backslash\{0\}$. However, now we have $y - \tilde{y} \in \mathrm{int}(C)$, which implies $\langle \mu, y - \tilde{y} \rangle < 0$. (In fact, the normality assumption was required for this inequality). By the same approach used in the proof of (13.22), we obtain a contradiction to $\langle \mu, y - \tilde{y} \rangle < 0$. The proof is complete. □

The following result uses the second-order lower Dini derivative and the notion of semi-differentiability:

Theorem 13.2.9. *Let X and Y be normed spaces, let $C \subset Y$ be a proper, pointed, closed, and convex cone, and let $F : X \rightrightarrows Y$ be a set-valued map. Let $(\bar{x}, \bar{y}) \in$*

graph(F) *and let* $(\bar{u}, \bar{v}) \in X \times Y$ *be arbitrary. Assume that* F *is second-order semidifferentiable at* $(\bar{x}, \bar{y})$ *in direction* $(\bar{u}, \bar{v})$. *Then (13.27) holds.*

Proof. Let $y \in D^2R(\bar{x}, \bar{y}, \bar{u}, \bar{v})(x)$. Then there exist sequences $\{t_n\} \subset \mathbb{P}$, $\{(x_n, y_n)\} \subset X \times Y$ such that $t_n \downarrow 0$, $(x_n, y_n) \to (x, y)$ and

$$a_n := \bar{y} + t_n\bar{v} + t_n^2 y_n \in R(\bar{x} + t_n\bar{u} + t_n^2 x_n). \tag{13.28}$$

Assume that $y \notin \mathrm{WMin}_C\, D^2F(\bar{x}, \bar{y}, \bar{u}, \bar{v})(x)$. Then there exists $\tilde{y} \in D^2F(\bar{x}, \bar{y}, \bar{u}, \bar{v})(x) = D_\ell^2 F(\bar{x}, \bar{y}, \bar{u}, \bar{v})(x)$ such that $y - \tilde{y} \in \mathrm{int}(C)$. Since $t_n \downarrow 0$ and $x_n \to x$, there exist $n_0 \in \mathbb{N}$ and $\tilde{y}_n \to \tilde{y}$ such that

$$b_n := \bar{y} + t_n\bar{v} + t_n^2\tilde{y}_n \in F(\bar{x} + t_n\bar{u} + t_n^2 x_n) \quad \text{for } n > n_0.$$

Since $y - \tilde{y} \in \mathrm{int}(C)$, we have $y_n - \widetilde{y_n} \in \mathrm{int}(C)$ for sufficiently large n. Moreover, because C is a cone, we obtain that

$$a_n - b_n = (\bar{y} + t_n\bar{v} + t_n^2 y_n) - (\bar{y} + t_n\bar{v} + t_n^2\tilde{y}_n) = t_n^2(y_n - \widetilde{y_n}) \in \mathrm{int}(C),$$

which, however, is a contradiction to (13.28). The proof is complete. □

Remark 13.2.10. The mapping Γ defined in Example 3.2 is second-order semi-differentiable at $(1, (1, 1))$ in the direction $(v_1 + v_2, (v_1, v_2))$.

For our next result, we consider the following parameter dependent vector-optimization problem

$$\mathrm{Min}(f(w, x), C) \quad \text{subject to } w \in H(x), \tag{13.29}$$

where H is a given set-valued map, and confine our discussion to the case when the set-valued map F is given by

$$F(x) := \{y \in Y \mid y = f(w, x) \text{ for some } w \in H(x)\}, \tag{13.30}$$

where X is the parameter space, $f : W \times X \to Y$ is a single-valued map, and $H : X \rightrightarrows W$ is a set-valued map.

For this situation, we have the following result.

Theorem 13.2.11. *Let* X *and* Y *be finite-dimensional normed spaces, let* $C \subset Y$ *be a proper, pointed, closed, and convex cone, and let* $F : X \rightrightarrows Y$ *be a set-valued map. Let* $(\bar{x}, \bar{y}) \in$ graph(F) *and let* $(\bar{u}, \bar{v}) \in X \times Y$ *be arbitrary. We make the following assumptions:*

1. f *is continuous and* C*-convex.*
2. H *is convex.*
3. F *is* C*-dominated by* Q *near* $\bar{x}$.

Then

$$\text{Min}(D^2F(\bar{x}, \bar{y}, \bar{u}, \bar{v})(x), C) \subset D^2Q(\bar{x}, \bar{y}, \bar{u}, \bar{v})(x).$$

Proof. For any $y \in \text{Min}(D^2F(\bar{x}, \bar{y}, \bar{u}, \bar{v})(x), C)$, we have $y \in D^2F(\bar{x}, \bar{y}, \bar{u}, \bar{v})(x)$, and hence there are sequences $\{t_n\} \subset \mathbb{P}$ and $\{(x_n, y_n)\} \subset X \times Y$ such that $t_n \downarrow 0$, $(x_n, y_n) \to (x, y)$, and $\bar{y}+t_n\bar{v}+t_n^2 y_n \in F(\bar{x}+t_n\bar{u}+t_n^2 x_n)$. Since F is C-dominated by Q near $\bar{x}$, for sufficiently large n, we have $\bar{y}+t_n\bar{v}+t_n^2 y_n \in Q(\bar{x}+t_n\bar{u}+t_n^2 x_n)+C$. Consequently, there exists a sequence $\{\tilde{y}_n\} \subset Y$ such that $\bar{y} + t_n\bar{v} + t_n^2 \tilde{y}_n \in Q(\bar{x} + t_n\bar{u} + t_n^2 x_n)$ and

$$y_n - \tilde{y}_n \in C. \tag{13.31}$$

Assume that the sequence $\{\tilde{y}_n\}$ is bounded. Since Y is finite dimensional, we may assume that $\tilde{y}_n \to \tilde{y}$ which implies $\tilde{y} \in D^2Q(\bar{x}, \bar{y}, \bar{u}, \bar{v})(x) \subset D^2F(\bar{x}, \bar{y}, \bar{u}, \bar{v})(x)$.

In view of (13.31), we have $y-\tilde{y} \in C$, and because $y \in \text{Min}_C\, D^2F(\bar{x}, \bar{y}, \bar{u}, \bar{v})(x)$, we must have $y = \tilde{y}$. This implies that $y \in D^2Q(\bar{x}, \bar{y}, \bar{u}, \bar{v})(x)$.

Therefore, to complete the proof it suffices to show that $\{\tilde{y}_n\}$ remains bounded. If possible, assume that this is not the case and $\|\tilde{y}_n\| \to \infty$. In view of (13.30), there are sequences $\{w_n\}$ and $\{\tilde{w}_n\}$ such that

$$\begin{aligned}
\bar{w} + t_n\bar{u} + t_n^2 w_n &\in H(\bar{x} + t_n\bar{u} + t_n^2 x_n) \\
\bar{y} + t_n\bar{v} + t_n^2 y_n &= f(\bar{w} + t_n\bar{u} + t_n^2 w_n, \bar{x} + t_n\bar{u} + t_n^2 x_n) \\
\bar{w} + t_n\bar{u} + t_n^2 \tilde{w}_n &\in H(\bar{x} + t_n\bar{u} + t_n^2 x_n) \\
\bar{y} + t_n\bar{v} + t_n^2 \tilde{y}_n &= f(\bar{w} + t_n\bar{u} + t_n^2 \tilde{w}_n, \bar{x} + t_n\bar{u} + t_n^2 x_n).
\end{aligned}$$

Due to the convexity of H, for some $0 \leq \alpha \leq 1$, we have

$$\begin{aligned}
\alpha(\bar{w} + t_n\bar{u} + t_n^2 w_n) + (1-\alpha)(\bar{w} + t_n\bar{u} + t_n^2 \tilde{w}_n) &= \bar{w} + t_n\bar{u} + t_n^2(\alpha w_n + (1-\alpha)\tilde{w}_n) \\
&\in H(\bar{x} + t_n\bar{u} + t_n^2 x_n).
\end{aligned}$$

Moreover, due to the assumption that f is C-convex, we obtain

$$\begin{aligned}
\bar{y} + t_n\bar{v} + t_n^2 y_n(\alpha) &:= f(\bar{w} + t_n\bar{u} + t_n^2(\alpha w_n + (1-\alpha)\tilde{w}_n), \bar{x} + t_n\bar{u} + t_n^2 x_n) \\
&\leq_C \bar{y} + t_n\bar{v} + t_n^2(\alpha y_n + (1-\alpha)\tilde{y}_n) \\
&\leq_C \bar{y} + t_n\bar{v} + t_n^2 y_n,
\end{aligned}$$

where we used the fact that

$$(\bar{y} + t_n\bar{v} + t_n^2 y_n) - (\bar{y} + t_n\bar{v} + t_n^2(\alpha y_n + (1-\alpha)\tilde{y}_n)) = t_n^2(1-\alpha)(y_n - \tilde{y}_n) \in C.$$

Since f is continuous, we have

$$\bar{y} + t_n\bar{v} + t_n^2 y_n(\alpha) \to \bar{y} + t_n\bar{v} + t_n^2 \tilde{y}_n \quad \text{as } \alpha \to 0$$
$$\bar{y} + t_n\bar{v} + t_n^2 y_n(\alpha) \to \bar{y} + t_n\bar{v} + t_n^2 y_n \quad \text{as } \alpha \to 1.$$

Since $\|\tilde{y}_n\| \to \infty$ and $y_n \to y$, by taking α_n appropriately close to 1, we have

$$\varepsilon t_n^2 \leq \|\bar{y} + t_n\bar{v} + t_n^2 y_n - (\bar{y} + t_n\bar{v} + t_n^2 y_n(\alpha_n))\| \leq t_n^2,$$

where $\varepsilon \in (0, 1)$. Setting $y_n(\alpha_n) = z_n$ we see that $\varepsilon \leq \|y_n - z_n\| \leq 1$.

Because $y_n \to y$, we deduce that (z_n) is bounded. Assume that $z_n \to z$, which implies $z \in D^2F(\bar{x}, \bar{y}, \bar{u}, \bar{v})(x)$. Since $\varepsilon \leq \|y_n - z_n\|$, we have $\varepsilon \leq \|y - z\|$, and hence $y \neq z$.

From the inequality

$$\bar{y} + t_n\bar{v} + t_n^2 z_n \leq_C \bar{y} + t_n\bar{v} + t_n^2 y_n,$$

we infer that $y_n - z_n \in C$, which implies $y - z \in C$. This, however, is a contradiction to the fact that $y \in \operatorname{Min}_C D^2F(\bar{x}, \bar{y}, \bar{u}, \bar{v})(x)$. Therefore, $\{\tilde{y}_n\}$ has to be a bounded. The proof is complete. □

Example 13.2.12. As an illustration of Theorem 13.2.11, let $X = W = (0, +\infty)$, $Y = \mathbb{R}^2$, $C = \mathbb{R}^2_+$, and define $H : X \rightrightarrows W$ by $H(x) = [x, +\infty)$ and $f : W \times X \to Y$ by $f(w, x) = (w, 1/w + x^2)$. In this example, the hypotheses of Theorem 13.2.11 are satisfied for any $\bar{x} > 0$. One can also see directly that the conclusion of Theorem 4.7 holds, since for each $x > 0$, we have

$$Q(x) = F(x) = \{(w, 1/w + x^2) \mid w \geq x\}.$$

To see what DF and D^2F look like in this example, consider the case where $\bar{x} = 1$ and $\bar{y} = (2, 3/2)$. Then

$$DF(1, (2, 3/2))(x) = \{(y, 2x - y/4) \mid y \in \mathbb{R}\}$$

and

$$D^2F(1, (2, 3/2), v_1, (v_2, 2v_1 - v_2/4))(x) = \{(y, 2x - y/4 + {v_1}^2 + {v_2}^2/8) \mid y \in \mathbb{R}\}.$$

□

13.3 Sensitivity Analysis in Set-Valued Optimization Using Coderivatives

In the previous sections, the notion of the contingent derivative was used to study the sensitivity properties of set-valued optimization problems. In this section, our objective is show that various coderivatives can also be used as the derivative concept to investigate the sensitivity properties of set-valued optimization problems. Numerous authors have used coderivatives in parametric optimization. One of the earlier works is the useful contribution by Levy and Mordukhovich [376], where the authors considered parametric families of constrained problems in mathematical programming and conducted a local sensitivity analysis for set-valued solution maps. More akin to this section is the work by Huy, Mordukhovich, and Yao [272] where the authors obtained formulas for estimating and computing the normal and mixed coderivatives of frontier and solution maps in parametric multiobjective optimization. Yao and Yen [604] employed coderivatives for some calculations related to a parametric affine variational inequality. Recently, Chung and Yao [98] used coderivatives of minimal point set-valued map in parametric vector optimization. In an interesting paper, Nam [438] provided precise formulae in terms of the initial data for the computation of the Fréchet and limiting normal cones to the normal cone mapping to a given convex polyhedron when the set of generating elements corresponding to the active constraints is linearly independent. Based on those results, he gave an explicit representation of the coderivatives of the normal cone mapping and computed the coderivative of solution maps to parametric variational inequalities. Our treatment of the subject in the following is based on the results derived by Chuong [97] where the author provides estimates for the Clarke coderivative of the efficient set map associated with a parametric vector optimization problem. These results are analogous to some of the results earlier obtained in [272].

We recall the setting of the model problem. Let P, X, and Y be Banach spaces, let $f : P \times X \to Y$ be a given single-valued map, and let $M : P \rightrightarrows X$ be a given set-valued map. Let $C \subset Y$ be a pointed, closed, and convex cone inducing a partial ordering in Y. We consider the following parametric vector optimization problem

$$\min_C \{f(p,x) |\ x \in M(p)\}, \tag{13.32}$$

depending on the parameter $p \in P$.

We define the associated set-valued map $F : P \rightrightarrows Y$ by

$$F(p) = (f \circ M)(p) := \{f(p, M(p))\} = \{f(p,x) |\ x \in M(p)\}.$$

Our objective is to investigate the differentiability properties of the following set-valued perturbation map $Q : P \rightrightarrows Y$ associated to the vector optimization problem (13.32):

$$Q(p) = \mathrm{Min}(F(p), C), \quad p \in P. \tag{13.33}$$

We begin by formulating and recalling a few concepts.

Definition 13.3.1. Let Y be a Banach space and Y^* be the topological dual of Y. Let $C \subset$ be a closed, convex, and pointed cone. The **uniformly positive polar** to the cone C is defined by

$$C^+_{UP} := \{y^* \in Y^* | \, \exists \, \beta > 0, \; \langle y^*, k \rangle \geq \beta \|k\|, \; \forall \, k \in C\}.$$

We recall that given Banach spaces X and Y, a single-valued map $g : X \to Y$ is called **strictly differentiable** at $\bar{x}$ if there exists a linear, continuous map $Dg(\bar{x}) : X \to Y$ such that

$$\lim_{x,z \to \bar{x}} \frac{g(x) - g(z) - \langle Dg(\bar{x}), x - z \rangle}{\|x - z\|} = 0.$$

We also recall that given Banach spaces X and Y and $S \subset X$, a single-valued map $L : S \to Y$ is said to be **local upper-Lipschitzian** at $\bar{x} \in S$ if there are numbers $\eta > 0$ and $\ell \geq 0$ such that

$$\|L(x) - L(\bar{x})\| \leq \ell \|x - \bar{x}\| \quad \text{for every } x \in B_\eta(\bar{x}) \cap S.$$

Definition 13.3.2. Let X and Y be Banach spaces, let $\mathbb{L} : D \rightrightarrows Y$ with $D = \operatorname{dom}(\mathbb{L})$ be a set-valued map, and let $(\bar{x}, \bar{y}) \in \operatorname{graph}(\mathbb{L})$. The map $\mathbb{L}$ is said to admit a **local upper-Lipschitzian selection** at $(\bar{x}, \bar{y})$ if there exists a single-valued mapping $L : D \to Y$ which is local upper-Lipschitzian at $\bar{x}$ satisfying $L(\bar{x}) = \bar{y}$ and $L(x) \in \mathbb{L}(x)$, for every $x \in D$.

We recall that given a Banach space, a set $S \subset X$, and $\bar{x} \in S$, the Clarke normal cone, the set of Fréchet ε-normals, the Fréchet normal cone, the Mordukhovich normal cone, are given by:

$$N_c(S, \bar{x}) := C(S, \bar{x})^- = \{x^* \in X^* | \, \langle x^*, x \rangle \leq 0, \;\; x \in C(S, \bar{x})\}.$$

$$\hat{N}_\varepsilon(S, \bar{x}) := \left\{ x^* \in X^* | \, \limsup_{x \xrightarrow{S} \bar{x}} \frac{\langle x^*, x - \bar{x} \rangle}{\|x - \bar{x}\|} \leq \varepsilon \right\}.$$

$$\hat{N}(S, \bar{x}) := \left\{ x^* \in X^* | \, \limsup_{x \xrightarrow{S} \bar{x}} \frac{\langle x^*, x - \bar{x} \rangle}{\|x - \bar{x}\|} \leq 0 \right\}.$$

$$N(S, \bar{x}) := \limsup_{x \xrightarrow{S} \bar{x}, \, \varepsilon \downarrow 0} \hat{N}_\varepsilon(S, \bar{x}).$$

We also recall that the following chain of inclusions is known to hold:

$$\hat{N}(S, \bar{y}) \subset N(S, \bar{y}) \subset N_c(S, \bar{y}). \tag{13.34}$$

The set S is called **Fréchet normally regular** at $\bar{x}$ if

$$\hat{N}(S,\bar{x}) = N_c(S,\bar{x}).$$

Finally, given Banach spaces X and Y with X^* and Y^* their duals, a set-valued map $F : X \rightrightarrows Y$ with $(\bar{x},\bar{y}) \in \text{graph}\,(F)$, the Clarke coderivative and the Mordukhovich coderivative are given by:

$$D_c^* F(\bar{x},\bar{y})(y^*) := \{x^* \in X^* |\, (x^*,-y^*) \in N_c(\text{graph}\,(F),(\bar{x},\bar{y})\},$$
$$D^* F(\bar{x},\bar{y})(y^*) := \{x^* \in X^* |\, (x^*,-y^*) \in N(\text{graph}\,(F),(\bar{x},\bar{y})\}.$$

It follows from above definition that for every $y^* \in Y^*$, the following inclusion holds:

$$D^* F(\bar{p},\bar{y})(y^*) \subset D_c^* F(\bar{p},\bar{y})(y^*). \tag{13.35}$$

We begin with the following auxiliary result:

Theorem 13.3.3. *Let P and Y be Banach spaces with P^* and Y^* their topological duals, let $G : P \rightrightarrows Y$ be a set-valued, and let $(\bar{p},\bar{y}) \in \text{graph}\,(G)$. Let $C \subset Y$ be a pointed, closed, and convex cone.*

1. Assume that $(G + C)$ is tangentially regular at $(\bar{p},\bar{y})$. Then for every $y^* \in Y^*$,

$$D_c^*(G + C)(\bar{p},\bar{y})(y^*) \subseteq D_c^* G(\bar{p},\bar{y})(y^*). \tag{13.36}$$

2. Assume that the following conditions hold:

a) The map G is Fréchet normally regular at $(\bar{p},\bar{y})$.
b) The map $H : P \times Y \rightrightarrows Y$ given by $H(p,y) := G(p) \cap (y - C)$ admits a local upper-Lipschitz selection at $(\bar{p},\bar{y},\bar{y})$.

Then for every $y^* \in C_{UP}^+$, the following inclusion holds:

$$D_c^*(G + C)(\bar{p},\bar{y})(y^*) \supseteq D_c^* G(\bar{p},\bar{y})(y^*). \tag{13.37}$$

Proof. By using the tangential regularity of $G + C$, we have

$$C(\text{graph}\,(G),(\bar{p},\bar{y})) \subseteq T(\text{graph}\,(G),(\bar{p},\bar{y})) \subset T(\text{graph}\,(G + C),(\bar{p},\bar{y}))$$
$$= C(\text{graph}\,(G + C),(\bar{p},\bar{y})),$$

which implies $N_c(\text{graph}\,(G + C),(\bar{p},\bar{y})) \subseteq N_c(\text{graph}\,(G),(\bar{p},\bar{y}))$, and consequently, for every $y^* \in Y^*$, we have $D_c^*(G + C)(\bar{p},\bar{y})(y^*) \subseteq D_c^* G(\bar{p},\bar{y})(y^*)$, which completes the proof of the first part.

For the second part, we fix an element $y^* \in C_{UP}^+$ and choose an arbitrary $p^* \in D_c G(\bar{p},\bar{y})(y^*)$. To obtain a contradiction, we begin by assuming that $p^* \notin$

$D^*_C(G+C)(\bar{p},\bar{y})(y^*)$. By (13.35), we deduce that $p^* \notin D^*(G+C)(\bar{p},\bar{y})(y^*)$, which implies that $(p^*,-y^*) \notin \hat{N}(\text{graph}\,(G+C),(\bar{p},\bar{y}))$.

Therefore, by using the definition of the Fréchet normal cone, there exists a sequence $\{(p_n,y_n)\} \subset \text{graph}\,(G+C)$ with $(p_n,y_n) \to (\bar{p},\bar{y})$ such that

$$\limsup_{n\to\infty} \frac{\langle (p^*,-y^*),(p_n,y_n)-(\bar{p},\bar{y})\rangle}{\|(p_n,y_n)-(\bar{p},\bar{y})\|} > 0. \tag{13.38}$$

From the assumptions that the map H admits a local upper-Lipschitzian selection at $(\bar{p},\bar{y},\bar{y})$ and that $\text{dom}\,(H) = \text{graph}\,(G+C)$, there are a number $\ell > 0$ and neighborhoods $U \times V \in \mathcal{U}(\bar{p},\bar{y})$ such that for each $(p,y) \in (U \times V) \cap \text{graph}\,(G+C)$, we can find $\hat{y} \in H(p,y)$ such that

$$\|\hat{y}-\bar{y}\| \le \ell \|(p,y)-(\bar{y},\bar{y})\|.$$

Due to the convergence $(p_n,y_n) \to (\bar{p},\bar{y})$, there exists $m \in \mathbb{N}$ such that $(p_n,y_n) \in U \times V$, for every $n \ge m$, and hence, for every $n \ge m$, there exists $\hat{y}_n \in H(p_n,y_n) = G(p_n) \cap (y_n - C)$ such that

$$\|\hat{y}_n-\bar{y}\| \le \ell \|(p_n,y_n)-(\bar{p},\bar{y})\|.$$

Using the definition of the map H, for each $n \ge m$, there exists $\hat{y}_n \in G(p_n)$ and $k_n \in C$ such that $\hat{y}_n = y_n - k_n$ which implies that there exists $(p_n,\hat{y}_n) \xrightarrow{\text{graph}\,(G)} (\bar{p},\bar{y})$ satisfying that

$$\|(p_n,\hat{y}_n)-(\bar{p},\bar{y})\| \le (\ell+1)\|(p_n,y_n)-(\bar{p},\bar{y})\| \quad \text{for every } n \ge m.$$

Since $y^* \in C^+_{UP}$, there exists $\beta > 0$ such that $\langle y^*,k\rangle \ge \beta\|k\| \ge 0$ for every $k \in C$, and consequently, for each $n \ge m$, we have

$$\begin{aligned}\langle (p^*,-y^*),(p_n,\hat{y}_n)-(\bar{p},\bar{y})\rangle &= \langle (p^*,-y^*),(p_n-\bar{p},y_n-k_n-\bar{y})\rangle \\ &= \langle (p^*,-y^*),(p_n-\bar{p},y_n-\bar{y})\rangle + \langle y^*,k_n\rangle \\ &\ge \langle (p^*,-y^*),(p_n-\bar{p},y_n-\bar{y})\rangle,\end{aligned}$$

which, by using (13.38), yields

$$\begin{aligned}\limsup_{n\to\infty} \frac{\langle (p^*,-y^*),(p_n,\hat{y}_n)-(\bar{p},\bar{y})\rangle}{\|(p_n,\hat{y}_n)-(\bar{p},\bar{y})\|} &\ge \limsup_{n\to\infty} \frac{\langle (p^*,-y^*),(p_n-\bar{p},y_n-\bar{y})\rangle}{\|(p_n,\hat{y}_n)-(\bar{p},\bar{y})\|} \\ &\ge \limsup_{n\to\infty} \frac{\langle (p^*,-y^*),(p_n-\bar{p},y_n-\bar{y})\rangle}{(\ell+1)(\|(p_n,y_n)-(\bar{p},\bar{y})\|)} \\ &> 0.\end{aligned}$$

Therefore,

$$\limsup_{(p,y)\overset{\text{graph}(G)}{\longrightarrow}(\bar{p},\bar{y})} \frac{\langle (p^*,-y^*),(p,y)-(\bar{p},\bar{y})\rangle}{\|(p,y)-(\bar{p},\bar{y})\|} > 0,$$

which implies that $(p^*,-y^*) \notin \hat{N}(\text{graph}(G),(\bar{p},\bar{y}))$, and by the fact that G is Fréchet normally regular, we deduce that $p^* \notin D_c^*G(\bar{p},\bar{y})(y^*)$, which is a contradiction. This completes the proof. □

Remark 13.3.4. Bao and Mordukhovich [26] proved an analogue of (13.36) for the Mordukhovich coderivative and an analogue of (13.37) for the mixed coderivative.

The following example shows that (13.37) does not hold without additional assumptions.

Example 13.3.5. For $P = Y = \mathbb{R}$ and $C = \mathbb{R}_+$, we define a set-valued map $G : \mathbb{R} \rightrightarrows \mathbb{R}$ by

$$G(p) = \begin{cases} \{0,2\} & \text{if} \quad p \geq 0, \\ \emptyset & \text{otherwise.} \end{cases}$$

We choose $(\bar{p},\bar{y}) = (0,2) \in \text{graph}(G)$ and $y^* = 2$. By a simple calculation, we obtain that

$$\hat{N}(\text{graph}(G),(0,2)) = N_c(\text{graph}(G),(0,2)) = \{(x,y) \in \mathbb{R}^2 |\ x \leq 0,\ y \in \mathbb{R}\},$$

and

$$D_c^*(G+C)(0,2)(2) = \emptyset,$$
$$D_c^*G(0,2)(2) = -\mathbb{R}_+,$$

showing that $D_c^*(G+C)(\bar{p},\bar{y})(y^*) \not\supset D_c^*G(\bar{p},\bar{y})(y^*)$. □

Remark 13.3.6. We note that for the data defined in Example 13.3.5, the map H does not admit a local upper Lipschitzian selection at $(0,2,2)$. Also note that in this example, the map G is Fréchet normally regular at $(0,2)$ and $y^* \in\in C_{UP}^+$.

The following example shows that the Fréchet normal regularity is essential for (13.37).

Example 13.3.7. We set $P = Y = \mathbb{R}$ and $C = \mathbb{R}_+$. We define a map $G : \mathbb{R} \rightrightarrows \mathbb{R}$ by $G(p) := \{|p|\}$.

We choose $(\bar{p},\bar{y}) = (0,0) \in \text{graph}(G)$ and $y^* = 1 \in C_{UP}^+$. By a simple calculation, we obtain that

$$\hat{N}(\text{graph}(G),(0,0)) = \{(x,y) \in \mathbb{R}^2 |\ y \leq x \leq -y,\ y \leq 0\},$$

$$N_c(\text{graph}\,(G),(0,0)) = \mathbb{R}^2,$$

and

$$D_c^*(G+C)(0,0)(1) = [-1,1],$$
$$D_c^*G(0,0)(1) = \mathbb{R},$$

showing that $D_c^*(G+C)(\bar{p},\bar{y})(y^*) \not\supset D_c^*G(\bar{p},\bar{y})(y^*)$. □

In the following result, we give an estimate for the Clarke coderivative of the composition of a single-valued map and a set-valued map.

Theorem 13.3.8. *Let P, X and Y be Banach spaces, let $f : P \times X \to Y$ be a single-valued map, and let $G : P \rightrightarrows X$ be a set-valued map. Let $\bar{p} \in P$ and $\bar{y} \in (f \circ G)(\bar{p})$. For $\bar{x} \in G(\bar{p})$ with $(\bar{p},\bar{x}) \in f^{-1}(\bar{y})$, let f be Fréchet differentiable at $(\bar{p},\bar{x})$ with $\nabla f(\bar{p},\bar{x}) := (\nabla_p f(\bar{p},\bar{x}), \nabla_x f(\bar{p},\bar{x}))$ as the derivative.*

1. *Assume that the composite map $f \circ G$ is tangentially regular at $(\bar{p},\bar{y})$. Then for every $y \in Y^*$,*

$$D_c^*(f\circ G)(\bar{p},\bar{y})(y^*) \subseteq \nabla_p f(\bar{p},\bar{x})^*(y^*) + D_c^*G(\bar{p},\bar{x})(\nabla_x f(\bar{p},\bar{x})^*(y^*)). \tag{13.39}$$

2. *Assume that the following conditions hold:*
   a) *The map G is Fréchet normally regular at $(\bar{p},\bar{x})$.*
   b) *The map $\tilde{G} : P \times Y \rightrightarrows X$ given by $\tilde{G}(p,y) := \{x \in G(p) |\ y = f(p,x)\}$ admits a local upper Lipschitzian selection at $(\bar{p},\bar{y},\bar{x})$.*

   Then for every $y^* \in Y^*$, the following inclusion remains valid:

$$D_c^*(f\circ G)(\bar{p},\bar{y})(y^*) \supseteq \nabla_p f(\bar{p},\bar{x})^*(y^*) + D_c^*G(\bar{p},\bar{x})(\nabla_x f(\bar{p},\bar{x})^*(y^*)). \tag{13.40}$$

Proof. We claim that for every $p \in P$, the following inclusion holds:

$$\{(p, \nabla f(\bar{p},\bar{x}))(p,x) | (p,x) \in C(\text{graph}\,(G),(\bar{p},\bar{x}))\} \subset C(\text{graph}\,(f\circ G),(\bar{p},\bar{y})). \tag{13.41}$$

To verify our claim, for $p \in P$, we pick $(p,x) \in C(\text{graph}\,(G),(\bar{p},\bar{x}))$. Then $(p,x) \in T(\text{graph}\,(G),(\bar{p},\bar{x}))$, and hence there are sequences $\{t_n\} \subset \mathbb{P}$ and $\{(p_n,x_n)\} \subset P \times X$ with $t_n \downarrow 0$ and $(p_n,x_n) \to (p,x)$ such that $\bar{x} + t_n x_n \in G(\bar{p}+t_n p_n)$ for every $n \in \mathbb{N}$. Therefore, $f(\bar{p}+t_n p_n, \bar{x}+t_n x_n) \in (f\circ G)(\bar{p}+t_n p_n)$ for every $n \in \mathbb{N}$, and by rearranging the above containment, for every $n \in \mathbb{N}$, we get

$$\bar{y} + t_n \frac{f(\bar{p} + t_n p_n, \bar{x} + t_n x_n) - f(\bar{p}, \bar{x})}{t_n} \in (f \circ G)(\bar{p} + t_n p_n).$$

In view of the Fréchet differentiability of the map f at $(\bar{p}, \bar{x})$, we get

$$\lim_{n \to \infty} \frac{f(\bar{p} + t_n p_n, \bar{x} + t_n x_n) - f(\bar{p}, \bar{x})}{t_n} = \nabla f(\bar{p}, \bar{x})(p, x),$$

and hence

$$(p, \nabla f(\bar{p}, \bar{x})(p, x)) \in T(\text{graph}\,(f \circ G), (\bar{p}, \bar{y})) = C(\text{graph}\,(f \circ G), (\bar{p}, \bar{y})), \tag{13.42}$$

where the last equality is due to the tangential regularity of $f \circ G$ at $(\bar{p}, \bar{y})$. This proves (13.41).

Since for each $(p, x) \in C(\text{graph}\,(G), (\bar{p}, \bar{x}))$ and $(u^*, -\nabla_x f(\bar{p}, \bar{x})^* y^*) \in N_c(\text{graph}\,(G), (\bar{p}, \bar{x}))$,

$$\langle (\nabla_p f(\bar{p}, \bar{x})^* y^* + u^*, -y^*), (p, \nabla f(\bar{p}, \bar{x})(p, x)) \rangle = u^*(p) - \nabla_x f(\bar{p}, \bar{x})^* y^*(x),$$

we deduce that for all $(p, x) \in C(\text{graph}\,(G), (\bar{p}, \bar{x}))$ and $(u^*, -\nabla_x f(\bar{p}, \bar{x})^* y^*) \in N_c(\text{graph}\,(G), (\bar{p}, \bar{x}))$,

$$\langle (-\nabla_x f(\bar{p}, \bar{x}))^* y^* + u^*, -y^*), (p, \nabla f(\bar{p}, \bar{x})(p, x)) \rangle \le 0. \tag{13.43}$$

Consequently, by using (13.41), we get

$$N_c(\text{graph}\,(f \circ G), (\bar{p}, \bar{y})) \subset \{(\nabla_p f(\bar{p}, \bar{x})^* y^* + u^*, -y^*) |\, (u^*, -\nabla_x f(\bar{p}, \bar{x})^* y^*) \in N_c(\text{graph}\,(G), (\bar{p}, \bar{x}))\}.$$

For $y^* \in Y^*$, let $p^* \in D_c^*(f \circ G)(\bar{p}, \bar{y})(y^*)$ be arbitrary. Then, $(p^*, -y^*) \in N_c(\text{graph}\,(f \circ G), (\bar{x}, \bar{y}))$, and in view of the above inclusion, there exists $(u^*, -\nabla_x f(\bar{p}, \bar{x})^* y^*) \in N_c(\text{graph}\,(G), (\bar{p}, \bar{x}))$ such that $(p^*, -y^*) = (\nabla_p f(\bar{p}, \bar{x})^* y^* + u^*, -y^*)$, which gives

$$(p^* - \nabla_p f(\bar{p}, \bar{x})^* y^*, -\nabla_x f(\bar{p}, \bar{x})^* y^*) \in N_c(\text{graph}\,(G), (\bar{p}, \bar{x}))$$

and consequently, we have $p^* \in \nabla_p f(\bar{p}, \bar{x})^* y^* + D_c^* G(\bar{p}, \bar{x})(\nabla_x f(\bar{p}, \bar{x})^* y^*)$. Therefore, we have shown that $p^* \in \nabla_p f(\bar{p}, \bar{x})^* y^* + D_c^* G(\bar{p}, \bar{x})(\nabla_x f(\bar{p}, \bar{x})^* y^*)$, and the proof of (13.39) is complete.

For the converse, let u^* be a fixed element such that $u^* \notin D_c^*(f \circ G)(\bar{p}, \bar{y})(y^*)$. We will show that

$$u^* \notin \nabla_p f(\bar{p}, \bar{x})^* y^* + D_c G(\bar{p}, \bar{x})(\nabla_x f(\bar{p}, \bar{x})^* y^*). \tag{13.44}$$

First note that $u^* \notin D_c^*(f \circ G)(\bar{p}, \bar{y})(y^*)$ implies $(u^*, -y^*) \notin \hat{N}(\text{graph}\,(f \circ G), (\bar{p}, \bar{y}))$. Thus,

$$\limsup_{(p,y)\overset{\text{graph}\,(f\circ G)}{\longrightarrow}(\bar{p},\bar{y})} \frac{\langle (u^*, -y^*), (p, y) - (\bar{p}, \bar{y}) \rangle}{\|(p, y) - (\bar{p}, \bar{y})\|} > 0,$$

implying that there exist $\{(p_n, y_n)\} \subset \text{graph}\,(f \circ G)$ and $\alpha > 0$ such that $(p_n, y_n) \to (\bar{p}, \bar{y})$ and

$$\langle u^*, p_n - \bar{p} \rangle \geq \langle y^*, y_n - \bar{y} \rangle + \alpha(\|p_n - \bar{p}\| + \|y_n - \bar{y}\|), \quad \text{for sufficiently large } n \in \mathbb{N}. \tag{13.45}$$

Since the map $\tilde{G}$ admits a local upper Lipschitzian selection at $(\bar{p}, \bar{y}, \bar{x})$, there exists a function $\ell : \text{dom}\,(\hat{G}) \to X$ which is local upper Lipschitzian at $(\bar{p}, \bar{y})$ and satisfies that $\ell(\bar{p}, \bar{y}) = \bar{x}$ and $\ell(p, y) \in \hat{G}(p, y)$ for all $(p, y) \in \text{dom}\,(\hat{G})$ sufficiently close to $(\bar{p}, \bar{y})$. Therefore, there exists $L > 0$ such that

$$\|x_n - \bar{x}\| \leq L(\|p_n - \bar{p}\| + \|y_n - \bar{y}\|) \tag{13.46}$$

for sufficiently large n, where $x_n := \ell(p_n, y_n) \in \tilde{G}(p_n, y_n)$. For such n, since $x_n \in \tilde{G}(p_n, y_n)$, $x_n \in G(p_n)$, and $y_n = f(p_n, x_n)$, it follows from (13.45) that

$$\begin{aligned}
\langle u^*, p_n - \bar{p} \rangle &\geq \langle y^*, f(p_n, x_n) - f(\bar{p}, \bar{x}) \rangle + \alpha(\|p_n - \bar{p}\| + \|f(p_n, x_n) - f(\bar{p}, \bar{x})\|) \\
&= \langle y^*, \nabla f(\bar{p}, \bar{x})(p_n - \bar{p}, x_n - \bar{x}) \rangle + o(\|p_n - \bar{p}\| + \|x_n - \bar{x}\|) + \alpha(\|p_n - \bar{p}\| \\
&\quad + \|f(p_n, x_n) - f(\bar{p}, \bar{x})\|) \\
&= \langle \nabla f(\bar{p}, \bar{x})^* y^*, (p_n - \bar{p}, x_n - \bar{x}) \rangle + o(\|p_n - \bar{p}\| + \|x_n - \bar{x}\|) \\
&\quad + \alpha(\|p_n - \bar{p}\| + \|f(p_n, x_n) - f(\bar{p}, \bar{x})\|),
\end{aligned}$$

and by using (13.46), we get that

$$\begin{aligned}
\langle u^*, p_n - \bar{p} \rangle &\geq \langle \nabla f(\bar{p}, \bar{x})^* y^*, (p_n - \bar{p}, x_n - \bar{x}) \rangle + o(\|p_n - \bar{p}\| + \|x_n - \bar{x}\|) \\
&\qquad + \frac{\alpha}{2}\|p_n - \bar{p}\| + \frac{\alpha}{2L}\|x_n - \bar{x}\| \\
&\geq \langle \nabla f(\bar{p}, \bar{x})^* y^*, (p_n - \bar{p}, x_n - \bar{x}) \rangle + o(\|p_n - \bar{p}\| + \|x_n - \bar{x}\|) \\
&\qquad + \hat{\alpha}\left(\|p_n - \bar{p}\| + \|x_n - \bar{x}\|\right),
\end{aligned}$$

where $\hat{\alpha} = \min\{\alpha/2, \alpha/(2L)\}$. Therefore,

$$\limsup_{(p,x)\overset{\text{graph}\,(G)}{\longrightarrow}(\bar{p},\bar{x})} \frac{\langle u^* - \nabla_x f(\bar{p}, \bar{x})^* y^*, p - \bar{p} \rangle - \langle \nabla_x f(\bar{p}, \bar{x})^* y^*, x - \bar{x} \rangle}{\|p - \bar{p}\| + \|x - \bar{x}\|} \geq \hat{\alpha},$$

which implies that $(u^* - \nabla_p f(\bar{p},\bar{x})^*(y^*), -\nabla_x f(\bar{p},\bar{x})^* y^*) \notin \hat{N}(\text{graph}(G), (\bar{p},\bar{x}))$. Then, by using the Fréchet normal regularity of G at $(\bar{p},\bar{x})$, we have $u^* - \nabla_p f(\bar{p},\bar{x})^*(y^*) \notin D_c^* G(\bar{p},\bar{x})(\nabla_x f(\bar{p},\bar{x})^* y^*)$, and hence (13.44) is verified. The proof is complete. □

The following example shows that the imposed conditions cannot be removed:

Example 13.3.9. We set $P = Y = \mathbb{R}$, $C = \mathbb{R}_+$, and $f(p,x) := x^2$. We define a set-valued map:

$$G(p) := \begin{cases} \{-\sqrt{p}, \sqrt{p}\} & \text{if} \quad p \geq 0, \\ \emptyset & \text{otherwise.} \end{cases}$$

Then,

$$\text{graph}(G) = \left\{(p,x) \in \mathbb{R}^2 \mid x^2 = p\right\},$$
$$\text{graph}(f \circ G) = \{(p,y) \in \mathbb{R}^2 \mid p \geq 0, \ y = p\}.$$

We choose $\bar{p} = \bar{x} = \bar{y} = 0$ and $y^* = 0$. By a simple calculation, we obtain that

$$\hat{N}(\text{graph}(G), (0,0)) = N_c(\text{graph}(G), (0,0)) = \{(x,0) \in \mathbb{R}^2 \mid x \in \mathbb{R}\},$$

and

$$D_c^*(f \circ G)(0,0)(0) = -\mathbb{R}_+,$$
$$D_c^* G(0,0)(0) = \mathbb{R},$$

proving that the inclusion (13.40) fails to hold. Note that the set-valued map $\tilde{G}$ does not admit any local upper Lipschitzian selection at $(0,0,0)$. □

We will now give some coderivative estimates for the perturbation map Q defined in (13.33). For this, we define two set-valued maps $T : P \times Y \rightrightarrows Y$ and $\tilde{M} : P \times Y \rightrightarrows X$ as follows:

$$T(p,y) := Q(p) \cap (y - C)$$
$$\tilde{M}(p,y) := \{x \in M(p) \mid y = f(p,x)\}.$$

Theorem 13.3.10. *Let $\bar{p} \in P$ and $\bar{x} \in M(\bar{p})$ be such that $\bar{y} := f(\bar{p},\bar{x}) \in Q(\bar{p})$. Let f be Fréchet differentiable at $(\bar{p},\bar{x})$ with the derivative $\nabla f(\bar{p},\bar{x}) := (\nabla_p f(\bar{p},\bar{x}), \nabla_x f(\bar{p},\bar{x}))$. Assume that the domination property holds for F around $\bar{p}$ and T admits a local upper Lipschitzian selection at $(\bar{p},\bar{y},\bar{y})$.*

1. *If F and $F + C$ are tangentially regular at $(\bar{p},\bar{y})$ and Q is Fréchet normally regular at $(\bar{p},\bar{y})$, then for every $y^* \in C^+_{UP}$,*

$$D_C^* Q(\bar{p}, \bar{y})(y^*) \subseteq \nabla_p f(\bar{p}, \bar{x})^* y^* + D_C^* M(\bar{p}, \bar{x})(\nabla_x f(\bar{p}, \bar{x})^* y^*). \tag{13.47}$$

2. *If $Q + C$ is tangentially regular at $(\bar{p}, \bar{y})$, F is Fréchet regular at $(\bar{p}, \bar{y})$, and M is Frećhet normal at $(\bar{p}, \bar{x})$ and admits a local upper Lipschitzian selection $(\bar{p}, \bar{y}, \bar{x})$, then for every $y^* \in C_{UP}^+$,*

$$D_C^* Q(\bar{p}, \bar{y})(y^*) \supset \nabla_p f(\bar{p}, \bar{x})^* y^* + D_C^* M(\bar{p}, \bar{x})(\nabla_x f(\bar{p}, \bar{x})^* y^*). \tag{13.48}$$

Proof. Since for every $p \in P$, the inclusion $Q(p) \subseteq F(p)$ holds and since the domination property holds for F around $\bar{p}$, there exists a neighborhood U of $\bar{p}$ such that

$$Q(p) + C = F(p) + C, \quad \text{for every } p \in U,$$

and consequently, for every $y^* \in Y^*$, we have

$$D_C^*(Q + C)(\bar{p}, \bar{y})(y^*) = D_C^*(F + C)(\bar{p}, \bar{y})(y^*). \tag{13.49}$$

Since the map Q is Fréchet normally regular at $(\bar{p}, \bar{x})$ and the map T admits a local upper Lipschitzian selection at $(\bar{p}, \bar{y}, \bar{y})$, it follows from (13.37) that for every $y^* \in C_{UP}^+$, we have

$$D_C^* Q(\bar{p}, \bar{y})(y^*) \subseteq D_C^*(Q + C)(\bar{p}, \bar{y})(y^*), \tag{13.50}$$

and, in view of the tangential regularity, we also have for every $y^* \in Y^*$,

$$D_C^*(F + C)(\bar{p}, \bar{y})(y^*) \subseteq D_C^* F(\bar{p}, \bar{y})(y^*). \tag{13.51}$$

Furthermore, using (13.39), for every $y^* \in Y^*$, we get

$$D_C^* F(\bar{p}, \bar{y})(y^*) \subseteq \nabla_p f(\bar{p}, \bar{x})^* y^* + D_C^* M(\bar{p}, \bar{x})(\nabla_x f(\bar{p}, \bar{x})^* y^*). \tag{13.52}$$

By combining (13.49) thorough (13.52) we finish the proof of the inclusion (13.47).

For (13.48), we begin by noticing that since the map $Q+C$ is tangentially regular at $(\bar{p}, \bar{y})$, it follows from Theorem 13.3.3 that for every $y^* \in Y^*$, we have

$$D_C^*(Q + C)(\bar{p}, \bar{y})(y^*) \supset D_C^* Q(\bar{p}, \bar{y})(y^*). \tag{13.53}$$

Define the set-valued map $H : P \times Y$ by

$$H(p, y) := F(p) \cap (y - C),$$

and note that $T(p, y) \subset H(p, y)$ for every $(p, y) \in P \times Y$, $\text{dom}\,(T) = \text{graph}\,(Q + C)$, and $\text{dom}\,(H) = \text{graph}\,(F + C)$. Therefore, the existence of a local upper Lipschitzian selection of T at $(\bar{p}, \bar{y}, \bar{y})$ implies the existence of local upper Lipschitzian selection of H at this point. Therefore, by the Fréchet normal regularity of F at $(\bar{p}, \bar{y})$, for every $y^* \in C^+_{UP}$, we get

$$D^*_c(F + C)(\bar{p}, \bar{y})(y^*) \supset D^*_c F(\bar{p}, \bar{y})(y^*). \tag{13.54}$$

Because M is Fréchet normally regular at $(\bar{p}, \bar{x})$ and because $\tilde{M}$ admits a local upper Lipschitzian selection at $(\bar{p}, \bar{y}, \bar{x})$, it follows from Theorem 13.3.3 that for every $y^* \in Y^*$, we have

$$D^*_c F(\bar{p}, \bar{y})(y^*) \supset \nabla_p f(\bar{p}, \bar{x})^* y^* + D^*_c M(\bar{p}, \bar{x})(\nabla_x f(\bar{p}, \bar{x})^* y^*). \tag{13.55}$$

Combining (13.49), (13.53), (13.54), and (13.55) completes the proof. □

The following examples depicts the importance of the domination property.

Example 13.3.11. We set $P = Y = \mathbb{R}$ and $C = \mathbb{R}_+$. We define $f(p, x) := x$ and

$$M(p) := \begin{cases} x \in \mathbb{R} | \; x \geq 0, & \text{if} \quad p = 0, \\ x \in \mathbb{R} | \; x > |p|, & \text{otherwise.} \end{cases}$$

Then,

$$\text{graph}\,(M) = \{(p, x) \in \mathbb{R}^2 | \; x > 0, 0 \; - x < p < x\} \bigcup \{(0, 0)\},$$
$$F(p) = F(p) + C = M(p) \quad \text{for every } p \in P.$$

We choose $\bar{p} = \bar{x} = 0$ and $y^* = 1 \in C^+_{UP}$. By a simple calculation, we obtain that

$$Q(p) := \begin{cases} \{0\}, \text{ if} & p = 0, \\ \emptyset, & \text{otherwise,} \end{cases}$$

confirming that $\bar{y} := f(p, \bar{x}) = 0 \in Q(\bar{p})$. We compute

$$D^*_c Q(0, 0)(1) = (-\infty, \infty)$$
$$\nabla_p f(0, 0)^*(1) + D^*_c M(0, 0)(1) = [-1, 1],$$

which shows that (13.47) does not hold.

Notice that the set-valued maps F and $(F + C)$ are tangentially regular at $(0, 0)$ and the map Q is Fréchet normally regular at $(0, 0)$. However, the domination property does not hold at F around $\bar{p} = 0$. □

Analogues of the above formulas for special constraints can be found in [98, 99, 272] and the cited references therein.

13.4 Sensitivity Analysis for Vector Variational Inequalities

Vector variational inequalities have been of great interest in the academic and professional communities ever since the path-breaking paper by Giannessi [199]. The interested reader can find some of the recent developments on this useful field in [4, 96, 118, 122, 177, 200, 203, 204, 377, 416, 424, 478] and the cited references therein. Vector variational inequalities provide an elegant and useful extension of variational inequalities. We remark that in recent years the theory of variational inequalities emerged as one of the most promising branches of pure, applied, and industrial mathematics, and provides us with a convenient mathematical apparatus for studying a wide range of problems arising in diverse fields such as structural mechanics, elasticity, economics, optimization, optimal control, inverse problems, financial mathematics, and others (see [30,33,121,139,201,202,221,238,289,320, 327,331,508]).

This section is devoted to differentiability and sensitivity properties of gap functions associated with vector-variational inequalities. Many authors have focused on studying sensitivity analysis of vector variational inequalities. For instance, Li, Yan, Chen [383] conducted a detailed study of differential and sensitivity properties of gap functions associated with vector variational inequalities and also gave optimality conditions for solutions of vector variational inequalities. This study was further enhanced by Meng and Li [418] who investigated the differentiability and sensitivity properties of gap functions associated with Minty vector-variational inequalities. Recently, Li and Li [378] presented the second-order differentiability and sensitivity properties of the gap functions for vector variational inequalities. Our treatment of the subject in this section follows the approach of [378, 383].

We begin by recalling the following notions of vector variational inequalities:

Definition 13.4.1. Let $C \subset \mathbb{R}^m$ be a pointed, closed, and convex cone, and let $K \subset \mathbb{R}^n$ be nonempty closed set. Let $F : K \to L(\mathbb{R}^n, \mathbb{R}^m)$ be a single-valued map.

1. The **vector variational inequality** seeks $\bar{x} \in K$ such that

$$\langle F(\bar{x}), x - \bar{x} \rangle \notin -C \setminus \{0\} \quad \text{for every } x \in K. \tag{13.56}$$

2. Given that C is solid, the **weak vector variational inequality** seeks $\bar{x} \in K$ such that

$$\langle F(\bar{x}), x - \bar{x} \rangle \notin -\text{int}(C) \quad \text{for every } x \in K. \tag{13.57}$$

The following definition recalls the notion of the gap functions associated to the above vector variational inequalities. See Chen, Goh, and Yang [90] and Meng and Li [418] for details.

Definition 13.4.2. Let $C \subset \mathbb{R}^m$ be a pointed, closed, and convex cone, and let $K \subset \mathbb{R}^n$ be nonempty closed set. Let $F : K \to L(\mathbb{R}^n, \mathbb{R}^m)$ be a given single-valued map.

1. A set-valued map $N : \mathbb{R}^n \rightrightarrows \mathbb{R}^m$ is said to be a gap function for vector variational inequality (13.56) if and only if the following conditions hold:
   a) $0 \in N(\bar{x})$ if and only if $\bar{x}$ solves vector variational inequality (13.56).
   b) $N(x) \cap (-C) = \{0\}$ for every $x \in K$.
2. Given that the cone C is also solid, a set-valued map $W : \mathbb{R}^n \rightrightarrows \mathbb{R}^m$ is said to be a gap function for weak vector variational inequality (13.57) if and only if the following conditions hold:
   a) $0 \in W(\bar{x})$ if and only if $\bar{x}$ solves weak vector variational inequality (13.57).
   b) $W(x) \cap (-\text{int}(C)) = \{\emptyset\}$ for every $x \in K$.

Given a compact set $K \subset \mathbb{R}^n$ and a single-valued map $F : K \to L(\mathbb{R}^n, \mathbb{R}^m)$, we introduce

$$G(x) := \bigcup_{z \in K} \{\langle F(x), x - z \rangle\} = \langle F(x), x - K \rangle. \tag{13.58}$$

Furthermore, using the above map G, we introduce the following set-valued maps:

$$N(x) = \text{Max}\,(G(x), C)\,, \tag{13.59}$$

$$W(x) = \text{WMax}\,(G(x), C)\,. \tag{13.60}$$

We have the following result regarding the above set-valued maps:

Theorem 13.4.3. *Let $C \subset \mathbb{R}^m$ be a pointed, closed, and convex cone, and let $K \subset \mathbb{R}^n$ be nonempty closed set. Let $F : K \to L(\mathbb{R}^n, \mathbb{R}^m)$ be a single-valued map. Then:*

1. *The set-valued map N is a gap function for vector variational inequality* (13.56).
2. *The set-valued map W is a gap function for weak vector variational inequality* (13.57).

Proof. Follows directly from the definition of the gap functions (see also [90]). □

In the following result, we study the differentiability properties of the set-valued map G:

Theorem 13.4.4. *Let $C \subset \mathbb{R}^m$ be a pointed, closed, and convex cone, let $K \subset \mathbb{R}^n$ be a compact set, and let $F : K \to L(\mathbb{R}^n, \mathbb{R}^m)$ be a continuously differentiable map. For $\bar{x}, u \in K$, let $\bar{y} = \langle F(\bar{x}), \bar{x} - u \rangle \in G(\bar{x})$, and*

$$\lim_{\|z\| \to \infty} \|\langle F(\bar{x}), z \rangle\| = \infty. \tag{13.61}$$

Then, for any $x \in \mathrm{dom}\,(DG(\bar{x}, \bar{y}))$*, the following identity holds:*

$$DG(\bar{x}, \bar{y})(x) = \langle \nabla F(\bar{x})(x), \bar{x} - u\rangle + \langle F(\bar{x}), x - T(K, u)\rangle,$$

where $DG(\bar{x}, \bar{y})$ *is the contingent derivative of* G *at* $(\bar{x}, \bar{y})$ *and* $\nabla F(\bar{x})$ *is the Fréchet derivative of* F*.*

Proof. For any $x \in \mathrm{dom}\,(DG(\bar{x}, \bar{y}))$, let $y \in DG(\bar{x}, \bar{y})(x)$. Then there are sequences $\{(x_n, y_n)\} \subset \mathbb{R}^n \times \mathbb{R}^m$ and $\{t_n\} \subset \mathbb{P}$ such that $(x_n, y_n) \to (x, y)$, $t_n \downarrow 0$, and $\bar{y} + t_n y_n \in G(\bar{x} + t_n x_n) = \langle F(\bar{x} + t_n x_n), \bar{x} + t_n x_n - K\rangle$ for every $n \in \mathbb{N}$. Therefore, for every $n \in \mathbb{N}$, there exists $\{z_n\} \subset K$ such that

$$\bar{y} + t_n y_n = \langle F(\bar{x} + t_n x_n), \bar{x} + t_n x_n - z_n\rangle,$$

which, after a rearrangement of terms, reads

$$\langle F(\bar{x}), \bar{x} - u\rangle - \langle F(\bar{x} + t_n x_n), \bar{x} - z_n\rangle = \langle F(\bar{x} + t_n x_n), t_n x_n\rangle - t_n y_n. \tag{13.62}$$

Using the Taylor expansion, we have

$$F(\bar{x} + t_n x_n) = F(\bar{x}) + t_n \nabla F(\bar{x})(x_n) + o(t_n x_n). \tag{13.63}$$

By substituting (13.63) into (13.62), and by dividing the resulting identity by t_n, we get

$$\left\langle F(\bar{x}), \frac{z_n - u}{t_n}\right\rangle = \left\langle \nabla F(\bar{x})(x_n) + \frac{o(t_n x_n)}{t_n}, \bar{x} - z_n\right\rangle + \langle F(\bar{x} + t_n x_n), x_n\rangle - y_n. \tag{13.64}$$

Since the set K is compact, we can assume that $z_n \to \tilde{x} \in K$. Since

$$\left(\left\langle \nabla F(\bar{x})(x_n) + \frac{o(t_n x_n)}{t_n}, \bar{x} - z_n\right\rangle + \langle F(\bar{x} + t_n x_n), x_n\rangle - y_n\right)$$
$$\to \langle \nabla F(\bar{x})(x), \bar{x} - \tilde{x}\rangle + \langle F(\bar{x}), x\rangle - y, \tag{13.65}$$

we deduce from (13.64) that $\left\{\left\langle F(\bar{x}), \frac{z_n - u}{t_n}\right\rangle\right\}$ must be a convergent sequence. We claim that $\left\{\frac{z_n - u}{t_n}\right\}$ remains bounded. Indeed, if this is not the case, then there is a subsequence $\left\{\frac{z_m - u}{t_m}\right\}$ such that $\left\|\frac{z_m - u}{t_m}\right\| \to \infty$. Then (13.61) yields $\left\|\left\langle F(\bar{x}), \frac{z_m - u}{t_m}\right\rangle\right\| \to \infty$, which is a contradiction of (13.65).

Therefore, we can assume that there exists $\beta > 0$ such that $\left\|\frac{z_n - u}{t_n}\right\| \leq \beta$, which, due to the fact that $t_n \downarrow 0$, implies that $\tilde{x} = u$. Since we can assume that $\frac{z_n - u}{t_n} \to \check{x}$, we get $\check{x} \in T(K, u)$. From (13.64) and the fact that $\tilde{x} = u$, we have $\langle F(\bar{x}), \check{x}\rangle = \langle \nabla F(\bar{x})(x), \bar{x} - u\rangle + \langle F(\bar{x}), x\rangle - y$, implying

$$y = \langle \nabla F(\bar{x})(x), \bar{x}-u\rangle + \langle F(\bar{x}), x-\check{x}\rangle \in \langle \nabla F(\bar{x})(x), \bar{x}-u\rangle + \bigcup_{z\in T(K,u)} \langle F(\bar{x}), x-z\rangle,$$

and since y was arbitrary, we have $DG(\bar{x}, \bar{y})(x) \subseteq \langle \nabla F(\bar{x})(x), \bar{x}-u\rangle + \langle F(\bar{x}), x - T(K, u)\rangle$.

For the converse, let $z \in T(K, u)$ be such that $y = \langle \nabla F(\bar{x})(x), \bar{x} - u\rangle + \langle F(\bar{x}), x - z\rangle$. Due to the containment $z \in T(K, u)$, there are sequences $\{z_n\} \subset K$, $\{t_n\} \subset \mathbb{P}$ such that $z_n \to u$, $t_n \downarrow 0$, and $\frac{z_n - u}{t_n} \to z$. We choose $\{x_n\} \subset \mathbb{R}^n$ such that $x_n \to x$ and define $\{y_n\} \subset \mathbb{R}^m$ by

$$y_n = \left\langle \frac{F(\bar{x} + t_n x_n) - F(\bar{x})}{t_n}, \bar{x} - u\right\rangle + \left\langle F(\bar{x} + t_n x_n), x_n - \frac{z_n - u}{t_n}\right\rangle.$$

Clearly, we have $y_n \to y$ and $\bar{y} + t_n y_n = \langle F(\bar{x} + t_n x_n), \bar{x} + t_n x_n - z_n\rangle \in G(\bar{x} + t_n x_n)$, which confirms that $y \in DG(\bar{x}, \bar{y})(x)$. This completes the proof. □

In the following result, we study the sensitivity properties of the gap function W:

Theorem 13.4.5. *Let $C \subset \mathbb{R}^m$ be a pointed, closed, convex and solid cone, let $K \subset \mathbb{R}^n$ be a compact set, and let $F : K \to L(\mathbb{R}^n, \mathbb{R}^m)$ be a continuously differentiable map. For $\bar{x}, u \in K$, we define $\bar{y} = \langle F(\bar{x}), \bar{x} - u\rangle \in W(\bar{x})$. Assume that G is derivable at $(\bar{x}, \bar{y})$ and* (13.61) *holds. Then for any $x \in \operatorname{dom}(DW(\bar{x}, \bar{y}))$,*

$$DW(\bar{x}, \bar{y})(x) \subset \operatorname{WMax}(DG(\bar{x}, \bar{y})(x), C).$$

Proof. For any $x \in \operatorname{dom}(DW(\bar{x}, \bar{y}))$, let $y \in DW(\bar{x}, \bar{y})(x)$. By the definition of W, we have $y \in DG(\bar{x}, \bar{y})(x)$. Assume that $y \notin \operatorname{WMax}(DG(\bar{x}, \bar{y})(x), C)$. Then, there exists $v \in DG(\bar{x}, \bar{y})(x)$ such that

$$v - y \in \operatorname{int}(C). \tag{13.66}$$

The containment $y \in DW(\bar{x}, \bar{y})(x)$ ensures that there exist sequences $\{(x_n, y_n)\} \subset \mathbb{R}^n \times \mathbb{R}^m$ and $\{t_n\} \subset \mathbb{P}$ with $(x_n, y_n) \to (x, y)$, $t_n \downarrow 0$ such that $\bar{y} + t_n y_n \in W(\bar{x} + t_n x_n)$ for every $n \in \mathbb{N}$.

Furthermore, from $v \in DG(\bar{x}, \bar{y})(x)$ and the fact that the map G is derivable at $(\bar{x}, \bar{y})$, there exists (u_n, v_n) with $(u_n, v_n) \to (x, v)$ such that $\bar{y} + t_n v_n \in G(\bar{x} + t_n u_n) = \langle F(\bar{x} + t_n u_n), \bar{x} + t_n u_n - K\rangle$, and consequently, there exists $\{z_n\} \subset K$ such that

$$\bar{y} + t_n v_n = \langle F(\bar{x} + t_n u_n), \bar{x} + t_n u_n - z_n\rangle. \tag{13.67}$$

Since the map F is continuously differentiable at $\bar{x}$, we also have

$$F(\bar{x}+t_n u_n) = F(\bar{x}) + t_n \nabla F(\bar{x})(u_n) + o(t_n u_n), \tag{13.68}$$

$$F(\bar{x}+t_n x_n) = F(\bar{x}) + t_n \nabla F(\bar{x})(x_n) + o(t_n x_n). \tag{13.69}$$

By rearranging the above equations, and using the notion $o(t_n) := \dfrac{o(t_n u_n) - o(t_n x_n)}{t_n}$, we get

$$\begin{aligned}\langle F(\bar{x}+t_n u_n), \bar{x}+t_n u_n - z_n\rangle &= \langle F(\bar{x}+t_n x_n), \bar{x}+t_n u_n - z_n\rangle \\ &\quad + \langle t_n \nabla F(\bar{x})(u_n - x_n) + o(t_n), \bar{x}+t_n u_n - z_n\rangle \\ &= \langle F(\bar{x}+t_n x_n), \bar{x}+t_n x_n - z_n\rangle \\ &\quad + \langle F(\bar{x}+t_n x_n), t_n(u_n - x_n)\rangle \\ &\quad + \langle t_n \nabla F(\bar{x})(u_n - x_n) + o(t_n), \bar{x}+t_n u_n - z_n\rangle.\end{aligned} \tag{13.70}$$

We set

$$\alpha(n) = \langle F(\bar{x}+t_n x_n), u_n - x_n\rangle + \left\langle \nabla F(\bar{x})(u_n - x_n) + \frac{o(t_n)}{t_n}, \bar{x}+t_n u_n - z_n\right\rangle.$$

Clearly $\alpha(n) \to 0$, and from (13.67) and (13.70), we get $\bar{y} + t_n v_n = \langle F(\bar{x} + t_n x_n), \bar{x} + t_n x_n - z_n\rangle + t_n\alpha(n)$ and hence $\bar{y} + t_n(v_n - \alpha(n)) \in G(\bar{x} + t_n x_n)$, which in view of the definition of W implies that $(\bar{y} + t_n(v_n - \alpha(n)) - (\bar{y} + t_n y_n) \notin \operatorname{int}(C)$ implying $v_n - \alpha(n) - y_n \notin \operatorname{int}(C)$, and hence $v - y \notin \operatorname{int}(C)$, a contradiction of (13.66). The proof is complete. □

Remark 13.4.6. Several related results can be found in [384]. Furthermore, differentiability and sensitivity of gap functions for the Minty vector variational inequalities can be found in [418].

In the remaining part of this section, we derive second-order analogues of the derivative formulas obtained above. We begin with the following result which was originally given in [378]:

Theorem 13.4.7. *Let $C \subset \mathbb{R}^m$ be a pointed, closed, and convex cone, and let $K \subset \mathbb{R}^n$ be a compact set. Let $F : K \to L(\mathbb{R}^n, \mathbb{R}^m)$ be a twice Fréchet differentiable map. Let $\bar{x}, u \in K$ be given elements and let $\bar{y} = \langle F(\bar{x}), \bar{x} - u\rangle \in G(\bar{x})$. Assume that* (13.61) *holds. Then, for any $x \in \operatorname{dom}(D^2G(\bar{x}, \bar{y}, \bar{u}, \bar{v}))$, we have*

$$\begin{aligned}D^2G(\bar{x}, \bar{y}, \bar{u}, \bar{v})(x) = \bigcup_{z\in T^2(K,\bar{x},w)} &[\langle F(\bar{x}), x - z\rangle + \langle \nabla F(\bar{x})(\bar{u}), 2(\bar{u} - w)\rangle] \\ &+\langle \nabla^2 F(\bar{x})(\bar{u}, \bar{u}), \bar{x} - u\rangle + \langle \nabla F(\bar{x})(x), \bar{x} - u\rangle,\end{aligned}$$

with w satisfying $\langle F(\bar{x}), w\rangle = -\bar{v} + \langle \nabla F(\bar{x})\bar{u}, \bar{x} - u\rangle + \langle F(\bar{x}), \bar{u}\rangle$.

Proof. For $x \in \operatorname{dom}(D^2G(\bar{x}, \bar{y}, \bar{u}, \bar{v}))$, let $y \in D^2G(\bar{x}, \bar{y}, \bar{u}, \bar{v})(x)$ be arbitrary. Then, there are sequences $\{(x_n, y_n)\} \subset \mathbb{R}^n \times \mathbb{R}^m$, $\{t_n\} \subset \mathbb{P}$ with $(x_n, y_n) \to (x, y)$ and $t_n \downarrow 0$ such that for every $n \in \mathbb{N}$,

$$\begin{aligned}\bar{y} + t_n\bar{v} + \frac{1}{2}t_n^2 y_n &\in G\left(\bar{x} + t_n\bar{u} + \frac{1}{2}t_n^2 x_n\right)\\ &= \left\langle F\left(\bar{x} + t_n\bar{u} + \frac{1}{2}t_n^2 x_n\right), \bar{x} + t_n\bar{u} + \frac{1}{2}t_n^2 x_n - K\right\rangle.\end{aligned}$$

Therefore, there exists $z_n \in K$ such that

$$\bar{y} + t_n\bar{v} + \frac{1}{2}t_n^2 y_n = \left\langle F\left(\bar{x} + t_n\bar{u} + \frac{1}{2}t_n^2 x_n\right), \bar{x} + t_n\bar{u} + \frac{1}{2}t_n^2 x_n - z_n\right\rangle, \tag{13.71}$$

and since K is compact, we have $z_n \to \breve{x} \in K$. The above equation then yields $\bar{y} = \langle F(\bar{x}), \bar{x} - \breve{x}\rangle$.

Since F is twice differentiable, we have

$$\begin{aligned}F\left(\bar{x} + t_n\bar{u} + \frac{1}{2}t_n^2 x_n\right) &= F(\bar{x}) + t_n\nabla F(\bar{x})(\bar{u}) + \frac{1}{2}t_n^2\nabla F(\bar{x})x_n + \frac{1}{2}t_n^2\nabla^2 F(\bar{x})\\ &\quad \left(\bar{u} + \frac{1}{2}t_n x_n, \bar{u} + \frac{1}{2}t_n x_n\right) + o\left(\left\|t_n\bar{u} + \frac{1}{2}t_n^2 x_n\right\|\right)^2.\end{aligned} \tag{13.72}$$

Using (13.71) and (13.72), we have

$$\begin{aligned}\bar{y} + t_n\bar{v} + \frac{1}{2}t_n^2 y_n &= \langle F(\bar{x}), \bar{x} - z_n\rangle + t_n\langle F(\bar{x}), \bar{u}\rangle + t_n\langle \nabla F(\bar{x})\bar{u}, \bar{x} - z_n\rangle + \frac{1}{2}t_n^2\langle F(\bar{x}), x_n\rangle\\ &\quad + \frac{1}{2}t_n^2\langle \nabla F(\bar{x})\bar{u}, 2\bar{u} + t_n x_n\rangle + \frac{1}{2}t_n^2\left\langle \nabla F(\bar{x})(x_n), \bar{x} + t_n\bar{u} + \frac{1}{2}t_n^2 x_n - z_n\right\rangle\\ &\quad + \frac{1}{2}t_n^2\left\langle \nabla^2 F(\bar{x})\left(\bar{u} + \frac{1}{2}t_n x_n, \bar{u} + \frac{1}{2}t_n x_n\right), \bar{x} + t_n\bar{u} + \frac{1}{2}t_n^2 x_n - z_n\right\rangle\\ &\quad + \frac{1}{2}t_n^2\left\langle \frac{o\left(\left\|t_n\bar{u} + \frac{1}{2}t_n^2 x_n\right\|\right)^2}{\frac{1}{2}t_n^2}, \bar{x} + t_n\bar{u} + \frac{1}{2}t_n^2 x_n - z_n\right\rangle,\end{aligned} \tag{13.73}$$

which yields

$$\left\langle F(\bar{x}), \frac{z_n - u}{t_n}\right\rangle \to -\bar{v} + \langle F(\bar{x}), \bar{u}\rangle + \langle \nabla F(\bar{x})\bar{u}, \bar{x} - \breve{x}\rangle. \tag{13.74}$$

We claim that the sequence $\left\{\frac{z_n-u}{t_n}\right\}$ is bounded. In fact, if this is not the case, then there exists a subsequence such that $\left\|\frac{z_n-u}{t_n}\right\| \to \infty$. By (13.61), we have $\left\|\left\langle F(\bar{x}), \frac{z_n-u}{t_n}\right\rangle\right\| \to \infty$, which contradicts (13.74). Therefore, we can assume that $\left\{\dfrac{z_n-u}{t_n}\right\}$ is bounded. This implies that $u = \check{x}$. Moreover, we can assume that $\dfrac{z_n-u}{t_n} \to w$ which implies that $w \in T(K,u)$. Furthermore,

$$\langle F(\bar{x}), w\rangle = -\bar{v} + \langle F(\bar{x}), \bar{u}\rangle + \langle \nabla F(\bar{x})\bar{u}, \bar{x}-u\rangle.$$

Using (13.73) and (13.74), we deduce that

$$\begin{aligned}\left\langle F(\bar{x}), \frac{\frac{z_n-u}{t_n}-w}{\frac{1}{2}t_n}\right\rangle &= -y_n - \left\langle \nabla F(\bar{x})(\bar{u}), \frac{z_n-u}{\frac{1}{2}t_n}\right\rangle + \langle F(\bar{x}), x_n\rangle \\ &+ \langle \nabla F(\bar{x})(\bar{u}), 2\bar{u} + t_n x_n\rangle \\ &+ \left\langle \nabla F(\bar{x})x_n, \bar{x} + t_n\bar{u} + \frac{1}{2}t_n^2 x_n - z_n\right\rangle \\ &+ \left\langle \nabla^2 F(\bar{x})\left(\bar{u}+\frac{1}{2}t_n x_n, \bar{u}+\frac{1}{2}t_n x_n\right), \bar{x} + t_n\bar{u} + \frac{1}{2}t_n^2 x_n - z_n\right\rangle \\ &+ \left\langle \frac{o\left(\left\|t_n\bar{u} + \frac{1}{2}t_n^2 x_n\right\|\right)^2}{\frac{1}{2}t_n^2}, \bar{x} + t_n\bar{u} + \frac{1}{2}t_n^2 x_n - z_n\right\rangle,\end{aligned}$$

and hence

$$\begin{aligned}\left\langle F(\bar{x}), \frac{\frac{z_n-u}{t_n}-w}{\frac{1}{2}t_n^2}\right\rangle &\to \langle F(\bar{x}), x\rangle + \langle \nabla F(\bar{x})\bar{u}, 2(\bar{u}-w)\rangle + \langle \nabla F(\bar{x})x, \bar{x}-u\rangle \\ &+ \left\langle \nabla^2 F(\bar{x})(\bar{u},\bar{u}), \bar{x}-u\right\rangle - y.\end{aligned}$$

By using the same arguments, we obtain $\frac{\frac{z_n-u}{t_n}-w}{\frac{1}{2}t_n^2} \to z$ with $z \in T^2(K,u,w)$. Therefore,

$$\begin{aligned}y \in \bigcup_{z\in T^2(K,\bar{x},w)} &[\langle F(\bar{x}), x-z\rangle + \langle \nabla F(\bar{x})(\bar{u}), 2(\bar{u}-w)\rangle] + \nabla^2 F(\bar{x})(\bar{u},\bar{u}), \bar{x}-u\rangle \\ &+\langle \nabla F(\bar{x})(x), \bar{x}-u\rangle.\end{aligned}$$

For the converse, assume that

$$y = \langle F(\bar{x}), x-z\rangle + \langle \nabla F(\bar{x})\bar{u}, 2(\bar{u}-w)\rangle + \langle \nabla F(\bar{x})(x), \bar{x}-u\rangle + \langle \nabla^2 F(\bar{x})(\bar{u},\bar{u}), \bar{x}-u\rangle$$

with $z \in T^2(K, u, w)$. Then there is a sequence $\{z_n\} \subset K$ with $z_n \to u$ and $t_n \downarrow 0$ such that $\frac{z_n - u}{t_n} \to w$ and $\frac{\frac{z_n-u}{t_n} - w}{\frac{1}{2}t_n} \to z$.

Since F is twice differentiable, we can take sequence x_n and y_n such that $x_n \to x$ and

$$y_n = \left\langle F(\bar{x} + t_n\bar{u} + \frac{1}{2}t_n^2 x_n), x_n - \frac{\frac{z_n-u}{t_n} - w}{\frac{1}{2}t_n} \right\rangle$$
$$+ \left\langle \frac{F(\bar{x} + t_n\bar{u} + \frac{1}{2}t_n^2 x_n) - F(\bar{x})}{t_n}, 2(\bar{u} - w) \right\rangle + \langle \nabla F(\bar{x}) x_n, \bar{x} - u \rangle$$
$$+ \left\langle \frac{F\left(\bar{x} + t_n\bar{u} + \frac{1}{2}t_n^2 x_n\right) - F(\bar{x}) - \nabla F(\bar{x})(t_n\bar{u} + \frac{1}{2}t_n^2 x_n)}{\frac{1}{2}t_n^2}, \bar{x} - u \right\rangle.$$

Therefore $y_n \to y$ and

$$\bar{y} + t_n(\langle \nabla F(\bar{x})(\bar{u}), \bar{x} - u \rangle + \langle F(\bar{x}), \bar{u} - w \rangle) + \frac{1}{2}t_n^2 y_n$$
$$= \left\langle F\left(\bar{x} + t_n\bar{u} + \frac{1}{2}t_n^2 x_n\right), \bar{x} + t_n\bar{u} + \frac{1}{2}t_n^2 x_n - z_n \right\rangle \in G\left(\bar{x} + t_n\bar{u} + \frac{1}{2}t_n^2 x_n\right).$$

Since $\bar{v} = \langle \nabla F(\bar{x})(\bar{u}), \bar{x} - u \rangle + \langle F(\bar{x}), \bar{u} - w \rangle$, we get $y \in D^2 G(\bar{x}, \bar{y}, \bar{u}, \bar{v})(x)$. The proof is complete. □

We now give a second-order sensitivity analysis for the weak vector variational inequalities. In the following result, we continue with the setting of Theorem 13.4.7.

Theorem 13.4.8. *Let $\bar{x} \in K$, $\bar{y} \in G(\bar{x})$ and $(\bar{u}, \bar{v}) \in T(\text{graph}\,(G), (\bar{x}, \bar{y}))$. Assume that G is second-order derivable $(\bar{x}, \bar{y})$ in the direction $(\bar{u}, \bar{v})$. Then for any $x \in \text{dom}\,(D^2 W(\bar{x}, \bar{y}, \bar{u}, \bar{v}))$, we have*

$$D^2 W(\bar{x}, \bar{y}, \bar{u}, \bar{v})(x) \subset \text{WMax}\left(D^2 G(\bar{x}, \bar{y}, \bar{u}, \bar{v})(x), C\right). \tag{13.75}$$

Proof. Let $y \in D^2 W(\bar{x}, \bar{y}, \bar{u}, \bar{v})(x)$. Therefore, $y \in D^2 G(\bar{x}, \bar{y}, \bar{u}, \bar{v})(x)$. If $y \notin \text{WMax}\left(D^2 W(\bar{x}, \bar{y}, \bar{u}, \bar{v})(x), C\right)$, then there exists $v \in D^2 G(\bar{x}, \bar{y}, \bar{u}, \bar{v})(x)$ such that

$$v - y \in \text{int}(C). \tag{13.76}$$

Since $y \in D^2 W(\bar{x}, \bar{y}, \bar{u}, \bar{v})(x)$, there are sequences $(x_n, y_n) \to (x, y)$ and $t_n \to 0$ such that

$$\bar{y} + t_n\bar{v} + \frac{1}{2}t_n^2 y_n \in W\left(\bar{x} + t_n\bar{u} + \frac{1}{2}t_n^2 x_n\right). \tag{13.77}$$

Analogously, since $v \in D^2G(\bar{x}, \bar{y}, \bar{u}, \bar{v})(x)$, there exists $(u_n, v_n) \to (x, v)$ such that

$$\bar{y} + t_n\bar{v} + \frac{1}{2}t_n^2 v_n \in G\left(\bar{x} + t_n\bar{u} + \frac{1}{2}t_n^2 u_n\right).$$

Therefore, there exists $\{z_n\} \subset K$ such that

$$\bar{y} + t_n\bar{v} + \frac{1}{2}t_n^2 v_n = \left\langle F\left(\bar{x} + t_n\bar{u}\frac{1}{2}t_n^2 u_n\right), \bar{x} + t_n\bar{u} + \frac{1}{2}t_n^2 u_n - z_n\right\rangle. \tag{13.78}$$

Since F is twice differentiable, we have

$$\begin{aligned} F\left(\bar{x} + t_n\bar{u} + \frac{1}{2}t_n^2 u_n\right) &= F(\bar{x}) + t_n \nabla F(\bar{x})\left(\bar{u} + \frac{1}{2}t_n u_n\right) \\ &\quad + \frac{1}{2}t_n^2 \nabla^2 F(\bar{x})\left(\bar{u} + \frac{1}{2}t_n u_n, \bar{u} + \frac{1}{2}t_n u_n\right) \\ &\quad + o\left(\left\|t_n\bar{u} + \frac{1}{2}t_n^2 u_n\right\|\right)^2. \end{aligned} \tag{13.79}$$

Using (13.72) and (13.79), we have

$$\begin{aligned} &\left\langle F\left(\bar{x} + t_n\bar{u} + \frac{1}{2}t_n^2 u_n\right), \bar{x} + t_n\bar{u} + \frac{1}{2}t_n^2 u_n - z_n\right\rangle \\ &= \left\langle F\left(\bar{x} + t_n\bar{u} + \frac{1}{2}t_n^2 x_n\right), \bar{x} + t_n\bar{u} + \frac{1}{2}t_n^2 x_n - z_n\right\rangle \\ &\quad + \frac{1}{2}t_n^2\left\langle F\left(\bar{x} + t_n\bar{u} + \frac{1}{2}t_n^2 x_n\right), u_n - x_n\right\rangle \\ &\quad + \frac{1}{2}t_n^2\left\langle \nabla F(\bar{x})(u_n - x_n), \bar{x} + t_n\bar{u} + \frac{1}{2}t_n^2 - z_n\right\rangle \\ &\quad + \frac{1}{2}t_n^3\left\langle \nabla^2 F(\bar{x})(\bar{u}, u_n - x_n), \bar{x} + t_n\bar{u} + \frac{1}{2}t_n^2 u_n - z_n\right\rangle \\ &\quad + \frac{1}{8}t_n^4\left\langle \nabla^2 F(\bar{x})(u_n, u_n) - \nabla^2 F(\bar{x})(x_n, x_n), \bar{x} + t_n\bar{u} + \frac{1}{2}t_n^2 u_n - z_n\right\rangle \\ &\quad + \frac{1}{2}t_n^2\left\langle \frac{o(t_n^2)}{\frac{1}{2}t_n^2}, \bar{x} + t_n\bar{u} + \frac{1}{2}t_n^2 u_n - z_n\right\rangle, \end{aligned}$$

where $o(t_n^2) = o\left(\left\|t_n\bar{u} + \frac{1}{2}t_n^2 u_n\right\|\right)^2 - o\left(\left\|t_n\bar{u} + \frac{1}{2}t_n^2 x_n\right\|\right)^2$.

We define

$$\begin{aligned}\alpha(n) = &\left\langle F\left(\bar{x} + t_n\bar{u} + \frac{1}{2}t_n^2 x_n\right), u_n - x_n\right\rangle \\ &+\left\langle \nabla F(\bar{x})(u_n - x_n), \bar{x} + t_n\bar{u} + \frac{1}{2}t_n^2 u_n - z_n\right\rangle \\ &+t_n\left\langle \nabla^2 F(\bar{x})(\bar{u}, u_n - x_n), \bar{x} + t_n\bar{u} + \frac{1}{2}t_n^2 u_n - z_n\right\rangle \\ &+\left\langle \frac{t_n^2}{4}\nabla^2 F(\bar{x})(u_n, u_n) - \frac{t_n^2}{4}\nabla^2 F(\bar{x})(x_n, x_n), \bar{x} + t_n\bar{u} + \frac{1}{2}t_n^2 u_n - z_n\right\rangle \\ &+\left\langle \frac{o(t_n^2)}{\frac{1}{2}t_n^2}, \bar{x} + t_n\bar{u} + \frac{1}{2}t_n^2 u_n - z_n\right\rangle,\end{aligned}$$

and note that $\alpha(n) \to 0$.

Furthermore

$$\begin{aligned}\bar{y} + t_n\bar{v} + \frac{1}{2}t_n^2[v_n - \alpha(n)] &= \left\langle F\left(\bar{x} + t_n\bar{u} + \frac{1}{2}t_n^2 x_n\right), \bar{x} + t_n\bar{u} + \frac{1}{2}t_n^2 x_n - z_n\right\rangle \\ &\in G\left(\bar{x} + t_n\bar{u} + \frac{1}{2}t_n^2 x_n\right),\end{aligned}$$

and it follows from the definition of the map W and (13.79) that

$$\left(\bar{y} + t_n\bar{v} + \frac{1}{2}t_n^2[v_n - \alpha(n)\right) - \left(\bar{y} + t_n\bar{v} + \frac{1}{2}t_n^2 y_n\right) \notin \operatorname{int}(C)$$

which implies that $v_n - \alpha(n) - y_n \notin \operatorname{int}(C)$. Therefore, $v - y \notin \operatorname{int}(C)$ which is a contradiction to (13.76). The proof is complete. □

Remark 13.4.9. We remark that numerous variants of the above results and illustrative examples can be found in [378]. All the results stated in this section are given in a finite-dimensional setting and it is natural to expect their extensions in general setting.

Chapter 14
Numerical Methods for Solving Set-Valued Optimization Problems

In this chapter we present solution procedures for solving set-valued optimization problems. In Sect. 14.1 a Newton method for solving general set-valued optimization problems is shown. For a special class of set-valued optimization problems where the objective map is polyhedral, convex and set-valued, we present an algorithm in Sect. 14.2.

Furthermore, in Sect. 15.4 we discuss the relationship between multiobjective optimization problems with uncertainties and set-valued optimization and present solution procedures for these special set-valued optimization problems (see Sect. 15.4.4).

14.1 A Newton Method for Set-Valued Maps

In Chap. 12 we have shown necessary optimality conditions for solutions of set-valued optimization problems. In algorithms for solving set-valued problems one has to study these optimality conditions, especially given in Theorem 12.8.9. This means that we have to solve certain inclusions involving set-valued maps.

It is well known from scalar optimization that the Newton method is one of the most powerful tools for solving systems of nonlinear equations. In the paper by Hoheisel, Kanzow, Mordukhovich, Phan [266] a new generalized Newton algorithm based on graphical derivatives is proposed and justified. The method in [266] reduces to the classical Newton method in the smooth case, being different from previously known versions of Newton's method in the case of Lipschitz continuous mappings. Based on advanced techniques of variational analysis and generalized differentiation, the well-posedness of the algorithm, its local superlinear convergence, and its global convergence of the Kantorovich type is derived in [266].

Set-valued generalizations of the Newton method are described in the paper by Dias and Smirnov [137]. In this section we follow the approach given in [137].

A.A. Khan et al., *Set-valued Optimization*, Vector Optimization,
DOI 10.1007/978-3-642-54265-7_14

The Newton method developed in [137] is a general tool to solve systems of nonlinear equations with geometric constraints and systems of nonlinear inequalities in a unified manner.

In the classical case these nonlinear equations are given for continuously differentiable functions $f : \mathbb{R}^n \longrightarrow \mathbb{R}^q$ (compare [446]) by

$$f(x) = 0 \tag{14.1}$$

and the Newton method generates the following sequence of iterates

$$x^{k+1} := x^k + \overline{x}^k, \qquad k = 0, 1, \cdots, \tag{14.2}$$

where $\overline{x}^k$ is a solution of the system of linear equations

$$\nabla f(x^k)\overline{x}^k = -f(x^k). \tag{14.3}$$

Dias and Smirnov [137] introduced a Newton-type method suitable to solve inclusions given for a set-valued map $F : \mathbb{R}^n \rightrightarrows \mathbb{R}^q$ by

$$0 \in F(x). \tag{14.4}$$

These Newton-type methods generate sequences of iterates (compare (14.2))

$$x^{k+1} := x^k + \overline{x}^k, \qquad k = 0, 1, \cdots,$$

where $\overline{x}^k$ is now a solution of an inclusion that is a generalization of the linear system (14.3). This generalized system involves set-valued map derivatives (see [16]). Namely, instead of (14.3) one solves an inclusion

$$-v^k \in \Lambda(x^k, v^k)(\overline{x}^k), \tag{14.5}$$

where v^k is a nearest to zero point belonging to the set $F(x^k)$, and the graph of the set-valued map $\Lambda(x^k, v^k)(\cdot)$ is a tangent cone in some sense to the graph of F at the point (x^k, v^k). In order to solve the linear system (14.3) and to prove convergence results it is essential to have the non-singularity of the matrix $\nabla f(x^k)$. This condition will be replaced by the condition of **metric regularity** (cf. [430, 499] and a survey in [21]). In the case of linear operators this method goes back to the open mapping theorem and in the case of smooth maps to the Lyusternik theorem. It was successfully applied to justify the well-posedness and convergence of the Newton method for nonsmooth equations by Hoheisel, Kanzow, Mordukhovich, Phan in [266]. Dias and Smirnov [137] show that the metric regularity of Lipschitzian set-valued maps is equivalent to the well-posedness of the Newton method for perturbed maps.

In the following we use the distance between a point $x \in \mathbb{R}^n$ and a set $S \subset \mathbb{R}^n$ denoted by

$$d(x, S) = \inf\{||x - s|| \mid s \in S\}.$$

The **projection of a vector** x **onto** S is defined by

$$P_S(x) := \{s \in S \mid ||x - s|| = d(x, S)\}.$$

In the formulation of the Newton method for set-valued maps we use the **contingent derivative** of $F : \mathbb{R}^n \rightrightarrows \mathbb{R}^q$ at the point $(\overline{x}, \overline{y}) \in \operatorname{graph} F$, i.e., the set-valued map $DF(\overline{x}, \overline{y}) : X \rightrightarrows \mathbb{R}^q$ defined by

$$\operatorname{graph} DF(\overline{x}, \overline{y}) = T(\operatorname{graph} F, (\overline{x}, \overline{y})),$$

where $T(\operatorname{graph} F, (\overline{x}, \overline{y}))$ is the **contigent cone** to the set graph F at the point $(\overline{x}, \overline{y}) \in \operatorname{graph} F$ (see Sect. 4.1.2). In other words, $y \in DF(\overline{x}, \overline{y})(x)$ if and only if $(x, y) \in T(\operatorname{graph} F, (\overline{x}, \overline{y}))$. The contingent cone to a set $S \subset X$ (X a real normed space) at a point $\bar{x} \in S$ is introduced in Sect. 4.1.2, Definition 4.1.11. We will use in the following the cone T_{13} studied in Theorem 4.1.12:

$$T(S, \bar{x}) := T_{13}(S, \bar{x}) = \left\{x \in X \mid \liminf_{t \to +0} \frac{d(\bar{x} + tx, S)}{t} = 0\right\}.$$

Furthermore, we use the coderivative by Mordukhovich $D^*F(\overline{x}, \overline{y}) : Y^* \rightrightarrows X^*$ of a set-valued map F at $(\overline{x}, \overline{y}) \in \operatorname{graph} F$ introduced in Definition 3.5.8.

Next, we recall the notion of metric regularity for set-valued maps and its coderivative characterization (see Mordukhovich [430], Rockafellar, Wets [499]), Durea, Strugariu [153, 154], Apetrii, Durea, Strugariu [9]). A set-valued map $F : \mathbb{R}^n \rightrightarrows \mathbb{R}^q$ is called **metrically regular around** $(\overline{x}, \overline{y}) \in \operatorname{graph} F$ if there exist $\varepsilon > 0$ and $\mu > 0$ such that

$$d(x, F^{-1}(y)) \leq \mu d(y, F(x)), \quad x \in \overline{x} + \varepsilon B_{\mathbb{R}^n},\ y \in \overline{y} + \varepsilon B_{\mathbb{R}^q}.$$

The following coderivative characterization of the metric regularity for set-valued maps is given in [430] and [499].

Theorem 14.1.1. *A set-valued map* $F : \mathbb{R}^n \rightrightarrows \mathbb{R}^q$ *with closed graph is metrically regular around* $(\overline{x}, \overline{y}) \in \operatorname{graph} F$ *if and only if the inclusion* $0 \in D^*F(\overline{x}, \overline{y})(y^*)$ *implies that* $y^* = 0$.

The **well-posedness of the Newton method** for perturbed set-valued maps $F(x) - \hat{y}$, where $F : \mathbb{R}^n \rightrightarrows \mathbb{R}^q$ is a Lipschitz continuous map in the sense of Definition 3.3.11 and the vectors $\hat{y} \in \mathbb{R}^q$ have a sufficiently small norm, can be used in order to characterize the metric regularity. Using the contingent derivative we say that the Newton method for perturbed set-valued maps F is well-posed around $(\overline{x}, \overline{y}) \in \operatorname{graph} F$ with modulus μ if there exists $\eta > 0$ such that for all $x \in \overline{x} + \eta B_{\mathbb{R}^n}$ and $\hat{y} \in \overline{y} + \eta B_{\mathbb{R}^q}$ there is $y \in P_{F(x)-\hat{y}}(0)$ satisfying the condition

$$DF^{-1}(y + \hat{y}, x)(-y) \cap \mu||y||B_{\mathbb{R}^n} \neq \emptyset. \tag{14.6}$$

Condition (14.6) implies that the largest possible Newton inclusion (14.5) for the perturbed map $x \longrightarrow F(x) - \hat{y}$ (the generalization of the Newton equation (14.3)) has at least one solution $x^0 \in \mu B_{\mathbb{R}^n}$.

The following theorem given by Dias and Smirnov [137, Theorem 3.1] explains under which conditions the generalized Newton method is well defined.

Theorem 14.1.2. *Assume that the set-valued map $F : \mathbb{R}^n \rightrightarrows \mathbb{R}^q$ with closed values is Lipschitz continuous with a constant $L_F > 0$ in a neighbourhood of a point $\overline{x}$ in the sense of Definition 3.3.11. Then the following conditions are equivalent:*

1. *The map F is metrically regular around $(\overline{x}, \overline{y}) \in \operatorname{graph} F$.*
2. *The Newton method for perturbed set-valued maps F is well posed around $(\overline{x}, \overline{y}) \in \operatorname{graph} F$.*

In the following we present a convergence analysis concerning the set-valued version of the Newton method. Consider a set-valued map $F : \mathbb{R}^n \rightrightarrows \mathbb{R}^q$ with closed values and $(x^k, y^k) \in \operatorname{graph} F$. The step-length in the algorithm is given by $t^k \in (0, 1]$. If $t^k = 1$ this method is called **Newton method** and if $t^k \in (0, 1]$, the method is called the **damped Newton method**. For example, the step-length t^k can be chosen from the condition

$$d(0, F(x^k + t^k\overline{x}^k)) = \min_{t \in [0,1]} d(0, F(x^k + t\overline{x}^k)).$$

For a given point $(x^k, y^k) \in \operatorname{graph} F$ we define the next iterate by

$$x^{k+1} := x^k + t^k\overline{x}^k, \qquad y^{k+1} \in P_{F(x^{k+1})}(0), \tag{14.7}$$

where $\overline{x}^k$ is a solution of the inclusion

$$-y^k \in \Lambda(x^k, y^k)(\overline{x}^k),$$

and $\Lambda(x, y) : \mathbb{R}^n \to \mathbb{R}^q$ is a positively homogeneous set-valued mapping satisfying the inclusion $\Lambda(x, y)(\overline{x}) \subset DF(x, y)(\overline{x})$ for all $(x, y) \in \operatorname{graph} F$ and $\overline{x}$. The vector $\overline{x}^k$ is chosen from the condition

$$\overline{x}^k \in P_{(\Lambda(x^k, y^k))^{-1}(-y^k)}(0). \tag{14.8}$$

The condition (14.8) uniquely defines $\overline{x}^k$ if $(\Lambda(x^k, y^k))^{-1}(-y^k)$ is convex.

Suppose $x \notin F^{-1}(0)$, $y \in P_{F(x)}(0)$ and $\overline{x} \in (\Lambda(x, y))^{-1}(\overline{y})$. Then there exists $p(t) \in \mathbb{R}^q$ such that

$$y + t\overline{y} + p(t) \in F(x + t\overline{x})$$

and

$$||p(t)|| = t\delta(x, y, \overline{x}, \overline{y}, t),$$

where

$$\delta(x, y, \overline{x}, \overline{y}, t) := t^{-1}d(y + t\overline{y}, F(x + t\overline{x})). \tag{14.9}$$

If F is Lipschitz continuous in the sense of Definition 3.3.11, then

$$\liminf_{t \to +0} \delta(x, y, \overline{x}, \overline{y}, t) = 0.$$

Furthermore, if $\overline{y} = -y$, then we have

$$d(0, F(x + t\overline{x})) \leq ||(1 - t)y + p(t)|| \leq (1 - t)||y|| + t\delta(x, y, \overline{x}, -y, t).$$

This means that the rate of convergence of the Newton method depends on the properties of the function δ. The following result concerning the rate of convergence is shown in [137, Theorem 4.1].

Theorem 14.1.3. *Suppose that the following conditions are satisfied:*

(i) *There exists $\mu > 0$ such that for all $x \in \mathbb{R}^n$, $y \in P_{F(x)}(0)$ and $\overline{y} \in \mathbb{R}^q$ the set $(\Lambda(x, y))^{-1}(\overline{y}) \cap \mu||\overline{y}||B_{\mathbb{R}^n}$ is non-empty.*
(ii) *There exists a monotone increasing function $\omega : [0, +\infty) \to [0, +\infty)$ such that $\lim_{\alpha \to +0} \omega(\alpha) = 0$ and $\delta(x, y, \overline{x}, -y, t) \leq ||y||\omega(t||y||)$ for all $x \in \mathbb{R}^n \setminus F^{-1}(0)$, $y \in P_{F(x)}(0)$ and $\overline{x} \in P_{(\Lambda(x,y))^{-1}(-y)}(0)$.*

Then for any initial point x^0 there exists a monotone non-decreasing sequence $t^k > 0$, such that $t^k = 1$ for large k, and the corresponding damped Newton method / Newton method, starting at x^0, converges to a point $x^\infty := \lim_{k \to \infty} x^k$ with $x^\infty \in x^0 + 2\mu d(0, F(x^0))B_{\mathbb{R}^n}$. If graph F *is closed, then $x^\infty \in F^{-1}(0)$ and the convergence is R-superlinear. If $\omega(\alpha) = O(\alpha)$, $\alpha \to +0$, then the convergence is R-quadratic. Furthermore, if F is additionally locally Lipschitz continuous, then the convergence is Q-quadratic.*

In the following, we discuss the set-valued version of the Newton method for systems of nonlinear equations with geometric constraints.

Suppose that $f : \mathbb{R}^n \times \mathbb{R}^l \longrightarrow \mathbb{R}^q$ is a continuously differentiable function. Its derivative is supposed to be Lipschitz continuous with a constant $L_{\nabla f}$. Let us consider the nonlinear equation

$$f(x, u) = 0. \tag{14.10}$$

We are looking for a solution (x, u) such that the variable u satisfies the geometric constraints

$$u \in U, \tag{14.11}$$

where $U \subset \mathbb{R}^l$ is a closed convex set. Let U_∞ be the **recession cone of** U, i.e.,

$$U_\infty := \{u \in \mathbb{R}^l \mid u + U \subset U\}.$$

Consider, for example, a system of equations and inequalities given for functions $g : \mathbb{R}^n \longrightarrow \mathbb{R}^{q-l}$ and $h : \mathbb{R}^n \longrightarrow \mathbb{R}^l$ by

$$g(x) = 0, \qquad h(x) \leq 0.$$

This system can be rewritten in the form (14.10) and (14.11) in the following way:

$$g(x) = 0, \qquad h(x) + u = 0, \qquad u \geq 0.$$

In this case it holds

$$U = U_\infty = \{u \in \mathbb{R}^l \mid u \geq 0\}.$$

The system (14.10) and (14.11) is equivalent to the inclusion

$$0 \in F(x) := f(x, U).$$

Let $y = f(x, u) \in P_{F(x)}(0)$. In order to apply the generalized Newton method to this inclusion, put for $\overline{x} \in \mathbb{R}^n$

$$\Lambda(x, u)(\overline{x}) := \nabla_x f(x, u)\overline{x} + \nabla_u f(x, u)U_\infty. \tag{14.12}$$

Let $(\hat{x}, \hat{u}) \in \mathbb{R}^n \times U$ such that $f(\hat{x}, \hat{u}) = 0$. Moreover, assume that

$$((\nabla_u f(\hat{x}, \hat{u}))^T)^{-1} U_\infty^+ \cap \ker \nabla_x f(\hat{x}, \hat{u}) = \{0\}.$$

Now, from Theorem 14.1.1 it follows that the set-valued map F is metrically regular around $(\hat{x}, 0)$. Furthermore, because of Theorem 14.1.2 the Newton method is well posed around $(\hat{x}, 0)$ and using the Lipschitz constant L_Λ we get

$$d(0, (\Lambda(x, u))^{-1}(\overline{y})) \leq L_\Lambda ||\overline{y}||,$$

whenever $(x, u) \in (\hat{x}, \hat{u}) + \eta B_{\mathbb{R}^{n+l}}$ and $\eta > 0$ is sufficiently small.

Let $y \in P_{F(x)}(0)$. Because of the well posedness of the Newton method, (14.5) and (14.12) there exist $\overline{x} \in \mathbb{R}^n$ and $\overline{w} \in U_\infty$ such that

$$-y = \nabla_x f(x, u)\overline{x} + \nabla_u f(x, u)\overline{w}$$

and

$$(||\overline{x}||^2 + ||\overline{w}||^2)^{\frac{1}{2}} \leq L_{\Lambda}||y||.$$

Because of $u + t\overline{w} \in U, t > 0$, we get (see (14.9))

$$\begin{aligned}
\delta(x, y, \overline{x}, -y, t) &= t^{-1}d(y - ty, F(x + t\overline{x})) \\
&\leq t^{-1}||y - ty - f(x + t\overline{x}, u + t\overline{w})|| \\
&\leq tL_{\nabla f}(||\overline{x}||^2 + ||\overline{w}||^2) \\
&\leq tL_{\nabla f}(L_{\Lambda}||y||)^2.
\end{aligned}$$

Applying Theorem 14.1.3 one gets assertions concerning the convergence of the generalized Newton method for solving the system (14.10) and (14.11). Especially, we can see that the Newton method, starting in a sufficiently small neighbourhood of $\hat{x}$, converges Q-quadratically to the set $F^{-1}(0)$.

14.2 An Algorithm to Solve Polyhedral Convex Set-Valued Optimization Problems

Kuroiwa [354] introduced an embedding approach for set-valued optimization problems (see Sect. 2.6.4) such that it is possible to use methods of vector optimization in order to solve set-valued problems. Furthermore, Löhne, Schrage [396] derived an algorithm for solving polyhedral convex set-valued optimization problems using a vectorial relaxation of the set-valued optimization problem. Very recently, Jahn [294] developed a derivative-free descent method for the solution of set optimization problems. In this section we present the results shown by Löhne, Schrage [396] for set-valued problems where the graph of the set-valued objective map is supposed to be given by finitely many linear inequalities. A minimal solution of the set-valued optimization problem is understood to be a set of elements in the domain satisfying two conditions: The attainment of the infimum and minimality with respect to a set order relation (see Sect. 2.6.2). The algorithm presented in this section is divided into two phases: In the first phase, a linear vector optimization problem, called the vectorial relaxation, is solved. The generated pre-solution yields the attainment of the infimum but, in general, not minimality. In the second phase of the algorithm, minimality is established by solving certain linear programs in combination with vertex enumeration of some values of the set-valued objective map.

Throughout this section we suppose that $C \subseteq \mathbb{R}^q$ is a proper pointed polyhedral convex cone with $\operatorname{int} C \neq \emptyset$. We present a solution procedure for set-valued optimization problems where the objective map $F : \mathbb{R}^n \rightrightarrows \mathbb{R}^q$ has a polyhedral convex graph. We will use a solution concept given by the order relation defined by (compare Sect. 2.6.2)

$$F(x) \preceq F(u) \Longleftrightarrow F(x) + C \supseteq F(u).$$

In the following $\mathscr{G}$ denotes the family of all closed convex subsets of R^q. The objective map $F : \mathbb{R}^n \rightrightarrows \mathbb{R}^q$ can be considered as a function from $\mathbb{R}^n$ into $\mathscr{G}$.

Furthermore, by $\mathscr{G}_C$ we denote the subfamily of those elements P of $\mathscr{G}$ having the additional property

$$P = P + C.$$

For $P, Q \in \mathscr{G}$ we use the lower set less order relation $\preceq^l_C$ (see Definition 2.6.9, for simplicity we denote this relation by $\preceq$ in this section):

$$P \preceq Q \Longleftrightarrow P + C \supseteq Q + C, \tag{14.13}$$

which can be equivalently written as (compare Sect. 2.6.2)

$$P + C \supseteq Q.$$

The order relation $\preceq$ is reflexive and transitive in $\mathscr{G}$ (preorder) and, additionally, antisymmetric in $\mathscr{G}_C$ (partial order). An **equivalence relation** is defined for $P, Q \in \mathscr{G}$ by

$$P \cong Q :\Longleftrightarrow P + C = Q + C. \tag{14.14}$$

Clearly, the quotient space $\mathscr{G}/\cong$ is isomorphic to $\mathscr{G}_C$ and thus $\preceq$ is a partial ordering in $\mathscr{G}/\cong$.

The partially ordered set $(\mathscr{G}/\cong, \preceq)$ provides a *complete lattice*, i.e., for every subset of $\mathscr{G}/\cong$ there exist the infimum and supremum (see Sect. 2.6.3) taking into account Proposition 2.6.25. To simplify the notation we express the infimum and supremum in terms of the (preordered) space $(\mathscr{G}, \preceq)$, where we have in mind that we are dealing with representatives of equivalence classes. Thus, for nonempty sets $\mathscr{P} \subseteq \mathscr{G}$ we have (compare Sect. 2.6.3, Proposition 2.6.48 and Löhne [395, Proposition 1.62])

$$\inf \mathscr{P} = \operatorname{cl}\operatorname{conv} \bigcup_{P \in \mathscr{P}} (P + C), \qquad \sup \mathscr{P} = \bigcap_{P \in \mathscr{P}} (P + C).$$

As usual, we set $\inf \emptyset = \emptyset$ and $\sup \emptyset = \mathbb{R}^q$.

14.2.1 Formulation of the Polyhedral Convex Set-Valued Optimization Problem

We consider an set-valued objective map $F : \mathbb{R}^n \rightrightarrows \mathbb{R}^q$ and the following set optimization problem (compare problem $(SP - \preceq)$ in Sect. 2.6.2):

$$\preceq -\text{minimize} \quad F(x), \quad \text{subject to} \quad x \in \mathbb{R}^n. \tag{$SP - \preceq$}$$

Problem $(SP - \preceq)$ is called **feasible** if dom $F \neq \emptyset$.

The objective map F can be considered as a map from $\mathbb{R}^n$ into the space $\mathcal{G}$ of all closed convex subsets of $\mathbb{R}^q$. With the above ordering relation (14.13), one obtains a complete lattice, i.e., the infimum $\inf\{F(x) \mid x \in \mathbb{R}^n\}$ in the sense of a greatest lower bound with respect to $\preceq$ always exists. Solution concepts for complete-lattice-valued problems are discussed in [257].

In this section we present an algorithm for solving $(SP-\preceq)$ under the assumption that $F : \mathbb{R}^n \rightrightarrows \mathbb{R}^q$ is a polyhedral convex set-valued map, this means that its graph graph F is a polyhedral convex set.

Furthermore, we assume throughout that

(A1) There is a matrix $Z \in \mathbb{R}^{q\times p}$ such that the **cone** C has the **representation**

$$C = \{y \in \mathbb{R}^q \mid Z^T y \geq 0\}, \tag{14.15}$$

(A2) There are $A \in \mathbb{R}^{m\times n}$, $B \in \mathbb{R}^{m\times q}$ and $b \in \mathbb{R}^m$ such that **the graph of F** has the **representation**:

$$\text{graph } F = \{(x, y) \in \mathbb{R}^n \times \mathbb{R}^q \mid Ax + By \geq b\}. \tag{14.16}$$

(A3) $(SP - \preceq)$ is **bounded** in the sense that

$$\exists v \in \mathbb{R}^q : \{v\} \preceq \inf_{x\in\mathbb{R}^n} F(x).$$

It is sufficient for the boundedness of $(SP - \preceq)$ in the sense of (A3) that dom F is a bounded set and

$$\forall x \in \mathbb{R}^n, \exists v \in \mathbb{R}^q : \{v\} \preceq F(x).$$

Obviously, the latter condition is fulfilled for a map F with bounded values $F(x)$.

Remark 14.2.1. The algorithm presented in Sect. 14.2.2 can verify whether the problem is bounded or not.

Remark 14.2.2. Because of assumption (A2) we consider problems with a polyhedral convex set-valued objective map.

The following solution concept is based on a combination of minimality and infimum attainment as these notions do no longer coincide in vector and set optimization. It is an adaptation of the concepts introduced in Sect. 2.6.3 and [257] to the present setting.

In order to express minimality in the following definition, we introduce for $P, Q \in \mathcal{G}$:

$$P \prec Q :\Longleftrightarrow P \preceq Q \quad \text{and not } P \cong Q.$$

Definition 14.2.3. A point $\overline{x} \in \operatorname{dom} F$ is said to be a **minimal solution** for $(SP-\preceq)$ if there is no $x \in \mathbb{R}^n$ with $F(x) \prec F(\overline{x})$. A finite set $\overline{S} \subseteq \operatorname{dom} F$ is called a **finite infimizer** for $(SP-\preceq)$ if the infimum is attained in $\overline{S}$, that is,

$$\inf_{x\in\overline{S}} F(x) = \inf_{x\in\mathbb{R}^n} F(x).$$

A finite infimizer $\overline{S}$ of $(SP-\preceq)$ is called a **solution set** of $(SP-\preceq)$ if it consists of only minimal solutions.

For deriving the algorithm we study the vectorial relaxation of the set-valued optimization problem $(SP-\preceq)$. Consider the linear function $f : \mathbb{R}^n \times \mathbb{R}^q \to \mathbb{R}^q$ defined by $f(x,y) := y$. Because of formal reasons we understand f as a set-valued map whose values are singleton sets, i.e.,

$$f : \mathbb{R}^n \times \mathbb{R}^q \rightrightarrows \mathbb{R}^q, \qquad f(x,y) = \{y\}.$$

The **vectorial relaxation of the set-valued optimization problem** $(SP-\preceq)$ is defined with $f : \mathbb{R}^n \times \mathbb{R}^q \rightrightarrows \mathbb{R}^q$ as:

$$\preceq -\text{minimize } f(x,y) \quad \text{subject to} \quad y \in F(x). \tag{VecRel}$$

Obviously, (VecRel) can be seen as a special case of a set-valued optimization problem, whence the above definitions apply also to (VecRel). Furthermore, (VecRel) is feasible if and only if $(SP-\preceq)$ is feasible. Since f is single-valued and the constraint $y \in F(x)$ can be expressed by finitely many linear inequalities, (VecRel) is (equivalent to) a linear vector optimization problem.

It holds

$$\inf_{x\in\mathbb{R}^n} F(x) = \inf_{x\in\mathbb{R}^n} \inf_{y\in F(x)} \{y\} = \inf_{x\in\mathbb{R}^n, y\in F(x)} f(x,y), \tag{14.17}$$

i.e., $(SP-\preceq)$ and (VecRel) have the same infima. This implies that $(SP-\preceq)$ is bounded if and only if (VecRel) is bounded. Equation (14.17) is a motivation for the following solution concept.

Definition 14.2.4. A finite set $\{x^i \in \mathbb{R}^n \mid i = 1, \cdots, k\}$ is called a **pre-solution set** of $(SP - \preceq)$ if there exist $y^i \in \mathbb{R}^q$, $i = 1, \cdots, k$, such that $\{(x^i, y^i) \in \mathbb{R}^n \times \mathbb{R}^q \mid i = 1, \cdots, k\}$ is a solution set of the vectorial relaxation (VecRel) of $(SP - \preceq)$.

It is important to mention that a pre-solution set of $(SP - \preceq)$ is in general not a solution set of $(SP - \preceq)$ (see Example 3.2 in [396]).

This discussion leads us to the algorithm described in the following section.

14.2.2 An Algorithm for Solving Polyhedral Convex Set-Valued Optimization Problems

Taking into account the assumptions (A1) and (A2), let a problem of type $(SP - \preceq)$ be given by $A \in \mathbb{R}^{m\times n}$, $B \in \mathbb{R}^{m\times q}$, $b \in \mathbb{R}^m$, $Z \in \mathbb{R}^{q\times p}$. The algorithm generates a solution set of $(SP - \preceq)$ if the problem is feasible and bounded. Otherwise it detects whether $(SP - \preceq)$ is infeasible or unbounded.

In the **first phase** of the algorithm, the vectorial relaxation (VecRel), that is (equivalent to) a linear vector optimization problem, is solved. A solution set of (VecRel) can be obtained, for instance, with Benson's algorithm, see e.g. [46, 160, 395, 528, 529]. We know that $(SP - \preceq)$ is bounded if and only if (VecRel) is bounded. However, Benson's algorithm is able to detect if (VecRel) is unbounded. It is important to mention that Benson's algorithm was extended for unbounded linear vector optimization problems in the book by Löhne in [395].

In the **second phase**, for every element x^0 belonging to the pre-solution set obtained in the first phase, one constructs a sequence $(x^0, x^1, x^2, \cdots, x^l)$ with $F(x^0) \succ F(x^1) \succ F(x^2) \succ \cdots \succ F(x^l)$ until a minimal solution x^l for $(SP - \preceq)$ in the sense of Definition 14.2.3 is obtained after finitely many steps.

For parameters $w \in C^-$ and $\overline{x} \in \operatorname{dom} F$, we consider the following scalar problem:

$$\text{maximize } w^T y \quad \text{subject to} \quad y \in F(x),\ F(x) \preceq F(\overline{x}). \qquad (P(w, \overline{x}))$$

As the $y = (y_1, \cdots, y_q)^T \in \mathbb{R}$ ($y_i \in \mathbb{R}$ for all $i = 1, \cdots, q$) are considered to be auxiliary variables, we use the following convention: $\hat{x}$ is said to be a solution of $(P(w, \overline{x}))$ if there exists $\hat{y} \in \mathbb{R}^q$ such that $(\hat{x}, \hat{y})$ is a solution of $(P(w, \overline{x}))$ in the ordinary sense. In practice this means that $(\hat{x}, \hat{y})$ is generated but only $\hat{x}$ is used. Of course, there is the following lower bound β for the optimal value α of $(P(w, \overline{x}))$:

$$\begin{aligned}\alpha(w, \overline{x}) &:= \sup\{w^T y \mid y \in F(x), F(x) \preceq F(\overline{x})\} \\ &\geq \sup\{w^T y \mid y \in F(\overline{x})\} =: \beta(w, \overline{x}).\end{aligned} \qquad (14.18)$$

In the following we will use certain **representations** of $F(\overline{x}) + C$. The **first representation** of $F(\overline{x}) + C$ is given for some $\overline{x} \in \operatorname{dom} F$, $w^j \in \mathbb{R}^q$, $\gamma_j \in \mathbb{R}$ $(j = 1, \cdots, r)$ by

$$F(\overline{x}) + C = \{y \in \mathbb{R}^q \mid (w^j)^T y \leq \gamma_j,\ j = 1, \cdots, r\}. \tag{14.19}$$

Furthermore, we use a **second representation** of $F(\overline{x}) + C$ given for $y^1, \cdots, y^s \in \mathbb{R}^q$ by

$$F(\overline{x}) + C = \operatorname{conv}\{y^1, \cdots, y^s\} + C. \tag{14.20}$$

It is well known that every non-empty polyhedral convex subset of $\mathbb{R}^q$ can be expressed as a (generalized) convex hull of finitely many points and finitely many directions belonging to $\mathbb{R}^q$. Note that the assumption that $(SP - \preceq)$ is bounded was used in (14.20), otherwise the cone on the right hand side can be a superset of C.

Remark 14.2.5. Numerical methods to compute a representation given by (14.20) from a representation given by (14.19) and vice versa are called vertex enumeration, see e.g. [31] and [80].

(14.18) yields the following sufficient optimality condition for minimal solutions of the set-valued optimization problem $(SP - \preceq)$ in the sense of Definition 14.2.3.

Lemma 14.2.6. *Let $(SP - \preceq)$ be feasible and bounded. Furthermore, for some $\overline{x} \in \operatorname{dom} F$ let the representation (14.19) of the set $F(\overline{x}) + C$ be given with the corresponding $w^j \in \mathbb{R}^q$, $\gamma_j \in \mathbb{R}$ $(j = 1, \cdots, r)$. If $\alpha(w^j, \overline{x}) = \beta(w^j, \overline{x})$ for all $j \in \{1, \cdots, r\}$, then $\overline{x}$ is a minimal solution for $(SP - \preceq)$.*

Proof. Suppose that $\overline{x}$ is not a minimal solution of $(SP - \preceq)$, i.e., there exists $x \in \mathbb{R}^n$ with $F(x) \prec F(\overline{x})$. Hence there is some $y \in F(x)$ and some $c \in C$ such that $y + c \notin F(\overline{x}) + C$. Because of $C + C = C$, we conclude that $y \notin F(\overline{x}) + C$. Thus there is some $j \in \{1, \cdots, r\}$ such that ${w^j}^T y > \gamma_j$ which implies $\alpha(w^j, \overline{x}) > \gamma_j \geq \beta(w^j, \overline{x})$, a contradiction to $\alpha(w^j, \overline{x}) = \beta(w^j, \overline{x})$ for all $j \in \{1, \cdots, r\}$. □

Using the representation (14.20) of $F(\overline{x}) + C$, problem $(P(w, \overline{x}))$ can be transformed into a linear program.

By (14.16), the first constraint $y \in F(x)$ in $(P(w, \overline{x}))$ can be expressed as $Ax + By \geq b$. The second constraint in $(P(w, \overline{x}))$ can be transformed as follows:

$$\begin{aligned} F(x) \preceq F(\overline{x}) &\Longleftrightarrow F(x) + C \supseteq F(\overline{x}) + C \\ &\Longleftrightarrow \forall i = 1, \cdots, s \ :\ y^i \in F(x) + C \\ &\Longleftrightarrow \forall i = 1, \cdots, s,\ \exists c^i \in C \ :\ y^i - c^i \in F(x) \qquad (14.21) \\ &\Longleftrightarrow \forall i = 1, \cdots, s,\ \exists c^i \in C \ :\ Ax - Bc^i \geq b - By^i. \end{aligned}$$

Thus, $(P(w, \overline{x}))$ is equivalent to the linear program:

$$\max w^T y \quad \text{subject to } (x, y, c^1, \cdots, c^s) \in S, \qquad (P(w; y^1, \dots, y^s))$$

where

$$S := \{(x, y, c^1, \cdots, c^s) \in \mathbb{R}^n \times \mathbb{R}^q \times \mathbb{R}^q \times \cdots \times \mathbb{R}^q \mid Ax + By \geq b,$$
$$Ax - Bc^i \geq b - By^i,\ Z^T c^i \geq 0\ (i = 1, \cdots, s)\}$$

which has $n + q(s + 1)$ variables and $m + ms + ps$ constraints. According to the above convention we will speak about a solution $x \in \mathbb{R}^n$ and do not mention the $q(s + 1)$ auxiliary variables $y, c^1, \cdots, c^s \in \mathbb{R}^q$.

The following algorithm for solving the set-valued problem $(SP-\preceq)$ is developed by Löhne, Schrage [396] under the assumptions (A1), (A2) and (A3).

Algorithm SetOpt

Input:
Representation of graph F according to (14.16): $A \in \mathbb{R}^{m\times n}$, $B \in \mathbb{R}^{m\times q}$, $b \in \mathbb{R}^m$;
Representation of the ordering cone C according to (14.15): $Z \in \mathbb{R}^{q\times p}$;
Output:
A solution set $\overline{X}$ of $(SP - \preceq)$ in the case that $(SP - \preceq)$ is feasible and bounded, $\overline{X} = \emptyset$ otherwise;
The solution status for $(SP - \preceq)$;

Phase 1: Solve the linear vector optimization problem (VecRel)
$\overline{X} \longleftarrow \emptyset$;
solve (VecRel);
if (VecRel) is infeasible **then** *status* $\longleftarrow$ "$(SP - \preceq)$ *is infeasible.*"; stop; **end**;
if (VecRel) is unbounded **then** *status* $\longleftarrow$ "$(SP - \preceq)$ *is unbounded.*"; stop; **end**;
store a pre-solution set $\{x^1, \cdots, x^k\}$ of $(SP - \preceq)$;

Phase 2: Computation of minimal solutions of $(SP - \preceq)$
for $i \longleftarrow 1$ **to** k **do**
flag $\longleftarrow 1$;
$K \longleftarrow \emptyset$;
while *flag* $= 1$ **do**

compute the representations (14.20) and (14.19) of $F(x^i) + C$:

$$F(x^i) + C = \text{conv}\{y^1, \cdots, y^s\} + C$$

$$F(x^i) + C = \{y \in \mathbb{R}^q \mid (w^j)^T y \leq \gamma_j,\ j = 1, \cdots, r\};$$

for $j \longleftarrow 1$ **to** r **do**
if $w^j/||w^j|| \notin K$ **then**
$K \longleftarrow K \cup \{w^j/||w^j||\}$;
solve $(P(w^j; y^1, \cdots y^s))$; $x_i \longleftarrow$ solution; $\alpha \longleftarrow$ optimal value;

$$\beta \longleftarrow \max\{(w^j)^T y^1, \cdots, (w^j)^T y^s\};$$

if $\alpha > \beta$ **then** break (i.e., exit the inner-most loop);
end;
if $j = r$ **then** *flag* $\longleftarrow 0$;
end;
end;
$\overline{X} \longleftarrow \overline{X} \cup \{x^i\}$;
end;
status $\longleftarrow$ *"(SP $-$ $\preceq$) has been solved."*;

14.2.3 Properties of the Algorithm

In [396, Lemmata 4.2 and 4.3] it is shown that the algorithm works correctly and is finite. This assertion is prepared by two lemmata.

Lemma 14.2.7. *Let* $(SP - \preceq)$ *be feasible and bounded and* $w \in \mathbb{R}^q$ *be fixed. Consider some* $\overline{x} \in \operatorname{dom} F$*, a halfspace* $H = \{y \in \mathbb{R}^q \mid w^T y \leq \gamma\}$ *containing the set* $F(\overline{x}) + C$ *and finitely many points* $y^1, \cdots, y^s \in \mathbb{R}^q$ *such that (14.20) holds. Then,*

(a) The linear program $(P(w; y^1, \ldots, y^s))$ *has a minimal solution;*
(b) The lower bound β *defined in (14.18) can be expressed as*

$$\beta(w, \overline{x}) = \max\{w^T y^1, \cdots, w^T y^s\}.$$

Proof. Let $\overline{y} \in F(\overline{x})$, such that $A\overline{x} + B\overline{y} \geq b$ because of (A2). By (14.20), for all $i \in \{1, \cdots, s\}$, we have $y^i \in F(\overline{x}) + C$, i.e., there is some $c^i \in C$ such that $y^i - c^i \in F(\overline{x})$, or equivalently, $A\overline{x} - Bc^i \geq b - By^i$. Hence, the point $(\overline{x}, \overline{y}, c^1, \cdots, c^s)$ is feasible for $(P(w; y^1, \ldots, y^s))$.

There exists $v \in \mathbb{R}^q$ such that $\{v\} + C \supseteq F(x)$ for all $x \in \mathbb{R}^n$ since $(SP - \preceq)$ is supposed to be bounded. Moreover, it holds $w \in C^-$ because H contains $F(\overline{x}) + C$. It follows that

$$\sup\{w^T y \mid x \in \mathbb{R}^n, y \in F(x)\} \leq w^T v + \sup\{w^T c \mid c \in C\} = w^T v,$$

which implies that $(P(w, \overline{x}))$ and hence $(P(w; y^1, \ldots, y^s))$ is bounded. So (a) is proven.

Statement (b) follows from (14.20) taking into account that $\sup_{c \in C} w^T c = 0$ for $w \in C^-$. □

Remark 14.2.8. The assumption that the halfspace $H = \{y \in \mathbb{R}^q \mid w^T y \leq \gamma\}$ contains the set $F(\overline{x}) + C$ implies $w \in C^-$ such that it is not necessary to suppose additionally $w \in C^-$.

Lemma 14.2.9. *Let $(SP - \preceq)$ be feasible and bounded and let $\overline{x} \in \operatorname{dom} F$. If $\hat{x} \in \mathbb{R}^n$ is a solution of $(P(w, \overline{x}))$ for some $w \in C^-$, then $\alpha(w, u) = \beta(w, u)$ for every $u \in \mathbb{R}^n$ with $F(u) \preceq F(\hat{x})$.*

Proof. It holds $\beta(w, u) \leq \alpha(w, u)$ (see (14.18). For $u \in \mathbb{R}^n$ with $F(u) \preceq F(\hat{x})$, we obtain

$$\begin{aligned}\alpha(w, u) &= \sup\{w^T y \mid y \in F(x),\ F(x) \preceq F(u)\} \\ &\leq \sup\{w^T y \mid y \in F(x),\ F(x) \preceq F(\hat{x})\} = \alpha(w, \hat{x}).\end{aligned}$$

Taking into account that $F(u) \preceq F(\hat{x})$ can be written as $F(u) \supseteq F(\hat{x}) + C$ and, as $w \in C^-$, we get

$$\beta(w, u) = \sup\{w^T y \mid y \in F(u)\} \geq \sup\{w^T y \mid y \in F(\hat{x})\} = \beta(w, \hat{x}).$$

The point $\hat{x}$ is supposed to be a solution of $(P(w, \overline{x}))$, this implies that there exists $\hat{y} \in F(\hat{x})$ such that $\beta(w, \hat{x}) \geq w^T \hat{y} = \alpha(w, \overline{x}) \geq \alpha(w, \hat{x})$. Altogether we get $\alpha(w, u) \leq \alpha(w, \hat{x}) \leq \beta(w, \hat{x}) \leq \beta(w, u) \leq \alpha(w, u)$, which yields the desired result. □

In [396, Theorem 4.4] Löhne and Schrage show that the algorithm **SetOpt** terminates after finitely many steps:

Theorem 14.2.10. *The algorithm* **SetOpt** *generates a solution set of $(SP - \preceq)$ whenever $(SP - \preceq)$ is feasible and bounded. Otherwise* **SetOpt** *states whether $(SP - \preceq)$ is infeasible or unbounded. If the representation*

$$F(\overline{x}) + C = \{y \in \mathbb{R}^q \mid (w^j)^T y \leq \gamma_j,\ j = 1, \cdots, r\}$$

(see (14.19)) computed in Phase 2 contains no redundant inequalities, **SetOpt** *terminates after finitely many steps. To be more precise, let the pre-solution set computed in the Phase 1 consists of k points and let l be the number of linear inequalities necessary to describe the polyhedral convex set* $\operatorname{graph} F + (0_{\mathbb{R}^n} \times C)$. *In Phase 2 of* **SetOpt**, *at most $l \cdot k$ linear programs have to be solved.*

Proof. If $(SP - \preceq)$ is infeasible or unbounded, so is (VecRel) and the algorithm terminates with the corresponding status. If $(SP - \preceq)$ is feasible and bounded, a pre-solution set $\{x^1, \cdots, x^k\}$, $k \geq 1$ of $(SP - \preceq)$ is obtained by solving (VecRel). For fixed $i \in \{1, \cdots, k\}$, denote by $\overline{x}^i$ the value of the variable x^i before the while loop has been entered and let $\hat{x}^i$ be the value after the while loop has been left.

Then, $\overline{X} := \{\overline{x}^1, \cdots, \overline{x}^k\}$ is the pre-solution set computed in Phase 1 and $\hat{X} := \{\hat{x}^1, \cdots, \hat{x}^k\}$ is the result of the algorithm. We will show that for all $i \in \{1, \cdots, k\}$,

$$F(\hat{x}^i) \preceq F(\overline{x}^i) \quad \text{and} \quad \nexists x \in \mathbb{R}^n \; : \; F(x) \prec F(\hat{x}^i). \tag{14.22}$$

In order to show (14.22), let $i \in \{1, \cdots, k\}$ be fixed. The constraint $F(x) \preceq F(x^i)$ of the equivalent formulation $(P(w^j, x^i))$ of the linear program $(P(w^j; y^1, \cdots, y^s))$ directly implies the first condition of (14.22). In the case where $(SP - \preceq)$ is feasible and bounded there always exists a minimal solution of $(P(w^j; y^1, \cdots, y^s))$ by Lemma 14.2.7. The while loop is left only after *flag* has been set to zero. This requires r iterations in the inner for loop, which occurs only if for every $j \in \{1, \cdots, r\}$, $\alpha = \alpha(w^j, \hat{x}^i)$ equals $\beta = \beta(w^j, \hat{x}^i)$. In case of $w^j/||w^j|| \in K$, this is known by Lemma 14.2.9, i.e., $(P(w^j; y^1, \cdots, y^s))$ does not need to be solved. But $\alpha(w^j, \hat{x}^i) = \beta(w^j, \hat{x}^i)$ for all $j \in \{1, \cdots, r\}$ implies that $\hat{x}^i$ is a minimal solution, compare Lemma 14.2.6. So the second condition in (14.22) is shown.

The first condition in (14.22) implies

$$\inf_{x \in \hat{X}} F(x) \preceq \inf_{x \in \overline{X}} F(x).$$

Since $\overline{X}$ is a pre-solution set of $(SP - \preceq)$, the infimum is attained in $\overline{X}$, that is,

$$\inf_{x \in \overline{X}} F(x) = \inf_{x \in \mathbb{R}^n} F(x).$$

It follows that the infimum is also attained in $\hat{X}$. The second condition in (14.22) states that $\hat{X}$ consists of only minimal solutions, whence $\hat{X}$ is a solution set of $(SP - \preceq)$.

To show finiteness, note first that there is a finite algorithm (such as Benson's algorithm) to solve (VecRel) in Phase 1. Consider the set-valued map $\tilde{F} : \mathbb{R}^n \rightrightarrows \mathbb{R}^q$, $\tilde{F}(x) := F(x) + C$. Of course, graph $\tilde{F}$ = graph $F + (0_{\mathbb{R}^n} \times C)$ is a polyhedral convex set. Thus it can be expressed as (compare (14.16))

$$\operatorname{graph} \tilde{F} = \{(x, y) \in \mathbb{R}^n \times \mathbb{R}^q \mid \tilde{A}x + \tilde{B}y \geq \tilde{b}\},$$

for some $\tilde{A} \in \mathbb{R}^{l \times n}$, $\tilde{B} \in \mathbb{R}^{l \times q}$, $\tilde{b} \in \mathbb{R}^l$. For every $x \in \operatorname{dom} F$, we have

$$F(x) + C = \{y \in \mathbb{R}^q \mid \tilde{B}y \geq \tilde{b} - \tilde{A}x\}.$$

Consequently, every representation of $F(x) + C$ given by (14.19) that contains no redundant inequalities consists of at most l inequalities. With other words, there are at most l different vectors $w^j/||w^j||$ for $j \in \{1, \cdots, k\}$ in the algorithm, and the proof is completed. □

Theorem 14.2.10 immediately implies the following existence result (see [396, Corllary 4.5]).

Corollary 14.2.11. *If* $(SP - \preceq)$ *is feasible and bounded, a solution set of* $(SP - \preceq)$ *exists.*

In order to reduce the computational effort of the algorithm for specific problems, an additional rule is suggested in [396]. Let $\{(x_1, y_1), \cdots, (x_k, y_k)\}$ denote the solution set of (VecRel) obtained in Phase 1 and consider iteration $i \in \{1, \cdots, k\}$ of the outer for loop in Phase 2:

If $y_j \in F(x_i)$ *for* j *with* $i < j \leq k$ *then skip all commands in iteration* j.

Obviously, this rule maintains the attainment of the infimum and thus the algorithm still works correctly.

Finally, we discuss the special case where $F : \mathbb{R}^n \longrightarrow \mathcal{G}_C$, i.e., we consider

$$F(x) = F(x) + C \quad \text{for all} \quad x \in \mathbb{R}^n.$$

In this case, (14.21) can be replaced by

$$\begin{aligned} F(x) \preceq F(\overline{x}) &\Longleftrightarrow F(x) \supseteq F(\overline{x}) \\ &\Longleftrightarrow \forall i = 1, \cdots, s \;:\; y^i \in F(x) \\ &\Longleftrightarrow \forall i = 1, \cdots, s \;:\; Ax \geq b - By^i. \\ &\Longleftrightarrow Ax \geq \max\{b - By^i \mid i = 1, \cdots, s\}. \end{aligned}$$

This means that the linear program $(P(w; y^1, \ldots, y^s))$ has only $n + q$ variables and $2m$ constraints. The problem to get a representation of graph$(F(\cdot) + C)$ given by (14.19) from a representation of graph F given by (14.16) seems to be difficult in practice where it is typical that n is much more larger than q. One way to obtain it is vertex enumeration of a polyhedral convex set in $\mathbb{R}^n \times \mathbb{R}^q$. In contrast, **SetOpt** involves vertex enumeration only in $\mathbb{R}^q$.

Chapter 15
Applications

15.1 Set-Valued Approaches to Duality in Vector Optimization

Applying the duality assertions for set-valued problems given in Sect. 8.3 we derive duality statements for vector optimization problems in this section.

In the literature there are several approaches to construct a dual problem for a given vector optimization problem. For instance, Luc [402] distinguishes between conjugation, Lagrangian and axiomatic Duality. However, there seems to be no unified approach to dualization in vector optimization. One of the difficulties is in the fact that the minimal solution in multi-objective optimization is not necessarily a single element, but in general becomes a subset of the image space. The definition of infimum (or supremum) of a set with partial order plays a key role in development of duality theory in multi-objective optimization. An interesting discussion of these aspects is given in the book by Pallaschke, Rolewicz [453] and in the paper by Nakayama [437]. There are at least three main ideas which are used for overcoming the difficulties that arise when generalizing well-known duality assertions from the scalar optimization theory to the vector-valued case.

The **first** one is the usage of scalarization in the formulation of the dual problem (see Schönfeld [525], Breckner [78], Jahn [290,292]). In this approach, scalarization concepts and corresponding duality assertions from real-valued optimization are often used in order to derive useful dual problems, to prove duality assertions or in order to solve the dual problem. In a lot of papers this procedure is used in the proofs too (see e.g. [52]). However, as shown in [234], this approach has the disadvantage that even in the case of linear vector optimization a duality gap may occur, although the usual assumptions are fulfilled.

A **second** category of dual problems is based on the observation that a dual vector optimization problem is naturally set-valued (see Tanino, Sawaragi [568], Kawasaki [316], Corley [111], Tanino [563, 566], Luc [402], Nakayama [437],

A.A. Khan et al., *Set-valued Optimization*, Vector Optimization,
DOI 10.1007/978-3-642-54265-7_15

Tammer [557], Dolecki, Malivert [145], Pallaschke, Rolewicz [453], Song [536], Hamel, Heyde, Löhne, Tammer, Winkler [234], Löhne, Tammer [397]). Duality assertions for vector optimization problems are shown without a scalarization "from the beginning". Instead, the dual problem becomes set-valued. For instance, in the paper by Tanino [566] the set-valued structure of the primal and dual vector optimization problem is taken into account: Embedding the primal problem into a family (depending from perturbation parameters) of set-valued optimization problems and applying an extension of Fenchel's inequality, Tanino derives a weak duality assertion and using the relationship between a map and its biconjugate he shows a strong duality statement. Furthermore, in the paper by Dolecki, Malivert [145] a set-valued approach in combination with Lagrangian techniques and perturbations of marginal relations is used in order to show duality assertions for general vector optimization problems where the solution concept is described by a transitive, translation-invariant relation.

A **third** type of dual problems is based on solution concepts with respect to the supremum and infimum in the sense of a vector lattice. In the book by Pallaschke and Rolewicz [453], special conditions concerning the order in the image space are supposed in order to prove duality assertions. A duality theory for objective functions with values in vector lattices is developed in [453] using corresponding notions of infimum and supremum in the sense of utopia minimum (maximum). However, these infima and suprema may not be solutions in the sense of vector optimization (Pareto) because they may not belong to the set of the objective function's values. In [453] notions of vector-convexity, vector-subgradients and vector duality are introduced. These notions are analogous to the scalar case taking into account the order in the vector lattice.

In the paper by Nieuwenhuis [442] solution concepts on the basis of infimal (supremal) sets are introduced because the assumption that the objective function has its values in a vector lattice is too restrictive for vector optimization. Nieuwenhuis [442] and Taninio [563, 566] derived duality assertions to these solution concepts. These infimal sets are closely related to weakly minimal points (see Definition 2.4.2). Dolecki and Malivert [145] extended these concepts to infimal sets being closely related to other kinds of efficiency, too.

In contrast to the these investigations the approach in this section (see Löhne, Tammer [397]) is characterized by an embedding of the image space of the vector-valued problem into a complete lattice without linear structure, namely a sublattice of the power set of the image space (compare Sect. 2.6.3). On the one hand, our primal and dual problems are set-valued and therefore related to the problems of the second type. The infimal (supremal) sets of Nieuwenhuis [442] are involved into the definition of the infimum (supremum) in the lattice which yields a relationship to the papers by Nieuwenhuis [442] and Taninio [563,566]. On the other hand, a consequent usage of the lattice structure yields that we can carry over the formulations, statements and proof techniques from the scalar optimization theory, even though our assertions are for solution concepts in the sense of vector optimization (weakly minimal points, see Definition 2.4.2). In [442, 563, 566] the construction of the dual problem is not completely analogous to the scalar case because the lattice structure is not taken into account.

Furthermore, we point out the relationship between duality in vector optimization and duality for set-valued problems based on the ordering relation "set inclusion", in the sequel called *set inclusion problems* (this approach is also based on the lattice structure of the image space), which were investigated in [393,394] (see Sect. 2.6.3).

In this section we discuss the relationship between a vector optimization problem and set-valued optimization problems with $\mathcal{F}$-valued and $\mathcal{I}$-valued objective maps (see Definitions 2.6.36 and 2.6.37). Since $\mathcal{F}$ and $\mathcal{I}$ are complete lattices (see Proposition 2.6.40) we can assign to a given problem a dual problem following the lines of scalar duality theory. In Sect. 8.3 we have shown that under the usual assumptions weak as well as strong duality assertions for $\mathcal{F}$-valued and $\mathcal{I}$-valued problems can be obtained. By the fact $\mathcal{F}$ and $\mathcal{I}$ being isomorphic and isotone (Proposition 2.6.39) it is clear that it is sufficient to prove the duality assertions just for one case, either for the $\mathcal{F}$-valued case or for the $\mathcal{I}$-valued case.

The **advantage of the $\mathcal{F}$-valued case** is that the operations, the ordering and the infimum and supremum have an easier structure, which is beneficial for proofs. In this case we speak about *set inclusion problems.*

The **advantage of the $\mathcal{I}$-valued case** is that it is closely related to vector optimization problems, therefore let us speak about *vector optimization problems* in this case.

This relationship can be seen as follows. Assume that $(Y, \leq)$ is a partially ordered linear topological space, where the order is induced by a proper pointed closed convex cone C with $\operatorname{int} C \neq \emptyset$. For an arbitrary set X, consider a set-valued objective map $F : X \rightrightarrows Y$. Using the notations introduced in Sect. 2.6.3, especially Definition 2.6.43, we study the vector optimization problem

$$\bar{P} := \operatorname{Inf} \bigcup_{x \in X} F(x). \qquad \text{(SVP)}$$

This means (compare Definition 2.6.31) that we are looking for the set of weakly minimal points of $\operatorname{Cl}_{+} \bigcup_{x \in X} F(x)$ with respect to C (see Definition 2.4.2). As usual, if putting $F(X) = \bigcup_{x \in X} F(x)$ we have the following relationships.

Because of Proposition 2.7.5 we have

$$\operatorname{Inf} F(X) = \operatorname{WMin}(F(X) + C, C)$$

under the assumption that $F(X) + C$ is closed.

Furthermore, under the assumption that $F(X) = \operatorname{cl}(F(X) + C)$ it holds (see Proposition 2.7.4)

$$\operatorname{WMin}(F(X), C) = \operatorname{WMin}(\operatorname{cl}(F(X) + C), C) = \operatorname{Inf} F(X).$$

Since $\operatorname{Inf}\bigcup_{x\in X} F(x) = \operatorname{Inf}\bigcup_{x\in X} \operatorname{Inf} F(x)$, (SVP) can be expressed as an $\mathcal{I}$-valued problem (see Sect. 2.6.3); without loss of generality we can assume that the sets $F(x)$ are self-infimal, i.e., $F : X \to \mathcal{I}$. Thus we consider the following problem (compare Definition 2.6.43)

$$\bar{P} = \operatorname{Inf} \bigcup_{x\in X} F(x) = \inf_{x\in X} F(x). \qquad (SP\text{--} \preceq_{\mathcal{I}})$$

We assign to $(SP\text{--} \preceq_{\mathcal{I}})$ a dual problem $(SD\text{--} \preceq_{\mathcal{I}})$: Let V be a set and $D : V \rightrightarrows \mathcal{I}$,

$$\bar{D} := \operatorname{Sup} \bigcup_{v\in V} D(v) = \sup_{v\in V} D(v). \qquad (SD\text{--} \preceq_{\mathcal{I}})$$

As usual (see Sect. 8.3.2), we speak about **weak duality** between $(SP\text{--} \preceq_{\mathcal{I}})$ and $(SD\text{--} \preceq_{\mathcal{I}})$ when $\bar{D} \preccurlyeq_{\mathcal{I}} \bar{P}$ and we speak about **strong duality** when $\bar{D} = \bar{P}$. In contrast to the primal problem the self-infimality of the values of the dual objective function $D(\cdot)$ plays an important role. If we replace a value $D(v)$ by $\hat{D}(v)$ with $\operatorname{Inf} D(v) = \operatorname{Inf} \hat{D}(v)$ but $\hat{D}(v)$ being not self-infimal, $\bar{D}$ might be changed. Therefore we shall understand $(SD\text{--} \preceq_{\mathcal{I}})$ itself as the dual problem to (SVP).

Remark 15.1.1. The vector optimization problem (SVP) means that we compute weakly minimal elements of the upper closure (elements belonging to the infimal set) of the image set with respect to the proper pointed closed convex cone C with nonempty interior (compare Sect. 2.6.3). We express the vector optimization problem by the $\mathcal{I}$-valued problem $(SP\text{--} \preceq_{\mathcal{I}})$ in order to use a complete lattice structure and to derive assertions analogously to the scalar optimization theory.

In the dual problem $(SD\text{--} \preceq_{\mathcal{I}})$ we study the problem to determine weakly maximal elements of the lower closure (elements belonging to the supremal set) of the dual image set with respect to the cone C.

Also in the case that the primal problem (SVP) is single-valued, the dual problem $(SD\text{--} \preceq_{\mathcal{I}})$ is always a set-valued problem. Indeed, by many authors (see Corley [111], Tanino [563,566], Luc [402], Dolecki, Malivert [145], Tammer [557], Pallaschke, Rolewicz [453]) it was observed that the dual of a vector optimization problem is “naturally” set-valued. Taking into account the space $\mathcal{I}$ and its lattice structure we give an explanation for this by our approach.

Remark 15.1.2. In case of $\bar{D}, \bar{P} \subset Y$, the weak duality inequality $\bar{D} \preccurlyeq_{\mathcal{I}} \bar{P}$ can be equivalently expressed by $(\bar{D} - \operatorname{int} C) \cap \bar{P} = \emptyset$, what is a well-known relation in vector optimization. Indeed, $\bar{D} \preccurlyeq_{\mathcal{I}} \bar{P}$ is equivalent to $\bar{P} \subset \operatorname{Cl}_{+}\bar{D}$ and by Proposition 2.6.33 (iv), (v), (vi) this is equivalent to $(\bar{D} - \operatorname{int} C) \cap \bar{P} = \emptyset$.

The reformulation of the vector optimization problem (SVP) as an $\mathcal{I}$-valued problem and the relationship to a set inclusion problem, in our setting an $\mathcal{F}$-valued problem (see Definition 2.6.36), is discussed and used for deriving Fenchel duality assertions in Sect. 15.1.1 (see Löhne, Tammer [397]). Furthermore, considering the reformulation of the given primal problem (SVP) as an $\mathcal{I}$-valued problem (see Definition 2.6.37) we derive corresponding Lagrange dual problems in Sect. 15.1.2.

Finally, in Sect. 15.1.3 we study the special case of linear vector optimization problems and apply our duality statements in order to show corresponding results for the linear case. So it is possible to understand better the structure of the dual linear vector optimization problem in comparison with results in the paper by Isermann [287].

15.1.1 Fenchel Duality for Vector Optimization Problems Using Corresponding Results for $\mathcal{F}$-Valued Problems

In this section we consider the vector optimization problem (SVP) for the special case $Y = \mathbb{R}^q$, partially ordered by a proper pointed closed convex cone $C \subset \mathbb{R}^q$ with $\operatorname{int} C \neq \emptyset$ and $X = \mathbb{R}^n$. We derive duality assertions for (SVP) using the duality statements that we have shown in Theorem 8.3.4 for $\mathcal{F}$-valued (compare Definitions 2.6.36) problems following the paper by Löhne and Tammer[397]. Since $\mathcal{F}$ is a complete lattice (Proposition 2.6.40) we can assign to a given problem a dual problem following the lines of scalar duality theory. We know that $\mathcal{F}$ and $\mathcal{I}$ being isomorphic and isotone (Proposition 2.6.39) and so it is sufficient to prove the duality assertions just for one case, either for the $\mathcal{F}$-valued case or for the $\mathcal{I}$-valued case. So we proceed as follows: We take the reformulation of (SVP) as $\mathcal{I}$-valued problem ($SP- \preceq_{\mathcal{I}}$), use the relationships between $\mathcal{I}$-valued and $\mathcal{F}$-valued problems and apply duality assertions for $\mathcal{F}$-valued problems given in Theorem 8.3.4. As already mentioned, it is beneficial for the proofs of the duality assertions that in the $\mathcal{F}$-valued case the operations, the ordering and the infimum and supremum have an easier structure.

So it is useful to consider a dual pair of $\mathcal{F}$-valued problems simultaneously. For $p : X \to \mathcal{F},\ p(x) := \operatorname{Cl}_+ F(x)$ and $d : V \to \mathcal{F},\ d(u) := \operatorname{Cl}_+ D(u)$ consider the problems

$$\bar{p} := \operatorname{cl} \bigcup_{x \in X} p(x) = \inf_{x \in X} p(x), \qquad (SP- \preceq_{\mathcal{F}})$$

$$\bar{d} := \bigcap_{u \in V} d(u) = \sup_{u \in V} d(u). \qquad (SD- \preceq_{\mathcal{F}})$$

We define **weak and strong duality** for the $\mathcal{F}$-valued problems ($SP- \preceq_{\mathcal{F}}$) and ($SD- \preceq_{\mathcal{F}}$) like in Sect. 8.3.1, i.e., by $\bar{d} \supseteq \bar{p}$ (that is $\bar{d} \preccurlyeq_{\mathcal{F}} \bar{p}$) and $\bar{\bar{d}} = \bar{p}$, respectively.

Of course, by Proposition 2.6.39 we have $\bar{P} = \operatorname{Inf} \bar{p}$, $\bar{D} = \operatorname{Inf} \bar{d}$, $\operatorname{Cl}_+ \bar{P} = \bar{p}$ and $\operatorname{Cl}_+ \bar{D} = \bar{d}$ and we have weak duality between ($SP- \preceq_{\mathcal{I}}$) and ($SD- \preceq_{\mathcal{I}}$) if and only if we have weak duality between ($SP- \preceq_{\mathcal{F}}$) and ($SD- \preceq_{\mathcal{F}}$), and we have strong duality between $(SP- \preceq_{\mathscr{I}})$ and $(SD- \preceq_{\mathscr{I}})$ if and only if we have strong duality between $(SP- \preceq_{\mathscr{F}})$ and $(SD- \preceq_{\mathscr{F}})$, i.e., we have

$$\bar{D} \preccurlyeq_{\mathscr{I}} \bar{P} \qquad \Longleftrightarrow \qquad \bar{d} \supseteq \bar{p},$$

$$\bar{D} = \bar{P} \qquad \Longleftrightarrow \qquad \bar{d} = \bar{p}.$$

In Sect. 8.3.1 we have shown weak and strong duality assertions for $\mathscr{F}$-valued problems (see Theorem 8.3.4). Now, we will apply these results for deriving duality assertions for vector optimization problems.

Let $F : X \rightrightarrows \mathbb{R}^q$ and $G : Z \rightrightarrows \mathbb{R}^q$ be two set–valued maps, $A : X \to Z$ is a linear map, where $X = \mathbb{R}^n$ and $Z = \mathbb{R}^m$ (as in Sect. 8.3.1). We now consider the following vector optimization problem

$$\bar{P} := \operatorname{Inf} \bigcup_{x \in X} \big(F(x) + G(Ax)\big). \tag{SVP_F}$$

In the same manner as at the beginning of this section, we do not loose generality if F and G are considered to be functions $F : X \to \mathscr{I}$, $G : Z \to \mathscr{I}$ and the vector optimization problem (SVP_F) can be equivalently expressed as an $\mathscr{I}$-valued problem

$$\bar{P} = \operatorname{Inf} \bigcup_{x \in X} \big(F(x) \oplus_{\mathscr{I}} G(Ax)\big). \tag{$SP_F- \preceq_{\mathscr{I}}$}$$

We define the conjugate (compare [566]) of a function $F : X \to \mathscr{I}$ (with respect to $c \in \mathbb{R}^q$) by

$$F_c^* : X^* \to \mathscr{S}, \qquad F_c^*(x^*) := \operatorname{Sup} \bigcup_{x \in X} \big(\langle x^*, x \rangle \{c\} - F(x)\big),$$

where $\mathscr{S}$ is the space of self-supremal sets with respect to C, which is defined analogously to $\mathscr{I}$. Following the lines of scalar optimization we consider the following dual problem:

$$\bar{D}_c := \operatorname{Sup} \bigcup_{u^* \in Z^*} \big(-F_c^*(A^T u^*) \oplus_{\mathscr{I}} -G_c^*(-u^*)\big). \tag{$SD_F^c- \preceq_{\mathscr{I}}$}$$

In view of Proposition 2.6.39 this is just a transformation of the dual pair of problems considered in Sect. 8.3.1 into the framework of functions with self-infimal values. Therefore we immediately obtain the same duality assertions.

Remark 15.1.3. Note that convexity of a function $F : X \to \mathscr{I}$ (see Definition 8.3.8) is equivalent to the so-called C-convexity (see e.g. Jahn [292, 293], Luc [402] and Sect. 8.3.2) of the corresponding set-valued map $\bar{F} : X \rightrightarrows \mathbb{R}^q$, $\bar{F}(x) := F(x)$ if $F(x) \subseteq \mathbb{R}^q$, $\bar{F}(x) := \emptyset$ if $F(x) = \{+\infty\}$, $\bar{F}(x) := \mathbb{R}^q$ if $F(x) = \{-\infty\}$, where C-convexity means that $\operatorname{graph}\big(\operatorname{cl}\big(\bar{F}(\cdot) + C\big)\big)$ is convex.

In the following theorem we will show duality results for the $\mathcal{I}$-valued problem $(SP_F- \preceq_{\mathcal{I}})$ (that is equivalent to the vector optimization problem (SVP_F) and the $\mathcal{I}$-valued dual problem $(SD^c_F- \preceq_{\mathcal{I}})$ using the results given in Theorem 8.3.4 for problems with $\mathcal{F}$-valued objective functions taking into account the relations between $\mathcal{F}$-valued and $\mathcal{I}$-valued problems (see Proposition 2.6.39).

Theorem 15.1.4 (Weak and Strong Duality Between $(SP_F- \preceq_{\mathcal{I}})$ and $(SD^c_F- \preceq_{\mathcal{I}})$). *For all $c \in \mathbb{R}^q$ it holds weak duality between $(SP_F- \preceq_{\mathcal{I}})$ and $(SD^c_F- \preceq_{\mathcal{I}})$, i.e., $\bar{D}_c \preccurlyeq_{\mathcal{I}} \bar{P}$. If F and G are convex, $0 \in \text{r-int}(\text{dom}\, G - A\, \text{dom}\, F)$ and $c \in \text{int}\, C$, then we have strong duality between $(SP_F- \preceq_{\mathcal{I}})$ and $(SD^c_F- \preceq_{\mathcal{I}})$, i.e., $\bar{D}_c = \bar{P}$; if additionally $\bar{P} \neq \{-\infty\}$, we have*

$$\bar{D}_c = \text{WMax}\Big(\bigcup_{u^* \in Z^*} \big(-F^*_c(A^T u^*) \oplus_{\mathcal{I}} -G^*_c(-u^*)\big), C\Big).$$

Proof. Consider the $\mathcal{F}$-valued dual pair $(SP- \preceq_{\mathcal{F}})$, $(SD_F- \preceq_{\mathcal{F}})$ from Section 8.3.1 and denote the corresponding $\mathcal{F}$-valued objective functions by p and d_c. The first part (weak and strong duality) follows from Theorem 8.3.4 and the considerations in Sect. 2.6.3, in particular Proposition 2.6.39. For the second part let $\bar{P} \neq \{-\infty\}$, hence $\emptyset \neq \text{Cl}_+ \bar{P} \neq \mathbb{R}^q$. Let $\bar{y} \in \text{Sup} \bigcup_{u^* \in Z^*} D_c(u^*) = \bar{D}_c = \bar{P} = \text{Inf}\, \bar{p}$. Then $\bar{y} \notin \bar{p} + \text{int}\, C$. By a separation theorem, there exists some $\bar{y}^* \in C^- \setminus \{0\}$ such that $\langle \bar{y}^*, \bar{y} \rangle \geq \sigma_{\bar{p}+\text{int}\, C}(\bar{y}^*) = \sigma_{\bar{p}}(\bar{y}^*) + \sigma_{\text{int}\, C}(\bar{y}^*) = \sigma_{\bar{p}}(\bar{y}^*)$. By (8.13) there exists some $\bar{u}^* \in Z^*$ such that $\langle \bar{y}^*, \bar{y} \rangle \geq \sigma_{d_c(\bar{u}^*)}(\bar{y}^*)$. Assuming that $\bar{y} \in d_c(\bar{u}^*) + \text{int}\, C$ we obtain $\langle y^*, \bar{y} \rangle < \sigma_{d_c(\bar{u}^*)}(\bar{y}^*)$ for all $y^* \in C^- \setminus \{0\}$, a contradiction. Hence $\bar{y} \notin d_c(\bar{u}^*) + \text{int}\, C$. On the other hand, $\bar{y} + \text{int}\, C \in \text{Inf}\, \bar{p} + \text{int}\, C = \bar{p} + \text{int}\, C \subseteq d_c(\bar{u}^*) + \text{int}\, C$. This yields that $\bar{y} \in \text{Inf}\, d_c(\bar{u}^*) = D_c(\bar{u}^*) \subseteq \bigcup_{u^* \in Z^*} D_c(u^*)$. Together we have $\bar{y} \in \text{WMax}(\bigcup_{u^* \in Z^*} D_c(u^*), C)$. □

Remark 15.1.5. Our dual problem $(SD^c_F- \preceq_{\mathcal{I}})$ differs from the dual problem given by Tanino [566] because of the formulation taking into account the complete lattice structure of the space $\mathcal{I}$ of self-infimal sets. So our formulation is completely analogous to that of scalar Fenchel duality theorems, that means our dual objective function is expressed in terms of the conjugates. Note further that the origin space of the dual problem is just the space Z^* instead of the space of linear continuous operators $\mathcal{L}(Z^*, \mathbb{R}^q)$.

Remark 15.1.6. In scalar optimization the primal and dual attainment of the solution are of interest in many cases. Even though we cannot expect that the infimum and supremum are attained by a single $x \in X$ and $u^* \in Z^*$, respectively, we speak about primal and dual attainment if we can replace the infimal set by the set of minimal (weakly minimal) elements in $(SP_F- \preceq_{\mathcal{I}})$ and the supremal set by the set of maximal (weakly maximal) elements in $(SD^c_F- \preceq_{\mathcal{I}})$, respectively. Concerning the primal attainment let us mention that even in scalar convex programming the infimum is not attained without additional assumptions. Therefore, also in the case of vector optimization one needs additional assumptions. In the paper by Dolecki, Malivert [145] and in the book by Pallaschke, Rolewicz [453] the submission (domination) property is supposed. This property guarantees the existence of minimal points. Using the infimal set in the primal problem, we do not need such a domination property.

As in the scalar theory the dual attainment in Theorem 15.1.4 follows without any additional assumption.

Remark 15.1.7. In many papers on duality for vector optimization problems (compare Jahn [292]) a strong direct and a strong inverse duality assertion is shown. In the formulation of the strong direct duality assertion the existence of a weakly minimal solution of the primal vector optimization problem is supposed and the existence of a weakly maximal solution of the dual problem with the same objective function value is shown. Conversely, in the strong inverse duality assertion a closedness condition concerning the primal problem and the existence of a weakly maximal solution of the dual problem are supposed in order to show the existence of a weakly minimal solution of the primal problem with the same objective function value. Such formulations like strong direct and inverse duality assertions are discussed in Sect. 15.1.3 (Corollaries 15.1.20, 15.1.21) for the case of linear vector optimization problems, but could be also obtained for the convex setting, as a conclusion of Theorem 15.1.4.

15.1.2 Lagrange Duality for Vector Optimization Problems Based on Results for $\mathscr{I}$-Valued Problems

In this section we apply the duality statements shown in Sect. 8.3.2 for deriving Lagrange duality assertion for set-valued vector optimization problems following Hernández, Löhne, Rodríguez-Marín, Tammer [247]. We study the primal and dual problem introduced by Li, Chen and Wu [379] and derive relationships to the dual problems discussed in Sect. 8.3.2 using the solution concepts of weakly minimal / maximal elements for vector optimization problems (see Definition 2.4.2).

Let X be a linear space, Y be a separated locally convex space partially ordered by a proper pointed closed convex cone C such that $\emptyset \neq \operatorname{int} C \neq Y$ and let Z be a separated locally convex space. Moreover, let $\langle Z, Z^* \rangle$ be a dual pair. Let $F : X \rightrightarrows Y$ be a set-valued map, let $G : X \rightrightarrows Z$ be a set-valued map with $\operatorname{dom} G = X$ and let $K \subseteq Z$ be a proper closed convex cone. We use the notations introduced in Sect. 2.6.3, especially in Definition 2.6.31.

We consider a set-valued vector optimization problem

$$\bar{P} := \operatorname{Inf} \bigcup_{x \in S} F(x) \qquad (SVP_S)$$

where

$$S = \{x \in X \mid G(x) \cap -K \neq \emptyset\}$$

(compare problem (SP) in Sect. 8.4.1).

From the relation

$$\operatorname{Inf} \bigcup_{x \in S} F(x) = \operatorname{Inf} \bigcup_{x \in S} \operatorname{Inf} F(x),$$

and setting

$$F_{\mathscr{I}} : X \to \mathscr{I}, \qquad F_{\mathscr{I}}(x) := \operatorname{Inf} F(x),$$

we can reformulate (SVP_S) as an $\mathscr{I}$-valued problem

$$\bar{P} := \inf_{x \in S} F_{\mathscr{I}}(x). \tag{P}$$

Denote by $\mathscr{L} := \mathscr{L}(Z, Y)$ the set of all linear continuous operators from Z to Y and by $\mathscr{L}_+$ the subset of all positive operators, that is, $\mathscr{L}_+ := \{T \in \mathscr{L} \mid T(K) \subseteq C\}$.

Thus, according to Sect. 8.3.2 we can associate to problem (P) two dual problems as follows: The first one is formulated as

$$\bar{D}_c := \sup_{u^* \in Z^*} \phi_c(u^*), \tag{SVD_L^C}$$

where

$$\phi_c(u^*) := \inf_{x \in X} L_c(x, u^*)$$

and (compare (8.18))

$$L_c(x, u^*) := F_{\mathscr{I}}(x) \oplus \inf_{u \in G(x)+K} \{\langle u^*, u \rangle \{c\} + \operatorname{bd} C\}.$$

Furthermore, the second one is given by

$$\tilde{D} := \sup_{T \in \mathscr{L}_+} \phi(T). \tag{SVD_L^T}$$

where

$$\phi(T) := \inf_{x \in X} L(x, T)$$

and (compare (8.22))

$$L(x, T) = F_{\mathscr{I}}(x) \oplus \inf_{u \in G(x)+K} \{T(u) + \operatorname{bd} C\}.$$

From now on without loss of generality we can assume that $X = \operatorname{dom} F_{\mathscr{I}}$.

Remark 15.1.8. We emphasize that problem (P) has been studied in several papers, for instance, Tanino [566], Song [535], Li et al [379] and Chen et al [89] to give conjugate duality. The authors consider an objective map F from X to $Y \cup \{+\infty\}$ but they work on $\mathrm{Dom}\, F = \{x \in X \mid F(x) \neq +\infty \text{ and } F(x) \neq \emptyset\}$. Note that the corresponding $\mathcal{I}$-valued map $F_{\mathcal{I}}$ defined in problem (P) satisfies that $\mathrm{Dom}\, F_{\mathcal{I}} = \{x \in X \mid F(x) \neq +\infty \text{ and } F(x) \neq \emptyset\}$. For a comparison of different approaches to duality in set-valued optimization see Sect. 8.4.

Li et al in [379] study problem (P) and the following dual problem

$$\bar{d}_L := \mathrm{WMax}(\{\mathrm{Inf} \bigcup_{x \in X} (F(x) + T(G(x))) \mid T \in \mathcal{L}_+\}, C) \tag{D$_L$}$$

(denoting Max instead of WMax).

By the previous notations, the problem (D$_L$) can be understood as the following $\mathcal{I}$-valued problem

$$d_L := \sup_{T \in \mathcal{L}_+} \mathrm{Inf} \bigcup_{x \in X} (F(x) + T(G(x))) .$$

Now we show the relationships between problem (D$_L$) and the dual problems studied in Sect. 8.3.2, especially the construction based on the Lagrangian map given in (8.22). The next result is easy to check taking account some properties established in Sect. 2.6.3.

Lemma 15.1.9. *For any $T \in \mathcal{L}_+$, the following conditions hold*

$$\inf_{u \in G(x)+K} \{T(u) + \mathrm{bd}\, C\} = \inf_{u \in G(x)} \{T(u) + \mathrm{bd}\, C\} = \mathrm{Inf}\, T(G(x)).$$

Therefore,

$$\inf_{x \in X} \{F_{\mathcal{I}}(x) \oplus \inf_{u \in G(x)+K} \{T(u) + \mathrm{bd}\, C\}\} = \inf_{x \in X} \{F(x) + T(G(x))\}$$
$$= \mathrm{Inf} \bigcup_{x \in X} (F(x) + T(G(x))) .$$

From Lemma 15.1.9 we obtain the following relationship between (SVD_L^T) and (D$_L$)

$$\bar{d}_L \subset d_L = \tilde{D}. \tag{15.1}$$

Consequently, the weak duality result presented in [379, Proposition 3.2] can be deduced from Theorem 8.3.19 taking into account Lemmata 8.3.7 and 15.1.9.

In order to establish relationships between the strong result given in [379] and the obtained results in Sect. 8.3.2 we need to consider the following stability notions (compare Sect. 8.4).

As in Sect. 8.4.2 let φ be a set-valued map from $X \times Z$ to Y defined by

$$\varphi(x,u) := \begin{cases} F(x) & \text{if } G(x) \cap (-K - u) \neq \emptyset \\ \emptyset & \text{else.} \end{cases}$$

Again, we use the **perturbation map** $W\colon Z \rightrightarrows Y^\bullet$ given by $W(u) := \operatorname{Inf} \bigcup_{x \in X} \varphi(x,u)$ with $W(0) = \operatorname{Inf} \bigcup_{x \in S} F(x) = \operatorname{Inf} F(S)$.

In [379, Definition 2.5] the following definition of positive subgradients is introduced in order to formulate the stability of the problem (SVP_S), compare Definition 8.4.5, where a corresponding formulation based on infimal sets is given. Furthermore, the positive subgradients in Definition 8.4.5 are vectors, where as the positive subgradients in the following definition are operators.

Definition 15.1.10. $T \in \mathscr{L}_+$ is a **positive subgradient** of W at $(\bar{u}, \bar{y}) \in \operatorname{graph} W$ (denoted by $T \in \partial^+ W(\bar{u}, \bar{y})$), if

$$T(\bar{u}) - \bar{y} \in \operatorname{WMax}\Big(\bigcup_{u \in Z} (T(u) - W(u)), C\Big).$$

The following definition for stable problems (SVP) is given in [379, Definition 5.2] (compare Definition 8.4.8).

Definition 15.1.11. The problem (SVP_S) is stable if $W(0) \neq \{+\infty\}$, $W(0) \neq \{-\infty\}$, $\partial^+ W(0, y) \neq \emptyset$ for all $y \in W(0)$.

Concerning the existence of positive subgradients in the sense of Definition 15.1.10 we get the following assertion (compare Lemma 8.4.7 for positive subgradients in the sense of Definition 8.4.5).

Lemma 15.1.12. *Let $y \in W(0) = \operatorname{Inf} F(S)$ and $T \in \mathscr{L}_+$. Then $T \in \partial^+ W(0, y)$ if and only if*

$$y \in \operatorname{Inf} \bigcup_{x \in X} (F(x) + T(G(x))).$$

Proof. Since $W(u) = \operatorname{Inf} \bigcup_{x \in S} \varphi(x,u)$ we have that $T \in \partial^+ W(0, y)$ if and only if

$$-y \in \operatorname{WMax}\Big(\bigcup_{u \in Z} (T(u) - W(u)), C\Big) = \operatorname{WMax}\Big(\bigcup_{u \in Z} \Big(T(u) - \operatorname{Inf} \bigcup_{x \in X} \varphi(x,u)\Big), C\Big)$$

or equivalently

$$y \in \operatorname{WMin}\Big(\bigcup_{u \in Z} \Big(\operatorname{Inf} \Big(\bigcup_{x \in X} \varphi(x,u)\Big) - T(u)\Big), C\Big).$$

By definition of $\varphi(x,u)$ we have

$$\begin{aligned}
&\mathrm{WMin}\Big(\bigcup_{u\in Z}\Big(\mathrm{Inf}\Big(\bigcup_{x\in X}\varphi(x,u)\Big) - T(u)\Big), C\Big)\\
&= \mathrm{WMin}\Big(\bigcup_{-u\in G(x)+K} \mathrm{Inf}\Big(\bigcup_{x\in X} F(x) - T(u)\Big), C\Big)\\
&= \mathrm{WMin}\Big(\bigcup_{-u=v\in G(x)+K} \mathrm{Inf}\Big(\bigcup_{x\in X} F(x) + T(v)\Big), C\Big)\\
&= \mathrm{WMin}\Big(\mathrm{Inf}\bigcup_{v\in G(x)+K}\Big(\bigcup_{x\in X} F(x) + T(v)\Big), C\Big)\\
&= \mathrm{WMin}\Big(\mathrm{Inf}\bigcup_{x\in X}\Big(F(x) + \bigcup_{v\in G(x)} T(v)\Big), C\Big)\\
&= \mathrm{WMin}\Big(\mathrm{Inf}\bigcup_{x\in X}(F(x) + T(G(x))), C\Big)\Big)\\
&= \mathrm{Inf}\bigcup_{x\in X}(F(x) + T(G(x)))\,.
\end{aligned}$$

□

Under certain convexity assumptions it is possible to show that the problem (SVP_S) is stable (compare Theorem 8.4.11).

Theorem 15.1.13. *If* $F \times G$ *is* $(C \times K)$*-convex,* $G(\operatorname{dom} F) \cap -\operatorname{int} K \neq \emptyset$ *and* $W(0) \neq \{-\infty\}$*, then* (SVP_S) *is stable.*

Proof. Because of $G(\operatorname{dom} F) \cap -\operatorname{int} K \neq \emptyset$ we have $W(0) \neq \{+\infty\}$. By Lemma 15.1.12 we have to show that for each $\bar{y} \in W(0) = \mathrm{Inf}\, F(S)$ there exits $T \in \mathscr{L}_+$ such that

$$\bar{y} \in \mathrm{Inf}\bigcup_{x\in X}(F(x) + T(G(x))).$$

If $\bar{y} = -\infty$ then $0 \in \partial^+ W(0)$.

Suppose that $\bar{y} \in Y$. Then $\bar{y} \in \mathrm{WMin}(\mathrm{Cl}_+(F(S)), C) \subset \mathrm{Cl}_+(F(S))$. By hypothesis the map $Q\colon X \rightrightarrows Y \times Z$ defined by $Q(x) := (F(x), G(x))$ is $(C \times K)$-convex. Thus, $Q(X) + (C \times K)$ is a convex set. Moreover, it is easy to check that (compare the proof of Theorem 8.4.11)

$$(Q(X) + (C \times K)) \cap \operatorname{int}(B \times (-K)) = \emptyset \tag{15.2}$$

where $B = \{\bar{y}\} - C$. Indeed, if there exist $x' \in X$ and (y,u) with

$$(y,u) \in \big((F(x'),G(x')) + (C \times K)\big) \cap \operatorname{int}(B \times (-K))$$

then $y \in (F(x')+C)\cap(\bar{y}-\operatorname{int} C)$ and $u \in (G(x')+K)\cap -\operatorname{int} K$. Thus, $y' = \bar{y}-c'$ where $y' \in F(x')$ and $c' \in \operatorname{int} C$ and $(G(x')+K)\cap(-K) \neq \emptyset$ (that is, $x' \in S$), a contradiction to $\bar{y} \in \operatorname{Inf} F(S)$.

By (15.2), applying the Hahn-Banach theorem there exists a pair $(\lambda,\mu) \in Y^* \times Z^* \setminus \{(0,0)\}$ such that

$$\langle \lambda, y\rangle + \langle \mu, u\rangle \le \langle \lambda, b\rangle + \langle \mu, -d\rangle \text{ for all } (y,u) \in Q(x)+(C\times K),\ x \in X,\ b \in B \text{ and } d \in K. \tag{15.3}$$

From this we deduce that $(\lambda,\mu) \in C^- \times K^- \setminus \{(0,0)\}$. In particular, by (15.3) we have

$$\langle \lambda, y\rangle + \langle \mu, u\rangle \le \langle \lambda, \bar{y}\rangle \text{ for all } (y,u) \in Q(x)+(C\times K),\ x \in X. \tag{15.4}$$

We show $\lambda \neq 0$. On the contrary, $\mu \neq 0$ and by (15.4), we have

$$\langle \mu, u\rangle \le 0 \text{ for all } u \in G(x),\ x \in X.$$

On the other hand, by hypothesis there exists $u' \in G(x)\cap(-\operatorname{int} K)$ then $\langle \mu, u'\rangle > 0$ since $\mu \in K^- \setminus \{0\}$. Hence, $\lambda \neq 0$.

Consider $T \in \mathscr{L}$ defined by

$$T(u) := -\langle \mu, u\rangle \cdot c_0 \quad \text{for } u \in Z$$

where $c_0 \in \operatorname{int} C$ and $\langle \lambda, c_0\rangle = -1$ (note that c_0 there exists since $\lambda \in C^- \setminus \{0\}$). It is clear that $T \in \mathscr{L}_+$. Moreover, we have

$$u \in G(x) \cap -K \Rightarrow T(u) = -kc_0 \text{ being } k \ge 0. \tag{15.5}$$

Case I. $\bar{y} \in F(S)$. Then there exists $\bar{x} \in S$ such that $\bar{y} \in F(\bar{x})$. By (15.4) we have $\langle \mu, u\rangle \le 0$ for all $u \in G(\bar{x})$. In addition, $\langle \mu, \bar{u}\rangle = 0$ for $\bar{u} \in G(\bar{x}) \cap -K$ since $\mu \in K^-$.

Therefore, if $\bar{u} \in G(\bar{x}) \cap -K$, by (15.4) and taking into account that $T(\bar{u}) = 0$ we have

$$\langle \lambda, y - T(u)\rangle \le \langle \lambda, \bar{y} - T(\bar{u})\rangle \text{ for all } (y,u) \in Q(x)+(C\times K),\ x \in X.$$

Since $T(\bar{u}) = 0$, applying [566, Proposition 2.7] we deduce

$$\bar{y} \in \operatorname{Inf} \bigcup_{x\in X} (F(x) + T(G(x)))$$

and we conclude the proof.

Case II. $\bar{y} \notin F(S)$. Since $\bar{y} \in \operatorname{Inf} F(S)$ we have $\bar{y} \notin F(S) + \operatorname{int} C$ and $\bar{y} + \operatorname{int} C \subset F(S) + \operatorname{int} C$. Thus, for each $\alpha > 0$ there exist $x' \in S$, $y' \in F(x)$ and $c' \in \operatorname{int} C$ such that $\bar{y} - c' = y' - \alpha c_0$.
By (15.3) we obtain

$$\langle \lambda, y \rangle + \langle \mu, u \rangle \leq \langle \lambda, \bar{y} - c' \rangle \text{ for all } (y, u) \in Q(x) + (C \times K),\ x \in X$$

or equivalently for any $\alpha > 0$

$$\langle \lambda, y \rangle + \langle \mu, u \rangle \leq \langle \lambda, y' - \alpha c_0 \rangle = \langle \lambda, y' \rangle + \langle \lambda, -\alpha c_0 \rangle$$
$$\text{for all } (y, u) \in Q(x) + (C \times K),\ x \in X. \tag{15.6}$$

Let $u' \in G(x') \cap -K$ (note that $x' \in S$). Taking into account (15.5) there exists $\alpha_0 > 0$ such that $T(u') = -\alpha_0 c_0$. From this and (15.6) we have

$$\langle \lambda, y \rangle + \langle \lambda, T(u) \rangle \leq \langle \lambda, y' \rangle + \langle \lambda, T(u') \rangle \text{ for all } (y, u) \in Q(x) + (C \times K),\ x \in X.$$

Since $\langle \lambda, T(u') \rangle = \langle \mu, u' \rangle$, by (15.4), we obtain $\langle \lambda, y' \rangle + \langle \lambda, T(u') \rangle \leq \langle \lambda, \bar{y} \rangle$. Hence,

$$\langle \lambda, y \rangle + \langle \lambda, T(u) \rangle \leq \langle \lambda, \bar{y} \rangle \text{ for all } (y, u) \in Q(x) + (C \times K),\ x \in X.$$

To conclude that $\bar{y} \in \operatorname{Inf} \bigcup_{x \in X} (F(x) + T(G(x))$ applying [566, Proposition 2.7] it is sufficient to show that

$$\bar{y} \in \operatorname{Cl}_+ \bigcup_{x \in X} (F(x) + T(G(x)))$$

which follows easily from $\operatorname{Cl}_+ F(S) \subset \operatorname{Cl}_+ \bigcup_{x \in X} (F(x) + T(G(x))$ (taking into account that $F(x) \subset F(x) + T(u) + \operatorname{int} C$ for $x \in S$ and $u \in G(x) \cap -K$) and $\bar{y} \in \operatorname{Cl}_+ F(S)$. □

Note that the assumptions given in [379, Proposition 5.3] are more stronger that those presented in the Theorem 15.1.13.

Consequently, we obtain the following result which improves [379, Theorem 5.2] and [535, Theorem 3.1]. Furthermore, we get relationships to the dual problems constructed in Sect. 8.3.2, especially to the dual problem $(SD_L^T - \preceq_{\mathscr{I}})$ with $\tilde{D} := \sup_{T \in \mathscr{L}_+} \phi(T)$ and the dual problem $(SD_L'^c - \preceq_{\mathscr{I}})$ with $\bar{D}_c := \sup_{u^* \in K^+} \phi_c(u^*)$ (see problems (SVD_L^T) and (SVD_L^C) in this section).

Theorem 15.1.14. *Let $c \in \operatorname{int} C$. If F is C-convex, G is K-convex, $G(\operatorname{dom} F) \cap -\operatorname{int} K \neq \emptyset$, then*

$$\bar{P} = \tilde{D} = \bar{D}_c = \bar{d}_L = d_L.$$

Proof. By Theorem 15.1.13 applying [379, Theorem 5.1] we have $\bar{P} = \bar{d}_L = d_L$. On the other hand, by Lemma 8.3.9 and Theorems 8.3.15 and 8.3.21 we have $\bar{P} = \bar{D}_c = \tilde{D}$. □

As conclusion, it is possible to simplify or to reduce the feasible set of several dual problems studied in the literature by considering problem (P) where the objective function is an $\mathscr{I}$-valued function.

15.1.3 Duality Assertions for Linear Vector Optimization Based on Lattice Approach

In this section we investigate the special case of linear vector optimization problems (see Löhne, Tammer [397], Heyde, Löhne, Tammer [258, 259]). We show that we can maintain a large part of the structures of scalar linear programming. In particular, we have no duality gap in the case that the right-hand side of the inequality constraints is zero, compare the discussion in [234]. On the other hand, our dual objective maps take their values in the space of self-infimal sets (compare Definition 2.6.37), in particular, we cannot expect a single-valued dual objective map. Nevertheless the values of the dual objective map have a simple structure, namely they are boundary points of translated cones, in some cases even hyperplanes. As usual, we use the abbreviation $f(S) = \bigcup_{x \in S} f(x)$. Suppose that $C \subset \mathbb{R}^q$ is a proper closed convex cone with nonempty interior.

Consider the following linear vector optimization problems using the notation given in Sect. 2.6.3, especially the infimal set $\operatorname{Inf} A$ of a set $A \subseteq \mathbb{R}^q$ (see Definition 2.6.31):

$$\operatorname{Inf} M(S), \qquad S := \{x \in \mathbb{R}^n \mid Ax \geq b\}, \tag{LP^1}$$

$$\operatorname{Inf} M(S), \qquad S := \{x \in \mathbb{R}^n \mid x \geq 0,\ Ax \geq b\}, \tag{LP^2}$$

$$\operatorname{Inf} M(S), \qquad S := \{x \in \mathbb{R}^n \mid x \geq 0,\ Ax = b\}, \tag{LP^3}$$

where $M \in \mathbb{R}^{n\times q}, A \in \mathbb{R}^{n\times m}, b \in \mathbb{R}^m$. We calculate the corresponding dual problems using the approach via conjugates presented in Sect. 15.1.1 and apply Theorem 15.1.4 following Löhne, Tammer [397]. Exemplarily we show the calculations starting with (LP^2). Starting with (LP^1) or (LP^3) we obtain the result similarly and even easier. We set

$$F(x) := \begin{cases} \{Mx\} + \operatorname{bd} C & \text{if} \quad x \geq 0 \\ \{+\infty\} & \text{else} \end{cases} \qquad \text{and} \qquad G(u) := \begin{cases} \operatorname{bd} C & \text{if} \quad u \geq b \\ \{+\infty\} & \text{else.} \end{cases}$$

For the choice $c \in \operatorname{int} C$, an easy calculation shows that

$$-F_c^*(x^*) = \operatorname{Inf}(M - cx^{*T})(\mathbb{R}^n_+),$$

$$-G_c^*(-u^*) = \begin{cases} \langle u^*, b\rangle \{c\} + \operatorname{bd} C & \text{if} \quad u^* \geq 0 \\ \{-\infty\} & \text{else.} \end{cases}$$

In order to obtain dual side conditions of a simple structure, it is useful to characterize the condition $\operatorname{Inf}(M - c\,x^{*T})(\mathbb{R}^n_+) = \{-\infty\}$. For this purpose consider the next two assertions.

Lemma 15.1.15. *Let $C_1, C_2 \subseteq \mathbb{R}^q$ be nonempty closed convex cones with* $\operatorname{int} C_2 \neq \emptyset$. *Then*

$$C_1 \cap -\operatorname{int} C_2 = \emptyset \iff C_1^- \cap (C_2^- \setminus \{0\}) \neq \emptyset.$$

Proof. Note that $y \in -\operatorname{int} C_2$, $z \in (C_2^- \setminus \{0\})$ implies that $\langle y, z\rangle > 0$. On the other hand, $y \in C_1$, $z \in C_1^-$ implies that $\langle y, z\rangle \leq 0$. This proves the implication "$\Leftarrow$".

If $C_1 \cap -\operatorname{int} C_2 = \emptyset$, by a separation theorem (e.g. [489, Theorem 11.3]), we obtain that there exists $y^* \in \mathbb{R}^q \setminus \{0\}$ such that $\inf_{y \in -C_2} \langle y^*, y\rangle = \sup_{y \in C_1} \langle y^*, y\rangle = 0$. Hence we have $y^* \in C_1^- \cap (C_2^- \setminus \{0\})$. □

Proposition 15.1.16. *Let $K \subseteq \mathbb{R}^n$ be a nonempty closed convex cone and let $H \in \mathbb{R}^{q \times n}$. Then, the following statements are equivalent:*

(i) $\operatorname{Inf} H(K) \neq \{-\infty\}$;
(ii) $\exists c^* \in C^- \setminus \{0\} : H^T c^* \in K^-$.

Proof. Set $L := H(K)$. Since $L \subseteq \mathbb{R}^q$ is a nonempty cone, we have

$$\begin{aligned} \operatorname{Inf} L \neq \{-\infty\} &\iff \exists y \in \mathbb{R}^q, \forall \alpha > 0 : \alpha y \notin L + \operatorname{int} C \ \wedge\ \alpha y + \operatorname{int} C \subseteq L + \operatorname{int} C \\ &\iff 0 \in \operatorname{Inf} L. \end{aligned}$$

By Lemma 15.1.15 we have

$$0 \in \operatorname{Inf} L \iff 0 \notin L + \operatorname{int} C \iff L \cap -\operatorname{int} C = \emptyset \iff L^- \cap (C^- \setminus \{0\}) \neq \emptyset.$$

Using the bipolar theorem [489, Theorem 14.1] we can conclude

$$\begin{aligned} L^- \cap (C^- \setminus \{0\}) \neq \emptyset &\iff \exists c^* \in C^- \setminus \{0\} : \forall l \in L : \langle c^*, l\rangle \leq 0 \\ &\iff \exists c^* \in C^- \setminus \{0\} : \forall x \in K : \langle c^*, Hx\rangle = \langle H^T c^*, x\rangle \leq 0 \\ &\iff \exists c^* \in C^- \setminus \{0\} : H^T c^* \in K^-. \end{aligned}$$

Together we obtain the desired assertion. □

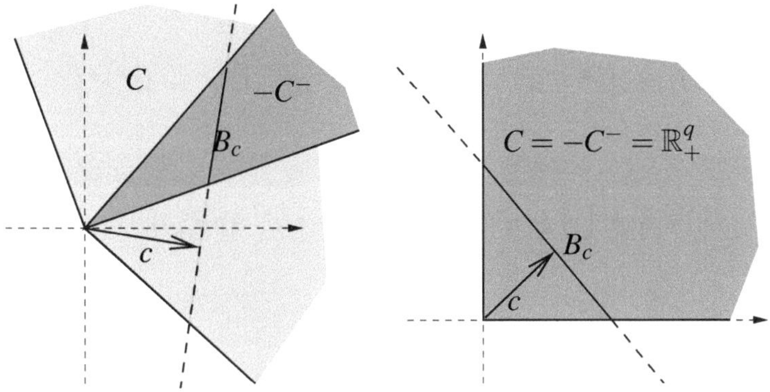

Fig. 15.1 Two examples for the set B_c

Applying the previous result for $c \in \operatorname{int} C$ we can write

$$-F_c^*(x^*) = \begin{cases} \operatorname{Inf}(M - cx^{*T})\left(\mathbb{R}^n_+\right) \subset \mathbb{R}^q & \text{if} \quad \exists c^* \in C^- \setminus \{0\} : x^* \leq M^T \frac{c^*}{c^T c^*} \\ \{-\infty\} & \text{else.} \end{cases}$$

In the following, the set $B_c := \left\{c^* \in C^+ \mid \langle c, c^* \rangle = 1\right\}$ is used to express the dual side conditions. In Fig. 15.1 we illustrate this set by two examples (with $\|c\| = 1$):

Proposition 15.1.17. *For $c \in \operatorname{int} C$, the set B_c is a compact convex subset of $\mathbb{R}^q$.*

Proof. Obviously, B_c is closed and convex. Thus, it remains to show that B_c is bounded. Assuming the contrary, we obtain a sequence $c_n^* \in B_c$ with $\|c_n^*\| \to \infty$. Without loss of generality we can assume that $c_n^* / \|c_n^*\| \to \bar{c}^* \in C^+ \setminus \{0\}$. Since $\langle c_n^* / \|c_n^*\|, c \rangle = 1/\|c_n^*\| \to 0$, it follows that $\langle \bar{c}^*, c \rangle = 0$. But $\bar{c}^* \in C^+ \setminus \{0\}$ and $c \in \operatorname{int} C$ implies that $\langle \bar{c}^*, c \rangle > 0$, a contradiction. □

According to problem $(SD_F^c - \preceq_{\mathscr{I}})$ in Sect. 15.1.1 we obtain the dual problem to (LP^2) with $c \in \operatorname{int} C$ as

$$\bar{D}_c = \operatorname{Sup} \bigcup_{u^* \in T_c} \left(c\, u^{*T} b \;+\; \operatorname{Inf}(M - c\, u^{*T} A)(\mathbb{R}^n_+)\right), \qquad (LD_c^2)$$

with

$$T_c := \left\{u^* \in \mathbb{R}^m \mid u^* \geq 0,\ \exists c^* \in B_c : A^T u^* \leq M^T c^*\right\}.$$

By a similar calculation we obtain the dual problems to (LP^1) and (LP^3) as

$$\bar{D}_c = \operatorname{Sup} \bigcup_{u^* \in T_c} \left(c\, u^{*T} b \;+\; \operatorname{Inf}(M - c\, u^{*T} A)(\mathbb{R}^n)\right), \qquad (LD_c^1)$$

with

$$T_c := \{u^* \in \mathbb{R}^m \mid u^* \geq 0,\ \exists c^* \in B_c : A^T u^* = M^T c^*\}$$

and

$$\bar{D}_c = \operatorname{Sup} \bigcup_{u^* \in T_c} \left(c\, u^{*T} b \;+\; \operatorname{Inf}(M - c\, u^{*T} A)(\mathbb{R}^n_+)\right), \tag{LD^3_c}$$

with

$$T_c := \{u^* \in \mathbb{R}^m \mid \exists c^* \in B_c : A^T u^* \leq M^T c^*\}.$$

In the next proposition we collect some properties of the dual feasible set T_c.

Proposition 15.1.18. *The set T_c in* (LD^1_c)-(LD^3_c) *(where $c \in \operatorname{int} C$) is always a closed convex subset of $\mathbb{R}^m$. Moreover, if C is polyhedral, then T_c is polyhedral, too.*

Proof. Exemplarily we give the proof for T_c in (LD^3_c). The set T_c can be expressed as $T_c = \{u^* \in \mathbb{R}^m \mid A^T u^* \in M(B_c) - \mathbb{R}^n_+\}$. Of course, $M(B_c) - \mathbb{R}^n_+$ is a convex set. By [489, Theorem 3.4], it follows that T_c is convex. Since B_c is compact (Proposition 15.1.17), the set $M(B_c) - \mathbb{R}^n_+$ is closed. As the map $A^T : \mathbb{R}^m \to \mathbb{R}^n$ is continuous, T_c is closed. If C is polyhedral, we conclude that C^-, B_c, $M(B_c)$ and $M(B_c) - \mathbb{R}^n_+$ are polyhedral, too. By [489, Theorem 19.3], it follows that T_c is polyhedral. □

From the Fenchel duality result in Theorem 15.1.4 and some additional considerations we obtain the following duality assertion.

Theorem 15.1.19 (Weak and Strong Duality Between (LP^{i}) **and** ($\mathrm{LD}^{\mathrm{i}}_c$) ($i = 1, 2, 3$))**.** *For $c \in \operatorname{int} C$ it holds weak and strong duality between* (LP^{i}) *and* ($\mathrm{LD}^{\mathrm{i}}_c$) *($i = 1, 2, 3$). More precisely we have*

(i) *$\bar{D}_c = \bar{P} \subseteq \mathbb{R}^q$ if $S \neq \emptyset$ and $T_c \neq \emptyset$, where "Sup" can be replaced by "Max" in this case,*
(ii) *$\bar{D}_c = \bar{P} = \{-\infty\}$ if $S \neq \emptyset$ and $T_c = \emptyset$,*
(iii) *$\bar{D}_c = \bar{P} = \{+\infty\}$ if $S = \emptyset$ and $T_c \neq \emptyset$.*

Proof. In the case where $S \neq \emptyset$ we obtain the strong duality by Theorem 15.1.4 taking into account Remark 8.3.6. In case (i), we have $\bar{P} \subseteq \mathbb{R}^q \cup \{-\infty\}$ and $\bar{D}_c \subseteq \mathbb{R}^q \cup \{+\infty\}$, hence $\bar{D}_c = \bar{P} \subseteq \mathbb{R}^q$. In case (ii) we have $\bar{P} = \bar{D}_c = \operatorname{Sup} \emptyset = \{-\infty\}$. In case (iii) we need some additional considerations. Let $\bar{u}^* \in T_c$ and $S = \emptyset$. Exemplarily we show the assertion for (LP^2) and (LD^2_c). By the Farkas Lemma there exists some $\hat{u}^* \in \mathbb{R}^m$ such that $\hat{u}^* \geq 0$, $A^T \hat{u}^* \leq 0$ and $b^T \hat{u}^* > 0$. Hence, for all $\alpha > 0$ we have $\bar{u}^* + \alpha \hat{u}^* \in T_c$. For $\alpha \to +\infty$ we have $(\bar{u}^* + \alpha \hat{u}^*)^T b \to +\infty$. Hence $\bar{D}_c = \{+\infty\} = \bar{P}$. □

Next we relate our result to classical formulations in vector optimization (see Remark 15.1.6), i.e., we formulate the duality assertions for weakly minimal/weakly maximal points introduced in Definition 2.4.2. Denoting the dual objective functions in (LD$_c^1$)-(LD$_c^3$) by $D_c : \mathbb{R}^m \to \mathscr{I}$, ($i = 1, 2, 3$), for all three pairs of dual problems we easily obtain the following conclusions from Theorem 15.1.19.

Corollary 15.1.20 (Direct Strong Duality). *For every $c \in \operatorname{int} C$ it holds*

$$\operatorname{WMin}(M(S), C) \subseteq \operatorname{WMax}(D_c(T_c), C).$$

Corollary 15.1.21 (Inverse Strong Duality). *If $M(S)$ is upper closed, then for all $c \in \operatorname{int} C$ it holds*

$$\operatorname{WMax}(D_c(T_c), C) \subseteq \operatorname{WMin}(M(S), C).$$

Finally, we give an example.

Example 15.1.22. (see Fig. 15.2) Let $q = m = n = 2$, $C = \mathbb{R}^2_+$ and consider the problem (LP^2) with the data

$$M = \begin{pmatrix} 1 & 0 \\ 0 & 1 \end{pmatrix}, \quad A = \begin{pmatrix} 1 & 2 \\ 2 & 1 \end{pmatrix}, \quad b = \begin{pmatrix} 2 \\ 2 \end{pmatrix}.$$

Let us calculate the dual problem for $c = (1, 1)^T \in \operatorname{int} \mathbb{R}^2_+$. For this choice, we set B_c is given by $B_c = \{c_1^*, c_2^* \geq 0 \mid c_1^* + c_2^* = 1\}$. The dual side conditions are as follows

$$u_1^*, u_2^* \geq 0, \qquad \exists c_1^* \geq 0 : \ u_1^* + 2u_2^* \leq c_1^*, \quad 2u_1^* + u_2^* \leq 1 - c_1^*.$$

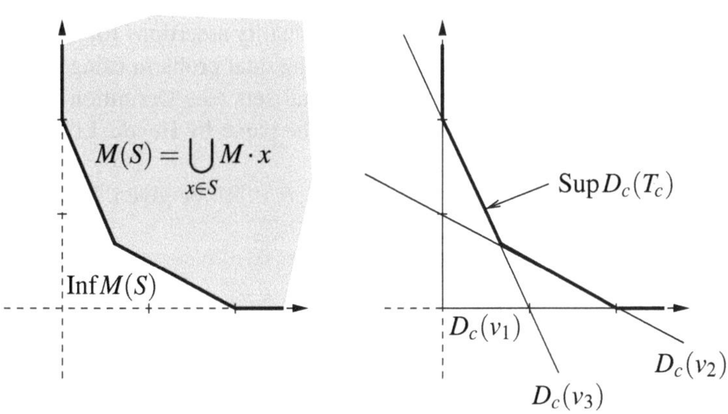

Fig. 15.2 The primal and dual values in Example 15.1.22

This can be equivalently expressed by $T_c = \{u_1^*, u_2^* \geq 0 \mid u_1^* + u_2^* \leq 1/3\}$. The vertices of T_c are the points $v_1 = (0,0)^T$, $v_2 = (1/3,0)^T$ and $v_3 = (0,1/3)^T$. The matrices $H_i := (M - c\, {v_i}^T A)$ can be easily computed as

$$H_1 = \begin{pmatrix} 1 & 0 \\ 0 & 1 \end{pmatrix}, \quad H_2 = \begin{pmatrix} \frac{2}{3} & -\frac{2}{3} \\ -\frac{1}{3} & \frac{1}{3} \end{pmatrix}, \quad H_3 = \begin{pmatrix} \frac{1}{3} & -\frac{1}{3} \\ -\frac{2}{3} & \frac{2}{3} \end{pmatrix}.$$

Hence we obtain

$$\begin{aligned} \operatorname{Inf} H_1(\mathbb{R}^2_+) &= \operatorname{bd} \mathbb{R}^2_+, \\ \operatorname{Inf} H_2(\mathbb{R}^2_+) &= \{y \in \mathbb{R}^2 \mid y_1 + 2y_2 = 0\}, \\ \operatorname{Inf} H_3(\mathbb{R}^2_+) &= \{y \in \mathbb{R}^2 \mid 2y_1 + y_2 = 0\}. \end{aligned}$$

Consequently, the values of the dual objective function at v_1, v_2, v_3 are $D_c(v_1) = \operatorname{bd} \mathbb{R}^2_+$, $D_c(v_2) = \{y \in \mathbb{R}^2 \mid y_1 + 2y_2 = 2\}$ and $D_c(v_3) = \{y \in \mathbb{R}^2 \mid 2y_1 + y_2 = 2\}$. We see that the three dual feasible points $v_1, v_2, v_3 \in T_c$ are already sufficient for strong duality.

15.1.4 Further Set-Valued Approaches to Duality in Linear Vector Optimization

In this section we will derive further duality statements for linear vector optimization problems where the solution concept is based on the concept of weakly minimal points (see Definition 2.4.2) using the lattice approach. We now discuss a set-valued dual problem to a linear vector optimization problem where also the scalarization parameter $c \in \mathbb{R}^q$ is considered as dual variable in difference to the construction of the dual problem in Sect. 15.1.3. We will show duality assertions for a primal linear vector optimization problem and a corresponding dual problem using a reformulation of the primal problem based on self-infimal sets (see Definition 2.6.37). The results presented in this section are derived in the paper by Heyde, Löhne, Tammer [258].

Let $m, n, q \in \mathbb{N}$ and $A \in \mathbb{R}^{m \times n}$, $M \in \mathbb{R}^{q \times n}$, $b \in \mathbb{R}^m$ be given. We consider the following vector optimization problem

$$\operatorname{WMin}((M(S) + \mathbb{R}^q_+), \mathbb{R}^q_+), \qquad S := \{x \in \mathbb{R}^n \mid Ax \geq b\}, \tag{LVP}$$

where $M(S) = \cup_{x \in S}\{Mx\}$. This means that we are looking for weakly minimal points of the set $(M(S) + \mathbb{R}^q_+)$ with respect to $\mathbb{R}^q_+$ in the sense of Definition 2.4.2. Furthermore, a point $x^0 \in S$ is called a **weakly efficient solution of (LVP)** iff

$$Mx^0 \in \operatorname{WMin}(M(S), \mathbb{R}^q_+) + \mathbb{R}^q_+ \quad \text{or equivalently} \quad Mx^0 \in \operatorname{WMin}(M(S), \mathbb{R}^q_+).$$

Note the a point x^0 is a weakly efficient solution of (LVP) if and only if it is a weakly efficient solution of the more common problem

$$\mathrm{WMin}(M(S), \mathbb{R}^q_+), \qquad S := \{x \in \mathbb{R}^n \mid Ax \geq b\},$$

even though the sets $\mathrm{WMin}(M(S), \mathbb{R}^q_+)$ and $\mathrm{WMin}((M(S) + \mathbb{R}^q_+), \mathbb{R}^q_+)$ can be different. The set $\mathrm{WMin}((M(S) + \mathbb{R}^q_+), \mathbb{R}^q_+)$ is closely related to the infimal set of $M(S)$ (compare Sect. 2.6.3, Definition 2.6.31).

Put $k := (1, 1, \ldots, 1)^T \in \mathbb{R}^q$. Consider the following set-valued dual objective map

$$\mathscr{H} : \mathbb{R}^m \times \mathbb{R}^q \rightrightarrows \mathbb{R}^q, \quad \mathscr{H}(u, c) := \left\{y \in \mathbb{R}^q \mid c^T y = b^T u\right\}.$$

The dual feasible set is given by

$$\mathscr{U} := \left\{(u, c) \in \mathbb{R}^m \times \mathbb{R}^q \mid (u, c) \geq 0,\ k^T c = 1,\ A^T u = M^T c\right\}.$$

We use the following notation

$$\mathscr{H}(\mathscr{U}) := \bigcup_{(u,c)\in\mathscr{U}} \mathscr{H}(u, c).$$

Since this dual objective map is set-valued we use a corresponding solution concept, namely, the concept introduced in Definition 2.6.5. As the dual problem to (LVP) we consider the problem

$$\mathrm{WMax}(\mathscr{H}(\mathscr{U}), \mathbb{R}^q_+). \tag{LVD}$$

This means the dual problem consists in determining weakly maximal points of the union of the hyperplanes $\mathscr{H}(u, c)$ defined by the points $(u, c) \in \mathscr{U}$ in the sense Definition 2.6.5.

Remark 15.1.23. The new idea in this approach compared with the results given in Sect. 15.1.3 (compare also [397] and [259]) consists in having a pair (u, c) of dual variables and having hyperplanes as values of the dual objective without making any assumptions on the rank of M.

A point $(u^0, c^0) \in \mathscr{U}$ is called a **weakly efficient solution of (LVD)** iff

$$\mathscr{H}(u^0, c^0) \cap \mathrm{WMax}(\mathscr{H}(\mathscr{U}), C) \neq \emptyset,$$

or equivalently,

$$\exists y^0 \in \mathscr{H}(u^0, c^0) : \forall (u, c) \in \mathscr{U} : \left(\{y^0\} + \mathrm{int}\, \mathbb{R}^q_+\right) \cap \mathscr{H}(u, c) = \emptyset. \tag{15.7}$$

This means that $((u^0, c^0), y^0) \in \mathscr{U} \times H(u^0, c^0)$ is a weak maximizer of (LVD) (compare Sect. 2.6.1, Definition 2.6.5). Subsequently, weakly efficient solutions of (LVP) or (LVD) are referred for simplicity as *solutions* of (LVP) or (LVD). The notion of a solution of problem (LVP) as a feasible point whose image is weakly minimal is common in vector optimization. We adapt this concept for the set-valued dual problem by defining solutions of (LVD) as feasible points whose image, which is a hyperplane, contains weakly maximal points (see Definition 2.6.5). Thus the solution concept for the dual problem (LVD) coincides with the concept based on vector approach introduced in Sect. 2.6.1.

In the following we prove weak and strong duality between the two problems directly. In the proofs the following pairs of dual scalar linear optimization problems depending on parameters $c, y \in \mathbb{R}^q$ play an important role.

$$c^T Mx \to \min \quad \text{s.t.} \quad Ax \geq b, \tag{$P_1(c)$}$$

$$b^T u \to \max \quad \text{s.t.} \quad u \geq 0,\ A^T u = M^T c, \tag{$D_1(c)$}$$

$$z \to \min \quad \text{s.t.} \quad Ax \geq b,\ Mx - kz \leq y, \tag{$P_2(y)$}$$

$$b^T u - y^T c \to \max \quad \text{s.t.} \quad u, c \geq 0,\ A^T u - M^T c = 0,\ k^T c = 1. \tag{$D_2(y)$}$$

The first pair of problems comes from classical linear scalarization and is mainly used for characterizing solutions of (LVD). The second pair of problems is very useful for characterizing weakly minimal and weakly maximal points in the image space $\mathbb{R}^q$. Similar problems also occur, for instance, in [286]. Note that the problem ($P_2(y)$) also provide a very common scalarization method in vector optimization, see e.g. [196, 454] and Chap. 5.

The following notion might also be useful for characterizing solutions of (LVP) and (LVD). A pair of points $(x, z) \in \mathbb{R}^n \times \mathbb{R}$ and $(u, c) \in \mathbb{R}^m \times \mathbb{R}^q$ is called **complementary** for the problems ($P_2(y)$) and ($D_2(y)$) if $u^T(Ax - b) = 0$ and $c^T(Mx - kz - y) = 0$.

Lemma 15.1.24. *If* $(x, z) \in \mathbb{R}^n \times \mathbb{R}$ *and* $(u, c) \in \mathscr{U}$ *are complementary points for* ($P_2(y)$) *and* ($D_2(y)$) *then* $z = b^T u - y^T c$.

Proof. If $(u, c) \in \mathscr{U}$ we have $k^T c = 1$ and $A^T u = M^T c$. Hence $c^T(Mx - kz - y) = 0$ and $u^T(Ax - b) = 0$ imply $z = c^T Mx - c^T y = u^T Ax - c^T y = u^T b - c^T y$. □

Subsequently, we use the following notation

$$\mathscr{M} := M(S) + \mathbb{R}^q_+ = \{y \in \mathbb{R}^q \mid \exists x \in S : Mx \leq y\}, \quad \mathscr{F}(u, c) := \mathscr{H}(u, c) \cap \mathscr{M}.$$

The following lemma can be interpreted as evidence of **weak duality**.

Lemma 15.1.25. *If* $(u,c) \in \mathscr{U}$ *and* $y \in \mathscr{M}$ *then* $c^T y \geq b^T u$.

Proof. Since $y \in \mathscr{M}$ there is some $x \in S$ such that $y \geq Mx$. Hence $(x,0)$ is feasible for $(\mathrm{P}_2(y))$. Duality between $(\mathrm{P}_2(y))$ and $(\mathrm{D}_2(y))$ implies $b^T u - y^T c \leq 0$. □

Remark 15.1.26. The preceding lemma implies **weak duality** in classical sense, i.e.,

$$x \in S, (u,c) \in \mathscr{U}, y \in \mathscr{H}(\mathscr{U}) \Longrightarrow Mx \not< y.$$

Indeed, if $x \in S$, $(u,c) \in \mathscr{U}$, $y \in \mathscr{H}(\mathscr{U})$ and $Mx < y$ would be true, then $c^T Mx < c^T y = b^T u$ but Lemma 15.1.25 implies $c^T Mx \geq b^T u$, a contradiction.

An interpretation of weak duality with the help of set relations is given later.

The next lemma states a sufficient optimality condition for (LVD), which is based on weak duality.

Lemma 15.1.27. *If* $(u^0,c^0) \in \mathscr{U}$ *and* $y^0 \in \mathscr{F}(u^0,c^0)$ *then* $y^0 \in \mathrm{WMax}(\mathscr{H}(\mathscr{U}), \mathbb{R}^q_+)$.

Proof. Let $(u^0,c^0) \in \mathscr{U}$ and $y^0 \in \mathscr{F}(u^0,c^0)$. Therefore we have $y^0 \in \mathscr{H}(\mathscr{U})$. We show that

$$\left(\{y^0\} + \operatorname{int}\mathbb{R}^q_+\right) \cap \mathscr{H}(u,c) = \emptyset \quad \text{for all} \quad (u,c) \in \mathscr{U}.$$

Assume on the contrary that there are $(u,c) \in \mathscr{U}$ and $y \in \mathscr{H}(u,c)$ with $y > y^0$. Since $c \geq 0, c \neq 0$ this implies $c^T y > c^T y^0 \geq b^T u = c^T y$, a contradiction. □

Remark 15.1.28. The assertion of Lemma 15.1.27 says that under the given conditions (u^0,c^0,y^0) is a weak maximizer of (LVD) in the sense of Definition 2.6.5.

The following theorem provides different characterizations of (weakly efficient) solutions of (LVD).

Theorem 15.1.29. *Let* $(u^0,c^0) \in \mathscr{U}$. *Then the following statements are equivalent.*

(i) (u^0,c^0) *is a solution of* (LVD),
(ii) u^0 *solves* $(\mathrm{D}_1(c^0))$,
(iii) *there exists some* $x^0 \in S$ *with* ${c^0}^T Mx^0 = b^T u^0$,
(iv) $\mathscr{F}(u^0,c^0)$ *is nonempty.*

Proof. (i)⇒(ii). Assume u^0 does not solve $(\mathrm{D}_1(c^0))$. Then there is some $u \in \mathbb{R}^m$ such that $(u,c^0) \in \mathscr{U}$ and $b^T u > b^T u^0$. But for each $y \in \mathscr{H}(u^0,c^0)$ we get

$$y + k(b^T u - b^T u^0) \in \left(\{y\} + \operatorname{int}\mathbb{R}^q_+\right) \cap \mathscr{H}(u,c^0)$$

contradicting (15.7), i.e., (u^0,c^0) being a solution of (LVD).

(ii)⇒(iii). If u^0 solves ($\mathrm{D}_1(c^0)$) then by duality between the problems ($\mathrm{P}_1(c^0)$) and ($\mathrm{D}_1(c^0)$) there is some $x^0 \in S$ such that ${c^0}^T M x^0 = b^T u^0$.

(iii)⇒(iv). If (iii) holds then $Mx^0 \in \mathscr{H}(u^0, c^0)$. Since $Mx^0 \in \mathscr{M}$ we have $Mx^0 \in \mathscr{F}(u^0, c^0)$.

(iv)⇒(i). By Lemma 15.1.27. □

Our aim is to show a strong duality theorem in the sense that the set of weakly minimal points for (LVP) and the set of weakly maximal points for (LVD) coincide. In the proof of the strong duality assertion in Theorem 15.1.46 for $\mathscr{I}$-valued problems we will essentially use the following first **strong duality result**.

Theorem 15.1.30. *The following four statements are equivalent.*

(i) $y^0 \in \mathrm{WMin}(\mathscr{M}, \mathbb{R}^q_+)$,
(ii) *there is some* $x^0 \in \mathbb{R}^n$ *such that* $(x^0, 0)$ *solves* ($\mathrm{P}_2(y^0)$),
(iii) *there is some* $(u^0, c^0) \in \mathscr{U}$ *with* $b^T u^0 = {y^0}^T c^0$ *solving* ($\mathrm{D}_2(y^0)$),
(iv) $y^0 \in \mathrm{WMax}(\mathscr{H}(\mathscr{U}), \mathbb{R}^q_+)$.

Proof. (ii)⇒(i). If $(x^0, 0)$ solves ($\mathrm{P}_2(y^0)$), then $x^0 \in S$ and $Mx^0 \leq y^0$, hence $y^0 \in \mathscr{M}$. Assume that there is some $y \in \mathscr{M}$ (i.e., there is some $x \in S$ with $Mx \leq y$) with $y < y^0$. Then there is some $z < 0$ such that $y \leq y^0 + kz$. This implies $Mx - kz \leq y - kz \leq y^0$, i.e., (x, z) is feasible for ($\mathrm{P}_2(y^0)$) and $z < 0$ contradicts the optimality of $(x^0, 0)$.

(i)⇒(ii). If $y^0 \in \mathrm{WMin}(\mathscr{M}, \mathbb{R}^q_+)$ then there exists some $x^0 \in S$ with $Mx^0 \leq y^0$, i.e., $(x^0, 0)$ is feasible for ($\mathrm{P}_2(y^0)$). Assume that there is some $(x, z) \in \mathbb{R}^{n+1}$ with $z < 0$ being feasible for ($\mathrm{P}_2(y^0)$). Let $y := y^0 + zk$ then $y < y^0$ and $Mx \leq y^0 + kz = y$, i.e., $y \in \mathscr{M}$ contradicting the weak minimality of y^0.

(ii)⇔(iii). By duality of ($\mathrm{P}_2(y^0)$) and ($\mathrm{D}_2(y^0)$).

(iii)⇔(iv). We have $y^0 \in \mathrm{WMax}(\mathscr{H}(\mathscr{U}), \mathbb{R}^q_+)$ iff

$$y^0 \in \mathscr{H}(\mathscr{U}) \tag{15.8}$$

and

$$y^0 \notin \mathscr{H}(\mathscr{U}) - \operatorname{int} \mathbb{R}^q_+. \tag{15.9}$$

Condition (15.8) is equivalent to

$$\exists (u^0, c^0) \in \mathscr{U} : {y^0}^T c^0 = b^T u^0, \tag{15.10}$$

and (15.9) is equivalent to

$$\forall (u, c) \in \mathscr{U} : {y^0}^T c \geq b^T u. \tag{15.11}$$

Since (iii) is equivalent to (15.10) together with (15.11), the statement follows. □

Now we are able to prove the following theorem which provides sufficient conditions for solutions of (LVP) and (LVD).

Theorem 15.1.31. *Let* $(u^0, c^0) \in \mathscr{U}$ *and* $x^0 \in S$ *be given. Then* x^0 *is a solution of* (LVP) *and* (u^0, c^0) *is a solution of* (LVD) *if one of the following equivalent conditions is satisfied.*

(i) $b^T u^0 = {c^0}^T M x^0$,
(ii) u^0 *solves* $(\mathrm{D}_1(c^0))$ *and* x^0 *solves* $(\mathrm{P}_1(c^0))$,
(iii) $(x^0, 0)$ *solves* $(\mathrm{P}_2(Mx^0))$ *and* (u^0, c^0) *solves* $(\mathrm{D}_2(Mx^0))$,
(iv) *for all* $y \in \mathbb{R}^q$ *there is some* $z^0 \in \mathbb{R}$ *such that* (x^0, z^0) *and* (u^0, c^0) *are complementary points for* $(\mathrm{P}_2(y))$ *and* $(\mathrm{D}_2(y))$.

Proof. First we show the equivalence of the four conditions.
(i)⇔(ii). By duality between $(\mathrm{P}_1(c^0))$ and $(\mathrm{D}_1(c^0))$.
(i)⇔(iii). By duality between $(\mathrm{P}_2(Mx^0))$ and $(\mathrm{D}_2(Mx^0))$.
(i)⇔(iv). If $(u^0, c^0) \in \mathscr{U}$, $x^0 \in S$ and

$$z^0 = b^T u^0 - y^T c^0 \tag{15.12}$$

then we have

$${u^0}^T (Ax^0 - b) = {c^0}^T (Mx^0 - kz^0 - y) = {c^0}^T Mx^0 - b^T u^0. \tag{15.13}$$

If (i) holds we define z^0 by (15.12) and then (i) and (15.13) imply (iv). If (iv) holds then (15.12) holds by Lemma 15.1.24 and then (iv) and (15.13) imply (i).

Now, sufficiency of these equivalent conditions for x^0 and (u^0, c^0) being solutions of (LVP) and (LVD) follows from Theorem 15.1.29 and Theorem 15.1.30. □

In the following we prove some statements showing the relationship between proper faces (in particular facets) of $\mathscr{M}$ and solutions of (LVD). Let us recall some facts concerning the facial structure of polyhedral sets. Let $\mathscr{A} \subseteq \mathbb{R}^q$ be a convex set. A convex subset $\mathscr{F} \subseteq \mathscr{A}$ is called a **face** of $\mathscr{A}$ iff

$$y^1, y^2 \in \mathscr{A}, \quad \lambda \in (0,1), \quad \lambda y^1 + (1-\lambda) y^2 \in \mathscr{F} \quad \Rightarrow \quad y^1, y^2 \in \mathscr{F}.$$

A face $\mathscr{F}$ of $\mathscr{A}$ is called **proper** iff $\emptyset \neq \mathscr{F} \neq \mathscr{A}$. A set $\mathscr{E} \subseteq \mathscr{A}$ is called an **exposed face** of $\mathscr{A}$ iff there are $c \in \mathbb{R}^q$ and $\gamma \in \mathbb{R}$ such that $\mathscr{A} \subseteq \{y \in \mathbb{R}^q \mid c^T y \geq \gamma\}$ and $\mathscr{E} = \{y \in \mathbb{R}^q \mid c^T y = \gamma\} \cap \mathscr{A}$. The proper $(r-1)$-dimensional faces of an r-dimensional polyhedral set $\mathscr{A}$ are called **facets** of $\mathscr{A}$. A point $y \in \mathscr{A}$ is called a **vertex** of $\mathscr{A}$ iff $\{y\}$ is a face of $\mathscr{A}$.

Theorem 15.1.32 ([599], Theorem 3.2.2). *Let* $\mathscr{A}$ *be a polyhedral set in* $\mathbb{R}^q$*. Then* $\mathscr{A}$ *has a finite number of faces, each of which is exposed and a polyhedral set. Every proper face of* $\mathscr{A}$ *is the intersection of those facets of* $\mathscr{A}$ *that contain it, and* r-bd $\mathscr{A}$

(the relative boundary of $\mathscr{A}$) is the union of all the facets of $\mathscr{A}$. If $\mathscr{A}$ has a nonempty face of dimension s, then $\mathscr{A}$ has faces of all dimensions from s to $\dim \mathscr{A}$.

Remark 15.1.33. If $\mathscr{M} \neq \emptyset$ then $\mathscr{M}$ is a q-dimensional polyhedral set, hence the facets of $\mathscr{M}$ are the $(q-1)$-dimensional faces of $\mathscr{M}$, i.e., the maximal (w.r.t. inclusion) proper faces. A subset $\mathscr{F} \subseteq \mathscr{M}$ is a proper face iff it is a proper exposed face, i.e., iff there is a supporting hyperplane $\mathscr{H}$ to $\mathscr{M}$ such that $\mathscr{F} = \mathscr{H} \cap \mathscr{M}$. We call a hyperplane $\mathscr{H} := \left\{y \in \mathbb{R}^q \mid c^T y = \gamma\right\}$ (i.e., $c \neq 0$) supporting to $\mathscr{M}$ iff $c^T y \geq \gamma$ for all $y \in \mathscr{M}$ and there is some $y^0 \in \mathscr{M}$ such that $c^T y^0 = \gamma$.

Lemma 15.1.34. *If* $\mathscr{H} = \left\{y \in \mathbb{R}^q \mid c^T y = \gamma\right\}$ *is a supporting hyperplane to* $\mathscr{M}$ *then* $c \geq 0$.

Proof. If $\mathscr{H}$ is a supporting hyperplane to $\mathscr{M}$ then there is some $y^0 \in \mathscr{M}$ with $c^T y^0 = \gamma$ and $c^T y \geq \gamma$ for all $y \in \mathscr{M}$. By definition of $\mathscr{M}$ we have $y^0 + w \in \mathscr{M}$ for all $w \in \mathbb{R}^q_+$, hence $c^T w \geq 0$ for all $w \in \mathbb{R}^q_+$. This implies $c \geq 0$. □

Lemma 15.1.35. *A set* $\mathscr{F} \subseteq \mathscr{M}$ *is a proper face of* $\mathscr{M}$ *if and only if there is a solution* $(u, c) \in \mathscr{U}$ *of* (LVD) *such that* $\mathscr{F} = \mathscr{F}(u, c)$.

Proof. '$\Longleftarrow$": If $(u, c) \in \mathscr{U}$ is a solution of (LVD) then there is some $x^0 \in S$ such that $Mx^0 \in \mathscr{H}(u, c)$, hence $Mx^0 \in \mathscr{F}(u, c)$. Moreover, if $y \in \mathscr{M}$ then $c^T y \geq b^T u$ by Lemma 15.1.25. Consequently, $\mathscr{H}(u, c)$ is a supporting hyperplane to $\mathscr{M}$ and $\mathscr{F}(u, c)$ is a proper face of $\mathscr{M}$.

'$\Longrightarrow$": If $\mathscr{F}$ is a proper face of $\mathscr{M}$ then there is some $c \in \mathbb{R}^q \setminus \{0\}, \gamma \in \mathbb{R}$ such that $\mathscr{H} := \left\{y \in \mathbb{R}^q \mid c^T y = \gamma\right\}$ is a supporting hyperplane to $\mathscr{M}$ and $\mathscr{F} = \mathscr{H} \cap \mathscr{M}$. By Lemma 15.1.34 we have $c \geq 0$. Since $c \neq 0$ we obtain $k^T c > 0$. Without loss of generality we can assume that $k^T c = 1$. Since $\mathscr{H}$ is a supporting hyperplane, we have $c^T y \geq \gamma$ for all $y \in \mathscr{M}$ and $c^T y^0 = \gamma$ for some $y^0 \in \mathscr{M}$. Hence there is some $x^0 \in S$ such that $c^T Mx^0 = c^T y^0 = \gamma$, i.e.,

$$\gamma = c^T Mx^0 = \min\left\{c^T Mx \mid x \in S\right\}.$$

By duality between $(\mathrm{P}_1(c))$ and $(\mathrm{D}_1(c))$, problem $(\mathrm{D}_1(c))$ has a solution u with $b^T u = \gamma = c^T Mx^0$. Thus $(u, c) \in \mathscr{U}$ is a solution of (LVD) by Theorem 15.1.29, and $\mathscr{H}(u, c) = \mathscr{H}$.

Hence $\mathscr{F} = \mathscr{F}(u, c)$. □

Corollary 15.1.36. *Each proper face of* $\mathscr{M}$ *is a subset of* $\mathrm{WMin}(\mathscr{M}, \mathbb{R}^q_+)$.

Proof. Let $\mathscr{F}$ be a proper face of $\mathscr{M}$. By Lemma 15.1.35 there is a solution $(u, c) \in \mathscr{U}$ of (LVD) such that $\mathscr{F} = \mathscr{F}(u, c)$. Let $y \in \mathscr{F} = \mathscr{F}(u, c)$, then $y \in \mathscr{M}$ (implying the existence of $x \in S$ such that $Mx \leq y$, i.e., $(x, 0)$ is feasible for $(\mathrm{P}_2(y))$) and $b^T u = c^T y$. Duality between $(\mathrm{P}_2(y))$ and $(\mathrm{D}_2(y))$ implies that (u, c) is optimal in $(\mathrm{D}_2(y))$ and $(x, 0)$ is optimal in $(\mathrm{P}_2(y))$, hence y is weakly minimal by Theorem 15.1.30. □

Corollary 15.1.37. $\mathrm{WMin}(\mathscr{M}, \mathbb{R}^q_+) \neq \emptyset$ *if and only if* $\emptyset \neq \mathscr{M} \neq \mathbb{R}^q$.

Proof. This is a direct consequence of Corollary 15.1.36, Theorem 15.1.32 and the fact that a nonempty set in $\mathscr{A} \subseteq \mathbb{R}^q$ has a nonempty boundary iff $\mathscr{A} \neq \mathbb{R}^q$. □

The following lemma shows that facets of $\mathscr{M}$ may be described by extreme solutions of (LVD) (i.e., solutions of (LVD) being a vertex of the feasible set $\mathscr{U}$).

Lemma 15.1.38. *If* $\mathscr{F}$ *is a facet of* $\mathscr{M}$ *then there is an extreme solution* (u^0, c^0) *of* (LVD) *such that* $\mathscr{F} = \mathscr{F}(u^0, c^0)$.

Proof. Let

$$\bar{\mathscr{U}} := \{(u,c) \in \mathscr{U} \mid \mathscr{F}(u,c) = \mathscr{F}\}.$$

By Theorem 15.1.29, all points of $\bar{\mathscr{U}}$ are solutions of (LVD) because $\mathscr{F}$ is nonempty as a facet of $\mathscr{M}$. Let $y \in \text{r-int}\,\mathscr{F}$ be arbitrary. Since $\mathscr{F}$ is a $(q-1)$-dimensional face we have $(u,c) \in \bar{\mathscr{U}}$ if and only if $(u,c) \in \mathscr{U}$ and $y \in \mathscr{H}(u,c)$, i.e., $b^T u = y^T c$. Hence $\bar{\mathscr{U}} = \mathscr{U} \cap \mathscr{H}_y$ with

$$\mathscr{H}_y := \left\{(u,c) \in \mathbb{R}^m \times \mathbb{R}^q \mid y^T c - b^T u = 0\right\}.$$

Since $y \in \mathrm{WMin}(\mathscr{M}, \mathbb{R}^q_+)$ by Corollary 15.1.36, Theorem 15.1.30 implies that $\mathscr{H}_y$ is a supporting hyperplane to $\mathscr{U}$, hence $\bar{\mathscr{U}}$ is a nonempty face of $\mathscr{U}$. Since $\bar{\mathscr{U}} \subseteq \mathscr{U} \subseteq \mathbb{R}^{m+q}_+$ contains no lines there is a vertex (u^0, c^0) of $\bar{\mathscr{U}}$ (see [489, Corollary 18.5.3]). Hence (u^0, c^0) is also a vertex of $\mathscr{U}$, i.e., an extreme solution of (LVD). □

We define the following sets.

$$\begin{aligned}
\mathrm{pFaces}(\mathscr{M}) &:= \{\mathscr{F} \subseteq \mathscr{M} \mid \mathscr{F} \text{ is a proper face of } \mathscr{M}\},\\
\mathrm{Facets}(\mathscr{M}) &:= \{\mathscr{F} \subseteq \mathscr{M} \mid \mathscr{F} \text{ is a facet of } \mathscr{M}\},\\
\mathrm{Sol(D)} &:= \{(u,c) \in \mathscr{U} \mid (u,c) \text{ is a solution of (LVD)}\},\\
\mathrm{ExtrSol(D)} &:= \{(u,c) \in \mathrm{Sol(D)} \mid (u,c) \text{ is a vertex of } \mathscr{U}\}.
\end{aligned}$$

Now we can extend the strong duality result in Theorem 15.1.30. Later we interpret the following result as the attainment of the supremum in the dual problem in extreme solutions.

Theorem 15.1.39. *We have the following chain of equalities*

$$\begin{aligned}
\mathrm{WMin}(\mathscr{M}, \mathbb{R}^q_+) = \mathrm{bd}\,\mathscr{M} &= \bigcup_{(u,c)\in \mathrm{ExtrSol(D)}} \mathscr{F}(u,c)\\
&= \mathrm{WMax}(\mathscr{H}(\mathrm{ExtrSol(D)}), \mathbb{R}^q_+) = \mathrm{WMax}(\mathscr{H}(\mathscr{U}), \mathbb{R}^q_+).
\end{aligned}$$

Proof. Theorem 15.1.32, Lemma 15.1.38, Lemma 15.1.35 and Corollary 15.1.36 imply the following chain of inclusions

$$\begin{aligned} \operatorname{bd}\mathscr{M} = \bigcup_{\mathscr{F}\in\operatorname{Facets}(\mathscr{M})} \mathscr{F} \subseteq \bigcup_{(u,c)\in\operatorname{ExtrSol}(\mathrm{D})} \mathscr{F}(u,c) \subseteq \bigcup_{(u,c)\in\operatorname{Sol}(\mathrm{D})} \mathscr{F}(u,c) \\ = \bigcup_{\mathscr{F}\in\operatorname{pFaces}(\mathscr{M})} \mathscr{F} \subseteq \operatorname{WMin}(\mathscr{M},\mathbb{R}^q_+) \subseteq \operatorname{bd}\mathscr{M}. \end{aligned}$$

Hence the first two equalities hold.

The equality $\operatorname{WMin}(\mathscr{M},\mathbb{R}^q_+) = \operatorname{WMax}(\mathscr{H}(\mathscr{U}),\mathbb{R}^q_+)$ was already shown in Theorem 15.1.30. Thus it remains to show that $\bigcup_{(u,c)\in\operatorname{ExtrSol}(\mathrm{D})} \mathscr{F}(u,c) = \operatorname{WMax}(\mathscr{H}(\operatorname{ExtrSol}(\mathrm{D})),\mathbb{R}^q_+)$.

If $y \in \bigcup_{(u,c)\in\operatorname{ExtrSol}(\mathrm{D})} \mathscr{F}(u,c)$ then there exists some $(u,c) \in \operatorname{ExtrSol}(\mathrm{D})$ such that $y \in \mathscr{F}(u,c) = \mathscr{H}(u,c) \cap \mathscr{M}$, i.e., $y \in \mathscr{H}(\operatorname{ExtrSol}(\mathrm{D}))$. Since (u,c) is a solution of (LVD) we have $(y + \operatorname{int}\mathbb{R}^q_+) \cap \mathscr{H}(\mathscr{U}) = \emptyset$, hence $(y + \operatorname{int}\mathbb{R}^q_+) \cap \mathscr{H}(\operatorname{ExtrSol}(\mathrm{D})) = \emptyset$ implying $y \in \operatorname{WMax}(\mathscr{H}(\operatorname{ExtrSol}(\mathrm{D})),\mathbb{R}^q_+)$.

On the other hand, if $y \in \operatorname{WMax}(\mathscr{H}(\operatorname{ExtrSol}(\mathrm{D})),\mathbb{R}^q_+)$ then $y \in \mathscr{H}(\operatorname{ExtrSol}(\mathrm{D}))$ and

$$y \notin \mathscr{H}(\operatorname{ExtrSol}(\mathrm{D})) - \operatorname{int}\mathbb{R}^q_+.$$

This is equivalent to

$$\exists(\bar{u},\bar{c}) \in \operatorname{ExtrSol}(\mathrm{D}) : y^T\bar{c} = b^T\bar{u} \tag{15.14}$$

and

$$\forall(u,c) \in \operatorname{ExtrSol}(\mathrm{D}) : y^T c \geq b^T u. \tag{15.15}$$

By Theorem 15.1.29, $\bar{u}$ solves $(\mathrm{D}_1(\bar{c}))$ hence $S \neq \emptyset$ by duality of $(\mathrm{P}_1(\bar{c}))$ and $(\mathrm{D}_1(\bar{c}))$.Thus the feasible set for $(\mathrm{P}_2(y))$ is nonempty as well. Since $(\bar{u},\bar{c}) \in \mathscr{U}$, i.e., $\mathscr{U} \neq \emptyset$, problem $(\mathrm{D}_2(y))$ has an optimal solution (u^0,c^0) being a vertex of $\mathscr{U}$. Optimality of (u^0,c^0) for $(\mathrm{D}_2(y))$ implies optimality of u^0 for $(\mathrm{D}_1(c^0))$ hence $(u^0,c^0) \in \operatorname{ExtrSol}(\mathrm{D})$ by Theorem 15.1.29. Now, (15.15) implies that $y^Tc^0 \geq b^Tu^0$. Moreover, optimality of (u^0,c^0) for $(\mathrm{D}_2(y))$ implies $b^Tu^0 - y^Tc^0 \geq b^T\bar{u} - y^T\bar{c} = 0$, i.e., $y^Tc^0 = b^Tu^0$. Consequently we have $y \in \mathscr{H}(u^0,c^0)$ and $y \in \operatorname{WMin}(\mathscr{M},\mathbb{R}^q_+) \subseteq \mathscr{M}$ by Theorem 15.1.30, i.e., $y \in \bigcup_{(u,c)\in\operatorname{ExtrSol}(\mathrm{D})} \mathscr{F}(u,c)$. □

In order to show weak and strong duality statements we give now a reformulation of the problems using the space $\mathscr{I}$ of self-infimal sets (compare Definition 2.6.37) and corresponding notations. Sequentially, it is important that $(\mathscr{I},\preceq_{\mathscr{I}})$ is a complete lattice (see Proposition 2.6.40).

We considered the linear vector optimization problem (LVP). It is easy to see that

$$\mathrm{WMin}((M(S)+\mathbb{R}^q_+),\mathbb{R}^q_+) = \mathrm{Inf}\, M(S) \iff S \neq \emptyset \text{ and } M(S)+\mathbb{R}^q_+ \neq \mathbb{R}^q.$$

Our aim is to reformulate problem (LVP) and its dual problem (LVD) as optimization problems with $\mathscr{I}$-valued objective function (compare Sect. 15.1.2). Consider the function

$$P : \mathbb{R}^n \to \mathscr{I}, \qquad P(x) := \mathrm{Inf}\,\{Mx\} = \{Mx\} + \mathrm{bd}\,\mathbb{R}^q_+.$$

It holds

$$\mathrm{Inf}\, M(S) = \mathrm{Inf} \bigcup_{x\in S} \{Mx\} = \mathrm{Inf} \bigcup_{x\in S} \mathrm{Inf}\,\{Mx\} = \inf_{x\in S} P(x).$$

Hence, we have

$$\inf_{x\in S} P(x) = \begin{cases} \{+\infty\} & \text{if} \quad S=\emptyset \\ \{-\infty\} & \text{if} \quad \mathscr{M} = \mathbb{R}^q \\ \mathrm{WMin}(\mathscr{M},\mathbb{R}^q_+) & \text{otherwise.} \end{cases}$$

Note that, by Corollary 15.1.37, $\mathrm{WMin}(\mathscr{M},\mathbb{R}^q_+) \neq \emptyset$ iff $\emptyset \neq \mathscr{M} \neq \mathbb{R}^q$. This means, if the set $\mathrm{WMin}(\mathscr{M},\mathbb{R}^q_+)$ is nonempty, it coincides with $\inf_{x\in S} P(x)$, otherwise if $\mathrm{WMin}(\mathscr{M},\mathbb{R}^q_+)$ is empty, we distinguish between two cases: $\inf_{x\in S} P(x) = \{+\infty\}$ if $S=\emptyset$ and $\inf_{x\in S} P(x) = \{-\infty\}$ otherwise. Thus, (LVP) is essentially equivalent to

$$\inf_{x\in S} P(x), \qquad S := \{x \in \mathbb{R}^n \mid Ax \geq b\}\,. \qquad (LVP')$$

Moreover, using the order relation $\preccurlyeq := \preccurlyeq_{\mathscr{I}}$ in $\mathscr{I}$ introduced in Sect. 2.6.3 it is easy to see that $x \in S$ is a (weakly efficient) solution of (LVP) if and only if

$$\left(x \in S,\ P(x) \preccurlyeq P(x^0)\right) \quad \Rightarrow \quad P(x) = P(x^0) \qquad (15.16)$$

The above considerations show the relationships between the solution concepts for (LVP) and (LVD) and lattice theoretical solution concepts for the primal problem.

We next want to reformulate the dual problem (LVD) using the supremum in $\mathscr{I}$. We first consider two auxiliary assertions.

Lemma 15.1.40. *The set $\mathscr{H}(\mathscr{U}) - \mathbb{R}^q_+$ is closed.*

Proof. Let $\{y_i\}_{i\in\mathbb{N}}$ be a sequence in $\mathscr{H}(\mathscr{U}) - \mathbb{R}^q_+$ converging to $\bar{y} \in \mathbb{R}^q$, thus for each i there is some $(u_i, c_i) \in \mathscr{U}$ with $y_i \in \mathscr{H}(u_i, c_i) - \mathbb{R}^q_+$, i.e., $y_i^T c_i \leq b^T u_i$. We have to show that there is some $(\bar{u}, \bar{c}) \in \mathscr{U}$ with $\bar{y}^T \bar{c} \leq b^T \bar{u}$.

Assume on the contrary that $\bar{y}^T c - b^T u > 0$ for all $(u, c) \in \mathscr{U}$. Since $\mathscr{U}$ is polyhedral there is some $\gamma > 0$ with $\bar{y}^T c - b^T u \geq \gamma$ for all $(u, c) \in \mathscr{U}$. Take $i_0 \in \mathbb{N}$ such that $\|y_{i_0} - \bar{y}\|_\infty < \gamma$, then

$$(\bar{y} - y_{i_0})^T c_{i_0} \leq \|y_{i_0} - \bar{y}\|_\infty \|c_{i_0}\|_\infty < \gamma$$

hence

$$\bar{y}^T c_{i_0} - b^T u_{i_0} < y_{i_0}^T c_{i_0} + \gamma - b^T u_{i_0} \leq \gamma,$$

a contradiction. □

Lemma 15.1.41. *It holds* $\mathrm{WMax}\left(\left(\mathscr{H}(\mathscr{U}) - \mathbb{R}^q_+\right), \mathbb{R}^q_+\right) = \mathrm{WMax}(\mathscr{H}(\mathscr{U}), \mathbb{R}^q_+)$.

Proof. We have

$$y \in \mathrm{WMax}(\mathscr{H}(\mathscr{U}), \mathbb{R}^q_+) \quad \Longleftrightarrow \quad \left(y \in \mathscr{H}(\mathscr{U}) \text{ and } y \notin \mathscr{H}(\mathscr{U}) - \operatorname{int} \mathbb{R}^q_+\right)$$

and

$$y \in \mathrm{WMax}\left(\left(\mathscr{H}(\mathscr{U}) - \mathbb{R}^q_+\right), \mathbb{R}^q_+\right) \Longleftrightarrow \quad \left(y \in \mathscr{H}(\mathscr{U}) - \mathbb{R}^q_+ \text{ and } y \notin \mathscr{H}(\mathscr{U}) - \operatorname{int} \mathbb{R}^q_+\right).$$

Thus it remains to show that

$$\left(y \in \mathscr{H}(\mathscr{U}) - \mathbb{R}^q_+ \text{ and } y \notin \mathscr{H}(\mathscr{U}) - \operatorname{int} \mathbb{R}^q_+\right) \quad \Longrightarrow \quad y \in \mathscr{H}(\mathscr{U}).$$

Indeed, $y \notin \mathscr{H}(\mathscr{U}) - \operatorname{int} \mathbb{R}^q_+$ implies $y^T c \geq b^T u$ for all $(u, c) \in \mathscr{U}$ and $y \in \mathscr{H}(\mathscr{U}) - \mathbb{R}^q_+$ implies the existence of some $(\bar{u}, \bar{c}) \in \mathscr{U}$ with $y^T \bar{c} \leq b^T \bar{u}$. Thus we obtain $y^T \bar{c} = b^T \bar{u}$, i.e., $y \in \mathscr{H}(\mathscr{U})$. □

Note that the hyperplane $\mathscr{H}(u, c) \subseteq \mathbb{R}^q$ is a self-infimal set, whenever $(u, c) \in \mathscr{U}$. Therefore the term $\sup_{(u,c) \in \mathscr{U}} \mathscr{H}(u, c)$ is well defined. The next lemma clarifies the relationship between this supremum and the solution concept of problem (LVD).

Lemma 15.1.42. *It holds*

$$\sup_{(u,c) \in \mathscr{U}} \mathscr{H}(u, c) = \begin{cases} \{-\infty\} & \textit{if} \quad \mathscr{U} = \emptyset \\ \{+\infty\} & \textit{if} \quad \mathscr{H}(\mathscr{U}) - \mathbb{R}^q_+ = \mathbb{R}^q \\ \mathrm{WMax}(\mathscr{H}(\mathscr{U}), \mathbb{R}^q_+) & \textit{otherwise.} \end{cases}$$

Proof. (i) If $\mathscr{U} = \emptyset$, we have $\sup_{(u,c) \in \mathscr{U}} \mathscr{H}(u, c) = \operatorname{Sup} \mathscr{H}(\mathscr{U}) = \operatorname{Sup} \emptyset = \{-\infty\}$, by definition.

(ii) The case $\mathscr{H}(\mathscr{U}) - \mathbb{R}^q_+ = \mathbb{R}^q$ follows from the definition of the supremal set.

(iii) Since $\mathscr{H}(\mathscr{U}) \subseteq \mathbb{R}^q$, we have

$$\sup_{(u,c)\in\mathscr{U}} \mathscr{H}(u,c) = \operatorname{Sup} \mathscr{H}(\mathscr{U}) = \operatorname{WMax}(\operatorname{cl}(\mathscr{H}(\mathscr{U}) - \mathbb{R}^q_+), \mathbb{R}^q_+),$$

by the definition of the supremal set. Lemma 15.1.40 and Lemma 15.1.41 yield that

$$\operatorname{WMax}(\operatorname{cl}(\mathscr{H}(\mathscr{U}) - \mathbb{R}^q_+), \mathbb{R}^q_+) = \operatorname{WMax}(\mathscr{H}(\mathscr{U}), \mathbb{R}^q_+).$$

The proof is completed. □

Remark 15.1.43. The preceding three lemmata remain valid if the set $\mathscr{U}$ is replaced by any finite or polyhedral subset.

Lemma 15.1.42 shows in fact the relationship between problem (LVD) and the following problem,

$$\sup_{(u,c)\in\mathscr{U}} \mathscr{H}(u,c), \qquad (LVD')$$

where

$$\mathscr{U} := \left\{(u,c) \in \mathbb{R}^m \times \mathbb{R}^q \mid (u,c) \geq 0,\ k^T c = 1,\ A^T u = M^T c\right\}.$$

Indeed, if $\operatorname{WMax}(\mathscr{H}(\mathscr{U}), \mathbb{R}^q_+)$ is nonempty, it coincides with $\sup_{(u,c)\in\mathscr{U}} \mathscr{H}(u,c)$ in problem (LVD′). Otherwise, if $\operatorname{WMax}(\mathscr{H}(\mathscr{U}), \mathbb{R}^q_+)$ is empty, we distinguish between the following two cases:

$$\sup_{(u,c)\in\mathscr{U}} \mathscr{H}(u,c) = \begin{cases} \{-\infty\} & \text{when } \mathscr{U} = \emptyset \\ \{+\infty\} & \text{otherwise.} \end{cases}$$

The solution concept for (LVD) as introduced in Sect. 2.6.3 can be expressed in terms of the ordering relation $\preccurlyeq := \preccurlyeq_{\mathscr{I}}$ in the complete lattice $\mathscr{I}$. This characterization is completely analogous to (15.16). So we obtain yet another motivation for this solution concept.

Lemma 15.1.44. *A point* $(u^0, c^0) \in \mathscr{U}$ *is a (weakly efficient) solution of* (LVD) *if and only if*

$$\Big((u,c) \in \mathscr{U},\ \mathscr{H}(u^0,c^0) \preccurlyeq \mathscr{H}(u,c)\Big) \quad \Rightarrow \quad \mathscr{H}(u^0,c^0) = \mathscr{H}(u,c). \tag{15.17}$$

Proof. Let $(u^0, c^0) \in \mathscr{U}$ be a solution of (LVD). Hence u^0 solves $(\mathrm{D}_1(c^0))$ by Theorem 15.1.29. Consider $(u, c) \in \mathscr{U}$ with $\mathscr{H}(u^0, c^0) \preccurlyeq \mathscr{H}(u, c)$. Then we have $c^0 = c$ and $b^T u^0 \leq b^T u$. Since $c^0 = c$, u is feasible for $(\mathrm{D}_1(c^0))$ hence $b^T u \leq b^T u^0$ and consequently $b^T u^0 = b^T u$. This means we have $\mathscr{H}(u^0, c^0) = \mathscr{H}(u, c)$.

Let $(u^0, c^0) \in \mathscr{U}$ be no solution of (LVD). By Theorem 15.1.29 there exists some $\bar{u} \geq 0$ with $A^T \bar{u} = M^T c_0$ and $b^T \bar{u} > b^T u^0$. Hence, we have $\mathscr{H}(u^0, c^0) \preccurlyeq \mathscr{H}(\bar{u}, c^0)$ but $\mathscr{H}(u^0, c^0) \neq \mathscr{H}(\bar{u}, c^0)$, i.e., (15.17) is not satisfied. □

As a consequence of the weak and strong duality assertion given in Lemma 15.1.25 and Theorem 15.1.30 and the above considerations, we present here duality assertions for vector optimization problems, formulated along the lines of the classical scalar duality theory. The complete lattice $(\mathscr{I}, \preccurlyeq)$ of self-infimal subsets of $\overline{\mathbb{R}}^q$ plays a key role in these results.

The first result shows that we have weak duality between the $\mathscr{I}$-valued problems (LVP') and (LVD').

Theorem 15.1.45 (Weak Duality). *Let $x \in S$ and $(u, c) \in \mathscr{U}$. Then it holds*

$$\mathscr{H}(u, c) \preccurlyeq P(x).$$

Proof. For all $y \in P(x) = \{Mx\} + \operatorname{bd} \mathbb{R}^q_+ \subseteq \mathscr{M}$, Lemma 15.1.25 yields $y^T c \geq b^T u$, hence $P(x) \subseteq \mathscr{H}(u, c) + \mathbb{R}^q_+$. This implies $\mathscr{H}(u, c) \preccurlyeq P(x)$. □

The next result shows strong duality between (LVP') and (LVD'). The following distinction between the three cases is well-known from scalar linear programming.

Theorem 15.1.46 (Strong Duality). *Let at least one of the sets S and $\mathscr{U}$ be nonempty. Then it holds strong duality between (LVP') and (LVD'), i.e.,*

$$V := \sup_{(u,c) \in \mathscr{U}} \mathscr{H}(u, c) = \inf_{x \in S} P(x).$$

Moreover, the following statements are true.

(i) *If $S \neq \emptyset$ and $\mathscr{U} \neq \emptyset$, then $\{-\infty\} \neq V \neq \{+\infty\}$ and*

$$V = \operatorname{WMax}(\mathscr{H}(\mathscr{U}), \mathbb{R}^q_+) = \operatorname{WMin}(P(S), \mathbb{R}^q_+) \neq \emptyset.$$

(ii) *If $S = \emptyset$ and $\mathscr{U} \neq \emptyset$, then $V = \{+\infty\}$.*
(iii) *If $S \neq \emptyset$ and $\mathscr{U} = \emptyset$, then $V = \{-\infty\}$.*

Proof. By the weak duality we have

$$\sup_{(u,c) \in \mathscr{U}} \mathscr{H}(u, c) \preccurlyeq \inf_{x \in S} P(x).$$

(i) If $S \neq \emptyset$ and $\mathcal{U} \neq \emptyset$, this implies that neither $\sup_{(u,c)\in\mathcal{U}} \mathcal{H}(u,c)$ nor $\inf_{x\in S} P(x)$ can be $\{-\infty\}$ or $\{+\infty\}$. Hence, Theorem 15.1.30 implies

$$\sup_{(u,c)\in\mathcal{U}} \mathcal{H}(u,c) = \mathrm{WMax}(\mathcal{H}(\mathcal{U}), \mathbb{R}^q_+) = \mathrm{WMin}(\mathcal{M}, \mathbb{R}^q_+) = \inf_{x\in S} P(x).$$

(ii) If $S = \emptyset$ and $\mathcal{U} \neq \emptyset$, we have $\inf_{x\in S} P(x) = \{+\infty\}$. Theorem 15.1.30 implies that

$$\mathrm{WMax}(\mathcal{H}(\mathcal{U}), \mathbb{R}^q_+) = \mathrm{WMin}(\mathcal{M}, \mathbb{R}^q_+) = \emptyset.$$

Since $\mathcal{U} \neq \emptyset$, we conclude $\mathcal{H}(\mathcal{U}) - \mathbb{R}^q_+ = \mathbb{R}^q$ and Lemma 15.1.42 yields

$$\sup_{(u,c)\in\mathcal{U}} \mathcal{H}(u,c) = \{+\infty\}.$$

(iii) If $S \neq \emptyset$ and $\mathcal{U} = \emptyset$, we have $\sup_{(u,c)\in\mathcal{U}} \mathcal{H}(u,c) = \{-\infty\}$. Theorem 15.1.30 implies that

$$\mathrm{WMin}(\mathcal{M}, \mathbb{R}^q_+) = \mathrm{WMax}(\mathcal{H}(\mathcal{U}), \mathbb{R}^q_+) = \emptyset.$$

Since $S \neq \emptyset$, we obtain $\mathcal{M} = \mathbb{R}^q$, hence $\inf_{x\in S} P(x) = \{-\infty\}$. □

In scalar linear programming, the attainment of the supremum of the problem in a vertex of the feasible set plays a key role in the simplex algorithm. It is therefore sufficient to search for a solution on a finite subset of the feasible set. The next result shows that we have a corresponding result for our dual problem. Typically, in our case, the supremum in (D′) is not attained in a single vertex, but in a finite number of vertices, namely, in the set of those vertices of $\mathcal{U}$ being solutions of (LVD), i.e., the set ExtrSol(D) of extreme solutions of (LVD).

Theorem 15.1.47 (Dual Attainment in Vertices). *Let $S \neq \emptyset$ and $\mathcal{U} \neq \emptyset$. Then the supremum in the dual problem (LVD') is attained in extreme solutions of* (LVD), *i.e.,*

$$\sup_{(u,c)\in\mathcal{U}} \mathcal{H}(u,c) = \sup_{(u,c)\in\mathrm{ExtrSol(D)}} \mathcal{H}(u,c).$$

Proof. Since $\mathcal{U} \neq \emptyset$ and $S \neq \emptyset$ we have

$$\sup_{(u,c)\in\mathcal{U}} \mathcal{H}(u,c) = \mathrm{WMax}(\mathcal{H}(\mathcal{U}), \mathbb{R}^q_+)$$

by Theorem 15.1.46. $\mathrm{WMax}(\mathcal{H}(\mathcal{U}), \mathbb{R}^q_+) = \mathrm{WMax}(\mathcal{H}(\mathrm{ExtrSol(D)}), \mathbb{R}^q_+)$ follows from Theorem 15.1.39.

It remains to show that

$$\mathrm{WMax}(\mathcal{H}(\mathrm{ExtrSol(D)}), \mathbb{R}^q_+) = \sup_{(u,c)\in \mathrm{ExtrSol(D)}} \mathcal{H}(u, c).$$

If $S \neq \emptyset$ and $\mathcal{U} \neq \emptyset$ then we conclude from Theorem 15.1.46 and Corollary 15.1.37 that $\emptyset \neq \mathcal{M} \neq \mathbb{R}^q$. Thus $\mathcal{M}$ has a facet and consequently $\mathrm{ExtrSol(D)} \neq \emptyset$ by Lemma 15.1.38. Moreover, $\mathcal{H}(\mathrm{ExtrSol(D)}) - \mathbb{R}^q_+ \subseteq \mathcal{H}(\mathcal{U}) - \mathbb{R}^q_+ \neq \mathbb{R}^q$. Hence the desired statement follows from Remark 15.1.43. □

15.2 Applications in Mathematical Finance

In this section we apply the duality statements for linear vector optimization problems presented in Sect. 15.1.4 for deriving corresponding duality assertions in portfolio management.

We consider a Markowitz-type bicriterial portfolio optimization problem (see Heyde, Löhne, Tammer [258]), where the expected return of the portfolio should be maximized and the risk of the portfolio, measured by the Conditional Value at Risk (see Example 5.1.4 and [191, Section 4.4, Definition 4.43]), should be minimized. For details about the Conditional Value at Risk (sometimes also called Average Value at Risk) see e.g. [497] or [191, Section 4.4].

We consider a market with n different financial instruments with returns r_j, $j = 1, \ldots, n$ being random variables combined in a random vector $r = (r_1, \ldots, r_n)^T$ with a given probability distribution $\mathbb{P}$. The decision vector $x \in \mathbb{R}^n$ represents a portfolio of these instruments, where the components x_j denote the fraction of the capital invested in instrument j. This yields the constraints

$$x \geq 0, \quad \sum_{j=1}^{n} x_j = 1.$$

The return of a portfolio x is given by $r^T x$ so the bicriterial optimization problem consists in minimizing the negative expected return, i.e., $-\mathbb{E}(r^T x)$ and the Conditional Value at Risk of the return (see Example 5.1.4 and [191, Section 4.4, Definition 4.43]), i.e., $CVaR_\beta(r^T x)$, for a given risk level $\beta \in [0, 1)$. We can approximate this problem by a linear one by sampling the probability distribution of r like it is done in [496]. If $r^1, \ldots, r^m$ denotes a sample of size m then

$$\mathbb{E}(r^T x) \approx \frac{1}{m} \sum_{k=1}^{m} {r^k}^T x$$

and

$$CVaR_\beta(r^T x) \approx \inf\left\{ \alpha + \frac{1}{(1-\beta)m} \sum_{k=1}^{m} z_k \mid \alpha \in \mathbb{R}, \forall k \in \{1, \ldots, m\} : \right.$$
$$\left. z_k \in \mathbb{R}_+, {r^k}^T x + \alpha + z_k \geq 0 \right\}.$$

Then the given problem accords essentially with the following linear vector optimization problem:

$$\text{WMin}(f(S) + \mathbb{R}^2_+, \mathbb{R}^2_+), \qquad (P_M)$$

where

$$S := \left\{ (x, z, \alpha) \in \mathbb{R}^n_+ \times \mathbb{R}^m_+ \times \mathbb{R} \mid \sum_{j=1}^{n} x_j = 1, \forall k \in \{1, \ldots, m\} : {r^k}^T x + \alpha + z_k \geq 0 \right\}$$

and

$$f(x, z, \alpha) = \begin{pmatrix} -\frac{1}{m} \sum_{k=1}^{m} {r^k}^T x \\ \alpha + \frac{1}{(1-\beta)m} \sum_{k=1}^{m} z_k \end{pmatrix}.$$

As already noted in Sect. 15.1.4, finding solutions of (P_M) is equivalent to finding weakly minimal solutions of the problem $\text{WMin}(f(S), \mathbb{R}^2_+)$.

We put

$$M := \begin{pmatrix} -\frac{1}{m}\mathbf{1}_m^T R^T & 0 & 0 \\ 0 & \frac{1}{(1-\beta)m}\mathbf{1}_m^T & 1 \end{pmatrix}, \quad A := \begin{pmatrix} I_n & 0 & 0 \\ 0 & I_m & 0 \\ \mathbf{1}_n^T & 0 & 0 \\ -\mathbf{1}_n^T & 0 & 0 \\ R^T & I_m & \mathbf{1}_m \end{pmatrix}, \quad b := \begin{pmatrix} 0 \\ 0 \\ 1 \\ -1 \\ 0 \end{pmatrix},$$

where

$$R := \begin{pmatrix} r_1^1 & \cdots & r_1^m \\ \vdots & \ddots & \vdots \\ r_n^1 & \cdots & r_n^m \end{pmatrix},$$

I_ℓ is the ℓ-dimensional identity matrix and $\mathbf{1}_\ell$ is the ℓ-dimensional vector with all components being 1. Then the problem ($\mathrm{P_M}$) is equivalent to

$$\mathrm{WMin}(M(S)+\mathbb{R}^q_+,\mathbb{R}^q_+), \qquad S := \left\{\bar{x}\in\mathbb{R}^{n+m+1} \mid A\bar{x}\geq b\right\},$$

a problem of type (LVP) in Sect. 15.1.4.

As the corresponding dual problem to ($\mathrm{P_M}$) we derive the following problem with a set-valued objective map as a special case of problem (LVD) in Sect. 15.1.4:

$$\mathrm{WMax}(\mathscr{H}(\mathscr{U}),\mathbb{R}^q_+), \tag{D_M}$$

where

$$\mathscr{U} = \left\{(\bar{u},c)\in\mathbb{R}^{n+2m+2}_+\times\mathbb{R}^2_+ \mid c_1+c_2=1,\ A^T\bar{u}=M^Tc\right\}.$$

In fact, we have

$$\mathscr{U} = \Bigg\{(w,p,v_1,v_2,u,c)\in\mathbb{R}^n_+\times\mathbb{R}^m_+\times\mathbb{R}_+\times\mathbb{R}_+\times\mathbb{R}^m_+\times\mathbb{R}^2_+ \mid c_1+c_2=1,$$

$$w+\mathbf{1}_n(v_1-v_2)+Ru=R\mathbf{1}_m\frac{-c_1}{m},\quad p+u=\mathbf{1}_m\frac{c_2}{(1-\beta)m},\quad \mathbf{1}_m^Tu=c_2\Bigg\}$$

and the set-valued objective map is given by

$$\mathscr{H}(w,p,v_1,v_2,u,c)=\left\{y\in\mathbb{R}^2 \mid c_1y_1+c_2y_2=v_1-v_2\right\}.$$

Interpreting w and p as slack variables and defining $v:=v_1-v_2$ we arrive at

$$\mathscr{U} = \Bigg\{(v,u,c)\in\mathbb{R}\times\mathbb{R}^m_+\times\mathbb{R}^2_+ \mid c_1+c_2=1,\quad \mathbf{1}_m^Tu=c_2,$$

$$1_nv+Ru\leq R\mathbf{1}_m\frac{-c_1}{m},\quad u\leq\mathbf{1}_m\frac{c_2}{(1-\beta)m}\Bigg\}$$

and

$$\mathscr{H}(v,u,c)=\left\{y\in\mathbb{R}^2 \mid c_1y_1+c_2y_2=v\right\}.$$

The following transformation of the dual variables results in dual variables being interpretable as probabilities. Note that $\mathscr{H}$ does not depend on u and for each $(v,u,c)\in\mathscr{U}$ there is

$$(v,q,c) \in \bar{\mathscr{U}} := \Bigg\{ (v,q,c) \in \mathbb{R} \times \mathbb{R}^m_+ \times \mathbb{R}^2_+ \mid c_1 + c_2 = 1, \quad \mathbf{1}_m^T q = 1,$$
$$\mathbf{1}_n v + Rqc_2 \leq R\mathbf{1}_m \frac{-c_1}{m}, \quad q \leq \mathbf{1}_m \frac{1}{(1-\beta)m} \Bigg\},$$

where q is given by $\frac{1}{c_2} u$ if $c_2 \neq 0$ and can be chosen as $q_k = \frac{1}{m}$ for all k if $c_2 = 0$. On the other hand for each $(v,q,c) \in \bar{\mathscr{U}}$ we have $(v, c_2 q, c) \in \mathscr{U}$, hence $\mathscr{U}$ can be replaced by $\bar{\mathscr{U}}$ and problem ($\mathrm{D_M}$) is equivalent to

$$\mathrm{WMax}(\mathscr{H}(\bar{\mathscr{U}}), \mathbb{R}^q_+), \tag{$\bar{D}_M$}$$

where

$$\bar{\mathscr{U}} = \Bigg\{ (v,q,c) \in \mathbb{R} \times \mathbb{R}^m \times \mathbb{R}^2 \mid c \geq 0, \quad c_1 + c_2 = 1, \quad \sum_{k=1}^m q_k = 1,$$
$$\forall k = 1, \ldots, m : \; 0 \leq q_k \leq \frac{1}{(1-\beta)m},$$
$$\forall j = 1, \ldots, n : \; v \leq -\sum_{k=1}^m \left(\frac{1}{m} r_j^k c_1 + r_j^k q_k c_2 \right) \Bigg\}.$$

Applying Theorem 15.1.29 we can characterize the solutions of ($\bar{\mathrm{D}}_\mathrm{M}$). A triple $(v^*, q^*, c^*) \in \bar{\mathscr{U}}$ is a solution of ($\bar{\mathrm{D}}_\mathrm{M}$) if and only if

$$v^* = \max \left\{ v \mid (v, q^*, c^*) \in \bar{\mathscr{U}} \right\} = \max \left\{ v \mid (v, q, c^*) \in \bar{\mathscr{U}} \right\},$$

i.e., if and only if

$$v^* = \min_{j=1,\ldots,n} -\sum_{k=1}^m \left(\frac{1}{m} r_j^k c_1^* + r_j^k q_k^* c_2^* \right) = \max_{q \in \mathscr{Q}} \min_{j=1,\ldots,n} -\sum_{k=1}^m \left(\frac{1}{m} r_j^k c_1^* + r_j^k q_k c_2^* \right)$$

with

$$\mathscr{Q} := \left\{ q \in \mathbb{R}^m \mid \sum_{k=1}^m q_k = 1, \quad \forall k = 1, \ldots, m : \; 0 \leq q_k \leq \frac{1}{(1-\beta)m} \right\}$$

being a nonempty set because of $\beta \geq 0$.

Since $q \geq 0$ and $\sum_{k=1}^m q_k = 1$, the numbers q_k may be interpreted as probabilities describing an alternative probability distribution $\mathbb{P}_q$ for the samples r^k. Then $\sum_{k=1}^m r_j^k q_k = \mathbb{E}_{\mathbb{P}_q}(r_j)$, the expectation of r_j under the alternative distribution

$\mathbb{P}_q$, and $\sum_{k=1}^{m} \frac{1}{m} r_j^k = \mathbb{E}_{\mathbb{P}}(r_j)$, the expectation of r_j under the given distribution $\mathbb{P}$. The numbers q_k are related to the dual description of the coherent risk measure **Conditional Value at Risk**. This dual description signifies that the Conditional Value at Risk of some financial position equals the worst case expected loss of this position under a certain set of alternative probability distributions (for details see e.g. [191, Theorem 4.47]). Moreover, the scalarization weights c_1 and c_2 describe the model uncertainty, i.e., c_1 can be interpreted as the probability for $\mathbb{P}$ being the right probability distribution and c_2 as the probability that $\mathbb{P}_q$ provides the appropriate distribution. Then $\mathbb{P}_{(c,q)} := c_1\mathbb{P} + c_2\mathbb{P}_q$ describes a probability distribution being a mixture of $\mathbb{P}$ and $\mathbb{P}_q$ and $\mathbb{E}_{\mathbb{P}_{(c,q)}}(r_j) = c_1\mathbb{E}(r_j) + c_2\mathbb{E}_{\mathbb{P}_q}(r_j)$. Hence, a solution for the dual problem consists of some (c^*, q^*) determining an alternative probability distribution $\mathbb{P}_{(c^*,q^*)}$ and a number $v^* = \min_{j=1,\ldots,n} -\mathbb{E}_{\mathbb{P}_{(c^*,q^*)}}(r_j)$ where the vector $q^* \in \mathcal{Q}$ must be chosen such that it maximizes $\min_{j=1,\ldots,n} -\mathbb{E}_{\mathbb{P}_{(c^*,q)}}(r_j)$ or minimizes $\max_{j=1,\ldots,n} \mathbb{E}_{P_{(c^*,q)}}(r_j)$, i.e., the largest expected return of the n given financial instruments, given the value of c^*. That means, (c^*, q^*) provides the worst case for the expected return of the "best" of the given financial instruments under the considered alternative probabilities $\mathbb{P}_{(c^*,q)}$.

Using the results of Sect. 15.1.4 we can see that a point $(x^*, z^*, \alpha^*) \in S$ is a solution of $(\mathrm{P_M})$ if and only if there is a solution (v^*, q^*, c^*) of $(\bar{\mathrm{D}}_{\mathrm{M}})$ such that

$$-\frac{c_1^*}{m}\sum_{k=1}^{m} {r^k}^T x^* + c_2^*\left(\alpha^* + \frac{1}{(1-\beta)m}\sum_{k=1}^{m} z_k^*\right) = v^* = \min_{j=1,\ldots,n} -\mathbb{E}_{\mathbb{P}_{(c^*,q^*)}}(r_j)$$

or equivalently if

$$-c_1^*\mathbb{E}^{appr}(r^T x^*) + c_2^* CVaR_\beta^{appr}(r^T x^*) = v^* = \min_{j=1,\ldots,n} -\mathbb{E}_{\mathbb{P}_{(c^*,q^*)}}(r_j), \qquad (15.18)$$

where $\mathbb{E}^{appr}$ and $CVaR_\beta^{appr}$ are the approximations of the expectation and the Conditional Value at Risk with the help of the samples. Thus one can find a solution of the portfolio optimization problem by first determining some "worst case" alternative probability $\mathbb{P}_{(c^*,q^*)}$ belonging to a solution (v^*, q^*, c^*) of $(\bar{\mathrm{D}}_{\mathrm{M}})$ and then searching for a portfolio x^* such that (15.18) is satisfied.

For vector optimization problems one often does not want to choose a scalarization in advance and prefers computing the whole set of minimal solutions. Concerning the dual problem, it might be also useful to compute all solutions of $(\bar{\mathrm{D}}_{\mathrm{M}})$ together with the corresponding efficient portfolios and to provide the decision maker (the investor) with this information because from solving the dual problem the investor gets an information about the relationship between the scalarization weights c^* and the "worst case" alternative probability scenario $\mathbb{P}_{(c^*,q^*)}$ taken into account under this scalarization.

15.3 Set-Valued Optimization in Welfare Economics

This section concerns applications of methods from variational analysis and generalized differentiation (especially Theorems 5.5.10 and 12.11.3) to nonconvex models of welfare economics (see [28] and [278]) with infinite-dimensional commodity spaces. Bao and Mordukhovich [28] established relationships between necessary conditions in set-valued optimization and appropriate extensions of the second fundamental theorem of **welfare economics** to nonconvex economies with general preference relations. The variational approach developed in the paper by Bao and Mordukhovich [28] (see Sect. 12.11) yields necessary conditions for various types of local optimal solutions of constrained set-valued problems such that it is possible to derive new versions of the second welfare theorem applied to Pareto as well as weak, strict, and strong Pareto optimal allocations of nonconvex economies (see Definition 15.3.2) under certain regularity conditions.

It is well known that vector optimization has its roots in economic modeling and general equilibrium theory. In our book we take attention to recent mathematical areas studying minimal points of sets and minimal solutions of set-valued optimization problems. We present some applications of set-valued optimization in welfare economics in this section following the presentation in the paper by Bao and Mordukhovich [28].

First, we formulate a conventional model of **welfare economics** that we will study in the following. For a given normed commodity space E, we consider an economy

$$Eco = (CS_1, \cdots, CS_n, PS_1, \cdots, PS_m, W) \qquad (m \in \mathbb{N}^*,\ n \in \mathbb{N}^*) \tag{15.19}$$

involving m firms with production sets $PS_j \subset E$ $(j = 1, \cdots, m)$, n customers with consumption sets $CS_i \subset E$ $(i = 1, \cdots, n)$, and a net demand constraint set W representing constraints related to the initial inventory of commodities in the economy Eco. We denote **production strategies** by $v = (v_1, \cdots, v_m) \in PS_1 \times \cdots \times PS_m$ and **consumption plans** by $y = (y_1, \cdots, y_n) \in CS_1 \times \cdots \times CS_n$. In the sequel we suppose that all the consumption sets CS_i $(i = 1, \cdots, n)$ are closed. Then the pair (v, y) is an **admissible state** of the economy Eco.

Furthermore, associate with each consumer a **preference / level set** $L_i(y)$ that consists of elements in CS_i preferred to y_i by this consumer at the consumption plan y. The corresponding **preference mappings** $L_i : Y \rightrightarrows E$ are set-valued with $Y := E^n$. By definition we have $y_i \notin L_i(y)$ for every $i = 1, \cdots, n$ and naturally suppose that $L_i(y) \neq \emptyset$ at least for some $i \in \{1, \cdots, n\}$. Put for convenience $\mathrm{cl}\, L_i(y) := \{y_i\}$ if $L_i(y) = \emptyset$.

Definition 15.3.1. An admissible state (v, y) of the economy Eco in (15.19) is called **feasible allocation** of Eco if

$$w := \sum_{i=1}^{n} y_i - \sum_{j=1}^{m} v_j \in W, \tag{15.20}$$

where W is the net demand constraint set representing constraints related to the initial inventory of commodities in the economy *Eco*.

The decision makers in welfare economics are looking for certain types of Pareto-type solutions (allocations) of the economy *Eco* in the following sense:

Definition 15.3.2 (Pareto-Type Optimal Allocations). Let $(\overline{v}, \overline{y}) \in E^m \times E^n$ be a feasible allocation of the economy *Eco*.

(i) $(\overline{v}, \overline{y})$ is called a **local weak Pareto optimal allocation of** *Eco* if there is a neighborhood $U \subset E^m \times E^n$ of $(\overline{v}, \overline{y})$ such that for every feasible allocation $(v, y) \in U$ we have $y_i \notin L_i(\overline{y})$ for some $i \in \{1, \cdots, n\}$.
(ii) $(\overline{v}, \overline{y})$ is called a **local Pareto optimal allocation of** *Eco* if there is a neighborhood $U \subset E^m \times E^n$ of $(\overline{v}, \overline{y})$ such that for every feasible allocation $(v, y) \in U$ either $y_i \notin \operatorname{cl} L_i(\overline{y})$ for some $i \in \{1, \cdots, n\}$, or $y_i \notin L_i(\overline{y})$ for all $i \in \{1, \cdots, n\}$.
(iii) $(\overline{v}, \overline{y})$ is called a **local strict Pareto optimal allocation of** *Eco* if there is a neighborhood $U \subset E^m \times E^n$ of $(\overline{v}, \overline{y})$ such that for every feasible allocation $(v, y) \in U$ with $y \neq \overline{y}$ we have $y_i \notin \operatorname{cl} L_i(\overline{y})$ for some $i \in \{1, \cdots, n\}$.
(iv) $(\overline{v}, \overline{y})$ is called a **local strong Pareto optimal allocation of** *Eco* if there is a neighborhood $U \subset E^m \times E^n$ of $(\overline{v}, \overline{y})$ such that for every feasible allocation $(v, y) \in U$ with $(v, y) \neq (\overline{v}, \overline{y})$ we have $y_i \notin \operatorname{cl} L_i(\overline{y})$ for some $i \in \{1, \cdots, n\}$.
(v) We omit the adjective "local" in (i) - (iv) or replace it by "global" if $U = E^m \times E^n$ above.

Remark 15.3.3. Of course, from the definitions given above it follows that (iv) $\Rightarrow$ (iii) $\Rightarrow$ (ii) $\Rightarrow$ (i) but not vice versa; the same implications hold with respect to the global version in (v). The notions of (both local and global) weak Pareto and Pareto optimal allocations are quite natural and often used in welfare economics. These notions correspond to similar Pareto-type concepts (weakly minimal and minimal solutions) for vector optimization problems based on preferences given by utility functions (see [214, 293, 402, 415]). The notion of strong Pareto optimal allocations (Definition 15.3.2 (iv)) was introduced by Khan [324] for models of welfare economics and later on intensively studied by Mordukhovich in [410,428,429,431]. Bao and Mordukhovich [28] established such relationships in a more general context. The notion of strict Pareto optimal allocations (Definition 15.3.2 (iii)) was first appeared in [431, Remark 8.15] as a modified version of the strong Pareto concept. Bao and Mordukhovich [28] show that local strict Pareto optimal allocations in welfare economics can be reduced in fact to the so-called fully localized minimizers (see Definition 2.6.49) in associated problems of constrained set-valued optimization.

In order apply the necessary optimality conditions derived in Theorem 12.11.3 to the general nonconvex model of welfare economics we formulate a constrained set-valued optimization problem which we associate with the welfare economic model introduced above.

We will formulate the welfare economic model as a set-valued optimization problem of the following type:

$$\prec -\text{minimize} \qquad F(x) \quad \text{subject to} \quad x \in S, \tag{15.21}$$

where $F : X \rightrightarrows Y$ is a set-valued cost mapping between Banach spaces, $S \subset X$ is a subset in X, and "$\prec$-minimization" in (15.21) is understood with respect to some *preference relation* $\prec$ on Y (compare Sect. 2.6.5 and [431, Subsection 5.3.1]): Given a subset $\mathscr{R} \subset Y \times Y$, we say that y_1 is preferred to y_2 ($y_1 \prec y_2$) if $(y_1, y_2) \in \mathscr{R}$.

We consider the **level-set mapping** $L : Y \rightrightarrows Y$ associated with the preference $\prec$ by

$$L(y) := \{u \in Y \mid u \prec y\} \tag{15.22}$$

introduced in Sect. 2.6.5.

In the following we study a set-valued optimization problem in form (15.21) with $X = E^{m+1}$, $x = (v, w)$, and $Y := E^n$ constructed upon the initial data of the welfare economy *Eco*:

$$\text{minimize } F(x) \quad \text{subject to} \quad x \in S, \tag{15.23}$$

where

$$F(x) := \{y \in Y \mid w = \sum_{i=1}^{n} y_i - \sum_{j=1}^{m} v_j\},$$

and

$$S := (\prod_{j=1}^{m} PS_j) \times W \subset X.$$

The "minimization" in (15.23) is understood with respect to the level-set mapping $L : Y \rightrightarrows Y$ (compare (15.22)) defined in the product form

$$L(y) := \prod_{i=1}^{n} L_i(y), \qquad y \in Y; \tag{15.24}$$

via the given preference mappings $L_i : Y \rightrightarrows E$ of the welfare economy *Eco*.

For deriving the extended second welfare theorem for local Pareto-type optimal allocations (see Definition 15.3.2) we recall at first that the notions of local weak Pareto, strict Pareto, and strong Pareto optimal allocations of the welfare economy (15.19) are equivalent to, respectively, the notions of fully localized weak minimizers, minimizers, and strong minimizers in the sense of Definition 2.6.49 for the set-valued optimization problem (15.23). So it is possible to derive a second welfare theorem for (15.19) from the corresponding necessary optimality conditions in Theorem 12.11.3 (see [28, Theorem 3.4]) for fully localized weak minimizers, minimizers, and strong minimizers.

The following result is shown in ([28, Theorem 4.1]).

Theorem 15.3.4 (Equivalence Between Local Pareto-Type Optimal Allocations in Welfare Economics and Fully Localized Minimizers in Set-Valued Optimization). *Let $(\overline{v}, \overline{y})$ be a feasible allocation of the welfare economy Eco in (15.19) with the preference sets $L_i(y)$, and let $\overline{x} := (\overline{v}, \overline{w})$ with $\overline{w} := \sum_{i=1}^{n} \overline{y}_i - \sum_{j=1}^{m} \overline{v}_j$.*

Then the following equivalence relationships hold:

(i) *$(\overline{v}, \overline{y})$ is a local weak Pareto optimal allocation of Eco if and only if $(\overline{x}, \overline{y})$ is a fully localized weak minimizer of the set-valued optimization problem (15.23) with respect to the preference mapping L defined in (15.24).*
(ii) *$(\overline{v}, \overline{y})$ is a local strict Pareto optimal allocation of Eco if and only if $(\overline{x}, \overline{y})$ is a fully localized minimizer of the set-valued optimization problem (15.23) with respect to the preference mapping L given in (15.24).*
(iii) *$(\overline{v}, \overline{y})$ is a local strong Pareto optimal allocation of Eco if and only if it is a fully localized strong minimizer of the set-valued optimization problem (15.23) with respect to the preference mapping L defined in (15.24).*

In Sect. 12.11 necessary optimality conditions for set-valued optimization problems of type (15.21) are shown. Now, we will use Theorem 12.11.3 to establish an extended version of the **second fundamental theorem of welfare economics** for local weak Pareto, strict Pareto, and strong Pareto optimal allocations of the economy *Eco* under the asymptotic closedness condition (see Definition 12.11.1) supposed at the corresponding optimal allocation on certain sets involved in the economy. The following theorem is shown in [28, Theorem 4.3] and ensures the existence of a marginal price supporting the local Pareto-type optimal allocations (see Definition 15.3.2) that are formalized via the limiting normal cone (3.25) computed at the optimal allocation in consideration. This theorem is derived directly from the corresponding first-order necessary conditions for fully localized minimizers of constraint set-valued problems presented in Theorem 12.11.3 (see [28, Theorem 3.4]) and based on the extended extremal principle in Theorem 5.5.10.

Theorem 15.3.5 (Extended Second Welfare Theorem for Local Pareto-Type Optimal Allocations). *Let $(\overline{v}, \overline{y})$ be a local optimal allocation of economy (15.19) in the senses listed below with respect to the preference sets $L_i(y)$ under the* ***local satiation requirement****:*

$$\overline{y}_i \in \operatorname{cl} L_i(\overline{y}) \quad \text{for all} \quad i = 1, \cdots, n. \tag{15.25}$$

Suppose that the commodity space E is Asplund, that the sets $PS_1, \cdots, PS_m$, and W are closed around the points in consideration, and that one of the sets

$$\operatorname{cl} L_i(\overline{y}),\ i = 1, \cdots, n; \quad PS_j, j = 1, \cdots, m; \quad and \quad W \tag{15.26}$$

is SNC at $\overline{y}_i, \overline{v}_j$ *, and* $\overline{w} = \sum_{i=1}^{n} \overline{y}_i - \sum_{j=1}^{m} \overline{v}_j$ *, respectively. Then there exists a marginal price* $0 \neq p^* \in E^*$ *satisfying the conditions*

$$-p^* \in N(\overline{x}_i; \operatorname{cl} L_i(\overline{y})),\ i = 1, \cdots, n,$$

$$p^* \in N(\overline{v}_j; PS_j),\ j = 1, \cdots, m,$$

$$p^* \in N(\overline{w}; W)$$

in each of the following cases of local optimal allocations of the economy Eco:

- $(\overline{v}, \overline{y})$ *is a local weak Pareto optimal allocation of Eco provided that all the sets* $L_i(\overline{y})$*,* $i = 1, \cdots, n$*, are asymptotically closed at the corresponding points* $\overline{y}_i$*;*
- $(\overline{v}, \overline{y})$ *is a local strict Pareto optimal allocation of Eco provided that there is an index* $i \in \{1, \cdots, n\}$ *such that the set* $\operatorname{cl} L_i(\overline{y})$ *is asymptotically closed at* $\overline{y}_i$*;*
- $(\overline{v}, \overline{y})$ *is a local strong Pareto optimal allocation of Eco provided that one of the sets in (15.26) is asymptotically closed at the corresponding point.*

Remark 15.3.6. In order to prove Theorem 15.3.5 one needs a specification of the asymptotic closedness property (compare Definition 12.11.1) for the product sets involved into the economy *Eco* (see [28, Proposition 4.2]).

Now, we consider economies with commodity spaces E ordered by proper closed cones $C \subset E$. We recall some conditions, well-recognized in welfare economics, that imply the asymptotic closedness property of the corresponding sets. The following result shows that the asymptotic closedness property holds in ordered Banach spaces E with generating proper closed cones C, i.e., $C - C = E$.

The next proposition is proven in [28, Proposition 4.4].

Proposition 15.3.7 (Asymptotic Closedness Property in Ordered Banach Spaces). *Let* E *be a Banach space with the generating proper closed ordering cone* $C \subset E$*. Assume that* A *is a closed subset of* E *satisfying the condition*

$$A - C \subset A, \tag{15.27}$$

and consider $\overline{y} \in \operatorname{bd} A$*. Then the set* A *is asymptotically closed at* $\overline{y}$*.*

According to the conventional terminology in welfare economics (compare Sect. 5.2.2), one says that the economy *Eco* (15.19) with the commodity space E, the generating proper closed ordering cone $C \subset E$ and the preference sets $L_i(y)$ satisfies:

- the **implicit free disposal of commodities** if

$$\operatorname{cl} W - C \subset \operatorname{cl} W \tag{15.28}$$

- the **free disposal of production** if

$$\operatorname{cl} PS_j - C \subset \operatorname{cl} PS_j \quad \text{for some} \quad j \in \{1, \cdots, m\}; \tag{15.29}$$

- the **desirability condition** if

$$\operatorname{cl} L_i(\overline{y}) + C \subset \operatorname{cl} L_i(\overline{y}) \quad \text{for some} \quad i \in \{1, \cdots, n\}. \tag{15.30}$$

The following corollary (see [28, Corollary 4.5]) is a consequence of Theorem 15.3.5 and employs the result of Proposition 15.3.7 providing in this way efficient implementations of the established extended version of the second welfare theorem for the cases of local strict Pareto and strong Pareto optimal allocations of nonconvex economies with ordered commodity spaces.

Corollary 15.3.8 (Extended Second Welfare Theorem for Local Strict Pareto and Strong Pareto Optimal Allocations with Ordered Commodities). *In addition to the assumptions of Theorem 15.3.5, assume that the commodity space E is generated by its proper closed ordering cone $C \subset E$. Then there is a positive marginal price $p^* \in C^+ \setminus \{0\}$ satisfying the conditions of the second welfare theorem (Theorem 15.3.5) in each of the following cases:*

- *$(\overline{v}, \overline{y})$ is a local strict Pareto optimal allocation of the economy Eco where the desirability condition (15.30) with respect to its preferences is fulfilled;*
- *$(\overline{v}, \overline{y})$ is a local strong Pareto optimal allocation of the economy Eco where either the implicit free disposal of commodities (15.28), or the free disposal of production (15.29), or the desirability condition (15.30) is fulfilled.*

15.4 Robustness for Vector-Valued Optimization Problems

In this section we apply solution concepts for set-valued optimization problems based on set approach, especially we use the set order relations introduced by Kuroiwa [348,349,351] (see Definition 2.6.9) in order to study robustness for vector-valued optimization problems.

Dealing with uncertainty in vector optimization is very important in many applications. On the one hand side, most real world optimization problems are contaminated with uncertain data, especially traffic optimization, scheduling problems, portfolio optimization, network flow and network design problems. On the other hand side, many real world optimization problems require the minimization of multiple conflicting objectives (see [543]), e.g. the maximization of the expected return versus the minimization of risk in portfolio optimization, the minimization of production time versus the minimization of the cost of manufacturing equipment, or the maximization of tumor control versus the minimization of normal tissue complication in radiotherapy treatment design.

This leads us to a concept of robustness for vector-valued optimization problems introduced and studied in a paper by Ehrgott, Ide and Schöbel [159]. In this section we follow the explanations in [159]. In [159] the authors extend the well-known concept of minmax robustness from single objective to multiobjective optimization. In single objective uncertain optimization this concept has been extensively studied, see e.g. [44, 173, 540] for older contributions and [43] for a recent survey. Detailed overviews of the existing robustness concepts for vector-valued optimization problems can be found in [541] and [130]. Furthermore, it is interesting to mention that another line of research investigates the use of multiobjective optimization techniques, especially scalarization methods, in order to model uncertain single objective optimization problems, see, e.g., [556] and [328].

For optimization problems with uncertain data it is typically that at the time when it is solved these data are not completely known. It is very important to estimate the effects of this uncertainty and so it is necessary to evaluate how sensitive an optimal solution is to perturbations of the input data. One way to deal with this question is **sensitivity analysis** (for an overview see [526]). Sensitivity analysis is a posteriori approach and provides ranges for input data within which a solution remains feasible or optimal. It does not, however, provide a course of action for changing a solution should the perturbation be outside this range. In contrast, **stochastic programming** (see Birge and Louveaux [50] for an introduction) and **robust optimization techniques** take the uncertainty into account during the optimization process. Depending from the concrete application one can decide whether robust or stochastic optimization is the more appropriate way of dealing with uncertainty.

For example, the emergency department with landing place for rescue helicopters in a ski resort should be chosen in such a way that the flight time to all ski slopes in the resort that are to be protected is minimized in the worst case, even though flight times are uncertain due to unknown weather conditions in the future. Similarly, if a aircraft-schedule of an airline is to be determined, one would want to be able to treat as many passengers as possible in a cost-effective manner, even though the exact number of passengers is not known at the time the schedule is fixed.

An important approach of dealing with such optimization problems is described in the concept of robustness: Instead of assuming that all data are known, one allows different scenarios for the input parameters and looks for a solution that works well in every uncertain scenario. Robust optimization is a growing up field of research, we refer to Ben-Tal, L. El Ghaoui, Nemirovski [43], Kouvelis and Yu [336] for an overview of results and applications for the most prominent concepts. Several other concepts of robustness were introduced more recently, e.g. the concept of light robustness by Fischetti and Monaci [186] or the concept of recovery-robustness in Liebchen et al. [388]. A scenario-based approach is suggested in Goerigk and Schöbel [212]. In all these approaches, the uncertain optimization problem is replaced by a deterministic version, called the *robust counterpart* of the uncertain problem.

For specific robustness concepts, the connection between uncertain scalar optimization problems and an associated (deterministic) multiple objective counterpart were observed by several authors. Kouvelis and Sayin [335, 522] use this relation to develop efficient solution methods for bi- and multiple objective discrete optimization problems based on algorithms that were originally developed to solve uncertain scalar optimization problems. They focus on two classical robustness concepts, see also [336].

Perny et.al. [476] use a multiple objective counterpart to introduce a robustness measure based on the Lorenz dominance rule in the context of minimum spanning tree and shortest path problems. From the stochastic programming perspective, a multiple objective counterpart for a two-stage stochastic programming problem was introduced in Gast [195] and used to interrelate stochastic programming models with the concept of recoverable robustness, see Stiller [544]. A critical analysis is given in Hites et.al. [264] who give a qualitative description of the similarities and differences between the two modeling paradigms.

In the literature, two main ways of dealing with optimization problems with uncertainty are suggested. In **stochastic optimization**, knowledge about the probability distribution of the uncertain data is assumed. Some authors use a probabilistic approach, for example, to minimize the probability that a solution is - in some sense - not good or not feasible. Other authors use the expected value of some cost function depending on the realization of the uncertain data as the objective function.

While in stochastic optimization a solution is desired which is good on average, a robust solution hedges against the worst case that may happen. Hence, different from stochastic optimization, no probability distribution is needed in **robust optimization**. There are different ways of defining robustness in the literature. Here, we focus on several models where uncertainties are assumed to occur in the objective function.

First, we formulate a **scalar optimization problem with uncertainties**. Throughout this section, let $\mathcal{U} \subseteq \mathbb{R}^m$ be the uncertainty set reflecting the potential scenarios that may occur. One could think of $\xi \in \mathcal{U}$ being real numbers or real vectors. Furthermore, let $f : \mathbb{R}^n \times \mathcal{U} \to \mathbb{R}$. Then an **uncertain scalar optimization problem** (uncertain OP) $P(\mathcal{U})$ is defined as a family of optimization problems

$$(P(\xi), \xi \in \mathcal{U}), \tag{15.31}$$

where for a given $\xi \in \mathcal{U}$ the optimization problem $(P(\xi))$ is given by

$$\begin{aligned} &f(x, \xi) \to \min \\ &\text{s.t. } x \in S \subseteq \mathbb{R}^n. \end{aligned} \tag{$P(\xi)$}$$

At the time the uncertain OP (15.31) has to be solved, it is not known which value $\xi \in \mathcal{U}$ is going to be realized.

An uncertain scalar optimization problem $P(\mathcal{U})$ is defined as a family of optimization problems with different objective functions defined by each scenario $\xi \in \mathcal{U}$. In order to clarify how it is possible to evaluate a feasible solution we are looking for feasible solutions $x \in S$ that are robust optimal solutions of the uncertain optimization problem. For scalar optimization problems, many different approaches to model robustness have been suggested in the literature. **Minmax robustness** (also called **strict robustness**) has been first mentioned by Soyster [540] and then formalized and analyzed by Ben-Tal, El Ghaoui, and Nemirovski in numerous publications, see e.g. [44] for early contributions and [43] for an extensive collection of results. The idea is that the worst possible objective function value is minimized in order to get a solution that is "good enough" even in the worst case scenario. Furthermore, constraints have to be satisfied for every scenario $\xi \in \mathcal{U}$. Then the **strictly robust counterpart** of the uncertain optimization problem $(P(\xi), \xi \in \mathcal{U})$ is defined by

$$\begin{aligned} \rho_{RC}(x) &= \max_{\xi \in \mathcal{U}} f(x, \xi) \to \min \\ \text{s.t. } & x \in S \subseteq \mathbb{R}^n. \end{aligned} \tag{RC}$$

In Ehrgott, Ide, Schöbel [159] the concept of minmax robustness is extended to **vector-valued optimization problems**. We derive corresponding results in more general settings (compare Ide, Köbis, Kuroiwa, Schöbel, Tammer [273]).

Unless otherwise mentioned, in the following we suppose that X is a linear space, Y a linear topological space and $C \subset Y$ is a proper closed convex and pointed cone.

In deterministic vector optimization one studies the problem (see Sect. 2.4)

$$\operatorname{Min}(f(S), C) \tag{VOP}$$

with a vector-valued objective function $f : X \to Y$, and a feasible set $S \subset X$.

In the following we assume that the objective function f may depend on scenarios ξ which are unknown or uncertain. As in uncertain single objective optimization, for an uncertainty set $\mathcal{U} \subseteq \mathbb{R}^m$, an **uncertain vector-valued optimization problem** $P(\mathcal{U})$ is given as the family

$$(P(\xi), \xi \in \mathcal{U})$$

of vector-valued optimization problems

$$\operatorname{Min}(f(S, \xi), C) \tag{$P(\xi)$}$$

with the objective function $f : X \times \mathcal{U} \to Y$, a feasible set $S \subseteq X$ and the notation (for $\xi \in \mathcal{U}$)

$$f(S, \xi) := \{f(x, \xi) \mid x \in S\}.$$

We call $\xi \in \mathcal{U}$ a scenario and $(P(\xi))$ an instance of $P(\mathcal{U})$.

Naturally, the issue of uncertain data affects these vector-valued optimization problems as it does in scalar optimization. Therefore, being able to find robust solutions would be very valuable in many applications. Research in this area deals with Branke's concept of robustness for single objective optimization [77]. Furthermore, Deb and Gupta [133] introduced certain concepts of robustness for vector-valued optimization problems.

Given an uncertain vector-valued optimization problem $P(\mathcal{U})$, the same question as in single objective optimization arises, namely, how to evaluate feasible solutions $x \in S$. In uncertain vector-valued optimization problems, we cannot evaluate solutions by just taking the worst case over all scenarios because we obtain a vector of objective values for each scenario. For each $x \in X$ the set of objective values of it is given by

$$F(x) := \{f(x,\xi) \mid \xi \in \mathcal{U}\} \subseteq Y. \tag{15.32}$$

Dealing with an uncertain vector-valued optimization problem $P(\mathcal{U})$ leads us to the following set-valued optimization problem with an objective map $F : X \rightrightarrows Y$ given in (15.32) and an order relation $\preceq$ (see Sect. 2.6.2):

$$\preceq -\text{minimize} \quad F(x), \quad \text{subject to} \quad x \in S. \tag{$SP - \preceq$}$$

Let $C \subset Y$ be a proper closed convex and pointed cone. Furthermore, we assume that F is compact-valued. In the following we will discuss several concepts of robustness for vector optimization problems based on the set order relations introduced in Sect. 2.6.2 (see Ide, Köbis, Kuroiwa, Schöbel, Tammer [273]).

15.4.1 $\preceq^u_C$-Robustness

With these notions we are able to apply the solution concept based on the upper set less order relation $\preceq^u_C$ introduced in Definition 2.6.9 for $A, B \subset Y$:

$$A \preceq^u_C B \Longleftrightarrow A \subseteq B - C.$$

Ehrgott, Ide, Schöbel [159, Definition 3.1] introduced robust efficient elements (minmax robust efficiency) for the case $X = \mathbb{R}^n$, $Y = \mathbb{R}^q$, $C = \mathbb{R}^q_+$: A feasible solution $\overline{x} \in S$ robust efficient, if $F(\overline{x}) - \mathbb{R}^q_+$ does not contain any other set $F(x)$ with $x \neq \overline{x} \in S$. Furthermore, Ehrgott, Ide, Schöbel formulated in [159, Definition 3.1] robust weakly efficient and robust strictly efficient elements. We will extend the concepts given in [159] to more general cones $C \subset Y$ using the upper set less order relation $\preceq^u_C$.

Definition 15.4.1 ($\preceq^u_C$-Robustness). Given an uncertain vector-valued optimization problem $P(\mathcal{U})$, a solution $x^0 \in S$ is called **$\preceq^u_C$-robust** for $P(\mathcal{U})$, if there is no $x \in S \setminus \{x^0\}$ such that

$$F(x) \preceq^u_C F(x^0).$$

The concepts of robust weak efficiency and robust efficiency introduced in Ehrgott, Ide, Schöbel [159, Definition 3.1] inspire us to extent the concept of upper set less order relation $\preceq^u_C$ introduced in Definition 2.6.9 for a proper closed convex and pointed cone $C \subset Y$ and $A, B \subset Y$ to the case of general nonempty sets $Q \subset Y$. Suppose that $Q \subset Y$ is a proper set with $C \subseteq \operatorname{cl} Q$ and $\operatorname{cl} Q \cap (-\operatorname{cl} Q) = \{0\}$. Under these assumptions we introduce an order relation with respect to Q analogously to the upper set less order relation $\preceq^u_C$ introduced in Definition 2.6.9:

$$A \preceq^u_Q B :\Longleftrightarrow A \subseteq B - Q. \tag{15.33}$$

If we are dealing with $Q = \operatorname{int} C$ we suppose $\operatorname{int} C \neq \emptyset$.

Definition 15.4.2 ($\preceq^u_{\operatorname{int} C}$-Robustness and $\preceq^u_{C\setminus\{0\}}$-Robustness). Given an uncertain vector-valued optimization problem $P(\mathcal{U})$ we call a solution $x^0 \in S$

- $\preceq^u_{\operatorname{int} C}$**-robust** , if $\operatorname{int} C \neq \emptyset$, $Q = \operatorname{int} C$ in (15.33) and there is no $x \in S \setminus \{x^0\}$ such that

$$F(x) \preceq^u_Q F(x^0);$$

- $\preceq^u_{C\setminus\{0\}}$**-robust** , if $Q = C \setminus \{0\}$ in (15.33) and there is no $x \in S \setminus \{x^0\}$ such that

$$F(x) \preceq^u_Q F(x^0);$$

Remark 15.4.3. In the case of $\preceq^u_C$-robust solutions in the sense of Definition 15.4.1 we have $Q = C$ in (15.33).

Remark 15.4.4. The motivation behind the concept of $\preceq^u_Q$-robustness is the following: When comparing sets with the upper set less order relation $\preceq^u_Q$ (with $Q = C$, $Q = \operatorname{int} C$ or $Q = C \setminus \{0\}$), the upper bounds of these sets, i.e., the "worst cases", are considered. Minimizing these worst cases is closely connected to the concept of minmax robust efficiency (see [159, Definition 3.1]) where one wants to minimize the objective function in the worst case. This risk averse approach would reflect a decision makers strategy to hedge against a worst case and is rather pessimistic.

Remark 15.4.5. The first scenario-based concept to uncertain multi-objective optimization, or *minmax*-robustness adapted to multi-objective optimization, has been introduced by Kuroiwa and Lee [355] and studied in [356]. In [355, 356] robust solutions of multi-objective optimization problems are introduced in the following way. The authors propose to consider the robust counterpart to $\mathcal{P}(\mathcal{U})$

$$\operatorname{Min}(f^{\mathcal{U}}_{RC}(S), \mathbb{R}^q_+), \tag{15.34}$$

$$f^{\mathcal{U}}_{RC}(x) := \begin{pmatrix} \max_{\xi_1 \in \mathcal{U}_1} f_1(x, \xi_1) \\ \cdots \\ \max_{\xi_q \in \mathcal{U}_q} f_q(x, \xi_q) \end{pmatrix}, \tag{15.35}$$

with functionals $f_i : \mathbb{R}^n \times \mathcal{U}_i \to \mathbb{R}$ for $i = 1, \dots, q$ and the convex and compact uncertainty sets $\mathcal{U} := (\mathcal{U}_1, \dots, \mathcal{U}_q)$ ($\mathcal{U}_i \subseteq \mathbb{R}^m$, $i = 1, \dots, q$). In [355], solutions to (15.34) are called *robust*.

Note that in [159] the authors pointed out that this concept differs from the concept of minmax robust efficiency.

In deterministic vector-valued optimization there is a relationship between weak minimality, minimality and strict minimality. This relationship also holds for robust minimality.

Lemma 15.4.6. *Let $P(\mathcal{U})$ be an uncertain vector-valued optimization problem, where $C \subset Y$ is a proper closed convex and pointed cone with* $\operatorname{int} C \neq \emptyset$. *Then we have:*

$$x \text{ is } \preceq^u_C\text{-robust} \;\Rightarrow\; x \text{ is } \preceq^u_{C\setminus\{0\}}\text{-robust} \Rightarrow\; x \text{ is } \preceq^u_{\operatorname{int} C}\text{-robust}\,.$$

Proof. The assertion holds because of $\operatorname{int} C \subset C \setminus \{0\} \subset C$. □

The robust counterpart of an uncertain vector-valued optimization problem is the problem of identifying all $x \in S$ which are $\preceq^u_{\operatorname{int} C}$-robust , $\preceq^u_{C\setminus\{0\}}$-robust , or $\preceq^u_C$-robust for $P(\mathcal{U})$.

The next lemma is shown by Ehrgott, Ide, Schöbel [159] for the case $X = \mathbb{R}^n$, $Y = \mathbb{R}^q$, $C = \mathbb{R}^q_+$ and gives important properties of the order relation that are used in order to derive solution procedures.

Lemma 15.4.7. *Consider the uncertain vector-valued optimization problem $P(\mathcal{U})$. Then it holds for all x', $\overline{x} \in S$ and for $Q = \operatorname{int} C$ ($Q = C \setminus \{0\}$, $Q = C$, respectively),*

$$F(x') \subseteq F(\overline{x}) - Q \iff F(x') - C \subseteq F(\overline{x}) - Q \tag{15.36}$$

$$F(x') \subseteq F(\overline{x}) - Q \iff \forall \xi \in \mathcal{U}\ \exists \eta \in \mathcal{U} :\ f(x', \xi) \in f(\overline{x}, \eta) - Q. \tag{15.37}$$

Proof. (15.36): (i) "$\Longrightarrow$": Suppose $F(x') \subseteq F(\overline{x}) - Q$. Then $f(x', \xi) \in F(\overline{x}) - Q$ for every $\xi \in \mathcal{U}$. This implies

$$f(x', \xi) - c \in F(\overline{x}) - Q$$

for every $\xi \in \mathcal{U}$ and for all $c \in C$ because of $C \subset \operatorname{cl} Q$. So we can conclude

$$F(x') - C \subseteq F(\overline{x}) - Q.$$

(ii) "$\Longleftarrow$": Suppose $F(x') - C \subseteq F(\overline{x}) - Q$. Because of $0 \in C$ we have

$$F(x') \subseteq F(\overline{x}) - Q.$$

(15.37): (i) "$\Longrightarrow$": Suppose the contrary. Then

$$\exists \xi \in \mathcal{U}\ \forall \eta \in \mathcal{U} : f(x',\xi) \notin f(\overline{x},\eta) - Q \Longrightarrow \exists \xi \in \mathcal{U} : \ f(x',\xi) \notin F(\overline{x}) - Q$$
$$\Longrightarrow F(x') \nsubseteq F(\overline{x}) - Q.$$

(ii) "$\Longleftarrow$": Suppose the contrary. Then

$$\exists \xi \in \mathcal{U} : f(x',\xi) \notin F(\overline{x}) - Q \Longrightarrow \exists \xi \in \mathcal{U}\ \forall \eta \in \mathcal{U} : f(x',\xi) \notin f(\overline{x},\eta) - Q.$$

□

Under the additional assumption that $X = \mathbb{R}^n$, $Y = \mathbb{R}^q$, $C = \mathbb{R}^q_+$ in Ehrgott, Ide, Schöbel [159, Lemma 3.4] the following result is derived:

Lemma 15.4.8. *Consider the uncertain vector-valued optimization problem $P(\mathcal{U})$ with $X = \mathbb{R}^n$, $Y = \mathbb{R}^q$, $C = \mathbb{R}^q_+$. Then it holds for all $x', \overline{x} \in S$ and for $Q = \operatorname{int} C$ ($Q = C \setminus \{0\}$, $Q = C$, respectively),*

$$F(x') \subseteq F(\overline{x}) - Q \Longrightarrow \sup_{\xi\in\mathcal{U}} f_i(x',\xi) \leq \sup_{\eta\in\mathcal{U}} f_i(\overline{x},\eta)\ \forall i \in \{1,\cdots k\}. \tag{15.38}$$

If $\max_{\xi\in\mathcal{U}} f_i(x,\xi)$ exists for all $x \in S$, then

$$F(x') \subseteq F(\overline{x}) - \operatorname{int} C \Longrightarrow \max_{\xi\in\mathcal{U}} f_i(x',\xi) < \max_{\eta\in\mathcal{U}} f_i(\overline{x},\eta)\ \forall i \in \{1,\cdots k\}. \tag{15.39}$$

Proof. (15.38): Let $F(x') \subseteq F(\overline{x}) - Q$. Suppose the contrary. Then there exists an $i \in \{1,\cdots q\}$ such that

$$\sup_{\xi\in\mathcal{U}} f_i(x',\xi) > \sup_{\eta\in\mathcal{U}} f_i(\overline{x},\eta).$$

Since the inequality is strict there exists a scenario $\xi^* \in \mathcal{U}$ such that $f_i(x',\xi^*) > \sup_{\eta\in\mathcal{U}} f_i(\overline{x},\eta)$. This means that for all $\eta \in \mathcal{U}$ it holds $f(x',\xi^*) \notin f(\overline{x},\eta) - Q$. This yields

$$F(x') \nsubseteq F(\overline{x}) - Q,$$

because of (15.37), a contradiction.

(15.39): Assume $\max_{\xi\in\mathcal{U}} f_i(x,\xi)$ exists for all $x \in S$ and let $F(x') \subseteq F(\overline{x}) - \operatorname{int} C$. Suppose the contrary. Then there exists an $i \in \{1,\cdots q\}$ such that

$$\max_{\xi\in\mathcal{U}} f_i(x',\xi) \geq \max_{\eta\in\mathcal{U}} f_i(\overline{x},\eta).$$

This inequality implies that there exists a scenario $\xi^* \in \mathcal{U}$ such that $f_i(x', \xi^*) \geq \max_{\eta \in \mathcal{U}} f_i(\overline{x}, \eta)$ and so we get for all $\eta \in \mathcal{U}$ that $f(x', \xi^*) \notin f(\overline{x}, \eta) - \operatorname{int} C$. This yields

$$F(x') \not\subseteq F(\overline{x}) - \operatorname{int} C,$$

because of (15.37), a contradiction. □

In the following we will see that for special structures of set-valued optimization problems we are able to use scalarization methods (compare Chap. 5) in order to derive solution procedures.

Lemma 15.4.7 is used in Ehrgott, Ide, Schöbel [159] for deriving solution procedures for uncertain vector optimization problems in the special case $X = \mathbb{R}^n$, $Y = \mathbb{R}^q$, $C = \mathbb{R}^q_+$ in (VOP). Using Lemma 15.4.7, we apply the knowledge of deterministic vector optimization (compare Sect. 2.4) and single objective robust optimization for deriving solution procedures for generating $\preceq^u_{C\setminus\{0\}}$-robust solutions of an uncertain vector optimization problem $P(\mathcal{U})$. The most common approach for computing properly minimal solutions for a deterministic vector optimization problem in finite dimensional image spaces is the weighted sum scalarization (see Definition 2.4.4, (a)). The general idea is to form a scalar optimization problem by multiplying each objective function with a positive weight and summing up the weighted objectives. The scalarized problem (P_{y^*}) for a given (deterministic) vector optimization problem $\operatorname{Min}(f(S), C)$ (see (VOP)) with $y^* \in C^+ \setminus \{0\}$ (see Sect. 2.4) is

$$\min_{x \in S} y^*(f(x)). \tag{P_{y^*}}$$

We now reduce the robust counterpart of an uncertain vector-valued optimization problem to a single objective uncertain optimization problem in order to be able to generate $\preceq^u_{C\setminus\{0\}}$-robust solutions by computing robust optimal solutions of the uncertain single objective problem applying Theorem 2.4.11, (b). To this end, we introduce the robust version of the weighted sum scalarization problem of an uncertain vector-valued optimization problem $P(\mathcal{U})$ as

$$\min_{x \in S} \sup_{\xi \in \mathcal{U}} y^*(f(x, \xi)). \tag{$P\mathcal{U}_{y^*}$}$$

The following result is shown by Ehrgott, Ide, Schöbel [159, Theorem 4.3] for the case $X = \mathbb{R}^n$, $Y = \mathbb{R}^q$, $C = \mathbb{R}^q_+$. We present the result in a more general setting where X is a linear space, Y is a linear topological space and $C \subset Y$ is a proper closed convex and pointed cone.

Theorem 15.4.9. *Consider an uncertain vector optimization problem $P(\mathcal{U})$. The following statements hold:*

(a) *If* $x^0 \in S$ *is a unique optimal solution of* $(P\mathcal{U}_{y^*})$ *for some* $y^* \in C^+ \setminus \{0\}$, *then* x^0 *is* $\preceq^u_C$*-robust for* $P(\mathcal{U})$.
(b) *If* $x^0 \in S$ *is an optimal solution of* $(P\mathcal{U}_{y^*})$ *for some* $y^* \in C^\#$ *and* $\max_{\xi \in \mathcal{U}} y^*(f(x,\xi))$ *exists for all* $x \in S$, *then* x^0 *is* $\preceq^u_{C\setminus\{0\}}$*-robust for* $P(\mathcal{U})$.
(c) *If* $x^0 \in S$ *is an optimal solution of* $(P\mathcal{U}_{y^*})$ *for some* $y^* \in C^+ \setminus \{0\}$ *and* $\max_{\xi \in \mathcal{U}} y^*(f(x,\xi))$ *exists for all* $x \in S$, *then* x^0 *is* $\preceq^u_{\operatorname{int} C}$*-robust for* $P(\mathcal{U})$.

Proof. Suppose that x^0 is not $\preceq^u_C$-robust ($\preceq^u_{C\setminus\{0\}}$-robust , $\preceq^u_{\operatorname{int} C}$-robust , respectively) for $P(\mathcal{U})$. Then there exists an element $\overline{x} \in S \setminus \{x^0\}$ such that

$$F(\overline{x}) \subseteq F(x^0) - Q, \tag{15.40}$$

for $Q = C$, ($Q = (C \setminus \{0\})$, $Q = \operatorname{int} C$, respectively). This implies

$$\forall \xi \in \mathcal{U}\ \exists \eta \in \mathcal{U} :\ f(\overline{x},\xi) \in f(x^0,\eta) - Q.$$

taking into account Lemma 15.4.7, (15.37).

Choose now $y^* \in C^+\setminus\{0\}$ for $Q = C$ ($y^* \in C^\#$ for $Q = C\setminus\{0\}$, $y^* \in C^+\setminus\{0\}$ for $Q = \operatorname{int} C$, respectively) arbitrary but fixed.

$$\Longrightarrow \forall \xi \in \mathcal{U}\ \exists \eta \in \mathcal{U} :\ y^*(f(\overline{x},\xi)) \leq\ (<, <, \textit{respectively})\ y^*(f(x^0,\eta))$$

$$\Longleftrightarrow \forall \xi \in \mathcal{U} :\ y^*(f(\overline{x},\xi)) \leq\ (<, <, \textit{respectively})\ \sup_{\eta' \in \mathcal{U}} y^*(f(x^0,\eta'))$$

$$\Longleftrightarrow \sup_{\xi' \in \mathcal{U}} y^*(f(\overline{x},\xi')) \leq\ (<, <, \textit{respectively})\ \sup_{\eta' \in \mathcal{U}} y^*(f(x^0,\eta')).$$

The last inequalities hold because for (b) and (c) $\max_{\xi' \in \mathcal{U}} y^*(f(\overline{x},\xi))$ exists. But this means that x^0 is not the unique optimal (an optimal, an optimal, respectively) solution of $(P\mathcal{U}_{y^*})$ for $y^* \in C^+ \setminus \{0\}$ ($y^* \in C^\#$, $y^* \in C^+ \setminus \{0\}$, respectively). □

Remark 15.4.10. In Theorem 15.4.9 (b) we consider $y^* \in C^\#$. Under our assumptions concerning the cone C and if we assume additionally $Y = \mathbb{R}^q$ we have $C^\# \neq \emptyset$ (compare [214, Theorem 2.2.12], [214, Example 2.2.16] and Sect. 2.2). Moreover, if Y is a Hausdorff locally convex space, $C \subset Y$ is a proper convex cone and C has a base B with $0 \notin \operatorname{cl} B$, then $C^\# \neq \emptyset$ (compare [214, Theorem 2.2.12]).

Based on Theorem 15.4.9, Ehrgott, Ide and Schöbel [159] derived the following algorithm for computing $\preceq^u_C$-robust ($\preceq^u_{C\setminus\{0\}}$-robust, $\preceq^u_{\operatorname{int} C}$-robust, respectively) solutions of the uncertain vector-valued optimization problem $P(\mathcal{U})$ for the special case $X = \mathbb{R}^n$, $Y = \mathbb{R}^q$ and $C = \mathbb{R}^q_+$. We present the algorithm for the more general case where X is a linear space, Y is a linear topological space and $C \subset Y$ is a proper closed convex and pointed cone.

Algorithm 1 for calculating ($\preceq^u_C \mid \preceq^u_{C\setminus\{0\}} \mid \preceq^u_{\text{int}\, C}$)-robust solutions of $P(\mathcal{U})$:

Input: Uncertain vector-valued problem $P(\mathcal{U})$, solution sets $\text{Opt}_{\preceq^u_C} = \text{Opt}_{\preceq^u_{C\setminus\{0\}}} = \text{Opt}_{\preceq^u_{\text{int}\, C}} = \emptyset$.

Step 1: Choose a set $\overline{C} \subset C^+ \setminus \{0\}$.

Step 2: If $\overline{C} = \emptyset$: **STOP. Output:** Set of $\preceq^u_C$-robust solutions $\text{Opt}_{\preceq^u_C}$, set of $\preceq^u_{C\setminus\{0\}}$-robust solutions $\text{Opt}_{\preceq^u_{C\setminus\{0\}}}$, set of $\preceq^u_{\text{int}\, C}$-robust solutions $\text{Opt}_{\preceq^u_{\text{int}\, C}}$ for $P(\mathcal{U})$.

Step 3: Choose $y^* \in \overline{C}$. Set $\overline{C} := \overline{C} \setminus \{y^*\}$.

Step 4: Find an optimal solution x^0 of $(P\mathcal{U}_{y^*})$.

a) If x^0 is a unique optimal solution of $(P\mathcal{U}_{y^*})$, then x^0 is $\preceq^u_C$-robust for $P(\mathcal{U})$, thus

$$\text{Opt}_{\preceq^u_C} := \text{Opt}_{\preceq^u_C} \cup \{x^0\}.$$

b) If $\max_{\xi\in\mathcal{U}} y^*(f(x,\xi))$ exists for all $x \in S$ and $y^* \in C^\#$, then x^0 is $\preceq^u_{C\setminus\{0\}}$-robust for $P(\mathcal{U})$, thus

$$\text{Opt}_{\preceq^u_{C\setminus\{0\}}} := \text{Opt}_{\preceq^u_{C\setminus\{0\}}} \cup \{x^0\}.$$

c) If $\max_{\xi\in\mathcal{U}} y^*(f(x,\xi))$ exists for all $x \in S$, then x^0 is $\preceq^u_{\text{int}\, C}$-robust for $P(\mathcal{U})$, thus

$$\text{Opt}_{\preceq^u_{\text{int}\, C}} := \text{Opt}_{\preceq^u_{\text{int}\, C}} \cup \{x^0\}.$$

Step 5: Go to Step 2.

The following algorithm is an interactive procedure for solving the uncertain vector-valued optimization problem $P(\mathcal{U})$. In this approach we reduce the scalarized problem to a 1-parametric optimization problem. The algorithm works in the case that X is a linear space, Y is a linear topological space and $C \subset Y$ is a proper closed convex and pointed cone.

Algorithm 2 for calculating a single desired ($\preceq^u_C \mid \preceq^u_{C\setminus\{0\}} \mid \preceq^u_{\text{int}\, C}$)-robust solution of $P(\mathcal{U})$:

Input: Uncertain vector-valued problem $P(\mathcal{U})$.

Step 1: Choose a nonempty set $\overline{C} \subset C^+ \setminus \{0\}$.

Step 2: Choose $\bar{y}^* \in \overline{C}$.

Step 3: Find an optimal solution x^0 of $(P\mathcal{U}_{\bar{y}^*})$.

a) If x^0 is a unique optimal solution of $(P\mathcal{U}_{\bar{y}^*})$, then x^0 is $\preceq^u_C$-robust for $P(\mathcal{U})$.
b) If $\max_{\xi\in\mathcal{U}} \bar{y}^*(f(x,\xi))$ exists for all $x \in S$ and $\bar{y}^* \in C^\#$, then x^0 is $\preceq^u_{C\setminus\{0\}}$-robust for $P(\mathcal{U})$.
c) If $\max_{\xi\in\mathcal{U}} \bar{y}^*(f(x,\xi))$ exists for all $x \in S$, then x^0 is $\preceq^u_{\text{int}\, C}$-robust for $P(\mathcal{U})$.

If x^0 is accepted by the decision-maker, then **Stop**. **Output:** x^0. Otherwise, go to Step 4.

Step 4: Put $k = 0$, $t_0 = 0$. Choose $\hat{y}^* \in \overline{C}$, $\hat{y}^* \neq \bar{y}^*$. Go to Step 5.

Step 5: Choose t_{k+1} with $t_k < t_{k+1} \leq 1$ and compute an optimal solution x^{k+1} of

$$\min_{x \in S} \sup_{\xi \in \mathcal{U}} (\bar{y}^* + t_{k+1}(\hat{y}^* - \bar{y}^*))(f(x,\xi)) \qquad (P\mathcal{U}_{\bar{y}^*+t_{k+1}(\hat{y}^*-\bar{y}^*)})$$

and use x^k as starting point. If an optimal solution of $(P\mathcal{U}_{\bar{y}^*+t_{k+1}(\hat{y}^*-\bar{y}^*)})$ cannot be found for $t > t_k$, then go to Step 2. Otherwise, go to Step 6.

Step 6: The point x^{k+1} is to be evaluated by the decision-maker. If it is accepted by the decision-maker, then **Stop**. **Output:** x^{k+1}. Otherwise, go to Step 7.

Step 7: If $t_{k+1} = 1$, then go to Step 2. Otherwise, set $k = k + 1$ and go to Step 5.

Furthermore, another approach based on scalarization by means of nonlinear functionals of type (5.5) is given in Ehrgott, Ide, Schöbel [159]. In [159] the ε-constraint method is used. In Chap. 5 we discussed the functional

$$\varphi_{A,k^0} : Y \to \overline{\mathbb{R}}, \quad \varphi_{A,k^0}(y) := \inf\{t \in \mathbb{R} \mid y \in tk^0 + A\},$$

where A is a nonempty subset of a real linear topological space Y and $k^0 \neq 0$ an element of Y (compare (5.5)).

The ε-constraint method for the case $X = \mathbb{R}^n$, $Y = \mathbb{R}^q$ and $C = \mathbb{R}^q_+$ was introduced by Haimes [230] (compare Eichfelder [161] and Ehrgott [158]). The idea of the ε-constraint method is to choose one of the functions f_i $(i \in \{1, \cdots q\})$ as objective function and to take for the other functions f_j $(j \neq i)$ upper bounds given by a vector $\varepsilon \in \mathbb{R}^q$. This ε-constraint method is a special case of this general nonlinear scalarization method by means of the functional (5.5) if we put

$$k^0 = (k^0_1, \ldots, k^0_q)^T, \quad \text{where } k^0_l := \begin{cases} 1 \text{ for } l = i \\ 0 \text{ for } l \neq i, \end{cases} \tag{15.41}$$

$$A := -\mathbb{R}^q_+ + \bar{b}, \text{ with } \bar{b} = (\bar{b}^1, \ldots, \bar{b}^q)^T, \ \bar{b}^l := \begin{cases} 0 & \text{for } l = i, \\ \varepsilon_l & \text{for } l \neq i. \end{cases} \tag{15.42}$$

For a deterministic vector-valued optimization problem (VOP), $i \in \{1, \cdots q\}$ and $\varepsilon \in \mathbb{R}^q$ the ε-constraint optimization problem is defined as

$$\min f_i(x) \qquad \text{s.t.} \quad f_j(x) \leq \varepsilon_j \ \ \forall j \neq i, \ x \in S. \qquad (P_{\varepsilon,i})$$

We now use the ε-constraint method to reduce a vector-valued uncertain optimization problem $P(\mathcal{U})$ to a single objective uncertain optimization problem. For doing this we define the robust ε-constraint version $(P\mathcal{U}_{\varepsilon,i})$ of $P(\mathcal{U})$, thus an ε-constraint problem for an uncertain vector-valued optimization problem:

$$\min \sup_{\xi \in \mathcal{U}} f_i(x, \xi) \qquad \text{s.t.} \quad f_j(x, \xi) \leq \varepsilon_j \quad \forall j \neq i, \ \forall \xi \in \mathcal{U}, \ x \in S. \qquad (P\mathcal{U}_{\varepsilon,i})$$

In the following theorem the relationships between solutions of $(P\mathcal{U}_{\varepsilon,i})$ and robust strict minimality (robust weak minimality) are presented (see [159]).

Theorem 15.4.11. *Consider an uncertain vector-valued optimization problem $P(\mathcal{U})$ with $X = \mathbb{R}^n$, $Y = \mathbb{R}^q$ and $C = \mathbb{R}^q_+$, then the following statements hold.*

(a) *If x^0 is the unique optimal solution of $(P\mathcal{U}_{\varepsilon,i})$ for some $\varepsilon \in \mathbb{R}^q$ and some $i \in \{1, \cdots, q\}$, then x^0 is $\preceq^u_C$-robust for $P(\mathcal{U})$.*
(b) *If $x^0 \in S$ is an optimal solution of $(P\mathcal{U}_{\varepsilon,i})$ for some $\varepsilon \in \mathbb{R}^q$ and $\max_{\xi \in \mathcal{U}} f_i(x, \xi)$ exists for all $x \in S$, then x^0 is $\preceq^u_{\text{int}\, C}$-robust for $P(\mathcal{U})$.*

Proof. (a): Suppose that x^0 is not $\preceq^u_C$-robust for $P(\mathcal{U})$. Then there exists an element $x' \in S$ with

$$F(x') \preceq^u_Q F(x^0),$$

where $\preceq^u_Q$ is given with respect to $Q = C = \mathbb{R}^q_+$.

With Lemma 15.4.7, (15.36) and (15.37), we get for $Q = C = \mathbb{R}^q_+$

$$\forall \xi \in \mathcal{U} \ \exists \eta \in \mathcal{U} : \ f(x', \xi) \in f(x^0, \eta) - \mathbb{R}^q_+.$$

This implies

$$\sup_{\xi \in \mathcal{U}} f(x', \xi) \leq \sup_{\eta \in \mathcal{U}} f(x^0, \eta)$$

and

$$\forall \xi \in \mathcal{U} \ \exists \eta \in \mathcal{U} : \quad f_j(x', \xi) \leq f_j(x^0, \eta) \leq \varepsilon_j \ (j \neq i).$$

This means that x' is feasible for $(P\mathcal{U}_{\varepsilon,i})$ and has an equal or better objective function value than x^0, in contradiction to the assumption that x^0 is the unique optimal solution of $(P\mathcal{U}_{\varepsilon,i})$.

(b): Suppose that x^0 is not $\preceq^u_{\text{int}\, C}$-robust for $P(\mathcal{U})$. Then there exists an element $x' \in S$ with

$$F(x') \preceq^u_Q F(x^0),$$

where $\preceq^u_Q$ is given with respect to $Q = C = \text{int}\, \mathbb{R}^q_+$.
Again, with Lemma 15.4.7, (15.36) and (15.37), we can conclude

$$\forall \xi \in \mathcal{U} \ \exists \eta \in \mathcal{U} : \ f(x', \xi) \in f(x^0, \eta) - \text{int}\, \mathbb{R}^q_+.$$

This yields

$$\max_{\xi \in \mathcal{U}} f(x', \xi) < \max_{\eta \in \mathcal{U}} f(x^0, \eta)$$

and

$$\forall \xi \in \mathcal{U} \; \exists \eta \in \mathcal{U} : \quad f_j(x', \xi) < f_j(x^0, \eta) \leq \varepsilon_j \; (j \neq i).$$

This means that x' is feasible for $(P\mathcal{U}_{\varepsilon,i})$ and has a better objective function value that x^0, in contradiction to the assumption that x^0 is the unique optimal solution of $(P\mathcal{U}_{\varepsilon,i})$. □

This leads to the following algorithm for computing $\preceq_C^u$-robust ($\preceq_{\text{int}\,C}^u$-robust) solutions of the uncertain vector-valued optimization problem $P(\mathcal{U})$ with $X = \mathbb{R}^n$, $Y = \mathbb{R}^q$ and $C = \mathbb{R}^q_+$.

Algorithm 3 for calculating ($\preceq_C^u \mid \preceq_{\text{int}\,C}^u$)-robust solutions of $P(\mathcal{U})$ based on ε-constraint method:

Input: Uncertain vector-valued problem $P(\mathcal{U})$, solution sets $\text{Opt}_{\preceq_C^u} = \text{Opt}_{\preceq_{\text{int}\,C}^u} = \emptyset$.
Step 1: Choose a set $\mathcal{E} \subset \mathbb{R}^q$.
Step 2: If $\mathcal{E} = \emptyset$: **STOP. Output:** Set of $\preceq_C^u$-robust solutions $\text{Opt}_{\preceq_C^u}$ and set of $\preceq_{\text{int}\,C}^u$-robust solutions $\text{Opt}_{\preceq_{\text{int}\,C}^u}$ for $P(\mathcal{U})$.
Step 3: Choose $\varepsilon \in \mathcal{E}$. Set $\mathcal{E} := \mathcal{E} \setminus \{\varepsilon\}$.
Step 4: For every $i \in \{1, \cdots q\}$ find an optimal solution $x^0(i)$ of $(P\mathcal{U}_{\varepsilon,i})$.

a) If $x^0(i)$ is a unique optimal solution of $(P\mathcal{U}_{\varepsilon,i})$, then $x^0(i)$ is $\preceq_C^u$-robust for $P(\mathcal{U})$, thus

$$\text{Opt}_{\preceq_C^u} := \text{Opt}_{\preceq_C^u} \cup \{x^0(i)\}.$$

b) If $\max_{\xi \in \mathcal{U}} f_i(x, \xi)$ exists for all $x \in S$, then $x^0(i)$ is $\preceq_{\text{int}\,C}^u$-robust for $P(\mathcal{U})$, thus

$$\text{Opt}_{\preceq_{\text{int}\,C}^u} := \text{Opt}_{\preceq_{\text{int}\,C}^u} \cup \{x^0(i)\}.$$

Step 5: Go to Step 2.

Remark 15.4.12. Further concepts for robustness and corresponding algorithms for solving uncertain vector-valued optimization problems $P(\mathcal{U})$ are developed in Ide, Köbis, Kuroiwa, Schöbel and Tammer [273].

Remark 15.4.13. The algorithms presented in this section can be considered as algorithms for solving special classes of set-valued optimization problems where the solution concept is given by the upper set less order relation (Definition 2.6.9).

15.4.2 $\preceq_C^l$-Robustness

In this section we use the l-type set-relation $\preceq_Q^l$ like in Definition 2.6.9 with $Q = C$, $Q = C \setminus \{0\}$ and $Q = \operatorname{int} C$, respectively, instead of $\preceq_C^l$:

$$A \preceq_Q^l B :\Longleftrightarrow A + Q \supseteq B,$$

where $A, B \subset Y$ are arbitrarily chosen sets. If we are dealing with $Q = \operatorname{int} C$ we suppose $\operatorname{int} C \neq \emptyset$. Consider again the set-valued map F given by (15.32). Using this notation we derive the new concept of $\preceq_Q^l$-robustness, defined analogously to $\preceq_Q^u$-robustness (Definitions 15.4.1 and 15.4.2):

Definition 15.4.14. Given an uncertain multi-objective optimization problem $P(\mathcal{U})$, a solution $x^0 \in S$ is called $\preceq_Q^l$-robust (with $Q = C$, $Q = C \setminus \{0\}$ and $Q = \operatorname{int} C$) for $P(\mathcal{U})$, if there is no $x \in S \setminus \{x^0\}$ such that

$$F(x) \preceq_Q^l F(x^0).$$

Remark 15.4.15. The $\preceq_Q^l$-robustness (with $Q = C$, $Q = C \setminus \{0\}$ and $Q = \operatorname{int} C$, respectively) can be interpreted as an optimistic approach. The $\preceq_Q^l$-robustness is an alternative tool for the decision maker for obtaining solutions of another type to an uncertain multi-objective optimization problem. This rather optimistic approach focuses on the lower bound of a set $F(\bar{x})$ for the comparison with another set $F(x^0)$. In particular, in the case $Q = C$, a point $x^0 \in S$ is called a $\preceq_C^l$-solution if there is no other point $\bar{x} \in S$ such that $F(x^0)$ is a subset of $F(\bar{x}) + C$. Contrary to the $\preceq_Q^u$-robustness approach, the $\preceq_Q^l$-robustness (with $Q = C$, $Q = C \setminus \{0\}$ and $Q = \operatorname{int} C$, respectively) is hence not a worst-case concept, thus the decision maker is not considered to be risk averse but risk affine. This optimistic concept thus hedges against perturbations in the best-case scenarios.

For calculating $\preceq_Q^l$-robust solutions again scalarization method is helpful, but in order to later on compute $\preceq_Q^l$-robust solutions to $P(\mathcal{U})$, we define a new weighted sum problem in a general setting:

Let $y^* \in C^+ \setminus \{0\}$ ($y^* \in C^\#$, respectively). Consider the weighted sum scalarization problem

$$\min_{x \in S} \inf_{\xi \in \mathcal{U}} y^*(f(x, \xi)). \tag{$P\mathcal{U}_{y^*}^l$}$$

Theorem 15.4.16. *Consider an uncertain vector optimization problem $P(\mathcal{U})$. The following statements hold:*

(a) *If x^0 is a unique optimal solution of $(P\mathcal{U}_{y^*}^l)$ for some $y^* \in C^+ \setminus \{0\}$, then x^0 is a $\preceq_C^l$-robust solution for $P(\mathcal{U})$.*

(b) *If* x^0 *is an optimal solution of* $(P\mathcal{U}^l_{y^*})$ *for some* $y^* \in C^\#$ *and* $\min_{\xi\in\mathcal{U}} y^*(f(x,\xi))$ *exists for all* $x \in S$*, then* x^0 *is a* $\preceq^l_{C\setminus\{0\}}$*-robust solution for* $P(\mathcal{U})$*.*

(c) *If* x^0 *is an optimal solution of* $(P\mathcal{U}^l_{y^*})$ *for some* $y^* \in C^+ \setminus \{0\}$ *and* $\min_{\xi\in\mathcal{U}} y^*(f(x,\xi))$ *exists for all* $x \in S$*, then* x^0 *is a* $\preceq^l_{\operatorname{int} C}$*-robust solution for* $P(\mathcal{U})$*.*

Proof. Suppose x^0 is not $\preceq^l_Q$-robust for $Q = C$ ($Q = C \setminus \{0\}$, $Q = \operatorname{int} C$, respectively). Consequently, there exists an $\bar{x} \in S \setminus \{x^0\}$ such that $F(\bar{x}) + Q \supseteq F(x^0)$ for $Q = C$ ($Q = C \setminus \{0\}$, $Q = \operatorname{int} C$, respectively). That is equivalent to

$$\begin{aligned}
&\forall \xi \in \mathcal{U}\ \exists \eta \in \mathcal{U} :\ f(\bar{x},\eta) + Q \ni f(x^0,\xi) \\
&\iff \forall \xi \in \mathcal{U}\ \exists \eta \in \mathcal{U} :\ f(\bar{x},\eta) \in f(x^0,\xi) - Q
\end{aligned} \tag{15.43}$$

Now choose $y^* \in C^+ \setminus \{0\}$ for $Q = C$ ($y^* \in C^\#$ for $Q = C \setminus \{0\}$, $y^* \in C^+ \setminus \{0\}$ for $Q = \operatorname{int} C$, respectively) arbitrary, but fixed. Hence, we obtain from (15.43)

$$\begin{aligned}
&\Longrightarrow \forall \xi \in \mathcal{U}\ \exists \eta \in \mathcal{U} :\ y^*(f(\overline{x},\eta)) \leq (<,<,\text{respectively})\ y^*(f(x^0,\xi)) \\
&\Longrightarrow \forall \xi \in \mathcal{U} :\ \inf_{\eta\in\mathcal{U}} y^*(f(\overline{x},\eta)) \leq (<,<,\text{respectively})\ y^*(f(x^0,\xi)) \\
&\Longrightarrow \inf_{\eta\in\mathcal{U}} y^*(f(\overline{x},\eta)) \leq (<,<,\text{respectively})\ \inf_{\xi\in\mathcal{U}} y^*(f(x^0,\xi)),
\end{aligned}$$

in contradiction to the assumptions. □

Based on these results, we are able to present the following algorithm that computes $(\preceq^l_C \mid \preceq^l_{C\setminus\{0\}} \mid \preceq^l_{\operatorname{int} C})$-robust solutions to $P(\mathcal{U})$:

Algorithm 4 for deriving $(\preceq^l_C \mid \preceq^l_{C\setminus\{0\}} \mid \preceq^l_{\operatorname{int} C})$-robust solutions of $P(\mathcal{U})$:

Input & Step 1-5: Analogous to Algorithm 1, only replacing $(P\mathcal{U}_{y^*})$ by $(P\mathcal{U}^l_{y^*})$ and replacing
$\max_{\xi\in\mathcal{U}} y^*(f(x^0,\xi))$ by $\min_{\xi\in\mathcal{U}} y^*(f(x^0,\xi))$.

The next algorithm computes $(\preceq^l_C \mid \preceq^l_{C\setminus\{0\}} \mid \preceq^l_{\operatorname{int} C})$-robust solutions via scalarization by means of linear continuous functionals:

Algorithm 5 for calculating a single desired $(\preceq^l_C \mid \preceq^l_{C\setminus\{0\}} \mid \preceq^l_{\operatorname{int} C})$-robust solution of $P(\mathcal{U})$:

Input & Step 1-7: Analogous to Algorithm 2, only replacing $(P\mathcal{U}_{\bar{y}^*})$ by $(P\mathcal{U}^l_{\bar{y}^*})$,
$\max_{\xi\in\mathcal{U}} y^*(f(x^0,\xi))$ by $\min_{\xi\in\mathcal{U}} y^*(f(x^0,\xi))$ and $(P\mathcal{U}_{\bar{y}^*+t_{k+1}(\hat{y}^*-\bar{y}^*)})$ by $(P\mathcal{U}^l_{\bar{y}^*+t_{k+1}(\hat{y}^*-\bar{y}^*)})$.

15.4.3 $\preceq^s_C$-Robustness

Now, we use the set less order relation $\preceq^s_Q$ with $Q = C$, $Q = C \setminus \{0\}$ and $Q = \operatorname{int} C$, respectively (compare Definition 2.6.7) for $A, B \subset Y$ arbitrarily chosen sets:

$$A \preceq^s_Q B :\Longleftrightarrow A \subseteq B - Q \text{ and } A + Q \supseteq B.$$

If we are dealing with $Q = \operatorname{int} C$ we suppose $\operatorname{int} C \neq \emptyset$.

We can now introduce the concept of $\preceq^s_Q$-robustness (with $Q = C$, $Q = C \setminus \{0\}$ and $Q = \operatorname{int} C$, respectively):

Definition 15.4.17. A solution x^0 of $P(\mathcal{U})$ is called $(\preceq^s_C \mid \preceq^s_{C\setminus\{0\}} \mid \preceq^s_{\operatorname{int} C})$-robust if there is no $\bar{x} \in S \setminus \{x^0\}$ such that

$$F(\bar{x}) \preceq^s_Q F(x^0)$$

for $Q = C$ ($Q = C \setminus \{0\}$, $Q = \operatorname{int} C$, respectively).

Remark 15.4.18. Note that a $\preceq^l_C$-robust solution is as well $\preceq^s_C$-robust by definition. The same assertion holds for a $\preceq^u_C$-robust solution.

The concept of $\preceq^s_C$-robustness can be interpreted in the following way: In a situation where it is not clear if one should follow a risk affine or risk averse strategy (e.g., the decision maker is not at hand or wants to get a feeling for the variety of the solutions) this concept might be helpful as it calculates solutions which reflect these different strategies. Therefore, this concept can serve as a pre-selection before deciding a definite strategy.

Computing $\preceq^s_C$-robust solutions is possible with the help of the following optimization problem:

$$h(x) := \begin{pmatrix} \inf_{\xi\in\mathcal{U}} & y^*(f(x,\xi)) \\ \sup_{\xi\in\mathcal{U}} & y^*(f(x,\xi)) \end{pmatrix} \to v - \min_{x\in S} \qquad (P\mathcal{U}^{biobj}_{y^*})$$

with $y^* \in C^+ \setminus \{0\}$ ($y^* \in C^\#$, respectively). For this bi-objective optimization problem we use the solution concept introduced in Definition 2.4.2. Especially, we say that $x^0 \in S$ is a weakly efficient element of the problem $(P\mathcal{U}^{biobj}_{y^*})$ if $h(x^0) \in \operatorname{WMin}(h(S), \mathbb{R}^2_+)$. Moreover, $x^0 \in S$ is called a strictly efficient element of the problem $(P\mathcal{U}^{biobj}_{y^*})$, if there is no $x \in S \setminus \{x^0\}$ such that $h(x) \in h(x^0) - \mathbb{R}^2_+$. We show the following theorem:

Theorem 15.4.19. *Consider an uncertain multi-objective optimization problem $P(\mathcal{U})$. The following statements hold:*

(a) *If x^0 is strictly efficient for problem $(P\mathcal{U}^{biobj}_{y^*})$ for some $y^* \in C^+ \setminus \{0\}$, then x^0 is $\preceq^s_C$-robust.*

(b) *If x^0 is weakly efficient for problem $(P\mathcal{U}_{y^*}^{biobj})$ for some $y^* \in C^+ \setminus \{0\}$ and* $\min_{\xi\in\mathcal{U}}\ y^*(f(x,\xi))$ *and* $\max_{\xi\in\mathcal{U}}\ y^*(f(x,\xi))$ *exist for all $x \in S$ and the chosen weight $y^* \in C^+ \setminus \{0\}$, then x^0 is $\preceq^s_{\operatorname{int} C}$-robust.*

(c) *If x^0 is weakly efficient for problem $(P\mathcal{U}_{y^*}^{biobj})$ for some $y^* \in C^\#$ and* $\min_{\xi\in\mathcal{U}}\ y^*(f(x,\xi))$ *and* $\max_{\xi\in\mathcal{U}}\ y^*(f(x,\xi))$ *exist for all $x \in S$ and the chosen weight $y^* \in C^\#$, then x^0 is $\preceq^s_{C\setminus\{0\}}$-robust.*

Proof. Let x^0 be strictly efficient (weakly efficient, weakly efficient) for problem $(P\mathcal{U}_{y^*}^{biobj})$ with some $y^* \in C^+ \setminus \{0\}$ ($y^* \in C^+ \setminus \{0\}$, $y^* \in C^\#$, respectively), i.e., there is no $\bar{x} \in S \setminus \{x^0\}$ such that

$$\inf_{\xi\in\mathcal{U}}\ y^*(f(\overline{x},\xi)) \le (<,\ <, \text{respectively}) \inf_{\xi\in\mathcal{U}}\ y^*(f(x^0,\xi))$$

$$\text{and} \sup_{\xi\in\mathcal{U}}\ y^*(f(\overline{x},\xi)) \le (<,\ <, \text{respectively}) \sup_{\xi\in\mathcal{U}}\ y^*(f(x^0,\xi)).$$

Now suppose x^0 is not $(\preceq^s_C \mid \preceq^s_{\operatorname{int} C} \mid \preceq^s_{C\setminus\{0\}})$-robust. Then there exists an $\bar{x} \in S \setminus \{x^0\}$ such that

$$F(\bar{x}) + Q \supseteq F(x^0) \text{ and } F(\bar{x}) \subseteq F(x^0) - Q$$

for $Q = C$ ($Q = \operatorname{int} C,\ Q = C \setminus \{0\}$). That implies

$$\exists \bar{x} \in S\setminus\{x^0\}:\ \forall \xi_1,\xi_2 \in \mathcal{U}\ \exists \eta_1,\eta_2 \in \mathcal{U}:\ f(\bar{x},\eta_1) + Q \ni f(x^0,\xi_1)$$
$$\text{and } f(\bar{x},\xi_2) \in f(x^0,\eta_2) - Q \tag{15.44}$$

for $Q = C$ ($Q = \operatorname{int} C,\ Q = C\setminus\{0\}$). Choose now $y^* \in C^+\setminus\{0\}$ ($y^* \in C^+\setminus\{0\}$, $y^* \in C^\#$) as in problem $(P\mathcal{U}_{y^*}^{biobj})$. We obtain from (15.44)

$$\exists \bar{x} \in S\setminus\{x^0\}:\forall \xi_1,\xi_2 \in \mathcal{U}\ \exists \eta_1,\eta_2 \in \mathcal{U}:\ y^*(f(\overline{x},\eta_1)) \le (<,\ <, \text{respectively})$$
$$y^*(f(x^0,\xi_1)) \text{ and } y^*(f(\overline{x},\xi_2)) \le (<,\ <, \text{respectively})\, y^*(f(x^0,\eta_2))$$
$$\Rightarrow \inf_{\xi\in\mathcal{U}}\ y^*(f(\overline{x},\xi)) \le (<,\ <, \text{respectively}) \inf_{\xi\in\mathcal{U}}\ y^*(f(x^0,\xi))$$
$$\text{and} \sup_{\xi\in\mathcal{U}}\ y^*(f(\overline{x},\xi)) \le (<,\ <, \text{respectively}) \sup_{\xi\in\mathcal{U}}\ y^*(f(x^0,\xi)).$$

The last two strict inequalities hold because the minimum and maximum exist. But this is a contradiction to the assumption. □

Based on these observations, we can formulate the following algorithm for computing $\preceq^s_C$-robust solutions to $P(\mathcal{U})$.

Algorithm 6 for computing $(\preceq^s_C \mid \preceq^s_{C\setminus\{0\}} \mid \preceq^s_{\text{int}\,C})$-robust solutions using a family of problems $(P\mathcal{U}^{biobj}_{y^*})$:

Input & Step 1-3: Analogous to Algorithm 1.
Step 4: Find a set of weakly efficient solutions $SOL_{we}(y^*)$ of $(P\mathcal{U}^{biobj}_{y^*})$.
Step 5: If $SOL_{we}(y^*) = \emptyset$, then go to Step 2.
Step 6: Choose $\bar{x} \in SOL_{we}(y^*)$. Set $SOL_{we}(y^*) := SOL_{we}(y^*) \setminus \{\bar{x}\}$.

a) If $\overline{x}$ is a strictly efficient solution of $(P\mathcal{U}^{biobj}_{y^*})$, then $\overline{x}$ is $\preceq^s_C$-robust for $P(\mathcal{U})$, thus

$$\text{Opt}_C := \text{Opt}_C \cup \{\overline{x}\}.$$

b) If $\overline{x}$ is weakly efficient for problem $(P\mathcal{U}^{biobj}_{y^*})$ and $y^* \in C^\#$ and $\min_{\xi\in\mathcal{U}} y^*(f(x,\xi))$ and $\max_{\xi\in\mathcal{U}} y^*(f(x,\xi))$ exist for all $x \in S$ and the chosen weight $y^* \in C^\#$, then $\overline{x}$ is $\preceq^s_{C\setminus\{0\}}$-robust for $P(\mathcal{U})$, thus

$$\text{Opt}_{C\setminus\{0\}} := \text{Opt}_{C\setminus\{0\}} \cup \{\overline{x}\}.$$

c) If $\overline{x}$ is a weakly efficient solution of $(P\mathcal{U}^{biobj}_{y^*})$ and $\max_{\xi\in\mathcal{U}} y^*(f(x,\xi))$ and $\min_{\xi\in\mathcal{U}} y^*(f(x,\xi))$ exist for all $x \in S$, then $\overline{x}$ is $\preceq^s_{\text{int}\,C}$-robust for $P(\mathcal{U})$, thus

$$\text{Opt}_{\text{int}\,C} := \text{Opt}_{\text{int}\,C} \cup \{\overline{x}\}.$$

Step 7: Go to Step 5.
Step 8: Go to Step 1.

In the following we present an algorithm that computes $\preceq^s_C$-robust solutions while varying the weights in the vector of objectives of problem $(P\mathcal{U}^{biobj}_{y^*})$.

Algorithm 7 for computing $(\preceq^s_C \mid \preceq^s_{C\setminus\{0\}} \mid \preceq^s_{\text{int}\,C})$-robust solutions using a family of problems $(P\mathcal{U}^{biobj}_{y^*})$:

Input & Step 1-3 & Step 5-7: Analogous to Algorithm 2, only replacing $(P\mathcal{U}_{\bar{y}^*})$ by $(P\mathcal{U}^{biobj}_{\bar{y}^*})$ and $(P\mathcal{U}_{\bar{y}^*+t_{k+1}(\hat{y}^*-\bar{y}^*)})$ by $(P\mathcal{U}^{biobj}_{\bar{y}^*+t_{k+1}(\hat{y}^*-\bar{y}^*)})$.
Step 4: Analogous to **Step 4** of Algorithm 5.

15.4.4 *Algorithms for Solving Special Classes of Set-Valued Optimization Problems*

In the following we will explain that the algorithms presented in Sect. 15.4 can be used for solving special classes of set-valued optimization problems.

Having a close look at all the concepts of robustness, we can see that in fact all of these are set-valued optimization problems.

Consider a set-valued optimization problem of the form

$$\preceq -\text{minimize} \quad F(x), \quad \text{subject to} \quad x \in S, \qquad (SP - \preceq)$$

with some given preorder $\preceq$ and a set-valued objective map $F : S \rightrightarrows Y$ we can see the following:

If the preorder $\preceq$ is given by $\preceq_C^l$, $\preceq_C^u$, or $\preceq_C^s$ with some proper closed convex pointed cone $C \subset Y$ and $F(x)$ can be parametrized by parameters $\xi \in \mathscr{U}$ with some set $\mathscr{U}$ in the way that

$$F(x) := \{f(x, \xi) \mid \xi \in \mathscr{U}\},$$

where $f : X \times \mathscr{U} \mapsto Y$, then the set-valued optimization problem $(SP- \preceq)$ is equivalent to finding $\preceq$-robust solutions to the uncertain vector optimization problem $P(\mathscr{U})$ and can therefore be solved by using one of the respective algorithms presented in Sect. 15.4.

We revealed strong connections between set-valued optimization and uncertain vector optimization. Furthermore, we derived the results in a more general setting as in [159] and [274]. In particular, we provided solution algorithms for a certain class of set-valued optimization problems. It seems possible to extend this class of problems to a more general one, but this is future work and of interest for the next steps in this area of research.

Moreover, Sect. 15.4 makes clear that finding robust solutions to uncertain vector optimization problems can be interpreted as an application of set-valued optimization. Thus, robust solutions to uncertain vector optimization problems can be obtained by using the solution techniques from set-valued optimization. Formulating concrete algorithms of this kind is another topic for future research.

Appendix

Very often, order structures are defined on subsets of linear spaces. We will consider only **real** linear spaces, so later on we shall omit the word real.

Let X be a linear space and $\emptyset \neq X_0 \subset X$; recall that X_0 is a **linear subspace** of X if $\lambda x + \mu y \in X_0$ for all $x, y \in X_0$ and $\lambda, \mu \in \mathbb{R}$. This shows that X_0 is stable w.r.t. addition and multiplication by scalars, and X_0 endowed with these operations is itself a linear space. Also recall that the nonempty set $M \subset X$ is **affine** (or a **linear manifold**) if $\lambda x + (1 - \lambda)y \in M$ for all $x, y \in M$ and $\lambda \in \mathbb{R}$; it is known that M is affine iff there exist $x_0 \in X$ and X_0 a linear subspace of X such that $M = x_0 + X_0$ (X_0 being called the parallel linear subspace to M).

Definition A.1. Let X be a nonempty set. We say that X is a **(real) linear space** or a **set equipped with linear structure** if an addition (that is, a mapping $+ : X \times X \to X$) and a multiplication by scalars (that is, a mapping $\cdot : \mathbb{R} \times X \to X$) are defined satisfying the following conditions:

(a) $\forall\, x, y, z \in X \;:\; (x + y) + z = x + (y + z)$ **(associativity)**,
(b) $\forall\, x, y \in X \;:\; x + y = y + x$ **(commutativity)**,
(c) $\exists\, 0 \in X,\ \forall\, x \in X \;:\; x + 0 = x$ **(null element)**,
(d) $\forall\, x \in X,\ \exists\, x' \in X \;:\; x + x' = 0$; we write $x' = -x$,
(e) $\forall\, x, y \in X,\ \forall\, \lambda \in \mathbb{R} \;:\; \lambda(x + y) = \lambda x + \lambda y$,
(f) $\forall\, x \in X,\ \forall\, \lambda, \mu \in \mathbb{R} \;:\; (\lambda + \mu)x = \lambda x + \mu x$,
(g) $\forall\, x \in X,\ \forall\, \lambda, \mu \in \mathbb{R} \;:\; \lambda(\mu x) = (\lambda\mu)x$,
(h) $\forall\, x \in X \;:\; 1x = x$ (**unity** element).

A linear space is often called a **vector space**. Defining $x + y := (x_1 + y_1, \dots, x_n + y_n)$ and $\lambda x := (\lambda x_1, \dots, \lambda x_n)$ for $x, y \in \mathbb{R}^n$ and $\lambda \in \mathbb{R}$, we have that $\mathbb{R}^n$ is a linear space.

In the sequel we shall use the following notation: Let A, B be nonempty subsets of a linear space X, $\emptyset \neq \Gamma \subset \mathbb{R}$, $x \in X$, and $\lambda \in \mathbb{R}$; then

$$A + B := \{x + y \mid x \in A,\ y \in B\},\ A + \emptyset := \emptyset + A := \emptyset,\ x + A := \{x\} + A,$$

A.A. Khan et al., *Set-valued Optimization*, Vector Optimization,
DOI 10.1007/978-3-642-54265-7

and

$$\Gamma \cdot A := \{\gamma x \mid \gamma \in \Gamma, x \in A\},\ \lambda A = \{\lambda\} \cdot A,\ \Gamma \cdot \emptyset = \emptyset.$$

In particular, $A - B = A + (-1)B = \{x - y \mid x \in A,\ y \in B\}$. If, moreover, I is a nonempty set and $\emptyset \neq A_i \subset X$ for every $i \in I$, we have

$$\lambda(A + B) = \lambda A + \lambda B,\ \lambda\left(\bigcup_{i \in I} A_i\right) = \bigcup_{i \in I} \lambda A_i,\ \lambda\left(\bigcap_{i \in I} A_i\right) = \bigcap_{i \in I} \lambda A_i.$$

Often it is useful to have also a topological structure on the set under consideration.

Definition A.2. Let X be a nonempty set. A **topology** τ on X is a family of subsets of X satisfying the following conditions:

(a) every union of sets of τ belongs to τ,
(b) every finite intersection of sets of τ belongs to τ,
(c) the empty set $\emptyset$ and the whole set X belong to τ.

The elements of τ are called **open** sets, and X equipped with τ is called a **topological space** and is denoted by (X, τ).

Important examples of topological spaces are furnished by metric spaces. Recall that a **metric** on a nonempty set X is a mapping $\rho : X \times X \to \mathbb{R}_+$ having the properties

$$\forall\, x, y \in X : \rho(x, y) = 0 \Leftrightarrow x = y,$$
$$\forall\, x, y \in X : \rho(x, y) = \rho(y, x),$$
$$\forall\, x, y, z \in X : \rho(x, z) \leq \rho(x, y) + \rho(y, z).$$

The set X equipped with the metric ρ is called a **metric space** and is denoted by (X, ρ).

Having the metric space (X, ρ), $x \in X$ and $r > 0$, consider the (closed) ball of center x and radius r :

$$B(x, r) := B_\rho(x, r) := \{y \in X \mid \rho(x, y) \leq r\}.$$

Define

$$\tau_\rho := \{\emptyset\} \cup \{D \subset X \mid \forall\, x \in D,\ \exists\, r > 0\ :\ B(x, r) \subset D\}.$$

It is easy to verify that τ_ρ is a topology on X, and so X becomes a topological space; a metric space (X, ρ) will always be endowed with the topology τ_ρ if not stated explicitly otherwise. Note that $B(x,r)$ is a closed set, while the open ball $\{y \in X \mid \rho(x,y) < r\}$ is an open set.

Taking $X = \mathbb{R}^n$ and $\rho(x,y)$ the usual Euclidean distance between $x \in \mathbb{R}^n$ and $y \in \mathbb{R}^n$, ρ is a metric on $\mathbb{R}^n$, and the open sets w.r.t. ρ are the usual open sets of $\mathbb{R}^n$. As in $\mathbb{R}^n$, in an arbitrary topological space (X,τ) we define other topological notions, that of neighborhood being very useful.

Definition A.3. Let (X,τ) be a topological space and $x \in X$. The subset U of X is a **neighborhood** of x (relative to τ) if there exists an open set $O \in \tau$ such that $x \in O \subset U$. The class of all neighborhoods of x will be denoted by $\mathscr{N}_\tau(x)$, or simply $\mathscr{N}(x)$. A subset $\mathscr{B}(x)$ of $\mathscr{N}_\tau(x)$ is called a **base of neighborhoods** (filter base) of x relative to τ if for every $U \in \mathscr{N}_\tau(x)$ there exists $V \in \mathscr{B}(x)$ such that $V \subset U$. One says that (X,τ) is **first-countable** if every element $x \in X$ has an (at most) countable neighborhood base.

For the topological space (X,τ) and a neighborhood base $\mathscr{B}(x)$ of x for every $x \in X$, the family $\{\mathscr{B}(x) \mid x \in X\}$ has the following properties:

(NB1) $\forall\, x \in X,\ \forall\, U \in \mathscr{B}(x) \ :\ x \in U,$
(NB2) $\forall\, x \in X,\ \forall\, U_1, U_2 \in \mathscr{B}(x),\ \exists\, U_3 \in \mathscr{B}(x) \ :\ U_3 \subset U_1 \cap U_2,$
(NB3) $\forall\, x \in X,\ \forall\, U \in \mathscr{B}(x),\ \exists\, V \in \mathscr{B}(x),\ \forall\, y \in V,\ \exists\, W \in \mathscr{B}(y) \ :\ W \subset U.$

Conversely, if for every $x \in X$ one has a nonempty family $\mathscr{B}(x) \subset \mathscr{P}(X)$ such that $\{\mathscr{B}(x) \mid x \in X\}$ satisfies the conditions (NB1)–(NB3), then there exists a unique topology τ on X such that $\mathscr{B}(x)$ is a neighborhood base for x, for every $x \in X$; moreover,

$$\forall\, x \in X \ :\ \mathscr{N}_\tau(x) = \{U \subset X \mid \exists\, V \in \mathscr{B}(x) \ :\ V \subset U\}.$$

In fact, the topology (X,ρ) of a metric space is constructed in this manner: The family $\{\mathscr{B}_0(x) \mid x \in X\}$ with $\mathscr{B}_0(x) := \{B_\rho(x,1/n) \mid n \in \mathbb{N} \setminus \{0\}\}$ satisfies the conditions (NB1)–(NB3) above; the topology defined by $\{\mathscr{B}_0(x) \mid x \in X\}$ is nothing else but the topology τ_ρ defined above. Since $\mathscr{B}_0(x)$ is countable for every $x \in X$, we obtain that every metric space is first-countable.

As seen above (property (NB2)), taking $\mathscr{B}(x)$ a neighborhood base of $x \in (X,\tau)$ and considering the relation $\succeq$ on $\mathscr{B}(x)$ defined by $U_1 \succeq U_2$ iff $U_1 \subset U_2$, then for all $U_1, U_2 \in \mathscr{B}(x)$ there exists $U_3 \in \mathscr{B}(x)$ such that $U_3 \succeq U_1$ and $U_3 \succeq U_2$. This is the prototype of a directed set. So, we say that the nonempty set I is **directed** by $\succeq$ if $\succeq$ is a partial order on I and

$$\forall\, i_1, i_2 \in I,\ \exists\, i_3 \in I \ :\ i_3 \succeq i_1 \text{ and } i_3 \succeq i_2.$$

Note that if $(I',\succeq')$ and $(I'',\succeq'')$ are directed sets, then $I' \times I''$ is directed by the relation $\succeq$ defined by $(i',i'') \succeq (j',j'')$ iff $i' \succeq' j'$ and $i'' \succeq'' j''$. Another important example of directed set is $\mathbb{N}$ endowed with its usual order $\geq$. A subset J of a directed set $(I,\succeq)$ is called **cofinal** if for every $i \in I$ there exists $j \in J$ such that $j \succeq i$; it follows immediately that $(J,\succeq)$ is directed if J is a cofinal subset of I.

The above discussion shows that $(\mathscr{B}(x), \succeq)$ is directed if $\mathscr{B}(x)$ is a neighborhood base for $x \in (X, \tau)$. This consideration is the base of the following definition.

Definition A.4. A mapping $\varphi : I \to X$, where $(I, \succeq)$ is a directed set, is called a **net** or a **generalized sequence** of X; generally, the net φ is denoted by $(x_i)_{i \in I}$, where $x_i := \varphi(i)$. When X is equipped with a topology τ we say that the net $(x_i)_{i \in I} \subset X$ **converges** to $x \in X$ if

$$\forall\, V \in \mathscr{N}_\tau(x),\ \exists\, i_V \in I,\ \forall\, i \succeq i_V\ :\ x_i \in V.$$

We denote this fact by $(x_i) \xrightarrow{\tau} x$ or simply $(x_i) \to x$, and x is called a **limit** of (x_i).

Lemma A.5. *Let (X, τ) be a topological space. Then every convergent net of X has a unique limit if and only if*

$$\forall\, x, y \in X,\ x \neq y,\ \exists\, U \in \mathscr{N}_\tau(x),\ V \in \mathscr{N}_\tau(y)\ :\ U \cap V = \emptyset, \tag{A.1}$$

i.e., (X, τ) is ***Hausdorff****.*

Proof. Suppose that (A.1) does not hold. Then there exist $x, y \in X$, $x \neq y$, such that

$$\forall\, U \in \mathscr{N}_\tau(x),\ \forall\, V \in \mathscr{N}_\tau(y),\ \exists\, x_{U,V} \in U \cap V.$$

As seen above, $I := \mathscr{N}_\tau(x) \times \mathscr{N}_\tau(y)$ is directed, and so $(x_{U,V})_{(U,V) \in I}$ is a net in X. It is obvious that $(x_{U,V}) \xrightarrow{\tau} x$ and $(x_{U,V}) \xrightarrow{\tau} y$, which shows that the limit of nets is not unique in this case.

Suppose now that (A.1) holds and assume that the net $(x_i)_{i \in I} \subset X$ converges to x and y with $x \neq y$. Take $U \in \mathscr{N}_\tau(x)$ and $V \in \mathscr{N}_\tau(y)$ such that $U \cap V = \emptyset$. Then there exist $i_U, i_V \in I$ such that $x_i \in U$ for $i \succeq i_U$ and $x_i \in V$ for $i \succeq i_V$. Because I is directed, there exists $i_0 \in I$ such that $i_0 \succeq i_U$ and $i_0 \succeq i_V$. We thus obtain the contradiction that $x_{i_0} \in U \cap V$. The proof is complete. □

A useful notion is that of subnet; so, we say that the net $(y_k)_{k \in K}$ is a **subnet** of the net $(x_i)_{i \in I}$ if there exists a mapping $\psi : (K, \succ) \to (I, \succeq)$ such that $y_k = x_{\psi(k)}$ for every $k \in K$ and for every $i \in I$ there exists $k_i \in K$ such that $\psi(k) \succeq i$ for $k \succ k_i$. If J is a cofinal subset of the directed set $(I, \succeq)$, then $(x_i)_{i \in J}$ is a subnet of the net $(x_i)_{i \in I}$. With the help of nets and subnets it is possible to characterize several topological notions as the closure of a set, the cluster points, the compact subsets, and the continuity of mappings between topological spaces in the same manner as these notions are characterized in $\mathbb{R}^k$ by using sequences and subsequences. For example, a subset A of the topological space (X, τ) is closed if and only if $x \in A$ whenever the net $(x_i)_{i \in I} \subset A$ converges to $x \in X$, but generally, taking sequences is not sufficient; we say that A is **sequentially closed** if $x \in A$ whenever the sequence $(x_n)_{n \in \mathbb{N}} \subset A$ converges to $x \in X$.

Recall that the function $f : (X, \tau) \to (Y, \sigma)$ is **continuous at** $x \in X$ if

$$\forall\, V \in \mathscr{N}_\sigma(f(x)),\ \exists\, U \in \mathscr{N}_\tau(x),\ \forall\, x' \in U \ :\ f(x') \in V;$$

f is **continuous** if f is continuous at every $x \in X$. When X is a linear space and τ is a topology on X, it is important to have compatibility between the linear and topological structures of X.

Definition A.6. Let X be a linear space endowed with a topology τ. We say that (X, τ) is a **topological linear space** or **topological vector space** (t.l.s. or t.v.s. for short) if both operations on X (the addition and the multiplication by scalars) are continuous; in this case τ is called a **linear topology** on X.

Since these operations are defined on product spaces, we recall that for two topological spaces (X_1, τ_1) and (X_2, τ_2), there exists a unique topology on $X_1 \times X_2$, denoted by $\tau_1 \times \tau_2$, with the property that

$$\mathscr{B}(x_1, x_2) := \{U_1 \times U_2 \mid U_1 \in \mathscr{N}_{\tau_1}(x_1),\ U_2 \in \mathscr{N}_{\tau_2}(x_2)\}$$

is a neighborhood base of (x_1, x_2) w.r.t. $\tau_1 \times \tau_2$ for every $(x_1, x_2) \in X_1 \times X_2$; $\tau_1 \times \tau_2$ is called the **product topology** on $X_1 \times X_2$. Of course, in Definition A.6 the topology on $X \times X$ is $\tau \times \tau$, and the topology on $\mathbb{R} \times X$ is $\tau_0 \times \tau$, where τ_0 is the usual topology of $\mathbb{R}$. It is easy to see that when (X, τ) is a topological linear space, $a \in X$ and $\lambda \in \mathbb{R} \setminus \{0\}$, the mappings $T_a,\ H_\lambda : X \to X$ defined by $T_a(x) = a + x$, $H_\lambda(x) := \lambda x$, are bijective and continuous with continuous inverses i.e., they are **homeomorphisms**). It follows, that $V \in \mathscr{N}_\tau(0)$ iff $a + V \in \mathscr{N}_\tau(a)$ for every $a \in X$ and every neighborhood of 0 is absorbing ($A \subset X$ is **absorbing** if for every $x \in X$ there exists $\delta > 0$ such that $[-\delta, \delta] \cdot x \subset A$, or equivalently, for every $x \in X$ there exists $\delta > 0$ such that $[0, \delta] \cdot x \subset A$). It is well known that the origin 0 of a t.v.s. (X, τ) has a neighborhood base formed by balanced absorbing sets ($\emptyset \neq A \subset X$ is **balanced** if $[-1, 1] \cdot A = A$). When X is a t.v.s., the class of all balanced neighborhoods of $0 \in X$ will be denoted by $\mathscr{N}_X$. Taking into account the formula

$$\operatorname{cl} A = \bigcap \{A + V \mid V \in \mathscr{B}\} \tag{A.2}$$

for the closure of a subset A of the t.l.s. (X, τ), where $\mathscr{B}$ is a neighborhood base for 0, one obtains that the origin of a topological linear space has a neighborhood base formed by closed, balanced, and absorbing sets.

Note also that a topological linear space (X, τ) is Hausdorff iff for every $x \in X \setminus \{0\}$ there exists $V \in \mathscr{N}_X$ such that $x \notin V$, or equivalently (see (A.2)), the set $\{0\}$ is closed.

Further examples of topological linear spaces are given in the next section.

References

1. Abadie, J.: On the Kuhn-Tucker theorem. In: Nonlinear Programming (NATO Summer School, Menton, 1964), pp. 19–36. North-Holland, Amsterdam (1967)
2. Abdouni, B.E., Thibault, L.: Optimality conditions for problems with set-valued objectives. J. Appl. Anal. **2**, 183–201 (1996)
3. Alonso, M., Rodríguez-Marín, L.: Set-relations and optimality conditions in set-valued maps. Nonlinear Anal. **63**(8), 1167–1179 (2005)
4. Altangerel, L., Boţ, R.I., Wanka, G.: Conjugate duality in vector optimization and some applications to the vector variational inequality. J. Math. Anal. Appl. **329**(2), 1010–1035 (2007)
5. Amahroq, T., Taa, A.: Sufficient conditions of optimality for multiobjective optimization problems with γ-paraconvex data. Studia Math. **124**(3), 239–247 (1997)
6. Amahroq, T., Thibault, L.: On proto-differentiability and strict proto-differentiability of multifunctions of feasible points in perturbed optimization problems. Numer. Funct. Anal. Optim. **16**, 1293–1307 (1995)
7. Anger, B., Lembcke, J.: Hahn-Banach type theorems for hypolinear functionals. Math. Ann. **209**, 127–151 (1974)
8. Anh, N.L., Khanh, P.Q., Tung, L.T.: Variational sets: calculus and applications to nonsmooth vector optimization. Nonlinear Anal. **74**(6), 2358–2379 (2011)
9. Apetrii, M., Durea, M., Strugariu, R.: On subregularity properties of set-valued mappings. Set Valued Var. Anal. **21**(1), 93–126 (2013)
10. Aragón Artacho, F.J., Mordukhovich, B.S.: Enhanced metric regularity and Lipschitzian properties of variational systems. J. Global Optim. **50**(1), 145–167 (2011)
11. Arrow, K.J., Hurwicz, L., Uzawa, H.: Studies in linear and non-linear programming. With contributions by H. B. Chenery, S. M. Johnson, S. Karlin, T. Marschak, R. M. Solow. Stanford Mathematical Studies in the Social Sciences, vol. II. Stanford University Press, Stanford, California (1958)
12. Artzner, P., Delbaen, F., Eber, J.M., Heath, D.: Coherent measures of risk. Math. Finance **9**, 203–228 (1999)
13. Attouch, H., Riahi, H.: Stability results for Ekeland's ϵ-variational principle and cone extremal solutions. Math. Oper. Res. **18**(1), 173–201 (1993)
14. Aubin, J.P.: Contingent derivatives of set-valued maps and existence of solutions to nonlinear inclusions and differential inclusions. In: Mathematical analysis and applications, Part A, Adv. in Math. Suppl. Stud., vol. 7, pp. 159–229. Academic Press, New York (1981)
15. Aubin, J.P., Cellina, A.: Differential inclusions. Set-Valued Maps and Viability Theory [Grundlehren der Mathematischen Wissenschaften (Fundamental Principles of Mathematical Sciences)], vol. 264. Springer, Berlin (1984)

A.A. Khan et al., *Set-valued Optimization*, Vector Optimization,
DOI 10.1007/978-3-642-54265-7

16. Aubin, J.P., Ekeland, I.: Applied nonlinear analysis. Pure and Applied Mathematics (New York). Wiley, New York (1984)
17. Aubin, J.P., Frankowska, H.: On inverse function theorems for set-valued maps. J. Math. Pures Appl. (9) **66**(1), 71–89 (1987)
18. Aubin, J.P., Frankowska, H.: Set-valued analysis. Systems & Control: Foundations & Applications, vol. 2. Birkhäuser Boston, Boston (1990)
19. Auslender, A., Cominetti, R.: A comparative study of multifunction differentiability with applications in mathematical programming. Math. Oper. Res. **16**(2), 240–258 (1991)
20. Avakov, E.R.: Necessary conditions for an extremum for smooth abnormal problems with constraints of equality and inequality type. Mat. Zametki **45**(6), 3–11, 110 (1989)
21. Azé, D.: A unified theory for metric regularity of multifunctions. J. Convex Anal. **13**(2), 225–252 (2006)
22. Baier, J., Jahn, J.: On subdifferentials of set-valued maps. J. Optim. Theory Appl. **100**, 233–240 (1999)
23. Banks, H.T., Jacobs, M.Q.: A differential calculus for multifunctions. J. Math. Anal. Appl. **29**, 246–272 (1970)
24. Bao, T.Q., Mordukhovich, B.S.: Existence of minimizers and necessary conditions in set-valued optimization with equilibrium constraints. Appl. Math. **52**, 452–562 (2007)
25. Bao, T.Q., Mordukhovich, B.S.: Variational principles for set-valued mappings with applications to multiobjective optimization. Control Cybern. **36**(3), 531–562 (2007)
26. Bao, T.Q., Mordukhovich, B.S.: Necessary conditions for super minimizers in constrained multiobjective optimization. J. Global Optim. **43**(4), 533–552 (2009)
27. Bao, T.Q., Mordukhovich, B.S.: Relative Pareto minimizers for multiobjective problems: existence and optimality conditions. Math. Program. **122**(2, Ser. A), 301–347 (2010)
28. Bao, T.Q., Mordukhovich, B.S.: Set-valued optimization in welfare economics. Adv. Math. Econ. **13**, 113–153 (2010)
29. Bao, T.Q., Tammer, C.: Lagrange necessary conditions for Pareto minimizers in Asplund spaces and applications. Nonlinear Anal. **75**(3), 1089–1103 (2012)
30. Barbagallo, A., Cojocaru, M.G.: Continuity of solutions for parametric variational inequalities in Banach space. J. Math. Anal. Appl. **351**(2), 707–720 (2009)
31. Barber, C.B., Dobkin, D.P., Huhdanpaa, H.: The quickhull algorithm for convex hulls. ACM Trans. Math. Software **22**(4), 469–483 (1996)
32. Barbu, V., Precupanu, T.: Convexity and optimization in Banach spaces, Mathematics and its Applications (East European Series), vol. 10, romanian edn. D. Reidel Publishing, Dordrecht (1986)
33. Bauschke, H.H., Combettes, P.L.: Convex analysis and monotone operator theory in Hilbert spaces, with a foreword by Hédy Attouch. CMS Books in Mathematics/Ouvrages de Mathématiques de la SMC. Springer, New York (2011)
34. Bazaraa, M., Goode, J., Shetty, C.: Constraint qualifications revisited. Manag. Sci. **18**, 567–573 (1972)
35. Bazaraa, M.S., Goode, J.J.: Necessary optimality criteria in mathematical programming in normed linear spaces. J. Optim. Theory Appl. **11**, 235–244 (1973)
36. Bazaraa, M.S., Goode, J.J., Nashed, M.Z.: On the cones of tangents with applications to mathematical programming. J. Optim. Theory Appl. **13**, 389–426 (1974)
37. Bazaraa, M.S., Shetty, C.M., Goode, J.J., Nashed, M.Z.: Nonlinear programming without differentiability in Banach spaces: necessary and sufficient constraint qualification. Applicable Anal. **5**(3), 165–173 (1975/76)
38. Bazarra, M., Shetty, C.: Foundations of Optimization. Springer, New York (1976)
39. Bednarczuk, E.M., Song, W.: Contingent epiderivative and its applications to set-valued optimization. Control Cybern. **27**, 375–386 (1998)
40. Bednarczuk, E.M., Zagrodny, D.: Vector variational principle. Arch. Math. (Basel) **93**(6), 577–586 (2009)
41. Bellaassali, S., Jourani, A.: Lagrange multipliers for multiobjective programs with a general preference. Set Valued Anal. **16**(2-3), 229–243 (2008)

42. Ben-Israel, A., Ben-Tal, A., Zlobec, S.: Optimality in Nonlinear Programming: A Feasible Directions Approach. Wiley, New York (1981)
43. Ben-Tal, A., Ghaoui, L.E., Nemirovski, A.: Robust Optimization. Princeton University Press, Princeton and Oxford (2009)
44. Ben-Tal, A., Nemirovski, A.: Robust convex optimization. Math. Oper. Res. **23**(4), 769–805 (1998)
45. Ben-Tal, A., Zowe, J.: A unified theory of first and second order conditions for extremum problems in topological vector spaces. Math. Program. Stud. (19), 39–76 (1982)
46. Benson, H.: An outer approximation algorithm for generating all efficient extreme points in the outcome set of a multiple objective linear programming problem. J. Global Optim. **13**, 1–24 (1998)
47. Benson, H.P.: On a domination property for vector maximization with respect to cones. J. Optim. Theory Appl. **39**(1), 125–132 (1983)
48. Bianchi, M., Kassay, G., Pini, R.: Ekeland's principle for vector equilibrium problems. Nonlinear Anal. **66**(7), 1454–1464 (2007)
49. Bigolin, F., Greco, G.H.: Geometric characterizations of C^1 manifolds in Euclidean spaces by tangent cones. J. Math. Anal. Appl. **396**(1), 145–163 (2012)
50. Birge, J.R., Louveaux, F.: Introduction to Stochastic Programming, 2nd edn. Springer Series in Operations Research and Financial Engineering. Springer, New York (2011)
51. Bitran, G.R., Magnanti, T.L.: The structure of admissible points with respect to cone dominance. J. Optim. Theory Appl. **29**(4), 573–614 (1979)
52. Boţ, R., Wanka, G.: An analysis of some dual problems in multiobjective optimization. I, II. Optimization **53**(3), 281–324 (2004)
53. Bonnans, J.F., Shapiro, A.: Perturbation analysis of optimization problems. Springer Series in Operations Research. Springer, New York (2000)
54. Bonnisseau, J.M., Cornet, B.: Existence of equilibria when firms follow bounded losses pricing rules. J. Math. Econ. **17**, 119–147 (1988)
55. Bonnisseau, J.M., Crettez, B.: On the characterization of efficient production vectors. Econ. Theory **31**, 213–223 (2007)
56. Borwein, J.M.: Multivalued convexity and optimization: A unified approach to inequality and equality constraints. Math. Program. **13**, 183–199 (1977)
57. Borwein, J.M.: Proper efficient points for maximizations with respect to cones. SIAM J. Control Optim. **15**(1), 57–63 (1977)
58. Borwein, J.M.: Tangent cones, starshape and convexity. Internat. J. Math. Math. Sci. **1**(4), 459–477 (1978)
59. Borwein, J.M.: Weak tangent cones and optimization in a Banach space. SIAM J. Control Optim. **16**(3), 512–522 (1978)
60. Borwein, J.M.: Stability and regular points of inequality systems. J. Optim. Theory Appl. **48**(1), 9–52 (1986)
61. Borwein, J.M.: Convex cones, minimality notions, and consequences. In: Recent Advances and Historical Development of Vector Optimization (Darmstadt, 1986), Lecture Notes in Economics and Mathematical Systems, vol. 294, pp. 62–85. Springer, Berlin (1987)
62. Borwein, J.M.: Epi-Lipschitz-like sets in Banach space: theorems and examples. Nonlinear Anal. **11**(10), 1207–1217 (1987)
63. Borwein, J.M., Lucet, Y., Mordukhovich, B.: Compactly epi-Lipschitzian convex sets and functions in normed spaces. J. Convex Anal. **7**(2), 375–393 (2000)
64. Borwein, J.M., O'Brien, R.: Tangent cones and convexity. Can. Math. Bull. **19**(3), 257–261 (1976)
65. Borwein, J.M., Strójwas, H.M.: The hypertangent cone. Nonlinear Anal. **13**(2), 125–144 (1989)
66. Borwein, J.M., Zhu, Q.J.: Techniques of variational analysis. CMS Books in Mathematics/Ouvrages de Mathématiques de la SMC, vol. 20. Springer, New York (2005)
67. Borwein, J.M.: Continuity and differentiability properties of convex operators. Proc. Lond. Math. Soc. (3) **44**(3), 420–444 (1982)

68. Borwein, J.M.: On the existence of Pareto efficient points. Math. Oper. Res. **8**(1), 64–73 (1983)
69. Borwein, J.M., Lewis, A.S.: Convex analysis and nonlinear optimization. Theory and examples. CMS Books in Mathematics/Ouvrages de Mathématiques de la SMC. 3. Springer, New York (2000)
70. Borwein, J.M., Strójwas, H.M.: Tangential approximations. Nonlinear Anal. **9**(12), 1347–1366 (1985)
71. Borwein, J.M., Zhuang, D.: Super efficiency in vector optimization. Trans. Am. Math. Soc. **338**(1), 105–122 (1993)
72. Boţ, R.I., Csetnek, E., Wanka, G.: Regularity conditions via quasi-relative interior in convex programming. SIAM J. Optim. **19**, 217–233 (2008)
73. Boţ, R.I., Grad, S.M., Wanka, G.: Duality in vector optimization. Vector Optimization. Springer, Berlin (2009)
74. Bouligand, G.: Sur l'existence des demi-tangentes à une courbe de Jordan. Fundamenta Math. **15**, 215–218 (1030)
75. Bouligand, G.: Introduction á la géométrie Infinitseminale Directe. Gauthier-Villars, Paris (1932)
76. Bourbaki, N.: Eléments de mathématique. XVIII. Première partie: Les structures fondamentales de l'analyse. Livre V: Espaces vectoriels topologiques. Chapitre III: Espaces d'applications linéaires continues. Chapitre IV: La dualité dans les espaces vectoriels topologiques. Chapitre V: Espaces hilbertiens. Actualités Sci. Ind., no. 1229. Hermann & Cie, Paris (1955)
77. Branke, J.: Creating robust solutions by means of evolutionary algorithms. In: Eiben, E., Bäck, T., Schenauer, M., Schwefel, H.P. (eds.) Parallel Problem Solving from Nature, Lecture Notes in Computer Science, pp. 119–128. Springer, Berlin, Heidelberg (1998)
78. Breckner, W.W.: Dualität bei Optimierungsaufgaben in halbgeordneten topologischen Vektorräumen. i. Math. Rev. Anal. Numér. Théor. Approx. **1**, 5–35 (1972)
79. Breckner, W.W., Kassay, G.: A systematization of convexity concepts for sets and functions. J. Convex Anal. **4**(1), 109–127 (1997)
80. Bremner, D., Fukuda, K., Marzetta, A.: Primal-dual methods for vertex and facet enumeration. Discrete Comput. Geom. **20**(3), 333–357 (1998)
81. Brumelle, S.L.: Convex operators and supports. Math. Oper. Res. **3**(2), 171–175 (1978)
82. Cambini, A., Martein, L., Vlach, M.: Second order tangent sets and optimality conditions. Math. Jpn. **49**(3), 451–461 (1999)
83. Canon, M.D., Cullum Jr., C.D., Polak, E.: Theory of Optimal Control and Mathematical Programming. McGraw-Hill, New York (1970)
84. Castellani, M.: A necessary second-order optimality condition in nonsmooth mathematical programming. Oper. Res. Lett. **19**(2), 79–86 (1996)
85. Castellani, M., Pappalardo, M.: First-order cone approximations and necessary optimality conditions. Optimization **35**(2), 113–126 (1995)
86. Castellani, M., Pappalardo, M.: Local second-order approximations and applications in optimization. Optimization **37**(4), 305–321 (1996)
87. Cesari, L., Suryanarayana, M.B.: Existence theorems for Pareto optimization; multivalued and Banach space valued functionals. Trans. Am. Math. Soc. **244**, 37–65 (1978)
88. Chai, Y., Liu, S., Li, Y.: The existence of contingent epiderivative for a set-valued mapping and vector variational-like inequalities. J. Inequal. Appl. **2013**(352), 13 (2013)
89. Chen, C.R., Li, S.J.: Different conjugate dual problems in vector optimization and their relations. J. Optim. Theory Appl. **140**(3), 443–461 (2009)
90. Chen, G.Y., Goh, C.J., Yang, X.Q.: On gap functions for vector variational inequalities. In: Vector variational inequalities and vector equilibria, Nonconvex Optim. Appl., vol. 38, pp. 55–72. Kluwer Academic, Dordrecht (2000)
91. Chen, G.Y., Huang, X., Yang, X.: Vector optimization. Lecture Notes in Economics and Mathematical Systems, vol. 541. Springer, Berlin (2005)

92. Chen, G.Y., Jahn, J.: Optimality conditions for set-valued optimization problems. Math. Methods Oper. Res. **48**(2), 187–200 (1998)
93. Chen, G.Y., Wang, Y.Y.: Generalized Hahn-Banach theorems and subdifferential of set-valued mapping. J. Syst. Sci. Math. Sci. **5**(3), 223–230 (1985)
94. Chen, X., Nashed, Z., Qi, L.: Convergence of Newton's method for singular smooth and nonsmooth equations using adaptive outer inverses. SIAM J. Optim. **7**(2), 445–462 (1997)
95. Chew, K.L.: Maximal points with respect to cone dominance in Banach spaces and their existence. J. Optim. Theory Appl. **44**(1), 1–53 (1984)
96. Chiang, Y., Yao, J.C.: Vector variational inequalities and the $(S)_+$ condition. J. Optim. Theory Appl. **123**(2), 271–290 (2004)
97. Chuong, T.D.: Clarke coderivatives of efficient point multifunctions in parametric vector optimization. Nonlinear Anal. **74**(1), 273–285 (2011)
98. Chuong, T.D., Yao, J.C.: Coderivatives of efficient point multifunctions in parameteric vector optimization. Taiwanese J. Math. **13**(6A), 1671–1693 (2009)
99. Chuong, T.D., Yao, J.C.: Generalized Clarke epiderivatives of parametric vector optimization problems. J. Optim. Theory Appl. **146**(1), 77–94 (2010)
100. Clarke, F.H.: Optimization and nonsmooth analysis. Canadian Mathematical Society Series of Monographs and Advanced Texts. Wiley, New York (1983)
101. Clarke, F.H., Stern, R.J., Wolenski, P.R.: Subgradient criteria for monotonicity, the Lipschitz condition, and convexity. Can. J. Math. **45**(6), 1167–1183 (1993)
102. Combari, C., Laghdir, M., Thibault, L.: Sous-différentiels de fonctions convexes composées. Ann. Sci. Math. Québec **18**(2), 119–148 (1994)
103. Cominetti, R.: Metric regularity, tangent sets, and second-order optimality conditions. Appl. Math. Optim. **21**(3), 265–287 (1990)
104. Constantin, E.: Higher order necessary and sufficient conditions for optimality. Panamer. Math. J. **14**(3), 1–25 (2004)
105. Constantin, E.: An application of higher order tangent cones to flow-invariance. Int. J. Appl. Math. Sci. **2**(1), 121–129 (2005)
106. Constantin, E.: Higher order necessary conditions in smooth constrained optimization. In: Communicating Mathematics, Contemp. Math., vol. 479, pp. 41–49. Amer. Math. Soc., Providence, RI (2009)
107. Constantin, E.: Second-order optimality conditions for problems with locally Lipschitz data via tangential directions. Comm. Appl. Nonlinear Anal. **18**(2), 75–84 (2011)
108. Conway, J.B.: A course in functional analysis, *Graduate Texts in Mathematics*, vol. 96, 2nd edn. Springer, New York (1990)
109. Corley, H.: Optimality conditions for maximization of set-valued functions. J. Optim. Theory Appl. **58**, 1–10 (1988)
110. Corley, H.W.: An existence result for maximizations with respect to cones. J. Optim. Theory Appl. **31**(2), 277–281 (1980)
111. Corley, H.W.: Duality theory for maximizations with respect to cones. J. Math. Anal. Appl. **84**, 560–568 (1981)
112. Corley, H.W.: Existence and Lagrangian duality for maximizations of set-valued functions. J. Optim. Theory Appl. **54**(3), 489–501 (1987)
113. Cornet, B.: Regularity properties of open tangent cones. Math. Program. Stud. (30), 17–33 (1987)
114. Correa, R., Jofré, A., Thibault, L.: Characterization of lower semicontinuous convex functions. Proc. Am. Math. Soc. **116**(1), 67–72 (1992)
115. Correa, R., Jofré, A., Thibault, L.: Subdifferential monotonicity as characterization of convex functions. Numer. Funct. Anal. Optim. **15**(5-6), 531–535 (1994)
116. Crespi, G.P., Ginchev, I., Rocca, M.: First-order optimality conditions in constrained set-valued optimization. Pac. J. Optim. **2**(2), 225–239 (2006)
117. Crespi, G.P., Ginchev, I., Rocca, M.: First-order optimality conditions in set-valued optimization. Math. Methods Oper. Res. **63**(1), 87–106 (2006)

118. Crespi, G.P., Ginchev, I., Rocca, M.: Existence of solutions of Minty type scalar and vector variational inequalities. Optimization **58**(7), 791–808 (2009)
119. Dancs, S.: Generalized tangent cone and an optimization problem in a normed space. J. Optim. Theory Appl. **67**(1), 43–55 (1990)
120. Dancs, S., Hegedüs, M., Medvegyev, P.: A general ordering and fixed-point principle in complete metric space. Acta Sci. Math. (Szeged) **46**(1-4), 381–388 (1983)
121. Daniele, P.: Dynamic networks and evolutionary variational inequalities. New Dimensions in Networks. Edward Elgar Publishing, Cheltenham (2006)
122. Daniele, P., Maugeri, A.: Vector variational inequalities and modelling of a continuum traffic equilibrium problem. In: Vector variational inequalities and vector equilibria, Nonconvex Optim. Appl., vol. 38, pp. 97–111. Kluwer Academic, Dordrecht (2000)
123. Daniilidis, A., Hadjisavvas, N.: Characterization of nonsmooth semistrictly quasiconvex and strictly quasiconvex functions. J. Optim. Theory Appl. **102**(3), 525–536 (1999)
124. Daniilidis, A., Pang, J.C.H.: Continuity and differentiability of set-valued maps revisited in the light of tame geometry. J. Lond. Math. Soc. (2) **83**(3), 637–658 (2011)
125. Das, P.C.: Application of Dubovitskii-Milyutin formalism to optimal settling problem with constraints. In: Optimization and optimal control (Proc. Conf., Oberwolfach, 1974), pp. 44–58. Lecture Notes in Math., Vol. 477. Springer, Berlin (1975)
126. Dauer, J., Saleh, O.: A characterization of proper minimal points as solutions of sublinear optimization problems. J. Math. Anal. Appl. **178**, 227–246 (1993)
127. Day, M.M.: Normed Linear Spaces, 3rd edn. Springer, New York (1973). Ergebnisse der Mathematik und ihrer Grenzgebiete, Band 21
128. De Blasi, F.S.: On the differentiability of multifunctions. Pac. J. Math. **66**(1), 67–81 (1976)
129. Dedieu, J.P.: Critères de fermeture pour l'image d'un fermé non convexe par une multiapplication. C. R. Acad. Sci. Paris Sér. A-B **287**(14), A941–A943 (1978)
130. Dellnitz, M., Witting, K.: Computation of robust Pareto points. Int. J. Comput. Sci. Math. **2**(3), 243–266 (2009)
131. Demyanov, V.F., Rubinov, A.M.: Constructive nonsmooth analysis. Approximation & Optimization, vol. 7. Peter Lang, Frankfurt am Main (1995)
132. Dentcheva, D.: Approximations, expansions and univalued representations of multifunctions. Nonlinear Anal. **45**(1, Ser. A: Theory Methods), 85–108 (2001)
133. Dep, K., Gupta, H.: Introducing robustness in multi-objective optimization. Evol. Comput. **14**(4), 463–494 (2006)
134. Di, S.: Classical optimality conditions under weaker assumptions. SIAM J. Optim. **6**(1), 178–197 (1996)
135. Di, S., Poliquin, R.: Contingent cone to a set defined by equality and inequality constraints at a Fréchet differentiable point. J. Optim. Theory Appl. **81**(3), 469–478 (1994)
136. Di, S., Poliquin, R.A.: Optimality conditions for max-type nonsmooth minimization problems. Optimization **41**(3), 219–236 (1997)
137. Dias, S., Smirnov, G.: On the Newton method for set-valued maps. Nonlinear Anal. **75**(3), 1219–1230 (2012)
138. Dien, P.H.: Locally Lipschitzian set-valued maps and generalized extremal problems with inclusion constraints. Acta Math. Vietnam. **8**(1), 109–122 (1984) (1983)
139. Djafari Rouhani, B., Khan, A.A.: On the embedding of variational inequalities. Proc. Am. Math. Soc. **131**(12), 3861–3871 (2003)
140. Dolecki, S.: Tangency and differentiation: some applications of convergence theory. Ann. Mat. Pura Appl. (4) **130**, 223–255 (1982)
141. Dolecki, S.: Tangency and differentiation: marginal functions. Adv. Appl. Math. **11**(4), 389–411 (1990)
142. Dolecki, S., Greco, G.H.: Towards historical roots of necessary conditions of optimality: regula of Peano. Control Cybern. **36**(3), 491–518 (2007)
143. Dolecki, S., Greco, G.H.: Tangency vis-à-vis differentiability by Peano, Severi and Guareschi. J. Convex Anal. **18**(2), 301–339 (2011)

144. Dolecki, S., Malivert, C.: Polarities and stability in vector optimization. In: Recent Advances and Historical Development of Vector Optimization (Darmstadt, 1986), Lecture Notes in Economics and Mathematical Systems, vol. 294, pp. 96–113. Springer, Berlin (1987)
145. Dolecki, S., Malivert, C.: General duality in vector optimization. Optimization **27**(1-2), 97–119 (1993)
146. Dontchev, A.L., Rockafellar, R.T.: Implicit functions and solution mappings. Springer Monographs in Mathematics. Springer, Dordrecht (2009)
147. Du, W.S.: The extension of continuity and generalized results for set-valued maps in convex analysis. Nonlinear Anal. **71**(7-8), 3176–3184 (2009)
148. Dubovickiĭ, A.J., Miljutin, A.A.: Extremal problems with constraints. Ž. Vyčisl. Mat. i Mat. Fiz. **5**, 395–453 (1965)
149. Durea, M., Dutta, J.: Lagrange multipliers for Pareto minima in general Banach spaces. Pac. J. Optim. **4**, 447–463 (2008)
150. Durea, M., Dutta, J., Tammer, C.: Bounded sets of Lagrange multipliers for vector optimization problems in infinite dimension. J. Math. Anal. Appl. **348**, 589–606 (2008)
151. Durea, M., Dutta, J., Tammer, C.: Lagrange multipliers for ϵ-pareto solutions in vector optimization with nonsolid cones in banach spaces. J. Optim. Theory Appl. **145**, 196–211 (2010)
152. Durea, M., Dutta, J., Tammer, C.: Stability properties of KKT points in vector optimization. Optimization **60**(7), 823–838 (2011)
153. Durea, M., Strugariu, R.: On parametric vector optimization via metric regularity of constraint systems. Math. Methods Oper. Res. **74**(3), 409–425 (2011)
154. Durea, M., Strugariu, R.: Openness stability and implicit multifunction theorems: applications to variational systems. Nonlinear Anal. **75**(3), 1246–1259 (2012)
155. Durea, M., Strugariu, R.: Calculus of tangent sets and derivatives of set-valued maps under metric subregularity conditions. J. Global Optim. **56**(2), 587–603 (2013)
156. Durea, M., Tammer, C.: Fuzzy necessary optimality conditions for vector optimization problems. Optimization **58**, 449–467 (2009)
157. Dutta, J., Tammer, C.: Lagrangian conditions for vector optimization in Banach spaces. Math. Methods Oper. Res. **64**, 521–541 (2006)
158. Ehrgott, M.: Multicriteria Optimization, 2nd edn. Springer, Berlin (2005)
159. Ehrgott, M., Ide, J., Schöbel, A.: Minmax robustness for multi-objective optimization problems. Eur. J. Oper. Res. **239**, 17–31 (2014)
160. Ehrgott, M., Löhne, A., Shao, L.: A dual variant of Benson's outer approximation algorithm. J. Glob. Optim. **52**(4), 757–778 (2012)
161. Eichfelder, G.: Adaptive scalarization methods in multiobjective optimization. Vector Optimization. Springer, Berlin (2008)
162. Eichfelder, G.: Optimal elements in vector optimization with a variable ordering structure. J. Optim. Theory Appl. **151**(2), 217–240 (2011)
163. Eichfelder, G.: Variable ordering structures in vector optimization. Habilitation Thesis, University Erlangen-Nürnberg (2011)
164. Eichfelder, G.: Variable ordering structures in vector optimization. In: Recent Developments in Vector Optimization, pp. 95–126. Springer, Berlin (2012)
165. Eichfelder, G.: Numerical procedures in multiobjective optimization with variable ordering structures. J. Optim. Theory Appl. (2013). DOI:10.1007/s10957-013-0267-y
166. Eichfelder, G.: Variable Ordering Structures in Vector Optimization. Springer, Berlin (2014)
167. Eichfelder, G., Gerlach, T.: Characterization of proper optimal elements with variable ordering structures. Preprint-Series of the Institute of Mathematics, Ilmenau University of Technology, Germany **No. M 14/01** (2014)
168. Eichfelder, G., Ha, T.X.D.: Optimality conditions for vector optimization problems with variable ordering structures. Optimization **62**(5), 597–627 (2013)
169. Eichfelder, G., Jahn, J.: Vector optimization problems and their solution concepts. In: Recent Developments in Vector Optimization, pp. 1–27. Springer, Berlin (2012)

170. Eichfelder, G., Kasimbeyli, R.: Properly optimal elements in vector optimization with variable ordering structures. J. Global Optim. (2013). DOI:10.1007/s10898-013-0132-4
171. Ekeland, I.: On the variational principle. J. Math. Anal. Appl. **47**, 324–353 (1974)
172. Ekeland, I., Temam, R.: Convex Analysis and Variational Problems. North-Holland, Amsterdam (1976)
173. El Ghaoui, L., Lebret, H.: Robust solutions to least-squares problems with uncertain data. SIAM J. Matrix Anal. Appl. **18**(4), 1035–1064 (1997)
174. El Maghri, M.: Pareto-Fenchel ϵ-subdifferential sum rule and ϵ-efficiency. Optim. Lett. **6**(4), 763–781 (2012)
175. El Maghri, M., Laghdir, M.: Pareto subdifferential calculus for convex vector mappings and applications to vector optimization. SIAM J. Optim. **19**(4), 1970–1994 (2008)
176. Ellaia, R., Hassouni, A.: Characterization of nonsmooth functions through their generalized gradients. Optimization **22**(3), 401–416 (1991)
177. Elster, K.H., Elster, R.: Vector variational inequality and geometric vector optimization. In: Variational Inequalities and Network Equilibrium Problems (Erice, 1994), pp. 55–67. Plenum, New York (1995)
178. Elster, K.H., Thierfelder, J.: On cone approximation of sets. In: Mathematical Methods in Operations Research, Lectures Presented at the Summer School of Opearetions Research, Primorsko, pp. 33–59 (1984)
179. Elster, K.H., Thierfelder, J.: The general concept of cone approximations in nondifferentiable optimization. In: Nondifferentiable Optimization: Motivations and Applications (Sopron, 1984), Lecture Notes in Econom. and Math. Systems, vol. 255, pp. 170–189. Springer, Berlin (1985)
180. Elster, K.H., Thierfelder, J.: Abstract cone approximations and generalized differentiability in nonsmooth optimization. Optimization **19**(3), 315–341 (1988)
181. Elster, K.H., Thierfelder, J.: On cone approximations and generalized directional derivatives. In: Nonsmooth Optimization and Related Topics (Erice, 1988), Ettore Majorana Internat. Sci. Ser. Phys. Sci., vol. 43, pp. 133–154. Plenum, New York (1989)
182. Federer, H.: Curvature measures. Trans. Am. Math. Soc. **93**, 418–491 (1959)
183. Fernández, F.R., Hinojosa, M.A., Puerto, J.: Set-valued TU-games. Eur. J. Oper. Res. **159**(1), 181–195 (2004)
184. Ferro, F.: An optimization result for set-valued mappings and a stability property in vector problems with constraints. J. Optim. Theory Appl. **90**(1), 63–77 (1996)
185. Ferro, F.: Optimization and stability results through cone lower semicontinuity. Set Valued Anal. **5**(4), 365–375 (1997)
186. Fischetti, M., Monaci, M.: Light robustness. In: Ahuja, R.K., Moehring, R., Zaroliagis, C. (eds.) Robust and Online Large-Scale Optimization, no. 5868 in Lecture Notes in Computer Science, pp. 61–84. Springer, Berlin/Heidelberg (2009)
187. Flett, T.M.: Differential Analysis. Cambridge University Press, Cambridge (1980)
188. Flores-Bazan, F.: Optimality conditions in non-convex set-valued optimization. Math. Methods Oper. Res. **53**, 403–417 (2001)
189. Flores-Bazán, F., Hernández, E., Novo, V.: Characterizing efficiency without linear structure: a unified approach. J. Global Optim. **41**(1), 43–60 (2008)
190. Föllmer, H., Schied, A.: Convex measures of risk and trading constraints. Finance Stoch. **4**, 429–447 (2002)
191. Föllmer, H., Schied, A.: Stochastic Finance. Walter de Gruyter, Berlin (2004)
192. Furukawa, N., Yoshinaga, Y.: Higher-order variational sets, variational derivatives and higher-order necessary conditions in abstract mathematical programming. Bull. Inform. Cybern. **23**(1-2), 9–40 (1988)
193. Gajek, L., Zagrodny, D.: Countably orderable sets and their applications in optimization. Optimization **26**(3-4), 287–301 (1992)
194. Gajek, L., Zagrodny, D.: Weierstrass theorem for monotonically semicontinuous functions. Optimization **29**(3), 199–203 (1994)

195. Gast, S.: Network flow problems with uncertain input data in the context of supply chain management applications. Ph.D. thesis, University of Wuppertal (2010)
196. Gerstewitz(Tammer), C.: Nichtkonvexe Dualität in der Vektoroptimierung. Wissenschaftliche Zeitschrift der TH Leuna-Merseburg **25**(3), 357–364 (1983)
197. Gerstewitz(Tammer), C., Iwanow, E.: Dualität für nichtkonvexe Vektoroptimierungsprobleme. Wiss. Z. Tech. Hochsch. Ilmenau **2**, 61–81 (1985)
198. Gerth, C., Weidner, P.: Nonconvex separation theorems and some applications in vector optimization. J. Optim. Theory Appl. **67**(2), 297–320 (1990)
199. Giannessi, F.: Theorems of alternative, quadratic programs and complementarity problems. In: Variational Inequalities and Complementarity Problems (Proc. Internat. School, Erice, 1978), pp. 151–186. Wiley, Chichester (1980)
200. Giannessi, F. (ed.): Vector variational inequalities and vector equilibria, mathematical theories. In: Nonconvex Optimization and Its Applications, vol. 38. Kluwer Academic, Dordrecht (2000)
201. Giannessi, F., Khan, A.A.: Regularization of non-coercive quasi variational inequalities. Control Cybern. **29**(1), 91–110 (2000)
202. Giannessi, F., Khan, A.A.: On the envelope of a variational inequality. In: Nonlinear Analysis and Variational Problems, Springer Optim. Appl., vol. 35, pp. 285–293. Springer, New York (2010)
203. Giannessi, F., Mastroeni, G., Yang, X.Q.: A survey on vector variational inequalities. Boll. Unione Mat. Ital. (9) **2**(1), 225–237 (2009)
204. Giannessi, F., Mastroeni, G., Yang, X.Q.: Survey on vector complementarity problems. J. Global Optim. **53**(1), 53–67 (2012)
205. Ginchev, I., Guerraggio, A., Rocca, M.: Geoffrion type characterization of higher-order properly efficient points in vector optimization. J. Math. Anal. Appl. **328**(2), 780–788 (2007)
206. Ginchev, I., Guerraggio, A., Rocca, M.: Higher order properly efficient points in vector optimization. In: Generalized Convexity and Related Topics, Lecture Notes in Econom. and Math. Systems, vol. 583, pp. 227–245. Springer, Berlin (2007)
207. Giorgi, G., Guerraggio, A.: On a characterization of Clarke's tangent cone. J. Optim. Theory Appl. **74**(2), 369–372 (1992)
208. Giorgi, G., Guerraggio, A.: On the notion of tangent cone in mathematical programming. Optimization **25**(1), 11–23 (1992)
209. Giorgi, G., Jiménez, B., Novo, V.: An overview of second order tangent sets and their application to vector optimization. Bol. Soc. Esp. Mat. Apl. SeMA (52), 73–96 (2010)
210. Girejko, E., Malinowska, A.B., Torres, D.F.M.: The contingent epiderivative and the calculus of variations on time scales. Optimization **61**(3), 251–264 (2012)
211. Girsanov, I.V.: Lectures on mathematical theory of extremum problems. Springer, Berlin (1972), edited by B. T. Poljak, Translated from the Russian by D. Louvish, Lecture Notes in Economics and Mathematical Systems, Vol. 67
212. Goerigk, M., Schöbel, A.: A scenario-based approach for robust linear optimization. In: Proceedings of the 1st International ICST Conference on Practice and Theory of Algorithms in (Computer) Systems (TAPAS), Lecture Notes in Computer Science, pp. 139–150. Springer, New York (2011)
213. Gong, X.H.: Optimality conditions for Henig and globally proper efficient solutions with ordering cone has empty interior. J. Math. Anal. **307**, 12–37 (2005)
214. Göpfert, A., Riahi, H., Tammer, C., Zălinescu, C.: Variational methods in partially ordered spaces. CMS Books in Mathematics/Ouvrages de Mathématiques de la SMC, vol. 17. Springer, New York (2003)
215. Göpfert, A., Tammer, C., Zălinescu, C.: On the vectorial Ekeland's variational principle and minimal points in product spaces. Nonlinear Anal. **39**(7, Ser. A: Theory Methods), 909–922 (2000)
216. Gorokhovik, V.V., Zabreiko, P.P.: On Fréchet differentiability of multifunctions. Optimization **54**(4-5), 391–409 (2005)

217. Gorokhovik, V.V, Gorokhovik, S.Ya.: A criterion of the global epi-Lipschitz property of sets (Russian). Izvestiya Natsional'no Akademii Nauk Belarusi. Seriya Fiziko-Matematicheskikh Nauk, pp. 118–120 (1995)
218. Götz, A., Jahn, J.: The Lagrange multiplier rule in set-valued optimization. SIAM J. Optim. **10**(2), 331–344 (1999)
219. Guerraggio, A., Molno, E., Zaffaroni, A.: On the notion of proper efficiency in vector optimization. J. Optim. Theory Appl. **82**, 1–21 (1994)
220. Gutiérrez, C., Jiménez, B., Novo, V.: A set-valued Ekeland's variational principle in vector optimization. SIAM J. Control Optim. **47**(2), 883–903 (2008)
221. Gwinner, J.: On a new class of differential variational inequalities and a stability result. Math. Program. **139**(1-2, Ser. B), 205–221 (2013)
222. Gwinner, J., Oettli, W.: Theorems of the alternative and duality for inf-sup problems. Math. Oper. Res. **19**(1), 238–256 (1994)
223. Ha, T.: Optimality conditions for several types of efficient solutions of set-valued optimization problems. In: Nonlinear Analysis and variational Problems, Springer Optim. Appl., vol. 35, pp. 305–324. Springer, New York (2010)
224. Ha, T.X.D.: A note on a class of cones ensuring the existence of efficient points in bounded complete sets. Optimization **31**(2), 141–152 (1994)
225. Ha, T.X.D.: On the existence of efficient points in locally convex spaces. J. Global Optim. **4**(3), 265–278 (1994)
226. Ha, T.X.D.: Lagrange multipliers for set-valued optimization problems associated with coderivatives. J. Math. Anal. Appl. **311**(2), 647–663 (2005)
227. Ha, T.X.D.: Some variants of the Ekeland variational principle for a set-valued map. J. Optim. Theory Appl. **124**(1), 187–206 (2005)
228. Ha, T.X.D.: Optimality conditions for several types of efficient solutions of set-valued optimization problems. In: Nonlinear Analysis and Variational Problems, Springer Optim. Appl., vol. 35, pp. 305–324. Springer, New York (2010)
229. Ha, T.X.D.: The Fermat rule and Lagrange multiplier rule for various efficient solutions of set-valued optimization problems expressed in terms of coderivatives. In: Recent Developments in Vector Optimization, pp. 417–466. Springer, Berlin (2012)
230. Haimes, Y., Lasdon, L.S., Wismer, D.A.: On a bicriterion formulation of the problems of integrated system identification and system optimization. IEEE Trans. Syst. Man Cybern. **1**, 296–297 (1971)
231. Halkin, H.: A satisfactory treatment of equality and operator constraints in the Dubovitskii-Milyutin optimization formalism. J. Optim. Theory Appl. **6**, 138–149 (1970)
232. Halkin, H.: Necessary conditions for optimal control problems with differentiable or non-differentiable data. In: Mathematical Control Theory (Proc. Conf., Australian Nat. Univ., Canberra, 1977), Lecture Notes in Math., vol. 680, pp. 77–118. Springer, Berlin (1978)
233. Hamel, A.: Remarks to an equivalent formulation of Ekeland's variational principle. Optimization **31**(3), 233–238 (1994)
234. Hamel, A., Heyde, F., Löhne, A., Tammer, C., Winkler, K.: Closing the duality gap in linear vector optimization. J. Convex Anal. **11(1)**, 163–178 (2004)
235. Hamel, A.H.: Equivalents to Ekeland's variational principle in uniform spaces. Nonlinear Anal. **62**(5), 913–924 (2005)
236. Hamel, A.H., Löhne, A.: Lagrange duality in set optimization. Submitted for publication (2012), http://arxiv.org/abs/1207.4433v1
237. Hamel, A.H., Tammer, C.: Minimal elements for product orders. Optimization **57**(2), 263–275 (2008)
238. Han, W.: A posteriori error analysis via duality theory, with applications in modeling and numerical approximations. In: Advances in Mechanics and Mathematics, vol. 8. Springer, New York (2005)
239. Hartley, R.: On cone-efficiency, cone-convexity and cone-compactness. SIAM J. Appl. Math. **34**(2), 211–222 (1978)

240. Hazen, G.B., Morin, T.L.: Optimality conditions in nonconical multiple-objective programming. J. Optim. Theory Appl. **40**(1), 25–60 (1983)
241. Henig, M.I.: Existence and characterization of efficient decisions with respect to cones. Math. Program. **23**(1), 111–116 (1982)
242. Henig, M.I.: Proper efficiency with respect to cones. J. Optim. Theory Appl. **36**(3), 387–407 (1982)
243. Henrion, R., Jourani, A., Outrata, J.: On the calmness of a class of multifunctions. SIAM J. Optim. **13**(2), 603–618 (electronic) (2002)
244. Hernández, E., Jiménez, B., Novo, V.: Benson proper efficiency in set-valued optimization on real linear spaces. In: Recent Advances in Optimization, Lecture Notes in Econom. and Math. Systems, vol. 563, pp. 45–59. Springer, Berlin (2006)
245. Hernández, E., Jiménez, B., Novo, V.: Weak and proper efficiency in set-valued optimization on real linear spaces. J. Convex Anal. **14**(2), 275–296 (2007)
246. Hernández, E., Khan, A.A., Rodríguez-Marín, L., Sama, M.: Computation formulas and multiplier rules for graphical derivatives in separable banach spaces. Nonlinear Anal. **71**, 4241–4250 (2009)
247. Hernández, E., Löhne, A., Rodríguez-Marín, L., Tammer, C.: Lagrange duality in vector optimization - a simplified approach based on complete lattices. Preprint (2009)
248. Hernández, E., Löhne, A., Rodríguez-Marín, L., Tammer, C.: Lagrange duality, stability and subdifferentials in vector optimization. Optimization **62**(3), 415–428 (2013)
249. Hernández, E., Rodríguez-Marín, L.: Duality in set optimization with set-valued maps. Pac. J. Optim. **3**(2), 245–255 (2007)
250. Hernández, E., Rodríguez-Marín, L.: Existence theorems for set optimization problems. Nonlinear Anal. **67**(6), 1726–1736 (2007)
251. Hernández, E., Rodríguez-Marín, L.: Lagrangian duality in set-valued optimization. J. Optim. Theory Appl. **134**(1), 119–134 (2007)
252. Hernández, E., Rodríguez-Marín, L.: Nonconvex scalarization in set optimization with set-valued maps. J. Math. Anal. Appl. **325**(1), 1–18 (2007)
253. Hernández, E., Rodríguez-Marín, L., Sama, M.: On solutions of set-valued optimization problems. Comput. Math. Appl. **60**(5), 1401–1408 (2010)
254. Hestenes, M.R.: Calculus of Variations and Optimal Control Theory. Wiley, New York (1966)
255. Hestenes, M.R.: Optimization Theory. Wiley-Interscience, New York (1975)
256. Heyde, F.: Coherent risk measures and vector optimization. In: Küfer, K.H., Rommelfanger, H., Tammer, C., Winkler, K. (eds.) Multicriteria Decision Making and Fuzzy Systems, pp. 3–12. SHAKER, Aachen (2006)
257. Heyde, F., Löhne, A.: Solution concepts in vector optimization: a fresh look at an old story. Optimization **60**(12), 1421–1440 (2011)
258. Heyde, F., Löhne, A., Tammer, C.: Set-valued duality theory for multiple objective linear programs and application to mathematical finance. Math. Methods Oper. Res. **69**(1), 159–179 (2009)
259. Heyde, F., Löhne, A., Tammer, C.: The attainment of the solution of the dual program in vertices for vectorial linear programs. In: Vincent, B. et al. (eds.) Multiobjective Programming and Goal Programming. Theoretical Results and Practical Applications. Lecture Notes in Economics and Mathematical Systems, vol. 618, pp. 13–24. Springer, New York (2009)
260. Heyde, F., Schrage, C.: Continuity concepts for set-valued functions and a fundamental duality formula for set-valued optimization. J. Math. Anal. Appl. **397**(2), 772–784 (2013)
261. Hiriart-Urruty, J.B.: New concepts in nondifferentiable programming. Bull. Soc. Math. France Mém **60**, 57–85 (1979)
262. Hiriart-Urruty, J.B.: Tangent cones, generalized gradients and mathematical programming in Banach spaces. Math. Oper. Res. **4**(1), 79–97 (1979)
263. Hiriart-Urruty, J.B.: A note on the mean value theorem for convex functions. Boll. Un. Mat. Ital. B (5) **17**(2), 765–775 (1980)
264. Hites, R., De Smet, Y., Risse, N., Salazar-Neumann, M., Vincke, P.: About the applicability of MCDA to some robustness problems. Eur. J. Oper. Res. **174**, 322–332 (2006)

265. Hoffmann, K.H., Kornstaedt, H.J.: Higher-order necessary conditions in abstract mathematical programming. J. Optim. Theory Appl. **26**(4), 533–568 (1978)
266. Hoheisel, T., Kanzow, C., Mordukhovich, B.S., Phan, H.: Generalized Newton's method based on graphical derivatives. Nonlinear Anal. **75**(3), 1324–1340 (2012)
267. Hu, S., Papageorgiou, N.S.: Handbook of multivalued analysis. Vol. I, Mathematics and Its Applications, vol. 419. Kluwer Academic, Dordrecht (1997)
268. Hu, S., Papageorgiou, N.S.: Handbook of multivalued analysis. Vol. II, Mathematics and Its Applications, vol. 500. Kluwer Academic, Dordrecht (2000)
269. Huang, H.: The Lagrange multiplier rule for super efficiency in vector optimization. J. Math. Anal. Appl. **342**, 503–513 (2008)
270. Huang, H., Li, R.: Global error bounds for γ-paraconvex multifunctions. Set Valued Var. Anal. **19**(3), 487–504 (2011)
271. Hurwicz, L.: Programming in linear space. In: Studies in Linear and Non-linear Programming. Stanford University Press, Stanford, California (1958)
272. Huy, N.Q., Mordukhovich, B., Yao, J.C.: Coderivatives of frontier and solution maps in parameteric multiobjective optimization. Taiwanese J. Math. **13**(6A), 1671–1693 (2009)
273. Ide, J., Köbis, E. Kuroiwa, D., Schöbel, A., Tammer, C.: The relationship between multicriteria robustness concepts and set-valued optimization. Fixed Point Theory Appl. **83** (2014)
274. Ide, J., Köbis, E.: Concepts of robustness for multi-objective optimization problems based on set order relations (2013)
275. Ioffe, A.D.: Calculus of Dini subdifferentials of functions and contingent coderivatives of set-valued maps. Nonlinear Anal. **8**(5), 517–539 (1984)
276. Ioffe, A.D.: Approximate subdifferentials and applications. II. Mathematika **33**(1), 111–128 (1986)
277. Ioffe, A.D.: Approximate subdifferentials and applications. III. The metric theory. Mathematika **36**(1), 1–38 (1989)
278. Ioffe, A.D.: Variational analysis and mathematical economics 1: Subdifferential calculus and the second theorem of welfare economics. In: Advances in Mathematical Economics, Vol. 12. Adv. Math. Econ., vol. 12, pp. 71–95. Springer, Tokyo (2009)
279. Ioffe, A.D., Tihomirov, V.M.: Theory of extremal problems. Studies in Mathematics and Its Applications, vol. 6. North-Holland, Amsterdam (1979)
280. Isac, G.: Sur l'existence de l'optimum de Pareto. Riv. Mat. Univ. Parma (4) **9**, 303–325 (1984) (1983)
281. Isac, G.: Pareto optimization in infinite-dimensional spaces: the importance of nuclear cones. J. Math. Anal. Appl. **182**(2), 393–404 (1994)
282. Isac, G.: The Ekeland's principle and the Pareto ε-efficiency. In: Multi-objective Programming and Goal Programming: Theories and Applications. Proceedings of the First International Conference, MOPGP (Portsmouth, UK, 1994), Lecture Notes in Econom. and Math. Systems, vol. 432, pp. 148–163. Springer, Berlin (1996)
283. Isac, G., Khan, A.: Second-order optimality conditions in set-valued optimization by a new tangential derivative. Acta Math. Vietnam. **34**, 81–90 (2009)
284. Isac, G., Khan, A.A.: Dubovitskii-Milyutin approach in set-valued optimization. SIAM J. Control Optim. **47**(1), 144–162 (2008)
285. Isac, G., Tammer, C.: Nuclear and full nuclear cones in product spaces: Pareto efficiency and an Ekeland type variational principle. Positivity **9**(3), 511–539 (2005)
286. Isermann, H.: Proper efficiency and the linear vector maximization problem. Oper. Res. **22**, 189–191 (1974)
287. Isermann, H.: On some relations between a dual pair of multiple objective linear programs. Z. Oper. Res., Ser. A **22**, 33–41 (1978)
288. Jachymski, J.R.: Caristi's fixed point theorem and selections of set-valued contractions. J. Math. Anal. Appl. **227**(1), 55–67 (1998)
289. Jadamba, B., Khan, A.A., Raciti, F.: Regularization of stochastic variational inequalities and a comparison of an L_p and a sample-path approach. Nonlinear Anal. **94**, 65–83 (2014)
290. Jahn, J.: Duality in vector optimization. Math. Program. **25**, 343–353 (1983)

291. Jahn, J.: Existence theorems in vector optimization. J. Optim. Theory Appl. **50**(3), 397–406 (1986)
292. Jahn, J.: Mathematical Vector Optimization in Partially Ordered Linear Spaces. Verlag Peter Lang, Frankfurt am Main (1986)
293. Jahn, J.: Vector Optimization. Theory, Applications and Extensions. Springer, Berlin (2004)
294. Jahn, J.: A derivative-free descent method in set optimization. Preprint-Reihe Angewandte Mathematik, Friedrich-Alexander-Universität Erlangen-Nürnberg **377** (2014)
295. Jahn, J., Ha, T.X.D.: New order relations in set optimization. J. Optim. Theory Appl. **148**(2), 209–236 (2011)
296. Jahn, J., Khan, A.A.: Existence theorems and characterizations of generalized contingent epiderivatives. J. Nonlinear Convex Anal. **3**(3), 315–330 (2002)
297. Jahn, J., Khan, A.A.: Generalized contingent epiderivatives in set-valued optimization: optimality conditions. Numer. Funct. Anal. Optim. **23**(7-8), 807–831 (2002)
298. Jahn, J., Khan, A.A.: The existence of contingent epiderivatives for set-valued maps. Appl. Math. Lett. **16**(8), 1179–1185 (2003)
299. Jahn, J., Khan, A.A.: Some calculus rules for contingent epiderivatives. Optimization **52**(2), 113–125 (2003)
300. Jahn, J., Khan, A.A., Zeilinger, P.: Second-order optimality conditions in set optimization. J. Optim. Theory Appl. **125**(2), 331–347 (2005)
301. Jahn, J., Rauh, R.: Contingent epiderivatives and set-valued optimization. Math. Methods Oper. Res. **46**(2), 193–211 (1997)
302. Jahn, J., Sachs, E.: Generalized quasiconvex mappings and vector optimization. SIAM J. Control Optim. **24**(2), 306–322 (1986)
303. Jameson, G.: Ordered linear spaces. Lecture Notes in Mathematics, Vol. 141. Springer, Berlin (1970)
304. Jiménez, B., Novo, V.: Second order necessary conditions in set constrained differentiable vector optimization. Math. Methods Oper. Res. **58**(2), 299–317 (2003)
305. Jiménez, B., Novo, V.: Optimality conditions in differentiable vector optimization via second-order tangent sets. Appl. Math. Optim. **49**(2), 123–144 (2004)
306. Jiménez, B., Novo, V.: Characterization of the cone of attainable directions. J. Optim. Theory Appl. **131**(3), 493–499 (2006)
307. Jofré, A., Penot, J.P.: Comparing new notions of tangent cones. J. Lond. Math. Soc. (2) **40**(2), 280–290 (1989)
308. Jourani, A.: Regularity and strong sufficient optimality conditions in differentiable optimization problems. Numer. Funct. Anal. Optim. **14**(1-2), 69–87 (1993)
309. Jourani, A.: Metric regularity and second-order necessary optimality conditions for minimization problems under inclusion constraints. J. Optim. Theory Appl. **81**(1), 97–120 (1994)
310. Jourani, A.: Open mapping theorem and inversion theorem for γ-paraconvex multivalued mappings and applications. Studia Math. **117**(2), 123–136 (1996)
311. Jourani, A.: Necessary conditions for extremality and separation theorems with applications to multiobjective optimization. Optimization **44**(4), 327–350 (1998)
312. Jourani, A., Thibault, L.: Chain rules for coderivatives of multivalued mappings in Banach spaces. Proc. Am. Math. Soc. **126**, 1479–1485 (1998)
313. Kalashnikov, V.V., Jadamba, B., Khan, A.A.: First and second order optimality condition in set-optimization. In: Dempe, S., et al. (eds.), Optimization with Multivalued Mappings, pp. 265–276. Springer, New York (2006)
314. Kasimbeyli, R.: Radial epiderivatives and set-valued optimization. Optimization **58**, 521–534 (2009)
315. Kawasaki, H.: Conjugate relations and weak subdifferentials of relations. Math. Oper. Res. **6**, 593–607 (1981)
316. Kawasaki, H.: A duality theorem in multiobjective nonlinear programming. Math. Oper. Res. **7**, 95–110 (1982)
317. Kawasaki, H.: An envelope-like effect of infinitely many inequality constraints on second-order necessary conditions for minimization problems. Math. Program. **41**(1, (Ser. A)), 73–96 (1988)

318. Khan, A.A., Raciti, F.: A multiplier rule in set-valued optimization. Bull. Aust. Math. Soc. **68**, 93–100 (2003)
319. Khan, A.A., Sama, M.: A multiplier rule for stable problems in vector optimization. J. Convex Anal. **19**(2), 525–539 (2012)
320. Khan, A.A., Sama, M.: Optimal control of multivalued quasi variational inequalities. Nonlinear Anal. **75**(3), 1419–1428 (2012)
321. Khan, A.A., Tammer, C.: Generalized Dubovitskii-Milyutin approach in set-valued optimization. Vietnam J. Math. **40**(2-3), 285–304 (2012)
322. Khan, A.A., Tammer, C.: Second-order optimality conditions in set-valued optimization via asymptotic derivatives. Optimization **62**(6), 743–758 (2013)
323. Khan, A.A., Ward, D.E.: Toward second-order sensitivity analysis in set-valued optimization. J. Nonlinear Convex Anal. **13**(1), 65–83 (2012)
324. Khan, M.A.: Ioffe's normal cone and the foundations of welfare economics: the infinite-dimensional theory. J. Math. Anal. Appl. **161**(1), 284–298 (1991)
325. Khanh, P., Tuan, N.: Higher-order variational sets and higher-order optimality conditions for proper efficiency in set-valued nonsmooth vector optimization. J. Optim. Theory Appl. **139**, 243–261 (2008)
326. Khanh, P., Tuan, N.: Variational sets of multi-valued mappings and a unified study of optimality condition. J. Optim. Theory Appl. **139**, 45–67 (2008)
327. Kinderlehrer, D., Stampacchia, G.: An introduction to variational inequalities and their applications. In: Classics in Applied Mathematics, vol. 31. Society for Industrial and Applied Mathematics (SIAM), Philadelphia, PA (2000). Reprint of the 1980 original
328. Klamroth, K., Köbis, E., Schöbel, A., Tammer, C.: A unified approach for different concepts of robustness and stochastic programming via non-linear scalarizing functionals. Optimization **62**(5), 649–671 (2013)
329. Klatte, D., Kummer, B.: Nonsmooth equations in optimization. Nonconvex Optimization and Its Applications, vol. 60. Kluwer Academic, Dordrecht (2002)
330. Klose, J.: Sensitivity analysis using the tangent derivative. Numer. Funct. Anal. Optim. **13**(1-2), 143–153 (1992)
331. Konnov, I.: Combined relaxation methods for variational inequalities. Lecture Notes in Economics and Mathematical Systems, vol. 495. Springer, Berlin (2001)
332. Kotarski, W.: Further generalization of the Dubovicki-Milutin theorem. J. Optim. Theory Appl. **54**(3), 565–573 (1987)
333. Kotarski, W.: Optimal control of a system governed by a parabolic equation with an infinite number of variables. J. Optim. Theory Appl. **60**(1), 33–41 (1989)
334. Kotarski, W.: On some specification of the Dubovicki-Milutin theorem for Pareto optimal problems. Nonlinear Anal. **14**(3), 287–291 (1990)
335. Kouvelis, P., Sayin, S.: Algorithm robust for the bicriteria discrete optimization problem. Ann. Oper. Res. **147**, 71–85 (2006)
336. Kouvelis, P., Yu, G.: Robust Discrete Optimization and Its Applications. Kluwer Academic, Dordrecht (1997)
337. Kowalewski, A., Kotarski, W.: Application of Milutin-Dubovicki's method to solving an optimal control problem for hyperbolic systems. Probl. Control Inform. Theory/Problemy Upravlen. Teor. Inform. **9**(3), 183–193 (1980)
338. Krabs, W.: Optimization and Approximation. Wiley, Chichester (1979)
339. Krasnosel′skiĭ, M.A.: Polozhitelnye resheniya operatornykh uravnenii. Gosudarstv. Izdat. Fiz.-Mat. Lit., Moscow (1962)
340. Krasnosel′skiĭ, M.A.: Positive solutions of operator equations. Translated from the Russian by Richard E. Flaherty; edited by Leo F. Boron. P. Noordhoff Ltd. Groningen (1964)
341. Křivan, V.: On the intersection of contingent cones. J. Optim. Theory Appl. **70**(2), 397–404 (1991)
342. Kruger, A.J., Morduhovič, B.Š.: Extremal points and the Euler equation in nonsmooth optimization problems. Dokl. Akad. Nauk BSSR **24**(8), 684–687, 763 (1980)

343. Kruger, A.Y.: On the extremality of set systems. Dokl. Nats. Akad. Nauk Belarusi **42**(1), 24–28, 123 (1998)
344. Kuk, H., Tanino, T., Tanaka, M.: Sensitivity analysis in parametrized convex vector optimization. J. Math. Anal. Appl. **202**(2), 511–522 (1996)
345. Kuk, H., Tanino, T., Tanaka, M.: Sensitivity analysis in vector optimization. J. Optim. Theory Appl. **89**(3), 713–730 (1996)
346. Kurcyusz, S.: On the existence and non-existence Lagrange multipliers in Banach spaces. J. Optim. Theory Appl. **20**(1), 81–110 (1976)
347. Kuroiwa, D.: Existence Theorems of Set Optimization with Set-Valued Maps. Manuscript Shimane University, Japan (1997)
348. Kuroiwa, D.: Lagrange duality of set-valued optimization with natural criteria. Sūrikaisekikenkyūsho Kōkyūroku (1068), 164–170 (1998)
349. Kuroiwa, D.: Natural Criteria of Set-Valued Optimization. Manuscript Shimane University, Japan (1068), 164–170 (1998)
350. Kuroiwa, D.: On natural criteria in set-valued optimization. RIMS Kokyuroku **1048**, 86–92 (1998)
351. Kuroiwa, D.: Some duality theorems of set-valued optimization with natural criteria. In: Tanaka, T. (ed.) Proceedings of the International Conference on Nonlinear Analysis and Convex Analysis, pp. 221–228. World Scientific, Singapore (1999)
352. Kuroiwa, D.: Existence theorems of set optimization with set-valued maps. J. Inf. Optim. Sci. **24**(1), 73–84 (2003)
353. Kuroiwa, D.: On derivatives of Set-Valued Maps in Set Optimization. Manuscript Shimane University, Japan (2008)
354. Kuroiwa, D.: On derivatives of set-valued maps and optimality conditions for set optimization. J. Nonlinear Convex Anal. **10**(1), 41–50 (2009)
355. Kuroiwa, D., Lee, G.M.: On robust multiobjective optimization. Vietnam J. Math. **40**(2-3), 305–317 (2012)
356. Kuroiwa, D., Lee, G.M.: On robust convex multiobjective optimization. Accepted to J. Nonlinear Convex Anal. (2013)
357. Kuroiwa, D., Nuriya, T.: A generalized embedding vector space in set optimization. In: Proceedings of the Fourth Intemational Conference on Nonlinear Analysis and Convex Analysis, pp. 297–304 (2006)
358. Kusraev, A.G., Kutateladze, S.S.: Subdifferentials: theory and applications. In: Mathematics and Its Applications, vol. 323. Kluwer Academic, Dordrecht (1995). Translated from the Russian
359. Laghdir, M.: Some remarks on subdifferentiability of convex functions. Appl. Math. E-Notes **5**, 150–156 (electronic) (2005)
360. Lalitha, C.S., Arora, R.: Conjugate maps, subgradients and conjugate duality in set-valued optimization. Numer. Funct. Anal. Optim. **28**(7-8), 897–909 (2007)
361. Lalitha, C.S., Arora, R.: Weak Clarke epiderivative in set-valued optimization. J. Math. Anal. Appl. **342**(1), 704–714 (2008)
362. Lasiecka, I.: Conical approximations of sets in optimization theory. Control Cybern. **4**(3-4), 39–58 (1975)
363. Laurent, P.J.: Approximation et Optimisation, Collection Enseignement des Sciences, No. 13. Hermann, Paris (1972)
364. Ledzewicz, U., Schaettler, H.: Second-order conditions for extremum problems with nonregular equality constraints. J. Optim. Theory Appl. **86**(1), 113–144 (1995)
365. Ledzewicz, U., Schättler, H.: A second-order Dubovitskii-Milyutin theory and applications to control. In: Advances in Nonlinear Dynamics. Stability Control Theory Methods Appl., vol. 5, pp. 179–192. Gordon and Breach, Amsterdam (1997)
366. Ledzewicz, U., Schättler, H.: High-order extended maximum principles for optimal control problems with non-regular constraints. In: Optimal control (Gainesville, FL, 1997), Appl. Optim., vol. 15, pp. 298–325. Kluwer Academic, Dordrecht (1998)

367. Ledzewicz, U., Schättler, H.: A high-order generalization of the Lyusternik theorem. Nonlinear Anal. **34**(6), 793–815 (1998)
368. Ledzewicz, U., Schättler, H.: High-order approximations and generalized necessary conditions for optimality. SIAM J. Control Optim. **37**(1), 33–53 (1999)
369. Ledzewicz, U., Schättler, H.: A high-order generalized local maximum principle. SIAM J. Control Optim. **38**(3), 823–854 (2000)
370. Ledzewicz, U., Walczak, S.: On the Lusternik theorem for nonsmooth operators. Nonlinear Anal. **22**(2), 121–128 (1994)
371. Ledzewicz, U., Walczak, S.: Generalizations of the Lusternik theorem and applications to nonsmooth and abnormal problems in optimization and optimal control. In: World Congress of Nonlinear Analysts '92, Vol. I–IV (Tampa, FL, 1992), pp. 2583–2594. de Gruyter, Berlin (1996)
372. Ledzewicz-Kowalewska, U.: On some specification of the Dubovitskiĭ-Milyutin method. Nonlinear Anal. **10**(12), 1367–1371 (1986)
373. Ledzewicz-Kowalewska, U.: Application of the method of contractor directions to the Dubovitskiĭ-Milyutin formalism. J. Math. Anal. Appl. **125**(1), 174–184 (1987)
374. Ledzewicz-Kowalewska, U.: Application of some specification of the Dubovitskiĭ-Milyutin method to problems of optimal control. Nonlinear Anal. **12**(2), 101–108 (1988)
375. Leung, S.F.: An economic application of the Dubovitskii-Milyutin version of the maximum principle. Optim. Control Appl. Methods **28**(6), 435–449 (2007)
376. Levy, A.B., Mordukhovich, B.S.: Coderivatives in parametric optimization. Math. Program. **99**(2, Ser. A), 311–327 (2004)
377. Li, J., Mastroeni, G.: Vector variational inequalities involving set-valued mappings via scalarization with applications to error bounds for gap functions. J. Optim. Theory Appl. **145**(2), 355–372 (2010)
378. Li, M.H., Li, S.J.: Second-order differential and sensitivity properties of weak vector variational inequalities. J. Optim. Theory Appl. **144**(1), 76–87 (2010)
379. Li, S.J., Chen, C.R., Wu, S.Y.: Conjugate dual problems in constrained set-valued optimization and applications. Eur. J. Oper. Res. **196**(1), 21–32 (2009)
380. Li, S.J., Meng, K.W., Penot, J.P.: Calculus rules for derivatives of multimaps. Set Valued Var. Anal. **17**(1), 21–39 (2009)
381. Li, S.J., Sun, X.K., Zhai, J.: Second-order contingent derivatives of set-valued mappings with application to set-valued optimization. Appl. Math. Comput. **218**(12), 6874–6886 (2012)
382. Li, S.J., Sun, X.K., Zhu, S.K.: Higher-order optimality conditions for strict minimality in set-valued optimization. J. Nonlinear Convex Anal. **13**(2), 281–291 (2012)
383. Li, S.J., Yan, H., Chen, G.Y.: Differential and sensitivity properties of gap functions for vector variational inequalities. Math. Methods Oper. Res. **57**(3), 377–391 (2003)
384. Li, S.J., Yang, X.Q., Chen, G.y.: Vector Ekeland variational principle. In: Vector Variational Inequalities and Vector Equilibria. Nonconvex Optim. Appl., vol. 38, pp. 321–333. Kluwer Academic, Dordrecht (2000)
385. Li, S.J., Zhu, S.K., Li, X.B.: Second-order optimality conditions for strict efficiency of constrained set-valued optimization. J. Optim. Theory Appl. **155**(2), 534–557 (2012)
386. Li, T., Xu, Y.: The strictly efficient subgradient of set-valued optimisation. Bull. Aust. Math. Soc. **75**(3), 361–371 (2007)
387. Li, Z.F., Chen, G.Y.: Lagrangian multipliers, saddle points, and duality in vector optimization of set-valued maps. J. Math. Anal. Appl. **215**(2), 297–316 (1997)
388. Liebchen, C., Lübbecke, M., Möhring, R.H., Stiller, S.: The concept of recoverable robustness, linear programming recovery, and railway applications. In: Ahuja, R.K.m Möhring, R., Zaroliagis, C. (eds.) Robust and online large-scale optimization. Lecture Note on Computer Science, vol. 5868. Springer, New York (2009)
389. Lin, L.J.: Optimization of set-valued functions. J. Math. Anal. Appl. **186**(1), 30–51 (1994)
390. Lin, L.J., Chuang, C.S.: Existence theorems for variational inclusion problems and the set-valued vector Ekeland variational principle in a complete metric space. Nonlinear Anal. **70**(7), 2665–2672 (2009)

391. Liu, C.G., Ng, K.F.: Ekeland's variational principle for set-valued functions. SIAM J. Optim. **21**(1), 41–56 (2011)
392. Liu, J., Song, W.: On proper efficiencies in locally convex spaces—a survey. Acta Math. Vietnam. **26**(3), 301–312 (2001)
393. Löhne, A.: Optimization with set relations. Dissertation, Martin-Luther-University Halle-Wittenberg (2005)
394. Löhne, A.: Optimization with set relations: Conjugate duality. Optimization **54(3)**, 265–282 (2005)
395. Löhne, A.: Vector Optimization with Infimum and Supremum. Springer, Berlin (2011)
396. Löhne, A., Schrage, C.: An algorithm to solve polyhedral convex set-valued optimization problems. Optimization **62**(1), 131–141 (2013)
397. Löhne, A., Tammer, C.: A new approach to duality in vector optimization. Optimization **56**(1-2), 221–239 (2007)
398. Löhne, A., Tammer, C.: Lagrange duality in vector optimization. In: Löhne, A., Riedrich, T., Tammer, C. (eds.) Festschrift zum 75. Geburtstag von Prof. Dr. Alfred Göpfert, pp. 17–33. Preprint Series Institut of Mathematics, Martin-Luther-University Halle-Wittenberg No. 07 (2009)
399. Long, X.J., Peng, J.W., Wong, M.M.: Generalized radial epiderivatives and nonconvex set-valued optimization problems. Appl. Anal. **91**(10), 1891–1900 (2012)
400. Lục, Ḏ.T.: Contingent derivatives of set-valued maps and applications to vector optimization. Math. Program. **50**(1, (Ser. A)), 99–111 (1991)
401. Luc, D.T.: An existence theorem in vector optimization. Math. Oper. Res. **14**(4), 693–699 (1989)
402. Luc, D.T.: Theory of vector optimization. Lecture Notes in Economics and Mathematical Systems, vol. 319. Springer, Berlin (1989)
403. Luc, D.T.: Theory of vector optimization. Lecture Notes in Economics and Mathematical Systems, vol. 319. Springer, Berlin (1989)
404. Lục, D.T.: Recession cones and the domination property in vector optimization. Math. Program. **49**(1, (Ser. A)), 113–122 (1990/91)
405. Luc, D.T., Malivert, C.: Invex optimization problems. Bull. Aust. Math. Soc. **46**, 47–66 (1992)
406. Luc, D.T., Swaminathan, S.: A characterization of convex functions. Nonlinear Anal. **20**(6), 697–701 (1993)
407. Luc, D.T., Tan, N.X., Tinh, P.N.: Convex vector functions and their subdifferential. Acta Math. Vietnam. **23**(1), 107–127 (1998)
408. Luenberger, D.G.: New optimality principles for economic efficiency and equilibrium. J. Optim. Theory Appl. **75**(2), 221–264 (1992)
409. Makarov, E.K., Rachkovski, N.N.: Unified representation of proper efficiency by means of dilating cones. J. Optim. Theory Appl. **101**(1), 141–165 (1999)
410. Malcolm, G.G., Mordukhovich, B.S.: Pareto optimality in nonconvex economies with infinite-dimensional commodity spaces. J. Global Optim. **20**(3-4), 323–346 (2001)
411. Malivert, C.: Fenchel duality in vector optimization. In: Advances in Optimization (Lambrecht, 1991). Lecture Notes in Econom. and Math. Systems, vol. 382, pp. 420–438. Springer, Berlin (1992)
412. Malivert, C., Penot, J.P., Théra, M.: Un prolongement du théorème de Hahn-Banach. C. R. Acad. Sci. Paris Sér. A-B **286**(3), A165–A168 (1978)
413. Martin, D.H., Gardner, R.J., Watkins, G.G.: Indicating cones and the intersection principle for tangential approximants in abstract multiplier rules. J. Optim. Theory Appl. **33**(4), 515–537 (1981)
414. Martin, D.H., Watkins, G.G.: Cores of tangent cones and Clarke's tangent cone. Math. Oper. Res. **10**(4), 565–575 (1985)
415. Mas-Colell, A.: Pareto optima and equilibria: the finite-dimensional case. In: Advances in Equilibrium Theory (Indianapolis, Ind., 1984). Lecture Notes in Econom. and Math. Systems, vol. 244, pp. 25–42. Springer, Berlin (1985)

416. Mastroeni, G.: On Minty vector variational inequality. In: Vector Variational Inequalities and Vector Equilibria. Nonconvex Optim. Appl., vol. 38, pp. 351–361. Kluwer Academic, Dordrecht (2000)
417. Megginson, R.E.: An introduction to Banach space theory. Graduate Texts in Mathematics, vol. 183. Springer, New York (1998)
418. Meng, K.W., Li, S.J.: Differential and sensitivity properties of gap functions for Minty vector variational inequalities. J. Math. Anal. Appl. **337**(1), 386–398 (2008)
419. Meng, Z.Q.: Hahn-Banach theorem of set-valued map. Appl. Math. Mech. **19**(1), 55–61 (1998)
420. Michel, P., Penot, J.P.: Calcul sous-différentiel pour des fonctions lipschitziennes et non lipschitziennes. C. R. Acad. Sci. Paris Sér. I Math. **298**(12), 269–272 (1984)
421. Minchenko, L., Volosevich, A.: Strongly differentiable multifunctions and directional differentiability of marginal functions. In: Quasidifferentiability and Related Topics. Nonconvex Optim. Appl., vol. 43, pp. 163–171. Kluwer Academic, Dordrecht (2000)
422. Minh, N.B., Tan, N.X.: On the C-Lipschitz continuities and C-approximations of multivalued mappings. Vietnam J. Math. **30**(4), 343–363 (2002)
423. Minh, N.B., Tan, N.X.: On the continuity of vector convex multivalued functions. Acta Math. Vietnam. **27**(1), 13–25 (2002)
424. Mishra, S.K., Noor, M.A.: On vector variational-like inequality problems. J. Math. Anal. Appl. **311**(1), 69–75 (2005)
425. Mordukhovich, B.S.: Maximum principle in the problem of time optimal response with nonsmooth constraints. Prikl. Mat. Meh. **40**(6), 1014–1023 (1976)
426. Mordukhovich, B.S.: Metric approximations and necessary conditions for optimality for general classes of nonsmooth extremal problems, (russian). Dokl. Akad. Nauk SSSR **254**, 1072–1076 (1980)
427. Mordukhovich, B.S.: Generalized differential calculus for nonsmooth and set-valued mappings. J. Math. Anal. Appl. **183**(1), 250–288 (1994)
428. Mordukhovich, B.S.: An abstract extremal principle with applications to welfare economics. J. Math. Anal. Appl. **251**(1), 187–216 (2000)
429. Mordukhovich, B.S.: Nonlinear prices in nonconvex economies with classical Pareto and strong Pareto optimal allocations. Positivity **9**(3), 541–568 (2005)
430. Mordukhovich, B.S.: Variational Analysis and Generalized Differentiation, Vol. I: Basic Theory, Vol. II: Applications. Springer, Berlin (2006)
431. Mordukhovich, B.S.: Variational analysis and generalized differentiation, Vol. II: Applications, Grundlehren der Mathematischen Wissenschaften [Fundamental Principles of Mathematical Sciences], vol. 331. Springer, Berlin (2006)
432. Mordukhovich, B.S., Shao, Y.H.: Nonsmooth sequential analysis in Asplund spaces. Trans. Am. Math. Soc. **348**(4), 1235–1280 (1996)
433. Mordukhovich, B.S., Treiman, J.S., Zhu, Q.J.: An extended extremal principle with applications to multiobjective optimization. SIAM J. Optim. **14**(2), 359–379 (electronic) (2003)
434. Mordukhovich, B.S., Wang, B.: Necessary optimality and suboptimality conditions in nondifferentiable programming via variational principles. SIAM J. Control Optim. **41**, 623–640 (2002)
435. Motreanu, D., Pavel, N.: Tangency, flow invariance for differential equations, and optimization problems. Monographs and Textbooks in Pure and Applied Mathematics, vol. 219. Marcel Dekker, New York (1999)
436. Motreanu, D., Pavel, N.H.: Quasitangent vectors in flow-invariance and optimization problems on Banach manifolds. J. Math. Anal. Appl. **88**(1), 116–132 (1982)
437. Nakayama, H.: Duality in multi-objective optimization. In: Gal, T., Stewart, T., Hanne, T. (eds.) Multicriteria Decision Making, pp. 3–29. Kluwer Academic, Boston/Dordrecht/London (1999)
438. Nam, N.M.: Coderivatives of normal cone mappings and Lipschitzian stability of parametric variational inequalities. Nonlinear Anal. **73**(7), 2271–2282 (2010)

439. Nehse, R.: The Hahn-Banach property and equivalent conditions. Comment. Math. Univ. Carolin. **19**(1), 165–177 (1978)
440. Németh, A.B.: A nonconvex vector minimization problem. Nonlinear Anal. **10**(7), 669–678 (1986)
441. Ng, K.F., Zheng, X.Y.: Existence of efficient points in vector optimization and generalized Bishop-Phelps theorem. J. Optim. Theory Appl. **115**(1), 29–47 (2002). Dedicated to Professor Wolfram Stadler
442. Nieuwenhuis, J.W.: Supremal points and generalized duality. Math. Operationsforsch. Statist. Ser. Optim. **11**(1), 41–59 (1980)
443. Nishnianidze, Z.G.: Fixed points of monotone multivalued operators. Soobshch. Akad. Nauk Gruzin. SSR **114**(3), 489–491 (1984)
444. Nurminskiĭ, A.E.: On the differentiability of set-valued mappings. Kibernetika (Kiev) (5), 46–48 (1978)
445. Oettli, W.: Optimality conditions for programming problems involving multivalued mappings. In: Modern Applied Mathematics (Bonn, 1979), pp. 195–226. North-Holland, Amsterdam (1982)
446. Ortega, J.M., Rheinboldt, W.C.: Iterative Solution of Nonlinear Equations in Several Variables. Academic Press, New York (1970)
447. Ovcharova, N., Gwinner, J.: On Moreau-Yosida approximation and on stability of second-order subdifferentials of Clarke's type. J. Convex Anal. **14**(2), 413–432 (2007)
448. Ovcharova, N., Gwinner, J.: Stability in generalized differentiability based on a set convergence principle. Set Valued Anal. **15**(4), 317–329 (2007)
449. Palata, J.: A survey of conical approximations used in the optimization. Optimization **20**, 147–161 (1989)
450. Páles, Z., Zeidan, V.: Nonsmooth optimum problems with constraints. SIAM J. Control Optim. **32**(5), 1476–1502 (1994)
451. Páles, Z., Zeidan, V.: Critical and critical tangent cones in optimization problems. Set Valued Anal. **12**(1-2), 241–258 (2004)
452. Páles, Z., Zeidan, V.: First- and second-order optimality conditions for a strong local minimum in control problems with pure state constraints. Nonlinear Anal. **67**(8), 2506–2526 (2007)
453. Pallaschke, D., Rolewicz, S.: Foundations of mathematical optimization. Mathematics and Its Applications, vol. 388. Kluwer Academic, Dordrecht (1997)
454. Pascoletti, A., Serafini, P.: Scalarizing vector optimization problems. J. Opt. Theory Appl. **42**, 499–524 (1984)
455. Pavel, N.: Second order differential equations on closed subsets of a Banach space. Boll. Un. Mat. Ital. (4) **12**(3), 348–353 (1975)
456. Pavel, N.H., Huang, Y.K., Kim, J.K.: Higher order necessary conditions for optimization. Libertas Math. **14**, 41–50 (1994)
457. Pavel, N.H., Ursescu, C.: Flow-invariant sets for autonomous second order differential equations and applications in mechanics. Nonlinear Anal. **6**(1), 35–74 (1982)
458. Peng, J., Lee, H.W.J., Rong, W., Yang, X.: A generalization of Hahn-Banach extension theorem. J. Math. Anal. Appl. **302**(2), 441–449 (2005)
459. Pennanen, T.: Graph-convex mappings and K-convex functions. J. Convex Anal. **6**(2), 235–266 (1999)
460. Penot, J.P.: L'optimisation à la Pareto: deux ou trois choses que je sais d'elle. In: Structures Economiques et Econométrie (Lyon, 1978), pp. 4.14–4.33. Lyon (1978)
461. Penot, J.P.: A characterization of tangential regularity. Nonlinear Anal. **5**(6), 625–643 (1981)
462. Penot, J.P.: Differentiability of relations and differential stability of perturbed optimization problems. SIAM J. Control Optim. **22**(4), 529–551 (1984)

463. Penot, J.P.: Variations on the theme of nonsmooth analysis: another subdifferential. In: Nondifferentiable Optimization: Motivations and Applications (Sopron, 1984). Lecture Notes in Econom. and Math. Systems, vol. 255, pp. 41–54. Springer, Berlin (1985)
464. Penot, J.P.: Erratum: "Differentiability of relations and differential stability of perturbed optimization problems" [SIAM J. Control Optim. **22**(4), 529–551 (1984); MR0747968 (85i:49041)]. SIAM J. Control Optim. **26**(4), 997–998 (1988)
465. Penot, J.P.: Second-order generalized derivatives: comparisons of two types of epi-derivatives. In: Advances in Optimization (Lambrecht, 1991), Lecture Notes in Econom. and Math. Systems, vol. 382, pp. 52–76. Springer, Berlin (1992)
466. Penot, J.P.: Second-order generalized derivatives: comparisons of two types of epi-derivatives. In: Advances in Optimization (Lambrecht, 1991), Lecture Notes in Econom. and Math. Systems, vol. 382, pp. 52–76. Springer, Berlin (1992)
467. Penot, J.P.: Optimality conditions in mathematical programming and composite optimization. Math. Program. **67**(2, Ser. A), 225–245 (1994)
468. Penot, J.P.: Second-order conditions for optimization problems with constraints. SIAM J. Control Optim. **37**(1), 303–318 (electronic) (1999)
469. Penot, J.P.: Recent advances on second-order optimality conditions. In: Optimization (Namur, 1998), Lecture Notes in Econom. and Math. Systems, vol. 481, pp. 357–380. Springer, Berlin (2000)
470. Penot, J.P.: Calculus without derivatives. Graduate Texts in Mathematics, vol. 266. Springer, New York (2013)
471. Penot, J.P., Sach, P.H.: Generalized monotonicity of subdifferentials and generalized convexity. J. Optim. Theory Appl. **94**(1), 251–262 (1997)
472. Penot, J.P., Sterna-Karwat, A.: Parametrized multicriteria optimization: continuity and closedness of optimal multifunctions. J. Math. Anal. Appl. **120**(1), 150–168 (1986)
473. Penot, J.P., Théra, M.: Semicontinuous mappings in general topology. Arch. Math. (Basel) **38**(2), 158–166 (1982)
474. Penot, J.P., Zălinescu, C.: Harmonic sum and duality. J. Convex Anal. **7**(1), 95–113 (2000)
475. Peressini, A.L.: Ordered Topological Vector Spaces. Harper & Row Publishers, New York (1967)
476. Perny, P., Spanjaard, O., Storme, L.X.: A decision-theoretic approach to robust optimization. Ann. Oper. Res. **147**, 317–341 (2006)
477. Plotnikov, A.V.: Differentiation of multivalued mappings. The T-derivative. Ukraïn. Mat. Zh. **52**(8), 1119–1126 (2000)
478. Popovici, N., Rocca, M.: Scalarization and decomposition of vector variational inequalities governed by bifunctions. Optimization **62**(6), 735–742 (2013)
479. Postolică, V.: Existence conditions of efficient points for multifunctions with values in locally convex spaces. Stud. Cerc. Mat. **41**(4), 325–331 (1989)
480. Postolică, V.: New existence results for efficient points in locally convex spaces ordered by supernormal cones. J. Global Optim. **3**(2), 233–242 (1993)
481. Pühl, H., Schirotzek, W.: Linear semi-openness and the Lyusternik theorem. Eur. J. Oper. Res. **157**(1), 16–27 (2004)
482. Qiu, J.H.: Dual characterization and scalarization for Benson proper efficiency. SIAM J. Optim. **19**, 144–162 (2008)
483. Qiu, J.H.: A generalized Ekeland vector variational principle and its applications in optimization. Nonlinear Anal. **71**(10), 4705–4717 (2009)
484. Qiu, J.H.: Set-valued quasi-metrics and a general Ekeland's variational principle in vector optimization. SIAM J. Control Optim. **51**(2), 1350–1371 (2013)
485. Raffin, C.: Sur les programmes convexes définis dans des espaces vectoriels topologiques. Ann. Inst. Fourier (Grenoble) **20**(fasc. 1), 457–491 (1970)
486. Ricceri, B.: Remarks: "On multifunctions with convex graph" [Atti Accad. Naz. Lincei Rend. Cl. Sci. Fis. Mat. Natur. (8) **77**(3-4), 64–70 (1984, 1985); MR0884937 (88d:54021)]. Arch. Math. (Basel) **52**(5), 519–520 (1989)

487. Rigby, L.: Contribution to Dubovitskiy and Milyutin's optimization formalism. In: Optimization Techniques Modeling and Optimization in the Service of Man. Part 2. Lecture Notes in Computer Science, vol. 41. Proceedings of the 7th IFIP Conference Nice, 8–12 September 1975, pp. 438–453. Springer, Heidelberg (1976)
488. Rocca, M.: Well-posed vector optimization problems and vector variational inequalities. J. Inf. Optim. Sci. **27**(2), 259–270 (2006)
489. Rockafellar, R.T.: Convex Analysis. Princeton Mathematical Series, No. 28. Princeton University Press, Princeton (1970)
490. Rockafellar, R.T.: Conjugate duality and optimization. Society for Industrial and Applied Mathematics. VI (1974), cBMS-NSF Regional Conference Series in Applied Mathematics, vol. 16. SIAM, Philadelphia
491. Rockafellar, R.T.: Clarke's tangent cones and the boundaries of closed sets in $\mathbb{R}^n$. Nonlinear Anal. Theory Methods Appl. **3**, 145–154 (1979)
492. Rockafellar, R.T.: Generalized directional derivatives and subgradients of nonconvex functions. Can. J. Math. **32**(2), 257–280 (1980)
493. Rockafellar, R.T.: The Theory of Subgradients and Its Applications to Problems of Optimization, *R & E*, vol. 1. Heldermann Verlag, Berlin (1981)
494. Rockafellar, R.T.: First- and second-order epi-differentiability in nonlinear programming. Trans. Am. Math. Soc. **307**(1), 75–108 (1988)
495. Rockafellar, R.T.: Proto-differentiability of set-valued mappings and its applications in optimization, analyse non linéaire (Perpignan, 1987). Ann. Inst. H. Poincaré Anal. Non Linéaire **6**(suppl.), 449–482 (1989)
496. Rockafellar, R.T., Uryasev, S.: Optimization of conditional value-at-risk. J. Risk **2**, 21–41 (2000)
497. Rockafellar, R.T., Uryasev, S.: Conditional value-at-risk for general loss distributions. J. Bank. Finance **26**, 1443–1471 (2002)
498. Rockafellar, R.T., Uryasev, S., Zabarankin, M.: Deviation measures in risk analysis and optimization. Preprint **7** (2002)
499. Rockafellar, R.T., Wets, R.J.B.: Variational Analysis. Grundlehren der Mathematischen Wissenschaften [Fundamental Principles of Mathematical Sciences], vol. 317. Springer, Berlin (1998)
500. Rodríguez-Marín, L., Sama, M.: About contingent epiderivatives. J. Math. Anal. Appl. **327**(2), 745–762 (2007)
501. Rodríguez-Marín, L., Sama, M.: (Λ, C)-contingent derivatives of set-valued maps. J. Math. Anal. Appl. **335**(2), 974–989 (2007)
502. Rodríguez-Marín, L., Sama, M.: Variational characterization of the contingent epiderivative. J. Math. Anal. Appl. **335**(2), 1374–1382 (2007)
503. Rodríguez-Marín, L., Sama, M.: Epidifferentiability of the map of infima in Hilbert spaces. J. Math. Anal. Appl. **342**(1), 371–385 (2008)
504. Rodríguez-Marín, L., Sama, M.: τ^w-contingent epiderivatives in reflexive spaces. Nonlinear Anal. **68**(12), 3780–3788 (2008)
505. Rolewicz, S.: On paraconvex multifunctions. In: Third Symposium on Operations Research (Univ. Mannheim, Mannheim, 1978), Section I, Operations Res. Verfahren, vol. 31, pp. 539–546. Hain, Königstein/Ts (1979)
506. Rolewicz, S.: On γ-paraconvex multifunctions. Math. Jpn. **24**(3), 293–300 (1979/80)
507. Rolewicz, S.: On $\alpha(\cdot)$-monotone multifunctions and differentiability of strongly $\alpha(\cdot)$-paraconvex functions, well-posedness in optimization and related topics (Warsaw, 2001). Control Cybern. **31**(3), 601–619 (2002)
508. Rouhani, B.D., Khan, A.A., Raciti, F.: Penalization and regularization for multivalued pseudo-monotone variational inequalities with Mosco approximation on constraint sets. J. Global Optim. **40**(1-3), 147–153 (2008)
509. Rubinov, A., Gasimov, R.: Scalarization and nonlinear scalar duality for vector optimization with preferences that are not necessarily a pre-order relation. J. Global Optim. **29**(4), 455–477 (2004)

510. Rubinov, A., Singer, I.: Topical and sub-topical functions. downward sets and abstract convexity. Optimization **50**, 307–351 (2001)
511. Rubinov, A.M.: Sublinear operators and their applications. Uspehi Mat. Nauk **32**(4(196)), 113–174, 287 (1977)
512. Ruszczyński, A., Shapiro, A.: Optimization of convex risk functions. Math. Oper. Res. **31**(3), 433–452 (2006)
513. Sach, P.: Nearly subconvexlike set-valued maps and vector optimization problems. J. Optim. Theory Appl. **119**, 335–356 (2003)
514. Sách, P.H.: Differentiability of set-valued maps in Banach spaces. Math. Nachr. **139**, 215–235 (1988)
515. Sach, P.H.: Sufficient conditions for generalized convex set-valued maps. Optimization **37**(4), 293–304 (1996)
516. Sach, P.H.: Another characterization of convexity for set-valued maps. Numer. Funct. Anal. Optim. **20**(3-4), 341–351 (1999)
517. Sach, P.H.: New generalized convexity notion for set-valued maps and application to vector optimization. J. Optim. Theory Appl. **125**(1), 157–179 (2005)
518. Sach, P.H., Yen, N.D.: Convexity criteria for set-valued maps. Set Valued Anal. **5**(1), 37–45 (1997)
519. Sama, M.: Some remarks on the existence and computation of contingent epiderivatives. Nonlinear Anal. **71**(7-8), 2997–3007 (2009)
520. Sama, M.: The role of directional compactness in the existence and computation of contingent epiderivatives. J. Math. Anal. Appl. **372**(1), 262–272 (2010)
521. Sawaragi, Y., Nakayama, H., Tanino, T.: Theory of multiobjective optimization. Mathematics in Science and Engineering, Vol. 176. Academic Press, Orlando (Harcourt Brace Jovanovich, Publishers). (1985)
522. Sayin, S., Kouvelis, P.: The multiobjective discrete optimization problem: A weighted min-max two-stage optimization approach and a bicriteria algorithm. Manag. Sci. **51**, 1572–1581 (2005)
523. Schinas, J., Boudourides, M.: Higher order differentiability of multifunctions. Nonlinear Anal. **5**(5), 509–516 (1981)
524. Schirotzek, W.: Nonsmooth Analysis. Universitext. Springer, Berlin (2007)
525. Schönfeld, P.: Some duality theorems for the non-linear vector maximum problem. Unternehmensforschung **14**, 51–63 (1970)
526. Scott, E.M., Saltelli, A., Sörensen, T.: Practical experience in applying sensitivity and uncertainty analysis. In: Sensitivity Analysis, Wiley Ser. Probab. Stat., pp. 267–274. Wiley, Chichester (2000)
527. Severi, F.: Sulla differenziabilità totale delle funzioni di più variabili reali. Ann. Mat. Pura Appl. **13**(1), 1–35 (1934)
528. Shao, L., Ehrgott, M.: Approximately solving multiobjective linear programmes in objective space and an application in radiotherapy treatment planning. Math. Methods Oper. Res. **68**(2), 257–276 (2008)
529. Shao, L., Ehrgott, M.: Approximating the nondominated set of an MOLP by approximately solving its dual problem. Math. Methods Oper. Res. **68**(3), 469–492 (2008)
530. Shi, D.S.: Contingent derivative of the perturbation map in multiobjective optimization. J. Optim. Theory Appl. **70,no.(2)**, 385–396 (1991)
531. Shi, D.S.: Sensitivity analysis in convex vector optimization. J. Optim. Theory Appl. **77**, 145–159 (1993)
532. da Silva, M.R.: Nonnegative order iterates of Bernstein polynomials and their limiting semigroup. Portugal. Math. **42**(3), 225–248 (1983/84)
533. Silverman, R.J., Yen, T.: The Hahn-Banach theorem and the least upper bound property. Trans. Am. Math. Soc. **90**, 523–526 (1959)
534. Simons, S.: Extended and sandwich versions of the Hahn-Banach theorem. J. Math. Anal. Appl. **21**, 112–122 (1968)

535. Song, W.: Conjugate duality in set-valued vector optimization. J. Math. Anal. Appl. **216**(1), 265–283 (1997)
536. Song, W.: A generalization of Fenchel duality in set-valued vector optimization. Math. Methods Oper. Res. **48**(2), 259–272 (1998)
537. Song, W.: Characterizations of some remarkable classes of cones. J. Math. Anal. Appl. **279**(1), 308–316 (2003)
538. Song, W.: Weak subdifferential of set-valued mappings. Optimization **52**(3), 263–276 (2003)
539. Sonntag, Y., Zălinescu, C.: Comparison of existence results for efficient points. J. Optim. Theory Appl. **105**, 161–188 (2000)
540. Soyster, A.: Convex programming with set-inclusive constraints and applications to inexact linear programming. Oper. Res. **21**, 1154–1157 (1973)
541. Steponavičė, I., Miettinen, K.: Survey on multiobjective robustness for simulation-based optimization. In: Talk at the 21st International Symposium on Mathematical Programming, August 19–24, 2012. Berlin, Germany (2012)
542. Sterna-Karwat, A.: Remarks on convex cones. J. Optim. Theory Appl. **59**(2), 335–340 (1988)
543. Stewart, T., Bandte, O., Braun, H., Chakraborti, N., Ehrgott, M., Göbelt, M., Jin, Y., Nakayama, H., Poles, S., Stefano, D.D.: Real-world applications of multiobjective optimization. In: Branke, J., Deb, K., Miettinen, K., Slowinski, R. (eds.) Multiobjective Optimization: Interactive and Evolutionary Approaches. Lecture Notes in Computer Science, vol. 5252, pp. 285–327. Springer, Berlin (2008)
544. Stiller, S.: Extending concepts of reliability: Network creation games, real-time scheduling, and robust optimization. Ph.D. thesis, TU Berlin (2009)
545. Treiman, J.S.: Shrinking generalized gradients. Nonlinear Anal. **12,no.(12)**, 1429–1450 (1988)
546. Studniarski, M.: Application of the Dubovitskii-Milyutin method to some locally convex extremal problems. Bull. Soc. Sci. Lett. Łódź **29**(6), 8 (1979)
547. Taa, A.: Necessary and sufficient conditions for multiobjective optimization problems. Optimization **36**, 97–104 (1996)
548. Taa, A.: Set-valued derivatives of multifunctions and optimality conditions. Numer. Funct. Anal. Optim. **19**(1-2), 121–140 (1998)
549. Taa, A.: Subdifferentials of multifunctions and Lagrange multipliers for multiobjective optimization. J. Math. Anal. Appl. **283**(2), 398–415 (2003)
550. Taa, A.: ϵ-subdifferentials of set-valued maps and ϵ-weak Pareto optimality for multiobjective optimization. Math. Methods Oper. Res. **62**(2), 187–209 (2005)
551. Taa, A.: On subdifferential calculus for set-valued mappings and optimality conditions. Nonlinear Anal. **74**(18), 7312–7324 (2011)
552. Taa, A.: Second-order conditions for nonsmooth multiobjective optimization problems with inclusion constraints. J. Global Optim. **50**(2), 271–291 (2011)
553. Tagawa, S.: Optimierung mit mengenwertigen Abbildungen. Dissertation, Univ. Mannheim, Mannheim (1978)
554. Tagawa, S.: Optimierung mit mengenwertigen Abbildungen. In: Third Symposium on Operations Research (Univ. Mannheim, Mannheim, 1978), Section I, Operations Res. Verfahren, vol. 31, pp. 619–629. Hain, Königstein/Ts (1979)
555. Takahashi, W.: Existence theorems generalizing fixed point theorems for multivalued mappings. In: Fixed point theory and applications (Marseille, 1989), Pitman Res. Notes Math. Ser., vol. 252, pp. 397–406. Longman Sci. Tech., Harlow (1991)
556. Tam, B., Ehrgott, M., Ryan, D., Zakeri, G.: A comparison of stochastic programming and bi-objective optimisation approaches to robust airline crew scheduling. OR Spectrum **33**(1), 49–75 (2011)
557. Tammer, C.: Lagrange duality in vector optimization. (Lagrange-Dualität in der Vektoroptimierung.). Wiss. Z. Tech. Hochsch. Ilmenau **37**(3), 71–88 (1991)
558. Tammer, C., Zălinescu, C.: Vector variational principles for set-valued functions. Optimization **60**(7), 839–857 (2011)

559. Tammer, C., Zălinescu, C.: Vector variational principles for set-valued functions. In: Recent Developments in Vector Optimization, pp. 367–415. Springer, Berlin (2012)
560. Tammer, C., Zălinescu, C.: Lipschitz properties of the scalarization function and applications. Optimization **59**, 305–319 (2010)
561. Tanaka, T.: Some minimax problems of vector-valued functions. J. Optim. Theory Appl. **59**(3), 505–524 (1988)
562. Tanaka, T., Kuroiwa, D.: The convexity of A and B assures $\operatorname{int} A + B = \operatorname{int}(A + B)$. Appl. Math. Lett. **6**(1), 83–86 (1993)
563. Tanino, T.: On supremum of a set in a multidimensional space. J. Math. Anal. Appl. **130**(2), 386–397 (1988)
564. Tanino, T.: Sensitivity analysis in multiobjective optimization. J. Optim. Theory Appl. **56**(3), 479–499 (1988)
565. Tanino, T.: Stability and sensitivity analysis in convex vector optimization. SIAM J. Control Optim. **26**(3), 521–536 (1988)
566. Tanino, T.: Conjugate duality in vector optimization. J. Math. Anal. Appl. **167**(1), 84–97 (1992)
567. Tanino, T.: Conjugate duality in vector optimization. J. Math. Anal. Appl. **167**(1), 84–97 (1992)
568. Tanino, T., Sawaragi, Y.: Duality theory in multiobjective programming. J. Optim. Theory Appl. **27**(4), 509–529 (1979)
569. Tanino, T., Sawaragi, Y.: Conjugate maps and duality in multiobjective optimization. J. Optim. Theory Appl. **31**(4), 473–499 (1980)
570. Tasset, T.N.: Lagrange multipliers for set-valued functions when ordering cones have empty interior. ProQuest LLC, Ann Arbor, MI (2010), thesis (Ph.D.)–University of Colorado at Boulder
571. Thibault, L.: Tangent cones and quasi-interiorly tangent cones to multifunctions. Trans. Am. Math. Soc. **277**(2), 601–621 (1983)
572. Thierfelder, J.: Nonvertical affine manifolds and separation theorems in product spaces. Math. Nachr. **151**, 329–344 (1991)
573. Thierfelder, J.: Separation theorems for sets in product spaces and equivalent assertions. Kybernetika (Prague) **27**(6), 522–534 (1991)
574. Treiman, J.S.: Characterization of Clarke's tangent and normal cones in finite and infinite dimensions. Nonlinear Anal. **7**(7), 771–783 (1983)
575. Turinici, M.: Maximal elements in a class of order complete metric spaces. Math. Jpn. **25**(5), 511–517 (1980)
576. Ursescu, C.: Multifunctions with convex closed graph. Czechoslovak Math. J. **25(100)**(3), 438–441 (1975)
577. Ursescu, C.: Tangent sets' calculus and necessary conditions for extremality. SIAM J. Control Optim. **20**(4), 563–574 (1982)
578. Ursescu, C.: On the calculus of some topologically open tangent sets. Boll. Un. Mat. Ital. A (7) **1**(2), 145–151 (1987)
579. Ursescu, C.: Adjacent and tangent regularity. Optimization **30**(2), 105–118 (1994)
580. Valadier, M.: Sous-différentiabilité de fonctions convexes à valeurs dans un espace vectoriel ordonné. Math. Scand. **30**, 65–74 (1972)
581. Vangeldère, J.: Sur une famille d'ensembles particuliers d'un espace vectoriel. Bull. Soc. Roy. Sci. Liège **38**, 158–170 (1969)
582. Varaiya, P.: Nonlinear programming in Banach spaces. SIAM J. Appl. Math. **15**, 284–293 (1967)
583. Vîrsan, C.: Necessary conditions of optimality of second order for the optimal control problem with functional constraints. Rev. Roumaine Math. Pures Appl. **18**, 767–791 (1973)
584. Vlach, M.: Approximation operators in optimization theory. Zeitschrift Oper. Res. **25**, 15–23 (1981)
585. Walczak, S.: Some properties of cones in normed spaces and their application to investigating extremal problems. J. Optim. Theory Appl. **42**, 561–582 (1984)

586. Wang, Q., Li, S., Teo, K.: Higher-order optimality conditions for weakly efficient solutions in nonconvex set-valued optimization. Optim. Lett. **4**, 425–437 (2010)
587. Wang, Q., Yu, G.: Second-order optimality conditions for set-valued optimization problems under Benson proper efficiency. Abstr. Appl. Anal. **Art. ID 432,963**, 16 (2011)
588. Wang, S.S.: A separation theorem for a convex cone on an ordered vector space and its applications. Acta Math. Appl. Sinica **9**(3), 309–318 (1986)
589. Wang, Y.T.: A characterization of paratingent cone and P-subderivative with applications in nonsmooth analysis. Acta Math. Sinica (N.S.) **7**(2), 181–192 (1991)
590. Ward, D.: Convex subcones of the contingent cone in nonsmooth calculus and optimization. Trans. Am. Math. Soc. **302**(2), 661–682 (1987)
591. Ward, D.: Isotone tangent cones and nonsmooth optimization. Optimization **18**(6), 769–783 (1987)
592. Ward, D.: The quantificational tangent cones. Can. J. Math. **40**(3), 666–694 (1988)
593. Ward, D.: Which subgradients have sum formulas? Nonlinear Anal. **12**(11), 1231–1243 (1988)
594. Ward, D.: Calculus for parabolic second-order derivatives. Set Valued Anal. **1**(3), 213–246 (1993)
595. Ward, D.: A chain rule for parabolic second-order epiderivatives. Optimization **28**(3-4), 223–236 (1994)
596. Ward, D.E.: A chain rule for first- and second-order epiderivatives and hypoderivatives. J. Math. Anal. Appl. **348**(1), 324–336 (2008)
597. Ward, D.E., Borwein, J.M.: Nonsmooth calculus in finite dimensions. SIAM J. Control Optim. **25**, 1312–1340 (1987)
598. Watkins, G.G.: Nonsmooth Milyutin-Dubovitskiĭ theory and Clarke's tangent cone. Math. Oper. Res. **11**(1), 70–80 (1986)
599. Webster, R.: Convexity. Oxford Science Publications. The Clarendon Press Oxford University Press, New York (1994)
600. Xia, Z.Q., Wang, M.Z., Zhang, L.W.: Directional derivative of a class of set-valued mappings and its application. J. Convex Anal. **10**(1), 211–227 (2003)
601. Xu, Y., Song, X.: The relationship between ic-cone-convexness and nearly cone-subconvexlikeness. Appl. Math. Lett. **24**(9), 1622–1624 (2011)
602. Yang, X.M., Li, D., Wang, S.Y.: Near-subconvexlikeness in vector optimization with set-valued functions. J. Optim. Theory Appl. **110**(2), 413–427 (2001)
603. Yang, X.Q.: A Hahn-Banach theorem in ordered linear spaces and its applications. Optimization **25**(1), 1–9 (1992)
604. Yao, J.C., Yen, N.D.: Coderivative calculation related to a parametric affine variational inequality. I. Basic calculations. Acta Math. Vietnam. **34**(1), 157–172 (2009)
605. Young, R.C.: The algebra of many-valued quantities. Math. Ann. **104**(1), 260–290 (1931)
606. Yu, P.: Cone convexity, cone extreme points, and nondominated solutions in decision problems with multiobjectives. J. Optim. Theory Appl. **14**, 319–377 (1974)
607. Zaffaroni, A.: Degrees of efficiency and degrees of minimality. SIAM J. Control Optim. **42**, 1071–1086 (2003)
608. Zagrodny, D.: General sufficient conditions for the convexity of a function. Z. Anal. Anwendungen **11**(2), 277–283 (1992)
609. Zălinescu, C.: The Fenchel–Rockafellar duality theory for mathematical programming in oreder-complete vector lattices and applications. Tech. Rep. 45, INCREST, Bucharest (1980)
610. Zălinescu, C.: Duality for vectorial nonconvex optimization by convexification and applications. An. Ştiinţ. Univ. Al. I. Cuza Iaşi Secţ. I a Mat. **29**(3, suppl.), 15–34 (1983)
611. Zălinescu, C.: Optimality conditions in infinite dimensional convex programming (Romanian). Ph.D. thesis, University Alexandru Ioan Cuza Iasi, Iasi, Romania (1983)
612. Zălinescu, C.: Stability for a class of nonlinear optimization problems and applications. In: Nonsmooth Optimization and Related Topics (Erice, 1988), Ettore Majorana Internat. Sci. Ser. Phys. Sci., vol. 43, pp. 437–458. Plenum, New York (1989)

613. Zălinescu, C.: A result on sets, with applications to vector optimization. Z. Oper. Res. **35**(4), 291–298 (1991)
614. Zălinescu, C.: Convex Analysis in General Vector Spaces. World Scientific, River Edge (2002)
615. Zălinescu, C.: Hahn-Banach extension theorems for multifunctions revisited. Math. Methods Oper. Res. **68**(3), 493–508 (2008)
616. Duality for vectorial convex optimization, conjugate operators and subdifferentials. The continuous case. Illmenau University, Eisenach (1984), Internationale Tagung Mathematische Optimierung – Theorie und Anwendungen
617. Zeidler, E.: Nonlinear Functional Analysis and its Applications. Springer, New York (1990)
618. Zheng, X.Y.: Proper efficiency in locally convex topological vector spaces. J. Optim. Theory Appl. **94**(2), 469–486 (1997)
619. Zheng, X.Y.: Scalarization of Henig proper efficient points in a normed space. J. Optim. Theory Appl. **105**(1), 233–247 (2000)
620. Zheng, X.Y., Ng, K.F.: The Fermat rule for multifunctions on Banach spaces. Math. Program. **104**(1, Ser. A), 69–90 (2005)
621. Zheng, X.Y., Ng, K.F.: The Lagrange multiplier rule for multifunctions in Banach spaces. SIAM J. Optim. **17**(4), 1154–1175 (electronic) (2006)
622. Zheng, X.Y., Ng, K.F.: Metric subregularity and calmness for nonconvex generalized equations in Banach spaces. SIAM J. Optim. **20**(5), 2119–2136 (2010)
623. Zhou, Z.A., Yang, X.M., Peng, J.W.: ϵ-strict subdifferentials of set-valued maps and optimality conditions. Nonlinear Anal. **75**(9), 3761–3775 (2012)
624. Zhu, Q.J.: Hamiltonian necessary conditions for a multiobjective optimal control problem with endpoint constraints. SIAM J. Control Optim. **39**(1), 97–112 (2000)
625. Zhu, Q.J.: Nonconvex separation theorem for multifunctions, subdifferential calculus and applications. Set Valued Anal. **12**(1-2), 275–290 (2004)
626. Zhu, S.K., Li, S.J., Teo, K.L.: Second-order Karush–Kuhn–Tucker optimality conditions for set-valued optimization. J. Global Optim. **58**(4), 673–692 (2014)
627. Zhuang, D.: Density results for proper efficiencies. SIAM J. Control Optim. **32**(1), 51–58 (1994)
628. Zowe, J.: A remark on a regularity assumption for the mathematical programming problem in Banach spaces. J. Optim. Theory Appl. **25**(3), 375–414 (1978)
629. Zowe, J.: Sandwich theorems for convex operators with values in an ordered vector space. J. Math. Anal. Appl. **66**(2), 282–296 (1978)

Index

A.A. Khan et al., *Set-valued Optimization*, Vector Optimization,
DOI 10.1007/978-3-642-54265-7

MIX
Papier aus verantwortungsvollen Quellen
Paper from responsible sources
FSC® C105338

If you have any concerns about our products,
you can contact us on
ProductSafety@springernature.com

In case Publisher is established outside the EU,
the EU authorized representative is:
Springer Nature Customer Service Center GmbH
Europaplatz 3, 69115 Heidelberg, Germany

Printed by Libri Plureos GmbH
in Hamburg, Germany